AUSTRALIA

130°E 140°E 150°E 160°E
10°S 20°S 30°S 40°S

ARAFURA SEA
CORAL SEA
PACIFIC OCEAN
TASMAN SEA
Gulf of Carpentaria
Great Australian Bight

NORTHERN TERRITORY
QUEENSLAND
SOUTH AUSTRALIA
NEW SOUTH WALES
VICTORIA
ACT
TASMANIA

Cape Van Diemen
Melville Island
Bathurst Island
Van Diemen Gulf
Wessel Islands
Cobourg Peninsula
Galiwinku
Nhulunbuy
Darwin
Jabiru
Arnhem Land
Alyangula
Groote Eylandt
Joseph Bonaparte Gulf
Katherine
Roper River
Ngukurr
Limmen Bight
Victoria R.
Kununurra
Lake Argyle
Borroloola
Mornington Island
Sturt Plain
Kalkarindji
Lake Woods
Barkly Tableland
Nicholson R.
Burketown
Tanami Desert
Tennant Creek
Camooweal
Mt Isa
Cloncurry
Davenport Range
Lake Mackay
Yuendumu
Mt Liebig 1274 m
Papunya
Macdonnell Ranges
Alice Springs
Hermannsburg
Lake Amadeus
Kaltukatjara
Uluru/Ayers Rock 863 m
Finke R.
Simpson Desert
Birdsville
Mt Woodroffe 1435 m
Musgrave Ranges
Great Victoria Desert
Lake Eyre
Cooper Cr.
Coober Pedy
Flinders Ranges
Andamooka
Leigh Creek
Lake Torrens
Lake Frome
Woomera
Hawker
Nullarbor Plain
Lake Everard
Ceduna
Lake Gairdner
Fowlers Bay
Streaky Bay
Port Augusta
Whyalla
Anxious Bay
Eyre Peninsula
Spencer Gulf
Port Pirie
Peterborough
Burra
Port Lincoln
Cape Spencer
Kangaroo Island
Adelaide
Murray Bridge
Victor Harbor
Bordertown
Kingston South East
Millicent
Mount Gambier
Thursday Island
Torres Strait
Cape York
Shelburne Bay
Weipa
Archer
Cape York Peninsula
Mitchell R.
Cooktown
Mossman
Mareeba
Cairns
Bartle Frere 1622 m
Innisfail
Tully
Normanton
Croydon
Gilbert River
Georgetown
Ingham
Townsville
Ayr
Home Hill
Bowen
Charters Towers
Richmond
Hughenden
Collinsville
Proserpine
Mackay
Sarina
Broad Sound
Great Barrier Reef
Great Dividing Range
Winton
Boulia
Longreach
Barcaldine
Emerald
Blackwater
Blackall
Yeppoon
Rockhampton
Gladstone
Biloela
Moura
Theodore
Monto
Bundaberg
Fraser Island
Maryborough
Gayndah
Gympie
Noosa
Nambour
Moreton Island
Augathella
Charleville
Quilpie
Lake Yamma Yamma
Sturt Desert
Mitchell
Roma
Kingaroy
Dalby
Toowoomba
Brisbane
Gold Coast
Thargomindah
Cunnamulla
St George
Goondiwindi
Warwick
Stanthorpe
Texas
Cape Byron
Lismore
Maclean
Grafton
Coffs Harbour
Moree
Inverell
Walgett
Bourke
Narrabri
Gunnedah
Armidale
Kempsey
Port Macquarie
Tamworth
Coonabarabran
Wilcannia
Cobar
Broken Hill
Menindee
Darling
Gilgandra
Dubbo
Scone
Taree
Maitland
Newcastle
Gosford
Sydney
Wollongong
Nowra
Parkes
Orange
Bathurst
Lithgow
Lord Howe Island
Renmark
Berri
Mildura
Lachlan
Griffith
Hay
Narrandera
Wagga Wagga
Goulburn
Canberra
Murray River
Albury
Wodonga
Shepparton
Horsham
Bendigo
Mt Bogong 1986 m
Mt Kosciuszko 2228 m
Bombala
Eden
Cape Howe
Ballarat
Melbourne
Geelong
Bairnsdale
Sale
Traralgon
Moe
Portland
Warrnambool
Colac
Cape Otway
Wilsons Promontory
Bass Strait
King Island
Furneaux Group
Cape Grim
Stanley
Burnie
Devonport
Launceston
St Marys
Mt Ossa 1617 m
Queenstown
Orford
Huonville
Hobart
South East Cape

400 600 km

118, 120, 124, 128, 130

76, 78

104, 166
68
70 74

76, 82, 84 88
90 92 96
100 108,

110, 114, 116
120
126, 128
130 134

138, 142, 146, 148, 152, 154

156, 158
164, 144, 168
170
172, 174
200 206 209
210, 212

222 248
250
252
254, 264

288

300
302, 304, 306, 308
310, 318 334
338

350

MICHAEL MORCOMBE

Field Guide to Australian Birds

MICHAEL MORCOMBE

Field Guide to Australian Birds

Steve Parish
PUBLISHING

First Published 2000 by
Steve Parish Publishing Pty Ltd
www.steveparish.com.au
PO Box 1058, Archerfield, Q 4108 Australia

National Library of Australia Cataloguing-in-Publication data:

Morcombe, Michael.
Field guide to Australian birds.

Bibliography.
Includes index.
ISBN 1 876282 10 X

1. Birds – Australia – Identification. 2. Bird watching – Australia. I. Title

598.0994

Front cover: Scarlet Honeyeater; Gouldian Finch; Orange Chat; Black-winged Monarch; Regent Bowerbird; White-winged Fairy-wren, race *leuconotus*

Half-title page: Budgerigar

Title page: Spotted Harrier hunting over spinifex plains of Pilbara. Acrylic on canvas, 74 × 50 cm

Pages 4 and 5: Gouldian Finch; Azure Kingfisher; Western Rosella; Comb-crested Jacana

Designed by Leanne Nobilio, SPP
Cover design by Linda Carling, SPP
Edited by Wynne Webber, SPP
Printed in China by South China Printing Company Ltd

Reprinted 2000, 2001, 2002 (twice)

ACKNOWLEDGEMENTS

The help I have received from a great many people extends beyond the fourteen years or thereabouts since this project began. This book drew heavily on my lifetime's experience in observing and photographing birds. So many have given me encouragement, assistance, company, often inspiration. This debt spans the years since my first book was published in 1966, through some 40 titles since as sole or joint author, and even back through my early years of competitive bird photography.

Photographs taken over four decades were a major source of information for the paintings. In order to photograph, I had to find the birds, often in remote regions, and, in many instances, I had to locate their nests. I have spent many hours in hides at nests or in other locations that attract birds. Observations of bird behaviour, and experience gained through an immense time in the company of birds across most of Australia, have been as valuable as the photographs obtained.

I wish to express my appreciation of the assistance given by Ron Johnstone, Assistant Curator of Ornithology at the Western Australian Museum, for access to the Museum's collection covering Australian and South-East Asian birds. On a great many occasions, he gave his time generously, answering my queries—giving his thoughts on the taxonomy of species or race, the colour of a plumage, or the distribution of birds, particularly those of the western half of the continent and the islands to the north. Space was provided where I could photograph, and make sketches of and notes about, birds from the museum's collection.

Over the years with each project—books on birds, wildflowers, national parks—I have tried to acknowledge my appreciation of all involved. Lack of space prevents a full list from all those past projects, but I am mindful that all the help, through all the years, has been part of the background that made this book possible. In addition to all those who have given generously of their time and knowledge during those past projects, this book has greatly benefited from the following, who have given generously of their knowledge, ideas, suggestions, and constructive criticism of ideas and the work in progress:

On many occasions I have taken advantage of visitors who, over the years, saw the work in progress, and gave their constructive comments, some of which were incorporated. For any advice not followed, I must take full responsibility. Others from whom I sought advice, including some we visited in our travels, gave their thoughts on my ideas in many useful discussions covering various aspects of the book's form and detail. Among these are Peter Brudenall, Alan Burbidge, Graeme Chapman, Stephen Davies, Stephen Debus, Ray Garstone, Alex George, Andrew Isles, Simon Nevill, Peter Slater, Tony Start and Donald Trounson. I give added thanks to those who loaned their own photos as references—Kevin Coate, John Estbergs and Max Howard.

For their company, encouragement and assistance on field trips, and for the pleasure of sharing common interests, I would like to thank Frank and Gwen Bailey, Della and John Davey, Judy and Andre DuPlessis, Elaine Hall, Pauline and Malcolm Lewis, David Morcombe, Raoul Slater, and Helen and John Start. To our family, from parents to children, thanks for their interest and support during this project and the many others preceding, for their company on trips, and for keeping things together in our absence.

I also thank Steve Parish for his enthusiasm and encouragement, and the publishing team with whom I have worked: Kate Lovett, Leanne Nobilio and Wynne Webber. I particularly thank Leanne, who polished my rough layout into a design, and set up the computer page shells into which I could write, paint and map.

Finally, I give special thanks to Irene for her company, support and encouragement through this long project. As with the many previous books, her contribution in managing photo and reference libraries, correspondence and other work has been in the background, but most valued and deeply appreciated.

CONTENTS

INTRODUCTION

The content of this field guide is confined to material that is likely to be used directly in the identification of birds or their nests. If such a book is to be used repeatedly in the field, its bulk and weight should not be increased by content that is not vital to its function, but would be more suited to books of general interest or reference.

This guide deliberately shows, on average, around 4 to 6 species on each colour page. This has the advantage, compared with 2 or 3 species to a page, that comparisons within a larger group of similar birds can be made more easily, on one or two pages. But this optimum number of species per page reduces the amount of text that can be placed directly opposite the illustrations.

In this book, maximum text for each species has been obtained by shifting most of the identification text to the right-hand page, among the illustrations, where it is most immediately useful. It is unnecessary to search text on the left-hand page for identifying features. These notes, in with the illustrations, typically include the overall character, typical posture and movements that make up the 'jizz' or essence of the bird. Notes that compare features of similar species may also be included on the right.

Further space opposite the illustrations has been saved by placing all breeding text in the illustrated nests-and-eggs guide (page 352). This material is not usually needed as urgently, but remains quickly accessible by the cross-reference page number at the end of the text for each species.

Space that was gained on the left-hand pages has allowed more comprehensive descriptions of behaviour, calls and song.

Distribution is covered by maps that include, for each species, both abundance and significant subspecies. For each species the map shows not only where that bird should reliably be found, but also where it is of extremely sparse, vagrant or doubtful occurrence, or areas where it may no longer exist. Also mapped are those small parts of the range where the species has been recorded in greatest numbers.

Breeding and nonbreeding ranges are not indicated. Unlike Europe and North America, information so far available is, as yet, far too sparse to allow accurate breeding range maps for many Australian species. There are few observers able to find and identify nests, particularly the small, well hidden nests, and atlas records are often gathered while travelling, when there is not time to search for nests.

This book includes not only the bird species that occur within Australia, but also many of the birds that have so far been recorded on its more distant islands: Norfolk and Lord Howe in the Pacific Ocean, Christmas and Cocos-Keeling in the Indian Ocean, Heard and Macquarie in the Southern Ocean, and the islands of Torres Strait.

These islands, especially Saibai and Boigu Islands, both very close to New Guinea, together with Christmas Island, close to Java, have been providing many of the species recently added to the combined Australian list.

There are also some recent discoveries for mainland Australia. While most are rare vagrants, reports from the greatly increasing number of bird observers that visit Australia's northern coastline are showing that some of these vagrants are of more frequent occurrence than was previously suspected.

The nests and eggs of most of the native species that breed within Australia are included. In some instances a distinctive nest may enable a bird of nondescript plumage to be identified. Though eggs are shown, the nest should not be approached closely just to see them. Do not touch either nest or eggs.

Often a nest will be noticed before a bird is seen, or a small bird will accidentally be flushed from a nest, at close quarters, giving opportunity for a brief look before the onlooker continues past. But detailed inspection should be made from some distance, using binoculars or a camera with a very long focal-length telephoto lens.

The taxonomy of Australian birds is in the midst of considerable change. Possible alterations are noted in the text. With few exceptions, names are from the current RAOU species list (Christidis and Boles, 1994).

While the sequence of families is based on this list, some birds are displayed where they can be compared with others of similar appearance, usually more convenient in the field. If the 'correct' place for a species is significantly different, this has usually been indicated.

Painting gives an opportunity to create bird images from any viewpoint, not easily possible using photography. The bird can be visualised, rotated, posed in the imagination and then on paper, to show it from above or below, front or side, with wings spread, or to emphasise plumage markings, typical postures or flight shapes that together create the character, or 'jizz', of the bird.

This, for every species, for many races, male and female, and at various ages, requires many thousands of tiny illustrations. Such coverage would be difficult, if not impossible, to obtain from available photographs, and the resulting book would probably be too bulky to be convenient as a guide.

Bird photographers well know how light influences their subject. Direct sunlight creates shadows with hard edges; these are often cast across a bird's body by its outstretched wings—especially when it glides or soars. In life these shadows are recognised as such, but, in photographs or paintings, they may appear to be plumage markings. To avoid misleading effects of light, all the illustrations in this guide are painted as if seen under the diffused light of a thinly overcast sky.

Photographers are also aware of the colour temperature of light. Greys become brownish greys in the warm glow of late afternoon sunshine, but this can become blue-grey when photographed in shade or with an electronic flash. Colour films have slight or significant colour bias—some warm-toned, others cooler. Photos can fade as they age, giving a colour error. Viewing bird skins under tungsten versus fluorescent light can also influence impressions of colour.

Any colour bias in lighting, although it will be most obvious when comparing whites and greys, will also alter all other colours. In this guide, the colour temperature is about midway between those extremes. When comparing living birds with illustrations, remember the influence of prevailing lighting, which may make the bird look different from art or photos, which in turn were influenced by the light prevailing at time of their creation.

COLOUR PAGE LAYOUT: 'FORMAL' AND 'INFORMAL'

Below: Robins, as an example of a formal page, where similar species are to be compared; then it may be helpful to have all similar images in columns for comparison. Males are in the first column down the left, then all females, all juveniles, all upper surface flights, finally underside flights. This avoids the need to search among birds scattered about the page, some not opposite their text, while trying to compare plumage details. The dotted blue dividing lines, in line with the relevant text, separate species illustrations, but only rarely subspecies.

Below: Rosellas, as an example of a less formal page layout; this page is typical of the display used when the species are not so closely alike. This allows a greater variety of postures and flight images and more artistic variation. However, as on the formal pages, the males or breeding plumages are at left, through to flights on the right. This guide uses the formal, informal and various intermediate page displays to best suit the group of birds being shown. Dotted blue horizontal dividing lines, lined up with text, always separate species illustrations, but rarely subspecies.

Birds are in line vertically for ease of comparison: breeding males in the first column, then, if unlike males, the females, juveniles; flight from above, beneath.

Flight comparisons, if any, are usually in the right-hand columns.

Blue lines between full species, rarely races

Juv.

Dash-dot line shows change of scale on the page

♂ ♀

Alternatively, the vertical columns may be used to compare breeding and nonbreeding plumages, and flight features, especially in waders and sea birds.

Pointers: These have no inherent identification significance on captions. They merely indicate the direction of wide-ranging notes that may, for example, give the bird's name, describe its plumage, the effects of light or compare it with a similar species.

Other links: On the left-hand page, each species entry begins with the bird's official common name (**bold type**), its scientific name (*italics*), the bird's size range and, if important, its wingspan, and, in brackets, the initials of island territories where it is found.

The family groups on the next two pages lead to the species pages (the main text, pages 14 to 351) via family colour markers, or using page numbers. In most cases the bird to be identified would be similar to, but not the same as, examples used on pages 8 and 9 to represent the groups of bird families.

1 *Sighting of bird of prey:* Remember general impressions, wing and tail shapes, positions, flight actions, perched posture, movements that combine to give overall character or jizz. Look for obvious markings in plumage, together with song, calls, habitat, location.

Field notebook: The following page of a notebook records a sighting of this raptor—sketches, comments, impressions of soaring bird; also of bird perched and at the nest, emphasising translucent white 'windows' near wingtips (very conspicuous against sun when underwings are dark, in shadow). At the nest (a sighting of adult, with large chick, sunning with outspread wings), the white is conspicuous on the upper wing surface.

2 *If the bird's identity is not obvious:* Record impressions of jizz, plumage pattern, calls, habitat and location in notes, backed by rough sketches or photos, even if distant. Notes and sketches of distinctive shapes, positions and movements also help in later identification, especially when there was no field guide available at time of sighting, or if another reference is consulted. Notes, sketches or photographs are necessary to verify a sighting of a rare vagrant species, or records of birds outside their usual range; nest notes, sketches or photos help support breeding records.

3 *Compare this bird with examples on pages 8–9:* Concentrate on overall similarities of wing shape, body contour and posture—all combine to give the character, or jizz. Best match in this example is group **6** (Osprey, kites, eagles, harriers, goshawks, falcons), leading by colour code to pages 80–95, with a full range of illustrations, where (perhaps with aid of notes and sketches) the example species is identified as the Black-breasted Buzzard.

Simplified family features and jizz: For purpose of selecting the family group, try to recognise basic characteristics, then use the more specific details (described page 11) to identify species in the main text pages. For raptors, basic points are 'fingered' spread wingtips of large raptors (usually more pointed in small species) and the similar shape and posture when perched. Finally, in close view, similar bills, eyes and talons make any member of this family very obviously a bird of prey (family group 6, page 9). A similar route can be followed to other family groups.

Non-passerines
Passerines

GUIDE TO FAMILY GROUPS

Select family example (nearest shape and character) or choose from the following list, then follow the page colour code:

PARTS OF THE PLUMAGE

At left: Photos of Blue-winged Kookaburra, Sacred Kingfisher and Yellow-billed Spoonbill. *Below:* Sketches of same species, with parts of plumage labelled. The exact numbers of feathers in tails, primaries, secondaries and other tracts vary between families, but in the field there is rarely an opportunity to count feathers.

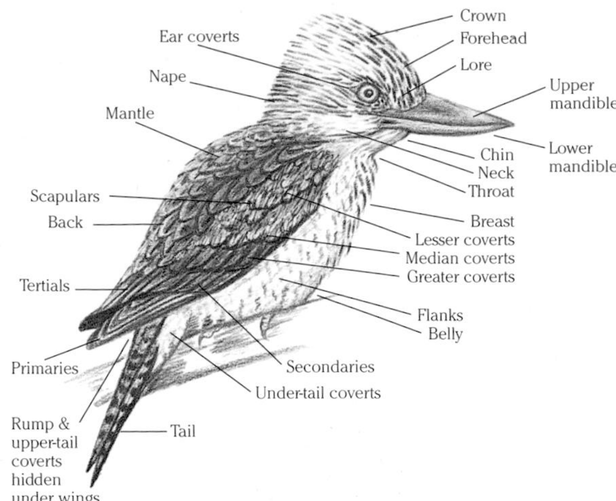

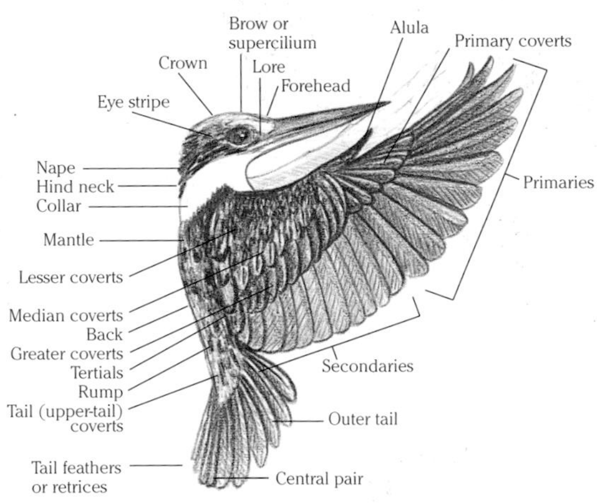

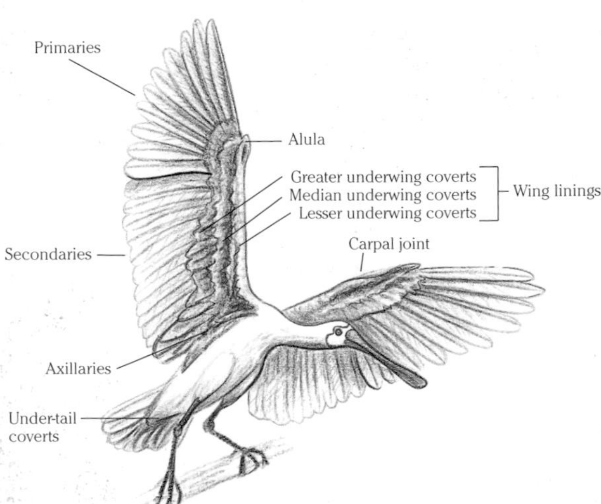

After using a bird's overall character or jizz to place it in a family group (leading to that family's pages in the main text), the species has to be identified. Many have a distinctive jizz—a combination of shape, behaviour, character—that may make them immediately recognisable, even in some instances when a bird may be but a far distant silhouette against the sky. But often the differences between species are slight. Usually, a number of characteristics must be compared: plumage markings, shape and colour of body, wings, head, tail: see the examples below with attached text labels. Also, in italics, are some common ornithological terms, linked to the illustration with pointers as is done in the body of the guide. A glossary is on page 434.

Plumages: The first true plumage, usually acquired in the nest (after the *down* of the hatchlings) is the *juvenile*, often followed by one or more *immature* stages, then, finally, *adult* plumage, in which the sexes may be the same or different. There may be seasonal changes, *breeding plumage* (usually summer), brighter than *nonbreeding* (winter). Wear may also change plumage colour and pattern.

Maps that show the distribution of each species are an essential and almost universal feature of modern field guides. Northern hemisphere guides often use colours to show the breeding range (usually summer range), winter range and where the two overlap, a year-round range. In Australia, with no large region of sub-zero winters, more resident birds are sedentary, or locally nomadic only. Breeding records of many species are sparse, especially those for small birds with well hidden nests and for birds of remote regions, some of which are very difficult to get to in the wet season when many tropical species breed. In this guide, density of map colour is used to show abundance within the known range, the colour depth corresponding with density of records of the species—the deepest tone is where the species has most often been recorded. Blue is used if there is but one race of the species in Australia, or usually for the nominate race when there are several, the others in brown, yellow and green.

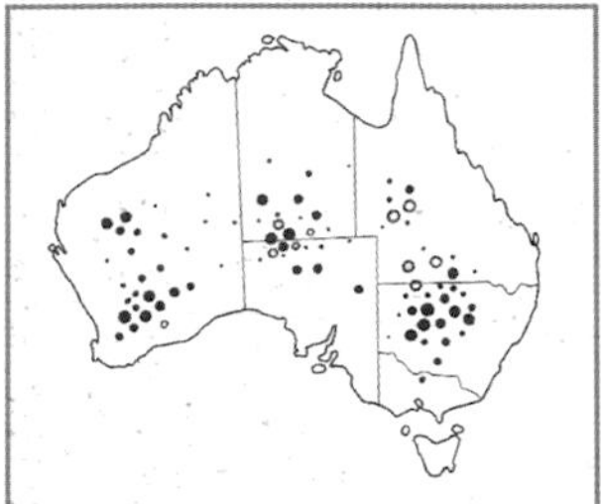

The *Atlas of Australian Birds* (Royal Australian Ornithologists Union, 1984) is a widely used source of bird distribution data. This book was the result of a 5-year survey of birds by some 3000 members and other 'atlassers'. As in the fictitious example at left, an open circle indicates, by its size, the number of sightings for each bird species; a solid circle marks where breeding was observed. Another survey is being undertaken between 1998 and 2002. The *Atlas of Victorian Birds* (Dept of Conservation, Forests, Lands, and RAOU, 1987) similarly shows the distribution of birds within Victoria.

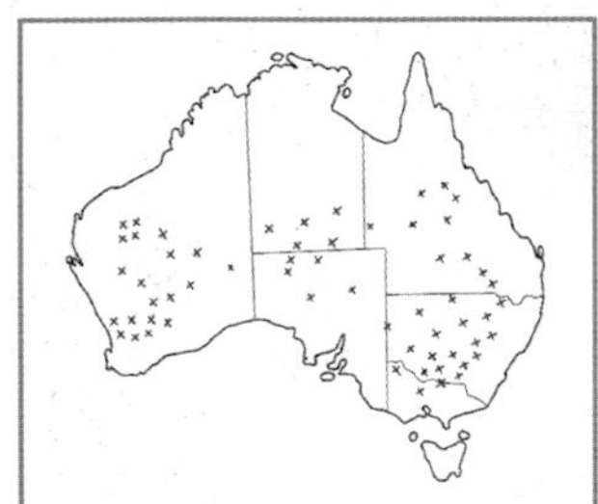

Early records, from around 1900 to the late 1960s (right) are also valuable references: they show the range of species from earliest observations and collections—valuable for species whose range has since contracted and for those so rare that there are few sightings. Among those used in this guide are Dr D.L. Serventy's maps in *Distribution of Birds on the Australian Mainland* (CSIRO, 1977), the Historical maps from the *Atlas of Australian Birds* (RAOU, 1984), and, based on many thousands of Western Australian Museum records, *Western Australian Birds*, volume 1, R.E. Johnstone and G.M. Storr (WA Museum, 1998).

Distribution data from these sources has been used to some extent by most guide and reference book authors, but with various interpretations, so that maps differ substantially from book to book. Some follow very closely the pattern displayed in the atlas type of bird distribution survey. Others are based on diverse references, both recent and historical. Often maps are partially based on the author's personal experience or research; others again might be based on, for example, museum records.

Some distribution maps include isolated and doubtful sightings far from the main range and fill the gaps between to present a broad interpretation of the range of the species. In parts of the range of that species, there will be very little chance of finding the species—but such areas of the map might not be differentiated from those parts where the bird is common.

In other maps the same bird records may have been interpreted much more conservatively, any sparse, vagrant or doubtful records being rejected to produce a map showing a far more restricted range. The bird would almost certainly occur throughout that range, but there is no indication that the species has, if only on rare occasions, or perhaps in early records, been observed far beyond the conservative range shown.

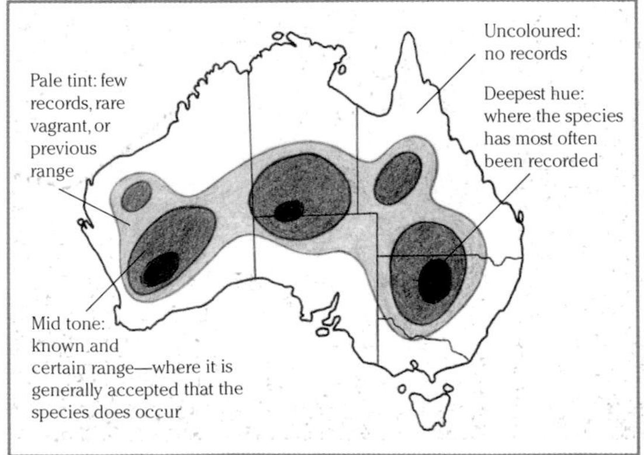

Multicolour, triple-density maps

1 The range in which a bird species could *reliably* be expected to occur—where the records are not extremely sparse or isolated, nor represent a part of the bird's range that is, apparently, no longer occupied—are coloured mid blue (or a mid tone of another colour for a subspecies or race). [Note that 'race' and 'subspecies' are used interchangeably.]

2 Regions where it is *unlikely*, but *possible*, that the bird could be seen are a paler tint. This pale tint shows the species' occurrence where it has been only very rarely recorded, perhaps being a rare vagrant, or a doubtful sighting, or of long-past occurrence.

3 Shown in the deepest hue are the areas where the species has most often been recorded: these are localities with the greatest number of sightings in the RAOU Atlas and other surveys, and in historical and other records. These may not be where the species actually occurs in greatest numbers, but rather show *where most bird observers have been successful* in recording the species. Often, as well as having a high population of the birds, the locality is easily accessible. This is where anyone who wants to see the bird is most likely to succeed—especially if time is limited.

Many Australian birds have several separated or merging populations, perhaps with hybrid forms where they meet. If these populations are sufficiently different, then each may be identified as a separate subspecies or race. On these maps, merging or intergradation of races is shown by the blending together of the colours.

A population that changes gradually across its range without any perceptible point of change is known as a 'cline', usually not a basis for subdivision into several races. Opinions differ widely on the number, name and distribution boundaries of the races of many species. Changes to species and races need to be soundly based and widely accepted—many proposed changes are not accepted, or are later reversed, causing public confusion.

HABITATS

Australia has greatly diverse habitats, each with its own landform, climate, vegetation and fauna. For each species, the habitat, or a diversity of habitats, provides what is needed for food, shelter and breeding. The list of habitats with each species account helps in identification and nest location. Vegetation is often the most important and obvious component of habitat—many habitats are named for their vegetation type. Others are dominated by a landform rather than by vegetation; these include islands, oceanic, coastal waters and lakes. Major habitat vegetation types are shown here.

Right: Rainforest
Left: Eucalypt forest
Below: Woodland
Below right: Mangrove

Above: Mallee *Right:* Acacia scrub
Below: Spinifex grasslands

Left: Saltbush *Above:* Coastal heath *Below:* Wetlands

Ostrich *Struthio camelus* 1.7–2.7 m

Largest living bird; much taller, more heavily built than the Emu. Australia's feral population originated from releases made when ostrich farms were abandoned in the 1920s. Now known in the ostrich industry as the Australian Grey to distinguish it from recently imported African black, red-necked and blue-necked forms. In the wild, the Ostrich is gregarious in small groups or flocks with females apparently outnumbering males. **Male:** body black, contrasting with white wings. In breeding season mature males develop red on bill, gape, legs. **Fem.:** similar, but brown with dull off-white wing plumes. **Juv.:** small downy chicks are buff with dark longitudinal striping; larger young are grey-brown, tipped white. **Variation:** other races or forms in Africa; some commercial imports since about 1995. **Voice:** usually silent. Males in breeding season give booming sounds—in distance an 'oo-oo-ooom', at close range likened to roaring, not as deep as the Emu's booming. **Similar:** Emu, smaller. **Hab.:** semi-arid grasslands, shrublands, very open woodlands. **Status:** rare in wild in Aust.; introduced.

Emu *Dromaius novaehollandiae* 1.5–2 m

Largest of Australian birds, widespread and familiar; extremely well adapted to the semi-arid regions. Feathers double-shafted, hair-like and loose; gives untidy, floppy, shaggy look. Emus are shy, fleeing at great speed if encountered, but have a curiosity that will often entice them very close. Live in pairs or family parties up to 12 or so birds, averaging 1–2 km apart, but much closer where conditions are favourable. They are migratory or nomadic, wandering in response to seasonal conditions and seeking the lush growth of vegetation that follows rain. Only rarely do they concentrate in large mobs where migrating birds are held back by a road or some other artificial barrier. The species is principally vegetarian, but takes some insects. Sexes similar, female larger. **Juv.:** chicks to about 3 months downy, buff with dark brown stripes. **Imm.:** smaller than adult, head black, blending on neck to the grey-brown of the body. **Voice:** usually silent, occasional low-intensity booming when approaching a strange object. In breeding season, deep thudding booming and drumming by female. Male gives a low growl to communicate with newly hatched chicks. **Hab.:** semi-arid grasslands, scrublands, open woodlands; less commonly heaths, alpine areas, tall dense forests. **Status:** common in preferred habitats. **Br.:** 352

Southern Cassowary *Casuarius casuarius* 1.5–1.75 m

Flightless, almost as tall as the Emu, but is more solid in build; has massive legs, coarse bristle-like black plumage, and the bare skin of head and neck are brightly coloured. A horn-surfaced casque on top of the head is internally spongy rather than bony, and possibly acts as a shock-absorber when the bird dashes through tangled rainforest thickets hitting the thick, woody stems and heavy, hanging loops of vines. Shy; most easily seen at edges of clearings, on tracks and in gardens in rainforest around dawn and dusk; feeds on fruits of rainforest trees. Solitary most of year; in pairs for a few weeks when breeding; family parties later. Cassowaries may be dangerous if provoked. **Fem.:** brighter colours of bare skin; larger. **Juv.:** heavily striped. **Imm.:** brownish plumage, smaller casque, brownish head and wattles. **Voice:** deep, thudding, resonant booming; abrupt rough grunts; hissing and roaring sounds when fighting. **Hab.:** rainforest, especially clearings, margins, vicinity of streams. Occasionally found in adjoining melaleuca swamps, eucalypt forest containing some rainforest fruiting trees and in gardens. **Status:** uncommon, though numerous in some optimum patches of rainforest. Threatened by loss of habitat, illegal hunting, attacks by dogs. Sedentary. **Br.:** 352

♂
Sparse, fine, soft, downy surfaced, pink tint when breeding
Soft, loose black body plumage
Large soft white plumes
Bare thighed, becoming pinkish or bluish in breeding season
Red edge to front of leg and edge of toes when breeding
♀
Pale sandy brown
Bare skin, blue to white
Loose shaggy hair-like appearance, individual feathers double-shafted
Elongated upper tail coverts form a mass of hanging, soft tail-like plumage.
♀
♂
Wings vestigial
Striping of downy young is gradually replaced by dark grey-brown juvenile plumage; no longer visible by age 4–5 months.
Very large, shaggy, loose plumage with prominent parting down the centre line of back; bounces as the bird runs. Females average larger and darker
Juv.
Downy young
Small casque; blue and pink expanding
Distinctive casque rising from crown is spongy rather than hard; may absorb shock of hitting heavy hanging loops of vines and low slender branches as the bird dashes head-down through the rainforest.
Immature: mix of brownish juvenile and black adult plumage
Coarse, drooping, bristle-like plumage
Bright blue and pink bare skin
Imm.
Pendulous pink wattles
Adult
Cinnamon
Massive, powerful; innermost of three toes has a very long claw that is used as a weapon.
Juv.
Downy young: dark brown and white stripes fade to merge into all-brown juvenile plumage by age 6 months.
No casque, first blue appears

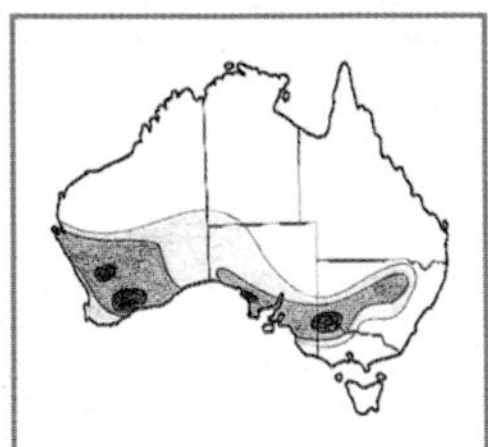

Malleefowl *Leipoa ocellata* 55–60 cm
Quiet, shy and wary. If disturbed, remains motionless and hard to detect, or walks silently, inconspicuously into concealing thickets. Will fly only if surprised or pursued. Most likely to be seen near an active nest mound; elsewhere difficult to observe for any length of time. Pairing permanent; both remain in home territory. Feeds by scratching about in leaf litter. **Voice:** male has a deep, double noted, booming territorial call; female a high crowing. Also various clucks, softer chuckles and grunts. **Hab.:** unburned mallee and woodland with abundant litter and low scrub. **Status:** reduced range; scarce, endangered. Sedentary. **Br.:** 352

Orange-footed Scrubfowl *Megapodius reinwardt* 40–50 cm (B, S)
When alarmed may scuttle away across the forest floor, or fly up to watch warily from a branch. Scrubfowls rake about in the forest litter, scattering twigs and leaves in search of food—a wide variety of fruits, seeds, insects or other tiny creatures that might be exposed. **Variation:** many races, probably 3 in Aust.; differences slight. **Voice:** noisy, calling both day and night, often many birds together—deep and powerful gurgling: 'ok-ok-owk-owwwk-owwwwk-ok-ok-ok-ok', and various chuckles and screams. **Hab.:** rainforests, riverine gallery forests, monsoon forest, vine thickets. **Status:** common; sedentary. **Br.:** 352

Australian Brush-turkey *Alectura lathami* 60–70 cm
Large black megapode, conspicuous and tame at tourist sites. Elsewhere wary; if disturbed, dashes into rainforest or may fly ponderously into a tree, jumping and flapping upwards to the safety of the high canopy. From below, looks like a vulture with black plumage and bare red head. Scratches about the forest floor in search of fallen fruits, seeds and invertebrates. **Variation:** race *purpureicollis*, white collar. **Voice:** vibrating low grunts. Loud 'gyok-gyok', probably only from male. **Hab.:** rainforest from temperate to tropical; densely vegetated gullies in wet eucalypt forest; monsoon and gallery forest; some drier scrubs further inland. **Status:** common, widespread, but population reduced by habitat destruction. **Br.:** 353

Wild Turkey *Meleagris gallapavo* 90–125 cm
A wild turkey descended from the North American Wild Turkey. Familiar large terrestrial game bird, usually in small flocks, male and several females. Forages on ground; runs or flies strongly if alarmed. **Voice:** territorial call of male is a loud gobbling, audible from afar, usually given early in the breeding season. Also yelps, clucks, croaks, hisses. **Hab.:** semi-domesticated birds on farmlands, grasslands, forest edges. **Status:** introduced, common on King Is. and Flinders Is.

Red Junglefowl *Gallus gallus* 40–70 cm
Birds similar to Asian Red Junglefowl were released in many sites around Aust., but only became established ferals on Heron and North-West Is., gradually reverting close to original junglefowl. **Fem.:** hen much smaller, less colourful, small comb. **Voice:** various clucks and cackles; male's territorial crows are like the domestic rooster, at higher pitch; noisiest at dawn. **Hab.:** rainforest and other dense closed forest; open understorey areas. **Status:** introduced, small colonies, declining.

Common Pheasant *Phasianus colchicus* 55–100 cm
Widely introduced in many parts of the world as a game bird; native to Asia. Only surviving populations now in S Tas., King Is. and Rottnest Is. Terrestrial and diurnal; usually in small to large parties or flocks; walks or runs if disturbed. A flock put to flight explodes from cover with much clapping of wings; strong swift flight. **Voice:** loud, hoarse crowing, 'korr-orr-ok', and many other noises including croakings, clucks and hisses. **Hab.:** in Aust. varied; imposed by site of release. **Status:** introduced, sedentary; scarce to common in small areas around site of introduction.

Indian Peafowl *Pavo cristatus* male 2–2.3 m; female 80–100 cm
Usually in small parties; male and five or so peahens. Typically conspicuous, noisy. Forages on the ground; strong flight despite length of train; roosts high in trees; seeks refuge on high perch if disturbed. **Voice:** from male an extremely loud, carrying, raucous 'heirr-elp' like a drawn out cry for 'help'. From both sexes various cluckings, shrieks and chuckles. **Hab.:** adaptable. In Aust., set by release site; diverse. **Status:** introduced, sedentary; common.

A large megapode with plumage coloured and intricately patterned to blend with the browns, fawns, black and white of stems and litter of bark, leaves and twigs in dappled sunlight and shadow typical of the dry woodlands and mallee scrubs.
Dark, very slight crest
Light chestnut; fades to buff tint.
Black line
Barred, streaked, blotched and variegated rufous, grey and brown, fringed and banded buff and white
Adult
Heavily built, powerful
Broad, with rounded tips
Runs to cover. If forced to fly, heavy laboured take off on patterned wings; soon drops to cover.
Long, barred
Outer tail dark
Adult
Deep dark chestnut
Small pointed crest
Small, dull ochre
Ground bird about size of domestic fowl. Appears almost black in gloom of rainforest, but has distinctive silhouette of oval body, small head with peaked crest, short thick orange legs.
Wings reach tail-tip.
Thick, powerful, orange
Ridge of dense black bristles
Bare red skin
Race *purpureicollis*, violet-tinged white collar
Distinctive long tail, feathers stacked in tall, flat-sided shape
Yellow wattle around base of neck; hangs lower on breast in breeding season.
Heavy laboured flight, legs dangling; all black except red head and neck, white or yellow collar
Newly hatched chicks are reddish brown. Immature: like adults but with more feathering on neck; colours are very dull.
Plumage of ferals may vary; is usually brown with iridescence giving tints of green, bronze and violet.
Bare grey and purplish grey skin; red wattles
Tuft of long feathers
Black band, tip rufous to buffy white
Long flame hackles
♂
Iridescent green or blue sheen over grey-black
♀
Red skin
Glossy iridescent green
Buff barred and mottled with dark brown
♀
Long, pointed, barred
♂
Gold and chestnut, extensively barred and spotted
Fan crests
♂
Grey-brown
♀
Body brilliant blue
Long train of greatly lengthened uppertail-coverts, lifted and spread over the back in display

California Quail *Callipepla californica* 24–27 cm

A species native to W parts of N America, the California Quail was brought into Aust. as a game bird in the mid 1800s and was released in the wild from Tas. to Qld. The mainland feral populations died out, but feral birds survived in Tas. until about 1950 and still remain on King Is. in Bass Str. The California Quail is only distantly related to Australian quails; differs in habitat and feeding requirements. Outside the breeding season coveys are formed—assemblies of up to 200 birds. Feeding is terrestrial, seeking seeds and edible vegetation. If disturbed, these quail run very fast, then all explode upward into flight, running again on landing to regroup. **Fem.:** dull colours, plain brown face, small head plume. **Juv.:** similar to adult but paler, duller. **Voice:** contact call when regrouping; loud territorial crowings in breeding season from males; many other calls from soft and plaintive to loud, staccato whistles. **Hab.:** in N America usually desert and semi-desert chaparral shrublands, thickets; avoids forests. On King Is. the California Quail has open country with bushes and shrubs. **Status:** introduced, extinct on Aust. mainland, remnant feral population on King Is.

Stubble Quail *Coturnix pectoralis* 17–20 cm

Large, heavily streaked quail with white eyebrow and cinnamon throat patch. Named for its liking for short grasses; often found in paddocks where low but dense cover of wheat-stalk stubble remains after harvesting. When disturbed, runs under cover or freezes; relies on camouflage colours until the intruder is very close; then flushes, bursting up on whirring wings, fast but low, and dropping abruptly. This is often the first indication that quail are present. In flight gives but a momentary glimpse of its colours and patterns. Usually gives no calls when flushed. In pairs when breeding; otherwise single birds or coveys of 20 or more. **Fem.:** white throat patch. **Imm.:** colours dulled. **Voice:** a clear, ringing 'cheery-wit' and 'too-too-weep', last note higher, louder, given repeatedly. **Similar:** Brown Quail, darker brown or reddish brown, lacks white brow line; wings broader and rounder at tips; makes whistling metallic rather than whirring sound when flushed. **Hab.:** both short and tall grasses, stubble, pasture, spinifex, saltbush and bluebush. **Status:** common; nomadic, irruptive. **Br.:** 353

King Quail *Coturnix chinensis* 13–15 cm

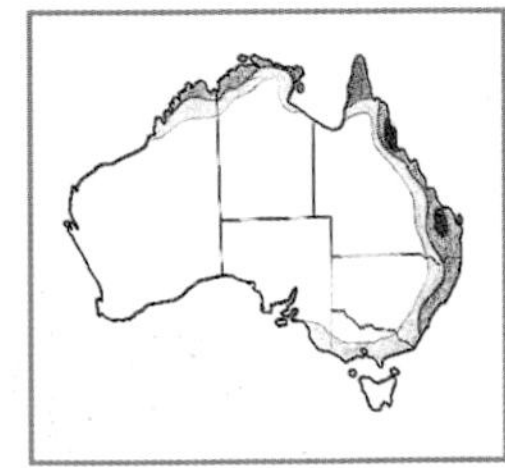

A small quail; male is colourful but secretive and elusive. In pairs or family groups. If disturbed, runs, keeping under cover; usually seen only if suddenly flushed, rising in comparatively weak flight with body tilted up, little wing noise. Usually drops to cover within a short distance, only rarely long flights; impression given is of a small, dark quail. **Juv.:** like female, but underparts are boldly streaked or spotted. **Voice:** a drawn out whistle, 2 or 3 notes, descending, husky and plaintive, 'kee-er, kee-er...' or 'kee-eee-er, kee-ee-er...'. Also quiet peeps, low husky growling. **Similar:** Red-backed and Little Button-quails. **Hab.:** prefers the dense wet vegetation of swampy grasslands, sedges and heaths; also found on drier crops and grasslands. **Status:** common in N, reduced in S. **Br.:** 353

Brown Quail *Coturnix ypsilophora* 17–20 cm

A large quail; overall brownish, finely streaked white above, barred beneath. Singly, in pairs or coveys of 10–30, which, when disturbed, dart along ground erratically. If close-pressed, burst upward, flushing explosively in all directions; a confusion of noisy wings and metallic whistling sound. Flight is low, fast, usually brief before plunging to cover again. Often gives alarm calls when flushed. Plumage colour of males is varied; grey to rufous-brown. **Fem.:** darker, less varied, rarely rufous, distinguished from males by large dark spots and bolder white streaks on the upper parts. **Juv.:** like female, but breast barring is indistinct. **Variation:** in Tas., the nominate race *ypsilophora*, known as the Swamp Quail, is larger and yellow eyed. On the mainland, the red-eyed race *australis* has a diversity and complexity of plumage between grey-brown and rufous-brown. **Voice:** a double noted, ascending whistle: 'pi-pieer', or 'tu-wieep'. Gives a sharp chirp in alarm. **Similar:** Stubble Quail. **Hab.:** grass, crops, heaths, rainforest edges, grassy and spinifex woodland; prefers damp, rank vegetation. **Status:** common. **Br.:** 353

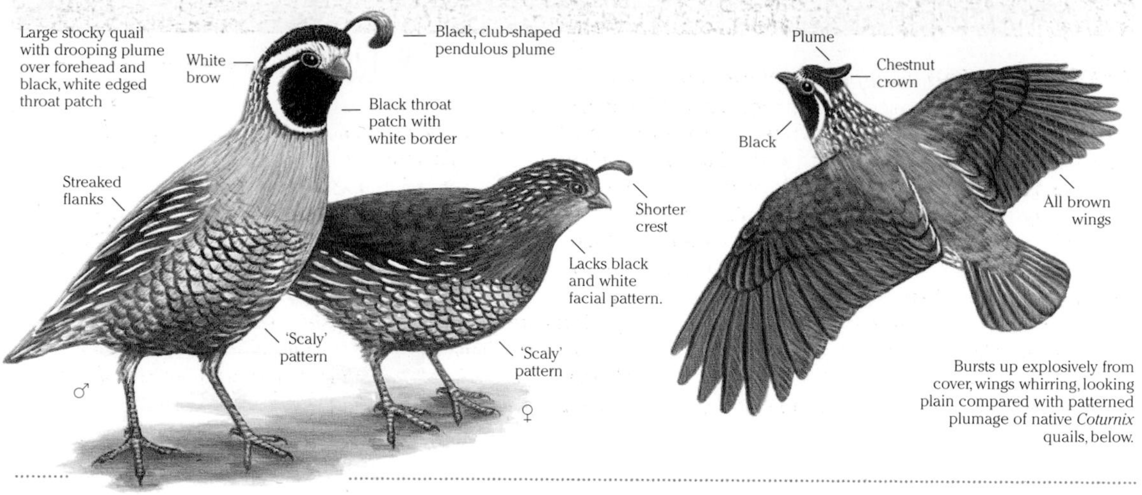

Large stocky quail with drooping plume over forehead and black, white edged throat patch
White brow
Black, club-shaped pendulous plume
Black throat patch with white border
Streaked flanks
Shorter crest
Lacks black and white facial pattern.
'Scaly' pattern
'Scaly' pattern
♂
♀
Plume
Chestnut crown
Black
All brown wings
Bursts up explosively from cover, wings whirring, looking plain compared with patterned plumage of native Coturnix quails, below.

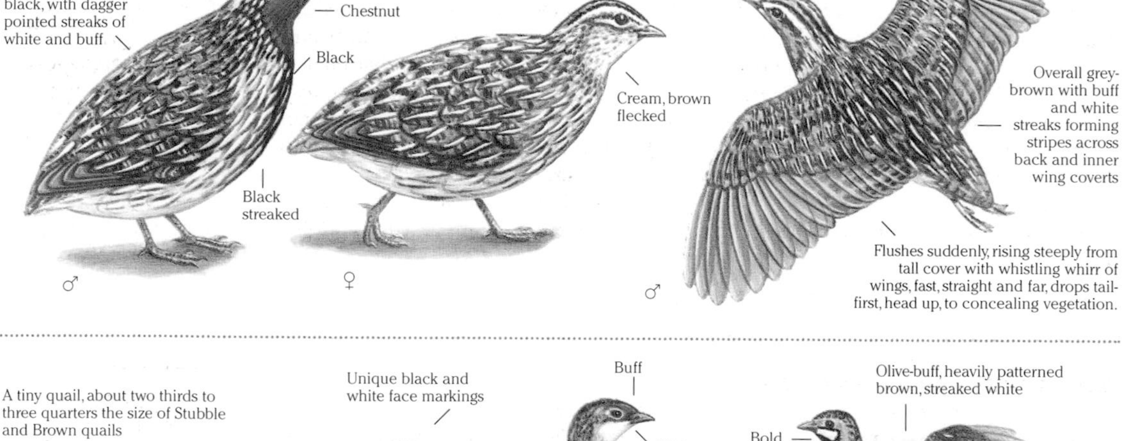

Long white brow
Buff, heavily patterned in dark browns and black, with dagger pointed streaks of white and buff
Chestnut
Black
Black streaked
♂
Untidy buffy brow
Cream, brown flecked
♀
Dark brown, streaked white
Overall grey-brown with buff and white streaks forming stripes across back and inner wing coverts
Flushes suddenly, rising steeply from tall cover with whistling whirr of wings, fast, straight and far, drops tail-first, head up, to concealing vegetation.
♂

A tiny quail, about two thirds to three quarters the size of Stubble and Brown quails
Unique black and white face markings
Deep chestnut
♂
Flanks and sides of neck blue-grey
Buff
White
Buff with bold dark barring
♀
Olive-buff, heavily patterned brown, streaked white
Bold pattern
♂
When flushed, rises steeply, with subdued whirr of wings; keeps low and tail-down; quickly drops again, tail-first, to cover.

The nominate race ypsilophora, the 'Swamp Quail' of Tas. (not illustrated) has yellow eyes, the only obvious and reliable difference between that race and the mainland race australis, the Brown Quail, shown below.
The Brown Quail of mainland Aust., race australis, has a continuous range of plumage tones from buff-brown to rufous-brown. There may also be a blue-grey form, or morph.
When flushed, birds explode into flight; scatter; dive headfirst to cover.
Race australis: both sexes and all colour forms have a red eye.
♀
Race australis, brown form: male is of similar colour but with finer markings.
♂ Race australis, rufous form
Race australis, ♂ rufous form

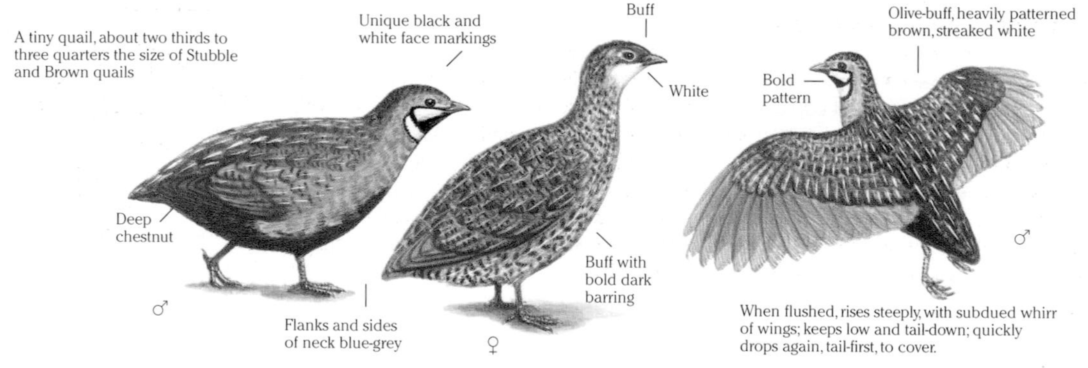

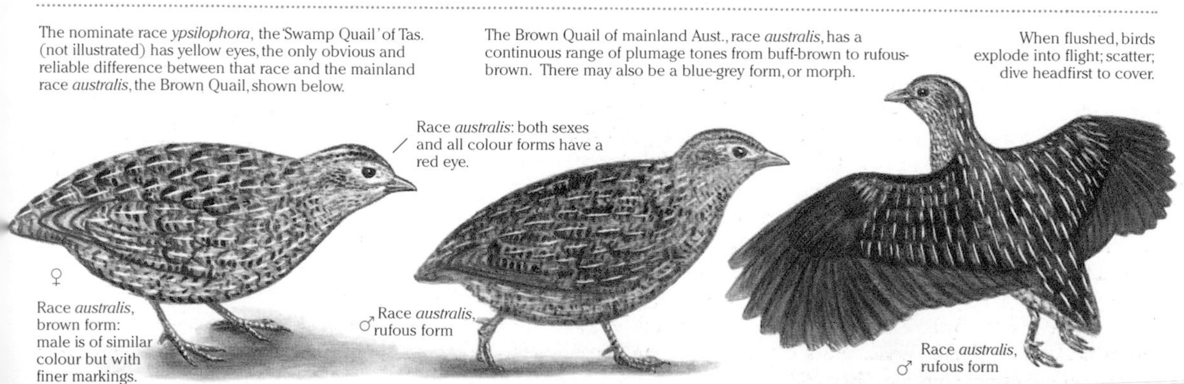

Magpie Goose *Anseranas semipalmata* 75–90 cm

Probably the most conspicuous, best-known water bird of tropical Aust.; in some places found in flocks of thousands. Bold pied plumage, pink legs and face make this large bird unmistakeable. Distinctive in several anatomical features: the looped windpipe of the male, the hooked bill, and the partly webbed feet. The bill and feet are thought to be adaptations to the habitat. In the dry season the geese dig up roots of rushes, wandering far over the mudflats of river floodplains to feed on plants of the seasonally inundated plains. Most of the foraging is on land and around billabong margins, and in shallows, head under water, probing for aquatic plants, up-ending vertically to reach the bottom. When disturbed, takes to air rather than to deeper water. The appearance of an eagle overhead can send a flock into the air with a great roar of wings. Flies heavily, long neck thrust forward; glides only towards landing; congregates on trees growing in swamps. **Juv.:** like adult, but entirely black on back; white areas mottled brownish. **Voice:** loud honking, that of male louder, higher. From a flock, a cacophony of nasal trumpeting. **Similar:** Black Swan, but longer neck, wholly black body, white wingtips, slower wingbeat. Straw-necked Ibis, but very long, downcurved bill; flaps and glides alternately. **Hab.:** tropical floodplains, shallows of dams, irrigated crops, swampy well-vegetated margins of deep waterways. **Status:** common or locally abundant across N Aust.; range reduced and declining numbers in SE; dispersive. **Br.:** 353

Cape Barren Goose *Cereopsis novaehollandiae* 75–90 cm

A large goose unique to Aust.; readily identified by predominantly grey plumage, deep pink legs and prominent yellow across the bill. Wary and difficult to approach, especially in the open environs in which it is usually likely to be encountered. Grazes on land, rarely takes to the water, but if harassed while with small young will take refuge in water, all swimming proficiently. Flight strong; rapid wingbeats interspersed with glides. In breeding season on islands, forms pairs, but gathers into small or large flocks through summer. In some regions, geese all leave the islands after breeding, grazing in large numbers on mainland pastures. On S coast of WA most keep to islands, which remain green through summer; few visit the mainland. **Imm.:** like adult, slightly paler. **Variation:** race *grisea*, SW of WA: larger, browner, more white on crown. **Voice:** often calls in flight. Male has a loud, harsh trumpet—a rapid 'ark-ark, ark-ark'. Both sexes give low grunts, softer honks, and hiss in aggression or when threatened. **Hab.:** offshore islands with scrub or pasture; ocean beaches, headlands, margins of lakes, swamps, farm pastures. **Status:** common within restricted range; locally dispersive or migratory. **Br.:** 354

Mute Swan *Cygnus olor* 1.3–1.6 m

Introduced swan; larger and more aggressive than the Black Swan. Entirely white except for orange bill, black lores and the knob over the base of the bill. The flight is powerful with audible sound of wind across the wings. The Mute Swan was first introduced into Aust. in 1886 and was subsequently released at many locations until about 1920. Some colonies still survive, including a well-known colony on the Avon R. at Northam, WA. **Juv.:** head and neck brownish grey; rest of upper parts mottled grey-brown. Underparts light grey-brown; underwing white. **Imm.:** white dulled by overall grey-brown mottling; bill grey. **Voice:** usually silent but can give soft, hoarse trumpeting in breeding site displays, and hisses of threat and defence. **Hab.:** lakes, rivers, dams, usually where release occurred. **Status:** introduced; localised; sedentary.

Black Swan *Cygnus atratus* 1.1–1.4 m

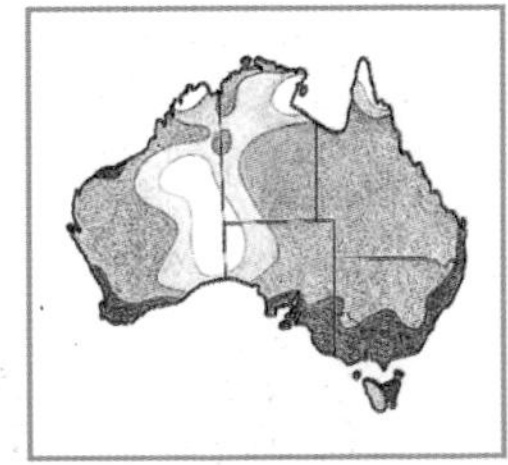

One of Australia's best-known birds. Easily identified; the only swan in the world that is so predominantly black. Most of white on wings is revealed only in flight, when the crisp, sharp white stands out against sky or other dark background colour, drawing attention to the powerful rhythmic beat of the wings. A long, splashing, frantically flapping take-off run is needed to get the heavy body airborne. Probably the only occasion a Black Swan could present an identification difficulty would be at a great distance with the bird in silhouette against sun, sunrise or sunset. Feeds on underwater vegetation, up-ending to reach deeper. **Juv.:** grey, mottled if seen close; flight feathers dull white, tipped black. **Voice:** a musical, clear bugling, carries far; probably a contact call given in flight and on the water. Loud hissing when defending the nest. **Similar:** Magpie Goose, but black flights, white body and, if all black in silhouette, a much shorter neck. Both Brolga and Black-necked Stork in distant silhouette, but, without plumage colour or pattern visible, should show long trailing legs, long bill. Any egret or heron flying with neck folded back would be obviously different in silhouette. **Hab.:** large areas of shallow water with aquatic vegetation; lakes, estuaries, flooded pastures. **Status:** common; nomadic. **Br.:** 354

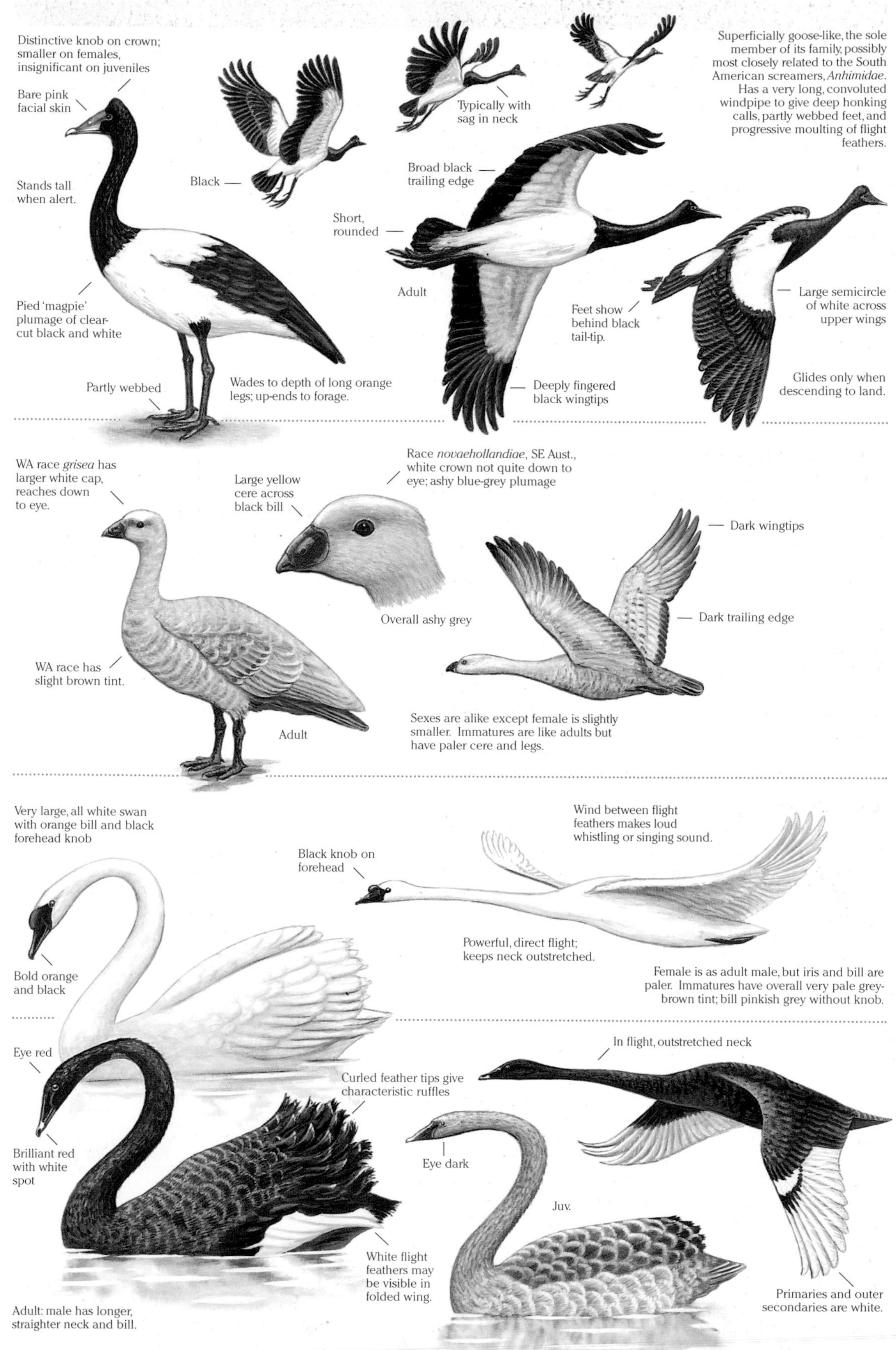

Distinctive knob on crown; smaller on females, insignificant on juveniles
Bare pink facial skin
Stands tall when alert.
Pied 'magpie' plumage of clear-cut black and white
Partly webbed
Wades to depth of long orange legs; up-ends to forage.
Black
Typically with sag in neck
Broad black trailing edge
Short, rounded
Adult
Feet show behind black tail-tip.
Deeply fingered black wingtips
Superficially goose-like, the sole member of its family, possibly most closely related to the South American screamers, *Anhimidae*. Has a very long, convoluted windpipe to give deep honking calls, partly webbed feet, and progressive moulting of flight feathers.
Large semicircle of white across upper wings
Glides only when descending to land.
WA race *grisea* has larger white cap, reaches down to eye.
Large yellow cere across black bill
Race *novaehollandiae*, SE Aust., white crown not quite down to eye; ashy blue-grey plumage
Overall ashy grey
Dark wingtips
Dark trailing edge
WA race has slight brown tint.
Adult
Sexes are alike except female is slightly smaller. Immatures are like adults but have paler cere and legs.
Very large, all white swan with orange bill and black forehead knob
Black knob on forehead
Wind between flight feathers makes loud whistling or singing sound.
Bold orange and black
Powerful, direct flight; keeps neck outstretched.
Female is as adult male, but iris and bill are paler. Immatures have overall very pale grey-brown tint; bill pinkish grey without knob.
Eye red
In flight, outstretched neck
Curled feather tips give characteristic ruffles
Brilliant red with white spot
Eye dark
Juv.
White flight feathers may be visible in folded wing.
Adult: male has longer, straighter neck and bill.
Primaries and outer secondaries are white.

Plumed Whistling-Duck *Dendrocygna eytoni* 40–60 cm
A goose-like duck with tall, upright stance and long, upswept plumes rising from the flanks. Grazes mostly at night on short vegetation of plains and margins of wetlands; rests in flocks by water most of the day. Will fly up to 30 km from water to graze. **Fem.:** slightly smaller, plumes tend to be shorter. **Imm.:** like adult, but slightly paler, muted colours, indistinct markings. **Voice:** a shrill whistling, given as a single sharp note, or a high, squeaky, wheezy 'tzwit-tzwit-tzwit-tzwit…'. Roosting flocks give an almost continuous 'jizzing-chittering' intermixed with high whistles; also from flocks when alarmed and in flight, and often heard at night. At the nest, soft whistles between pair. A different whistling in flight from wind across wings. **Similar:** Wandering Whistling-Duck, shorter plumes, lower stance. **Hab.:** grasslands not far from water; near swamps, lakes, floodwaters, dams. **Status:** common; migratory or nomadic to regions of optimum conditions. **Br.:** 353

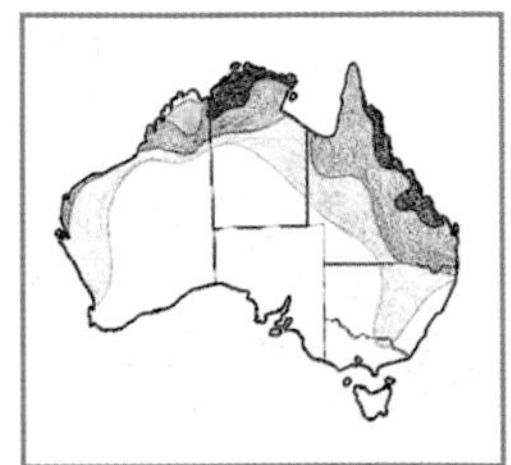

Wandering Whistling-Duck *Dendrocygna arcuata* 55–61 cm
Identifiable as a whistling-duck by its twittering, whistling voice, dense flocking, lowered head in flight. Differs from the Plumed Whistling-Duck in its far greater use of water, where it always feeds; swims and dives for aquatic plants, mostly at night. Often flies or swims in huge close-packed flocks. **Imm.:** colours paler, duller. **Voice:** a high, whistled, rather tremulous 'wit-wit-wit…', slightly slurred. From flocks, a confusion of whistling and twittering. **Similar:** Plumed Whistling-Duck. **Hab.:** in the dry season, wetlands with permanent water and aquatic vegetation such as river billabongs, floodplain pools and tidal creeks. Extends further afield in wet season with spreading floodwaters. **Status:** common, locally abundant; seasonally migratory or nomadic. **Br.:** 353

Spotted Whistling-Duck *Dendrocygna guttata* 43 cm
A rare vagrant to Australia's N coast; probably from NG where it is one of the most abundant ducks. **Imm.:** white streaks on flanks. **Voice:** differs from resident Australian whistling-ducks; not as high or twittering. Usually a nasal 'wheeow' or 'zziow', which may continue as a sequence, 'whu-wheow-whee' repeatedly. Flocks give a varied confusion of sound; most excitedly if put to flight. **Similar:** Wandering Whistling-Duck; on water is like immature of the Spotted, but in flight shows longer, more drooped neck. **Hab.:** in NG confined to the lowlands, rivers, margins of freshwater swamps, lakes and marshes. **Status:** widespread, Philippines through NG to E Indonesia. First record for Aust. at Weipa Sewerage Farm, Mar.–Apr. 1995 and Dec. 1995, when 39 birds were reported. Flocks have more recently been sighted in coastal swamps of E Cape York, where the species may now breed.

Blue-billed Duck *Oxyura australis* 36–44 cm
A small compact duck. Male is deep chestnut with blue 'scooped' bill. Performs elaborate courtship displays, bobbing with bill dipping into water. Tends to be shy, secretive; if disturbed may dive to depart under water rather than fly. Well adapted to aquatic life, rarely coming on to land, where it walks slowly, clumsily. Can fly strongly, though take off appears laboured. **Voice:** usually silent, but in display male makes a soft, throbbing, increasingly rapid 'dunk, dunk, dunk-dunk-dunk…' sound, and a shrill but soft 'chi-chi-chi-…'. Female, rebuffing male, soft 'tet-tet-tet…'. **Hab.:** breeding, deep, permanent, densely vegetated freshwater lakes, swamps and dams; winters on more open waters. **Status:** uncommon; sedentary. **Br.:** 354

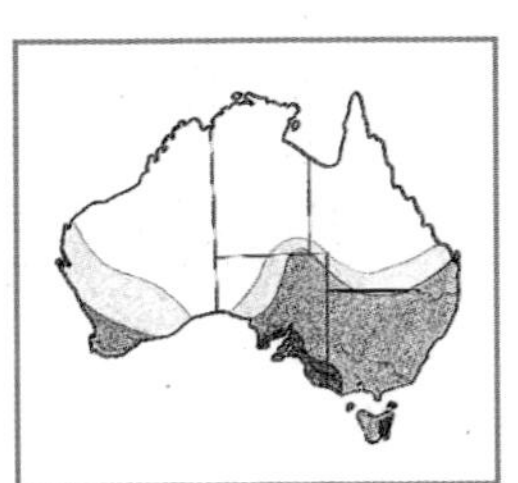

Musk Duck *Biziura lobata* male 60–70 cm; female 47–55 cm
A dark, low-floating duck with broad bill; male has a large leathery lobe beneath the bill. Attracts attention through breeding season with lengthy periods of display. Almost invariably seen on water, to which it is highly adapted, submerging with scarcely a ripple to feed or escape. Rarely seen on land, but occasionally emerges; waddles clumsily. In flight has rapid, shallow wingbeats; can travel swiftly and far, though take offs appear laborious and landings are clumsy. **Voice:** the male in display gives a deep grunt and loud, piercing whistle. At the same instant the drake's feet, or perhaps its wings, make a jetting splash with a deep 'plonk' sound. **Similar:** female and immature are like those of Blue-billed and Freckled Ducks. **Hab.:** deep permanent lakes, swamps and dams having areas of dense reedbeds and open waters. **Status:** sedentary; common. **Br.:** 354

Mid to light cinnamon-olive crown, nape and back of neck
Bill pink, mottled black on upper surface
Head low in flight
Pale
Tall, upright posture; plumage in warm cinnamon and rufous tones
Arising from flanks are unique long, lanceolate, upswept plumes; the tallest rise above the level of the back.
Slight crest
Barred
Feet trail behind short tail.
Plumes
Pink, webbed
Sexes alike, but female tends to have slightly shorter plumes. Immatures are slightly paler, duller, with faint barring on breast.
Black
Rich chestnut
Usually holds body lower, neck more curved than does the tall, erect Plumed Whistling-Duck.
Humpbacked, low necked
Plumes
Scalloped brown and deep rufous
Dark back of neck
Feet trail.
Plumes
Black
Sexes similar; juveniles are dull with pale centre to abdomen.
Slight crest
Broad, dark wings
All dark
Grey face
A NG species; overall darker, shorter necked, especially in flight.
White
Spotted flanks
Dark, edged rufous
Chestnut, spotted white
Paler, dull colours, indistinct spotting, white streaks on flanks
Feet trail behind tail.
Adult
Juv.
Adult
Neck is shorter, less drooped in flight than Aust. whistling-ducks.
Rounded, black
Alert posture, head high, tail cocked
Deep chestnut with bright highlights
Blue
Raised, fanned in display or alarm
♂
♂
♀
'Dished' bill profile
Typically head low, tail on water
Compact 'stiff-tailed' duck; usually floats low with tail trailing under water.
♀
Male in nonbreeding plumage is like female; darker with blotchy, dark green bill.
Runs across water to take flight; wingbeats rapid, shallow; flight fast and direct.
Almost straight from side
Stiff, triangular tail is trailed in water or raised forward over back.
Entirely dark grey-brown above, fine bars visible only in close view
Leathery lobe
♂
♂
Female smaller, often paler, no bill lobe
Rarely seen in flight; occasional short flights low over water; rapid shallow wingbeats; clumsy splashdown
♀
Dark, short necked, low-floating duck with heavy broad bill. A large, pendulous leathery lobe hangs from bill to water, expanded in display.
♀

Australian Wood Duck *Chenonetta jubata* 45–60 cm

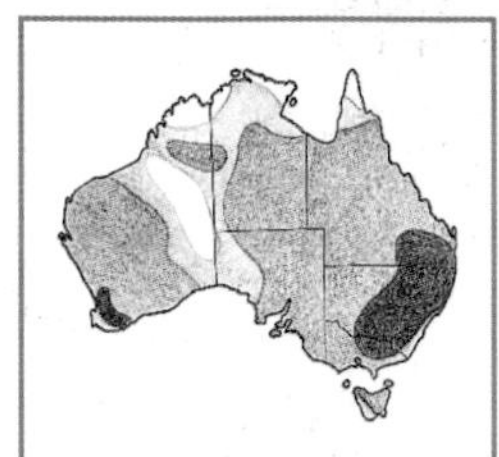

Must be seen at close range for its delicate beauty to be appreciated. Has gained from clearing, provision of dams and pastures. Water provides sanctuary and a safe camp for the flock, but the pasture nearby is vital, providing most of the grazing. Also dabbles and up-ends in the shallows as it searches for aquatic plants. Flight swift, low; small parties swerve among trees. **Voice:** female a drawn out, querulous, nasal 'grouwwk', beginning low, rising. Male a similar, but higher, more abrupt 'nowk!'. From flock, rapid clucking, louder when agitated. **Hab.:** open woodlands, farmlands, flooded pastures; vicinity of dams, lakes, estuaries, ponds. **Status:** abundant, sedentary or dispersive. **Br.:** 355

Pink-eared Duck *Malacorhynchus membranaceus* 38–45 cm

Usually in small parties to large flocks; keeps to water or loafs on logs and limbs rising from water; seldom on land. Groups work through shallows, filtering muddy water and soft bottom sediments for microscopic aquatic plants and animals, often working in stirred-up water from the bird ahead. More easily approached than most ducks. If disturbed, usually do not fly far, the flock circling to land again nearby. **Imm.:** as adult but paler. **Voice:** flocks noisy in flight and on the water with rapid, whistled twittering; also a sharp 'ti-wit, ti-wit, ti-wit' alarm call and a drawn out 'wheeii-ooo' in display. **Hab.:** shallow, open, muddy wetlands and temporary floodwaters. **Status:** widespread, nomadic; common. **Br.:** 355

Grey Teal *Anas gracilis* 42–45 cm

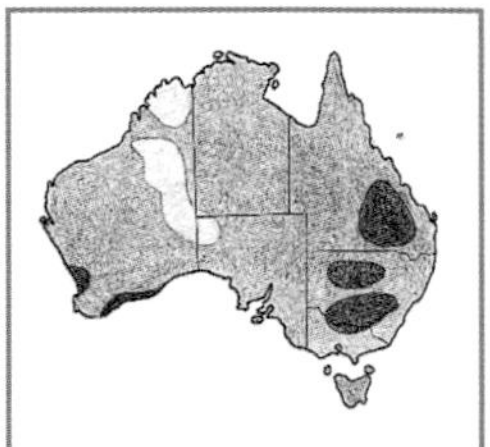

In pairs, small to huge flocks; swims buoyantly, head often high emphasising slender-necked appearance. Takes flight with explosive upward leap and splash; rises swiftly. Flight fast; shows white centrally on under and upper wings. Mainly aquatic; feeds among floating aquatic vegetation, usually in shallows. Highly mobile and opportunistic; quickly reaches temporary floodwaters on distant inland plains. **Voice:** male gives a sharp whistle with soft low grunt and loud, whistled 'gedg-ee-oo'. Female has a loud, chuckled, descending series of quacks; also a slow, harsh, drawn out 'que-aark'. **Hab.:** varied, almost any wetlands, uses temporary floodwaters of interior. **Status:** abundant; nomadic and dispersive. **Br.:** 355

Chestnut Teal *Anas castanea* 38–48 cm

Feeds mostly at dusk and dawn, dabbling and up-ending in the shallows; small flocks loaf much of the day beside the water. **Voice:** almost indistinguishable from Grey Teal except for minor differences of pitch; male's slightly deeper. **Similar:** Grey Teal most like female and juvenile of Chestnut, but latter have slightly darker, warmer tone overall with heavier mottling. **Hab.:** wetlands; prefers salt and brackish coastal estuaries, lakes, salt marshes, tidal mudflats and coastal islands. **Status:** abundant Tas., common elsewhere in SE and extreme SW of WA; usually sedentary, but some wander far N and well inland. **Br.:** 355

Garganey *Anas querquedula* 37–41 cm

A distinctive teal-sized duck; the male has a long, downcurving, white eyebrow line. Gregarious; individuals or pairs are usually in company with other ducks. Aquatic, rarely further onto land than waterline. Dabbles and up-ends to feed in shallow water. **Nonbr.:** like female, but coarser streaking on head, whiter throat, grey on shoulders. **Imm.:** like female, paler. **Voice:** male has a distinctive strident rattle, but unlikely to be heard from birds wintering in Aust. as it is given during courtship. **Similar:** female Pacific Black Duck most like female and nonbreeding Garganey; has bolder facial stripes and much larger. Grey Teal, but lacks facial stripes. **Hab.:** freshwater wetlands, swamps, shallow lakes, flooded grasslands, floodplains, sewage farms. **Status:** uncommon annual migrant to coastal N Aust.; rare vagrant further S.

Northern Pintail *Anas acuta* 50–65 cm including tail

Large, slender dabbling duck; male in breeding plumage is distinctive. Swift flight with rapid wingbeats. Floats high, neck straight and tall when alert, up-ending rather than diving when feeding. **Voice:** usually silent; male has a nasal warning call, female a rather hoarse quack or croak. **Similar:** female superficially like females and juveniles of local *Anas* species. **Hab.:** in native range, shallow open wetlands and nearby pastures, lakes, estuaries. **Status:** highly migratory species that, in Asia, winters S to Burma; vagrant to Indonesia and NG. In Aust., recorded NE of NSW and SW of WA.

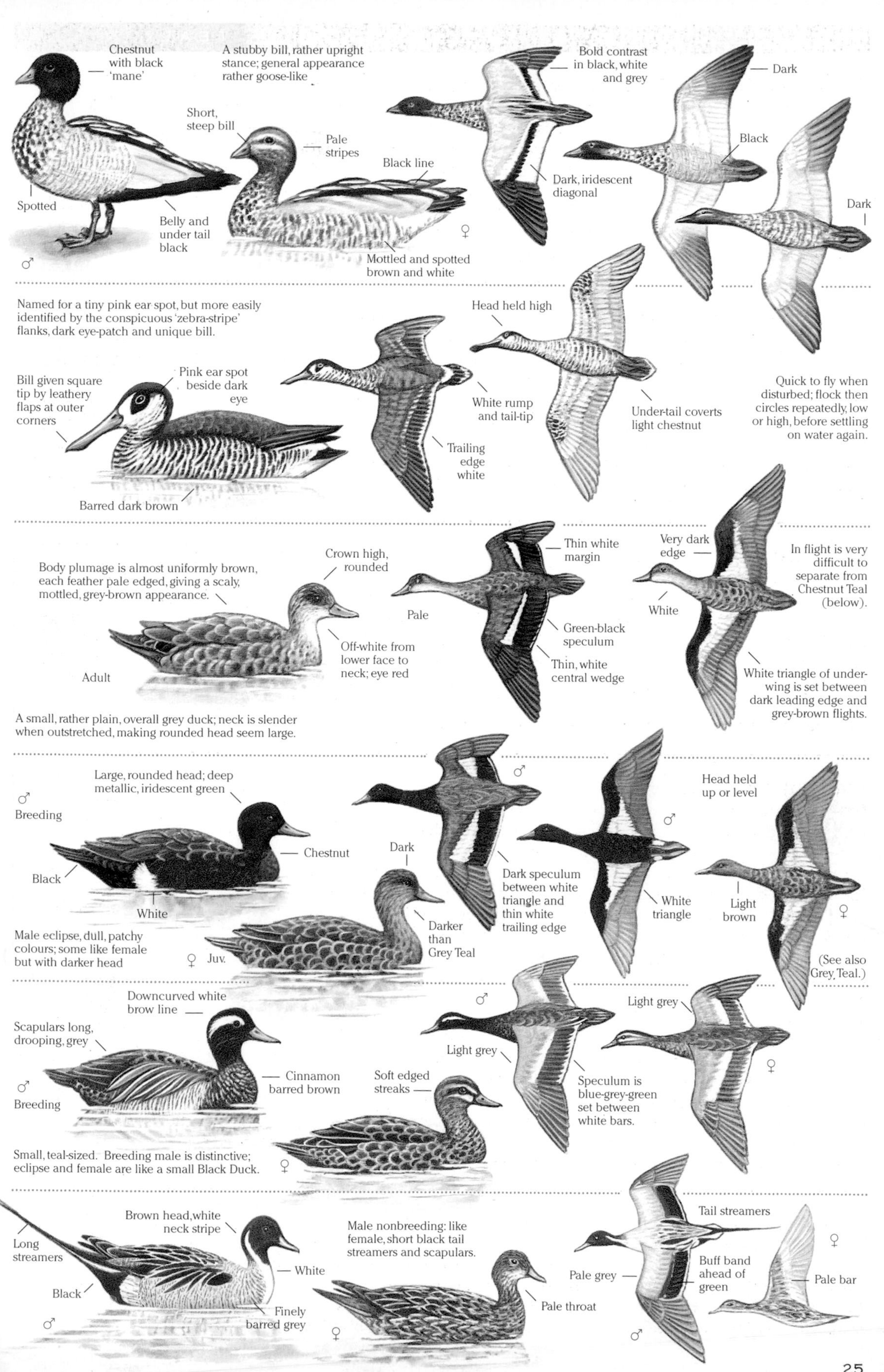

Chestnut with black 'mane'
Spotted
Belly and under tail black
♂
A stubby bill, rather upright stance; general appearance rather goose-like
Short, steep bill
Pale stripes
Black line
Mottled and spotted brown and white
♀
Bold contrast in black, white and grey
Dark, iridescent diagonal
Dark
Black
Dark
Named for a tiny pink ear spot, but more easily identified by the conspicuous 'zebra-stripe' flanks, dark eye-patch and unique bill.
Bill given square tip by leathery flaps at outer corners
Pink ear spot beside dark eye
Barred dark brown
White rump and tail-tip
Trailing edge white
Head held high
Under-tail coverts light chestnut
Quick to fly when disturbed; flock then circles repeatedly, low or high, before settling on water again.
Body plumage is almost uniformly brown, each feather pale edged, giving a scaly, mottled, grey-brown appearance.
Crown high, rounded
Off-white from lower face to neck; eye red
Adult
A small, rather plain, overall grey duck; neck is slender when outstretched, making rounded head seem large.
Pale
Thin white margin
Green-black speculum
Thin, white central wedge
Very dark edge
White
In flight is very difficult to separate from Chestnut Teal (below).
White triangle of underwing is set between dark leading edge and grey-brown flights.
♂
Breeding
Large, rounded head; deep metallic, iridescent green
Chestnut
Black
White
Male eclipse, dull, patchy colours; some like female but with darker head
Dark
Darker than Grey Teal
♀ Juv.
♂
Dark speculum between white triangle and thin white trailing edge
♂
White triangle
Head held up or level
Light brown
♀
(See also Grey Teal.)
Downcurved white brow line
Scapulars long, drooping, grey
Cinnamon barred brown
♂
Breeding
Soft edged streaks
Small, teal-sized. Breeding male is distinctive; eclipse and female are like a small Black Duck.
♀
♂
Light grey
Speculum is blue-grey-green set between white bars.
Light grey
♀
Long streamers
Brown head, white neck stripe
White
Black
Finely barred grey
♂
Male nonbreeding: like female, short black tail streamers and scapulars.
Pale throat
♀
Tail streamers
Pale grey
Buff band ahead of green
♂
♀
Pale bar

Australian Shelduck *Tadorna tadornoides* 56–73 cm

An impressively large and richly coloured shelduck, the female even more so than the male. Dominantly terrestrial in habits: swims frequently, preens and mates on the water; but feeds on pasture and, to a lesser extent, aquatic vegetation obtained in shallow wetlands. Very alert, quick to take flight. When travelling, flocks may form either long lines or 'V' formation. **Nonbr.:** breast of male becomes paler, white collar less distinct. **Imm.:** as adults but paler, duller plumage. **Voice:** loud calls, often in flight. Male gives a deep, harsh, nasal, buzzing grunt or honk; female has a higher call, though with similar nasal tone, 'ank-aank' or 'an-ganker'. **Hab.:** congregates in large flocks, sometimes a thousand birds or more, on extensive fresh, brackish or saline expanses of water such as estuaries and large lakes. In breeding season pairs disperse, many moving inland to nesting territories centred on a swamp, lake, stream or dam to which the ducklings can be taken after leaving the nest. **Status:** locally common in SE and SW Aust., otherwise uncommon. **Br.:** 354

Radjah Shelduck *Tadorna radjah* 49–61 cm

A large shelduck; dumpy, head low unless alert, appears black and white at distance unless strong light reveals dark areas as being deep chestnut. Usually in pairs or flocks; feeds around margins of wetlands and in shallows, resting on banks or mudflats, or perched on limbs over water. Flight swift and powerful; dashes fast and low between trees. **Voice:** very vocal, calling in flight and on water or land. Male has a rattling whistle; female a deeper, slower, more harsh rattling. **Hab.:** in tropical dry season flocks congregate on coastal waters, mangrove-lined river channels, tidal mudflats and beaches, or remain inland on permanent lagoons and pools along major rivers and where deeper swamps persist on river floodplains. In wet season moves from littoral habitat to the shallow margins of the expanding wetlands. **Status:** common in optimum habitat areas of NT, scarce to rare elsewhere; sedentary with local seasonal movements. **Br.:** 354

Freckled Duck *Stictonetta naevosa* 50–59 cm

A dark duck with peaked crown and scooped bill. In flight, rapid wingbeats, head low on outstretched neck, rather humpbacked. Feeds by dabbling in shallows. **Voice:** silent in flight. Both sexes give a low, resonant, querulous groan, often becoming drawn-out and peevish. Male gives an 'axe-grind' sound, a low buzz, ending in a sharp squeak, soft and not audible at a distance. Female has a throaty chuckle, louder defending nest or ducklings. **Similar:** Blue-billed Duck, female and eclipse, but usually swims with tail in water; different bill and crown contours. **Hab.:** breeds on densely vegetated freshwater lakes, swamps, creeks and floodwaters with thickets of melaleuca, casuarina, leptospermum, or in the interior with canegrass or lignum. After breeding moves to open waters. **Status:** rare, possibly endangered; nomadic. **Br.:** 354

Hardhead *Aythya australis* 45–60 cm

Australia's only representative of the true diving ducks that feed by deep diving. Steep take off; swift, direct flight with rapid wingbeats and whirring sound. **Voice:** male gives a soft, wheezy whistle and 'whirr'; female a loud, harsh rattle, the first note loudest, then faster and trailing away, 'gaaak,gaak,gak,gak-gak-gakgak'. **Similar:** Blue-billed Duck if females of both species are seen on the water with differences of underparts not visible; Blue-billed holds its stiff tail flat on the water. **Hab.:** prefers large, deep lakes and swamps with abundant aquatic vegetation, but also on smaller creeks, flooded crops, shallow floodplain pools; only rarely on coastal lagoons, mangrove swamps and salt lakes. **Status:** common in SE and SW Aust., nomadic or dispersive to areas of heavy rains across most of interior and far N. **Br.:** 355

Pacific Black Duck *Anas superciliosa* 48–60 cm

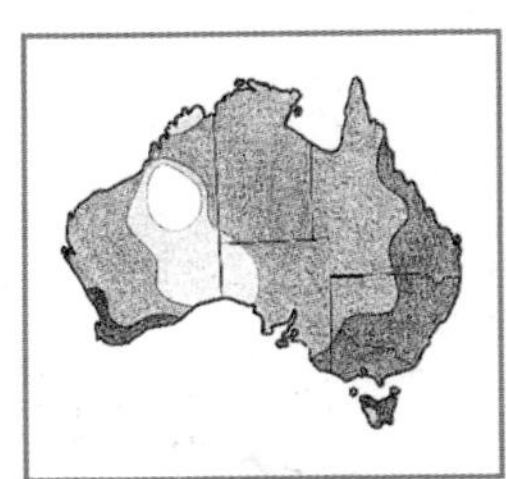

Well-known; common both in public parks and in the wild. **Voice:** male, a quick 'rhaab-rhaab' of varying strength; in warning a loud, extended 'rhaaaeeb'; in display a high pitched whistle immediately followed by a deep grunt. Female, typical loud duck quackings, single quacks of varying strength and a long sequence, descending and fading, 'quaak, quaak, quak, quak-quakquak'. **Similar:** Mallard, female, but single paler stripe through eye, orange on bill and legs. **Hab.:** almost any and every wetland habitat throughout Aust. from small freshwater ponds to sheltered marine waters; preference is for shallow, well-vegetated swamps. **Status:** common; sedentary and nomadic. **Br.:** 355

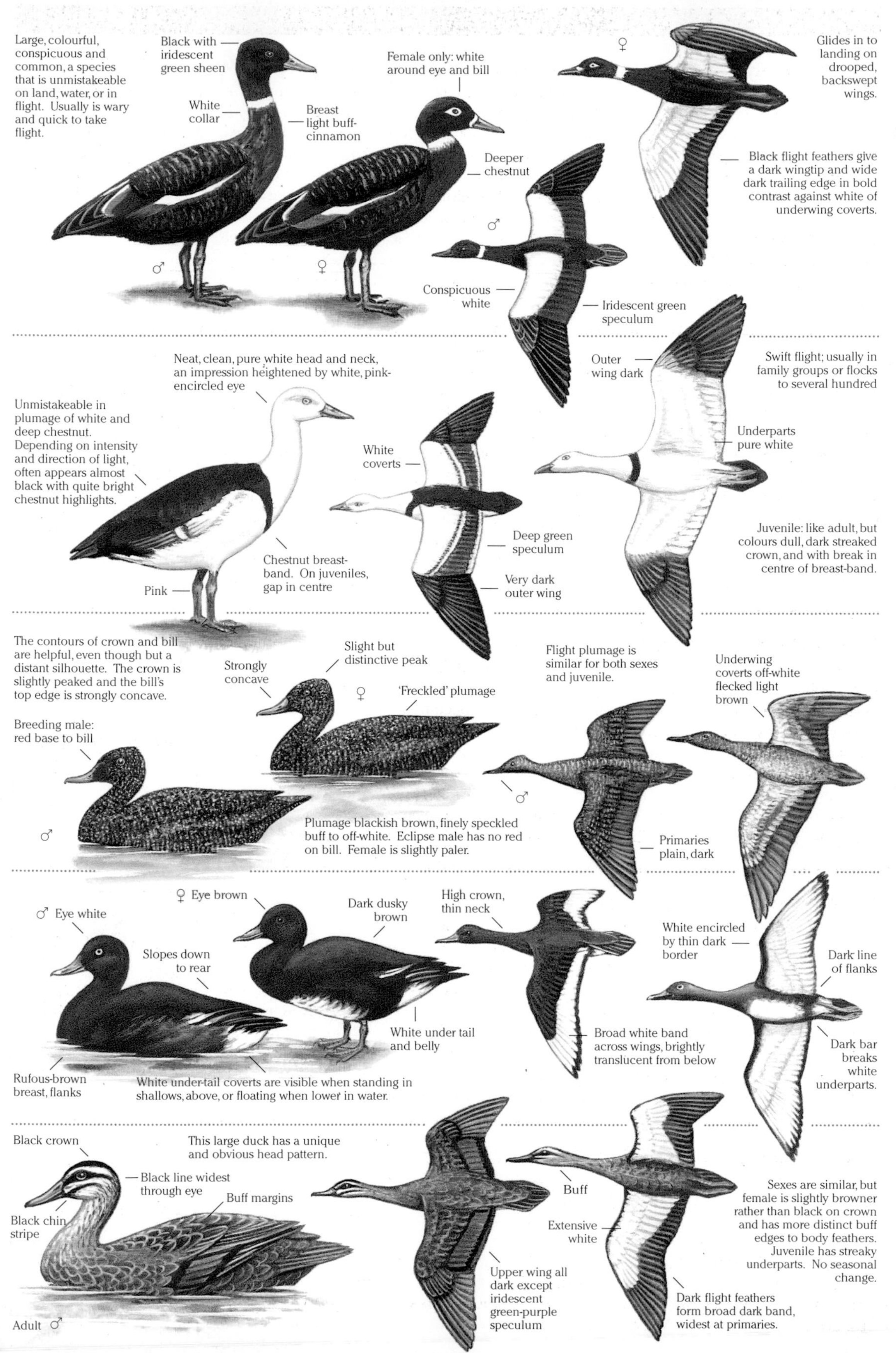
Large, colourful, conspicuous and common, a species that is unmistakeable on land, water, or in flight. Usually is wary and quick to take flight.
Black with iridescent green sheen
White collar
Breast light buff-cinnamon
Female only: white around eye and bill
Deeper chestnut
♂
♀
♀
Glides in to landing on drooped, backswept wings.
Black flight feathers give a dark wingtip and wide dark trailing edge in bold contrast against white of underwing coverts.
♂
Conspicuous white
Iridescent green speculum
Neat, clean, pure white head and neck, an impression heightened by white, pink-encircled eye
Unmistakeable in plumage of white and deep chestnut. Depending on intensity and direction of light, often appears almost black with quite bright chestnut highlights.
Pink
Chestnut breast-band. On juveniles, gap in centre
White coverts
Deep green speculum
Very dark outer wing
Outer wing dark
Swift flight; usually in family groups or flocks to several hundred
Underparts pure white
Juvenile: like adult, but colours dull, dark streaked crown, and with break in centre of breast-band.
The contours of crown and bill are helpful, even though but a distant silhouette. The crown is slightly peaked and the bill's top edge is strongly concave.
Strongly concave
Slight but distinctive peak
♀
'Freckled' plumage
Breeding male: red base to bill
♂
Plumage blackish brown, finely speckled buff to off-white. Eclipse male has no red on bill. Female is slightly paler.
Flight plumage is similar for both sexes and juvenile.
Underwing coverts off-white flecked light brown
♂
Primaries plain, dark
♂ Eye white
♀ Eye brown
Slopes down to rear
Dark dusky brown
White under tail and belly
Rufous-brown breast, flanks
White under-tail coverts are visible when standing in shallows, above, or floating when lower in water.
High crown, thin neck
Broad white band across wings, brightly translucent from below
White encircled by thin dark border
Dark line of flanks
Dark bar breaks white underparts.
Black crown
This large duck has a unique and obvious head pattern.
Black line widest through eye
Buff margins
Black chin stripe
Adult ♂
Upper wing all dark except iridescent green-purple speculum
Buff
Extensive white
Dark flight feathers form broad dark band, widest at primaries.
Sexes are similar, but female is slightly browner rather than black on crown and has more distinct buff edges to body feathers. Juvenile has streaky underparts. No seasonal change.

Green Pygmy-Goose *Nettapus pulchellus* 30–36 cm

A tiny iridescent goose-like duck; swims and dives strongly; rarely on land or perch; reluctant to fly, but then is swift in flight. **Nonbr.:** like breeding male, but white cheek patches are blotched brown. **Juv.:** like adult female, but green dull. **Voice:** male has a sharp, high, whistled contact call, given almost continuously in flight or on the water, fluctuating in pitch, strength and rapidity, 'chi-wip, chi-wip, chiwip…'; from the female, a whistle of declining pitch and intensity, 'phee-oo'. On coming in to land, a deeper, softer 'kuk-ka-kadu'. Alarm call from both sexes is a sharp, whistled 'whit!' or 'whit-whit'. **Similar:** Cotton Pygmy-Goose, some likeness between juveniles. **Hab.:** deep permanent freshwater lakes, dams and lagoons with abundant aquatic vegetation, especially waterlilies; birds move into floodplain swamps as they fill in the wet season. **Status:** locally common; sedentary and locally nomadic. **Br.:** 355

Cotton Pygmy-Goose *Nettapus coromandelianus* 34–38 cm

A miniature, goose-like duck; almost exclusively aquatic, rarely on land. Usually out on deeper water; perches on logs; flies readily. **Voice:** in flight male gives a continuous, rapid, high, nasal, chattering quack like a rapid, sharply metallic, nasal frog's croak, 'car-car-carwak'; a cacophony of sound when from many birds. The female has a much softer quacking. **Similar:** juvenile like that of Green Pygmy-Goose. **Hab.:** coastal wetlands, preferring deep permanent pools and swamps with abundant aquatic grasses; moves out to floodplain swamps and pools as these fill in the wet season. **Status:** uncommon or rare vagrant over most of range; sedentary with some local seasonal movements. **Br.:** 355

Mallard *Anas platyrhynchos* 50–70 cm

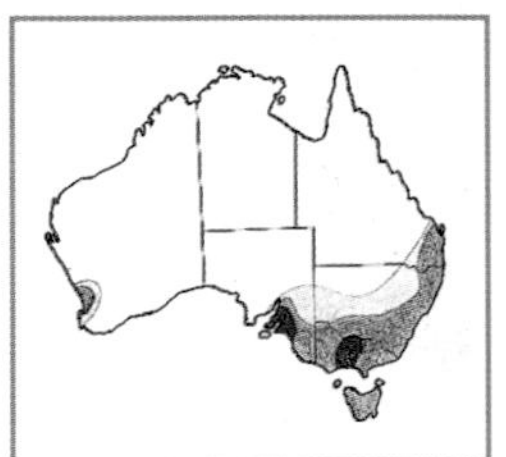

A boldly plumaged introduction from N hemisphere. Male unmistakeable; has white collar separating green head from chestnut of breast, and orange legs. Female has dark eye stripe, orange legs. Forages by dabbling and up-ending in shallows. Alert and wary, but often tame in public parks. **Voice:** probably indistinguishable from Pacific Black Duck. **Male:** usually a quick 'rhaab-rhaab'; in warning a loud 'rhaaaeeb'; in display a whistle and deep grunt. **Fem.:** typical duck quackings—loud single quacks and, in long sequences, a descending 'quaak, quaak, quak-quak-quak'. **Similar:** female and juvenile of Pacific Black Duck. **Hab.:** freshwater lakes, dams, ponds. **Status:** introduced; widespread and common around some towns and farms, but not rare in the wild. Hybridises with Pacific Black Duck.

Australasian Shoveler *Anas rhynchotis* 45–54 cm

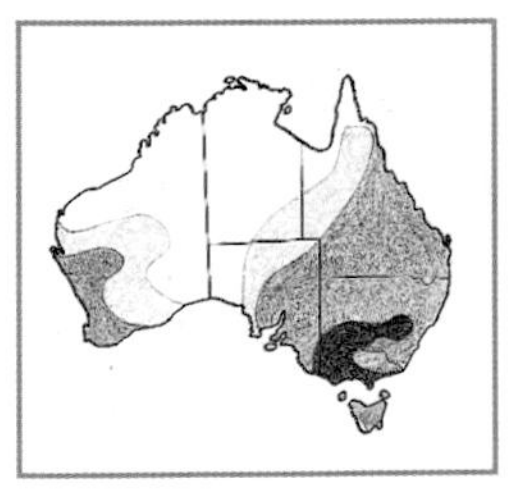

The massive bill is used to feed on small creatures that are filtered from the water through lamellae on the bill's fringe. Often in close-packed flocks; ducks swim in circles, head partly submerged, bill moving rapidly in dabbling action. A semi-nocturnal feeder; during the day often loafs or floats among other ducks far out on open water. Extremely wary; takes off steeply; flight swift and direct with a whirring sound that is unique among Australian ducks. **Voice:** usually silent, but courting male gives a soft, double noted 'took-ook…, took-ook…', given at widely spaced intervals, but sometimes rapidly. Female has a rapid sequence of soft but sharp quacks, becoming faster, lower and softer, 'quaaak…, quaak, quak, quak-quakquak.' **Similar:** Pink-eared Duck, but has zebra stripes, black eye-patch; Northern Shoveler female, but paler, outer edge of tail white. **Hab.:** a wide variety of wetlands with preference towards large permanent lakes or swamps that have abundant shrubbery cover of melaleuca and cumbungi. Only an infrequent user of mangrove swamps, floodwaters, sewage farms. **Status:** generally rare, but quite common around SW coast of WA; nomadic and dispersive. **Br.:** 355

Northern Shoveler *Anas clypeata* 45–55 cm

A large, heavily built duck with a huge bill; a filter-feeder found singly or in groups; holds head low, bill dabbling; rarely dives, but up-ends to reach bottom mud. Seldom leaves water, then waddles clumsily. Flight strong and swift, wings set well back. **Voice:** usually silent; calls weak. Male gives a soft hoarse 'took' or 'took-took, took-took', distinguishable from calls of Australasian Shoveler. **Similar:** Australasian Shoveler. **Hab.:** in Aust. recorded on freshwater billabongs, lagoons, swamps. **Status:** a N hemisphere species; migrates S to SE Asia, the Philippines and Borneo; a rare vagrant to Aust., very few records.

Large white cheek patch
Iridescent green highlights across greenish black; mostly black under dull light
Iridescent green
♂
♂
White panel across dark green-brown wings
♀
White brow
♀ White
Bright white panels in dark underwings
A tiny duck superficially like a miniature goose, largely due to the stubby goose-like bill
Flanks of both sexes are barred with dark greenish brown lines following the shapes of feathers.
Fast direct flight low over water. Highly aquatic; feeds and rests on water, rarely on land or perch.
White face and neck
Dark iridescent green, largely black in dull light
♂
Dark green cap
White neck with dark band
White
Brownish green
♂
White bar across width of wings
Thin white rear edge is also present on upper wing.
Small goose-like duck with white neck and dark breast-band; dark green back
Flanks of both sexes are white very finely barred with grey-buff.
♀
Juvenile, like female, has dull brown upper parts.
♀
Pale yellow
Dark green highlights over black
Large introduced duck, male distinctive. Female and hybrid between Mallard and black duck are more difficult to identify.
Mid brown eye stripe
Light grey-brown coverts
♂
Grey-brown, finely vermiculated
Brown with buff margins
♀
♂
Chestnut breast with white collar
Face pattern more like Black Duck
The Mallard is closely related to the Pacific Black Duck, and interbreeds; the hybrid is a paler version of the Black Duck.
Blue violet between thin white bars
Head dark blue-grey with vertical crescent-shaped facial streak
♂
Long, pointed, white streaked scapulars
Long bill blends to sloping forehead.
Coverts pale blue
Triangular white bar between coverts and iridescent green speculum
♀
Underwing same for both sexes
♂
White
White flank patch
Chestnut, dark spotted flanks
The shoveler is a richly plumaged duck that needs to be seen in sunlight to reveal its full intensity of colour; otherwise, in dull light, appears very dark.
Nonbreeding male: loses white flank patch; colours overall rather faded. Juvenile is like the female with paler, finely marked underparts.
♀
Head is deep green; usually appears black with green highlights.
White
Green
Female and nonbreeding male are similar to those of Australasian Shoveler, but with some white towards outer edges of tail, and orange along the lower bill. Head of nonbreeding male is darker than that of the female. Juvenile male is like adult female, but darker and with underparts streaked rather than spotted.
White breast
Bright chestnut flanks
White

Australasian Grebe *Tachybaptus novaehollandiae* 23–25 cm

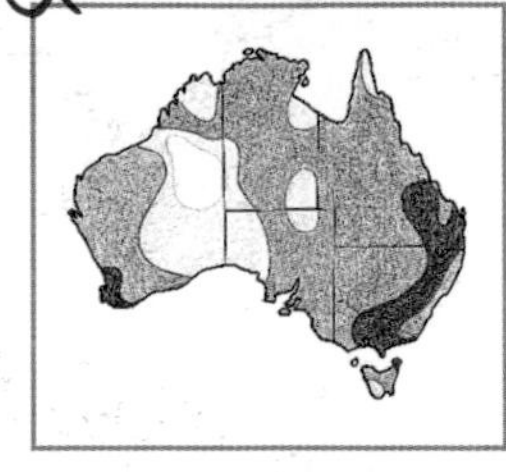

A common small waterbird of lakes, swamps and farm dams. Floats high and buoyantly, often with feathers fluffed out to give a rotund shape. Dives with forward plunge, headfirst; causes but a slight ripple on the water. Spends much of the day hunting small freshwater crayfish and other aquatic life at depths of 3 m or more; brings prey to surface, shaking it violently, probably breaking off the hard claws. Occasionally makes short pattering flights across the water surface on rapidly whirring wings, displaying white upper wing panels. **Voice:** noisy in breeding season and in defending territory against rivals. Both sexes give a fast, loud, high and metallic chittering; each burst of sound is continuous for 3–5 sec, vibrating rather than trilling. Also gives harsher versions in threat or aggression. **Similar:** Hoary-headed Grebe in nonbreeding and juvenile plumage, but dark cap of Hoary-headed Grebe extends lower to bottom of the eye. **Hab.:** greatly diverse wetlands—lakes, swamps, small farm dams—usually fresh water with abundant aquatic vegetation. Sometimes after breeding is found in large congregations on open waters, including saline estuaries. **Status:** widespread; common. **Br.:** 356

Little Grebe *Tachybaptus ruficollis* Rare vagrant to N coast: Darwin, Sep. 1999; possibly Derby, 1961.

Hoary-headed Grebe *Poliocephalus poliocephalus* 29–30 cm

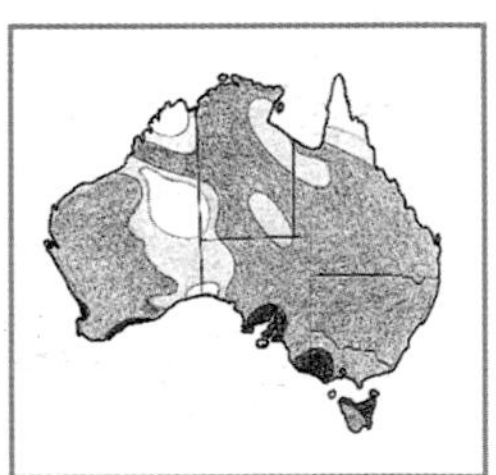

A small grebe with distinctively streaked head. If disturbed, will often fly away across the surface, a reaction different from that of the Australasian Grebe, which, when alarmed, dives, surfacing at a distance, screened by vegetation. Feeds almost entirely by diving; gives a little jump forwards and up for momentum to dive headfirst. Prey includes crustaceans such as small yabbies and jilgies, bugs and beetles, freshwater snails, dragonflies, spiders, small fish and some water plants. **Voice:** usually silent, gives very muted churring and low guttural sounds in courtship and at the nest, but audible only over a very short distance. Have been reported as giving a single 'hrow' sound when breeding. **Similar:** Australasian Grebe. **Hab.:** wetlands, both fresh water and estuarine; coastal and throughout the interior. Usually on large open areas of water, either permanent or temporary flooding. In SE and SW Aust. occurs on lakes, river pools, reservoirs, swamps and farm dams. Recorded coastal habitats include inlets and bays, lagoons, saltfields and mangrove swamps. In semi-arid regions, waterholes and river pools, dams and floodwaters. When it is breeding, prefers wetlands with abundant vegetation. **Status:** common or, in favourable conditions, locally abundant, but scarce in far N; nomadic and dispersive. **Br.:** 356

Great Crested Grebe *Podiceps cristatus* 47–61 cm

A large, long necked grebe; distinctive in breeding or other plumage. A specialised aquatic species, it never emerges onto land; it is sleek bodied with large-lobed feet set far back for optimum propulsion; an excellent swimmer both on the surface and under water. When threatened, vanishes with scarcely a ripple, travelling far under water to surface in the distance or behind screening reeds. Rarely seen flying, and then only in short, low flights across water. In flight it holds its neck outstretched, back humped, feet trailing. White panels on the rapid, shallow beating wings are obvious. Probably longer flights between lakes are made at night. Usually seen in pairs during the breeding season. Males vigorously defend nest territories from rivals, contesting any territorial trespass with aggressive displays, head lowered, crests and ruffs extended. Fights occur and can be quite vicious. Rivals may be pecked and held under water. Crested Grebes have elaborate courtship rituals that include a fascinating 'penguin-dance'. After breeding the grebes often gather in groups or large flocks on open areas of water. **Fem.:** as male, slightly smaller. **Nonbr.:** Aust. race *australis* retains breeding plumage throughout the year, although crests and ruffs become reduced; in N hemisphere the birds attain a dull winter plumage. **Imm.:** as adult, crest much smaller, no ruff; overall much paler, especially in winter. **Voice:** usually silent; calls when breeding and when nest is threatened. Calls include a throaty, gurgling, rolling 'quarr-r, quarr-r, quarr-r', a rapid, rattling, frog-croaking or barking 'rhag-rhag-rhag', a guttural, growling 'ghor-rr' and various other noises associated with display and territory. **Hab.:** prefers well-vegetated margins and reedbeds in channels near open waters of large lakes and reservoirs; also on sheltered estuaries and bays. Rarely seen on small stock dams or ponds. **Status:** uncommon and localised in SE and SW; irregular and scarce vagrant to N and semi-arid regions following flooding rains. **Br.:** 356

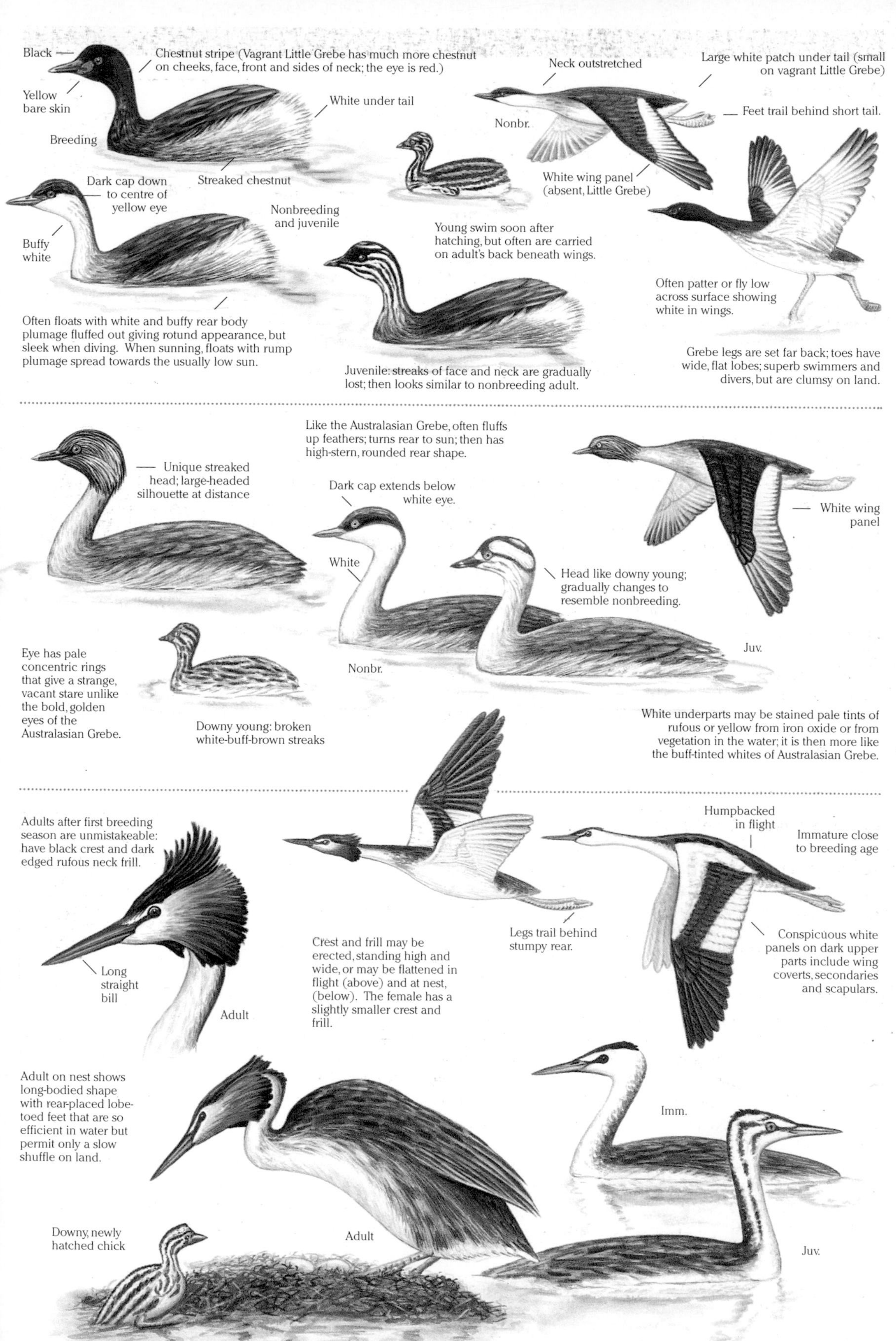
Black
Chestnut stripe (Vagrant Little Grebe has much more chestnut on cheeks, face, front and sides of neck; the eye is red.)
Yellow bare skin
White under tail
Breeding
Streaked chestnut
Dark cap down to centre of yellow eye
Buffy white
Nonbreeding and juvenile
Often floats with white and buffy rear body plumage fluffed out giving rotund appearance, but sleek when diving. When sunning, floats with rump plumage spread towards the usually low sun.
Neck outstretched
Large white patch under tail (small on vagrant Little Grebe)
Nonbr.
Feet trail behind short tail.
White wing panel (absent, Little Grebe)
Young swim soon after hatching, but often are carried on adult's back beneath wings.
Often patter or fly low across surface showing white in wings.
Juvenile: streaks of face and neck are gradually lost; then looks similar to nonbreeding adult.
Grebe legs are set far back; toes have wide, flat lobes; superb swimmers and divers, but are clumsy on land.
Like the Australasian Grebe, often fluffs up feathers; turns rear to sun; then has high-stern, rounded rear shape.
Unique streaked head; large-headed silhouette at distance
Dark cap extends below white eye.
White wing panel
White
Head like downy young; gradually changes to resemble nonbreeding.
Juv.
Nonbr.
Eye has pale concentric rings that give a strange, vacant stare unlike the bold, golden eyes of the Australasian Grebe.
Downy young: broken white-buff-brown streaks
White underparts may be stained pale tints of rufous or yellow from iron oxide or from vegetation in the water; it is then more like the buff-tinted whites of Australasian Grebe.
Adults after first breeding season are unmistakeable: have black crest and dark edged rufous neck frill.
Humpbacked in flight
Immature close to breeding age
Legs trail behind stumpy rear.
Conspicuous white panels on dark upper parts include wing coverts, secondaries and scapulars.
Long straight bill
Adult
Crest and frill may be erected, standing high and wide, or may be flattened in flight (above) and at nest, (below). The female has a slightly smaller crest and frill.
Adult on nest shows long-bodied shape with rear-placed lobe-toed feet that are so efficient in water but permit only a slow shuffle on land.
Imm.
Downy, newly hatched chick
Adult
Juv.

Macaroni (Royal) Penguin *Eudyptes chrysolophus* 65–75 cm (H,M)

A large penguin with massive red bill, black or white face, and long yellow crest plumes sweeping back and down. During breeding season feeds in seas around the colony, principal food being small krill taken in dives at depths of 20–100 m. After breeding disperses widely, on rare occasions being reported from Australia's S coasts. **Juv.:** lacks extensive crest of long plumes; brows are few fine yellow lines. **Imm.:** similar to adult, but shorter plumes; overall colours dull. **Variation:** the Royal Penguin with white face (formerly classified as a separate species *E. schlegeli*) is now regarded as being a form of the dark faced Macaroni Penguin. **Voice:** noisy with loud trumpetings of greeting and recognition at the nest. The raucous, rattling, braying 'kaaa-kaa-aargh' calls, coming from the huge number of birds that make up a colony, are a continuous, harsh, loud, vibrating buzz. **Hab.:** Antarctic and sub-Antarctic waters, islands. **Status:** abundant. **Br.:** in colonies, some with hundreds of thousands of pairs, on Antarctic islands.

Rockhopper Penguin *Eudyptes chrysocome* 45–60 cm

A small crested penguin with unique, strongly diverging plumes that begin behind bill and do not meet in centre. Often observed in small groups at sea; swims with head and part of back showing. At speed can clear the surface in porpoising leaps. On land, walks rather insecurely, or, moving faster, jumps along, both feet together as if tied. Also jumps feet first from rocks into sea rather than diving headfirst. **Fem.:** bill not quite as deep and heavy. **Variation:** 3 subspecies, of which *moseleyi* is most numerous in southern Australian waters. Race *filholi* is an occasional visitor to southern coasts; race *chrysocome* is a very rare straggler to the Australasian region. **Voice:** at sea, abrupt barks or croaks as contact calls; in colonies is noisy with loud, raucous braying. **Similar:** other crested penguins. **Hab.:** marine, Antarctic and sub-Antarctic waters; feeds in pelagic waters; breeding colonies are on rugged islands. **Status:** several races are moderately common stragglers to Aust. waters. **Br.:** in colonies on rocky islands of S oceans.

Fiordland Penguin *Eudyptes pachyrhynchus* 55–65 cm

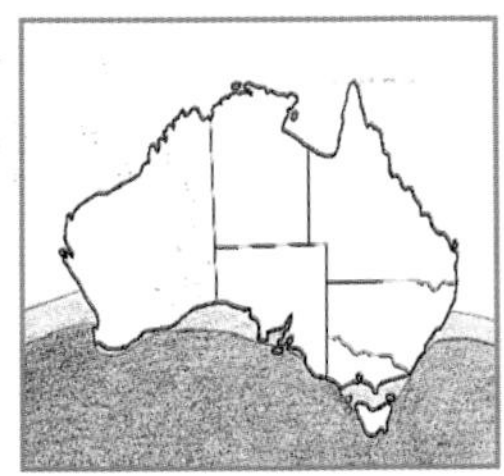

A medium size crested penguin similar to the Rockhopper; heavier bill without fleshy membrane at base; white bases to cheek feathers usually show as streaks. Breeds around S of NZ, then wanders W to Aust. coasts where most records have been of young birds coming ashore to moult. At sea, feeds in small groups or alone; porpoises from waves when travelling fast. Usually walk when on land, but hop with feet together when hurried. **Fem.:** bill not quite as deep and heavy. **Voice:** in breeding colonies, loud, harsh, deep; less calling at sea. **Similar:** Erect-crested and Snares Penguins. **Hab.:** when breeding, temperate rainforests and coastal waters of NZ; otherwise absent from coasts, presumably pelagic. **Status:** straggler to Aust. coasts, mostly one- to two-year-old immatures. Tas. to SW of WA.

Erect-crested Penguin *Eudyptes sclateri* 65–70 cm

Black headed with the densest and bushiest of all crests—golden, upswept unless wet. May porpoise above surface when swimming fast. On land, walks or uses two-legged hop to move more rapidly. **Voice:** rattling and shrill sounds interspersed with deep, throbbing grunts; calling is loud and persistent in breeding colonies, far more subdued when moulting on land, silent at sea. **Similar:** other crested penguins. Juveniles difficult to separate at sea. **Hab.:** when breeding, rocky islands and surrounding waters; otherwise open ocean, but little is known of distribution out of breeding season. **Status:** rare vagrant to coasts of Tas., Vic., SA and S coast of WA.

Snares Penguin *Eudyptes robustus* 50–60 cm (M)

Breeds on Snares Islands S of NZ. Little is known of its winter range away from the Snares, but immatures appear to disperse more widely than adults. Close to the Fiordland Penguin in appearance and these two are sometimes regarded as one species—but although the two intermingle at the Snares, hybrids are unknown. **Voice:** in colony, loud, harsh, ratchetty squawkings intermixed with barks, deep croaks, throbbing grunts. Noisiest early in breeding cycle; far less noisy when moulting or at sea. **Similar:** Fiordland Penguin, which lacks pink at base of bill, has white cheek stripes, different plume shape. **Hab.:** marine, pelagic waters. **Status:** recorded rarely from Aust. coasts, usually Tas., but recorded W to SA.

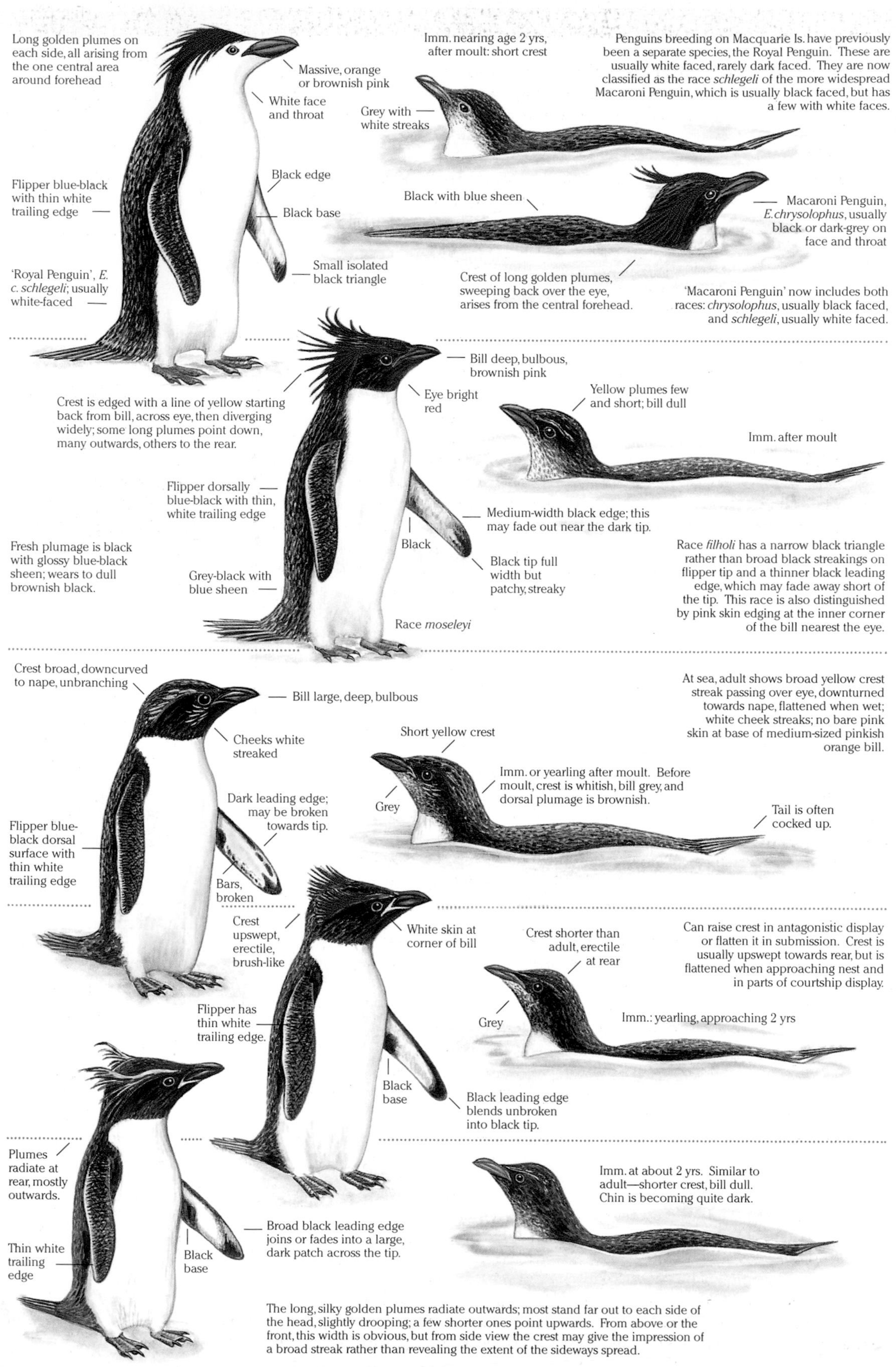

Long golden plumes on each side, all arising from the one central area around forehead
Massive, orange or brownish pink
White face and throat
Flipper blue-black with thin white trailing edge
Black edge
Black base
Small isolated black triangle
'Royal Penguin', *E. c. schlegeli*; usually white-faced
Imm. nearing age 2 yrs, after moult: short crest
Grey with white streaks
Penguins breeding on Macquarie Is. have previously been a separate species, the Royal Penguin. These are usually white faced, rarely dark faced. They are now classified as the race *schlegeli* of the more widespread Macaroni Penguin, which is usually black faced, but has a few with white faces.
Black with blue sheen
Macaroni Penguin, *E. chrysolophus*, usually black or dark-grey on face and throat
Crest of long golden plumes, sweeping back over the eye, arises from the central forehead.
'Macaroni Penguin' now includes both races: *chrysolophus*, usually black faced, and *schlegeli*, usually white faced.
Bill deep, bulbous, brownish pink
Eye bright red
Crest is edged with a line of yellow starting back from bill, across eye, then diverging widely; some long plumes point down, many outwards, others to the rear.
Yellow plumes few and short; bill dull
Imm. after moult
Flipper dorsally blue-black with thin, white trailing edge
Medium-width black edge; this may fade out near the dark tip.
Black
Black tip full width but patchy, streaky
Fresh plumage is black with glossy blue-black sheen; wears to dull brownish black.
Grey-black with blue sheen
Race *moseleyi*
Race *filholi* has a narrow black triangle rather than broad black streakings on flipper tip and a thinner black leading edge, which may fade away short of the tip. This race is also distinguished by pink skin edging at the inner corner of the bill nearest the eye.
Crest broad, downcurved to nape, unbranching
Bill large, deep, bulbous
Cheeks white streaked
At sea, adult shows broad yellow crest streak passing over eye, downturned towards nape, flattened when wet; white cheek streaks; no bare pink skin at base of medium-sized pinkish orange bill.
Short yellow crest
Imm. or yearling after moult. Before moult, crest is whitish, bill grey, and dorsal plumage is brownish.
Grey
Dark leading edge; may be broken towards tip.
Tail is often cocked up.
Flipper blue-black dorsal surface with thin white trailing edge
Bars, broken
Crest upswept, erectile, brush-like
White skin at corner of bill
Crest shorter than adult, erectile at rear
Can raise crest in antagonistic display or flatten it in submission. Crest is usually upswept towards rear, but is flattened when approaching nest and in parts of courtship display.
Grey
Imm.: yearling, approaching 2 yrs
Flipper has thin white trailing edge.
Black base
Black leading edge blends unbroken into black tip.
Plumes radiate at rear, mostly outwards.
Imm. at about 2 yrs. Similar to adult—shorter crest, bill dull. Chin is becoming quite dark.
Thin white trailing edge
Black base
Broad black leading edge joins or fades into a large, dark patch across the tip.
The long, silky golden plumes radiate outwards; most stand far out to each side of the head, slightly drooping; a few shorter ones point upwards. From above or the front, this width is obvious, but from side view the crest may give the impression of a broad streak rather than revealing the extent of the sideways spread.

King Penguin *Aptenodytes patagonicus* 85–95 cm (H,M)
Very tall; colourful; second largest of all penguins. Highly gregarious at island breeding colonies. At sea, solitary or in small groups. When 'porpoising', launches from waves and re-enters with a smooth, graceful action rather than the leaping and splashing return displayed by some smaller species. On land progresses with a waddle rather than a two-footed hopping action. **Voice:** in colonies and at sea both sexes give cooing contact calls that can be audible more than a kilometre away. Many other calls given on land as part of courtship and nesting behaviour. **Similar:** Emperor Penguin, but larger with ear patch more rounded, unlike the vertical oval of the King Penguin. **Hab.:** marine, pelagic. **Status:** vagrant to Aust. seas, Tas. to WA. Juveniles especially wander far. **Br.:** on islands of S Atlantic and Indian Oceans.

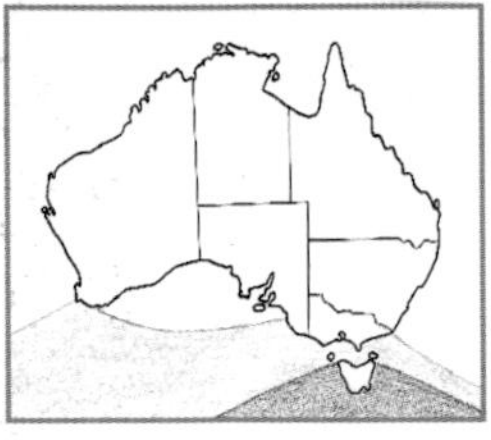

Gentoo Penguin *Pygoscelis papua* 70–80 cm
A medium sized penguin with a distinctive white triangle over each eye, joining narrowly across the crown in an hourglass shape from above. **Variation:** 2 subspecies, *papua* in Australasian waters, *ellsworthii* in seas between Antarctica and S America. **Voice:** calls at sea are unknown. In colonies has various loud trumpetings, harsh croaking, vibrant, harsh, throaty raspings, 'aargh', 'aa-aa-argh', and hissing sounds. **Hab.:** marine, often on Antarctic pack-ice; usually avoids continental coasts. **Status:** rare vagrant to Aust. coasts; very few records, all from Tas. Abundant. **Br.:** in colonies, Antarctic Peninsula and sub-Antarctic islands.

Adelie Penguin *Pygoscelis adeliae* 70 cm (H,M)
A highly gregarious penguin, both on land and at sea. Has displays used within the huge colonies. A slight crest gives a peaked contour to the crown and is used in some displays, including threat and courtship, when crest is raised, neck upstretched and flippers beaten. Conversely, in submissive behaviour, crest is flattened. **Voice:** contact call is a sharp barking given at sea and elsewhere away from the colony. Calls in display include thumping, rasping sounds. **Hab.:** seas in vicinity of pack-ice. **Status:** occurs Heard Is. and Macquarie Is.; rare vagrant to Tas., dispersing from huge Antarctic colonies. **Br.:** around coasts of Antarctica.

Chinstrap Penguin *Pygoscelis antarctica* 70–76 cm (H,M)
A distinctive black and white penguin that appears to have its black cap held on by a thin, black 'chinstrap' line. **Voice:** known calls are from nest site and include a loud, cackling 'arh-kauk-kauk', soft humming and hissing sounds. Calls at sea and on islands outside breeding season are not known. **Hab.:** when breeding, forages in shallower waters around islands; at other times, Antarctic seas, often around icefloes and among light pack-ice. **Status:** on rare occasions vagrants in Aust. waters; all records from Tas. Common in seas around breeding grounds. **Br.:** on Antarctic Peninsula and islands of Antarctic and sub-Antarctic seas.

Magellanic Penguin *Spheniscus magellanicus* 70 cm
Often in large parties when feeding at sea. In travelling fast may 'porpoise' above the waves; sometimes long lines of birds break from the water, diving forward. Although timid on land, shows no fear of humans in water; aggressive and can give a serious bite with the massive bill. **Voice:** in colony, loud staccato braying calls. **Hab.:** when breeding, coasts, islands and offshore waters near colony; otherwise oceanic. **Status:** abundant and apparently increasing within usual range; normally sedentary or short nomadic movements. Accidental to Aust.; one specimen from Phillip Is., Vic. **Br.:** coasts and islands around southern S America.

Little Penguin *Eudyptula minor* 35–45 cm (L)
The smallest of all penguins; also known as the Fairy Penguin. White undersurfaces blend softly into steely blue-grey upper parts without the sharp-edged contrast evident in the plumage pattern of other penguins. **Juv.:** as adult, smaller bill. **Variation:** other races around NZ coast. **Voice:** noisy ashore at night with sharp yapping, high trilling, a resonant 'quarr', vibrating, deep grunts and braying. Short, sharp yaps at sea. **Hab.:** usually inshore waters of mainland coast and islands. **Status:** common around S coast of Aust. **Br.:** 356

Distinctive inverted, teardrop shaped orange or deep yellow auricular patch is narrowly connected to similar tint on upper breast.
Orange or pinkish orange streak
Back blue-grey, silvery towards nape and shoulders
Wide dark edge tapers towards tip.
Large dark tip, but some entirely white
Adult
Pink to ivory
Pale lemon tint
Imm.
Adult
Swims head high, orange ear patch well above water.
Triangular white
Orange with black ridge
Has unique white triangle over each eye; these join narrowly across crown to form an hourglass shape when seen from above.
Flipper has broad white trailing margin, very thin white leading edge.
Thin black leading edge that often fades to a break midway down the flipper.
White triangle over eye is smaller than on adult, and usually not quite reaching down to the eye.
Black triangle
Conspicuous bright orange
Juv.
Adult
Erected crest makes pointed crown.
White eye-ring
Bill looks stubby, partly covered by feathers, dull red and black.
Medium sized penguin with white eye-ring, black hooded head and small, stubby, dark bill. Plumage of head at times forms a peaked ridge towards rear of crown. The shape of the crown profile, or crest, is an important component of displays, both in courtship and when defending vicinity of nest from other penguins or predators such as skuas.
Crest is often lowered, giving a sleek rounded crown contour to both adults and juveniles.
Imm.
Tapering dark edge
Blue-black
Small dark tip
Adult
Has appearance of having a black cap held on by a slender strap under the chin.
Dark grey
'Chinstrap'
On land has a waddling gait holding flippers out and slightly back, or toboggans on belly across ice or snow. Very aggressive; will deliberately attack intruders including humans and dogs.
Fine black leading edge
Imm.
Grey face
Fine white line along trailing edge
Pale grey tip
Bare pink skin around eye
White band encircling face
Brownish black; more brown in worn plumage
Black band on throat and upper breast
Adult
Flipper has thin white leading edge and broad white trailing edge.
Strangely and distinctively banded penguin with eye-ring of bare pink skin and deep, heavy, dark bill
White with scattered small black feathers
This is the species that is so well known as the 'Penguin Parade' tourist attraction at Phillip Is., Vic. It is common and breeds on other coastal islands from SW of WA to mid-NSW coast.
White with silvery sheen
Dark silvery blue-grey or with slight brownish tone in worn plumage
Lacks crest, bright colours or bold pattern, but this plain look is in itself unique, and together with the very small size, makes the Little Penguin easy to identify.
Adult
Light flesh pink

Southern Fulmar *Fulmarus glacialoides* 45–50 cm; span 115–120 cm (H, M)
Distinctive in southern oceans, where most petrels are dark. At sea a gregarious, quarrelsome scavenger; hangs about fishing boats, trawlers; otherwise rarely follows ships. Long periods of graceful wheeling and gliding, but settles on sea to feed, rest; glides low to snatch or make shallow plunges. Often active at night when small squid are near surface. **Juv.:** as adult, smaller bill. **Voice:** noisy at sea when quarrelling over food, and in nest colonies. Various fowl-like cluckings, croakings and short, sharp, harsh cacklings. **Similar:** White-headed and Grey Petrels. **Status:** rare visitor to S coasts. **Br.:** Antarctic mainland and islands.

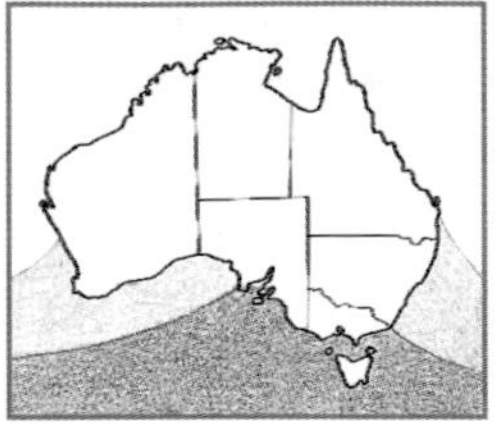

Antarctic Petrel *Thalassoica antarctica* 40–45 cm; span 110–120 cm (H,M)
Usually a bird of the pack-ice in Antarctic seas. In flight keeps well above waves. Stiff-winged glides are interspersed with bursts of rapid, shallow wingbeats; sometimes hovers before plunging into sea to depths of 1.5 m; occasionally scavenges around ships. Swims well; often floats with wings outstretched. In small to large flocks at sea. **Voice:** usually silent at sea; in colonies, rapid, harsh, abrupt 'argh-argh-argh'; some notes are deep, croaked, drawn out like creaky door. **Similar:** Cape Petrel. **Status:** abundant in Antarctic seas; very rare vagrant to coastal seas of SE Aust. **Br.:** huge colonies on cliffs of Antarctic and islands.

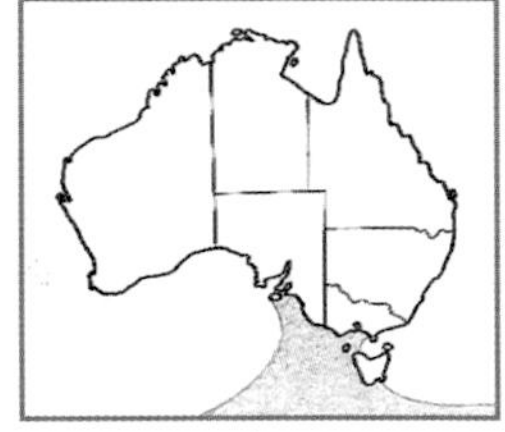

Cape Petrel *Daption capense* 35–42 cm; span 80–90 cm (H, M, L, N)
In flight displays typical fulmar pattern of brief glides interspersed with bursts of shallow fluttering flight. Tends to fly much higher than most other petrels and shearwaters, especially in strong winds, when flight is buoyant, swooping and soaring. Dives for food from a height or while floating. Swims well, floating high. Often gathers in large, noisily quarrelsome flocks around fishing boats. **Voice:** noisy when breeding and in squabbling flocks at sea; has a rapid, hoarse 'arghargh-arg', a sharper 'cac-cak-cak', and an excited but softer, almost musical 'cook-cook-cook'. **Hab.:** Antarctic and sub-Antarctic seas. Aust., inshore and shelf-edge waters. **Status:** common to rare around Aust. coastline; abundant in S seas.

Snow Petrel *Pagodroma nivea* 30–40 cm; span 75–85 cm (H, M)
The only small, all-white petrel. Flight erratic and fluttering; rapid, shallow wingbeats; occasional brief glides, coming close to water, almost hovering, but rarely alighting. Does not usually follow or forage around ships. **Variation:** two subspecies, both from Antarctic. **Voice:** usually silent at sea, but varied 'teck-teck' and other calls swinging from guttural to high-pitched. **Similar:** White Tern, albino forms of other petrels. **Hab.:** usually confined to the Antarctic pack-ice and nearby seas. **Status:** rare vagrant to warmer seas; may be accidental to S Aust. coast; several doubtful records. **Br.:** in colonies around Antarctica.

Blue Petrel *Halobaena caerulea* 26–32 cm; span 66–71 cm (H, M)
Small blue-grey petrel; unique square-cut tail with white tip is visible at a distance and in poor light. Flight fast and buoyant; frequently glides low on stiff, bowed wings, dipping between waves, uplifting over crests; high and wheeling in strong winds, shows alternately the white underparts then the steely grey upper pattern. Gregarious, small scattered flocks, often with other species. **Voice:** silent at sea; in courtship a soft 'ku-ku-ku-COO-COO'. **Similar:** Gould's Petrel, but has no white tail-tip. **Hab.:** usually pelagic, occasionally over shallow waters. **Status:** uncommon but regular winter visitor to S Aust. waters. **Br.:** islands around Antarctica.

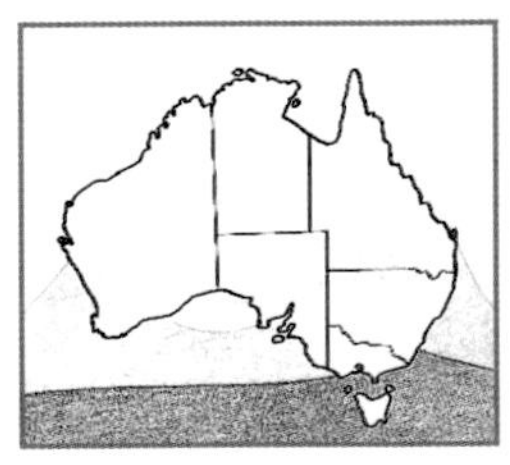

Mottled Petrel *Pterodroma inexpectata* 33–35 cm; span 85 cm (L, M, N)
Medium sized gadfly petrel; bold black diagonals and dark outer primaries form an 'M' shape across upper surface of outstretched wings. Flight swift with brief bursts of quick, shallow wingbeats and long, sweeping glides. Sweeps high in the wind; turns in great arcs; again swoops towards the waves. Ignores ships. **Voice:** silent at sea; in colonies, rapid, vibrating, metallic 'ki-ki-ki-' and various growling and crooning sounds. **Similar:** Soft-plumaged Petrel, but underwing darker. **Hab.:** open seas and Antarctic iceberg belt. **Status:** accidental or rare vagrant to coastal waters of SE Aust. **Br.:** islands near NZ, including Snares Is.

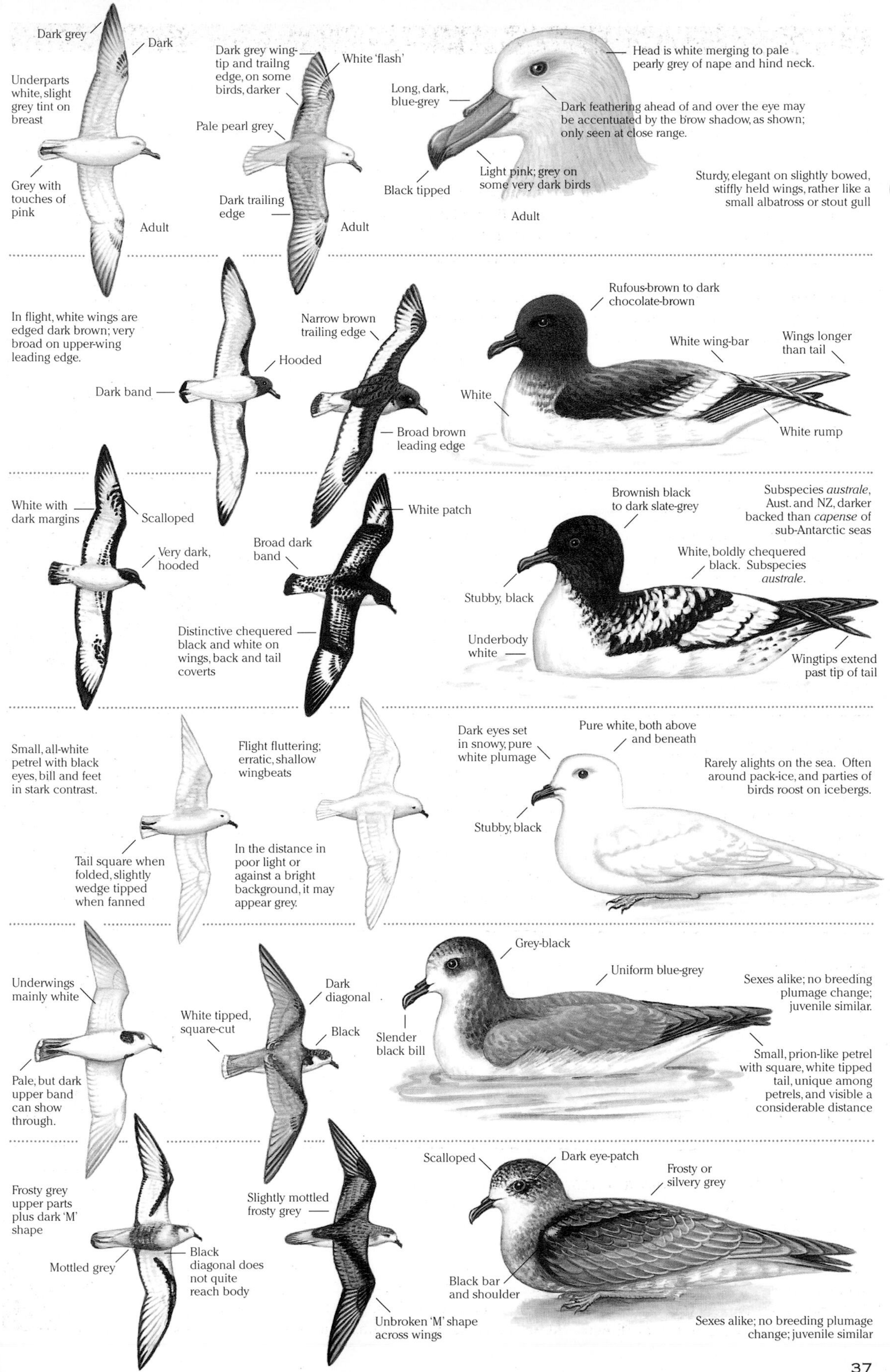
Dark grey
Dark
Underparts white, slight grey tint on breast
Grey with touches of pink
Adult
Dark grey wing-tip and trailng edge, on some birds, darker
White 'flash'
Pale pearl grey
Dark trailing edge
Adult
Long, dark, blue-grey
Black tipped
Light pink; grey on some very dark birds
Adult
Head is white merging to pale pearly grey of nape and hind neck.
Dark feathering ahead of and over the eye may be accentuated by the brow shadow, as shown; only seen at close range.
Sturdy, elegant on slightly bowed, stiffly held wings, rather like a small albatross or stout gull
In flight, white wings are edged dark brown; very broad on upper-wing leading edge.
Dark band
Hooded
Narrow brown trailing edge
Broad brown leading edge
Rufous-brown to dark chocolate-brown
White wing-bar
Wings longer than tail
White
White rump
White with dark margins
Scalloped
Very dark, hooded
White patch
Broad dark band
Distinctive chequered black and white on wings, back and tail coverts
Brownish black to dark slate-grey
Subspecies *australe*, Aust. and NZ, darker backed than *capense* of sub-Antarctic seas
White, boldly chequered black. Subspecies *australe*.
Stubby, black
Underbody white
Wingtips extend past tip of tail
Small, all-white petrel with black eyes, bill and feet in stark contrast.
Flight fluttering; erratic, shallow wingbeats
Tail square when folded, slightly wedge tipped when fanned
In the distance in poor light or against a bright background, it may appear grey.
Dark eyes set in snowy, pure white plumage
Pure white, both above and beneath
Rarely alights on the sea. Often around pack-ice, and parties of birds roost on icebergs.
Stubby, black
Underwings mainly white
Pale, but dark upper band can show through.
White tipped, square-cut
Dark diagonal
Black
Grey-black
Uniform blue-grey
Slender black bill
Sexes alike; no breeding plumage change; juvenile similar.
Small, prion-like petrel with square, white tipped tail, unique among petrels, and visible a considerable distance
Frosty grey upper parts plus dark 'M' shape
Mottled grey
Black diagonal does not quite reach body
Slightly mottled frosty grey
Unbroken 'M' shape across wings
Scalloped
Dark eye-patch
Frosty or silvery grey
Black bar and shoulder
Sexes alike; no breeding plumage change; juvenile similar

Black-winged Petrel *Pterodroma nigripennis* 28–30 cm; span 64–71 cm (L, N)
A small species with distinctive markings. Flight is fast, powerful; at times dashes and weaves low across the sea. More often, in light winds, follows a leisurely, undulating path; brief bursts of rapid wingbeats alternate with short glides. With the power of a strong wind added, flight gains vigour; wheels, soars high, swoops low. **Voice:** in flight over colony a rapid, high 'peet-peet-peet'. **Similar:** Cook's Petrel. **Hab.:** marine, pelagic; usually comes no closer to land than the edge of the continental shelf, but sometimes driven into coastal waters by storms. **Status:** records along E coast increasing. **Br.:** colonies, Lord Howe and Norfolk Is. Page 356

Gould's Petrel *Pterodroma leucoptera* 30 cm; span 68–71 cm
Flight is slower than similar petrels, less high towering; banks, weaves and dips, glides on stiff wings. **Variation:** race *leucoptera* breeds Aust.; *caledonicus*, rare vagrant. **Voice:** usually silent at sea. At breeding grounds calls in flight and on the ground. After dark, circles above colony giving a high, metallic ticking, 'kik-kik-kik-kik' or 'zit-zit-', more rapidly and excitedly in aerial pursuits. Usual call from birds on ground is a shrill 'pee-pee-peeoo' or a tremulous growling. **Similar:** Cook's, Black-winged Petrels. **Hab.:** offshore waters, oceans. **Status:** population of Aust. race small, about 500 pairs breeding on Cabbage Tree Is., NSW. **Br.:** 357

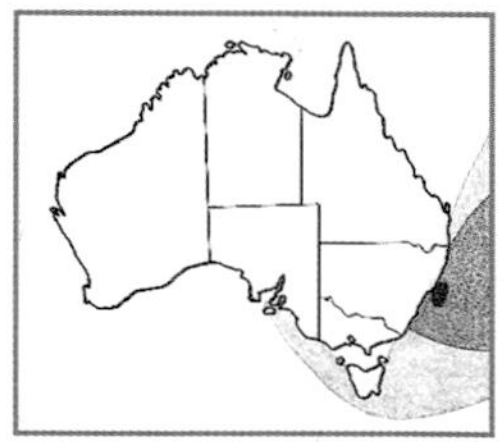

Cook's Petrel *Pterodroma cookii* 25–30 cm; span 66 cm
A gadfly petrel with black underwing diagonals that stop well short of the body. Flight leisurely and rolling in light winds, or vigorous with rapid, jerky wingbeats and erratic banking and weaving that can reveal diagnostic markings of both under and upper plumage. **Voice:** usually silent at sea and on ground. Calls in flight over colonies varied: includes high, nasal, rather duck-quacked 'kwek-kwek-kwek', soft, or loud and harsh; like call of Mottled Petrel. **Similar:** Black-winged Petrel. **Hab.:** pelagic; subtropical and, rarely, sub-Antarctic waters. **Status:** rare vagrant offshore E coast of Aust. **Br.:** breeds on islands off N NZ.

Juan Fernandez Petrel *Pterodroma externa* 40–43 cm; span 96–98 cm
Larger than most 'Cookilarias'—the petrels that are similar to Cook's Petrel. Flight similar to but less energetic than that of White-necked and Barau's, appearing strong, graceful and effortless; wheels in high arcs. **Variation:** previously a race (nominate) of the White-necked Petrel. **Voice:** possibly similar to that of White-necked Petrel. **Similar:** Barau's, Gould's and White-necked Petrels. **Hab.:** pelagic across tropical and subtropical parts of central and E Pacific. **Status:** extremely rare vagrant or accidental visitor to offshore waters of E coast of Aust. **Br.:** breeds Dec.–Jun. on Juan Fernandez group of islands, Chile.

Barau's Petrel *Pterodroma baraui* 38 cm; span 96 cm (H)
Large, dark capped petrel. Flight slower, not as agile or aerobatic as most other petrels of same range. **Voice:** probably silent at sea, but noisy over breeding colonies; calls are described as a high 'ti-ti-ti-ti-' and lower 'kek-kek-kek-kek-'. **Similar:** unique in its usual range; elsewhere, Juan Fernandez, Atlantic, Herald and Gould's Petrels. **Hab.:** appears to use areas of warm surface waters across the tropical and subtropical parts of the Indian Ocean. **Status:** occurs W half of Indian Ocean, vagrant further E, and only extremely rare accidental occurrences in coastal waters of WA. **Br.:** summer, islands near Madagascar.

White-necked Petrel *Pterodroma cervicalis* 40–43 cm; span 1 m (L, N)
Flight leisurely yet strong, a gentle 'roller-coaster' travel; in strong winds wheels and arcs high. Feeds by dipping and snatching from the surface; solitary, occasionally in loose flocks. **Variation:** previously a race of the species *Pterodroma externa*; common name unchanged. **Voice:** silent at sea, noisy in the colonies during the breeding season. In flight, a loud, harsh 'ka-ka-ka-ka' and a softer 'tse-tse-tse-tse'. Also in flight and from ground, a long 'kukoowik-ka-ka'. **Similar:** white collar around back of neck is unique. **Hab.:** oceans, edge of continental shelf. **Status:** rare or uncommon vagrant; sightings usually well offshore, occasionally from land, NSW and Qld, summer. **Br.:** Nov.–May, Kermadec Islands, NZ.

Broad black bar almost to body
Dark, wedge tip
Small gap
Broad black diagonals meet across rump.
Wedge tip
Wing grey, darker with wear, reducing clarity of diagonal bar
Dark grey eye-patch
Stubby, black
Small gap
Back and collar grey; contrasts more with wings as these are darkened by wear.
Underparts white except collar
Black face
Dark, short, rounded
Tapered point towards body
Narrow dark edge
Dark hood comes over side of face and neck.
Dark grey-brown
Broad black diagonal and dark outer primaries form a broad 'M' joining across the rump.
Hood extends broadly through eye, over face and side of neck.
Pale
Fades to fine point well out from body.
Dark tipped, outer edges white
Crown and back pale grey
Dark, wide 'M' shape from wingtip to wingtip, conspicuous against surrounding lighter greys
White brow line
Pale grey
Black, long, slender
Pale grey in strong contrast to dark wing plumage
Pale partial collar
Laterals, tail feathers white on outer edge; this is obvious only on spread tail.
Narrow dark rim around tail-tip
Small dark carpal patch with only a short, fine trace of diagonal bar
Dark eye-patch
Narrow black trailing edge
Dark outer wings
Dark head blends into grey of hind neck and back without white collar.
Black 'M' across wings
Dark head blends into grey of hind neck and back without rear white collar..
Only faint trace of grey collar
Tail long, dark; outer feathers are edged white.
Long, narrow straight-winged jizz
Faint
Slender body, narrow wedge tail. Similar to Juan Fernandez and White-necked Petrels
Diagonal bar extends just over halfway from carpal to body.
Dark edge
Slender, wedge tipped
Dark cap
Wide 'M' pattern wingtip to wingtip. But in worn plumage most of wing becomes darker grey-brown; 'M' is less obvious.
Dark
Black behind eye
White
Grey back, darker in worn plumage
Faint
Tail dark; white on outer pair of feathers
Glides with long wings slightly flexed back.
Dark trailing edge
Wide
Dark edged
Dark wing band crosses rump with rounded point towards tail.
Black 'M' band joins across wings.
White
Black cap; extends below eyes.
Broad white hind-neck collar is a diagnostic feature.
Silvery mid grey
Pale

Great-winged Petrel *Pterodroma macroptera* 38–43 cm; span to 1.02 m (H, M, L)
A dark, thickset, long-winged petrel. Flight in light winds usually a lethargic, buoyant, meandering glide; rises and falls in wide arcs with occasional wingflaps. In direct travel, strong, steady wingbeats, each with high, gull-like upstroke; brief pause; deep, strong downbeat. In high winds flight is powerful, soaring high, plunging down to sweep across the waves. **Voice:** at nest and display, rapid, squeaky, screeched 'kee-kee-kee', gruff 'carrrk' or 'quawer'. **Similar:** Providence, Kermadec, Kerguelen Petrels. **Hab.:** oceans, coastal waters. **Status:** common; usually race *macroptera* WA to SA, both *macroptera* and *gouldi* further E. **Br.:** 356

Providence Petrel *Pterodroma solandri* 40 cm; span about 1 m (C, L)
A heavily built gadfly petrel; flight relaxed with slow, easy wingbeats; but in strong winds is fast, dashing and weaving close to the waves, then lifts to soar high in great arcs. **Voice:** in flight over colonies a loud, rapid, harsh 'kak-kak-kaaak-kak-kak' or 'kir-rer-rer'; on ground a deeper, trilled 'kerr-rer, kuk-kuk-kuk, ker-er-er.' **Similar:** Great-winged Petrel, dark morphs of Kermadec and Herald. **Hab.:** marine, pelagic; the E edge of Australia's continental shelf is possibly a favoured feeding ground for Lord Howe birds during winter. **Status:** migratory, dispersive; spreads into N Pacific after breeding. **Br.:** Lord Howe Is. and Phillip Is.

White-chinned Petrel *Procellaria aequinoctialis* 50–58 cm; span 1.3–1.4 m (H)
Heavily built dark petrel; usually has a tiny patch of white under chin. Shares with following two species a distinctive 'albatross pattern' of flight: close to the waves, wingbeats slow, deliberate and graceful, interspersed with long glides; but also soars high. **Variation:** near Aust., only nominate occurs. Race *conspicillata*, breeding S Atlantic, not yet recorded in Aust. seas; may be a full species. **Voice:** usually silent at sea, but, around trawlers, high trilling and aggressive, harsh, clucking 'chek-chek' or 'kek-kek-'. **Similar:** Black and Westland Petrels. **Hab.:** marine, pelagic; Antarctic to subtropical waters, often over shallower coastal water. **Status:** uncommon in Aust. seas; elsewhere common. **Br.:** Antarctic and sub-Antarctic seas.

Westland Petrel *Procellaria westlandica* 50–55 cm; span 1.3–1.4 m
Flight appears rather laboured in light winds; long glides close to water broken by occasional leisurely beating of wings. Action becomes more vigorous in the lift of stronger winds: weaves and wheels on strongly flexed wings; low between crests, then rising albatross-like to soar and bank. Usually solitary, but congregates around fishing trawlers. Feeds by snatching from surface or by shallow surface diving. **Voice:** silent at sea; noisy visiting nest colonies at night; various harsh cackles. **Similar:** Black Petrel and dark form of White-chinned Petrel. **Hab.:** when breeding, offshore waters; after breeding, marine dispersing into Pacific pelagic waters. **Status:** rare visitor in SE Aust., usually in summer. **Br.:** South Is., NZ.

Black Petrel *Procellaria parkinsoni* 46 cm; span 1.1 m
Usually indistinguishable at sea from the larger Westland unless bill detail is visible. Lighter build, slender wings; more graceful, easy, buoyant flight. Glides with carpals forward, outer wings backswept, pointed. **Voice:** in colonies, harsh, sharp clacking, 'ak-ak-ak-ak-', and a throaty squawk, 'argk, argk, argk'. **Similar:** Westland Petrel, but heavier build; larger, black tipped bill. White-chinned, but less buoyant flight; lacks white chin. **Hab.:** marine, preferring shelf-edge pelagic seas; usually avoids inshore waters. **Status:** rare but probably regular visitor to SE Aust., usually in summer. Small NZ population. **Br.:** summer, on NZ islands.

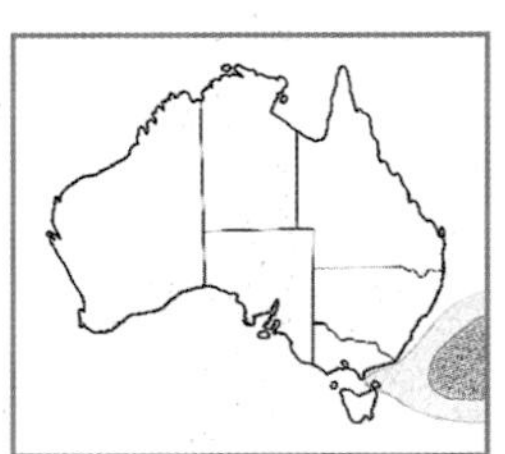

Kerguelen Petrel *Lugensa brevirostris* 33–36 cm; span 80–82 cm (H, M)
Blunt bodied yet slender winged silhouette. Plumage is glossy; silvery flashes from underside of flight feathers catch the sunlight. This reflective quality alters the plumage pattern with changing intensity of light. In light winds it has a gently weaving flight; glides interspersed with brief bursts of bat-like fluttering. In strong winds soars and circles, tern-like, very high. **Voice:** calls at sea unknown, possibly silent. In colonies it calls mainly at dusk, a wheezy 'ch-chee-chee-chay'. **Similar:** Great-winged and Providence Petrels. **Hab.:** highly pelagic; rarely near land unless driven inshore or wrecked by gales. **Status:** uncommon winter visitor to S coasts of Aust. **Br.:** in colonies on sub-Antarctic islands.

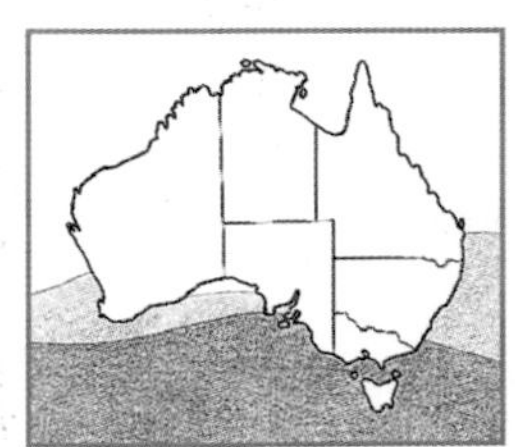

Long, narrow, finely pointed, zigzag wing silhouette
Almost uniformly dark, above and beneath
Stocky, 'bull necked' jizz
Wedge shaped tail
Pointed when folded
Subspecies *gouldi*
Deep, stubby
White, forehead to chin
Subspecies *macroptera*, grey around face
Long wings extend past tail-tip.
White under primaries
White coverts
Forehead 'scaly'
White
Mottled slaty grey-brown
Worn plumage much warmer brown
Stout, black
A large, heavily built petrel, identified by white beneath outer wing.
Upper parts almost uniformly dark brown
In shadow, almost black
Slaty grey-brown
Long brown tail and wings like similar species
Large, very dark petrel
Sooty black
Feet show just slightly beyond tail-tip.
Any white under chin is visible only at close range.
Wedge tipped tail, often partly fanned
Small head, hunched close
White chin varies in size and shape, quite large to very small or absent. S Atlantic race *conspicillata* has white lines from chin up in front of eye and onto forehead, and from chin up behind eye and over crown.
Variable patch of white under the chin
Large, thick necked, deep chested bulky body tapers to tail.
Large, heavily built brownish black to dusky black petrel with bill that is ivory to buff, bulbous and black tipped.
Lighter silvery grey highlights
Under full sun upper surfaces appear more brown than black.
Plumage wears to dark mottled brown, paler on outer wings.
Bill large, deep, all white to off-white; bulbous tip, prominent tube nostrils
Tips of black feet show behind tail.
Wedge tipped
Underwing sooty black
Colour intensity varies with lighting; usually pale cream, may appear ivory to buff.
Bulbous, entirely black
Like above species but smaller, lighter build.
Black undersurfaces
Wedge tipped
Tips of feet behind tail
Sooty black in fresh plumage, wearing to blackish brown; full sunlight tends to emphasise the warmer browns.
Upper edge grey to buff, tip black
Slightly less bulbous and hooked
Greenish buff to pale grey-buff; varies with lighting and distance.
Medium sized, all dark petrel with large, high head
White or pale
Plumage glossy so that sun on pale or white areas under the wing can flash brightly, silvery white.
Silvery grey
Wedge tipped
Grey-brown
Forehead high, steep
Black on forehead, lores and around eye
Slaty dark grey
Small, stubby black bill
Underbody is slate grey, slightly paler than upper.

Kermadec Petrel *Pterodroma neglecta* 38 cm; span 93 cm (N)
Stocky body, long pointed wings that are held forward to the carpel joint, outer wing backswept, short tail; these are part of the overall impression, or 'jizz', of the large gadfly petrels of the genus Pterodroma. Identification of the species is much more difficult, with sightings at sea often brief and distant. Adding to the complexity in recognising this species, the Kermadec has three recognised colour phases, but actually a continuous grading through from light to dark. Fortunately all retain a common plumage feature that identifies them as Kermadec: a pale or white area beneath the outer wing at the base of the primaries. On the upper wing, the same patch shows a white 'flash'—white streaks formed by the reflective, glossy white shafts of the primaries. The pattern of flight also helps identification. Most of its time in the air is spent gliding, effortlessly riding on the wind, wheeling and banking, assisted occasionally by a few deep, leisurely, gull-like beats of the wings. In strong winds the flight pattern changes; becomes more dashing, towering up against the wind, wheeling in wide arcs, diving lower in long glides across the waves. At sea this species is solitary; feeds by dipping and seizing to take sea life such as small squid and crustaceans; ships are ignored. **Variation:** two subspecies. The nominate *neglecta* breeds on islands across the SW Pacific, while the slightly larger birds of subspecies *juana* breed on islands along the W coast of S America. Both subspecies have the range of colour morphs, light through to dark. **Voice:** apparently silent at sea. In breeding colonies, calls in flight and on ground, very noisy in evenings as thousands of birds return and circle the nesting grounds. The call is varied, but usually a loud 'yuk-ker-a-wooo-WUK'—the first part a hoot, the final loud and abrupt. **Similar:** Herald Petrel, but always has darker head and face; White-headed, but differs from pale morph of Kermadec in having dark eye-patch. **Hab.:** oceanic, pelagic. **Status:** extremely rare vagrant or accidental visitor to E coast NSW. **Br.:** in colonies of S Pacific, including Ball's Pyramid near Lord Howe Is.

Grey Petrel *Procellaria cinerea* 50 cm; span 1.1–1.3 m (H)
A distinctive large petrel plumaged ashy grey on the upper parts, softly blending to white on underbody. In flight, wings are often held rather straight and stiff, which, together with a rather tubby body and slender wings, imparts a rather shearwater-like jizz. Appears swift and strong in flight; long graceful glides intersperse with periods of powered flight with distinctive wing action, quick and powerful with a characteristic jerkiness. Often plunges into the sea from considerable heights, the momentum of the dive carrying it well under water where it propels itself forward with its wings; may stay under water a minute or two in pursuit of small fish or other marine prey. On emerging, shakes water from its plumage. In strong winds the energy in the flight reflects the power of the wind with albatross-like soaring, wheeling and gliding; wings are held more flexed than in the stiff, straight-out position held for lighter airs. Accompanies and scavenges around ships and fishing boats, often following for long periods. At sea may be solitary or in small groups, occasionally gathering in large rafts. **Voice:** apparently silent at sea. Noisy in colonies, especially early in breeding cycle; cackles, moans, resonant alarm calls. **Similar:** White-headed Petrel, but no white head with black eye-mark or white tail. **Hab.:** marine, mainly pelagic; tends to keep offshore, not often sighted from land. **Status:** rare vagrant to S Aust., elsewhere common in Antarctic and sub-Antarctic seas. **Br.:** winter breeder, colonies on islands of S oceans.

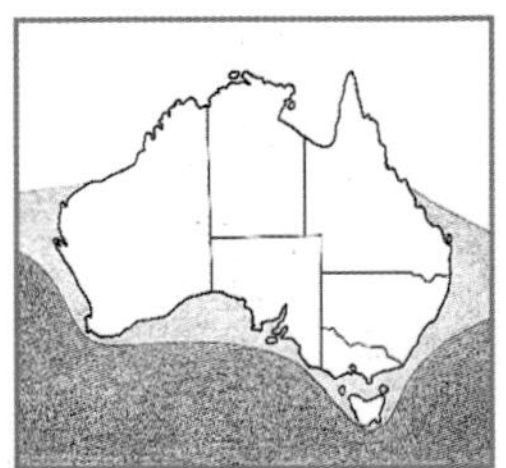

Soft-plumaged Petrel *Pterodroma mollis* 32–36 cm; span 84–95 cm (M, H)
The two morphs or colour forms have in common a white underbody with grey breast-band, grey underwings with wide, dark, diagonal band, and grey upper parts. Flight fast and with rapid wingbeats, alternating with glides on angled wings, travelling in wide arcs; zigzag progression, rarely soars very high. Gregarious, often found in small parties travelling low and fast, occasionally in loose flocks of up to a thousand birds. Occasionally follows ships. **Variation:** several subspecies and rare dark morph. The latter varies from an overall very dark grey-brown on which the breast-band and underwing band are barely discernible to individuals with a wide dark breast-band and heavily streaked underparts. **Voice:** usually silent at sea; possibly some communication within large flocks. At colony calls only in flight—a low, wavering, musical wail, repeated several times, often ending with a sharp upwards 'whik'; has several other sounds including sharp squeaks and trills. **Similar:** White-headed Petrel, but dark eye-patch; Mottled Petrel but dark belly and dark underwing-bar. **Hab.:** marine, pelagic; Antarctic to subtropical waters. **Status:** may be common in seas S of Aust.; a regular and quite common visitor to NW of WA; less common or scarce on SE coasts at any time of year—some beach-wrecked birds after winter gales. Dark morph occurs in S Atlantic; not yet recorded in Aust. **Br.:** in colonies on islands of S oceans.

White shafts on upper wing primaries catch sunlight as a white flash.
Light morph
The Kermadec shows more variation of plumage than any other 'gadfly' petrel. Within this species, the 'light morph' indicates the palest plumage, while the dark morph represents the other extreme.
White strip along humerus
White
Light morph
White with buff-brown tint
Pale
Some white under wing primaries
Intermediate morph
Between light and dark morphs is a continuous range of colour forms. This plumage, at about mid-range, further illustrates the series, but represents only one stage.
Mid grey-brown
Intermediate morph
Grey-brown
White
The dark morph is entirely brownish grey-black. All morphs have the same identifying markings—the pale patch on underwing primaries, and fine white lines of primary shafts on the upper wing.
Dark morph
Stubby black bill
Dark morph
Dark brown
Primaries black with glossy white shafts
White
Narrow, finely pointed wings
Grey-black
Dark slaty grey gives dark headed appearance.
Slender, pale greenish yellow
Short, wedge tipped, dark
Silvery or ash-grey in fresh plumage wearing to a slightly brownish grey
White with faint grey wash mainly towards tail coverts in stark contrast to near black wings
Feet visible
White with touches of faint blue-grey wash; brilliant white in sunlight
Primary undersides reflective, flash in sunlight.
Large, bulky bodied with proportionately small headed appearance
Tail dark grey; undertail coverts very light blue-grey
Dark tips
Speckled dark grey
White brow
Eye is set in a dark patch that merges with the grey nape; this dark area is emphasised by the white brow above and made darker by the shadowing beneath the brow.
Broad dark diagonal
Grey-black diagonal almost lost on dark underwing
Soft grey
Dark eye
Dark grey
Light morph
Indistinct 'M' band across upper wings
Identified by a combination of grey upper parts, grey underwing with darker diagonal band and grey breast-band across white underside of body
Light morph
Dark morph
Light morph

Tahiti Petrel *Pseudobulweria rostrata* 38–40 cm; span 85–95 cm (N)

Typical gadfly petrel jizz; stocky, bull necked, tapering silhouette; dark hood; large bill; long wings often held relatively straight; pale reflective underwing-bar. In light weather, leisurely gliding with occasional languid wingflaps; keeps close to the surface. With strong winds, a more purposeful, albatross-like soaring and swooping. Usually solitary, at times in loose flocks. **Variation:** several subspecies. **Voice:** silent at sea; in colony at night a long whistled sequence. **Hab.:** pelagic; Aust. sightings usually beyond continental shelf. **Status:** an uncommon, perhaps regular, visitor; recorded E, N and NW Aust. seas. **Br.:** Pacific islands.

Atlantic Petrel *Pterodroma incerta* 43 cm; span 1–1.1 m

A large petrel with stocky jizz; dark brown head and upper breast. Wings long, uniform dark brown beneath, held forward to the carpel then backswept. Flight described as swift and 'careening'; often follows ships. **Similar:** Tahiti Petrel, but has a pale central line along underside of each wing. **Hab.:** pelagic waters of S Atlantic, and rarely the W parts of Indian Ocean. **Status:** common in central S Atlantic, and is one of the most abundant petrels; but, as yet, no more than a possibility of being a very rare vagrant to Aust. waters. **Br.:** Tristan da Cunha group, S Atlantic.

White-headed Petrel *Pterodroma lessonii* 40–46 cm; span 1.1 m (H,M)

Dark eye-patch conspicuous against white of head. In light winds keeps close to the waves with short glides and periods of steady, shallow, stiff-winged beats. In strong winds, little flapping, wheels in high arcs, catching the lift of wind over waves. **Voice:** usually silent at sea. In colonies, noisy after sunset and again before sunrise with much calling from birds overhead. Flight call, 'wit-wit-wit' and 'tiew-ee, tiew-ee-' with an occasional gruff 'ooo-err'. **Similar:** light phase of Soft-plumaged Petrel, Grey Petrel. **Hab.:** Antarctic and sub-Antarctic oceans, cold seas as far S as the edge of pack-ice. Except near breeding grounds, avoids inshore waters. **Status:** uncommon winter visitor to seas along S coast of Aust.—edge of continental shelf unless blown in by storms. **Br.:** on islands of sub-Antarctic seas.

Herald Petrel *Pterodroma arminjoniana* 35–39 cm; span 1 m

Stocky build, typical gadfly petrel in overall impression or jizz. Distinctive underwing pattern with white patch at base of primaries. Usually glides low with leisurely wing-flaps; in strong winds, higher banking and wheeling. **Variation:** subspecies *heraldica* extends from E coast of Aust. across Pacific towards S America; the nominate subspecies occurs in the Atlantic and Indian Ocean towards Madagascar. Both races have light to dark plumage morphs or phases. The small restricted Aust. population belongs to subspecies *heraldica*; it appears that most are of the light phase. **Voice:** apparently silent at sea. At the island colony, pairs and individuals call in display flights, night or day, rarely from the ground. Usual call is a high chattering variously described as 'hi-hi-hi-' or 'chi-chi-chi-'. **Similar:** Kermadec Petrel, but much larger white patch at base of primaries; Providence Petrel, but less under-primary white than all but darkest Herald Petrels. **Hab.:** pelagic, i.e. uses the ocean surface and the air close above. Tropical and subtropical waters are preferred with most Aust. sightings being beyond the edge of the continental shelf. **Status:** other than on breeding island, rarely seen; disperses from colony away from coastal waters; a very few vagrant and beach-washed records for E coast of Aust. **Br.:** 356

Bulwer's Petrel *Bulweria bulwerii* 26–27 cm; span 68–72 cm

Small dark petrel with slender body tapering to fine tail; long narrow wings with carpal 'wrist' joints carried well forward, outer wing backswept. In flight zigzags low, usually within several metres of the waves. Prion-like twisting and turning, each brief burst of quick wingbeats followed by a short glide on bowed wings; tail fans briefly to manoeuvre. When feeding, circles low, dipping and snatching at surface prey. Solitary or in small groups, occasionally resting on surface of calm seas. **Voice:** silent at sea. **Similar:** Matsudaira's Storm-Petrel, but deeply forked tail and pale fore-wing patch; Wedge-tailed Shearwater, dark morph, but larger, longer bill and head, so appears longer necked. **Hab.:** marine, pelagic; typically over warmer waters. **Status:** may be quite common Sep.–Apr. in NW Aust. waters. Elsewhere around Aust. coast a rare vagrant; records NE Qld, Vic. **Br.:** oceanic islands.

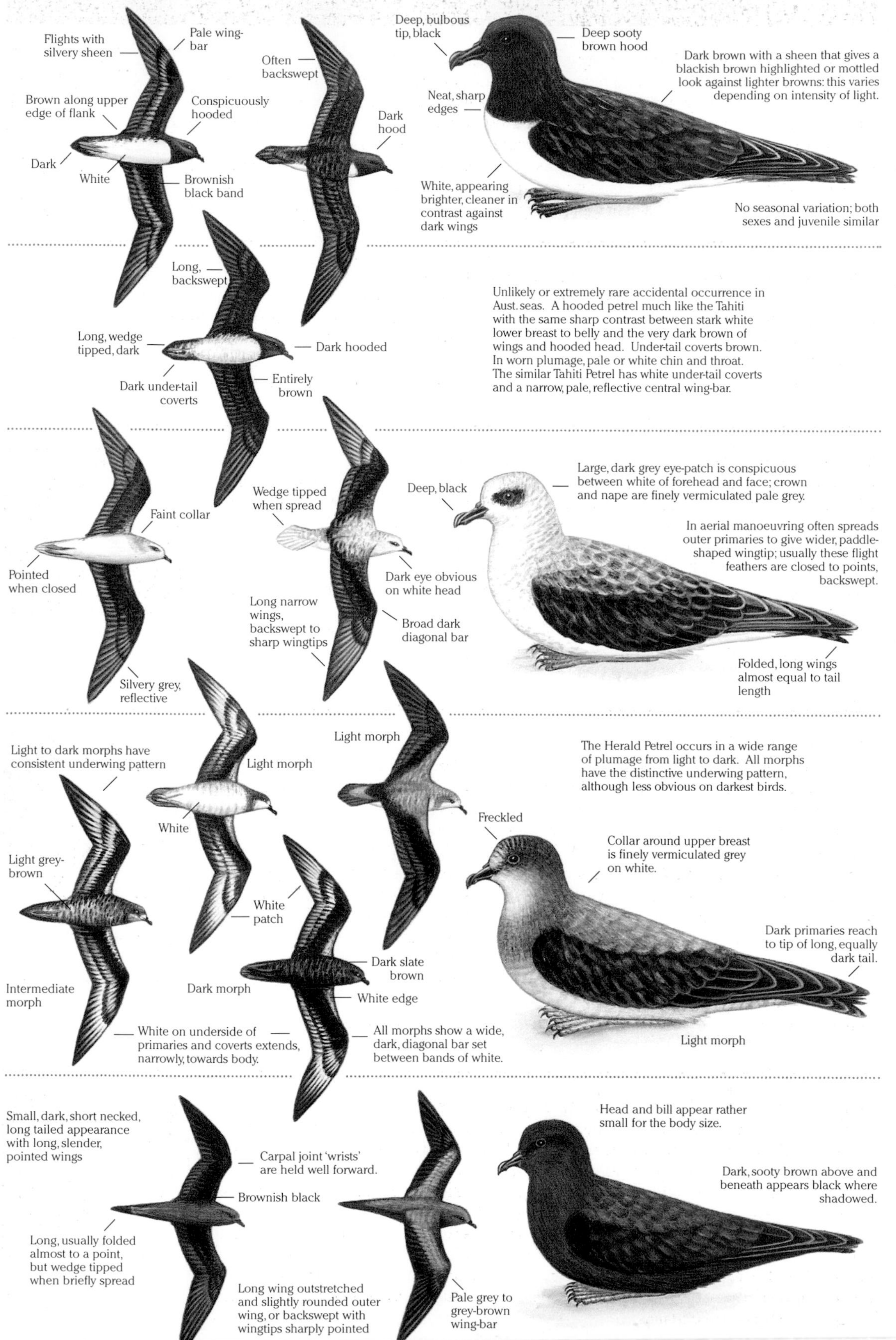

Flights with silvery sheen
Pale wing-bar
Brown along upper edge of flank
Conspicuously hooded
Dark
White
Brownish black band
Often backswept
Dark hood
Deep, bulbous tip, black
Deep sooty brown hood
Dark brown with a sheen that gives a blackish brown highlighted or mottled look against lighter browns: this varies depending on intensity of light.
Neat, sharp edges
White, appearing brighter, cleaner in contrast against dark wings
No seasonal variation; both sexes and juvenile similar
Long, backswept
Long, wedge tipped, dark
Dark hooded
Dark under-tail coverts
Entirely brown
Unlikely or extremely rare accidental occurrence in Aust. seas. A hooded petrel much like the Tahiti with the same sharp contrast between stark white lower breast to belly and the very dark brown of wings and hooded head. Under-tail coverts brown. In worn plumage, pale or white chin and throat. The similar Tahiti Petrel has white under-tail coverts and a narrow, pale, reflective central wing-bar.
Faint collar
Pointed when closed
Silvery grey, reflective
Wedge tipped when spread
Dark eye obvious on white head
Long narrow wings, backswept to sharp wingtips
Broad dark diagonal bar
Deep, black
Large, dark grey eye-patch is conspicuous between white of forehead and face; crown and nape are finely vermiculated pale grey.
In aerial manoeuvring often spreads outer primaries to give wider, paddle-shaped wingtip; usually these flight feathers are closed to points, backswept.
Folded, long wings almost equal to tail length
Light to dark morphs have consistent underwing pattern
Light grey-brown
Intermediate morph
Light morph
White
White patch
Dark morph
Light morph
Dark slate brown
White edge
White on underside of primaries and coverts extends, narrowly, towards body.
All morphs show a wide, dark, diagonal bar set between bands of white.
The Herald Petrel occurs in a wide range of plumage from light to dark. All morphs have the distinctive underwing pattern, although less obvious on darkest birds.
Freckled
Collar around upper breast is finely vermiculated grey on white.
Dark primaries reach to tip of long, equally dark tail.
Light morph
Small, dark, short necked, long tailed appearance with long, slender, pointed wings
Carpal joint 'wrists' are held well forward.
Brownish black
Long, usually folded almost to a point, but wedge tipped when briefly spread
Long wing outstretched and slightly rounded outer wing, or backswept with wingtips sharply pointed
Pale grey to grey-brown wing-bar
Head and bill appear rather small for the body size.
Dark, sooty brown above and beneath appears black where shadowed.

Buller's Shearwater *Puffinus bulleri* 46–47 cm; span 97–99 cm

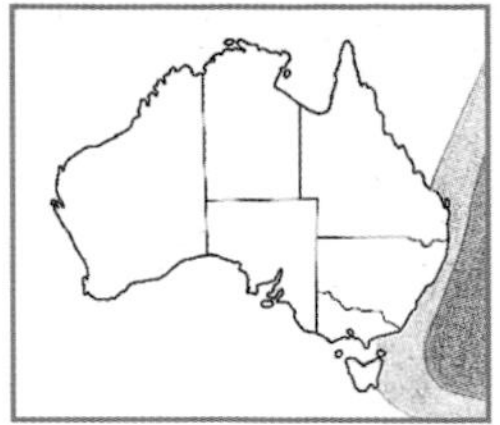

Flight leisurely, buoyant, banking, gliding; slow, deliberate wingbeat; arcs higher in strong winds. Forages low over surface; hunts by low shallow plunges. **Voice:** silent at sea; noisy in colonies after dark and pre-dawn; varied wailing, screaming, howling sounds, like other shearwaters. **Similar:** pale morph of Wedge-tailed Shearwater; Juan Fernandez, White-necked, Barau's Petrels. **Hab.:** tropical seas; preference for areas of strong upwelling along continental shelves and the food-rich waters where warm and cooler currents meet. **Status:** regular but scarce summer visitor to SE Aust. **Br.:** coastal islands off North Is., NZ.

Wedge-tailed Shearwater *Puffinus pacificus* 38–45 cm; span 1 m (C, K, L, N)

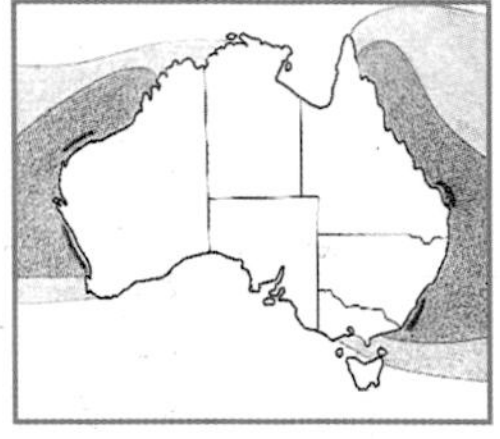

The common 'Muttonbird'. Flight is slow and leisurely; seems to drift buoyantly close to the water, wrist-like carpels well forward, tail usually tapering to a long point. **Voice:** silent in flight; noisy in colonies at night, a wailing 'ka-wooo-ah', repeatedly, becoming faster, louder, rising to an almost hysterical climax. **Similar:** Flesh-footed Shearwater. **Hab.:** tropical and subtropical seas; pelagic, frequenting and feeding across the ocean surface, often at junction of warm and nutrient-rich cool waters, and also where inshore waters meet deep oceanic waters. **Status:** common in coastal and oceanic waters of E and W Aust., most being dark plumages; light morph rare in E Aust., but about 25% of population in WA. **Br.:** 357

Fleshy-footed Shearwater *Puffinus carneipes* 40–45 cm; span 1–1.5 m (L, M, N)

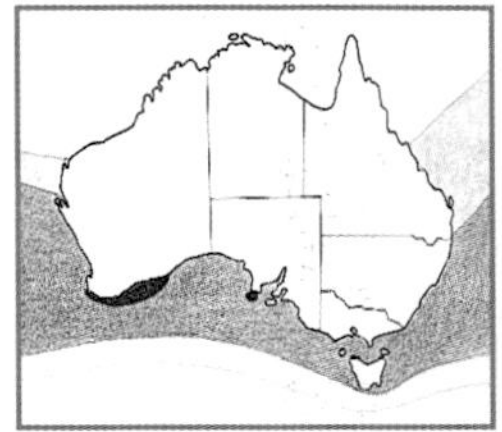

A large, burly, brownish black shearwater with characteristic lethargic flight; long slow glides, banking with one wingtip then the other towards the waves, interspersed with deep, slow wingbeats. In strong winds the glides become more swift and purposeful, banking, lifting and swooping between the swells, but without the erratic energy of the smaller and lighter shearwaters. Solitary or flocks; follows boats. **Voice:** usually silent at sea; noisy in colonies with its repeated hoarse 'ku-KOOO-uhg', rising to a strident scream. **Hab.:** pelagic, usually beyond edge of continental shelf. **Status:** common S Australian seas. **Br.:** 357

Sooty Shearwater *Puffinus griseus* 40–46 cm; span 1 m (H, L, M, N)

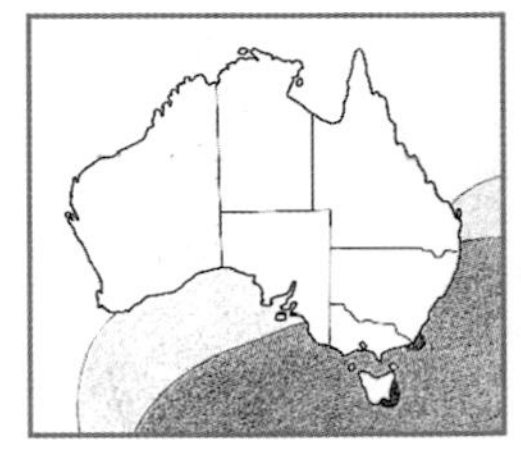

Flight fast and direct: mixes bursts of rapid, stiff-winged beats with long banking glides when one wingtip brushes the water's surface. In strong winds, arcs higher on flexed wings. **Voice:** usually silent at sea, but noisy on arrival and prior to the pre-dawn departure. The usual call is a loud, regularly repeated, rather dove-like, wailing coo; begins slowly, softly, then rises through the sequence, building to a loud frenzy of excitement before dying away again: 'awook, awook, awow, awow-AWOW-awow-awook, awook'. **Similar:** Short-tailed and Flesh-footed Shearwaters. **Hab.:** pelagic in Antarctic to subtropical seas; occasionally coastal waters. **Status:** common to abundant around S coasts. **Br.:** 357

Short-tailed Shearwater *Puffinus tenuirostris* 40–45 cm; span about 1 m (M)

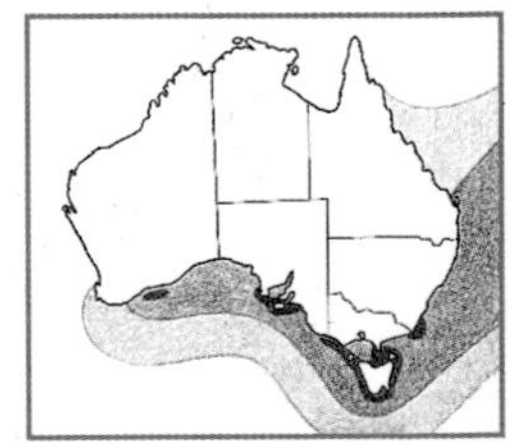

The 'muttonbird' harvested on islands of Bass Str. At sea around colonies may be seen in huge flocks, some estimated to contain millions of birds. These may form large rafts on the sea surface near breeding islands in late afternoon; birds await the cover of darkness before approaching nests. Flight is slow, appears leisurely; floats over ocean on stiffly held wings. In stronger winds the gliding birds skim the waves, dipping into the hollows and lifting over the crests. **Similar:** Sooty Shearwater. **Hab.:** marine, pelagic; usually over continental shelf waters. **Status:** The Tas. population has been estimated at about 16 million birds. **Br.:** 357

Pink-footed Shearwater *Puffinus creatopus* 48 cm; span 1.1 m

A large shearwater with grey-brown plumage; variable extent of mottling on flanks and underwing. Buoyant, languid flight with rather heavy, laboured flapping action alternating with long glides on downcurved wings. In strong winds the flight becomes invigorated with higher wheeling and banking. **Voice:** not recorded. **Variation:** light, intermediate and dark morphs; intermediate most common. **Similar:** light morph of Wedge-tailed Shearwater, but has longer, more pointed tail; other minor differences. **Hab.:** a preference for shallower waters of continental shelf seas. **Status:** rare vagrant to Aust.; several possible sightings.

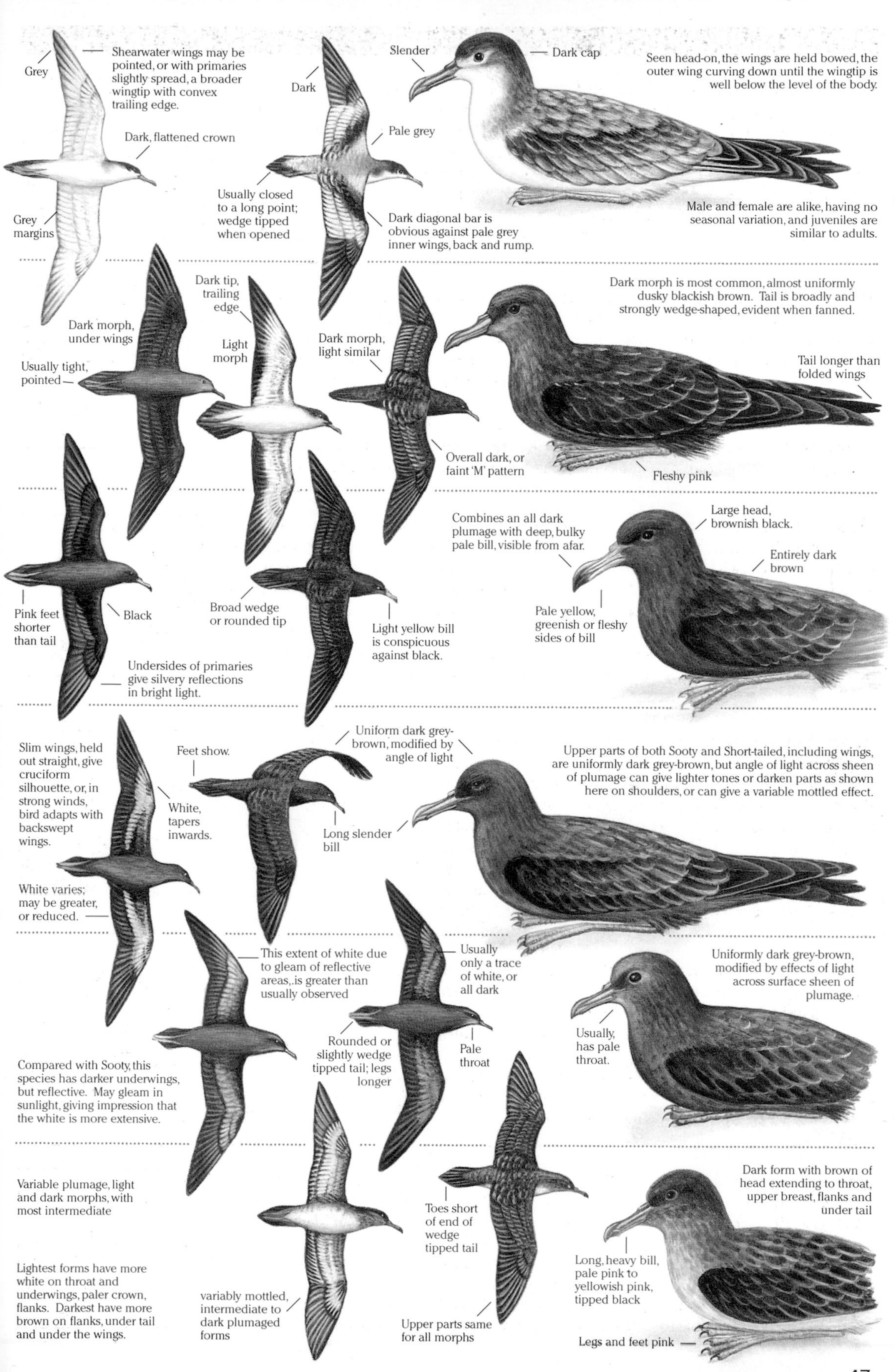

Grey
Shearwater wings may be pointed, or with primaries slightly spread, a broader wingtip with convex trailing edge.
Dark, flattened crown
Grey margins
Dark
Usually closed to a long point; wedge tipped when opened
Slender
Pale grey
Dark diagonal bar is obvious against pale grey inner wings, back and rump.
Dark cap
Seen head-on, the wings are held bowed, the outer wing curving down until the wingtip is well below the level of the body.
Male and female are alike, having no seasonal variation, and juveniles are similar to adults.
Dark morph, under wings
Usually tight, pointed
Dark tip, trailing edge
Light morph
Dark morph, light similar
Overall dark, or faint 'M' pattern
Dark morph is most common, almost uniformly dusky blackish brown. Tail is broadly and strongly wedge-shaped, evident when fanned.
Tail longer than folded wings
Fleshy pink
Pink feet shorter than tail
Black
Undersides of primaries give silvery reflections in bright light.
Broad wedge or rounded tip
Light yellow bill is conspicuous against black.
Combines an all dark plumage with deep, bulky pale bill, visible from afar.
Large head, brownish black.
Entirely dark brown
Pale yellow, greenish or fleshy sides of bill
Slim wings, held out straight, give cruciform silhouette, or, in strong winds, bird adapts with backswept wings.
White, tapers inwards.
White varies; may be greater, or reduced.
Feet show.
Uniform dark grey-brown, modified by angle of light
Long slender bill
Upper parts of both Sooty and Short-tailed, including wings, are uniformly dark grey-brown, but angle of light across sheen of plumage can give lighter tones or darken parts as shown here on shoulders, or can give a variable mottled effect.
This extent of white due to gleam of reflective areas,.is greater than usually observed
Compared with Sooty, this species has darker underwings, but reflective. May gleam in sunlight, giving impression that the white is more extensive.
Usually only a trace of white, or all dark
Rounded or slightly wedge tipped tail; legs longer
Pale throat
Uniformly dark grey-brown, modified by effects of light across surface sheen of plumage.
Usually, has pale throat.
Variable plumage, light and dark morphs, with most intermediate
Lightest forms have more white on throat and underwings, paler crown, flanks. Darkest have more brown on flanks, under tail and under the wings.
variably mottled, intermediate to dark plumaged forms
Toes short of end of wedge tipped tail
Upper parts same for all morphs
Dark form with brown of head extending to throat, upper breast, flanks and under tail
Long, heavy bill, pale pink to yellowish pink, tipped black
Legs and feet pink

Great Shearwater *Puffinus gravis* 43–51 cm; span 1–1.2 m
Large with clearcut dark cap. Flight swift and deliberate with powerful, rapid beating of stiff, straight wings; glides and banks close to the waves, plunging into the water or rising 4 to 10 m to dive at a shoal of fish or other marine life. Usually solitary, often follows ships. **Voice:** noise like fighting cats when squabbling around trawlers. **Similar:** Streaked, Grey, Buller's Shearwaters, Grey Petrel. **Hab.:** open seas; rarely inshore. **Status:** rare vagrant, several sightings of a single bird at sea, Robe, SA, Jan.–Feb. 1989.

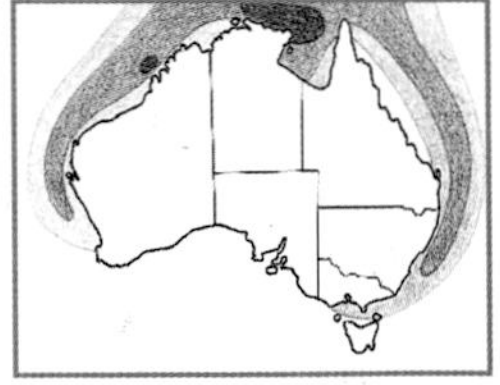

Streaked Shearwater *Calonectris leucomelas* 47–49 cm; span 1.22 m
Large shearwater; face and crown white with black streaks. Flight is slow and graceful in leisurely arcs; glides on bowed wings, occasional deep, rather heavy flaps. In strong winds, flaps less; lifts and banks higher, faster. Follows fishing boats, solitary or in small groups, or mixed with other seabirds. **Voice:** usually silent at sea. **Similar:** Wedge-tailed Shearwater, light morph. **Hab.:** pelagic seas, shelf waters and further out; unusual inshore. **Status:** common summer–autumn visitor to N, W and E coasts of Aust. **Br.:** on islands offshore Japan and Korea.

Manx Shearwater *Puffinus puffinus* 30–38 cm; span 78–88 cm
Medium sized; black upper parts in sharp contrast against white underparts. Glides and banks low above the waves with brief flurries of rapid, shallow, stiff wingbeats. Against stronger winds, lifts, banks and glides with little flapping. **Voice:** silent at sea, but calls at colonies. **Similar:** Fluttering and Hutton's Shearwaters. **Hab.:** favours edge of continental shelf; infrequent further out; rarely over inshore waters. **Status:** extremely rare vagrant to Aust. seas; one dead on beach, SA, several possible sightings SE Aust. **Br.:** N hemisphere.

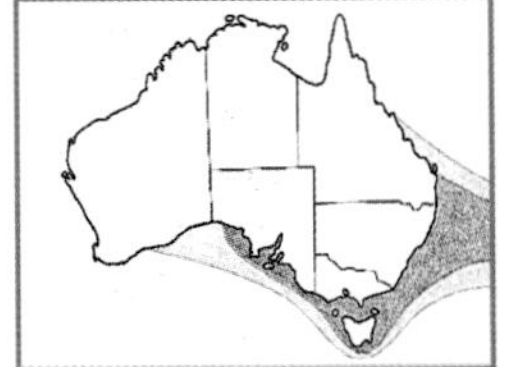

Fluttering Shearwater *Puffinus gavia* 32–37 cm; span 75–77 cm
Dark brown upper parts, white beneath. Flight is swift and low, short glides on stiff wings. Slightly slower, deeper wing action than other small shearwaters, which gives a fluttering effect. Its flight follows a slightly less direct course with more banking, lifting and falling. **Voice:** silent at sea, noisy over island colonies. **Similar:** Hutton's almost identical; Little Shearwater smaller. **Hab.:** more frequently seen inshore in estuaries and harbours than are other shearwaters. **Status:** common Aust. inshore waters Apr.–Oct. **Br.:** NZ coastal islands.

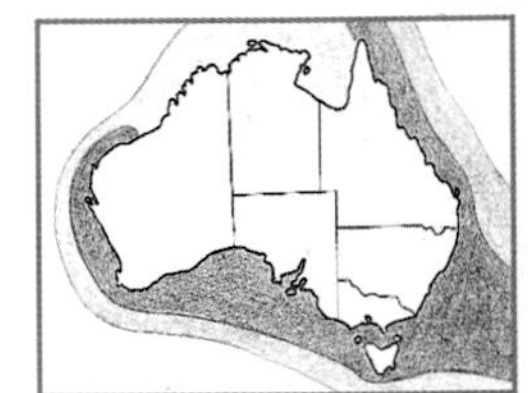

Hutton's Shearwater *Puffinus huttoni* 36–38 cm; span 72–78 cm
Flight like Fluttering Shearwater; slightly longer necked, shorter tailed appearance. Often in flocks, small to huge. Floats high when swimming. **Voice:** usually silent at sea, but occasionally some cackling sounds. Very noisy in breeding colonies; a juddering, squawking 'kouw-kouwkee-kee-aah' with variations. **Hab.:** prefers waters of continental shelf; at times comes inshore into estuaries, bays and channels. **Status:** migratory around most of Aust. coast, mainly birds under 2 yrs. Adults probably sedentary near NZ breeding grounds.

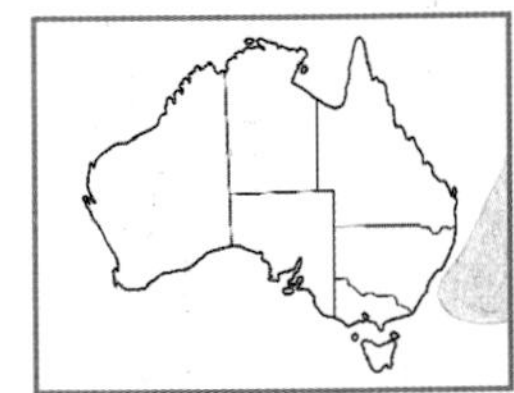

Audubon's Shearwater *Puffinus lherminieri* 27–33 cm; span 64–74 cm
Small, stocky; long tail with black under-tail coverts and underside of tail, which is often spread. In light winds it glides more than other fluttering shearwaters; flight looks smoother. In strong winds has high, wheeling glides. **Voice:** noisy in breeding colonies; usual call 'shooo-kree', either rasping or sharply screeched. **Similar:** Fluttering and Little Shearwaters. **Hab.:** tropical and subtropical oceans; occasionally inshore bays, estuaries. **Status:** several possible sightings off E coast. **Br.:** colonies on islands of tropical oceans.

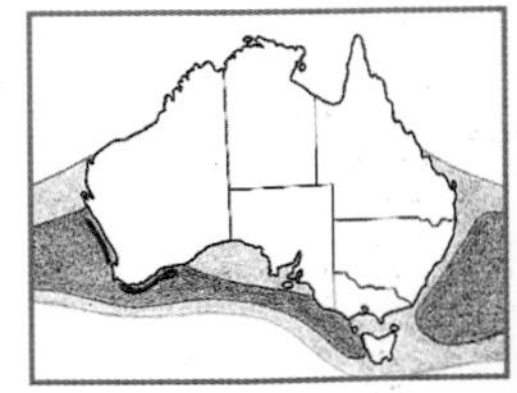

Little Shearwater *Puffinus assimilis* 25–30 cm; span 58–67 cm (L, M, N)
Smaller, slenderer than Audubon's Shearwater. Flight is 'flutter and glide'—shallow, whirring wingbeats followed by brief low glides; flat, not banking or twisting. In strong winds, long banking glides over waves and troughs, fluttering to catch the wind on crests. Alights on water; swims under water. **Variation:** two Aust. races, *assimilis* in E and *tunneyi* in W. **Voice:** silent at sea; breeding, a growling, sobbing 'wah-i-wah-i-wah-ooo'. **Hab.:** oceanic; continental shelf waters. **Status:** common offshore S of WA; uncommon in SE Aust. **Br.:** 357

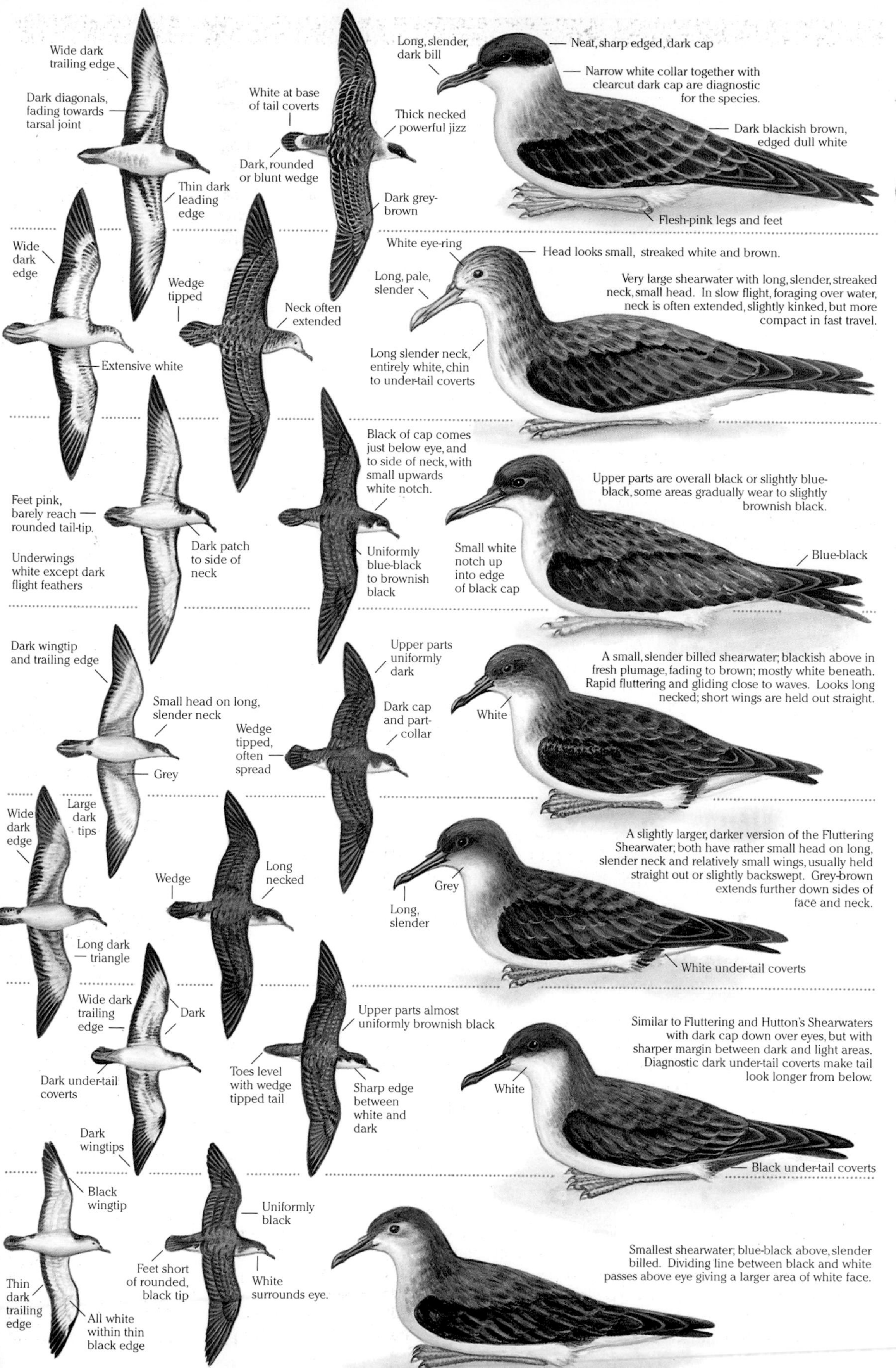
Wide dark trailing edge
Dark diagonals, fading towards tarsal joint
Thin dark leading edge
White at base of tail coverts
Dark, rounded or blunt wedge
Thick necked powerful jizz
Dark grey-brown
Long, slender, dark bill
Neat, sharp edged, dark cap
Narrow white collar together with clearcut dark cap are diagnostic for the species.
Dark blackish brown, edged dull white
Flesh-pink legs and feet
Wide dark edge
Extensive white
Wedge tipped
Neck often extended
White eye-ring
Long, pale, slender
Head looks small, streaked white and brown.
Very large shearwater with long, slender, streaked neck, small head. In slow flight, foraging over water, neck is often extended, slightly kinked, but more compact in fast travel.
Long slender neck, entirely white, chin to under-tail coverts
Feet pink, barely reach rounded tail-tip.
Dark patch to side of neck
Underwings white except dark flight feathers
Black of cap comes just below eye, and to side of neck, with small upwards white notch.
Uniformly blue-black to brownish black
Small white notch up into edge of black cap
Upper parts are overall black or slightly blue-black, some areas gradually wear to slightly brownish black.
Blue-black
Dark wingtip and trailing edge
Small head on long, slender neck
Grey
Wedge tipped, often spread
Upper parts uniformly dark
Dark cap and part-collar
White
A small, slender billed shearwater; blackish above in fresh plumage, fading to brown; mostly white beneath. Rapid fluttering and gliding close to waves. Looks long necked; short wings are held out straight.
Wide dark edge
Large dark tips
Wedge
Long necked
Long dark triangle
Long, slender
Grey
A slightly larger, darker version of the Fluttering Shearwater; both have rather small head on long, slender neck and relatively small wings, usually held straight out or slightly backswept. Grey-brown extends further down sides of face and neck.
White under-tail coverts
Wide dark trailing edge
Dark
Dark under-tail coverts
Toes level with wedge tipped tail
Upper parts almost uniformly brownish black
Sharp edge between white and dark
White
Similar to Fluttering and Hutton's Shearwaters with dark cap down over eyes, but with sharper margin between dark and light areas. Diagnostic dark under-tail coverts make tail look longer from below.
Black under-tail coverts
Dark wingtips
Black wingtip
Thin dark trailing edge
All white within thin black edge
Uniformly black
Feet short of rounded, black tip
White surrounds eye.
Smallest shearwater; blue-black above, slender billed. Dividing line between black and white passes above eye giving a larger area of white face.

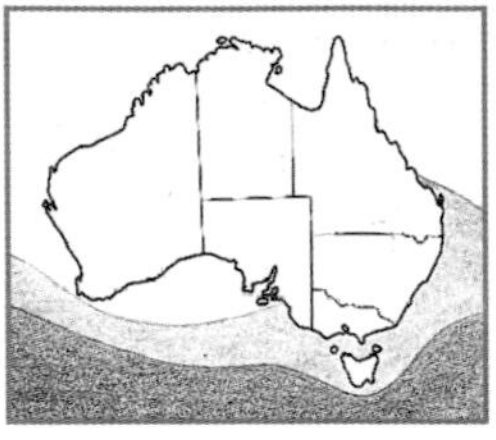

Broad-billed Prion *Pachyptila vittata* 28 cm; span 60–62 cm
Largest of prions, extremely wide bill. Flight slower; skims low across sea; slower wingbeats and more gliding, less erratic, steep banking than smaller narrow-billed prions. Wings held well forward; appears short necked, long tailed; massive bill and high forehead, looking front-heavy in flight. **Voice:** noisy at night in colonies; gives a rough, rasping 'ku, ku-aah, kuk'. **Similar:** Salvin's, Antarctic Prions. **Hab.:** feeds in dense flocks; usually deep pelagic seas, inshore waters in storms. Subtropical, tropical and sub-Antarctic waters. **Status:** rare vagrant; some beach-cast on SE and SW coasts. **Br.:** islands of southern oceans.

Salvin's Prion *Pachyptila salvini* 25–30 cm; span 58 cm
Is sometimes included within Broad-billed Prion species as race *salvini*, and is probably indistinguishable at sea from that species and the Antarctic Prion. However, recent studies show *vittata* to be a distinct species, more closely related to Antarctic Prion; flight is similar to both related species. Gregarious, often in dense flocks. **Voice:** not well known; probably calls only in colonies, loud 'KA-kaka-DU'. **Hab.:** sub-Antarctic, extending into adjacent Antarctic and subtropical waters. **Status:** common visitor to Aust. offshore waters Jun.–Oct.; presence revealed by many beach-wrecked specimens. **Br.:** islands of S Indian Ocean.

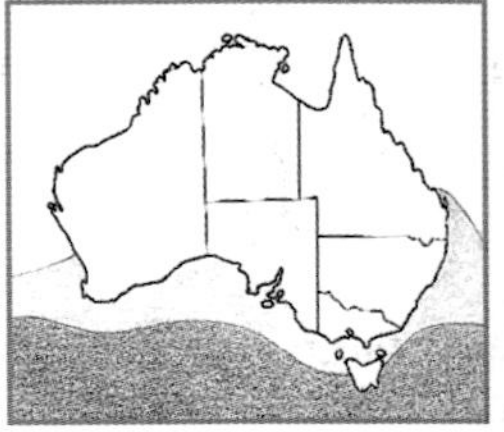

Antarctic Prion *Pachyptila desolata* 25–28 cm; span 62–65 cm (H, M, L)
At sea, similar in appearance and flight to above species; has same feeding tactics: dense flocks work together across the water, into the wind, wings outstretched, feet paddling—shallow plunges just beneath the surface to snatch small marine creatures. **Voice:** silent at sea; noisy in colonies with its throaty, dove-like cooing: 'uk-coo-uk-cooo-uk-uk-u-cooo'. **Hab.:** Antarctic to sub-Antarctic, favouring cold waters just N of the zone of pack-ice. **Status:** moderately common in S Aust. seas; often beach-cast winter to spring; considerable numbers are wrecked after storms. **Br.:** islands of sub-Antarctic seas.

Slender-billed Prion *Pachyptila belcheri* 25–26 cm; span 56 cm (H, M)
Flight fast, erratic, buoyant, keeping close to the waves; lifts over crests with bursts of rapid wingbeats and through troughs with long, weaving, twisting glides. Feeds largely on small crustaceans. When feeding, glides and flutters just above the waves, pattering the water occasionally for added lift, often snapping with slender bill at the surface. **Voice:** silent at sea; in colonies usually a harsh cooing, occasional squawks and trills. **Similar:** Fairy and Fulmar Prions. **Hab.:** sub-Antarctic and Antarctic seas S to edge of pack-ice; shallower shelf waters. **Status:** common SW coast; uncommon in SE. **Br.:** islands of sub-Antarctic seas.

Fairy Prion *Pachyptila turtur* 24–28 cm; span 56–60 cm (M)
Smallest prion; keeps close to the waves, often huge flocks weaving and banking in unison, displaying blue-grey then white surfaces. Flight is buoyant and erratic; feeds by dancing lightly across the surface, wings fluttering and feet pattering, dipping head to surface, then plunging beneath it. **Voice:** noisy at night in breeding grounds. **Hab.:** sub-Antarctic seas and islands while breeding, then wanders to subtropical seas; rarely close inshore except when sheltering from storms. **Status:** commonly seen; usually well offshore. In the SE this is the most abundant of the prions; breeding colonies hold up to 25 000 pairs. **Br.:** 357

Fulmar Prion *Pachyptila crassirostris* 28 cm; span 60 cm (H)
A rare species closely resembling the much more abundant Fairy Prion; the two species are usually indistinguishable at sea. Flight similar; Fulmar more erratic and often performs an acrobatic, high, up-and-over loop that returns it to the original course. Strongly gregarious, feeds in flocks. Females slightly smaller. **Voice:** noisy at night in colonies; gives stuttering, low, harsh croakings. **Hab.:** marine, pelagic. **Status:** vagrant; beachwashed specimen from Tas., unconfirmed records from Vic. and S coast NSW. May visit more often but very difficult to confirm from sightings at sea. **Br.:** sub-Antarctic islands S of NZ and Heard Is.

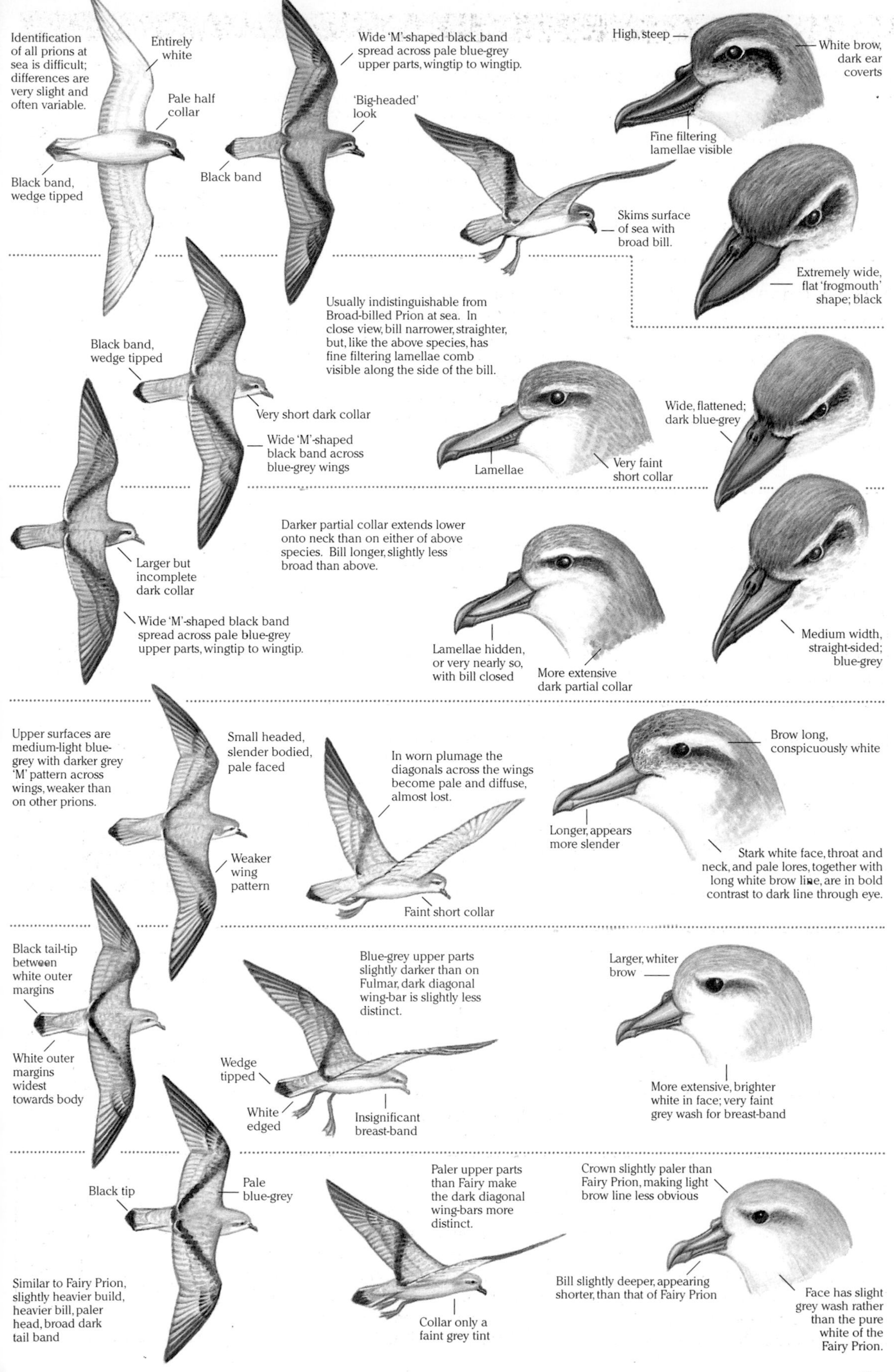
Identification of all prions at sea is difficult; differences are very slight and often variable.
Entirely white
Pale half collar
Black band, wedge tipped
Black band
Wide 'M'-shaped black band spread across pale blue-grey upper parts, wingtip to wingtip.
'Big-headed' look
High, steep
White brow, dark ear coverts
Fine filtering lamellae visible
Skims surface of sea with broad bill.
Extremely wide, flat 'frogmouth' shape; black
Usually indistinguishable from Broad-billed Prion at sea. In close view, bill narrower, straighter, but, like the above species, has fine filtering lamellae comb visible along the side of the bill.
Black band, wedge tipped
Very short dark collar
Wide 'M'-shaped black band across blue-grey wings
Lamellae
Very faint short collar
Wide, flattened; dark blue-grey
Darker partial collar extends lower onto neck than on either of above species. Bill longer, slightly less broad than above.
Larger but incomplete dark collar
Wide 'M'-shaped black band spread across pale blue-grey upper parts, wingtip to wingtip.
Lamellae hidden, or very nearly so, with bill closed
More extensive dark partial collar
Medium width, straight-sided; blue-grey
Upper surfaces are medium-light blue-grey with darker grey 'M' pattern across wings, weaker than on other prions.
Small headed, slender bodied, pale faced
In worn plumage the diagonals across the wings become pale and diffuse, almost lost.
Brow long, conspicuously white
Longer, appears more slender
Weaker wing pattern
Faint short collar
Stark white face, throat and neck, and pale lores, together with long white brow line, are in bold contrast to dark line through eye.
Black tail-tip between white outer margins
White outer margins widest towards body
Blue-grey upper parts slightly darker than on Fulmar, dark diagonal wing-bar is slightly less distinct.
Wedge tipped
White edged
Insignificant breast-band
Larger, whiter brow
More extensive, brighter white in face; very faint grey wash for breast-band
Black tip
Pale blue-grey
Paler upper parts than Fairy make the dark diagonal wing-bars more distinct.
Crown slightly paler than Fairy Prion, making light brow line less obvious
Similar to Fairy Prion, slightly heavier build, heavier bill, paler head, broad dark tail band
Collar only a faint grey tint
Bill slightly deeper, appearing shorter, than that of Fairy Prion
Face has slight grey wash rather than the pure white of the Fairy Prion.

Southern Giant-Petrel *Macronectes giganteus* 85–100 cm; span 1.8–2.1 m (H, M)
Largest of petrels; superficially albatross-like, but lacking their graceful shape and flight. Massive, ugly, yellowish green bill with conspicuous large nasal tube extending more than halfway along top edge. Dimorphic, either dark grey-brown with whitish head and neck, or overall white with scattered dark feathers. Undersurface leading edge of inner wing is pale. In active flight, heavy, humpbacked, laboured. The stiff-winged glides lack the graceful skill of an albatross. In strong winds the gliding improves, with some wheeling and weaving. Gathers at carrion, offal, sewage outlets. **Juv.:** brownish black, scattered white about face. **Imm.:** dark morph is like adult Northern Giant-Petrel, blackish brown, mottled white about face; imm. white morph is like adult. **Voice:** usually silent at sea except when squabbling over carrion or other food. At the breeding grounds is noisy throughout year; various sounds, mostly unpleasant. A deep, stuttered sequence of groaning sounds, 'ur-ur-ur-ur', beginning very deep and resonant, becoming high and rapid. Also deep, throaty, croaking groans, loud creaky door noises, braying, gurgling and cat-like mewing. **Similar:** Northern Giant-Petrel is like the dark morph of this species and even more like the juvenile; at close range, bill colours differ. **Hab.:** marine, over open seas and inshore waters; favours edge of continental shelf and edge of pack-ice. **Status:** common around S coast of Aust., predominantly immatures. **Br.:** colonies on Antarctic and sub-Antarctic islands, and Antarctic mainland.

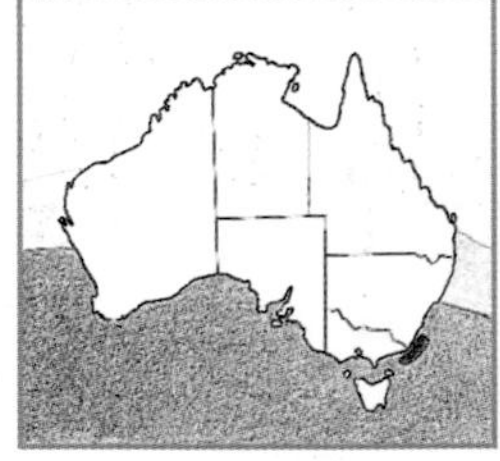

Northern Giant-Petrel *Macronectes halli* 81–94 cm; span 1.8–2 m (H, M)
Flight and behaviour similar to the preceding species. Bill is pale yellow, usually reddish tipped. Leading edge of inner-wing undersurface is dark near body. **Voice:** usually silent at sea except in a squabbling mob near carrion or refuse. Generally like calls of Southern Giant-Petrel; display calls possibly lower and harsher; some deep: 'argh-argh-argh-'; or rising to a whistling, screaming crescendo. **Similar:** dark phase juveniles of Southern Giant-Petrel, but bill details differ. Superficially similar to sooty albatrosses, juveniles of Wandering Albatross. **Hab.:** marine, temperate and sub-Antarctic seas, frequenting inshore and pelagic seas out from edges of continental shelves. On average tends to prefer slightly warmer seas than Southern Giant-Petrel, temperate to sub-Antarctic. **Status:** common on coasts and offshore waters of S Aust., May–Oct. **Br.:** as single pairs on islands of sub-Antarctic seas.

Sooty Albatross *Phoebetria fusca* 85–90 cm; span 2–2.5 m (M)
This small sooty brown Albatross gives an all dark impression at sea; tapering streamlined shape has little detail to interrupt sleek contours. Flight swift, appearing graceful and effortless, especially in strong winds. Aerobatic gliding, soaring and stooping on wings that are always slightly flexed; rarely resorts to flapping; uses the energy of waves, and the winds above, to a height of 10 or 15 m. Swims in upright posture, often follows ships. **Fem.:** slightly smaller. **Voice:** silent at sea except for threat call when feeding—a harsh, throaty 'ghaaaow'; bill clappering in display. Calls at breeding colony are similar to those of Light-mantled Sooty Albatross. **Similar:** Light-mantled Sooty, but has a light grey neck, mantle and back. Also, immature of Wandering Albatross and giant-petrels. **Hab.:** uses the pelagic environs, the oceans beyond the shallower continental shelf waters, their surface waters and resources. **Status:** regular visitor in small numbers to S Aust., usually well out beyond the continental shelf, Jan.–Nov. **Br.:** on islands of S Indian Ocean and S Atlantic.

Light-mantled Sooty Albatross *Phoebetria palpebrata* 80–90 cm; span 2 m (H, M)
Similar to Sooty Albatross, but upper body, from nape to upper tail coverts, is pale ash-grey in contrast to dark, sooty brown head. Flight graceful, elegant; demonstrates complete mastery of forces of wind and waves. Soars and glides with very little flapping. **Juv.:** in fresh plumage lacks white-shafted primaries; eye crescent and sulcus line along bill are grey. **Voice:** usually silent at sea except for a harsh 'ghaaaa!' threat call, usually while squabbling for food. At nest a loud 'PEEE-iew, PEEE-aagh'. **Hab.:** S to Antarctic pack-ice in summer; N to subtropical waters in winter. On average, prefers colder seas than does the Sooty Albatross, but overlapping. Usually over deep pelagic and shelf-edge seas; occasionally inshore waters. **Status:** rare winter–spring visitor to SE Aust. **Br.:** islands of sub-Antarctic oceans.

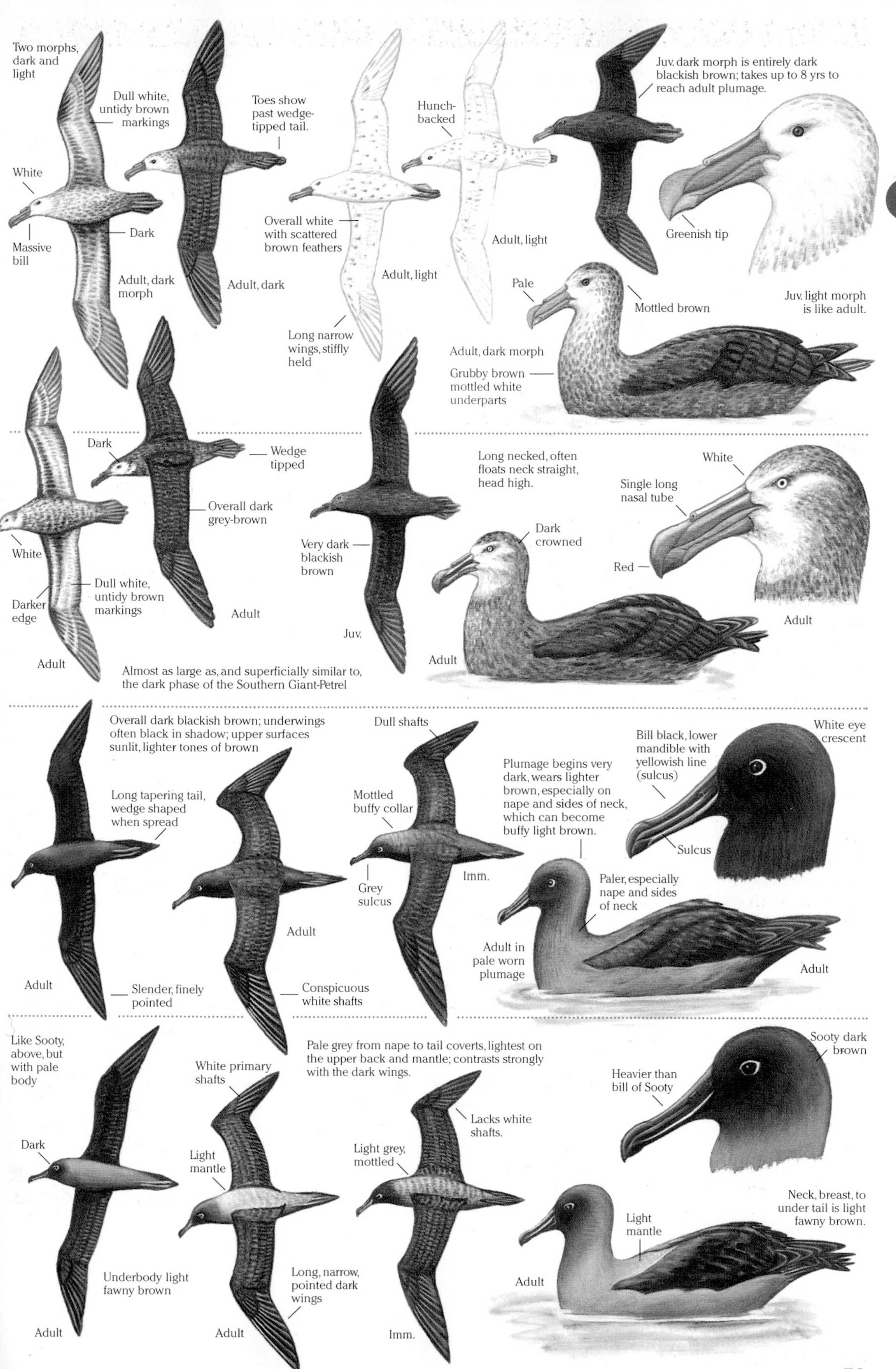

Two morphs, dark and light
Dull white, untidy brown markings
White
Massive bill
Dark
Adult, dark morph
Adult, dark
Toes show past wedge-tipped tail.
Overall white with scattered brown feathers
Long narrow wings, stiffly held
Adult, light
Hunch-backed
Adult, light
Juv. dark morph is entirely dark blackish brown; takes up to 8 yrs to reach adult plumage.
Greenish tip
Pale
Mottled brown
Juv. light morph is like adult.
Adult, dark morph
Grubby brown mottled white underparts
Dark
Wedge tipped
Overall dark grey-brown
White
Darker edge
Dull white, untidy brown markings
Adult
Adult
Very dark blackish brown
Juv.
Almost as large as, and superficially similar to, the dark phase of the Southern Giant-Petrel
Long necked, often floats neck straight, head high.
Dark crowned
White
Single long nasal tube
Red
Adult
Adult
Overall dark blackish brown; underwings often black in shadow; upper surfaces sunlit, lighter tones of brown
Long tapering tail, wedge shaped when spread
Adult
Slender, finely pointed
Adult
Conspicuous white shafts
Dull shafts
Mottled buffy collar
Grey sulcus
Imm.
Plumage begins very dark, wears lighter brown, especially on nape and sides of neck, which can become buffy light brown.
Bill black, lower mandible with yellowish line (sulcus)
White eye crescent
Sulcus
Paler, especially nape and sides of neck
Adult in pale worn plumage
Adult
Like Sooty, above, but with pale body
Dark
Underbody light fawny brown
Adult
White primary shafts
Light mantle
Adult
Long, narrow, pointed dark wings
Pale grey from nape to tail coverts, lightest on the upper back and mantle; contrasts strongly with the dark wings.
Lacks white shafts.
Light grey, mottled
Imm.
Heavier than bill of Sooty
Sooty dark brown
Light mantle
Neck, breast, to under tail is light fawny brown.
Adult

Wandering Albatross *Diomedea exulans* 1.1–1.2 m; span 2.5–3.5 m (H, L, M, N)
Very large, extremely long wings, usually stiffly outstretched; graceful, dynamic gliding and soaring, using the updraft from waves to sustain flight for long periods with little need for wing flapping. In strong winds, wings held bowed and flexed. Usually within 15 m above the sea, the bird often rests on sea in calm conditions—requires long take-off run; follows ships. **Male:** when mature, at 8–10 years or more, is approaching the maximum extent of white in the plumage; retains black only on the primaries and a narrow margin along the secondaries, both above and beneath. **Fem.:** seldom acquires as much white on upper wing as old males. **Juv.:** plumage initially dark brown except for conspicuous white face and white underwing, gradually showing increasingly white upper parts over the following 7 to 9 years. The white of the upper wing begins along the centre-line and spreads forward and back. No seasonal changes in plumage. **Voice:** at sea usually silent, but occasionally makes gurgling, croaking sounds. In colonies, groaning, yapping, croaking interspersed with harsh braying and shrill shrieking. **Variation:** S race *chionoptera* is snowy white at maturity; N race *exulans* is slightly smaller, usually retains slightly darker plumage. Those around SE Aust. coasts are usually of the race *exulans*, but most reaching the SW of WA seem to be of the race *chionoptera*, which has one of its breeding sites on Macquarie Is. A few of the race *gibsoni* (sometimes not separated from *exulans*) reach the SW corner of the continent. Another population, previously within the nominate race *exulans*, breeds on Amsterdam Is. in the S Indian Ocean; this may be a separate species, which would be known as the Amsterdam Albatross, *Diomedea amsterdamensis*. These birds permanently retain the dark plumage typical of the juvenile Wandering, and have a unique bill, pink like that of the Wandering, but with brownish green tips and dark cutting edge. **Similar:** Royal Albatross, but which are white-bodied, with whitening of upper wing gradually spreading back from the leading edge, rather than from the centre-line of the wing. Mature birds with white of upper parts fully developed are most difficult to separate, except on small details: black cutting edge to bill indicates a Royal, and only the Wandering has any black to tip or edges of tail. Sooty albatrosses and giant petrels are superficially similar to young brown-bodied Wandering Albatrosses, but lack the great overall size and white underwing of the Wandering Albatross. **Hab.:** marine, open oceans, edge of pack-ice; feeds over both the deep pelagic and the shallower continental shelf waters. **Status:** common visitor to offshore and continental waters of SE Aust., and regular around S coast to SW of WA; more often sighted winter to early spring; most belong to the race *exulans*. **Br.:** Antarctic and sub-Antarctic islands.

Royal Albatross *Diomedea epomophora* 1.1–1.2 m; span 3–3.5 m (H, M)
Huge, long winged, heavy bodied albatross with massive, bulbous tipped bill that has a black cutting edge, visible as a fine, dark line along the edge of the upper mandible. In flight, graceful and skilled in using the uplifts of wind across waves. The soaring, banking turns and long, sweeping glides are only infrequently interrupted by any flapping of wings. But the flight becomes laboured in calm conditions when albatrosses often resort to resting, or 'loafing', on the surface. Food is taken by snatching from the water or shallow plunging into the sea. Solitary or gregarious, it follows fishing boats, squabbling over scraps. **Fem.:** on average slightly smaller, darker. **Juv.:** head and body white, upper wing surfaces black, almost joined by dark, mottled plumage of back. White gradually increases on the wings, beginning closest to the body at the leading edge of each wing and spreading outwards in successive years. On adult birds (about 9 years of age) the entire upper wing is white except for the primaries and a narrow black trailing edge, both of which always remain black. **Voice:** at sea, occasional harsh squawks. Noisy in breeding colonies; calls are similar to, perhaps not as harsh as, those of the Wandering Albatross. These include loud ratchetting sounds, slow, deep, resonant, gurgling clucks and low vibrant groans. **Variation:** two subspecies, markedly different. The nominate race *epomophora*, as described above, gradually changes from the dark-winged juvenile to the white-winged adult, but the race *sanfordi* retains the all black upper surface to the wings with the body, including back and tail, entirely white. **Similar:** Wandering Albatross, but juvenile initially has dark body and, as that becomes white in following years, the wings also become white, but spreading outwards from the centre strip. At closer range, tail is black tipped or slightly black edged for all but a few old birds, and the bill lacks the black cutting edge along its upper mandible. Birds of race *sanfordi* are always distinctly different from Wandering Albatross. **Hab.:** subtropical and sub-Antarctic oceans; occasionally further S into the Antarctic. In Aust., occurs over both open ocean and shallower inshore waters. **Status:** commonly and regularly seen in Aust. waters: race *epomophora* is predominant in the SE; race *sanfordi* is more common off the South Australian coast. **Br.:** NZ mainland and nearby islands.

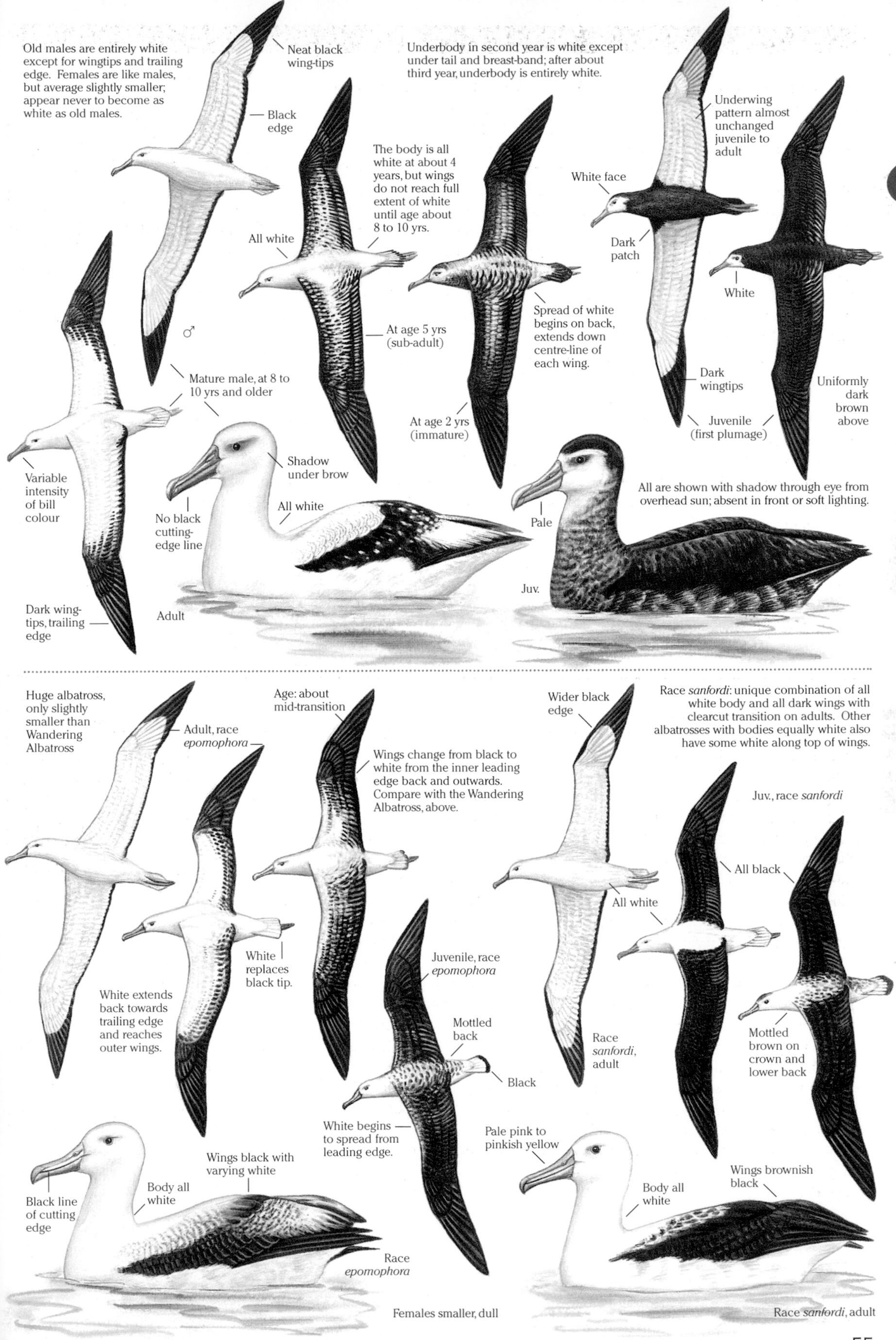
Old males are entirely white except for wingtips and trailing edge. Females are like males, but average slightly smaller; appear never to become as white as old males.
Neat black wing-tips
Black edge
♂
Underbody in second year is white except under tail and breast-band; after about third year, underbody is entirely white.
The body is all white at about 4 years, but wings do not reach full extent of white until age about 8 to 10 yrs.
All white
At age 5 yrs (sub-adult)
Spread of white begins on back, extends down centre-line of each wing.
At age 2 yrs (immature)
Underwing pattern almost unchanged juvenile to adult
White face
Dark patch
White
Dark wingtips
Juvenile (first plumage)
Uniformly dark brown above
Mature male, at 8 to 10 yrs and older
Variable intensity of bill colour
Shadow under brow
No black cutting-edge line
All white
Adult
Dark wing-tips, trailing edge
Pale
Juv.
All are shown with shadow through eye from overhead sun; absent in front or soft lighting.
Huge albatross, only slightly smaller than Wandering Albatross
Adult, race *epomophora*
Age: about mid-transition
Wings change from black to white from the inner leading edge back and outwards. Compare with the Wandering Albatross, above.
White replaces black tip.
White extends back towards trailing edge and reaches outer wings.
Juvenile, race *epomophora*
Mottled back
Black
White begins to spread from leading edge.
Wider black edge
Race *sanfordi*: unique combination of all white body and all dark wings with clearcut transition on adults. Other albatrosses with bodies equally white also have some white along top of wings.
Juv., race *sanfordi*
All black
All white
Race *sanfordi*, adult
Mottled brown on crown and lower back
Black line of cutting edge
Body all white
Wings black with varying white
Race *epomophora*
Females smaller, dull
Pale pink to pinkish yellow
Body all white
Wings brownish black
Race *sanfordi*, adult

Black-browed Albatross *Diomedea melanophris* 80–90 cm; span 2.2–2.5 m (H, M, N)
A medium-sized albatross with a unique combination of characteristics. A short dark brow, conspicuous against the white head, overhangs and shadows the eye, giving a frowning, penetrating or angry expression, while the bill is a bright orange-yellow with a waxy sheen. The underwing shows broad black margins, leaving a comparatively narrow zigzag white panel up the centre of each wing. In flight, this albatross is light and graceful. In strong winds, soars seemingly effortlessly, wheeling high on rigid wings, but it is forced to flap in light winds; often waits out calm periods on the water. **Juv.:** grey collar, narrowest or broken at the centre of the breast; very dark underwing. **Variation:** race *impavida*; brow darker and underwing has reduced area of white; pale iris. **Voice:** at sea it is usually silent except for harsh croaks when squabbling over food; noisy in colonies. **Hab.:** a wide range of marine habitats from inshore shallows, bays and channels to the edge of the continental shelf and beyond to pelagic ocean environs. **Status:** common from May to Aug. around Australia's southern coasts. **Br.:** islands of sub-Antarctic seas.

Buller's Albatross *Diomedea bulleri* 75–80 cm; span 2–2.15 m (M)
A small, slender albatross with graceful and apparently effortless flight; soars high on rigid wings, wheeling in wide arcs, gliding low into wave troughs. In light airs often rests on calm water, floating high, upright posture emphasising slender necked, small headed appearance. Solitary, or in small groups or mixed gatherings. **Juv.:** bill dark tipped grey-brown; plumage slightly browner; faint smudgy collar. **Variation:** race *platei*, pale forehead. **Voice:** silent at sea; in colony low, nasal, grating croaks like a crow, 'argh-argh-argh', abrupt or drawn out. Males give longer croaks, also wail and groan. **Similar:** Grey-headed Albatross, but has wider black leading edge to underwing and darker crown; Yellow-nosed, but narrow yellow edge only to top of bill, head and neck white. **Hab.:** subtropical to sub-Antarctic; shelf edge and pelagic, preferring warmer waters, or in S seas where currents from N are warmer. Uses inshore waters; visible from land when sheltering from rough conditions on open seas. **Status:** regular but uncommon visitor to SE Aust.; usually offshore Apr.–Jul. **Br.:** race *bulleri* in colonies on Snares Is. and Solander Is., NZ; race *platei* on Chatham Is. and Three Kings Is., NZ.

Grey-headed Albatross *Diomedea chrysostoma* 70–85 cm; span 1.8–2.3 m (H, M)
Medium sized, has unique characteristics—entirely grey head and neck; black bill with a yellow line going along top and bottom edges; broad black leading edges underwing. Soars, glides and wheels effortlessly in broad arcs; makes only the slightest adjustments to the stiffly outstretched wings to use the uplift of wind across waves. In light or calm conditions often rests on the surface. **Juv.:** head and neck darker grey; bill dark grey; underwing black and grey. **Imm.:** gradually more like adult with underwings developing a white central strip, head becoming paler, bill developing some dull yellow. **Voice:** silent in flight. At breeding grounds, vibrating, nasal croaking, a drawn out 'aarrrgh, aarrrgh' like a heavy, creaking door, and an abrupt, sharper 'a-a-a-a-a'. **Similar:** Buller's and Yellow-nosed Albatrosses, but have narrower black leading edge to the underwing. **Hab.:** marine: Antarctic to sub-Antarctic in summer; subtropical waters in winter, but seeking colder waters of southerly currents. Feeds over deep seas beyond the shelf-edge shallows; only seeks refuge inshore in rough conditions. **Status:** regular winter visitor to Aust.; offshore and coastal waters from SE and Tas. to SW of WA. **Br.:** colonies on islands of sub-Antarctic oceans.

Yellow-nosed Albatross *Diomedea chlororhynchos* 70–80 cm; span 1.8–2 m (K)
A small, slender 'mollymawk' with yellow-orange top edge to the long, slender, glossy black bill. Flight is like that of other small albatrosses—elegantly aerobatic in medium to strong winds, but often floats on the surface when it is too calm to get wind uplift from the waves. Solitary, small groups, or mixed with other seabirds around trawlers or other food sources. **Variation:** race *bassi*, 'Indian Yellow-nosed Albatross', possibly a full species. **Voice:** usually silent at sea except when fighting over food. Noisy in colonies with a rapid, high, vibrating, nasal yapping, 'hek–ek–ek-ek-ek-ekek-eg-eg-eeg-eeeg'. **Hab.:** shallower waters of continental shelves, vicinity of upwellings and confluence of oceanic currents. Subtropical to sub-Antarctic seas; most abundant over warmer waters. **Status:** race *bassi* a regular and common visitor from southern coasts of Aust. to SE Qld; race *chlororhynchos* rare on Aust. seas. **Br.:** in pairs and colonies on islands of S oceans.

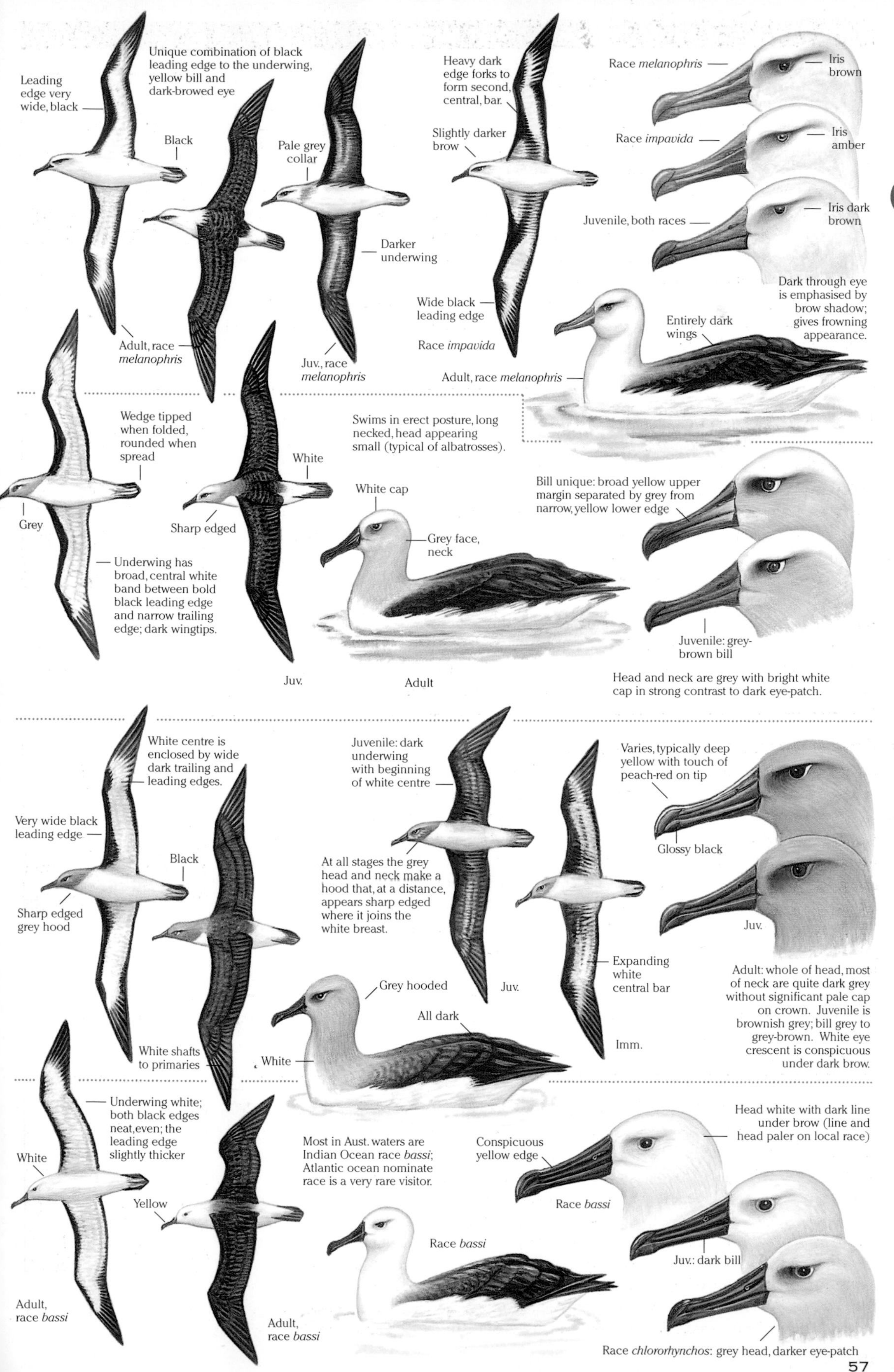
Leading edge very wide, black
Unique combination of black leading edge to the underwing, yellow bill and dark-browed eye
Black
Pale grey collar
Darker underwing
Adult, race melanophris
Juv., race melanophris
Heavy dark edge forks to form second, central, bar.
Slightly darker brow
Wide black leading edge
Race impavida
Race melanophris
Iris brown
Race impavida
Iris amber
Juvenile, both races
Iris dark brown
Dark through eye is emphasised by brow shadow; gives frowning appearance.
Entirely dark wings
Adult, race melanophris
Wedge tipped when folded, rounded when spread
Grey
White
Sharp edged
Underwing has broad, central white band between bold black leading edge and narrow trailing edge; dark wingtips.
Juv.
Swims in erect posture, long necked, head appearing small (typical of albatrosses).
White cap
Grey face, neck
Adult
Bill unique: broad yellow upper margin separated by grey from narrow, yellow lower edge
Juvenile: grey-brown bill
Head and neck are grey with bright white cap in strong contrast to dark eye-patch.
White centre is enclosed by wide dark trailing and leading edges.
Very wide black leading edge
Black
Sharp edged grey hood
White shafts to primaries
Juvenile: dark underwing with beginning of white centre
At all stages the grey head and neck make a hood that, at a distance, appears sharp edged where it joins the white breast.
Juv.
Grey hooded
All dark
White
Expanding white central bar
Imm.
Varies, typically deep yellow with touch of peach-red on tip
Glossy black
Juv.
Adult: whole of head, most of neck are quite dark grey without significant pale cap on crown. Juvenile is brownish grey; bill grey to grey-brown. White eye crescent is conspicuous under dark brow.
Underwing white; both black edges neat, even; the leading edge slightly thicker
White
Yellow
Adult, race bassi
Adult, race bassi
Most in Aust. waters are Indian Ocean race bassi; Atlantic ocean nominate race is a very rare visitor.
Race bassi
Conspicuous yellow edge
Race bassi
Head white with dark line under brow (line and head paler on local race)
Juv.: dark bill
Race chlororhynchos: grey head, darker eye-patch

Shy Albatross *Diomedea cauta* 1 m; span 2–2.5 m (M)
The largest of the black backed 'mollymawk' albatrosses, but still a half-metre to a metre smaller in wingspan than the two great albatrosses, the Wandering and the Royal. Easily identified by its size and the extent of white under the wings, and, at closer range, by the small black 'notch' on the black leading margin of each wing close to the body. It is the only albatross to breed in Australian waters and yet breed only within the Australasian region. It is proportionately very long winged, yet rather stout-bodied. The massive bill adds to an overall heavier, less agile jizz than have the other smaller, black backed mollymawks. In flight this species's casual effortless grace resembles the larger, white backed Wandering Albatross; it soars and wheels high above the horizon on rigid but slightly drooped wings. In light airs, without the lift and energy of wind across the tall waves of open ocean, the Shy Albatross must use deep, slow, heavy beats of its long wings, a laboured flight that usually keeps close to the sea. During calm, it often settles on the sea, at times forming rafts of many birds. Take off requires a long run across the surface, wings beating and feet kicking. Feeds by snatching from the surface and shallow diving at the surface. **Juv.:** dark to mid grey breast-band and head; dark grey bill with black tip. **Imm.:** pale grey head; narrow pale breast-band; bill pale grey-buff with dark tip. **Variation:** two other races, *salvini*, the 'Grey-backed Albatross', and *eremita*, the 'Chatham Island Albatross'. **Voice:** usually silent at sea, but harsh croaking when squabbling with other birds for food, especially around fishing boats. Noisy in colony, usually a deep, vibrating, nasal croaking, 'argk-argk-argk-argk'; also a harsh, strident wailing. **Similar:** size, mainly white underwing and the black notch at the base of the leading edge of wing separate this from other black backed albatrosses. **Hab.:** subtropical to sub-Antarctic oceans with preference for shallower shelf waters around islands and continental shelves. Comes close inshore, entering bays and harbours, and offshore out to the shelf edge; further out, scarce over pelagic depths. **Status:** nominate race *cauta* breeds in Aust. waters; disperses around Aust. coast from Carnarvon on W coast of WA to SE Qld; common throughout the year, but is more often seen in winter. Race *salvini* is a rare visitor to SW and SE Aust.; race *eremita* is a rare vagrant to SE Aust. **Br.:** 357

Laysan Albatross *Diomedea immutabilis* 80–82 cm; span 1.95–2.05 m (N)
A possible recent addition to the Australian bird list with reported sightings on NSW coast and at Norfolk Is. Has white head with extensive dark lores extending back through eyes to give a very dark browed look, then fading towards nape and down onto face. Underbody white, underwings white with black margins and variable dark streaks across the underwing coverts. Upper surfaces of wings are dark brown, continuous across the back. Tail dark, feet projecting beyond tail-tip. **Hab.:** marine, pelagic. **Status:** abundant across N Pacific having recovered from destruction of some colonies in early part of 20th century and during World War II. Breeding had been known only from islands of the NW Hawaiian chain, but has expanded its breeding range to the NW Pacific, closer to the NE Australian coast. An increased W Pacific breeding population may also result in expansion of the overall nonbreeding range in that region, thus increasing the chance of sightings of vagrant individuals in E Australian seas. In Aust. there has been one sighting off the coast near Wollongong, NSW. **Br.:** on islands of Hawaiian chain and Bonin Is. in NW Pacific, S of Japan.

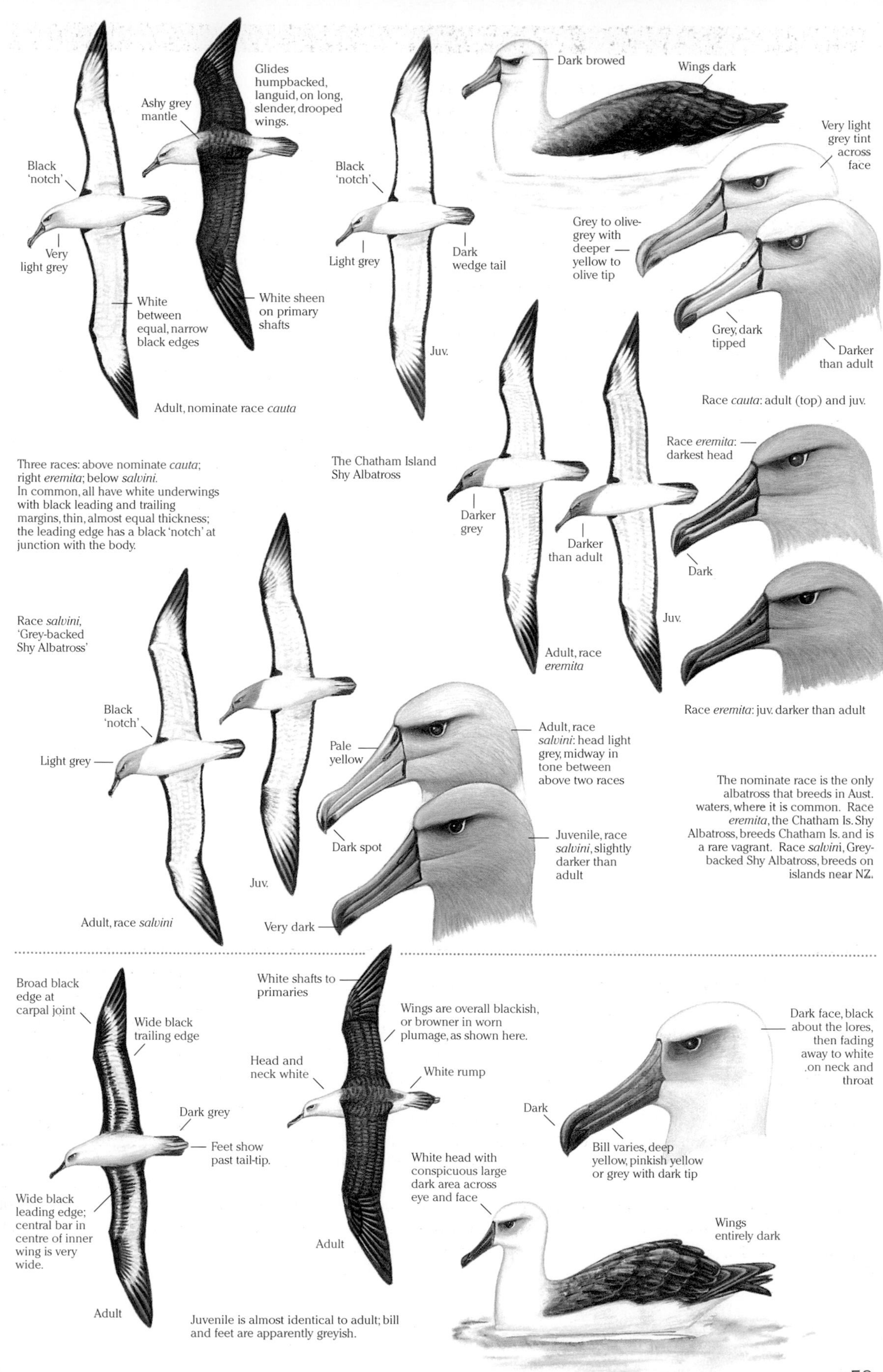
Glides humpbacked, languid, on long, slender, drooped wings.
Ashy grey mantle
Black 'notch'
Very light grey
White between equal, narrow black edges
White sheen on primary shafts
Adult, nominate race *cauta*
Black 'notch'
Light grey
Dark wedge tail
Juv.
Dark browed
Wings dark
Very light grey tint across face
Grey to olive-grey with deeper yellow to olive tip
Grey, dark tipped
Darker than adult
Race *cauta*: adult (top) and juv.
Three races: above nominate *cauta*; right *eremita*; below *salvini*. In common, all have white underwings with black leading and trailing margins, thin, almost equal thickness; the leading edge has a black 'notch' at junction with the body.
The Chatham Island Shy Albatross
Darker grey
Darker than adult
Adult, race *eremita*
Juv.
Race *eremita*: darkest head
Dark
Race *eremita*: juv. darker than adult
Race *salvini*, 'Grey-backed Shy Albatross'
Black 'notch'
Light grey
Adult, race *salvini*
Juv.
Pale yellow
Dark spot
Very dark
Adult, race *salvini*: head light grey, midway in tone between above two races
Juvenile, race *salvini*, slightly darker than adult
The nominate race is the only albatross that breeds in Aust. waters, where it is common. Race *eremita*, the Chatham Is. Shy Albatross, breeds Chatham Is. and is a rare vagrant. Race *salvini*, Grey-backed Shy Albatross, breeds on islands near NZ.
Broad black edge at carpal joint
Wide black trailing edge
Dark grey
Feet show past tail-tip.
Wide black leading edge; central bar in centre of inner wing is very wide.
Adult
White shafts to primaries
Head and neck white
Wings are overall blackish, or browner in worn plumage, as shown here.
White rump
Adult
Juvenile is almost identical to adult; bill and feet are apparently greyish.
Dark face, black about the lores, then fading away to white .on neck and throat
Dark
Bill varies, deep yellow, pinkish yellow or grey with dark tip
White head with conspicuous large dark area across eye and face
Wings entirely dark

Common Diving-Petrel *Pelecanoides urinatrix* 20–25 cm; span 33–38 cm (H, M)
In common with other diving-petrels, this is a small, stocky petrel with short winged, short tailed, tubby, neckless jizz. When floating on the surface, it looks like a miniature penguin. Flight is distinctive, unlike any other type of seabird. Always low, fast and direct with quail-like whirring of wings, clipping wave tops. May fly directly into waves to burst out and into flight at the other side; dives deeply, using wings in flight-like action in pursuit of small fish. Usually in small flocks. **Variation:** race *exsul*, 'Kerguelen Diving-Petrel', vagrant to SW of WA, has a broader bill than nominate race of SE Aust. **Voice:** apparently silent at sea, but very noisy at night in breeding colonies. Male gives a slow 'koo-ah', rising in pitch, and 'koo-ah-ka, koo-ah-ka-', also described as 'ku-ku-miaw'. Female has variations like 'koo-aka-did-a-did'. **Similar:** South Georgian Diving-Petrel, almost impossible to separate at sea, rarely identified in Aust. waters. **Hab.:** waters of continental shelf, inshore waters, bays, open oceans. **Status:** common in and near Bass Str., rare in coastal waters as far N as SE Qld and westwards to SW of WA. **Br.:** 357

South Georgian Diving-Petrel *Pelecanoides georgicus* 18–21cm (H, M)
At sea, indistinguishable from Common Diving-Petrel, but so rare that any diving-petrel seen in Aust. waters is almost certainly the Common. Positive identification requires that the bird be in-hand to examine details. Previously it was thought that the distribution of the various species did not overlap, but it has been found that both species share many of the island breeding grounds where there are differences in calls and colour of downy chicks. Chicks are initially grey; those of the Common Diving-Petrel are white. **Voice:** silent at sea. From burrows, a series of 5–10 varied squeaks; in flight a squeaked 'ku-eeek' at intervals of several seconds. Calls differ from those of Common Diving-Petrel. **Hab.:** coastal and offshore waters around islands of Antarctic and sub-Antarctic seas; probably also over open ocean, but identification for confirmation of range is difficult at sea. **Status:** extremely rare vagrant to SE Aust. **Br.:** islands of S and NZ seas.

Wilson's Storm-Petrel *Oceanites oceanicus* 15–19 cm; span 38–42 cm
When travelling, flight is swift, swallow-like with bursts of rapid wingbeats interspersed with short glides on upswept wings close to the waves. When feeding, switches to a slow, fluttering flight; works across the surface against the wind, skimming along the windward slopes of wave troughs. Often hovers with rapid, shallow, fluttering beats of uplifted wings, feet pattering the waves, head bobbing down to snatch small fish, crustaceans or carrion on or close beneath the surface. **Variation:** subspecies *exasperus* has slightly longer wings and tail, more southerly range. **Voice:** at sea is silent except for occasional querulous, sparrow-like chattering within feeding flock; chirping calls at night. In colonies, churring and peeping sounds. **Similar:** Grey-backed Storm-Petrel, dark morph of White-bellied Storm-Petrel. **Hab.:** deep pelagic seas, shelf slopes and shallower shelf and inshore waters; ranges from Antarctic pack-ice to subtropical. Aust. records usually from edge of continental shelf. **Status:** widespread and abundant. Migrates from Antarctic breeding grounds to winter across tropical and N seas, passing northwards along W and E coasts of Aust. in autumn; remains around N coast, returns S in spring. **Br.:** Antarctic coasts and surrounding islands.

Leach's Storm-Petrel *Oceanodroma leucorhoa* 19–22 cm; span 45–48 cm
A rare vagrant; similar to Wilson's Storm-Petrel; overall jizz slender, graceful; flight distinctive. When travelling is fast and direct; sudden changes of direction, bursts of powered flight with rapid, shallow but strong wingbeats alternating with zigzag glides on bowed wings. When foraging, the flight is slow and buoyant; flutters with wings slightly raised, almost hovering against the wind, moving gradually forward, feet paddling the surface, snatching with the bill at small surface creatures. Does not usually follow ships, but occasionally attracted to boats during fishing. **Variation:** nominate race, NE Pacific, shows less white on rump southwards until birds of southernmost colonies have entirely dark rumps. **Voice:** calls at sea while in flight as well as over the colony. Most common call is a staccato ticking that ends with a slurred trill, and chuckling sounds. **Similar:** White-rumped. **Hab.:** offshore waters. **Status:** rare vagrant to Aust. Several records from SW of WA, and an old record from W Vic. **Br.:** N Pacific and Atlantic, colonies of millions on islands offshore Nova Scotia, Maine, Labrador and Newfoundland, with smaller colonies in N Scotland.

Tubby, short necked; flight fast and low with rapid, whirring wingbeats; fly into waves; patter along surface.
Scapular feathers lightly tipped white in fresh plumage
Pale to mid grey
Feet cobalt blue
Brown edges to primaries
Race *exsul* has broader bill; may also have darker breast-band.
Darker
Cross-divider set to rear.
Nostrils are divided in two by a longitudinal ridge or septum with a second, transverse divider towards rear of nostrils.
Glossy black
Both Common and South Georgian become slightly browner as plumage wears.
Dive while swimming as well as by aerial plunges.
Scapular tips white in fresh plumage
In hand, small details of bill give safer identification.
Divider set to centre
Nostril subdivision differs from the Common (above).
White to very pale grey
Ear coverts, sides of face are slightly paler than Common Diving-Petrel.
Pale grey inner edges to primaries
Slightly smaller and paler than Common Diving-Petrel
Fresh plumage brownish black; wing-bar less obvious
White may show.
Bright white
Yellow feet trail behind tail.
Long thin legs often dangling to patter across water surface
Wings short, rounded
Wing-bar is much lighter in worn plumage.
Worn plumage is paler, browner with wing-bar more conspicuous.
Conspicuous white crescent
Steep, high forehead characteristic of all storm-petrels
Tail's upper surface often dished, which may make the square tip look slightly rounded or concave
White of rump extends partly under tail.
Slender, short bill
Fresh dark plumage
Legs, toes grey-black; webs yellow
Fresh plumage is overall dark, sooty, blackish brown.
Pale brown wing-bar
White rump may show
White rump is divided by dark line or point.
Forked
Feet black, shorter than tail
Underparts are brownish black except for edges of white rump.
Wings long, slender, backswept
Worn plumage is paler with more obvious wing-bar.
High steep forehead, slender hooked bill

Matsudaira's Storm-Petrel *Oceanodroma matsudairae* 24–25 cm; span 54–58 cm

In flight appears slender with long, pointed wings. Compared with smaller species, flight is relaxed; slower wingbeats, more time spent in flapping flight, fewer erratic changes of direction, but capable of bursts of speed. Glides low, sometimes tipping the water. To feed, alights on the surface holding wings up with an occasional flap for stability as waves pass, or in calm conditions feeds with wings folded; follows ships. **Hab.:** tropical and subtropical pelagic waters; possibly at upwellings of cooler water. **Status:** migratory; travels from NW Pacific to Indian Ocean via NW Aust. with records of sightings from Kimberley coast to Indonesia and Christmas Is., Jul.–Nov. **Br.:** S of Japan on N Volcano Is. Jan.–Jun.

White-bellied Storm-Petrel *Fregetta grallaria* 18–22 cm; span 46–48 cm

Occurs in two colour morphs, light and dark, with full range of intermediate plumages. Distinctive flight using the wind—glides into the wind with uplifted wings, touching the surface with dangling feet, hugging the wave contours—brief bursts of fluttering, shallow wingbeats. **Variation:** four subspecies. Probably only the Australasian-breeding nominate *grallaria* occurs in Aust. seas, but it has variable plumage from light through to dark phases. **Voice:** silent at sea and in flight over colonies; calls are usually from within nest burrow, long series of high 'pew-pew-pew-pew' notes. **Similar:** light morph plumage pattern similar to the Black-bellied Storm-Petrel, but without the distinctive longitudinal black breast-band. **Hab.:** in Aust., along edge of continental shelf and further out; only occasionally over inshore waters. **Status:** after breeding disperses across Pacific and over Tasman Sea. Possible Aust. range from S Tas. N to Coral Sea. Most claimed sightings from S coast NSW to S coast Qld. **Br.:** Lord Howe Is., Kermadecs, and islands of S Atlantic and S Pacific.

Grey-backed Storm-Petrel *Garrodia nereis* 16–19 cm; span 38–40 cm

Travelling flight is swift and direct with continuous rapid fluttering—rather bat-like wingbeats uninterrupted by glides. Feeding flight is much slower; patters across the surface using feet to kick and bounce across the water. **Voice:** silent at sea; in colonies, a wheezy, scratchy chirping or twittering. **Similar:** White-faced and Black-bellied Storm-Petrels. **Hab.:** observations around SE Aust. show species is most common along the shelf edge and further out; rarely closer inshore. **Status:** birds from the NZ population are regular visitors to the Aust. coast in winter. **Br.:** Oct.–May, islands SE of NZ; S Atlantic and Indian Oceans.

Black-bellied Storm-Petrel *Fregetta tropica* 20 cm; span 46 cm

In flight is distinguished from beneath by rough edged black band down centre of white belly; band harder to see from side view. Keeps low with erratic changes of direction, skimming and hugging contours of the waves, legs dangling and feet dipping, occasionally plunging breast against water then pushing clear with feet and wings. **Variation:** race *melanoleuca*, S Atlantic. **Voice:** silent at sea; in colonies makes a distinctive, whistled 'hieuuuuw' and softer pipping. **Hab.:** Antarctic and sub-Antarctic in summer; tropical and subtropical seas in winter, usually beyond the shelf-edge; rarely enters inshore waters. **Status:** uncommon winter visitor to W, S and E coasts of Aust. **Br.:** islands of sub-Antarctic seas, Nov.–Apr.

White-faced Storm-Petrel *Pelagodroma marina* 18–21 cm; span 42–43 cm

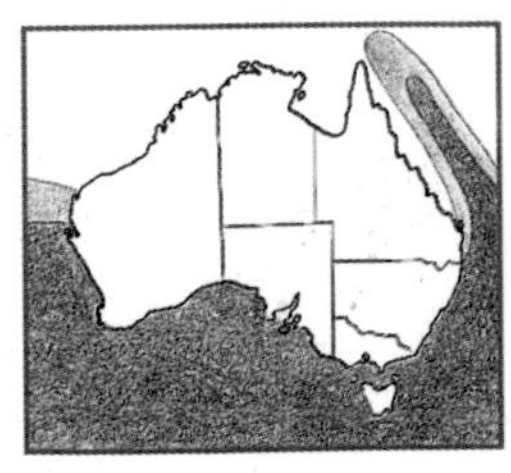

The only storm-petrel breeding in Aust. waters; distinctive facial pattern; common. Flight when travelling is erratic and weaving—low, level glides on rigid down-bowed wings, steep banking turns, brief bursts of jerky wingbeats. In low foraging flight, glides slowly into the wind, wings outstretched parallel to waves, long legs dangling; but, unlike some other storm-petrels, hits surface with feet together, bounding rather than striding. **Variation:** numerous subspecies around world; 3 in Australasian region. Race *dulciae* common; breeds around S and SE coasts of Aust. **Voice:** in breeding colony a soft, repetitive 'peeoo-peeoo' or 'woooo, wooo'. **Hab.:** uses continental shelf waters when breeding, but is usually well out of sight of land. Otherwise much further out over deep oceanic waters. **Status:** common; breeding, migratory. **Br.:** 357

A large storm-petrel of tropical seas, entirely dark brown with long, pointed wings.
White primary shafts form a white spot, obvious at distance
Forages across waves with wings uplifted, feet and bill dipping the surface, often settling briefly on water with wings up.
Pointed wings curve back at carpal joints
Tail deeply forked; trailing feet do not show beyond tail-tip
Buffy brown coverts form a pale curved band across the wings
Long, forked
Uniformly brownish black
Underwing pattern of broad black leading edge and wide dark trailing edge is conspicuous on light phase, much less obvious on intermediate and darker forms.
Black hood
Conspicuous white crescent
Black chin
Feet not visible
In fresh plumage, inner coverts and scapulars have white edging.
High, steep forehead
Black hooded
Dark morph
Black
Brownish black with white fringing
Sharp edge to hood higher than on Black-bellied Storm-Petrel
White between wide dark edges
Folded wings longer than square-cut black tail
Dark hooded
Rounded outer wing
Back feathers ashy grey and pale edged from mantle through scapulars and rump to tail coverts; gives slightly scaly pattern.
Grey from back to tail without any white on rump
Black
Wide black
Feet show
Sharp edged
Broad black band
Pale diagonal across coverts
Under tail white
Square tip
Black upper wing
Black
White crescent
Black
Head to upper breast always black, but belly varied; most have central black band.
Feet trail behind tip of tail that is square to slightly rounded.
Wide, black
Head to upper breast always black, while throat usually pale or mottled
Underside of body extremely varied. The central band (shown left) may be wide, narrow or spread to cover most of underbody black. The black band may be broken, consisting of small flecks of black, or absent (at right).
Whitish, often not visible
Faint coverts diagonal in fresh plumage
Black band
Whitish spot
Curves back.
Brow and face white; crown and through eye, dark grey
Grey, slightly browner in worn plumage
Feet trail
Rump and upper tail coverts pale grey
Tail square tipped or slightly forked
Very thin black edge
Grey
White
Curved back
Grey, webs pale yellow
Feet extend well to rear of tail in travelling, or down to touch water when feeding

Red-tailed Tropicbird *Phaethon rubricauda* 45–50 cm; tail streamers 40 cm (C, K, L, N)

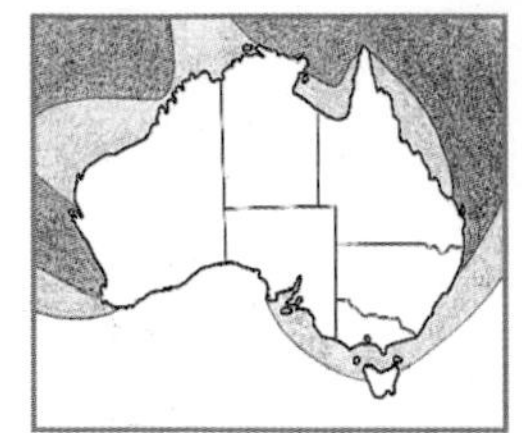

White with a satiny sheen and often a faint glow of pink; conspicuous against blues of sea and sky; less obvious at a distance are the long, red tail streamers. In flight usually keeps quite high; 'butterfly' flight of fluttering wingbeats, alternating glides. Dives vertically, plummeting entirely under water; takes flying fish and small squid. **Voice:** loud, sharp, harsh, cackling calls during aerial displays; squawks and purring at nest. **Hab.:** tropical and subtropical seas; pelagic, often far from land, closer around colonies. **Status:** breeding W coast WA; otherwise regular to rare accidental visitor to parts of Aust. coastline within its range. **Br.:** 358

White-tailed Tropicbird *Phaethon lepturus* 38–40 cm; tail 35–40 cm (C, K, L, N)

Small, slender and graceful; has white tail streamers. At sea is usually solitary; flies high and strongly with rapid tern-like wingbeats. Takes fish and other marine life in vertical dives from 10 to 20 m height; plunges deep below the surface. Often hovers before descending. May swoop low across surface to take flying fish. **Voice:** in courtship aerial displays, a 'kek-kek-kek' call while flying in high circles. **Similar:** Red-tailed Tropicbird. **Hab.:** oceanic, probably pelagic; rarely seen over inshore waters; comes to land only on returning to nest colony. **Status:** vagrant or accidental to both NW and E coasts of Aust. **Br.:** 358

Masked Booby *Sula dactylatra* 75–85 cm; span 1.6–1.7 m (K, L, N)

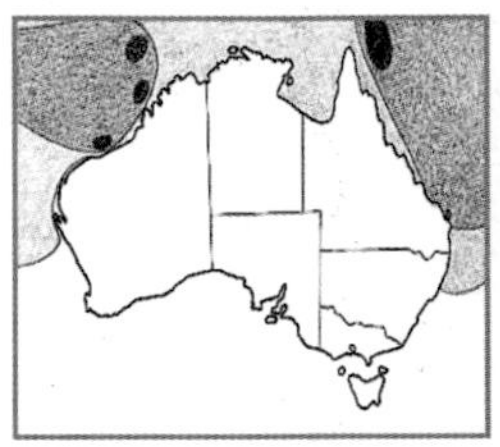

Flies fast and high; strong, steady wingbeats alternate with glides; solitary or in loose groups when returning to colony. Hunts by vertical dives from heights of 15 to 100 m; plunges several metres beneath surface. Boobies are built for such high plunges; the tapered bill breaks the surface, and their velocity and sleek lines carry them deep under water. **Juv.:** distinctly different; ashy brown head, neck and wing coverts, and mottled with white across the back. **Voice:** usually silent at sea; noisy in breeding colonies. Male gives a high, descending, whistled 'whieeoooo', female a loud, harsh, honking 'aarh-aarh-a-yah'. **Similar:** Gannets and white morph of Red-footed Booby. See also Abbott's Booby, Christmas Is. **Hab.:** marine, pelagic; often far from land beyond the shelf edge over deep water. Breeding birds will travel far out from island colonies to reach deep seas. **Status:** moderately common offshore. **Br.:** 358

Red-footed Booby *Sula sula* 70–80 cm; span 1.4 m

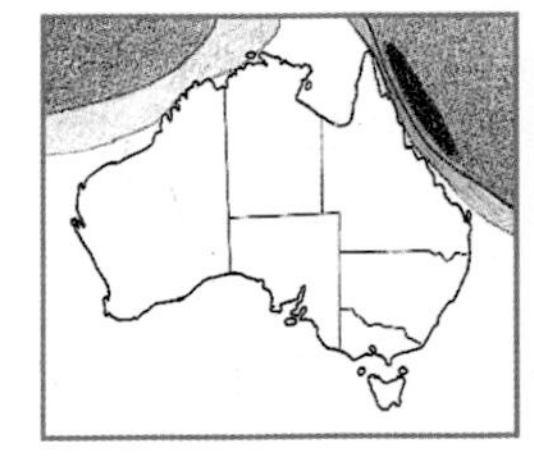

Smallest booby; flies swiftly in long glides that skim the waves; travels far out from land. Has agility to catch flying fish in aerial pursuit, but generally feeds by plunge-diving vertically from 8 m and higher. Feeds in flocks; often follows boats. **Variation:** plumage extremely varied. Two basic colour morphs: white and white tailed brown. The former is entirely white except for the flight feathers and their coverts; white may have slight or strong golden tint. The white-tailed brown morph is extremely variable; the browns vary from dark grey-brown to pale gingery brown. **Voice:** silent at sea except for alarm call, a loud, grating 'karrak, karrak'. Noisy at breeding colony, a grating 'kurr-uk, kurr-uk'; in flight over colony, a loud 'rarh, rarh'. **Hab.:** tropical and pelagic; often travels far from breeding island to deepwater feeding grounds, but also feeds in shallows of lagoons around islands and reefs. **Status:** common, widespread; breeds only on Qld offshore islands—not often recorded around other parts of Aust. coast. **Br.:** 358

Brown Booby *Sula leucogaster* 65–75 cm; span 1.3–1.5 m (C, K, L, N)

Sleek, slender booby with clearcut brown and white plumage; sharp demarcation between brown neck and upper breast and white belly. Flies low in bursts of smooth, easy wingbeats alternating with low glides; alone or in flocks. **Voice:** calls at sea when squabbling and fishing. Noisy in colonies: male has soft hiss; female a harsh, deep honk or quack. **Similar:** juv. Masked Booby, but has a white collar and breast, and back and wings mottled brown; juv. Red-footed, but entirely brown. **Hab.:** marine, largely tropical; deep waters and inshore shallows. **Status:** common N coast of Aust. from North West Cape in WA to SE Qld. Breeds on islands off NW of WA, and on Barrier Reef. **Br.:** 358

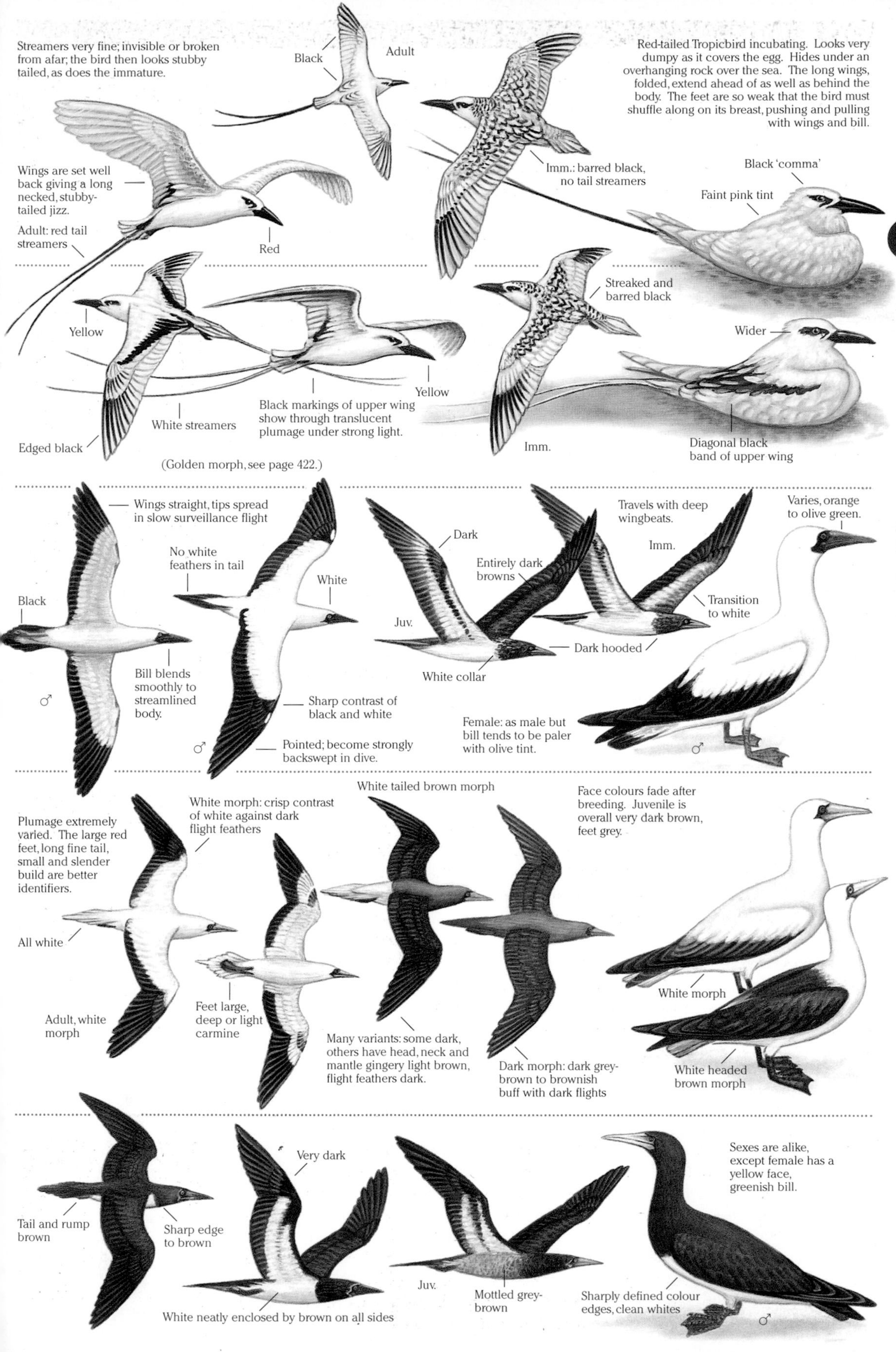

Streamers very fine; invisible or broken from afar; the bird then looks stubby tailed, as does the immature.
Black
Adult
Red-tailed Tropicbird incubating. Looks very dumpy as it covers the egg. Hides under an overhanging rock over the sea. The long wings, folded, extend ahead of as well as behind the body. The feet are so weak that the bird must shuffle along on its breast, pushing and pulling with wings and bill.
Wings are set well back giving a long necked, stubby-tailed jizz.
Imm.: barred black, no tail streamers
Black 'comma'
Faint pink tint
Adult: red tail streamers
Red
Streaked and barred black
Yellow
Wider
Yellow
White streamers
Black markings of upper wing show through translucent plumage under strong light.
Edged black
Imm.
Diagonal black band of upper wing
(Golden morph, see page 422.)
Wings straight, tips spread in slow surveillance flight
Travels with deep wingbeats.
Varies, orange to olive green.
Dark
No white feathers in tail
Entirely dark browns
Imm.
White
Black
Transition to white
Juv.
Dark hooded
Bill blends smoothly to streamlined body.
White collar
♂
Sharp contrast of black and white
Female: as male but bill tends to be paler with olive tint.
♂
Pointed; become strongly backswept in dive.
♂
White tailed brown morph
White morph: crisp contrast of white against dark flight feathers
Face colours fade after breeding. Juvenile is overall very dark brown, feet grey.
Plumage extremely varied. The large red feet, long fine tail, small and slender build are better identifiers.
All white
White morph
Feet large, deep or light carmine
Adult, white morph
Many variants: some dark, others have head, neck and mantle gingery light brown, flight feathers dark.
Dark morph: dark grey-brown to brownish buff with dark flights
White headed brown morph
Very dark
Sexes are alike, except female has a yellow face, greenish bill.
Tail and rump brown
Sharp edge to brown
Juv.
Mottled grey-brown
Sharply defined colour edges, clean whites
♂
White neatly enclosed by brown on all sides

Darter *Anhinga melanogaster* 85–90 cm; span 1.2 m

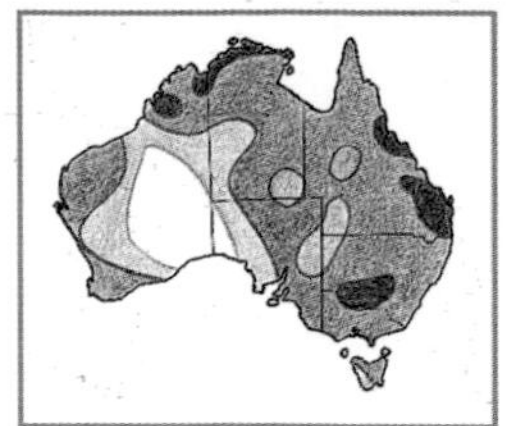

Long, sinuous neck merges to small, sleek head and straight, dagger-like bill. Floats very low, often body submerged, only head and neck visible, snake-like. Slips beneath surface with scarcely a ripple. Hunts under water with a spearing action as the kinked neck suddenly straightens. **Voice:** a harsh 'kar, kar, ka, ka-ka-kakaka', slow and loud, becoming more rapid, fading. At nest, loud, brassy cacklings by male; clicking sounds. **Hab.:** most wetlands, fresh or brackish, of 0.5 m or more depth that have trees, logs, limbs, well-vegetated banks. Also in estuaries, sheltered bays. **Status:** common in suitable habitat; vagrant Tas. **Br.:** 359

Great Cormorant *Phalacrocorax carbo* 80–85 cm; span 1.3–1.5 m

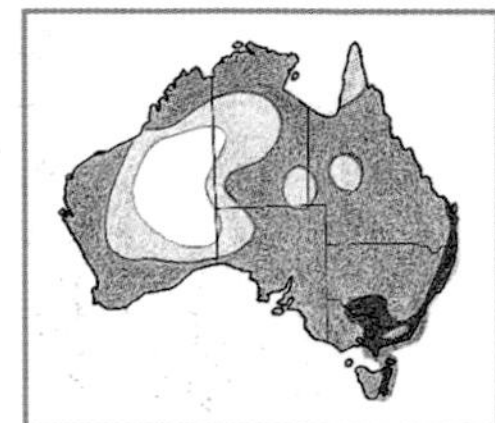

Large and black with yellow throat and face. When breeding has white on face and flanks. Flies with rapid wingbeats between glides. Solitary or small groups; holds wings out to dry. **Voice:** usually silent. Breeding male gives croaking, stuttering groans like a heavy, creaking door, and raucous, barking threat calls; hoarse hissing from female. **Similar:** Little Black Cormorant. **Hab.:** large expanses of fresh or salt water; most abundant on coastal estuaries, bays, lagoons, inland on deep rivers, lakes, swamps and floodwaters; uncommon on small or shallow waters. **Status:** common almost throughout Aust. wherever habitat exists. **Br.:** 359

Little Black Cormorant *Phalacrocorax sulcirostris* 55–65 cm; span 1 m (L, N, T)

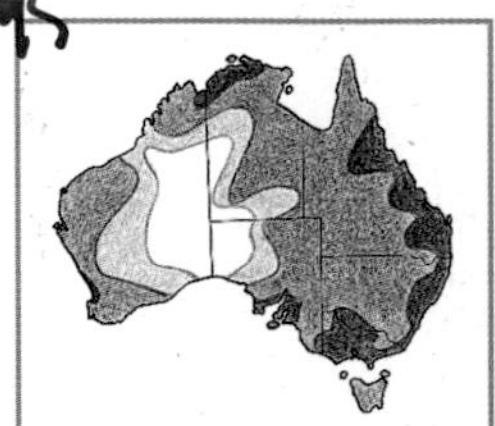

Appearance is superficially like Great Cormorant. Uses smaller bodies of water such as farm dams. Flight is strong; rapid wingbeats alternate with glides; flocks take up 'V' formation. Highly gregarious; large flocks fish cooperatively. **Voice:** usually silent, but some ticking and croaking among birds in fishing flocks. Guttural croaks and tickings at nest. **Hab.:** on mainland, most common on inland waters—lakes, river pools, deep open swamps, estuaries, lagoons; more coastal in Tas. **Status:** common; sedentary and nomadic. **Br.:** 359

Little Pied Cormorant *Phalacrocorax melanoleucos* 55–60 cm (L, N, T)

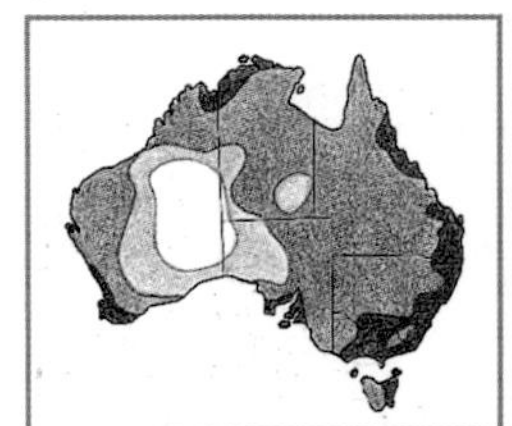

Small, pied, stubby yellow bill, long tail. Flight strong; can take off and rise more steeply than other cormorants. **Voice:** near the nest, cooing 'keh-keh-keh-' and a harsh, deep 'uk-uk-urk' by male arriving with nest materials or food for young. A sharp croak is used as an alarm call. **Similar:** Pied Cormorant. **Hab.:** sheltered coastal lagoons, harbours, bays. Inland, uses small lakes, dams, billabongs, swamps, floodwaters. Follows creeks well inland. Uses dams in forest and farm country to advantage and feeds on introduced carp and redfin. **Status:** common almost throughout Aust., including interior after heavy rains. **Br.:** 359

Pied Cormorant *Phalacrocorax varius* 70–80 cm; span 1–1.3 m

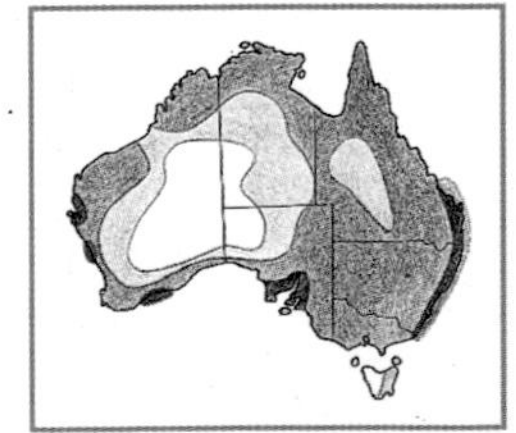

May be solitary, in small groups or in thousands. Flight strong, direct; slower wingbeats than Little Pied; head high, neck often kinked. When travelling in numbers forms 'V' pattern. Rests, roosts and nests in trees; often perches with wings extended. **Voice:** usually silent, but in breeding colony various cacklings, loud ticking, deep guttural grunting. **Hab.:** waters both marine and inland—lakes, rivers, billabongs, open and deep areas of swamps, coastal lagoons, estuaries. **Status:** moderately common to abundant all states. **Br.:** 359

Black-faced Cormorant *Phalacrocorax fuscescens* 60–70 cm; span 1 m

Facial skin black. In flight holds head level or lower than body axis; does not perch on trees. **Breeding:** back of neck is finely streaked white. **Nonbr.:** without plumes. **Voice:** usually silent, but, within the breeding colony, loud, guttural croakings from the male, soft, hoarse hissing from the female. **Similar:** Pied and Little Pied Cormorants. **Hab.:** marine, feeding predominantly over inshore waters and reefs, around offshore islands; occasionally enters estuaries, tidal reaches of large rivers. **Status:** common to abundant, S coasts of mainland and Tas., SE of NSW around to SW of WA. **Br.:** 359

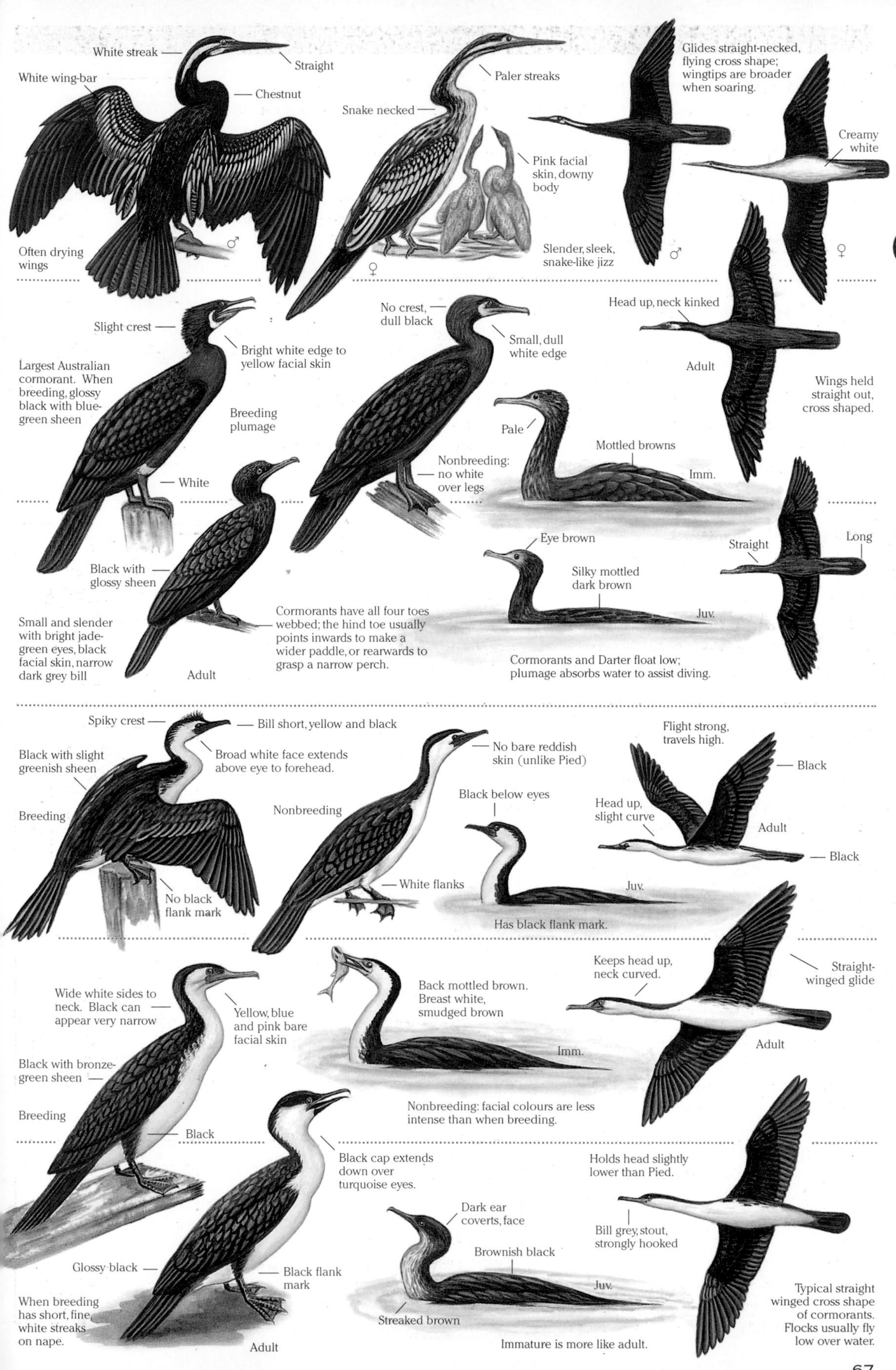

White streak
Straight
White wing-bar
Chestnut
Often drying wings
Paler streaks
Snake necked
Pink facial skin, downy body
Slender, sleek, snake-like jizz
Glides straight-necked, flying cross shape; wingtips are broader when soaring.
Creamy white
Slight crest
Bright white edge to yellow facial skin
Largest Australian cormorant. When breeding, glossy black with blue-green sheen
Breeding plumage
White
No crest, dull black
Small, dull white edge
Nonbreeding: no white over legs
Head up, neck kinked
Adult
Wings held straight out, cross shaped.
Pale
Mottled browns
Imm.
Black with glossy sheen
Small and slender with bright jade-green eyes, black facial skin, narrow dark grey bill
Adult
Cormorants have all four toes webbed; the hind toe usually points inwards to make a wider paddle, or rearwards to grasp a narrow perch.
Eye brown
Silky mottled dark brown
Juv.
Straight
Long
Cormorants and Darter float low; plumage absorbs water to assist diving.
Spiky crest
Bill short, yellow and black
Black with slight greenish sheen
Broad white face extends above eye to forehead.
Breeding
No black flank mark
Nonbreeding
No bare reddish skin (unlike Pied)
White flanks
Black below eyes
Juv.
Has black flank mark.
Flight strong, travels high.
Black
Head up, slight curve
Adult
Black
Wide white sides to neck. Black can appear very narrow
Yellow, blue and pink bare facial skin
Black with bronze-green sheen
Breeding
Black
Back mottled brown. Breast white, smudged brown
Imm.
Nonbreeding: facial colours are less intense than when breeding.
Keeps head up, neck curved.
Straight-winged glide
Adult
Black cap extends down over turquoise eyes.
Glossy black
Black flank mark
When breeding has short, fine white streaks on nape.
Adult
Dark ear coverts, face
Brownish black
Juv.
Streaked brown
Immature is more like adult.
Holds head slightly lower than Pied.
Bill grey, stout, strongly hooked
Typical straight winged cross shape of cormorants. Flocks usually fly low over water.

Australian Pelican *Pelecanus conspicillatus* 1.6–1.8 m; span 2.3–2.5 m (T)

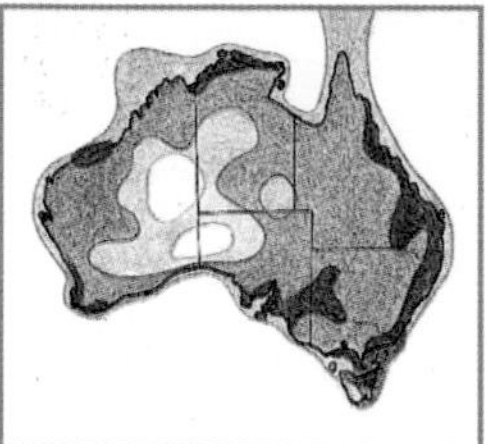

Familiar huge bird with massive bill. Head tucks back in flight giving tubby shape. Soars on flat wings; circles in thermals to great heights, then travels far in long glides, helped by occasional slow flapping. Flocks travel in formation, often 'V' shaped. Gregarious, pairs to large flocks in flight, loafing or fishing. At times work cooperatively to round up fish. **Voice:** in display or squabbling, deep resonant croaks and guttural grunting. **Similar:** at height, possibly White-bellied Sea-Eagle. **Hab.:** diverse large or small areas of water from sheltered coastal bays to temporary pools in desert. **Status:** common; nomadic, dispersive. **Br.:** 358

Cape Gannet *Morus capensis* 85–92 cm; span 1.7–1.8 m

First recorded in Aust. 1980; a single bird found nesting, mated to an Australasian Gannet in a small colony on Wedge Light, Port Phillip Bay, Vic. Another, banded in S Africa, was recaptured off Cape Leeuwin, WA. **Status:** accidental to Aust.

Australasian Gannet *Morus serrator* 85–90 cm; span 1.7–1.9 m (L, M, N)

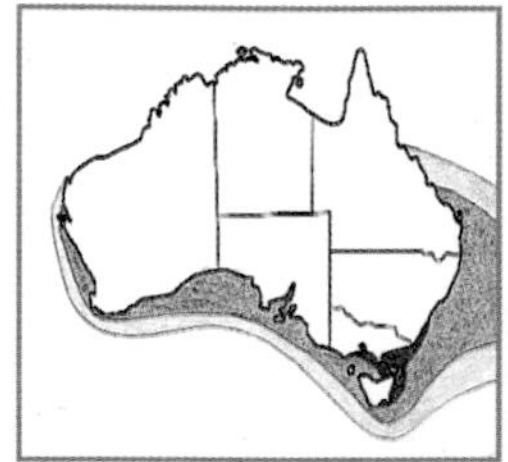

Sleek, with pointed bill, wingtips, tail. Hunts by plunge-diving; groups or large flocks wheel 20–30 m above schools of fish, plunging bill first in spectacular dives on half-closed wings. Gregarious, lines of birds travel close to waves; casual flap-and-glide flight; gather in 'rafts' on calm seas. **Voice:** noisy in colonies; raucous, sharp squawks intermingled with deep, barked 'urragh-urragh'. **Hab.:** marine, usually continental shelf waters. Prefers seas near coastal islets, capes, channels. Enters estuaries, bays, harbours to shelter from rough seas. **Status:** common mid WA coast to mid Qld coast; far larger NZ population. **Br.:** 358

Great Frigatebird *Fregata minor* 85–105 cm; span 2.1–2.3 m (C, K, L, M, N)

Large; predominantly black with long pointed wings, deeply forked tail. Wings typically held far forward, strongly bowed. Soars high on thermal uplifts or on updrafts above cliffs of oceanic islands; uses only occasional deep wingbeats. Feeds by snatching flying fish and squid from surface of sea, and, to some extent, by harassing other seabirds arriving from fishing grounds until they drop food for the frigatebird to catch. **Variation:** five subspecies: *F. m. minor* breeds Christmas Is.; *F. m. palmerstoni* in W Pacific and Coral Sea. **Voice:** usually silent in flight; when landing at nest, male gives a repetitive, yelping 'tjiew-tjiew-'; from nest, call a braying 'wah-hoo-hoo-hoo'. **Similar:** Christmas and Lesser Frigatebirds. **Hab.:** tropical seas, pelagic; occasionally inshore shelf waters. **Status:** regular around N coasts from Pt Cloates, WA, to N Stradbroke Is., Qld. **Br.:** 358

Lesser Frigatebird *Fregata ariel* 70–80 cm; span 1.8–1.9 m (C, K, L, N)

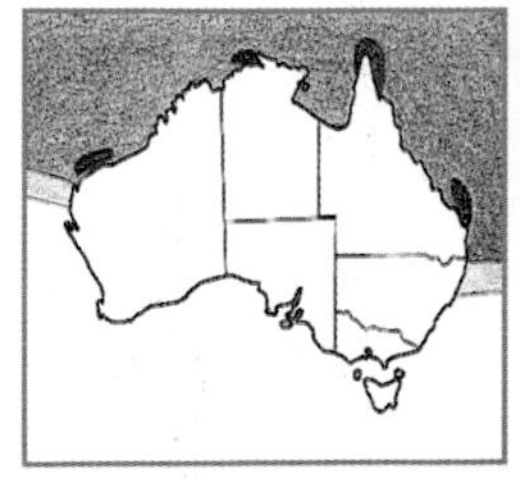

Flight graceful; soars, glides, dives; deep, easy wingbeats; can stay airborne all day. Feeding and harassing behaviour similar to other frigatebirds. Perches on trees, rarely on ground. **Male:** like other frigatebirds, has inflatable scarlet throat sac. **Fem.:** extent and shape of white on breast separates the three species. **Voice:** both sexes give landing calls—the male a whistled 'wheees-wheees-', the female a high 'chip-ah, chip-ah' followed by sharp shrieks. **Hab.:** marine; airspace over tropical seas, usually pelagic. Often far from land, but also over shelf waters, in places close inshore, and inland over continental coastlines. **Status:** common N Aust. Breeding species. Sites include Barrier Reef islands and Ashmore Reef, WA. **Br.:** 358

Christmas Frigatebird *Fregata andrewsi* 0.9–1 m; span 2.1–2.3 m (C, K)

Breeds only on Christmas Is., a small population with restricted range. Both sexes have greater areas of undersurface white than either of the other species. Like others, avoids settling in low sheltered situations or on water—it has difficulty taking off again except from trees, cliffs or into wind. **Voice:** calls restricted to vicinity of nests in breeding season. Male has a landing call, a descending 'i-eer, i-eer'; female a squeaky chatter. **Hab.:** pelagic tropical seas, favouring warm waters of S equatorial current. **Status:** confined to NE Indian Ocean around breeding site on Christmas Is. Probably sedentary. **Br.:** Christmas Is.

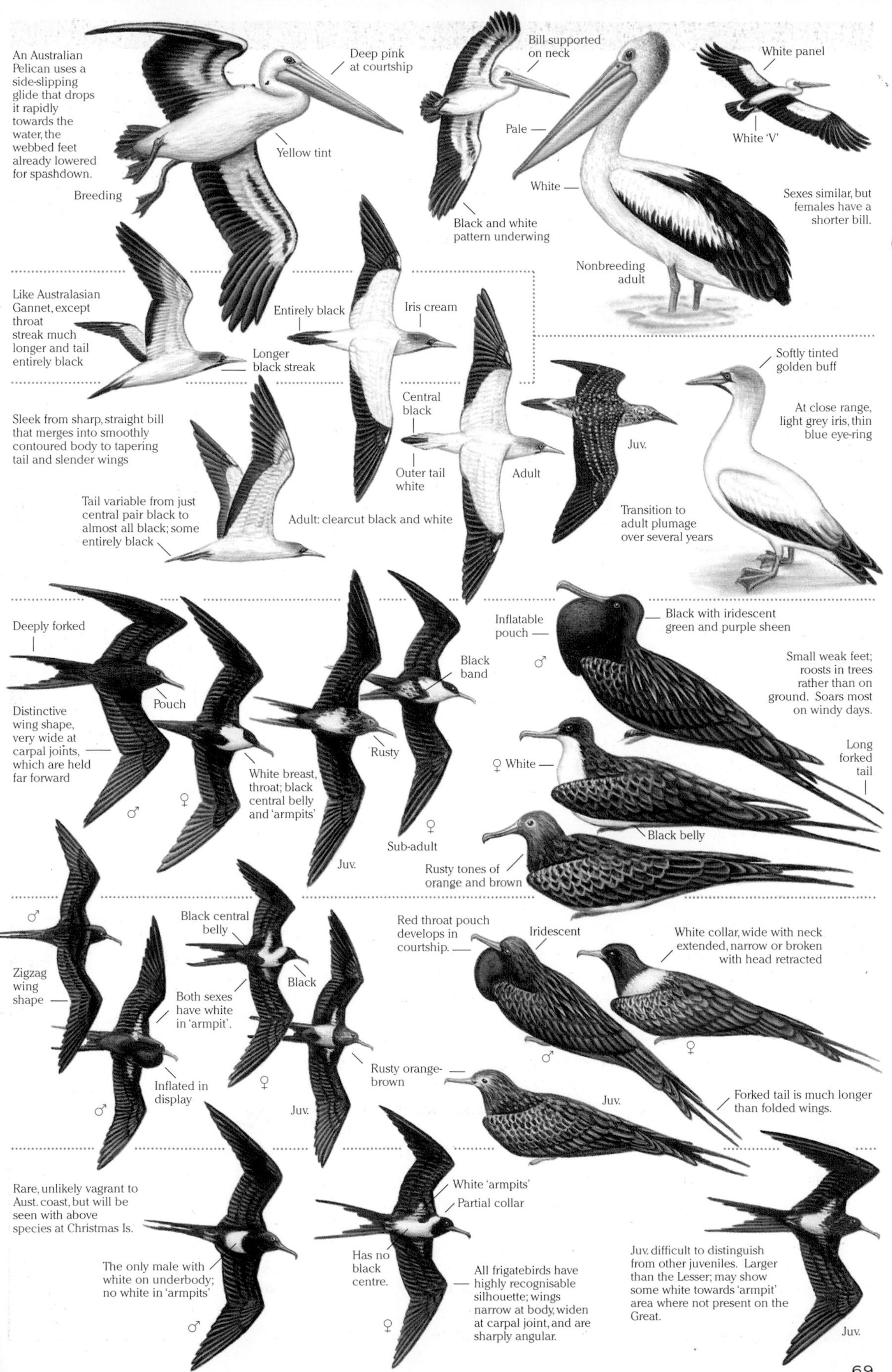
An Australian Pelican uses a side-slipping glide that drops it rapidly towards the water, the webbed feet already lowered for spashdown.
Breeding
Deep pink at courtship
Yellow tint
Bill supported on neck
Pale
Black and white pattern underwing
White
White panel
White 'V'
Sexes similar, but females have a shorter bill.
Nonbreeding adult
Like Australasian Gannet, except throat streak much longer and tail entirely black
Longer black streak
Entirely black
Iris cream
Central black
Outer tail white
Adult
Juv.
Softly tinted golden buff
At close range, light grey iris, thin blue eye-ring
Sleek from sharp, straight bill that merges into smoothly contoured body to tapering tail and slender wings
Tail variable from just central pair black to almost all black; some entirely black
Adult: clearcut black and white
Transition to adult plumage over several years
Deeply forked
Pouch
Distinctive wing shape, very wide at carpal joints, which are held far forward
♂
♀
White breast, throat; black central belly and 'armpits'
Juv.
Rusty
Black band
♀
Sub-adult
Inflatable pouch
♂
Black with iridescent green and purple sheen
Small weak feet; roosts in trees rather than on ground. Soars most on windy days.
♀ White
Long forked tail
Black belly
Rusty tones of orange and brown
♂
Zigzag wing shape
Black central belly
Black
Both sexes have white in 'armpit'.
Inflated in display
♂
♀
Juv.
Rusty orange-brown
Red throat pouch develops in courtship.
Iridescent
White collar, wide with neck extended, narrow or broken with head retracted
♂
♀
Juv.
Forked tail is much longer than folded wings.
Rare, unlikely vagrant to Aust. coast, but will be seen with above species at Christmas Is.
The only male with white on underbody; no white in 'armpits'
♂
White 'armpits'
Partial collar
Has no black centre.
♀
All frigatebirds have highly recognisable silhouette; wings narrow at body, widen at carpal joint, and are sharply angular.
Juv. difficult to distinguish from other juveniles. Larger than the Lesser; may show some white towards 'armpit' area where not present on the Great.
Juv.

Great Egret *Ardea alba* 0.85–1.05 m (C, K, L, M, N)

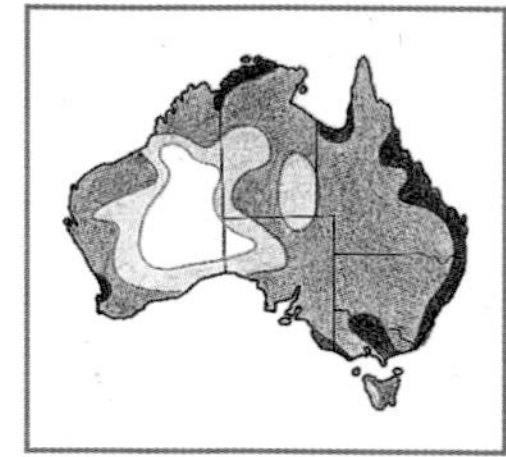

Tall, graceful, long legs, long kinked neck, snowy white plumage. In flight, slow, deliberate wingbeats, neck folded. Hunts with slow, stealthy movements; often poises motionless, neck extended, but kinked ready for a thrust of the straight, spearing bill. Solitary or small groups; rarely large flocks. **Voice:** alarmed, taking flight, low, hollow croaking 'argh-argh-arrgh-aargh'. Disturbed at nest, abrupt guttural croaks, 'grok-grok-grok-grok-grok'. Greeting call on landing at nest, a slow succession of deep, hollow, rasping croaks, 'gor-rork, grok-gro-grkgrok', fading. **Hab.:** wetlands, flooded pastures, dams, estuarine mudflats, mangroves and reefs. **Status:** common, very widespread in any suitable permanent or temporary habitat. **Br.:** 360

Intermediate Egret *Ardea intermedia* 55–70 cm (C, T)

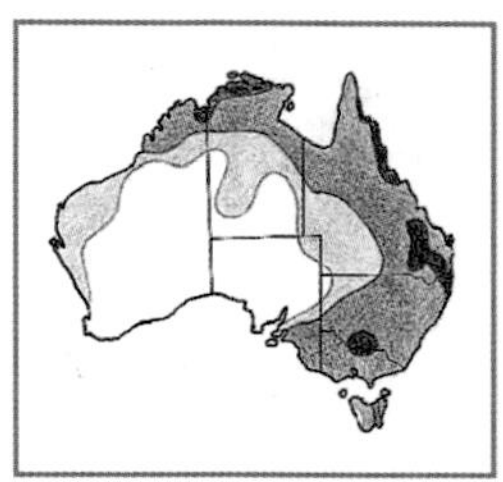

Long plumes in fine filaments hang in a lacy veil from the breast and back, cascading across folded wings and over the tail; enhanced in courtship and greeting displays, when the plumes are lifted and spread. Colours of bare parts intensify during courtship, fading by time eggs are laid. In flight, head is drawn back, feet extend beyond tail-tip. **Voice:** least vocal of the white egrets; silent away from colony. If startled at nest, alarm call is a hollow croak, 'glok-glok-glok'; if threatened, a loud, throaty 'kroooo-krooo'. Greeting at nest is a soft, rasping croak, 'grrrawk-grrrawk'. **Similar:** other white egrets. **Hab.:** freshwater wetlands, especially lake margins, billabongs and swamps with abundant emergent vegetation; also occasionally mangrove swamps, tidal mudflats. **Status:** common in N; uncommon in SE; vagrant to Tas. **Br.:** 360

Cattle Egret *Ardea ibis* 48–53 cm (C, K, L, M, N)

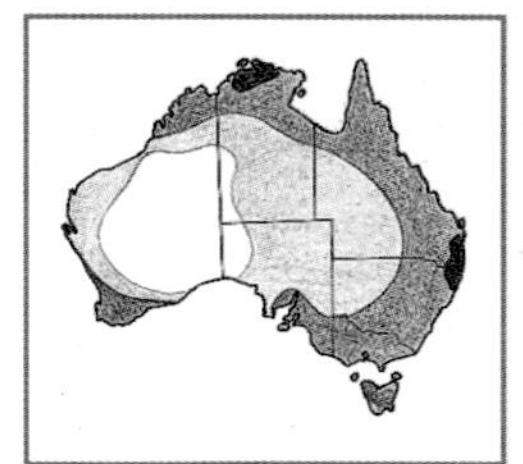

Small, squat egret, often in flocks with livestock. Posture is usually hunched; walks with forward–back swing of head; darts forward, neck outstretched, to take prey. Flight swift with rapid wingbeats. Highly sociable, small groups to very large flocks. Arrived in N Aust. about 1950; has now spread to the SW, the SE and Tas. The Cattle Egret's association with livestock, the advantages derived from feeding close to grazing cattle added to habitat adaptability have helped the species spread. **Voice:** in colonies, a very deep croak, 'krok, krok', and aggressive, harsh 'krow'. **Similar:** Intermediate and Little Egrets are like nonbreeding Cattle Egret, but stand tall. **Hab.:** moist pastures with tall grass; shallow open wetlands and margins, mudflats. **Status:** common N Aust., uncommon in S Range and population expanding. **Br.:** 360

Little Egret *Ardea (Egretta) garzetta* 55–65 cm (C, K, T, N)

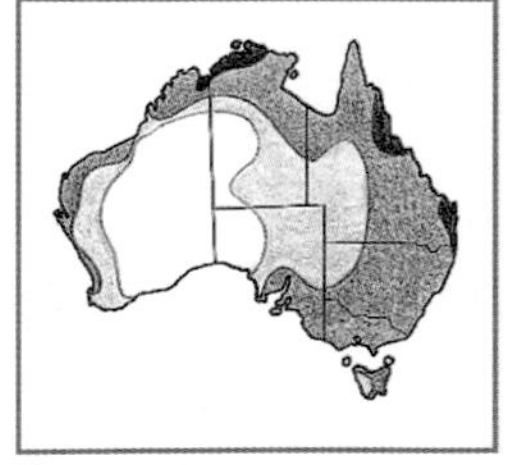

Light, delicate build, evident in its sleek, slender body and emphasised by a long neck. Flight is light and buoyant, uninterrupted by glides. Dashes about in erratic pursuits; lifts wings to startle water creatures. **Breeding:** long, slender plumes curve from rear of crown; fine filamentous plumes from breast and over the back to tail-tip. **Voice:** silent away from colonies except for a croak of alarm. Noisy when breeding; harsh croaking 'argk-argk-argk', squawked 'kiaw', 'kurik-kurik', gurgling sounds. **Similar:** other all-white egrets and white morph of Eastern Reef Egret. **Hab.:** wetlands, both fresh and marine. Usually forages in shallows of open waters—swamps, billabongs, floodplain pools, mudflats and mangrove channels. **Status:** common around N coasts, uncommon to rare in S of range; nomadic or migratory. **Br.:** 360

Eastern Reef Egret *Ardea (Egretta) sacra* 60–65 cm (C, K, T)

Medium-sized heron of the shoreline. Shape distinctive: long necked but short legged without visually balanced neck-to-leg ratio typical of other egrets. This is emphasised by the thick legs, and is most obvious when it stands alert, neck upstretched. **Variation:** two colour morphs: white and very dark grey. **Voice:** gives a harsh, abrupt 'yowk, yowk' in alarm. At nest during courtship, a deep, abrupt, guttural, frog-like 'yrok, yroak' or 'yok-yok'. **Similar:** dark morph is unique—no significant white in plumage; white morph is superficially like other white egrets, which all have longer legs, different bill colours. **Hab.:** shorelines of mainland coasts and islands; estuarine mudflats; inshore reefs. **Status:** common N and NE coasts; uncommon to rare in SE and SW; now absent or vagrant to most of Vic. and Tas. **Br.:** 360

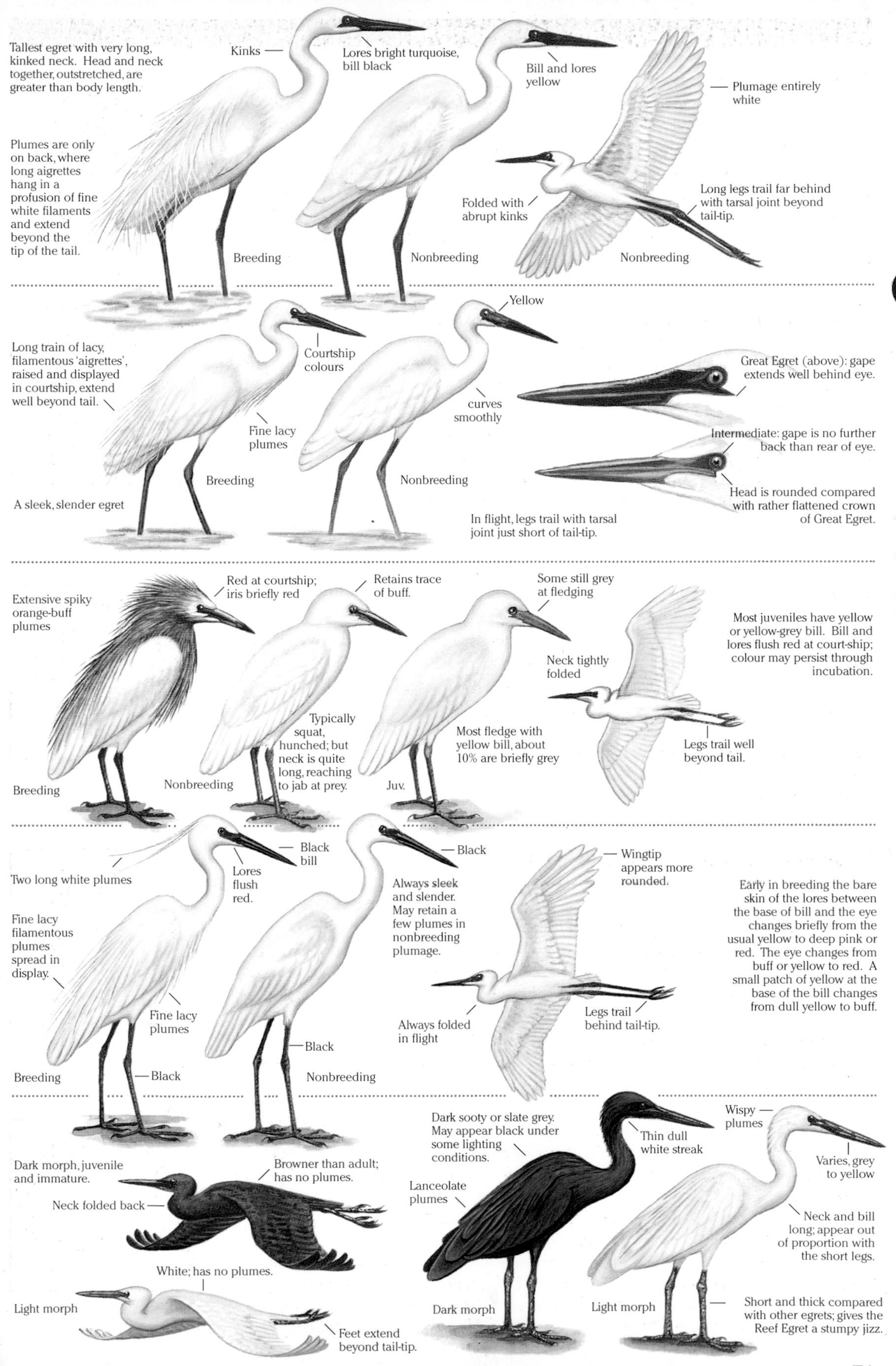

Tallest egret with very long, kinked neck. Head and neck together, outstretched, are greater than body length.
Kinks
Lores bright turquoise, bill black
Bill and lores yellow
Plumage entirely white
Plumes are only on back, where long aigrettes hang in a profusion of fine white filaments and extend beyond the tip of the tail.
Folded with abrupt kinks
Long legs trail far behind with tarsal joint beyond tail-tip.
Breeding
Nonbreeding
Nonbreeding
Yellow
Long train of lacy, filamentous 'aigrettes', raised and displayed in courtship, extend well beyond tail.
Courtship colours
Great Egret (above): gape extends well behind eye.
curves smoothly
Fine lacy plumes
Intermediate: gape is no further back than rear of eye.
Breeding
Nonbreeding
Head is rounded compared with rather flattened crown of Great Egret.
A sleek, slender egret
In flight, legs trail with tarsal joint just short of tail-tip.
Red at courtship; iris briefly red
Retains trace of buff.
Some still grey at fledging
Extensive spiky orange-buff plumes
Most juveniles have yellow or yellow-grey bill. Bill and lores flush red at court-ship; colour may persist through incubation.
Neck tightly folded
Typically squat, hunched; but neck is quite long, reaching to jab at prey.
Most fledge with yellow bill, about 10% are briefly grey
Legs trail well beyond tail.
Breeding
Nonbreeding
Juv.
Black bill
Black
Two long white plumes
Lores flush red.
Wingtip appears more rounded.
Always sleek and slender. May retain a few plumes in nonbreeding plumage.
Early in breeding the bare skin of the lores between the base of bill and the eye changes briefly from the usual yellow to deep pink or red. The eye changes from buff or yellow to red. A small patch of yellow at the base of the bill changes from dull yellow to buff.
Fine lacy filamentous plumes spread in display.
Fine lacy plumes
Always folded in flight
Legs trail behind tail-tip.
Black
Breeding
Black
Nonbreeding
Dark sooty or slate grey. May appear black under some lighting conditions.
Wispy plumes
Thin dull white streak
Varies, grey to yellow
Dark morph, juvenile and immature.
Browner than adult; has no plumes.
Lanceolate plumes
Neck folded back
Neck and bill long; appear out of proportion with the short legs.
White; has no plumes.
Light morph
Dark morph
Light morph
Short and thick compared with other egrets; gives the Reef Egret a stumpy jizz.
Feet extend beyond tail-tip.

Great-billed Heron *Ardea sumatrana* 1–1.1 m
Very large, heavily built, wary, solitary; stalks with slow deliberate movements. Elusive, rarely in open; retreats into cover of mangrove or paperbark swamp while any observer is still distant. Low, heavy flight; slow, deep beats of dark wings; keeps below the tall canopy of mangroves and follows the water's edge and airspace above channels in swamps; rarely seen overhead. **Voice:** a slow, hollow, drawn out, croaking 'a-arr-argk'. Also attributed with a deep, rumbling 'bull-roar' or 'crocodile roar', usually at night. **Similar:** juvenile Black-necked Stork, but it flies neck extended, has longer legs, white trailing edge to wings. **Hab.:** dense, gloomy swamp-edge forest of tropical coastal estuaries; mudflats, river-edge paperbark thickets, billabong swamps; well inland on major rivers. **Status:** uncommon; sedentary. **Br.:** 360

White-faced Heron *Ardea novaehollandiae* 66–69 cm (C, L, M, N, T)
Common and well-known small heron with white face and throat, and yellow legs. **Breeding:** long lanceolate plumes on nape, mantle and back; shorter lanceolate plumes on lower foreneck and breast. **Voice:** calls at nest, in flight, landing at roost, in courtship flights, and in contact or alarm. Various croaking, grunting sounds: 'urrk-urrk-urrk'; 'arrrgh, arrrgh, arrrgh'; 'graaow'; grunted 'urgk-urg-urgh'. **Similar:** White-necked Heron. **Hab.:** diverse, widespread; usually shallow wetlands, margins of swamps, dams and lakes, damp or flooded pastures; also salt and brackish shallows of estuaries, mudflats, mangroves, saltpans, reefs, beaches, dunes. **Status:** abundant almost throughout Aust. in permanent or temporary habitats. Favoured by increased open wet pasture, irrigated land, ditches and dams. **Br.:** 360

White-necked Heron *Ardea pacifica* 75–105 cm (N, T)
Large; has white neck, dark wings, bold white spot to edge of upper wing. Folds neck back in flight; slow, deep wingbeats. Solitary or in small to large flocks; soars high on thermals. **Breeding:** lanceolate maroon plumes; white neck. **Nonbr.:** dark spots down centre may fade or disappear with wear. **Juv.:** neck spots spread wider; no plumes. **Imm.:** like non-breeding adult, but slight grey cap, no plumes. **Voice:** a loud single or double croak in alarm or flight: 'argh, aarrgh'. At nest gives a deep, loud 'oomph!', a raucous cackle at changeover. **Similar:** Pied Heron, White-faced Heron, Grey Heron. **Hab.:** shallow wetlands, swamps, floodwaters, wet grasslands, shallows of lakes; mainly fresh water, but occasionally coastal mudflats. **Status:** common, widespread, migratory; dispersive or irruptive after widespread interior rains. **Br.:** 360

Pied Heron *Ardea picata* 43–52 cm (T)
Small, dainty, dark capped heron with neat, pied plumage. In breeding season has long, dark plumes at nape, mantle; white plumes around base of neck. Forages actively, dashing after prey; hunts and travels in flocks. **Voice:** in flight, an abrupt, deep, rough croak, 'orrrk!', given as alarm or when feeding in groups; at nest, soft cooing. **Similar:** White-necked Heron; superficially like immature and nonbreeding Pied Heron, but much larger; has white wing spots, grey or black bill and, usually, dark neck spots. **Hab.:** coastal wetlands of N Aust., including floodplain pools, billabong shallows, swamps, wet pastures, tidal and mangrove mudflats. **Status:** across N, abundant only in areas of extensive wetlands, dispersive. **Br.:** 360

Rufous (Nankeen) Night Heron *Nycticorax caledonicus* 55–65 cm (C, K, L, T)
Predominantly nocturnal, most likely to be seen if flushed from dense foliage of its daytime roost trees. Takes off with a croak of alarm; circles overhead with rapid wingbeats, neck hunched back, flight feathers translucent red under sunlight. Solitary or groups; often large numbers roost together. **Voice:** harsh croak if disturbed feeding at night or flushed from nest; also on arrival and departure from roosts or wetland feeding sites at night. The alarm is a hoarse, but not deep, croak, 'ow-uk' or 'qwu-ok'; at roost and nest a more abrupt, nasal 'auk-auk-ak'. **Similar:** adults are unique. Juveniles show some likeness to Striated Heron, which is smaller, and Australasian Bittern (much larger, browner). **Hab.:** most wetlands, billabongs, flooded grasslands, damp fields; also estuarine environs such as mangroves, tidal channels. Prefers sites with some cover of tall vegetation and dense trees nearby for roosting. **Status:** common in SE and SW; scarce in interior. **Br.:** 360

Usually a humpbacked silhouette
Small crest
White
Massive dark bill with yellowish lower base
Dark grey-brown with long silvery grey plumes
Plumes longer when breeding
Breeding
Overall more brown than adult with rufous streaks or edging to much of the plumage
Streaked buff and rufous
Feet extend beyond tail.
Juv.
A huge heron, standing almost 1.2 m tall. Flight typical of herons; keeps neck folded back. Most likely to be confused with juvenile Black-necked Stork, which extends neck in flight.
Grey Heron, Ardea cinerea
Unsubstantiated. Unlikely, but not impossible, vagrant to N Aust. from SE Asia
Black stripes
Adult
Plumes on back and nape in breeding season
White
Plumes pale brown
Breeding
White obvious
Feet extend beyond tail.
Contrast between dark flight feathers and pale underwing coverts; but shadow may make underwing appear darker.
Slaty blue-black
Long legs trail far behind short tail.
Conspicuous white spots at carpal joints
Juv.
Long neck kept folded back. Extensive spotting to sides of neck indicates juvenile.
Long lanceolate plumes with iridescent maroon and green sheen
White neck
Breeding
Imm.
Immature and nonbreeding adult have up to 4 rows of dark spots down centre of neck. Imm. may show some grey on crown, lost by wear of feather tips. Nonbreeding adults may retain some back plumes after breeding.
Scale change
Long nuchal crest
Dark cap extends below eyes.
Dark blue-grey plumes cascade down the back in breeding season
White plumes
Breeding
Nonbreeding adult lacks crest, other plumes are reduced.
White, no dark cap
Imm.
Immature, having no black cap, is similar to the White-necked Heron, but much smaller.
Slaty black
Erratic darting flight; rapid wingbeats. Flocks fly in untidy clusters.
In flight, keeps neck folded back.
Feet trail well behind tail.
Colourful stocky heron; head hunched down onto shoulders in daytime roosting pose
Black crown with several long white plumes
Green
Cinnamon-rufous, bright in sunlight
Delicate tints of cinnamon blend into pure white.
Breeding
Legs are thick and quite long (as is necessary for a wading hunter).
Streaked on juvenile, black for immature
White, streaked brown
Juv.
When hunting, neck is extended ready to spear bill into water; hunched posture is replaced by alert stance.
Wing linings are white, but shadowed under sun can look dark, while translucent flight feathers become brighter rufous.
Neck folded
Feet trail behind short tail.

Striated Heron *Butorides striatus* 45–50 cm (C,T)

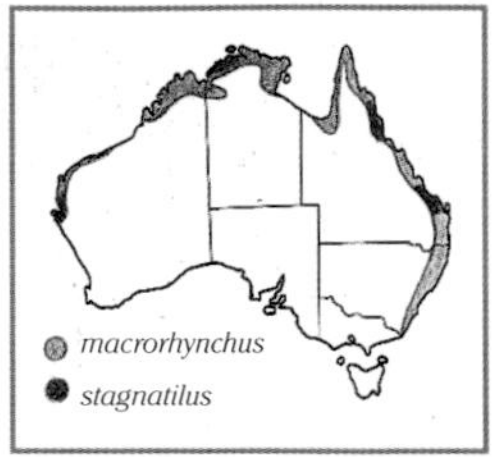

Small stocky bittern-like heron. Secretive—skulks in shadowy mangroves, emerging at low tide to forage along pools and channels of mudflats. In flight, keeps low with fast beating, rounded wings; folds neck, legs trailing. **Variation:** race *macrorhynchus*, E coast. Race *stagnatilis* has a dark grey form on grey, mud subtrate and is rufous where the coast sand or mud is reddish, mostly the Pilbara region of WA. **Voice:** in courtship display, a scratchy, sneeze-like 'tsch-aar' and abrupt, loud 'hooh!'. If flushed or disturbed at nest, sqawks loudly; stays in vicinity making 'chuk-chuk-' cluckings. At other times of year, silent unless alarmed or flushed; harsh, scratchy screech, 'tchew-chit-chit'. **Hab.:** tidal strip around coast; adjoining wetlands. Typically in mangroves, estuaries, tidal rivers. **Status:** common in N; uncommon to vagrant in SE. **Br.:** 361

Black Bittern *Ixobrychus flavicollis* 55–65 cm (T)

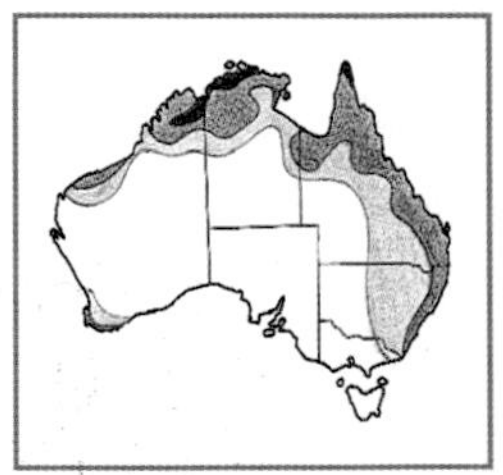

Usually seen when accidently put to flight; may perch before dropping down to cover. Predominantly black streaked yellow and white from chin to breast. **Fem.:** as male but black replaced by dusky brown; other colours dull; back feathers edged buff. **Imm.:** like female; yellow of neck very pale, back feathers edged rufous. When alarmed, may freeze with long neck and bill extended skywards, mimicking vertical stems of reeds; otherwise skulks away or takes to the air with powerful flight. Occurs singly or in pairs, hunting in dense water-edge vegetation. **Voice:** presence may be revealed by loud, deep, drawn out, booming 'whOOOm, whOOm' at intervals of 10–20 sec; often answered by similar calls from a distance. Also soft moaning sounds, soft, low hissing, and a repetitive 'e-eh, e-eh' at the nest. **Hab.:** diverse wetlands, estuarine and littoral. Requires dense water-edge vegetation, even if only a narrow fringe. Habitats include dense surrounds of freshwater springs and billabongs, and tidal reaches of creeks and rivers. **Status:** moderately common around freshwater swamps (N coast); suffers loss of habitat where cattle have destroyed the sheltering water-edge vegetation. **Br.:** 361

Little Bittern *Ixobrychus minutus* 25–35 cm

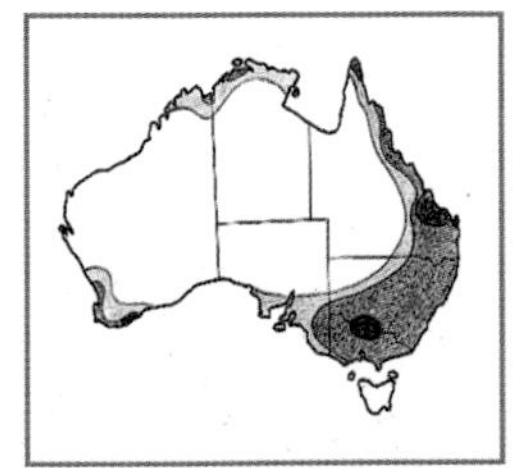

A very small, colourful, but extremely secretive bittern, whose presence may not be expected until it is flushed or glimpsed momentarily at the edge of a reedbed; conspicuous. Black crown, chestnut face; neck streaked chestnut and buff; large cinnamon shoulder patches. Usually adopts a rail-like crouch as it skulks through dense vegetation at water's edge; rarely emerges into the open. When disturbed, stands with neck vertical, bill skywards, mimicking the surrounding reeds; is assisted by vertical streakings that match the chestnut and buff tones of dead, dry reeds. **Voice:** when flushed from nest, a sharp 'cra-aa-ak, khok-kuk-kuk-kuk-'. In breeding season, male has a monotonously repetitive, resonant 'ook-ook-ook-ook-ook-' or 'corr-orr-orr-orr-', the notes at 0.5 sec intervals, the sequence lasting about 10 sec. At the nest, female gives soft crooning sounds. **Similar:** Yellow Bittern, Striated Heron. **Hab.:** freshwater swamps, lakes and rivers with dense reedbeds, tall sedges and well-vegetated margins; also in brackish–saline mangroves, saltmarsh and coastal lagoons. **Status:** localised, probably more common than sightings suggest; apparently migratory. **Br.:** 361

Yellow Bittern *Ixobrychus sinensis* 30–40 cm

Occurs S Asia, Indonesia, NG. One confirmed Aust. record—accidental, wind-blown immature after a cyclone, 1967. Several very old unconfirmed records. **Voice:** deep croak; also possibly a softer 'kak-ak, kak-ak' in flight. **Similar:** Little Bittern. **Hab.:** within usual range, reedbeds, densely vegetated margins of lakes, mangroves, ricefields.

Australasian Bittern *Botaurus poiciloptilus* 65–75 cm

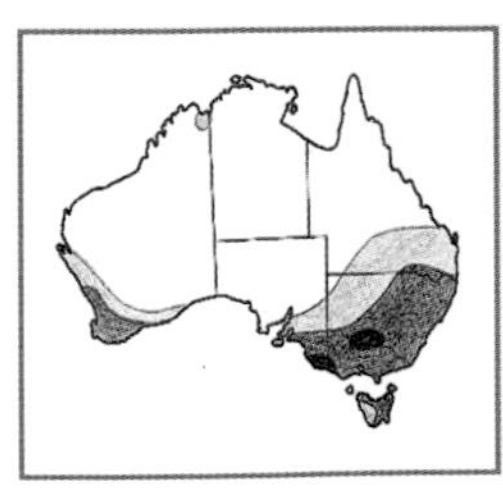

Large, powerfully built bittern. If startled into flight, rises heavily on broad wings, initially neck outstretched, legs dangling. On longer flights uses slow, steady, shallow wingbeats; holds neck hunched back and appears short; feet trail behind tail-tip; usually keeps low, but may circle high. Forages in shallows or hunts in deeper water from platforms of bent-over reeds. **Voice:** in alarm, flushed, an abrupt, harsh 'craaak!'. In spring and summer, usually at night, males give a deep, resonant, double noted, booming 'oo-OOM, oo-OOM', repeatedly; audible more than a kilometre away. **Similar:** juvenile of Nankeen Night Heron, but it is smaller, more distinctly streaked and spotted. **Hab.:** freshwater wetlands, occasionally estuarine; prefers heavy vegetation—shrubbery, reedbeds, sedges. **Status:** uncommon to rare. Migratory, dispersive, irruptive. **Br.:** 361

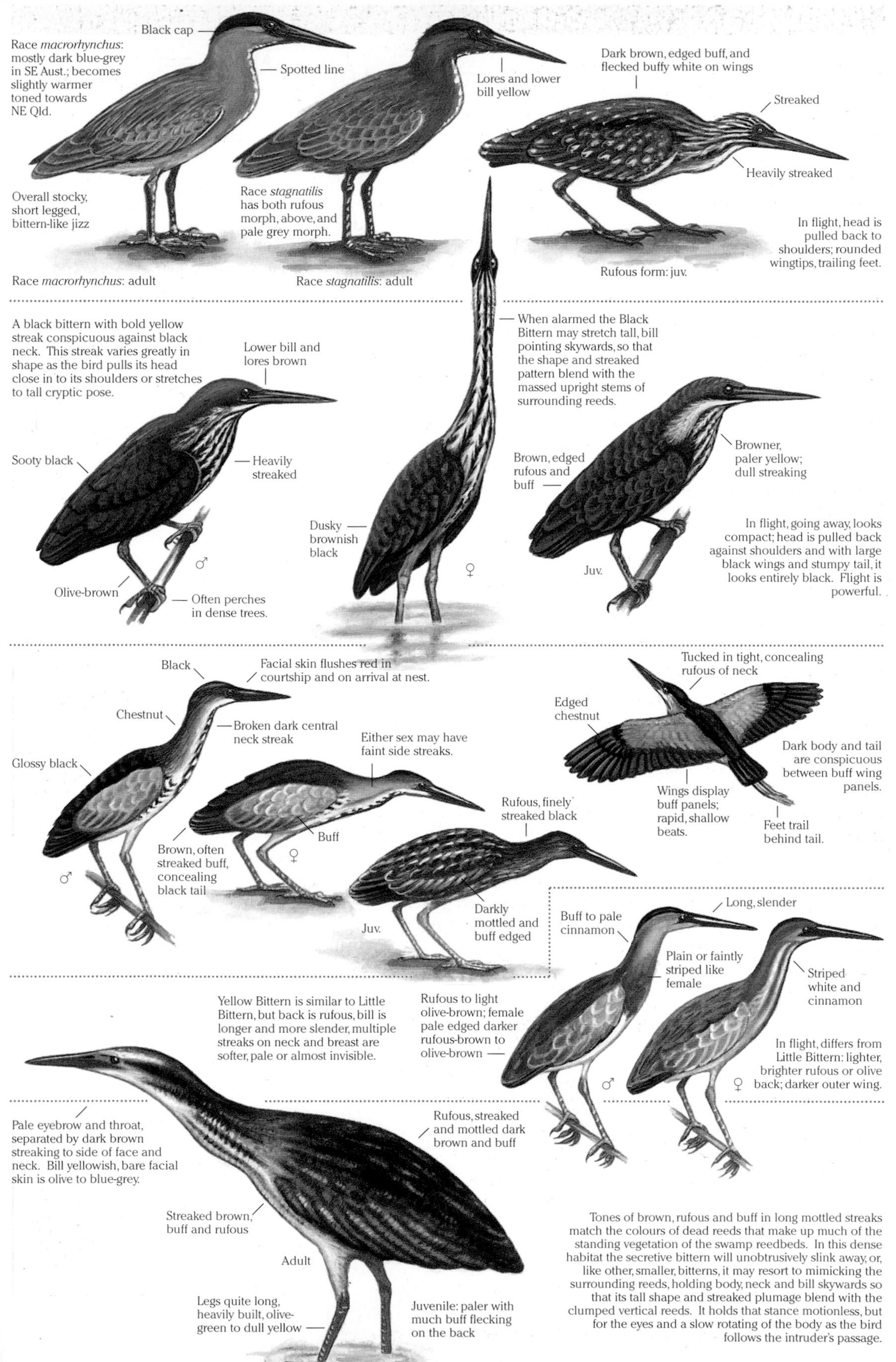
Black cap
Race *macrorhynchus*: mostly dark blue-grey in SE Aust.; becomes slightly warmer toned towards NE Qld.
Spotted line
Lores and lower bill yellow
Dark brown, edged buff, and flecked buffy white on wings
Streaked
Heavily streaked
Overall stocky, short legged, bittern-like jizz
Race *stagnatilis* has both rufous morph, above, and pale grey morph.
In flight, head is pulled back to shoulders; rounded wingtips, trailing feet.
Race *macrorhynchus*: adult
Race *stagnatilis*: adult
Rufous form: juv.
A black bittern with bold yellow streak conspicuous against black neck. This streak varies greatly in shape as the bird pulls its head close in to its shoulders or stretches to tall cryptic pose.
Lower bill and lores brown
When alarmed the Black Bittern may stretch tall, bill pointing skywards, so that the shape and streaked pattern blend with the massed upright stems of surrounding reeds.
Sooty black
Heavily streaked
Brown, edged rufous and buff
Browner, paler yellow; dull streaking
Dusky brownish black
In flight, going away, looks compact; head is pulled back against shoulders and with large black wings and stumpy tail, it looks entirely black. Flight is powerful.
♂
♀
Juv.
Olive-brown
Often perches in dense trees.
Black
Facial skin flushes red in courtship and on arrival at nest.
Tucked in tight, concealing rufous of neck
Chestnut
Broken dark central neck streak
Edged chestnut
Either sex may have faint side streaks.
Glossy black
Dark body and tail are conspicuous between buff wing panels.
Rufous, finely streaked black
Wings display buff panels; rapid, shallow beats.
Buff
Feet trail behind tail.
Brown, often streaked buff, concealing black tail
♀
♂
Darkly mottled and buff edged
Juv.
Long, slender
Buff to pale cinnamon
Plain or faintly striped like female
Striped white and cinnamon
Yellow Bittern is similar to Little Bittern, but back is rufous, bill is longer and more slender, multiple streaks on neck and breast are softer, pale or almost invisible.
Rufous to light olive-brown; female pale edged darker rufous-brown to olive-brown
In flight, differs from Little Bittern: lighter, brighter rufous or olive back; darker outer wing.
♂
♀
Pale eyebrow and throat, separated by dark brown streaking to side of face and neck. Bill yellowish, bare facial skin is olive to blue-grey.
Rufous, streaked and mottled dark brown and buff
Streaked brown, buff and rufous
Adult
Legs quite long, heavily built, olive-green to dull yellow
Juvenile: paler with much buff flecking on the back
Tones of brown, rufous and buff in long mottled streaks match the colours of dead reeds that make up much of the standing vegetation of the swamp reedbeds. In this dense habitat the secretive bittern will unobtrusively slink away, or, like other, smaller, bitterns, it may resort to mimicking the surrounding reeds, holding body, neck and bill skywards so that its tall shape and streaked plumage blend with the clumped vertical reeds. It holds that stance motionless, but for the eyes and a slow rotating of the body as the bird follows the intruder's passage.

Royal Spoonbill *Platalea regia* 75–80 cm (L,T)

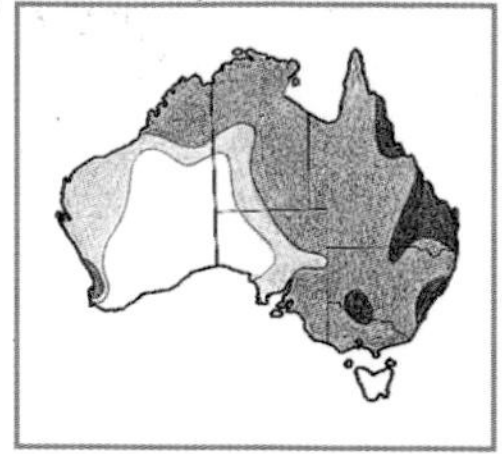

Tall, white plumaged; long bill tapers to distinctive wide flat tip that gives the two species of spoonbill unique use of wetlands food resources. These birds stride through the shallows sweeping the slightly opened bill in broad arcs side to side. Any small creature—tiny fish, crustacean or insect—that touches the inside of the broad tip, triggers it to shut instantly. Birds usually form small parties flying in lines or 'V' formation. **Breeding:** white plumes from nape. **Voice:** silent except at nest site when breeding; it gives soft, low grunts, groans and hisses. Also sounds non-vocally in displays: a loud 'whoof' made by the wings and soft bill-snapping. **Similar:** Yellow-billed Spoonbill, but this has dull yellowish bill and legs. **Hab.:** shallow wetlands and margins of deeper waters—fresh or saline swamps and flooded pastures, either open water or vegetated; also coastal lagoons, mangroves. **Status:** common except in arid parts of range; rare Tas. **Br.:** 361

Yellow-billed Spoonbill *Platalea flavipes* 75–90 cm (L,N)

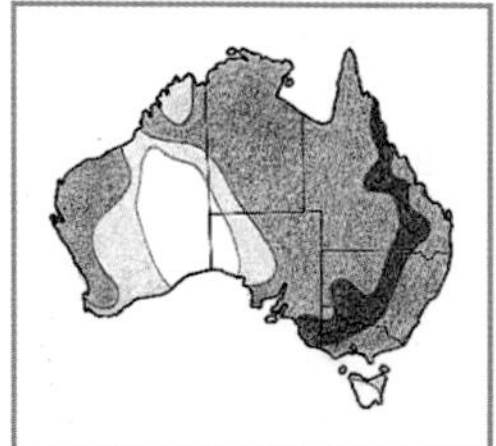

Distinguished from Royal by its dull yellow to off-white bill and legs. Like the Royal, feeds by sweeping slowly side to side with bill slightly open, but also feeds by moving rapidly forward, thrusting, probing, stirring, chasing. Flies with neck outstretched, bill level; steady, shallow wingbeats broken by short glides; travels in flocks; soars very high. **Voice:** usually silent, but in threat displays gives soft, nasal coughs or grunts, clappers bill. **Similar:** Royal Spoonbill (black bill and legs). **Hab.:** shallow swamps, fresh or brackish waters, flooded pastures; small pools and dams; rarely, tidal. **Status:** common on coast, inland after heavy rains. **Br.:** 361

Australian White Ibis *Threskiornis molucca* 65–75 cm

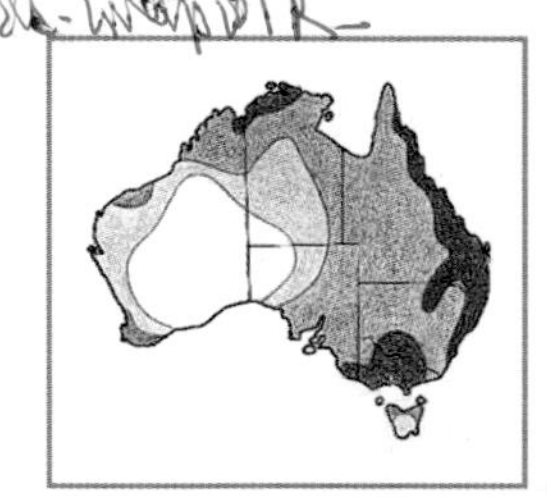

Bare black skin of head, neck, legs; lacy black of tertials in clearcut contrast to white plumage; long downcurved bill is distinctive. Flocks circle, soar, travel in undulating lines or rough 'V' formation; appear clean white against deep blue sky. At close range, plumage and skin are often stained muddy grey. **Voice:** usually silent but for deep, grunted 'urrrk'; noisy in colonies and when settling to roost—deep croaked and grunted honkings. **Similar:** at a distance, Straw-necked Ibis or spoonbills. **Hab.:** shallow fresh and tidal wetlands; and pastures. **Status:** common to abundant N, SE and SW; sedentary, dispersive. **Br.:** 361

Straw-necked Ibis *Threskiornis spinicollis* 60–70 cm

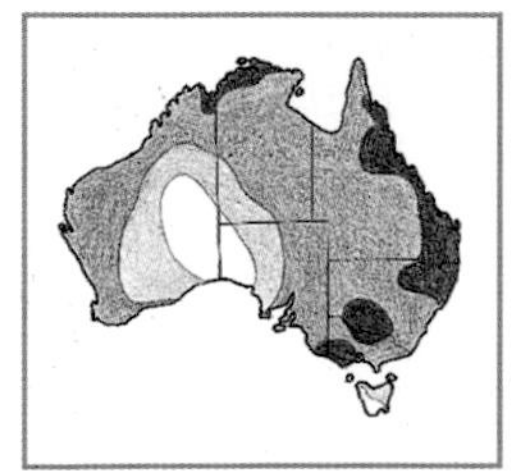

Glossy iridescent black wings, back and collar, most of underparts white, making this species distinctive even at distance; straw-like neck feathers are unique. In small groups or flocks of hundreds, work in lines across pasture, or rise in updraft of thermals, wings rigid, showing black and white. When travelling, flocks maintain 'V' formation, each holding position with deep, steady wingbeats and brief glides. Typical of ibis, neck is extended, toes trail behind tail. Flocks often perch conspicuously on dead trees. **Voice:** croaks when intending to fly; louder if alarmed; repeatedly on take off. During flight, occasional long, harsh croak. When joining flock, announces arrival with rapid series of croaks. **Hab.:** grasslands, wet or dry; prefers cultivated and irrigated pastures. Occasionally uses shallows of wetlands; rarely arid or marine. Large flocks wherever food such as grasshoppers abound. **Status:** common, locally abundant, nomadic. **Br.:** 361

Glossy Ibis *Plegadis falcinellus* 50–54 cm (K,T)

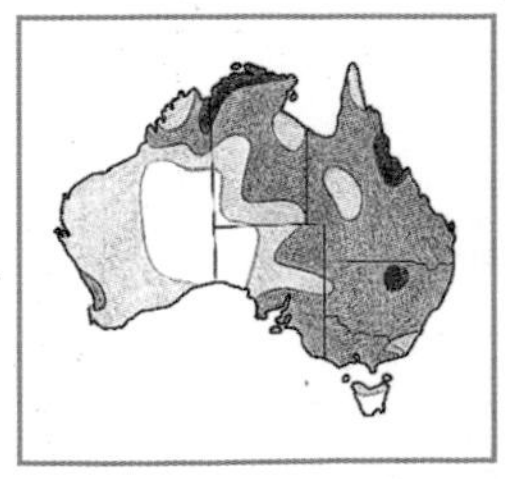

Small dark ibis with glossy iridescence and highlights of bronze, green or purple sheen, most obvious in breeding season. Travels in clusters, 'V' formation or spread out in wavering lines. Flocks congregate and roost on dead trees near water. When feeding, bird walks or wades with slow, deliberate movements, probing mud with long bill. **Voice:** in flight, occasional deep, grunted croaks; if startled, loud hoarse croaks. In colonies gives grunts and croaks. **Similar:** curlews have long downcurved bills, but lighter, patterned plumage. Little Black Cormorants at distance—flock jizz similar but different bill, legs. **Hab.:** shallows of swamps, floodwaters, sewage ponds, flooded or irrigated pastures; occasionally feeds on moist pastures, sheltered marine habitats. **Status:** common across coastal N; elsewhere generally uncommon, nomadic. **Br.:** 361

Wide-tipped black 'spoon' bill
Breeding: white plumes, yellow over each eye, red forehead spot
Black wingtips, juveniles only
Flies with neck and legs extended, long bill horizontal; rapid wingbeats interspersed with glides.
Lacks plumes, facial marks; has shorter bill.
Breeding: buff wash on breast (hidden)
Feet trail behind tail-tip.
Juv.
Nonbreeding: lacks plumes and buff tint on breast.
Feeds in shallow waters.
Sexes are alike except female is slightly smaller.
Black
Wide tipped, pale yellow or creamy white
Breeding: fine black plumes; black edging around face
Nonbreeding and juveniles: have no plumes on lower back or breast.
Breeding: buff breast plumes emerge or lengthen.
Dull greenish or creamy yelow
Breeding
Nonbr.
Long, downcurved bill typical of all ibis
Black, unfeathered
Black tipped
Bare pink skin follows line of wing bones, becomes scarlet in breeding season.
When breeding has sparse, stiff, pale yellow plumes hanging from lower neck.
Lacy black plumes
The White Ibis, feeding in mud and swamp water, often has neck, breast and belly plumage stained brown or grey.
Dark grey with smudge of red
Adult, nonbreeding
Feet trail behind tail.
In flight, neck is always extended.
Long downcurved bill
Bare grey-black skin
Black with iridescent green and violet sheen
Wings black with conspicuous, stark white, wing linings
Straw-like spiny feathers partly cover black neck band that is unbroken on female but has central gap on male.
Breeding adult
Imm.
In flight, neck is held extended.
Red
Female and immatures have slightly shorter bills.
Black
Tail and rump white
Feet slightly behind tail-tip
Brown with iridescent green and violet
White flecks or streaks on head and neck
Adult nonbreeding is like juvenile, but without white down the fore-neck.
Shorter than bills of Straw-necked and White Ibis
Sooty rather than iridescent brown
Underwing entirely dark
Entirely dark brown with a glossy iridescence that gives a sheen or burnishing of shifting greens, violet and magenta across the plumage, especially on the wings.
Bill, neck and feet sag slightly in flight.
Adult, breeding
Juv.
Feet trail well behind tail-tip.
The iridescent colours vary greatly with changes of lighting; at times this ibis can appear entirely black.

Black-necked Stork *Ephippiorhynchus asiaticus* 1.1–1.3 m; span 1.9–2.2 m (T)

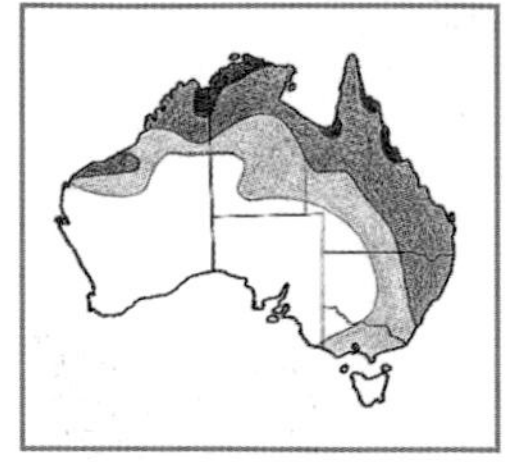

This tall, stately stork with massive black bill is commonly known as the 'Jabiru'. Head and neck are highlighted by an iridescent shimmering green and purplish sheen. Movements are slow and deliberate as it stalks the shallows of tropical swamps; suddenly it unleashes a spearing jab of the powerful bill or dashes after prey with long strides and flapping wings. Most small creatures of swamp and grassland environs are prey: fish, frogs, eels, turtles, small crabs, snakes. File Snakes are on record as victims, probably also small Olive Pythons; both are aquatic. These are rendered harmless, broken and limp by lengthy jabbing and thrashing of the heavy bill, then are swallowed whole. Black-necked Storks fly with necks fully extended, long legs trailing well behind stubby tail-tips; wings beat slowly but powerfully. They soar on long, broad wings, using thermals to reach high altitudes before setting off on long glides that are broken irregularly by languid wing flapping. **Male:** eye brown. **Fem.:** as male, but eye yellow. **Nonbr.:** no seasonal difference in plumage. **Juv.:** as adult, but plumage has less contrast (dark brown where adult is black, and the white is mottled with light grey-buff). **Imm.:** probably a gradual change from juvenile to adult through a series of immature plumage stages. **Variation:** race *australis* over whole of Aust. and NG; nominate race in India and SE Asia. **Voice:** clacking and sharp snap sounds are made with the bill; sometimes the bird gives guttural grunts, usually in threat or dancing displays; there are doubtful reports of booming calls. **Similar:** unmistakeable on the ground, but perhaps less distinctive in flight, especially at a distance, as a silhouette or in other difficult lighting conditions. Australian Pelican, but folds neck back so that only the head and large pale bill protrude ahead of the wing; Great-billed Heron, but folds neck back, bill much smaller. **Hab.:** wetlands and vicinity; prefers open freshwater environs, including the margins of billabongs, swamps, shallow floodwaters over grasslands, wet heathlands, watercourse pools, sewage farms, dams, adjacent grasslands and savannah woodlands. Less often at inter-tidal shorelines, margins of mangroves, mudflats and estuaries. **Status:** moderately common across coastal N Aust.; becomes increasingly scarce southwards on both E and NW coasts until it becomes a rare vagrant at southern and inland extremities of its range. **Br.:** 362

Brolga *Grus rubicunda* 0.8–1.3 m; span 1.7–2.4 m

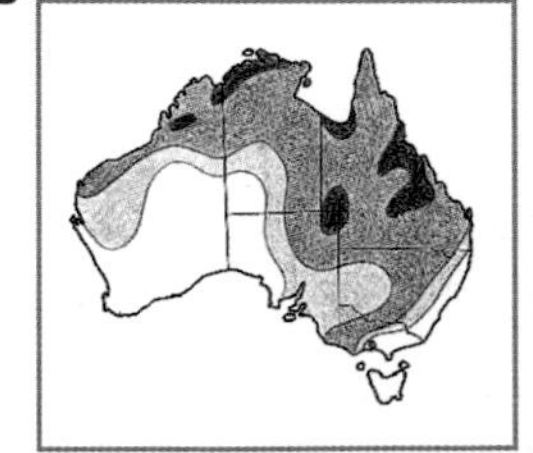

Tall, elegant, graceful in almost every movement; well-known dancing displays, bugling call. When encountered, usually walks steadily away unless forced to fly. Flies with neck outstretched; broad wings have conspicuously fingered tips, slow, deliberate beat; black legs trail well behind the tail-tip. Likely to be encountered in diverse habitats, but usually near water in the dry season, moving out over greater areas of wet, green grassland and shallow floodwater. Usually in flocks, some of thousands through the dry season when the Brolgas congregate on the last shrinking wetlands. When the wet arrives, flocks disperse and pairs move inland, probably to traditional breeding sites. Elaborate dancing displays as pairs form; groups and pairs leap gracefully, wings outspread, head thrown back, bill skywards to give the wild, carrying, bugling call of the Brolga. Sexes similar; female smaller. **Imm.:** only trace of pink at back of crown; no dewlap. **Voice:** best known is the loud bugling, given as a duet by a pair; carries far, audible 2 km away; often first indication of Brolgas' presence. Stimulates other distant pairs to call. Contact call a low, purring 'grruw'; alarm call, a high 'ga-r-r-oo'. In flight, a hoarse, grating 'graough'. **Similar:** Sarus Crane. **Hab.:** freshwater swamps, flooded grasslands, margins of billabongs, lagoons, dry grasslands, floodplains, irrigated pastures; occasionally estuaries, mangroves. **Status:** widespread, common N and NE Aust.; uncommon in S. Dispersive and seasonally migratory. **Br.:** 362

Sarus Crane *Grus antigone* 1.2–1.5 m; span to 2.4 m

So closely resembles the Brolga that, for many years, it occurred as an established breeding species in NE Qld without being differentiated. Further sightings indicated that the species had an expanding range across most of N Aust. Although occupying the same territory and habitats, Sarus and Brolga do not interbreed. In behaviour there appears little difference. The Sarus tends to be wild, wary; flies from disturbance where the Brolga would likely walk. In pairs, family parties or larger groups; roosts in large numbers in swamps. **Imm.:** trace of pink over head and neck. **Voice:** like call of Brolga, but slightly higher pitch. **Hab.:** in the dry, the Sarus occupies drier habitat than the Brolga provided there is water in the area for drinking and bathing, and secure roosting sites. Dry season habitats include tall grassland, grassy paddocks, open woodlands, tidal flats, edges of billabongs and dams. In the wet, is most common in woodlands around flooded areas. **Status:** first recorded Normanton, 1966. Found to occur more widely. Expanding range, occasionally recorded from Kimberley. Most abundant NE Qld—large flocks occur on Atherton Tableland. **Br.:** 362

Although named 'Black-necked', dark parts of the plumage, especially the neck, have a changing iridescent green sheen, with highlights of purple and blue.
Male has brown eye.
Massive bill
Black with iridescent colours
Square-tipped, deeply fingered
Black panel visible both sides of long white wings
Female has yellow eye.
Legs trail far behind.
Distinctive silhouette is apparent at distance: extended neck, domed crown and massive bill with straight upper edge.
Deep pink
Adult ♂
Brown without iridescence; fades gradually towards pale belly.
Juvenile brown
Juv.
In transition to adult pattern and colour
Brown, imm. primaries
Immature birds leap in exuberant play or perhaps greeting display. Juveniles change plumage gradually over period of about four years, acquiring glossy adult colours first on head and neck, last on primaries.
Imm.
Imm.
Adult leg colour developing
Head bare, grey, with red across face, bill to nape
Dark 'dewlap' bulge under chin
Blue-grey with pale edging
'Bustle' of dense elongated grey tertial feathers down over primaries and rump
Legs dark grey
Black, square tipped
Downbeats are slow and powerful; upstrokes end with a quick flicking action.
Dark trailing edge
Neck always outstretched, often sagging in centre, drooped when landing
Dark legs trail far behind end of tail.
Scarlet on head extends to upper neck.
Lacks dark 'dewlap'; pale grey under chin
Black, deeply fingered
May have some white feathers in 'bustle'.
Secondaries are entirely pale without any dark trailing edge.
Sarus is slightly taller than Brolga; both share similar dancing displays, gregarious flocking when roosting and feeding.
Red legs trail far behind tail.
Dull pink or dusky red

Osprey *Pandion haliaetus* male 50 cm; female 65 cm; span to 1.7 m (T)

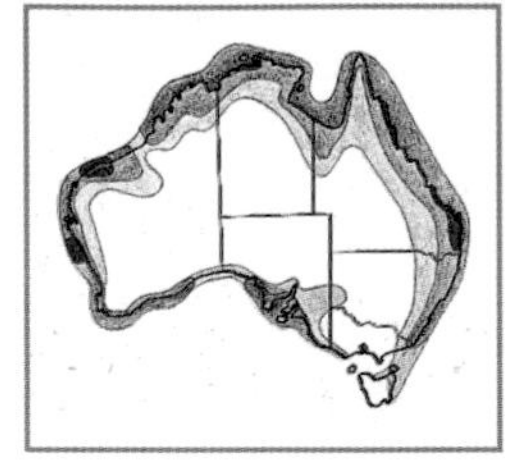

Usually seen near water; distinctive in flight and when perched. Wingbeats are slow, leisurely, shallow but steady; powerful enough to cause noticeable undulating or bobbing movement of the body. When soaring or slow gliding, the wings taper significantly, widest near body, narrow towards backswept wingtips. When hunting, the Osprey may soar high to spot schools of fish, gliding lower before folding back the wings to plunge into the sea. Unlike the White-bellied Sea-Eagle, the Osprey often submerges, catching fish up to 1 m beneath the surface. At other times it may make a shallow strike, immersing only the legs to snatch at fish close to the surface. With reversible outer toes and rough surfaces to the feet, the Osprey can securely grip the slippery fish, which can be held head forward in the direction of the bird's movement through the water as it returns to the surface. Then the Osprey must take the full weight of the fish and struggle heavily into the air. **Fem.:** has a dark-streaked neckband. **Juv.:** similar to adult, but heavier dark streaking on nape, crown; wide dark band around the breast; most of the feathers of the upper parts are white edged, giving a scaly effect that diminishes as the edges wear. **Voice:** often silent, but noisy near the nest in breeding season. Usual call is a drawn out, plaintive or peevish whistled 'pee-ieer', harsh screams and an anxious, sharp 'tchip-tchip'. Alarm calls are harsher, especially when diving at an intruder near the nest. **Hab.:** coastal waters and estuaries, beaches, islets and reefs—but usually not far out to sea except on islets or exposed reefs. Follows major rivers and wetlands far inland from the coast to large river pools, even to arid regions where large pools occur in gorges hundreds of kilometres inland. **Status:** common around N coast, especially on rocky shorelines, islands and reefs; uncommon to rare or absent from closely settled parts of SE Aust. **Br.:** 362

Brahminy Kite *Haliastur indus* male 45 cm; female 50 cm; span 1.2–1.3 m (T)

Unique, chestnut and white; adult unmistakable. Usually glides with the broad wings held flat or slightly drooped, bowed, wingtips upcurved. When soaring to gain height, wings are still bowed, but slightly uplifted in a very shallow dihedral. Primaries are spread to form broader, more rounded wingtips. In faster glide the primaries are backswept, wrists forward. Occasional casual, 'rowing' wingbeats intersperse with gliding or soaring. Glides and soars low along shoreline, shallows, mangroves and mudflats. Scavenges for carrion along shoreline and shallows, but also an opportunistic predator of small fish and other marine creatures of reefs and beaches—insects and small reptiles from adjoining land can be taken in a shallow swoop. Will perch on dead trees by the shoreline, from which vantage point it scans the water's edge. **Juv.:** brownish, gradually acquires adult plumage over several years, chestnut and white feathers appearing among the juvenile browns and fawns. Flight exposes pale bar along underwing; widest at base of dark flight feathers. **Voice:** a high, harsh, wheezy, drawn out, descending 'pei-ir-ah' is mainly heard in spring breeding season. At the nest, harsh, wheezy whistling sounds. **Similar:** adults are distinctive, but other species are similar to the juvenile Brahminy—the Whistling Kite, Little Eagle (light morph) and juvenile White-bellied Sea-Eagle. **Hab.:** usually the more sheltered waters of tropics and subtropics; prefers coasts with islands, mangroves, estuaries, mudflats, harbours, coastal towns; penetrates far inland along rivers. **Status:** common around Australia's N coastline; uncommon in S of range; widespread through SE Asia. **Br.:** 362

Black Kite *Milvus migrans* male 50 cm; female 55 cm; span 1.2–1.5 m

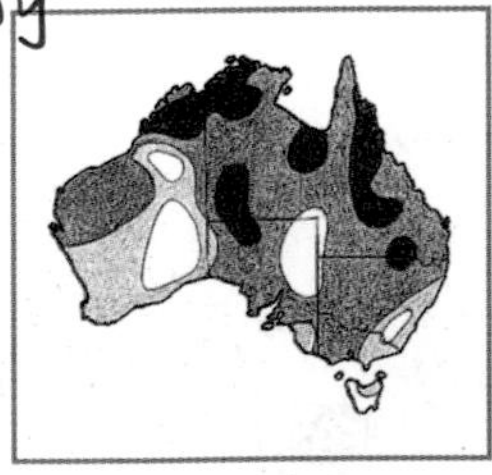

Overall blackish brown; slow floating glide, occasional lazy wing flaps, widely fingered wingtips, forked tail. Leisurely gliding and soaring with tail often fanned and twisted to gain lift in the breezes and updrafts. Wings are usually held slightly bowed, highest at the carpals, outer wings slightly drooped. But often, possibly more in calm airs, holds wings almost flat. When soaring and gaining height, may lift the profile to a very shallow 'V' dihedral. Often glides just above or between treetops; skilfully maintains a glide in still airs with only an occasional flap of the wings. Upright posture on perch; looks very long winged, the tips of the folded wings reaching almost to the tip of the long tail. Rarely solitary, often in gatherings of hundreds. Principally scavenges carrion and rubbish around coastal towns and stations. **Voice:** a plaintive, peevish, descending, quavering 'kwe-ee-ier'; also a sharp, staccato 'kee-ee-ki-ki-ki'. **Similar:** Square-tailed Kite, but it has unforked tail, white face, very long fingered primaries; Little Eagle (dark phase), but shorter tail, longer feathered legs, soars in circles on flat wings. **Hab.:** woodlands, scrublands, tree-lined watercourses, mangroves, mudflats, swamps; often around homesteads, grass fires. **Status:** common coastal N Aust.; elsewhere uncommon, nomadic or vagrant. **Br.:** 363

Dark line
Slight crest
Dark brown
Wing is long, extends beyond tail.
Male white or lightly streaked
♂
Brown
Brown
White edged upper parts
Juv.
Wings strongly kinked back in fast glide
White
♂
Barred
Always dark
Long, tapered wings, widest at body; slightly kinked back when soaring
Always very dark
Dark collar, female; darker on juvenile
♀
Soars and glides on distinctively up-kinked and drooped wings.
Barred
Plunges headlong with feet forward beneath sea or snatches from surface.
Toes opposed, fish is carried head forward.
♀
White
Chestnut
Short, unfeathered
Adult
Brown, streaked cream
onger
han tail
Juv.
Bowed
Upturned
Glide and soar
Slow glide
Pale tipped, short
Black tips
Fast glide
Held well forward
Pale
Strongly 'S'-curved trailing edge
Juv.
Soaring
Gently S-curved trailing edge
Rounded when spread
Dark behind eye
Dusky blackish brown
Adult
Long tail, usually forked
Juv.
Soars and glides on slightly drooped and arched wings.
Tail often twisted
Pale streaks
Fork of tail is not always visible when kite is standing on ground or moulting, or when feather tips are worn.
Conspicuously 'fingered'
Pale, lightly barred
S-curved trailing edge
When spread, makes square-tipped triangle; more distinctly forked when partly closed.

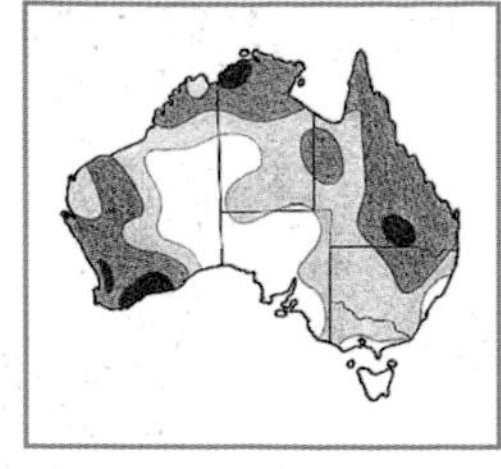

Square-tailed Kite *Lophoictinia isura* 50–55 cm; span 1.3–1.45 m
A colourful endemic; rare over most of its range. Conspicuous white face; large wings, when folded, extend well beyond the tip of long, square tipped tail. In flight, wide tipped wings with long, deeply fingered primaries derive lift from slightest updraft or thermal; needs only an occasional slow, casual wingbeat. Often glides around treetop level, or low over heath or grassland—harrier-like with similar buoyant, hesitant, almost hovering quartering of the ground in search of small birds or other prey. In slow glide has a slight sideways rocking motion. The male brings nestlings of small birds, often complete with nest, to the brooding female. **Voice:** loudest is a yelped call, clear, plaintive and rather musical, 'yip-yip-yip-', rising in pitch to an excited 'yeep-yeep-yeep'; or a hoarse, yelped 'airk-ek-k'. Food begging call by female is a high, whistled 'wheee' and from male a slightly lower version; female on nest also gives a brief, shrill chittering. **Hab.:** in S Aust., eucalypt woodland, open forest and heath-woodland; in N Aust., more diverse, dominated by eucalypts, pandanus, gallery forest, heath. **Status:** rare, scattered, sedentary or partly migratory; moderately common in hills near Perth. **Br.:** 362

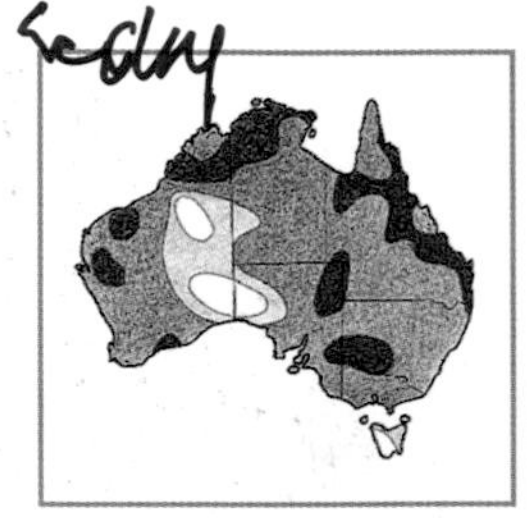

Whistling Kite *Haliastur sphenurus* 50–60 cm; span 1.2–1.5 m (T)
Scruffy looking, gingery brown kite. Flight buoyant; easy, languid soaring and gliding on wings slightly drooped, held far forward to give short necked, long tailed appearance. Soars in haphazard, untidy circles; widely fingered wingtips, fanned tail. In fast glide, wingtips back, tail narrowed. Often glides low around treetops; deep, slow wingbeats with undulation of body. **Voice:** in flight and at nest, a rapid, sharp, penetrating, carrying call; first a drawn out, descending 'peee-arrgh' followed by a burst of short, sharp, harsh, upward notes, 'ka-ke-ki-ki'. The full call: 'peee-aa-rgh, ka ke-ki-kiki... peee-arrgh, ka ke-ki-ki-ki'; sequence repeated several times. Harsh 'eeargh' in defending nest; a loud 'kaairr' as warning near nest. Solitary or gregarious; often at carrion, roadside kills; takes some live prey. **Hab.:** varied, often wetlands, but also arid regions (especially in good seasons) and near watercourses; open woodlands, scrublands, farmlands, estuaries, littoral mudflats. **Status:** common in N, rare in S of range; sedentary and migratory, part of N population moves S in summer. **Br.:** 363

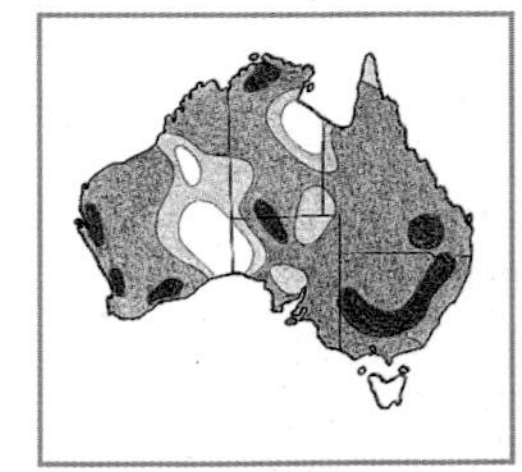

Little Eagle *Hieraaetus morphnoides* 45–55 cm; span 1.1–1.35 m
One of Australia's two species of true eagle, or three if Gurney's Eagle is now counted in addition to the Wedge-tailed. These are booted or trousered eagles whose legs are heavily feathered to the feet, adding to the impression of solid build, especially of legs and talons, compared with the slender bare legs of other Aust. raptors. The Australian Little Eagle looks a true eagle: stocky build; compact and powerful; head with slight crest to the nape; long legs sleekly feathered in buff or tawny brown; feet and talons large and powerful—an eagle also in behaviour, posture and presence, in ability to hunt large, live prey, and in the undulating display flight and ringing calls that attract attention kilometres away. The Australian Little Eagle in flight is distinctive, usually seen soaring in tight circles that reveal the flat wing position. From directly beneath, the overall shape is compact and neat, the wings long and broad, of even width, no taper. Rounded wingtips are deeply fingered; pattern of underwing is distinctive. To hunt, soars, circling; dives steeply to strike prey on ground or in trees. Prey is mainly rabbits, but also birds as large as ducks, and possums, cats and foxes. Although itself 650 g to 1 kg, the Little Eagle kills prey to 1.5 kg and can lift and fly with 500 g. **Light morph:** head, neck cinnamon-buff, underbody buffy white finely streaked cinnamon; underwings have a pale buff diagonal band to a large pale window at the base of flight feathers. **Dark morph:** white and buff replaced by light reddish brown streaked dark brown; underwing pattern remains, but darker. Northern birds average smaller than southern. **Imm.:** a plainer, more rufous plumage within each phase. **Voice:** a musical, yelping whistle in distinctive double or triple note sequence: a very rapid 'chik-a-chuk' or 'chik-a-chuk, kuk'; the first 'chik' strong, sharp and high, the '-a-chuk' softer, lower, mellow, musical; sometimes a soft low fourth note, 'kuk', at the end. Emphasis can vary. **Similar:** Whistling Kite, but different pattern on underwing; longer, rounded tail. Black-breasted Buzzard has some similarity to dark phase Little Eagle, but has bold white 'bullseye' in wingtips and shorter tail; to a lesser extent, juvenile of Brahminy Kite. **Hab.:** favours hilly country where it often soars on the updrafts generated by wind deflected up the slopes. Forests, woodlands, open scrublands, tree-lined watercourses of interior. Most abundant where open country intermixes with wooded or forested hills, as in farmland, irrigated lands. Avoids dense forest, but will use clearings and margins of dense eucalypt and rainforest. **Status:** widespread but uncommon; vagrant to Tas. Adults sedentary; young dispersive. **Br.:** 363

Dark grey brown
White
Pale wingbar. Sunlit, may show whitish highlights.
Rufous, heavily streaked brown
Adult
Longer than tail
Cinnamon -rufous
Lightly streaked
Short legs
Juv.
Soaring: wings are held well forward.
Broad, dark barring
Off-white panel
Long, square tipped
Long, deeply fingered; makes wide wingtips.
Circling low, Kite shows widespread primaries.
White often visible from afar
Gliding
Sunlit white face can be conspicuous.
Soaring
From behind, gliding away, wingtips appear quite pointed.
Sandy fawn, pale streaks
Adult
Tail is longer than wings.
Darker than adult; buff or white spots
Heavier streaks
Juv.
Legs are short, pale, lower part unfeathered.
Soars and glides on arched, slightly drooped wings.
Throws head back to give whistling call.
Long, rounded, plain, sandy, pale tipped
Distinctive pale M-shaped pattern under wings
Black, deeply fingered
May display small dark crest.
Adult: light morph
Buffy white, finely streaked cinnamon
Wingtips almost to end of tail
Legs are always feathered over their full length.
Juvenile: light morph
Soars and glides on flat wings.
Glide: wingtips backswept
Juvenile: light morph
Adult: light morph
Barred
Diagonal pale bar across underwings
Mid brown streaked darker
Adult: dark morph
Legs fully feathered
Bright deep rufous
White
Rounded when spread, soaring
Square when folded
Many intermediate plumages between lightest and darkest morphs. Except for the darkest birds, is distinguished by pale underwing M-pattern. Dark birds are similar to Buzzard, but latter species has much brighter white outer wing patch; soars and glides on upswept wings.
Juvenile: dark morph
Adult: dark morph
Juvenile: dark morph
Glides with wingtips back, tail folded.

Black-breasted Buzzard *Hamirostra melanosternon* 50–60 cm; span 1.4–1.55 m

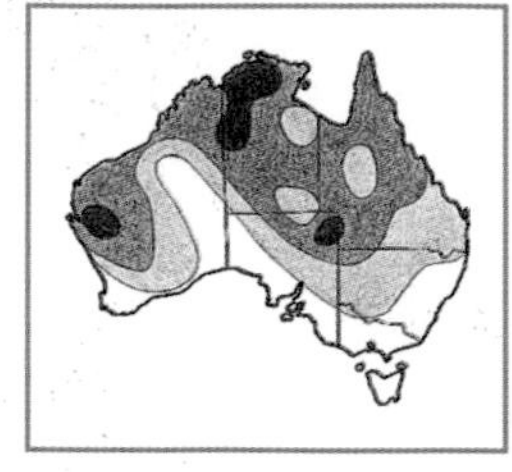

Medium sized, large winged, robust; richly plumaged in black and deep rufous; white bullseye patches at base of unbarred black primaries. Soars high; also glides low, rather like a harrier, across the plains seeking rabbits, small reptiles, ground birds and carrion. Also takes eggs, breaking them with stones thrown from bill. When sailing low, often sways or tilts side to side; in flapping flight uses deep, deliberate, powerful wingbeats. Usually solitary or in pairs, unlikely to be in a group larger than family party. **Voice:** usually silent but quite vocal near nest. On return to nest, excited yelping, 'kyik-kyik-kyik'; as alarm, high, long 'screee'. Also a variety of harsh, scratchy, grating sounds. **Similar:** Wedge-tailed, but much larger; long pointed tail; Square-tailed Kite, but long, barred, deeply fingered primaries make outer part of wings the widest; Black Kite, but longer, larger tail, lacks the prominent white wing patches, largely a gregarious scavenger. **Hab.:** mainly in semi-arid and arid regions; nests in large trees along inland watercourses and hunts out over surrounding scrub or grassland plains. Also uses open woodlands, floodplains; avoids forests and other dense vegetation. **Status:** widespread but scattered; all mainland states, most common through semi-arid inland and in N. Sedentary. **Br.:** 363

Wedge-tailed Eagle *Aquila audax* 0.9–1.1 m; span 1.8–2.5 m

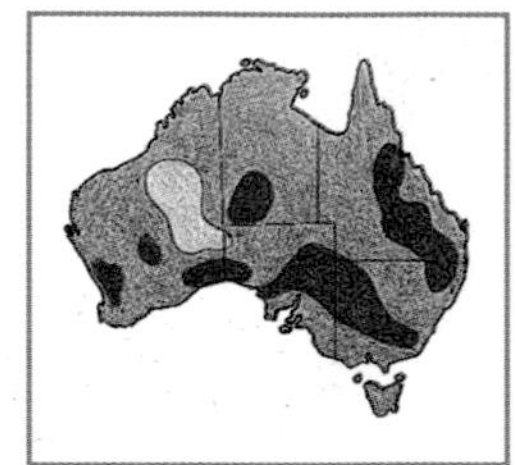

Largest Aust. raptor; dark, legs heavily feathered to feet; long tail overall diamond shaped, the tip a pointed 'wedge'. Often seen at roadkill; takes flight heavily with slow, deep, powerful wingbeats. Once aloft, skilfully uses updrafts of thermals or hillslopes; rises effortlessly, rarely needing to flap the huge wings. Soars very high on upswept wings in great circles, showing widely spread flight feathers of deeply fingered wingtips; at times exceeds 2000 m above ground. When high enough, the eagle may set the wings to a flatter dihedral, outer wings backswept, and depart on a long, down-sloping glide, soon vanishing in the far distance. Active, flapping flight used only close to ground; deep, slow powerful beats interspersed with flat glides. **Fem.:** larger, averages 4.2 kg to male's 3.2 kg. **Imm.:** dark brown with rufous and cinnamon about head and neck. **Variation:** race *fleayi*, Tas.; adults retain pale nape. **Voice:** often silent, but vocal in breeding season. Usually a wavering, high, rising whistle leads into a short, lower, descending note, 'tsIET-you, tsIET-you'. Also a hoarse, drawn out yelp that rises to a short, scratchy squeal, and a long scream, perhaps of alarm. **Similar:** Black-breasted Buzzard, but smaller, short blunt tail, stark white underwing patches; juvenile of White-bellied Sea-Eagle, but stubby tail, narrow outer wings; Gurney's Eagle, but rounded tail-tip. **Hab.:** diverse climatic and vegetative types; usually hunts over open country but nests in forests: arid scrublands, alpine, mallee, coastline, wetlands, forested areas with farm clearings. **Status:** common except in closely settled regions and extensive heavy forests. **Br.:** 363

White-bellied Sea-Eagle *Haliaeetus leucogaster* 75–85 cm; span 1.8–2.2 m (C,T)

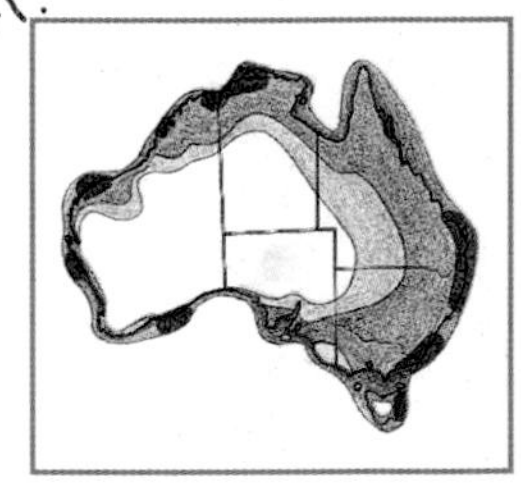

A large, neatly plumaged eagle with white head and body, contrasting dark grey wings, short white tipped tail. Often conspicuously perched on a high limb that gives a view over coast or billabong. Soars skilfully; circles on rising air columns above sun-heated islands or headlands, or above coastal ranges or cliffs where the wind is deflected upwards. When soaring, wings are held in an upcurved dihedral; short tail is slightly wedge tipped, or rounded when fully spread. Sometimes it soars in company with the similarly sized Wedge-tailed Eagle. **Voice:** harsh, nasal, carrying, goose-honked 'ank,ak-ak,ank,ak-ak-', or two birds in chorus—a confused, rapid 'ank-ank-arkakak-ank-akakak-ak'. **Similar:** adults unmistakeable. Juveniles may be confused with juveniles and immatures of the Wedge-tailed Eagle, but these have the longer diamond-shaped tail, fully feathered legs. **Hab.:** usually coastal, over islands, reefs, headlands, beaches, bays, estuaries, mangroves, seasonally flooded inland swamps, lagoons and floodplains; often far inland on large pools of major rivers. **Status:** established pairs usually sedentary, immatures dispersive. Common around most of the coastline, scarce near major coastal cities. **Br.:** 363

Gurney's Eagle *Aquila gurneyi* 74–87 cm; span 1.7–1.9 m (T)

Large solidly built blackish brown eagle; fully feathered legs, heavy bill. In flight, primaries are deeply fingered. Soaring and gliding it looks slow and lethargic, but has powerful, deep wingbeats in active flight. When gliding, wings are almost level, but are lifted slightly higher to a shallow V dihedral when soaring or in fast glide. Soars high, hunts low over forest and open ground. **Voice:** nasal, downward, piping whistle. **Similar:** the Wedge-tailed Eagle, immature White-bellied Sea-Eagle. **Hab.:** lowland rainforests, adjacent coasts and islands. **Status:** in Australia, reported from Boigu Is., Torres Str.; also NG and the Moluccas.

Red, crested
Black
Rufous feathered
Adult
Wings longer than tail
Short tail
Rufous, lightly streaked
Short, partly feathered
Juv.
Stark bright white
Black
Short, square
Stark bright white
Gliding
Soaring and slow, sailing glide
When in this glide, may rock slowly side to side.
Rufous
Colours of underwing are here as if seen well lit by low sun; much darker if shadowed by high sun.
Juv.
Old males are black with rufous hackles on nape, tawny across wing.
Adult
Long, powerful fully feathered legs
Juv.
Shorter than tail
Tail diamond-shaped, 6 pairs of feathers
Proportionately long necked, small headed
Long 'fingered' tips flex upwards
Trailing edge S-shaped
Gliding away, long diamond-shaped tail is narrowed, more pointed; in rear view the backswept wingtips appear pointed.
Wedge-tailed Eagle soars on upswept wings; may rise to 2000 m or more.
Wedge-tailed Eagle in glide; wingspan to 2.3 m.
Sea-eagle, soaring.
Sea-eagle in glide; span to 2.2 m
Entirely white
Heavy bill
Grey
Yellow, unfeathered
Adult
Wings longer than tail
Brown, streaked cream
Brown, edged buff
Juv.
White, long necked
Adult has unique white and black underwing pattern.
In slow glide, primaries are slightly backswept.
Short, grey and white, wedge tipped when partly folded
Juv.
Buffy white
Dark
Dark band on buffy white
Juv.
Overall blackish brown
Adult
Large, grey
Dark, fully feathered
Tawny rufous
Feathered; yellow feet
Juv.
Soar and glide
Wings broad, rounded, dark.
Rufous-buff
Pale
Faintly barred
Rounded, dark
Soars with S-curved trailing edge, narrow at body.
Dark band; lightly barred, rounded
Juv.

Spotted Harrier *Circus assimilis* 50–60 cm; span 1.2–1.45 m

A large, slender raptor with long, broad wings; blue-grey upper parts; colourful chestnut and white underparts; chestnut face with rather owl-like facial disc. In flight has a slim body, long barred tail; wings display conspicuously widely fingered black tips to the long primaries. Often seen gliding low over grasslands, spinifex or crops; wings upswept, long tail fanned widely, and with characteristic slow, side to side rocking motion. In faster gliding, wings are held lower, slightly bowed. Occasionally hovers briefly, just above the vegetation, with long legs dangling to snatch at a ground bird, small mammal, reptile or other prey. The soaring or gliding may be assisted by a few deep, slow flaps of the wings. Most commonly seen over open country with low dense cover, often harrying with trailing talons almost brushing the tops of grass or shrubs. Sometimes will soar high, wings held up in a wide V; then, when high enough, travels fast on long, downward glide with wings held lower, tips backswept and pointed. The Spotted Harrier is unique among harriers—it nests in trees, only very rarely on the ground. Other harriers nest on the ground in tall grass or crops, or over water in reedbeds. The Spotted Harrier may have adapted to a continent that lacks extensive wetlands where ground-nesting birds can be protected by water, and has taken to tree nesting. But it seems to retain some instinct for the character of the ground vegetation, for it often places the tree nest in a clump of mistletoe or among dense regrowth at the end of a broken limb, where the density of the vegetation holds and conceals the nest as would dense reedbeds or tall grass. **Juv.:** conspicuously different, with much chestnut and buff, especially on underparts. **Voice:** usually silent, but near the nest gives a high pitched chipping sound, a 'see-eep' food call and a soft 'kitter-kitter-kitter' between the pair. **Similar:** Swamp Harrier male has similar flight, size and silhouette, but has white rump, unbarred tail, streaked rather than spotted from throat to belly, less colourful underparts. The Swamp Harrier is more like the juvenile of the Spotted Harrier, which also has a pale rump, but is distinguished by a boldly barred tail. Square-tailed Kite also has colourful underwings, but the tail is proportionately shorter; wingtips are extremely widely fingered so that outer wing is the widest part; tips are barred rather than black; short legs and white face. **Hab.:** open grasslands, spinifex, open shrublands, saltbush, very open woodlands, crops and similar low vegetation that allows the low 'harrying' mode of hunting. **Status:** most common in drier inland areas; nomadic, part migratory or dispersive; movements linked to abundance of prey species. Much of the range of the Spotted Harrier is across semi-arid country, including vast areas of spinifex, relatively undeveloped. Also in regions of broadacre wheat farming. **Br.:** 365

Swamp Harrier *Circus approximans* 50–60 cm; span 1.2–1.5 m (L, M, N, T)

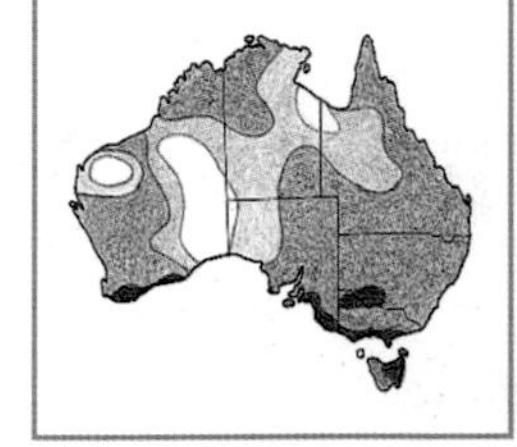

A typical harrier in its mode of hunting; glides low over reedbeds, open water of swamps and reedy lakes on long, broad, upswept wings. Flight action is buoyant, but not as light and floating as that of the Spotted Harrier. With that species it shares a rocking, slow glide, harrying of prey; hovers with slow, powerful wingbeats, long, thin legs dangling with talons ready to strike at any small creature hiding in the vegetation. Often harries waterbirds, forcing ducklings to dive repeatedly as it glides and circles low overhead. May be seen perched low on clumped reeds, stumps, mudflats; rarely high on trees. Distinctive features are white rump, brown facial disc, streaked breast, unbarred tail. Males are paler, greyer, but both sexes become more grey with age. Light blue-grey gradually replaces brown, especially on tail and flight feathers. Prey includes any small animal of the wetlands and vicinity—mammals up to rabbit size, birds, frogs, reptiles, fish, large insects. **Voice:** occasionally a threatening, chattered 'kik-kik-kik' in competition for food. In the breeding season, calls are given near the nest or while soaring high over the nest territory, a loud, brief, whistled 'ki-yoo'. **Similar:** Spotted Harrier. **Hab.:** wetlands, including swamps and lakes, vegetated or with open waters, mangroves, salt marshes, temporary floodwaters. Less typically, marine waters, heathlands, grasslands and pastures. **Status:** common; sedentary and nomadic or migratory—some birds wander N to wetlands in summer, return S to breeding wetlands by winter. **Br.:** 365

Papuan Harrier *Circus spilonotus* 50–55 cm; span 1.3–1.4 m

Male distinctive; black-hooded head, back, shoulders and black-streaked breast in bold contrast to white undersurfaces, streaked breast. In flight the pattern of bold black and white is distinctive. **Fem.:** Brown; pale upper-tail coverts; bold banding on tail; pale streakings on neck, breast and, often, the head. **Juv.:** overall dark brown; pale nape, and, sometimes, crown and rump. **Voice:** usually silent; a shrill 'whieeeuw' has been recorded from captive birds. **Similar:** Swamp Harrier, female and juvenile; Spotted Harrier, 'dark-hooded' first immatures. **Hab.:** NG grasslands, floodplains and swamp margins. **Status:** probably quite common in NG; unconfirmed records for Aust.

Grey, barred darker
Chestnut face bordered grey, forms a facial 'disc'
Chestnut, spotted white
Adult
Long, barred
Long, yellow
Widely fingered
Adult
Long, barred, rounded
Soaring, and slow, sailing gliding on upswept wings
Cinnamon-buff streaked brown
Brown, edged buff
Dark edged facial disc
Widely fingered black wingtips are conspicuous.
Backswept in fast glide
Juv.
Juv.
Long with rounded tip
Chestnut shoulder
Barred
Lowered outer wing in fast glide
Patrols low over grasslands in slow, sailing glide, wings in upswept V, often with slight side to side rocking, occasional slow flap of wings.
'Harrying', almost hovering over low vegetation; long wiry legs ready to reach into vegetation to snatch at prey.
Pale-edged facial disc
White rump, obvious in flight, is a feature of species; on dark-plumaged juvenile, rump is pale rufous.
Buff streaked
Pale
Juv.
Soar and slow glide
♂
♀
Dark brown
Barred
Long, grey; plain or very faintly barred
Juv.
Long, slender, pale yellow.
White rump, adults; pale rufous, juvenile
♀
♂
White, barred
Glides with backswept wingtips.
Hovering, harrying prey.
Barred, rounded
Male has black-hooded appearance.
Streaked white
Soar and slow glide
Pale wing panel
Black streaked
Gliding
Appears hooded.
Barred primaries
Black
Pale, unbarred
More heavily barred than Swamp Harrier
♂
♀
Juv. (not shown) is dark brown like juv. Swamp Harrier, but with larger white patch on nape.
♂

Black-shouldered Kite *Elanus axillaris* 35–38 cm; span 80–95 cm

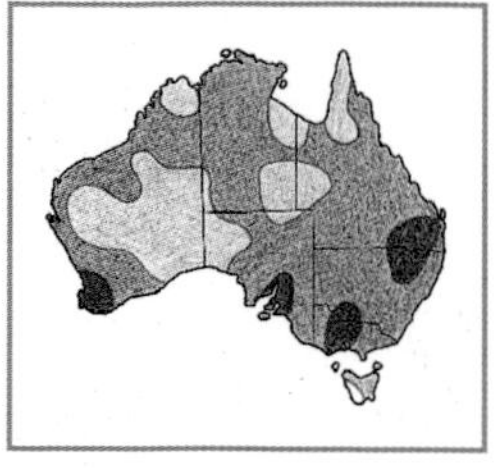

Elegant, small, pale grey hawk with white head and black shoulder patch. In flight it shows all-white underparts except for a small black patch at each 'wrist' joint and dark wingtips; superficially gull-like. Hovers frequently, skilfully while hunting, wings moving through wide arc; bird drops to take mouse, small lizard or ground bird. Soars on strongly upcurved wings; primaries slightly spread, blunt wingtips, tail widely fanned, tip rounded. Often hunts at daybreak and dusk when mice most active. **Voice:** short, plaintive, worried, piping 'siep', repeated regularly at intervals of about 5 sec; a drawn out, wheezy, husky or scraping 'scrair' at intervals of 5–10 sec. Also a 'chek-chek-chek' contact call and a sharp 'kik-kik-kik' distress call, given aggressively when defending nest. **Similar:** Letter-winged Kite. **Hab.:** natural grasslands and farmland stubble with height to harbour mice or other prey; heaths, saltbush with scattered trees. **Status:** common coastal Aust.; scarce semi-arid and absent arid regions, vagrant Tas. Irruptive after rodent plagues. **Br.:** 364

Letter-winged Kite *Elanus scriptus* 30–36 cm; span 85–95 cm

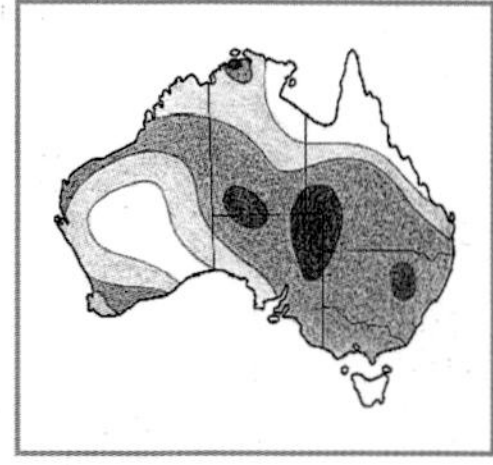

Similar to Black-shouldered Kite, but in flight shows bold black bar across underwings in shape of shallow 'M' or 'W'; wings otherwise white with translucence that may further brighten parts of wings. Black in front of eye does not extend behind eye; gives slightly owl-like face. Largely nocturnal; roosts by day, often in flocks; most hunting at night, seeking Long-haired Rat, house mice, small native mice and marsupials. When soaring or gliding, wing dihedrals are similar to Black-shouldered Kite, but active flight appears more erratic, indirect with changes of direction interspersed with brief glides and downward swoops; wing action is deep, loose; longer, slower sweeps back and forth while hovering; a slower, more agile flight that may be advantageous in nocturnal hunting. Gregarious, often in flocks; nests in colonies. **Voice:** sharp, clear, penetrating, whistled 'pseep', like that of the Black-shouldered but louder. Also a harsh rasping, louder and more penetrating than from the Black-shoudered Kite, rather like a distant cockatoo screech. **Similar:** Black-shouldered Kite. **Hab.:** semi-desert and desert along tree-lined creeks; hunts over grasslands and other low vegetation. **Status:** usually rare, at times locally common. Stronghold and breeding range the grasslands of Barkley Tableland and the Diamantina, Georgina and Cooper Creeks NE of Lake Eyre. Numbers explode in good seasons, then disperse widely. **Br.:** 364

✓**Pacific Baza** *Aviceda subcristata* 35–45 cm; span 0.8–1.1 m Mamungwala

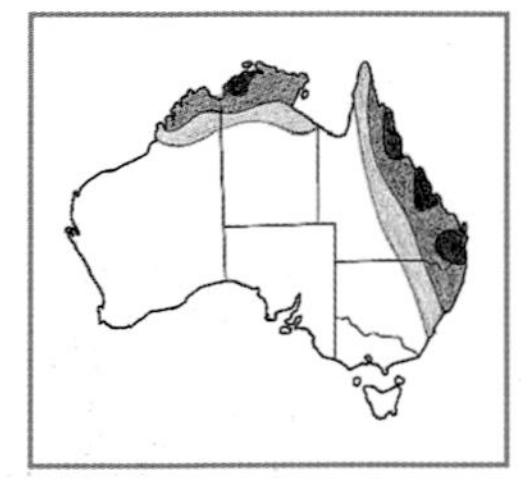

A slender, colourful, medium-sized hawk with bold dark bars across white belly and flanks. The small grey head has a small crest, deep yellow eyes. Perches in rather upright posture, wings almost reaching tip of long, broad tail. In flight, wings are uniquely shaped paddles—outer wing is widest; inner wing is narrow near body. Wingtips are rounded, deeply fingered; primaries deeply barred. Soars and glides with wings held almost flat; weaves and circles around the treetops, often in family parties, sometimes in flocks of 30 or more; plunges into canopy foliage, snatching at stick insects, mantids, frogs or small reptiles on branches, foliage or, at times, the ground. Often hunts from a perch concealed under forest canopy; prefers forest edges or gaps where foliage does not impede visibility or flight toward prey. Gliding and soaring flight is graceful, effortlessly floating on long, wide wings. Active flight with deep, relaxed wingbeats, the action hesitating above back level before downbeat; becomes an overarm rowing action at speed. Generally rather unobtrusive and seems to cause little alarm among small birds. **Fem.:** slightly less colourful. **Juv.:** crown and upper parts browner, breast light cinnamon barred darker. **Voice:** quiet, unobtrusive and secretive in vicinity of nest, though reported to be noisy during nest-building. Calls during spectacular aerial display flights. In its undulating display the bird climbs with deep wing flapping, plunges downward with wings held in steep dihedral showing colour under wings; then swoops up to climb and repeats the whole sequence, all the while whistling loudly. This characteristic call is a distinctive, rising and falling, musical 'whiech-yoo, whiech-yoo, whiech-yoo'; the first part scratchy yet musical, rising strongly; the final 'yoo' low and mellow. This call is repeated many times in succession, the calls a few seconds apart. Also a rather mellow and musical 'kaka-kaka-kak-a kak-ak-ak', at times by both birds together, very rapid. **Similar:** Brown Goshawk and Collared Sparrowhawk, but both have finer barring, lack crest, longer yellow legs and different wing shape in flight. **Hab.:** margins and spaces of gallery forests, monsoon forests, swamp forests, rainforest–woodland margins, tropical and subtropical open forests and woodlands, preferably adjoining closed forest areas, and often near water; occasionally suburban gardens in leafy tropical suburbs. **Status:** uncommon across N and NE Aust., rare in NSW. **Br.:** 364

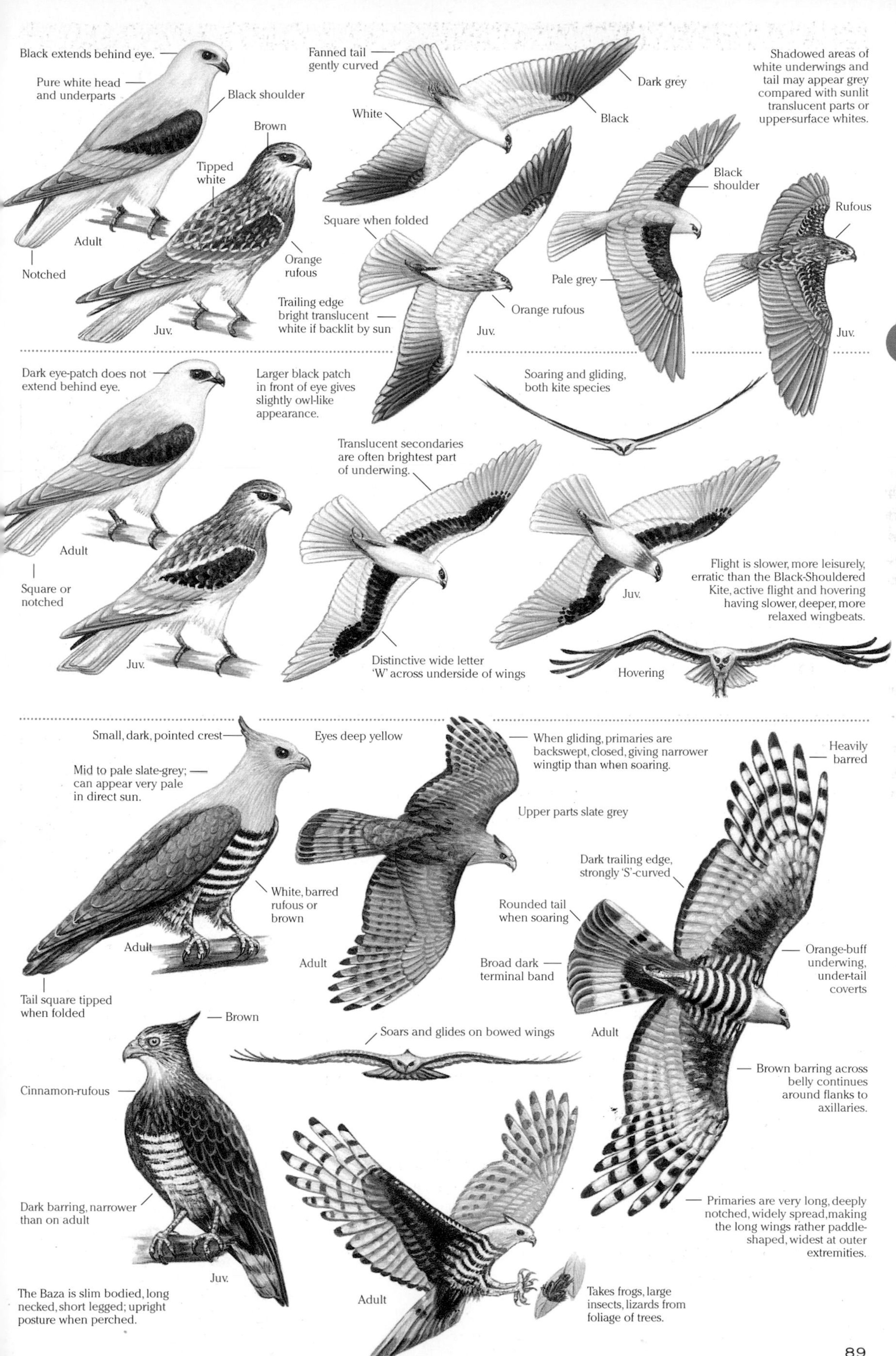
Black extends behind eye.
Pure white head and underparts
Black shoulder
Brown
Tipped white
Adult
Notched
Orange rufous
Juv.
Fanned tail gently curved
White
Dark grey
Black
Square when folded
Trailing edge bright translucent white if backlit by sun
Orange rufous
Juv.
Shadowed areas of white underwings and tail may appear grey compared with sunlit translucent parts or upper-surface whites.
Black shoulder
Pale grey
Rufous
Juv.
Dark eye-patch does not extend behind eye.
Larger black patch in front of eye gives slightly owl-like appearance.
Soaring and gliding, both kite species
Translucent secondaries are often brightest part of underwing.
Adult
Square or notched
Juv.
Distinctive wide letter 'W' across underside of wings
Juv.
Flight is slower, more leisurely, erratic than the Black-Shouldered Kite, active flight and hovering having slower, deeper, more relaxed wingbeats.
Hovering
Small, dark, pointed crest
Mid to pale slate-grey; can appear very pale in direct sun.
White, barred rufous or brown
Adult
Tail square tipped when folded
Eyes deep yellow
When gliding, primaries are backswept, closed, giving narrower wingtip than when soaring.
Upper parts slate grey
Adult
Heavily barred
Dark trailing edge, strongly 'S'-curved
Rounded tail when soaring
Broad dark terminal band
Orange-buff underwing, under-tail coverts
Adult
Brown
Soars and glides on bowed wings
Cinnamon-rufous
Brown barring across belly continues around flanks to axillaries.
Dark barring, narrower than on adult
Juv.
Primaries are very long, deeply notched, widely spread, making the long wings rather paddle-shaped, widest at outer extremities.
The Baza is slim bodied, long necked, short legged; upright posture when perched.
Adult
Takes frogs, large insects, lizards from foliage of trees.

Collared Sparrowhawk *Accipiter cirrhocephalus* 30–40 cm; span 55–80 cm

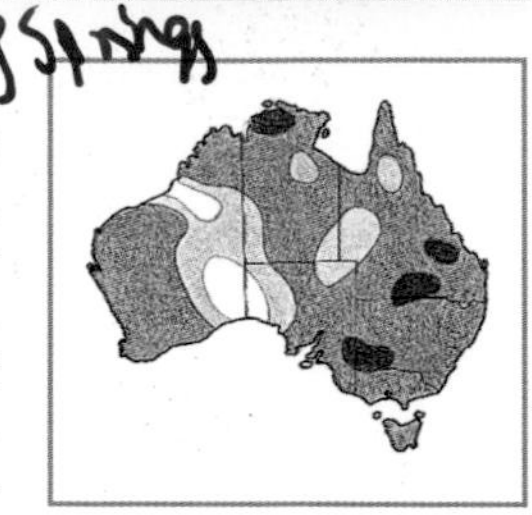

Small, lightly built hawk; slender, square tipped or notched tail; long, thin, wiry yellow legs; rufous collar and yellow eyes. Rapid tail-waggle on landing. Prey is usually small passerines, occasionally to rosella size, usually taken in flight. Regularly patrols territory in fast, low flight, suddenly encounters or flushes prey that is then taken after twisting, turning pursuit. Also waits in foliage, launching attacks at small birds passing by. In low flight, agile bursts of rapidly flickering wingbeats. Gliding and soaring shows a rounded wing silhouette. **Voice:** near nest and in display, a shrill, chattered 'ki-ki–ki-ki' and a slower, piercing 'kwiek-kwiek-kwieek-'. Calls are similar to Brown Goshawk, but higher pitched, weaker, more rapid. **Similar:** Brown Goshawk. **Hab.:** forests and woodlands almost throughout Aust. from tree-lined watercourses of arid interior to wet coastal forests. **Status:** widespread; probably quite common. **Br.:** 364

Brown Goshawk *Accipiter fasciatus* 40–50 cm; span 75–95 cm (C, N)

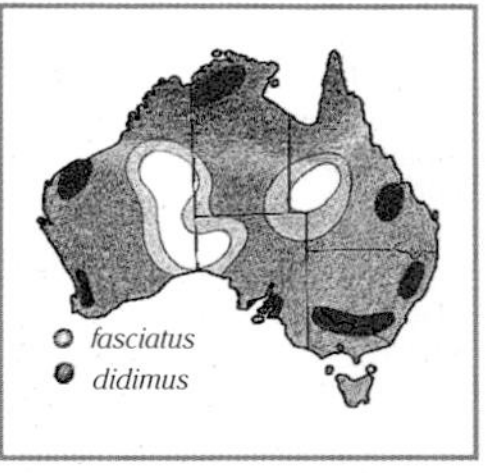

Fast, agile, powerful hunter of forests and woodlands. Often hunts from perch, waiting half concealed to dash or dive suddenly at still or moving prey; less frequently pursues very small birds in flight. Takes much more of its prey from the ground than does the Sparrowhawk, especially rabbits where these are numerous. On landing, waggles tail side to side. **Juv.:** much heavier streaking and barring. **Imm.:** finer barring, lacks rufous collar; adult plumage in third year. **Variation:** race *didimus*, tropical N Aust. is smaller, paler with whitish throat; may differ sufficently to be recognised as a separate species. In SW of WA, race *fasciatus* may differ slightly from those of the SE, including heavier barring of underside of tail. **Voice:** calls loudly in vicinity of nest when breeding, a high 'keek-keek-keek', rising in pitch. Also a rapid, excited, descending 'kik-kik-ki-ki-kikik' possibly in defence of nest site. At times uses a slow, drawn out 'ee-you-wick, ee-you-wick'; female deeper than male. **Similar:** Collared Sparrowhawk. Similar plumage but has smaller, thinner legs with elongated middle toe; tail square cut or slightly notched when folded and more rapidly waggled on landing; proportionately longer wings. **Hab.:** forests and woodlands, dry scrublands, farms. **Status:** common; sedentary and part migratory. **Br.:** 364

Grey Goshawk *Accipiter novaehollandiae* 40–55 cm; span 70–110 cm

Sleek, heavily built goshawk plumaged in softly blending greys; cere and legs deep orange, and eyes crimson. White morph is even more dramatic with these colours set in snowy white. Skulks amid foliage, bursting out with speed and aerial agility to take prey, whether a bird in flight or a rabbit. Powered flight is fast and direct; strong, deep wingbeats alternating with glides; makes short flights among the trees or soars just above the canopy. **Voice:** male a slow, high, piercing 'kieek-kieek-kieek-kieeek-', 10–20 times; female's call similar but slower, a drawn out, mellow 'kuuwieek-kuuweik-kuuweik-.' **Similar:** Grey Falcon, but has pointed, dark tipped wings. **Hab.:** rainforests, gallery forest, tall mangroves and eucalypt forests, woodlands, river edge forest; prefers mature forest with overhead canopy and open understorey that suits hunting technique. **Status:** uncommon, patchy; in many places rare due to clearing. **Br.:** 364

Red Goshawk *Erythrotriorchis radiatus* 45–60 cm; span 1.1–1.35 m

Largest of Aust. goshawks and unique in many respects. Appears to have features similar to both the goshawks and the harriers; face harrier-like. In flight, more like kite than goshawk—long winged when soaring, or, with wings back in fast glide, like a rufous-toned and swifter Brown Falcon. Soaring often interspersed with strong, deep wing-flaps—gives powerful, fast flight. Probably always rare or uncommon, this species was not studied in detail until recent years. When soaring shows long, heavily barred wings and tail; rounded wingtips; holds wings slightly raised. Usually hunts from foliage-screened perches—a typical goshawk attack, fast, and from ambush. Alternatively, will fly low and fast with deep powerful beats and brief flat-winged glides along creeks, around swamps and lagoons, all places where larger numbers of birds occur, its very long, large and powerful legs and feet capable of snatching and holding prey that might be flushed from undergrowth or water. Recorded prey includes birds up to Brush Turkey size. **Voice:** unlike other goshawks. Female has harsh, strident, crowing call, 'arhk, arhk, awk', repeated for up to 30 min., especially the month before laying. Male's call higher, yelped but similar; various other noises. **Similar:** many raptors that have reddish juvenile plumage. **Hab.:** undisturbed forest or woodland with mosaic of mixed vegetation, especially areas which include bits of river, billabong or swamp wetlands with large bird populations. **Status:** scarce even in far N; rare and declining in S part of range. **Br.:** 364

Rufous half collar
Brown streaked neck, grey-brown back
Barred with broad, dark 'V' markings
Finely barred
In slow glide, wingtips are closed, slightly backswept.
Long, slender, wiry yellow legs, very long middle toe
♀
Juv.
Broad bulge at secondaries
Soars with primaries spread, rounded wingtips.
Folded in glide, has long, slightly notched or square tip
Soaring
Gliding
Males small, slender
♂
Partly spread when soaring
Heavier hooded brow gives eyes a menacing glare not evident in Sparrowhawk.
Second-year plumage darker, heavily barred
Rufous half-collar, may extend around neck as a complete collar.
Male similar, smaller
Race didimus, tropical N, smaller, paler
♀
Outer pair often very lightly barred
Broad bulge along secondaries
Browner, darker
Rather than a single plumage feature that will separate Goshawk from Sparrowhawk, often the difference is a combination of size, shape, colour, build and movement that make up the jizz of each species.
♀
Very long, usually with rounded tip
Long, bare yellow legs: often perches with legs slightly splayed outwards; middle toe is not greatly longer than others.
Juv.
Darker, wider, 'V' pattern barring
Glides with wings flat or very slightly drooped; raised to a shallow dihedral when soaring.
Red eye
Deep yellow cere
Wings are uplifted slightly when soaring.
Grey morph: grey upper surfaces, entirely white beneath
Glides on bowed wings.
Finely barred
Slow glide: rounded wingtips
White morph: entirely white plumage
Adult
Wings are very broad at secondaries, narrowed at body.
Juvenile: iris brown, yellow after first year, then red. Juvenile of grey morph is slightly darker than adult.
Large, powerful, yellow or orange legs and feet
Medium length, wide, square cut
Adult
Dark streaked
Eye orange; immatures rufous-orange
Bright rufous
Slow glide, with primaries partly closed and backswept
Large, powerful tropical goshawk with upright stance; female of size similar to Australian Little Eagle
Heavily barred
Trailing edge bulges at secondaries.
Dark grey and brown, scalloped rufous
Gently rounded when spread
Tail folded, square tip
Rufous feathering over thick, powerful yellow legs
Soars with wings slightly uplifted.
♀
Rather falcon-like shape when in fast glide
Backswept in fast glide
Glides with wings flat or slightly bowed.

Nankeen Kestrel *Falco cenchroides* 30–35 cm; span 60–80 cm (C, L, T)

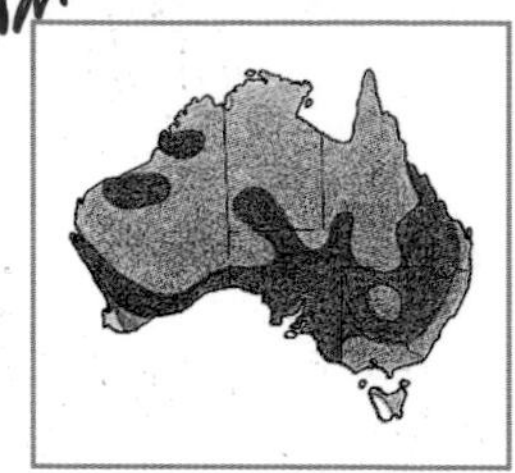

A small falcon with rufous upper parts, boldly contrasting black flight feathers, underparts whitish, conspicuous dark tail band. Flight compared with other falcons is lighter, wandering, often changing direction, stopping to hover; indirect flight has rapid wingbeats, sweeping glides. Soars on flat, blunt tipped wings, widespread tail. Hovers skilfully into wind with quick, shallow wingbeats, or hangs with wings flexed, uplifted; drops suddenly to take prey in groundcover. **Voice:** usually silent, but quite noisy in early breeding season. The sharp, high, almost metallic 'ki-ki-ki' is used with many variations: territorial defence, displays, fighting, approaching nest with food: ranges from fast, shrill, chattered 'kikikikik-' to slow 'kee-kee-kee' and very slow, metallic, tapping 'kik, kik, kik'. Also has a drawn out, screaming, rising 'keeeiir, keeiir' at food exchange and copulation. **Hab.:** open woodlands, grasslands, mulga and other sparse scrublands, heathlands, farmlands, roadsides, coastal dunes and heaths. **Status:** common mainland Aust.; nonbreeding visitor to Tas. **Br.:** 365

Australian Hobby *Falco longipennis* 30–35 cm; span 70–90 cm

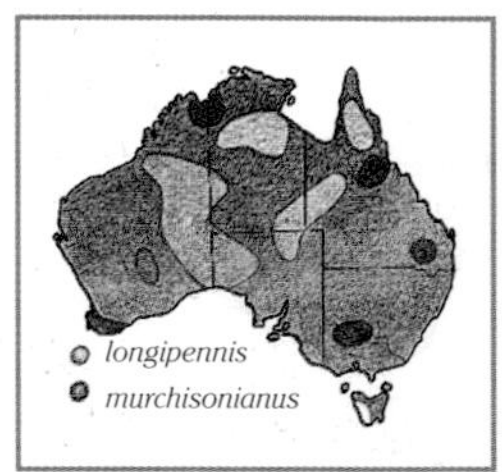

A small, fast, black faced falcon about Kestrel size but darker; buff collar around to nape, underbody rufous, streaked brown; when perched, finely pointed wingtips reach at least to tail-tip. Flight swift, direct with rapid, shallow wingbeats; occasional brief glides on flat or slightly lowered wings. In pursuit of birds the wing action becomes vigorous, deeper; the wing silhouette curves back, becomes scythe shaped, almost swift-like. Fleeing birds are overtaken by superior speed and aerial agility, are struck down in fast stooping dives from behind. When soaring, the wings (which are slightly angled back in glides) are held fully outstretched so that the trailing edge is almost straight; the tail is fanned, rounded. Aerobatic skill enables the Australian Hobby to take, in flight, prey ranging from large insects to bats, and birds from swallows to parrots. **Juv.:** darker, richer plumage colours; legs and feet pale greyish yellow. **Variation:** two races in Aust. Race *longipennis* occurs E, SE and SW Aust. Race *murchisonianus*, distinguished by much paler colour above and beneath, through inland, NW, N and drier parts of NE Aust.; boundaries indefinite, often wide overlap of these races. **Voice:** sharp, harsh, metallic 'kiek-kiek-kiek-kiek-', accelerating to 'kir-kie-kie-kie' or 'kikikikikiki' with greater anger or anguish. Given in territorial defence, or defending nest, or by male approaching nest with food. Calls are similar to those of Kestrel, but more harsh and metallic; also similar to calls of Peregine Falcon, but not as deep or powerful. **Similar:** Peregrine Falcon, Kestrel. **Hab.:** woodlands and open forests, surrounds of lakes and swamps, watercourse trees of interior scrublands, heathlands and farmlands with scattered trees, garden suburbs. **Status:** uncommon; scarce in Tas.; widely distributed. **Br.:** 365

Peregrine Falcon *Falco peregrinus* 35–50 cm; span 85–100 cm

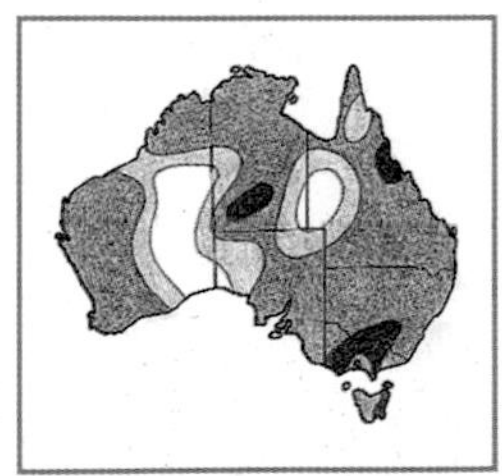

A large, powerfully built, black hooded, dark backed falcon with chin, throat and underbody white; breast to under tail has fine, transverse dark barring. Wings are long, pointed, almost reach tip of tail when perched. Entire crown, including forehead, is black, extending down onto cheeks as a broad triangular patch. In flight, the Peregrine is powerful, fast, agile; active flight with stiff, shallow, rapid wingbeats interrupted by short glides. In pursuit, the wingbeats become deeper, powerful, lashing strokes, soon overtaking most prey. Often hunts from above, stooping with tremendous velocity on half-closed wings; hits the prey, perhaps a flying duck or pigeon, with tremendous impact. In powered flight and fast glide, the outer wings are backswept, pointed, held flat or very slightly drooped; tail is folded. When soaring, circling high, the silhouette changes—wings are held straight out with the trailing edge almost a straight line wing to body; tail fanned, rounded. **Voice:** calls higher from male; deeper and very harsh from the larger female. Noisy in breeding season and when nest is approached with variations of the 'hek-hek' or chek-chek' call, depending on the excitement or agitation of the birds. This call is used by the male returning to nest with food, or when an intruder is first sighted from the nest, then becoming a harsh but drawn out, loud 'kiek-kiek-kiek-kiek-' or 'hiek-hiek-hiek', about 2 or 3 per second when danger is close. Peregrines may swoop aggressively at intruders, and calls will become extremely loud, grating, harsh, staccato; much more rapid, about 5 per second, 'kek-kek-kek-' or 'hek-hek-hek-' almost continuously; rises to a machine-gun-fire crescendo in diving attack, 'ekekekekekek-'; slows and fades as attacking birds circle and climb for another attack. Also quiet sounds between pair at nest, or female to young—frog-like croaks, plaintive 'chip-chip-' and 'chirrup' sounds. **Similar:** Hobby, but smaller with differences of plumage and flight silhouette. Brown Falcon, but very different voice, flight and plumage details. **Hab.:** diverse, from rainforest to arid scrublands, from coastal heath to alpine. Requires abundant prey, secure nest sites, lack of human interference. Also hunts over rainforest canopy, estuaries, offshore island seabird colonies. **Status:** widespread; generally uncommon to rare. **Br.:** 365

♂ Light grey
♀ Rufous
Dark grey 'tearmark'
Chestnut or mid-rufous
White, finely streaked
Spread when hovering, soaring. Pointed in fast glide
Landing or hovering, a contrast of rufous against black
Female, juvenile: barred rufous
♀
Soars and glides almost flat.
Heavier streaking
Grey
Juv.
Backswept and pointed in fast glide
Blunt tipped compared with Australian Hobby
Adult wingtips usually almost to tail-tip
Collar reaches nape
Overall darker, brown and rufous
Kestrel has conspicuous black tail band; tail spread in slow glide, soaring, hovering.
Blue-grey to grey
Narrow black point
Face markings remain black although most of plumage of this race is much paler.
Folded tail is square tipped, gently rounded when spread in slow glide.
Juv.
Race *murchisonianus*: plumage pale
Slow glide, with wings slightly backswept. Straighter and broader when soaring
Wingtip slightly longer than tail
Square tip is not evident in side view.
Soars and glides with wings slightly bowed and lowered.
Carpel joints are held forward.
Primaries are backswept and pointed in fast glide.
Wing scythe shaped, trailing edge curved
Heavily barred
Buff-fronted form, SW corner of WA
Black hood, yellow cere and eye-ring
Overall darker, brown and rufous
Broad black point
Beginning a dive in display or attack: carpal joints forward, outer wings backswept, pointed.
Juv.
Heavily built, deep-chested appearance
Deep yellow, powerful legs, feet and talons
White
Fast glide, wings pointed, trailing edge curved
Square tipped
Sexes alike, except female is larger
Trailing edge curved in glide
In soaring configuration, wings are held straight out; trailing edge then almost straight; primaries are slightly spread in blunt wingtips; tail is very widely fanned.
Wing dihedral for both soar and glide
Square tipped tail

Brown Falcon *Falco berigora* 40–50 cm; span 90–120 cm (L)

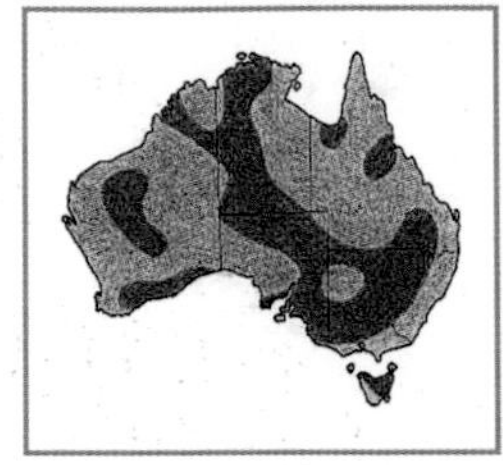

One of Australia's most common birds of prey, it can be seen across the continent, and is conspicuous throughout most of the interior on roadside perches. Once its unique, loud, carrying, raucous, crowing and cackling calls are known, any traveller or camper will be aware of its presence where it would otherwise be unnoticed. Unlike Australia's other large falcons, the Brown Falcon feeds mostly on the ground; hunts reptiles, grasshoppers, beetles and mice, and also feeds on carrion. Roads offer surveillance of long expanses clear of any concealment; any small creature making its way across the road is prey to the falcon as it glides down for an easy catch. This versatile raptor is well equipped for terrestrial hunting: its long, heavily scaled legs and short talons give it the ability to run, leap and snatch at such agile small prey as mice, dragon lizards, snakes up to a metre in length and small ground birds. The Brown Falcon's flight, speed and agility, while not comparable to a Peregrine's, are adequate to catch small birds that it may take by surprise as it glides low across scrub or grasslands. While it usually hunts from perches, the Brown also quarters open country at 5–20 m height, hovering low in search of prey. In active flight the Brown Falcon is unmistakeable once its peculiar wing action is recognised: a deep, laboured, overarm rowing action, interrupted by brief, wobbling, side slipping glides, meanwhile attracting attention with loud, raucous cacklings, the whole performance unique among Aust. raptors. Often calls kilometres from the nest and continues until in the nest tree; together with equally noisy behaviour at the nest, this can make the presence of a nest, even if well concealed, impossible to overlook. **Colour morphs:** plumage ranges from almost uniform brownish black to white fronted with sandy brown back and to some that are much more rufous. **Voice:** loud raucous cacklings somewhat like a laying hen but louder and harsher, 'karairk-kuk-kukkuk', the first part raucous, rising, the following 'kuk-kuk-' as a low clucking. Also 'karark', 'kar-r-rak', 'kairrrk' as single calls a few seconds apart. **Similar:** Black Falcon, most like dark morph of Brown Falcon; Kestrel most like light phase; Red Goshawk, some similarity with reddish morph of Brown Falcon. **Hab.:** widespread in most open habitats—woodlands, lightly treed farmlands, mulga scrub, watercourse tree lines, alpine woodlands and meadows, heathlands, coastal dunes, farmlands. **Status:** common; sedentary, at times irruptive. **Br.:** 365

Black Falcon *Falco subniger* 45–55 cm; span 95–110 cm

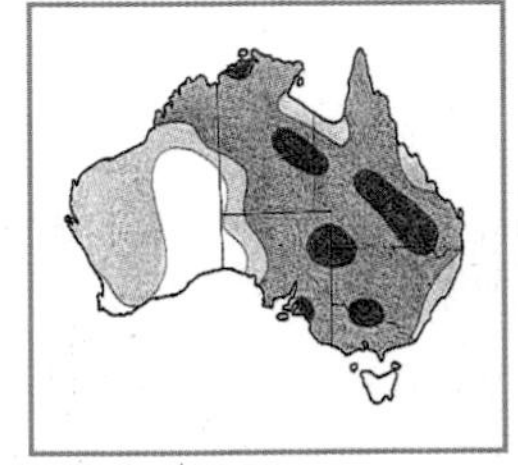

Large, sooty black, broad shouldered falcon. Flight variable, leisurely with relaxed crow-like wingbeats, occasional glides on slightly drooped wings. Accelerates with deep, powerful, rapidly flickering wingbeats to pursue avian prey. **Voice:** usually silent, but several reports of loud screams when attacking prey. Call like that of Peregrine, but deeper, slower, 'gaak-gaar-gaak-', becoming a more excited 'gak-gak-gak-' if an intruder is near the nest tree. In sudden alarm a single 'gaaark!'. Also gives a call quite unlike other falcons in courtship and display flights—a loud, high, sharp, scratchy 'eeik…eeik…' every 3 to 5 sec. **Similar:** Brown Falcon, juveniles and darkest adults. **Hab.:** usually across the semi-arid and arid interior; uses tree-lined watercourses and isolated stands of trees; hunts out over the low vegetation of surrounding plains, grasslands, saltbush and bluebush. Also often hunts over wetlands, temporary waters or bore drains in the arid regions, taking advantage of the abundant birdlife attracted to the water. **Status:** uncommon; migratory. Main stronghold and breeding region appears to be interior of SE Qld and NW Vic. After breeding, migrates and disperses across most of E Aust. and NW across NT to Kimberleys; occasional vagrants to S interior of WA. **Br.:** 365

Grey Falcon *Falco hypoleucos* 30–45 cm; span 85–95 cm

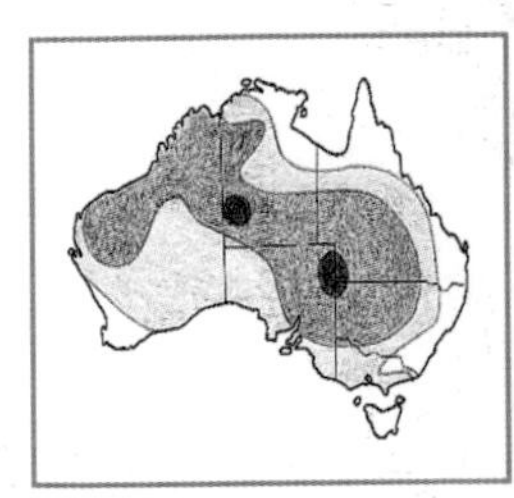

A sleek, grey plumaged falcon of the interior plains; flight swift and direct; patrols low over groundcover below treetop level, propelled by easy, shallow wingbeats and brief glides, when dark wingtips are most obvious. Soars with wings held close to level, dark tipped primaries slightly spread to give a blunt wingtip. When gliding, outer wing is backswept. Small birds flitting between trees or flushed from the ground are likely to be caught by surprise; the falcon takes them in a brief burst of speed generated by deep, strong wingbeats. As well as catching birds in flight, usually close to the ground, it will catch small mammals, reptiles and grasshoppers on the ground. **Imm.:** mottled darker above, heavier streaking on breast. **Voice:** usually silent, but when breeding gives a loud, slow 'kek-kek-kek' or 'kak-ak-ak-ak'. Similar to the call of the Peregrine Falcon but slower, deeper, harsher. **Similar:** Black-shouldered and Letter-winged Kites, but these have black shoulder patches, very different silhouette and flight; Grey Goshawk, but different habitat, silhouette, flight. **Hab.:** usually in lightly timbered country, especially stony plains and lightly timbered acacia scrublands. **Status:** rare; resident or nomadic to most of semi-arid interior. **Br.:** 365

Plumage varies from pale rufous to near black. All colour morphs have double dark vertical streaks enclosing paler cheeks.
Dark
Pale
Dark vertical streak
Adult
Dark
Juv.
Dark
Brown morph
Light morph
Dark thigh patches
Juveniles are darker versions of each colour phase; all have the distinctive facial markings.
Fast glide
Brown morph
When soaring, wings are held straight out, trailing edge is a long gentle curve; primaries spread enough to give a rounded wingtip.
Fast glide
Rufous morph
Heavily barred
Adult
Brown Falcons attract attention in noisy display flights, holding wings steeply upwards, swaying and tumbling towards nest tree, all the while cackling loudly.
Dark 'trousers', on all morphs, but harder to see on darkest forms
Rounded
Juv.
Wings slightly short of tail-tip
Dark morph
Soaring and gliding: wings are held in a moderately uplifted dihedral.
Soar and glide
Similar: dark morph of Brown Falcon
Largest Australian falcon: wide shoulders are obvious from front, perched—the bulky shoulders make the head look small.
With age, streaky white throat
Blue-grey to white
Glide
Fast glide
Strongly curved trailing edge, straight only when soaring
Uniformly sooty brown to sooty black
Linings darkest
Slightly lighter
Tips of folded wings much shorter than tail
Juv.
Fast flight, active or gliding on strongly backswept, pointed wings
Tends to be slightly darker than adult.
Folded, except when soaring, then fanned with notched corners
Legs partly feathered; large, powerful grey feet
Grey
Yellow
Dark wingtips are conspicuous and diagnostic.
Grey with fine brown streaks
Glide
Rounded or square tipped
Mottled, streaked darker
Curved trailing edge
When soaring, wings are held out straight and well forward so that trailing edge is straight; primaries spread to make a rounded wingtip; tail is fanned and gently rounded.
Very pale grey with fine dark streaks
Glide
Juv.
Adult
Deep yellow; large, powerful
Wings reach tail-tip.
Soaring and gliding
Distinctive dark tips
Barred

Dusky Moorhen *Gallinula tenebrosa* 35–40 cm (L)

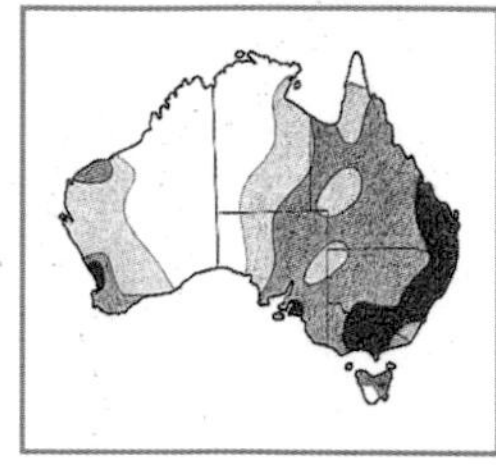

Widespread, common, often around lakes of suburban parks. Runs and swims well; always wary, quick to dash to cover. Swims jerkily, paddling with jerks of head and flicks of tail. On land forages with slow deliberate step; in alarm runs with flapping wings. In flight holds neck extended. Legs and feet dangle initially, trail behind tail-tip while travelling in direct, strong flight. **Voice:** as a territorial call, an abrupt, raucous 'krurk!' or 'krruk-uk-uk'. Also a repeated, resonant 'krek' or 'krok' and loud, harsh screeches. **Similar:** Purple Swamphen, which is much larger, blue breast; Eurasian Coot, but has bill and shield white. **Hab.:** diverse wetlands; prefers open waters with well vegetated margins; usually fresh water, some brackish including swamps, lakes, estuaries. **Status:** common in SE and SW Aust.; uncommon to vagrant in N. **Br.:** 366

Purple Swamphen *Porphyrio porphyrio* 45–50 cm (L,T)

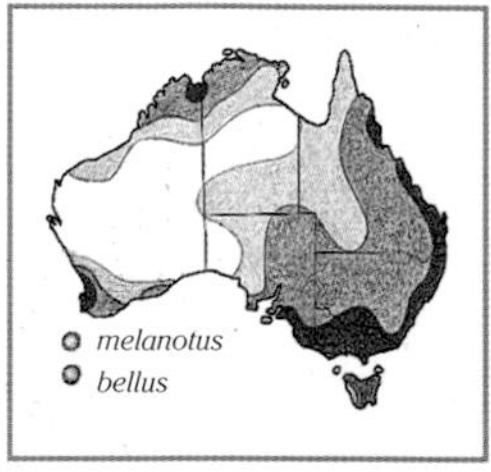

Large, colourful, common waterhen. Massive scarlet bill and shield. Deep tones of plumage need full sunlight to bring out intensity of colour. Breast can appear dark slaty blue-grey in dull light, but direct sunlight transforms this to a bright, intense blue with azure sheen, or, on the SW race, reveals an emerald hue down the central breast. Male and female are alike; full colour is kept through the year. Aggressive and bullying towards other waterbirds; kills ducklings. Has strength to pull up reeds as food. Clumsy, legs-dangling, crashlanding flight. **Variation:** nominate race outside Aust. Race *melanotus* in N and E Aust.; race *bellus* in the SW of WA. **Voice:** great variety of sounds. Harsh, abrupt 'kak, kak' rising to sharp, grating 'kiark, ki-aark'. Also querulous, grating 'qua-ark' and loud, harsh squawks of warning when with small chicks. At night often gives wild shrieks and boomings, perhaps basis of bunyip stories. Deep thudding sounds from beating wings against body. **Hab.:** diverse wetlands, typically swamps, well vegetated lake and river margins, adjacent grasslands, agricultural land, lawns; also estuarine wetlands. **Status:** widespread and common in E, N and SW Aust. **Br.:** 366

Black-tailed Native-hen *Gallinula ventralis* 32–38 cm

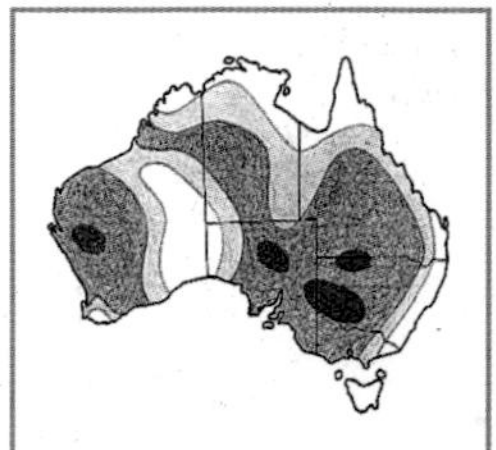

Highly adapted to an arid continent; responds to rain in semi-desert by breeding rapidly to large populations. Where there were few, large new flocks suddenly forage around any pool or green patch of ground. Returning drought causes them to vanish as quickly as they arrived. Runs swiftly; swims; flies with rapid, shallow wingbeats, feet trailing. Sexes alike; juvenile as adult; no seasonal plumage change. **Voice:** usually silent, even in large flocks; but at times a sharp 'kak', perhaps in alarm, or a rapid, continuous 'kak-kak-ak-ak-ak-'. **Hab.:** permanent or temporary wetlands of semi-arid regions, fresh or saline; often around shallow claypan pools, saltbush surround of lakes. **Status:** common; dispersive, irruptive. **Br.:** 366

Tasmanian Native-hen *Gallinula mortierii* 42–50 cm

A large, heavily built native-hen, almost as large as the well-known Purple Swamphen. One of only three flightless Australian birds, the others being the Emu and Cassowary. Usually in small family parties: breeding groups of 2 or 3 adults together with young in their first year or two. Typical breeding group: two males and one female. **Voice:** drum-like grunts, probably contact calls, higher pitched in aggression or alarm. Males and female combine in see-sawing, harsh, rasping sounds, rising to crescendo in pitch and loudness. **Similar:** Black-tailed Native-hen, but unlikely in Tas. **Hab.:** swamps, marshes, lakes; feeds out across surrounding grasslands, farmlands. **Status:** endemic to Tas.; common; sedentary. **Br.:** 366

Eurasian Coot *Fulica atra* 35–38 cm

A common, well-known and easily recognised waterbird with dumpy, dark, slate-grey body, conspicuous white forehead shield and bill. Highly aquatic: its flattened, lobed toes (an effective alternative to webbed feet) give strong propulsion under water and on the surface. The Coot forages in shallow to deep water where it up-ends or dives for plant material. At times it emerges onto land; then looks rather clumsy and unbalanced. The legs are set far back to give better thrust in water, but the Coot actually walks quite well; individuals and groups may emerge to feed on grassland near water. **Voice:** most often heard is an abrupt 'krek,' and sharper 'krik'; also a grating, sharp 'kiek-kiek-kiek'. **Hab.:** diverse wetlands that include rivers, lakes, swamps; also, rarely, marine wetlands such as estuaries. **Status:** principally coastal regions, but has been widely recorded on temporary lakes and floodwaters in the semi-arid interior; common. **Br.:** 366

Bright red
Bill bright red, tipped yellow
Dark, slaty grey, fresh plumage wears to slightly paler, brownish grey.
White
Olive, some smudged black, dull red; smaller shield
Nonbreeding in browner, paler, worn plumage
White
Older males appear to retain all or some of bill and shield's breeding colours throughout the year.
Breeding
Bare orange skin shows through sparse, wiry black plumage.
Red, tipped yellow
Downy young
In flight plumage is all dark except under tail edges. Legs trail behind tail.
Juvenile brownish grey, very small shield, retains red and yellow on bill from downy young stage.
Massive, scarlet
Black with blue sheen
Race bellus, turquoise centre of deep blue underparts; obvious only in direct sunlight
Powerful legs, long toes
White
Downy young: black with fine, white filament tips
Rufous skin visible
Flight is awkward: labours into air, legs dangling; strong in fast flight, neck extended, legs trailing. Clumsy crashlandings into vegetation or water.
Many displays including showing of white undertail in a 'wing-up, tail-up' posture, swimming or on land. Note slight blue edge to outer primaries is just visible.
Yellow eye has direct, penetrating stare.
Green shield above green and red bill
The large, black tipped tail is usually held steeply upwards, varying from vertical to horizontal, and tends to be roughly square cut.
Shown in alert, tall, upright stance. Often with back horizontal, head level with tail, hen-like shape
Intense carmine-pink
If disturbed, quick to dash for cover, half running, half flying, across pools or open ground
Long, pointed, white tipped feathers on flanks
Olive-brown
Large, triangular, greenish yellow bill; bright red eye.
Tail black; may be steeply upright, but often lower
Adult
White patch on flanks
Blue-black
Blue-grey
Juvenile is like adult but paler brown above, and brownish grey rather than blue-grey underparts; bill grey; eye brown.
Downy young
The Tasmanian Native-hen has the distinctions of being flightless and is one of the species found only in Tas.
Eye brilliant red in black surrounds
Bill and shield white
Rounded, plump body shape tapers to pointed wisp of tail.
Adult
Toes with flattened fringing lobes rather than webbed
Sparse wispy feathers reveal red skin
Downy young
Juvenile is pale grey beneath; throat whitish; dull brown above; iris brown.
Swims high on water; head jerks to match thrusting of feet.
Adult
Wings entirely dark
Adult
Chases, patters and skitters low across surface, leaving long trail of splashes; often in aggressive pursuits.

Corncrake *Crex crex* 26–30 cm
Extremely shy; keeps out of sight in low, dense vegetation; most likely to be revealed by call or accidental flushing; then shows bright chestnut wings, bold pattern of back. **Voice:** in breeding season, a persistent, rasping 'crrek, crrek, crrek'. **Hab.:** grassland—usually meadows, hayfields, irrigated lush pastures. **Status:** two very old records, none since 1944.

Bush-hen *Amaurornis olivaceus* 24–26 cm
Noisy, secretive. Tends to emerge from cover early and late or when heavily overcast. **Juv.:** paler, whitish throat, bill brown. **Voice:** varied, includes loud, shrieked, cat-like 'ki-arr-rk, ki-arr-rk', often a long succession, fading away, or with two birds in duet; a clear, piping 'keek, keek, keek', all at same pitch, for long periods; sometimes drawn out, higher 'kee-eek, kee-eek'; soft ticking sounds. **Hab.:** dense margins of freshwater wetlands, rainforest margins and regrowth, dense tall crops, mangrove edges. **Status:** sedentary or part-nomadic. **Br.:** 366

White-browed Crake *Porzana cinerea* 18–20 cm
Unlike other secretive crakes of dense reedbeds, the White-browed Crake is a bird of the waterlily pads and other floating vegetation where it forages across the lilies, helped by exceptionally long toes. It flutters low across intervening water. Most active after sunrise, near sunset and beneath overcast skies. **Voice:** calls most heard are fast, chattering 'kiak-kiak-kiak', 'chika-chika', a frog-croaked 'kak', and a soft 'charr-ar, charr-ar' in danger. **Hab.:** wetlands with waterlilies or other floating plants, such as billabongs, swamps, floodwaters; also surrounding forest or woodland. **Status:** common in optimum habitat. **Br.:** 366

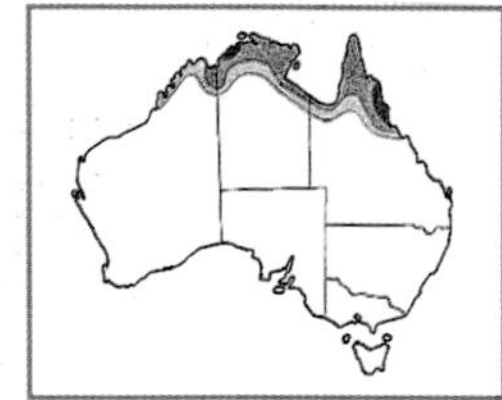

Spotless Crake *Porzana tabuensis* 17–20 cm
A plain, dark crake with bright red eyes, pink legs. Rarely out of dense cover; forages in shallows, wading, swimming, climbing over fallen and floating vegetation. Rarely seen in flight, but in any brief glimpse would appear dark with pink dangling or trailing legs. **Voice:** sharp, harsh 'chaik'; also fast, rattling, descending and fading 'chak, chak-chakchak-'; bubbling and trilling whistling. **Hab.:** reedbeds and other dense aquatic vegetation around lakes, swamps, saltmarshes, mangroves. **Status:** uncommon; nomadic, migratory. **Br.:** 367

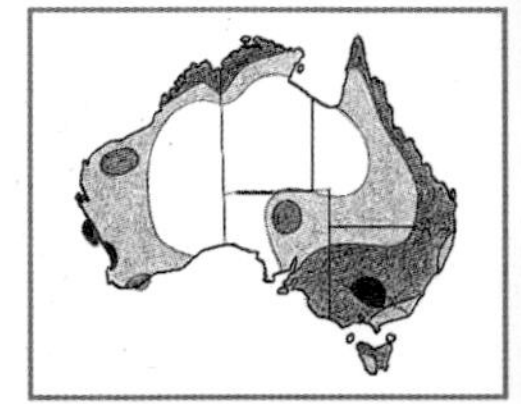

Baillon's Crake *Porzana pusilla* 15–16 cm
Smallest of Aust. rails; secretive, but like other crakes may forage out of cover early and late in the day; probably less shy than other species. Forages over floating vegetation, tail upright, constantly flicked. Weak, fluttering flight among reeds, legs dangling. **Voice:** loud, sharp, very rapid, ratchetting 'kar-r-r-r-r-ak'; also a sharp 'chak' or 'krek' in alarm. **Hab.:** permanent or temporary fresh or saline wetlands, swamps, vegetated lake margins on floating vegetation in deep and shallow water. **Status:** not often seen but probably common; migratory. **Br.:** 367

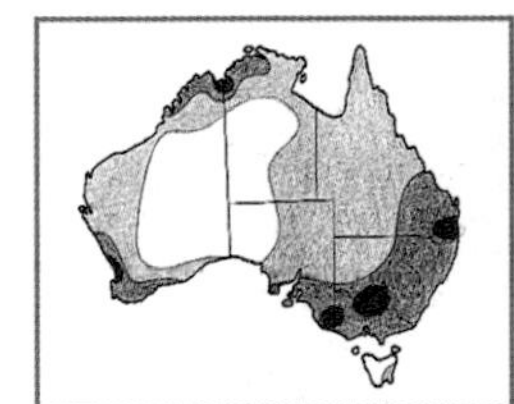

Australian Spotted Crake *Porzana fluminea* 19–22 cm
An inhabitant of dense reedbeds, but early and late in the day will venture out onto nearby shallow open water, mudflats or floating vegetation. Forages slowly; constantly flicks tail. Swims across intervening water; rarely flies. **Voice:** the usual call is an abrupt, sharp, metallic 'chaik-chaik, chaik-chaik', more musical than harsh; also a rapid descending sequence, 'chak-ak-ak-ak-akakakak'. **Hab.:** dense cover, fresh or salt wetlands, lakes, swamps, saltmarsh; at times far from water. **Status:** common SE and SW; nomadic. **Br.:** 367

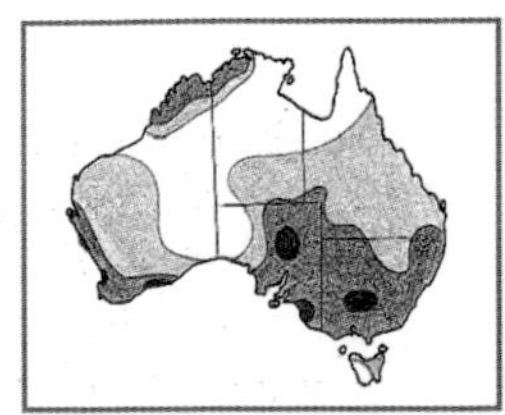

Lewin's Rail *Dryolimnas pectoralis* 21–25 cm
A plump rail with long magenta-pink bill and extensive barring. Secretive and difficult to sight even momentarily in the swamps where it forages; tends to come into the open less than other rails. **Voice:** a sharp, loud 'krek' given as a burst of 10 to 20 calls; becomes faster and louder, then fades away. Also a whistle, very loud, in a series of up to 50. **Similar:** Buff-banded Rail. **Hab.:** swamps, lakes, tidal creeks, saltmarsh, lush wet pasture, or paperbark and other swamp woodlands. **Status:** uncommon; patchy, nomadic. **Br.:** 366

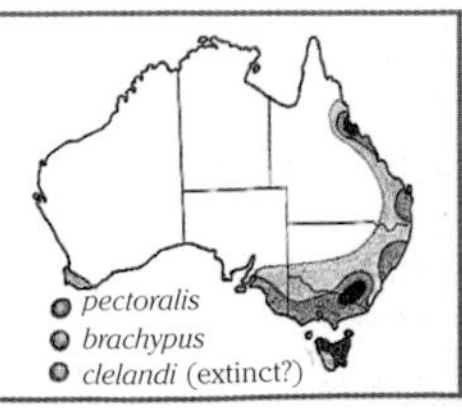

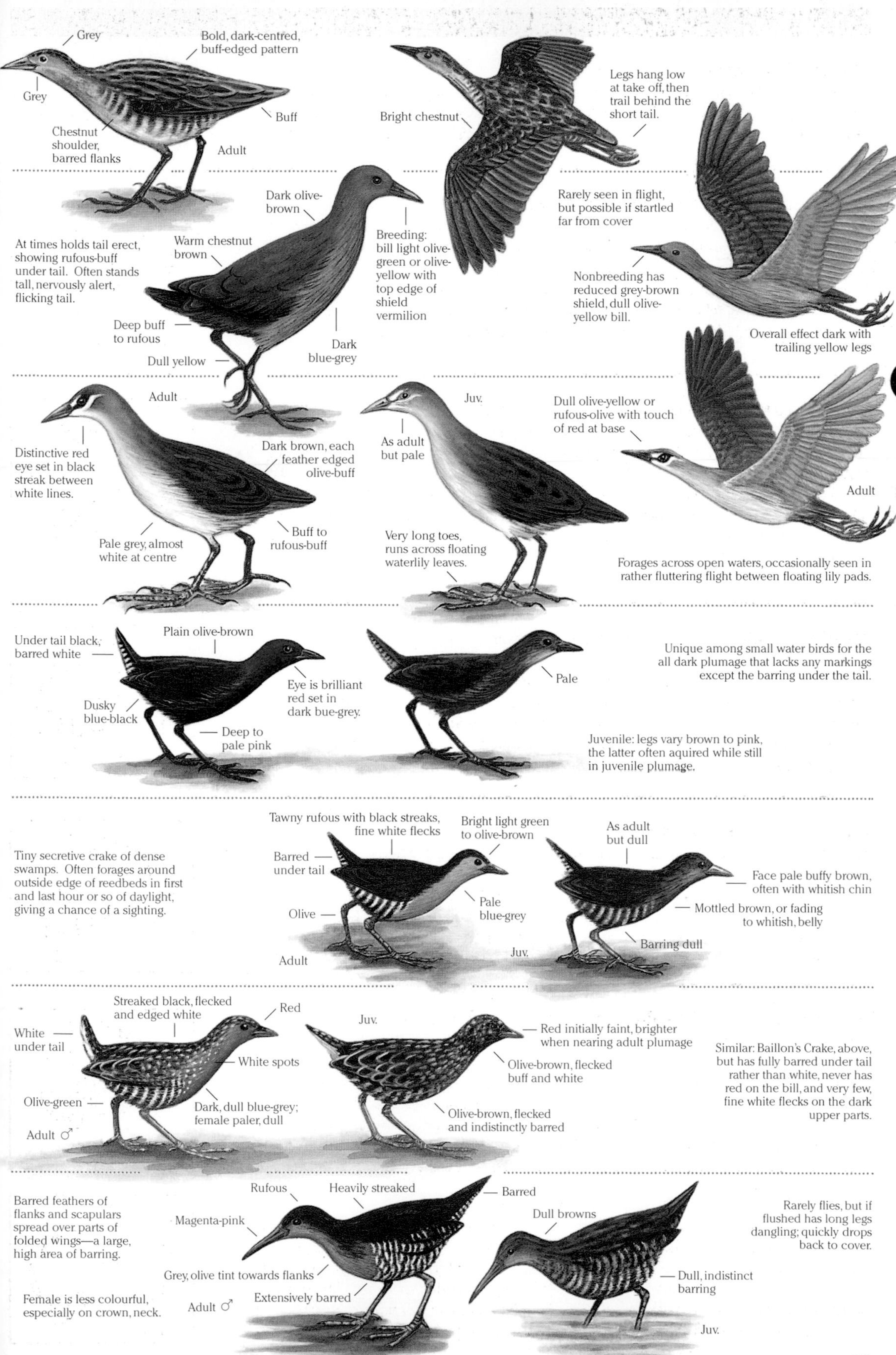
Grey
Bold, dark-centred, buff-edged pattern
Grey
Buff
Chestnut shoulder, barred flanks
Adult
Bright chestnut
Legs hang low at take off, then trail behind the short tail.
Rarely seen in flight, but possible if startled far from cover
Dark olive-brown
At times holds tail erect, showing rufous-buff under tail. Often stands tall, nervously alert, flicking tail.
Warm chestnut brown
Breeding: bill light olive-green or olive-yellow with top edge of shield vermilion
Nonbreeding has reduced grey-brown shield, dull olive-yellow bill.
Deep buff to rufous
Dark blue-grey
Dull yellow
Overall effect dark with trailing yellow legs
Adult
Juv.
Dull olive-yellow or rufous-olive with touch of red at base
Distinctive red eye set in black streak between white lines.
Dark brown, each feather edged olive-buff
As adult but pale
Adult
Pale grey, almost white at centre
Buff to rufous-buff
Very long toes, runs across floating waterlily leaves.
Forages across open waters, occasionally seen in rather fluttering flight between floating lily pads.
Under tail black, barred white
Plain olive-brown
Eye is brilliant red set in dark bue-grey.
Pale
Unique among small water birds for the all dark plumage that lacks any markings except the barring under the tail.
Dusky blue-black
Deep to pale pink
Juvenile: legs vary brown to pink, the latter often aquired while still in juvenile plumage.
Tawny rufous with black streaks, fine white flecks
Bright light green to olive-brown
As adult but dull
Tiny secretive crake of dense swamps. Often forages around outside edge of reedbeds in first and last hour or so of daylight, giving a chance of a sighting.
Barred under tail
Face pale buffy brown, often with whitish chin
Olive
Pale blue-grey
Mottled brown, or fading to whitish, belly
Barring dull
Juv.
Adult
Streaked black, flecked and edged white
Red
Juv.
White under tail
Red initially faint, brighter when nearing adult plumage
White spots
Olive-brown, flecked buff and white
Similar: Baillon's Crake, above, but has fully barred under tail rather than white, never has red on the bill, and very few, fine white flecks on the dark upper parts.
Olive-green
Dark, dull blue-grey; female paler, dull
Olive-brown, flecked and indistinctly barred
Adult ♂
Rufous
Heavily streaked
Barred
Barred feathers of flanks and scapulars spread over parts of folded wings—a large, high area of barring.
Magenta-pink
Dull browns
Rarely flies, but if flushed has long legs dangling; quickly drops back to cover.
Grey, olive tint towards flanks
Dull, indistinct barring
Female is less colourful, especially on crown, neck.
Adult ♂
Extensively barred
Juv.

Red-necked Crake *Rallina tricolor* 24–29 cm

Distinctive; often heard, but secretive, difficult to sight or observe. Forages along edges and shallows of rainforest streams and pools, and in nearby rainforest leaf litter; most active and calling around dawn and dusk. Often flicks tail; stands tall, alert; quick to dash for cover. **Voice:** a harsh, sharp 'kark' or 'karrark' as a single note; often a sequence, a rasping, aggressive 'karr-ak, karrrk, kak, kak, kak-kakakakak'. Its monotonous 'klok, klok, klok' may continue for hours, often at night. **Similar:** Red-legged Crake. **Hab.:** vicinity of streams and swamps in or near dense thickets of rainforest and vine scrub. **Status:** sedentary; possible migration between Cape York and NG. Has lost much of its lowland rainforest habitat. **Br.:** 367

Red-legged Crake *Rallina fasciata* 20–25 cm

Rare vagrant; keeps to cover; partly nocturnal. **Voice:** noisy, often calls at night; screams, grinding sounds; loud, regular 'kek'. **Similar:** Red-necked Crake. **Hab.:** in normal range, wet parts of rainforest, regrowth, swamps, paddyfields. **Status:** in Aust. known only from a single weak male found on a lugger in Broome, 1958. Widespread India to Lesser Sundas; Aust. record may be storm-blown or off-course migrant. A possible vagrant to NW coast of Aust. **Br.:** India to Indonesia.

Buff-banded Rail *Gallirallus philippensis* 28–32 cm (L, K)

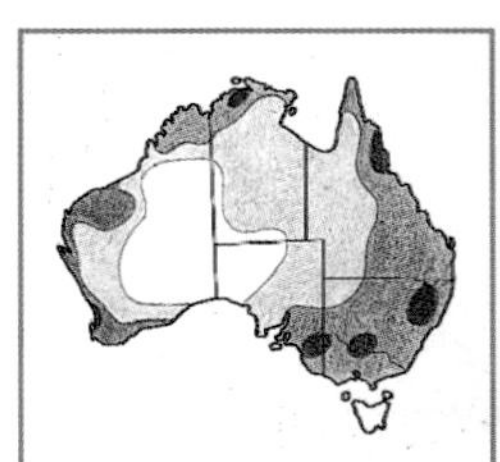

Colourful; wary; easily recognised. Often emerges from dense cover at dawn to feed before full sun reaches exposed mudflats or open marshy ground. Feeds again towards dusk. **Voice:** common call a squeaky 'swiit' and loud, carrying, creaky, harsh 'kiek' or 'priep'; often answered by others. Also reported are triple noted 'tchuk-e teika' and braying sounds. Makes soft grunting sounds to chicks and growling hisses when chicks are threatened. **Similar:** Lewin's Rail, but it lacks buff breast-band and eyebrow, and has much longer, pink bill. **Hab.:** dense, damp vegetation around swamps, lakes, creeks, coastal lagoons, tidal mudflats, rainforest margins, moist paddocks and sewage farms. **Status:** probably more common on coast than sightings would suggest; sedentary, nomadic, dispersive. **Br.:** 367

Chestnut Rail *Eulabeornis castaneoventris* 44–52 cm

A very large rail, predominantly chestnut, infrequently sighted; loud calls may reveal its presence. Struts across mangrove mudflats with tail flicked nervously; runs swiftly through tangled vegetation rather than flying. **Voice:** diverse descriptions, but seems to include pig-like grunts or drumming intermixed or alternating with louder, raucous screeches, 'whuh-WHAIKA, whu-WHAIKA', repeated very many times to a steady rhythm. Apparently a single bird makes the noise, but sounds like two birds in a duet. **Similar:** no others within Aust. of comparable size and colour. **Hab.:** bays and estuaries lined with mangroves; usually large blocks of mangrove swamp across broad estuarine tidal mudflats rather than narrow fringing strips. Prefers dense areas of mangrove cover—a closed canopy—especially where nesting. Occasionally ventures out of mangroves to search adjacent mudflats or grassy woodlands; feeds principally on crabs and other small creatures probed from the mud. **Status:** population scattered rather sparsely along a slender line of habitat around far NW coast; difficulty of access both protects habitat and makes population assessment difficult. **Br.:** 367

Australian Bustard *Ardeotis australis* 0.8–1.3 m; span to 2 m

Tall, stately bird of open inland plains. Stands neck upstretched, bill uplifted and freezes if an intruder approaches; then slowly walks away, speeding up if followed, or takes off with deep, slow, powerful wingbeats. **Br.:** remarkable courtship display: foreneck and gular sac are inflated; long feathers of upper breast spread in fan shape; tail lifts and fans over the back; male begins strutting and booming. This continues for much of the day. **Voice:** loud calls in breeding season, deep booming, something like the roar of a distant lion, rising then falling. Closer, the sounds include an abrupt, hoarse exhalation 'huhhh!', often leading into a hoarse throaty growling—in all, 'huhh!, huhh! -aa-a-r-r-rgh, aa a-r-r-rrrgh'. A sharper barking call is directed at rivals and given in alarm; the female gives a similar but higher call. **Hab.:** grasslands, especially tussock grasses like speargrass, Mitchell grass; spinifex; arid scrub with saltbush, bluebush; open dry woodland of mulga, mallee, heath. **Status:** nomadic; remains common away from more heavily settled regions. **Br.:** 367

Uniform grey-brown
Bright chestnut
Green
Tail is often raised, repeatedly flicked.
Indistinct pale barring
Olive
Adult
Olive-brown
Juv.
Paler face
Olive-green
If startled, rather than flying, will usually run, head down, wings spread or flapping, to cover of undergrowth.
Olive-brown
Entire underwing barred
Feet trail behind tail.
South-east Asian species, one record for Australia, but always a possibility in suitably swampy parts of the Kimberley and NT coast.
Chestnut
Deep olive-brown
Tail is often raised, frequently flicked.
Entire undersurface, wings and body, except neck, heavily barred
Bold, obvious black and white barring
Elegant, alert, tall posture displays deep buff breast crescent.
Distinctive face: long white brow, chestnut eye streak, grey lower face and throat, streaked chestnut crown
Deep buff
Barred
Heavily barred
Barred under tail
Adult
Paler ochre-brown crown, eye streak
Juv.
Faint buff tint
Bars pale, indistinct
Juv.
Chicks hatch clad in dense black down; leave nest within a day. Down is replaced by juvenile feathering at about two weeks of age.
Grey
Yellowish
Large rail, back colour variable, but head always grey, bill olive-yellow, underparts always chestnut.
Chestnut
Colour of upper parts varies: the chestnut-backed form in Qld and NT; the olive-backed form in WA; some NT birds intermediate in colour.
Olive backed form
Pale under chin
Chestnut
Pointed tail, often raised, often flicked.
Thick, powerful legs
Underwing all brown
Head partly tucked in
Feet trail behind tail-tip.
Black
Chestnut, finely barred darker
Broad wings, widely fingered
Runs into wind for heavy take off; slow, powerful wingbeat.
Feet trail behind short tail.
♀
Female lacks black breast-band.
Juv. has adult face, but with heavier barring around neck; dull colours.
If threathened, juv. may freeze beside spinifex or other grass clumps, bill skywards, easily overlooked, while adults lead predator away.
Juv.
Black and white patterned wing coverts
♂ Breast-band
In breeding display male has inflated, white feathered breast sac hanging almost to the ground.

Red-backed Button-quail *Turnix maculosa* 12–15 cm (T)

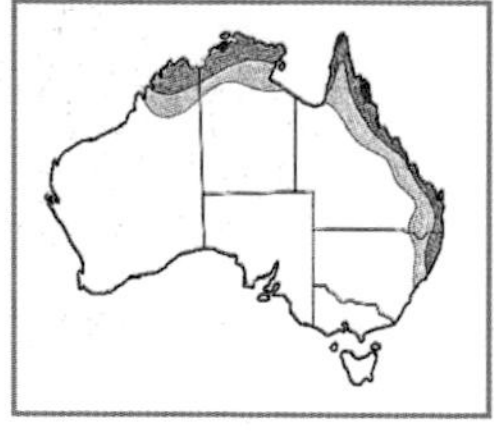

Lives in dense vegetation in small coveys; difficult to flush—they sit tight, then explode into fast flight; drop again at a distance. **Voice:** a soft, tremulous 'oo, oo; oo,...' or 'whoo, whoo, whoo,...', each note ending with a slight lift. Given as a long sequence—begins very softly, resonant and musical; repeats many times; rises, strengthens, until finally quite loud, clear, rather vibrant. **Similar:** Red-chested Button-quail, but has lighter back and rufous collar. **Hab.:** tussock grasslands of blacksoil plains and clay flats, grassy tropical woodlands, rainforest margins, spinifex, crops, sedges—usually near water, especially in breeding season. **Status:** common in some parts of tropical N; scarce or vagrant in SE. **Br.:** 368

Chestnut-backed Button-quail *Turnix castanota* 15–18 cm

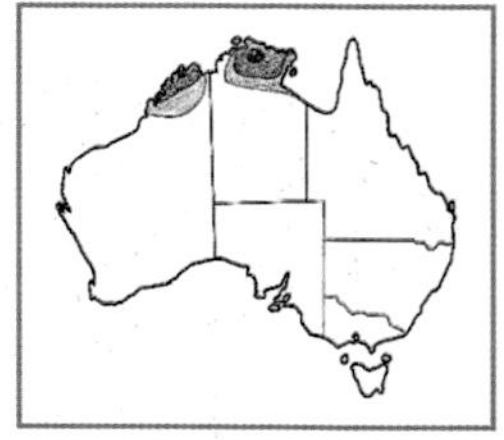

If alarmed, runs with head high. Difficult to flush; then whole group bursts up, but drops back to cover very quickly. Has darker rufous undertone to the upper parts than Buff-breasted, and has more black markings, which sometimes extend over mantle. Can have fine black barring on the rump. **Voice:** low, moaning 'oom', perhaps a contact call when a group becomes separated. **Similar:** within range, closest are Little Button-quail (which is smaller; also white streaking on upper parts extends to rump) and Brown Quail, but its wings and back are almost uniformly dark in flight. **Hab.:** usually dry, elevated sandstone terrain—ridges, escarpments, plateaus with eucalypt woodlands and open forests, groundcover of spinifex, sparse grass. **Status:** sedentary; population uncertain. **Br.:** 368

Buff-breasted Button-quail *Turnix olivii* 18–22 cm

Can remain hidden even in its typically sparsely covered habitat; difficult to sight. Prefers to walk or run from danger. Finally flushes when hard-pressed; it rises steadily on whirring wings rather than exploding upwards; usually travels 50–100 m before dropping to cover. **Voice:** female gives a deep, slow booming. Begins softly, a long sequence with increasing intensity and frequency. Male gives a softly whistled 'chiew-chiew'. **Similar:** within range, only the Painted Button-quail, but its back is heavily marked. **Hab.:** open tropical woodlands with sparse tussocky grass. **Status:** rare; locally and seasonally nomadic. **Br.:** 368

Little Button-quail *Turnix velox* 13–16 cm

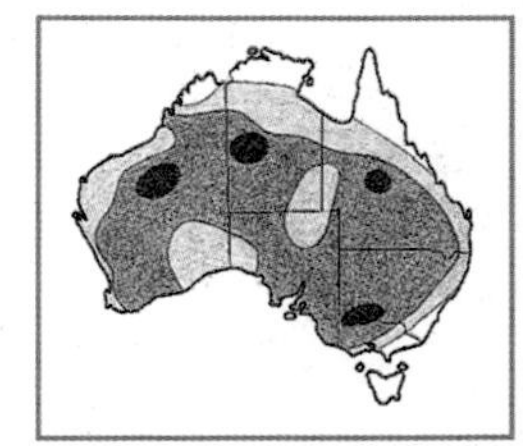

Small, rufous and cinnamon, dark winged quail with short, deep grey bill. Occurs in small coveys or pairs. When disturbed, squats or scuttles through grass, or rises on whirring wings—keeps low, often turns to show white flanks before it drops to hide in distant cover. **Voice:** soft, high resonant musical 'whoo, whoo, whoo...'. Squeaky chatter when flushed. **Similar:** in flight differs from other small quail in strong contrast between cinnamon-rufous panel of secondary coverts and dark grey of outer wing and secondaries. **Hab.:** diverse grassland and open woodland country; includes mallee, mulga, tussock grasslands, spinifex, crops. **Status:** common; population density varies locally; irruptive, nomadic. **Br.:** 368

Red-chested Button-quail *Turnix pyrrhothorax* 12–16 cm

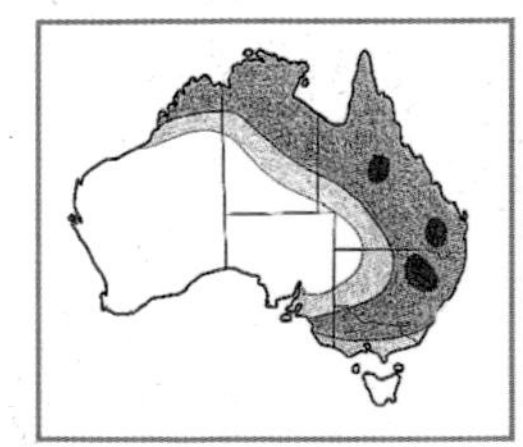

Common in E and N Aust. Populations fluctuate greatly with varying climatic conditions: abundant in good seasons, then forced to disperse widely, often to greener coastal regions, when conditions deteriorate. Some birds stay throughout year, others are seasonal or vagrant. Occurs in pairs or small coveys. When approached, tends to squat rather than fly; eventually explodes from cover; flies fast, low and far before dropping to ground and running. **Male:** typical of button-quails—smaller, paler and duller than the female in keeping with its role as incubator and parent in the first few days after chicks hatch. **Fem.:** larger, brightly coloured, even more so in breeding season. Female apparently chooses nest site and initiates construction. Male probably does most of the work. After laying eggs, hen may lay clutches in nests prepared with other males. Breeding habits may depend on the season—polyandrous when food is in plenty, monogamous in leaner years. **Voice:** soft, quite high, booming 'oom, oom' at 1 sec intervals; notes slightly slurred and rising through a sequence of 20–30 calls. **Hab.:** grasslands on blacksoil plains and other flat, heavy soil country, grassy woodlands, rainforest margins, spinifex, crops; in breeding season, grasslands and sedgelands near water. **Status:** common; sedentary or nomadic. **Br.:** 368

Small button-quail with slender yellow bill, rufous rear collar
Narrow dull collar
Pale buff
♂
Male small, more buff or ochre than rufous
Grey and bright rufous, barred and heavily blotched with black
Rufous collar, shoulders
Yellow eye
Fine, yellow
Rufous
Buff
♀
Buff panels across wing coverts in sharp contrast to dark flight feathers
Dark outer wing
♀
Rufous or cinnamon-rufous rump usually plain, sometimes fine barring
Face finely speckled grey
Pale olive-grey
♂
Male smaller, lighter colours
Bill short, deep; grey-brown
Deep rufous
Speckled dark grey
Yellow eye
Grey, pale spotted
♀
Female larger, in stronger colours than male, especially when breeding
Grey between dark lines
Black
Plain rufous rump
Dark
♀
Largest button-quail; plain buff breast, upper parts pale cinnamon-buff only lightly patterned and mottled with rufous, black and white. In flight, dark flight feathers contrast with plain, pale cinnamon rump.
Male is smaller, overall lighter, markings less distinct; pale forehead.
Rump plain or lightly marked pale cinnamon-buff
Pale cinnamon, plain or only lightly marked
Chestnut streak
Yellow eye
Dark grey forehead
Large, heavy, drooped olive-grey bill
Off-white
♀
Buff breast, unmarked by any obvious spotting or barring
Bursting into flight, it is similar to Chestnut-backed, above, but relatively pale and plain. Gives a fleeting glance of pale cinnamon-buff underlying colour, lightly marked rufous and black, and white on inner wings, all in contrast to dark grey or grey-brown outer wings.
♂
White flanks
Dark-scalloped breast margins
Male and juv. are darker than female on crown, and sides of breast have small dark scallops
Plain cinnamon-rufous
Crown usually lightly mottled
Eye pale
Thick bill
♀
Cinnamon, brighter in breeding season
Chestnut, streaked white
Bold contrast of rufous coverts against darker flight feathers
♀
♂
Buff
Flanks rufous-buff
Face speckled grey and white
Deep orange-rufous
Rufous
♀
Spotted
Back, rump and tail coverts grey-buff streaked white and mottled with black
Grey-buff coverts make pale inner-wing panel.
♀
Dark flight feathers
Occurring within the range of the Red-chested Button-quail is the Plains-wanderer. It has similar colour on breast, and is superficially quail-like, but is not related to quails. See page 104.
Red-chested Button-quail in flight, shows buffy brown wing coverts as pale panels on otherwise dark wings. Is separated from Little Button-quail by the grey-buff rather than rufous back and wing panels, and by rufous-buff in place of pale buffy white flanks.

Painted Button-quail *Turnix varia* 17–23 cm

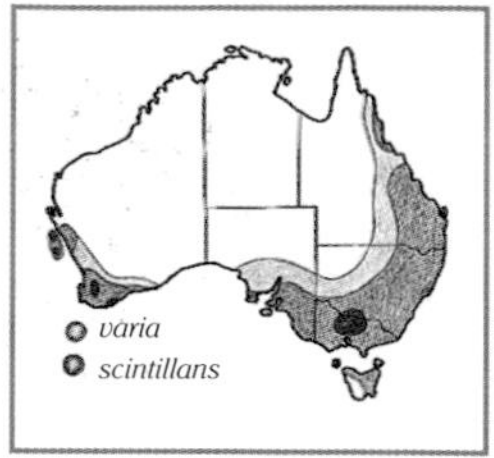

Often in small coveys, pairs or alone. Usually walks away when approached; then runs with head high, or squats and freezes. Will finally explode into flight. Like many button-quails, is active day or night; forages among litter for seeds and insects, often leaving small round bare patches. **Voice:** in long series of even pitched booms from female (up to 30 secs) with notes more rapid towards the end. **Hab.:** open forests and woodlands; includes banksia woodland, mulga and brigalow, mallee. Prefers stony ridges, abundant leaf litter but sparse grass. **Status:** widespread and common, but habitat reduced. **Br.:** 368

Black-breasted Button-quail *Turnix melanogaster* 17–19 cm

Conspicuously distinctive: female's head is black hooded; both sexes' breasts are white, spotted black. Usually in pairs or small parties of about five. Squats and freezes, or runs when approached. Flies only if forced; takes off in a sudden whirring. Flight is short and fast; glides to landing. Feeds on seeds and insects, scratching about in litter, leaving small, bare circular depressions. **Voice:** from female a low drumming or booming 'oo-ooom, oo-ooom', quickly repeated many times, rising and falling through the series, difficult to establish direction of calls. Also a quick 'ook' when disturbed. **Similar:** female unique in appearance and drumming sound. Male is less distinctive—male Painted Button-quail also has a spotted breast, but grey rather than black, and has more chestnut on the shoulder. **Hab.:** usually low canopy, closed rainforest or monsoon forests, vine thickets, and drier shrubby scrubs such as hoop pine, brigalow, belah and bottletree thickets where there is a deep leaf-litter layer. Also in eucalypt forests such as spotted gum, especially where there is a dense understorey such as lantana with grass groundcover. **Status:** sedentary; formerly common, but populations reduced or eliminated by habitat destruction or alteration—now rare, localised and vulnerable. **Br.:** 368

Plains-wanderer *Pedionomus torquatus* 17–18 cm

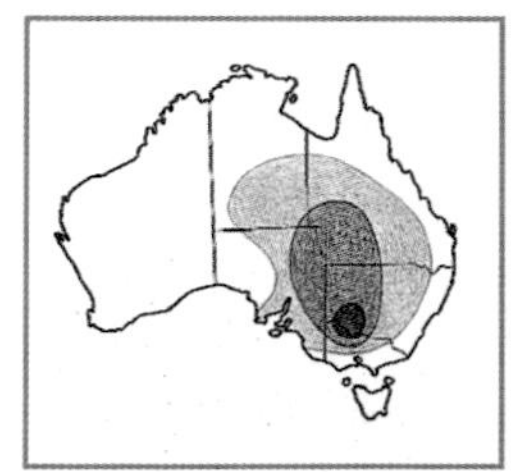

Distinctive quail-like ground bird, the sole member of its family, which is closer to the waders than to quail. Use of similar habitat seems to be reflected not only in appearance, but also in polyandrous breeding, evidently more effective within these habitat and climatic parameters. **Voice:** repetitive, low, hollow 'coo', given only in spring, perhaps only by female. Described as dove-like, or, more prosaically, like the moo of a distant cow. Calls chicks with a soft 'tchuk!'. **Similar:** quail, but they lack the tall, narrow necked, angular headed stance and silhouette of the Plains-wanderer. The spotted neck and wing pattern of the Plains-wanderer in flight is unlike any quail. **Hab.:** natural open grasslands, treeless with patches of open ground; may be lightly grazed. Avoids country where grass is too tall or dense, or too sparse, low or heavily grazed. Feeds on seeds on bare ground at base of grass clumps. A heavy sward of grass would conceal seeds and hinder escape from predators; on the other hand, extremely low, sparse or heavily grazed grass offers little concealment. Infrequent reports from saltbush and similar low sparse shrublands, stubble or other low crops. **Status:** most common in the Riverina of NSW and in NW Vic. Elsewhere sparse, patchy occurrence, generally rare, vulnerable. Sedentary until forced to move or disperse by change such as heavy grazing or dense grass growth. **Br.:** 368

Ruddy Turnstone *Arenaria interpres* 22–24 cm (C, K, L, N, T, M)

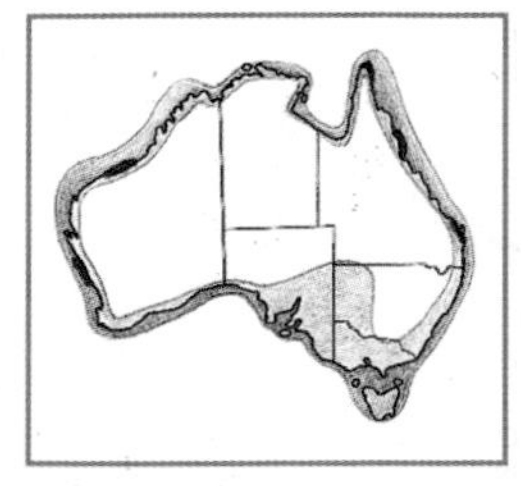

Named for its foraging technique of turning over beach stones, seaweed and other objects to feed on tiny crustaceans, sand-hoppers and other small creatures hiding beneath. Gregarious, usually in small groups that rush about prodding, probing and turning. Short legs give the impression that leads to this wader being described as tubby, dumpy or stocky. The body is actually quite slender, tapering towards the tail. The species' breeding plumage is easily recognised—the bold pattern of the head and across the wings and back is unique and distinctive when in flight. Nonbreeding plumage is only slightly less distinctive. The juvenile's indistinct and rather variable pattern may cause difficulty. Plumage transitions from juvenile to adult are rather indistinct and varied, particularly from 'immature nonbreeding' to 'first breeding'. The plumage contains mixtures of heavily worn juvenile plumage and new chestnut areas of adult plumage. Head, neck and upper breast of juvenile can be almost all mottled brown, or can have pale areas that, with passing moults, follow more or less closely the pattern of the adult. **Voice:** a rapid, irregular, weak chattering and fast twittering in short bursts; a high 'trit-tit-tit-tit-tit...' or 'trititititi...', fast enough to be a rippling trill. Also gives a clear, whistled 'kiew'. **Hab.:** beaches and coasts with exposed rock, stony or shell beaches, mudflats, exposed reefs and wave platforms. **Status:** common; entire Aust. coastline.

Black, grey streaked, spotted white
Entire upper parts rufous, heavily barred black, streaked grey and white
Eye red
White spots on grey breast
Shoulders chestnut
Buff
♀
Off-white
Male is smaller, paler dull colours; juv. is small, lacks chestnut shoulder.
Grey streak
Rufous body and coverts, heavily black and grey patterned, less contrast with dark outer wings than in other species
Dark leading edge
In flight, rump and tail appear rufous-grey.
When flushed, bursts up with whirr of wings; darts away, keeping a low, weaving course; drops to cover again at distance; then runs.
♂
Black with white spots along brow; slight chestnut tint on crown and nape
Heavily spotted white
Black-hooded head, yellow eye
Faintly barred
♂
Breast black, heavily spotted white each side
Chestnut, heavily patterned black and grey, white streaks
Female's breast heavily spotted with white crescent markings leaving an almost solid black streak down the centre. Breast pattern of male is similar, but comparatively dull.
Mid grey
♀
♀
Extensive dark mottling over rufous body and coverts reduces contrast with dark outer wings.
Slender
Outer wing pattern unlike any quail
About size of Stubble Quail, superficially quail-like, but more closely related to waders.
Often adopts tall upright stance showing thin neck
Black spotted
Diagnostic pattern and colour of neck and, if visible, rufous breast-band
Rufous breast-band
White base to primaries continues as a pale diagonal across secondaries.
Pale trailing edge
Legs longer than tail; hang low at take off and on landing approach.
♂
Legs are longer than those of quail.
♀
♀
Erratic pattern of black and white bands
Chestnut
Rufous and white bars converge towards rump
Black
♂ Breeding
Orange
Nonbr.: lacks rufous in wings; head brownish; basic patterns remain.
Complex, distinctive pattern of black banding passes through eye and around neck to a broad patch across breast.
While wintering in Aust., many begin to show some rufous-orange.
Unique pattern of upper parts; in nonbreeding plumage, brown replaces rufous.
Slightly uptilted, narrowly wedge shaped
Brownish nonbreeding head tones in transition to clear black and white breeding pattern
Nonbreeding adult in pre-breeding moult with partial breeding colours, March–May, prior to return journey to N Asia
Juv.
Juv.: variable version of nonbreeding; often more extensive brown over head and shoulders

Latham's Snipe *Gallinago hardwickii* 27–30 cm (L, M,N)

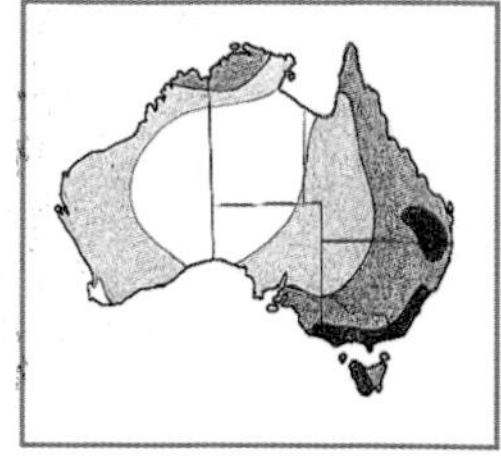

As a group, the three species of *Gallinago* snipe are easily recognisable, having extremely long straight bills and a distinctive high head with a steep forehead, flattened crown and eyes set further up and back than usual to give a wide field of vision (probably almost as much to the rear and above as to the front). While these snipe differ distinctly from other birds, differences among the three species are so slight that positive identification is usually based on details like tail structure and requires an in-hand specimen. Latham's Snipe is the largest; wings and tail appear even longer proportionately, giving a tapering, pointed rear end. Snipe are active around dusk and dawn. They remain concealed through the day and are seen only if flushed. They burst up to fly a fast zigzag course and then drop to cover again. **Fem.:** slightly larger, longer billed. **Voice:** harsh rasp: 'kzek-kzek-kzek'; usually on take off and in flight. **Similar:** Pin-tailed and Swinhoe's Snipes. **Hab.:** low vegetation around wetlands in shallows, sedges, reeds, heaths, salt marsh, irrigated crops. **Status:** regular summer migrant; NE to SE Aust.; rare visitor to SW of WA. Locally common in optimum habitats; breeds Japan.

Pin-tailed Snipe *Gallinago stenura* 25–27 cm

Smallest of the three species recorded for Aust.; it looks stumpy tailed. The short tail is no longer than the tips of the folded wings, which, in turn, are about equal to tips of tertials. Projection points—wingtips, tertials, tail-tip—do not usually fall neatly together as one point, but spread apart to make a truncated, squat rear end in comparison with the other two species, whose longer, projecting tails give a single, finely tapered point. In flight the short tail also looks stubby, and the wings are shorter, more rounded. Usually sighted when flushed from cover unexpectedly. Explodes into flight, usually fast, low, direct. May drop quickly to cover, or circle higher overhead before descending steeply to the ground. **Voice:** calls when put to flight, a rather high, startled sound, unlike usual flight calls, which tend to be more throaty or nasal, 'tchaa' or 'tchet'. **Hab.:** in Aust., the few records have been from coastal freshwater wetlands; includes swamps, river pools, sewage ponds, usually with grass. **Status:** uncommon; probably regular winter visitor. **Br.:** N Siberia.

Swinhoe's Snipe *Gallinago megala* 27–29 cm

Extremely difficult to separate from above species, especially the Pin-tailed. In large areas of Aust., only one species has so far been found, which makes it easier for observers; however, the overlap will probably increase with further observations; almost any sighting could potentially be any of the three species. When flushed, Swinhoe's dashes away in characteristic zigzag flight; sometimes keeps low, but may circle high overhead before diving into low, dense vegetation. All these snipe use their unusual tails in display flights over their N hemisphere breeding grounds. The narrow outer feathers are strong and stiff, and, when the tail is fully fanned, stand out at right angles from each side of the body. In display the bird dives, wings half folded, so that air blasts across the series of slots between the stiff feathers to make various unusual sounds—whistling, humming, winnowing—perhaps supplemented by vocal sounds. Such display is most unlikely ever to be seen in Aust. **Voice:** when flushed, the call is an abrupt, rasping 'skaik!'; has other calls in display. **Hab.:** in Aust., around billabongs, swamps, flooded grasslands, sewage ponds, claypans. **Status:** scattered records in W and NW coastal Aust.; regular visitor to Kimberley region; rare vagrant to SW Aust.

Painted Snipe *Rostratula benghalensis* 23–26 cm (L)

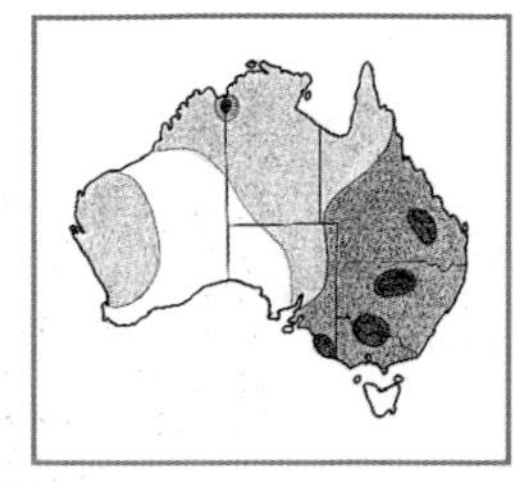

Bold and colourful plumage, but rarely seen—extremely secretive, keeping to dense vegetation of swamps, emerging only in subdued light of dawn and dusk. Silent except when breeding. Freezes when approached. Finally, with intruder very close, will flush, but flies only a short distance. Airborne, has broad, rounded wings and slow, erratic wingbeats, legs initially dangling. The Painted Snipe is one of only two species of the family Rostratulidae, which appears more closely related to jacanas than to *Gallinago* snipe. But, to enable convenient comparisons for identification, the Painted Snipe is presented here with its nearest look-alikes—it shares the name 'snipe' because of that likeness. As with both jacanas and snipe, the female is larger and more brightly coloured than the male, and the roles are reversed. The male incubates while the polyandrous female may lay several clutches in other nests to be attended by other males. **Voice:** in breeding season, 'advertisement' calls of female, typically around dusk, a long series of 'kot, kot, kot...' and soft, resonating 'whoo' sounds. When flushed, an abrupt, harsh 'krek!'. **Hab.:** surrounds and shallows of wetlands that are well vegetated with dense low cover. **Status:** breeding resident of inland SE Aust.; uncommon. **Br.:** 367

Long, straight bill and similar facial markings in all three species
Toes about equal tip of long tail
Eye streak becomes a double line behind the eye.
Long winged; tail projects even further, giving a rear end tapered smoothly to a long slender point.
Back boldly patterned with dark brown markings and whitish lines
Irregular dark scallops or broken brown barring
Tertials
Primaries
Projecting rufous tail
In the folded wing, the dark brown primaries are scarcely longer than the buff-edged tertials, and are often hidden by them. Both fall well short of tip of the long, bright, rufous tail.
Olive-grey to olive
Spreads tail on landing to show dark laterals.
Toes not projecting
Tail feathers: usually 18 or fewer
Tertials, primaries and tail all about equal length; primaries may be hidden under tertials.
Toes extend behind tail.
Square cut, stubby rear end
Back markings less conspicuous, with rufous-buff replacing white
Tail: usually 26; varies 24 to 28 retrices with outer 6 to 8 pairs fine, pin-like.
Grey-green to brownish green
Dark-speckled rather than scalloped or in broken transverse bars
Feet trail well behind short tail; those of Swinhoe's, below, do not extend as far.
With tail longer than folded wings, Swinhoe's has a rear end that tapers to a fine point rather than being stubby.
In folded wing, dark brown primaries extend beyond buff-edged tertials; bright rufous tail projects even further.
Dark facial stripes
Dark brown surrounds paler diagonal panel across inner wing.
Feet trail behind tail-tip.
More white shows on tail than with Pin-tailed Snipe.
Irregular mottling or indistinct chevrons rather than bars or speckling
Buff-edged tertials
Tail feathers usually total 20; vary 18 to 26 with outer 6 on each side narrow, but not fine pins like Pintail, above.
Buffy to pale grey-brown coverts; plumage may gradually fade or wear paler.
Tail, rufous
Plain brown primaries
Olive-yellow
Distinctive eye-patch shape
Slight buff tint
Rufous
Diagnostic bold white band from breast curves back and, finer and buff, down the back.
White underwing coverts
♀
Dark brown
♂
Bill down-curved, with slightly swollen tip
Legs dangle low or trail behind tail.
Backswept white and buff band common to both sexes and juvenile
Banded with large buff spots
Olive-yellow to greyish olive
Rounded wings, heavily banded
♂
Imm. female: wings similar to adult female; head is more like adult male.
Imm. ♀
Body is bobbed up and down while bird walks and forages.

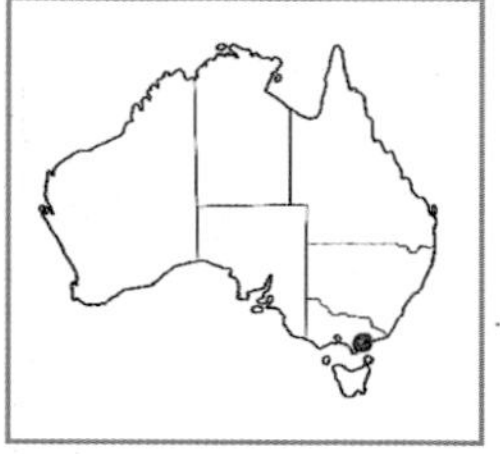

Short-billed Dowitcher *Limnodromus griseus* 25–30 cm
Dowitchers are migratory shorebirds with long, almost straight snipe-like bills with a slightly bulbous, sensitive tip. With these they probe deeply into soft mud with rapid vertical vibrations, seeking small invertebrates. **Voice:** usually a mellow 'tsew-tsew-'. **Similar:** Long-billed Dowitcher, but (nonbreeding) has breast almost plain grey; flanks and under tail with bold bars, lower back white; in flight, lacks fine white wingbars. Asian Dowitcher, larger and longer, slightly downcurved bill. **Hab.:** in Aust., lagoons, pools with soft mud, moist grass and semi-aquatic vegetation. **Status:** recorded in adult nonbreeding plumage at the Corner Inlet, Vic., 1995, and initially identified as being the similar Long-billed Dowitcher.

Asian Dowitcher *Limnodromus semipalmatus* 33–36 cm
Distinctive, long, heavy bill that is slightly downcurved and thickened near the tip; used in rapid-action probing just ahead of feet as bird moves forward. White of barred tail and faintly barred rump extends in a triangle between dark wings, reaching mid-back. White edged secondary coverts form a pale band across wings. **Br.:** rich bright chestnut on head, neck and breast; back feathers are dark, edged rufous. **Nonbr.:** crown dark capped; white brow line; back dark grey with white margins; breast white, mottled grey-brown. **Juv.:** as nonbreeding with buff and brown tints. **Voice:** contact calls are a yelping 'chiewp' and softer 'kriow'. **Similar:** Bar-tailed Godwit. **Hab.:** beaches, mudflats, sewage ponds, saltfields. **Status:** a regular visitor in small numbers to NW Aust.; elsewhere a rare vagrant.

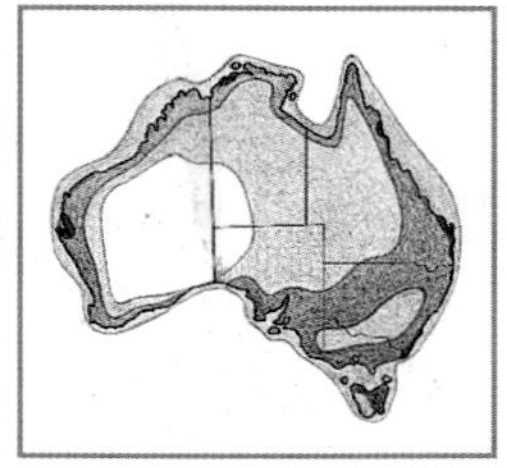

✓ **Bar-tailed Godwit** *Limosa lapponica* 37–39 cm (C, L, M, N) Cairns
This godwit is one of the most common Australian migratory waders; occurs in huge numbers at wintering grounds with counts as high as 60 000 at Broome where waders rest and feed. Breeding colours appear during the period in Aust.; are displayed on many of those passing northwards in March and April. **Br.:** head, neck and underbody deep rufous or chestnut. **Nonbr.:** grey-brown above; underparts white with buffy brown mottling and streaking in breast. **Juv.:** overall warmer colours; fine buff barring on dark back feathers. **Voice:** in flocks, have calls of contact and alarm. The former is a sharp 'kak' or 'kerk'; the alarm 'kirrik' or 'kirrark'. **Similar:** Whimbrel; Black-tailed Godwit, which is more slender, longer legged, unbarred tail. **Hab.:** coastal mudflats, sandbars, shores of estuaries, salt marsh, sewage ponds. **Status:** scattered records most of Australian coast; common. **Br.:** far N Asia, across N Russia to Siberia and Alaska.

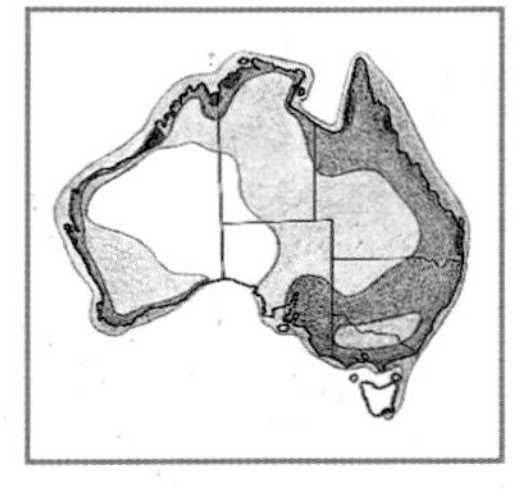

Black-tailed Godwit *Limosa limosa* 36–44 cm (L, T)
Compared with Bar-tailed, smaller and more slender. Longer neck, slimmer body and longer legs that trail substantially behind the tail-tip combine to give a more slender, delicate jizz. Bill is straighter, very slightly bulbous-tipped. Female is slightly larger than male with longer bill. **Br.:** male has extensive rufous on head, neck, breast; female not as bright. **Voice:** quite vocal with a tuneful song at its northern-hemisphere breeding grounds, but gives only contact and alarm calls in Aust. Some soft chattering sounds when feeding and some calling from flocks when flushed. Sounds recorded include sharp 'witta-wit', a rather harsh, strident 'wieka-wiek-wieka', and soft 'kek' or 'kuk'. **Similar:** Hudsonian Godwit, but it has black underwing coverts; Bar-tailed Godwit (barred underwing and tail). **Hab.:** usually coastal—estuaries, sheltered bays, lagoons with extensive tidal mudflats or sandbars—shores and islets of large, ephemeral inland lakes; infrequently on rocky coasts, islets; sewage farms. **Status:** common; most abundant on N coast E of Darwin, progressively less common further E and S. **Br.:** across N Eurasia.

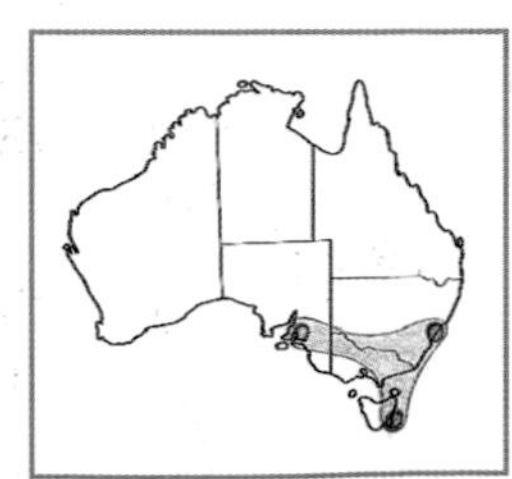

Hudsonian Godwit *Limosa haemastica* 38–41 cm (N)
An extremely rare visitor to Aust. Breeds N America, but very small numbers visit NZ regularly. Very few accepted records—SA and NSW, usually June–Sept. **Br.:** head and neck grey, breast to belly chestnut, barred black. **Nonbr.:** darker than above species, especially on breast; flanks barred. **Juv.:** like nonbreeding, but warmer tone overall. **Voice:** usually silent away from the N American breeding grounds except for the alarm call, a sharp 'weit-weit-weit', higher pitched than call of the Bar-tailed Godwit. **Similar:** Bar-tailed, but lacks its white underwing and longer white bar along upper wing. **Hab.:** in Aust. on coastal lagoons, estuaries, freshwater lakes, salt ponds. **Status:** accidental or rare vagrant; NSW 1982, SA 1986.

The bill is long, but slightly shorter than Long-billed and Asian Dowitchers. It is almost straight, unlike the gently downcurved bill of the Asian Dowitcher, or the gently upcurved, tapering bills of the godwits.
Lower breast faintly streaked or mottled
Long, straight, but shorter than bill of Asian Dowitcher
Like non-breeding, grey with wide white brow
Toe tips project.
Darker underwing than Asian Dowitcher, finely barred, appearing grey
White
Thin white edge to secondary coverts
Lightly barred
Greenish yellow
Faintly barred. (Long-billed Dowitcher has heavier bars.)
Nonbreeding
Juv.
Juv. as nonbreeding, but back is slightly rufous, underparts have buff tint.
Faintly speckled white triangle from rump to back
White edge
Tail barred
Nonbreeding
Like a smaller version of Bar-tailed Godwit but has swollen, slightly drooped bill tip.
Black
White brow
Heavily streaked
White
In all plumages, tail is barred. Lighter barring extends up the pale rump, which, with whitish lower back, forms a pale wedge between dark wings.
Wing and tail-tips equal length
Bright chestnut
Bars
White with barred flanks
Spotted or mottled in loosely vertical columns
Buff tint
Upper parts dark grey-brown
Breeding
Nonbreeding
Juv.
Bill is very long, slightly downcurved and swollen towards the tip.
Tail barred; finer bar on rump
Large wader, often standing tall, but rather heavy build; bill upturned, tail barred
Rufous or chestnut
Long, gently upcurved; pink with dark tip; bill of female is longer than that of male.
Dark
Heavily streaked cap on juv.
Plain
Breeding female less colourful
White triangle into upper back
Nonbreeding
Barred tail; feet visible
Dark centres, barred buff
Blotched rufous on buff
Barred
Plain chestnut
Tail barred (like breeding)
♀ Non-breeding
No obvious wing-bar
♂ Breeding
Black
Juv.
Barred underwing
Similar to above species, but has plain black rather than barred tail; longer bill that is straight or has a very slight gradual upturn.
Rufous-chestnut
Very long, nearly straight, pink with dark tip
White
Long white wing-bar
White rump, dark back
Black
Barred
Black
Cinnamon, buff tinted, plain
White
Long, trailing
Tail plain black
White, barred
Tail black, without bars or other markings
Thin black
Bold white bar across dark flight feathers
Long black legs
Breeding
Non-breeding
Juv.
Legs extend further behind tip of tail than on other godwits.
Extremely rare vagrant; shorter legs, overall darker in all plumages than the more common godwits.
Grey
Grey
Slightly upturned, finer point
White face, brow, dark eye streak
Steeper, higher forehead, rounded crown
Black
White rump, black tail and lower back
Chestnut, barred black; female has white as well as black bars.
White, barred black and rufous
Barred flanks
Darker grey-brown
White central underwing
Wide black
Wide part of white wing-bar is very short.
Legs shorter than Bar- and Black-tailed
Nonbreeding
Feet trail.
♂ Male breeding.

Little Curlew *Numenius minutus* 30–36 cm (C,K,L)

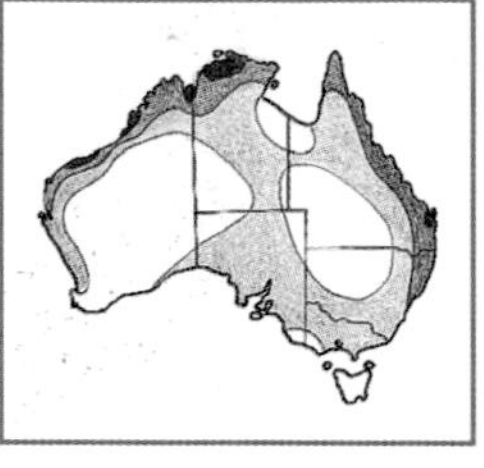

Flocks of hundreds, even thousands, of birds can be found around the extensive swamps and billabongs of the coastal blacksoil plains in N Aust. Stands erect, alert; forages busily, actively probing at ground, usually in groups. Flies with easy, relaxed wing action; shows brownish rump, whitish shafts to outer primaries, but no wing bars; holds wings up briefly on landing. Tips of toes just visible beyond tail-tip. **Voice:** usually three notes in flight—sharp, rising 'kee-kee-kee', like, but without the clarity of, the Greenshank's call. In alarm, husky, high 'tchiew-tchiew-tchiew' and rasping 'kwiekek'. **Hab.:** dry grasslands of clay and blacksoil plains, river floodplains, woodlands with grassy understorey, around billabongs and freshwater swamps, also similar artificial environs—pasture, airfields, sports fields, lawns. Often forages over recently burnt grassland or open woodland. **Status:** abundant across coastal N; scattered through inland to southern regions; vagrant to Tas.

Whimbrel *Numenius phaeopus* 39–44 cm (C,K,L,N,T)

A medium sized curlew with twin dark streaks along the crown and mid length bill. In flight, usually has a white wedge from rump to lower back; underwings are entirely barred, brown on buff. Wingbeats seem more rapid than those of curlews. Gregarious; feeds in small flocks or mixed with other waders on coastal mangrove and estuary mudflats, beaches and reefs. Takes molluscs and crustaceans; includes berries in parts of its migration route. Feeds with energetic dashing about—short runs, rapid jabbing of bill—rather than deep probing. **Variation:** most migrating to Aust. are of Siberian race *variegatus* with white to light brown barred wedge from rump to back; a few are of N American race *hudsonicus* with entirely dark rump and back. **Voice:** in Aust. the distinctive call is a rather musical, even pitched 'ti-ti-ti-', but so rapid that the notes run together in almost unbroken tittering, 'ti-tititititititi…', usually given in flight. Also a high, clear 'keer-keer-keer' and slow, tremulous 'ke-ee-r'. **Similar:** Little Curlew (smaller, dark rump, partially barred underwing); Eastern Curlew (larger, extremely long downcurved bill and dark rump); Eurasian Curlew (white underwings). **Hab.:** mudflats of estuaries, lagoons, preferably with mangroves; less often sandy beaches, reefs, salt lakes. **Status:** common across N Aust. coast; uncommon to rare further S. **Br.:** N Europe, Russia, Siberia, Alaska, Canada.

Eastern Curlew *Numenius madagascariensis* 60–65 cm (L,N,T)

Largest wader in Aust., over 0.5 m in length including long downcurved bill. In small groups or alone, or sometimes very large flocks; much more wary and quick to take flight than other waders; most need a short take-off run; wings beat slowly compared with the rapid action typical of most waders. Feeds by probing deeply into mud or sand. **Fem.:** slightly larger, longer bill. **Juv.:** like adult but paler with neater, clearer edges and notchings to feathers, finer streaking on breast; bill is initially much shorter, but then slowly grows to full adult length. **Br.:** slightly more rufous, more contrast between darks and lights of plumage, compared with nonbreeding. **Voice:** beautiful, haunting, melancholic yet melodious sound used as contact call, given in flight or from ground; a high, drawn out, two-part call, the second lifting higher, attenuated, 'coor-lee' or 'cur-eek'; also a rapid, softer, melodious 'curee-cree-cureecuree', and a more strident version as alarm. **Similar:** Eurasian Curlew, which has white wedge from rump to back and white underwings; Wimbrel, but it is much smaller with short bill, usually a white rump. **Hab.:** tidal mudflats, sand spits of estuaries, mangroves, lake shores, ocean beaches. **Status:** common migrant to N, NE and SE, Tas.; occasional to SW of WA and SA. **Br.:** Russia, NE China.

Eurasian Curlew *Numenius arquata* 50–60 cm

Extremely rare vagrant or accidental visitor; probably birds that travelled too far S or joined with Eastern Curlews that travel through to Aust. In flight, wingbeats are slow, deliberate, rather gull-like, similar to action of Eastern Curlew. **Juv.:** similar to nonbreeding adults, but more buff about neck and breast. Uses the long downcurved bill to probe deep into mud of tidal flats on estuaries; seeks crustaceans, molluscs, worms. Birds from north-central Russian breeding grounds travel to locations as far apart as southern Africa, Japan, Malaysia and the Philippines. **Variation:** two races: *arquata*, Europe to Urals; *orientalis* from the Urals through central Russia; intermediate forms. **Voice:** loud, carrying, ringing 'cour-liu, cour-liu', and, in alarm, 'tiu-yiu-yiu-yiu-yiu'. **Similar:** Eastern Curlew, which has longer bill, dark rump and back, barred underwing; Whimbrel, which is much smaller, shorter bill. **Hab.:** estuarine mudflats, beaches. **Status:** only two reports for Aust.—these not accepted as certain—near Darwin, 1948; coast near Perth, 1969. Normally winters southwards to Africa, SE Asia, Malaysia, Borneo. **Br.:** Europe to Siberia.

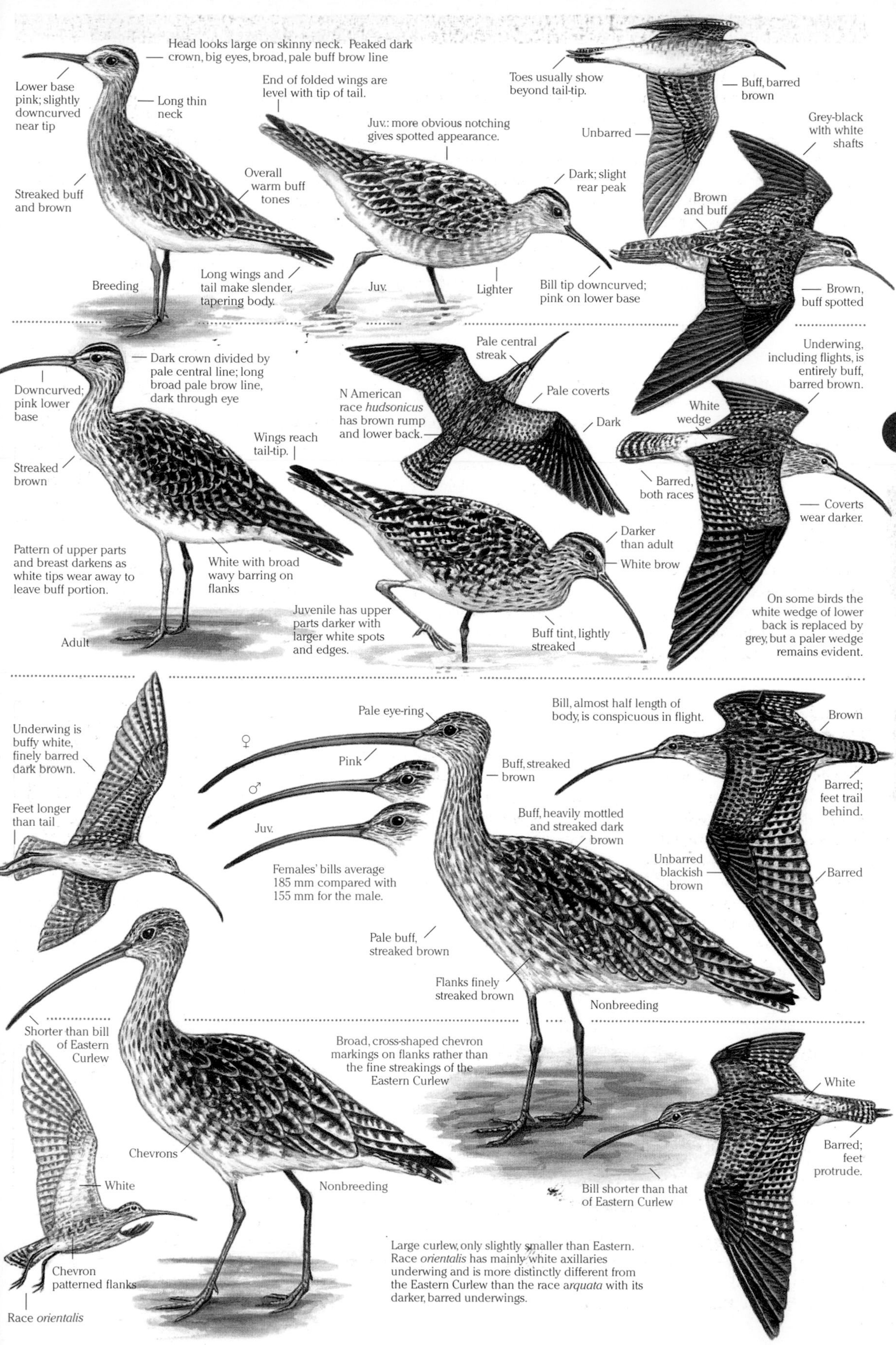

Head looks large on skinny neck. Peaked dark crown, big eyes, broad, pale buff brow line
Lower base pink; slightly downcurved near tip
Long thin neck
Streaked buff and brown
Overall warm buff tones
Long wings and tail make slender, tapering body.
Breeding
End of folded wings are level with tip of tail.
Juv.: more obvious notching gives spotted appearance.
Dark; slight rear peak
Juv.
Lighter
Bill tip downcurved; pink on lower base
Toes usually show beyond tail-tip.
Buff, barred brown
Unbarred
Grey-black with white shafts
Brown and buff
Brown, buff spotted
Downcurved; pink lower base
Dark crown divided by pale central line; long broad pale brow line, dark through eye
Streaked brown
Wings reach tail-tip.
Pattern of upper parts and breast darkens as white tips wear away to leave buff portion.
White with broad wavy barring on flanks
Adult
N American race *hudsonicus* has brown rump and lower back.
Pale central streak
Pale coverts
Dark
Juvenile has upper parts darker with larger white spots and edges.
Darker than adult
White brow
Buff tint, lightly streaked
Underwing, including flights, is entirely buff, barred brown.
White wedge
Barred, both races
Coverts wear darker.
On some birds the white wedge of lower back is replaced by grey, but a paler wedge remains evident.
Underwing is buffy white, finely barred dark brown.
Feet longer than tail
Pale eye-ring
♀
Pink
♂
Juv.
Buff, streaked brown
Females' bills average 185 mm compared with 155 mm for the male.
Buff, heavily mottled and streaked dark brown
Pale buff, streaked brown
Flanks finely streaked brown
Nonbreeding
Bill, almost half length of body, is conspicuous in flight.
Brown
Barred; feet trail behind.
Unbarred blackish brown
Barred
Shorter than bill of Eastern Curlew
Broad, cross-shaped chevron markings on flanks rather than the fine streakings of the Eastern Curlew
Chevrons
White
Nonbreeding
Chevron patterned flanks
Race *orientalis*
Bill shorter than that of Eastern Curlew
White
Barred; feet protrude.
Large curlew, only slightly smaller than Eastern. Race *orientalis* has mainly white axillaries underwing and is more distinctly different from the Eastern Curlew than the race *arquata* with its darker, barred underwings.

Spotted Redshank *Tringa erythropus* 29–32 cm

A tall, elegant wader; long necked with long, slender legs. The full length of feet usually trail behind the tail-tip. Bill is distinctive: fine and straight with a slight but noticeable downturn almost at the tip. This is an extremely rare vagrant to Aust. with few accepted records. Shy and wary, it frequently bobs head and body. Forages by delicately probing mud, sometimes working slightly deeper water where it swims and up-ends to reach bottom. Flight is fast and direct; on longer travels may tuck legs up into plumage. The few seen in Aust. have been alone, but in regular wintering regions occurs in small groups to large flocks. Male and female similar in breeding and nonbreeding plumages. **Juv.:** overall slightly darker brown tone; more heavily spotted on darker body. **Voice:** distinctive, flute-like 'tchuet'. **Similar:** Common Redshank, which has a wide white panel along the upper wing's trailing edge and a straight-tipped bill. **Hab.:** prefers salt marshes, shallow freshwater swamps and lagoons. **Status:** sightings, all from Oct. to Apr., come from NW of WA, Top End of NT, Hunter R. district of NSW, Seaford, Vic.

Common Redshank *Tringa totanus* 27–29 cm (C)

Often intermixed with other waders; forages briskly; bobs head and body when suspicious; always alert and quick to give alarm calls; if alarmed, departs with strong but erratic flight. **Variation:** many different subspecies or races, and several colour forms. Those visiting NW Aust. are probably race *ussuriensis*, which breeds in Mongolia, Manchuria and E Russia. The plumage on the back includes cinnamon-rufous heavily patterned with black, and some tips of light cinnamon, buff and white. **Juv.:** as adult nonbreeding, but warmer tones, buff streaking on crown, buff edgings to feathers of upper parts; legs and base of bill orange rather than red. **Voice:** noisy any time of year. In flight, usually two rapid, high, ringing notes, a third extremely abrupt: 'kier-kier-kp, kier-kier-kp, kier-kier-kp', the three-note sequence repeated several times. Alarm call on the ground is a piercing 'kieer'. **Similar:** distinctive with combination of broad, brightly translucent white trailing edge and white wedge from rump to back. **Hab.:** coastal wetlands, including estuaries and lagoons where there are open areas of shallows with mudflats and sandbars. **Status:** uncommon but probably regular summer visitor to scattered sites.

Lesser Yellowlegs *Tringa flavipes* 23–25 cm

Tall, graceful stance and movement on long legs. The few records in Aust. are of single birds, but in more popular wintering grounds it gathers in small or large flocks, sometimes of hundreds of birds. Moves about mudflats and shallows with brisk movements, jabbing and probing the surface. Shares the wary, nervous bobbing of head and tail typical of the *Tringa* species; noisy, excitable. If put to flight, has easy, relaxed, rather slow wing action, but can travel fast; legs trail by almost the full length of the foot. **Voice:** usual flight call a whistled 'tiew-tiew-'; the 'ti-' lifting, the final 'ew' dropping in pitch; louder in alarm. On ground, loud 'tiew', each note of long series accompanied by bobbing of head and tail. **Juv.:** like nonbreeding, but slightly darker, warmer with extensive clear, pale buff, feather-edge notches and spots; darker tone to head and upper body; crown and breast more heavily streaked. **Similar:** Wood Sandpiper, especially juvenile, but that species is not as slender, folded wingtips are shorter, bill is shorter, brow line also extends behind eye, legs are greener, has different call. Marsh Sandpiper (same differences except bill is longer and finer than Lesser Yellowlegs). **Hab.:** margins of muddy wetlands, marshy edges of swamps, mudflats. **Status:** rare vagrant, very few sightings.

Greater Yellowlegs *Tringa melanoleuca* 29–33 cm

Possible vagrant or accidental to N Aust. coast; breeds Alaska and Canada, winters S of North America and to southernmost coasts of South America. Vagrant to Europe; records from South Africa, Hawaii, possibly Cook Islands. Tall, elegant wader with bright yellow or orange legs; bill slightly upcurved and more than 1.5 times length of head from base of bill through to nape. At usual wintering grounds occurs in small flocks or singly, probing mud, and, like Greenshank, wading quite deeply; often the first species in mixed gatherings to give alarm calls and take flight. **Voice:** almost identical to Greenshank, perhaps higher pitched and faster, a very rapid 'kier-kier-kier, kier-ker-kier-kier-'. **Juv.:** like nonbreeding, but warmer tones to upper parts, which are more extensively spotted and notched with buff. **Similar:** Lesser Yellowlegs, which is smaller with shorter bill relative to head length, lacks spots on flight feathers, and has slightly flatter, less resonant call; Greenshank, which has pale greenish yellow legs, upper body streaked rather than spotted, bill finer, and obvious white wedge extending from rump to upper back. **Hab.:** coastal mudflats, salt marsh; possibly inland on fresh or brackish muddy lakes. **Status:** apparently not yet a substantiated record for Aust., but a possibility.

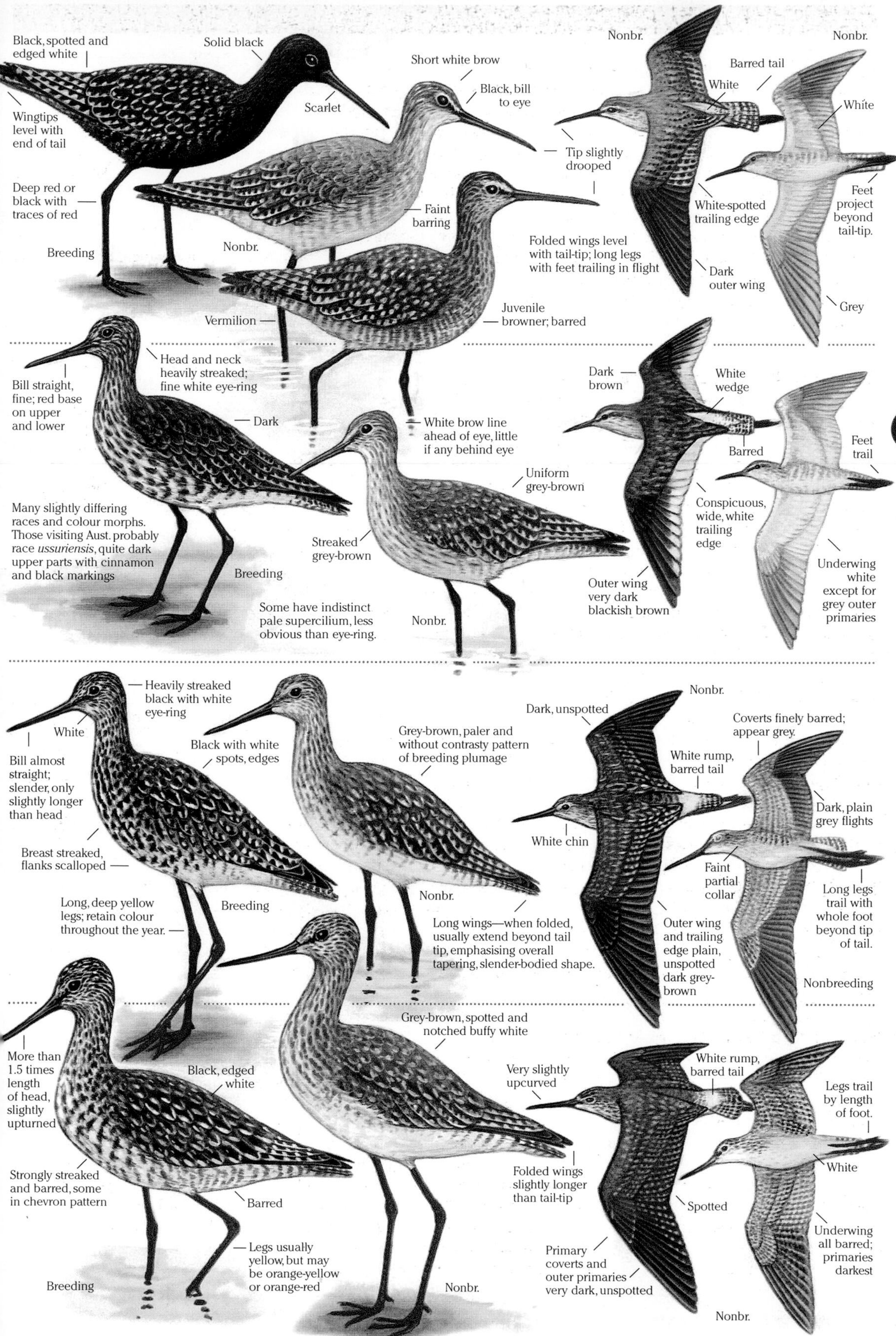
Black, spotted and edged white
Solid black
Wingtips level with end of tail
Scarlet
Short white brow
Black, bill to eye
Deep red or black with traces of red
Faint barring
Breeding
Nonbr.
Vermilion
Juvenile browner; barred
Nonbr.
Nonbr.
Barred tail
White
White
Tip slightly drooped
Feet project beyond tail-tip.
White-spotted trailing edge
Folded wings level with tail-tip; long legs with feet trailing in flight
Dark outer wing
Grey
Bill straight, fine; red base on upper and lower
Head and neck heavily streaked; fine white eye-ring
Dark
White brow line ahead of eye, little if any behind eye
Uniform grey-brown
Many slightly differing races and colour morphs. Those visiting Aust. probably race *ussuriensis*, quite dark upper parts with cinnamon and black markings
Streaked grey-brown
Breeding
Some have indistinct pale supercilium, less obvious than eye-ring.
Nonbr.
Dark brown
White wedge
Barred
Feet trail
Conspicuous, wide, white trailing edge
Outer wing very dark blackish brown
Underwing white except for grey outer primaries
Heavily streaked black with white eye-ring
White
Bill almost straight; slender, only slightly longer than head
Black with white spots, edges
Grey-brown, paler and without contrasty pattern of breeding plumage
Breast streaked, flanks scalloped
Long, deep yellow legs; retain colour throughout the year.
Breeding
Nonbr.
Long wings—when folded, usually extend beyond tail tip, emphasising overall tapering, slender-bodied shape.
Nonbr.
Dark, unspotted
Coverts finely barred; appear grey.
White rump, barred tail
Dark, plain grey flights
White chin
Faint partial collar
Long legs trail with whole foot beyond tip of tail.
Outer wing and trailing edge plain, unspotted dark grey-brown
Nonbreeding
More than 1.5 times length of head, slightly upturned
Black, edged white
Grey-brown, spotted and notched buffy white
Very slightly upcurved
White rump, barred tail
Legs trail by length of foot.
White
Strongly streaked and barred, some in chevron pattern
Barred
Folded wings slightly longer than tail-tip
Spotted
Underwing all barred; primaries darkest
Legs usually yellow, but may be orange-yellow or orange-red
Primary coverts and outer primaries very dark, unspotted
Breeding
Nonbr.
Nonbr.

Common Greenshank *Tringa nebularia* 30–35 cm (C, K, L, M, T)

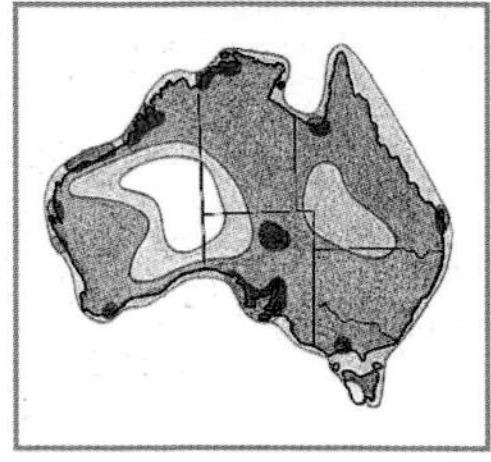

Stands tall, erect; bobs head in alarm; leaps into flight with ringing calls; is very timid and wary. Occurs alone, or in small or large flocks, often with other waders. Active and excitable; forages briskly with sudden dashes, erratic changes in direction. **Voice:** the loud, ringing alarm call is one of the most familiar sounds in wader habitat—a clear, very rapid 'tiew–tiew–tiew' with a slight drop in pitch on the last note, typically given as three quick, evenly spaced notes that together occupy less than a second. Notes in a series may range from 1 to 6 with a pause of 1 sec before the next set. **Similar:** Marsh Sandpiper, but it is much smaller, slender build, fine straight bill, dark streaked crown, very long legs with feet trailing much further in flight. **Hab.:** diverse inland and coastal spots. Away from the coast uses both permanent and temporary wetlands—billabongs, swamps, lakes, floodplains, sewage farms and saltworks ponds, flooded irrigated crops. On the coast uses sheltered estuaries and bays with extensive mudflats, mangrove swamps, muddy shallows of harbours and lagoons, occasionally rocky tidal ledges. This species generally prefers wet and flooded mud and clay rather than sand. **Status:** widespread common migrant Sep.–Apr.; a few remain through winter. Breeds Eurasia, Britain to Kamchatka Peninsula to NE of Japan.

Nordmann's Greenshank *Tringa guttifer* 30–33 cm

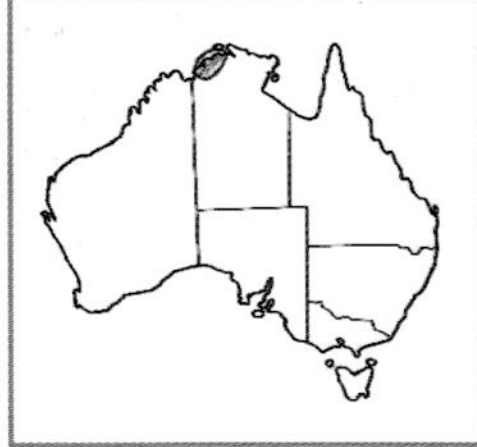

Not only rare in Aust., where it remains but a possible visitor, but also uncommon with very restricted breeding grounds on Sakhalin Is., NE Japan. The small numbers that migrate and the scarcity of records from along its migration route suggest that the species is in decline. Migration follows a long, south-westward path via Japan, Korea, Hong Kong, Malaysia, then NW to Burma and India's E coast. Any that may stray further S to Aust. would be easily overlooked because its appearance is so like the Common Greenshank. Unique is the slight webbing between the toes, but identification requires a bird in hand. **Voice:** distinctly different from the Common Greenshank; a less musical, loud, sharp, piercing 'kiyiew'. **Similar:** Common Greenshank, Marsh Sandpiper. **Hab.:** estuarine mudflats, coastal lagoons. **Status:** rare possible vagrant; uncertain sight record in coastal NT.

Marsh Sandpiper *Tringa stagnatilis* 22–26 cm (C, L, T)

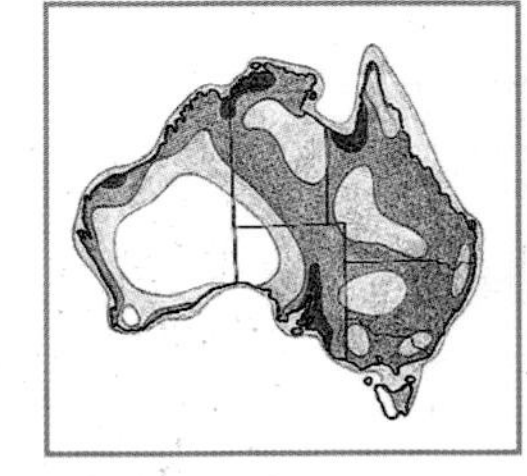

Tall, elegant, long necked and very long legged sandpiper, like a greenshank but much smaller, more slender. Also distinctive in flight; long legs trail so that all toes are completely behind the tail-tip, and the deep white wedge that extends from rump far up the back is clearly visible. Occurs alone, in small groups or with other waders. Forages in shallows; has rather upright stance, often with quick dashes to take some small creature; wades out to depth limit of long legs, searching and feeling for prey; probes mud and around marshy vegetation. **Voice:** usual call a quick, soft, mellow, musical 'kier' or 'teoo'; alarmed, excited or put to flight, gives short, very high, thin, rather metallic and rapid 'kier, kier-kp, kier-kp, kier-kier', becoming a confused mass of chittering from even small flocks. **Similar:** Greenshank, which has much larger, heavier, upturned bill, and relatively shorter and thicker legs. Wood Sandpiper, which has much shorter legs and bill, and darker, spotted upper parts. Wilson's Phalarope (nonbreeding) has shorter legs, no white wedge on back. **Hab.:** coastal and inland wetlands, salt or fresh; typically estuarine and mangrove mudflats, beaches, shallows of swamps, lakes, billabongs, temporary floodwaters, sewage farms and saltworks ponds. **Status:** regular summer migrant to Aust., Aug.–May. Moderately common across far N; more scattered around other coastal parts; sparse through inland regions. Breeds across N Asia; those that reach Aust. are probably coming from central Siberia.

Upland Sandpiper *Bartramia longicauda* 26–28 cm

A strange, rather plover-like wader with tall upright posture; often holds long slender neck vertically. Head is small for bulk of bird, yet looks large, rounded and distinct compared with the thin neck. Unlike most waders, it prefers drier grasslands—the prairies of N America and similar environs. Behaviour, posture and feeding actions are more like a plover; runs and stops abruptly to jab at small prey. If alarmed, has sandpiper-like bobbing action of rear end. Flight is swift with steady, easy wingbeats; feet do not reach beyond long tail. Often perches on posts or power poles; holds wings up briefly after landing. **Voice:** wide variety of sounds—flight call a piping 'quip-eip-eip', and lower, liquid 'pull-ip' or 'quie-lip'. **Hab.:** open grasslands or prairies, commonly on farmlands, pasture, stubble or airfields; likely to use similar habitats on migration. **Status:** single confirmed record near Sydney, 1848; breeds Alaska, Canada, USA.

Heavily streaked
Long, slightly upturned
Darker than nonbreeding
Barred, pale
Grey
Chestnut, heavily marked with black central streaks—some whole feathers are black, most have buff-white margins and notches.
Dark
Darker streaking than on nonbreeding
Bold black chevrons
Juv.
Project slightly
Grey-brown, finely edged white and notched black
Folded wings extend slightly beyond tail-tip.
Pale, finely streaked
Dark under eye-ring
Obvious white rump and back, tail lightly barred
Dark
Long, greenish yellow to grey-green
Breeding
Very dark, plain
Grey
Pale streaking extends down to sides of breast
Black
Nonbr.
Light grey-brown
Unbarred, whiter than common greenshank
Yellow, thick based
Like nonbreeding, but slightly warmer tone
Dark
Lightly streaked
Slightly smaller than than Common Greenshank. Legs obviously short above the joint.
Greenish or olive-yellow
Short legs only just reach tip of tail.
Short upper leg
Slight toe webbing is invisible at usual observation distances.
Nonbr.
Juv.
Very slightly upcurved
Streaked brown
Fine straight black bill tapers evenly from base to tip.
Grey-brown
Upper parts are overall pale cinnamon sparsely patterned with dark brown centres, fine buffy grey or white margins.
Dark bar
White
Long white wedge
Heavier 'V' bars, spots and streaks
Cap streaked grey-brown
Breeding
Soft grey-brown, pale margins
Very long legs
Tail-tip
White
White
Trailing feet
Browner, darker
The stilt-like, long, thin legs are usually dull yellowish green, some brighter yellow.
Faint grey-brown
Dark, plain
Nonbr.
White
Very dark, plain
Juv.
Grey-green
Large dark eyes, crown dark with pale central line
Thin neck
Rounded triangular head
Buff, heavily barred and streaked dark brown, edged buffy white
Slightly drooped
Dark brown, edged pale buff
Black flights contrast with lighter coverts.
Pale buff with dark brown bars and chevrons
Barred
Adult
Heavy streaking and barring
Legs yellow ochre to greyish olive; rather short
Wings shorter than barred tail
Juv.
Pale buff, streaked
Legs are shorter than the long, barred tail.

Wood Sandpiper *Tringa glareola* 20–22 cm (C,T)

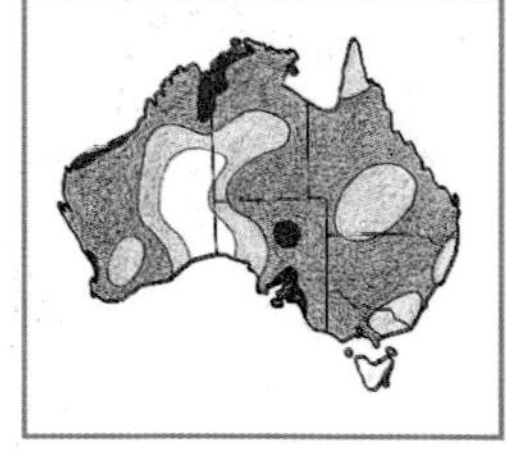

A graceful, active, slender wader, more often seen in the far N than the interior or S, inland rather than coastal; prefers shallows of wooded lakes or swamps with living or dead trees, such as river gums or paperbarks; often forages among fallen limbs and vegetation. Occurs singly, in pairs or small parties, occasionally large flocks, often with other waders. Wary, quickly agitated; then it holds its head high on the long, slender neck and bobs lower body. When flushed, swiftly rises high with loud, shrill cries; circles with bursts of quick wingbeats. When feeding, busy but graceful with high-stepping movements. **Voice:** sharp, high, piercing, rapid 'chi-chi-chip', 'chi-chi-chi-chip', the last 'chip' slightly lower, abrupt. **Similar:** Green Sandpiper, Lesser Yellowlegs. **Hab.:** freshwater swamps, lakes, flooded pasture; less frequently on brackish waters, occasionally in mangroves. Often uses artificial wetlands such as large farm dams. **Status:** an uncommon migrant.

Terek Sandpiper *Tringa (Xenus) cinereus* 22–24 cm (C,L,N)

A small, pale, rather dumpy wader; distinctive long, upcurved, smoothly tapered bill; rather short orange legs. 'Teeter's, with nervous, exaggerated, bobbing action. The Terek is a lively, active feeder, dashing erratically about the mudflats with head down, pecking and probing in mud and sand, chasing or wading through shallows with sideways sweeping action of bill. Flies with a flickering wing action; strong and direct. **Voice:** alarm call a musical 'pee-peeweer, peeweer, peewit' or alternatively 'teeu-duey, duey, wi-wi-wi-yu'. Probably remembered more for the pleasant, yet haunting, flute-like quality rather than the precise pattern of notes. **Hab.:** coastal mudflats in sheltered estuaries and lagoons as well as sandbars, reefs, coastal swamps, saltfields. **Status:** common summer migrant on N coasts, scarce to rare in S.

Common Sandpiper *Tringa (Actitis) hypoleucos* 19–21 cm (C,K,L,N,T)

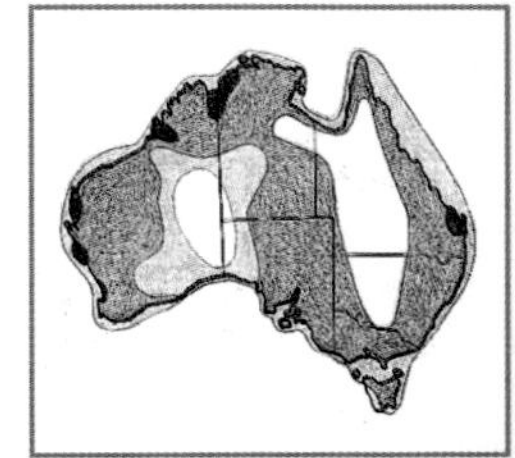

Almost constant teetering and bobbing of tail. Put to flight, darts away level and close to the water with distinctive spasmodic wing action—wingbeats hesitate a fraction of a second every beat or so; glides to land and teeters vigorously. **Voice:** in flight, fine squeaks, very abrupt, penetrating; may be given as two rapid squeaks, the second fractionally lower, 'tsie-tsiep', or as a longer series—four, six, or perhaps more, at six or so per second at even, high pitch, but drops slightly in the last note or two, 'tsie-tsie-tsie-tsie-tsiep-tsiep'. **Hab.:** varied coastal and interior wetlands—narrow muddy edges of billabongs, river pools, mangroves, among rocks and snags, reefs or rocky beaches; avoids wide open mudflats. Perches on branches, posts, boats. **Status:** widespread, scattered, generally uncommon.

Green Sandpiper *Tringa ochropus* 21–24 cm

Possible rare migrant; like a larger, plumper Wood Sandpiper with shorter greenish legs, longer heavier bill, white eyebrow only ahead of eye. In flight, white rump patch contrasts boldly with dark wings. Often adopts horizontal, rather hunched posture. **Juv.:** darker greenish brown above, heavily spotted dull buff; white brow more obvious. **Voice:** a sharp, ringing 'tlee-it-wit-wit', usually on being put to flight. **Similar:** Wood Sandpiper, but it has some white under the wings, warmer browns with heavier white flecking, long white brow; Common Sandpiper, but it has obvious wingbar, dark-centred rump, white peak ahead of wings, warmer browns. **Hab.:** freshwater wetlands. **Status:** unconfirmed NT sightings.

Buff-breasted Sandpiper *Tryngites subruficollis* 18–21 cm (L)

An unusual, plover-like sandpiper that, like the Upland Sandpiper, is usually a bird of drier grasslands. The upright stance when alert combines with the rounded head on long, slender neck and high-stepping walk to make this species distinctive. May crouch in the grass with compact, hunched posture when feeding. Approached, may crouch, freeze or run. In flight, looks long winged; keeps a low, zigzag path. **Nonbr.:** paler, dull buff edging on upper parts, dark centres less pronounced. **Voice:** usually silent; flight call a soft 'tchu', or 'prreei-t'. **Similar:** Ruff (female) but it is considerably larger, longer neck, heavier bill; Upland Sandpiper, but larger, different underwing. **Hab.:** dry open grasslands, pastures, sometimes in vicinity of swampy areas. In Aust., usually in short sparse grass, lawn, salt marsh or low samphire close to estuaries, lagoons, swamps. Only rarely on mudflats. **Status:** rare; perhaps regular migrant or vagrant to SE and S Aust.

Long white brow and forehead
Dark, streaked white
Straight, fine, black; medium length
Brown and black, boldly speckled white
Heavily streaked
Legs moderately long, yellow or greenish yellow
Wingtips level with tail
Breeding
White
White
Grey-brown and black, notched and edged white
Brown tint, light streaks
Nonbr.
Dark brown, broadly edged buff
Buff wash, streaked brown
Juv.
Long white brow line
Square white patch on lower rump; thin bars on tail
Pale flecked coverts
Outer wing dark
White shaft
Feet project well beyond tail-tip.
White
Slender white body
White, finely barred
High crown and steep forehead
A rather dumpy small wader, deep chested shape
Long, upcurved; black with dull orange base
Grey-brown with dark line along scapulars
Dusky grey-brown wash each side of breast, streaked
Folded wings just reach tail-tip.
Legs rather short, orange
Breeding
Plain, paler, dark shoulder
Brownish sides to breast
Nonbr.
Grey-brown with white brow, forehead, throat
Juvenile is like adult but darker with cinnamon fringes to back feathers; some dark scapulars.
Tips of toes protrude
White trailing edge to secondaries and inner primaries
Blackish brown
Nonbr.
Dark cap
White eye-ring
Olive-brown with faint black streaks and barring
Brown, streaked darker, a clearly defined bib-like breast-band
Tail longer than folded wing
Breeding
Straight, fine, evenly tapering
Indistinct brow line
Colours more uniform, dull
Divided
Wing
Tail
Characteristic white peak or hook shape
Juv.: similar but upper parts are barred dark brown, edged and spotted buff.
Nonbr.
Barred white outer edge
White trailing edge
Plain dark brown
Conspicuous white wing-bar
Nonbr.
Short white brow
White brow does not extend behind eye.
Bill straight, greenish base, medium length
Heavily streaked
Breeding
Plain light olive-brown
Olive-brown, sparsely flecked white
White; in flight a strong contrast against all dark underwings
Short, dull, grey-green
Nonbr.
Large, square white rump
Toes protrude slightly from barred tail-tip.
Dark, plain olive-brown without any wing-bar
Dark olive-brown
Underwing is entirely dark brown in strong contrast to white undersurface of body and plain white under tail.
Often stands tall, walks with head jerking; plover-like.
Large dark eye in plain pale face
Dark streaked
Dark centred pattern
Primaries are level with tail-tip.
Deepest buff from face to breast
Distinctive in overall buff tones, stronger about the face, sides of neck and breast
Breeding
With head tucked in, can look very tubby.
Juv.
Often almost white on rear underparts
Dark band
Dark mottled trailing edge
Dark
Dark
Underwing white
Feet just visible
Dark edge; no wing-bar
Dark plain outer wing

Sharp-tailed Sandpiper *Calidris acuminata* 17–22 cm (C, K, L, N)

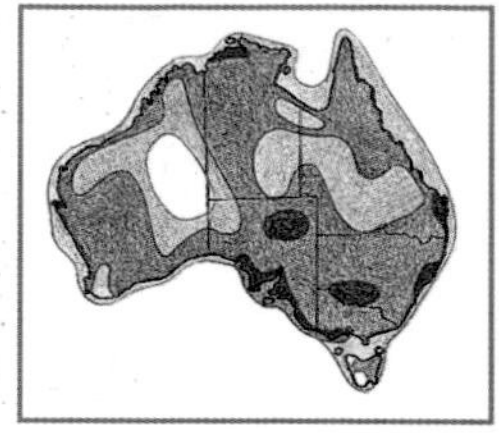

Plump, medium sized sandpiper; in flight, white sides to dark centre of rump. Thin wing stripe. **Voice:** when flushed, a quick 'pliep'; also rapid, high, scratchy, squeaky trills like the highest of fairy-wren trills; chatterings with intermixed soft, low and very high, squeaky sounds like Welcome Swallow's chatter. **Similar:** Pectoral Sandpiper, but it has slimmer look, longer neck, shorter legs, more upright stance and longer, more slender downcurved bill. **Hab.:** fresh or salt wetlands—the muddy edges of lagoons, swamps, lakes, dams, soaks, sewage farms, temporary floodwaters. **Status:** abundant in SE, common elsewhere.

Pectoral Sandpiper *Calidris melanotos* 18–24 cm (L)

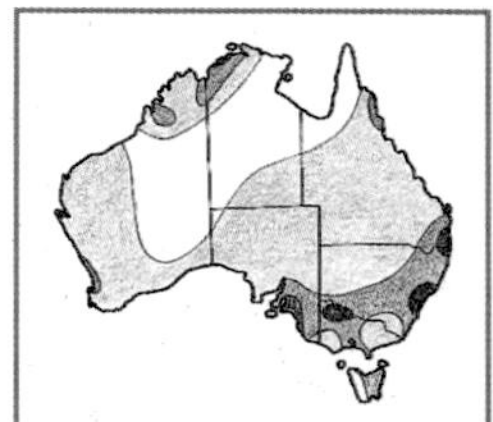

Medium-sized sandpiper. All plumages have heavily streaked breast sharply demarcated from clean, white belly. Often adopts upright stance, but also hunches low, compact. When flushed, flies a fast, low, twisting course. Forages in shallows and soft mud; solitary or small groups, or mixed with other waders. **Voice:** when flushed, repeated loud, harsh 'tirrit' or 'prrip'. **Similar:** Sharp-tailed Sandpiper. **Hab.:** usually coastal wetlands, both fresh and saline, but also inland on permanent and temporary wetlands. Uses sites with mudflats, fringing vegetation, swamps with heavy overgrowth of vegetation. **Status:** regular uncommon visitor to Aust., mostly to SE, but scattered almost throughout, coastal and inland.

Curlew Sandpiper *Calidris ferruginea* 18–23 cm (C, L, T)

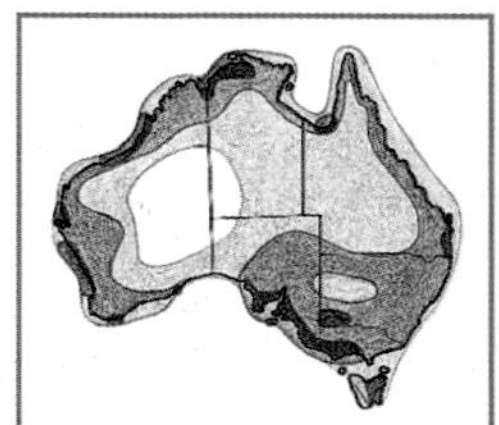

Medium sized sandpiper, slender if standing tall, plump if hunched down or resting; wings extend beyond tail-tip giving long, slender rear profile. Females have longer bill; in fresh breeding plumage have more pale tips and dark barring. Male solid, plain red. **Juv.:** upper parts warmer brown tones, buff tinted breast. **Voice:** contact call, repeated at intervals of about 1 sec, a mellow, rippling 'chirrip, chirrip, chirrup' like some Budgerigar calls; given by flock becomes a pleasant confusion of twitterings. **Similar:** Red Knot, but it is larger; has short straight bill. **Hab.:** inter-tidal mudflats of estuaries, lagoons, mangrove channels; around lakes, dams, floodwaters, flooded saltbush surrounds of inland lakes. **Status:** widespread, common summer migrant to Australian coastal sites; some scattered across suitable interior sites.

Stilt Sandpiper *Micropalama himantopus* 19–23 cm

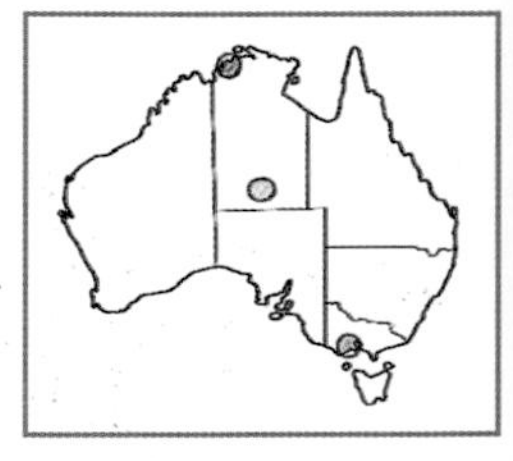

Medium sized sandpiper with long, thin stilt-like legs that seem to hold the rather slender body strangely high above the ground. The neck is long—is stretched up if alert, well forward when feeding, or pulled compactly onto shoulders in a stubby, hunched shape. The effect of posture on all species should be taken into account when assessing 'slim', 'plump' descriptions. Usually feeds in shallows, often to the extent of the long legs; feeds with rapid jabbing action, sometimes completely immersing head when reaching down into deeper water. Flies powerfully; long pointed wings, long bill and trailing legs give a distinctive silhouette. Usually occurs as a solitary vagrant in company with Sharp-tailed or Curlew Sandpipers or other waders. **Voice:** flight call a soft, rather rattling, trilled 'kirrt' and clearer 'whiuw'. **Hab.:** usually shallow waters of estuaries, tidal mudflats, swamps, ponds of saltworks and sewage farms; at times among flooded or swampy vegetation. **Status:** rare vagrant; very few sightings in NT and Vic., all on either sewage or saltworks ponds.

Ruff *Philomachus pugnax* male 26–32 cm; female 20–25 cm

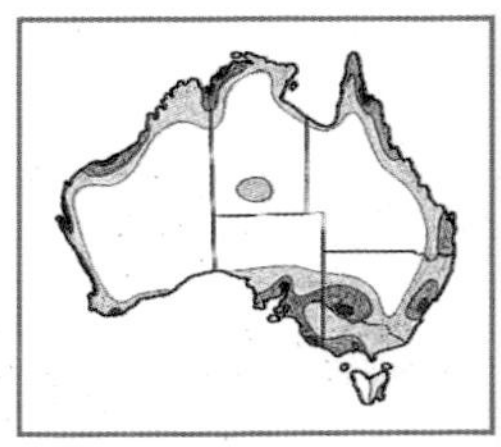

The name 'Ruff' is used for the male, which is not only substantially larger, but has remarkable breeding plumage and displays. The smaller female is the Reeve, and, while it acquires distinctive breeding plumage, it lacks the adornments of the male. Males in breeding season develop large ear tufts and exaggerated, colourful ruffs about neck and shoulders; colours vary; hardly any two birds are the same. Some have extensive black contrasting against faces of bare red skin; others have ruffs of snowy white, rufous or barred brown, and facial skin of ochre, orange or red. Ruffs congregate on display grounds or 'leks' to display, are then chosen by a watching female, and the pair departs to mate. The female alone builds the nest and raises the young. Breeding is in N Europe and Asia—most arrive in Aust. in dull nonbreeding plumage. A few may arrive with remnants of breeding plumage, or may start to acquire breeding colours before they leave in the autumn. **Voice:** usually silent, or low grunts. **Hab.:** mudflats and sedges around fresh or saline lakes, tidal pools, estuaries. **Status:** rare but recurrent visitor to coast, but few interior sightings.

White eye-ring and long white brow that is widest behind the eye
Downcurved
Rufous; white margins
Rufous
White
Rump white each side of a dark central stripe
Feathers pointed; edges rufous
Pointed
Orange-buff tint
Fine wing-bar
Primaries level with tail-tip
Breast tinted grey with fine streaks
Underwing predominantly white
Always has streaked breast; rather 'potbellied'.
Olive
Faint dull streaks
Juv. bright, rufous and buff
Dark
Nonbr.
Breeding
Long neck, short legs; often upright stance; breast always heavily streaked
Brown, streaked
Rufous margins
In flight, toes do not extend beyond tip of white-edged and dark-tipped tail.
Longer, more downcurved than Sharp-tailed
White
Edged rufous-buff
Primaries level with tail-tip
Buff, streaked brown
Faint, fine white wing-bar
Grey-buff breast heavily streaked; clearcut lower junction with white belly
Neat sharp edge between streaked breast and clean white of belly
Underwing mainly white
Dull olive to yellow
Nonbr.
Juv.
Dark wings
Breeding
When breeding, much bright chestnut-red in plumage, patterned black, fringed white
White
White
Bold white rump, dark tail-tip
Downcurved, quite long
Pale mottled
Dark grey
White coverts
Wings slightly longer than tail-tip
White
White wing-bar
In fresh plumage, pale tips create barring and fringing that will disappear as tips wear, darkening plumage.
♀ In fresh plumage: some barring and fringing. Male plain, solid, red-chestnut
Long, black
Buff, only slightly streaked
In flight, feet extend beyond tip of tail.
Nonbr.
Juv.
Nonbr.
Pale grey
Tall with distinctive, long, tapering, slightly drooped bill
Rufous
Downturned, with slightly swollen tip
Distinct long pale brow
Tip slightly drooped, bulbous
Wings long and pointed
Dark brown, edged rufous and white
Thick base
Plain grey-brown
Fringed rufous
Rump white, tail plain grey
Buff tint, faint streaks
Barred
Entire length of foot projects behind tail.
Long, spindly, stilt-like, greenish yellow legs
Grey streaked neck, breast, flanks
Wingtips extend beyond tip of tail.
Wing-bar very thin, faint
White
Breeding
Nonbr.
Juv.
Ruff, moulting. A patchy mix of breeding and nonbreeding plumages, showing remnants of the red and black breeding plumage
Dark, buff edged
Buff
Ruff, nonbreeding, at about same scale as above species. Reeve nonbreeding is similar but about 75% of size.
Head looks small. Bill is slightly bulbous tipped.
Narrow white wing-bar
Pale
Whole foot trails.
Humpbacked
White ovals each side of rump and tail
Wings longer than tail-tip
White
Long, yellow to orange and red
Long necked, potbellied shape
Long winged; dark outer wing and secondaries
♂ x 1/2
Adult nonbreeding
Juv. ♂ x 1/2 scale

Red Knot *Calidris canutus* 23–25 cm (L, M, N)

Breeding plumage is extensively rufous-chestnut. Most have lost this colour by August or September when they reach wintering grounds in Aust., and wear dull grey nonbreeding plumage. However, a few reach N Aust. in breeding colours. In the southern autumn, when waders begin their migration back to N breeding grounds, some knots leave in fresh breeding colours. **Fem.:** slightly uneven chestnut, less extensive. **Voice:** soft calls from feeding flocks—low, throaty, harsh 'knut' or 'knot'; from flocks on migration, 'nyup-nyup' sounds. In alarm, whistled 'qwik-ick', 'twit-wit' or 'twit-twit-twit'. **Similar:** in breeding plumage, Curlew Sandpiper, Asian Dowitcher, Great Knot. **Hab.:** sheltered coasts on mudflats and sandbars of estuaries, harbours, lagoons; occasionally on beaches, reefs. **Status:** summer migrant to Aust.; abundant in N Aust., less common in S.

Great Knot *Calidris tenuirostris* 26–28 cm

Medium-large wader; bright in breeding plumage, otherwise dull; long bill with thick base, very slight droop towards tip. Highly gregarious, flocks in large single species or mixed species groups, sometimes of hundreds or thousands of birds. Forages on inter-tidal flats, usually in shallows and at waterline, moving forwards slowly and deliberately, thoroughly probing the mud. Flies with slow beats of long wings; loose flocks. **Fem.:** averages slightly larger; less chestnut shows in scapulars of breeding plumage. **Juv.:** as nonbreeding, but upper parts darker with contrasting white fringes to feathers; brownish buff wash to breast. **Voice:** usually silent, but has a rapid, mellow, hollow 'krok-kok' or 'knut-nut'; the first note longer, rising, the second abrupt, slightly lower. Taken up by a flock, becomes a continuous, rapid babble of 'krok-knut-kok-nut-nok' sounds. **Similar:** in breeding plumage, the size and heavy black markings make this species unique. In nonbreeding, some similarity to nonbreeding Red Knot and tattlers. **Hab.:** sheltered coastal mudflats of estuaries, inlets, harbours, lagoons, mangrove swamps. Also on sandy bars and beaches near mudflats. Occasionally seen in salt lakes, lagoons and saltworks ponds, but only rarely on inland lakes or swamps. **Status:** abundant across N, less common in S.

Wandering Tattler *Heteroscelus incanus* 26–28 cm (L, N)

Tattlers are easily separated from other shorebirds, the upper parts having entirely plain grey plumage, both in nonbreeding and breeding; flight shows darker grey underwings. Distinguishing between the two species in nonbreeding plumage is far more difficult. A tattler whose breeding plumage has any bars on the rear central belly or under-tail coverts is most likely a Wandering. Usually solitary when feeding or resting. Feeds almost exclusively on rocky coastline; sneaks about rocks, probing, teetering. **Voice:** flight call a sharp rippling trill of 5 to 10 piercing notes lasting 0.5–1 sec, evenly pitched, accelerating but fading slightly in strength. Also a flute-like alarm call of just one or two notes. **Hab.:** almost entirely confined to rocky shorelines—wave-washed tidal platforms and exposed reefs around headlands or high islands. Likely to use these sites, or occasionally jetties, to loaf and roost. **Status:** uncommon but probably regular summer migrant to coastal NE Aust.

Grey-tailed Tattler *Heteroscelus brevipes* 24–27 cm (C, L, N, T)

Far more common than the Wandering Tattler; in comparisons best not to rely on any one feature, but rather the whole range of observable plumage and behaviour characteristics. Some features can, alone, be unreliable. The name refers to the slightly paler grey rump, where, in fresh plumage, feathers are pale fringed, but not obviously paler, tending to wear to a grey similar to the upper parts. The extent to which folded primary or flight feathers of the wings project beyond the tail-tip differs between the two tattlers. The Wandering Tattler's wingtips extend further on average than do the Grey-tailed's. But there is some overlap of such individual measurements; wingtip wear and moult of outer primary feathers can further reduce the reliability of this feature. Many other features such as the length of brow (supercilium) and the length of nasal groove are difficult to see. The Grey-tailed Tattler darts about mudflats, sandbars and beaches, bobbing and teetering between dashes. **Voice:** flight call is distinctive—fluid, musical, but slightly mournful, drawn out 'too-weet', initially falling in pitch, then rising sharply in the final 'eet'. Also as a rapid sequence, slightly sharper, rising in pitch and accelerating, 'weit-weit-weet-weet-weetweetweet'. **Hab.:** coastal; forages in inter-tidal pools, shallows, soft surfaces of mudflats and sand beaches as well as rock ledges, reefs. Often perches on branches, posts or jetties; roosts in groups, same or mixed species. **Status:** common summer migrant to coastal N Aust.; uncommon in S.

Heavily patterned
Bill not longer than head
Indistinct pale brow line
Darker through eye
Rump lightly barred
White
Plain
Tapered, almost straight
Scalloped
Feet not projecting
Legs quite short, grey-green
Rufous-chestnut
Narrow wing-bar
Breeding ♂
Wings slightly longer than tail
Pale brow
Dull white
Nonbr.
Buff tint
Juv.
A few arrive in spring still displaying this chestnut plumage. Some acquire fresh breeding colours before flying N in autumn.
Like Red Knot; longer bill and bulkier body make head look small.
Overall streaked
Bill is longer than head; tapers from heavy base, very slightly downcurved in outer half.
White, lightly streaked
Feet not projecting
Rufous shows on scapulars.
Grey-brown with dark streaks, pale margins
Dark streaked; little if any pale brow line
Dark
Wings fold longer than tail-tip.
Dark
Sparse black arrowhead markings
Breast heavily spotted and scaled black; centre may be totally black.
Pale
Wing-bar thin, weak
Streaked brownish grey
Breeding
Dark grey-brown to olive-grey
Grey arrowheads
Comparisons are between this species and the following, very similar Grey-tailed Tattler.
Heavier bill
Slightly darker on crown and back
Upper tail coverts are of a grey uniform with rest of upper parts.
Darker grey
Pale brow, not extending much behind eye
Underwing dark grey
Wings usually extend well beyond tail-tip.
Throat white
Breast darker grey
Slightly darker, plain grey upper parts
Usually has more grey on the flanks than does the Grey-tailed Tattler.
Barred
Often a small unbarred area
Always heavily barred from neck to belly or further back
White, belly to under tail
Nonbr.
Breeding
Nonbr.
Paler rump when fresh, but narrow pale fringes soon wear to grey similar to back
Nonbr.
Primaries usually level with, or only slightly longer than tail-tip
Pale row extends behind eye.
Underwing dark grey
Grey-tailed slightly larger, heavier build.
Plain light to mid-grey
Plain grey
Juvenile: spotted edges
Pale grey
White
Narrow barring
Always white back to tail
Juv.
Nonbr.
Breeding
Yellow tint at all ages

Sanderling *Calidris alba* 19–20 cm (C,K)
This common, small, pale wader of sandy beaches runs behind receding waves, pecking and chasing, darting on twinkling legs up the beach as each wave breaks. Flies with flickering wing action, emphasised by wide, bold, white bar across dark wings. Congregates in small or large flocks, sometimes hundreds on favoured beaches, often with other waders. **Voice:** in flight, a sharp, quick, liquid 'tlick-tlick'; also a very soft twittering within flocks. **Similar:** Red-necked and Little Stints (both smaller). **Hab.:** open sandy beaches washed by ocean swells. **Status:** regular summer migrant to most of the coast; abundant on some beaches.

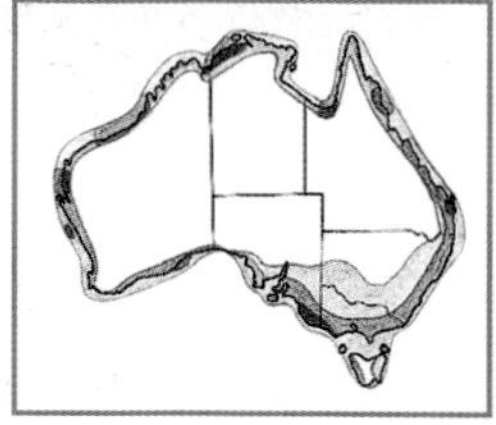

Red-necked Stint *Calidris ruficollis* 13–16 cm (C,L,M,N,T)
An extremely small wader, highly sociable in flocks and intermixed with other waders; darts about the mud or sand stopping to peck and probe. Flocks frequently burst into flight, swift on long wings, white undersurfaces flashing against sea or sky. **Voice:** a fast, extremely high, disyllabic 'chirit' or 'chrit', often so abrupt, a fraction of a second, that it sounds more like a single 'chit' or 'prip'. **Hab.:** diverse—tidal and inland on mudflats, salt marshes, beaches, saltfields, temporary floodwaters. **Status:** common migrant in huge numbers—up to 100 000 birds on most favoured sites, and scattered widely elsewhere.

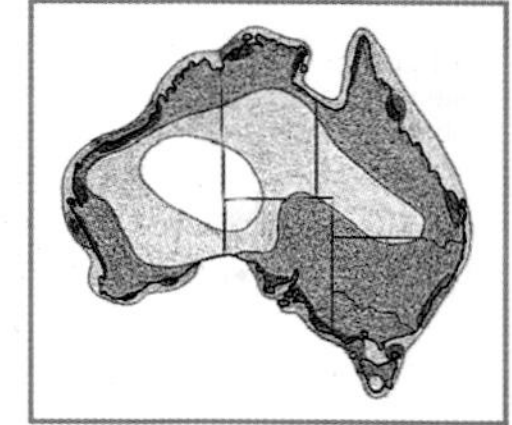

Little Stint *Calidris minuta* 12–14 cm
A tiny wader, rarely sighted; always as individuals in flocks of other small waders. Foraging and general behaviour is similar to the Red-necked Stint; scurries about the sand or mud, pecking and probing; may stand more upright when alarmed. Flight action like Red-necked Stint. **Voice:** extremely high, quick 'chit' or 'stit', more abrupt, higher than the 'chrit' of the Red-necked; also a quieter, high 'tsee-tse'. **Similar:** Red-necked Stint, Western Sandpiper. **Hab.:** mudflats, salt marshes, beaches, saltfields, most places where Red-necked Stints are found. **Status:** rare; few sightings scattered across coastal S Aust. and NT.

Long-toed Stint *Calidris subminuta* 13–15 cm (C)
Appears to be less gregarious than other stints; seen in pairs, singly or in flocks at favoured sites. Alert, secretive; crouches low, hunched; pecks crake-like around vegetation. At times stands tall, neck extended. Perches on logs, low branches close mud or water. If flushed, may rise high or zigzag low, calling. **Voice:** a trilled 'chirrip', so quick that the sound is more like 'trrp', 'prrp' or 'chrrp'; slightly lower, less metallic than other stints. **Hab.:** prefers shallow, fresh water and brackish swamps, lakes with muddy edges; often among low vegetation rather than on open mudflats. **Status:** regular visitor but scarce; commonest in WA.

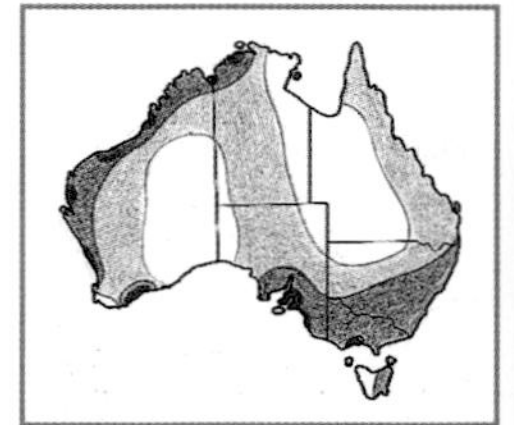

Temminck's Stint *Calidris temminckii* 13–15 cm
One of just two Aust. stints with yellowish legs. Tail is unique among stints—has pure white outer feathers, slightly longer than folded wingtips, and longer than the trailing feet in flight. White of tail is conspicuous on take off and landing, and with sudden changes of direction. **Juv.:** upper parts lightly scaled with brown edges and dark subterminal lines. **Voice:** a distinctive, high-pitched, thin, cricket-like trilling, 'tirr-r' and 'trrr-it'; as flight call, 'tir-ir-ir-irir'. Also has a display song, not likely to be given in Aust. **Similar:** Long-toed Stint. **Hab.:** freshwater wetlands—muddy, partly vegetated margins of swamps and lakes, flooded irrigated crops. **Status:** breeds N Europe, Asia; winters S to Borneo; a possible vagrant to Aust.

Western Sandpiper *Calidris mauri* 15–17 cm
Often in company with other waders, but may be solitary. Not greatly wary or timid; more approachable than most other waders. Feeds with jabbing and probing actions; may wade quite deep when feeding. Females tend to have a proportionately longer, more drooped bill tip, and average slightly larger. **Juv.:** clear white brow contrasting with dark eye stripe; has strong chestnut fringes to mantle and upper scapulars. **Voice:** a thin, sharp 'jeet' or 'cheit'. **Similar:** Dunlin is larger; nonbreeding Red-necked Stint has straighter and shorter bill, grey tone each side of breast. **Hab.:** mudflats and shallow wetlands. **Status:** possible vagrant; unconfirmed sight records on NSW coast; if this species occurs in Aust., it is very rare.

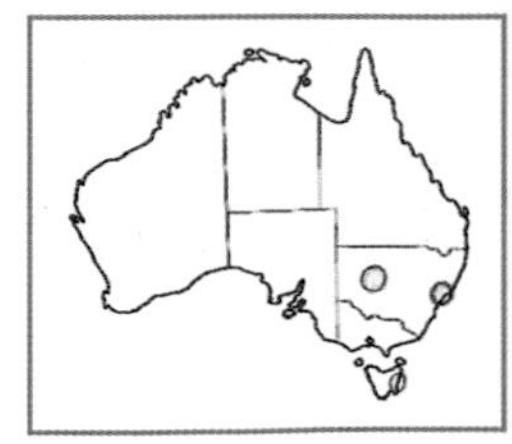

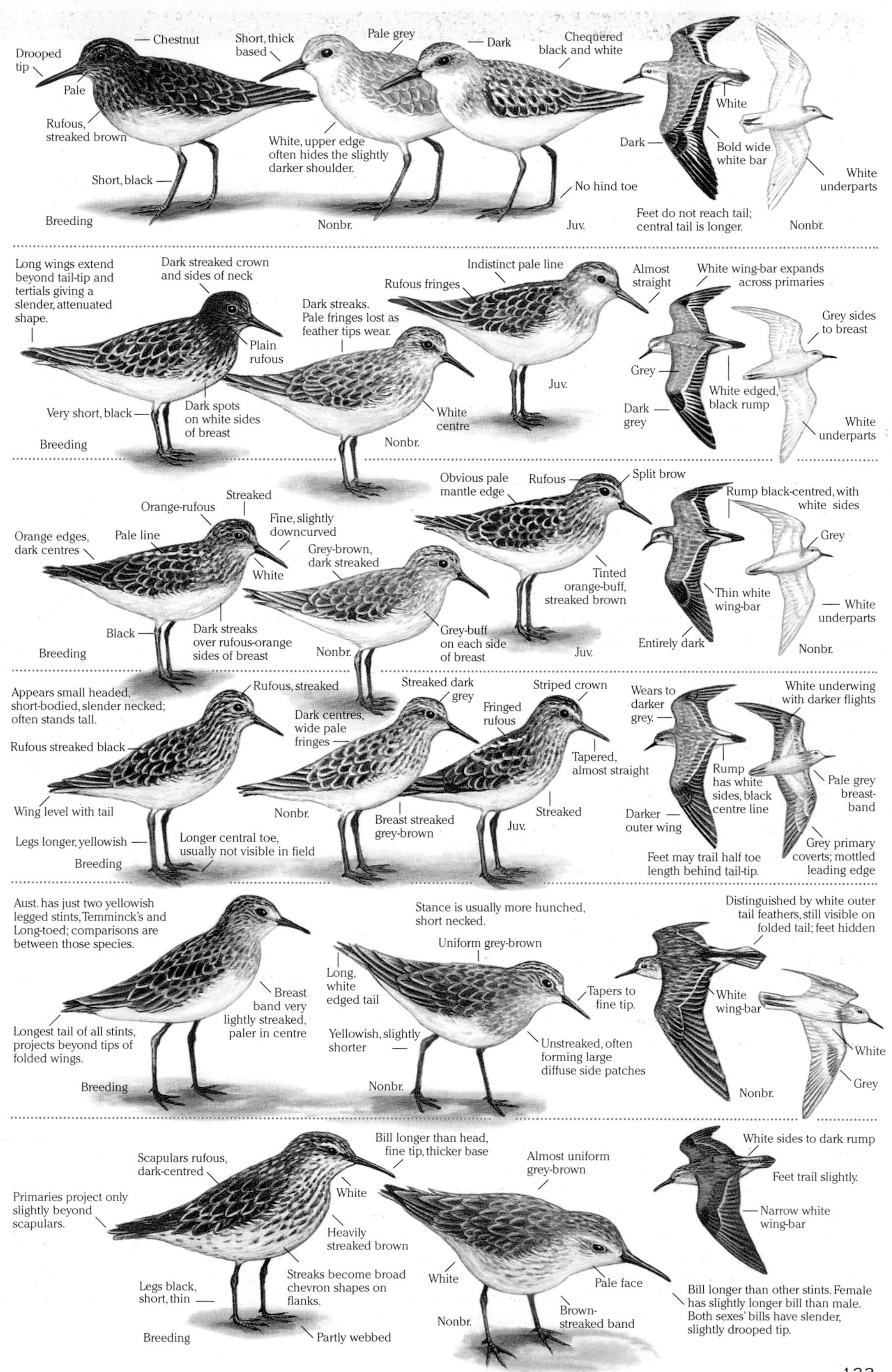
Drooped tip
Chestnut
Pale
Rufous, streaked brown
Short, black
Breeding
Short, thick based
Pale grey
White, upper edge often hides the slightly darker shoulder.
Nonbr.
Dark
Chequered black and white
No hind toe
Juv.
White
Dark
Bold wide white bar
White underparts
Feet do not reach tail; central tail is longer.
Nonbr.
Long wings extend beyond tail-tip and tertials giving a slender, attenuated shape.
Dark streaked crown and sides of neck
Plain rufous
Very short, black
Dark spots on white sides of breast
Breeding
Dark streaks. Pale fringes lost as feather tips wear.
White centre
Nonbr.
Indistinct pale line
Rufous fringes
Juv.
Almost straight
White wing-bar expands across primaries
Grey sides to breast
Grey
Dark grey
White edged, black rump
White underparts
Orange-rufous
Streaked
Orange edges, dark centres
Pale line
Fine, slightly downcurved
White
Black
Dark streaks over rufous-orange sides of breast
Breeding
Grey-brown, dark streaked
Grey-buff on each side of breast
Nonbr.
Obvious pale mantle edge
Rufous
Split brow
Tinted orange-buff, streaked brown
Juv.
Rump black-centred, with white sides
Grey
Thin white wing-bar
Entirely dark
White underparts
Nonbr.
Appears small headed, short-bodied, slender necked; often stands tall.
Rufous, streaked
Rufous streaked black
Wing level with tail
Legs longer, yellowish
Breeding
Longer central toe, usually not visible in field
Dark centres, wide pale fringes
Nonbr.
Streaked dark grey
Breast streaked grey-brown
Fringed rufous
Striped crown
Tapered, almost straight
Streaked
Juv.
Wears to darker grey.
Rump has white sides, black centre line
Darker outer wing
Feet may trail half toe length behind tail-tip.
White underwing with darker flights
Pale grey breast-band
Grey primary coverts; mottled leading edge
Aust. has just two yellowish legged stints, Temminck's and Long-toed; comparisons are between those species.
Breast band very lightly streaked, paler in centre
Longest tail of all stints, projects beyond tips of folded wings.
Breeding
Stance is usually more hunched, short necked.
Uniform grey-brown
Long, white edged tail
Tapers to fine tip.
Yellowish, slightly shorter
Unstreaked, often forming large diffuse side patches
Nonbr.
Distinguished by white outer tail feathers, still visible on folded tail; feet hidden
White wing-bar
White
Grey
Nonbr.
Scapulars rufous, dark-centred
Bill longer than head, fine tip, thicker base
White
Primaries project only slightly beyond scapulars.
Heavily streaked brown
Streaks become broad chevron shapes on flanks.
Legs black, short, thin
Breeding
Partly webbed
Almost uniform grey-brown
White
Pale face
Brown-streaked band
Nonbr.
White sides to dark rump
Feet trail slightly.
Narrow white wing-bar
Bill longer than other stints. Female has slightly longer bill than male. Both sexes' bills have slender, slightly drooped tip.

Dunlin *Calidris alpina* 16–22 cm

Small, dumpy, hunched. When feeding, moves slowly, head down, jabbing rapidly, wading; upright posture in alarm. **Variation:** many races; *sakhalina* and *pacifica* most likely in Aust. Considerable variation in length of legs and bills. **Voice:** flight call a harsh, reedy or buzzing 'chzee', chzeep', 'trzeep' and 'treep', higher than call of Curlew Sandpiper. While feeding or roosting in flocks, birds make faint, soft titterings. **Similar:** Broad-billed Sandpiper, but it has a distinctive bill; Curlew Sandpiper, but white rump in flight; Western Sandpiper, but much smaller. **Hab.:** sheltered coasts around estuaries, lagoons, mudflats, sandbars; inland on edges of lakes, floodwaters. **Status:** rare vagrant; most sightings in Aust. are uncertain.

Broad-billed Sandpiper *Limicola falcinellus* 16–18 cm

Uncommon small sandpiper; superficially like Curlew Sandpiper, but smaller with distinctive heavy bill: greater than the length of head; thick at base; initially straight and tapering, then downturned and flattened, but remains quite broad towards the tip. From the side this downturn is obvious; the width of bill is much less apparent. Generally stint-like behaviour; more deliberate and persistent as its long bill probes vertically in mud and shallow water. Seems less timid than most waders; tends to crouch when disturbed. Alone or in small loose flocks, often accompanies Red-necked Stints or Curlew Sandpipers. **Voice:** flight call a high, buzzing 'chzeeip' and an abrupt 'tzit'. **Similar (nonbr.):** Curlew Sandpiper, Dunlin—both noticeably larger. **Hab.:** sheltered coastal estuaries, lagoons with soft inter-tidal mudflats; muddy coastal creeks, swamps, sewage ponds; occasionally reefs. **Status:** migrant; generally uncommon, more frequent on N coast, rare inland.

White-rumped Sandpiper *Calidris fuscicollis* 15–17 cm

This small wader has to travel further than most others to reach Aust.—it breeds only in N Canada and Alaska. The species is known for its wanderings, which extend to Europe, South Africa and South America. The little bird's long wings, when folded, extend beyond the tail-tip and give the species a slender, tapering shape. **Voice:** the flight call is a high, thin squeak, more like that of a bat, but with a slight vibrant quality, 'tzreit', also described as 'jeeet' or 'eeet'. **Similar:** Baird's Sandpiper, distinguished in flight by its dark centred, grey tail and thin white wing-bar. **Hab.:** inland rather than marine wetlands—margins of swamps, lagoons, lakes, muddy pools; only occasionally tidal mudflats. **Status:** a rare vagrant to Aust.; only a few accepted records among sightings, mostly from SW and SE Aust.

Baird's Sandpiper *Calidris bairdii* 14–16 cm

Similar to White-rumped, but even more slender with long, tapering rear profile; usually a horizontal stance. Tends to be alone or in small groups. When alarmed, either crouches or stands tall; feeds briskly, darting about, probing mud, wading and sometimes dunking head to reach deeper. In flight, the long wings give slower wing action than stints. **Voice:** flight call rather low pitched with touch of harshness, a rolling trill, 'prreet' or 'kyrrrp'; also a sharp 'tsiek'. **Hab.:** coastal, margins of freshwater and brackish wetlands; often feeds on outer edge in drier vegetation as well as damp areas and mudflats; rarely in dense swamp vegetation. **Status:** rare vagrant in Aust.; a few sightings scattered around coast in most states and NT; lone birds in with other small waders.

Cox's Sandpiper *Calidris paramelanotos* 19 cm

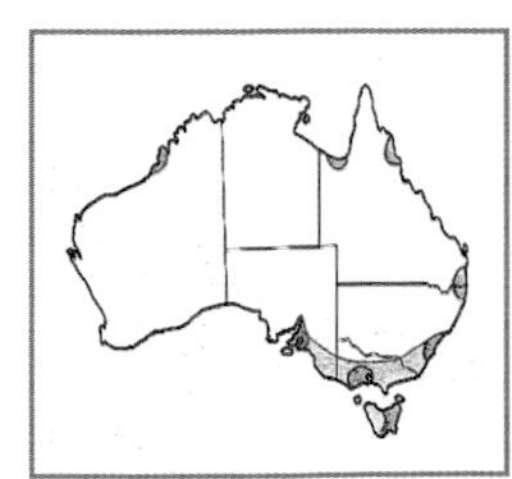

Described as a full species in 1982 on the basis of a collection of specimens from Aust., but not known from any likely breeding areas of N hemisphere; intermingles with other waders. In flight, thin white wing-bar; tail and central rump darker with white edges. DNA testing now shows it to be a natural hybrid, probably with Curlew Sandpiper and Pectoral Sandpiper as parent species. **Br.:** not known except from birds partially moulted into a pre-breeding plumage. **Voice:** like Pectoral, but higher, shriller 'trilt'. **Similar:** Curlew Sandpiper, Pectoral Sandpiper, Sharp-tailed Sandpiper. **Hab.:** fresh and marine wetlands, including tidal mudflats, salt farms, sewage ponds. **Status:** rare summer migrant; now regarded a hybrid.

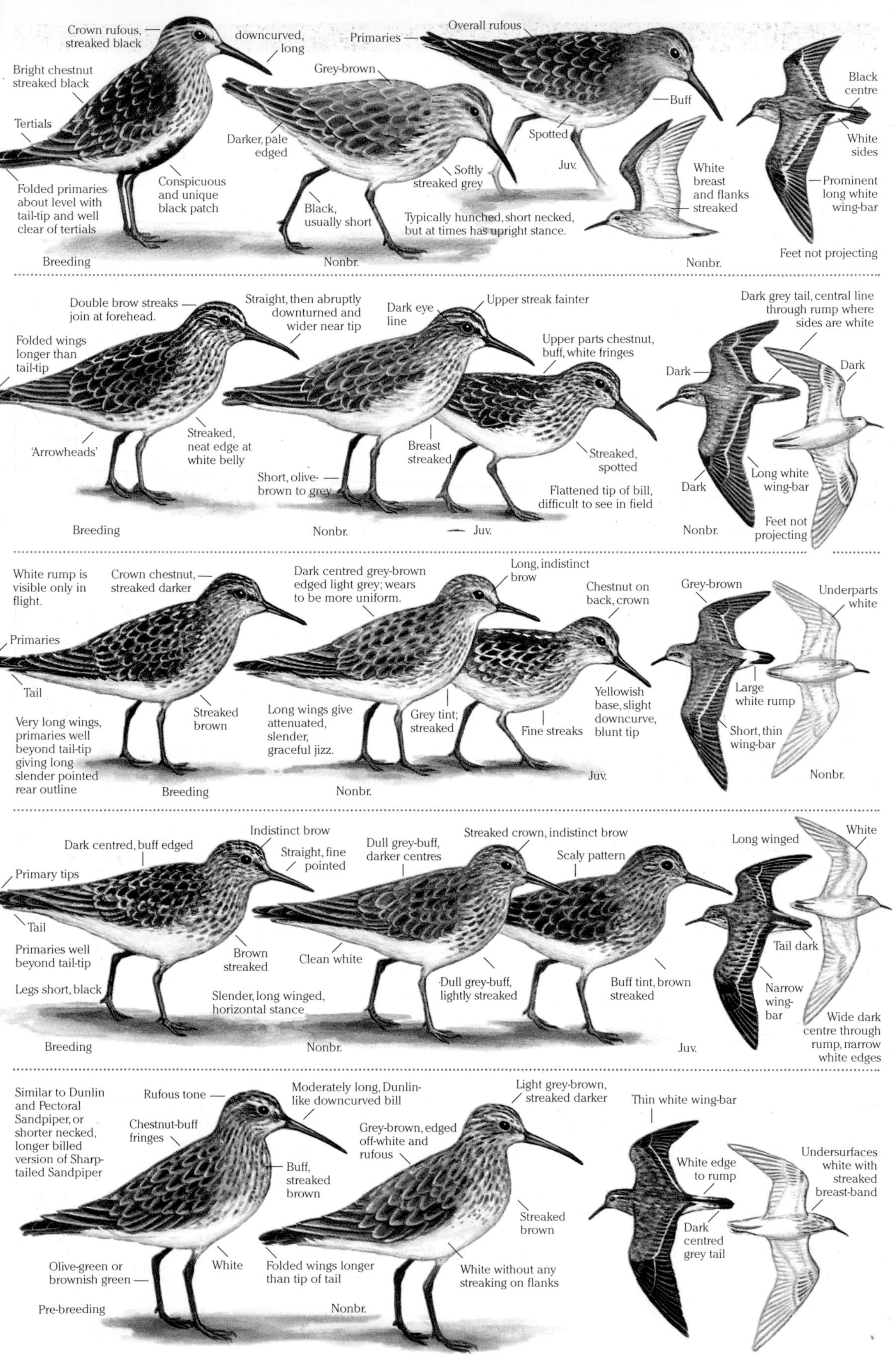

Crown rufous, streaked black
downcurved, long
Bright chestnut streaked black
Tertials
Folded primaries about level with tail-tip and well clear of tertials
Conspicuous and unique black patch
Breeding
Grey-brown
Darker, pale edged
Black, usually short
Softly streaked grey
Nonbr.
Primaries
Overall rufous
Buff
Spotted
Juv.
Typically hunched, short necked, but at times has upright stance.
White breast and flanks streaked
Nonbr.
Black centre
White sides
Prominent long white wing-bar
Feet not projecting
Double brow streaks join at forehead.
Folded wings longer than tail-tip
'Arrowheads'
Streaked, neat edge at white belly
Breeding
Straight, then abruptly downturned and wider near tip
Dark eye line
Upper streak fainter
Short, olive-brown to grey
Breast streaked
Nonbr.
Upper parts chestnut, buff, white fringes
Streaked, spotted
Flattened tip of bill, difficult to see in field
Juv.
Dark grey tail, central line through rump where sides are white
Dark
Dark
Dark
Long white wing-bar
Feet not projecting
Nonbr.
White rump is visible only in flight.
Crown chestnut, streaked darker
Primaries
Tail
Streaked brown
Very long wings, primaries well beyond tail-tip giving long slender pointed rear outline
Breeding
Dark centred grey-brown edged light grey; wears to be more uniform.
Long, indistinct brow
Long wings give attenuated, slender, graceful jizz.
Grey tint; streaked
Nonbr.
Chestnut on back, crown
Yellowish base, slight downcurve, blunt tip
Fine streaks
Juv.
Grey-brown
Underparts white
Large white rump
Short, thin wing-bar
Nonbr.
Dark centred, buff edged
Indistinct brow
Straight, fine pointed
Primary tips
Tail
Brown streaked
Primaries well beyond tail-tip
Legs short, black
Breeding
Dull grey-buff, darker centres
Clean white
Slender, long winged, horizontal stance
Dull grey-buff, lightly streaked
Nonbr.
Streaked crown, indistinct brow
Scaly pattern
Buff tint, brown streaked
Juv.
Long winged
White
Tail dark
Narrow wing-bar
Wide dark centre through rump, narrow white edges
Similar to Dunlin and Pectoral Sandpiper, or shorter necked, longer billed version of Sharp-tailed Sandpiper
Rufous tone
Chestnut-buff fringes
Moderately long, Dunlin-like downcurved bill
Buff, streaked brown
White
Olive-green or brownish green
Pre-breeding
Light grey-brown, streaked darker
Grey-brown, edged off-white and rufous
Streaked brown
Folded wings longer than tip of tail
White without any streaking on flanks
Nonbr.
Thin white wing-bar
White edge to rump
Dark centred grey tail
Undersurfaces white with streaked breast-band

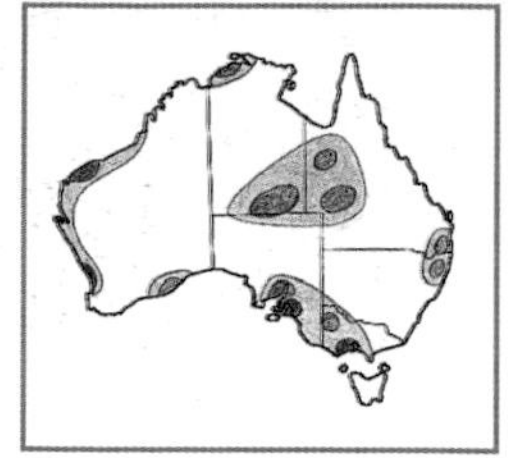

Red-necked Phalarope *Phalaropus lobatus* 18–19 cm (M)
Gregarious; congregates on seas rich in plankton; occasionally blows ashore or shelters from gales on near-coastal wetlands. Often swimming, floats very high, lightly, on water or wades; infrequently comes onto land. Feeds by pecking at small creatures on water's surface. Male smaller than female and paler in breeding plumage. **Voice:** in flight, soft 'chek' and 'chwik'. **Similar:** Grey Phalarope in nonbreeding plumage, but it has a larger, heavier build, eye-patch is square ended rather than slender and downturned, and upper parts usually plainer, paler grey. **Hab.:** coastal and near-coastal on waters and muddy margins of brackish or saline lagoons. Most spend nonbreeding period at sea. **Status:** rare but probably regular vagrant.

Wilson's Phalarope *Steganopus tricolor* 22–24 cm
Often with other waders—very active, forages in shallows, swinging bill through water. Runs erratically; stalks insects. **Male:** darker, brown and dull orange replace the female's lighter, brighter grey and rufous. **Voice:** usually silent, occasionally a soft, nasal, grunted 'aangh'. **Similar:** in nonbreeding plumage, Red-necked Phalarope, Marsh Sandpiper. **Hab.:** less aquatic than other phalaropes—marine situations in sheltered tidal estuaries, lagoons; more often around swamps, shallows of lakes. Often forages standing in shallows and around waterline rather than predominantly by swimming. **Status:** extremely rare, uncertain vagrant.

Grey Phalarope *Phalaropus fulicaria* 20–22 cm
A large, plump phalarope, appears to have long slender neck, small head, bold white wing-bar, extensive white underwing. This species and the Red-necked both spend much time at sea while escaping the N winter. Strays or shelters from time to time on lagoons or other protected coastal waters. Swims quite high on water, perhaps with tail held higher than the Red-necked, and, like it, often spins on the water surface, pecking at surrounding floating items. In flight, erratic over short distances, or has a distinctive, side to side 'jinking'. **Voice:** usually quiet, but, as flight and contact call, a sharp 'whit' or 'twit'. In alarm, a more musical 'zhwhit'. **Similar:** Red-necked Phalarope. **Hab.:** Aust. records from coastal sites, typically brackish and saline lagoons surrounded with salt marsh or wet grassland and with exposed mud margins. **Status:** rare vagrant or migrant; few confirmed records; most sightings from SE and SW Aust.

Comb-crested Jacana *Metopidius (Irediparra) gallinacea* 20–27 cm
Highly adapted to a life entirely on tropical lagoons. Not only feeds but nests on the floating lily leaves, building a flimsy nest of fine green water-plant stems, largely supported by the lilies. The nests are easily damaged, even if accidentally, by large wildlife, or flood as lagoons rise after heavy rain. Jacanas may shift the eggs to another nest, carried tucked under the chin, or pushed, floating, across the water. Chicks, already with long toes at hatching, are quickly able to avoid predators that congregate at wetlands such as raptors and water pythons. At any alarm, probably a parent's call, chicks tumble into the water, hiding beneath floating leaves, just the tip of a bill occasionally showing. Small chicks are often carried tucked up under the wings. Adult Jacanas can dive, but usually just fly a short distance to another part of the lagoon. **Voice:** not noisy in frequency or loudness, but calls quite often and regularly with diverse sounds that have been grouped into two broad types. The first group is a regularly repeated series of notes typically given in confrontation with other Jacanas; the second is of twittering, chittering, piping notes, usually given while in flight or standing tall in upright, alert pose. Sequences may blend from one to the other. Also has a sharp, nasal alarm call. **Similar:** quite unique in breeding plumage, only likely similar species is Pheasant-tailed Jacana in nonbreeding plumage. **Hab.:** lakes, swamps and dams where there are waterlilies or other extensive cover of floating vegetation. The Jacana forages nimbly across such waters, walking or running on the leaves, its weight widely spread by the long toes, turning, probing and searching for small water creatures and insects. **Status:** locally abundant; sedentary on large, rich, undamaged wetlands in remote or protected far N coastal Aust. Increasingly uncommon to scarce towards the SE extremity of its range and where lagoon habitat is damaged by stock or other activity. **Br.:** 367

Pheasant-tailed Jacana *Hydrophasianus chirurgus* 30–40 cm—illustration far right
Br.: very long dark tail, spectacular white, black tipped wings, gold on back of neck. **Nonbr.:** considerably larger than Jacana without any wattle rising above crown; obvious white wings above and beneath; long gold and black line from brow to shoulder. **Status:** nearest usual occurrence Java, Timor, NG; an extremely rare vagrant to Aust.; only a few confirmed sight records or possible sightings in the Kimberley and Pilbara regions of WA.

In breeding plumage females are more brightly coloured than males.
As female, but paler
Dark with white brow spot
Needle-like, slightly downturned at tip
Chestnut, extends high to ear coverts
♂
Breeding ♀
Buff scapulars, mantle edge fade with wear.
Eye-patch dark, downturned
Grey, dark centres, white fringes
Nonbr., both sexes
Juv.
Pale buff, mantle, scapulars
Juveniles have yellow legs, later olive-grey to dark grey.
White sides
White wing-bar
Lobed feet are usually hidden.
Nonbr.
Grey from crown to upper back
Broad black eye stripe continues down neck.
Rufous
Legs dark when breeding
Pale grey
Dark line
Legs yellow, nonbreeding; extend behind in flight.
Faint white wing-bar
White fringes on fresh plumage
Dark eye streak on white face
Fine straight bill
Moult from juvenile to nonbreeding: has some darker feathers intermixed on coverts, tertials.
Face oval, white
Black cap
Extensive chestnut-red
Yellow
Breeding ♀
Nonbr.
Upper parts scaled with white margins in fresh plumage, then more uniform pale grey
Smudgy grey
Streaked
Colours dull
May be white.
Olive to yellow
Breeding ♂
Male may be almost as bright, or very dull.
Fringed buff
Pinkish buff tint
Juv.
White
Wide white bar along wings
In flight, legs do not extend beyond tip of tail.
Tail well short of folded wings
Dark olive-brown, lighter iridescent highlights
Extremely long toes enable these birds to walk on lily leaves. The slender claws are enormously elongated, especially that of the hind toe.
Deep pink, rather glossy surfaced fleshy wattle
Golden-buff
Olive-green, olive-grey to dark grey
Adult
Rufous
Small, pale wattle and comb
White, no buff
Juv.
Dark outer wing and trailing edge
Comb-crested Jacana: in flight, at distance, the olive-brown wing coverts and back, brighter under sun, form an obvious lighter panel surrounded by blackish wings, mantle and tail.
In flight, long legs and toes trail conspicuously, far behind.
Underwing is entirely black.
Underside of body is white with a bold black breast-band.
Pheasant-tailed Jacana, nonbreeding. In breeding season, tail equals head and body length.
Tail and folded wing are of equal length
White wings
Eyebrow, side of neck, golden
Legs and toes proportionately shorter than Comb-crested
Unique with white, black-tipped wings

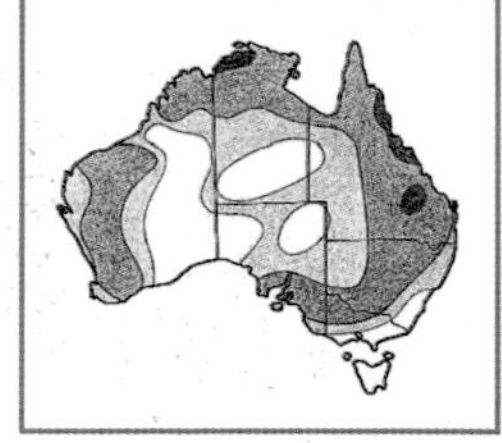

Bush Stone-curlew *Burhinus grallarius* 55–60 cm

The wailing call is one of the most characteristic sounds of the bush at night. By day the secretive, camouflaged birds, though large, are hard to see; they freeze flat on the ground or skulk away. **Voice:** calls are eerie, a drawn out, mournful—'wee-ier, wee-ieer, whee-ieeer, whee-ieer-loo'. Each call rises, strengthening, faster, building to a climax, then trails away. Calls are often, probably typically, given by several birds. One starts, another joins in; calls overlap, building the intensity of sound. Sometimes other groups will join in or answer from afar. At night, these spine-tingling sounds will carry far across lonely bush and paddocks. **Similar:** generally unmistakeable; nearest perhaps is the Beach Stone-curlew. **Hab.:** open woodlands, lightly timbered country, mallee and mulga—anywhere with groundcover of small sparse shrubs, grass or litter of twigs. Avoids dense forest, closed canopy habitats. **Status:** common across N and NE; uncommon to rare in SE and SW. **Br.:** 368

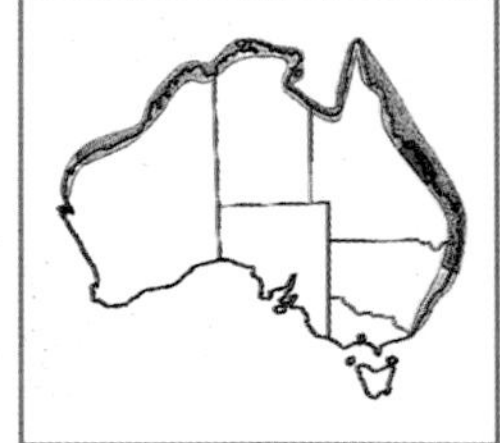

Beach Stone-curlew *Esacus neglectus* 54–56 cm

Less strictly nocturnal than the Bush Stone-curlew; forages with slow, deliberate, heron-like actions. Usually wary; flies ahead showing distinctive wing pattern if disturbed. **Voice:** as alarm call, a quick 'chwip' repeated at intervals of 1 sec, perhaps also as 'chwip-chwip'. Territorial calls are given at night, like those of the Bush Stone-curlew, but harsher and at higher pitch, a 'weer-liew' repeated about eight times, each higher and faster. **Hab.:** confined to the marine tidal zone—mudflats, mangroves, sandy, stony and rocky shores, sheltered or exposed to ocean breakers. **Status:** considered vulnerable due to coastal disturbance, more secure in remote parts of N coast. **Br.:** 368

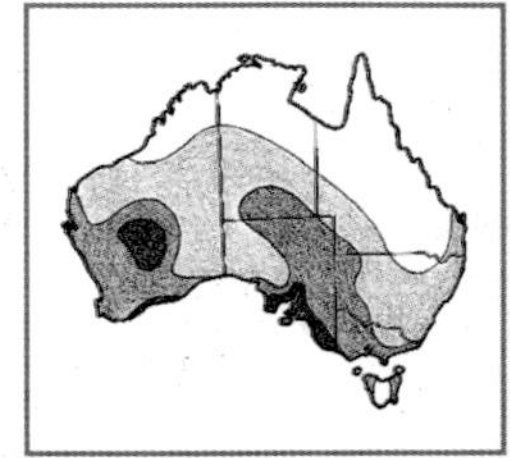

Banded Stilt *Cladorhynchus leucocephalus* 35–43 cm

Unique endemic wader, most likely to be seen on ocean beaches after dispersal from breeding. **Voice:** from flock, musical resonant yapping; varied, deeper, more mellow than Black-winged Stilt. Individual notes range from abrupt 'ohk' to slightly longer 'chowk' and 'chowk-ok'. In a large assembly, intermingled calls of hundreds of birds become a pleasant babble of 'ohk-chowk-ok-chok-ohk', a mixture of high and low, loud and soft. **Hab.:** salt lakes of coast and inland: in large flocks on temporarily flooded saltpan lakes; also on marine beaches of estuaries and inter-tidal flats. **Status:** common on parts of range; congregate to breed on interior salt lakes in good conditions; disperse widely. **Br.:** 369

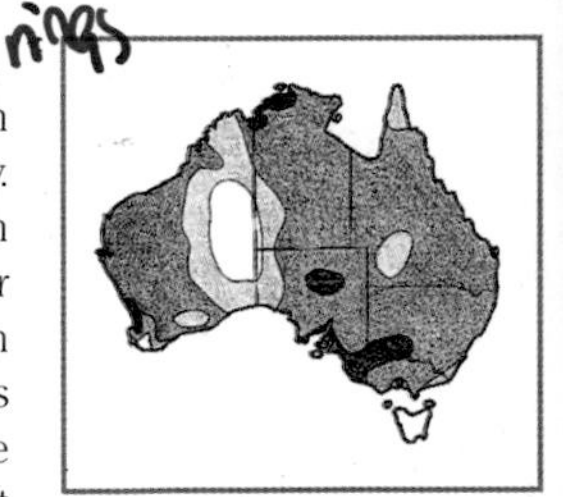

Black-winged Stilt *Himantopus himantopus* 33–37 cm (C, L, M, N, T)

Slender, elegant, gregarious wader with incredibly long legs. Struts gracefully through shallows; bobs head and calls when alarmed. Flies swiftly; long legs trail; wings beat strongly. Aust. race *leucocephalus* of widespread species. **Voice:** a regular high, nasal, yapping with slight variations of pitch and strength, like a toy trumpet, 'ap, ak, ap-ap, ak, ap, ap, ak-ap'; higher pitch and more nasal than the mellow notes of the Banded Stilt; slight variations from individuals of group. **Similar:** Banded Stilt, especially nonbreeding and juvenile, but it has plump bellied, round backed, shorter legged shape, and its head and back are always pure white. **Hab.:** diverse shallow freshwater wetlands—interior claypans, flooded paddocks, salt lakes. **Status:** widespread; mostly common; absent in sandy deserts; rare in Tas.; dispersive or nomadic. **Br.:** 369

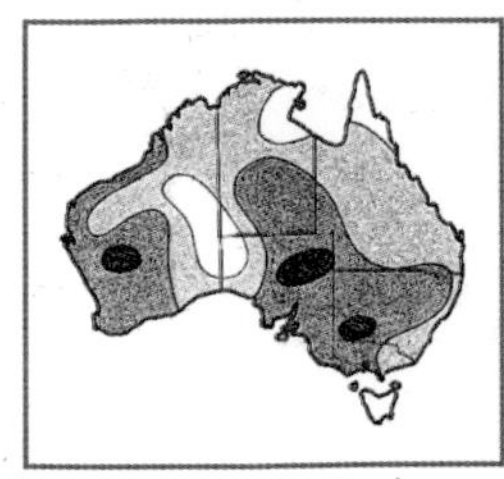

Red-necked Avocet *Recurvirostra novaehollandiae* 40–48 cm

Unique to Aust.—graceful, colourful, large wader with long, slender, strongly upcurved bill. Feeds by wading through shallow water; sweeps the submerged bill from side to side, slightly open, just above and parallel to the muddy or sandy bottom. Probably relies on touch to identify prey. Almost always in large flocks or groups, often with stilts. Feeds and roosts on water a safe distance from shore. **Voice:** often silent; usually calls when disturbed—a yapping similar to that of the Black-winged Stilt, but less abrupt, higher and more metallic or nasal, 'aik, airk airk, airk, aik, aik' with slight variations of pitch, strength and length of notes. Probably more variety from a group of birds than from one. **Hab.:** both salt and freshwater wetlands. Large numbers tend to congregate on large, shallow salt lakes, particularly as salinity increases through evaporation; then feeds on brine shrimp. Also use shallow freshwater swamps and lakes, temporary floodwaters, claypans, dams, saltworks. **Status:** most of continent; scarce near N and NE coasts, vagrant Tas. **Br.:** 369

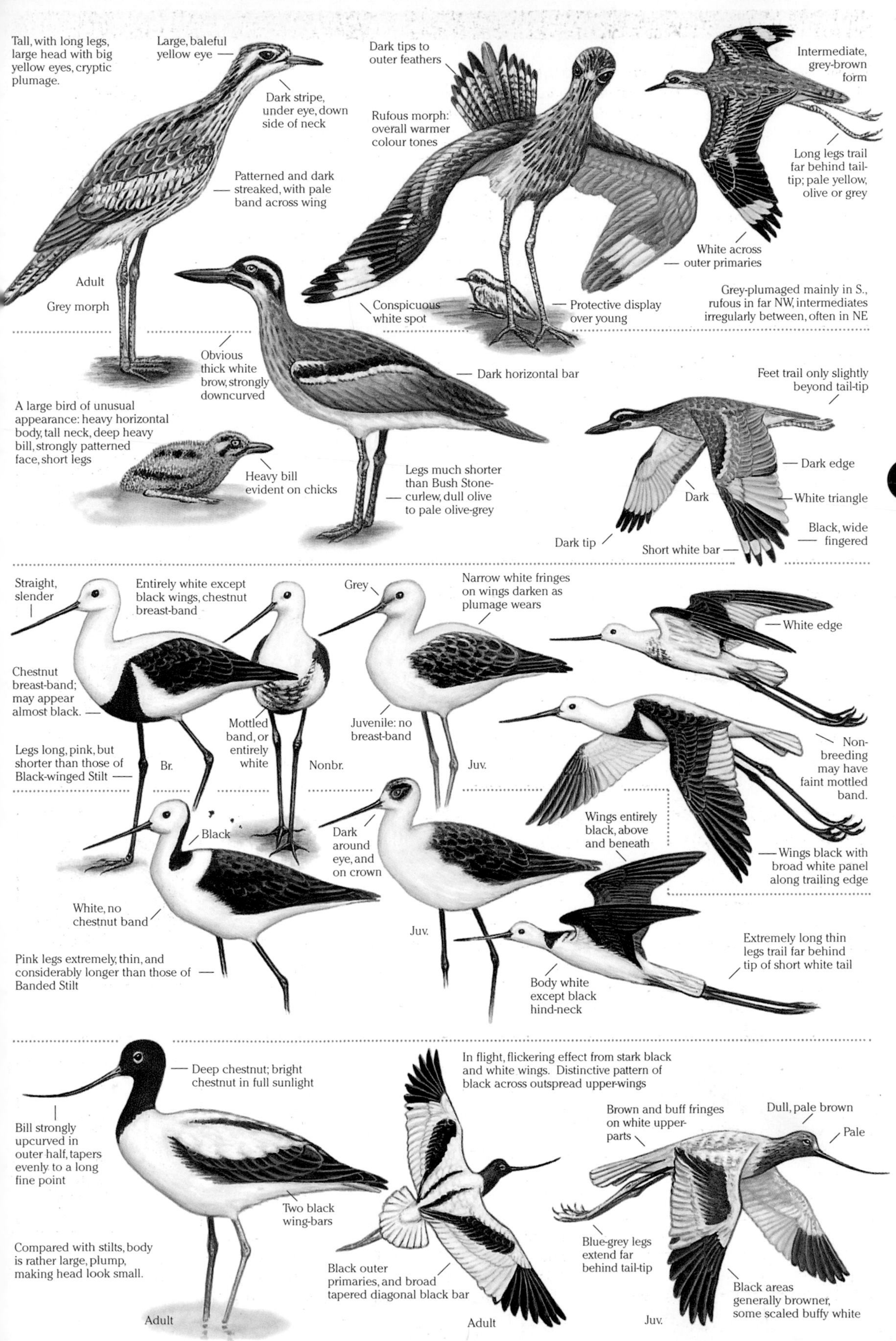
Tall, with long legs, large head with big yellow eyes, cryptic plumage.
Large, baleful yellow eye
Dark stripe, under eye, down side of neck
Patterned and dark streaked, with pale band across wing
Adult
Grey morph
Dark tips to outer feathers
Rufous morph: overall warmer colour tones
Conspicuous white spot
Protective display over young
Intermediate, grey-brown form
Long legs trail far behind tail-tip; pale yellow, olive or grey
White across outer primaries
Grey-plumaged mainly in S., rufous in far NW, intermediates irregularly between, often in NE
Obvious thick white brow, strongly downcurved
Dark horizontal bar
A large bird of unusual appearance: heavy horizontal body, tall neck, deep heavy bill, strongly patterned face, short legs
Heavy bill evident on chicks
Legs much shorter than Bush Stone-curlew, dull olive to pale olive-grey
Feet trail only slightly beyond tail-tip
Dark edge
Dark
White triangle
Black, wide fingered
Dark tip
Short white bar
Straight, slender
Entirely white except black wings, chestnut breast-band
Chestnut breast-band; may appear almost black.
Mottled band, or entirely white
Legs long, pink, but shorter than those of Black-winged Stilt
Br.
Nonbr.
Grey
Narrow white fringes on wings darken as plumage wears
Juvenile: no breast-band
Juv.
White edge
Non-breeding may have faint mottled band.
Wings black with broad white panel along trailing edge
Black
White, no chestnut band
Pink legs extremely, thin, and considerably longer than those of Banded Stilt
Dark around eye, and on crown
Juv.
Wings entirely black, above and beneath
Body white except black hind-neck
Extremely long thin legs trail far behind tip of short white tail
Bill strongly upcurved in outer half, tapers evenly to a long fine point
Deep chestnut; bright chestnut in full sunlight
Two black wing-bars
Compared with stilts, body is rather large, plump, making head look small.
Adult
In flight, flickering effect from stark black and white wings. Distinctive pattern of black across outspread upper-wings
Black outer primaries, and broad tapered diagonal black bar
Adult
Brown and buff fringes on white upper-parts
Dull, pale brown
Pale
Blue-grey legs extend far behind tail-tip
Black areas generally browner, some scaled buffy white
Juv.

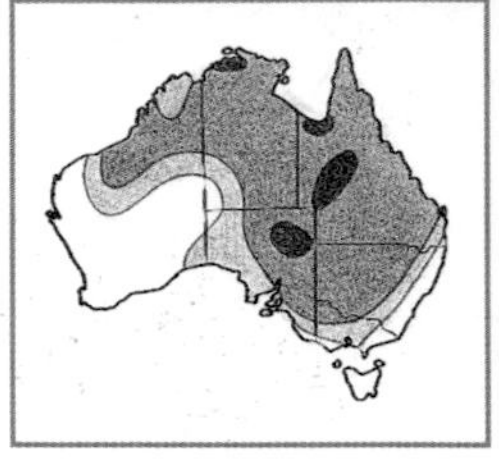

Australian Pratincole *Stiltia isabella* 22–24 cm (C,L,T)
Graceful, slender pratincole; stands tall and erect on long legs. Hawks for insects swallow-like on backswept, finely pointed wings. Long legged, often dashes about on ground after insects and other small prey; bobs head. Sandy rufous plumage matches bare ground typical of habitats. **Nonbr.:** almost identical to breeding adult; hard to separate reliably in the field; individual variations and effects of plumage wear may be as much as seasonal changes. Lores and flanks probably paler, plumage generally dull; may have a few additional dark throat spots. **Voice:** undulating, pleasant, cheery whistle as flight call, a varied 'whit-WEIT', 'WEET-whit-whitwhit' and 'wirr-ie-WEIT'; as alarm calls, similar but sharp, urgent, loud. Flying birds' calls attract attention before they are seen; calls in thunderstorms. **Hab.:** treeless and sparsely wooded plains, grasslands—areas of sparse vegetation like claypans, gibberstone; never far from water of lagoon or livestock dam. **Status:** widespread; common in spots through N and NE; occurrence varies, unpredictable. **Br.:** 369

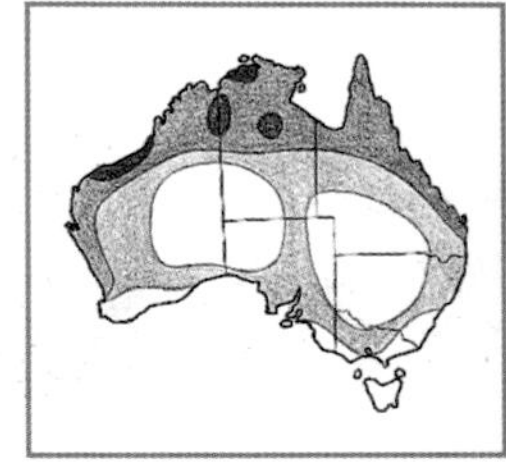

Oriental Pratincole *Glareola maldivarum* 23–24 cm (C,L)
Elegant in flight; has long, pointed wings, deeply forked tails; hawks swallow-like for insects, especially around dawn and dusk, and frequently in large flocks. During heat of day, rests on areas of flat, open land not far from water—paddocks, airfields, mudflats, roads. Tends to squat low, well concealed by their colour in rough ground or small depressions, even the impressions from stock hooves. **Voice:** compared with Australian Pratincole, a rather rough, but not unpleasant, 'chak-a-chak'; started by one or a few birds, taken up by flock. With sharper 'cha-rik' or 'krik' sounds intermixed, becomes an almost continuous confusion of 'char-rakichikiakak...' noise. **Similar:** Australian Pratincole shares long winged, sleek shape, short, thick, downcurved bill, but has much longer legs, finer wingtips, warmer rufous-buff tones; is more terrestrial in habits. **Hab.:** often hawk over wetlands where clouds of insects accumulate; at times around bushfires; on open plains, open areas around tidal flats, beaches, wetlands. **Status:** migrant from India, SE Asia; arrives N Aust. with tropical wet season, Nov.–Jan.; disperse, nomadic; flocks range widely seeking most productive sites. Leave Aust. to return to SE Asia between Feb. and April.

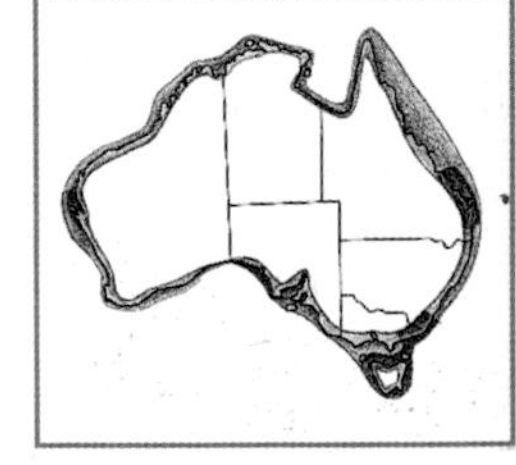

Pied Oystercatcher *Haematopus longirostris* 42–50 cm
With confirmation of presence on central E coast of Aust. (and on Norfolk Is.) of New Zealand's South Island Pied Oystercatcher, Aust. now has three species of oystercatchers. These are all large, heavily built waders with bills suited to prising molluscs from rocks and probing into sand. As the name indicates, oysters and other shellfish are a major part of the diet, the chisel-like bill prising open, or breaking into the shell. Various other invertebrates are also taken, such as the large marine worms. These oystercatchers have many characteristics in common: all stand and forage with hunched posture, bill downwards, or, when alert, adopt a tall, upright stance. Flight is fast and direct with rapid, shallow wingbeats; keeps along the beach or out to nearby reefs and islets. Usually solitary, in pairs or in small family groups, spaced out along the beach, each lot defending its territory. **Voice:** high, not harsh or sharp, but a ringing, mellow, resonant 'quip-quip-quip-quip-quip', very even in pitch. In flight, 'quip-a-peep, quipapeep, quipapeep'. In alarm, loud, sharp, high pitched 'kervee-curvee-curvee', increasing in pitch and rapidity as the nest is approached; lures intruders away with distraction display, pretending s wing is broken. **Hab.:** beaches and mudflats of inlets, bays, ocean beaches and offshore islets; less often rocky coasts, headlands. **Status:** moderately common; vulnerable to disturbance, in parts of SE becoming uncommon. **Br.:** 369
South Island Pied Oystercatcher *H. finschi.* Shorter legs, more white in wings. NZ vagrant (Island Birds).

Sooty Oystercatcher *Haematopus fuliginosus* 40–52 cm (C)
Large, stocky oystercatcher. Scarlet bill, eyes and legs are conspicuous against the black plumage. Several birds group together in flocks of up to 20 or even 50. When disturbed, walk away rapidly; fly if close pressed—low to waves, straight ahead or circling back 50 m or so. **Variation:** race *ophthalmicus*, NW, N and NE coasts; larger bill and wider, perhaps slightly more orange, eye-ring. **Voice:** high, clear, piping calls; sharper, more piercing and quicker than the Pied; usually 'kier-kier-kier-kier-kier...'. May develop into a double noted 'kwi-keer, kwikeer, kwikeer'; again sharper than similar calls of the Pied Oystercatcher. Often given in flight. Also gives loud whistling calls before taking flight, and piercing calls if an intruder approaches nest. **Hab.:** marine, usually rocky shoreline, high rocky islets, boulders below cliffs, wave-cut platforms and reefs. Also visits and forages on sandy beaches and coves between rocky headlands. **Status:** sedentary; generally uncommon—scarce on disturbed coastlines, common on parts of N coasts. **Br.:** 369

Upright stance, streamlined, rakish jizz
Red
Sandy rufous
Streaked
Buff edged
Pale
Pale
Dark chestnut flanks meet in centre.
Very long, crossed
Long legs
Breeding
Juv.
Adult
Tail short, square, white with black subterminal band
Breeding
Extremely long, slender, backswept wings
Black underwings
Feet extend behind tail.
Chestnut flanks
White
Dark chestnut flanks can appear almost black at distance, meet to varying extent across belly.
Nonbreeding similar, usually reduced flank patches, dull colour
Long, fine points to wings
Black line around white edged buff throat
Juv.: mottled and streaked
Dark primaries
Tertials
Tail
Breeding
Folded wing is much longer than tail, which reaches midway between black primary tips and olive-brown tertials.
Streaked
Breast and upper parts dull olive-brown
Short legs
Nonbr.
White, dark tip
Forked
Streaks
Dull olive-brown
Wingtips not as finely pointed as Australian Pratincole
Underwing coverts chestnut
Nonbr.
Scale change
The common, widespread black and white oystercatcher
Alert, upright posture
Eye-ring, iris and bill are all brilliant scarlet.
Brown eye
White point or hook between breast and wing
Dark brown with orange-buff tips and notches, perhaps lost with wear
Orange-red with dark tip
White of wing stipe barely visible; much larger on South Island Oystercatcher.
White
Adult
Juv.
Legs pink, stout. South Island Pied has legs 1/3 shorter, giving an obviously low, dumpy, short legged appearance; bill is longer.
Downy young, striped
Under-surface of flight feathers dark, white confined to coverts. South Island Pied has underwing all white except thin dark edge line.
Black
On inner wing, white bar narrows to a fine point.
Sharp edge to black breast
Black upper parts, white wing-bars tapering inwards to a point. South Island Pied has bar wide towards body, and extending to the trailing edge of inner secondaries.
Black tail, white rump
Scarlet eye-ring and iris
Stocky build; plumage entirely sooty black
Nominate race, small eye-ring and bill
Scarlet-pink bill
Shorter than legs of Pied Oystercatcher; deep crimson-pink
Adult
Race *ophthalmicus*, N Aust., has larger, more conspicuous eye-ring and longer bill. May be only one race, as variation, at least on W coast, appears gradual, or clinal.
Adult
In flight, entirely black except bill, eyes and legs, which are shorter than tail-tip
Blunt, rounded wingtips
Nonbreeding: colour of legs, eyes, bill is slightly dulled.
Juvenile is as adult but has dull brownish black body with buffy tips on back, bill grey with faint orange on base of lower bill, orbital ring and legs grey.

Pacific Golden Plover *Pluvialis fulva* 24–26 cm (C K,L, N)

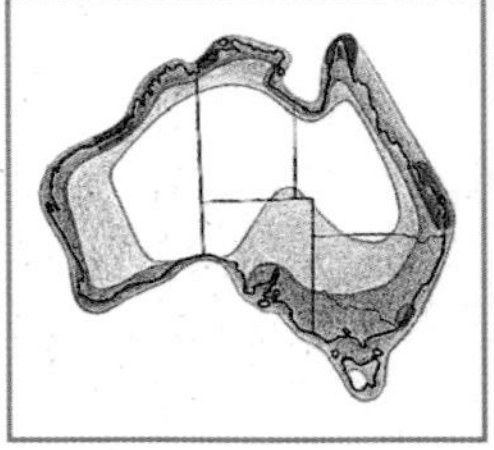

Slender, upright, shy, wary; feeds in typical stop–start plover manner. **Variation:** previously a race of American Golden Plover, *P. dominica*. **Voice:** usually silent on ground; calls in alarm, at take off, in flight and when landing. In flight a high, clear, musical 'tiu-ee' or 'tchu-eet', the second note higher. Similar calls, more subdued, when feeding or resting birds are put to flight and on coming in to land. In alarm, calls are more urgent, sharper. **Similar:** other golden plovers (more easily separated in breeding plumage); *fulva*'s upper parts tend to be lighter and brighter. Differences in folded wingtip length, primary projection beyond tertials, foot projection, colours underwing and other details are useful. **Hab.:** mainly coastal; usually in small parties or quite large flocks on estuaries, inter-tidal mudflats, beaches, reefs, salt marshes, offshore islands; only rarely far inland. **Status:** common migrant from Arctic; arrives Aug.–Sep.; disperses to suitable habitat around coast; leaves Apr.–May when some will have already moulted into breeding plumage. A few remain through southern winter.

American Golden Plover *Pluvialis dominica* 24–28 cm

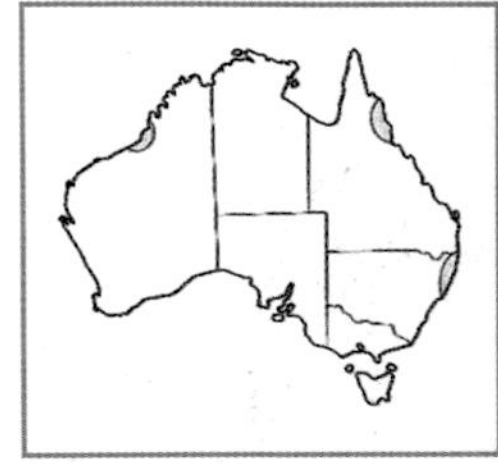

Very difficult to distinguish between this species and Pacific Golden Plover. The American is, on average, larger with shorter legs, heavier bill. Tips of folded wings protrude further from under gold-spangled tertials—at least 4 and up to 6 primary tips show. The Pacific's wingtip protrudes less; no more than two or three primary feathers show. The American's longer wingtips also extend noticeably beyond the tip of the tail, while the Pacific's tend to be level with the tail-tip. In flight, both have grey underwings, which helps distinguish them from the Eurasian and Grey Plovers. The Pacific usually has toes showing beyond the tail-tip, while the American's shorter legs fail to reach the tip. **Nonbr.:** slightly more grey than other golden plovers in both nonbreeding and juvenile plumages. **Voice:** similar to that of Pacific Golden Plover—loud, clear, quick, double noted call lifting to a higher second note: 'keer-eet, keer-eet' or 'tiu-eet'. Call changes when birds are put to flight and become much sharper, more urgent in alarm. **Hab.:** coastal inter-tidal mudflats; out of the breeding season may use varied habitats. In usual wintering range makes more use of inland habitats than does the Pacific Golden Plover, a tendency yet to be observed for Aust. **Status:** breeds Arctic N America; migrates to S America; some vagrant to Europe, Africa and possibly Aust. Unconfirmed records of sightings on NSW coast and NW of WA.

Eurasian Golden Plover *Pluvialis apricaria* 26–28 cm

Larger than the other golden plovers; slightly smaller than the Grey Plover. Within usual range tends to be very alert, shy, easily put to flight, when it may circle warily for some time before landing. **Fem.:** in breeding plumage, so like male that difference is only likely to be noticed in comparisons within breeding pairs, which show that females tend to have face and underparts browner than true black and often have scattered specks of white across dark underparts. **Variation:** two races—northern (*P. a. altifrons*) from N Scandinavia into Arctic Russia, and southern (*P. a. apricaria*) from S Scandinavia to Ireland. If this species is eventually added to the Aust. bird list, it would seem that the northern race would be the more likely—that population is more highly migratory. **Voice:** use of calls may differ in Aust. if species does occur; a clear, liquid or yodelled flight call, 'tloo-ee'; clear and loud enough to be heard before the bird is seen. **Hab.:** in N hemisphere uses permanent grassland, farmlands, inter-tidal mudflats, salt marsh, lake shores. **Status:** common in usual range; no accepted records for Aust.; several possible sightings.

Grey Plover *Pluvialis squatarola* 28–30 cm (C, K, L, M)

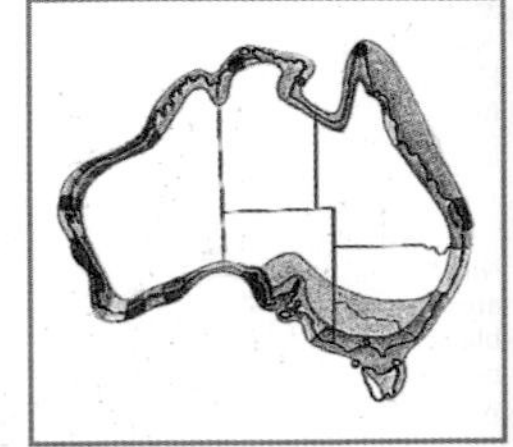

Medium sized; large head and heavy bill; long legs; often stands hunched, body horizontal and head pulled down, but also stands tall, alert. Solitary or in small groups, at times with golden plovers. Shy, tends to stay far out on shallows or flats. Powerful, swift, graceful flight, loose flocks in irregular lines. Typical plover behaviour—stop–start running, pecking, often out into shallows; probes into sand or mud, occasionally pasture, for molluscs, marine worms, crustaceans. **Fem.:** breeding, as male except black underparts with some flecks of white. **Voice:** flocks often silent; may flush without calling; distinctive flight call—loud, triple noted whistle, plaintive, undulating 'whee-oo-eeir, whie-oo-eeir', rather like beginning of Bush Stone-Curlew's call. Can sound agitated, giving similar call more rapidly. **Similar:** Pacific Golden Plover, but it lacks diagnostic black armpits. **Hab.:** coastal, usually marine shores of estuaries or lagoons on broad, open mudflats, sandy bars or beaches, rock platforms and reef flats of rocky coasts; inland but near the coast on margins of salt lakes and swamps. **Status:** common migrant from Arctic; arrive N Aust. in Sep.; maximum numbers in southern Aust. in Dec.; leave April. Occurs almost all parts of coast; common W and S coasts, abundant in Kimberley, scarce in Tas.

Bold white line from brow through flanks to under tail
Black
Golden spangled
Often stands upright.
Primaries
Tertials
Moult
White flecked
Juv.
Wingtips just beyond tail-tip
Breeding
Retains some gold spangling.
Tertials
2 or 3 dark primaries visible
Golden buff flecked brown
Buff brow
Longer legs
Adult: nonbr.
Darker streaks
Lacks wing-bar.
Feet just exceed tail-tip.
Outer wing, secondaries, dark brown
Underwing grey-brown
Nonbr.
Toes do not project.
Nonbr. grey-brown, little gold
White
No wing-bar
Grey-brown
Proportionately longer wings than other golden plovers
White bands down neck; then, as bold wide patches, almost meet in centre of breast.
Dull grey-brown, little gold spangling
Nonbr.
Grey-buff
Brightly gold and white spangled
Long primaries
Dark, pale flecked
White band from brow to sides of breast
Long projection of black primaries, 4–6 showing beyond golden tertials
White
Relatively short legs
Juvenile has more yellow-buff than nonbreeding; breast is streaked, flanks are barred brown.
Flanks black
Under-tail is solid black.
Wings extend well beyond barred tail-tip.
Breeding
Broad white band extends from forehead to barred flanks and under-tail coverts.
Almost as breeding
Juvenile: finely barred flanks and belly
Indistinct brow line
Only species with entirely white underwings. Pacific and American, above, are grey-brown, while Grey Plover, below, has black axillaries ('armpits')
Spangled with gold spots and notches
Nonbr.: only slightly less colourful than breeding
White
Nonbr.
Gold spangled
White bands almost meet in centre, leaving a narrow vertical black centre line.
Barred
Mottled gold, brown
Black
White
Upper, all plumages, thin white wing-bar, spreading on primaries as white streaks
Nonbr.
♂ Breeding
Wings and tail almost equal
Larger than golden plovers; plump, tends to stand in hunched posture.
Heavy bill
Black white pattern
Darker, bolder, white and yellow markings
Grey-brown, spotted and notched dull off-white
Black
Nonbr.
Neck and breast: black centre between wide white bands
Juv.
White, barred
Feet and tail equal
White
Wings slightly longer than tail
Plain grey-brown, spotted and edged white
In flight shows broad black band from neck to belly, then spreading wide across flanks onto axillaries, all combining as a black cross.
Dark legs; only *Puvialis* with hind toe
Nonbr.
Wing-bar spreads as white streaks on primaries.
Larger and bulkier than golden plovers; similar patterns, but without gold
Nonbr.
♂ Breeding

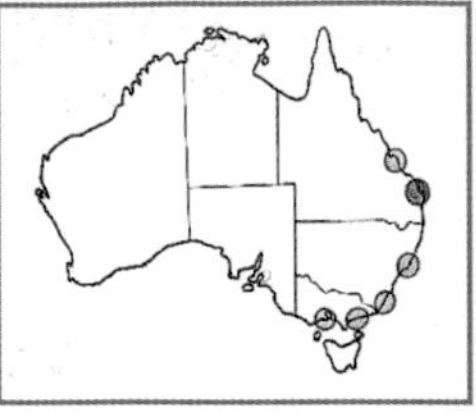

Ringed Plover *Charadrius hiaticula* 18–20 cm
Small robust wader; only solitary sightings in Aust. Has a wary, quick, scurrying action when foraging; pauses and stands tall. Glides to land; holds wings up on alighting. **Juv.:** breast-band reduced or broken. **Voice:** distinctive contact call, mellow or fluty, two notes, second higher, 'too-lee, too-lee'; also a low 'too-weep'; sharper versions in alarm. **Hab.:** in Aust., tidal mudflats and sand of estuaries, salt marsh, saltpans, sewage ponds. **Status:** rare migrant or accidental visitor; sightings quite widespread—Qld, NSW, Vic., SA, NT.

Little Ringed Plover *Charadrius dubius* 14–17 cm (C)
Small, slender plover; busy, erratic, wary. In flight, backswept pointed wings. **Juv.:** obscure head markings, breast-band usually broken in centre. **Voice:** a descending, clear 'peeeooo', usually as a flight call—contrasts to the rising call of the Ringed Plover; in alarm, an abrupt, rapidly repeated 'pip-pip!, pip!, pip-pip'. **Similar:** Ringed, which lacks eye-ring, has wing-bar, orange bill, different call. **Hab.:** muddy edges or mudflats of tidal or freshwater wetlands including estuaries, lakes, lagoons, dams, ponds. **Status:** rare but regular visitor.

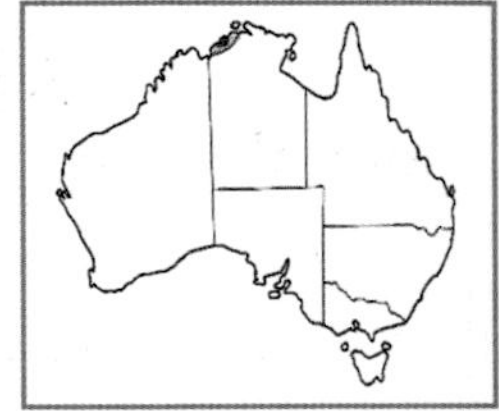

Kentish Plover *Charadrius alexandrinus* 15–17 cm
A small, widespread plover; slightly larger than Red-capped; behaviour typical of small plovers. **Variation:** about 6 subspecies, all zoogeographical regions except Aust. American races are known as 'Snowy Plover' and are very pale. **Voice:** flight call is a sharp but not loud 'twit' or 'wit' and rougher 'pirrr'. **Similar:** Red-capped Plover. **Hab.:** in usual range, usually marine beaches, tidal flats, rarely inland; in Aust., a record from coastal mangrove mudflats. **Status:** rare vagrant; one verified occurrence on NT coast, probably N Asian race *dealbatus*.

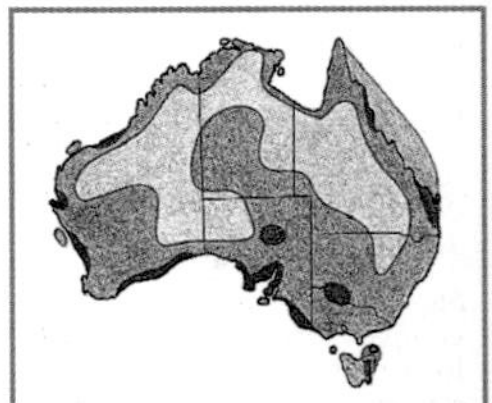

Red-capped Plover *Charadrius ruficapillus* 14–16 cm
Busy, gregarious—darts along water's edge alone or in flocks of hundreds: darts forward; stops abruptly; bobs head nervously. Flocks fly fast, erratically, displaying dark then light surfaces. **Voice:** as contact, faint 'wit, wit, wit' or trilled 'twitwit twitwit' or 'trrrrrit'; frantic chirring in distraction display; alarm call a plaintive 'twink'. **Hab.:** greatest numbers occur inland on salt lakes, salty edges of waterways, brackish pools, claypans. Coastal on sheltered estuaries, salt marsh lagoons. **Status:** common; sedentary or nomadic. **Br.:** 369

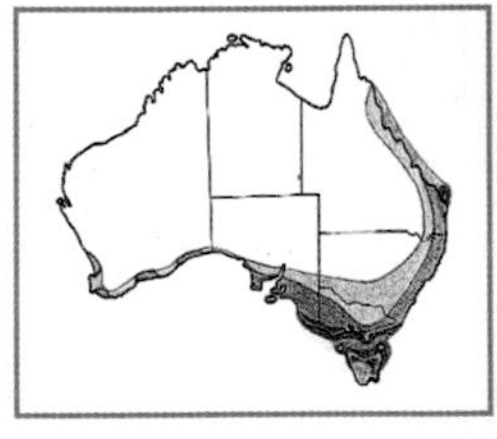

Double-banded Plover *Charadrius bicinctus* 18–21 cm (L, N)
Unique double breast-band, or upper band plus remnants of lower band as side tabs, or two remnant tabs or smudges each side. Usually in groups or flocks; behaviour typical of small plovers: abrupt stop–start, run–pause movements. Tends to stand more upright than other small plovers. **Voice:** a loud 'pit', usually high pitched, but varied—sometimes a double 'tink-tink'. Flocks in flight maintain contact with constant musical tinkling of many individual 'tink' and 'chip' calls. **Similar:** in breeding plumage is unique in region; in other plumages is like Mongolian and Large Sand Plovers. **Hab.:** tidal mudflats, beaches, exposed reefs, salt marshes, freshwater wetlands, inland salt lakes, short grass of golf courses, airfields. **Status:** widespread, common migrant to SE Aust., autumn to spring, with scattered records from as far as NE Qld and W coast of WA.

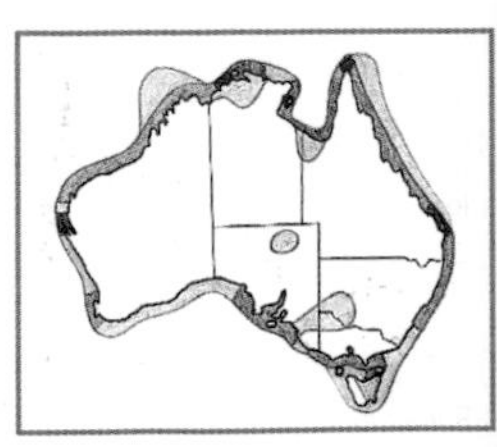

Lesser Sand Plover *Charadrius mongolus* 18–21 cm (C, L, N, T)
Small plover seen both in bright breeding and dull nonbreeding plumages while in Aust. Many moult to breeding colour before departing in autumn. **Fem.:** black of male replaced by brown. **Juv.:** buff-brown breast patches and buff fringing to tertials and scapulars. **Variation:** numerous races; most likely to reach Aust. are *mongolus* from interior of E Russia and *atrifrons* from Tibet and Himalayas. **Voice:** differs slightly from that of Greater Sand Plover; usual call a quick, hard 'chrik' and 'chrik-it' on taking flight. As flight call, soft musical 'tirrrit-tirrirt trritt'. **Similar:** Large Sand Plover, but it has longer, heavier bill, larger, rounded head and longer, paler legs that trail behind tail-tip; Double-banded Plover, but it has double breast-bands or side tabs. **Hab.:** inter-tidal sandflats and mudflats, beaches, estuary mudflats and sandbars; reef flats. **Status:** nonbreeding migrant Sep.–Apr.; abundant Qld, uncommon further SW and S; some remain through winter in N.

Short white brow behind eye
Black bands
Brown
Orange
Wide
Breeding
Very small, compact; black breast-band, black face, black across base of bill and forehead.
Orange legs
Longer
Short, blunt, less orange
Brown breast patches
Nonbr.
Underside of wings and body white, except breast-band
Black
White
White
Encircling white collar
Long dusky brow
Less orange
Paler
Outer wing almost black
White wing-bar spreads onto primaries.
Nonbr.
Long white line
Two white bands
Upperparts mid grey-brow; narrow white collar
Fine sharp bill
Thin black breast-band
Slender rear body
Long legs, pinkish to orange
Breeding
Faint brow
Pale eye-ring
Brown
Often horizontal body position
Nonbr.
Darker
Underwing coverts and axillaries white
Dark
Collar
Faint brow
White on sides of rump
Compared to Ringed Plover, uniformly dark, insignificant wing-bar
None of these small plovers have feet that show behind the tail-tip.
Nonbr.
Olive-brown with white hind-neck collar
Larger, darker
Breeding
Black or grey or olive-grey, long
Like Red-capped Plover, slightly larger; comparisons are between these species.
Complete collar
Longer, thicker
Nonbr.
Collar
Longer brow line
White sides to rump; visible on spread outer tail
Wing-bar spreads wide on primaries
Nonbr.
Shape can mislead; a bird hunched, head pulled in, feathers fluffed out, may look large, plump. When active, alert, plumage is tight, sleek; neck outstretched, standing upright, it will look tall, slender, long necked.
Bright chestnut, no rear collar
Conspicuous chestnut on crown, neck, shoulders
Short fine
Small lateral breast patches
Short rear end gives compact shape; looks long-legged; often stands tall-upright
Chestnut-brown
Small side patches
Nonbr.
Brownish
Juv.
Narrow wing-bar expands as white streaks on very dark primaries.
White
Pale upper parts
Black
Wing-bar widest on primaries
White coverts
The only plover in Aust. region with two breast-bands, or two tabs each side on nonbreeding. More upright stance than most small plovers.
Black mask
Black
Black
Chestnut
♂ Breeding
Breeds in NZ, winters in Aust. Most seen here are in dull grey-brown nonbreeding plumage.
Brown mask
Faint brow
Brown, often broken
Nonbreeding, both sexes, no black or chestnut
Lower tab is a brown smudge each side.
Nonbr.
Brown
Dull
♀ Breeding
Dark
White
Dark outer wing, narrow wing-bar. Underwing white
Some breeding females may be almost as strongly coloured as the males; others are like nonbreeding.
Bright chestnut
White, split by vertical line
Similar to Large Sand Plover, but smaller, rounded head, shorter legs and bill.
Dark
♂ Breeding
Race mongolus
Black
Bright chestnut
♂ Br.
Race atrifrons
Brown lateral breast tabs
Tabs may join centrally.
Nonbr.
Buff tint to brown of lateral tabs
Juv.
White
White
White sides of rump and tail show when tail is spread, landing. In flight, toes are about equal to tail-tip.
Underwing coverts white
Narrow wing-bar, spreads wider on dark inner primaries
♂ Br.

Greater Sand Plover *Charadrius leschenaultii* 22–25 cm (K, C, L, T)

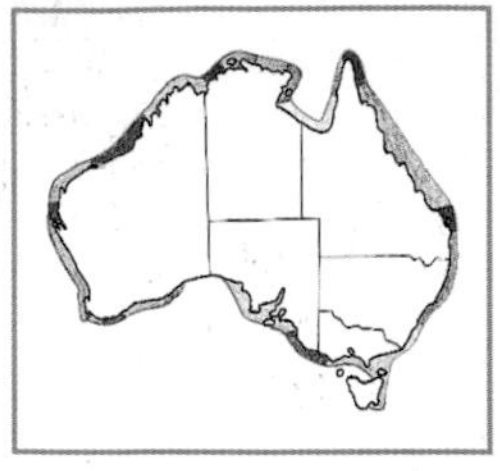

Relaxes in hunched, horizontal stance. Alert, is more upright, neck extended. Gregarious, often in mixed flocks. **Fem.:** in breeding like male, but black often replaced by dark brown. **Juv.:** buff tint to breast patches and buff edging to brown upper parts. **Voice:** rather quiet calling from flocks feeding and in flight—a short rippling 'drrit' or 'trreet'; extended as 'trrrri-trrrri-trrri-trrri', so rapid individual notes are not discernible, but at times becoming a rattling sound. Almost unique; with familiarity calls can assist identification. **Similar:** Lesser Sand Plover, Oriental Plover. **Hab.:** coastal, inter-tidal mudflats and sandbanks of sheltered bays and estuaries; sandy cays of coral reefs, reef platforms; less often coastal salt marsh, brackish and, very rarely, freshwater wetlands. **Status:** migrant Aug.–May; common in N; uncommon in S.

Caspian Plover *Charadrius asiaticus* 18–20 cm

Although different from the Oriental Plover in breeding plumage, these two are almost identical in other plumages and have at times been combined as one species. The Caspian breeds around the Caspian Sea in W Asia. Most migrate to S and E Africa where they winter inland on dry plains, not unlike the Pine Creek area of inland NT where Australia's sole specimen was shot in 1896. Behaviour is similar to the Oriental Plover. **Juv.:** similar to nonbreeding, but with buff tint to breast-band; upper surface feathers are dark centred with rufous or buff fringes. **Voice:** usual call a loud, sharp 'tyip' or 'tyik' and more rapidly as a rattling 'tip-tp-tp-tptptp'; also a loud, shrill, whistled 'kwheeeit'. **Similar:** Oriental Plover, which is larger with longer legs and neck, entirely dark underwing and which has very pale head in breeding plumage. Sand plovers, but they are more compact and rounded, and folded wings only just reach tail-tip. **Hab.:** differs substantially from similar plovers in using much drier inland habitats, grasslands, lake-edge mud, temporary floodwaters. Sole Aust. specimen Pine Creek, 1896; more recent sightings have been coastal NT and NE Qld. **Status:** very rare vagrant or accidental; several records only.

Oriental Plover *Charadrius veredus* 21–25 cm (C, K, L)

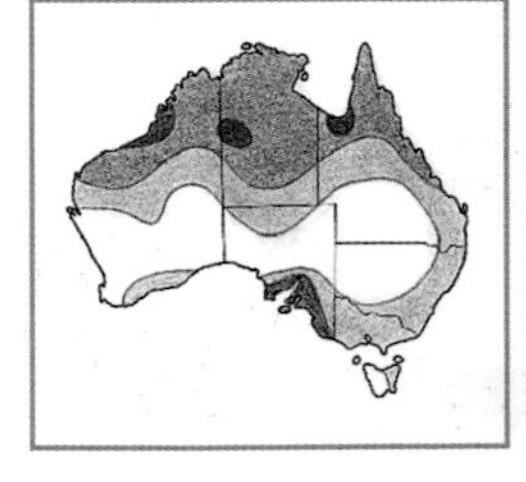

Long-legged plover of inland regions. On migration in N Aust. gathers in flocks on open, thinly vegetated, grassland plains; these flocks reach hundreds, even thousands, of birds. Often forage at night, roost by day with other waders on beaches or mudflats. May be seen in company with Inland Dotterels or pratincoles. Foraging is typical—stop–start, running, bobbing down to peck at prey. Tend to be rather wary, bobbing head when alarmed; runs rapidly. If put to flight, has powerful action; fast and high while twisting and turning erratically, the many birds of the flock seem always to change direction in unison. **Voice:** flight call is a soft, very quick 'tik' or 'tink' repeated irregularly, especially among birds of a flock; also has a sharp, rapid trill. **Similar:** male in breeding plumage is distinctive; especially the combination of chestnut breast-band with thin black lower edge. In nonbreeding and juvenile plumages, similar species are Caspian Plover, Large Sand Plover, Lesser Sand Plover. **Hab.:** usually inland in semi-arid regions on open grasslands, claypans or gibberstone plains; less often on the marine sites such as tidal mudflats typically used by other plovers. Also, occasionally, where dense vegetation of spinifex, heathland or similar has recently been burnt. **Status:** migrant from Mongolia and Russia; arrives N Aust. in large numbers, Sep.–Nov., then disperses southwards; returns northwards about March.

Inland Dotterel *Peltohyas (Charadrius) australis* 19–22cm

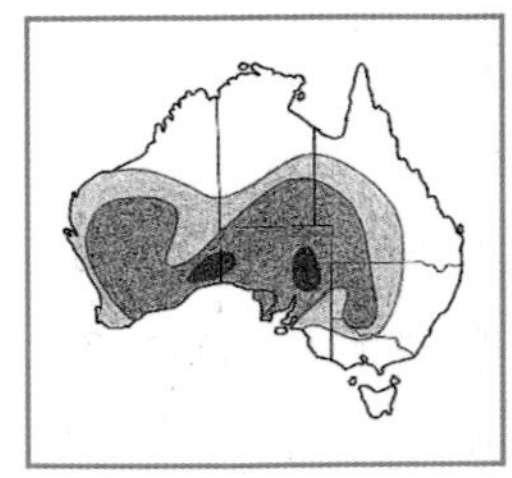

Unique wader now adapted to arid habitat. Gregarious, small to large flocks; mainly nocturnal; runs to escape—takes flight as a last resort, showing long wings, buff-brown patterning. **Nonbr.:** probable that all birds in pale plumage with indistinct dark markings are juveniles. Nonbreeding adults may show some loss of colour intensity, but the distinctive dark markings, the vertical eye stripe and the 'Y' shape on the breast will always be present. **Voice:** quiet outside breeding season unless disturbed. Then, as contact by birds separated from flock, a sharp, metallic 'kwik' or 'quoik' given as a single or double noted call, or as a longer sequence with notes at 0.5 sec intervals. Calls persist while flock is scattered, running; has been heard from flocks feeding at dusk. Maintains contact and keeps flock together; given loudly as alarm at intrusion or threat, or in anxiety by birds separated from main group. **Similar:** adult markings are unique; juveniles superficially similar in colour to Buff-breasted Sandpiper and pratincoles. **Hab.:** semi-arid and arid plains, open flat country with sparse vegetation—gibber plains, claypans, gravelly flats, open bare clay soils, all with expansive bare ground between the scattered tufts of vegetation. At times seen near water of ephemeral creeks on claypans, but at times far from any water. **Status:** endemic species, probably nomadic or locally migratory. **Br.:** 369

Chestnut, crown grey-brown
Some have fine black edge; this is absent from breeding female.
Like Mongolian Sand Plover but larger; appears big headed, long legged.
Large eye set in black mask.
Chestnut
Long
Often stands with head hunched down, making neck thick, increasing apparent head size, head then looking much too large for body.
Head can look too large for body; bill is large for head.
Large, heavy, black bill
Wingtips slightly longer than tail
Nonbr.
Lateral breast patches may join centrally.
Olive-grey, straw, or grey-brown
Feet trail behind tail.
White
Dark
Feet show behind tail.
White
Nonbr.
Grey-brown
Dark
White wing-bar spreads wide in inner primaries
Dark
Long white brow
Rarely sighted in Aust. Distinctive only in breeding plumage, otherwise is difficult to separate from more common Oriental Plover.
Clean white
Bright chestnut
Breeding ♂
Dark border
Nonbr.
Long, olive or grey
Female breeding: breast-band is mostly brown or has a few scattered chestnut feathers; head more like nonbreeding.
Head pattern is like breeding, but dull brown.
Buffy white brow, wide behind eye
Long, attenuated rear body
Mottled grey-brown
Folded wings extend well beyond tail-tip
Nonbr.
Brown rump and tail
Narrow wing-bar leads out to flash of white on inner primaries
Feet extend slightly beyond tail tip
Underwing coverts white
Very pale headed; brown of crown and through eye pale, soft edged
Chestnut with black edge
Legs long, varied, greenish, buff, dull orange, buffy pink
Breeding female has breast-band grey-brown; may have some chestnut intermixed.
Breeding
Long, wide, indistinct eyebrow line
Folded wings extend beyond tail-tip.
Mottled buff and pale brown, as complete or partial breast-band
Nonbr.
Underwing is entirely brown, against which the white belly stands out in contrast.
Dark rump with narrow white sides; dark tail with with fine white tip
Toes visible
Nonbr.
Only very fine wing-bar with no white streaks to bases of inner primaries.
Long, dark, pointed wings
A dark line through the eye is common, but this small plover is the one species with an obvious vertical, a line through each eye and meeting as a band across the crown
Crown and eye streak
'Y' shaped breast-band
Wingtips are about level with tail tip.
White
Chestnut flanks
Adult
In adult plumage, sexes alike, slight seasonal variation
Faint streak through eye
White
Long legs, upright alert stance
Faint beginning of of 'Y' breast pattern
Juv.
Underlying overall colour a sandy buff, heavily streaked dark brown. The resultant combination of colour and camouflage pattern, together with the disruptive face and breast markings, merges with the colours and textures of the gibber plain environment.
Twin encircling black bands
Deep buff
Darker
Toes protrude beyond dark tail-tip.

Black-fronted Dotterel *Charadrius (Elseyornis) melanops* 16–18 cm

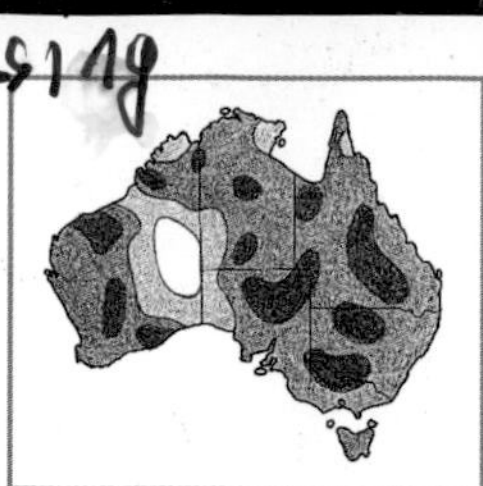

Small, slender plover, distinctive plumage pattern. Gregarious, gatherings from few to many birds. When foraging, holds body horizontal; bobs head; runs on twinkling legs; stops abruptly; pecks; runs again. In flight, keeps low with deep flickering wingbeats. **Voice:** contact call a regular, sharp 'tip-tip-tip-' at intervals of a second or two; becomes much louder, sharper in alarm. **Hab.:** usually freshwater wetlands, shallow, muddy bottomed swamps, billabongs, lake margins, temporary claypan pools; only rarely coastal saline waters, tidal mudflats or other shoreline sites. **Status:** common and widespread; sedentary, dispersive. **Br.:** 369

Hooded Plover *Charadrius (Thinornis) rubricollis* 19–23 cm

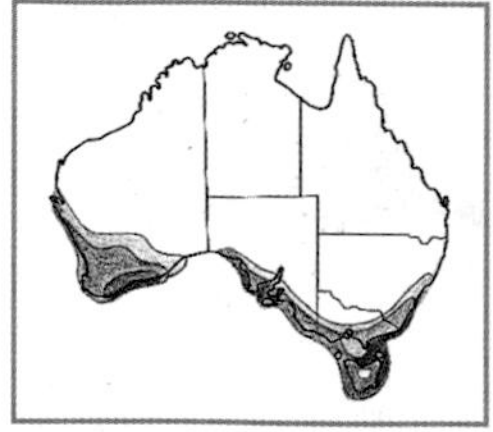

A small plover with black hood and bars that disrupt the shape, making it inconspicuous when still. Usually in pairs or small parties; dashes about near the water's edge, pecking, bobbing, often darting down the beach as waves recede. In flight shows boldly patterned wings. **Voice:** a husky, very abrupt, almost barking 'kep, kep, kep' or 'kue, kue, kue' at intervals of 1 or 2 sec; also a similar, soft, flute-like 'quiep, quiep…' and a higher piping. **Hab.:** sandy beaches of ocean, estuaries, coastal lakes and inland salt lakes. **Status:** endemic species vulnerable to human disturbance of beaches; declining in those locations. **Br.:** 369

Red-kneed Dotterel *Erythrogonys cinctus* 17–19 cm (T)

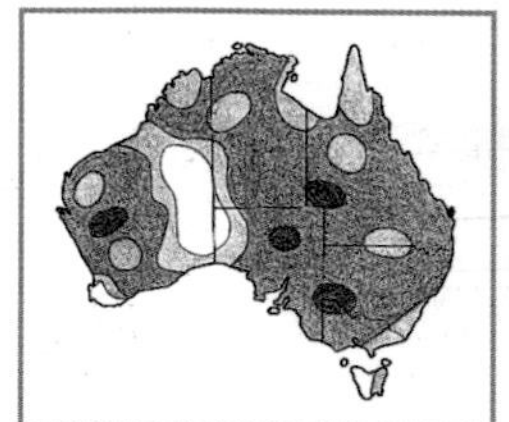

A medium sized, long legged, rather plump plover; stands with body level or with breast low, tail high. Often stands motionless, then bobs head and dashes forward at prey. Probes around muddy shoreline; wades and sometimes swims while feeding. In flight, birds twist and turn; travel fast with shallow, flickering wingbeats. **Voice:** flight and alarm call is a rather sharp, double 'tet-tet' or 'chit-chit'; gives a musical 'prit-prit-pri-t' trill on being put to flight. **Hab.:** uses well-vegetated freshwater wetlands—swamps, lakes, billabongs, interior claypans, sewage ponds, overflows from bores, windmills—working the shallows among emergent vegetation. **Status:** resident; nomadic in response to rainfall; common. **Br.:** 369

Masked Lapwing *Vanellus miles* 35–39 cm (C, L, N, T)

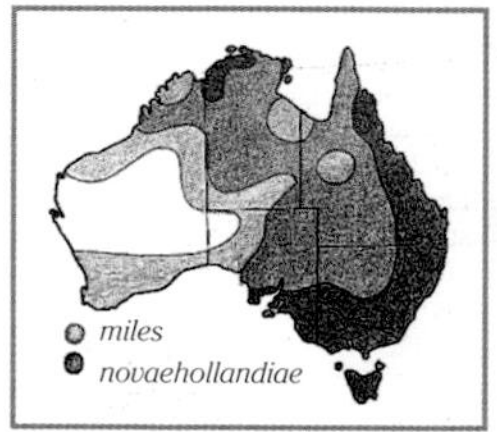

Includes Spur-winged Plover, race *novaehollandiae*. A large, conspicuous, noisy and often aggressive plover; well known in those parts of Aust. where it occurs. Bold enough to claim suburban parks and gardens as its territory. Noisy early in nesting season; strongly defends nest site; dives at intruders, at times striking with wing spurs. In pairs or small family groups during breeding season; at other times in large flocks that may number several hundred. When feeding, stalks slowly, deliberately, body horizontal, dipping and stabbing at prey; when alert, stands upright. **Variation:** two subspecies or races that overlap and interbreed so that many birds have intermediate characteristics with wattles between the extremes illustrated and with varying amounts of black on the sides of the neck. **Voice:** loud, penetrating, insistent; not easily ignored when nearby. As alarm and in threat, noisy 'karrek-karrak-karrak'; rolls the 'r's. If the young are approached, the noise lifts in pitch and tempo, 'karri-karrik-karrik' and 'kek-kekkekekeke'. **Hab.:** varied open, short-grassed sites, both natural and modified, often beside water of swamps, lagoons, salt marsh. **Status:** breeding resident of N and E Aust.; may have benefited from clearing, provision of dams. **Br.:** 370

Banded Lapwing *Vanellus tricolor* 25–29 cm (L)

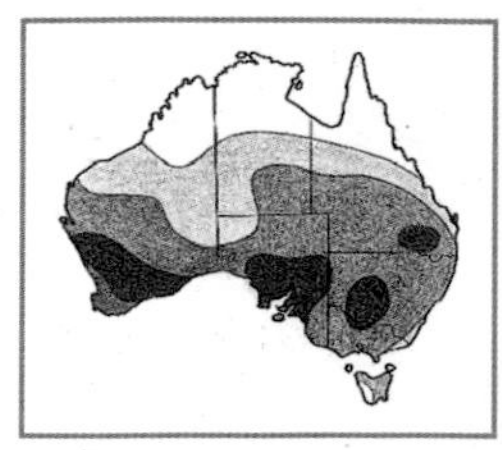

Lapwings are named for their flight; the broad-ended wings flap with a distinctive, hesitant action. Banded Lapwings feed among short grass, often on ploughed, harvested or grazed paddocks. Strongly territorial when nesting; at other times highly gregarious. Stands very upright, wary and alert; walks steadily away if disturbed; quick to take flight with abrupt jerky wingbeats. **Voice:** calls of similar character to those of Masked Lapwing. Usually a grating, strident, resonant, very abrupt 'kerr-kerr-kerr-kerr-' given in rapid-fire bursts. Also a higher, sharper 'quirrrk-quirrrk-quirrk-', at times with a ringing effect. **Hab.:** open country, short grasses, low sparse shrubbery, open acacia or eucalypt woodland where trees are sparsely scattered. Often on agricultural land—ploughed paddocks, grazed pasture, emergent crops or mown grass such as golf courses. Tends to be in general vicinity of water in drier parts of interior. **Status:** endemic species; common and widespread across S Aust.; vagrant, nomadic or dispersive. **Br.:** 370

Red ringed eyes in broad black band
Dull pink eye-ring in grey-brown band; beginning of 'Y' band shows at sides of breast.
Dark grey brown
Scapulars deep chestnut
Collar
Dark, white edges
Bill red, black-tipped
Pale coverts
Deep 'V', black breast-band
Both adult and juv. have a dark horizontal shoulder bar of purplish brown scapulars alongside pale coverts.
White with black, conspicuous 'V'
Very dark outer wings
Dull pink
Adult
Juv.
Adult
Grey hood
Bright red eye-ring
Collar
Light brownish grey
Black-tipped red
Black hood
Tail to rump is black centred with white outer edges.
Collar and side breast bars like adult but smudgy grey. These and the hood wear, become faded, patchy.
Black hind neck collar forks onto breast.
Orange
Adult
Juv.
White, dark tipped
Back of neck black without any white collar
Red bill
Toes beyond tail-tip
White edges to rump and tail
Imm. plumage shows beginning of breast-band and flank streaks. The preceding juvenile plumage has little if any breast-band.
Black-edged white
Wide white trailing edge
Flanks grade from black to chestnut.
At all ages has clearcut contrast between white throat and black or brown hood.
Wide
Red 'knee'
Legs down to 'knee' red; lower leg is blue-grey.
Black primaries
Adult
Juv.
Small black cap
Masked Lapwing, race *miles*: large yellow facial wattles, all-white neck
Race *novaehollandiae*, black hind neck
The Masked Lapwing has two subspecies or races that were long regarded as two distinct species; both have a prominent sharp spur at the carpal joint of each wing and yellow facial wattles.
Complete white collar
Feet trail behind tail-tip.
White
Spurwinged Plover, race *novaehollandiae*: small wattles, black hind neck
Spur
Black
Small wattle
Black of crown continues down sides of neck.
Spur
Flight feathers black, wingtips broad
Fringed buff
Adults
Juv.
Long red legs
Brown, speckled buff
Red
White eye stripe
Black, tipped white
Yellow
Plain grey-brown
Broad buff feather edges throughout
Rump and tail white except broad black subterminal band
Eye-ring yellow
Dark brown with buff fringes to feathers
Bold, heavy, black vertical line from bill down sides of neck to lower breast; from the front view this is a deep, rounded 'U' shape enclosing the pure white throat.
Black
Wing-bar an obvious white diagonal across each wing
Pinkish grey
Legs pink above joint, pinkish grey below
Underwing coverts and axillaries are white in contrast against black wingtips.
Juv.
Adult

Great Skua *Catharacta skua* 51–66 cm (H, M)

A very large, powerful, predatory and scavenging seabird. Usually keeps to the open oceans, but found occasionally over inshore waters. Floats high on surface; looks bulky. Powerful flight; uses its great agility and acceleration in aerial pursuit of other seabirds for their food; in breeding colonies a predator and scavenger among smaller nesting birds. **Juv.:** like adult but plain dark brown; lacks the fine, golden shaft streaks. **Voice:** silent except in breeding colonies. **Hab.:** open ocean, continental shelf slopes; less often over inshore waters; rarely bays or estuaries. **Status:** regular but uncommon winter visitor to southern Aust. seas.

South Polar Skua *Catharacta maccormicki* 53 cm (M, N)

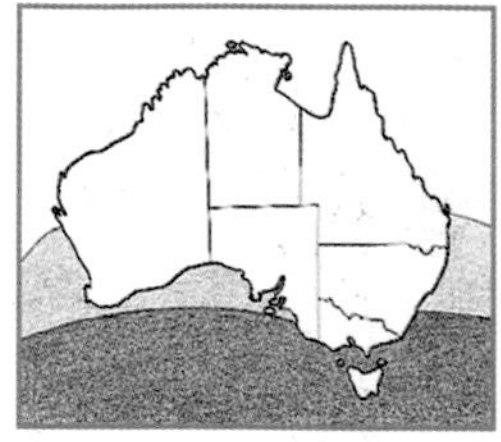

Large gull-like predator; forages over open sea; plummets from heights up to 6 m to take prey on or just beneath the surface. **Variation:** light to dark morphs with continuous colour gradation between. **Voice:** silent at sea; squeals and screams at Antarctic breeding grounds. **Similar:** light morph is distinctive; dark morph similar to Great Skua and juv. of Pomarine Jaeger. **Hab.:** usually Antarctica, its islands and iceshelf; few records away from those habitats, then usually offshore, only occasionally close inshore, rarely in bays or harbours. **Status:** few confirmed records; scattered sightings around much of the southern coastline.

Arctic Jaeger *Stercorarius parasiticus* 41–46 cm (C)

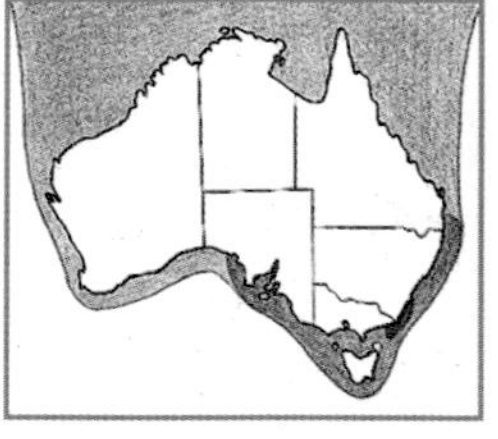

Medium sized with slender bill, long narrow wings; falcon-like, rapid, jerky wingbeats in attack; impressive acceleration, speed and aerial agility. Aggressive; feeds often by aerial piracy, attacking and robbing other birds, harassing a gull or tern until it disgorges its food, only then letting the smaller bird escape. Other times flies with alternating periods of quick wingbeats and glides. **Variation:** light and dark morphs, intermediates. **Voice:** silent away from the breeding grounds except for some high, nasal squealing at others of its kind when squabbling over food. **Similar:** Pomarine Jaeger, but it has heavier build, slower wingbeat; Long-tailed Jaeger, which has smaller, more slender, fine-tipped streamers. **Hab.:** in nonbreeding months, subtropical and sub-Antarctic seas, inshore waters, shallower waters of the continental shelf, and into bays, estuaries, harbours. **Status:** arrives Aust. waters Oct.–Nov.; departs late Apr.–early May. **Br.:** N hemisphere, circumpolar.

Pomarine Jaeger *Stercorarius pomarinus* 66–78 cm (L, M)

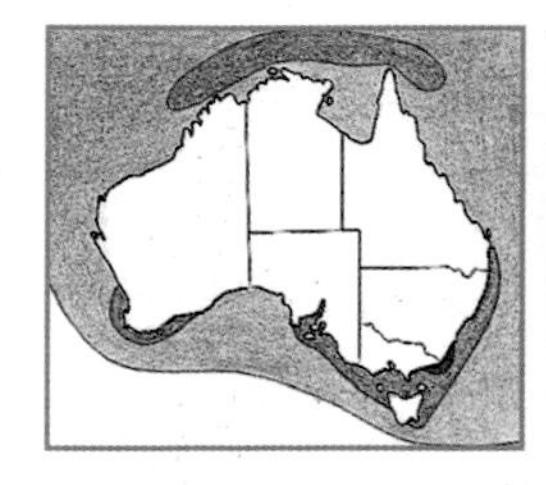

A more heavily built jaeger, rather large headed, barrel chested. In flight, gives more powerful impression than other jaegers; deep wingbeats like large skua or gull. Makes direct attacks on chosen seabird victims, flying with great agility through dives and tight turns until the harassed bird, which may be up to small albatross size, gives up its catch. **Variation:** light and dark morphs; the light form is thought to outnumber the dark by a ratio of about 20:1. **Voice:** at sea, is known to call when it is competing for scraps around fishing trawlers. **Similar:** Arctic Jaeger, which is smaller with more slender, rapid, falcon-like flight; Long-tailed Jaeger (when breeding, it has fine-pointed tail streamers). **Hab.:** during migration it wanders tropical and subtropical seas; occurs in Aust. coastal waters Oct.–May. Most common along edge of continental shelf; scarce closer inshore, and only rarely into harbours and estuaries. **Status:** uncommon, nonbreeding migrant visitor; recorded most frequently off Qld coast. This is the most common species of jaeger off the NSW coast, and there are many records around most of the coast of Vic., northern Tas., SA, WA and NT.

Long-tailed Jaeger *Stercorarius longicaudus* 50–55 cm (L)

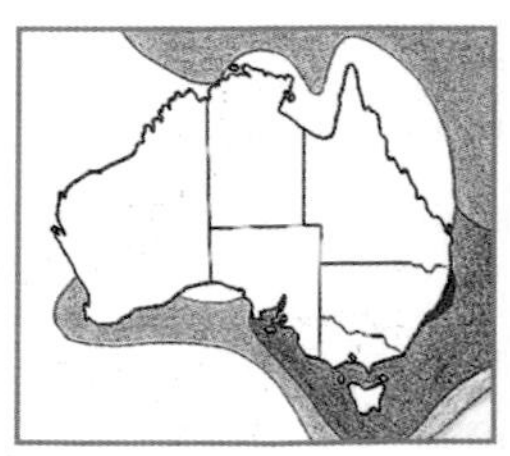

In flight lighter, more elegant and tern-like than other jaegers; the buoyancy imparted by undulations of body with each wingbeat. Patrols low; almost stalls, then often dips to surface. Robs other seabirds, harassing with quick, agile flight; but often feeds tern-like at sea surface. **Variation:** adults polymorphic, but almost all are light phase. Dark phase is very rare; intermediate phase, if it exists, is extremely rare. But juveniles occur in light, dark and intermediate phases, all barred, especially under-tail and underwing coverts, although this is not obvious on the darkest; all juveniles have short tail streamers. Apparently intermediate plumaged juveniles become light adults, as do light juveniles. **Voice:** rarely if ever call away from breeding grounds. **Similar:** other jaegers. **Hab.:** in nonbreeding months wanders oceans; tends to use edge of continental shelf where most Aust. sightings occur, rarely inshore waters. **Status:** regular but uncommon, Oct.–May.

Wings broad, pointed
Brown, streaked and mottled buff and rufous
Tail a rounded wedge, little or no projection of central tail feathers
White bases and shafts form conspicuous white wing 'flash'.
Massive with wingspan to 1.5 m, stout gull-like shape
Almost uniformly dark brown, finely streaked pale golden around the neck.
Bill short, heavy, strongly hooked
Plumage varies: some have body brown, streaked and margined light rufous; others have more extensive pale markings; lighter coloured birds retain darker crown. Plumage darkest after moult, gradually wears paler:
Massive, dark, heavy, gull-like shape
White
Dark except head
Dorsal uniform brown
Light morph
Pale buff head, body
White
Tail a short dark wedge
Central retrices project
Pale neck
Buffy brown
Intermediate morph
Always white, all morphs
Pale nape
Dark morph
Pale around bill
Less white
Dark grey, fringed and streaked buff; collar grey-buff
Juv.
Long tail streamers
Creamy collar
Long narrow wings
Breeding, light morph
Dark cap
Partial band
Tip may be broken.
Breast-band streaked
Palest birds have breast-band reduced to dark tab each side.
Nonbreeding, light phase. Juv. similar, but barred underwing
Darker
Complete band
Nonbreeding, intermediate morph
Juvenile, intermediate morph
All have white flash on both sides of outer primaries, less bright on dark phase.
Pale
All dark
Breeding, dark morph
Has deep chested, large headed, broad winged and powerful jizz.
White shafts
Black cap, buff tint on neck
Tail has central pair as long, twisted, spoon tipped streamers.
Breeding, light morph
Streamers may be broken but still appear broad; tail otherwise rather short, wide.
Breeding, light
Breast-band: some may have only small side patches.
White flash
Streamers shorter, less twisted, still with blunt tips
Loses buff tint.
Barred
Nonbr., light morph
All have white flash on underside of primaries, streaks of white shafts on upper side.
Rounded tips of short streamers
Juvenile, light morph
Spoon tipped streamers
Knob tipped, twisted streamer
Breeding, dark morph
Dark cap
Lighter rufous
Often feeds on water; dives to retrieve food.
Red to grey-brown
Black
Yellow tint
Lightly built, slim, long winged, rather tern-like
Black trailing edge
Long, fine tail streamers while in breeding plumage
White, yellow tinted
Only two white shafts
Breeding, light morph
Black tip
Underwing plain dark grey with only outer 2 or 3 shafts finely white
Black cap
Underbody white with some grey towards tail, or light grey from breast back
Breeding, light morph
Barred
Nonbreeding, light morph
Outer 2 or 3 whiter
Central tail longer than on other juveniles
Juv.
Intermediate, barred
Dark morph is extremely rare, more slender than above.
Has only 2–3 fine white primary shafts.
Black
Long, fine
Black
Black
Dark morph, Br.

Pacific Gull *Larus pacificus* 50–66 cm

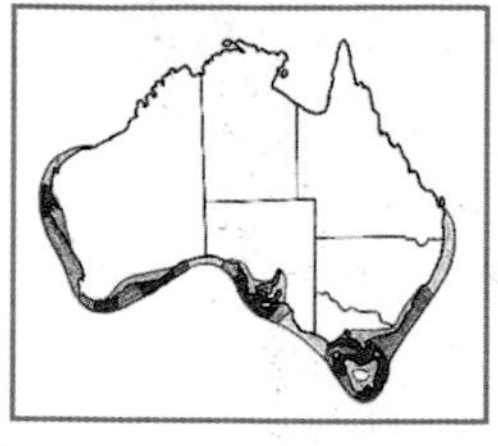

Very large gull; patrols water's edge; has heavy lumbering flight interspersed with glides. Gregarious—in flocks, small groups, alone. **Imm.:** recognisable up to the third or fourth year, but aging is complicated by the individual variation in progress toward adult plumage. **Br.:** slightly brighter colours of bill, orange legs and orbital ring compared with nonbreeding. **Variation:** two subspecies, nominate *pacificus* in SE Aust. and *georgii* in SA and WA; some of race *georgii* have red iris, appear dark eyed. **Voice:** in alarm, mournful 'ohk-ohk'; shorter and sharper with greater anxiety, 'owk-owk-owk'; barked 'ok-ok-ok' flying over intruders. Long, peevish, thin, piping 'airrk, airrk'. **Similar:** Kelp Gull, but it has smaller bill, adults have red only on lower mandible. **Hab.:** coastal with tendency for E race to prefer sheltered beaches; those in W also occur on coasts exposed to open oceans. Often on coastal islands; occasionally on inland rivers, coastal lakes. **Status:** endemic to Aust.; common on W and S coasts; scarce parts of SE where it is being displaced by the Kelp Gull. **Br.:** 370

Kelp Gull *Larus dominicanus* 50–62 cm (H, L, M)

Slightly smaller than Australia's endemic Pacific Gull; bill not as massive. The combination of black wings, all-white tail, white tipped primaries and wider white trailing edge to wings makes identification of adult birds easy. Usually in pairs, alone or in small flocks, or flocks in breeding colonies on small coastal islands; flocks at some breeding strongholds are very large. More gregarious than Pacific Gull and will defend nest sites vigorously. In flight has deep, slow, strong wingbeats alternating with glides like the Pacific Gull. **Nonbr.:** during S summer some of white primary tips reduced or lost. **Voice:** most common throughout year is a long call that starts softly, followed by long sequence of loud notes: 'huh, huh, ee-arh-har-har-har-har' or 'yho-yho-yho-'; usually around 15 notes in series. Used territorially and as warning. Various other calls, most relating to interaction between birds when breeding. **Similar:** Pacific Gull. **Hab.:** in Aust., almost entirely coastal; usually within sight of land; rarely more than 10 km offshore. Uses the more protected parts of coast—estuaries, bays, mudflats, offshore islands. Also on wetlands near coast. **Status:** not recorded for Aust. until 1943; increasing; now common in parts of E coast, and is breeding in WA, NSW and Tas. **Br.:** 370

Silver Gull *Larus novaehollandiae* 38–42 cm (L, N)

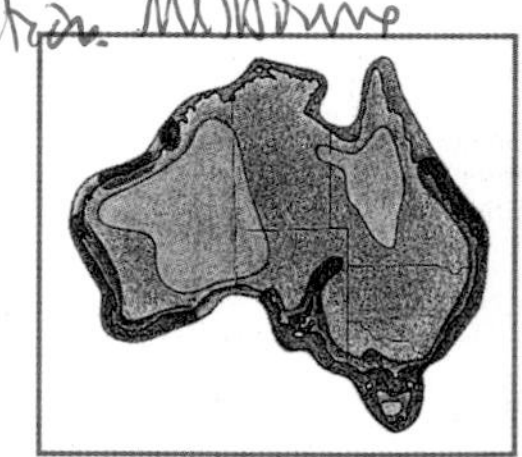

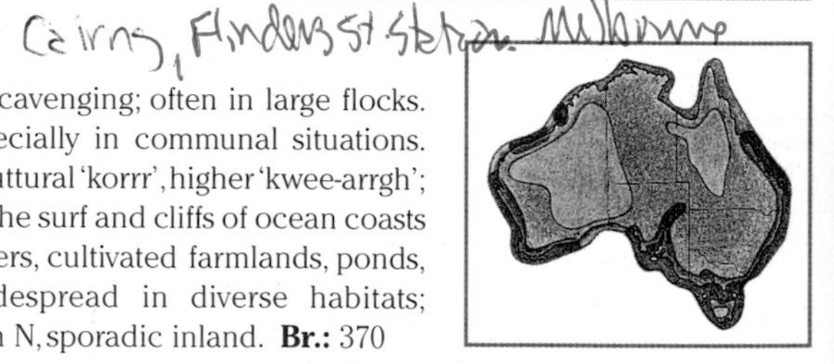

Well known, in many places tame; noisy, bold, gregarious, scavenging; often in large flocks. Flight direct, unhurried. **Voice:** great variety of calls, especially in communal situations. Include an aggressive, rough 'karrgh-karrgh-karrgh', deeper, guttural 'korrr', higher 'kwee-arrgh'; wheezy sounds from juveniles. **Hab.:** diverse, ranging from the surf and cliffs of ocean coasts to offshore islands, inland rivers, lakes, temporary floodwaters, cultivated farmlands, ponds, coastal towns, rubbish dumps. **Status:** extremely widespread in diverse habitats; opportunistic and adaptable; abundant in S, less common in N, sporadic inland. **Br.:** 370

Lesser Black-backed Gull *Larus fuscus* 50–60 cm

A very large gull that may occur in Aust waters. This species migrates from Iceland, N Europe and N Russia, not normally as far as Aust. **Adult:** wing has quite broad white trailing edge, narrow white leading edge, white primary tips with white 'window' spots on outermost feathers and white tail. **Juv.:** tail very white at base with dark band; darker grey coverts form dark line near trailing edge. **Hab.:** in normal range, coasts and inland, inshore waters, estuaries, lagoons. **Status:** occurrence on Aust. seas as yet unproven, but the number of claimed sightings suggests that an acceptably confirmed sighting may yet occur.

Black-tailed Gull *Larus crassirostris* 45–48 cm

Grey backed, long winged, short legged, medium sized gull. Adult has a broad black tail band, plain dark wingtips, yellow bill with red and black tip. Stands slightly taller than the common Silver Gull, even though the legs are relatively short. Flight is similar to that of Silver Gull, but slower wing action. **Voice:** a hard bass 'kaou, kaou' or 'yaou'; also plaintive mewing. **Similar:** immatures could be confused with similar stage of Kelp Gull. **Hab.:** in normal range, found on sandy or rocky coasts, inshore waters, estuaries, harbours, inshore islands. **Status:** the several vagrants to Aust. have been from NT and Vic. The one accepted record was from Port Phillip Bay, a single bird with Silver Gulls feeding on fish at the bay's entrance.

Wings are entirely black except for narrow white inner wing trailing edge.
Black band
White tip
Adult
Developing white rump, black band
Whiter
Darker, nearer to blackof adult
Imm., yr 3
Upper parts dark brown, scaled dull buffy white
Brown, paler tip
Narrow edge
Juv.
Dark brown, paler on face
Massive, pinkish base
Dark brown, edged pale buff
White
Nominate race pacificus
Red
Race georgii
Juv.
White
Massive, yellow, red tipped bill
Black
White
Folded, plain dark primaries extend far beyond tip of black banded tail.
Adult
All white
Narrow white leading edge
Adult
Wide white
Mottled brown
Dark tip
Mottled brown
White 'mirror' on first primary; white tips to outer six
Brownish black
Imm., about yr 2
Partly barred
Dark brown, pale streaked
Juv.
Wider edge
Bill less massive than Pacific Gull; red spot only on lower tip
White
Slaty black
Bold white crescents
White tips
All white tail, no band
Long wings, when folded, extend not far past tail-tip; compare with Pacific Gull.
Legs bright yellow when breeding, or yellow, grey-green
Darker
Black
Upper parts dark brown fringed dull off-white
Off-white, streaked brown
Juv.
Red only on lower tip
Black, with white 'windows' on outer three primaries
White
Red
Thin subterminal band
Dark edge
Wings mottled grey and brown
Brown tint
Juv.
White tips, spots on outer 3
Red
Eye white
White
Pale grey
White tipped
Adult
Red
Smudged brown
Pale grey mottled brown and buff
White tips
Subterminal bar
Brown
Juv.
White
Dark slate grey
Broad white trailing edge
White edge
Black
White tips, 'windows'
Adult, br.
Brownish
Grey
Dark
White edge
Entirely dark
Imm., yr 2
Streaked
White
Scaly pattern
Streaked
Juv.
Streaked brown
Br.: head and neck white
Adult, nonbreeding
Yellow legs
Edge white
Black
White tail edge
Mid grey
Plain black
Adult, breeding
Dark
Dark
Adult, nonbreeding
White edge
Plain, black
Black, tipped white
Wings black; exceed tail length.
Mottled light grey-brown
Mid grey
Bill yellow, black subterminal band, red tip, red spot on lower
Nonbreeding adult. Breeding similar except head all white
Short, greenish yellow
Streaked grey brown
Pale, tipped black
Brown, edged off-white
Juv.
Primaries protrude well beyond tail-tip giving attenuated rear end
Black
Red
Red

Franklin's Gull *Larus pipixcan* 32–35 cm

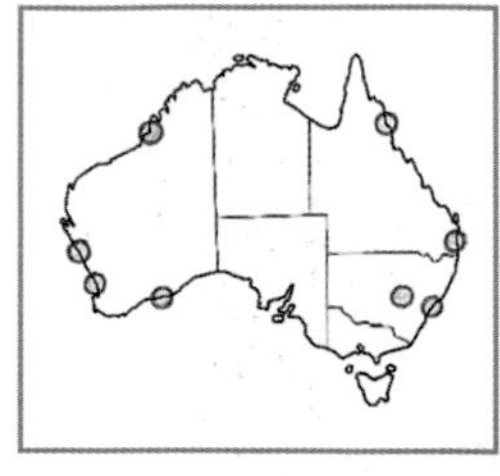

A small distinctive gull, most like Laughing Gull. Breeding grounds are in prairie marshes of N America, foraging out over surrounding grasslands and cropped land. On migration uses both coastal and inland habitats. Along main routes down W coast of S America occurs in very large flocks. Perched, has a rather low, hunched horizontal posture. In flight is agile, light and buoyant with paddling beats of slightly rounded wings. Feeds by dipping and snatching at surface. **Br.:** black hood comes down to upper neck, white eye crescents almost join at rear of eye; large white spots along folded primaries; white spotted, black wingtip is isolated from grey upper wing by white that joins white trailing edge; legs red. **Nonbr.:** hood reduced to distinctive half hood, the rear half of the breeding plumage hood; similar on immature, but even less clearly defined. **Voice:** silent outside breeding colonies, where call is a soft 'krruk'. **Similar:** Laughing Gull. **Hab.:** in Aust., sightings have been on sandy beaches, mudflats, rocky spits and near coastal wetlands, all on sheltered parts of the coast in estuaries and bays. **Status:** widespread sightings, although few verified and accepted; rare vagrant, early spring to autumn.

Laughing Gull *Larus atricilla* 36–40 cm

A small, slender, sleek gull, appearance emphasised by long wings that, when folded, extend well beyond the tail-tip, and by large, slightly downcurved bill that blends smoothly to a gently sloping forehead. Small details include white crescents above and below eye, conspicuous on the black heads of birds in breeding plumage, and black encircling the tip of the red bill. While only occasional solitary vagrants have been sighted in Aust., this gull in its usual range is gregarious. Flight is slower—deeper wingbeats, more leisurely—than Silver Gull. Readily and skilfully snatches food from sea surface. On land stands tall, upright on long legs when alert, or more horizontal when relaxed. **Voice:** a strident, high 'ca-ha', often prolonged to a frenetic laughing: 'ha-ha-ha-ha-haah-haah'. **Similar:** Franklin's Gull, but it has slightly smaller, shorter wings and legs, smaller bill, black hood tends to come further down on fore-neck. **Hab.:** coastal—would be likely in any situation favoured by the Silver Gull; scavenges in harbours, along tide line, over coastal reaches of rivers; follows fishing boats. **Status:** a number of accepted records from Tas., Eyre Bird Observatory (S coast WA), SE Vic., Cairns in NE Qld.

Sabine's Gull *Larus sabini* 27–31 cm

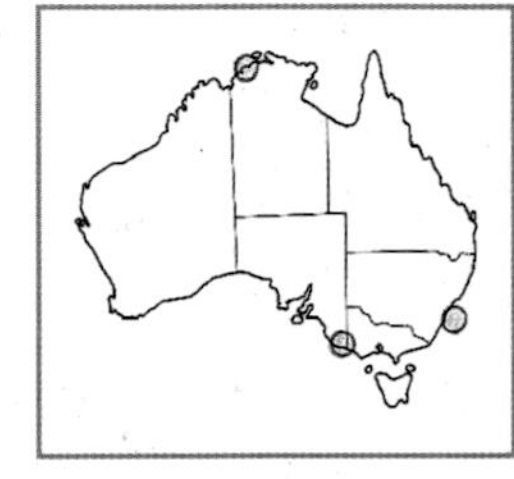

Small, slender bodied; long pointed wings accentuated by distinctive upper wing pattern of grey, white and black; the tail has a shallow fork, unique in Aust. region. This species differs from most other gulls in retaining juvenile plumage until its arrival in southern wintering areas. It has a complex pattern of moults—6 plumages in the 2.5 years to reach adult breeding plumage. Breeds arctic parts of Russia, Alaska, Canada. Migrates S down coasts to S Africa and S America, with vagrants to Japan and Indonesia; a very few as far as Aust. In usual wintering regions is gregarious in small to large flocks; in Aust. has only occurred as solitary vagrants. Flight tern-like, light and buoyant, deep wingbeats, little gliding. Feeds tern-like, picking up items from water or land. Usually avoids ships, but will approach to inspect gatherings of other seabirds around fishing boats. **Voice:** harsh grating, like that of Arctic Tern. **Hab.:** in the nonbreeding season is mainly pelagic, favouring deep waters along edge of S continental shelves, but also occasionally comes over shallower shelf waters. May shelter on protected inshore waters during gales. **Status:** few sightings, very widely spaced, in NT, NSW and SA; all occurring autumn to winter months.

Black-headed Gull *Larus ridibundus* 36–42 cm

A small, slender gull, thin long-pointed wings, rapid wingbeats; in general appearance like Silver Gull, but slightly smaller. Diagnostic wing pattern of white triangle is conspicuous between black margins of wingtip. Contrary to name, hood is not black but brown—dark in fresh plumage, fading to mid or even light brown through the season; contrast against bright white of neck and nape ensures that the hood usually looks near black at a distance. **Voice:** vocal throughout the year: varied screams; harsh, quite high pitched scolding 'karrgh' and 'kraaak'; various other calls—harsh screams, a dtawn out 'kreeoo' and 'kekk' in alarm. **Similar:** distinctive in breeding plumage; otherwise Laughing, Franklin's. **Hab.:** in usual range, varied shallow wetlands, coasts and inshore waters, bays, estuaries, salt marsh, wet grasslands and ploughed fields. In Aust., recorded at sewage ponds. **Status:** an abundant, widespread, gregarious species; breeds Iceland, Europe, through Asia to Japan, China, Kamchatka Peninsula. Migrates to Africa, SE Asia; reaches Indonesia, NG, and, as a rare vagrant, to N Aust. A confirmed sighting in breeding plumage at Broome.

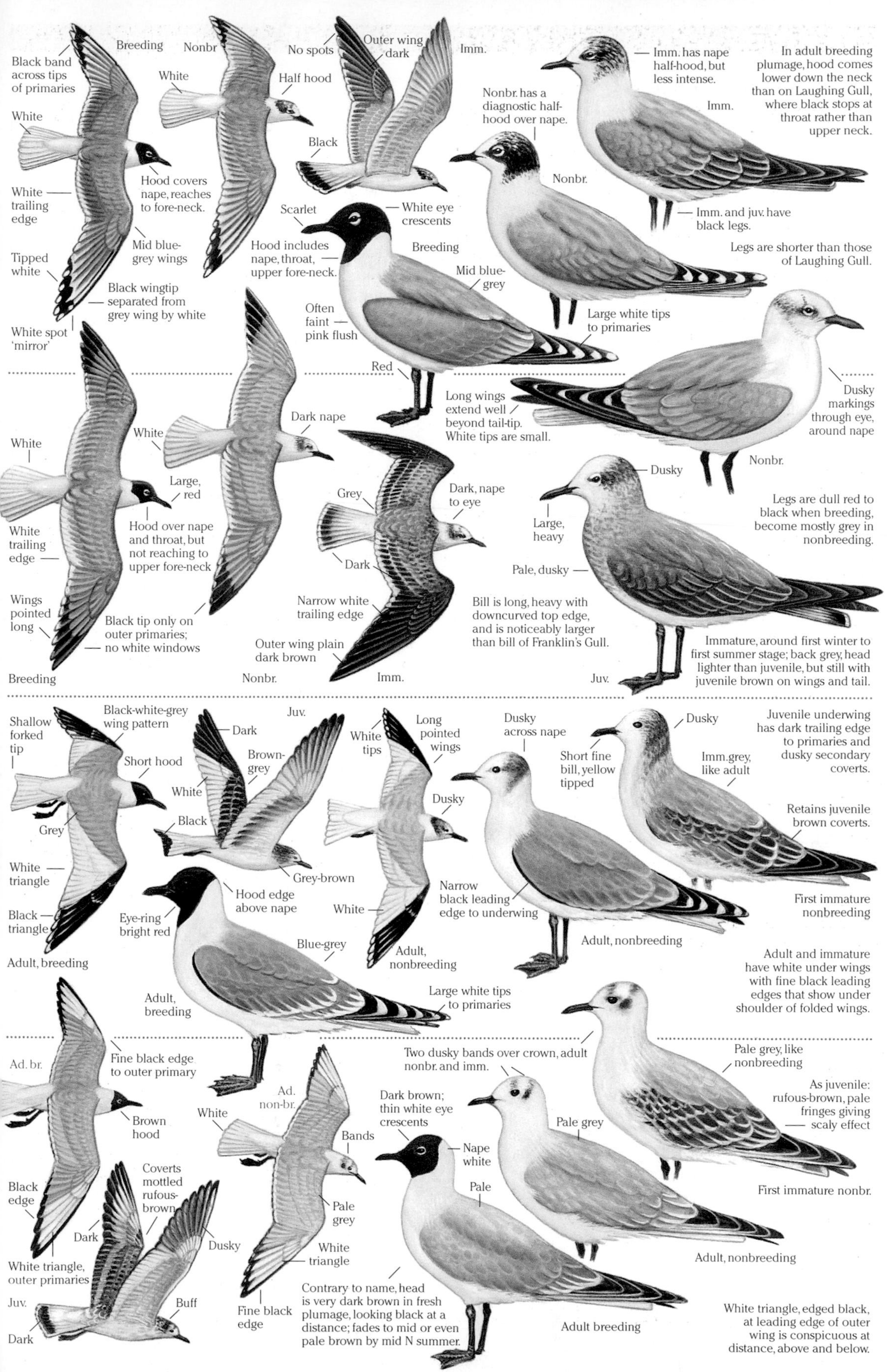

Breeding
Black band across tips of primaries
White
White trailing edge
Tipped white
Hood covers nape, reaches to fore-neck.
Mid blue-grey wings
Black wingtip separated from grey wing by white
White spot 'mirror'
Nonbr
White
No spots
Half hood
Outer wing dark
Black
Imm.
Scarlet
White eye crescents
Hood includes nape, throat, upper fore-neck.
Breeding
Mid blue-grey
Often faint pink flush
Red
Nonbr. has a diagnostic half-hood over nape.
Nonbr.
Large white tips to primaries
Imm. has nape half-hood, but less intense.
Imm.
Imm. and juv. have black legs.
In adult breeding plumage, hood comes lower down the neck than on Laughing Gull, where black stops at throat rather than upper neck.
Legs are shorter than those of Laughing Gull.
White
White
White trailing edge
Large, red
Hood over nape and throat, but not reaching to upper fore-neck
Wings pointed long
Black tip only on outer primaries; no white windows
Breeding
Dark nape
Nonbr.
Grey
Dark
Dark, nape to eye
Narrow white trailing edge
Outer wing plain dark brown
Imm.
Long wings extend well beyond tail-tip. White tips are small.
Dusky markings through eye, around nape
Nonbr.
Dusky
Large, heavy
Pale, dusky
Bill is long, heavy with downcurved top edge, and is noticeably larger than bill of Franklin's Gull.
Juv.
Legs are dull red to black when breeding, become mostly grey in nonbreeding.
Immature, around first winter to first summer stage; back grey, head lighter than juvenile, but still with juvenile brown on wings and tail.
Shallow forked tip
Black-white-grey wing pattern
Short hood
Grey
White triangle
Black triangle
Adult, breeding
Juv.
Dark
Brown-grey
White
Black
Grey-brown
Hood edge above nape
Eye-ring bright red
Blue-grey
Adult, breeding
Large white tips to primaries
White tips
Long pointed wings
Dusky
Narrow black leading edge to underwing
White
Adult, nonbreeding
Dusky across nape
Adult, nonbreeding
Short fine bill, yellow tipped
Dusky
Imm. grey, like adult
Juvenile underwing has dark trailing edge to primaries and dusky secondary coverts.
Retains juvenile brown coverts.
First immature nonbreeding
Adult and immature have white under wings with fine black leading edges that show under shoulder of folded wings.
Ad. br.
Fine black edge to outer primary
Brown hood
Black edge
White triangle, outer primaries
Coverts mottled rufous-brown
Dark
Dusky
Juv.
Buff
Dark
Ad. non-br.
White
Bands
Pale grey
White triangle
Fine black edge
Two dusky bands over crown, adult nonbr. and imm.
Dark brown; thin white eye crescents
Nape white
Pale
Pale grey
Contrary to name, head is very dark brown in fresh plumage, looking black at a distance; fades to mid or even pale brown by mid N summer.
Adult breeding
Adult, nonbreeding
Pale grey, like nonbreeding
As juvenile: rufous-brown, pale fringes giving scaly effect
First immature nonbr.
White triangle, edged black, at leading edge of outer wing is conspicuous at distance, above and below.

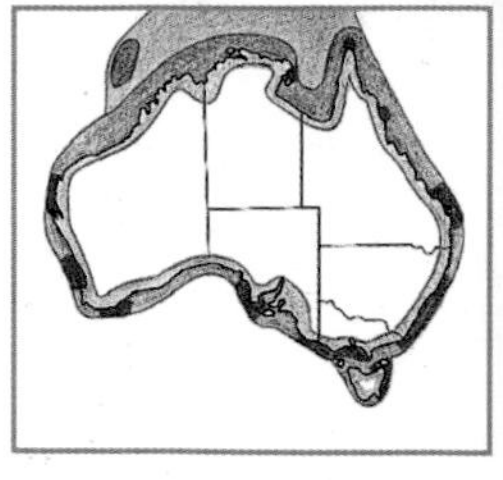

Crested Tern *Sterna bergii* 43–48 cm (C, L, T)
A large, black-capped tern with long, slightly downcurved yellow bill, long wings and forked tail. The common, familiar large tern of bays and harbours, it often roosts on boats and jetties. Flight is powerful, swift; long pointed wings angled back; deep beats. May be in mixed flocks with other terns and gulls. Forages by plunging from several metres to take prey just under the surface. **Variation:** numerous subspecies; one in Australasia, the race *cristata*, with pale plumage. **Voice:** noisy, especially at breeding colonies, often at night and in dawn flights. Intruders near nests or chicks cause much screeching from birds overhead. Alarm call an abrupt 'wep'. Common advertising call a raucous 'graaak' or 'kirrak'. In threat and alarm, used in response to aerial predators, fighting, duelling in colony, a deep, throaty 'korr-korr-korr-' and 'krow-krow'. **Similar:** Lesser Crested Tern, but it is considerably smaller, has slimmer, shorter legs and crest, straighter and more orange bill. **Hab.:** coastal—ocean beaches, offshore islands, extending out to deeper pelagic waters; inshore on estuaries, bays, harbours, coastal lagoons; inland on major rivers, occasionally on saline lakes, salt ponds near coast. **Status:** widespread and common around Aust. coast; breeding resident. **Br.:** 370

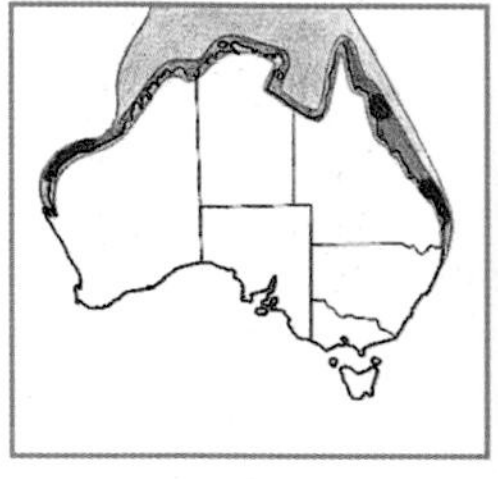

Lesser Crested Tern *Sterna bengalensis* 38–43 cm (C)
Closely resembles the Crested Tern, but is smaller, more slender, paler grey. In flight shows shorter but slim wings giving more delicate, compact jizz. If the two species are seen together the size difference is obvious. **Variation:** three subspecies, birds in Australasian region being race *torresi*. **Voice:** like Crested but less harsh; a grating 'kik-kerek' and ratchetting 'gr-a-a-a-k', each 'a' hard, metallic, abrupt; also a slower, harsh 'grruk-uk-uk'. **Similar:** Crested Tern, but it has a larger, greenish yellow bill and a wider white gap between black of cap and base of bill. The long shaggy crests of the two species of crested tern separate them from all other Aust. terns; the Caspian has a much shorter crest and massive red bill. **Hab.:** coastal seas, using shores of sandy beaches, coral cays, exposed reefs, islands. On parts of coast on mudflats of estuaries, creek channels. **Status:** breeding resident, most common NE Qld, breeding on reef islands S to Capricorn Group. Most colonies have a few to several hundred pairs; a few have above a thousand pairs. Also breeds NW of WA, Adele Is. and Bedout Is. **Br.:** 370

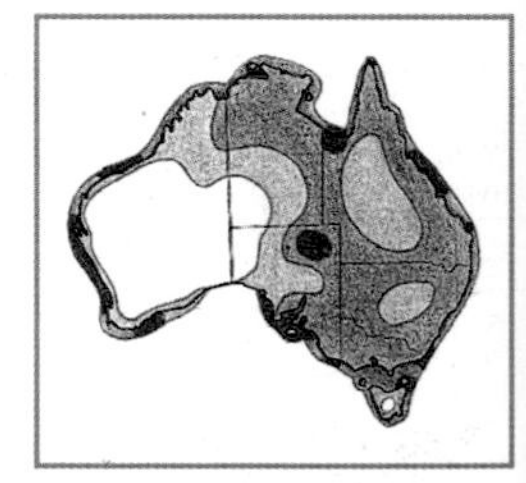

Caspian Tern *Sterna caspia* 48–54 cm (T)
A huge gull-like tern with massive red bill and black capped, angular head. Looks very large on beach among other terns, usually with body horizontal, head pulled down tightly, or alert, standing tall. Widely distributed in Europe, Asia, N America, Africa, Australasia. Flight like that of large gulls—powerful deep, regular, wingbeats, yet swift and graceful on long, slender, backswept wings. Patrols the surf line and inshore waters, bill typically downwards; often pauses, turns, or briefly hovers heavily, before plunging into the water. **Voice:** as an alarm call, the most common call and often given while hovering overhead, a loud, deep, rasping, abrupt 'owgk, owgk' or 'kowk, kowk, kowk' at irregular intervals, and, varied slightly, 'urgk, urgk', or, becoming agitated, more rapidly as 'urgk-urk-uk-uk-uk-'. Also, drawn out, harsh, aggressive sounding 'ar-ar-rr-rk', or 'kraark', and higher, squealed 'ai-air-arrk'. **Hab.:** usually coastal, with preference for sheltered estuaries, inlets, bays, harbours, lagoons with muddy or sandy shores. Keeps close inshore, not usually out beyond reef line. Also extends well inland on fresh or salt lakes, temporary floodwaters, large rivers, reservoirs, sewage ponds. **Status:** breeding resident species; nest sites usually coastal; uses offshore islands. Common, though scattered rather than concentrated; usually sedentary, although those using temporary wetlands of interior then disperse; such movements cause numbers to fluctuate seasonally. **Br.:** 370

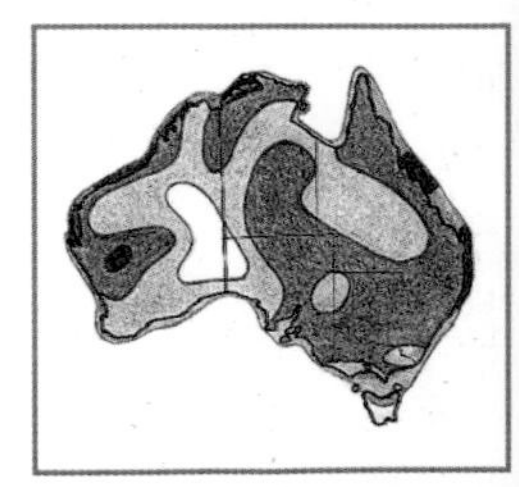

Gull-billed Tern *Sterna nilotica* 36–42 cm (T)
Occurs across every continent as an inland species, only rarely over the ocean. Breeds in colonies on small islands of inland shallow lakes. At a distance looks like a large white gull but has a bulkier body, broader wings and is more rounded than is typical of terns. Close up the heavy gull-bill may be noticed. **Nonbr.:** differs in having a dark patch through each eye, or, reduced even further, just a grey streaked area around the ear coverts. **Variation:** many subspecies; in Aust. most are race *macrotarsa*, larger and paler; some in N Aust. may be summer migrants from Asia of race *affinis*, slightly smaller. **Voice:** a hoarse, nasal, quavering call, 'ar-ark, ar-ark, ar-arrk'. **Hab.:** unusual in using inland waters, fresh or saline, for nesting. Often uses temporary waters on mudflats or claypans, saltpans, salt marsh, open floodplains in arid regions where heavy rains of a cyclone or other rare downpour causes extensive shallow flooding. Out of the breeding season, shows some preference for lagoons and salt marshes near the coast. **Status:** nomadic; highly dispersive. **Br.:** 370

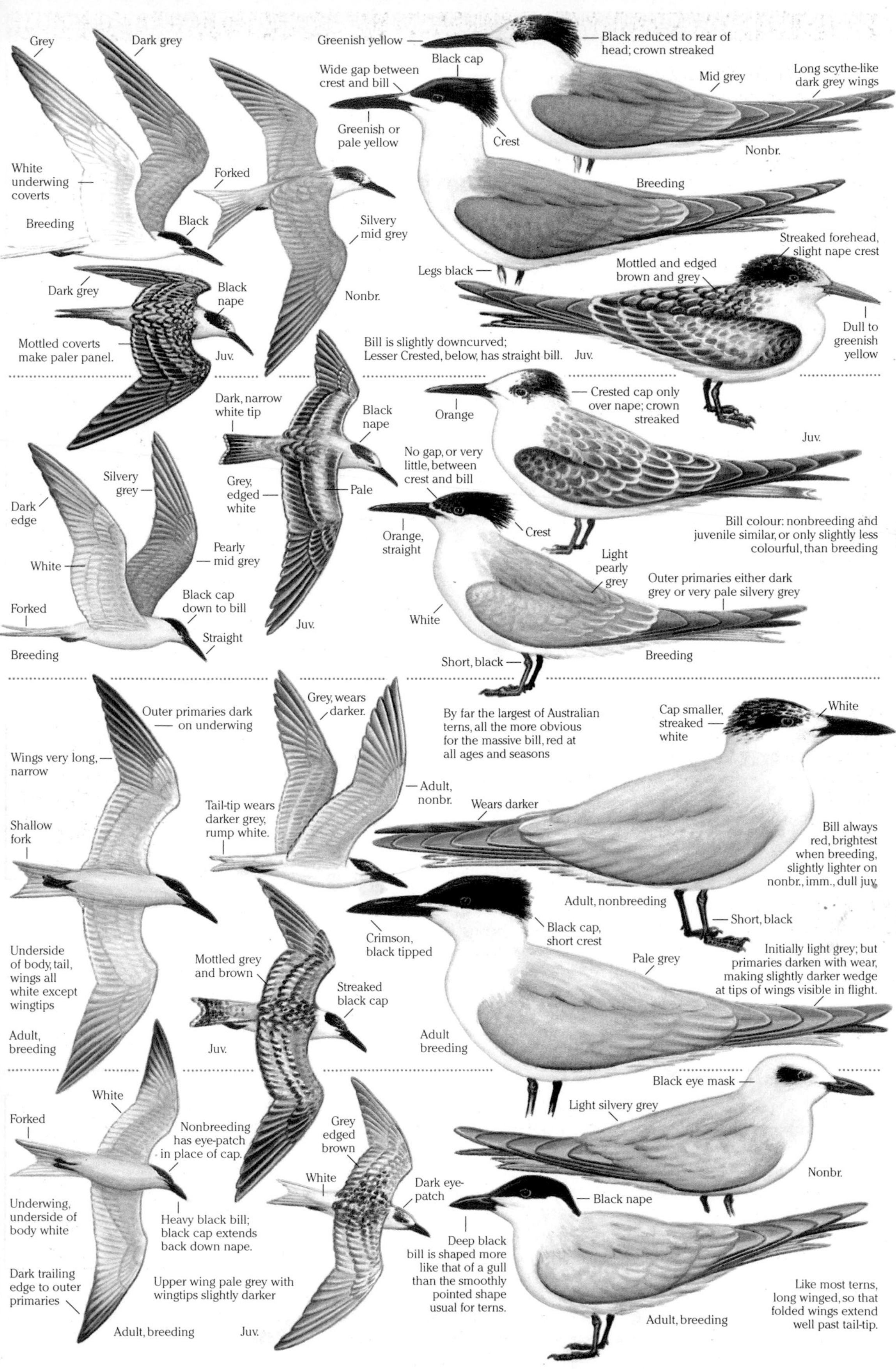
Grey
Dark grey
White underwing coverts
Breeding
Forked
Black
Silvery mid grey
Nonbr.
Dark grey
Black nape
Mottled coverts make paler panel.
Juv.
Greenish yellow
Black reduced to rear of head; crown streaked
Black cap
Wide gap between crest and bill
Greenish or pale yellow
Crest
Mid grey
Long scythe-like dark grey wings
Nonbr.
Breeding
Legs black
Streaked forehead, slight nape crest
Mottled and edged brown and grey
Dull to greenish yellow
Bill is slightly downcurved; Lesser Crested, below, has straight bill.
Juv.
Dark, narrow white tip
Black nape
Orange
Crested cap only over nape; crown streaked
Juv.
Silvery grey
Dark edge
Grey, edged white
Pale
No gap, or very little, between crest and bill
Orange, straight
Crest
Bill colour: nonbreeding and juvenile similar, or only slightly less colourful, than breeding
White
Pearly mid grey
Light pearly grey
Outer primaries either dark grey or very pale silvery grey
Forked
Black cap down to bill
Straight
Breeding
Juv.
White
Short, black
Breeding
Outer primaries dark on underwing
Grey, wears darker.
By far the largest of Australian terns, all the more obvious for the massive bill, red at all ages and seasons
Cap smaller, streaked white
White
Wings very long, narrow
Adult, nonbr.
Wears darker
Tail-tip wears darker grey, rump white.
Shallow fork
Bill always red, brightest when breeding, slightly lighter on nonbr., imm., dull juv.
Adult, nonbreeding
Short, black
Crimson, black tipped
Black cap, short crest
Initially light grey; but primaries darken with wear, making slightly darker wedge at tips of wings visible in flight.
Pale grey
Underside of body, tail, wings all white except wingtips
Mottled grey and brown
Streaked black cap
Adult, breeding
Juv.
Adult breeding
Black eye mask
Light silvery grey
White
Forked
Nonbreeding has eye-patch in place of cap.
Grey edged brown
Nonbr.
White
Dark eye-patch
Black nape
Underwing, underside of body white
Heavy black bill; black cap extends back down nape.
Deep black bill is shaped more like that of a gull than the smoothly pointed shape usual for terns.
Dark trailing edge to outer primaries
Upper wing pale grey with wingtips slightly darker
Like most terns, long winged, so that folded wings extend well past tail-tip.
Adult, breeding
Juv.
Adult, breeding

White-fronted Tern *Sterna striata* 35–43 cm (N)

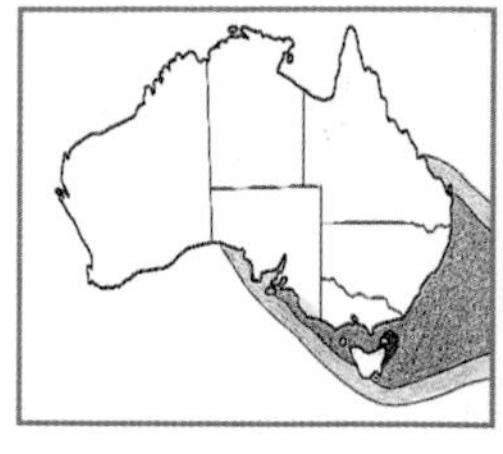

A medium sized, robust tern with rather large head; a steep forehead rather abruptly levels to a comparatively flat crown. The black cap does not touch the bill, but remains separated by a white band, not wide but usually quite obvious between black of bill and black of cap; this is the 'White Front' of its name. On a perched bird, the cap turns down quite sharply, extending down nape to upper part of hind neck. In flight, with head blending smoothly to body, the cap becomes a neat, long, black oval. **Voice:** when disturbed or in flight, a sharp, squeaky, rasping 'ki-erk', beginning sharply, becoming a vibrating, rasping sound, still quite high; the whole call quite quick. Also gives a more abrupt 'kiek' as a contact call, and an angry scream at aerial predators. From colony or flock, a confusion of sharp squeaks intermingle with lower, but not deep, rasping sounds. **Similar:** nonbreeding Roseate, but it has a rounded crown, finer bill, small slim build. Also Common and Arctic, but these are absent in winter when White-fronted Terns visit. **Hab.:** marine, usually exposed ocean coastline, beaches, rocky headlands, offshore islets; less often in sheltered estuaries, harbours. **Status:** most breed in NZ, visit SE Aust. in substantial numbers through winter, including more juveniles than adults; small breeding colony SE Bass Str. **Br.:** 370

Black-naped Tern *Sterna sumatrana* 30–32 cm (L)

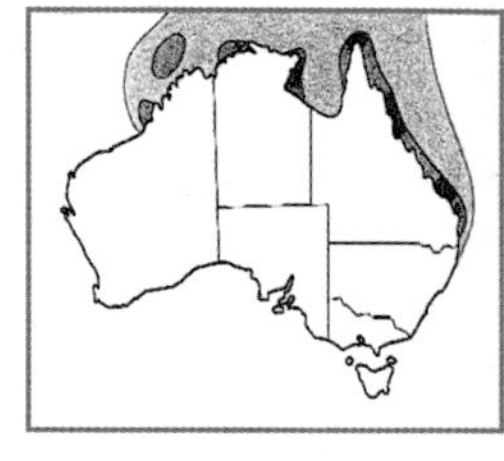

Distinctive, a very pale tern with a conspicuous, broad, black band encircling the back of the neck, then tapering forward to terminate in a point ahead of the eye. This feature, retained throughout the year, is similar on immatures, though somewhat blurred by a streaked crown, and still recognisable in the indistinct speckled brown juvenile version. Although pale grey above, it can look white in strong light. Tends to rest and roost on sandspits, beaches, rocks at water's edge; forages both inside barrier reefs and over open seas beyond the outer reef. Direct, fast flight; short, quick wingbeats; often skims low across the water or swoops to snatch small fish from just beneath the surface. **Voice:** sharp, scratchy, abrupt 'chaik-chaik, chaik, chaik' at irregular intervals; also harsh scolding sounds, 'karrk-karrk-'. **Similar:** nonbreeding and juveniles of Little Tern, Fairy Tern, Roseate Tern. **Hab.:** offshore coral cays, inner-reef rocky islands; rarely continental coasts, estuaries, harbours. **Status:** resident, breeding; most common along Great Barrier Reef. **Br.:** 371

Antarctic Tern *Sterna vittata* 32–36 cm (H, M)

Larger and looking bulkier than Arctic and Common Terns, the Antarctic Tern is in breeding plumage through the summer, Oct.–Apr., when those northern terns are in nonbreeding plumage. Flight like White-fronted Tern—steady, direct, regular, deep wingbeats. Patrols some 5–10 m above the sea; hovers then plunges; also shallow dips to surface. **Voice:** a high 'chirr-chirr-chirrah'; a rattling alarm call, and a scolding, squawking noise when nest site threatened. **Similar:** Arctic Tern, but it is smaller, more slender and shorter with a more slender bill, longer and more deeply forked tail, outer primaries more translucent and darker, more obvious outer edges to tail; in breeding plumage, lighter grey on underparts. **Hab.:** usually sub-Antarctic seas, coasts and islands, some moving over oceans out of breeding season. Most feeding is on shallower inshore waters, plunging below surface. **Status:** widespread in southern oceans, usually N as far as S tip of NZ. Rare vagrant to Aust. waters where there have been two specimens: one SW of WA; one from near Kangaroo Is., SA.

Roseate Tern *Sterna dougallii* 31–38 cm

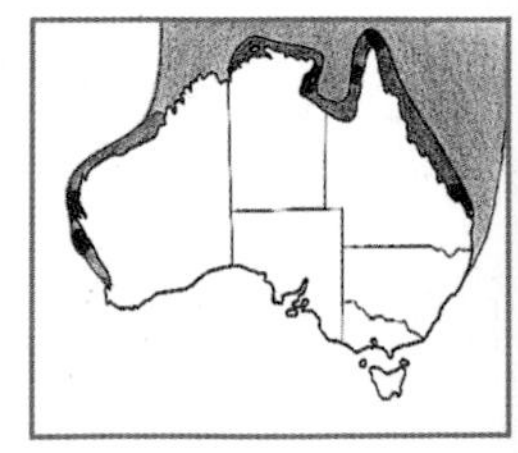

This is the smallest and palest of the group of terns known as the 'commic' terns, that is, all similar to the Common Tern. The slender body and long tail streamers give a delicate appearance in flight; wingbeats are shallow; flight is graceful; pale upper parts look almost white, and underparts are gleaming white. The faint pink blush for which the bird is named is evident only while breeding. On upper parts, contrast is lacking between wings and white rump. The flight is distinctive: direct and fast, not as wavering or buoyant as other commic terns. Tends also to feed over deeper waters; hovers then dives almost vertically from 2 to 10 m above the water, plunging to take small fish close to the surface. **Voice:** usual flight call is a rising 'ch-vriek' or clear 'chu-ick'. Also gives a deep, grating 'krarrk'. **Similar:** usually most difficulty arises from nonbreeding Common Tern, Arctic Tern, Black-naped Tern. **Hab.:** marine, coastal, in tropical and subtropical regions. Often around coral reefs, foraging over reef, reef lagoons and surrounding sea. Usually keeps away from mainland shoreline, but may use shallow waters just a hundred or so metres offshore. **Status:** common around N Aust., breeding on islands of both NE and W coasts, perhaps expanding southwards in WA where it is now a casual visitor, or a vagrant further S, as in NSW. **Br.:** 371

Dark edge
Breeding
Nonbr.
Mottled brown
Streaked crown
Brown ear coverts, nape
Black around nape
Heavy dark markings
Black
Brownish
Juv.
Grey
White, blue-grey in shadow
Black cap
Streaked
Peaked cap
Nonbreeding
Slightly forked
Red
Dark grey
White cheeks or 'whiskers'
Dark
White cheeks
Dark edge
Short
Pale
Dark edge
Dark
Black coverts
Dark outer primaries
White
Darker
Breeding: dark grey beneath
White
White
White collar
Streaked
Tail shorter than wings
Dark
White
Black
Breeding
White
Head, body, coverts, black
Dark edge and saddle
Nonbr.
Non-breeding
Red
Black head and body
Juv.
White wings conspicuous, surrounded by black
Pale
Dark
White vent to tail
Grey
Mid grey
Br.
Black, long
Black ear
Pale grey
Mottled black and brown
White
Black
White
Dark side-breast mark
Dark grey
Mottled grey-brown
Dark
Pale grey
Juv.
Head and body black, with folded wings mid grey
Black over eye and ear coverts
White, in contrast to black, from vent back to under tail, as with both above species
Nonbr. has similar head pattern, grey back, wings
Nonbr.
Juv.
Breeding
Red
Dark mark
Flight feathers and tail are translucent, transmitting sunlight from above, brilliant white.
Thicker structures not translucent, darker
Juv.
Steep high forehead
High rounded crown
Small dark patch at front of eye, dark eye-ring and dark eyes combine to give big-eyed appearance.
Brown nape
Long, finely tapered
Grey, mottled brown
Adult; no seasonal plumage differences
Fine black shafts to outer primaries and outer tail
Darker
Fine dark shafts
Dark eye and bill
Adult
Forked
Dark behind eye, brown nape
Short, black
Dark
Tail forked, falls short of tips of very long wings
Entirely white; grey is underwing shadow where not translucent.
Dark shafts to outer primaries
Brownish, wears grey
Bill black, conspicuous against white plumage; long, evenly tapered to a fine point; straight except slight upcurve of top edge towards the steep, high forehead.
Adult
Dark remiges in strong contrast with nearly white coverts
Upper wing both morphs grey; dark wingtips
Grey-brown
Pale blue-grey
Dark patch at front edge of eye makes dark eye, set in almost-white head, appear much larger.
Slightly downturned
Large dark eyes
Dark underwing coverts
Black with pinkish webs
Adult, pale morph
Medium forked
White underwing coverts
Bill slender, slightly downcurved
Mottled grey-brown, wearing towards grey
Dark morph
Medium-dark edge
Slightly mottled brownish grey, probably wears towards grey
Pale morph
Upper wing, both morphs: pale grey, darker grey-brown wingtips
Juv.
Juv.

Sooty Tern *Sterna fuscata* 38–45 cm (C, K)

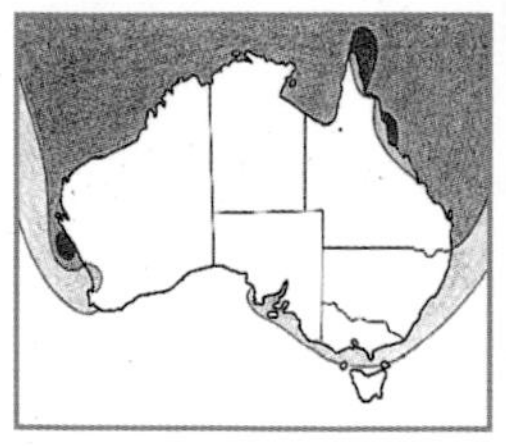

Black and white oceanic tern, feeds in mixed flocks, usually far out from land. Rarely alights on water; powerful flight, deep wingbeats; wheels and soars, dives close to water, skimming the surface for small fish and crustaceans; some nocturnal species also taken. **Voice:** does not often call at sea, but noisy in colonies with almost constant yapping calls, rather like yapping of a small dog, 'ker-waka-wak', from which comes the alternative common name, 'Wide-awake'. As alarm, a harsh, drawn out 'kraaak'. In colonies the incessant cacophony of hundreds of 'wak-awak' and similar calls intermingle, so that it is difficult to single out the calls of any individual birds. **Similar:** Bridled Tern, but more slender, longer legs and finer bill, very light grey rather than white underparts, brown back, longer eyebrow line. **Hab.:** oceanic, tropical and subtropical Indian and Pacific Oceans. Frequents and feeds over pelagic waters far from land, returns to land only rarely to breed or escape rough weather. Occasionally seen close inshore, on coast or inland after gales at sea. **Status:** breeding species off coasts of WA and Qld, occasional sightings further S; some colonies have thousands of birds. **Br.:** 371

Bridled Tern *Sterna anaethetus* 35–41 cm (K, L, N)

Superficially similar to the Sooty Tern, but distinctive in many details. In distance, lacks the strongly contrasting, crisp black and white pattern of the Sooty; instead has brown upper parts with dulled, faintly grey-tinted whites of underbody and underwing coverts. Closer, this tern has a narrow white forehead band and sharply pointed, narrow, triangular eyebrow lines that extend back behind the eyes, quite unlike the Sooty's broad white forehead patch, which does not extend behind the eyes. **Voice:** like yapping of a small dog—similar to the yap of a Black-winged Stilt; very abrupt, nasal 'ak, ak-ak, ak-ak-aak' or 'wep-wep, wep-wep-wup', and harsher, scolding 'airrrgh, arrrrgh'. Becomes very noisy if nesting colony is approached; much calling at night in breeding season. **Similar:** Sooty Tern. **Hab.:** tropical and subtropical seas, often far from land; usually forages on open seas, but frequents breeding islands, reefs and, occasionally, inshore waters. More likely to be seen from land than is Sooty Tern. **Status:** common around NW to NE Australian coasts; rare to absent in SE. **Br.:** 371

Common Noddy *Anous stolidus* 40–45 cm (C, K, L, N)

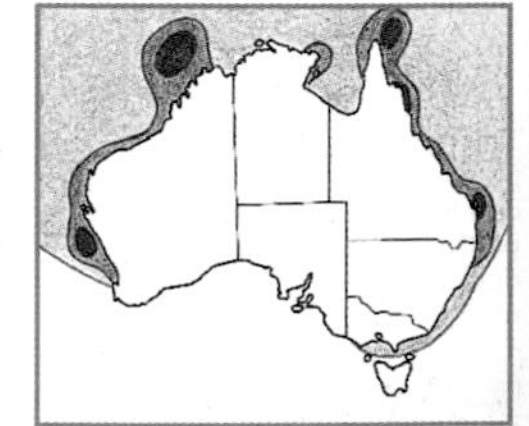

Large brown noddy with ashy white cap; in flight, paler grey under wing, two-toned brown on upper wing. Gregarious—often dense flocks attack shoals of fish, usually far out at sea. Large body, slower wingbeats give heavier jiz than other noddies. **Voice:** a grating or guttural 'urk-urk-rrk-rrk-rrkrrk' that combines in a colony as constant, murmuring background noise. In alarm or threat, harsh, guttural squawks: 'kar-rrark'; 'kraaaa'. **Juv.:** plain brown head, fine pale fringes to upper-wing covert. **Similar:** other noddies. Juveniles similar to those of Sooty and Bridled Terns. **Hab.:** when breeding, coastal waters in vicinity of colony islands; otherwise oceanic. **Status:** resident in large colonies on islands of mid west WA and Qld. **Br.:** 371

Black Noddy *Anous minutus* 35–40 cm (L, N)

Medium-sized, slender; all brownish black except for silvery white cap and pale central tail; reflective highlights along rear of underwing. Usually in flocks over schools of fish; hovers before dropping to the surface. Flight lighter, faster wingbeats, more fluttering than Common Noddy. **Voice:** harsh, fast, rattling, 'ak-ar-r-r-r-k, ak-ar-r-r-r-k', 'ak-ak-ak-akak', 'ak-aairak-ak-air-ark'; screeches in alarm. A colony vibrates with rattling noise. **Juv.:** as adult, larger white cap to nape, faint pale edges on upper feathering. **Hab.:** in Aust., the islands of the offshore reefs of Qld and the Kimberley coast of WA, and the surrounding seas and reefs. **Status:** breeding resident; common in region around colonies; casual visitor further S. **Br.:** 371

Lesser Noddy *Anous tenuirostris* 30–34 cm

A delicately plumaged, graceful noddy confined to the Indian Ocean. In flight, more like Black Noddy than larger Common Noddy. **Voice:** a purring sound, churring or soft rattling; similar to, but probably softer than, sounds of other noddies; a colony of thousands described as sounding like a 'vast purring chorus'. Also has louder rattle of alarm. **Hab.:** in Aust., low, flat coral–limestone isles in Abrolhos Group; reefs and lagoons with dense, low, fringing mangroves. Forages around islands, reefs and lagoons, and further out to sea. **Status:** rare in Aust., although locally abundant at breeding colonies; probably sedentary. **Br.:** 371

Adult
White, mottled grey tint
White edge
Tail black with white outer edges; deeply forked; adults have long streamers.
Grey, mottled brownish
Lacks tail streamers; fork is not as deep as that of adult.
Juv.
White
Black
Wide white forehead; white brow does not extend behind the eye.
Black
Nonbreeding may differ in having white streaked crown, and often shows a white hind-neck collar.
Long streamers
Adult
Black
White scalloped
Bill quite long, rather heavy
White
Predominantly dark sooty brown with white-scalloped back and wings, white vent to under-tail coverts
Juv.
Adult, fresh plumage
Dark brown, wears paler
Light grey
Forked, with streamers
Wears to thin collar
White edge
Under the body and under wings very light grey, palest on the breast, becoming lighter with age. Also with age and wear, a narrow white collar appears around the hind neck.
Linings pale grey, mottled
Shorter, forked
Juv.
Dull white
Streaked
Coverts edged buffy white
Narrow white band across forehead, with fine white point extending behind eye.
Pale brown
Brown upper parts
Adult
Streaked brown
Brown, edged white
Dark through eye, ear coverts
Juvenile lacks any tail streamers.
Juv.
Bill shorter than adult's, which is long, stout, straight, dark
Tail long, pointed if closed; if partly open, notched
Lighter dusky grey
White
Adult
Juv.
Juv.: pale coverts. On adult, paler with wear
Dark crown
Shallow tail notch
Largest noddy: white forehead, ashy grey crown merging to dark brown body
Ashy white forehead blends smoothly to grey-brown nape and darker neck
Face dark brown, meeting pale forehead in clear-edged black line through lores
Bill thicker than those of other noddies; slightly downcurved
Dark grey-brown
Adult
Partial eye-ring; thin, only below eye
Adult
Plumage more evenly black than that of other noddies; whiter cap in greater contrast, centre of tail pale
Glossy highlights
Brown-black, no pale bar
Pale
White
Compared with Common Noddy, above, Black Noddy is smaller, more slender, darker sooty black; has finer, straighter bill.
Adult
Brownish black
Bright silvery white cap
Long, slender, almost straight
Partial eye-ring, above as well as beneath eye
Lores, between eye and base of bill black, in stark contrast to the silvery white cap
Brownish black
Face and neck brownish black
Adult
Shallow fork; tip rounded when spread
Long, narrow wings
Adult
Light grey of crown blends softly, evenly, into blue-grey neck and darker body
Almost white, pearly grey forehead; dark oval in front of eye
Bill long, fine, very slightly downcurved
Lores pale grey; continuous with soft merging of forehead and lower face
Partial eye-ring above and beneath eye
Grey-brown
Adult
Juveniles are like adults, but whiter cap with sharper transition to nape; pale markings on mantle and scapulars
Primaries blacker
Tip of shallowly forked tail level with ends of folded wings

Banded Fruit-Dove *Ptilinopus cinctus* 33–35 cm

Identified by obvious black breast-band. Occurs in small groups, pairs or alone in pockets of rainforest; feeds on figs or other fruits early and late; shelters in dense foliage through heat of day. Shy and often well-concealed in foliage; never far from water—visits waterholes at least once a day; sudden take off causes loud wing-clap. Flight powerful; clearly audible, whistling wingbeats. **Voice:** advertising call a deep booming 'coo' repeated 6–8 times at intervals of several seconds. **Hab.:** rugged sandstone escarpment among gorges and cliffs in patches of monsoon forest with fruiting trees. **Status:** scarce; restricted, but secure. **Br.:** 372

White-headed Pigeon *Columba leucomela* 38–40 cm

Large; entire head and underparts white; back and wings black. One of the shyest and wariest of pigeons. Often the first sign of its presence is a loud clatter of wings as a flock panics into flight close overhead. **Voice:** a soft, high-low 'whOO-wuk, whOO-wuk'; the first half is loud, with emphasis on the resonant, musical 'OO', the second half, '-wuk', much softer, and fading away. **Hab.:** rainforest, favouring edges and regrowth; cleared land with abundant Camphor Laurel; suburban gardens. **Status:** scarce; common NE NSW; dispersive. **Br.:** 372

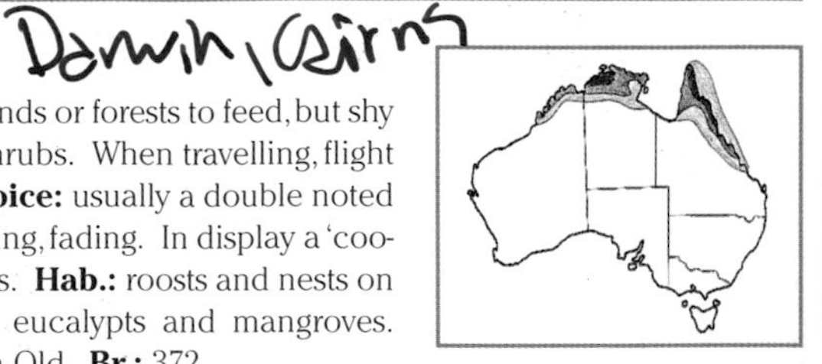

Pied Imperial-Pigeon *Ducula bicolor* 38–42 cm (C,T)

Large pigeon, conspicuous in big flocks travelling to roosting islands or forests to feed, but shy and wary in small parties feeding in foliage of fruiting trees or shrubs. When travelling, flight is direct, swift with powerful, steady uninterrupted wingbeats. **Voice:** usually a double noted 'ook! –whuuu'; the first part abrupt, the second drawn out, moaning, fading. In display a 'coo-whoo-hoo'; soft then loud, synchronised with bowing movements. **Hab.:** roosts and nests on offshore islands; commutes to mainland rainforests, adjacent eucalypts and mangroves. **Status:** sedentary or migratory; common, probably increasing in Qld. **Br.:** 372

Elegant Imperial-Pigeon *Ducula concinna* 44–48 cm

A recent vagrant to N Aust. from Indonesia. A new addition to the Aust. bird list; seen in Darwin, but just as likely to be encountered elsewhere along the N coast. **Voice:** a deep, gruff, growling 'urr-wooo' and a shorter, abrupt, almost barked version; the two intermixed or given by two birds—one growling, the other grunting. **Hab.:** within usual range, small tropical islands, in canopy of monsoon forest, forest edges and lightly wooded cultivated land. In NT, a sole bird occupied urban Darwin with remnant rainforest species, especially dense canopy of Banyan Fig, where it foraged, also feeding on Carpentaria Palm and other fruiting trees. **Status:** rare vagrant; usually nomadic; elsewhere common.

Collared Imperial-Pigeon *Ducula mullerii* 40–43 cm (T)

Unusual and distinctive plumage colour: maroon and soft mauve-pink set against dark slate-grey with diagnostic bold black neckband. A NG species that is poorly known; juvenile not as yet described. Solitary birds or groups cross narrow channel to Boigu Is., an Australian territory in Torres Str., to forage in canopy of fruiting trees. Flight strong, fast and direct. **Voice:** not well-known, but includes a rising sequence of about five cooing calls. **Hab.:** dense lowland rainforests, swamp forest or mangroves, usually near water of rivers or lakes. Feeds in fruiting trees in small parties, pairs or alone; usually seen only when flying across open space of waterways or above the forest canopy. **Status:** the species occurs as two populations on lowlands of S and N NG, separated by the mountainous interior, but linked by narrow coastal corridor; these are separate subspecies, the southern being the nominate race.

Topknot Pigeon *Lopholaimus antarcticus* 40–45 cm

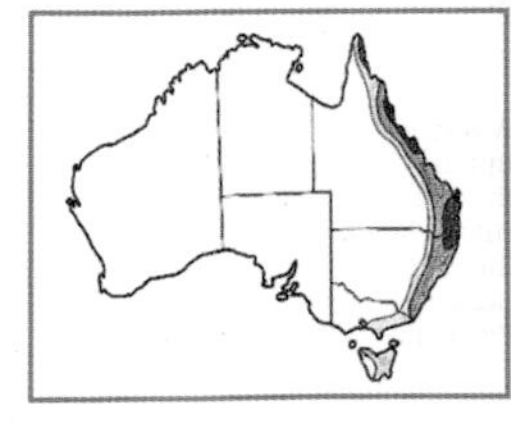

The unique backswept crest makes this species easily identified at close range; even far off, the bulky crest makes a distinctive silhouette. But the Topknot can be recognised by its flight; flocks of these very large birds fly across open country, travelling far and fast in search of rainforest trees with ripening fruit. Flocks of thousands were said to be common before extensive clearing and uncontrolled hunting. Flocks are now much smaller, but still wander the forests: are seen winging their way, powerful, fast and straight; swooping in long sloping glides on flat wings from ridge-top through to valley; or circling high and easily over forests. **Voice:** often silent; has a low, resonant, abrupt grunt 'whug, whug, whug'; in flock, feeding or fighting, short sharp screeches. **Hab.:** rainforests, remnant and regrowth, and nearby eucalypt forest; forages in high canopy, occasionally in shrub layer, but not on ground. **Status:** nomadic; common in Qld, N NSW; rare further S. **Br.:** 372

A large fruit-dove with head, neck and upper breast white
Greenish, yellow tip
Sooty black or slaty blue-black
Faint creamy tint
Black band
Grey
Adult ♂
Bronzed
White with very pale grey tint
Paler breast-band
Pink
Juv.
Similar: Pied Imperial-Pigeon, White-headed Pigeon. Both larger, all white under-body
Wings broad, rounded
Flight feathers, or remiges, glossy grey with dark shafts
Grey-brown outer wing
Underwing coverts grey
Adult female similar but with a green sheen on breast-band
Grey
Black tail
White tip
Red, tip yellow
Red eye-ring
White of head, neck and underparts in strong contrast to back and wings, which are black with a slight iridescent sheen of green and violet. Red accent on bill, eye-ring, feet
Adult
Dark, plain
White from head to under-tail coverts
Adult
In flight, bold visual contrast of pure white body carried powerfully and swiftly between entirely dark wings
On females, the white is a very light, slightly mottled, uneven grey; whiter with age. Juveniles even darker, and with dark cap
White, often with faint cream or grey tint on head and neck
Pale yellow
Black of folded wing makes a distinctive tapered curve of black band along the otherwise white body
Primaries, their coverts, and most secondaries are slaty black.
Remiges black; all coverts, axillaries white
A large pigeon. The pattern of black areas on wings and tail is diagnostic in flight and when folded. Juvenile differs—white parts are very light grey and buff tint to underbody is strongest on under-tail coverts, where black bars are only faint.
Black spots
White rump
Adult
Black
Black bars
White edged black
Barred
Adult
Iris yellow, rarely red
Faint pink tint
Underside of flight feathers black
Also known as the 'Yellow-eyed Imperial-Pigeon'. Birds in E of usual range have more coppery upper parts; western populations have more pink on nape. Juvenile is possibly similar but dull colours.
White at base of bill
Iridescent dark green or blue-green with coppery sheen under some lighting conditions
Green sheen
In length slightly exceeds well-known Pied Imperial-Pigeon, above. The dark plumage is iridescent, varying with lighting; dark green with purple or bronze sheen, or dull black.
Pale silvery grey
Iridescent greenish or purplish black
Black collar, broadest at nape, completely encircles the neck.
Variable, pink to deep crimson
Upper mantle glossy maroon or wine red
Chestnut
Throat, face, nape band are white or silvery grey.
Black
Chestnut
In flight, from beneath, pale grey body in strong contrast to black of wings and tail
Blue-black with iridescent green or purplish sheen
White band
Mauve-pink
Backswept grey fore-crest
Rusty rufous hind-crest
Smaller, often tinged rufous
Crest short, only rear crown
Breast feathers look spiky.
Slaty grey
Breast streaky, mottled rather than spiky appearance of adults
Slaty grey
Grey-brown
A very large pigeon with unique crests rising grey from the forehead and drooping, rufous or gingery, down over the nape to the back of the neck, forming a colourful and distinctive 'topknot'
Flight feathers grey-black
Grey wing linings
Back of tail darker, band wider, stronger contrast
White subterminal band on grey-black
Tail band dull, blurred
Adult
Juv.

Flock Bronzewing *Phaps histrionica* 27–31 cm

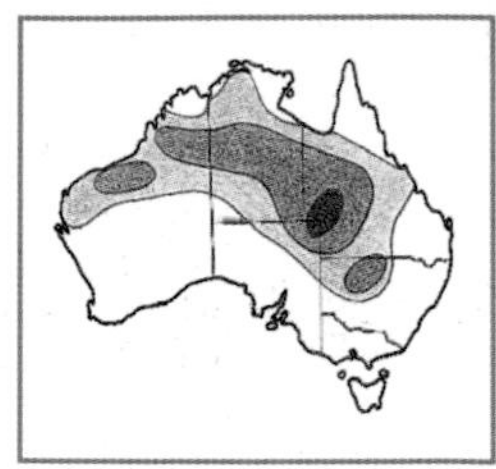

Nomadic inhabitant of grassland plains of N interior; formerly in immense flocks, still occurs in thousands. Dispersed on reddish ground, feeding birds are inconspicuous. Morning and evening visits to water can be spectacular; they arrive in tight flocks, crowding the water's edge; then burst up with a muffled roar of wings, displaying fanned, white-tipped tails. Wings create a loud clap on the first beat—really large flocks can sound like thunder or ocean breakers. Flight is fast and direct with deep, strong action of pointed, backswept wings. **Voice:** no audible calls at water or in flight. Birds in feeding flocks give a soft, low, drawn out moaning at even pitch and volume; in display a 'wook' call. **Hab.:** open grassland plains, clumped grasses, small shrubs with open spaces. **Status:** common; erratic occurrence. **Br.:** 372

Crested Pigeon *Ocyphaps lophotes* 31–35 cm

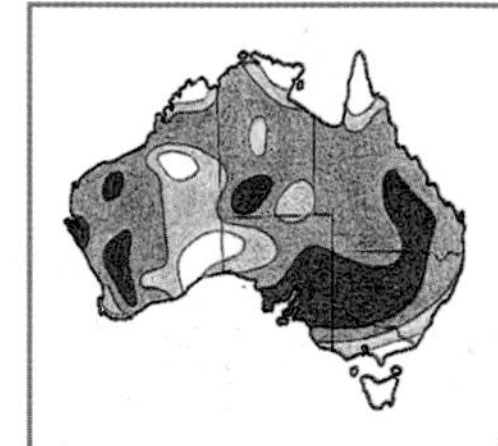

Usually in small flocks; feeds on ground. In flight, accelerates with vigorous bursts of whistling wingbeats interspersed with fast, direct glides with wings held stiff and flat; on landing, tips forward, long tail lifting vertically. **Juv.:** short crest, dull colours. **Voice:** a rather musical 'whoo'; starts softly, lifts up and strengthens, ending abruptly, 'whoo,– whoo,– whoo–'; also a very quick 'woop!'. **Hab.:** open woodlands, acacia scrublands, farmlands, roadside tree lines, homesteads and yards, always near water; avoids dense or wet coastal habitats. **Status:** very widespread and common; expanding range as heavier vegetation is cleared. **Br.:** 372

Spinifex Pigeon *Geophaps plumifera* 20–24 cm

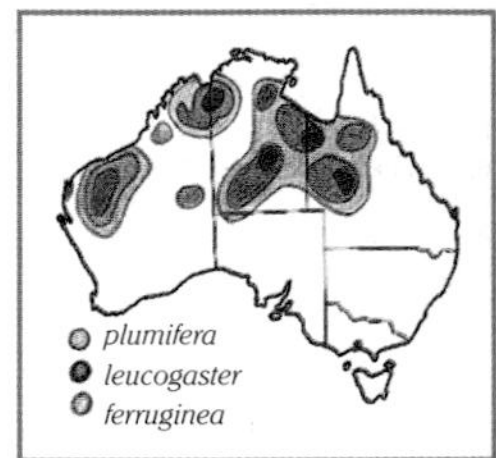

Unique little reddish, plump, upright, red-crested, ground-feeding pigeon—unmistakeable. Small parties live on rocky ground; blends with red rock and gold spinifex. Runs erratically, stopping to stand tall, alert; flushes abruptly, quail-like, with loud whirr of wings; glides away on stiff downcurved wings, giving a brief glimpse of rufous back and wings. **Juv.:** paler, markings similar to adult but weaker; orbital skin khaki; iris pale yellow. **Variation:** many slightly differing populations; opinions differ whether variation is clinal, or justifies subspecies. **Voice:** a soft, husky, throaty yet musical 'coo, coo-roo!' with final 'roo!' uplifting and clearer, the sequence repeated for long periods. **Hab.:** usually in spinifex of rocky ranges, gorges, less often in sandy country; never far from water. **Status:** sedentary; common. **Br.:** 372

Common Bronzewing *Phaps chalcoptera* 30–36 cm

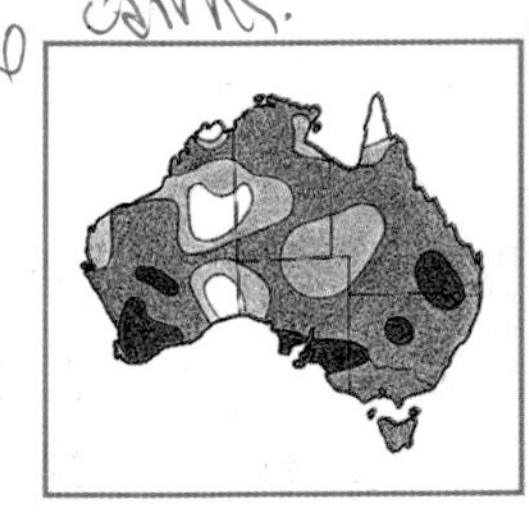

Large, bulky bodied, small-headed, ground-foraging pigeon. Solitary or in small parties; feeds by pecking about on ground for seeds. If flushed, takes off with a clatter of rufous-lined wings; often flies up into tree foliage and watches warily. Flight very fast, direct with strong, deliberate wingbeats. **Juv.:** similar to adults; dull colours, little iridescence. **Variation:** no subspecies, but considerable clinal variation: in Tas. is larger with brighter chestnut underwings; Central Aust., paler with cinnamon underwings; Kimberley, smaller and paler. **Voice:** advertising call a far-carrying, mournful, slow, resonant and deeply vibrating 'whooo'; begins softly, is carried through at very even, low pitch, dropping slightly to finish; lasts almost a second, and is repeated at several second intervals. The display call is very different—a short, soft 'whoo-, hoo-hoo'. **Similar:** Brush Bronzewing. **Hab.:** extremely diverse, covering most of the continent: coastal forests, woodlands, arid scrubs, mallee, heaths, alpine woodland, farmland. **Status:** endemic; locally nomadic; usually common, scarce in very arid regions. **Br.:** 372

Brush Bronzewing *Phaps elegans* 25–33 cm (L)

Similar to but slightly smaller and more colourful than the Common Bronzewing, with more rufous to sides of neck and breast. Forages on ground beneath shrubbery. More secretive and wary than Common Bronzewing; flushes suddenly with loud wing-clapping. Flight fast, direct; may give glimpse of chestnut shoulders, rufous underwing. **Juv.:** like adult but dull, dark; face grey with only a trace of rufous on throat; darker plumage with paler edges. **Variation:** in SW of WA (possibly a race, '*occidentalis*') forehead may be paler buff, back more olive; in Tas., darker. **Voice:** higher, shorter, faster than call of Common Bronzewing, lacking the slight downward slur; a repeated 'hoo, hoo, hoo, hoo', each call seeming to grow slightly higher and quicker with building excitement. **Hab.:** heathlands and habitats with dense groundcover dominated by shrubs and scattered trees of banksia, leptospermum, casuarina. Also in shrubby undergrowth of open woodlands, mallee thickets, heath. **Status:** uncommon, preferring undisturbed habitats; affected by grazing, burning. **Br.:** 372

Nape to tail coverts and inner wing coverts rufous-brown in fresh plumage, wearing or fading to a much paler sandy or cinnamon brown
Black head, rather comical face; white around bill and white line swirling back around ear coverts
Long, sandy rufous upper-tail coverts hide folded tail or centre of fanned tail.
Blue-grey
Female face pattern like male, but often with much less black
In distant flocks, the rufous and blue-grey helps identify the species.
Primaries white tipped
Some females have more black on the face, and are closer to male's head pattern.
♂
Juv.: like paler faced females; muted colours, with pale scallops on breast, back, rump, and wing coverts
Head grey with fine pointed, long, dark, usually upright, crest
♀
Feet dusky pink
Innermost flight feathers darkest
Tail grey, darker towards white tip
Orbital ring red, thick; iris orange
Soft fawny brown; dark bars across wing coverts
Speculum iridescent green, violet and gold, edged with white
Adult
Feet deep pink
Race *plumifera,* white-bellied, Victoria R., NT, and E Kimberley region
Upright crest, rusty buff like body
Bare red skin around red eye
Race *leucogaster* like *plumifera*, but upper parts slightly more rufous in Central Aust. or paler, sandy, buff-brown towards N Qld; intermediate but rather variable extent of white on belly
Race *ferruginea*, Pilbara, WA; entirely red-bellied, and overall brightest rufous of the races
Face pattern is the same for all races.
Three subspecies of Spinifex Pigeon recognised, but hybridisation occurs between adjoining populations. A general trend in colour from brighter rufous in the W to paler sandy tones in the NE.
White
Black
Variable extent of white
White, variable extent, medium to large area
Entirely bright rufous
Nominate race *plumifera,* large white patch on belly, narrow breast-bands of both white and black. Overall plumage tone is light rufous-brown.
Adult
Adult
Adult
Buff
Pale blue-grey
White cheek stripe curves back and wider, down towards blue-sided neck.
Blue of neck blends into pinkish grey breast.
Less blue down sides of neck
Underwing rufous with darker trailing edge. When flushed, takes off with clatter of wings lifted high, giving a brief glimpse of colour that is usually hidden.
Dark olive-brown with multicoloured iridescence across all wing coverts
Iridescent colours of wing coverts vary with direction of light: green, bronze, violet, magenta or black.
♂
♀
Blue-grey, tipped white
♂
Blue-grey
Cinnamon or chestnut, a deeper colour than on Common Bronzewing
Light blue-grey
Rufous underwing
White cheek line merges back and down into pale blue-grey of face
Chestnut
Rich chestnut, from nape down sides of neck, wrapping around to sides of breast
Throat patch very small, pale rufous
Soft greyish blue extends from face to fore neck and entire underbody.
Changing green, bronze, violet iridescent colours in two bands across coverts
Sides of neck olive-brown rather than the chestnut of the male
Deep olive-brown
Deep, dull pink
Upper tail brown, dark band near tip of outer feathers
Blue-grey
♂
♀
♂
White tips under tail

Squatter Pigeon *Geophaps scripta* 26–30 cm

A heavily built, ground-dwelling pigeon, usually in small flocks or pairs; responds to disturbance by 'freezing' or running, darting erratically among grass tussocks. If pushed too closely, a flock will burst into flight with a clapping of wings; birds scatter widely to trees or back to ground. Flight fast—flapping interspersed with swift glides on stiff, downcurved wings. **Variation:** N form, central Qld to Cape York Peninsula, is probably a race, *peninsulae*, but variation may prove clinal. **Voice:** low, deep murmuring, seeming to come from deep within, muffled or throaty, yet rather musical; sequence is often varied—fast, slow, slow *or* slow, fast, fast—'coo-coo, coore, coore, coo-coo, cooor'. Also higher, uplifting 'coo-hooop!'. **Similar:** Partridge Pigeon, similar colour and flank markings, but different head pattern. **Hab.:** open grassy woodlands; prefers sandy soils interspersed with low gravelly ridges. These poorer soils have more open, shorter grass cover that allows easier, faster movement than in the densely matted grass of rich blacksoil country. Birds are never far from water. **Status:** widespread NE Aust.; uncommon to locally common; rare in S parts of range; sedentary, partly nomadic. **Br.:** 372

Partridge Pigeon *Geophaps smithii* 25–28 cm

This species is the NW replacement for the Squatter Pigeon of NE Aust. and shares its choice of habitat; both are terrestrial, feeding, nesting and usually roosting on the ground, colours blending well with environs. Will squat or stand motionless if approached, then runs erratically through the grass or leaps into flight with loudly whirring wings. Other behaviour shared by both species includes a breeding season that extends throughout the year with maximum activity in the midwinter dry season. Like the Squatter, a gregarious species; usually in small parties; occasionally forms flocks of many hundreds. **Variation:** race *blaauwi*, N of WA. **Voice:** similar to calls of Squatter, but slightly more husky and querulous; slow, varying in strength—some strong, some softly fading and very low, often just single notes or a series of notes more widely spaced, 'khwoor, khwuoor, kwoo-kwoo-kwooor, khwuoor'; higher, clearer 'oo-poo-poor' in alarm. **Hab.:** open tropical woodlands, often on sandy, stony ground with short, rather sparse grass. **Status:** endemic; locally common; sedentary. **Br.:** 373

White-quilled Rock-Pigeon *Petrophassa albipennis* 28–30 cm

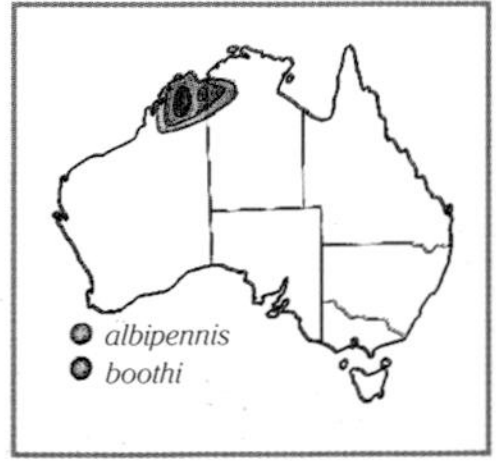

A bulky, small-headed, dark brown pigeon with horizontal, droop-winged stance that gives a low, humpbacked, short-legged shape. With the attenuated rear end and tail and the dark, scaly looking plumage, this and the similar Chestnut-quilled Rock-Pigeon appear almost reptilian. Lives in small flocks, pairs or alone in rugged sandstone escarpment country; usually seen perched on huge boulders or ledges of cliffs where they run with agility despite their low, short-legged appearance. From these heights the pigeons launch themselves into flight, gliding down to search for seeds on the ground among boulders or out on grassy, open woodland nearby. If startled, they take off with a loud clapping of wings, flying up to the sanctuary of the cliffs where they characteristically run a few metres across the rock after landing, sometimes hiding in rock crevices. **Variation:** nominate race *albipennis* (NW Kimberley from N of Derby to Kununurra in E Kimberley and NE of Victoria R., NT) always has variable white wing patches and dark brown to reddish brown plumage; race *boothi* (NW of NT, from Stokes Ra. to Waterloo Stn) has reddish brown plumage and little, if any, white across outer wings. **Voice:** starts with rough, low, grating 'grr-'; rises to an abrupt, high, sharp '-oook'; repeated at regular intervals, 'grr-oook, grr-oook'. Also has a softer, more husky or throaty 'coo, car-ook'. **Similar:** Chestnut-quilled Rock-Pigeon, but wing patches are chestnut, and much further E on Kakadu escarpments. **Hab.:** rugged sandstone ranges, gullies, gorges in pockets of monsoon forests, figs and adjoining eucalypt woodlands. **Status:** locally common in suitable habitat; sedentary. **Br.:** 373

Chestnut-quilled Rock-Pigeon *Petrophassa rufipennis* 29–32 cm

The Chestnut-quilled, like the White-quilled Rock-Pigeon, is a bird specialised for the habitat provided by a restricted type of environment—the cliffs and gorges of rugged sandstone escarpments scattered across N Aust. The very limited habitat would always restrict the total population, while the intervening flat country breaks these sedentary rock-pigeons into separated populations, facilitating the evolution of two species, and of subspecies also among the widespread populations of White-quilled Rock-Pigeons. In flight, behaviour and use of habitats, this species is like the White-quilled. **Voice:** a loud, musical, rapid 'cook-ar-ook' like the Aboriginal name, Kukarook. The display call is a low, grinding, grating 'owrrgh-owrrgh-owrrgh', uttered at the bottom of each bow. **Hab.:** like that of the White-quilled. Neither species will be far from water, usually in pools of gorges. **Status:** restricted range, but locally common. **Br.:** 373

Distinctive head pattern: twin white vertical stripes curve down and are made conspicuous by black surrounds.
Race *scripta* is identified by bare blue skin around the eye.
Race *peninsulae* is identified by bare red skin around the eye.
The two forms (races, or representing clinal variation, and differing mainly in colour of facial skin) appear to meet in a broad but poorly understood hybrid zone in central Qld around the Einasleigh uplands and N Burdekin lowlands.
Pale edges give wings a scaly pattern.
Plumage of both races identical, dominated by dull olive-brown
Blue-grey breast tapers downwards to a triangular point when seen from front.
Iridescent green and violet spots on secondary coverts form a band across brown folded wings, both races.
White flanks curve up around wing.
Adult: nominate race *scripta*
Adult: race *peninsulae*
Bare orbital skin around eye is red.
Fine white and black brow lines follow top edge of coloured skin; on race *blaauwi*, white brow may be only back to the eye, then black.
Race *blaauwi*, orbital skin deep yellow
Dark brown cheek patch outlined in white
Squatter and Partridge Pigeons are very much alike, both in appearance and in their use of habitat, but each occupies a separate, non-overlapping portion of tropical N Aust.
The similarity between the two species is heightened because each has a subspecies that differs in the colour of bare skin forming the mask around the eyes, but which differs only slightly, if at all, in plumage detail.
Dull, greyish olive-brown
Small blue-grey patch scalloped black
In flight, short rounded wings
White flanks curve up in front of wing.
Identifying features of plumage are same as described for nominate race.
Tail is almost hidden under long coverts, brown with dark band across all but central feathers.
Adult: nominate race *smithii*
Feet dull pink
Adult: race *blaauwi*
Rock-Pigeons are large-bodied, small-headed, ground-dwelling birds: two species, both plumaged dark brown with small identifying differences.
White line under eye, downturned at ear coverts
Short, broad, rounded wings
Large white patch across primaries
Nominate, darker race *albipennis*, W Kimberley, has large white wing patches; its E Kimberley form is smaller, reddish, but still has quite large white patches across primaries.
Attenuated rear body: back, rump and tail coverts have dark, scaly feather pattern.
Tail long, rounded
In flight, reddish race *boothi* of NW of NT would show little, if any, white on outer wings. Body colour may be linked to rock colour in the habitat of a population. Juveniles of both races are almost identical to adults.
Adult: nominate race *albipennis*
White 'quill', the white base of outer primaries
Adult: race *boothi*
Race *boothi* has plumage lighter brown and very small, if any, patch of white on primaries; white edge very small or absent in folded wing.
Bright chestnut across the outer wing
Head and neck flecked with white
When in flight, the reddish or white wing patches of both species are conspicuous, but vanish as wings fold on landing so that these dark birds appear to vanish; this, combined with the habit of running after landing, could mislead any pursuing predator.
Thin but prominent line under eye, back around ear coverts
Sooty black speckled silvery grey
Entirely sooty brown without lighter or darker bands or tips
Dark sooty, blackish brown with paler brown scale-like feather edges
Sooty, brownish black feathers edged paler brown give scaly look to the long expanse of rump and upper-tail coverts.
Edge of outer primaries may show as an orange-chestnut streak low on the folded wing.

Wompoo Fruit-Dove *Ptilinopus magnificus* 38–48 cm

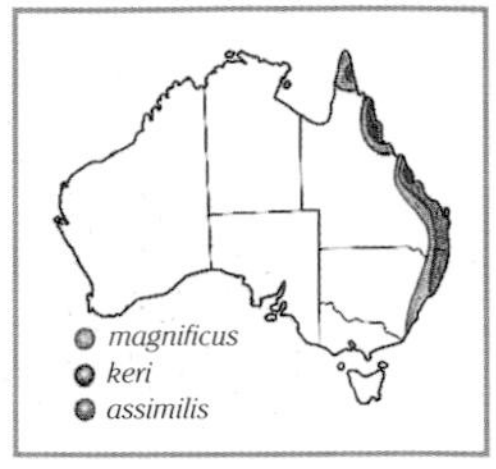

Despite their size, these birds are hard to find in the rainforest foliage canopy; they often stay in one big tree through the day, climbing in agile, parrot-like manner among slender outer branchlets to reach fruit. Presence sometimes revealed by sounds of falling fruit, or by early morning calls, more often in the breeding season. Sexes similar, but male slightly larger. **Voice:** the advertising call is very loud, audible up to 1 km away. Repeated lengthily, but hard to determine its direction. This call is powerful, deep, reverberating, much as in the name 'wom-poo', but the first part like the deep, inward 'plonk' of a rock dropped in deep water: 'g'lonk-ooo', 'g'lonka-ooo' or 'wolloka-ooo'. **Similar:** no other species has similar colours and such large size. **Hab.:** tropical and subtropical rainforest of lowland and ranges; also monsoon forests and closed gallery forests of Cape York; occasionally temperate rainforests of SE Qld and NSW; wet eucalypt forests near rainforests. Needs large blocks of undisturbed forest. **Status:** sedentary, localised nomadic wanderings in search of fruiting trees. Quite abundant in NE Qld; declining southwards—rare in NE of NSW, almost extinct S of Sydney. **Br.:** 373

Superb Fruit-Dove *Ptilinopus superbus* 22–24 cm

A small fruit-dove. The male is resplendent in bright colours, yet is often difficult to sight; small parties and pairs feed high in the canopy of fruiting rainforest trees. Flight fast, direct, through canopy spaces with whistling sound from wings. **Voice:** a series of mellow, musical calls, beginning softly, slowly, rising in pitch and volume until a loud and clear 'whoop, whoop, whooop'; also low 'oom', in steady sequence. **Hab.:** rainforest and similar closed forests—monsoon forest, regrowth, lantana thickets, woodland adjoining rainforest at all altitudes. Most foraging is done within rainforest, usually high, but lower when shrubbery carries fruits. **Status:** quite common in N parts of range; considered endangered in NSW. **Br.:** 373

Rose-crowned Fruit-Dove *Ptilinopus regina* 22–24 cm (N)

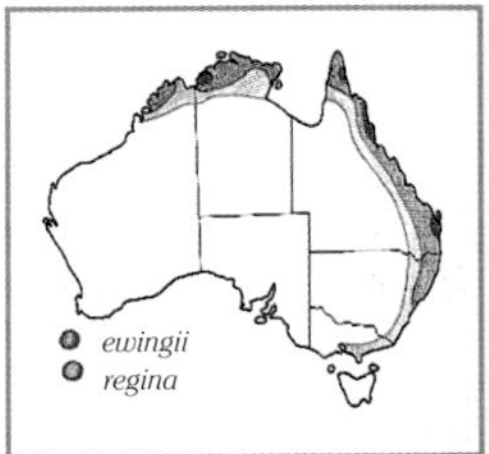

Similar to Superb, but different enough for adult males to be easily separated. **Fem.:** Rose-crowned female only slightly different from male; slightly greener hind neck, weaker yellow and pink on underbody. **Juv.:** entirely green and yellow, belly yellow. **Variation:** nominate race *regina*, E coast of Qld and NSW; race *ewingii*, NT and WA. **Voice:** noisy, loud calls given in long series: 'whup-whooo, whp-whooo, whp-whooo'; the first part low and abrupt, followed by a drawn out 'whooo'. Also as 'whu-whoo, whu-whoo-whuk-whukwuk...', becoming faster and fading away. **Hab.:** rainforest, monsoon forest, vine scrubs, mangroves, swampy woodland. **Status:** quite common; migratory, nomadic, dispersive. **Br.:** 373

Emerald Dove *Chalcophaps indica* 23–27 cm (C, L, N, T)

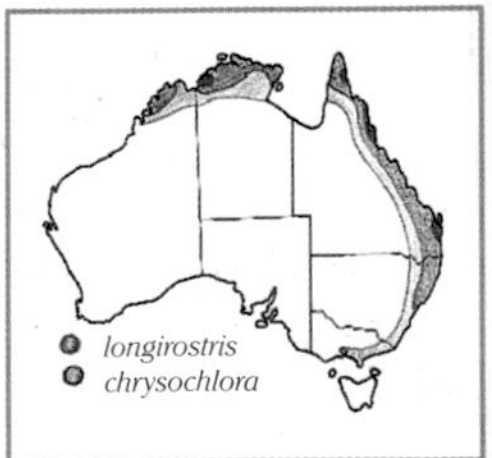

Small, plump, green and brown dove widespread through SE Asia and Aust. Typically seen in pairs, rarely in larger groups; forages on ground or, at times, feeds on seeds or fruits accessible in trees. **Variation:** in Aust. 2 races: *longirostris*, coastal NT, Kimberley; *chrysochlora* (*rogersi*) E coastal Qld, NSW, and Vic. **Voice:** as advertising call gives a series that begins with an extremely low purring-moaning that gradually, with each call, rises higher: 'p-ur-r-r-oom, p-r-r-ooom, p-rrooom, p-rooom, prooom'. **Hab.:** rainforests, monsoon forests, wet eucalypt forests, melaleuca woodlands, lantana thickets, regrowth scrub along creeks. **Status:** abundant N and NE Aust., declining towards SE end of its range. Dispersive, nomadic. **Br.:** 373

Brown Cuckoo-Dove *Macropygia amboinensis* 40–45 cm

Unmistakeable large, long, slender, pigeon-like dove; its call is one of the most common rainforest sounds. Typically a bird of the forest edges; it is seen along tracks, in clearings and regrowth in small parties or pairs, feeding in the vegetation or on the ground. Put to flight, it vanishes swiftly into forest with easy, strong wingbeats. **Variation:** in Aust., 2 or 3 races (or races plus clinal variations). From S to N are probably *phasianella*, *quinkan* and *robinsoni*. **Voice:** a loud and almost ringing call, which at a distance sounds like a single, clear, rising 'waaalk'; starts low, rises in pitch and volume; changes from soft, low and slightly husky to loud, clear and penetrating: a sound that seems to resonate through the forest. Close at hand, two soft, quiet, low preceding notes may be heard, like 'cook-coo' but more abrupt, giving the complete call as 'kuk-k-wawrk' or 'cuk-c-waaalk'—the whole about one full second from low, softly stuttered beginning to loud, clear, ringing end. A call given in display is similar, but more harsh and gruff. **Hab.:** rainforest, highland as well as lowland, and wet eucalypt forest, with a preference for the margins; brigalow scrubs, regrowth, thickets of lantana, wild tobacco. **Status:** common, ranging from abundant in some northern areas to scarce in SE. Locally, seasonally migratory or nomadic. **Br.:** 373

Head and neck white and pale grey merging gradually, softly into light yellowish green
Bill red, tipped yellow; eye red
Deep purple of breast tapers to narrow point at throat.
Bright light green with broken golden wing-bar
Adult
Deep golden yellow
Largest Australian Fruit-dove, including long tail, almost half a metre long, around double the length of the small fruit-doves below, and similar or slightly longer than the Brown Cuckoo-Dove, but much more bulky
Greenish grey
Blotched green
Edge of gold of underwing, juvenile or adult
Yellow spots fewer, paler
Juv.
Dulled yellow
Grey, both juv. and adult
Flight feathers grey tipped, chestnut towards base
Underwing linings deep golden yellow, conspicuous in flight
Purplish black in shadow
Race *keri*, NE Qld, slightly darker green, breast more maroon than purple; race *assimilis*, Cape York, breast paler purplish maroon
Vermilion
Purple
Male distinctive; a rainbow of bright colours
Golden green, blue-black centres
Streaky grey
Blue-black
♂
Deep pink
Brilliant light green
Purple spot
Patchy green band
♀
Light leaf green
Mottled green-grey-white
Gold edged wing coverts
Juv.
Grey undertail
Short, white tip
♂
Grey, darkest on flight feathers
Yellow line
Deep violet to rose pink
Dark blue-green with yellow edges
Rough grey
Rose
♂ Golden
Rough texture of grey central breast is created by pointed, slightly protruding feathers.
Paler pink
Yellow above weak rose belly
♂ Race *ewingii*
Underwings darker against the light
Deep yellow
Yellow
♂ Race *regina*
In flight overhead, any yellow confirms fruit-dove to be Rose-crowned, not Superb. Note that the Wompoo also has yellow underbody, but also has extensive yellow under wings, dark purple from breast to throat; much larger.
Iridescent emerald green in colourful contrast against adjoining warm brown tones of the body
Orange to red
Pinkish or slightly violet-tinted brown
Twin pale grey bars
White shoulder
Pink
When spread, black band across all but central feathers
Adult
Brown, extensively barred darker
Iridescent green on wings edged with chestnut
Juv.
Dark brown, with black band
Adult
Underwing cinnamon or chestnut with dark trailing edge
Male has slight iridescence on neck and mantle, reflecting tints of violet or green depending on the light direction.
Fine white line under eye
Coppery rufous
Cinnamon-buff underparts
Wings dark brown
♂
General impression: very long bodied, slender, small headed. Upper parts coppery or rufous-brown
Extremely long, tapering tail
♀
Visible only at close range is a small area of blue-grey skin around the eye and the red orbital ring.
Fem: rufous cap
Scalloped pattern to sides of neck
Rufous-brown
Brighter rufous than adult male
Fine dark barring to throat, neck
Dark scalloping around neck, unlike smooth-necked appearance of adult
Wing coverts have brighter rufous fringes.
Juv.

Wonga Pigeon *Leucosarcia melanoleuca* 35–40 cm

A large, plump grey and white pigeon with distinctive markings. Forages on ground, not often seen in flight unless flushed. Takes off with loud wing-claps; soon lands, may lift tail showing black flecked under-tail coverts. Glides on flat wings with occasional quick flicking wing-flaps. **Juv.:** like adult, slightly browner, dull, frontal 'V' shapes not quite as bold. **Voice:** long series —quite high, rapid, very penetrating 'whoik-whoik-whoik-whoik-'—repeated lengthily at constant pitch becoming monotonous. **Hab.:** rainforest, vine thickets, eucalypt forests and woodlands. Inland can be found in brigalow forest, tea-tree thickets, deciduous vine scrubs. **Status:** sedentary; uncommon, although can be locally abundant in areas of most favourable habitat. **Br.:** 373

Feral Pigeon *Columba livia* 33–35 cm (L, N)

Other names: Domestic Pigeon, Rock Dove, Feral Pigeon, Homing Pigeon. **Variation:** huge diversity of cultivated strains. **Voice:** cooing. **Hab.:** towns, other built environs and surrounding country. **Status:** introduced soon after European settlement; now widespread, mainly in the vicinity of towns and farms.

Peaceful Dove *Geopelia placida (striata)* 20–24 cm (L, T)

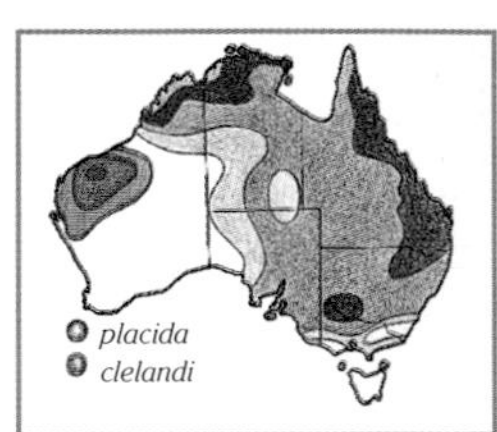

Small dove; forages on ground; roosts, rests and nests in shrubs or trees. Undulating flight gives glimpses of rufous wing linings, dark flights and pale tips to dark outer tail. **Juv.:** like adults, but barring less distinct; pale tips to some of upper parts. **Variation:** in Pilbara, race *clelandi*, very slightly warmer tone overall. **Voice:** a pleasant, musical, lilting, loud and clear 'coo-wi-ook, coo-wiook, coo-iook, cooiook'; also a throaty but musical 'quo-r-r-r-r'. **Hab.:** open forest, woodland, tall shrubland (usually of acacias), watercourse tree lines through arid country. **Status:** very common in better watered regions of N; sedentary. **Br.:** 373

Diamond Dove *Geopelia cuneata* 20–24 cm

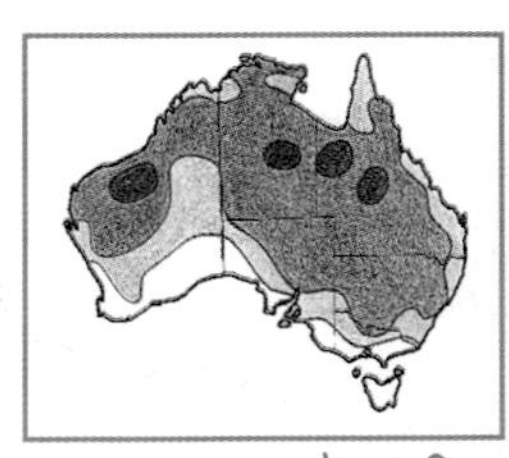

Tiny red-eyed dove; usually small parties feed on the ground, running fast if disturbed, or flying with whistling sound from wings. Flight swift, undulating yet direct. **Voice:** two versions given, each of four notes, all of clear, pleasant, musical character with slight but distinctive reverberation. One call has the four notes alternating low-high, low-high 'coor-cooo, coor-cooo'; alternatively, the four notes are quite level in pitch, 'coo-cooo, coo-cooo'; also gives a drawn out 'quorrr'. **Hab.:** widespread in dry areas—grassy woodlands, semi-arid grasslands, spinifex, dry mulga and similar dry scrublands—but never far from water. **Status:** common, but fluctuating, nomadic populations that move to coastal SE when interior is in drought. **Br.:** 373

Bar-shouldered Dove *Geopelia humeralis* 27–30 cm (T)

Medium size, long tailed; feeds on open ground near cover; roosts and nests in shrubs or trees. **Variation:** gradually paler SW from NT; opinions differ whether this is clinal (no races) or several races. **Voice:** a double-noted, cheery 'cookaw-cookor', 'cookaw-cookor'; first part is higher, stronger and clearer, the second slightly lower, fading. Also, long bubbling, laughing, descending 'cook-aw-cookawcookawcookaw–'. **Hab.:** usually near wetlands, especially inland; woodlands, forests, monsoon and vine scrubs, mangroves, brigalow, spinifex scrubs. **Status:** very common across N Aust.; sedentary; uncommon or vagrant in S. **Br.:** 373

Spotted Turtle-Dove *Streptopelia chinensis* 30–32 cm (L)

Introduced around 1860–1920; two subspecies that have since interbred in some localities. **Voice:** a high, triple-noted, musical cooing, 'cook-oo-ook', 'coo-coo-crooo' with other variations. **Hab.:** mainly suburban; parks, gardens, remnant vegetation, farms, plantations. Rarely in undisturbed natural bushland. **Status:** common, and spreading to country towns.

Laughing Turtle-Dove *Streptopelia senegalensis* 25–26 cm

Small introduced dove; in flight shows conspicuous white outer tail feathers. **Voice:** laughing, undulating or bubbling, mellow sequence like 'did-you-see-a-cuckooo', the last word loudest, clearest: 'quook-kuk-a-kuk-KUKooo'. **Hab.:** city, suburbs, country towns, parks, schoolyards, rail yards, farms, occasionally pine plantations; rarely in undisturbed natural bushland. **Status:** introduced to Perth around 1898; now common through most of SW wheatbelt, towns and farms.

White forehead and face
Double grey 'V'
Double white 'V'
Distinctive double 'V' pattern in grey or white with all markings from throat and sides of breast pointing to the centre of the breast
Wings grey-brown
Underside of body and tail coverts heavily spotted.
Adult
Dark slate grey
Base of bill red
Wings grey-brown
Grey and white 'V' markings all point to the centre of the breast in diagnostic pattern.
White with large, dark pointed or crescent-shaped spots
Adult
Feet pink to red
The Feral Pigeon (far right) is an introduced species, now widespread and common.
Breeding of the Feral Pigeon (or Rock Dove) for show and racing has created an almost infinite variety of colours and fancy ruffs or wattles; these types continue to be released or escape into the wild.
Short dark bill, large cere
Iridescent sheen, purple and green
The feral population has diverse colours, but the plumage illustrated has some commonly seen features and approximates the ancestral type.
Upper parts blue-grey, dark bars across wings
Underparts often paler
Red
Tail black or black tipped
In flight, rufous underwing coverts and grey flight feathers show clearly.
Fine dark cross-barring
Eye-ring, lores, cere and iris are all blue-grey; bill darker blue-grey.
White
Scalloped black and white
Heavily barred
Faint tint of pink
Feet deep pink
Long tapering tail
Clear white spots
Bright red eye and orbital ring
In flight the strongly tapered tail is held closed as a long, narrow trailing point. When fanned out at take off or landing, shows dark outer feathers with white tips.
Brown
Soft pale blue-grey delicately merging into off-white underparts
Long, tapered. If spread, dark outer tail is white tipped.
Dark barred and white tipped
Finely barred
Adult ♂
Juv.
Adult female like male, very slightly browner tone to upper parts' plumage; orbital ring slightly dulled red
Smallest Aust. dove; distinctive with bright red eye, exaggerated greatly by thick red eye-ring, conspicuous against blue-grey
Possible races are *huneralis* in E; *inexpectata* from N Qld to Kimberley; and palest, *headlandi* in Pilbara region of WA.
Rufous coppery with dark barring
Dull pink, bare, orbital skin is set in soft blue-grey plumage that extends from face to fore neck, upper breast, and dark barred crown.
Sexes alike, orbital skin pinkish in breeding season, otherwise greyish. Juvenile dull, without coppery hind neck, dark barring on fore neck. Plumage becomes gradually paler from NT through inland Kimberley and into Pilbara
Tail grey-brown with white tips to all but the central feathers; white is usually obvious only when tail is spread in flight.
Blue-grey, not barred
In flight, entire underwing shows cinnamon or chestnut with dark trailing edge; strongly contrasting colour flashes against white flanks. Upper wing also has a panel of chestnut, similar to that of hind neck but smaller, across primaries and primary coverts, and conspicuous in flight.
Adult
Feet deep pink or pinkish brown
Introduced species, now widespread and common
In flight, spread tail is dark brown with large white tips to the outer four feathers on each side; these are hidden under central two pairs of plain brown tips when the tail is folded. Outer wings and secondaries are plain dark brown, both on upper and lower surfaces. The spotted dark nape is usually visible.
Pale grey
Neck patch black with white spots
Upper breast is a lightly spotted coppery tone on male, less on female, plain on juvenile.
Head and neck pale brown with violet-pink tint
Central tail feathers are brown, hiding white-tipped outer tail feathers.
Light pinkish brown
Blue-grey shoulders are a diagnostic feature when wings are folded or in flight.
Adult
Adult
Feet pink
White

Palm Cockatoo *Probisciger aterrimus* 55–65 cm
Largest Australian cockatoo; found singly, in pairs, small parties. Roosts high in rainforest; each morning groups congregate, display, disperse to feeding trees. Likely to be seen in flight over breaks in forest canopy; big broad wings have steady, deep action; glides on downcurved wings. Feeds on seeds, nuts, berries; able to crack extremely hard nuts of pandanus palms. **Voice:** call of two notes; low then rising through a rasping beginning to a sharp, abrupt screech: 'aar-rraiik!' **Hab.:** lowland rainforest, margins adjoining eucalypt, swamp woodlands; rarely ventures far from rainforest. **Status:** common in restricted range. **Br.:** 374

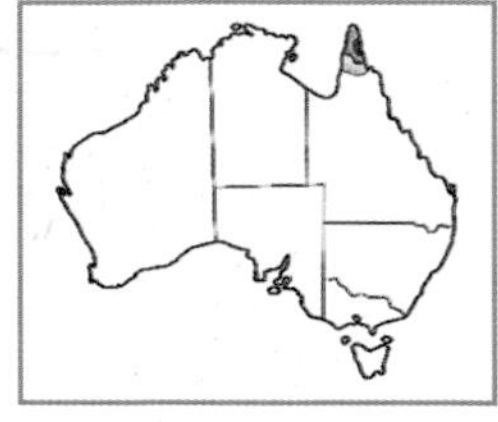

Red-tailed Black-Cockatoo *Calyptorhynchus banksii* 50–65 cm
A magnificent, conspicuous, noisy cockatoo. Travels to water morning and evening in noisy, often very large flocks. **Male:** bulky black crest overhanging heavy bill; unbarred scarlet panels in tail. **Fem.:** spotted and edged yellow; tail yellow to orange, barred black. **Imm.:** as female. **Variation:** race *banksii* largest, female has orange-yellow tail; race *macrorhynchus* large, female with pale yellow tail. Races *samueli* and *naso* have smaller bodies, crests and bills, and females have more orange in their tails; *naso* also has somewhat rounded crest. Race *graptogyne* smallest, females orange-red tailed. **Voice:** contact call, usually in flight, a grating, metallic 'karraak', 'karrark' and 'airrk'; some calls hard and grating, others husky or squeaky, whistling 'kreeeeik'. **Similar:** Glossy, but smaller, browner, very short low crest. **Hab.:** diverse; tall wet mountain forests of SE to open tropical forests of far N, tree-lined rivers of interior. **Status:** nomadic, migratory; common in N, SW; rare, threatened in SE. **Br.:** 374

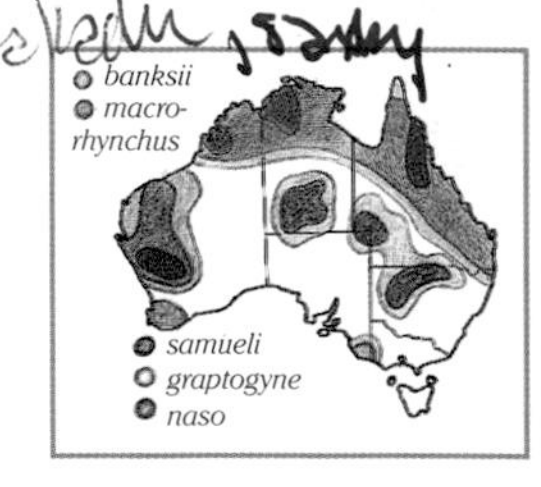

Glossy Black-Cockatoo *Calyptorhynchus lathami* 46–50 cm
Highly specialised; reliant on casuarina seeds and requires habitats that include these trees. Not as conspicuous as the Red-tailed; usually not in large flocks. Small parties, commonly up to ten birds, spend most of the day quietly feeding in the foliage of casuarina trees, the only sound being the busy clicking of bills as they demolish the hard, woody seed capsules. **Variation:** three races. **Voice:** weaker, higher and more wheezy than that of the Red-tailed. Varied, comparatively soft, wheezing and grating 'kee-aiirrk', 'airr-riiek', 'airrk', 'airrek 'arr-errk'. **Similar:** Red-tailed. **Hab.:** forests and woodlands with abundant casuarina trees; any of about five species. **Status:** uncommon, perhaps declining; limited by remaining habitat. **Br.:** 374

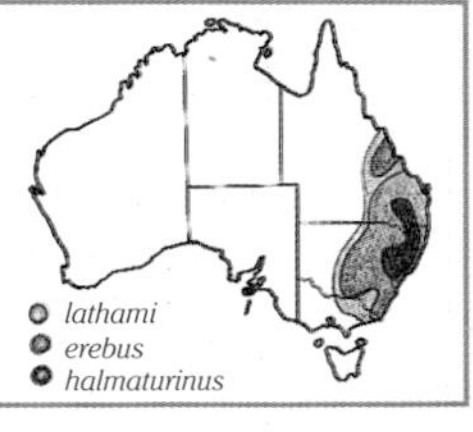

Yellow-tailed Black-Cockatoo *Calyptorhynchus funereus* 58–65 cm
Flight buoyant: effortless wheeling among the treetops with slow, deep wingbeats and floating glides. In pairs or small parties; forms large flocks in winter. **Variation:** number of races 2, perhaps 3. **Voice:** far-carrying 'why-eeela, weee-la'; in alarm, harsh screeches. Grinding noises when feeding. **Hab.:** diverse: coastal, inland and alpine; eucalypt forests, woodlands, rainforests. Feeds largely on seeds of native trees and shrubs including eucalypts, banksias, hakeas, xanthorrhea. Pines in plantations are worked for both pine seeds and for wood-boring insects. **Status:** nomadic or locally migratory; moderately common. **Br.:** 374

Short-billed Black-Cockatoo *Calyptorhynchus latirostris* 55–60 cm
Closely related to, and previously a race of, the Yellow-tailed; similar flight and calls; often in large wandering flocks. **Voice:** loud, querulous, high, drawn out, wailing, wheezy, complaining 'aa-ieeer-la' or 'wy-ieee-la'; the emphasis on the middle 'ieeer'. **Similar:** Long-billed Cockatoo. **Hab.:** forests, woodlands, heathlands, farms; feeds on banksias, hakeas, dryandras—often on ground; also exploits pine plantations. **Status:** declining due to clearing and resultant loss of food plants and large hollow nest trees; local migratory and nomadic movements. **Br.:** 374

Long-billed Black-Cockatoo *Calyptorhynchus baudinii* 55–60 cm
Except for bill, calls and behaviour, almost identical to the Short-billed. **Voice:** compared with above species, less of the long, drawn out, wheezy wailing; a more clear, whistled sound, rising clear and high then falling away, not harsh, 'wiee-ier', 'whyie-ier', or 'whyie-rrk'; deeper querulous corella-like sounds. **Hab.:** forests, farm trees; feeds primarily on seed from large woody capsules of marri, a common SW eucalypt; also strips bark from dead trees in search of wood-boring insects, occasionally damages apple and pear crops. **Status:** locally nomadic; secure while sufficient old-growth forests remain. **Br.:** 374

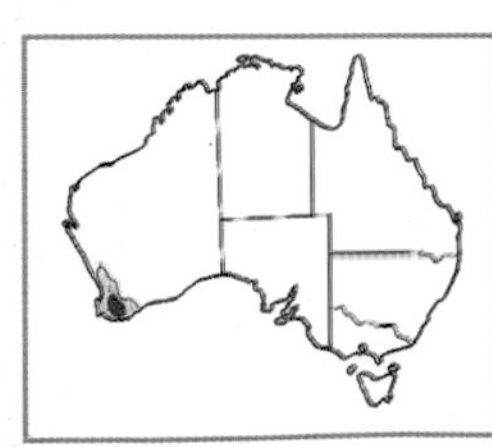

Crest feathers long, back-curved
Bare facial skin, bright orange-scarlet; it turns deeper crimson with excitement.
Bill huge, dark grey; female has smaller face patch and bill.
Overall slaty black
Adult
Juvenile has pinkish grey facial skin and yellow-edged feathers on underparts and wing linings.
Black without any colour panels
In flight, crest folds back; bill may be tucked in to breast.
Flight appears heavy, laboured with deep, slow beats of broad wings; glides on downcurved wings.
From beneath, entirely black except red face
Wings almost square tipped
Adult
Massive, rounded, helmet-like crest overhangs far forward of the heavy grey bill.
Crest down
Plumage overall black, brownish on primaries
♂
Female and imm. are spotted and barred yellow.
♀ race *macrorhynchus*
Tail panels red, translucent, brilliant
Tail panel pale yellow; other races yellow-orange to orange-red
With sun overhead, translucent panels become fiery red.
Buoyant, languid flight, slow, deep wingbeats; travels high; often glides; lands with tail spread, displaying red.
♂
Brownish
Bill dark, rounded, massive. Crest small, low, usually flat following head contour
Brownish black, not obviously glossy
♂
Red tail panels
Low flat crown; very small crest at front, both sexes; usually folded down
Female has tail panels red, edged yellow
Variable irregular patches of yellow around neck and face; male may have a few sparse yellow spots.
♀
Brilliant red translucent panels
Flight appears lazy, buoyant, languid; flocks drift along, travelling high; often glide and land with tails spread, displaying red or orange.
Two, perhaps three races: *funereus*, E coast S to Gippsland, and *xanthanotus* in Tas.; those in W Vic. and SE of SA are either the same as those in Tas., or make up a third race, *whitei*.
Yellow ear patch
Red eye-ring
Eye-ring grey, bill pale
Yellow tail panels
Bill dark
Tail panels yellow
♂
Yellow margins
♀
♀
Yellow-tailed and Short-billed species have similar wide, short-tipped bills.
The Short-billed Black-Cockatoo is more closely related to the Yellow-tailed, above, than to the other white-tailed species, the Long-billed, below. The differences between the two white-tailed species are more significant.
♂
Conspicuous large white tail panels
Pale edged feathers
Long-billed: long slender tip
The Long-billed and the Short-billed are similar except for the length of the tip of the bill.
Male: red eye-ring, black bill, dusky ear-patch
Female has grey eye-ring, pale bill, white ear-patch.
♀
Both Short-billed and Long-billed Black-Cockatoos have long tails with obvious white panels.
Flight is leisurely, floating, with slow wingbeats; drifts lazily, undulating. Coming to land in a tree, swoops upwards to alight, displaying the spread tail.
♂
Both white-tailed species also have white ear patch.
Long-tipped bill
Although the ranges of the two white-tailed species overlap, there are differences in habitat and feeding: the fine bill helps this species exploit some seed sources.

Gang-gang Cockatoo *Callocephalon fimbriatum* 33–36 cm

Distinctive dark grey cockatoo with wispy, fine-plumed crest. Bulky head and shoulders taper to short tail. Once settled in a tree and feeding on seeds, nuts and berries, Gang-gangs can be hard to see: when feeding or resting in dense foliage, they remain silent except for soft growling sounds. Sometimes reveal their presence by dropping debris as they rip seed capsules apart, but remarkably placid and tolerant of close approach. Its long wings and deep powerful wingbeats make the Gang-gang a strong flier; often travels tree to tree by swooping low then rising steeply to land. **Voice:** as contact call an incredibly drawn out, creaky rasp, 'gr-raer-iriek!'; beginning as a rough croak, ending with a squeak like a rusty hinge. Makes soft growling sound while feeding. **Hab.:** dense, tall, wet forests of mountains and gullies, alpine woodlands. **Status:** common in prime habitat of S NSW and NE Vic., but becoming rare where habitat is degraded. **Br.:** 374

Long-billed Corella *Cacatua tenuirostris* 38–41 cm

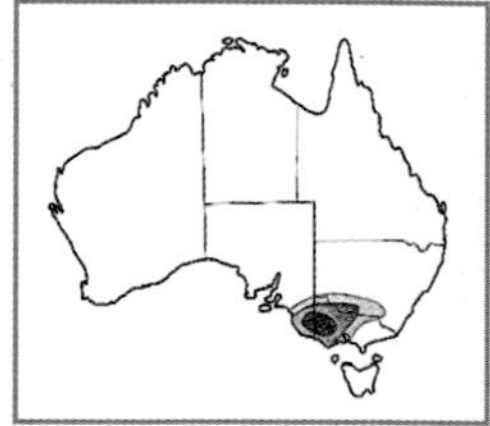

A white cockatoo with crimson or salmon-pink splashed about the face and neck; crest often folded and not noticeable. Usually in pairs or alone in the spring breeding season; at other times in small to large flocks. **Voice:** the usual contact call in flight is a wavering, nasal, falsetto, triple-noted 'ar-aer-ek, ar-aer-aerk', which, from a flock, becomes a raucous din. When disturbed makes very loud, harsh screeches. The long bill is used to dig up roots, bulbs and corms, the plumage becoming stained with red dust. **Hab.:** savannah woodland, open forest, grassland with scattered or watercourse trees, roadside and paddock trees; close to water. **Status:** now generally uncommon; aviary escapee colonies near Sydney, Brisbane, Adelaide, Perth, Hobart. **Br.:** 375

Little Corella *Cacatua sanguinea* 36–39 cm

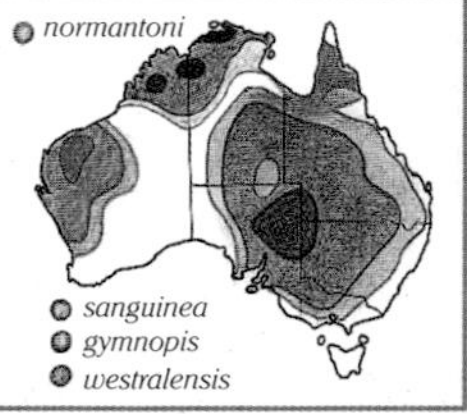

White with pink patch at lores small or absent; feeds in flocks on ground; congregates on trees to strip the leaves. Flight swift with glides. **Variation:** *sanguinea*, Kimberley and Top End NT; race *gymnopis*, E interior; *race westralensis*, NW of WA; *normantoni*, NE Qld. **Voice:** very noisy in flight and perched. Call varies in length—a harsh, resonant, nasal, brassy 'air-er-ek' that also varies from sharp to guttural—'aier-ek, aier-rr-k, aer-rk, errk, urrk, aiirk'; also loud screeches. Flocks of thousands create a crescendo of screeches and hollow croaks. In breeding season flocks are smaller, pairs keep closer to their nest sites. **Hab.:** tree-lined watercourses and adjacent plains; savannah woodland, mulga, mallee. **Status:** generally abundant. **Br.:** 375

Western Corella *Cacatua pastinator* 36–39 cm

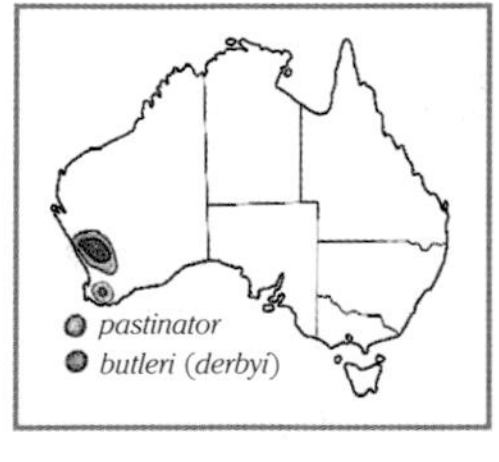

Similar to Long-billed Corella, with red or deep pink to lores and throat; differs in having an often erect crest, long rounded wings, longer tail. The Western Corella uses its long bill to dig for roots and corms. The plumage often becomes reddish, stained by soil. **Variation:** two subspecies. Opinions differ on naming of N population, whether *butleri* or *derbyi*. Prior to settlement there was a wide gap between this N race and the S race *pastinator*. With clearing of wheatbelt woodlands and forests, the N race has extended further S while the S race has contracted southwards to the more heavily forested SW corner. **Voice:** like that of Little Corella. **Similar:** Little Corella, but lacks pink on throat, weaker pink at lores; Long-billed Corella, but separated by Nullarbor. **Hab.:** open forest, woodlands, farmlands with abundant trees in shelter belts, road reserves, paddocks; SW of WA. **Status:** S race constricted, locally quite common, vulnerable; N race common. **Br.:** 375

Cockatiel *Nymphicus hollandicus* 31–33 cm

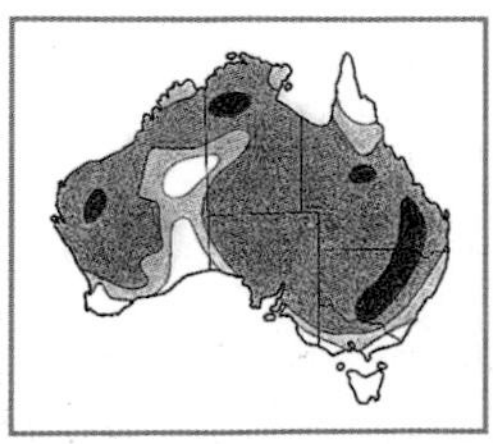

A graceful small cockatoo with slender silhouette. Flight swift on slender, flickering, backswept wings, long tail trailing; travels direct and without the undulations typical of parrots; small flocks keep together with precision formation flying. The Cockatiel's classification is still uncertain, it being the sole member of its genus. Until quite recently, it was placed in the parrot family. Evidence now suggests it is a cockatoo, and it is placed in that family, to which it has anatomical similarities and behavioural links in breeding rituals and the begging calls of the young. Biochemical comparisons support this classification. **Voice:** a distinctive, loud, clear, flight call, a slightly husky, mellow, rolling yet sharply penetrating 'whee-it, wheeit' or 'querr-eel', which gives the name 'Quarrion'. **Hab.:** most open country, woodlands, scrub or grasslands with scattered trees that is near water. **Status:** common, abundant in some northern inland regions; highly nomadic—follows rain into drier regions or moves towards coast during inland droughts. **Br.:** 375

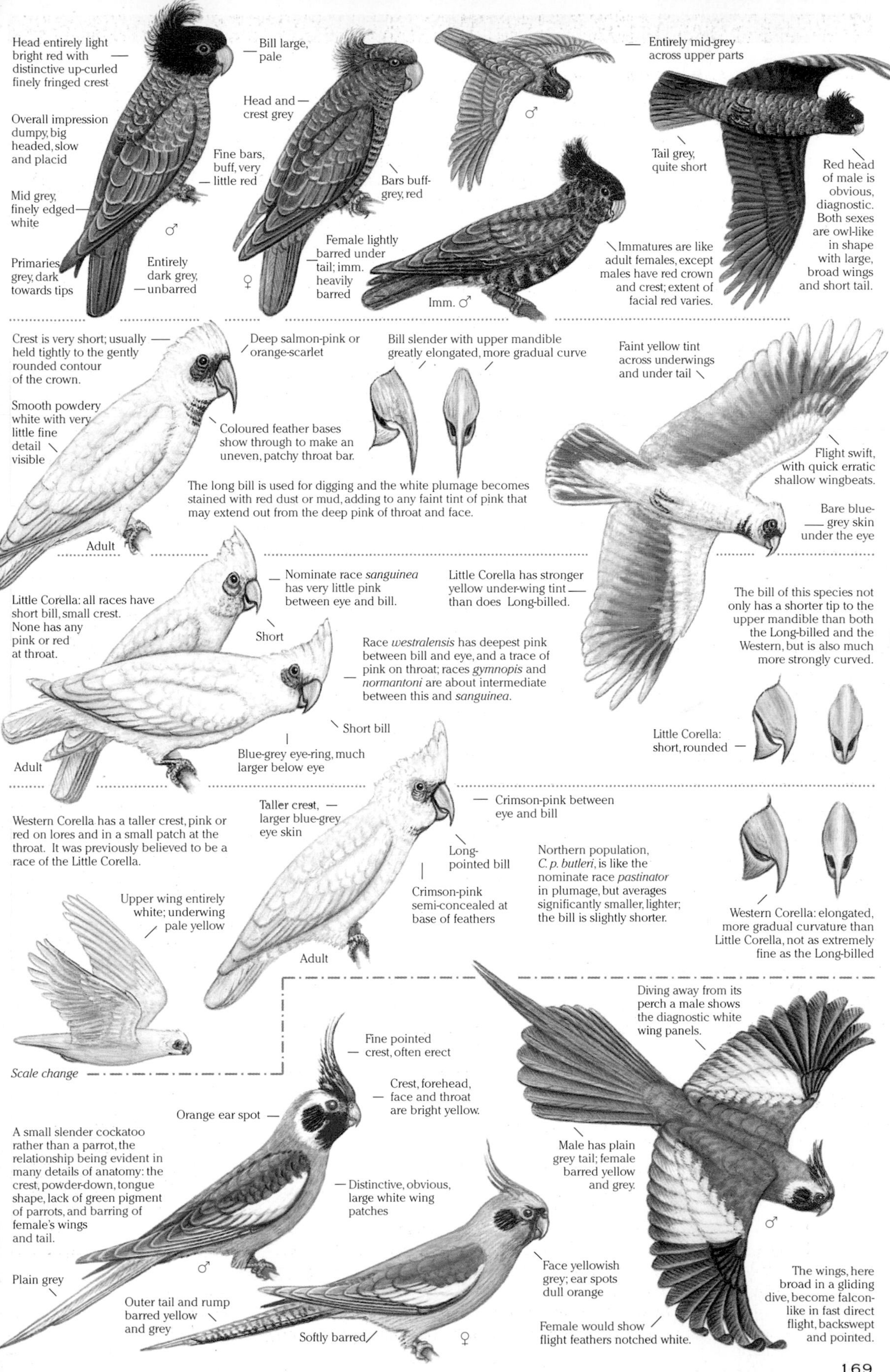

Head entirely light bright red with distinctive up-curled finely fringed crest
Overall impression dumpy, big headed, slow and placid
Mid grey, finely edged white
Primaries grey, dark towards tips
Entirely dark grey, unbarred
♂
Bill large, pale
Head and crest grey
Fine bars, buff, very little red
Bars buff-grey, red
♀
Female lightly barred under tail; imm. heavily barred
Imm. ♂
♂
Entirely mid-grey across upper parts
Tail grey, quite short
Red head of male is obvious, diagnostic. Both sexes are owl-like in shape with large, broad wings and short tail.
Immatures are like adult females, except males have red crown and crest; extent of facial red varies.
Crest is very short; usually held tightly to the gently rounded contour of the crown.
Deep salmon-pink or orange-scarlet
Smooth powdery white with very little fine detail visible
Coloured feather bases show through to make an uneven, patchy throat bar.
Bill slender with upper mandible greatly elongated, more gradual curve
The long bill is used for digging and the white plumage becomes stained with red dust or mud, adding to any faint tint of pink that may extend out from the deep pink of throat and face.
Adult
Faint yellow tint across underwings and under tail
Flight swift, with quick erratic shallow wingbeats.
Bare blue-grey skin under the eye
Little Corella: all races have short bill, small crest. None has any pink or red at throat.
Nominate race *sanguinea* has very little pink between eye and bill.
Short
Little Corella has stronger yellow under-wing tint than does Long-billed.
Race *westralensis* has deepest pink between bill and eye, and a trace of pink on throat; races *gymnopis* and *normantoni* are about intermediate between this and *sanguinea*.
Short bill
Blue-grey eye-ring, much larger below eye
Adult
The bill of this species not only has a shorter tip to the upper mandible than both the Long-billed and the Western, but is also much more strongly curved.
Little Corella: short, rounded
Western Corella has a taller crest, pink or red on lores and in a small patch at the throat. It was previously believed to be a race of the Little Corella.
Taller crest, larger blue-grey eye skin
Crimson-pink between eye and bill
Long-pointed bill
Crimson-pink semi-concealed at base of feathers
Adult
Northern population, *C. p. butleri*, is like the nominate race *pastinator* in plumage, but averages significantly smaller, lighter; the bill is slightly shorter.
Western Corella: elongated, more gradual curvature than Little Corella, not as extremely fine as the Long-billed
Upper wing entirely white; underwing pale yellow
Scale change
Fine pointed crest, often erect
Crest, forehead, face and throat are bright yellow.
Orange ear spot
A small slender cockatoo rather than a parrot, the relationship being evident in many details of anatomy: the crest, powder-down, tongue shape, lack of green pigment of parrots, and barring of female's wings and tail.
Distinctive, obvious, large white wing patches
Plain grey
♂
Outer tail and rump barred yellow and grey
Softly barred
♀
Face yellowish grey; ear spots dull orange
Female would show flight feathers notched white.
Diving away from its perch a male shows the diagnostic white wing panels.
Male has plain grey tail; female barred yellow and grey.
♂
The wings, here broad in a gliding dive, become falcon-like in fast direct flight, backswept and pointed.

Sulphur-crested Cockatoo *Cacatua galerita* 45–50 cm

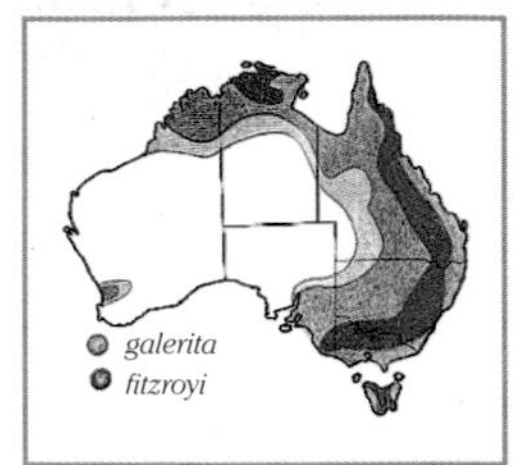

A big, noisy white cockatoo with yellow crest. In S Aust., forms huge flocks with regular roosts and midday shelters on tree-lined watercourses; the flocks fly out to open country to feed on the ground. In tropics, flocks are small; the birds are more arboreal, feeding on seeds, berries, and flowers of trees and shrubs. Has stiff-winged, irregular flap and glide flight. **Voice:** loud, raucous, unpleasant screeches, usually an intermix of harsh and sharp sounds, varying from deep, grinding and guttural to powerful, piercing screeches of 'ear-splitting' intensity: 'airrrik, aarrrk, ahrk, aieirrk, aieirieik!'. **Hab.:** diverse, ranging from high rainfall forests—eucalypt, rainforest, coastal mangroves—to semi-arid inland regions, watercourse trees and partly cleared farmlands. **Status:** sedentary; common to abundant. **Br.:** 375

Major Mitchell's Cockatoo *Cacatua leadbeateri* 35–40 cm

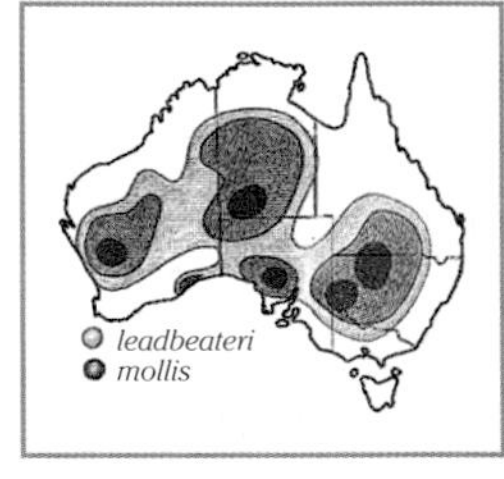

One of the most beautiful Australian cockatoos, plumaged in delicate tints of pink and white with a display of fiery colour when the crest is raised; colour most effective in flight, showing pink underwings, or at landing, when the wings are lifted and the crest is spread to show the colour bands. Shallow, quick, erratic wingbeat action is interrupted by brief glides. Usually travels by brief low flights, often landing in trees to break the journey into shorter stages. In pairs, small family parties, rarely large flocks; at times intermingles with Galahs or Little Corellas. **Imm.:** like adults but the iris is brown instead of red. **Voice:** contact call, given frequently while in flight, is a thin, querulous, drawn out, wavering screech, the sound stuttering or undulating, hoarse and scratchy, 'ar-ai-ar-a-ar-iagh, ai-ra-a-iagh'. In alarm, several loud, harsh, abrupt screeches. **Similar:** Galah, but grey back and wings, deeper pink underbody, low, pale pink crest. **Hab.:** open sparsely timbered grasslands, drier farmlands with well-treed paddocks, mulga and similar open scrublands, open mallee country, callitris and casuarina country, watercourse trees; never far from water. **Status:** sedentary; generally uncommon and of patchy occurrence, but may be locally common; has probably declined in southern farming regions. **Br.:** 375

Galah *Cacatua roseicapilla* 35–38 cm

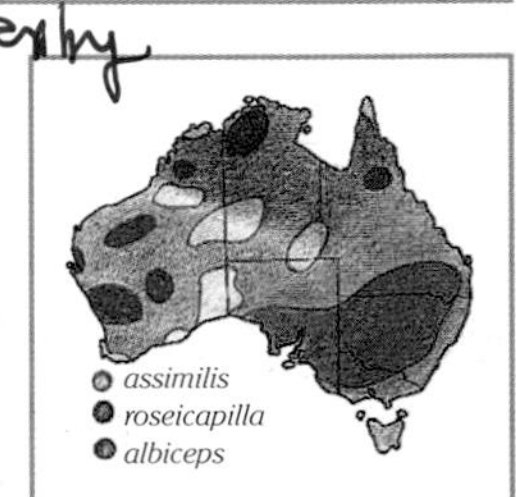

Familiar pink and grey. In flight, deep, abruptly varied wingbeats, steady and direct or a wild, erratic, crazy route across the sky. Often big flocks wheel in unison, showing massed pink then grey. **Variation:** race names may not be settled until origin of the type specimen of *roseicapilla* can be determined. **Voice:** rather harsh, metallic and abrupt, yet not unpleasant, 'chirrink-chirrink, chirrink-chirrink' and variations: 'chzink-chzink', 'czink-czink-czink-czink'; in alarm, harsh, scolding, rasping screeches. Feeds on ground or low shrubs in small parties or flocks. **Hab.:** diverse open country, typically open spaces of interior, but increasingly into coastal areas opened by clearing. **Status:** abundant; range and numbers increased. **Br.:** 374

Eclectus Parrot *Eclectus roratus* 40–43 cm (T)

Magnificent bulky parrot; sexes vastly different, both spectacular. Male looks best in flight or displaying underwings and flanks. Very noisy, conspicuous; small flocks congregate to roost in tall rainforest trees. In morning, small groups, occasionally large numbers, noisily move out to feed on big fruiting trees. Eclectus travel high above the rainforest canopy—heavy, slow and direct, deep purposeful strokes—the wings do not lift above shoulder height and the action is interspersed with glides on down-bowed wings. **Voice:** a frequently repeated harsh screech with the recurrent 'r' strongly 'rolled' to give heavily vibrating throaty roughness, 'arrrk-arrrk-arrrk' and 'airrk'. While feeding, an occasional wailing cry or soft mellow whistle. **Hab.:** usually in rainforest, but ventures into adjoining eucalypt woodlands and travels through woodlands to reach other areas of rainforest. **Status:** quite common within optimum rainforest; also in Solomons, NG, Indonesia. **Br.:** 376

Red-cheeked Parrot *Geoffroyus geoffroyi* 21–24 cm

Small bright parrot of Cape York rainforests. Uses forest foliage canopy; feeds on seeds, fruits, berries, usually in pairs or small flocks. Difficult to see high in the leaves, but noisy, which helps locate where birds are feeding. May be glimpsed fluttering high in the branches, dropping fruit and debris. The flight is distinctive, swift and direct, swerving and twisting so that bright blue of underwings shows; wingbeats are shallow and rapid without gliding. **Voice:** loud, metallic, piercing 'airk, airk, airk' and long rapid series at even pitch and strength: 'haik-haik-haik-haik...'; all piercing, ringing notes. **Hab.:** dense tropical rainforests and margin of adjoining woodland. **Status:** common within their small area of habitat. **Br.:** 376

High, deep yellow crest curving forward
Characteristic flight pattern of brief glides interspersed with bursts of shallow, rapid wingbeats. Travels high, then drops suddenly to treetop level in spiralling glide; displays crest at landing.
Bill dark grey to black
Nominate race has bare whitish skin around eye.
Yellow tint on ear coverts
Crest folds almost flat, usually with slight upwards curve.
Nominate race galerita, Cape York to Tas. Race fitzroyi of NT and Kimberley has bluish eye ring, little if any yellow on ear coverts.
Yellow tint on underside of wings and tail
Crimson and yellow
On landing, the Major Mitchell lifts its wings high displaying the deep pink underwings that are usually hidden, and displays its brightly coloured crest.
Race mollis, no yellow
The Major Mitchell flies with irregular, shallow, almost fluttering wingbeats interspersed with glides on downcurved wings.
Male has brown eyes; female has red eyes and a wider yellow band through crest.
Pink
Pink
Underwing deep salmon pink, white edges
Very light pinkish white
Race assimilis, grey or white eye-ring
Male brown iris, female red
Wing and tail-tips may be darker
Rump and tail very pale grey
Vent to under tail, very pale grey
Nominate race roseicapilla has red eye-ring, deep pink body and very pale crown.
Deep rose pink
Rump and tail very pale grey
Brilliant green with red eye, yellow-orange bill
Crimson; yellow eyes, black bill
In flight the brilliant red and green of the male's plumage is no less spectacular than the crimson and deep blue-violet of the female.
Glimpses of scarlet flanks
Dark flight feathers enhance brilliance of both sexes.
Scarlet underwing coverts and flanks
Crimson body broken by blue-violet band
Violet-blue
Cheeks rose-red, extending to forehead and ear coverts
Greenish yellow
Olive-brown
Bronzed
Green
Light leaf green
Males are slow to gain their red-cheeked plumage. Initially green headed, then brown like the female, finally reach full male colours after two years.
Dark blue-grey
Underwing coverts bright blue
Brown to grey-brown
Imm.

Rainbow Lorikeet *Trichoglossus haematodus* 26–31cm (T)

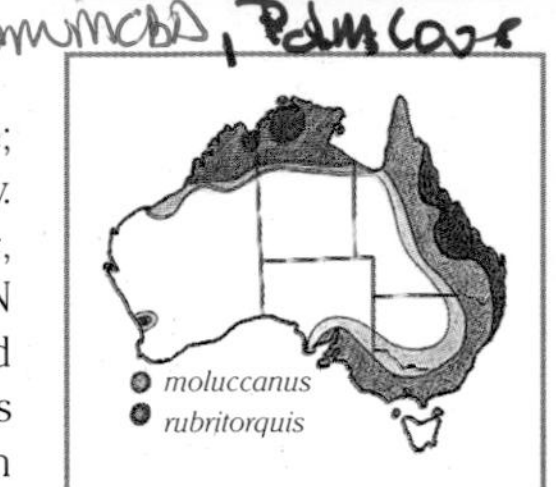

A large, brightly coloured, bold, noisy and well-known species. Distinctive flight silhouette; flocks dart, twist and wheel among trees, or fly straight and direct above the forest canopy. Feeds on fruit, nectar, blossoms, seeds, berries and orchard fruit. **Variation:** race *moluccanus*, Rainbow Lorikeet, occupies the E coastal strip from Cape York to SA; it shows considerable N to S variation, apparently mostly clinal, but those of NE Qld are sometimes listed as a third mainland race, *septenrionalis*. The second race, *rubritorquis*, 'Red-collared Lorikeet', extends through the Top End of NT to Kimberley. A third race, *caeruleiceps*, occurs on the Australian island of Sabai in N Torres Str. **Voice:** in flight gives frequent, quite pleasant, softly rasping or vibrating musical screeches; softer mellow chattering and subdued screeches while feeding; quiet twittering while resting. **Hab.:** diverse; rainforests, eucalypt forests, woodlands, riverside trees, farmlands with remnant stands of trees, paperbark and heath woodlands, mangroves, gardens. **Status:** common in N; uncommon to rare in SE; vagrant to Tas. **Br.:** 375

Scaly-breasted Lorikeet *Trichoglossus chlorolepidotus* 22–24 cm

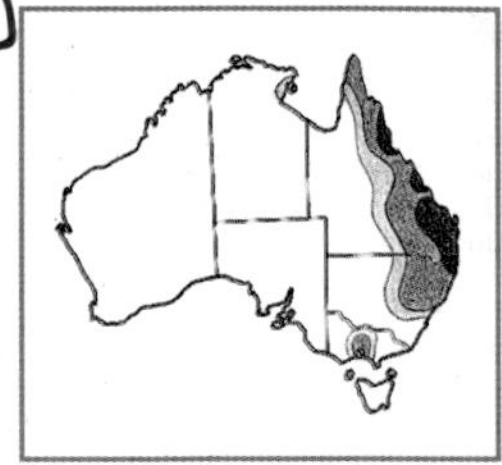

A rather plain green lorikeet with yellow margins to plumage of neck and breast, giving scaly appearance. When perched, the only strongly contrasting colour is the vermilion of bill and eye, and sometimes a small glimpse of red at the shoulder. In flight this plain green parrot is transformed. The red underwing coverts, broad band of orange-red across dark primaries and bright green leading edge combine to create spectacular underwings that, in the brief glimpse given as the birds dash past, compensate for the comparatively unexciting image presented with wings folded. Scaly-breasted Lorikeets feed on nectar, flowers, seeds, berries and fruits from grevilleas such as the silky oak, and from banksias, erythrina, xanthorrhoeas and eucalypts. **Voice:** sharp, short, clear screeches, something like the Rainbow Lorikeet, but usually sharper. A 'chewip', often in flight, is varied from extremely high and sharp to mid range and rather mellow, or slightly rasping. Also gives scolding 'charr!' and soft chattering while perched. **Similar:** other green parrots with red underwings, including larger Rainbow Lorikeet, which has greatly different head and breast colours, as well as similar-sized Swift Parrot, and smaller Purple-crowned Lorikeet. **Hab.:** diverse; most vegetation types along lowlands of E coast where there are flowering trees, including eucalypt forests, woodlands, heath-woodlands, paperbarks. **Status:** common; nomadic. **Br.:** 375

Purple-crowned Lorikeet *Glossopsitta porphyrocephala* 15–16 cm

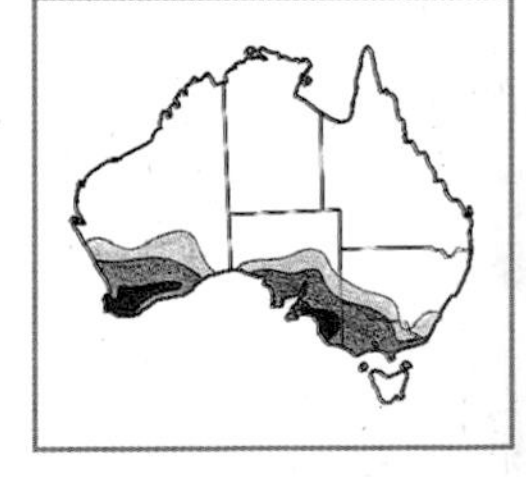

More often heard than seen; almost continuous sharp, high screeches attracting attention as pairs or flocks dash through the treetops, the sunlight at times catching the crimson beneath their wings. In the tops of flowering trees, whether coastal forest giants or mallee scrub of the drier interior, the green plumage of these tiny parrots allows them to merge into the foliage as they climb among the leaves and flowers in their search for nectar and pollen. The Purple-crowned Lorikeet is highly nomadic, gathering in large numbers where a species of eucalypt, whether wetland or dryland, is in flower; it then vanishes, not to be seen in such abundance in that locality for many years. Nesting is in spring, occurring wherever the wandering flocks of lorikeets have found the most favourable conditions; that is, wherever a species of forest or woodland is flowering strongly. The following year they may nest hundreds or thousands of kilometres away where some different eucalypt species has a strong nectar flow. On these occasions the nesting can become concentrated at the richest sites; almost every tree that has a hollow may have one or more nesting pairs, so that there is constant chatter and whirring of small wings as parties of lorikeets dart away to raid flowers in the countryside surrounding. Such nesting can extend over many kilometres and can include thousands of birds, but so small and inconspicuous are they that the whole busy scene may well go unnoticed. **Voice:** calls almost continuously in flight with a high, slightly metallic or vibrating 'tziet, tziet', less sharply screeched than the Little Lorikeet. When feeding in treetops, soft chattering mixes with louder flight calls. **Similar:** three others with red underwing: Rainbow Lorikeet, but it is much larger and has blue and orange on head and body; the Scaly-breasted Lorikeet, but it is larger with red on the flights as well as the wing-lining and has plain green head plumage; Swift Parrot, but larger with bright red under-tail coverts and red around the bill. **Hab.:** although primarily an inland species, this lorikeet also visits coastal forests when these are in flower, and will, on occasions, be heard (more than seen) in the crowns of the karri forests, the tallest and wettest of SW WA. It is also present where more open woodland or mallee occurs on drier parts of the coast. Inland mallee and woodland are its most characteristic habitat, but it also raids orchards at times to feed on ripe fruit. **Status:** moderately common; nomadic, erratic, locally abundant to rare; the only naturally occurring lorikeet in SW Aust. **Br.:** 376

The Rainbow and Red-collared have in the past been separate species, then combined as races of the one species. These may again be split into two species. Whether one or two species, the common names reflect the obvious difference between the forms.
Race *moluccanus*, Rainbow Lorikeet
Collar greenish yellow
Bill and eye bright red
Head streaky blue-violet
Blue-violet
Yellow, mottled green
Adult
If primaries are spread, yellow spots on their inner webs begin to show as a thinner dotted replica of the yellow bar that is prominent across the underwing.
Pale yellow bar across primaries of both races, but larger on the Red-collared
Both races have bright red underwing linings.
Collar orange-red
Orange, lightly edged with red
Dark blue-black
Race *rubritorquis*, 'Red-collared Lorikeet'
Plain greenish lorikeet that is transformed in flight by display of underwing colour
Two-toned red underwing: scarlet inner-wing lining; orange-salmon band across dark brown flight feathers
Green body, tail; orange outer wing, red inner
Edge of red underwing coverts may be visible.
Adult
Upper surfaces, body and wings, light bright green
Yellow-green outer edges
Bill and eye light, bright vermilion
As primaries spread wider, pink spots along their inner webs may show as a thin dotted wingbar.
Bill and iris brown
Imm.
Head plain green unmarked by other colour or pattern
Green with yellow scaly pattern
Lateral feathers of tail greenish yellow
Red underwing coverts curve around to the leading edge and slightly onto the upper wing at the bend, sometimes visible at shoulder; more likely to be seen from front than side view.
Purple crown
Orange ear coverts
Orange-red
Pale blue
Underwing coverts crimson
Yellow-green
Yellow inner webs
In small parties or flocks, clusters of small specks dashing across the sky, their identity revealed by distinctive screeches.
Underwing coverts crimson
In fast flight, wings backswept
Turquoise
Upper parts bright green, smudge of rufous-bronze across nape and mantle
At bend of wing, touch of bright blue. This is more evident from front.
Landing, or in short fluttering flights between canopy perches, wings are out wide, tail is spread.
Inner web of each flight feather grey-brown
Purple crown
Blue leading edge
Underwing coverts crimson
Juvenile has only slight hint of purple.
Pale yellowish forehead, ear coverts
Bronzed nape, mantle, adult and juvenile
Preening gives glimpse of crimson underwing.
Imm.
Adult
At a distance, wings folded, this lorikeet appears rather plain, but close up in good light, if wings are lifted to stretch, preen or come in to land, it is quite spectacular. Usually there are only fleeting glimpses of these colours as the birds dash through the treetops.
Scale change
Similar: Swift Parrot page 187 23–26 cm
Red
Red
Size similar to Scaly-breasted Lorikeet; much larger than Purple-crowned

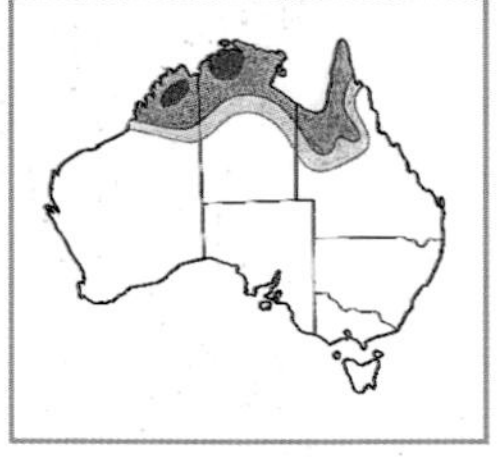

Varied Lorikeet *Psitteuteles versicolor* 18–20 cm
Multicoloured, yellow-streaked with light grey or white goggle-like eye-ring. Small parties or large flocks feed on flowers or fruit, favouring nectar of bloodwoods and melaleucas. Flight swift, direct in tight formation. **Variation:** little geographical variation, but breast colour varies greatly from pink to dull mauve. **Voice:** contact call, constant while in flight, is a thin, shrill, metallic screech, noticeably higher pitched but not as loud as the calls of the Rainbow Lorikeet. While feeding, chatters busily—varied, quick, sharp little screeches, intermingling with softer scoldings, some quietly low and slightly rasping, others soft or higher, all abrupt. **Similar:** only lorikeet across most of its range showing green underwings in flight; Rainbow (the Red-collared race in NT and WA), has orange-red underparts. **Hab.:** tropical forests and eucalypt or melaleuca woodland, wherever trees or lower shrubs like grevilleas are flowering; often heavier vegetation near watercourses. **Status:** moderately common; probably more abundant in tropical NT than elsewhere in range. **Br.:** 375

Musk Lorikeet *Glossopsitta concinna* 21–23 cm
Predominantly bright leaf green with splashes of bright crimson, yellow and blue, but not easily seen dashing across the sky in screeching flocks or climbing among dense foliage and flowers of tall eucalypts. Often intermingles with Purple-crowned or Little Lorikeets, the Musks being obviously larger. Flight very swift, direct with audible whirr of rapidly vibrating wings. The Musk Lorikeet is nomadic, perhaps more seasonal and predictable than other lorikeets in its search for nectar. **Fem:** crown, which on male is blue, is dull turquoise. **Voice:** in flight a shrill metallic screech in contact; feeds in treetops with much soft chattering, the varied notes intermingling, sharp, husky and querulous. **Similar:** Little and Varied Lorikeets also have green underwing coverts. **Hab.:** woodlands, open forests, mallee, cleared country with trees along watercourses and roads. **Status:** common to uncommon; locally abundant where flowering is heavy. **Br.:** 375

Little Lorikeet *Glossopsitta pusilla* 15–16 cm
A tiny emerald green lorikeet with crimson surrounding its black bill. Flight swift, direct; rapidly whirring small wings give a fleeting glimpse of yellowish green underwing linings; widely fanned tail reveals orange-red colour that is usually hidden. Feeds on nectar, pollen, fruits, berries, seeds; approachable while feeding. **Voice:** sharp, short screeches, slightly undulating or warbling, 'zrrit, zrit' or 'chirrit, chrrit'. While feeding, almost constant, pleasant, subdued, chatter-like screeches, sharp or mellow, with intermingled brief, harsh scoldings. **Similar:** Musk Lorikeet, but larger with some red behind eye but not on throat, yellow flank patches. **Hab.:** forests, woodlands; favours open country—trees along watercourses and paddock trees. **Status:** nomadic; common in Qld, NSW, less common Vic., uncommon SA, vagrant Tas. **Br.:** 376

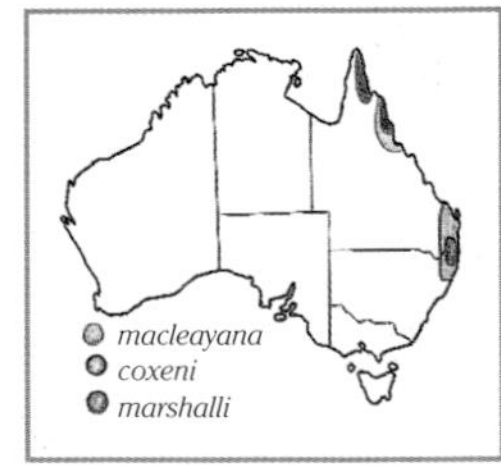

Double-eyed Fig-Parrot *Cyclopsitta diophthalma* 13–15 cm
An extremely small, short-tailed parrot of tropical rainforests. The name 'Double-eyed' comes from a NG race that has a dark mark near the eye. Fig-parrots feed on native figs, rainforest fruits and nectar. They are well adapted to a life in the rainforest canopy, scurrying mouse-like along branches and among foliage. Flight is swift, direct with short bursts of wingbeats, only slightly undulating; usually travels above rather than through the rainforest canopy. **Voice:** contact call is a sharp, penetrating 'tseit-tseit-tseit-', usually in flight and on landing; alarm call is a shrill screech. **Similar:** Little Lorikeet, but slender shape, and with facial red entirely in front of eye, around base of bill. **Hab.:** rainforest, varying from the low riverine monsoon forest types of Cape York to the lowland and high-altitude tall rainforests of the Cooktown–Ingham coast and high ranges; lowland rainforests of SE Qld and NE NSW. These rainforest blocks are isolated, surrounded by eucalypt forest or woodland into which the fig-parrots venture for only very short distances. **Variation:** NG is home to the nominate race and many subspecies. Australia's small outlying representation of the species now comprises isolated races in each of three tracts of rainforest down the E coast. These races have in times past been described as full species with common names. Race *macleayana*, NE Qld, occurs in both lowland and highland rainforests from about Mackay to Cooktown, and is quite common around Cairns where it may be seen in, and occasionally nests in, densely vegetated gardens of the foothills; this would be the most easily sighted of the races, occurring in many easily accessible national parks. Race *marshalli* occurs further N in lowland rainforests of the NE coast of Cape York Peninsula, S to about Rocky R.; most sightings have been where access is easiest: Iron Range, Lockhart R., Claudie R. The S race *coxeni* is confined to lowland rainforests of SE Qld from about Maryborough S to vicinity of Macleay R. in NSW. **Status:** N races moderately common and secure, S race *coxeni* now rare and endangered, having lost much of its lowland rainforest habitat. **Br.:** 376

Scarlet
Whitish eye-ring
Pink to mauve, streaked yellow
Tiny lorikeet of varied, streaked colours; male brighter than female
Dull, breast only slightly mauve
Imm.
The varied colours from nape to back and belly have yellow-shafted feathers, giving a heavily streaked appearance unique among lorikeets.
Adult ♂
Overhead, gives glimpse of entirely bright yellowish green underwing linings, under tail and body, except for a smudge of mauve or pink on the breast.
Darting flight on backswept wings
♂
Scarlet cap
Underwing linings bright lime green
Pink to dull mauve
Under tail has yellow inner margins.
Forehead through to ear coverts bright crimson; crown blue to turquoise
Red tipped, dark base
Upper parts bright leaf green with bronzed tone on nape and mantle
Adult ♂
Female has much less blue on crown; immature has dull reds, brown bill.
Diving away it shows colours of tail hidden when perched: yellowish olive with lateral feathers towards the base edged orange-red.
Underwing coverts yellowish green
Yellow patches on flanks, largely hidden under wings when perched
Above green, with streak of scarlet around the blue crown
Flight swift, body tapering to pointed tail, backswept wings
Tiny emerald lorikeet with scarlet surrounding a black bill, yellowish green under wings, hidden colour in tail
Vivid scarlet of forehead, lores and throat encircles black bill.
Nape and upper mantle bronzed
Wings and tail vivid emerald green
Immatures have paler orange face, brown iris and bill.
Ear coverts faintly, sparsely streaked, not usually noticeable
Light yellow-green
From above, bright emerald
Swift flight on pointed, backswept wings
Diving down and away it displays colours usually hidden beneath the tail, yellow-green with orange-red inner web towards base of all but central feathers.
Orange-red blends through yellow to green.
Yellowish green
Male: blue encircles eye, red cheek streak underlined with dark blue.
Red, both sexes
Red
Blue
Underwing, (all races, both sexes) has pale yellow bands across base of flight feathers and across greater coverts; remainder of wing linings are bright yellowish green.
Smallest Australian parrot, three subspecies. Race macleayana occurs in NE Qld from Cooktown south to Mackay.
♀
Race macleayana
♂
Race macleayana 'Red-browed or Blue-faced Fig-Parrot'
Race marshalli has larger red cheek patch, small blue brow line, very narrow blue edge under red at throat.
Race marshalli 'Marshall's Fig-Parrot'. NE Cape York Peninsula
Blue with grey cheeks
♀
Blue forehead
♂
Red edge to the innermost of the wing-coverts, all races
Race coxeni 'Blue-browed or Coxen's Fig-Parrot'
♂
Small area of patchy red set in large area of deep blue.
Outer edges of flight feathers blue, all races
Race coxeni
Very small area of pale patchy red above blue lower cheek streak; small patch of blue on forehead.
All races
Yellow edges to breast, yellow flanks largely hidden under folded wing; all races.
♀
Race coxeni is found only in SE Qld and NE NSW where it is rare and seriously endangered.
Diving away it shows stumpy tail, yellow wing-bars.

Australian King-Parrot *Alisterus scapularis* 42–44 cm

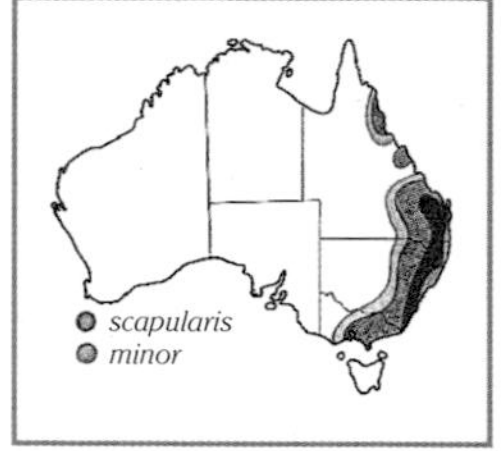

In pairs, small flocks; flight direct, heavy with full, regular wingbeats. In the wild it is wary; flies far if disturbed. **Imm.:** like female, but brown eyes. **Variation:** two Aust. subspecies; *scapularis* (*alisterus*) in SE, north to Cardwell, Qld; *minor*, smaller, Cardwell N to Cooktown. **Voice:** in flight gives abrupt, sharp, clear 'krassiek', 'k-wiek', 'chriek' or 'charrak' with slight variations; or rapid 'chrak-chrak-chrak-'. Alarm call is a harsh, metallic, screeched 'karrark!'. Males give a sequence of high sharp whistles, 'chreip, peeip, peip', while perched. **Similar:** Crimson Rosella, but wings and cheek patch blue. **Hab.:** keeps to heavier coastal and mountain forests in breeding season, including rainforests, eucalypt forests, palm forests, dense river-edge forests and closely adjoining eucalypt woodlands. After nesting, King-Parrots wander further to lowlands, farmlands, shelter-belts, parks, gardens. **Status:** sedentary, dispersive; usually common where habitat remains. **Br.:** 376

Red-winged Parrot *Aprosmictus erythropterus* 31–32 cm

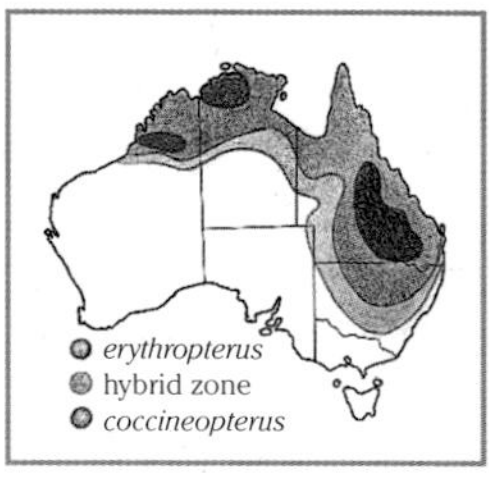

An unmistakeable parrot clad in brilliant green; crimson wings set in dark surrounds add visual impact. Typically seen in small parties or pairs feeding on fruits, seeds, nectar and insects in foliage of trees and shrubs; rarely on ground. Tend to be wary, easily put to flight; rise with much calling. **Variation:** probably 2 Aust. races: *erythropterus*, NSW–Qld; and *coccineopterus* of NT–Kimberley. A broad intermediate hybrid zone (Cape York) suggests clinal variation **Voice:** flight call a sharp, metallic 'chrrik-chrrik, chrrik-chrrik' or 'crillik, crillik'; in distance sounds like a Budgerigar chattering. Alarm call is a series of harsher, more abrupt screeches, 'chak-chak-chak chrak-k-kak'. **Hab.:** open eucalypt forests and woodlands, typically with open grassy understorey, fringes of rainforest or monsoon forest; further inland, mulga, brigalow, casuarina and callitris; mangroves along the far N coast. **Status:** nomadic, dispersive, common through most of its range. **Br.:** 376

Princess Parrot *Polytelis alexandrae* 35–45 cm

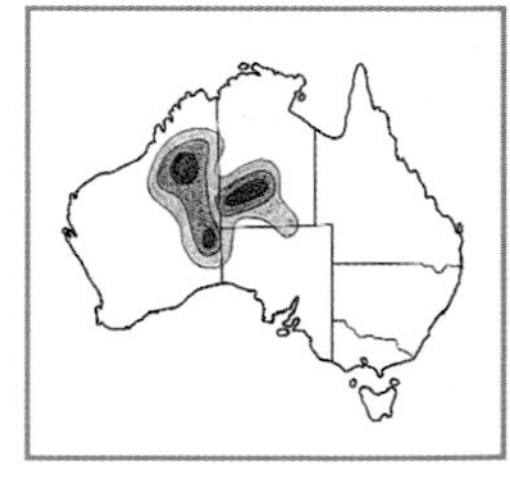

One of Australia's most beautiful parrots, but probably better known in aviculture than in the wild because its natural range covers some of the most arid country in this continent. May be sighted in pairs or small parties, travelling with slightly irregular wingbeats and undulating flight; drops slowly to the ground with fluttering, almost hovering wing action. Much time is spent on the ground searching for seed; spinifex seed is probably a major part of the diet—observers have noted that this parrot is seldom found far from spinifex. **Voice:** rather quiet, occasionally a rather loud, unmusical 'kee-ahrk-carruk'. **Hab.:** arid regions with sparse trees, eucalypts, casuarinas, acacias, spinifex; also vicinity of salt lakes with succulents and saltbush groundcover. **Status:** highly nomadic; these birds may appear after an absence of years, even decades, perhaps after flooding rains of cyclones or thunderstorms, when fresh vegetation occurs in some part of their range. **Br.:** 377

Superb Parrot *Polytelis swainsonii* 37–42 cm

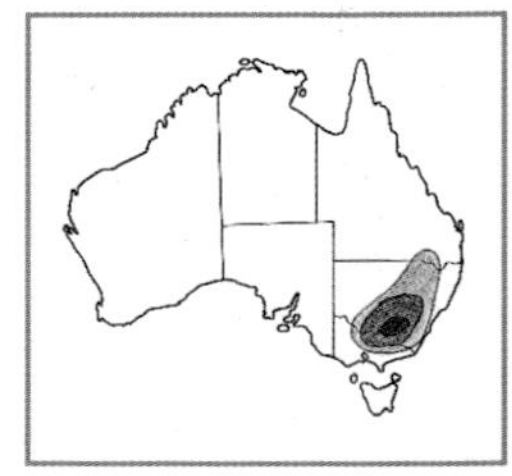

A swift, long tailed, slender, green and gold parrot; usually seen in small parties or flocks. **Voice:** varied calls; commonly a strong, penetrating, rather rough yet musical 'querr-ieek, querrieek' with final 'ieek' loud and sharp; or 'krak-karrark'. Other calls include a sharp, penetrating, whistled 'whiek, whiek, whiek' at regular intervals, and harsh, deep, rough scolding, 'quarrarrk'. **Hab.:** River Red Gum, box and similar forests, river-edge forest, nearby mallee, native cypress, farmlands. Migrates from SW Riverina to north-central NSW along Namoi and Macquarie rivers for winter. **Status:** has declined with habitat loss; species now common only locally, mainly in a few preserved habitats such as Barmah Forest. **Br.:** 376

Regent Parrot *Polytelis anthopeplus* 38–40 cm

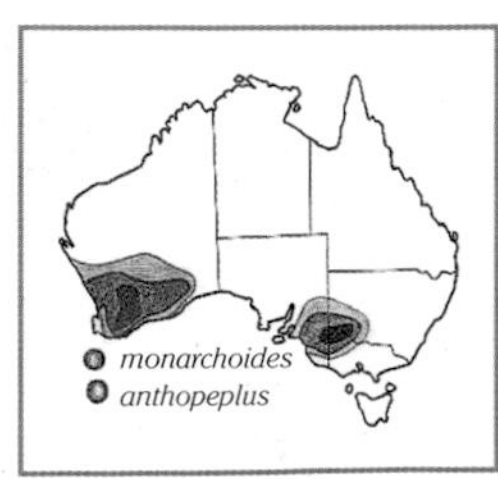

Colourful, gregarious large parrot with a long tail; flies swiftly on backswept wings. **Variation:** nominate race *anthopeplus*, plumage rather dull olive-green, is found over much of SW of WA; race *monarchoides*, with brighter plumage, as illustrated, occurs in a restricted area around junction of Vic., SA and NSW. **Voice:** distinctive, mellow, rolling, deep-in-the-throat 'quarrak-quarrak-quarrark'; scoldings and chattering. **Hab.:** eastern population inhabits floodplain woodlands, mainly of river gum, *Eucalyptus camaldulensis*, which provide big old trees for nesting. The WA race uses open forests and woodlands and appears to be increasing in range and numbers, while the eastern race, although quite common within its restricted range, may be less secure. **Status:** moderately common; nomadic. **Br.:** 376

A large, brilliantly plumaged parrot, usually in forests. A flock of these birds winging their way between soaring trees beneath the high dark canopy of foliage brings a fleeting, spectacular splash of colour to those gloomy spaces. But that colour will be seen only in full brilliance if the birds pass through sunlight beneath a break in the overhead foliage.

Green Rosella *Platycercus caledonicus* 32–38 cm

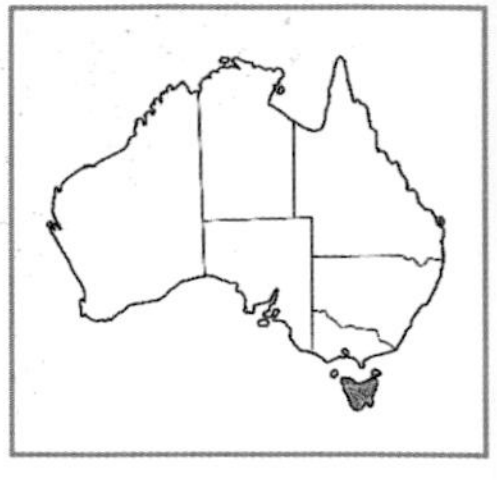

Found only in Tas. and some islands of Bass Str., this species is the largest rosella. Usually seen in flocks or small parties, feeding on the ground or among the foliage of trees where they can be well hidden by their greenish plumage. In flight they often draw attention with a distinctive 'kzink' or 'kussink'. Their colour is most conspicuous in flight, overhead, when the yellow undersurfaces can be caught in sunlight, bright against blue or green of sky or forest. Their flight is strong, swift, with only slight undulations; bursts of quick wingbeats are broken by brief glides. **Variation:** King Is. population may be a subspecies, *brownii*. **Voice:** a very high, ringing call; three piercing whistles, the second note the highest, 'whee-whieit-whee'. Another common call is a harsh, sharp, metallic 'k-ziek, kziek-kziek; kziek, kziek', the last '-iek' a brief piercing squeak, with erratic breaks, some but a fraction of a second, others several seconds. From pairs, parties or flocks, typical rosella chattering, but more harsh, squeaky, than is usual for rosellas. **Hab.:** the Green Rosella utilises most Tasmanian habitats, except treeless moorlands and farmlands cleared of trees. It is most common in heavy rainfall districts, dense forests and mountains. **Status:** abundant in most parts of Tas.; sedentary but with some localised wanderings. **Br.:** 377

Crimson Rosella *Platycercus elegans elegans* 32–37 cm (N)

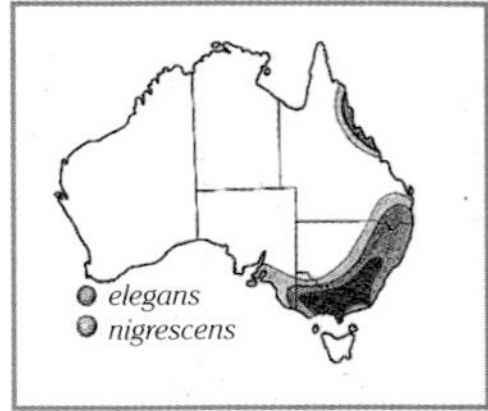

Plumaged in crimson and dark blue, this rosella would no doubt be far more appreciated were it rare, but it is quite common and its beauty is all too often taken for granted. Flies strongly, weaving swiftly among massed forest trees, displaying bright red or orange. The Crimson Rosella species now includes two other rosellas that were, for a very long time, considered separate full species; they have distinctively different plumage colours and long-accepted common names. For this reason, the 'Adelaide Rosella' and the 'Yellow Rosella', follow below, as if full species, under their well-known common names, but with their present classification as subspecies recognised. **Variation:** as well as the yelow-orange *flaveolus* and *adelaidae,* below, there are two crimson races similar to the crimson nominate race elegans: *nigrescens*, confined to Queensland's north-eastern coast, is like the nominate race *elegans*, but smaller and darker; race *melanoptera*, Kangaroo Is., SA, is like *elegans* but larger, darker. **Voice:** a clear, ringing 'k-teee-tip, k-tee-tip', the central 'teee' loud, very high, penetrating, but more musical than squeaky; strong; variations include 'k-tee-it-tip', kteeeit-tip' and 'tip-teee', always with the 'teee' as the strongest part. Also has various harsh scoldings: 'chiak-chiak', 'chak-chak'. **Similar:** Australian King-Parrot. **Hab.:** principally the heavy humid forests of the eastern coast and ranges: rainforests, tall, dense wet eucalypt forests, timbered watercourses and farmlands near forests, parks and gardens. **Status:** abundant throughout, including suburbs of cities. **Br.:** 377

'Yellow Rosella' *Platycercus elegans flaveolus* 31–35 cm

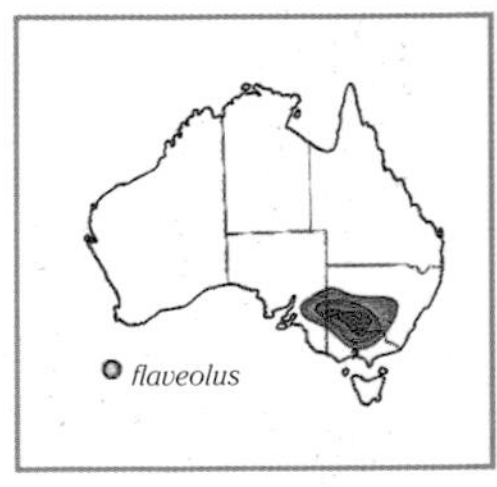

A distinctive subspecies of the Crimson Rosella. In its natural habitat, among foliage of river gums, the yellow and green of this parrot can be difficult to sight, but, in flight, from beneath, the entirely yellow underbody looks bright against the dark tones of wings and tail; quite spectacular in low-angled sunlight against blue of the sky. Flight appears more direct, less undulating than that of the Crimson Rosella. Usually in pairs or small parties; feeds more among foliage, less on the ground than other rosellas; rather shy. **Juv.:** upper parts dull olive-yellow with little black, underparts greenish yellow, rump olive. **Voice:** similar to calls of Crimson, but not as high or clear; includes a 'chwit, chwit-chwit, chwit-' and very rapid, rather budgie-like chattering. **Similar:** Adelaide Rosella, but adults more overall orange; immatures with more extensive red surrounding the blue cheek patch. **Hab.:** a restricted, specialised habitat—parts of major rivers, their tributaries and, in places, broad floodplains with river-edge and floodplain woodlands of river red gums and box trees; nearby roadside trees, paddock tree belts, mallee. **Status:** sedentary; quite common within its restricted range. **Br.:** as Crimson Rosella, 377

'Adelaide Rosella' *Platycercus elegans adelaidae* 34–35 cm

A distinctive subspecies of the Crimson Rosella. Common around Adelaide, often seen in suburbs and parks, and quite abundant in parts of nearby ranges. Flight and feeding habits are similar to those of the Crimson Rosella. After the breeding season the adults remain in pairs or small parties while the immatures gather into flocks. **Variation:** appears to have roughly clinal variation from N to S. Opinions differ on races: probably breaks into yellowish *subadelaidae*, orange-yellow *adelaidae* and orange-red *fleurieuensis*. **Voice:** pleasant and musical, clear, ringing yet abrupt 'chwik, chwik-chwik, chwik'; also, in groups, a confusion of typical rosella chatter. **Similar:** no similar species, but yellowest of 'Adelaide Rosella' similar to subspecies known as 'Yellow Rosella', and orange-red individuals like the nominate race *elegans* known as 'Crimson Rosella'. **Hab.:** Adelaide Hills and Mt Lofty Ranges S to the Murray R: tall, dense forest of ranges, lines of trees along watercourses, across plains, river flats and nearby mallee. **Status:** sedentary; common. **Br.:** as Crimson, 377

Edged blue
This and the Crimson Rosella make up the 'Blue-cheeked Rosella Complex. The Yellow and Adelaide Rosellas were previously full species, but now races of Crimson Rosella; they have well established common names. and distinctive plumages.
Greenish yellow
Blue
Golden
Red frontal band
Mottled green and black
Brownish black edged dark blue
Although known as the Green Rosella, the yellow of crown and breast can appear quite golden, especially under the warmth of direct sunlight.
Blue cheeks
Greener than adult
Olive-green
Adult
Greenish yellow
Juvenile greener overall compared with adult
Central tail bronzed green; outer tail blue, edged white
Blue
Adult
Simple bold plumage: overall deep crimson with dark blue on cheeks, wings and tail
Olive green
Outer web of primaries edged blue
Patchy crimson and green
Extensive blue on upper coverts, secondaries
Green
Long tail tipped pale blue
Crimson rump
Green
Deep blue
Blue, often with greenish or brown tone
Dives away through the narrow spaces of dense forests, turning and twisting skilfully between trunks, limbs and foliage.
Adult
Juv..
Adult
Adult
Red frontal band
Black
Under tail white tipped
Underwing coverts blue
Head and body almost entirely bright yellow with feathers of nape, mantle and scapulars black centred.
Bend of wing blue
Black centred, broadly edged yellow
Obvious blue cheeks
Outer webs of primaries blue
Secondaries blue
Older birds have some red on throat
Black-blue band across yellow from shoulder to wingtips
Yellow, throat to under tail
Lateral feathers blue tipped white; central pair greenish towards yellow rump
Adult
Forehead and crown red blending through orange to the deep yellow edgings of the black upper parts
Cheek patch deep blue
Grey-black
A deeper coloured southern form of Adelaide Rosella (race *fleurieuensis*)
Varies, orange to yellow.
Blue; often with greenish tint towards rump
Outer tail feathers blue, tipped white
A paler, yellowish, northern form of Adelaide Rosella (race *subadelaidae*). There is some interbreeding with the Yellow Rosella where the races meet around Mannum and Morgan in SW SA.
The immature of the Adelaide Rosella is olive-green above, yellowish green beneath, with crown, face and throat red.
Blue-grey, white tipped

Western Rosella *Platycercus icterotis* 25–28 cm

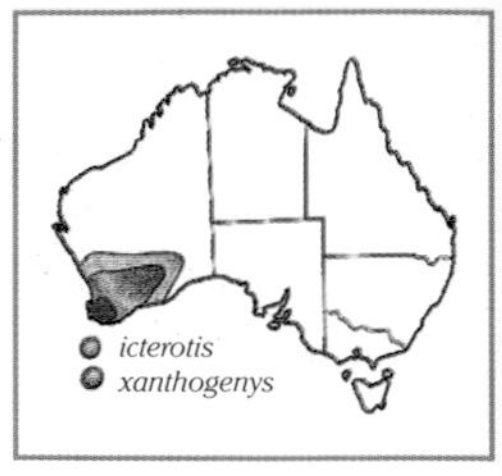

A quiet, unobtrusive bird; usually in pairs or small family parties rather than in flocks; feeds quietly on the ground or in foliage. The flight is lighter, more fluttering with only slight undulations. This small, predominantly red rosella, like many birds of the isolated SW corner of the continent, has developed distinctive plumage and behaviour. These are sufficient that, in grouping similar rosellas into three 'complexes', blue-cheeked, white-cheeked and yellow-cheeked, the Western Rosella stands alone as the sole species of yellow-cheeked rather than being lumped into one of the other groups. Apart from its slightly smaller size, greater difference between sexes and unique yellow cheeks, the Western Rosella behaves differently; it is more placid, less wary and timid, and less aggressive in aviaries. **Variation:** nominate race inhabits W coast and more heavily forested ranges; race *xanthogenys* lives in the semi-arid woodlands from W edge of wheatbelt out through salmon-gum country towards the Nullarbor. Across intervening cleared farmlands and open woodlands, the two forms merge gradually. **Voice:** varied calls, clear, musical, high-pitched, ringing 'quink-quink-quink-quink' and slightly softer 'whip-a-wheee'. **Hab.:** diverse, from tall wet karri to dry woodland well inland towards the Nullarbor. Most common in woodlands of salmon gum and wandoo, farmlands; less common in heavy wet karri and jarrah; scarce on most of sandy W coastal plain. **Status:** generally quite common; probably gaining from land clearing, provision of water. **Br.:** 377

Eastern Rosella *Platycercus eximius* 29–33 cm

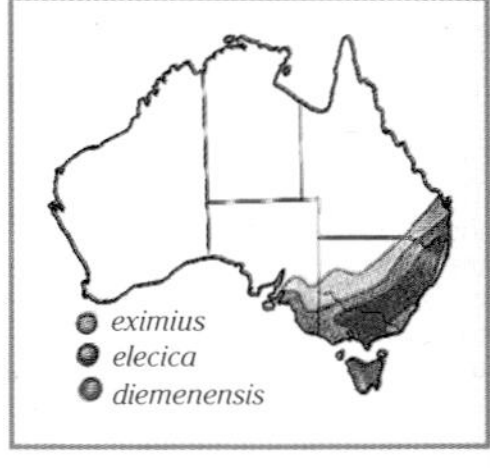

The brightly coloured Eastern Rosella is familiar in small flocks around towns and along roadsides; feeds mainly on the ground. Flight is undulating when flying tree to tree—dives down, swoops up—but travels high and level over longer distances. **Variation:** three races: nominate race in SE NSW; race *elecica*, 'Golden-mantled Rosella', which has more extensive gold fringing on the back, on tablelands, NE NSW and SE Qld. Tasmanian race, *diemenensis*, has darker reds, larger white cheek patches. **Voice:** in flight, brisk, sharp, clear, rapid 'quink-quink, quink-quink' and even more rapid 'whit-whit-whit-whit-'. Slower, more drawn out, is a high, clear, ringing 'pee-pt-eee'; there is also much rapid, confused chattering within groups. **Hab.:** woodlands with scattered trees but mainly grassy groundcover, farmlands, watercourse trees, croplands, parks, gardens; generally more open environs than Crimson Rosella, usually below 1200 m altitude. **Status:** abundant through most of its range where suitable habitat exists; has benefited from the opening up of heavier forest. **Br.:** 377

Pale-headed Rosella *Platycercus adscitus* 28–32 cm

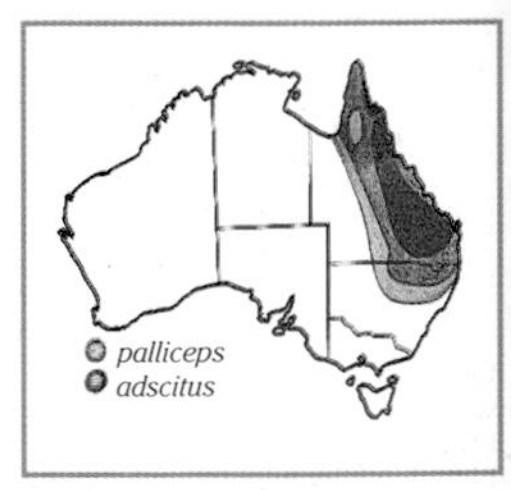

Feeds in pairs or small flocks on ground or among foliage; flight undulating, obviously pale head. **Fem:** like male, but with wing stripe that is absent on males. **Imm.:** slightly duller than females, often some grey or red on head; has wing stripe. **Variation:** nominate race *adscitus*, 'Blue-Cheeked Rosella', Cape York S to Cairns, then intergrades with southern race *palliceps*, 'Pale-headed Rosella', which extends S to northern NSW. These were once considered to be two separate species. **Voice:** in flight, abrupt 'czik-czik-, czik-czik-'; from trees, high but soft, thin and slightly tremulous 'fee-e-fee-e-fe-e' or 'fwe-we-we-wee'. Also typical confusion of typical rosella chattering by several birds. **Hab.:** grassy woodlands, farmlands with scattered trees, lines of trees along watercourses and roads, dry scrubby ridges, but usually in lowlands rather than on the highest parts of ranges. **Status:** abundant, sedentary; has benefited from thinning of heavier forests. **Br.:** 377

Northern Rosella *Platycercus venustus* 28–30 cm

The only rosella found in far NW Aust.; easily identified by bold plumage pattern. Usually small parties of five or ten birds feed on the ground in the cooler morning and evening, or shelter in leafy treetops during the heat of the day, where they may attract attention with their sharp 'whit-whit-' calls or typical rosella chatter accompanied by much tail wagging. The flight is undulating, often swooping close to the ground, then turning steeply upwards, tail and wings spread, to land. Feeds on seeds of grasses, shrubs and trees, principally those of eucalypts and melaleucas, but also berries and flowers. **Imm.:** duller than adults; may have some red flecks on head or breast; greener central tail feathers; usually has a faint wing stripe. **Voice:** calls high, very clear and ringing; includes an extremely rapid, high, piercing 'whit-whit-whit-whit-whit-whi-', so rapid that the ten or twenty notes are given within two or three seconds and at an unwavering, even pitch. Quieter, more husky, mellow, is an erratic 'chak, chak-chak, chakchakchak, chak'. **Hab.:** grassy open forests and woodlands, especially those dominated by eucalypt or melaleuca trees; further inland, where woodlands give way to more open country, the tree belt lining watercourses. Also visits coastal swamp-forests and mangroves to feed. **Status:** sedentary; uncommon, but not scarce in sites of most suitable habitat; population appears to be declining. **Br.:** 377

The Western Rosella has at times been included as one of the white-cheeked group, but is now considered to be a quite distinct species.
Green edged on west coastal race *icterotis*
Cheek patch yellow,
Mottled red and green
Blue, tipped white
Blue
Central pair green
Red from throat to under-tail coverts
Green
♂
Deeper crimson
Mantle and upper parts scalloped red and black
Dull red mottled with green
There are two subspecies of Western Rosella: the nominate green-backed race in the wetter west coastal forests, and the red-backed race, *xanthogenys*, in the semi-arid eastern part of the species' range, with intermediate forms between.
♀
Red-backed inland race *xanthogenys*
Rump slightly more olive grey-green
Race *icterotis*
Central pair mainly blue
The best known of the three species grouped together here, conveniently known as the 'white-cheeked rosellas'. This grouping is more in recognition of affinities rather than of importance of the white cheeks. This species hybridises with the Pale-headed Rosella in the N of its range; some consider the Pale-headed to be a subspecies of the Eastern Rosella.
Scarlet
Upper plumage black edged with light green or yellow creating a boldly scalloped pattern; female and immatures may have slightly greener, finer edgings.
Deep blue
Black
Patchy yellow to nape
♂
Bright greenish yellow or blue-green
Tails relatively broad, not extremely long
Rump light, bright green
Scarlet
White cheeks
♀
Yellow
♂
Central tail feathers blue-green, laterals blue tipped white
♂
Primaries edged blue
Upper-wing and underwing coverts deep blue
Blue-cheeked race
Blue
Deeper golden yellow compared with the pale, thin, slightly green-tinted feather edging of the blue-cheeked nominate race
White with yellow tint
Red
Yellow
Cheeks white with yellow tint
Central tail feathers blue with slight green tint towards coverts; outer tail blue tipped white
Yellow
Cheek patch white with blue-violet lower edge
Blue extends to upper breast or throat.
Adult
Nominate race *adscitus*, 'Blue-cheeked Rosella'
Blue, with green tint towards rump
Race *palliceps*, Pale-headed Rosella
The third species, with the above two, form the 'White-Cheeked Rosella' complex.
White cheeks conspicuous; partly blue in W of range
Pale yellow; black edging may be finer in W of range
Pale yellow margins to black centred feathers of mantle, back, wings
Light blue wing-bar
Yellow rump
Scarlet under-tail coverts
Scarlet
Pale blue fading to whitish tips; these are more obvious on upper surface of tail.
Bold conspicuous contrast of white cheeks and bill against black head
Central tail feathers blue-green, lateral feathers blue with white tips
Black
Dark blue
Deep blue
A subspecies, *hilli*, is sometimes recognised. It has a more violet cheek patch, finer or no black fringes to yellow of underparts and a larger bill.
Adult
Adult
Outer webs blue

Red-capped Parrot *Purpureicephalus spurius* 34–37cm

A large parrot, strongly coloured almost to the point of being gaudy. A male in full colour has areas of solid, rich, bright or deep colour: crimson, bright greenish yellow, deep blue-violet. The rump is bright, slightly greenish yellow, conspicuous as the birds fly up and away, and head and breast are often not visible. The female can be almost as colourful as the male; only the immature is rather nondescript—still recognisable as a Red-cap, although the cap is reduced to a small red forehead-bar. The flight is swift, undulating as bursts of wingbeats are interspersed with glides; travels level rather than swooping low to the ground like many other parrots. A distinctive feature of the Red-capped Parrot is the bill with its long slender tip. This appears to be an adaptation for feeding on the large, very hard and woody seed capsules of the marri, *Eucalyptus calophylla*, one of the most common tree species throughout the range of the Red-capped Parrot in the SW of WA. As the capsule of the gumnut matures, many months after flowering, the seeds that develop within can only be reached by other parrots and cockatoos by laboriously cutting away the wood of the capsule. But the Red-cap can insert the fine tip of the upper mandible into the capsule's small valve openings through which seed would ultimately be shed and prise out the seeds—an easier, quicker way of exploiting this major food source. The range of the marri approximately matches that of the Red-cap, and the marri occurs intermixed with other eucalypts such as karri, jarrah and wandoo throughout the SW. On most farms of the western wheat belt and the coastal plains, marri trees have been retained as paddock shelter trees, ensuring food for these parrots, which can, however, make use of other seed-bearing and fruiting trees and can feed on the ground on fallen seed. **Voice:** in flight a distinctive, rolling 'kchurrrink!'. From trees, abrupt, rough 'chrrek!' and rapid 'chirek-achek'. **Hab.:** likely to be found in almost any habitat in the SW corner of WA: tall wet karri and heavy jarrah forests, their intermixture with marri and wandoo, banksia heathlands, mallee, and farmland with remnants of any of these vegetations. **Status:** sedentary or with small local movements; common. **Br.:** 377

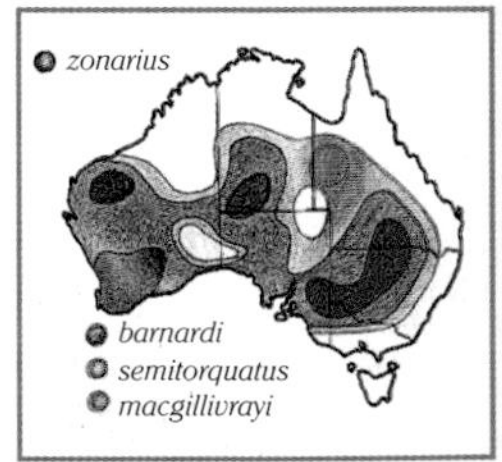

Australian Ringneck *Barnardius zonarius* 34–38 cm

At present all these parrots with yellow 'ringneck' hind-collars are races of a single species, the Australian Ringneck. Previously they have been grouped as two or four separate species; all have long-established common names still used in aviculture and as local common names. The races of Australian Ringneck differ not just in appearance, but in calls and other details. The subspecies interbreed within transitional zones. Flight is undulating with deep wingbeats in bursts; in between, wings remain extended in short glides.The races are:

'Port Lincoln Parrot' *Barnardius zonarius zonarius.* The nominate race of the Australian Ringneck. With black head, golden collar at nape, bright yellow belly, blue in wings and tail, and extensive green, this is the most boldly coloured of the races. Often feeds on the ground on native seeds and plants, or on spilled grain in paddocks or on roadsides. Flies up from ground to tree directly, tail spread showing colours. Travels with strong, undulating flight, swooping low then up to land in a tree, long tail spread on landing. **Voice:** ringing double-noted 'klingit, klingit, klingit' and rapid, sharply ringing 'kling-kling-kling-'; in groups, much noisy chattering. **Hab.:** the drier regions: woodlands, mallee, mulga, spinifex, especially tree-lined watercourses, roadside and farm trees. The slightly smaller, paler race *occidentalis,* Pilbara of WA, has no separate common name. **Status:** common in E; abundant in WA.

'Twenty-eight Parrot' *Barnardius zonarius semitorquatus.* Underside of body almost uniformly deep green; also slightly larger than nominate race. **Voice:** triple-noted call, more mellow than sharply ringing, 'teu-wit, teoo', or 'twen-ty, eight'. Other calls more like those of Port Lincoln. **Similar:** 'Port Lincoln Parrot'; between these black-headed races there is a broad hybrid zone where the abdominal patch becomes yellower further inland, changing gradually from green to yellow, possibly contributing to the local use of 'Twenty-eight' for all WA ringnecks, including the Pilbara race *occidentalis.* **Hab.:** heavier forests of the wet SW, the jarrah and karri, and intermixed wandoo and marri. **Status:** Clearing of E and coastal plains margins of the SW heavy forest block has brought the Port Lincoln race further into the previously exclusive habitat of the green-bellied 'Twenty-eight' race, which is being reduced in range as a distinct race without any yellow underneath.

'Mallee Ringneck' *Barnardius zonarius barnardi.* A colourful race with greenish rather than black head and with red over the bill. Feeds both in trees and on the ground on seeds, and flowers. In flight, brief bursts of deep wingbeats interspersed with short glides that give an undulating motion. Travels low, fast and near the ground; swoops up with tail spread to land. **Voice:** calls in flight, a high, ringing 'kling-kling-kling'; much varied chattering while feeding in foliage, but silent on ground; harsher alarm call. **Hab.:** mallee, mulga scrub, eucalypt woodlands, river gums along watercourses, farmlands with scattered trees. **Status:** sedentary or locally nomadic; common but possibly reduced in some areas by clearing, farming; in this respect differs from the nominate race, which thrives on farmlands in the wheat belt of WA.

'Cloncurry Ringneck' *Barnardius zonarius macgillivrayi.* This pale race is confined to a relatively small area between Cloncurry and Mt Isa, and W slightly into NT. **Hab.:** arid, hilly, open forest and grassland of the Selwyn Ranges; especially the vicinity of large eucalypts along watercourses. **Status:** this race appears to be holding its numbers near Cloncurry and around the upper reaches of the Diamantina R. **Br.:** all races 377

Cap bright red
Bright greenish yellow rump
Back and wing coverts — bright rich green
— Yellow
♂
Blue, tipped white
As the Red-capped Parrot flies away it displays a bright yellow rump that can be a helpful identifier if the other bright colours are not visible.
Deep purplish blue
Spectacular large parrot, unique colour pattern, strong bold and bright colours
♂
Red of crown is dull, may be streaked green.
Small red band
Thighs to under-tail coverts bright red
Long hooked bill
Greenish yellow
Faint wing stripe, never on males
Blue, paler tips
Greenish yellow
Only immatures likely to cause any difficulty in identification
Dull mauve-blue
Central pair bronze green
Red with flecks of green
♀
Central pair of tail feathers dark green
Juv.
The Australian Ringneck species is made up of dark headed W races and paler headed E races; all have yellow collar.
Bright yellow collar
Black head with dark blue-violet cheeks
Black and dark blue-violet
Red frontal band
Nominate race *zonarius*, 'Port Lincoln Parrot'
White tipped
Yellow
Entirely green
Edged blue
SW race *semitorquatus*, 'Twenty-eight'
This race confined to the forested SW corner of WA; eastwards, increasingly yellow bellied, merging with Port Lincoln.
Race *semitorquatus*, red frontal band
Females have slightly softer colours; immatures dull, brown head.
Yellow collar
Red frontal band
Yellow collar less obvious against pale, indistinct, overall colour pattern
Dusky blue-grey
Australian Ringneck, nominate race *zonarius*, locally 'Port Lincoln Parrot', bright yellow belly with sharp edge to dark green breast
Pale yellow
Race *barnardi*, 'Mallee Ringneck'
Small orange patch
Soft blue-grey
Bright blue underwing coverts
Race *macgillivrayi*, 'Cloncurry Ringneck'
Large yellow abdominal patch
Lateral feathers pale blue, edged white
Ringnecks have long tails, tapering when folded, the green central feathers almost hiding the blue and white of laterals.
Outer margins and tips pale blue, central pair dark

Red-rumped Parrot *Psephotus haematonotus* 26–28 cm

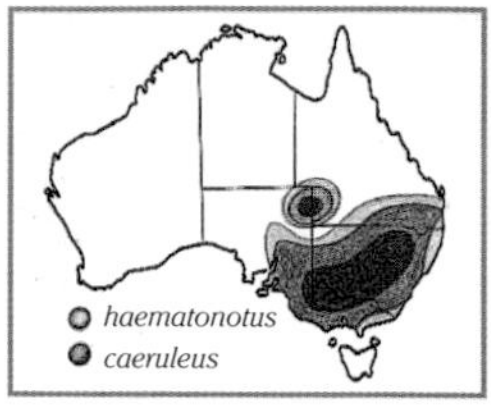

Often feeds on ground in pairs to large flocks. Flight swift, slightly undulating, often travels high. **Variation:** race *caeruleus*, Lake Eyre basin, has paler plumage. **Voice:** sharp, metallic, scratchy, squeaky, abrupt 'chwie-chwiep, chwie-chwiep-'; long squeaked 'chwieee'. Also lower, rather husky, harsh 'chwier-querrk'; harsh scolding, 'querrk, querrk', intermixed with the fine squeaks. **Similar:** Mulga Parrot. **Hab.:** open grassy and lightly timbered plains, timbered watercourses, mallee, farmlands. **Status:** sedentary, rarely dispersive; common. **Br.:** 378

Mulga Parrot *Psephotus varius* 26–30 cm

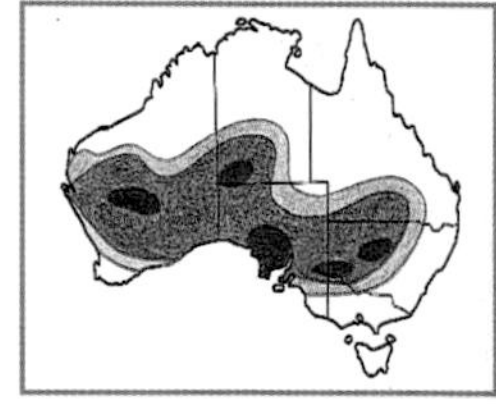

Undulating low flight that displays intense colours of male; generally quiet and unobtrusive. **Imm.:** like adults, but orange-red of thighs and belly is absent or reduced on young males. **Voice:** flight call a distinctive, slightly husky yet still sharp and strong 'zwit-zwit, zwit-zwit' or 'chwit-chwit'; from trees, a very rapid, sharp 'wit-wit-witwitwit', and softer fast chattering. **Hab.:** semi-arid regions; mulga, mallee, saltbush with scattered small trees, lines of big trees along watercourses, drier farmlands; never far from water. **Status:** moderately common; abundant in good seasons, at other times may be scarce. **Br.:** 378

Paradise Parrot *Psephotus pulcherrimus* 27–30 cm

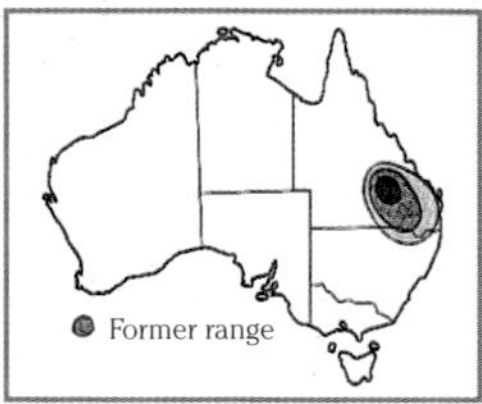

One of the most beautiful parrots, it lived in small family parties or pairs; fed on ground. **Imm.:** like female, dull colours. **Voice:** series of soft, rather musical whistles; in alarm, a sharp metallic call. **Similar:** Blue Bonnet, NE race *haematorrhous*, shares brown upper parts, red shoulder patch, scarlet under-tail coverts, belly; but has brown and yellow breast and rump, and blue face. **Hab.:** open grassy woodlands and scrubby grasslands on broad river valleys and plains where termite mounds are common. **Status:** almost certainly extinct. **Br.:** 378

Golden-shouldered Parrot *Psephotus chrysopterygius* 25–26 cm

A rare and beautiful parrot. In small parties or pairs it usually feeds on ground; not timid, flies up to nearby trees, soon returns to feed when disturbance has ceased. Flight swift with only slight undulation. During heat of day, rests in shady foliage; visits waterholes early in the morning, occasionally during day. **Voice:** flight call a sharp, scratchy, metallic yet musical, pleasant 'chwit, chwit' and 'chirrit, chirrit'; also sharp, quiet, varied chattering from treetops. Silent on ground. **Similar:** Hooded, but widely separated range. **Hab.:** grassy woodlands, with scattered eucalypts, paperbarks, abundant large termite mounds; occasionally river-edge mangroves. **Status:** endangered, probably by altered habitat, changes in burning, regrowth. **Br.:** 378

Hooded Parrot *Psephotus dissimilis* 25–26 cm

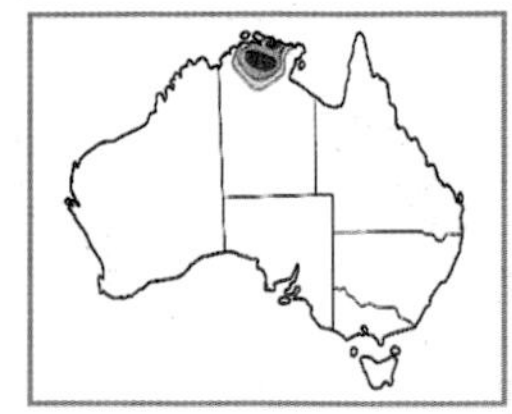

Bold strong colours make the male spectacular, and even more so in flight; golden wing panels flicker with the action of outspread dark wings. Travels swiftly, slightly undulating. Not timid, flies up to nearby trees if flushed; often feeds along roadsides. **Voice:** sharp, very high, thin 'chseit, chseit' in flight and sharp, metallic 'chsink, chsink'; slightly harsh or scolding 'charrak, charrak'. In trees there is quiet, varied, squeaky, budgie chatter. **Hab.:** dry savannah woodlands and open forest with grass or spinifex groundcover, termite mounds, plains or stony ridges; not far from water. **Status:** sedentary; uncommon, localised. **Br.:** 378

Blue Bonnet *Northiella haematogaster* 27–34 cm

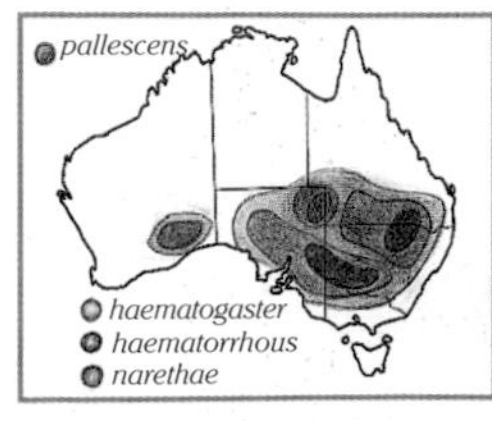

Blue-faced parrot; usually feeds on the ground in pairs or small parties; often in shade by roadside seeking seeds, flying up to nearby tree when disturbed. Also feeds on fruits, berries, flowers. Roosts very quietly in shady foliage through heat of day. Flight undulating, rather erratic, dropping low on leaving a tree, swooping up to land in another, tail spread showing white-tipped blue outer feathers. **Variation:** four races, three well known by common name; a fourth, from the Lake Eyre Basin, is a pale form of the nominate race. **Voice:** contact call given in flight is harsh, nasal with metallic, scratchy quality, very abrupt, 'chrak, chrak' with variations; shorter 'chak'; in alarm, rapid 'chzak-chzak-czakczakczak'; other calls include longer 'chrraak', soft whistles and 'chrik' chattering. The Naretha race has a slightly softer call. **Hab.:** lightly timbered grasslands, mulga, mallee, she-oak, watercourse and paddock trees. **Status:** common; Naretha form remote, possibly uncommon. **Br.:** 378

Often forages in busy flocks on the ground, males showing red rump moving away or flying up and away.
Dull blue-green
Dull olive-brown
Red of rump patch is displayed between blues of wings as birds fly up from ground or as they land on a branch, wings outspread.
Head emerald green, dark bill
Olive-green
Rump red
Yellow
Small yellow patch at shoulder
Central pair blue-green
Lateral feathers blue-green tipped white
White under-tail coverts
♂
♀
Outer webs deep blue
Secondaries blue, their coverts turquoise
Yellow forehead, red crown
Intense bright emerald green upper parts with brick-red band across tail coverts
♀
Also known as 'Varied', or 'Multi-coloured Parrot', and that is the impression given by the male: intense green, splashed crimson, red, yellow, orange and with much blue on the wings
Green rump
Dull brick red on shoulder, crown
♂
Orange-red
Light green
Extensive deep blue on outer wing
Deep golden
Beautiful small parrot, now almost certainly extinct
Dark brown cap
Yellow around eyes
Bright blue
Blue, tipped white
Light buffy to brownish olive
Deep blue
Red
Emerald
♂
Scarlet
Scarlet
♂
♀
Pale blue
In flight, from above, Paradise has bright blue rump in strong contrast against dark browns.
In flight, this parrot, and even more the Hooded with its larger, brighter golden wing patches, is a spectacular bird with flickering gold set in dark wings beside brilliant turquoise and scarlet.
The Golden-shouldered and Hooded Parrots are so much alike that in the past they were combined as a single species, the Hooded being the subspecies.
Black cap
Yellow
Bronze-olive
Green tint
Golden median wing coverts
Turquoise-blue
Turquoise or pale blue
Median wing-coverts golden
Olive tint
Scarlet, edged yellowish
Blue tint
Turquoise
On wings of the Hooded Parrot, the golden yellow patches are larger, extending back onto the next row, the greater coverts.
Back darker, blackish brown; black cap down over the eyes
Pale grey-green
The Hooded Parrot has larger golden wing patches set in darker brown upper parts and a larger black cap, but a smaller area of orange-scarlet beneath.
Orange-scarlet edged yellowish
Salmon, yellow
When alert, like many other species, holds neck extended; looks small headed, slender. At other times, feeding on ground or relaxed, resting, head is lowered, neck is shorter, bird does not look so slender.
Edged deep blue
Rusty red wing coverts; olive or yellow on other races
Face, forehead both deep blue
Coverts olive and scarlet
Rusty red
Olive-gold
Rump olive-orange
Forehead patch larger, brighter blue; cheeks deep mauve-blue
Rump olive-ochre
Red
Red
Yellow
Red
Pale yellow
Red does not extend to thighs.
'Red-vented' race *haematorrhous*
Race *haematogaster*, 'Yellow-vented Blue Bonnet'
Race *haematorrhous*, 'Red-vented Blue Bonnet'
Race *narethae*, 'Naretha Blue Bonnet'

Bourke's Parrot *Neopsephotus bourkii* 19–22 cm

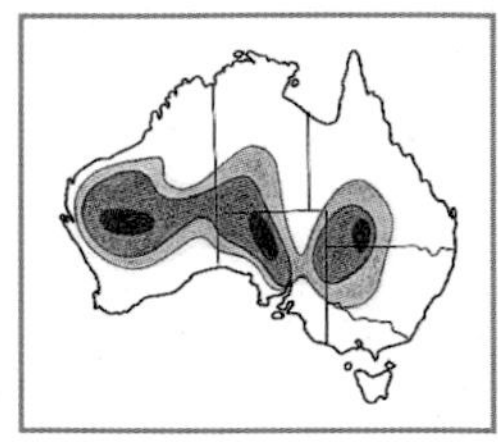

With its soft colours and quiet behaviour, this small parrot is easily overlooked. But a watch at a waterhole at sundown is likely to reveal its presence as pairs and small flocks fly in to drink. This habit of coming to water well after sunset has earned this bird the name 'Night Parrot', confusing it with the true Night Parrot, *Pezoporus occidentalis*. In the soft dim light after the sun has dipped below the horizon, these birds, while on the ground, blend with the red earth that is typical of much of the interior, inconspicuous as they approach and drink at the edge of the pool. During the day when birds feed under mulga bushes, these plumage colours blend well with greys and browns of weathered dead limbs and shadowy ground. Usually quiet, approachable. Flight is direct, swift, low with audible whirring of wings; shows pink and blue underparts, tail prominently tipped white if spread. **Voice:** flight is call not extremely high, scratchy or harsh, but is mellow, pleasantly musical yet quite sharp, penetrating and clear: 'chiew-eet, chw-iet, chew-iet'. In trees, soft musical twittering. Alarm call a shrill double note. **Hab.:** mulga and similar acacia scrublands, tree-lined watercourses in mulga country, semi-arid woodlands. **Status:** nomadic; generally quite common but varies locally by season—abundant or scarce. **Br.:** 379

Budgerigar *Melopsittacus undulatus* 17–20 cm

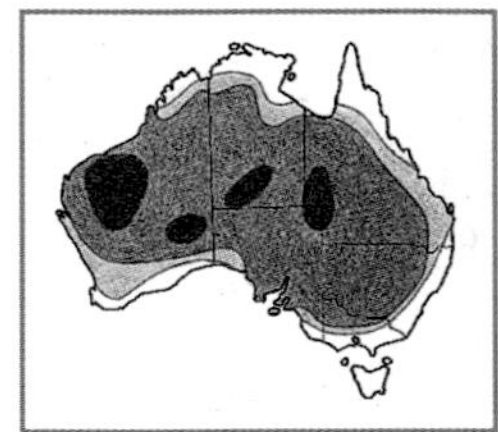

Familiar small bright green parrot, usually in small parties, but at times in huge flocks that dart across the sky, twisting and turning in unison. **Voice:** in flight, warbling, musical 'tirrrit', 'tir-rit, tirit'; on taking flight, a rapid, rasping or scolding 'tzzit-tzit-tzit, tzt-zt-zt', fading away. From a large assembly of Budgerigars there is an overwhelming wave of sound. **Hab.:** spinifex, saltbush and grassland of plains and ranges, treeless or with scattered trees or shrubs. Lines of trees along watercourses and stands of open eucalypt woodland scattered through open country are centres for these birds, providing nesting hollows and water from creek-bed pools. Although Budgerigars can survive for some time without water, they are rarely found far from it; the fast-flying flocks can quickly cover considerable distances to water. **Status:** highly nomadic, responding to heavy rains by congregating where lush grass ensures abundant seed, quickly raising several broods, each of up to eight young. The resultant population explosion forces the birds to disperse as soon as arid conditions return. Budgerigars occasionally wander nearer the coasts, avoiding only the most heavily forested regions. Abundant. **Br.:** 379

Swift Parrot *Lathamus discolor* 23–26 cm

Aptly named for swift flight; flocks dart across sky, weaving fast and low among trees with audibly whirring wings and clear, sharp calls. Often associates with lorikeets; feeds on nectar, scale insects, fruits. **Voice:** different from lorikeets; lacks their screeching, scratchy, harsh sounds; rather has pleasant, musical, high notes, sharp but clear. In flight, contact call is 'chi-wit, chiwit, chiwit'. Also has a great variety of musical chattering and trills, intermixed deep, mellow sounds and clear, sharp notes. **Similar:** Musk Lorikeet, Purple-crowned Lorikeet, Little Lorikeet, but all with different placing of red in plumage; smaller, stubby build, short tails. **Hab.:** forests and woodlands with flowering trees. **Status:** sparse, endangered. **Br.:** 378

Ground Parrot *Pezoporus wallicus* 28–30 cm

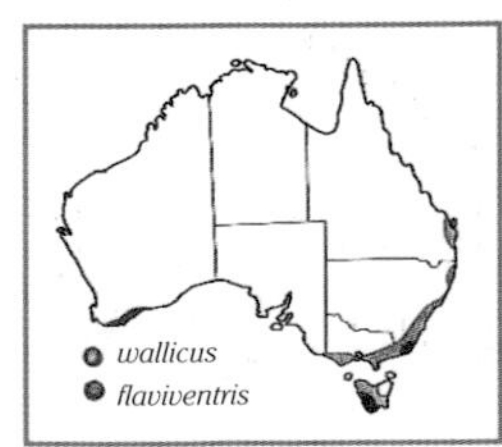

The Ground Parrot is a shy, elusive bird, usually not seen unless flushed, or at twilight when calls are heard and birds can be seen fluttering low over the heath. **Voice:** calls at dawn and dusk, a series of piercing, ringing, resonating whistles, rising in steps, each note flowing on almost unbroken, but abruptly higher than the preceding; or lower notes at more even pitch. Cheerful budgie-like warbling and sharp, rapid trills. **Hab.:** heathlands of coasts, ranges, swampy heaths, drier ridges in swamps, occasionally grasslands near heathlands; in Tas., the stronghold of this species, it favours button-grass plains. **Status:** uncommon to rare; endangered in many parts of range by loss of habitat in excessively frequent fires. **Br.:** 379

Night Parrot *Pezoporus occidentalis* 22–24 cm

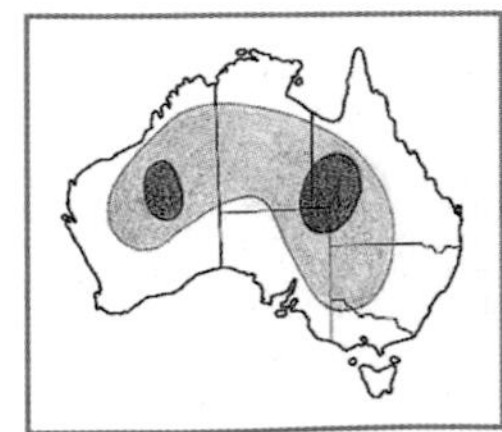

Nocturnal—hides in dense spinifex by day, drinks after dark, then feeds among spinifex. **Voice:** old records are of sharp squeak when flushed; contact call in flight a short, sharp whistle, several times, rapidly. A low, drawn out, double-noted whistle when coming to water. In alarm, a harsh, almost croaking sound. **Hab.:** appears often to have been associated with spinifex, or among samphire bushes on margins of salt lakes. **Status:** rare, thought probably extinct until discovery of a dried specimen on a Qld roadside in 1990. **Br.:** 379

Clad in soft pastel pinks and blues, the little Bourke's Parrot blends surprisingly well into the colours of its semi-desert habitat.
Dusky mid blue-violet
Outer feathers tipped white
Underwing coverts are mid dusky blue-violet, deeper when shadowed, as is usual when seen from beneath, or brighter if the underside of a wing catches direct sunlight.
Deep rose-pink
♂
Female and juveniles have no blue on forehead.
Bright blue of under tail extends to edges of upper-tail coverts.
Female is only very slightly less colourful.
Only the male has a touch of blue on forehead.
Juveniles have dull colours.
Male does not have the stripe that crosses underwing of female and immatures.
Central feathers dark grey-brown; outer feathers dull blue-grey edged white and with large white tips
Pale blue
♂
Immatures are like female, but less pink; have underwing stripe, but very faint on immature males.
Juveniles and some females have dull plumage.
One of best-known Australian birds; many colour forms have been bred from the green and yellow wild stock.
Male feeds female on a perch near the nest hollow.
Pale wing-bar, more yellow on underside of wing
Underwing coverts vivid green
Pale yellow bar across underwing
Yellow head, heavily barred except forehead to throat
Bright green
Barred black and yellow
Juveniles are dull with barring extending down forehead; may not have dark spots on throat.
Brown, barred black and yellow
Cheek spots
Edged blue-green
Sexes similar but male has darker blue-grey cere; the pale cere of the female becomes brown in the breeding season.
Yellow band across tail
Tail blue-green
Bright green
Crown dark blue
Deep emerald green
Spectacular in flight with scarlet underwing coverts in bold contrast against dark brown of flight feathers
♂
Bend of wing scarlet
Swift Parrot superficially resembles lorikeets (p. 173)
Forehead to throat, crimson
Small splashes of scarlet on inner edges of tertials and on shoulder
Sexes similar, female slightly duller, has creamy underwing bar. Juvenile like female, but dark iris, paler red on throat and under tail.
Scarlet under-tail, and spots along flanks
Tail finely pointed, unusual dusky maroon colour
Scarlet
Scale change
Green heavily mottled with yellow and black
Small red band
Outer feathers brighter yellow, lightly barred
Ground dwelling bird of heathlands
Yellow, strongly barred brown
Yellow wing-bar across brown, green edged primaries
Tail very long, green, with cross barring of yellow, brighter at outer margins
Race *flaviventris*, SW of WA, has brighter yellow underparts with lighter, broken barring.
Flight low, twisting, with bursts of wingbeats; glides on stiffly downcurved wings; appears bright green.
Looks rather large headed.
Green, mottled black and yellow
The thickset, dumpy Night Parrot is a terrestrial, nocturnal and secretive species. By day it keeps hidden in dense spinifex or other cover where it may use concealing passageways through the vegetation. If flushed, flutters away erratically, low, soon dropping to the ground, then scurrying into hiding beneath the concealing spinifex.
Outer feathers of tail barred black
Plump, stocky bodied
In flight, primaries, secondaries and primary coverts show as dark brown edged olive-green, and with yellowish wing stripe across primaries near base, the whole outer wing quite dark and in strong contrast to bright green, heavily barred patterning of rest of plumage.
Sexes alike; immatures reputed to be dull: plain, yellowish throat, neck

Blue-winged Parrot *Neophema chrysostoma* 20–22 cm
Golden olive above, light, bright yellow beneath with the largest, deep blue shoulder patches found among the neophemas. Flight is swift, direct with little undulation. Usually in pairs or small flocks; forages on ground, early and late in day. **Voice:** extremely high, thin tinkling, in bursts, fast then slow 'tsiwee-tsiwee-tsiweet, tsi-weet, tsi-weet'; more like highest squeaks of a thornbill or fairy-wren than a parrot. **Similar:** Rock, Elegant Parrots, Orange-bellied Parrots. **Hab.:** forests and alpine to grasslands, mulga, saltbush, coastal dunes. **Status:** abundant in Tas.; common in Vic. and SE of SA; elsewhere uncommon to rare. **Br.:** 379

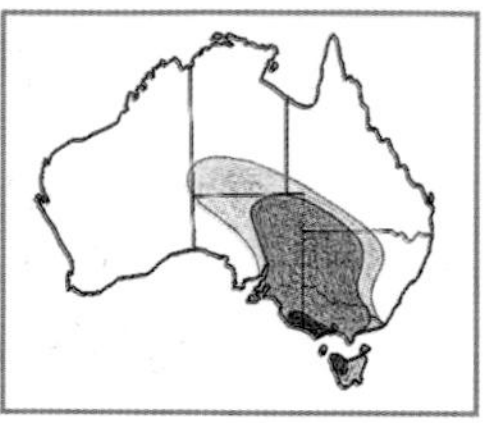

Scarlet-chested Parrot *Neophema splendida* 18–21 cm
Scarlet chest identifies male, but female, lacking red, is more like other neophema parrots. Although so colourful, they are inconspicuous in their natural habitat: calls are soft; feed on ground; fly low, keeping close to cover. In pairs or small parties, rarely large flocks. May remain hidden unless accidentally flushed by close approach. The flight is swift and rather erratic or fluttering. **Voice:** mellow, abrupt, chattering, twittering 'tooweet', 'chwit', 'chweet'; does not carry far. **Similar:** Turquoise Parrot, but range differs; Blue-winged; Elegant like female Scarlet, but far less blue on head. **Hab.:** open woodlands of eucalypts, she-oak, mulga with spinifex, saltbush. **Status:** nomadic; irruptive—scarce, but at times locally common. **Br.:** 379

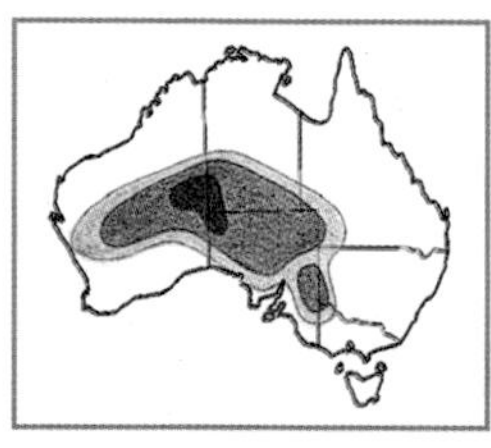

Rock Parrot *Neophema petrophila* 21–23 cm
Olive and deep yellow blend well with low shoreline vegetation and lichen-covered rocks. Usually not seen unless flying or flushed or attention drawn by the tinkling calls. In flocks, at times quite large. If flushed from dense low cover, birds dash away with a flurry of quick wingbeats and burst of calls. Erratic flight with jerky wingbeats, flocks rising high at times. **Voice:** sharp, thin tinkling calls, usually given in flight, 'tsee, tzit-tseit'; but, from flock passing overhead, draws attention. **Hab.:** coastal: shoreline, dunes and heaths, usually within sound of the sea, rarely venturing inland; offshore islands. **Status:** sedentary; common. **Br.:** 379

Orange-bellied Parrot *Neophema chrysogaster* 20–22 cm
One of Aust.'s rarest birds; only 150–200 left in the wild. In flight, blue underwing coverts, blue secondaries and margins of primaries. Most of day spent hidden under low vegetation; feeds on ground around dawn and dusk. Wary, easily flushed, then flies high with loud calls; travels far before dropping again to ground. **Voice:** in alarm, rapid, buzzing 'zzt-zzt-zzt-zzt'; tinkling contact calls in flight. **Hab.:** breeds summer, SW Tas., buttongrass and swampy sedgeland plains. Winters S coast of Vic. and SE of SA; uses tidal flats, salt marsh and heath, islets, pasture close to shore. **Status:** migratory; endangered, possibly still declining. **Br.:** 379

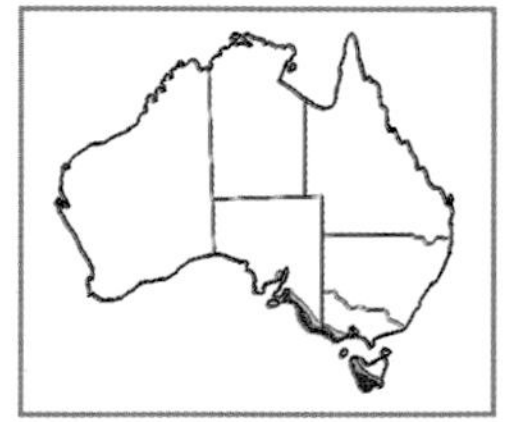

Turquoise Parrot *Neophema pulchella* 19–21 cm
Feeds inconspicuously on ground; flies up with outspread wings displaying extensive blue, red and green on inner wings, outspread tail blue edged and tipped yellow. Overhead, shows yellow beneath body and tail, and blue underwing coverts. Flight is generally swift, erratic, fluttering wing action; brief glides on downcurved wings; the tail is spread on take off and landing, displaying extensive yellow. **Voice:** high, weak, musical, tinkling 'tzeit-tzeit, tzeit-tzeit'; faint high twittering when feeding or at water. **Hab.:** woodland and open grassland, natural or partly cleared. **Status:** rare, may be locally common; semi-nomadic. **Br.:** 379

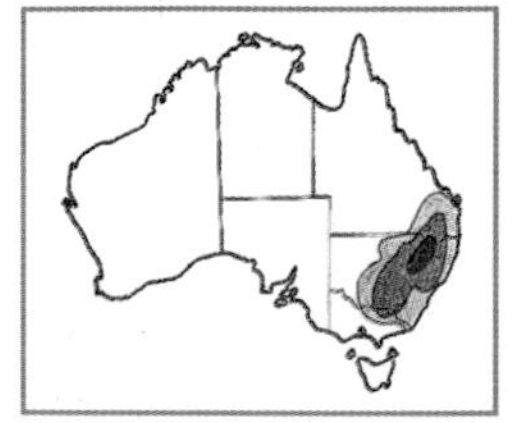

Elegant Parrot *Neophema elegans* 22–23 cm
Out of breeding season, small to large flocks feed on ground. **Voice:** in flight, sharp, squeaky 'chwit' or 'tzit'; usually silent while feeding on ground, but may give occasional faint, sharp, squeaky twitters, like some of the squeakiest sounds from fairy-wrens. **Similar:** Blue-winged Parrot, but broad blue wing patch not as golden beneath, blue of forehead not extending back behind eyes. **Hab.:** woodlands, lightly timbered grasslands, partly cleared farmlands, margins of clearings in heavy forest, tree-lined watercourses, mallee, mulga. **Status:** locally nomadic with localised movements, mainly at fringes of range; generally common. **Br.:** 379

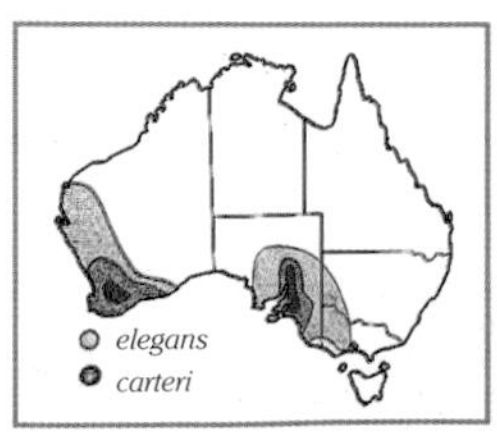

Of all neophemas, the shoulder patches of this species are the largest and the deepest blue.
Crown dull golden, merging to dull golden olive-green on back, wing coverts, rump
Frontal band blue almost back to the eyes
Lores, and around eyes yellow
Displays extensive deep blue of both upper and under-wing coverts in flight.
Light, bright golden yellow
Blue-green
Immature and female have dull olive-green crown; overall slightly less colourful.
Juveniles and female have blue of shoulders reduced, mottled green
Juv.
Juveniles have no blue frontal band above bill; female has a smaller, paler blue band than that of male.
Brilliant scarlet chest set between bright blue head, and deep yellow underbody
Cobalt blue
Scarlet
Quiet and unobtrusive; easily overlooked even though so brightly coloured
♂
Yellow upper margins; underside yellow
Extensive light cobalt blue
Wings and tail spread as 'air-brakes' on landing
Wing coverts pale blue
♀
Juv. like female, but bill and cere orange; males have darker blues.
Underside of outer tail feathers deep golden yellow
Flight feathers edged deep blue
Small parrot of shoreline environs; has dull, brownish, olive-green upper parts. Race *zietzi*, SA, has slightly paler upper parts, slightly more yellow underparts, but, with plumage wear, colours are often no different from the nominate race of SW of WA.
Dull olive-green with slight brown tone
Deep blue frontal band extends back to encircle the eye.
Bend of wing deep blue, outer primary coverts light bright blue. Underwing coverts deep blue
Flight feathers brown edged blue
Underparts deep yellow, dulled by slight olive-green or olive-brown cast towards upper breast
Olive-green becoming blue towards the tip
Female has slightly dulled colours; immature lacks blue frontal band.
Has orange belly patch, but this may be difficult to see, especially when parrots are feeding on the ground.
'Grass' green
Frontal band blue; does not extend behind eye.
Yellowish green, face to breast
Wings emerald green
Bend of wing deep blue
Flight feathers brown, outer margins blue
♂
Orange patch, lower belly through to vent
Juvenile lacks blue forehead band.
Slightly dulled greens
Often feathers of breast are fluffed out to overlap and conceal bend of wing, so that blue is hidden.
Orange patch is smaller, paler.
Imm.
Turquoise male is spectacular in flight with outspread wings and yellow-tipped tail.
Red band across inner-wing coverts
Extensive turquoise blue
Extensive deep blue and turquoise on flights, coverts
♂
Older birds of either sex may have orange tone to lower centre of belly.
Turquoise face divided by white lores
Lacks red band, but extensive deep blue, turquoise
♀
Central pair green, all others dull blue with broad yellow tips and outer margins
Yellow, throat to under tail
Yellow, belly to under tail
Head golden olive with two-tone blue frontal band. Olive-green on back, wings, rump.
Blue behind eye
Immature: dull colours compared with adults
Deep blue as a rather narrow band from shoulder back to margins of flight feathers
♂
Juvenile
Similar to adult, but plain olive-green about head; lacks blue forehead band.
A rather plain golden olive parrot with rather narrow edge of blue on shoulder, blue margined flight feathers and narrow two-toned blue frontal band.
Bright yellow, on some birds has a touch of orange.
Under tail and under-tail coverts deep yellow
Female (not shown): colours dulled compared with male, blue frontal band reduced
In flight from above, shows golden rump and blue tail, with deep yellow outer half to outer tail feathers; outer wing blue. Beneath, breast to under tail yellow, underwing coverts blue

Oriental Cuckoo *Cuculus saturatus* 28–33 cm (C,L,T)

Large, underparts boldly barred; long pointed wings; swift, slightly undulating, dashing flight, like a small falcon. **Variation:** races visiting Aust. are mostly *optatus*, from NE Asia, but possibly also slightly smaller nominate race, *saturatus*, of S Asia. Often solitary, dispersed; some congregate at points of departure for the return migration. Always shy and elusive, quiet; slips away unobtrusively if approached. Hunts from perch, darting out to take insects from foliage, ground. **Voice:** usually silent; may give several, up to about six, whistled notes, 'kee-kee-kee', and harsh 'graak-graak-gak-ak-ak'. **Similar:** Pallid Cuckoo, but the adult has no underbody barring; juv. and immature differ in many details including pattern of underbody barring; Barred Cuckoo-shrike, but black lores, plain tail, dark grey breast. **Hab.:** rainforest margins, monsoon forest, vine scrubs, riverine thickets, wetter, densely canopied eucalypt forests, paperbark swamps, mangroves. Typically in denser vegetation types of more closed canopy than is usually occupied by the Pallid Cuckoo. **Status:** breeds across Eurasia while in northern hemisphere; parasitic, laying in nests of northern warblers. Migrates, in northern autumn, as far S as Indonesia, NG and N Aust. Departs Aust. in autumn; some remain through Australian winter. Uncommon.

Pallid Cuckoo *Cuculus pallidus* 28–34 cm (C,L,N)

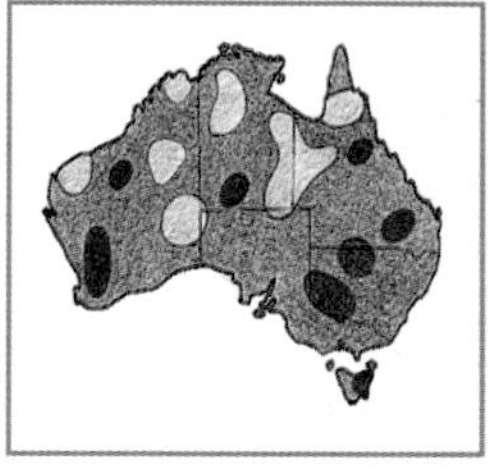

Large, long tailed, superficially falcon-like in flight, but feeds on large insects from ground and foliage; eats hairy caterpillars, avoided by most birds. **Variation:** sexes very different. Male in light grey and dark grey morphs with intermediates. Females in light rufous and dark rufous morphs. There are no subspecies or seasonal variation in these plumages. **Juv.:** first plumage is often seen when plaintive begging calls lead to a recently fledged cuckoo, or one that is the sole, very large, occupant of the nest of a host species. Upper parts are heavily blotched, streaked dark brown and white, probably an effectively disruptive, bark-like camouflage worn while it overflows the nest of its host parents, and for a time after it has left the nest. **Imm.:** similar to adults in their respective morphs, and may for a while be recognised by a scattering of retained juvenile feathers, usually darker, more heavily mottled or notched. **Voice:** in spring, male advertises with repeated series of slightly husky, mellow whistles on a rising scale, 'quip-peer-peer-peeer-peeer-peer-', 'quip-quip-pip-pip-pieeer'. Female responds with a long, drawn out, husky, sharp 'queeeep!'. **Hab.:** open country; avoids dense, closed vegetation types. **Status:** migratory in S; elsewhere only attract attention by calls in spring; common. **Br.:** 380

Brush Cuckoo *Cacomantis variolosus* 22–26 cm (L,T)

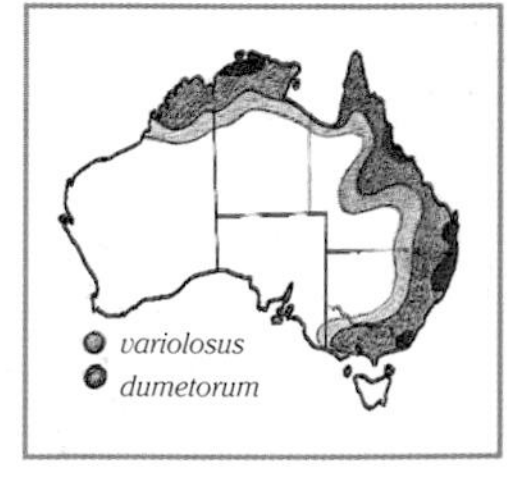

Takes its name from use of 'brush' for rainforest; inhabits dense, often closed canopy habitats. There it tends to sit silent and motionless for long periods, occasionally darting out to take caterpillars or other insects from vegetation; can be difficult to sight unless it moves. Only in the breeding season does this species become conspicuous; males then very active and noisy, chasing other males, pursuing females. The flight is swift on pointed, backswept wings, slightly undulating and showing pale wing-bar. **Variation:** many subspecies through islands to N of Aust., Malaysia to NG and Solomon Is. In Aust., probably two races. In the SE nominate race, *variolosus*, which, in mid Qld, has a wide intergrade zone with N race *dumetorum*, the latter slightly smaller but otherwise little different. **Voice:** a long series of descending, clear, mellow yet piercing calls, 'feee-ip, feeeip, feeeip-'. A ringing, metallic 'pee-whip-ee, pee-whip-ee' may be either slow, gently rising and quite mellow, or rapid, rising in pitch, ever quicker and higher, until continuous—an excited, frenzied, piercing sound. **Hab.:** rainforests, monsoon and gallery forests, stream thickets, paperbark swamps, eucalypt forests and woodlands. **Status:** common across N Aust.; migrant to SE Aust., where uncommon. Some winter on northern islands. **Br.:** 380

Black-eared Cuckoo *Chrysococcyx osculans* 19–21 cm

Solitary, inconspicuous; hunts from perch in shrub or tree, dropping to ground to take insects, including hairy caterpillars, which are disliked by most other birds. **Voice:** usually quiet, but in breeding season males call from a prominent open perch, giving a piercing, drawn out, slightly descending whistle repeated at regular intervals, each call of the same length and each at the same penetrating high pitch as the previous: 'feeeieuw, feeeieuw, feeeiew'. Also a loud but less piercing 'fee-ew-it, fee-ew-it' or 'fee-ew-eer'. Occasionally an abrupt, quieter 'feeit'. **Hab.:** occurs across most of the Australian mainland, avoiding only the wet, heavily forested east coast and similar SW corner of WA. Occurs in most drier habitats: open woodlands, mulga and mallee; sparsely vegetated arid country with spinifex, grasslands or salt marsh; widely scattered trees and shrubs; lines of vegetation along watercourses. **Status:** migrates into SE and SW for the summer, only rarely reaching Tas.; central Aust. movements are uncertain. Present across N Aust throughout the year; probably some travel to NG. Generally uncommon, but is easily overlooked and is probably occasionally locally common. **Br.:** 380

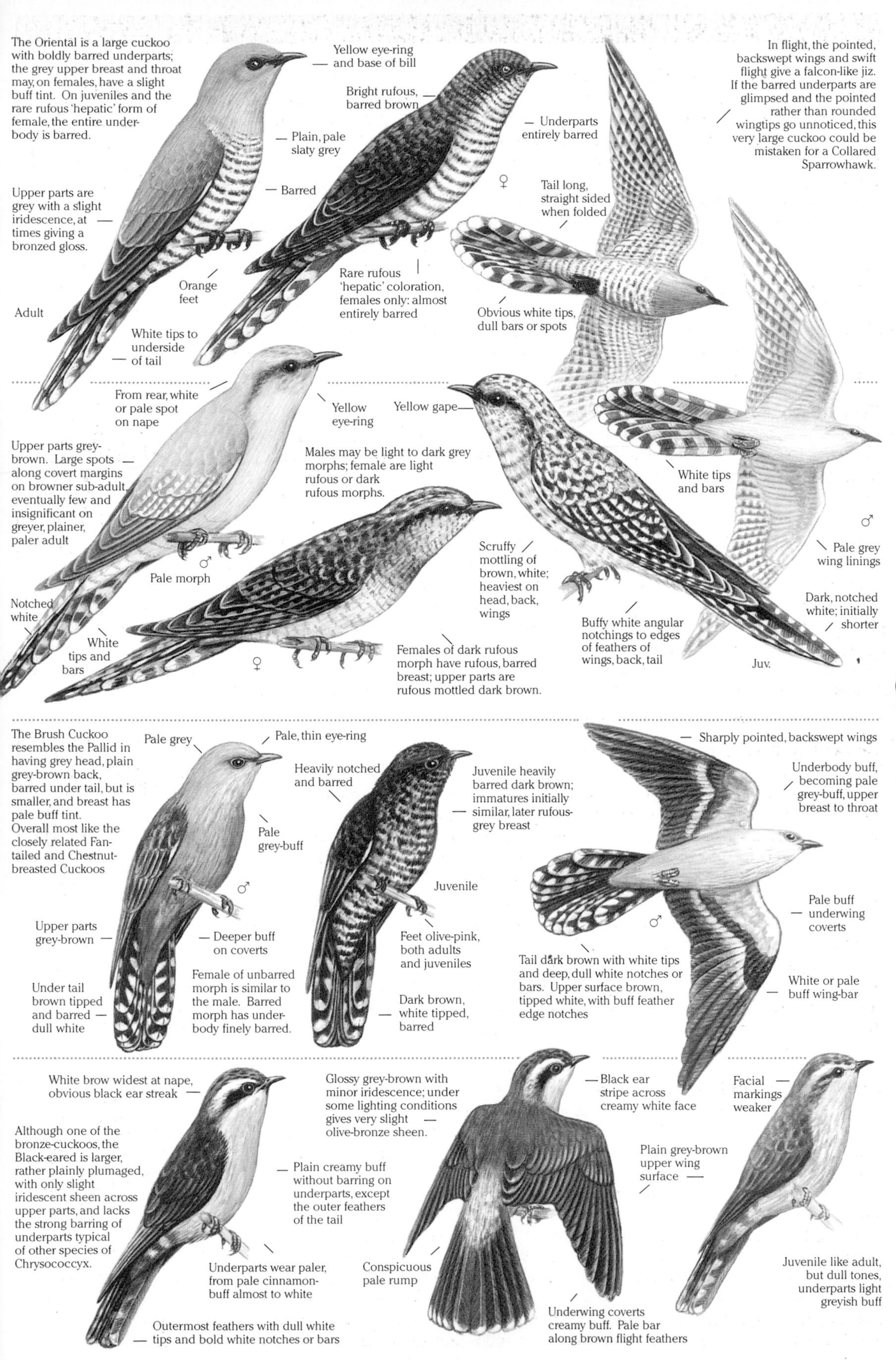
The Oriental is a large cuckoo with boldly barred underparts; the grey upper breast and throat may, on females, have a slight buff tint. On juveniles and the rare rufous 'hepatic' form of female, the entire underbody is barred.
Yellow eye-ring and base of bill
Bright rufous, barred brown
Plain, pale slaty grey
Underparts entirely barred
In flight, the pointed, backswept wings and swift flight give a falcon-like jiz. If the barred underparts are glimpsed and the pointed rather than rounded wingtips go unnoticed, this very large cuckoo could be mistaken for a Collared Sparrowhawk.
Barred
♀
Tail long, straight sided when folded
Upper parts are grey with a slight iridescence, at times giving a bronzed gloss.
Orange feet
Rare rufous 'hepatic' coloration, females only: almost entirely barred
Adult
Obvious white tips, dull bars or spots
White tips to underside of tail
From rear, white or pale spot on nape
Yellow eye-ring
Yellow gape
Upper parts grey-brown. Large spots along covert margins on browner sub-adult, eventually few and insignificant on greyer, plainer, paler adult
Males may be light to dark grey morphs; female are light rufous or dark rufous morphs.
White tips and bars
♂
Scruffy mottling of brown, white; heaviest on head, back, wings
Pale grey wing linings
Pale morph
♂
Dark, notched white; initially shorter
Notched white
Buffy white angular notchings to edges of feathers of wings, back, tail
White tips and bars
♀
Females of dark rufous morph have rufous, barred breast; upper parts are rufous mottled dark brown.
Juv.
The Brush Cuckoo resembles the Pallid in having grey head, plain grey-brown back, barred under tail, but is smaller, and breast has pale buff tint. Overall most like the closely related Fan-tailed and Chestnut-breasted Cuckoos
Pale grey
Pale, thin eye-ring
Heavily notched and barred
Juvenile heavily barred dark brown; immatures initially similar, later rufous-grey breast
Sharply pointed, backswept wings
Underbody buff, becoming pale grey-buff, upper breast to throat
Pale grey-buff
♂
Juvenile
Pale buff underwing coverts
Upper parts grey-brown
Deeper buff on coverts
Feet olive-pink, both adults and juveniles
♂
Tail dark brown with white tips and deep, dull white notches or bars. Upper surface brown, tipped white, with buff feather edge notches
Female of unbarred morph is similar to the male. Barred morph has underbody finely barred.
White or pale buff wing-bar
Under tail brown tipped and barred dull white
Dark brown, white tipped, barred
White brow widest at nape, obvious black ear streak
Glossy grey-brown with minor iridescence; under some lighting conditions gives very slight olive-bronze sheen.
Black ear stripe across creamy white face
Facial markings weaker
Although one of the bronze-cuckoos, the Black-eared is larger, rather plainly plumaged, with only slight iridescent sheen across upper parts, and lacks the strong barring of underparts typical of other species of Chrysococcyx.
Plain creamy buff without barring on underparts, except the outer feathers of the tail
Plain grey-brown upper wing surface
Underparts wear paler, from pale cinnamon-buff almost to white
Conspicuous pale rump
Juvenile like adult, but dull tones, underparts light greyish buff
Outermost feathers with dull white tips and bold white notches or bars
Underwing coverts creamy buff. Pale bar along brown flight feathers

Chestnut-breasted Cuckoo *Cacomantis castaneiventris* 22–25 cm

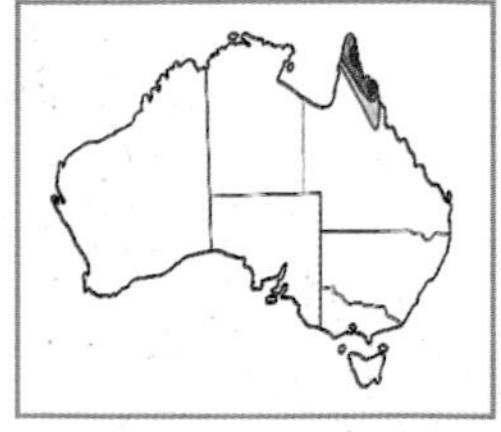

Secretive bird of understorey and mid levels of tropical rainforests; watches from perch then drops to ground, or flutters around foliage to take insects or other small prey. Sits still for long periods; inconspicuous unless calling. **Variation:** nominate race in Aust. and NG; two other races in NG. **Voice:** a wavering rather than descending, rattling, musical trill that is sharply penetrating if close; similar to descending trill of Fan-tailed Cuckoo; some similarity to call of Yellow-billed Kingfisher. Also a mournful, husky, 'wheeer-wheeer-'. **Similar:** Fan-tailed and Brush Cuckoos, but both much paler under body, slightly larger. **Hab.:** rainforests, monsoon forests, mangroves, river-edge thickets. **Status:** uncommon; sedentary or migratory. **Br.:** 380

Fan-tailed Cuckoo *Cacomantis flabelliformis* 25–27 cm (L)

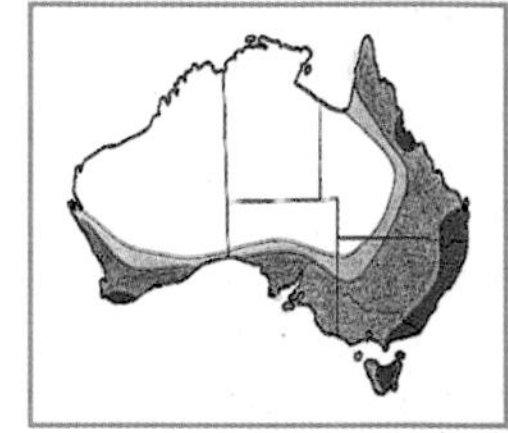

Adopts a rather upright posture, tail downwards, often keeping to a perch for long periods while calling. Flight undulating; elevates tail on landing. Hunts from perch; waits patiently, drops to take insects from ground, foliage. **Voice:** a descending, rattling, mellow, musical trill, quite loud if close; similar to trill of Chestnut-breasted, which undulates rather than descends, and may be slightly sharper. Also a musical, almost level, vibrating 'pheeweer'; high, thin, whistled 'fweeit'. **Similar:** Brush Cuckoo, Chestnut-breasted Cuckoo. **Hab.:** wet eucalypt forests, open forests and rainforest margins, but, on average, less dense than habitats preferred by Brush and Chestnut-breasted. **Status:** common; partly locally migratory. **Br.:** 380

Horsfield's Bronze-Cuckoo *Chrysococcyx basalis* 14–17 cm (C)

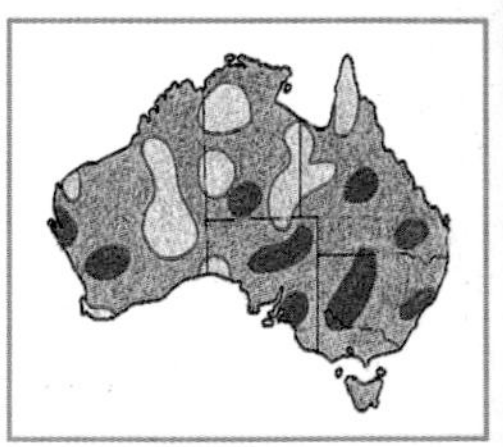

A common cuckoo throughout Aust. in almost all habitats except the densest and wettest vegetation types. Hunts from a perch; darts to ground to take insects. Flight swift; glimpse of pale under-wing-bar. **Imm.:** colours dull; underbody plain except for faint short bars along flanks. **Voice:** sharp, piercing, descending whistle, 'tsieeew, tsieeew, tsieeew'; the initial 'tsie' an extremely high, thin squeak of sound, then descending and strengthening through the final 'eew'. Also a quick, cheery chirrup like that of Budgerigar. **Similar:** Shining, Gould's and Little Bronze-Cuckoos, but all have unbroken barring up to throat. **Hab.:** open forests, woodlands, roadside trees and farm shelter belts. **Status:** common; migrant to S. **Br.:** 380

Shining Bronze-Cuckoo *Chrysococcyx lucidus* 16–18 cm (L, N)

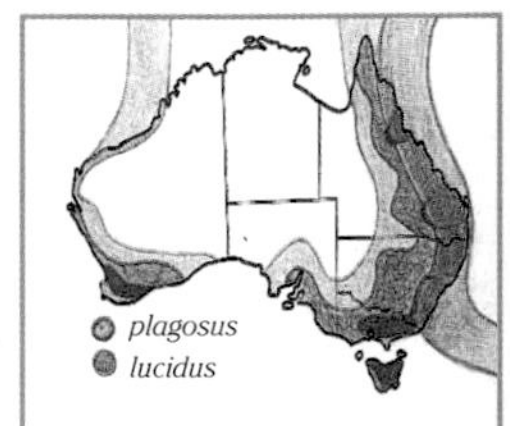

Hunts from perch; darts out to take insects from foliage and ground. **Imm.:** upper parts grey-brown; underbody dull white, faintly barred. **Variation:** race *plagosus* breeds Aust.; *lucidus*, nominate race, breeds in NZ, passes through E coast of Aust. to and from wintering grounds on islands to N of Aust. **Voice:** several different sequences, all very high pitched, metallic. As a very long series, piercingly high, at even pitch: 'wheee-wheee-wheee...'. Another call begins high and sharp; ends wavering downwards: 'phee-ieer, phee-ierr, phee-ier'. **Hab.:** mid to upper strata, wet dense rainforests, eucalypt forests, woodlands. **Status:** common. **Br.:** 380

Little Bronze-Cuckoo *Chrysococcyx minutillus* 14–15 cm (T)

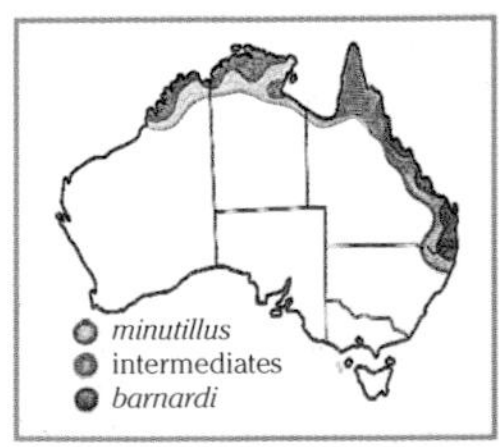

Small inconspicuous in dense habitats unless it is calling. Hunts from perch; sallies out to take insects on wing as well as from ground and foliage. **Variation:** races *barnardi*, mid E; *russatus*, NE Qld. **Voice:** a long, descending, rippling trill, loud, clear and ringing; also gives series of 3 to 6 descending notes like a slower, drawn out version of the trill, 'tiew-tiew-tiew-tiew-'. **Hab.:** vicinity of varied dense, wet vegetation types: rainforest edges, tropical monsoon forest, paperbark swamps, mangroves, woodlands, lush gardens; often near water. **Status:** sedentary, but race *barnardi* migratory from SE to NE of range; common. **Br.:** 380

Gould's Bronze-Cuckoo *Chrysococcyx russatus* 14–15 cm

Very small, more rufous than any other Bronze-cuckoo. Hunts insects in flight or on foliage or the ground. **Variation:** the taxonomy of Little-Bronze-Cuckoo group remains confused; currently the Gould's and Little Bronze-Cuckoos are considered separate species, but apparently with interbreeding in NE Qld. They may be combined as a single species, with Gould's Bronze then being race *russatus* of *C. minutillus*. **Voice:** like Little Bronze, a descending trill; also a slower, descending 'tiew-tiew-'. **Hab.:** rainforests, monsoon and dense river-edge forest, mangroves, paperbark swamps. **Status:** sedentary; common. **Br.:** 380

Beautiful small cuckoo confined to Cape York; upper parts dark slaty blue-grey with metallic sheen
Yellow eye-ring
Deep rufous chestnut, chin to under-tail coverts
Adult
Immature, like the Black-eared, but unlike other immatures, does not have barred underbody.
dull greyish cinnamon
Imm.
Dull mid brown
Plain, dull, pale cinnamon-buff
Pale wing-bar across primaries, secondaries
Partly fanned tail shows white tips and notches.
Rufous
Adult
Mid slaty grey
Yellow eye-ring
Adult
Male underparts entirely cinnamon; female on breast only, remainder whitish
Long tail tipped and barred dull white
Brown, edged rufous, giving finely scalloped appearance
Thin white eye-ring
Dull dark brown
Barred brown, lighter towards tail
Tail blackish brown, barred light brown
Juv.
In flight, Fan-tailed Cuckoo is similar to Chestnut-breasted, above, but with the rufous-chestnut replaced by pale cinnamon-buff.
Long white eyebrow curves down.
Dark line through eye
Brown with bronzed green iridescence
Bright rufous edges to outer feathers of tail
Adult
White tips to inner webs
White notches along outermost feathers only. Hidden when tail fully folded
Throat plain or streaked, not barred
Barring incomplete down centre-line
Adult
Rufous bases to under-tail feathers
Pointed, backswept wings
Pale broad wing-bar across primaries and secondaries
Tail feathers tipped white; outermost pair notched white along margins
Adult
Crown metallic green
From front, white specks up forehead
Race *lucidus*, brighter glossy iridescent green
In Aust., most are race *plagosus*, 'Golden bronze-Cuckoo'. Nominate race, *lucidus*, 'Shining Bronze-Cuckoo', is an uncommon transient visitor to E coast.
Adult
Bronze-green bars, unbroken, up to chin
White tips hidden when tail is folded
Coppery bronze
Race *plagosus*, bronzed, glossy iridescent green
Unbroken barring
Broad pale bar across primaries and secondaries
Bars bronze in sunlight
Adult
The Little Bronze-Cuckoos, and Gould's, below, are closely alike.
Dark cap, obvious white brow
Male: red eye-ring and iris
Green of plumage has a less rufous tone than has Gould's.
Metallic, iridescent olive-green
Underparts white without any rufous tint to breast, and entirely covered with short dark scallops
Underside of tail rufous, except for outer tail feathers
♂
Eye-ring pale creamy green to pale fawn; iris brownish
♀
Upper surface of tail is plain olive-green on central pair, others increasingly barred and rufous edged outwards; only visible if tail is spread. Undersurface of tail is barred and tipped white.
Gould's Bronze-Cuckoo has at times been placed with the Little Bronze Cuckoo, as a subspecies; these may again be combined.
Iridescent rufous-bronzed olive green
Rufous tint or band
Green plumage has strongly bronzed iridescence, wing covert margins bronze.
Compared with Little Bronze-Cuckoo, tail has more intense and more extensive rufous tone.
Underparts are entirely covered with short, dark bronzed scallops, not complete transverse bars.
♂
Eye-ring pale grey to tan
Rufous tint on breast, both sexes
Scallops broken, rather than the long, neat bars of above species
In flight, both this and the Little Bronze show a pale bar across the underside of primaries and secondaries.
♀

Common Koel *Eudynamys scolopacea* 40–46 cm (T)

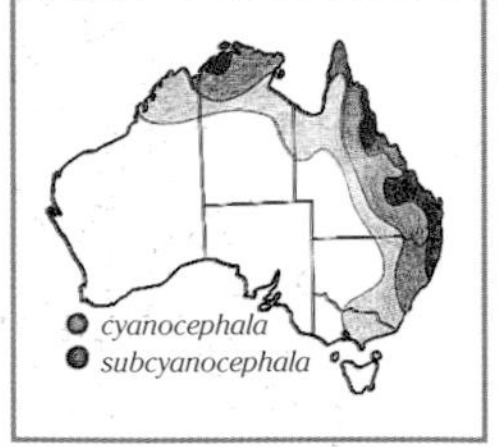

A large parasitic cuckoo well known in coastal N and E Aust. for its far-carrying 'koo-eel'; in breeding season males often call for long periods, day and night, advertising their territories. The male's blue-black iridescent plumage differs substantially from the female, immature and juvenile plumages. The Koel usually keeps to rainforest margins and other vegetations mainly comprised of dense leafy trees, where it can be difficult to sight unless it is calling from a perch unscreened by intervening foliage. May be in pairs, small groups or alone; usually wary and elusive, although males display noisily and chase females throughout the breeding season. The Koel feeds on native fruits, but often extends its selection to cultivated fruits in suburban gardens. **Variation:** many subspecies, about 17 or 18, from India to NG, China, Solomon Is. and Aust., where there are two. The race *cyanocephala* occurs down the E coast of Aust.; the smaller race *subcyanocephala* across N Aust. **Voice:** the male has a mellow and rather musical, ringing, loud 'quow-eel, quow-eel, quowee, quowee'; initially the 'quo' quite deep and mellow, the final 'ee' lifting in pitch. The sequence of perhaps five to ten calls starts with slow calls with long pauses of several seconds between, but gradually becomes quicker and higher. Also has a similar sequence but more rapid; begins quite deep and mellow yet with a resonant, ringing quality; the calls come quicker, the series rising, 'quowil-quoil-quoil-quoi-quoi'. The female gives a series of four or so shrieking whistles, 'quieek-quieek-quieek-quieek'. Generally silent outside the breeding season. **Similar:** Spangled Drongo, but its outcurving tail feathers make a wide 'fish-tail' tip. **Hab.:** rainforests, monsoon forests and dense wet eucalypt forests, especially the margins; leafy trees of river edges, farmlands, woodlands and gardens. **Status:** occurs widely from India through SE Asia to S China, Malaysia, Indonesia, NG and, as a summer migrant, into N and E Aust. **Br.:** 380

Channel-billed Cuckoo *Scythrops novaehollandiae* 58–65 cm (N,T)

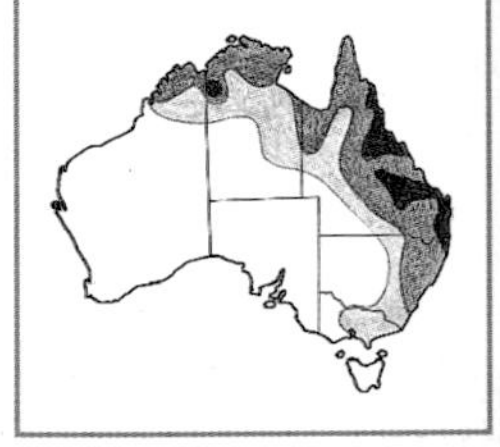

A huge cuckoo, largest of all parasitic birds, with huge bill giving more the appearance of a hornbill. This is a migratory species; breeds in Aust. in spring and summer and migrates to NG, Indonesia and other northern islands for the winter. On arrival in Aust., about Sep.–Oct., the Channel-billed Cuckoo makes its presence obvious with loud raucous calls, activity which, coming shortly before the onset of the wet season, has given this species a variety of alternative local names such as Floodbird, Rainbird and Stormbird. So large a cuckoo requires a large host, and the Channel-billed Cuckoo uses currawongs, magpies and crows. With the huge bill extending the body length forward, in flight this species looks from below like a flying cross; the body is well over half a metre long and the wingspan approaches a metre. The flight is strong, wings pointed and at times partly backswept, giving a falcon-like although rather long bodied silhouette. Has strong, deep wingbeats and often travels high, conspicuously. Channel-billed Cuckoos feed principally on fruit, but also take insects, especially large insects such as locusts; they have also been reported to feed on eggs and nestlings of other birds. During breeding season they are usually in pairs or solitary; at other times they may be in small flocks. **Voice:** loud, raucous, maniacal crowing and squawking: 'awrrk, aworrk, oirrk, oik-oik-oik' with many variations—'aiirrk-aiirrk-ark-ark-urk' and crow-like 'arrk, arrk, arrk, airrk, urrgk urrgk'. **Hab.:** rainforest, monsoon forest, eucalypt forest and woodlands, river-edge thickets, swamp woodlands. **Status:** breeding migrant arriving Aug. in N Aust., Oct. in SE NSW. Reaches the Kimberley, where it is uncommon. Departs Jan.–Feb. from SE, and Mar.–Apr. from N Aust. **Br.:** 380

Pheasant Coucal *Centropus phasianinus* 60–75 cm (T)

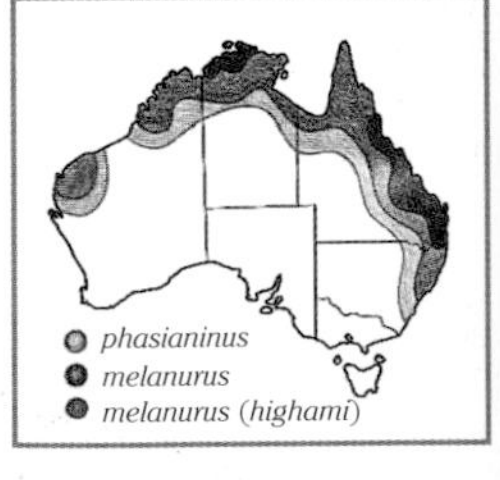

A large, superficially pheasant-like member of the cuckoo family; unlike all other Australian cuckoos, it is a ground-dweller rather than arboreal, inhabiting tall grass and other dense groundcover. Its barred and intricately patterned plumage matches the colours and detail of ground debris—leaf litter, grass and twigs. With short, rounded wings the flight is weak, typically moving higher into shrubbery or trees in a series of flapping leaps higher among the branches, eventually returning to ground in a steep glide, terminating in a rather clumsy plunge into the grass; it does, however, fly quite high and far at times. The Pheasant Coucal also differs from other Australian cuckoos in building its own nest. Insects and other small creatures form the major part of the diet, but it apparently takes small birds, frogs, rodents and reptiles. **Variation:** two subspecies in Aust.: nominate race *phasianinus*, E coast of Aust. N to the Burdekin R. Race *melanurus* occurs from the Burdekin R. around the N coast to the Kimberley. An isolated population in the Pilbara is sometimes listed as race *highami*, but differences seem not sufficiently significant. Race *thierfelderi* occurs in Torres Str. Other races NG to Timor. **Voice:** a long, rapid series of resonant notes; initially slow, accelerating, slowing and falling away, or, at other times, beginning abruptly, rapid notes sustained, then falling away: 'coop, coop, cook-kook-kook-kook...'. Sometimes accompanied by a second bird at different pitch. Also a metallic, tapping 'chak-chak-chok, chok, chowk, chowgk'. **Hab.:** dense, tall grass of river floodplains and tropical woodlands, watercourse scrub, canefields, lantana thickets, dense roadside and riverbank undergrowth, mangrove fringes, swampy heathlands. **Status:** sedentary; common. **Br.:** 380

Eyes bright red
Very large cuckoo, male black with iridescent blue-green sheen
Crown black
Blackish brown, spotted white
Buffy, streaky, black edged throat
White to deep buff, barred
Immature female is like adult female, but eye may remain brown into second summer.
Dark line through eye
Juvenile very different. Colours rather like light female, but has creamy white cap with dark centre line, dark eye stripe. Eye is brown.
♀ Light form
♂
♀ Dark
Wings are often held low to each side of tail.
Rufous-brown barred white and buff
Tail long, straight sided, round tipped
Long, tail, rounded, barred buffy white
Adult female variable. Dark form black capped, dark brown, white spotted upper parts. Light form has rufous crown; chestnut replaces dark brown on back: has many intermediate forms.
Dark brown barred and spotted white
♀
♂
Mature male entirely black; sub-adult males may still have remnants of rufous barring.
Long, round tipped tail, black with blue-green sheen
Scale change
Red bare skin eye-ring
Huge pale-bodied cuckoo with massive bill; contrasting dark wings and tail
Pale grey
Bill varies; always very large, and often enormous. Appears disproportionately massive even for so large a bird.
Dark tipped wings
Grey with each feather broadly tipped black
Scalloped buff
Bare skin of lores grey
Bill smaller than that of adult
Pale buff
Pale grey
Broad black subterminal band
Adult
Juv.
White tipped
Adult
Central pair has very narrow white tip; outer feathers have large white tips.
Black subterminal band and white tips on both surfaces of tail; underside also barred
Heavy black subterminal band
A very large, superficially pheasant-like bird, in size surpassed only by the Channel-billed Cuckoo, which, although of similar total length, has a shorter tail and larger and heavier body
Bright rufous-chestnut intricately barred black and buff
Bill black when breeding
Head, neck and entire underparts black with fine, glossy white feather-shaft streaks
Tail very long, dark brown heavily barred black. Buff or buffy white barring heaviest on inner feathers
Flight appears weak and clumsy on short, stubby wings.
Wings and back remain similar to breeding plumage
Breeding plumage
Cinnamon streaked rufous and buff
Towards tips of outer feathers the brown is almost obscured by black; buff barring is reduced.
Nonbreeding
When moulting may have plumage intermediate to breeding and nonbreeding. Juveniles like nonbreeding, but paler with more fawn and buff

Rufous Owl *Ninox rufa* 45–55 cm

Large rufous-fronted owl, most likely to be seen roosting in dense foliage of a rainforest tree. A powerful hunter, its prey is often large: Brush Turkey, Scrubfowl, Sugar Glider, flying-fox. **Fem.:** smaller, usually darker. **Juv.:** white, downy head, front. **Variation:** race *rufa*, Kimberley and NT; race *queenslandica*, NE Qld, slightly darker; race *meesi* also darker, and smaller. **Voice:** usually a slow, deep 'whooo-hoo'; the first 'whoo' drawn out the second note shorter, same or slightly lower pitch. Also, low moaning sounds. When both sexes call, male's hoots slightly deeper; courting male may give up to seven rapid hoots. **Hab.:** roosts in dense vegetation of rainforest, also river-edge gallery forest, monsoon scrubs, swamps, mangroves; hunts through adjoining eucalypt woodlands. **Status:** probably rare across most parts of its wide range. **Br.:** 381

Powerful Owl *Ninox strenua* 60–65 cm

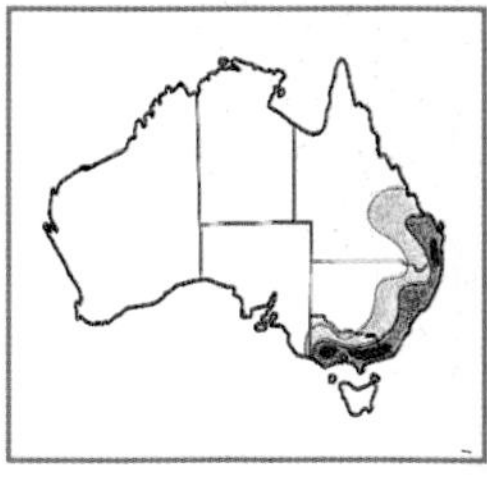

Would be very difficult to detect on its daylight roost but for its use of limbs that, although shaded by a dense overhead canopy, are usually clear of concealing low foliage. The prey includes possums, gliders, roosting birds and some terrestrial animals, usually rabbits and small marsupials. **Voice:** in breeding season, male's territorial call is a clear, almost ringing, far carrying single or double 'whoo'; begins softly and rises clear and loud; may be heard a kilometre away. When a pair is calling, the male's call is deeper; female's higher hooting seems to be a response or contact call. Various totally different sounds are given near the nest hollow. **Hab.:** eucalypt forests, preferring tall wet forests of ranges where the territories centre on densely vegetated gullies. Also found marginally in lower or drier forest that holds both prey and large hollows. **Status:** sedentary; uncommon. Requires large territories, 300–1500 ha, to support each breeding pair. **Br.:** 381

Brown Hawk-Owl *Ninox scutulata* 29–30 cm

Deep golden eyes emphasised by surrounding of dark brown. Often takes large flying insects on the wing at dusk, like a small, stout hawk. **Variation:** Aust. record was race *japonica*, a winter migrant to Indonesia from SE Asia. **Voice:** mellow, double noted, whistled 'cooo-hoo', the second note higher, shorter. **Hab.:** forests, woodlands. **Status:** rare vagrant.

Barking Owl *Ninox connivens* 35–45 cm

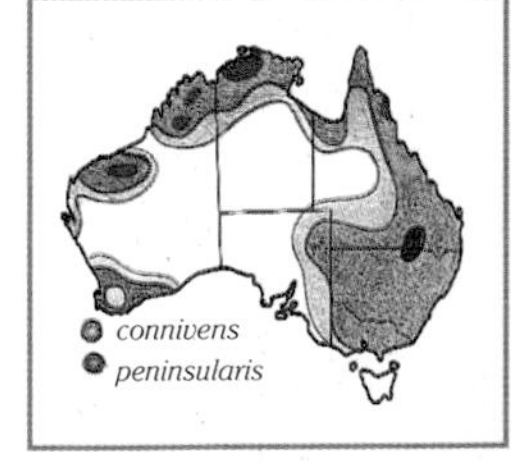

A medium-sized owl, superficially resembling the well-known Boobook, but considerably larger, taller, more upright with brilliant yellow eyes; upper parts grey to grey-brown. The presence of these owls is usually revealed by their barking calls, sometimes given from daytime roosts, and, if followed back to their source, may give a sighting of a roosting bird staring down with piercing yellow eyes. Often the owl will fly when its roosting place is approached, its silent flight betrayed by noisy attention from small birds that give chase and reveal where the owl has landed by their mobbing, harassing behaviour. Male and female similar. **Variation:** gradually becomes darker and smaller northwards, possibly a subspecies, *N. c. peninsularis*, with the larger, paler nominate race *N.c. connivens* across southern Aust. **Voice:** a rapid 'wook-wook', clear and very carrying; pleasant and rather musical, especially when a pair is calling, the male slightly deeper, the female responding with slightly higher, clearer notes. Close by, a soft, gruff 'arr' may be heard, leading in to the loud, clear barks, becoming 'arr-wook-wook'. At times there is a brief flurry of calls, a fast, excited, higher 'wook-wook,-wook-wook-'; at other times the calls die away to a soft, low, relaxed 'wuf-wuf, wuf-wuf'. Also, rarely, perhaps only at the beginning of the breeding season, a chilling 'screaming woman' call, a long, wavering 'a-aii-eer-lp'. **Hab.:** open country with stands of trees, tree-lined watercourses, paperbark swamps in N and NW Aust. **Status:** common NW and N; now uncommon in SW and SE. Sedentary. **Br.:** 381

Southern Boobook *Ninox novaeseelandiae* 25–35 cm (N)

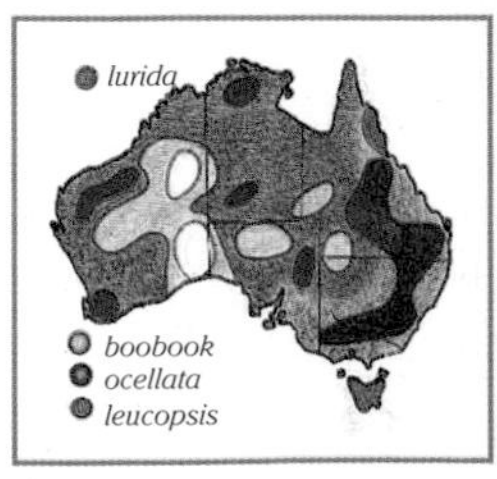

Small brown owl with a dark patch around each eye, most conspicuous on the paler western-inland race, obscure on dark races. Preys mostly on insects and arthropods, but small birds and mouse-sized mammals are also taken; insects are sometimes caught in flight. Roosts by day in dense foliage. **Variation:** four Aust. races. **Voice:** a mellow, musical, double hoot, 'kook-kook' or 'book-book' at varied pitch; the second note is lower, throaty and vibrating in the deepest calls. The sound carries far and may be repeated monotonously; often a pair, calling back and forth in duet, add diversity and excitement to this call—slow then rapid, with variations from low and tremulous to high and clear, or a deep, throaty, rapid, vibrating 'quorrr-quorrr-quorrr' and a high, tremulous, descending whistle. **Hab.:** almost anywhere with trees, especially open eucalypt forests and woodlands. Except for NE Qld race *lurida*, is rare in dense rainforests. **Status:** common; sedentary. **Br.:** 381

White, heart-shaped facial disc with fine rim of of fawn and brown
Upper back pale pearly grey
Upper parts beautifully plumaged in soft intermixed buff, chestnut and grey with dark-tipped white spots
Underparts white with sparse dark spots
In flight, feet trail only slightly behind tip of tail
Legs moderately long, but much shorter than those of Grass Owl; lightly feathered, almost to the foot
Feet and talons quite lightly built
Adult
Facial discs of members of the barn owl family are shaped to act as sound reflectors, concentrating sounds accurately calculating direction and distance, and giving the ability to hunt in darkness.
Often adopts a hunched posture, with head pulled down onto shoulders, or thrust forward. The Tasmanian Masked Owl is large, females' bodies up to 570 mm long, weight 1.25 kg, compared with 460 mm and 800 g on the Aust. mainland.
Deep wine-red shading to facial disc
Chestnut and buff, heavily patterned with dark brown or black
Dusky rufous
Dark Morph
♀
Lightest forms are buff marked with pale grey.
Intermediate form: rufous tint on face and breast
Mask broad, rounded, with darker border
Three colour morphs: pale, dark, intermediate, usually in each race. The small northern race, *kimberli*, is the palest: some are white with pale bufffy grey upper parts; but has few if any dark morphs.
Pale morph: white underparts
In any race or locality the female is larger, often much darker, than male.
Females of the large Tas. race *castanops* are very dark, but some males are of paler intermediate morph
Feathered legs, powerful feet
Facial disc varies, brownish buff tint or almost white; chestnut 'tear' marks
Crown dark brown
Back patchy brown, buff and rufous, unlike the pale grey of the Barn Owl
Often hunts near dusk: slow leisurely flapping flight interspersed with slow glides, long legs trailing, or dangling low ready to snatch a rodent or other prey hidden in grass or other cover.
Wings barred, more heavily on upper surface
In flight, feet trail well beyond tail-tip. At take off and at landing, long legs dangle far below body. But Barn owls also occasionally roost in grass and the legs, although much shorter, hang down in a similar manner.
In flight, the Grass Owl, like the Barn and Masked, has a flat or blunt-faced, rather large-headed appearance.
Underparts vary from entirely white to partly or entirely pale, or quite deep rufous-buff; males white, females darker and larger.
Long legs, sparsely feathered on lower half
♂
Tail barred
♀
Owls have opposable outer toes, often pointed sideways for wider grab at prey, or to rear when perched; sometimes more forward than sideways on flat surfaces.
Facial is disc broadly oval with soft grey-brown streaks radiating outwards from large black eyes.
Sooty dark grey with large silvery white spots and overall dense, fine white speckling that makes plumage appear paler
Breast is silvery grey with fine, dark, broken barring, fading to silvery white towards under tail area.
The Lesser Sooty is a small, speckled grey owl of northern rainforests. By day it roosts in tree hollows, the chimneys of strangler figs or dense foliage such as large epiphytic fern clumps.
Often hunts in darkness using low perches in rainforest; locates prey by sound, dropping on terrestrial marsupial-mice and rodents.
In flight, under-wing is very pale, almost white, with grey barring.
Back and wings are sparsely spotted white with little of the very fine speckling of the Lesser Sooty.
Huge black eyes are set in very dark centres of wide facial disc.
Upper breast dark grey with white spots
Broad wings fold to give stubby rear body shape.
Massive, fully white-feathered legs, powerful feet and long talons can kill animals to the size of the Greater Glider, potoroos or rabbits.
Sooty Owl shows great sexual dimorphism, the female being much larger, weighing around 875 g compared to 600 g for the male, and with much more massive talons and bill that allows hunting of larger prey. The Lesser Sooty, however, as well as being far smaller, shows little difference of size between the sexes.

Tawny Frogmouth *Podargus strigoides* 34–52 cm

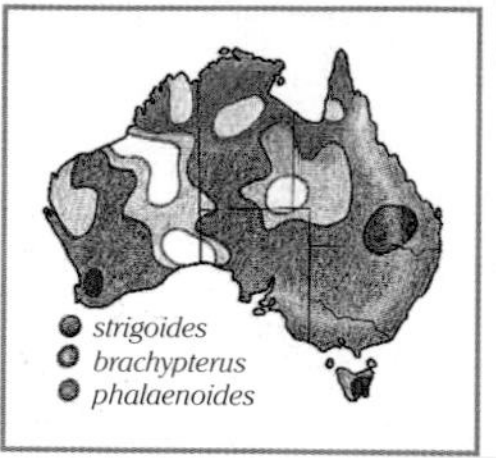

Probably the best known Australian nocturnal bird; occasionally seen in camouflage pose on an exposed limb, stiffly posed to mimic a broken branch. The streaked and mottled plumage looks like old weathered wood or bark, the bill and bristles like the jagged end of a broken branch, and the untidy white spots and dark streaks are like lichens and sap stains on old timber. Yellow eyes look through narrowed slits; the head turns almost imperceptibly to follow an intruder's movements. At night one may be lit by headlights while perched wide-eyed on a roadside post, not the broken-limb pose used by day. Hunts mostly ground-dwelling creatures, large insects, spiders, frogs, small mammals and ground birds. Watches and listens from a low perch, gliding down on broad, silent wings to take prey on ground. **Variation:** three mainland Aust. races, all with grey males and grey, rufous or chestnut plumaged females. E Race *strigoides* largest, darker patterned grey males; females, grey or chestnut. Race *phalaenoides* is smaller, males paler grey; females grey or rufous. Those on Groote Eylandt, Gulf of Carpentaria, are small and palest, silvery grey; sometimes listed as a fourth race, *lilae*. **Voice:** a muffled, low, resonant, rapid 'ooo-oom-oom-oom'; from males, a similarly rapid 'ar-oom, aroom, aroom-aroom'. **Similar:** Marbled Frogmouth, but eyes orange-red; Papuan Frogmouth, eyes orange. Both usually in dense closed habitats, have restricted range. **Hab.:** woodlands, open eucalypt forests and wide variety of other vegetation types; least often seen in treeless deserts, rainforest and wet, dense, tall eucalypt forests. **Status:** sedentary; common. **Br:** 382

Marbled Frogmouth *Podargus ocellatus* 37–46 cm

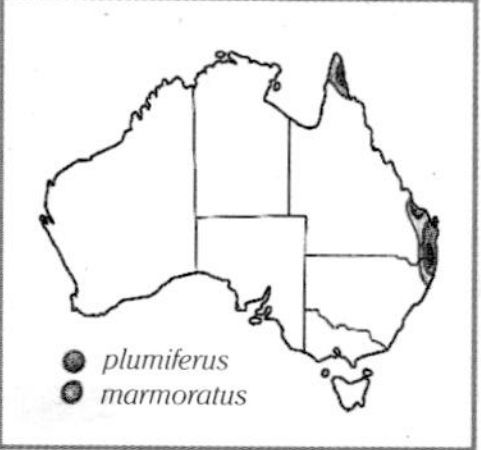

Beautifully plumaged frogmouth with deep orange or reddish eyes and extensive mottling or marbling of white, greys and browns that gives the look of rough bark, complete with pale patches of lichens, fissuring and weathering. When disturbed, like other frogmouths, the Marbled will freeze motionless, at times with head lowered rather than in the high, upward-pointing, straight broken-branch pose used by the Tawny Frogmouth. The marbled plumage is so effective a camouflage that this frogmouth is still almost as difficult to sight as with bill upwards. Hunts from low branches, dropping down to take small creatures from the ground. **Male:** marbled greys and white, often with slight brown wash. **Fem.:** usually slightly smaller with softer, less contrasting plumage pattern and, especially in northern and arid regions, tends towards a warmer rufous or more brownish tone than the male. **Variation:** two very widely separated populations: race *marmoratus*, the 'Marbled Frogmouth' of Cape York Peninsula, has bristles reaching to the tip of the bill; race *plumiferus*, 'Plumed Frogmouth' of SE Qld and NE NSW, is slightly larger, and has slightly shorter bristles and more heavily barred wings. **Voice:** usually a double noted, rapid, musical 'whooo-whoop, whooo-whoop', the second part uplifted; a guttural, growling 'quorr-orr, quorr-orr', which is deep, throaty, vibrating; a rapid, descending bubbling sound. **Similar:** Tawny Frogmouth, but paler eyes, bristles not barred; Papuan Frogmouth, huge size, red eyes. **Hab.:** race *marmoratus* of NE Qld lives in river-edge gallery forest, monsoon forest, rainforest, adjoining eucalypt or melaleuca forests and woodlands; race *plumiferus* is in patches of subtropical lowland rainforest at altitudes up to about 800 m. **Status:** Marbled on Cape York probably uncommon but secure; habitat of Plumed on mid E coast is greatly reduced; race now apparently rare, perhaps endangered. **Br:** 382

Papuan Frogmouth *Podargus papuensis* 50–60 cm

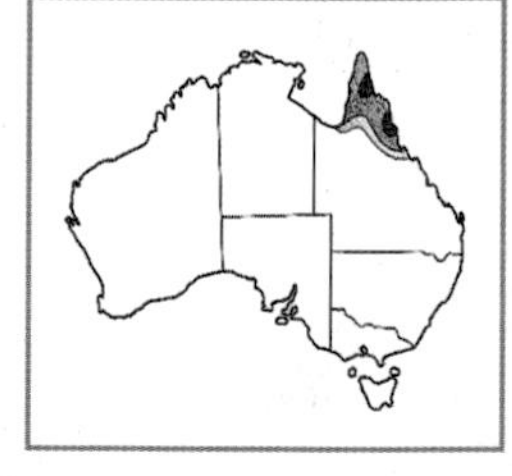

Frogmouths are much lighter than owls of similar size; the soft open feathering makes their slender bodies look much larger. The long, broad wings give greater bulk when perched and a light, buoyant flight. At 60 cm long, this is a huge frogmouth, with massive head adding to the bulky appearance. But its size is exaggerated by its loose feathering: although nearly the size of a Rufous Owl, it is only one-third the weight. With small, thin, weak feet and slow flight, frogmouths are no match for the Rufous Owl, often becoming its prey. Male and female differ considerably: males have much heavier bills; the females are slightly smaller and usually browner. A rare rufous plumage seems to occur only on females. **Variation:** on Cape York, race *rogersi* is larger and paler; S of Cooktown, race *baileyi* is smaller, darker. **Voice:** soft, low, steady 'orrm-orrm-orrm-orrm', slower and deeper than calls of either smaller frogmouth and with a slight but discernible rough or harsh quality; also a descending sequence of bubbly sounds, often ending with a 'clak', perhaps a snapping of the bill. Also a harsh, long, scolding, rasping scream: 'haaarrrrck!'. This screaming screech is often intermixed with the harshest, loudest and most unpleasant of frog-croak noises. **Hab.:** edges of dense closed forests such as rainforest and monsoon forest; seems to avoid the interior of these closed forests, preferring to hunt out through more open tropical woodlands, but using the denser vegetation for daytime concealment. In these daylight roosts they are rather sociable, often in pairs or small family parties. Roost sites tend to have a dense overhead canopy and, at times, as in mangroves, are over water. Hunting is from perches; birds regularly visit the many perches through their territory, listening and watching, then gliding down to take prey—grasshopper, beetle, lizard, small rodent or even occasionally a small ground bird. Most hunting is early in the evening or before dawn while there is faint light. **Status:** common in northern part of range; scarce in S. **Br:** 382

Well-known Frogmouth or 'Morepork' is often seen during the day imitating a broken limb—bill is uplifted, feathers are tight against the stiffly outstretched body, colour and pattern are like bark or weathered wood. Follows movements through narrowed eyes.
Bristles extend past bill tip.
Grey morph female would have brown on shoulders and malar streak.
Eyes pale to deep yellow. If seen in lights at night, is transformed: eyes wide, posture relaxed.
Upper surface has large white spots that form uneven barring.
Male always grey, streaked and mottled darker and paler greys or browns; all races. Grey morph females similar
Long dark streaks
♂
♀
Black streaks, throat to breast
Wings long and broad, giving light, buoyant flight
Rufous birds, and the rare E coast chestnut phase, are always females.
Nestling
Race *strigoides*: females grey or chestnut; males always grey. Both darkest in humid coastal forests
Long barred bristles
Pale brow
Eye deeper orange-yellow to vermilion
The Marbled Frogmouth occurs in two very widely separated populations: race *marmoratus*, 'Marbled', on Cape York; race *plumiferus*, 'Plumed', in SE Qld and NE NSW.
Dark streak down each side of throat
Bristles, or 'plumes' are slightly shorter than those of N race *marmoratus*.
Heavily 'marbled' in greys and white; males are more heavily marked than females.
Sides of breast, wings 'marbled' with mottled white and buff
If disturbed by day, adopts broken-branch pose—bill up, plumage held tight and sleek against body.
Race *plumiferus*, Plumed Frogmouth, has larger, more heavily barred wings.
Barred
♀
Race *marmoratus*, 'Marbled Frogmouth'
Males usually have a bolder pattern, greyer tone, and are slightly larger.
♂
Females slightly browner, with softer plumage pattern
♀
Tails all long, graduated or tapering
Will often freeze in postures less upright, less like a straight broken branch, relying more on colour than shape for concealment.
Eyes, both sexes, are typically red when identifying the species; but some orange-red, like Marbled Frogmouth, above.
Massive bill
Brow buff to white
Plumage varies greatly between individuals. Males overall dark grey to dark brown on upper parts with greater contrast between dorsal and ventral surfaces. Detail of mottling and streaking varies, but gives general impression of large, whitish spots or blotches in rows along secondary coverts and scapulars.
Buff or off-white edge to scapulars
Soft fawn to buff, spotted and mottled white, finely streaked chestnut and brown
Females are slightly browner to deep rufous; more uniform above and beneath. There are clinal and individual variations, but no colour morphs. Grey males and rufous females seen together may give impression of morphs.
♂
Nestling
Flight feathers and tail rufous or brown, barred black and white
Feet small, comparatively weak
♀
Shoulders are mottled deeper russet or brown with extremely variable blotches of white, grey or buff across the coverts.
Adult: huge, 'frog', bill

Large-tailed Nightjar *Caprimulgus macrurus* 25–28 cm (T)
Tropical nightjar with prominent white on tips of tail and wing feathers; usually found near rainforest. Hawks for insects with lighter, more fluttering, gliding, moth-like flight than other large nightjars. **Juv.:** when first hatched, has short pinkish buff down, soon replaced by first rufous-brown plumage; lacks the white throat band and has dull buff wing spots; tail initially very short. **Voice:** in distance, hollow 'quok-quok-quok-' or 'chop-chop-chop-'; close at hand the call is a quick 'quorrok-quorrok-quorrok-'—usually three to six calls, brief pause, then repeat. Each 'quorrok' is very abrupt. In the distance sounds more like 'quok' or 'chop'. **Similar:** Spotted and White-throated Nightjars, but neither has white in both tail and wings. **Hab.:** margins of rainforests, monsoon and vine forests, mangroves. These nightjars are rarely deep in the rainforest, but use the heavy layer of leaf litter on the ground near the edge or just outside closed forests where they are close to dense cover for daytime concealment and have open woodlands for night hunting nearby. **Status:** common in undamaged habitat; vulnerable to stock damage in small patches of monsoon and similar vegetation across N Aust. **Br:** 382

Spotted Nightjar *Eurostopodus argus* 29–32 cm
Most colourful of nightjars; a bird of drier and more open country where the ground is red or brown, littered with stones, grey and brown twigs, and dry leaves of yellow, ochre and rufous. The intricately patterned plumage mimics these colours so that a Spotted Nightjar, roosting or brooding on the ground, is almost invisible. Hawks aerially for insects with fluttering, bat-like flight; most active just after dark, and again towards dawn. If flushed from nest by day, the nightjar is likely to fly a short distance before dropping to ground; may then give a distraction display, wings and tail spread, mouth wide open, hissing. **Juv.:** extensive cinnamon-russet on upper parts. **Variation:** dull, pale, sandy buff to rich, rufous-red, tending to match colours of local earth or rock; no subspecies. **Voice:** a beautiful call that carries far on still nights in open country. Starts with slow notes, accelerates, becomes rapid and loud, then dies away in a long series of bubbling sounds: 'coooor, cawoor, cawow-caWOW-caWOW-tooka-tooka-tooka-tooka.....'. **Habitat:** eucalypt, acacia, mulga and other open woodland like mallee and spinifex; favours stony ridges. **Status:** quite common, especially in more remote regions; part migratory. **Br:** 382

White-throated Nightjar *Eurostopodus mystacalis* 32–37 cm
A large, dark nightjar that, if flushed, shows no large white patches on wingtips or tail. In headlights or spotlight beam its eyes return a fiery red reflection. Like other species, the disruptive plumage patterns and colours of this nightjar are almost impossible to detect on the ground among leaves, twigs and rocks; its darker overall colours match rich grey-black soils and litter of a more coastal range than of that other widespread species, the Spotted Nightjar. Flight swift and agile, twisting and turning at will; insects are taken on the wing. These nightjars are most active around dusk and dawn when their long winged, hawk-like silhouette may be seen against a darkening sky. **Voice:** attractive long sequence of musical, bubbling sounds; begins with very slow notes that become more rapid, rising to a crescendo, then abruptly switching to a long cascade of bubbling notes: 'awwk, awwk, awk, awk-AWk-AWk-AWk-quok-quok-quok-quok-quok-'. This latter sound, given very rapidly, can be like the initial chuckle of a Laughing Kookaburra before it launches into full raucous laughter, or like small, low-pitched motorcycle engine. Also has a sharp, penetrating 'aik!, aik!, aik!'. **Hab.:** forests, woodlands, heathlands, often on the more open, drier ridges among rocks and fallen timber. **Status:** common. **Br:** 382

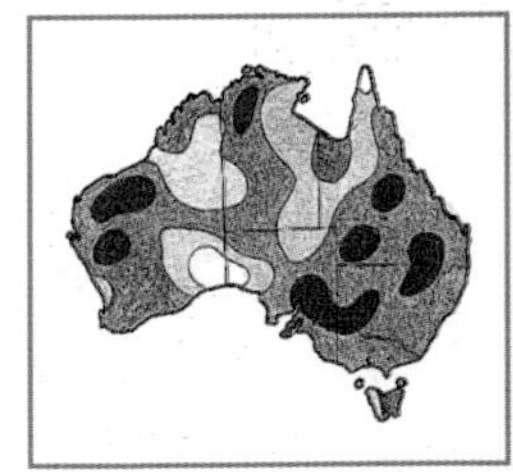

Australian Owlet-nightjar *Aegotheles cristatus* 21–25 cm
A tiny owl-like bird with proportionately huge eyes; the head markings give a remarkable likeness to the Sugar Glider, a common small marsupial. This is especially so when only the face can be seen peering down from a hollow. Captures insects in flight or on the ground at night. **Voice:** a distinctive, pleasant churring, one of the most common calls of the night in the Australian bush, especially in the arid interior where even the smallest stand of stunted trees seems to have its resident Owlet-nightjars. At a distance the call is a pleasant, vibrating, musical churring, ending slightly sharper. Closer, the churring is recognisable in the first, loudest part of the call, like a rolling of r's that gives a throaty 'quar-rr-rgh-'. It then blends to a higher, sharper, but not as loud, '-a-kak'—the whole call an undulating 'quar-rr-rgh-a-kak, quar-rr-rgh-a-kak'. Also gives a single, sharp, penetrating, metallic 'aeiiirk!'. Calls throughout the year, and often in the day, when it may then be seen sitting, sunning, at the entrance to its hollow. Flushes easily; darts immediately to another hollow, seemingly familiar with all the hollows within its territory. **Hab.:** diverse: coastal and mountain rainforests, eucalypt forests, interior woodlands, tree-lined watercourses, mulga, spinifex with scattered clumps of trees. **Status:** common; widespread. **Br:** 382

This tropical northern nightjar is unique in having large white patches in both the wingtips and the outer tail.
Rufous face underlined by by two white streaks
With wings folded, white spots of wing patches are often visible on edges of primaries.
Folded wings much shorter than tail
Roosts and nests on leaf litter.
White spots across wing coverts
Long wings and light body give buoyant, quiet, fluttering, butterfly-like flight.
Twin thin white streaks
In flight, unique combination of white wing patches and very large white tips to outer tail feathers
Large white wing-bar, narrower on female
Large, obvious white tips to outer tail
Extensively spotted and barred buff and rufous; female may be slightly paler.
Underparts relatively plain rufous
Tail rufous or buff, heaviest dark barring on outer feathers
Tail without white tips
White wing patch
In flight identified by presence of white wing patches together with lack of white patches on tail
Roosts on ground among leaf litter and rocks in open exposed sites. Relies on extremely effective camouflage patterns for concealment.
Ruby-red eye reflection in spotlight
Rufous collar
Back grey to grey-brown
Folded wings shorter than tip of tail
White around throat
Underparts tawny, finely spotted and barred darker
Edges of white wing spots not usually visible
Large, dark, long-winged nightjar with white throat band but without white in tail and with very little in outer wing; distinctive call
Juvenile: overall rufous tone, dull compared with adult
Erratic stiff 'double-beat' wing action; long glides on uplifted wings
Grey brow line, very dark face
Dark back
Scapulars pale grey, tipped black
Tips of long wings almost reach end of tail
Small white spots across wingtips
Identified in flight by absence of white in tail together with small white wingtip spots in place of white bars
Adult; sexes alike
Undersurfaces buff, barred brown
Huge eyes, each set in a dark line from base of bill to back of crown; a third, finer, central dark line extends from bill to crown.
Soft fringe-edged flight feathers give quiet, fluttering, moth-like flight.
Tail and wings with softly speckled barring
Adult, rufous-grey morph
Colour morphs are:
dark grey—coastal SE and SW Aust.;
light grey and rufous-grey—throughout range;
rufous—N and arid regions.
Sexes are alike.
Tasmanian Owlet Nightjar is smaller, but otherwise no significant plumage difference—not darker, nor more sooty. Is probably a race, *tasmanicus*, but is sometimes included within *cristatus*.
Eyes, unlike those of true nightjars (above) do not glow brightly in a spotlight beam.
Most strongly rufous of the morphs is found in the Pilbara region of NW Aust. and across arid parts of the interior; this colour form appears to occur in both sexes.
Juveniles similar to adults, but with less strongly marked face, and shorter facial bristles and tail.
Adult, rufous morph

Glossy Swiftlet *Collocalia esculenta* 9–11.5 cm

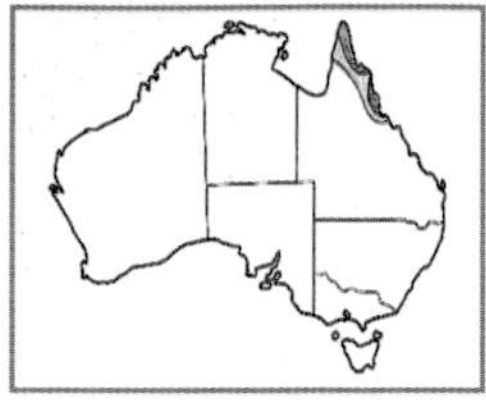

Smallest Aust. swift, no white on rump, underside all dark except contrasting white belly. Erratic flight, on shallow-fluttering downcurved wings; abrupt turns, stalls, glides. Sweeps very low through forest spaces, often over watercourses, or, in faster flight, high above canopy. **Variation:** race *nitens*, upper parts glossed bluish, vagrant to NE Qld; race *esculens*, glossed greenish and with more white in fanned tail, breeds on Christmas Is. **Voice:** silent, or weak, sharp twitter. **Hab.:** airspace above and through forests. **Status:** occasional vagrant.

White-rumped Swiftlet *Aerodramus spodiopygius* 11–12 cm

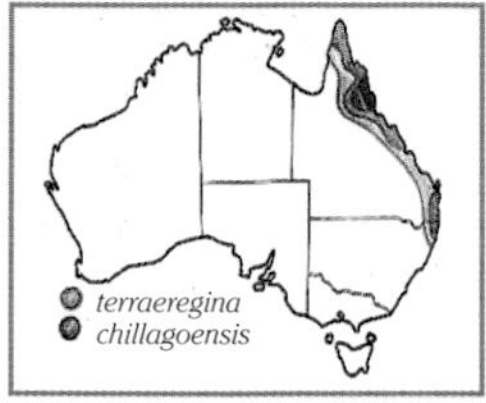

Small, dark, grey-brown swiftlet with dull white rump. Small parties to large flocks; flight erratic, twisting and turning in pursuit of flying insects, occasionally quite low, at other times high. Usually seen in vicinity of breeding caves. **Variation:** in Aust., race *terraereginae* and, at Chillagoe caves, *chillagoensis*. **Voice:** high pitched twittering, squeaking; in caves, echolocation by metallic clicking. **Hab.:** skies over coastal ranges, gorges, islands, woodlands to 1000 m altitude. **Status:** sedentary; common within restricted range. **Br:** 382

Uniform Swiftlet *Collocalia vanikorensis* 11–12 cm

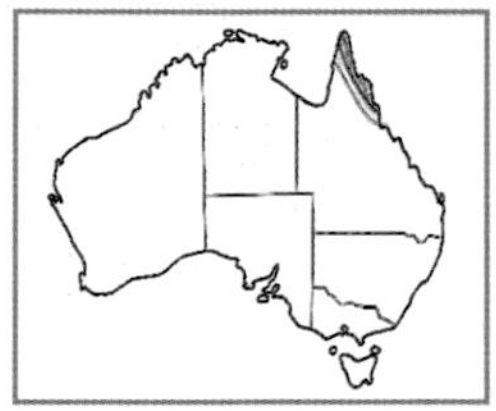

As the name indicates, almost uniform in its dusky dark colour. Typical erratic swiftlet flight, hawking for insects in large flocks. **Voice:** when flying low and close, scratchy, high twittering may be heard. **Similar:** other all-dark NG swiftlets. **Hab.:** in NG, mostly over lowland rainforest but will use airspace over almost any terrain, including coastal and offshore islands. In open country, flocks will hawk for insects almost to ground level; at other times, just above forest canopy, or very much higher. **Status:** common and widespread across N islands, many races; few sightings in Aust., only one confirmed record, from Cape York.

House Swift *Apus affinis* 13–14 cm

A small, rather stocky swift. Large, clear-cut white rump patch wraps around, usually visible from below. Flight not as fast, powerful or graceful as larger swifts; more fluttering, weaker, often gliding. **Voice:** harsh, shrill, rippling or rattling trill; occasionally high, squeaky shrieks. **Similar:** Fork-tailed Swift, but tail more deeply forked; White-throated, but has white under-tail coverts, not rump. **Hab.:** airspace above most environs within its range, and around coastal cliffs, caves, buildings. **Status:** common, Indonesia to Asia, Africa; rare vagrant to NE Qld.

Fork-tailed Swift *Apus pacificus* 17–18 cm (C, L, M, N, T)

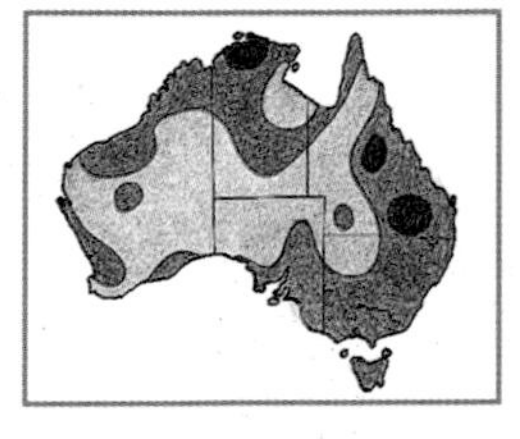

A medium-large swift with white throat and rump, and deeply forked tail that forms a single fine point when fully closed, or spreads into a wide 'V' shape in slow aerial manoeuvres. In flocks, or rarely, one or two individuals, it hawks for insects, at times very high and sometimes in mixed flocks with the larger White-throated Needletail. Often, when hawking for food, the flocks make circling sweeps, gathering flying insects in wide-gaping bills; they sometimes scoop water on the wing from inland lakes or small pools. Flight buoyant, more erratic and fluttering than the Needletail; stays on the wing day and night, sleeping in high, circling flocks. **Voice:** high-pitched buzzing, screeching, lorikeet-like, shrill, excited twittering. **Hab.:** low to very high airspace over varied habitat, rainforest to semi-desert; most active just ahead of summer storm fronts. **Status:** summer migrant, Oct.–Apr.

White-throated Needletail *Hirundapus caudacutus* 19–21 cm (K, L, M, N)

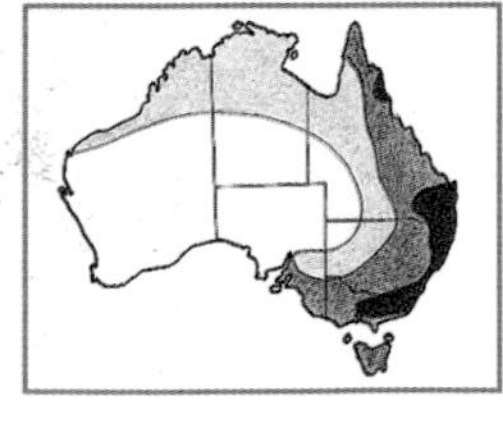

Australia's largest swift; its rather heavy looking body merges to a broad, short, square-cut tail; long winged, powerful; one of fastest of all birds; able to fly very high, but at times dives low in pursuit of flying insects. Soars, making slower aerial manoeuvres with wings forward and broadly flared; accelerates powerfully, falcon-like with outer wings backswept, pointed. To dive steeply and very fast, the wings curve back sleekly beside the tail. **Voice:** sharp, musical, uneven twittering with high to very high squeaks, very rapid, some clear, some scratchy, 'chi-chiet-chit-chi–chit-chiet-chi-'. **Similar:** Fork-tailed Swift, but body tapers to finely pointed or sharply forked tail. Australian Hobby, but only in distance or poor light. **Hab.:** high open spaces of sky above almost any habitat, including oceans; at times gathers over ranges, headlands, often in humid, unsettled weather preceding thunderstorms. **Status:** breeds northern Asia, migrates S to reach Aust. early Oct., and the SE corner by Dec.; the return begins Mar. and most have departed by end of April; moderately common, locally common E coast and ranges.

White inner edge spots visible only if tail fully spread in a slow, banking turn
Downcurved
Dark grey
White
Tail almost square when spread; downcurve may make slight fork from some angles.
Upper parts black with blue gloss
When closed, tail tip forms a shallow notch.
Scythe-shaped wings with backswept tips; become straighter, broader tipped in slow manoeuvres.
Translucent
In fast flight, wings backswept, narrowed
Tail black
Breast mottled dark grey, face and throat dark
Contrasting white belly
Black
Typical martin shape
Going away, downcurved wings appear very thin.
Narrow, dull white
Race *terraereginae*: nape and back dark
Glides on stiffly down-curved wings
Wingtips slightly rounded
Translucent, paler inner edge
Grey-brown: race *terraereginae*
Wings broad in slow glides
Narrow in fast flight
Paler nape; broader, whiter rump
Race *chillagoensis*: chin to belly paler; whiter patch on rump; paler nape
Wide white rump
Wing shape varies, aerobatics, slow, or fast flight.
Small, almost uniformly dark swiftlet
Shape changes often in insect-hawking acrobatics
Slightly forked
Rump dark
Rear edge is more square or slightly convex as tail is spread wider.
Upper parts dark grey-brown, with greenish bronze sheen
Underparts almost uniformly brownish or dusky grey, very slightly paler at throat. Other SE Asian swiftlets are also uniformly dark, and probably not able to be separated in the field. These include Black-nest Swiftlet, *Aerodramus maxima*, Edible Nest Swiftlet, *A. fuciphaga*, and Mountain Swiftlet, *A. hirundinacea*.
Wing and tail shapes become more slender, pointed, backswept in faster, higher flight.
Wings forward, broadly flared, in slow stalls and turns
Small, dark, stocky swift identified by combination of white rump and very shallowly forked tail
White rump patch extends to sides, may be glimpsed from below.
White throat
Compared with swiftlets, House Swift is more heavily built, has white throat as well as rump, and stronger flight.
Very shallow fork, becomes square-tipped if only partially spread.
White throat
Closed, has small notch
As with other swifts, sunlight through thin, translucent inner flight feathers can make them much brighter than opaque parts of the wing.
White edges of rump may show from below.
Large swift whose body tapers to a long, forked or tapering tail
Long, sickle-shaped, pointed wings
White rump
Tail deeply forked
Black, or brownish black in worn plumage
Long tail usually narrowly forked
White throat
Tapering body
Pale undersurface scalloping is not obvious at distance.
Wings vary greatly: straight out and broad, or backswept, narrow and finely pointed.
Tail forked when spread, but usually as a narrow, deep fork, or as a single fine point when tightly closed in fast flight with wings backswept
White forehead
White throat
Upperparts greenish-black; centre of back paler brown; white on inner edges of tertials
Rear body is thickset, not obviously tapering towards tail.
White inner edge to tertials
Short, broad tail tipped with protruding spines, but these not usually visible at distance
Appearance, and jizz, can in some wing shapes be falcon-like
Long inner primaries give distinctive wing shape.
White of under-tail coverts extend forward to flanks.
Tail usually with square-cut tip
In powerful acceleration, wings are backswept; rapid, shallow wingbeats.
Wings backswept, sharply pointed, tail narrowed in fast flight, dives
Soaring in high circles, gliding, or fluttering among insect swarms–holds wings well forward, broadly flared, tail widely fanned.

Azure Kingfisher *Alcedo azurea* 17–19 cm

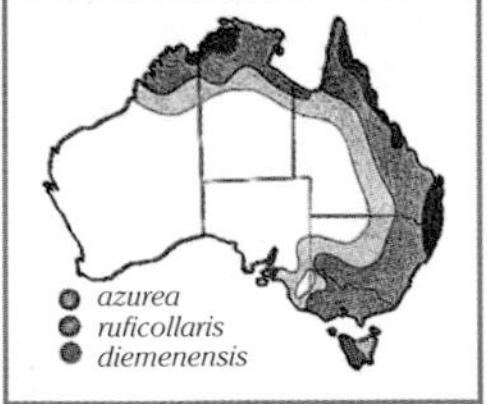

Named for the azure blue that dominates the plumage, in sunlight a deep blue with violet sheen. Often overlooked unless a high squeak draws attention as the kingfisher darts across a river pool, skimming low across the water, a momentary glimpse of blue vanishing around a river bend or into shadowy overhanging vegetation. When perched can be hard to detect, but regularly lifts tail, occasionally bobs head. Most likely to draw attention by dropping to the water with a small splash, to take yabbies or small fish. **Variation:** nominate *azurea* E and SE Aust.; race *ruficollaris* across N Aust., Cape York to Kimberley; *diemenensis* in Tas. **Voice:** often silent, but frequent sharp squeaking when breeding; the high, thin 'eeeet' given in flight is easily audible if bird darts past very close, but at any distance needs nearly perfect high-frequency hearing. **Similar:** Little and Forest Kingfishers, but white underparts. **Hab.:** well-vegetated banks of creeks, swamps, lakes, mangroves. **Status:** common where habitat remains suitable, mainly in far N, otherwise uncommon; most sedentary, some migratory. **Br:** 382

Little Kingfisher *Alcedo pusilla* 12–13 cm (T)

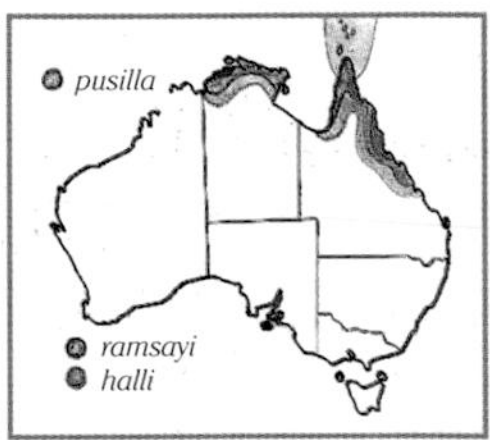

Even smaller than the Azure Kingfisher and not often seen. Sits quietly, motionless except for an occasional bob of the head, on perches low over water. This species is wary and timid, often seen only when darting away into the gloomy channels of mangroves or rainforest. From its perch it dives to take small fish, crustaceans, water beetles and other aquatic creatures; occasionally it hovers briefly, low over the water, before plunging to take its prey. The Little Kingfisher's tiny body is neatly divided, bright blue above and white beneath, the blue a cooler cerulean hue compared with the warmer violet-blue of the Azure Kingfisher. The flight is fast and direct, skimming the surface, to a perch rarely more than a few metres above the water; it is unlikely to be seen over land. The Little and Azure are Australia's only representatives of the *Alcedo* kingfishers, specialists at exclusivly deep-diving into water for their prey. They are further distinguished in having only three toes, just two pointing forward. **Juv.:** blue is dusky, the breast lightly scalloped with brown. **Variation:** 3 Australian subspecies. The nominate *pusilla,* a NG species, extends S through islands of Torres Str., possibly to the NE tip of Cape York. Race *ramsayi*, coastal NT and E Cape York, has blue side patches extending partway in towards the centre of the breast; its blue is lighter, and with a touch of turquoise. Race *halli,* of coastal NE Qld, has only small inwards patches of blue each side of the breast. **Voice:** extremely high-pitched squeak, 'tzeeet' or 'szeeit', often given in flight; sometimes a rapid, pipping 'tseet-tseet-tseet-tseet'. These calls are even higher-pitched and weaker than those of the Azure, and probably inaudible to many observers if the bird is not very close. **Hab.:** densely vegetated channels of mangroves, swamps, creeks, rainforest streams; occurs on coastal lowlands and altitudes to 750 m in mountain rainforests. **Status:** generally uncommon, but probably more common in remote northern rainforests and mangroves; sedentary. **Br:** 382

Buff-breasted Paradise-Kingfisher *Tanysiptera sylvia* 30–35 cm (T)

In flight through the rainforest, this bird is spectacular, its long, white tail plumes streaming behind, showing orange-buff undersurfaces of wings and body. In the gloom of the forest, these colours may be bright only if the bird happens to fly through a patch of sunlight. This kingfisher keeps to the mid level and lower canopy of the rainforest where it tends to perch in dense foliage, remaining almost motionless and silent. Unless it calls or drops to the ground to take some small lizard or frog, its presence might easily remain undetected. The habit of occasionally flicking the tail upwards while perched helps to reveal its presence. It is best seen soon after its arrival from NG, early in Nov., when it is in its brightest plumage, the long tail plumes not yet stained or damaged by nest digging. At that time also, frequent territorial calling makes location easier; later, with eggs or young in the nest, the birds are more secretive. **Voice:** a rapid succession of notes, not high-pitched but rather metallic with a touch of harshness, resonates and carries far through the rainforest: 'karow-karow-karow-karow-' or 'charow-tcharow', the 'ow' slightly uplifting, stronger, higher and emphasised. Each call is rapid, the usual series of six to ten calls taking only 3 or 4 seconds. It also gives a mellow, musical trill that begins strongly and continues for 3 or 4 seconds, fading away in pitch and volume. The alarm, if the nest is disturbed, is a high screech. **Hab.:** rainforests of coastal lowlands and lower slopes of ranges; riverine thickets, tropical gardens. Prefers rainforest with open lower levels, being usually a terrestrial hunter. **Status:** locally common in restricted range and habitat; breeding migrant Nov. to Mar. **Br:** 383

Common Paradise-Kingfisher *Tanysiptera galatea* 33–43 cm

Common in lowland rainforests of NG. Possible vagrant to NE Qld or Australian islands of Torres Str.; one doubtful record. **Voice:** mournful notes lead into a rapid, loud trill. **Hab.:** in NG, lowland rainforest and monsoon forest.

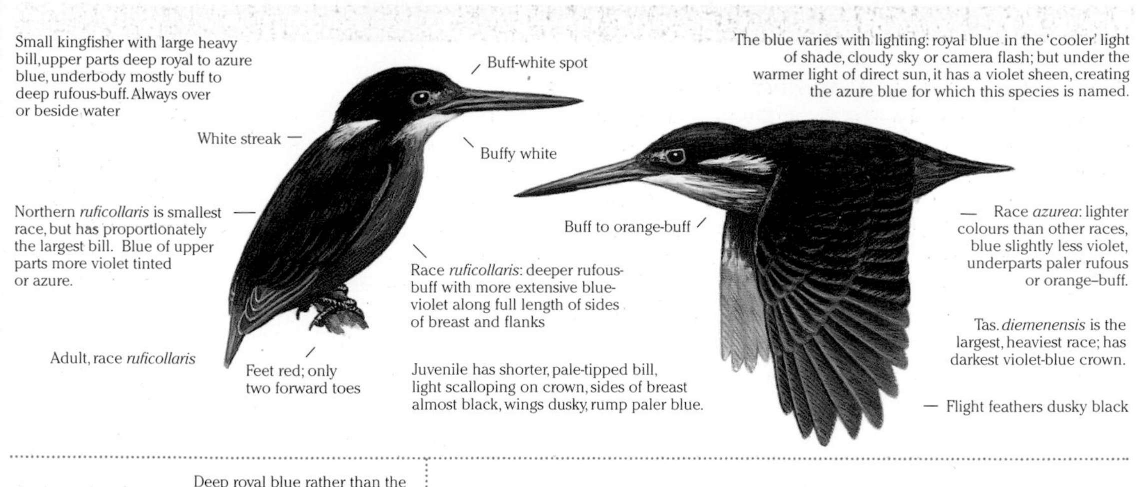

Small kingfisher with large heavy bill, upper parts deep royal to azure blue, underbody mostly buff to deep rufous-buff. Always over or beside water
Buff-white spot
White streak
Buffy white
The blue varies with lighting: royal blue in the 'cooler' light of shade, cloudy sky or camera flash; but under the warmer light of direct sun, it has a violet sheen, creating the azure blue for which this species is named.
Northern *ruficollaris* is smallest race, but has proportionately the largest bill. Blue of upper parts more violet tinted or azure.
Buff to orange-buff
Race *azurea*: lighter colours than other races, blue slightly less violet, underparts paler rufous or orange–buff.
Race *ruficollaris*: deeper rufous-buff with more extensive blue-violet along full length of sides of breast and flanks
Tas. *diemenensis* is the largest, heaviest race; has darkest violet-blue crown.
Adult, race *ruficollaris*
Feet red; only two forward toes
Juvenile has shorter, pale-tipped bill, light scalloping on crown, sides of breast almost black, wings dusky, rump paler blue.
Flight feathers dusky black

White spot
Deep royal blue rather than the violet-tinged blue of the Azure Kingfisher
Eastern race *halli* has only a small inwards expansion of blue each side of white breast
Adult, race *halli*
Race *ramsayii* is a paler turquoise blue
White streak
Long fine black bill
Race *ramsayi*: the blue patches each side of breast are larger, closer to forming a breast-band; this is even more noticeable on the juveniles.
Feet dark brown

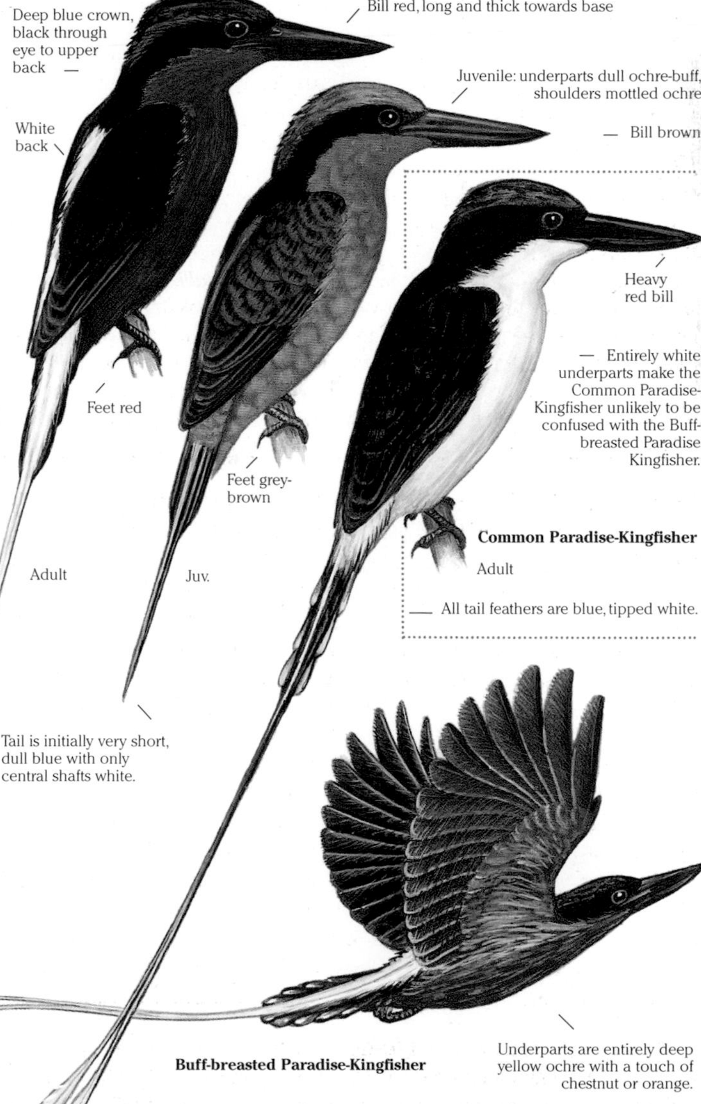

One of the most spectacularly beautiful of Australian birds, best seen on its first arrival from NG in Nov., before the new breeding plumage has become muddied and perhaps damaged in the process of digging a nest tunnel into a rainforest termite mound.
Deep blue crown, black through eye to upper back
Bill red, long and thick towards base
Juvenile: underparts dull ochre-buff, shoulders mottled ochre
White back
Bill brown
Heavy red bill
Entirely white underparts make the Common Paradise-Kingfisher unlikely to be confused with the Buff-breasted Paradise Kingfisher.
Feet red
Feet grey-brown
Buff-breasted Paradise-Kingfisher
Common Paradise-Kingfisher
Adult
Centre of the tail is white, continuing up to rump and back where it is often partly covered by folded wings, so that there appear to be two smaller patches of white, the upper back, and the rump.
Adult
Juv.
All tail feathers are blue, tipped white.
Tail is initially very short, dull blue with only central shafts white.
The tail, with two greatly elongated central feathers, makes up about half the total length of the Buff-breasted Paradise-Kingfisher. The outer feathers of the tail are deep royal blue, the central pair white.
Many paradise-kingfishers, including the Common Paradise-Kingfisher, have large spatulate tips to the tail, but the Buff-breasted has only a slight enlargement of the tail tip.
Buff-breasted Paradise-Kingfisher
Underparts are entirely deep yellow ochre with a touch of chestnut or orange.

Yellow-billed Kingfisher *Syma torotoro* 18–20 cm

Although brightly coloured, can be difficult to sight; sits still in rainforest, merging into surrounding foliage where orange and red of large deciduous or dying leaves and colourful new growth spread through the greenery. The unique calls are most likely to reveal its presence, but seem to echo through the rainforest so that the direction of their origin can be hard to judge. Sits very still, often quite low; occasionally drops to ground. **Voice:** gives a loud, ringing, musical trill, at first quite rapid, but rising steeply in pitch and tempo until it is a shrill, vibrating whistle, undulating, finally fading away to be repeated after a pause. Often given repeatedly with increasing intensity, especially when answered from an adjoining territory. Scolds and screeches when defending the nest site. **Hab.:** rainforest, monsoon forest, mangroves; favours edges, often hunts and nests in adjoining tropical woodlands. **Status:** sedentary; common in very restricted range. **Br:** 383

✓ **Forest Kingfisher** *Todiramphus macleayii* 17–23 cm (T) Darwin (Bot. Gard.)

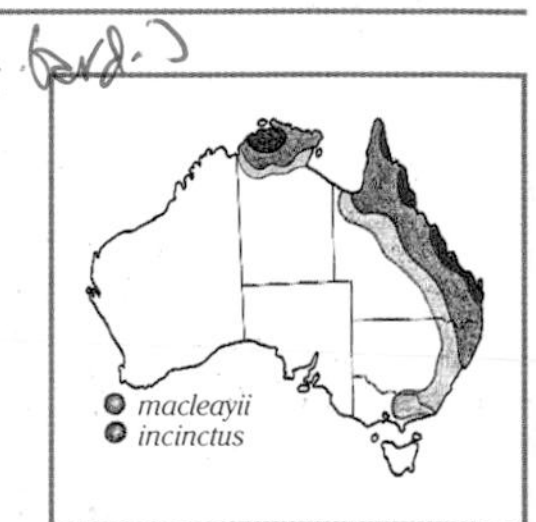

Deep royal blue above, pure white undersurfaces in neat bold contrast. In flight, white patch in outer wing. Hunts from exposed perches in fairly open country; dives to ground or shallow water to take its prey; quite conspicuous, noisy in breeding season. **Variation:** 2 Aust. races, a third confined to NG. Nominate race *macleayii* found only in NT; race *incinctus* from Cape York to SE Aust. **Voice:** sharp, piercing, very rapid 'kik-kik-kik-kikkikikikik-' in long sequences; much faster rate than quite similar call of the Sacred Kingfisher. **Similar:** Little Kingfisher, but far smaller; Sacred and Collared but greener; none has white wing spot. **Hab.:** open forests, woodlands; favours vicinity of rivers, margins of swamps, billabongs; also mangroves, farmlands. **Status:** nominate race *macleayii* sedentary; common. Eastern *incinctus* part migratory—some winter in NG, at which time absent from SE Aust. and less common in NE. **Br:** 383

Red-backed Kingfisher *Todiramphus pyrrhopygia* 20–24 cm

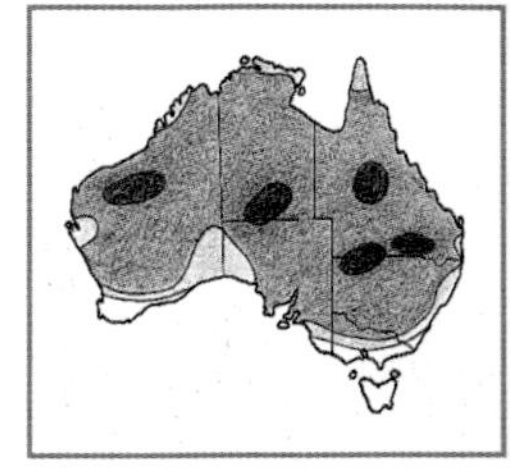

Hunts from perch in open; drops to ground to take small reptiles, occasionally mouse- sized mammals, large insects. **Fem.:** crown more heavily streaked white, wings dull turquoise, more buff in whites. **Juv.:** like female but dull grey-green mantle and back; buff fringes to wing coverts. **Voice:** a mournful 'kee-ip', the last '-ip' ending abruptly, given in a long, slow series at intervals of several seconds: 'kweee-ip, kweee-ip, kweee-ip'. Also uses a slightly deeper, nasal rather than clear, 'quaeirr-ip, quaeirr-ip', which has an even more melancholic quality than the higher call. **Similar:** Sacred Kingfisher. **Hab.:** throughout the drier regions, usually inland, in semi-arid woodlands, mulga and mallee scrublands, spinifex and other almost treeless country, often far from water. **Status:** moderately common; nomadic and migratory. **Br:** 383

Collared Kingfisher *Todiramphus chloris* 24–28 cm (C,T)

Large, stocky, heavy billed; a 'tree' kingfisher, it hunts over low-tide mud and puddles rather than deep-diving. **Voice:** calls like the Sacred, but slower, stronger, sharp and penetrating, at greater intervals: 'kik-, kik-, kik-', perhaps closer to 'kiek' than the quick 'kik' of the Sacred; also a rattling, piercing 'kr-r-riek-kr-r-riek-'. **Similar:** Sacred Kingfisher, but it has buff-tinted whites, smaller, with lighter bill. **Hab.:** in Aust., coastal mangroves; favours mangrove creeks and along seaward side of the mangrove belts, but also hunts out over nearby open mudflats and beaches. **Status:** sedentary in N, migratory in S of range; quite common. **Br:** 383

Sacred Kingfisher *Todiramphus sanctus* 20–23 cm (C, L, N, T)

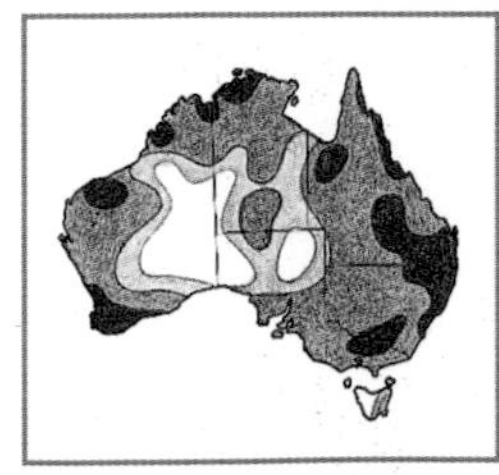

Well-known, widespread, common kingfisher; migrates into southern Aust. for summer breeding; attracts attention with persistent calling. Hunts on dry land for small reptiles and large insects, but also uses margins of wetlands. **Variation:** nominate race on mainland Aust., other races Norfolk and Lord Howe Is. **Voice:** early in breeding season, a loud, sharp, penetrating, far-carrying 'kik-kik-kik-kik-', usually of 3 to 5 notes, occasionally 'kek-kek-kek-kek'; these calls repeated almost incessantly through the day. Between a pair near the nest, a clear, shrill 'kieer-kieer-kieer-', often when about to fly to nest or mate with food. On arrival, 'kek-krreee-krreee-kreee-kreee', a vibrant and rising sound. In defending nest site, dives with loud harsh scream. **Similar:** Collared, but has pure white underparts and collar, heavier bill, larger and more stocky build. **Hab.:** open forests, woodlands, semi-arid scrublands, mangroves. **Status:** sedentary in N, summer migrant to S. **Br:** 383

Head rich rufous-buff; black nape spots, like eyes from behind
Black top edge, outer half of bill
Black 'eyes' on nape
Female has black centre to crown.
Eye encircled by black ring; partly black lores
Top of outer half of bill is black. Cutting edge is finely serrated along outer part (both sexes).
Wing coverts blue-green to olive-green
Nominate race is in New Guinea; Aust. race is *flavirostris*
Flight feathers black
Buffy white
Paler than male
Deep blue
Some birds have dark upper surface to outer bill, others have entirely yellow bills.
Rump light blue
In flight, underwing rufous-buff
White spot each side of forehead
Buff
Neat plumage pattern of deep blue, white and black. Nominate race *macleayii* in N Aust., *incinctus* coastal NE to SE.
Female blue; male mottled blue and white
Female: blue
Deep to mid sky blue: *macleayii*
Male: white
Coverts edged buff
Turquoise tint: race *incinctus*
Buff tint
Race *macleayii*: large patch of white curving around base of primaries; on race *incinctus*, this patch averages smaller.
NW race *macleayii* has slight buff tint on flanks.
Adults white
Race *incinctus*
Juv.
Crown bluish or greenish grey heavily streaked with white
Female similar, more white on crown, slightly greyed colours
In flight, all show white wing linings, white patch on underside of primaries.
Black facial mask encircles nape
Complete white collar, both sexes
Mantle dark grey-green
Shoulder (coverts) blue
Dark olive or brownish green
Long, very heavy bill
Underparts white
Mid back to tail coverts rufous
Wide white collar
Rufous
White; lacks buff tint of Sacred.
Australia has 2, perhaps 3 subspecies of the Collared Kingfisher. Race *sordidus*, illustrated, occurs on eastern and northern coasts; the colour of its upper parts changes from olive green in SE of its range, gradually browner N to Cape York, and W through NT to W Kimberley. It is possible that *sordidus* may divide into 2 races, the greener SE population then becoming a third Aust. race, *colcloughi*. Race *pilbara*, confined to the NW Pilbara coast, is smaller and paler with brownish green head and back. Vagrants to Christmas Is. are probably of the Indonesian race *palmeri*.
Not only larger than Sacred, but of broader, more stocky build
Bronzed olive-green
The Collared (Mangrove) Kingfisher has about 50 subspecies widely distributed from India in W through SE Asiato N Aust. and Pacific islands.
Flight feathers edged deep blue
Female slightly browner-olive on back, blues dull. Juvenile has markings like juv. Sacred.
Black mask
Collar buff-tinted
Buff lores
White
Flight feathers deep blue
Faint buff tint
Crown turquoise green
Green
Blue-green wing coverts
Buff, mottled brown
Pinkish-buff base to lower mandible
Buff tint strongest on lower flanks
Upper-wing coverts fringed buff
Blue
Female greener, duller than male, less buff beneath
Juv.
In flight, the Sacred Kingfisher shows rufous-buff underwing.

Blue-winged Kookaburra *Dacelo leachii* 38–42 cm

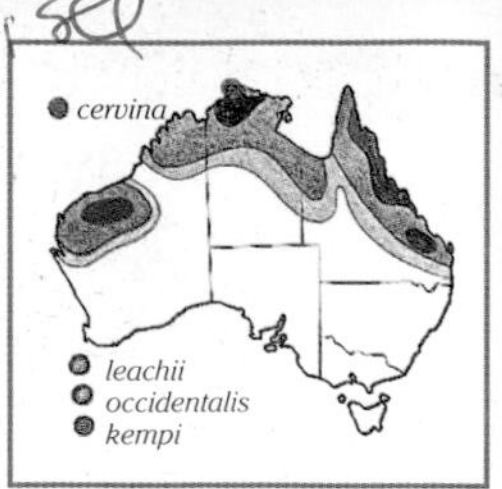

More colourful than Laughing Kookaburra but has rather unpleasant, staring white eye. Calls described as maniacal, insane. Small family parties, up to ten or so birds. **Juv.:** similar to adult female, paler crown, darker scalloping of underparts. **Variation:** number of races uncertain; differing opinions. Insufficiently distinctive may be *kempi* and *cervina*, probably better included within *leachii*. Race *occidentalis* (*cliftoni*), Pilbara, may likewise be within *leachii*. **Voice:** a raucous cacophony, begins slowly, builds louder: 'kok-kok-kok-QUOR-R-RORK-QUOR-R-RORK-QUOR-R-RORK-car-r-aark-craark-crak-qwok-quok'. One bird starts calling, followed by others in unsynchronised fashion, building up, rising and falling in cacophony. **Similar:** Laughing Kookaburra, but less blue in wings, dark eye streak, pleasant laughter. **Hab.:** open forest, woodland, tropical woodlands, tree-lined rivers. **Status:** sedentary or local movements; moderately common. **Br:** 383

Laughing Kookaburra *Dacelo novaeguineae* 40–47 cm

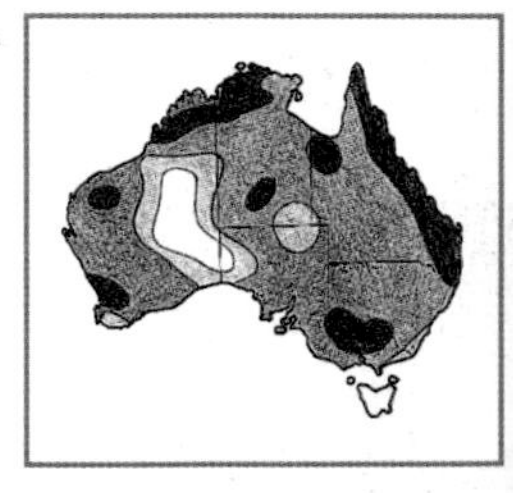

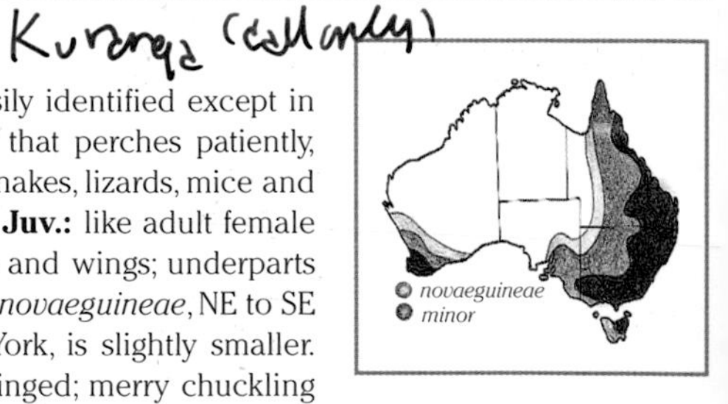

Perhaps Australia's best-known bird, famed for its 'laughing' call; easily identified except in separating this species from the Blue-winged. A terrestrial hunter that perches patiently, gliding down to ground to take small, but occasionally quite large, snakes, lizards, mice and other small mammals, large insects, small birds and their nestlings. **Juv.:** like adult female with more pronounced pale or white edging to feathers of mantle and wings; underparts more heavily barred, bill shorter, all dark. **Variation:** nominate race *novaeguineae*, NE to SE Aust. (introduced to both Tas. and SW Aust.). Race *minor*, Cape York, is slightly smaller. **Voice:** a far more pleasant, jovial 'laughter' than that of the Blue-winged; merry chuckling rising to raucous laughter, fading away again to a slow throaty chuckle: 'chok, kok, kok-kak-KAK-KAK–KAK-KOK-KAK-KOK-kook-kook-kok, kok, kok', often with other birds joining in, giving greater volume to the laughter. Also some low, grinding 'growk-growk' noises. **Similar:** Blue-winged Kookaburra, but it has a lightly streaked, whitish head, lacks dark band through eye, and has much more blue in wings. **Hab.:** diverse habitats: include open forest, woodlands, partly cleared farmlands, watercourse trees of semi-arid inland, parks and gardens. **Status:** sedentary; common. **Br:** 383

Rainbow Bee-eater *Merops ornatus* 23–27 cm (L,T)

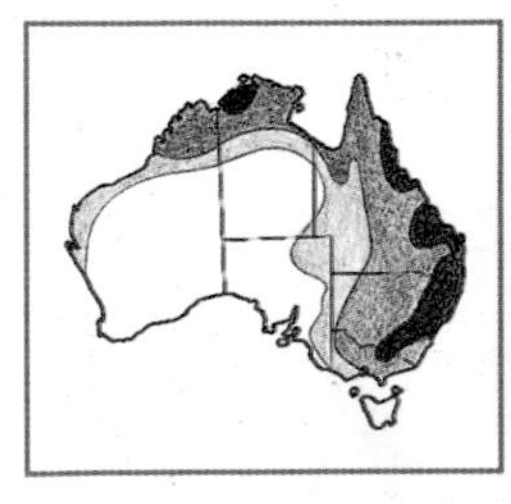

Orange wings flash against blue sky as a Rainbow Bee-eater twists and turns in pursuit of fast-flying insects; then glides back to its perch to beat the captured bee or dragonfly against the wood. The plumage combines delicate pastel tints with strong, bright hues; wings are long and pointed, tail usually has long central streamers. **Voice:** a loud, clear, sharp, cheerily vibrating burst of 'pir-r-r' calls is often given in flight, carrying far and attracting attention. Delivered with characteristic rapid rattling or vibrating; a few or very many calls in quick succession; these fluctuate slightly in pitch and volume: 'pirr-pirrrp-pirr-pirrr-pirrp-pirr-pirrr-pirr-pirrp-'. Perched birds exchange a similar, softer, slower call with less vibrancy, 'quiera-quiera-quirea-'. **Similar:** kingfishers are generally rather similar but plumper with very large, thick, straight bills; do not take prey in flight. **Hab.:** open country of woodlands, open forest, semi-arid scrub, grasslands, clearings in heavier forests, farmlands; avoids heavy forest that would hinder its aerial pursuit of insects. In breeding season, requires also an open clearing or paddock with loamy soil soft enough for nest tunnelling, yet firm enough support the tunnel. **Status:** common; regular summer migrants to southern Aust., Sep. to Apr. In the N they are resident and remain to breed. **Br:** 383

Dollarbird *Eurystomus orientalis* 25–29 cm (C,L,N)

Australia's sole representative of the roller family; a regular breeding migrant to Australia, winters on northern islands. Conspicuous and noisy early in breeding season with its diving, rolling display flights. Hunts from high bare limbs; takes all prey in flight in powerful pursuits and acrobatic aerial manoeuvring, snapping up large flying insects such as cicadas, mantids, and beetles in the broad bill. Tends to hunt very early morning and late into twilight. Its deep, relaxed, strong wingbeats and glides display the large, rounded crescents of white across the base of the primaries on both upper and lower surfaces. On the very dark, almost black violet-blues and greens of the wings, these white patches are very conspicuous, the origin of 'dollar' bird. **Voice:** usually a rapid cackle—an abrupt, grinding, harsh, metallic, jarring, buzzing, rattle, 'kzak-kzak-kzak-kzak-kzak-kzak-' or 'kak-kak-kak-'. Sometimes it gives similar, slightly slower, stronger 'kzaak!' calls at wide erratic intervals. Other calls include a deeper, slower, growling 'kzarrk-' and a higher, raucous but not grating 'kraak-kriak-kraak-'. **Hab.:** woodlands, edges of heavier forests, open country with widely scattered trees, inland watercourse trees, farmlands, city suburbs. **Status:** summer breeding migrant to Aust.; common in N, sparse in drier and SE regions. **Br:** 383

Eye white
Head off-white, streaked dark brown
Massive bill, dark above, pale creamy pink beneath
Off-white to buff, variable fine, broken, dark barring or scalloping
Extensive bright blue in wings
Race leachii
♂
Bright pale blue
Lighter streaking, head almost white
Light rufous-buff; female deeper
Brown, pale-tipped
♂
Race occidentalis
Rump light blue
White-streaked head, without any dark streak through eye
Blue patch on wings is always much larger, brighter than on Laughing Kookaburra.
♀
Females, both races, have red tails.
Massive deep bill with pale underside
Brown, feathers pale-edged; white margins more evident on juveniles
Tips of wing coverts very pale blue or often almost white
Feet relatively small, weak
Both sexes have rufous barred black tail, tipped and edged white.
Dark nape crescent joins eye-streaks
Adults
Flight feathers grey-brown
Extensive blue in coverts and flight feathers
Rufous, pale tipped
♀ Blue-winged
Both species have white bar at base of primaries.
Scale change
Juvenile: greener crown
Eye brown
Bright red eye set in long, wide, black eye stripe.
Lacks throat bands
Orange crown blends softly downwards into yellow-green, which in turn changes gradually to pale and azure blues.
Black crescent
Orange underwings seem often to catch the sun, often against deep blue sky or cool greens of background trees; the wings seem to give out a fiery inner glow.
Eye bright red
Long bill, evenly tapered, only slightly downcurved
Black edge
♀
Juv.
Paler blue
Tail without long streamers
Female has shorter, thicker streamers.
♂
Male has very long, fine tail streamers with small clubbed tips, often worn or broken shorter.
Dark brown; may appear almost black.
Throat blue-violet
Flight feathers dark blue-violet
Brown of back merges into deep viridian on wings.
Bill fleshy pink to red, tipped black
Dark, square tip
Flight feathers violet-blue
Feet red to dull orange
Adult; colours of female are very slightly softer than those of male.
In their breeding season, Dollarbirds engage in aerial displays: several birds circle, dive steeply, then, on pulling out of the dive, tip and roll side-to-side, displaying their bold, white 'dollar' wing patches.
Dark-tipped
Large white wing patches give name 'dollar' bird.
Juvenile has very dull colours, brownish bill and feet, lacks blue at throat.

Red-bellied Pitta *Pitta erythrogaster* 17–19 cm

Most spectacularly colourful Australian pitta, but surprisingly inconspicuous in its dimly lit rainforest habitat, the plumage matching similar colours in leaf-litter and foliage. Usually alone; forages on the forest floor, scratching about with strong feet to expose insects and other small creatures concealed under the moist leaf-litter layer. Flicks wings and tail; probes into mud; breaks snails on an 'anvil' rock. **Variation:** twenty-six races; only one, *macklotii*, in Aust. **Voice:** calls from ground and from forest canopy perches. The call is slow, drawn out and mournful, beginning so low that it is almost a growl, rising steadily in volume and pitch, mellow and tremulous at first, finishing quite high and clear. The second note follows at slightly lower pitch and volume, softer and more mellow, not tremulous; the whole call a long, slow 'gr-r-r-r-rork, groir'; may be enticed with call imitations. **Hab.:** rainforest, monsoon forest. **Status:** considered to be a summer breeding migrant from NG; arrives about Nov., departs Apr. Probably uncommon, although the remote and difficult country of its range and its shy and inconspicuous behaviour probably mean that sightings are few in proportion to population. **Br:** 384

Blue-winged Pitta *Pitta moluccensis* 18–21 cm (C)

Rare vagrant. All records are from north-west WA: Derby; near Broome; Burrup Peninsula near Dampier; Christmas Is. Widespread in Asia, notably SE Asia and southern China. Moves southwards in northern winter, commonly to Borneo, and occasionally to Sumatra, Java, Bali, Christmas Is. and, rarely, into NW Aust. **Voice:** the call is similar to that of the Noisy Pitta, but more husky, slightly lower and with different emphasis and timing. The double-noted call of the Noisy Pitta rises sharp and querulous at the end of the second note, seemingly delivered quite rapidly and cheerfully, but the call of the Blue-winged is a more uniform, scratchy 'quaraerk, quaraerk' with a longer pause between notes and a barely discernible uplift in the second note. The calls have also been described as 'taew-laew', 'pu-wi-u, pu-wi-u' and 'pu-wiu'. The Blue-winged Pitta also has a loud 'skieew!' alarm call. **Similar:** Noisy Pitta, but black throat and black on lower belly. **Hab.:** in SE Asia, rainforests, mangroves; in Aust., much drier NW coast, but vagrants could arrive anywhere in the NW, perhaps occasionally in the remote mangroves, monsoon scrubs and vine thickets of the rugged W Kimberley coast, where unlikely to be noticed. **Status:** common in usual range; rare vagrant to Aust.

Noisy Pitta *Pitta versicolor* 18–20 cm

Big colourful pitta, widespread; usually forages in leaf litter of forest floor; favours snails, which are broken on a rock or hard wooden 'anvil'. It is shy and usually silent except in the breeding season,, when pittas call almost continuously by day and into the night from the ground and from perches high in the forest canopy. **Fem.:** similar to male; pinky rather than red under tail. **Juv.:** duller, lacks red and black under tail to belly. **Variation:** uncertain: *versicolor*, mid E coast of NSW N to mid E Qld; *simillima* on Cape York; a possible race, *intermedia*, between. **Voice:** loud, brisk, ringing, metallic, double noted, mellow whistles, 'quarreek-quarrik, quarreek-quarrik', commonly described as 'walkto-work'; carries far in the rainforest. A single, harsh, mournful 'kieow', sometimes given at night, may be a warning or alarm call. **Similar:** Blue-winged Pitta also has yellowish buff underparts, but much more blue in wings, white throat, red instead of black stripe up lower belly. **Hab.:** tropical and subtropical rainforests, monsoon forests, wet dense eucalypt forests, mangrove forests. **Status:** common. Race *simillima* is migratory: winters in southern NG; breeds in Cape York rainforests in summer. Race *versicolor* is locally migratory; moves down from mountain forests to those of coastal lowlands in winter. **Br:** 384

Rainbow Pitta *Pitta iris* 16–19 cm

The Rainbow Pitta, confined to the Top End of the NT and the Kimberley in WA, is the sole Aust. pitta that does not also occur in NG or other islands to the N. Typical of pittas, it has a rather dumpy body, large head, very short tail, broad wings and colourful plumage, and forages in the litter layer, scratching about with long legs and large strong feet. The Rainbow inhabits vegetation that has some rainforest characteristics such as a relatively closed canopy and a dense layer of litter. There it is inconspicuous for most of the year, but, in the breeding season, it attracts attention with its distinctive loud calls. Usually solitary, it feeds on invertebrates from ground debris and soil, and snails that are broken against a rock. **Variation:** race *iris*, coast of NT. The Rainbow Pitta population of the of the NW Kimberley appears sufficiently distinctive to justify a proposed second race, *johnstoneiana*. **Voice:** clear, rather sharp, metallic. Differs from other Aust. pittas in having a preliminary note, then the main call rapidly in three parts, a quieter 'kwe', followed by a call similar to that of the Noisy, but somewhat sharper: 'kwe, kweeik-a-kweek'. **Hab.:** monsoon rainforests or scrubs, vine scrubs, thickets of cane-like native bamboo. **Status:** sedentary; moderately common in restricted pockets of suitable habitat. **Br:** 384

The spectacular Red-bellied Pitta, a breeding summer migrant from New Guinea, unique among Australian pittas for its blue upper breast, and the bright red from lower breast to under tail.
Chestnut
Grey-brown
Pale blue, bordered above and below by black bands
Blackish flight feathers under deep blue wing coverts
Extensive bright red
Adult
Extensive blue on wings
Chestnut
Although brightly coloured, can blend into background of the dim, damp rainforest among rust and gold of fallen leaves, greens of foliage, reds of new growth, brownish sticks and wet, black decaying leaves.
Female similar, but usually chestnut nape is not as bright. Juvenile is brownish on back, breast buff mottled darker and with the rest of the underparts brownish or pinkish buff.
Very short stumpy tail; looks even more so with tail folded—in flight looks almost tailless.
White bars across blackish primaries
Rare vagrant, extensive cobalt blue on wing coverts, bold buff-black-white facial pattern.
Intense green
Dull golden buff brow lines, black centre of crown
White
In folded wings, white of primaries is usually concealed.
Under-tail coverts to central lower belly, pinkish red; this narrow central strip often invisible from side view.
Golden buff with line of deep pink extending forward onto lower belly, visible from front if pitta stands tall.
Adult
Black central streak, often extends back to join black nape band.
Deep cobalt blue of lesser coverts merges into turquoise along inner greater coverts.
Large white patch extends to trailing edge.
Rump bright cobalt blue
Deep dusky blue
Largest of Aust. pittas. Olive-green mantle merges into bright leaf green of back and much of visible wings.
Deep chestnut band encircles black centre.
Very broad black facial band extends down as black bib.
White spots on primaries are usually hidden.
Deep yellow-ochre-buff
Black streak
Red on male or deep pink on female
Long legs, quite large, strong feet pink or brownish pink
Deep chestnut band encircles central black streak.
Size decreases clinally S to N. White wing-bar and green tail-tip also reduced S to N; this occurs most noticeably between Gympie and Rockhampton.
Glossy iridescent light blue with slight turquoise tint
White spots across blackish primaries
Glossy bright pale blue
Upper tail blackish, widely tipped green
Long chestnut line over eye
Bright glistening green becoming darker, more olive, further down on back and wings
Short white bar across primaries
Shoulder patch iridescent, deep blue with highlights of bright light blue
Black face and breast
Chestnut encircles black centre of crown.
Race johnstoneiana has a bolder rufous brow line and larger white bar across primaries.
Primaries blackish but almost entirely hidden under the green of tertials and secondaries
Unique among Australian pittas for the black that extends from head to mid belly (on race johnstoneiana, reaches under-tail coverts).
Bright red
Brownish
Rump and tail green separated by narrow edge of blue on tail coverts
Green-edged inner flight feathers
Sexes similar, juvenile dull with little blue on shoulder

Superb Lyrebird *Menura novaehollandiae* male 80–95 cm; female 75–85 cm

Since it was discovered the Superb Lyrebird has fascinated people with the structure of its tail, the strength, quality and mimicry of its song, and the frantic activity and beauty of its displays. **Male:** the long, ornate tail is used in elaborate displays, a shimmering veil as he competes for the attention of any female in the vicinity. The male constructs a mound of litter from the forest floor; he displays and sings from this elevated platform. If the male's display is persistent enough and the presentation is attractive to a female, she may approach his display mound and allow mating; later she takes on all the labour of nest construction, incubation of eggs and feeding of chicks. The male stays in attendance at his mound throughout the winter to early spring breeding season. Females tend to establish their nesting territories around or near the display territory of a mature male that has developed a sufficiently convincing song, tail structure and display performance. **Fem.:** differs from the male most obviously in having a shorter tail of rather ordinary, long, brownish feathers; she also has a brownish or rufous tint on the throat that is not evident on older males. **Juv.:** similar to female, but rufous on throat and forehead, and the plumage is slightly fluffy. **Imm.:** in first year, similar to juvenile, but less fluffy plumage. In second year, rufous tone is confined to the throat; on males this gradually reduces and is entirely lost by the end of the third year. Males develop the long and complex tail from the second year; it lengthens with each moult until the magnificence of the mature male's tail is reached by the sixth year. **Variation:** the northern race *edwardi* inhabits timbered country, often rocky ranges and gullies, through the New England district of N NSW and into SE Qld. A second race, *novaehollandiae*, occurs in rainforests and dense wet eucalypt forests through most of the southern range of the species, from the mid NE coast and ranges through SE NSW into eastern Vic., where it may be distinctive enough to be recognised as a third race, *victoriae*. **Voice:** powerful, far-carrying; can reproduce faithfully a huge range of calls from tiny, squeaky voiced birds to the harsh growlings of Gang-gang Cockatoos. The sounds of Eastern Whip-bird, Kookaburra, Rosella and the rasping churring of the Satin Bowerbird intermingles with the thin, high notes of the smallest thornbills and robins. These are intermixed with strange complex sounds of the Lyrebird's own creation, often mechanical, rattling, whirring in nature, and thudding noises. Sounds of saws and other machinery will be faithfully reproduced if heard in nearby forestry or farming activities. **Similar:** Albert's Lyrebird, but shorter tail, lyrates without notches, more chestnut than grey overall colour tone, rufous throat and under-tail coverts on both sexes. **Hab.:** nominate race in rainforests and wet eucalypt forests with a preference for valleys of dense undergrowth, often treeferns. The northern race uses more open forests and woodland, often granite country with sandy soils, boulder-capped hills, rocky gullies, sand-heath undergrowth. **Status:** common throughout most of natural range where habitat remains; rare in Tas.. **Br:** 384

Albert's Lyrebird *Menura alberti* male 80–90 cm; female 65–75 cm

This lyrebird is not well known, confined to a far smaller range than the Superb Lyrebird, and occupying rugged country far from the large SE cities. Although unmistakably a lyrebird, and superficially like the Superb, it differs in many details of appearance and behaviour. The tail is proportionately shorter, being little more than half the total length of 85 cm, whereas the Superb has a tail that is around two-thirds of a total length of almost a metre. The tail is less ornate, the outermost lyrates being very plain and short compared with those of the Superb. The body, however, is more colourful: the upper parts have much chestnut on wings and back, brighter rufous around rump and tail coverts, throat rufous-orange, and remainder of underparts creamy buff. Feeds by scratching about in the leaf litter and into rotting logs and loose topsoil, finding small creatures such as worms, insects, snails. Runs with fluttering wings to escape danger, but can fly quite well and roosts in trees. The large, obviously quite deep and vigorous scratchings produced by their powerful, large-clawed feet mark their presence and help to locate these birds at times in the year when they are not often calling. The displays are similar to those of the Superb, but the shorter tail filaments may not so completely envelop the bird. Uses a flat piece of ground from which all debris has been raked for its stage rather than a mound of debris. These display sites are often in dense undergrowth beneath cover such as lawyer vine; the agitated movements of the bird often send vines shivering as it brushes against their stems. **Fem.:** has a different tail structure from the male—has no fine, filamentous plumes, but rather ordinary tail feathers that, though long, are plain, tapering, dull grey-brown. **Juv.:** initially has rather downy plumage; both the juvenile and the later immature plumage have a bright rufous throat. **Voice:** both this species and the Superb have powerful, flexible voices and use a mixture of their own calls and mimicry of other species in long unbroken passages of song. But compared with the Superb, the Albert's Lyrebird restricts its mimicry to a few species. The voice can create sounds at one moment deep and resonant, switch to high thin squeaks and sharp trills, then change again to harsh noises. Some of the passages of song begin with a soft, mellow sound that rises clearer and louder, and which has been likened to the howl of a dingo. Males call many hours daily in the peak of the winter breeding season; quiet at other times. The Green Catbird and Satin Bowerbird feature strongly in its imitations. In alarm gives a shrill shriek. Even when calling strongly, this shy and elusive species is not easily sighted in the dense tangled vegetation of its habitat with dim light and the birds wary. **Similar:** Superb Lyrebird. **Hab.:** subtropical rainforests, in areas of dense thickets and undergrowth. **Status:** sedentary; restricted occurrence, but can be locally quite common. **Br:** 384

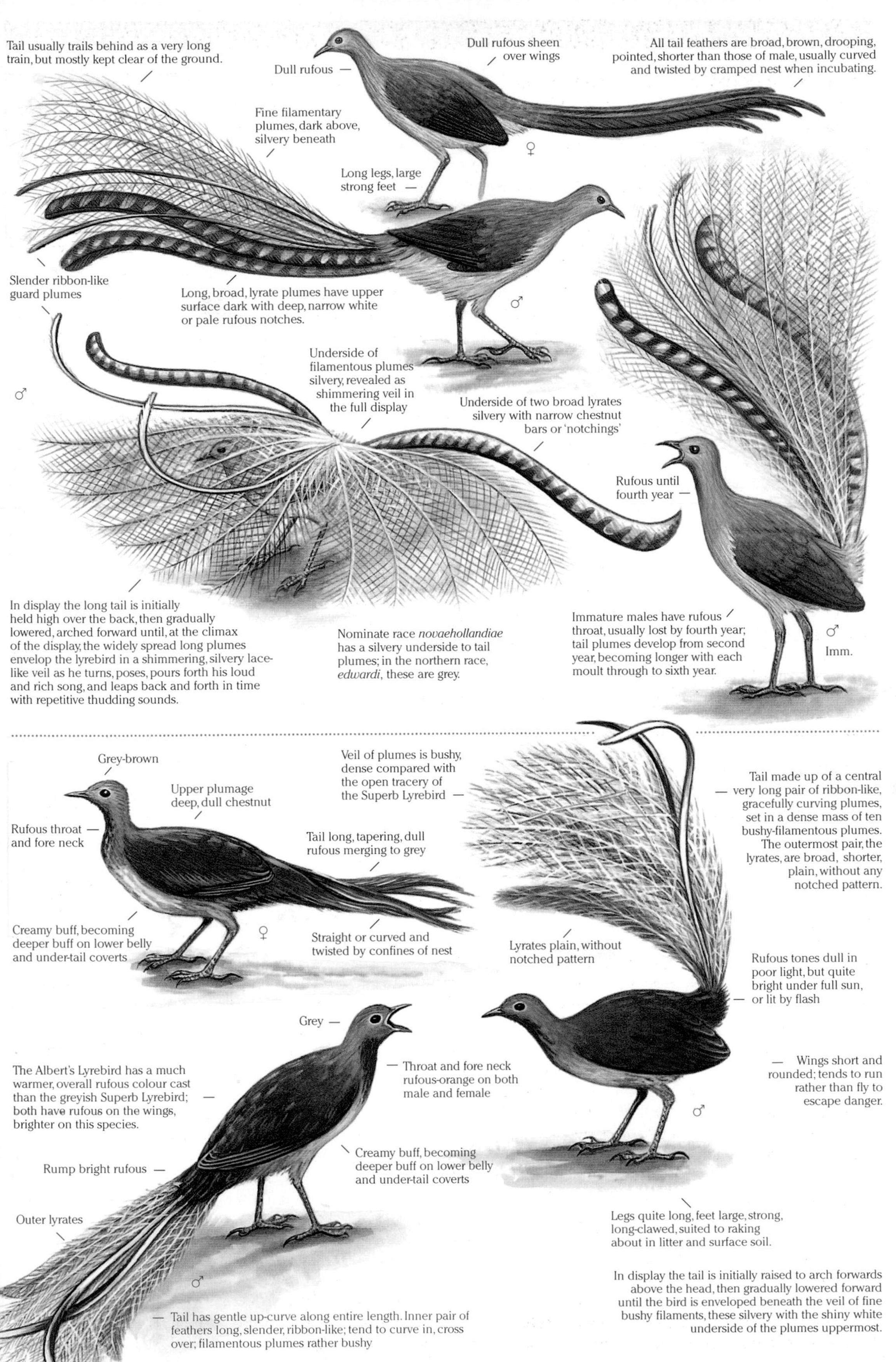

Tail usually trails behind as a very long train, but mostly kept clear of the ground.
Dull rufous
Dull rufous sheen over wings
All tail feathers are broad, brown, drooping, pointed, shorter than those of male, usually curved and twisted by cramped nest when incubating.
Fine filamentary plumes, dark above, silvery beneath
♀
Long legs, large strong feet
Slender ribbon-like guard plumes
Long, broad, lyrate plumes have upper surface dark with deep, narrow white or pale rufous notches.
♂
♂
Underside of filamentous plumes silvery, revealed as shimmering veil in the full display
Underside of two broad lyrates silvery with narrow chestnut bars or 'notchings'
Rufous until fourth year
In display the long tail is initially held high over the back, then gradually lowered, arched forward until, at the climax of the display, the widely spread long plumes envelop the lyrebird in a shimmering, silvery lace-like veil as he turns, poses, pours forth his loud and rich song, and leaps back and forth in time with repetitive thudding sounds.
Nominate race *novaehollandiae* has a silvery underside to tail plumes; in the northern race, *edwardi*, these are grey.
Immature males have rufous throat, usually lost by fourth year; tail plumes develop from second year, becoming longer with each moult through to sixth year.
♂
Imm.
Grey-brown
Upper plumage deep, dull chestnut
Veil of plumes is bushy, dense compared with the open tracery of the Superb Lyrebird
Tail made up of a central very long pair of ribbon-like, gracefully curving plumes, set in a dense mass of ten bushy-filamentous plumes. The outermost pair, the lyrates, are broad, shorter, plain, without any notched pattern.
Rufous throat and fore neck
Tail long, tapering, dull rufous merging to grey
Creamy buff, becoming deeper buff on lower belly and under-tail coverts
♀
Straight or curved and twisted by confines of nest
Lyrates plain, without notched pattern
Rufous tones dull in poor light, but quite bright under full sun, or lit by flash
Grey
The Albert's Lyrebird has a much warmer, overall rufous colour cast than the greyish Superb Lyrebird; both have rufous on the wings, brighter on this species.
Throat and fore neck rufous-orange on both male and female
Wings short and rounded; tends to run rather than fly to escape danger.
♂
Creamy buff, becoming deeper buff on lower belly and under-tail coverts
Rump bright rufous
Outer lyrates
Legs quite long, feet large, strong, long-clawed, suited to raking about in litter and surface soil.
♂
In display the tail is initially raised to arch forwards above the head, then gradually lowered forward until the bird is enveloped beneath the veil of fine bushy filaments, these silvery with the shiny white underside of the plumes uppermost.
Tail has gentle up-curve along entire length. Inner pair of feathers long, slender, ribbon-like; tend to curve in, cross over; filamentous plumes rather bushy

Rufous Scrub-bird *Atrichornis rufescens* 17–19 cm
Famed for powerful, varied calls; secretive and confined to small parts of coastal ranges. Occurs in areas of dense groundcover where sighting is very difficult, even when the bird comes very close, often calling from a few metres away yet staying out of sight. Fossicks for insects and other small animal life in deep layers of litter and debris. **Voice:** the powerful, ringing calls can be heard as rather harsh and metallic from far away; at close range they are extremely, almost painfully, loud—a slow, deliberate 'cheip' with the central 'ei' strongly emphasised. Calls include a 'cheip, cheip, cheip, cheip', a more abrupt, equally loud, 'chep-chep-chep-', a sharper 'chwiep-chwiep-chwiep' and a whip-cracked 'chweeip-chweeip-'. Other sounds are only audible at close range: soft mellow cheeps, trills, scoldings. **Hab.:** temperate rainforest, adjacent eucalypt forest, about 600 to 1000 m altitude; previously in lower habitat, now cleared, with tall, dense undergrowth and tussocky rushes under breaks in the canopy. **Status:** fragmented population on range-tops; sedentary, rare. **Br:** 385

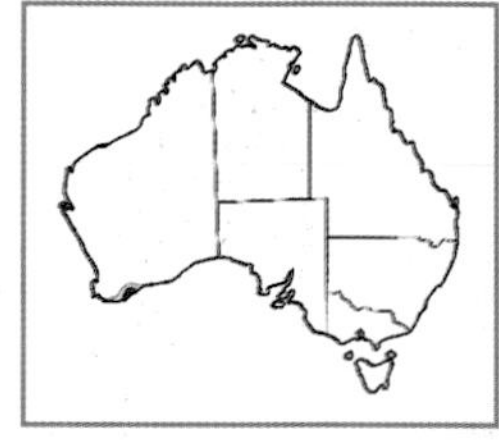

Noisy Scrub-bird *Atrichornis clamosus* 21–23 cm
One of Australia's rarest birds; long thought extinct, rediscovered in 1961. Its powerful calls attract attention, but birds are elusive, difficult to sight even when close; they scuttle mouse-like under very dense, low groundcover. **Voice:** loud, clear, ringing territorial calls of male can be heard up to a kilometre away in winter breeding season. Call is a series of notes, initially slow, sharp, becoming lower, richer, faster, increasingly powerful, almost painfully loud if the bird comes close under low cover. The sound is not unpleasant, is even musical, yet has something of the metallic screech of a dry bearing forced to turn, a short, sharp, loud squeak at each revolution, faster and louder: 'chwiep, chwiep, chwip-CHWIP-CHWIP-CHIWIEP-chwp-chwp'. **Similar:** Western Bristlebird, but has a scaly-patterned breast without black markings. **Hab.:** dense heath, rushes, tall sedges under stunted shrubby trees; some in this type of dense low vegetation in hillside gullies, others in similar thickets around sandplain swamps. **Status:** rare, restricted habitat; now re-introduced to other S-coast sites. **Br:** 385

Eastern Bristlebird *Dasyornis brachypterus* 20–22 cm
Secretive; moves quickly through dense heath and tussocky vegetation, tail horizontal, occasionally uplifted, sometimes fanned; flight low and brief. Presence revealed in spring by loud clear calls. **Variation:** population around NE of NSW and SE Qld is a possible second race, *monoides*. **Voice:** songs of varied cheerful, musical notes, some extremely high and weak, others more mellow, clear and strong; some sequences with slight whip-crack finish. Repeats a sequence several times, then adds variations: 'wit-wit wheerie-whEIT'; 'chi-chi-cheery-whEIT'; 'whit-wheeeir-weeit-weeit-whit'; 'queeity-queeeir'. In alarm, a scolding 'chzzzt-chzzzt'. **Hab.:** ranges, sedge-tussock gullies, gaps and edges of rainforests, wet eucalypt forests; also dense coastal heaths. **Status:** once common, now scattered remnants; rare, vulnerable; sedentary. **Br:** 388

Western Bristlebird *Dasyornis longirostris* 18–21 cm
Rare with very restricted range; shy, difficult to sight; song often the only sign of its presence. **Voice:** similar to Eastern Bristlebird; has slightly sad quality, reminiscent of lonely moonlight singing of Willie Wagtail. Song at the eastern end of its range differs in detail from song of separated populations further west; variations by individual birds. Intermixed sharp and mellow notes form descending sequences: 'tchip-eeee, quit-a-weer'; 'tchip-it-er-pieerit'; and 'quiee-ity-pieer, quieeity-pieer'. **Similar:** Noisy Scrub-bird. **Hab.:** dense, low closed heaths, open heaths if near refuge clumps of taller dense shrubbery or dense watercourses thickets. **Status:** rare, reduced to scattered remnant populations; vulnerable. **Br:** 388

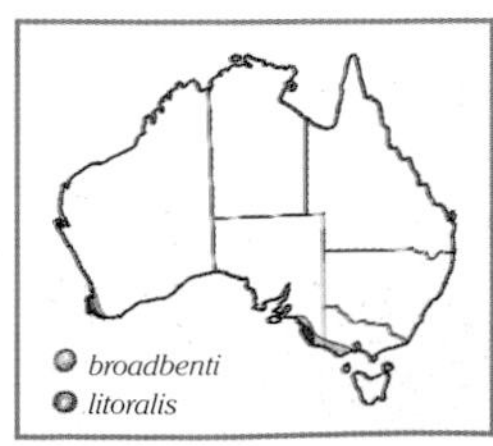

Rufous Bristlebird *Dasyornis broadbenti* 24–27 cm
Large bristlebird with rufous crown and ear coverts. Elusive in typically dense habitat, but often inquisitive; scurries through low vegetation, seldom flying, occasionally raising and spreading the long tail. **Voice:** a clear, ringing whistle; if recognised, helps locate these shy birds. Usually a series of clear, sharp notes lead into a much stronger, quick, double whip-crack; these last notes are often repeated by the female: 'chip, chip, chip, chup, CHEE-O-WHIP'. **Variation:** WA race *littoralis* probably extinct. Otway Range, Vic., may have a third race, *caryochrous*. **Hab.:** dense coastal heaths and thickets, wiregrass; in Otway Range, in dense undergrowth of wet forests. **Status:** sedentary; eastern race moderately common. **Br:** 388

Tail carried low, at times lifted, as when calling
White streak down sides of throat
Upper parts rufous, finely barred black
When male gives his powerful calls, the feathers of throat and breast stand out giving a roughly barred effect as the darker bases of the feathers are revealed.
Rufous underparts fluffed out making calling bird appear larger
At peak of powerful, carrying, territorial call, the male scrub-bird holds bill high, tail uplifted, wings drooped.
♂
Both sexes have head slightly paler buffy brown with fine, dark brown speckling and barring.
Female has lighter buffy rufous throat and breast; male has mottled blackish throat with conspicuous white edges.
Both sexes have upper body, wings and tail rufous-brown, closely and finely barred black.
♀
Tail moderately long, rufous-brown, fine dark barring
Eye dark
Brown, very finely barred black, giving impression of blackish brown in distance; wings slightly more rufous
Tapering black line from breast-band to chin, set between bold white streaks
Legs and feet large, pinkish brown
Under-tail coverts brighter rufous-buff
Rarely flies, scuttles under dense low undergrowth; wings short and rounded, feet large and strong.
Plain grey-brown without barring
Triangular bill, giving pointed head shape
Lacks black; side streaks diffuse, pale.
Pale buffy cinnamon, greyish belly
Tail long, broad, round-tapered tip, barred, often cocked slightly or steeply upwards
Bristlebirds belong in the family Pardalotidae, but are shown here for convenient comparison with the scrub-birds. The two groups are similar not only in appearance, but also have similar extremely elusive behaviour within dense low vegetation, and both have clear, ringing calls. Some bristlebirds have a distribution that coincides closely with that of one of the two species of scrub-bird.
Eye bright red
Pale around eye, lores
Visible at close range, strong 'bristles'
Light scaly pattern, in parts may form indistinct rough bars.
Rump, tail coverts rufous-brown
Tail long; faint fine bars barely visible
Wings short, rounded, chestnut
Flanks and belly dull grey-brown
A rather weak and reluctant flier; if flushed, skims away over the low heath shrubbery, long tail fanned, before dropping back to cover.
Crown, mantle, grey-brown with pale feather tips giving light flecked or dappled effect
Indistinct pale brow, red eye
Undersurface feathers dark tipped giving mottled or unevenly scalloped rather than barred appearance
Tail quite long, yet slightly shorter than those of the other species of bristlebird; usually held low, trailing as the bird runs through and under the low scrub, but elevated in alarm, or trailing and fanned in flight.
Rump, tail-coverts rufous-brown
Wings short, rounded, strongly rufous
Sexes alike; immatures plain olive-brown without paler spottings or scallopings
The Western Bristlebird has within its restricted range, the even smaller range of the Noisy Scrub-bird. The Two Peoples Bay conservation reserve has both these similar, and rare species side by side, their clear calls often intermingling.
Bright rufous crown and ear coverts; not as bright at east end of range
Chin and throat white, finely scalloped and flecked with grey
Sexes alike, except female is slightly smaller. Immatures are probably similar to adults.
Wings short, rounded, dull chestnut
Long tail usually trailing, occasionally raised and spread widely, as when alarmed
Underparts grey-brown, pale edged feathers give strongly scalloped pattern.
The slightly smaller western race *littoralis*, with lighter, brighter plumage, formerly occurred in coastal heathland habitat of extreme SW corner of WA between Cape Naturaliste and Cape Leeuwin. It has not been seen for decades. Proposed race *caryochrous*, Otway Range, has overall darker plumage, inhabits heavier forest.
Largest bristlebird, length 27 cm

White-browed Treecreeper *Climacteris affinis* 14–15 cm
Predominantly arboreal; spirals upwards searching crevices of bark and limbs for insects; also descends to search fallen logs, ground. In pairs or small family parties, occasionally solitary. **Voice:** sharp, abrupt, very metallic and penetrating, 'chrip, chrip, chrip' and as a double noted 'chip-chip, chip-chip'. Also a rapid trill, a ripple of sharp, metallic sound lasting several seconds. **Hab.:** semi-arid interior, including mulga, mallee, desert oak, callitris, spinifex with scattered trees, watercourse eucalypts. **Status:** sedentary; uncommon. **Br:** 385

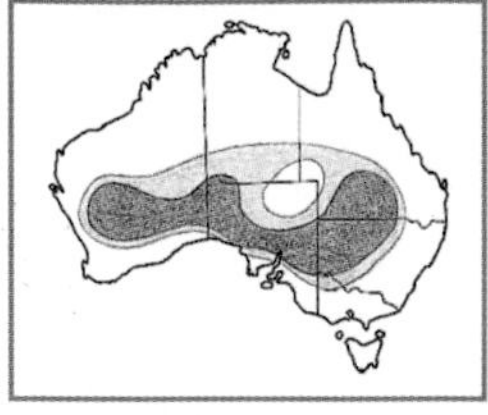

Brown Treecreeper *Climacteris picumnus* 16–18 cm
A large, well-known, widespread treecreeper; long, pale buff eyebrow separated from pale face and breast by dark line through eye, slight cresting of crown. Feeds spiralling up trees, and often on fallen timber or ground where it tends to hop with head high. Flight swift, undulating with glides that display colour of wing-bar. Gregarious, often in family groups of 4 to 5, but a territory may be defended by just 1 or 2 birds. **Voice:** a clear, high 'whit, whit, whit, whit'; each note has a strong sharp finish. **Hab.:** eucalypt forests and woodlands, scrubs of the drier areas, river-edge trees, timbered paddocks. **Status:** common; sedentary. **Br:** 385

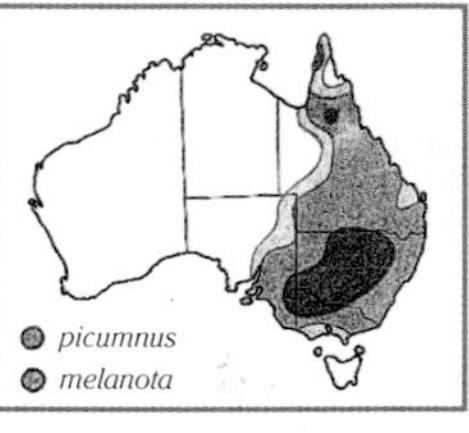

Rufous Treecreeper *Climacteris rufa* 15–17 cm
Easily identified by its orange-rufous underparts. Forages spiralling up tree trunks and limbs, using jerky two-footed jumps, then gliding down to base of next tree. Over greater distances, uses low undulating flight; pale buff wing bands are conspicuous. Much time spent among fallen timber and often on open ground. Birds nearer W coast have more intense colours than those of drier interior. **Voice:** sharp, ringing 'chip, chip, chip', widely and regularly spaced, 2–5 seconds apart. The trill is very high and sharp, descending and slowing, lasting 2 or 3 seconds. **Hab.:** wide range; jarrah and wandoo forest near Perth, dense wet SW coastal karri to dry woodland, mallee on Nullarbor, farmland trees. **Status:** moderately common. **Br:** 385

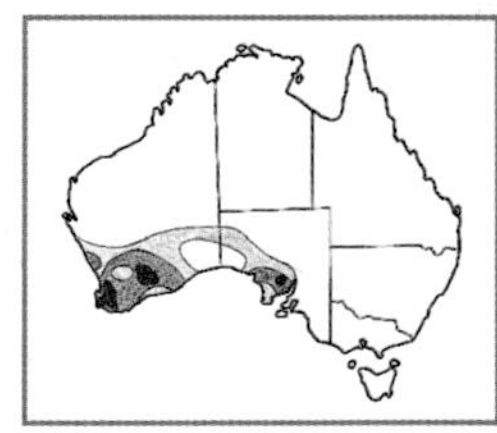

Black-tailed Treecreeper *Climacteris melanura* 17–20 cm
Through most of its range in N this is the only treecreeper. Spirals up trees; often feeds on ground. Flight direct, only slight undulation; shows buff wing-band. **Similar:** Brown Treecreeper, dark race *melanota*, slight overlap in central N Qld, but similar only in rear view. **Variation:** race *wellsi*, an isolated population in Pilbara region of WA. **Voice:** sharp, regular 'chip, chip, chip' at intervals of about a second, similar to call of Rufous; also as a rapid descending trill, 'chip-chip-chipchipipipip'. **Hab.:** tropical open forests, woodlands, spinifex with scattered trees, tree-lined rivers in arid regions. **Status:** common; sedentary. **Br:** 385

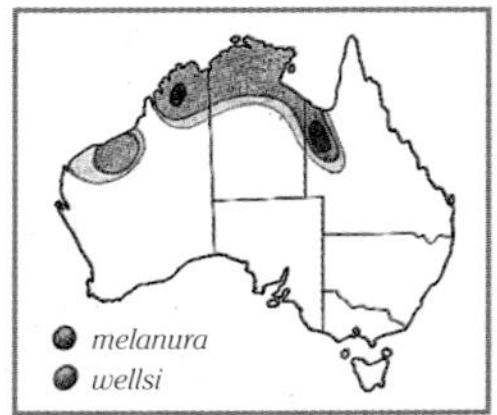

Red-browed Treecreeper *Climacteris erythrops* 14–15.5 cm
Distinctive dark treecreeper with red brow, white throat. Favours tall, smooth-barked trees of wet eucalypt forests; forages among long strips of bark. Flight swift, undulating; shows grey-buff wing-band. Gregarious, usually in small family parties that appear to contain only one female. **Similar:** White-throated Treecreeper, has larger white throat patch, to upper breast; no brow line. **Voice:** squeaks at very high pitch, weak, thin 'iep, iep, iep'. Also a fast, scratchy trill; begins extremely high and squeaky, ends with several stronger, harsher squeaks: 'iep-ip-ip-ip-i-i-i-ip–chzip-chziep-chziep'; in the distance sounds more like the trill of a fairy-wren than of a treecreeper. **Hab.:** forested ranges with big trees; rainforest margins. **Status:** uncommon; sedentary. **Br:** 385

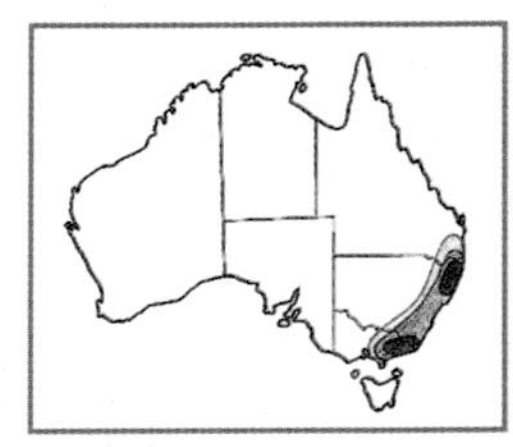

White-throated Treecreeper *Cormobates leucophaeus* 13–15 cm
Differs enough to be in a separate genus. Searches bark vigorously, spiralling rapidly up trunks and branches, swooping down to begin low on the next tree; rarely hunts on ground. Longer flights are fast, direct, undulating. In pairs or family parties only in breeding season; else usually solitary. **Variation:** subspecies include *minor*, 'Little Treecreeper' of NE Qld. The SE race *leucophaeus* may be split into several races. **Voice:** loud, ringing, clear, musical 'whit-whit-whit-whit-', rippling 'quit-quit-quit-' and more mellow, musical 'twiet-twiet-twiet-'. Also a loud, rippling trill. **Hab.:** rainforest, eucalypt forest and woodland; forested river margins of drier regions. Also lower scrubs of banksia woodland, mallee and brigalow. **Status:** common; sedentary. **Br:** 385

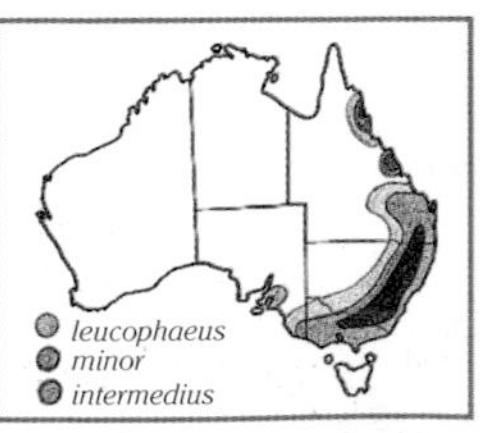

Crown grey-brown, darker than crown of Brown Treecreeper. Ear coverts have obvious black and white streaks.
Brow crisp white
Races intergrade N of Spencer Gulf, SA. Nominate race affinis has grey-brown rump, mid brown back.
Male has plain grey throat and upper breast
Lower breast and belly are white with black-edged streaks.
Tail grey-brown with black band
Wing linings pale buff
Buff band across flight feathers, conspicuous in flight from both above and beneath
Female rufous streaked. Remainder of underparts streaked like male
Western race superciliosus has rufous-brown back, rump.
Black band on all but central pair
Ear coverts softly streaked
Long pale brow
Nominate race picumnus light to mid grey-brown. Buff edge of wing bands may be visible.
Male: fine black lines
Female: fine rufous lines
Race melanota, has upper parts blackish brown
Long brow line, off-white, conspicuous
Female has rufous streaks, male black.
Dark tail band not evident
Race melanota, 'Black Treecreeper', occurs NE Qld; nominate race southwards.
Dark band on all except central pair
Most colourful of treecreepers, with much bright rufous in plumage
Bright rufous-buff band across wing
Brow, face, underparts bright rufous; male has black-edged, pale streaks on breast.
Grey from crown to mantle, merging into olive-grey to brownish grey lower on the back
In flight, dark, blackish flight feathers emphasise rufous wing band; seen from below, sunlight through translucent band gives it a rufous-orange glow.
Dark subterminal band, except on central pair
Wings grey-brown with rufous edges of primaries showing
Feet large, strong, enabling bird to hop up vertical surfaces.
Pale streaks
Barred coverts
Largest treecreeper; darkness of upper parts rivalled only by the dark race, melanota, of the Brown Treecreeper
More marked sexual dimorphism than other treecreepers: male has black, white-streaked throat; female has plain white throat.
Race melanota: crown to tail blackish brown
Pilbara race wellsi is overall lighter in tone, and brighter rufous.
Dark brown, or deep chestnut in direct sunlight
In flight, buffy white wing-bar
Under-tail coverts barred
Red brow, white throat and lines of white streaks make this a distinctive and attractive species.
Large rufous brow patch encircles eye
In flight, shows wing-bar of pale buff across both upper wing and underwing.
Male has plain white throat.
Female has rufous streaks on upper breast.
Back, wings, tail dark brown
Olive-grey, each feather with white streak and black border, forming long lines of streaks
Immature lacks red of brow and around eye; under-parts grey, mottled rather than streaked.
Dark subterminal band, all except central pair
In flight, prominent pale buff wing-bar, edges visible on folded wing
Female has orange spot.
'Little Treecreeper' of NE Qld
Within nominate race leucophaeus may be recognised races metastasis (N NSW into SE Qld) and race grisescens (Mt Lofty Ra., SA).
Mid to very dark grey-brown upper parts, and large area of white at throat.
Throat white, pale central underparts
Race minor, a distinctive small, very dark race
Dark band
Barred
Flanks white-streaked
Rump grey
Juvenile females, all races, have rufous rump, otherwise like adults.

Purple-crowned Fairy-wren *Malurus coronatus* 14–16 cm

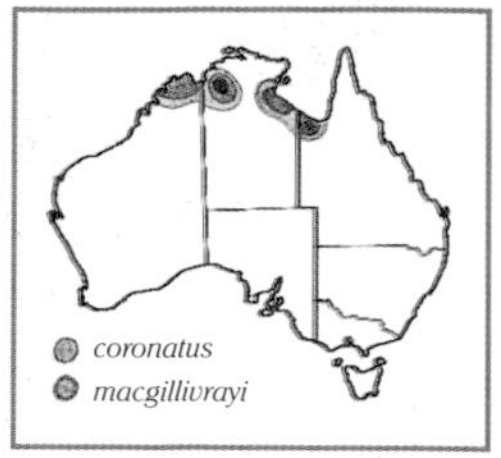

Plumage pattern of this fairy-wren is unlike any other; the identity of a male in breeding plumage is obvious. The eclipse male and the female are also quite distinctive with broad rufous or brown ear patches. Occur in pairs or small family parties; move quickly through their territory; short, low direct flights; pause to hop about ground and low shrubbery after insects. **Variation:** two subspecies: nominate race in Kimberley of WA and adjoining NW of the NT; greyer backed race *macgillivrayi* is much further E around the northern part of the NT–Qld border. **Voice:** distinctive, not the high and often weak trill of other fairy-wrens, but a strong, vigorous, reeling sequence with some almost harsh notes. Commonly a vigorous, rollicking, scrub-wren-like 'zeepa-zeipa-zeipa' or 'cheipa-cheipa-'. Many variants: 'tcheip-tcheip-tcheip-' and 'tzwip-tzwip-tzwip-'. **Hab.:** fringing vegetation of rivers and creeks, especially pandanus and paperbark clumps with dense undercover of tall cane grass, occasionally in mangroves. **Status:** moderately common in remnant suitable habitat; at many sites the essential water's edge vegetation has been trampled, often destroyed, by cattle. **Br:** 386

White-winged Fairy-wren *Malurus leucopterus* 12–13 cm

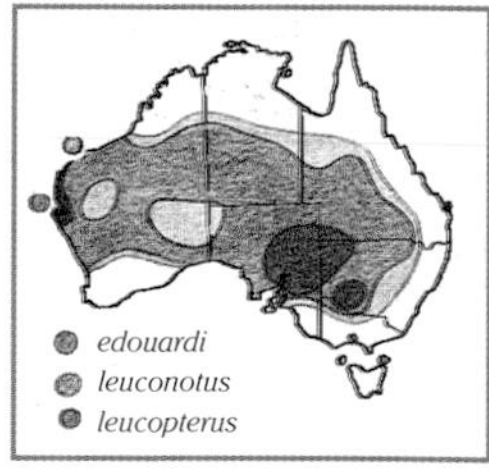

Widespread and conspicuous in inland and arid coastal regions. The black and white form was named by French zoologists Quoy and Gaimard after landing on Dirk Hartog Is. in 1818; the mainland race *leuconotus* was not described until 1865. A third race, *edouardi*, with black and white plumage occurs on Barrow Is. **Voice:** an undulating, musical, reeling trill that carries some distance in open country and which has a metallic, mechanical quality to it. Undulating pitch and volume give an impression of uneven mechanical rotation, like a rapidly retrieving fishing reel or a sewing machine at fluctuating speed: 'chirr-IRR-irr-IRIRit-chirr-IRRirit-irirIRIRRIT-chirrir-IRRITirrit-', a pleasant, rollicking sound that is sustained and repeated. Also has an abrupt, high, weak, insect-like 'prrit!'. **Similar:** male in breeding plumage is unique. Female and immature of Red-backed both have a very plain face and shorter, brownish tail. **Hab.:** nominate race found in dense, low scrub of Dirk Hartog Is.; blue and white mainland race found on open plains of low saltbush, bluebush, spinifex, marshy margins of salt lakes, coastal heaths, cane grass. **Status:** locally nomadic; common. **Br:** 386

Splendid Fairy-wren *Malurus splendens* 13–14 cm

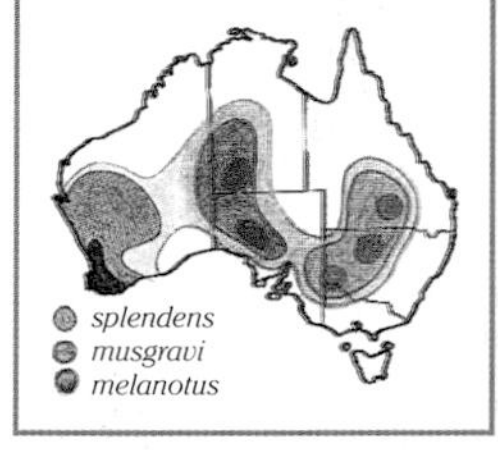

Spectacular brilliant blue, obvious, active, usually easily observed; nominate race common around Perth. **Variation:** nominate race *splendens*, southern WA; race *musgravi*, E of WA through NT and SA; race *melanotus*, SE of SA, interior of Vic., NSW, Qld. The N part of Qld population, near Longreach, has been proposed as a race, *emmotorum*; it has sky-blue crown and mantle. **Voice:** song a very rapid tremulous trill, the 'reeling' much less pronounced than in the call of the White-winged, but gives the Splendid's song an undulating quality. The notes forming the 3 or 4 seconds duration of the song vary; usually high, squeaky sounds intermingle with softer or mellow notes, or begins with very rapid, high, insect-like squeaks for a second or two, then moves to lower, stronger or whistled notes, but usually with a reeling, rhythmic pattern: 'trit-triiit-trit-tirreet-tirreet-trit-tirreet-trit-tirreet'. Other sounds include extremely high, short, vibrating trills, a loud, whistled warning trill, frequent weak 'trrrip' contact squeaks and a plaintive, soft 'sreee' seeking contact. **Similar:** breeding males unique; females and eclipse males have lighter, brighter blue in tails than in the otherwise similar females and eclipse of the Superb Fairy-wren. **Hab.:** race *splendens*, shrubby low undergrowth of wet SW forests, open forests and woodlands, semi-arid woodlands, mallees and scrubs; races *melanotus* and *musgravi* are both found in semi-arid woodlands, mallee, mulga and belar. **Status:** sedentary or locally nomadic; patchy variable occurrence, often locally quite common. **Br:** 386

Superb Fairy-wren *Malurus cyaneus* 13–14 cm

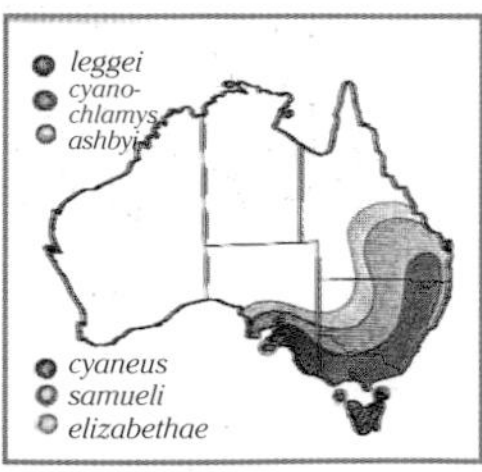

The familiar blue wren of SE Aust; the range takes in the major cities from Brisbane SE to Adelaide. Often in family parties—a dominant male, the female, and brownish young. **Male:** distinctive; upper body and breast are entirely bright, deep blues with black banding, underparts dusky white. **Variation:** nominate *cyaneus* in Tas.; *cyanochlamys* widespread on mainland. **Voice:** song a vigorous trill, beginning squeakily, but quickly strengthening into a strong, downward cascade of louder, less sharp musical notes. Foraging party maintains contact with weak, sharp 'trrit' or plaintive 'treee' from those falling behind; strident version in alarm, 'terrrrit!'. **Similar:** Splendid, but breeding males entirely blue and black, eclipse brighter blue in tails and some blue on wings; females brighter blue on tails. Variegated, but breeding male has chestnut shoulders, eclipse has white tips on blue-grey tail; female has darker chestnut lores. **Hab.:** dense undergrowth of grass, bracken, shrubbery in forests, heaths, gardens, roadsides and inland watercourses. **Status:** common. **Br:** 386

Race coronatus: cinnamon-brown back, wings
Deep lilac crown underlined by black face mask and nape, and with black centre spot
Blue of tails may have turquoise tint, perhaps more on eastern race macgillivrayi than on western nominate race.
Race coronatus occurs in Kimberley and NW of NT; race macgillivrayi NW Qld and NE of NT.
Tail deep blue; all except central pair tipped white
Ear coverts deep chestnut
Crown grey
Fine white eye-brow, eye-ring
Race macgillivrayi grey-brown on back, wings
Throat, breast off-white, flanks buff
Off-white, gradually deepening to buff on flanks
♂
Male in nonbreeding eclipse plumage is like the female, but ear coverts are grey rather than chestnut.
♀
Throat and breast white, flanks tinted buff
♂
Pale, soft blue-grey, long with tapered tip
Plain faced, no markings on eye or lores
Scapulars stark white set in cobalt blue
Similar: female Red-backed Fairy-Wren, but shorter brownish tail
♀
Flickering wings of males are like white butterflies in distance.
Plumage of female lacks distinguishing markings; light sandy brown upper parts merge softly into off-white underparts.
Nominate race leucopterus, Dirk Hartog and Barrow Is.: glossy black in place of the blue of the mainland race.
♂
Primaries brown; white spreads further out on secondaries of older males.
Race leuconotus, widespread across mainland; under some lighting conditions may appear near or partly black.
Tail medium length, intense cobalt blue
Race melanotus, 'Black-backed Wren'
Overall pale to deep cobalt, less violet
Tail bright blue
Black lower back
Upper parts blue-violet
Ear coverts sky blue
Light bright blue
Lores dark
No black across lower back
Thin band
Overall purplish sheen in sunlight, cobalt-turquoise under colder light of shade, or flash
♂
Nonbreeding or eclipse males are like females, but some blue on wings.
Violet gloss over deep cobalt
♂
Blues silvery, except breast deep blue-violet
Deep blue-violet, like breast
Broad black band
Race musgravi, 'Turquoise Wren'
Black
Lores pale rufous
♂
♀
Race splendens, 'Splendid Fairy-wren'
Wide breast-band; black scapulars often hidden under mantle
Wings brownish
♂
Dark dull blue
Pale blue
Mid grey-brown
Lores and eye-ring rufous
Wings grey-brown
Grey-brown
Deep blue, almost black in shadow
Off-white
Pale grey, almost white in direct sun
Eclipse male is like female, but tail is dull blue, bill black, lacks rufous lores.
Light grey-brown
♂
♂
♀
In flight away, mantle and crown are conspicuous bright, light blue.
Nominate race cyaneus in Tas.; larger and darker. Up to 5 other races on islands of SA, Bass Str. and the Aust. mainland
Tail long, dark dull blue

Variegated Fairy-wren *Malurus lamberti* 13–14 cm

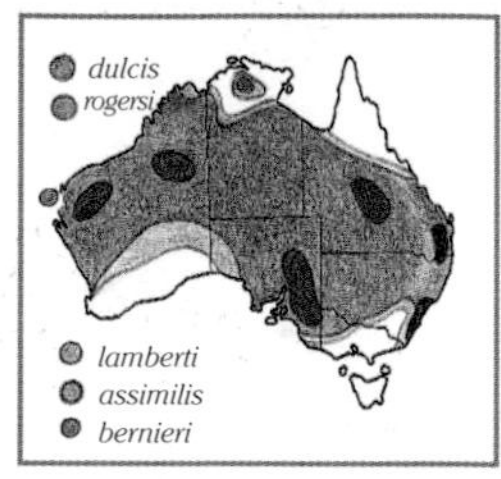

By far the most widespread chestnut-shouldered fairy-wren; has been split into four species, which are now usually described as races. Across most of the inland and drier regions (most of the continent), one or another of the four races is the sole chestnut-shouldered fairy-wren. Usually in pairs or small family parties of up to five birds. Tends to be shy, inclined to stay in undergrowth in presence of intruders. **Variation:** five subspecies. Nominate race *lamberti*, 'Variegated Fairy-Wren', mid-coastal E Aust. Race *assimilis*, 'Purple-backed Fairy-wren', across most of interior and semi-arid Aust. Race *rogersi*, 'Lavender-flanked Fairy-wren', Kimberley and NE corner of NT. Race *dulcis*, also known as 'Lavender-flanked', Arnhem Land, NT. Race *bernieri*, confined to Bernier Is., W coast WA. **Voice:** sings only in the breeding season, a high, metallic, squeaky and rather clockwork or mechanical rattling trill at fairly uniform pitch, 'tririt-tirirrit-tirit-trit-tirrririt-trit-tirrit'. The most common call, for much of the year the predominant sound, is a hard, staccato, erratic and abrupt 'trrt-trtt, rrrrt-trrt-trt', strong rather than high or squeaky, and apparently in warning, attracting others of the party to take up the call, and to dive into cover if there is sufficient urgency in the sound. Varied contact calls are given while foraging; these change from race to race—a rapid, sharp 'trrit' from east-coastal nominate race, and a thinner, squeakier, plaintive 'sreee' from inland and northern races. **Similar:** Lovely Fairy-wren, but has large conspicuous white tips to tail, broad pale ear coverts; Red-winged, but extremely dark blue rather than black breast, silvery blue crown, distinctive song, wetter habitat; Blue-breasted, but deep royal blue breast that is obviously blue rather than black in good light, violet-blue crown. **Hab.:** in such a wide-ranging species the habitat is extremely varied. Along the east coast the nominate race *lamberti* inhabits forest undergrowth, coastal heaths and dune vegetation, and roadside and garden thickets of native and introduced shrubbery. The inland 'Purple-backed' race *assimilis* occurs in drier country with sheltering shrubbery—mulga and mallee country, scattered through spinifex and along watercourses beneath larger trees. The 'Lavender-flanked' races *rogersi* and *dulcis* inhabit rugged country where spinifex and low shrubbery grows among the boulders of ranges and gorges, and along the pandanus and melaleuca thickets of watercourses. **Status:** usually sedentary; moderately common. **Br:** 386

Lovely Fairy-wren *Malurus amabilis* 13–14 cm

Distinctive, upright tail often briefly fanned and flicked, conspicuous white tips may serve as a visual signal; together with contact calls, keeps parties together in gloomy habitats. Forages more in shrubbery and lower tree foliage than on ground. **Nonbr.:** older dominant males appear to remain in full colour throughout the year without any intervening female-like eclipse plumage. Younger males may moult to eclipse plumage for several months or to scruffy intermediate plumage. **Variation:** in Atherton–Cairns region blues are bright; slightly deeper on Cape York where females also are darker. **Voice:** sings at any time of year, launching directly into a slightly descending, rippling trill. Contact call a high yet rather soft 'streee' given continually to keep family party together; churring in alarm. **Similar:** other chestnut-shouldered species. **Hab.:** thickets of rainforest margins, monsoon and vine scrubs; forages short distances into closed forest and adjoining open woodland; also swamp woodlands, mangroves. **Status:** sedentary; common. **Br:** 386

Blue-breasted Fairy-wren *Malurus pulcherrimus* 14–15 cm

Now confined to SW of WA and Eyre Peninsula in SA. Although separated by the arid country of the Nullarbor, there appears to be no difference between birds of the two populations, which suggests that the break is comparatively recent. Rather shy and secretive, it forages under cover, usually on ground or very low in shrubbery. **Voice:** song begins with extremely high, squeaky, thin notes leading in to several seconds of lower pitched, very rapid, hard, mechanical, vibrating trilling. More often heard are rattling, mechanical alarms, 'trrrt-trt-trtt-trt-trrrt'. **Habitat:** dry and semi-arid woodlands, mallee, wattle with shrubby understorey, open sand-plain heaths. **Status:** sparse; locally common in remnant habitat. **Br:** 386

Red-winged Fairy-wren *Malurus elegans* 14–16 cm

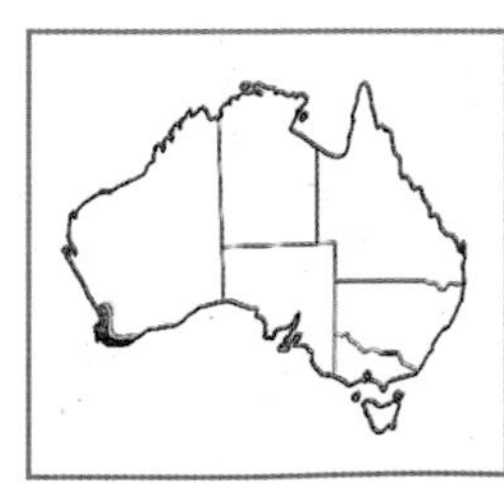

An inhabitant of the wetter SW corner of WA. It can usually been separated from the Blue-breasted and Variegated by its location, habitat and song. **Voice:** song begins with several squeaky notes followed by a rather rattling, uniform trill; quite unlike song of Splendid, the usual overlapping species, and separable from call of Blue-breasted, which is usually in drier inland habitats. Alarm call a strident, rapid 'trrit-trrit-trrit'. **Similar:** Blue-breasted, but deeper crown colour, lighter breast, further inland. **Hab.:** undergrowth of tall wet SW karri forests, dense coastal heaths, further north in drier jarrah–wandoo forests, the lush vegetation along creeks, around swamps. **Status:** sedentary; common, especially in SW of range. **Br:** 386

Silvery azure crown merges into similar colour ear coverts.
Bright chestnut
Nominate race *lamberti*, Variegated Fairy-wren
♂
Black, not dark blue, in direct sunlight
Tail grey-brown with faint blue tint
Grey-brown
Lores and eye-ring deep chestnut
Off-white
♀
Females of races *lamberti* and *assimilis* alike
Mantle and back purplish cobalt
Crown deep violet-cobalt; ear coverts lighter, brighter blue
♂
Race *assimilis*, 'Purple-backed Fairy-wren'
Tail longer, fine white tips
Race *rogersi*, wide white tips to tail
Crown deep purple-blue, ear coverts lighter, brighter, blue
Deep-chestnut scapulars
Breast black
Males of race *dulcis* have narrow white tips to tails, otherwise like race *rogersi*.
♂
Lavender wash along flanks
Blue-grey
Lores and eye-ring light chestnut
♀ Female, race *rogersi*
Lores and eye-ring white
Dark blue-grey
Eclipse males, all races, like females but retain black bill, have dark lores.
Tail blue with fine white tips
Female, race *dulcis*, 'Lavender-flanked Fairy-wren'
♀
Some males briefly, often only partially, go to eclipse plumage similar to year-round female pattern.
Large white tips
Silvery azure, but slightly deeper azure blue on Cape York population
Broad ear patch
Black
♂
White outer edge
Dusky grey-blue upper parts, becoming darker in N of range
White lores and eye-ring
Ear coverts pale blue
Grey-brown
♀
Crown and back light cerulean blue in southern part of range
♂
Tails in this species relatively slightly shorter, rounded
Conspicuous large white tips
Tail dusky blue-grey with narrow white tips
Bright chestnut
Crown and ear coverts both a similar mid violet-blue
Deep royal blue in sunlight; can appear black in shade
♂
Blue-breasted is separated from western race of Variegated by ear coverts the same deeper blue-violet as the crown.
Dusky blue-grey
Back and wings grey-brown; slight blue tint to grey of head
Greyish white with very faint buff tint
Chestnut lores, eye-ring
♀
Tail dusky blue; white tips are very narrow and often worn away.
Shoulders rusty chestnut
Silvery, very light blue on crown, ear coverts, back
Very dark navy blue in sunlight, darker than on Blue-breasted; appears black in shade.
♂
A long tail helps make this one of the largest species of fairy-wren.
Rufous tint
Lores, eye-ring deep chestnut
Black bill unusual among females of chestnut-shouldered species
♀
The Red-winged has pale, silvery blue crown and ear coverts, much lighter than those of the Blue-breasted, while the breast is even darker blue. Breast of Variegated is black in direct sunlight.

Red-backed Fairy-Wren *Malurus melanocephalus* 10–13cm

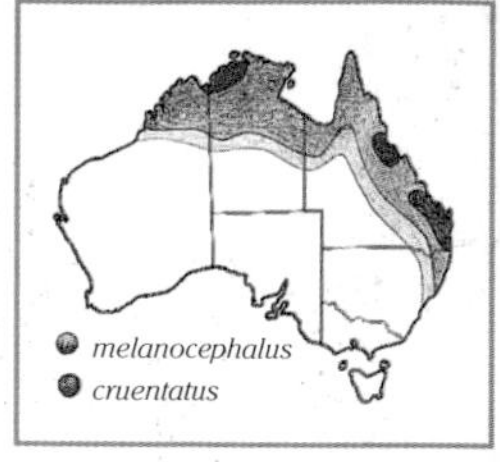

Males of this species display spectacularly. The expanded scarlet plumage creates a rounded ball of fiery colour as they make long, rising flights over the vegetation, the effect emphasised by the red set against black of wings and head against the cool background colours of sky and foliage. Between flights they bounce excitedly from perch to perch through shrubbery; each male uses the impact of his vivid colour to intimidate his rival. May be seen in pairs, small family groups or larger parties in which fully coloured males are greatly outnumbered by brownish birds, including females, immatures and males in eclipse plumage. In autumn males moult to eclipse plumage similar to that of females; this is retained for five or six months. But older, dominant males moult direct to their new colourful breeding plumage, completely missing the dull eclipse stage. **Fem.:** in common with males, females have no blue in plumage. Even the tail is without a trace of the blue that lightly tints the tails of most other fairy-wren females. **Voice:** begins with squeaks so high and weak that they would be heard only when very close, followed by a trill beginning with high squeaky notes, usually switching to lower, louder, rattling sounds: 'sreee-sreee-trichee-trichee-tchakka-tchakka-tchakka'. This is usually given by males in breeding season. Contact call within foraging group is a soft 'tsiet'; the alarm call a louder, sharper 'trrrrit!'. **Hab.:** most abundant in tall rank grasses with scattered trees and thickets of wattle or other shrubbery, coastal and inland along watercourses, but also in heaths, rainforest margins, swamp woodlands, spinifex. **Status:** common; sedentary. **Br:** 386

Southern Emu-wren *Stipiturus malachurus* 15–19 cm; includes tail 11–12 cm

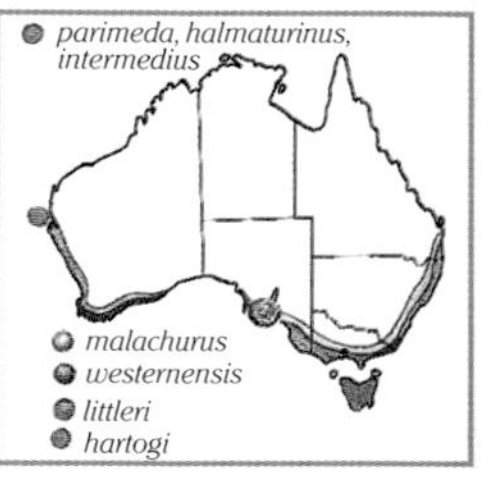

Tiny with a long tail of filamentous feathers that superficially resemble those of the Emu. Runs mouse-like through dense low vegetation; bounces across open spaces, climbing briefly to shrub-top vantage points. Usually stays under cover; occasional flights are insect-like, seemingly slowed by length of tail streaming behind. **Voice:** similar to fairy-wrens but higher, weak, scratchy, yet it carries well over open heath, giving the best chance to locate birds; may be beyond upper limit of hearing of some observers. The squeaky calls have an undulating, rollicking variation of volume and pitch, 'siee-seit-tzeet-tzeet-siee-seeit-tzeeit-tzeet-seee-seeit'. Contact call is a faint squeak, 'tsiee'; alarm a sharp 'trrrit!'. **Variation:** many races, slight variations: *malachurus* occurs S Vic to SE Qld; *littleri* in Tas.; *westernensis* in SW of WA, *hartogi* on Dirk Hartog Is.; *parimeda* on S Eyre Peninsula; *halmaturinus* on Kangaroo Is.; race *intermedius* of Mt Lofty Ra., SA. Subspecies list not finally resolved; may also include a population SE of SA–SW Vic. as a new race, *polionotum*. **Hab.:** diverse, regional differences. All in dense, low cover, damp heaths, sedges, sand-dune and sandplain heaths, spinifex shrublands, buttongrass plains. **Status:** common in extensive habitat; generally secure, but some populations endangered. **Br:** 386

Mallee Emu-wren *Stipiturus mallee* 13–15 cm; includes tail 8–9 cm

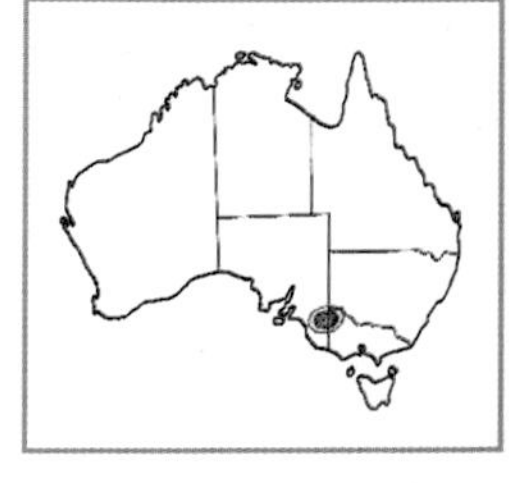

A tiny, shy emu-wren of very restricted range. When first disturbed, may momentarily climb above the dense, spiny clumps of spinifex. It is difficult to locate again; elusive, darts through spinifex; flutters low across open ground to keep behind cover; vanishes silently. **Voice:** like Southern Emu-wren, but weaker. The song is a feeble trill of squeaky, weak, insect-like sounds lasting 2 or 3 seconds, usually from males in breeding season. More frequently given is the high, metallic squeak of the contact call, a thin, drawn out 'sreeep'. Although weak, this sound carries some distance, and, together with the stronger alarm call, gives the best chance of locating these birds that so persistently stay out of sight. The alarm is a sharper, harsh, abrupt 'tirrit'. **Hab.:** low sand dunes where old spinifex forms large clumps under low mallee and native cypress; also tall heaths with tea-tree and broom-bush. **Status:** rare; in localised colonies vulnerable to fire. **Br:** 386

Rufous-crowned Emu-wren *Stipiturus ruficeps* 12–13 cm; includes tail 6.5–7.5 cm

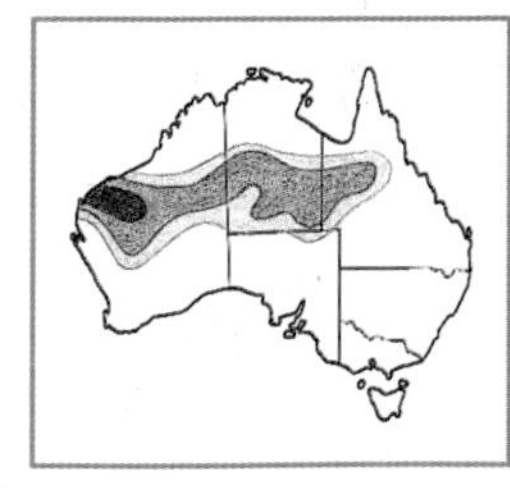

The smallest of the emu-wrens; male very colourful with deep violet-blue set against bright rufous. But for the long tail, would be almost the smallest of Australian birds. Remote, harsh habitat and its secretive behaviour make it difficult to find. Often uses a small shrub as a vantage point, briefly in clear sight before again dropping into the concealing spinifex. In early morning activity often makes short flights; flutters low over the spinifex, long tail trailing. Tends to remain silent and concealed later in the day. **Voice:** similar to other emu-wrens, but extremely high, weak. Contact call a fine, penetrating, cricket-like, squeaked 'tseeet-tseeet-tzeeet-'. Song, usually given by males in breeding season, is a fast succession of scratchy, squeaky notes, seeming jumbled rather than trilled, at varied strengthes and pitches. Like that of Southern Emu-wren but weaker, higher; most calls high to extremely high, often beyond observers' hearing unless very close. **Hab.:** spinifex on dunes, ranges or plains, typically with a scattering of small shrubs. **Status:** probably quite common across the vast spinifex expanses of NW and interior—calls are often heard when birds would otherwise be undetected. **Br:** 386

Tail square tipped, relatively short, making this species the smallest fairy-wren in overall length
Nominate race: orange-scarlet
♂
Tail pale brown like rest of upper-parts, and relatively short. Females of otherwise similar White-winged Fairy-wren have longer, blue-grey tails.
Face plain, lacks identifying markings.
♀
In display, the male of this species fluffs out his crimson plumage in spectacular contrast against surrounding black, becoming a ball of fire when in flight or bouncing between twigs and branches.
♂
♂
Race *cruentatus*, back deeper crimson, tail even shorter
Crimson race occurs across tropical northern Australia.
♂
Patchy, scruffy male in transition to eclipse plumage
Entirely, uniquely, black and red—no trace of blue in plumage
Six long, finely filamented plumes, the few fine barbs lacking the interlocking hooks typical of most feathers
Upper parts buff with soft-edged streakings of black.
Lores buff, finely streaked
Underparts tawny buff with paler central to lower belly
♀
Streaked
Lavender-blue 'bib'
Tail often held upwards when overhead clearance allows
♂
Underparts deep tawny buff-white on lower belly where they are often not visible.
Nominate race has plain rufous-buff forehead
Back rufous-buff, streaked black
In flight is like a large insect, slowed by long trailing tail.
♂
Males of SW race have greyer back; dark streaks extend to forehead, and brow has slightly more blue.
Back and wings olive-grey with feathers dark streaked and edged buff
Crown plain rufous-buff without streakings
Mid sky-blue, including ear coverts
Underparts pale buff
Smaller with shorter, even finer tail than the widespread Southern Emu-wren
♂
Legs relatively short
Mid olive-grey, streaked darker
Olive-rufous, finely streaked darker
Face and ear coverts finely streaked white and dark blue-grey
♀
Rufous-ochre streaked brown
Crown deep rusty-rufous, quite bright in sunlight, lacks dark streaking
Deep sky-blue face, throat and ear coverts, which are finely streaked black and white
Tawny buff
♂
Tails often trail behind to allow passage through dense vegetation.
Tail shorter, 1.2 to1.5 times body length
♀
Grey-brown edged rufous-ochre
Tail feathers have some of the fine interlocking barbs typical of most feathers, but absent in other emu-wrens.
♂
Occasional brief flights; keeps very low; wings short and rounded.
The Rufous-crowned Emu-wren is the lightest bodied Australian bird, weighing 5 g or slightly less; its long tail deprives it of any chance of being the smallest in total length.
The tail feathers, while very slender, are slightly more dense, less lacy, filamentous, than those of other emu-wrens, yet are far less substantial than the tails of the fairy-wrens.

Black Grasswren *Amytornis housei* 19–21 cm

A rare Australian bird confined to extremely rugged, remote parts of the NW Kimberley. First discovered in 1901, it was not seen again until 1968, and not until 1998 was a nest with eggs discovered. In pairs or parties of three or four in the breeding season; at other times, larger groups of six to nine birds forage in and around the large clumps of spinifex that cover the ground among the huge tumbled boulders of ranges and gorges. Not secretive, it often flies onto vantage points on rocks or cliffs when moving through its foraging territory. Shelters in the shade of ledges or crevices in the hottest part of the day. **Voice:** calls are distinctive and quite loud, making any search for these birds easier. The song is harsher, stronger than that of any of the much smaller fairy-wrens of the region. Gives a rather metallic, slow, rattling sequence, into which are inserted low, husky and high, scratchy, squeaky sounds and contrasting brief, clear, mellow, musical, rising whistles: 'chaka-chuk-seeip-chuk-zwiEEEP, chak-chukchuk-chewiEEEP! chuk-chuk-seeip'. **Hab.:** rough ranges, gorges, huge sandstone boulders, cliffs with spinifex, scattered trees, shrubs. **Status:** restricted range, sedentary; in places quite common **Br:** 387

White-throated Grasswren *Amytornis woodwardi* 20–22 cm

Largest grasswren; usually runs rather than flies; scurries among boulders, through spinifex; occasionally bounds to a high point to call. Although the voice is loud, which greatly helps to find it, the rugged terrain limits the range of the call. **Voice:** foraging flocks keep together with sharp, scratchy 'chwiep' contact calls; break into harsher 'tzzrrrt!' alarm calls at the slightest sign of danger. The song, perhaps given only in the breeding season, is loud, rich, and varied, rather like that of the Reed Warbler, an intermix of sharp and deep rich, musical notes, 'chziep-chiep-quorrop, chirrip-chok, chwiep-quarriep-'. **Hab.:** rough sandstone ranges and escarpment valleys with spinifex, sparsely scattered shrubs and trees. **Status:** sedentary or locally nomadic; restricted range, patchy occurrence, but quite common in places. **Br:** 387

Carpentarian Grasswren *Amytornis dorotheae* 16–17 cm

Occupies terrain consisting of rocky low cliffs, ledges and boulders with sparse vegetation only. Here these grasswrens can be extremely elusive, bouncing away to disappear behind rocky bluffs or into shadowy crevices. Often take flight on the first encounter; flutter away in slow, emu-wren-like flight, long tail trailing, to drop back to cover and vanish some 20 to 40 m away. **Voice:** contact within groups is maintained with a high, cricket-like 'tseeit'. In warning a harsher, buzzing 'tzzeit, tzzeit'. The song is a rapid, cheerful, canary-like warbling, a musical intermingling of squeaky, high trills and deeper, stronger, prolonged churring sounds: 'chrrip-chewip-chur-r-r-r-r-r-p'. **Hab.:** sandstone ranges and ridges broken into ledges and boulders sparsely clad with spinifex and sparse, stunted trees. **Status:** sparsely scattered colonies. **Br:** 387

Grey Grasswren *Amytornis barbatus* 17.5–20 cm

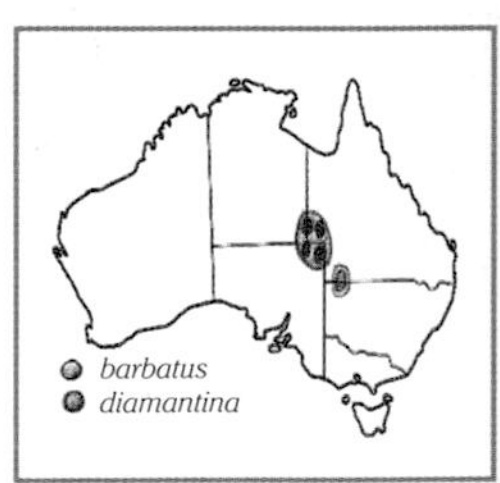

The striped plumage pattern makes these birds hard to see among tangled stems of swamp lignum and cane-grass thickets. The best times are early morning and towards dusk when they venture out onto open mudflats or dunes to feed. Live in parties of up to 15 birds, or in pairs in breeding season. Often flies to cross floodwaters or mudflats, long tail streaming behind. **Voice:** contact between birds is maintained by very high-pitched, quick 'tsip' calls, given by one, answered by others in the group—an almost unbroken, twittering rippling of metallic, squeaky 'tsit-tsit-tsit, tsit-tzit, tsit' calls that continues for several seconds. The alarm is a high, squeaked 'tseeep'. **Hab.:** periodically flooded river overflow swamps; dense clumps of lignum and swamp cane grass, sedges, saltbush. **Status:** very restricted; but may be locally common. **Br:** 387

Eyrean Grasswren *Amytornis goyderi* 15–16 cm

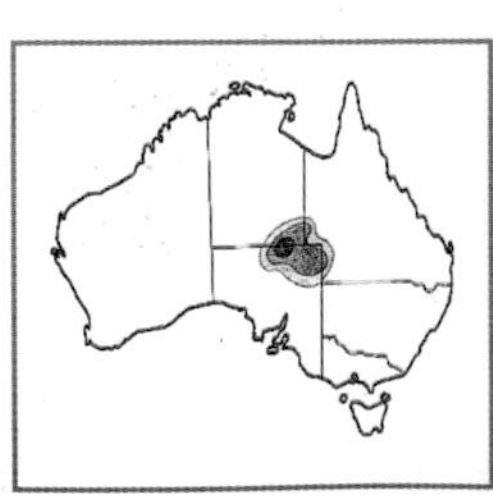

Confined to cane grass on dunes; bounces rapidly across the sand, at great speed when disturbed, wings perhaps half-spread, but rarely flying, then only fluttering low and for a short distance before dropping down into concealing cane grass. In pairs or small parties up to some dozen birds. **Voice:** contact calls are extremely high-pitched, weak, insect-like squeaks, probably inaudible unless close by: 'tseee' and tinkling, squeaked 'tsip-tsip-tseee-tsip'. Song begins with these same high squeaks, changing abruptly to a fast, louder, lower, rattling, musical chatter, 'tsee-tseee-chakachakachaka'; alarm a sharp, strong 'tzeeet'. **Similar:** Striated Grasswren, darker beneath, narrow bill. **Hab.:** dunes with metre-high clumps of cane grass and, less often, spinifex at the base of cane-grass dunes. **Status:** perhaps quite common through remote habitats. **Br:** 387

Darkest of grass-wrens, yet colourful in full sunlight; confined to NW Kimberley
Rich rufous-chestnut
Head and breast black, heavily streaked white
Belly to under tail unstreaked black
♂
Tail and underbody black, but the whole bird can look near black when sheltering under shadowed rock ledges and crevices.
Black, with finer streaking
Chestnut with fine pale streaks
♀
Light rufous-chestnut
Back and wing coverts are intense rufous-chestnut.
Head of male is very heavily streaked white; female has finer white streaks.
♂
Inner flight feathers and tertials edged chestnut
Wings short, broadly rounded, blackish brown
Tail long, rounded, blackish with barely discernible, fine barring
Largest grasswren. Similar to Black Grasswren, but with obvious white throat patch; the two species are widely separated.
Chestnut brown; appears very dark unless in direct sunlight.
White throat, upper breast
Black breast-band is heavily streaked white.
♀
Female is rich chestnut beneath.
Dark blackish brown; may trail low or be held steeply upwards.
Black, streaked white except on lower margin, which has a solid black malar 'whisker' streak
Deep chestnut brown
Male paler tawny buff beneath
♂
Moderately long, tapered
The Carpentarian Grasswren is smaller and paler than the White-throated. Like that species and the Black, it is isolated from all other grasswrens in its rough, northern sandstone ranges.
Rufous-chestnut with fine white, dark-edged streaks
Crown black, streaked; brow rufous
White extends to belly; no dark breast-band.
Male tawny buff on flanks, belly
♂
Upper parts, forehead to tail, similar for both sexes
Crown black, finely streaked; brow rufous
Bold black streak from bill back to edge of breast, gives a strongly defined edge; both sexes.
Female is chestnut on flanks, belly.
♀
The unique facial pattern of the Grey Grasswren simplifies identification; male and female differ only in minor details.
Upper parts cinnamon-fawn to rufous with fine white streaks; these between fine or heavy dark streaks
Face white with clear black markings in a pattern quite unlike that of any other grasswren.
Race barbatus more cinnamon fawn, nape to tail and on wing coverts, with heavier dark streaks; race diamantina is larger, cinnamon-rufous, with lighter, finer dark streaks, both on upper parts and breast.
Proportionately the longest-tailed grasswren; feathers brownish, pale edged
♂ ♀
Female is like male, only very slightly duller, less distinct markings on breast. Flanks are pale buff like male, unlike some other desert grasswrens in which female has deeper rufous flanks.
Cinnamon
A reddish backed inhabitant of cane grass on sand dunes of Simpson and Strzelecki deserts.
Thick
Face finely streaked black and white
White
The bill is distinctive: a deep, stubby, finch-bill shape, perhaps for crushing the large seeds of cane grass.
Underparts, chin to belly, are white except for pale buff flanks on male.
♂
Presence of this grasswren may sometimes be detected by paired footprints in the soft sand of the dunes around clumps of cane grass. These are slightly larger than would be left by the local White-winged Fairy-wrens, and typically have one foot of the pair slightly ahead of the other.
Brown, edged buff
Underparts white, throat to belly; flanks deep rufous, much richer than the pale buff of the males
♀

Striated Grasswren *Amytornis striatus* 15–18.5 cm

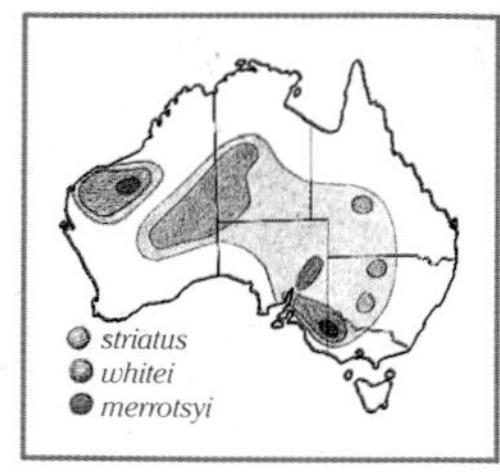

Like other grasswrens, skilled at hiding within low dense cover; contrive always to keep a clump of spinifex between themselves and any observer. Should they choose to keep low and silent, the chance of a sighting would be slight, but, in the spring breeding season, perhaps at other times, they will occasionally climb into taller, more open shrubs to call loudly and give bursts of strong, rippling song before dropping down again to vanish in the spinifex. In a sudden encounter they are likely to bounce away in different directions; in their short flights, low above the spinifex, with fluttering wings and long tails streaming behind, they are rather like large, dark emu-wrens. They are most active very early, bouncing onto exposed rocks and up into more open shrubs; later, through the heat of the day, they are likely to be silent and hidden. Even when feeding young their movements back and forth with food are all but invisible as they scurry through and behind the spinifex clumps. At other times of the year when not confined to the environs of a nest, and less inclined to climb above the spinifex to sing, locating these birds is far more difficult. **Variation:** most widespread of all grasswrens, broken range through much of the interior. Probably three subspecies. Nominate race *striatus* has widest range, from reddish birds of sandy deserts of interior of WA through to greyer birds of the mallee of NW Vic. The most colourful, *whitei*, occurs in the Pilbara region, NW of WA; its bright rusty rufous and deep buff tones match the red rock and spinifex environs. A previous subspecies, *merrotsyi*, differs sufficiently to be recognised as a full species. **Voice:** the contact calls are high, scratchy squeaks and trills, 'tsee, tsee, tzee, chweep-chweep-chweep'; in alarm a harsher, louder 'tzirrr!'. The song begins with similar high squeaks, quickly followed by rapid, loud, clear, rippling, musical notes, 'tsee-tsee, piew-piew-piew'. **Similar:** Eyrean Grasswren, but has deep finch-like bill and white from breast to belly; Dusky and Thick-billed Grasswrens, but lack black moustache line, dull colours. **Hab.:** the nominate race occurs in spinifex, preferring big old clumps on sand dunes; in E part of range large spinifex clumps under mallee. Races *whitei* and *merrotsyi* are in spinifex on rocky ranges. **Status:** widespread, patchy, in places quite common, but difficult to find. **Br:** 387

Short-tailed Grasswren *Amytornis merrotsyi* Previously included within Striated Grasswren; this form occurs in Flinders Ra. and Gawler Ra., SA; is now considered sufficiently distinctive to be given full species status. **Br:** 387

Thick-billed Grasswren *Amytornis textilis* 15–20 cm

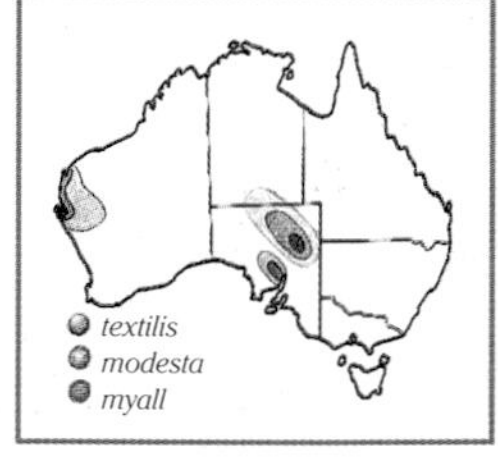

Extremely shy and elusive; hides under shrubs; darts across ground if disturbed. Rarely flies, then low, fluttering. Feeds in open, very early morning, often in pairs or family groups. Races differ significantly; once a separate species, western birds are larger, long tailed—male's tail is longer than female's—and the darker plumage is heavily streaked. The central eastern race is short tailed, male and female equally so, with paler, greyer, lightly streaked plumage. **Variation:** nominate race *textilis*, 'Western Grasswren', Dirk Hartog Is., Peron Peninsula and adjoining parts of central W coast WA (formerly through southern WA into SE of SA); race *modesta*, 'Thick-billed', southern NT to NW of NSW; race *myall*, Gawler Ra., SA. **Voice:** western race has quite a strong call, usually a persistent, rather treecreeper-like, whistled 'cheep-cheep-cheep' that may be extended into a brief musical, rattling trill, 'chip-chip, chew-wekaweka'; eastern race extremely high, soft 'see-see-see' and brief, squeaky trill, 'see-see, tsewit-tsewit'. **Similar:** Dusky, but different habitat, range. **Hab.:** the only grasswren reliant on a shrubby habitat. In west, race *textilis* on sandplain with sparse scrub of acacia and other shrubs. Elsewhere, treeless depressions with dense saltbush, blue-bush, cottonbush; some central Aust. sandy river channels in cane grass, saltbush, spinifex. **Status:** lost from much of former range with grazing of low shrubbery; locally common in a few places. **Br:** 387

Dusky Grasswren *Amytornis purnelli* 15–18 cm

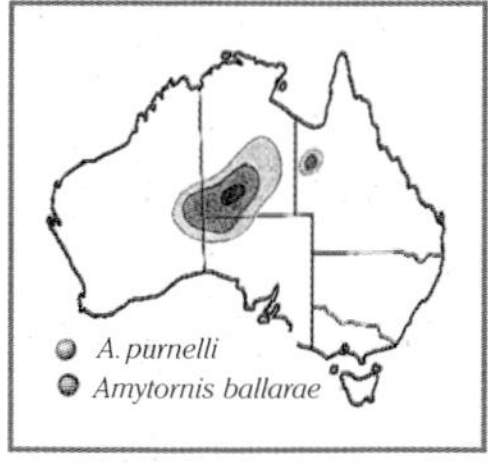

Less timid, more inquisitive and easily seen than most grasswrens. In family parties of ten or twelve birds; pairs when breeding. Usually noticed bouncing among boulders and spinifex, tail held steeply upwards, but often scuttling, tail trailing low, into spaces under and between boulders; very rarely flies. More likely to be seen around dawn or dusk; shelters through hottest hours in shadowed crevices. **Variation:** race *ballarae* is no longer included as a race of the Dusky; possibly a race in the southern Kimberley. **Voice:** high rippling trill, the first several notes faint squeaks, the rest an undulating low–high sequence, 'see-see, tchoo-tchoo-chee'; contact calls, weak twitterings, 'tsee-tsee-tsee-'; harsher, staccato 'tchik-tchik!' in alarm. **Similar:** Thick-billed Grasswren. **Hab.:** rough, boulder covered ranges, gorges; favours large, high clumps of spinifex. **Status:** one of the more common grasswrens, easily seen in accessible ranges and gorges of central Aust. **Br:** 387

Kalkadoon Grasswren *Amytornis ballarae* (Previously race *ballarae* of Dusky.) Selwyn Ra. near Mt Isa, NW Qld; widely separated from similar Dusky Grasswren. Now considered sufficiently distinctive to be a full species. **Br:** 387

Tail quite long, rounded, dusky grey-brown edged buff
Dull cinnamon-brown with white streaks edged dark
Race *striatus*, grey-brown coloration of SE Aust; male
Flanks pale grey-buff
♂
Tail dusky grey-brown, edged rufous
Deep rust-red
Edge of breast lightly streaked
Flanks, lower breast and belly golden buff; female has darker rufous patch on flank.
♂
Race *whitei*, Pilbara region, NW WA, deep rusty rufous tones; slightly larger than other races
Race *striatus*, grey-brown coloration of SE Aust.; sexes similar except flanks
Brow bright rufous
♀
Bold black moustache line separates streaked face from white throat, all races.
Wings grey-brown, buff edged
Female has rufous flanks
Nominate race *striatus* varies, rather dull grey-brown backed in the mallee country of SE Aust. (above), gradually brighter, more rufous-cinnamon towards the sandhill deserts of central WA (below).
Central Australian reddish toned form of race *striatus*
♂
Females similar, but with rufous flank patches
Back bright rust red with fine streaking
Bill long and deep
Tail short relative to similar grasswrens
Black facial line broken with white
Female has rufous flanks; males are pale buff on flanks.
♀
Short-tailed Grasswren *Amytornis merrotsyi*
N Flinders Ra.: rusty backed, long legged, shorter tailed
Drab grey-brown upper parts, heavily streaked
Race *textilis*, 'Western Grasswren'
Uniformly mid sandy brown, only slightly paler than upper parts and white streaked
♂
Eastern race *modesta* has shorter tail, dull grey-brown upper parts with dark-edged white streaks.
Bills of both races rather blunt and finch-like although shape not as deep as that of the Eyrean Grasswren
Very pale fawn, finely streaked
♂
Heavy bill
Nominate race *textilis*, mid W of WA, has long tail; that of male longer.
Race *myall* similar to *textilis*; may prove to be one form if pre-settlement contact occurred.
Female has underparts sandy fawn with deeper chestnut patch on flanks.
♀
Race *modesta*: female pale fawn with rufous flanks
♀
Dull red-brown
Russet brown, white streaked on breast
♂
Lores rufous
Tail quite long, dark brown with buff edges
Female has underparts buffy brown, streaked breast, rufous flanks.
Dusky Grasswren, *A. purnelli*, dusky grey-brown, found in rocky ranges of Central Australia
♀
Bright reddish brown
Sandy, rufous streaked throat to upper breast; has unusual stiff barb-tipped feathering.
Head and face darker brown with clear white streaks
Males, plain grey flanks and belly
♂
Bright reddish brown with dark edged streaks
Kalkadoon Grasswren *Amytornis ballarae*
Confined to rough, spinifex-clad Selwyn Ra., NW Qld. 'Kalkadoon' is derived from name of the Aboriginal group from this locality.
Female: rufous flanks, dark flecked belly
♀

Spotted Pardalote *Pardalotus punctatus* 8–10 cm

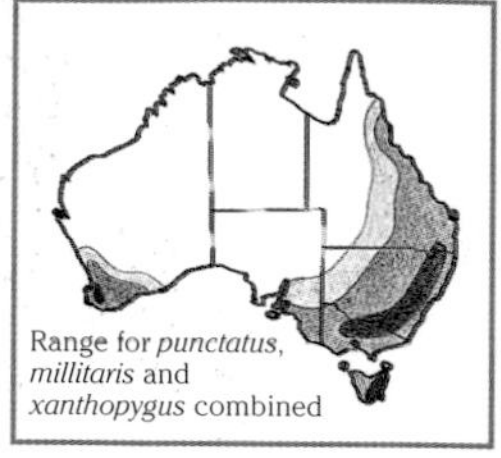
Range for *punctatus*, *millitaris* and *xanthopygus* combined

Sharply defined, clear white spots on black crown and wings give both the 'Spotted' and the 'Diamondbird' descriptive names. Bright yellow throat, under tail and reddish rump make this a most colourful bird. One of the smallest of Australian birds, it feeds in tree foliage where, leaf-sized, it is inconspicuous; but it often draws attention by persistent soft calls and occasional clicks of the bill as it takes lerps and other insects from leaves and flowers of eucalypts. The beauty of these pardalotes is usually not apparent when high in trees, but they tend to tolerate close approach—occasionally, when they feed in low foliage or descend to their ground nest tunnels, they can be watched more closely and their neat colourful pattern seen in detail. In pairs while breeding, at other times in family groups, occasionally large foraging parties. **Variation:** race *punctatus*, Spotted Pardalote, E, SE and SW Aust.; race *millitaris*, NE Qld. The 'Yellow-rumped Pardalote', race *xanthopygus*, at times considered a separate species, is now included within Spotted Pardalote as a subspecies. **Voice:** contact call is repeated regularly for long periods; an even, musical, soft, whistled 'weep-weeip'. The song is a sequence of clear whistled notes, 'whee, whee-bee' or 'sleep, may-bee'. The sound of the yellow-rumped race has a quiet, rather mournful quality: 'chnk-whee-a-bee', soft, high, then falling away. **Similar:** Forty-spotted Pardalote, but olive brownish crown and rump; only in Tas.; Red-browed, but wings brownish edged yellow without white spots. **Hab.:** eucalypt forests, woodlands, parks, gardens, watercourse vegetation; race *xanthopygus* is found in sandplain mallee and drier open woodland; the two races tend to be kept apart by these preferences. **Status:** seasonally nomadic or migratory; common. **Br:** 388

Forty-spotted Pardalote *Pardalotus quadragintus* 9–10 cm

One of Australia's rarest birds, evidently a declining population now fragmented into several colonies. Confined to SE Tas.; Maria and South Bruny Is. are strongholds, but occasionally wanders to Hobart suburbs. Pairs or small parties forage in foliage of forest trees; rather slowly and methodically searches the leaves for lerps, spiders and other small creatures. **Voice:** similar to that of Spotted Pardalote, perhaps harsher, double-noted 'wheet-whoo'; a louder, nasal 'twnnt' in the breeding season. **Similar:** Spotted Pardalote female, but has spotted crown, chestnut rump. **Hab.:** eucalypt forests, usually relatively dry forest dominated by the manna gum, *Eucalyptus viminalis*; foraging is almost exclusively confined to this species of tree. **Status:** sedentary or locally nomadic; restricted range, rare and endangered. **Br:** 388

Red-browed Pardalote *Pardalotus rubricatus* 10–12 cm

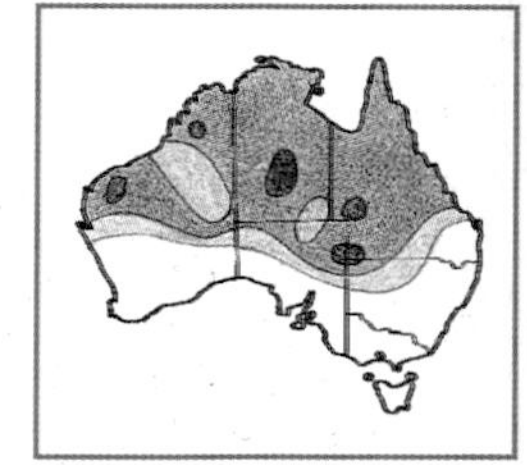

Unusual pale-eyed, large-billed pardalote. Feeds in foliage of trees and shrubs; solitary or in pairs, often high. **Variation:** two subspecies, very slight differences. Race *rubricatus* has grey-brown upper parts and pale yellow on the breast, and occupies all this species's range, except for Cape York. There its place is taken by race *yorki* with upper parts olive-brown and deep yellow breast spot. **Voice:** distinctive, pleasant song quite unlike that of other pardalotes; a loud, mellow whistle of five or six notes, one or several slow and rising, three to five higher, quicker, the result like a softly muted Crested Bellbird or rosella. **Hab.:** drier woodlands, watercourse trees, mulga. **Status:** nomadic; common in NW, N, uncommon in E of range. **Br:** 388

Striated Pardalote *Pardalotus striatus* 9.5–11.5 cm

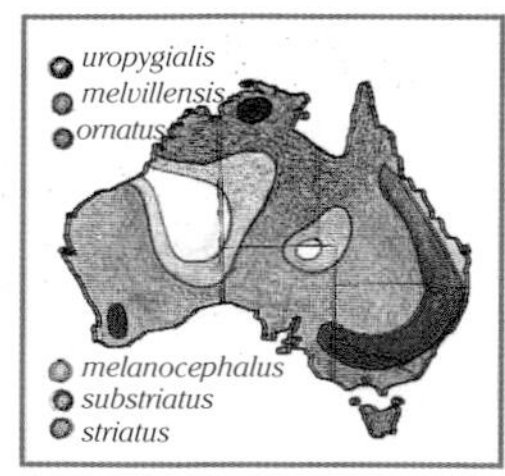

Widespread pardalote; the five races cover almost the entire continent. All have black caps and conspicuous yellow and white brow line. All have a wing spot, either yellow or red, and broad or narrow white wing stripe. Females almost identical to males, slightly less colourful; juveniles have softer, dull colours, indistinct facial markings, pale crown, lightly scalloped. May be seen in pairs, alone or, except when breeding, small parties to large flocks. Forages in foliage, taking insects, especially lerps, and other tiny creatures in the leaves, bark and flowers. **Variation:** previously four full species with well-established common names and differing appearance; now combined as races of Striated Pardalote. These include the nominate race *striatus*, previously 'Yellow-tipped Pardalote', in Tas. and migratory to SE mainland; race *substriatus*, 'Striated Pardalote', throughout interior to central and SW coast of WA; race *ornatus*, 'Eastern Striated Pardalote', mid coastal E Aust.; race *melanocephalus*, 'Black-headed Pardalote', NE NSW to NE Qld; race *uropygialis*, also known as 'Black-headed Pardalote', Kimberleys, Top End of NT and NW Qld. **Voice:** clear, sharp, musical 'witta-witta', the second part slightly lower, repeated regularly at several second intervals for long periods; also gives soft, low trills. **Hab.:** diverse, from wet eucalypt forests, rainforests, mangroves of coastal populations, to semi-desert watercourse trees and arid scrubland habitat used by those in the much more arid interior. **Status:** common; some races migratory, others locally nomadic or sedentary. **Br:** 388

Most colourful of the tiny birds whose clear white spots give this genus the alternative name 'Diamondbird'
White brow
Crown black, spotted buffy white
Dull chestnut
Deep golden yellow
Like female, but with brownish head, indistinct buffy brow and spots; overall rather dull.
Rump and upper-tail coverts bright chestnut-red
Female lacks yellow throat, but almost as bright as male on rump and beneath tail
Juv.
♂
Yellow
♀
White spotted
Yellow rump paler than that of the male
Underparts as for red-rumped race of Spotted Pardalote
♂
Spotted Pardalote now includes former species *xanthopygus*, 'Yellow-rumped Pardalote'
Yellow rump
Wings blue-black, flight feathers and most coverts tipped with large white spots
Females, both races, plain buffy white from throat to belly; pale yellow under tail
♀
Tail usually shows small white tips
Rump dull olive
Dull greenish brown
Lacks the bold brow line of other pardalotes.
Face pale olive-yellow merging into surrounding greenish brown
The Forty-spotted retains the spotted wing pattern of the common, widespread Spotted Pardalote, but, in its long isolation in Tas., has acquired a rather drab olive tone on crown and rump. This species may be the remnant population from a previous invasion into Tas. of Spotted Pardalotes; now it competes with three other pardalote species.
Under tail dull yellow
Underparts white with greyish yellowish tints
Wings black, prominently spotted white on flight and most of covert tips
Adult
As with other pardalotes, no seasonal variation; female like male. Immatures are slightly less colourful than adults.
The Red-browed Pardalote combines the crown pattern of the Spotted Pardalote with the striped wing of the Striated.
Black; white-spotted
Long wide brow line yellow with red patch at top edge just ahead of eye; brows join above bill.
Crown mottled grey-brown
Brow line pale, indistinct
Only pardalote with a pale eye; bill longer, deeper; does not have the distinctively stubby billed appearance of other pardalotes.
Wings edged white and pale yellow
Buffy white
Wings edged yellow and white
Underparts buffy white, variable amount of pale yellow on breast
Two races, many other slight variants, some listed as races in the past
Adult
Juv.
Stubby bill
Crown, face, streaked
Crown, face, streaked
Grey-brown
Grey
Race *substriatus*, 'Striated Pardalote'
Pale markings
Red spot, wide wing stripe
Yellow wing spot, thin white wing stripe
Juvenile, race *substriatus*
Nominate race *striatus*, 'Yellow-tipped Pardalote'
All races share prominent yellow and white brow; variable extent of yellow at throat. Females are slightly less colourful.
Black crown, ear coverts, nape, not streaked
Rump mustard-yellow
Streaked
Grey
Rump light brown
Red spot, wide wing stripe
Red spot, wide white wing-stripe, buff flanks
Red spot, thin stripe: race *ornatus*, 'Eastern Striated'
Rump dull ochre
Race *melanocephalus*, 'Black-Headed Pardalote'
Buffy grey flanks
Race *uropygialis*, yellow rumped form of 'Black-headed Pardalote'

Pilotbird *Pycnoptilus floccosus* 16.5–17 cm

Often forages with Superb Lyrebird, taking small creatures exposed when litter and soil are raked about; its far-carrying calls 'piloted' bushmen searching for lyrebirds. Usually keeps on or near the ground; if alarmed runs swiftly on long, strong legs. **Voice:** the male's loud call described as having 'piercing sweetness': 'whit-wheet-WHEE-a-wer' or fast 'whit-whit-WHEER, whit-whit-WHEER'; these and other variations have the clear musical quality of bristlebird calls. Females also may give these calls and softer 'whitta-whittee' and 'too-whit-too-whittee' responses. **Hab.:** wet eucalypt and temperate rainforest, alpine and coastal woodland in dense undergrowth with abundant debris. **Status:** sedentary; common. **Br:** 388

Rockwarbler *Origma solitaria* 13–14 cm

Best known for using caves, often behind waterfalls, for nest sites. Mainly terrestrial, it hops about on sandstone ledges, clings to the rock face, and takes insects from crevices and the margins of streams, only occasionally foraging on bark of tree trunks or limbs. **Voice:** a clear 'chweep-chweep' with a slightly mournful quality to the sound. Also a sharper 'chwik-chwik', and harsh, scolding 'tzzt, tzzt'. **Similar:** Pilotbird, but larger, broader, upwards flicked tail; forages on leaf litter rather than rock faces. **Hab.:** forested sandstone and limestone gullies, caves, cliffs; often with streams, waterfalls. **Status:** sedentary; locally common. **Br:** 388

Atherton Scrubwren *Sericornis keri* 13–14 cm

In behaviour and voice similar to the White-browed Scrubwren, but differs from that species in its preference for gloomy depths of rainforest that the White-browed does not usually penetrate. Tends to forage on the ground or a few metres above it, very rarely high; often alone or in pairs, or, after breeding, in small family groups of up to four birds. Rather shy, unobtrusive, almost mouse-like in movements. **Voice:** sharp, scratchy, deliberate 'chi-wiep, chi-wiep,-'; rapid 'chiep-chiep-chiep'; musical 'chippity-wiep, chippity-wiep'. **Similar:** Large-billed Scrubwren, but paler, more arboreal in habits, different calls. **Hab.:** rainforests above about 600 m altitude. **Status:** sedentary; restricted range, uncommon. **Br:** 389

Large-billed Scrubwren *Sericornis magnirostris* 12–13 cm

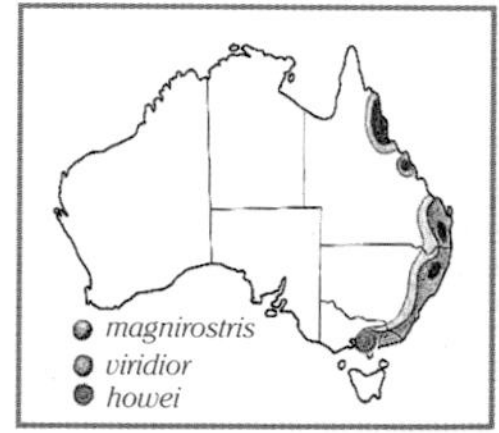

Plain scrubwren with rather long, slender bill that appears very slightly uptilted. The large-billed Scrubwren is more gregarious than the Atherton. Usually found in small flocks, actively foraging in mid levels of the rainforest—on trunks and branches, over bark and among debris suspended in vines and epiphytes, and often, briefly, hovers and flutters around high foliage. **Voice:** cheerful, sharp, loud 'chiWIP-chiWIP-chiWIP'; faster, scratchy 'chizip-chzipchzip'; piercing 'chweeip-chweeip'; scolding 'kzzip-kzzip-kzzip'; mimics other small birds. **Hab.:** tropical rainforests up to 1500 m altitude. **Status:** sedentary; quite common. **Br:** 389

Tropical Scrubwren *Sericornis beccarii* 11–12 cm

Small scrubwren, closely related to Large-billed Scrubwren. Forages on tree trunks, branches, vines and foliage of lower to mid level shrubs and trees, and occasionally feeds among ground leaf litter. Pairs or small family groups rather than flocks. **Imm.:** plainer version of adult, slight rufous tint to flanks. **Variation:** race *minimus* at tip of Cape York merges with plainer plumaged race *dubius* in vicinity of Iron Ra. This species could be combined within the Large-billed Scrubwren as races *S. m. beccarii* and *S. m. dubius*. **Voice:** similar to that of widespread White-browed Scrubwren, a very rapid, undulating 'chippity-wippit, chippity-wippit–'; quick, scolding 'squarr-squarr-squarr'. **Hab.:** tropical rainforests, monsoon forests and river-edge vine scrubs. **Status:** sedentary; quite common. **Br:** 389

Scrubtit *Acanthornis magnus* 11–12 cm

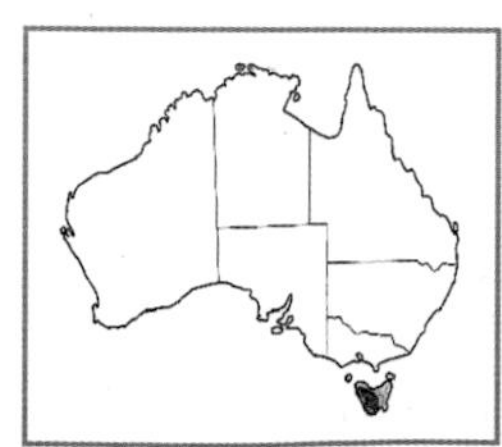

Shy, inconspicuous in the rainforest's dim light; scuttles mouse-like beneath undergrowth, searching with treecreeper-like hopping action on logs and lower trunks. Fans tail in display. **Voice:** piercing whistle, extremely rapidly repeated, sharp 'chi-chi-' followed by louder, clearly whistled 'chiew' or 'chiewit', sequence varied: 'chi-chi-chi-chi-CHIEW, chi-chi-ch-chi-CHEWIT'; 'chiewit-chiewit, chi,chi,chi'; 'chiewit, chiewit, chiewit'. Also low, scolding chatter, 'chak-chak-chak'. **Hab.:** dense, tangled undergrowth with much debris in wet temperate rainforests, fern gullies. **Status:** sedentary; common SW Tas., uncommon elsewhere. **Br:** 389

Also in this family, after pardalotes, are usually placed the three species of bristlebirds. In this guide, these are with the scrub-birds on page 216, for ease of comparison between similar birds. The bristlebirds are much larger than all others of the 'pardalote family,' and more like the scrub-birds in appearance, and preference for low dense habitat. Within the Pardalotidae, their nearest likeness is the Pilotbird, smaller, and occupying different habitat.

The Western Bristlebird and Noisy Scrub-bird so closely share the Two Peoples Bay reserve in WA that the calls of the two may be heard intermingling. The Eastern Bristlebird occurs within the same restricted range as the Rufous Scrub-bird, and shares some conservation reserves with that species.

The Rockwarbler is unusual in building a nest that is suspended from the roof of a cave or the underside of a rock ledge, often near a waterfall and frequently in semi-darkness.

Female is similar to male; immatures have slightly dulled colours.

Although the Atherton and Large-billed Scrubwrens are similar, they build very different nests, the Atherton's being a well-concealed domed nest on the ground, the Large-billed building a rather obvious, suspended nest well above ground in foliage or vines.

Sexes alike; immature has more rufous upper parts, slightly stronger yellow tint beneath.

The two similar, rather plain brown rainforest scrub-wrens of north-eastern Qld, the Large-billed and Atherton, can be identified using a combination of appearance, habits and voice. Their nests differ greatly and help confirm identity; that of the Large-billed is quite conspicuous and more likely to be seen.

The Scrubtit, inhabitant of dense wet Tasmanian forests, is sufficiently distinctive to be placed in a genus of its own, *Acanthornis*. It closely resembles both the Tasmanian Scrubwren and the Brown Thornbill, which share the same environs and have similar calls.

Fernwren *Oreoscopus gutturalis* 13–15 cm

Confined to undergrowth of highland rainforests, sufficiently different from scrubwrens to be in a separate genus. Although the bird may be hard to see, the dark olive-brown blending into dark tones of moist leaf litter as it sneaks through dense ferns and accumulated forest-floor debris, even a brief frontal glimpse should show the conspicuous white throat and brow pattern. **Voice:** a high, fine, 'pee-pee-PEEEE', the last note strongly whistled, musical; another call is softer, brief, more even: 'pip-pip-pip-peeeee'. These calls are rather like display calls of Pied Honeyeater. Also a scolding, metallic, buzzing 'pzeep-pzeep-pzzzeep'. **Hab.:** rainforests above 600 m altitude. **Status:** sedentary; moderately common. **Br:** 389

Yellow-throated Scrubwren *Sericornis citreogularis* 12–14 cm

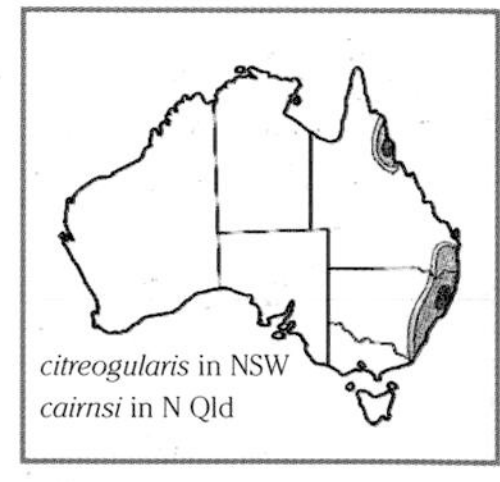

Distinctive, bold, black mask bordered with bright yellow. These birds mostly forage on the ground—solitary or in pairs, they methodically hop across litter, mossy logs, occasionally into lower shrubbery, merging surprisingly well into the background of gold, chestnut and black litter of fallen leaves and moist, decaying debris. **Variation:** probably 2, perhaps 3 races. In NE Qld, race *cairnsi* is isolated. In NSW, nominate race *citreogularis* may be split, intergrading in NE NSW with race *intermedius*, which extends into SE Qld. **Imm.:** like female; dull, fawny underparts. **Voice:** varied melodious, cheery, whistled calls, rather like those of Mistletoebird. Commonly a sequence of very sharp notes intermixed with slightly lower, more mellow notes, 'sieeip-chzweep-chip-sieep-chzeep'. An accomplished mimic of the calls of other birds. Contact are calls sharp, abrupt: 'chiep, chiep, chiep'. Scolds harshly: 'chzzak-chzzak-chzzak'. Birds of far northern population can have more chattering, less whistling in repertoire. **Similar:** White-browed Scrubwren, but throat white or buffy white. **Hab.:** gloomy gullies and undergrowth, ferns and debris beneath a dense canopy; in NE Qld, highland rainforests at 600–1000 m altitude; southern population in both mountainous and coastal forests. **Status:** sedentary; common. **Br:** 389

White-browed Scrubwren *Sericornis frontalis* 11–13 cm

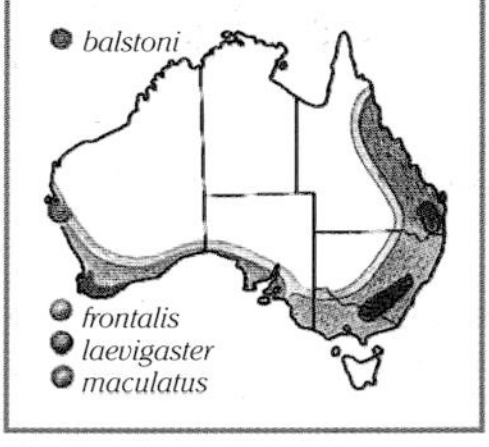

One of the more widespread and common small birds; only its rather secretive habits and preference for dense cover keep it from being more widely recognised. Across its wide range from NE Qld to SW WA, this scrubwren uses any dense, scrubby cover—the moist, low undergrowth of forests along coasts and ranges, the creekside thickets of drier woodlands, and dense low sandplain heaths. Always alert, scrubwrens usually give their harsh scolding, chattering and churring alarm calls before they are sighted, then may be glimpsed climbing about in the undergrowth. Very inquisitive—they may be called up with squeaky noises that will often elicit their responding chatter from dense thickets that look to be likely scrubwren habitat. Foraging is typically on or near the ground in territories held by a breeding pair; these dominant birds are often accompanied by several juveniles. **Variation:** 4 races, perhaps divided in up to 11 races. Race *laevigaster* from NE Qld S to N NSW (where there may be an intemediate race, *tweedi*). The nominate race *frontalis* from NE NSW S and E to Mt Lofty Ra., SA, where there may be a separate local race, *rosinae*. On Flinders Is., race *flindersi*; on King Is., *tregellasi*. Race *maculatus* extends W from Adelaide into WA: it may may be further split—those from about Adelaide W to Esperance, WA, as *mellori*; those on Kangaroo Is. as race *ashbyi*. Race *balstoni* covers a population around Shark Bay, mid west coast of WA. The Tasmanian Scrubwren may be returned to this species (as race *S. f. humilis*). **Voice:** song is a high, clear, varied whistle: 'ch-weip, ch-weip, ch-weip'; 'chi-wipip, chi-wipip'; 'chp-wiep, chp-wiep'; 'cheweep-chip'; and a very high 'ts-eeer, ts-eeer', the second syllable usually higher. In alarm, scolds with sharp, hard, metallic, rattling 'chk-chk-ckchkchk' or buzzing 'tzzzt-tzzt-tzzt'. As contact between members of the group while foraging, softer, more intimate chittering and churring sounds. Often includes mimicry of other species. **Similar:** Yellow-throated Scrubwren. **Hab.:** coasts, ranges in dense undergrowth of forests, along creekbanks, in sandplain heaths, swamp thickets, mangroves, gardens. **Status:** sedentary; generally common. **Br:** 389

Tasmanian Scrubwren *Sericornis humilis* 12–14 cm

Keeps on or near the ground, skulking beneath dense thickets; rarely flies more than a few metres; often not noticed unless it starts up its staccato alarm calls, similar to those of White-breasted Scrubwren. **Voice:** calls 'chzit-chzit', very scratchy, sharp; the song is described as like a squeaking wheel, a scratchy, squeaky 'chzeit-chzeit-chzitty-chzitty-chzitty'. The alarm call is a scolding, rasping chatter. **Hab.:** undergrowth of forests, dense thickets of creek-edges and ravines of ranges. **Status:** a larger, dull plumaged, indistinctly patterned version of the White-browed Scrubwren. Was previously race *humilis* of that species—it may, yet again, be returned to subspecies status confined to Tas. and nearby islands; common. **Br:** 389

Overall dark; appears even darker on the moist leaf litter of the gloomy rainforest floor.
Long, gently downcurved, thin white brow line
Upper parts entirely, almost uniformly, rich dark brown
The prominent white throat, sweeping back to match brow line, encloses blackish face underlined by black bib.
Mid olive-brown; in shade often appears almost as dark as the upper parts.
Adult
Dark brown
Facial markings in pale and dark browns, little different from body colour
Overall paler than adult
Lacks white and black of adult's throat and breast.
Juv.
Red eye
Long brow, white and yellow
Primaries edged yellow
Face markings paler than on male
Red eye set in black mask, bright yellow throat
Underparts rather patchy olive-buff and creamy white
♂
♀
Largest and most brightly coloured of scrubwrens; easily identified by its brightly coloured black and yellow face markings.
This species's presence is often revealed by the conspicuous nests before the birds themselves are sighted.
Builds big greenish black nest, usually hanging above creek-beds in rainforest
Yellow eye
Long all-white brow line
Upper parts rich dark brown
Black lores, grey ear coverts
Paler, less contrast in facial markings
♀ Race *frontalis*
Nominate race, *frontalis*, streaky grey throat; dull, pale, grey-buff flanks
♂
All races have white tipped greater coverts visible at the shoulder.
Immatures are rather duller versions of the adults of the various races.
Tail of most races is variably, indistinctly, darker towards the tip.
Mask black, set between clear white streaks
Underparts buffy yellow; brightest in N of range, gradually more olive, duller, southwards
♂
In SE Qld and NE NSW, the 'Buff-breasted' race has dark, grey-brown breast and olive-buff flanks.
♂
Previously a separate species ('Buff-breasted Scrubwren'), *S.f. laevigaster* is now a race, the most brightly coloured, of the White-browed Scrubwren.
All races have a tiny white line under eye.
Pale tipped
Race *maculatus*, 'Spotted Scrub-wren', dark spots, throat to breast
♂
All races have brow line low across top of yellow eye, giving an intense staring, almost scowling, expression.
Clear white brow line extends back to eye then fades away.
Grey-brown
Olive-brown
Female similar to male; immatures like adults, but dulled colours, indistinct facial pattern
Dull greyish white
Tail dull rufous-brown
Primary coverts are lightly tipped white.
Flanks dull olive-brown
The Tasmanian Scrubwren has also been known as the Brown Scrubwren, and at another time as a race of the White-browed Scrubwren.
Long legs, large strong feet
Adult

Chestnut-rumped Heathwren *Hylacola pyrrhopygia* 13–14 cm
Wren-like with uplifted tail, but has rather stout legs and feet, and earthy colours that match the environs of leaf litter and twigs. The chestnut rump catches attention as the heathwren bounces or flutters away through low scrubby vegetation; forages on or near the ground. Usually shy and secretive, except in breeding season, when it sings from higher perches in bushes; but even then keeps partly concealed. **Voice:** renowned for its spirited song into which is blended passages of mimicry of other birds in high, clear, pure, silvery notes: 'chweep-tweep-toowheet, tweet-wheeit-wheeit, chirreeep, twitchy-twitchy-quorrop-quorreep', and similar notes in seemingly endless variety. Harsh chatting and scolding, 'chzzt-chzzzt!', in alarm. **Hab.:** heaths, low dense thickets in woodlands, forests. **Status:** sedentary; uncommon. **Br:** 389

Shy Heathwren *Hylacola cauta* 12–14 cm
Similar to Chestnut-rumped, but with bolder, more contrasty pattern, more fiery rufous rump and tail coverts; probably not exceptionally shy. Pairs or small parties forage on ground among low vegetation and debris. Flight reluctant, low and brief; gives but a fleeting glimpse of the right rufous rump and white wing patches. **Variation:** many isolated populations with only slight differences: possible races may be *cauta* (SA–Vic.); *macrorhynchus* (NSW); *halmaturinus* (Kangaroo Is.); *whitlocki* (WA). **Voice:** song a clear, sharp, musical sequence, not as sustained or varied as that of the Chestnut-rumped. It begins with several high, thin squeaks, then stronger, a touch of buzzing rattle that fades away: 'see-sree, chweip-chzeip-quirrip-quip-chip-chiep'. In alarm, harsh scolding. **Similar:** Chestnut-rumped Heathwren. **Hab.:** mallee and coastal thickets with dense low cover; grass tussocks on sandplain. **Status:** sedentary; uncommon, often overlooked. **Br:** 389 .

Striated Fieldwren *Calamanthus fuliginosus* 13–14 cm
Heavily streaked, overall buffy toned fieldwren; tail usually carried steeply upwards. Usually in pairs, solitary or in small family parties; easily overlooked, flying briefly only if suddenly startled or forced. Its streaked plumage, shy, retiring nature and ability to scurry, hop and bound away behind cover make this, like the Rufous Fieldwren, visible only in brief glimpses. **Variation:** uncertain, differences slight; possibly 3 races. **Voice:** the male climbs to the top of a shrub to deliver his song, tail waving sideways as he pours forth a lively, although subdued, succession of trills and chatterings: 'whit-whit wirree, whit-whit-whitCHIER, chrrit-chrrit-chrrit'. The first part is in notes clear, spirited and musical, leading to a louder 'chier', then quickly followed by a slightly harsh, chattered 'chrrit', repeated several times. **Hab.:** swampy alpine and coastal heathlands, tussocky grasslands, low shrubby vegetation, margins of swamps. **Status:** sedentary; locally common. **Br:** 389

Rufous Fieldwren *Calamanthus campestris* 12–13 cm
Although secretive, in the breeding season males sing from bare perches with a clear view over the shrubbery, giving the chance of a distant sighting, but then quickly drops into cover. **Variation:** has many small populations scattered across mostly drier parts of southern and western Aust. Their differences have been considered clinal or to have insufficient difference between them to justify races; an alternative view is that these represent many races, perhaps 7, as well as a second species ('Western Fieldwren', *C. montanellus*, SW of WA). **Voice:** similar to that of Striated, 'chit-whit-whit whirrrrrreet', with the first short notes whistled clear and high, the last long passage a rattled chattering. **Similar:** Striated Fieldwren, but heavy streakings, range further to SE in wetter habitats. **Hab.:** inland and dry country with groundcover of saltbush, bluebush, scattered low shrubs on sandplain, gibber, saltmarsh. **Status:** sedentary; has lost much habitat. **Br:** 389

Speckled Warbler *Chthonicola sagittata* 12–13 cm
Has a boldly patterned face, heavily streaked, yellow tinted underparts, and tail that is held horizontally rather than cocked upright. Quiet and unobtrusive, yet more visible in its open habitat than fieldwrens and heathwrens; flies readily when disturbed; perches in trees easily visible; soon returns to the ground where it forages, often in company with thornbills and other small birds. **Voice:** musical song with extensive use of mimicry of other species. A rather undulating, cheery, canary-like, rapid, rambling mix of clear, sharp whistles and deeper, more mellow notes, 'chwiep-cheerip-chip-weip-weip-weip-cheerip-weip-chip', with many variations. **Hab.:** open eucalypt woodlands with rocky gullies, ridges, tussocky grass groundcover, scattered logs, sparse shrubbery. **Status:** patchy distribution; scarce to moderately common. **Br:** 389

Brow line dull white
Unstreaked grey-brown
Central feathers chestnut at base, darkening towards tip
Of the two similar species of heathwren, this, the Chestnut rumped, is slightly less colourful and has less contrast in its plumage pattern, than does the Shy Heathwren, below.
Throat and breast dull white, streaked brown
Tail brown; becomes blackish brown in outer part, and with white tips is conspicuous in flight.
In flight, the wings of the Chestnut Heathwren are comparatively plain brown, lacking the conspicuous pattern evident in the outer wings of the Shy Heathwren.
Tail coverts bright chestnut
Wings plain brown without noticeable pale tips to coverts
Rear of flanks mid grey-buff
Legs and feet quite large, sturdy
♂
Dark coverts
White patch around base of primaries
Eye yellowish brown
Long clean white brow across dark crown
Rump and tail-coverts bright rufous-chestnut
Heavily streaked dark brown on white
Flanks plain, white or pale grey
White bases of primaries form obvious white spot on folded wing
♂
Tail darkens outwards from bright chestnut base to blackish brown towards the white tips.
♂
Male has brow and lores white.
Very heavy, wide streakings across entire upper parts, except tail
Black centres to tertials
White tips to outer feathers
Blackish brown subterminal band
Female has eyebrow, lores and throat tawny-buff.
Buffy white with broad black striations
Flanks buff, heavily streaked
♀
White tips to outer tail
Wings grey-brown, edged buff
Immatures are like females, but duller markings.
♂
This species is an inhabitant of drier country to the north and west of the range of the similar Striated Grasswren, above.
Slight rufous-buff tint to forehead, crown; striations very fine
Cinnamon-rufous
Brow and lower rim of eye white
White tips and dark subterminal band, except central pair, which are only slightly darker brown towards tip
Grey brown, streaked
Creamy white with narrow black striations
Buffy white, finely striated
Upper parts change gradually from grey in some southern localities, through grey-brown to cinnamon-rufous in NW of WA. Sexes are similar.
Legs quite large, sturdy, in keeping with terrestrial habits
Often climbs to an exposed perch above shrubbery to sing, principally in breeding season.
Broad, dull white brow line extends down to include lores and lower rim of eye.
Male has black upper margin to brow.
Female has rufous upper edge to brow
Back, wings, tail grey-brown, softly streaked darker.
Holds tail horizontally; white tips of outer tail may be hidden under longer, dark-tipped inner feathers.
♂
Yellow tint
In flight, spread tail shows a wide black band above a broken band of white-tips of outer tail feathers.
Face off-white, streaked on ear coverts with buffy brown; large white patch at sides of neck towards nape
Plumage overall heavily streaked rather than speckled; these streaks are more conspicuous on white undersurface.
♀
Previously with scrubwrens in *Sericornis*, but now the sole species of *Chthonicola*.

Redthroat *Pyrrholaemus brunneus* 11–12 cm

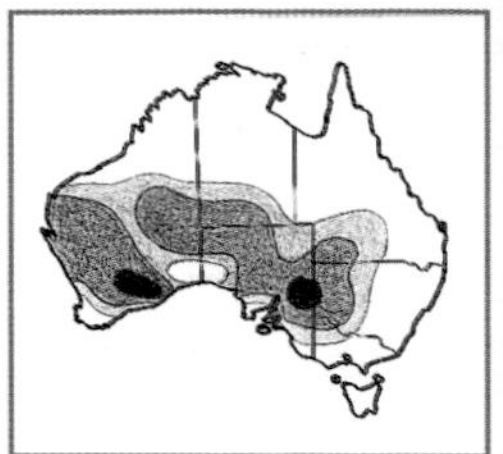

Small, grey-brown; sole claim to distinction a patch of bright colour, more orange than red, at throat of male; the female is more difficult to identify unless in company with the male. Forages briskly on the ground in the usually rather sparse low shrubbery of its semi-arid habitat. Undulating low flight with frequently fanned, white tipped tail. **Voice:** a distinctive intermix of sharp, clear, whistled notes alternate with harsher, chattered sounds: 'whit-whit-chee-chee-quorr-quorr, whit-whit-chee-chee-chak-chak'; 'chi-chi-chi-chi, quirrrr'. **Hab.:** semi-arid open eucalypt woodland, mulga and mallee with scattered low shrubs, open areas with saltbush and bluebush. **Status:** sedentary; uncommon to moderately common. **Br:** 389

Western Gerygone *Gerygone fusca* 10–11 cm

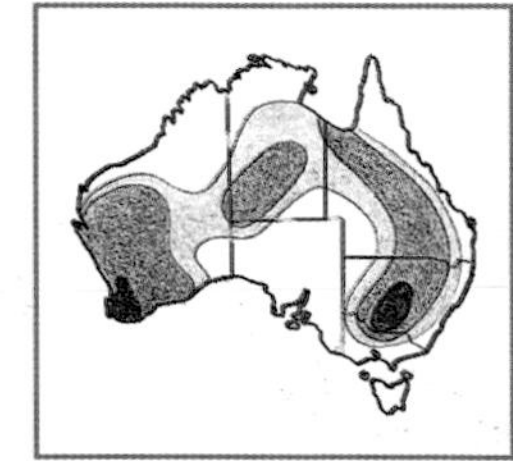

A small, plain species; forages in the tree canopy; often hovers about outer foliage; solitary or in pairs, rarely in small mixed or same-species parties. **Variation:** across N Aust., race *mungi*. In SW of WA and inland E Aust., race *fusca*. Alternatively, those E of Spencer Gulf, SA, may form a third race, *exsul*. **Voice:** pensive, whistled tune; high, clear violin notes, descending and rising, meandering, in slow, drawn out, wistful, sweet notes, but ceasing abruptly before the tune seems complete. **Similar:** other small, grey-brown gerygones; compare tail patterns. **Hab.:** drier country W of Great Dividing Range in open forests, woodlands, mulga, mallee; in SW WA it extends closer to coast, and is also found in heavier jarrah and wet karri forests. **Status:** sedentary; common. **Br:** 390

Brown Gerygone *Gerygone mouki* 9–11 cm

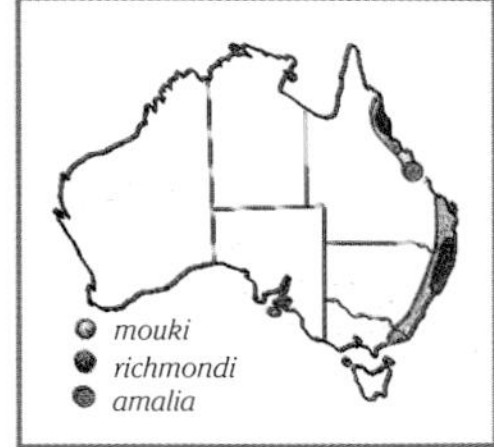

An inhabitant of dense wet forests, noticeable for its almost constant twitterings and brisk activity; often in small parties or larger flocks. Forages for insects and other small creatures on foliage and bark of forest trees; moves rather methodically among leaves, inspecting mossy limbs; often hovers over and under leaves of the canopy. **Variation:** three isolated populations: *mouki*, *richmondi*, *amalia*. **Voice:** brisk, sharp, more vibrant and insect-like than songs of other gerygones: 'wit-chippitee-wit-chippitee-chippitee-chipitee', 'whitchip-whitchip-whitchip' and other variations; scolding 'tzip-tzip-tzip'. Song differs slightly in NE Qld; notes are given in an ascending sequence. **Similar:** Large-billed and Mangrove Gerygones. **Hab.:** rainforests, adjoining dense wet eucalypt forests and mangroves. **Status:** sedentary; common **Br:** 390

Mangrove Gerygone *Gerygone levigaster* 10–11 cm

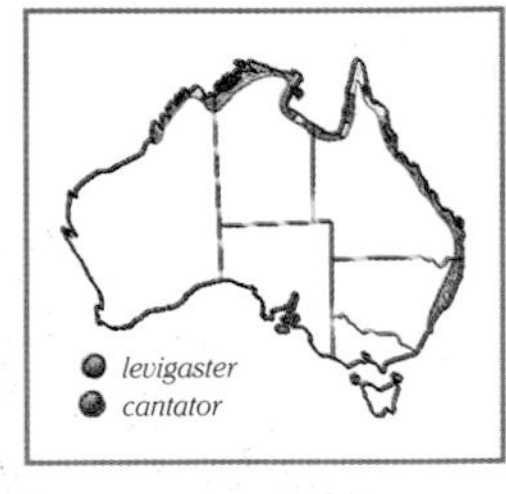

This gerygone's distribution follows the narrow strip of mangroves extending around most of northern and eastern Aust. Here it forages through the foliage canopy of the mangroves, occasionally hovering to search the outermost leaves; solitary, in pairs, infrequently larger parties. Has a reputation for being rather tame; often allows a close approach when in low foliage. Identification is complicated by the occasional use of mangroves by Brown and Large-billed Gerygones, and by some overlap with the Dusky on the W Kimberley coast. **Variation:** nominate race *levigaster* across the north and Kimberleys, the race *cantator* down the E coast. **Voice:** song similar to that of White-throated and Western Gerygones, but probably superior. Clear notes, some strong and piercing, others thin, tenuous, clear, like miniature violin, the tune falling wistfully, then lifting strongly, undulating. **Hab.:** usually tidal mangroves, but out of breeding season ventures into adjacent rainforests, swamps, woodlands. **Status:** sedentary or locally nomadic; common within usual habitat. **Br:** 390

Dusky Gerygone *Gerygone tenebrosa* 11–12 cm

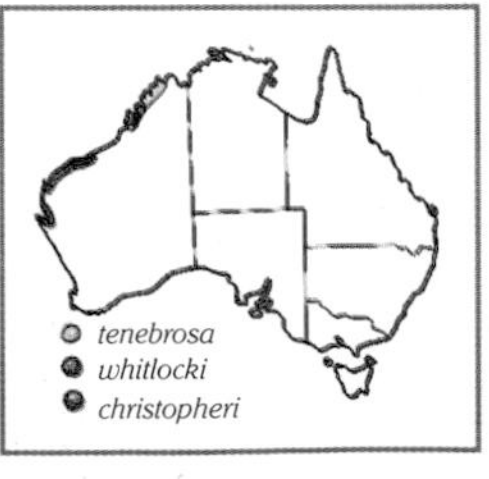

Tail markings are indistinct, unlike other gerygones that share its range and habitat. Forages on foliage of mangroves; in Kimberley tends to be found along the seaward side of wider mangrove belts, the Large-billed Gerygone usually on the landward side. **Variation:** breaks in the coastal mangrove belt have isolated 3 populations. Opinions differ: some argue that these represent only a clinal variation, thus just 1 race, *tenebrosa*. Others argue each population should be a race. The third view has *tenebrosa* used only for the Kimberley population and *whitlocki* for those in both Pilbara and Shark Bay. **Voice:** plaintive, mellow, notes, like those of a whistler, although softer, quite unlike the lively, high-pitched, thin, at times sharp, songs of most other gerygones, but rather a slow, rather hesitant variations 'queeit, toowheet, queeit, quit, queeit, whit'. **Hab.:** mangroves, occasionally adjoining dryland scrub. **Status:** sedentary; common **Br:** 390

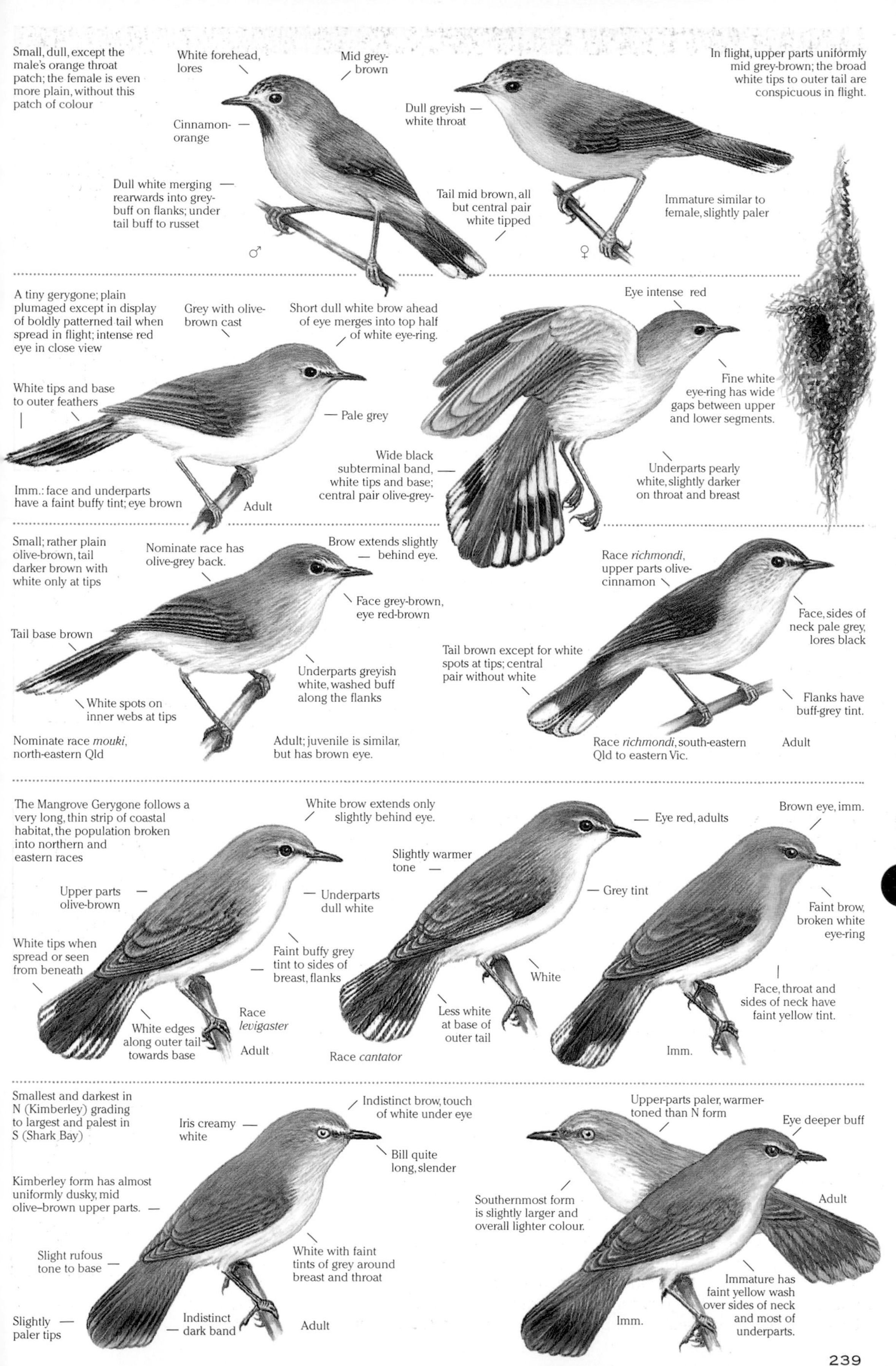
Small, dull, except the male's orange throat patch; the female is even more plain, without this patch of colour
White forehead, lores
Mid grey-brown
Cinnamon-orange
Dull white merging rearwards into grey-buff on flanks; under tail buff to russet
♂
Dull greyish white throat
Tail mid brown, all but central pair white tipped
In flight, upper parts uniformly mid grey-brown; the broad white tips to outer tail are conspicuous in flight.
Immature similar to female, slightly paler
♀
A tiny gerygone; plain plumaged except in display of boldly patterned tail when spread in flight; intense red eye in close view
Grey with olive-brown cast
Short dull white brow ahead of eye merges into top half of white eye-ring.
Eye intense red
White tips and base to outer feathers
Pale grey
Fine white eye-ring has wide gaps between upper and lower segments.
Wide black subterminal band, white tips and base; central pair olive-grey-
Underparts pearly white, slightly darker on throat and breast
Imm.: face and underparts have a faint buffy tint; eye brown
Adult
Small; rather plain olive-brown, tail darker brown with white only at tips
Nominate race has olive-grey back.
Brow extends slightly behind eye.
Face grey-brown, eye red-brown
Tail base brown
Underparts greyish white, washed buff along the flanks
White spots on inner webs at tips
Nominate race *mouki*, north-eastern Qld
Adult; juvenile is similar, but has brown eye.
Race *richmondi*, upper parts olive-cinnamon
Face, sides of neck pale grey, lores black
Tail brown except for white spots at tips; central pair without white
Flanks have buff-grey tint.
Race *richmondi*, south-eastern Qld to eastern Vic.
Adult
The Mangrove Gerygone follows a very long, thin strip of coastal habitat, the population broken into northern and eastern races
White brow extends only slightly behind eye.
Eye red, adults
Brown eye, imm.
Upper parts olive-brown
Underparts dull white
Slightly warmer tone
Grey tint
Faint brow, broken white eye-ring
White tips when spread or seen from beneath
Faint buffy grey tint to sides of breast, flanks
White
Face, throat and sides of neck have faint yellow tint.
White edges along outer tail towards base
Race *levigaster*
Adult
Race *cantator*
Less white at base of outer tail
Imm.
Smallest and darkest in N (Kimberley) grading to largest and palest in S (Shark Bay)
Iris creamy white
Indistinct brow, touch of white under eye
Bill quite long, slender
Upper-parts paler, warmer-toned than N form
Eye deeper buff
Kimberley form has almost uniformly dusky, mid olive-brown upper parts.
Southernmost form is slightly larger and overall lighter colour.
Adult
Slight rufous tone to base
White with faint tints of grey around breast and throat
Immature has faint yellow wash over sides of neck and most of underparts.
Slightly paler tips
Indistinct dark band
Adult
Imm.

Large-billed Gerygone *Gerygone magnirostris* 10–11 cm (T)

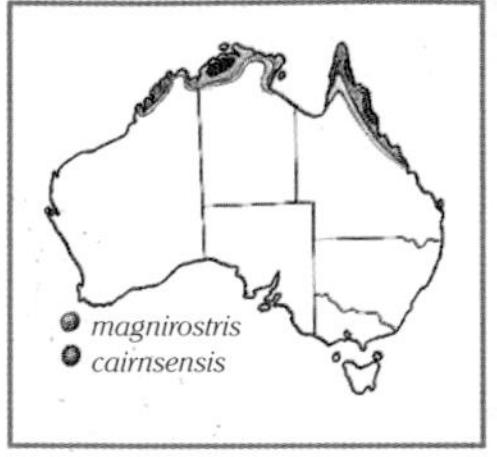

Plain plumaged gerygone distinguished by superb voice. Forages in canopy vegetation, often hovering about outer foliage. Usually in pairs, at times singly or in small parties of four or five birds; busily seeks insects on leaves and flowers. Presence of this species is often indicated by slender nest suspended over water. **Imm.:** like adult except eye is dull brown, bill is brownish. **Variation:** race *magnirostris* mainly near coast of Kimberley and NT; race *cairnsensis*, Cape York and southwards to vicinity of Mackay. A third race, *brunneipectus*, occurs on Sabai and Boigu Is., Torres Str. **Voice:** powerful, rich, whistled notes in diverse sequences, some rapid, others slow and deliberate; high, clear, or low and mellow. Will often create a tune, repeat it faithfully a few or many times; then switch abruptly to a new or similar version: 'wheecheeapip-wheecheeapip-wheecheeapip; cheeapip-cheeapip-cheeapip; wichee-wichee-wichee-wichee'. **Hab.:** mangroves, paper-bark swamps and around streams in dense forests. **Status:** sedentary; common. **Br:** 390

Green-backed Gerygone *Gerygone chloronotus* 9–10 cm

Distinctive green overtone to back, wings, rump; otherwise very plain with insignificant white markings to face and tail. Inconspicuous, blends into foliage; solitary or in pairs, occasionally in small groups; vigorously forages in foliage for insects. As with some other gerygones, the nest can be a useful aid to identifying the very plain little birds that may be in attendance. **Variation:** in Kimberley, yellowish cast; possible race *darwini*. **Voice:** clear and ringing, but lacks the diversity of the best of the gerygone songsters; calls repeated in sustained fast bursts: 'CHIRReep-CHIRReep-CHIRReep-' with emphasis on the first clear, sharp 'CHIRR'; at times delivered extremely rapidly 'CHIRRiepCHIRRiepCHIRRiep-' in long bursts like the reeling trill of a White-winged Fairy-wren. **Similar:** White-throated Gerygone, but yellow underparts; Lemon-bellied Flycatcher, race *tormenti*, but olive-brown back, more white on brow, dark eye, breast light grey. **Hab.:** mangroves, monsoon scrubs, paperbark swamps, vegetation along river gorges well inland. **Status:** sedentary; common. **Br:** 390

Fairy Gerygone *Gerygone palpebrosa* 10–11 cm

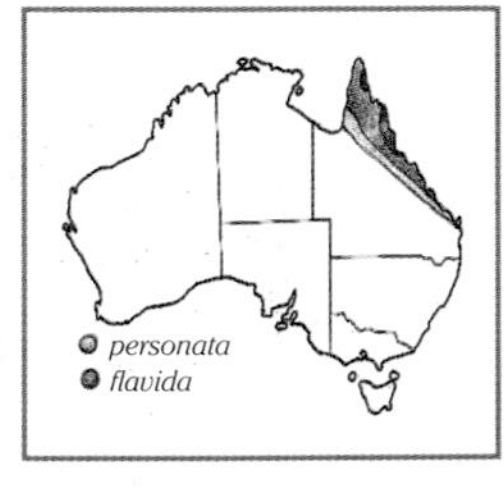

The two races of the species *palpebrosa* were previously separate species, the Black-throated and the Fairy Gerygones. These quite distinctive birds are now combined under the name Fairy Gerygone. Males of the two races differ greatly in appearance, but where their ranges overlap in the Cairns–Atherton region, there are intermediate amounts of black on face and throat. Forage in foliage canopy where individuals, pairs or small parties busily seek insects and other tiny creatures; often hover, attracting attention with scratchy chatter and song. **Imm.:** of both races are like adult females, but lack white forehead spots and white throat, have noticeably pale eye-ring. **Variation:** race *palpebrosa* and others in NG; northern race *personata* (Black-throated) occurs Cape York S to Cairns–Ingham district, where it merges with the southern race *flavida* (Fairy), which then extends S to about Fraser Is. **Voice:** typically lively and cheerful, not very high-pitched, but quite strong and penetrating, rather scratchy, metallic, repetitive: 'whit-WHITch-e-tew, whit-WHITch-e-tew, whit-WHITch-e-tew'; 'whit-a-whit-a-WHITchu, whit-a-whit-a-WHITchu'. In these and similar tunes it lacks the variety and creativity of the best of the gerygone songsters. **Similar:** White-throated Gerygone is like southern race of Fairy Gerygone, *flavida*, but has white spots in tail, more clearly defined white patch at throat and under tail, stronger yellow to remainder of underparts, never any black under chin; juveniles are difficult to separate. **Hab.:** lowland rainforests, monsoon, riverine forests, mangroves extending into nearby eucalypt woodlands and watercourse vegetation. **Status:** sedentary; moderately common. **Br:** 390

White-throated Gerygone *Gerygone olivacea* 10–11 cm

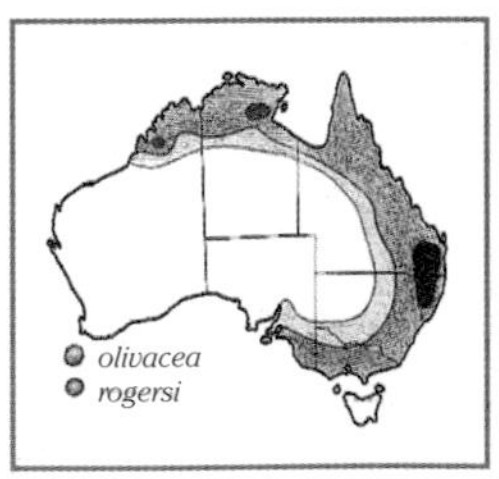

Wide range, conspicuous bright yellow underparts and exquisite song combine to make this one of the best-known gerygones. Forages busily in trees' outer foliage, darting, hovering about leaves and flowers to take insects, quickly moving on to other trees. In spring and summer the loud, clear, unmistakeable song attracts attention at a distance. At other times of year is often silent, or gives only a few brief notes. In small family parties, pairs or solitary. **Variation:** race *olivacea* from SE Aust. to NW Qld; race *rogersi* from NW Qld to W Kimberley. The population on Cape York may be separated as race *cinerascens*. **Voice:** loud, clear, carrying. Usually begins with several loud, piercing, high notes immediately followed by pure, high, clearly whistled, violin-like notes that descend in an undulating, silvery, sweet cascade, at times lifting briefly, only to resume the downward, tumbling momentum. Abruptly returns to the initial louder, sharper notes to repeat the whole sequence, often with slight variations. **Similar:** Fairy Gerygone, Weebill, Yellow Thornbill. **Hab.:** woodlands, open forests, watercourse vegetation. **Status:** south-eastern birds migrate to join the sedentary northern population. **Br:** 390

Two populations: nominate race magnirostris across coastal Kimberley and NT; race cairnsensis along the NE coast of Qld
Eye red
Bill slightly longer and heavier than those of most other gerygones
White throat
Warmer olive-brown on back, rufous margins to wings
Back olive; warmer brown on wings
Faint buff tint at edge of breast
Deeper buff on breast, flanks
Grey-brown with indistinct darker band, dull spots showing in wide-spread tail
Underparts white
Race cairnsensis, slightly smaller than nominate race
Adult
Adult
Race magnirostris
Underside of tail has dull, pale spots on outer tips, indistinct dark subterminal band.
Larger white tips to outer tail feathers
Head grey or grey-brown
Distinguished by distinctly green tone to mantle, back and rump, and edging the flight feathers. Lacks the white brow and tail markings of most other gerygones. A little-known species that also occurs in New Guinea.
Face almost plain grey with barely apparent pale brow and eye-ring
Eye red
Brown with overlay of metallic olive-green
Nest is a distinctive globular shape with a long slender tail suspended among foliage, often over or near water of mangroves or swamps.
White with touch of dull buff along the flanks
Flight feathers brown, edged light olive-green
Juv: Brownish head, eyes and bill
Tail plain grey-brown with slight greenish overtone
Adult
Race personata, 'Black-throated Gerygone', has a dark head, is boldly-patterned from face to upper breast.
White spot each side
Pale underwing coverts
Blackish brown
Olive-grey
White streak each side
Flight feathers, and underside of tail, grey to grey-brown.
♂
Underparts soft yellow
Black streak
♂
Females have faint white forehead spots.
♂
Race flavida, 'Fairy Gerygone', overall pale, colours of head pattern merge softly.
White throat merges into pale yellow underparts; some males have small black streak under chin.
Little difference between females of the two races
Females of both races lack the black facial markings of the respective males.
♀
Ranges widely across northern and eastern Aust. The White-throated has two races: the nominate race olivacea in the east; race rogersi across northern Aust.
Eye red
Upper parts grey-brown with olive tone to back
White spot each side of forehead
Eye brown, pale eye-ring
White forehead spots reduced or absent
Overall dull coloration
Race olivacea has a white patch at the base of the tail; northern race rogersi has this white absent or dull greyish
Clearly defined white throat patch
Yellow
White
Underparts entirely pale yellow
Tail grey-brown with a large white spot at the tip of each tail feather. These spots are not readily visible on upper surface of fully folded tail.
Adult
Imm.
Immatures are similar to both Weebill and Yellow Thornbill; different shape, behaviour and song.

Weebill *Smicrornis brevirostris* 8.5–9.5 cm
Tiny, greenish, with loud, clear, distinctive 'wee-bill' calls, otherwise inconspicuous in mid to upper canopy where pairs or small parties flutter about the foliage seeking small insects. **Variation:** 3 races, or 4 if Pilbara population is a separate race, *ochrogaster*. **Voice:** strong far-carrying whistled call, slightly husky compared with the clear, pure notes of gerygones. Often a single, loud 'wheetiew'; longer sequences are 'whit, whit, WHEETiew' and 'whit, whit, WHEET-a-whit'. **Similar:** Yellow Thornbill, streaked face, red eye. **Hab.:** dry open eucalypt forests, woodlands, mallee. **Status:** sedentary, locally nomadic; common. **Br:** 390

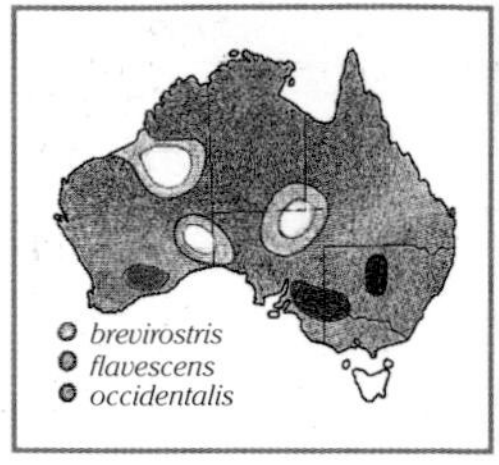

Yellow Thornbill *Acanthiza nana* 9–10 cm
Strictly arboreal; small parties, pairs or solitaries unobtrusively work through mid to upper foliage; when high, easily overlooked but for calls. Always busily active and on the move, fluttering and hovering about foliage or flowers, and then often displaying the dark, pale-tipped tail. **Voice:** a harsh, scratchy rather than high, insect-like sound, erratic 'chzip-chzzzip, chzip-chzip, chzeeep, chzip-chzzip-chzzzip'. Also gives a strong, harsher, scolding 'kirrzz, kirrrzz-kirrrzz'. **Similar:** Weebill, Buff-rumped Thornbill, gerygones with yellowish underparts, especially immatures. **Hab.:** open forests, woodlands of acacia, casuarina, melaleuca, brigalow, in preference to eucalypts. **Status:** sedentary; common. **Br:** 391

Striated Thornbill *Acanthiza lineata* 10 cm
Like the Yellow Thornbill, the Striated is strictly arboreal, but concentrates its foraging on eucalypts. After the breeding season Striated Thornbills often gather in larger flocks; hunts energetically through foliage; briefly hovers to take insects from leaves. **Voice:** soft, insect-like, 'tzzt, tzzit, tzzt-tizzit', at times run together rapidly to form a short tune, almost trilled, 'tzt-tzt-tizzit-tizzit-tzt-tzt-tizzit-tizzit-'. **Similar:** in the same habitat, Yellow, Brown and Buff-rumped Thornbills, Weebill. **Hab.:** usually high-rainfall, heavy eucalypt forests, but also drier open forests, woodlands and coastal mangroves. **Status:** sedentary; common. **Br:** 391

Buff-rumped Thornbill *Acanthiza reguloides* 10.5–11.5 cm
Usually seen on ground, sometimes in foliage. Often in small parties; forages busily, restlessly; departs with bouncy flight displaying yellow rump, rather like the Yellow-rumped Thornbill. **Variation:** up to 4 races; includes *reguloides*, 'Buff-rumped', and *squamata*, 'Varied Thornbill'. **Voice:** lively, sharp, twittering calls, 'tsip-twit-titch-tseip, twit-ti-twit-twit-tseip, tseip-tseip'. Song is very much more rapid, an unbroken, tinkling 'chipitit-chipitit-chipitit-chipitit'. **Hab.:** open dry eucalypt forests, woodlands, heaths; forages in low shrubs and across debris-littered ground; northern race lives in woodlands with grassy groundcover, foraging higher in foliage of large shrubs and trees. **Status:** sedentary; moderately common. **Br:** 391

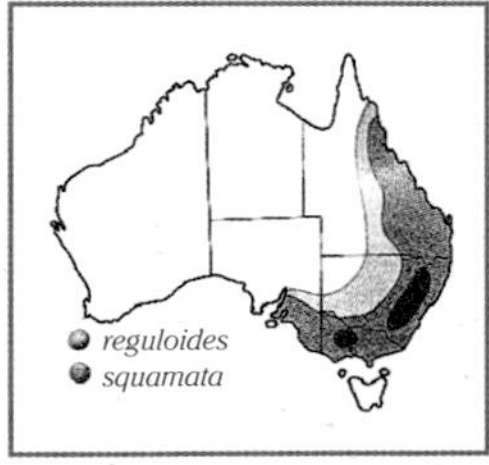

Yellow-rumped Thornbill *Acanthiza chrysorrhoa* 11–12 cm
One of the most familiar small birds. Usually forages on ground, often in small flocks; they work across grass or leaf litter with a jerky, hop-and-peck action. **Variation:** variation slight, appears largely clinal; possibly several races. **Voice:** cheery, undulating, chittering, tinkling song; lasts just 3 or 4 seconds, usually ending with two clear whistled notes, the second descending: 'chip-chip-chippity-cheepity-chippity-cheepity-wheit-wheehoo'. Whole song is then repeated almost identically, a few or many times. **Similar:** Buff-rumped Thornbill. **Hab.:** grassy woodlands, scrublands, farms, gardens. **Status:** sedentary; abundant. **Br:** 391

Western Thornbill *Acanthiza inornata* 10 cm
Busy, restless small bird of tree foliage and undergrowth; often on the ground. In pairs, or small flocks, at times with other species. **Voice:** weak, erratic with broken sequences of scratchy chittering, 'tchip-tsip' sounds, with an occasional louder 'choweep', mellow and rather musical: 'tchip-tseeip, cheewip, tseep-tsip, choweep, tsip-cheowip, tsip, seep, choweep, tsip-'. Includes mimicry of other birds. **Similar:** Slender-billed Thornbill, but usually further inland in different habitat. **Hab.:** tall, wet, dense eucalypt forest and coastal heaths to rather arid open woodlands and dry sandplain heaths. **Status:** sedentary; common. **Br:** 391

Forehead lacks any speckling
Eye creamy white
Bill short, thick, light brown
Olive
Some may have slight dark streaks.
Race *brevirostris*
Imm. (all races): colours dull, iris grey
Yellowish olive
Plain pale yellow
Underparts pale creamy buff
Race *flavescens*: brighter yellow underparts
Grey-brown
Face grey-brown with faint rufous tint to ear coverts, sides of neck
Grey to grey-brown
Back deeper olive-green — to brownish olive
Dark tail band
White tips to tail, except central pair of feathers, which are entirely dark
Race *occidentalis*
Tiny thornbill, varies from pale inland to deeper golden near coast, brightest in far N
Olive
— Brow short, pale buff, not conspicuous
Cheeks to ear coverts grey, streaked white
Adult
Some forms slightly deeper ochre tone on chin and throat
Broad dark tail band
Inland form (*modesta*) smaller, paler; underparts soft creamy buff
Faint pale tips
In south-east, deeper olive-green to brownish olive (race *nana*, 'Little Thornbill')
Bill longer, finer than that of Weebill, above
Birds of isolated Atherton population have brightest overall yellowish coloration (race *flava*, 'Yellow Thornbill')
Named for the extensive streaking—black or dark brown on the breast, fine white on the dark rufous crown, and more heavily white across the dark cheeks and ear coverts. Immatures are like adults.
Shows some variation across range, darkest in SE; bright and obviously streaked around Mt Lofty Ra., SA; bright and more lightly streaked towards N. Possibly 4 races, *alberti* in SE Qld; *lineata* through E NSW and S Vic.; *clelandi* in SE of SA; *whitei* on Kangaroo Is.
Eye grey-brown —
— Crown rufous-brown, finely streaked white
Olive-green to olive-brown —
Cheeks and ear coverts dark, heavily streaked white
Tail olive-brown with dark subterminal band
Throat white, grading to yellow, dark streaked
Crown olive-brown, merging — into rufous tint of forehead
— pale tipped feathers give finely scalloped effect.
Nominate race *reguloides*, — olive-brown to olive-grey back, brownish wings
Throat, breast, flecked grey
Rump pale pinkish — buff
Yellowish buff
Very pale buff
Pale creamy white iris
Throat pale lemon
Race *squamata*, brighter yellow underparts, greener back
Pale tips
The pale nominate race *reguloides* occurs across much of SE Aust., while the brighter yellow race *squamata* ('Varied Thornbill') is confined to coastal NE Qld. Within these races, several others have at times been proposed.
Female is similar to male, perhaps slightly paler in northern race. Immatures are like adults, but have dark eye and rather dull plumage.
Best-known and most widespread of the thornbills; likely to be seen anywhere from suburban gardens to central semi-desert plains. Renowned for its huge, double-storey nest.
Olive-brown to olive-grey
— Forehead black, spotted white
Eye pale fawny brown
Wings mid brown —
Rump light, bright yellow —
Off-white; olive-buff along flanks
Bright yellow obvious as bird flies up from — ground
Black with very — narrow white tips
Imm: like adult, but with darker, brownish eyes
In flight, has a rapid undulation that shows the yellow rump as the bird departs.
The Western is the plainest thornbill, plumaged in greys with only the faintest tints of olive and buff. Plumage pattern is confined to a slight speckling about the face.
Eye off-white to greyish white —
— Forehead browner with tiny pale-edged feathers giving a slightly scalloped effect
Back olive-grey to — brownish olive
Face and ear coverts are lightly flecked buffy white.
Pale brownish buff rump —
Broad dark tail band, pale fawn tip —
Underparts are greyish white with a faint buff tint, strongest towards the rump.
Some variation from N to S. Western thornbills from near the S coast are darker with an olive-brown cast to the upper parts and more buff beneath. Those further N are paler, greyer. Juveniles are like adults, but facial markings are even less obvious.

Slender-billed Thornbill *Acanthiza iredalei* 9–10 cm

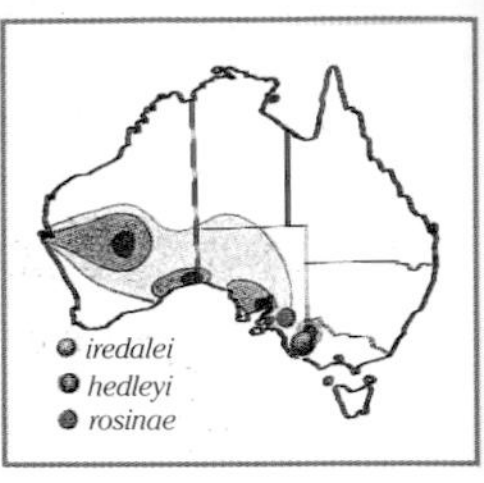

Variable; quite colourful to very pale. Forages in low shrubs, on ground. In pairs or small parties, seeking insects, spiders and other small prey from the foliage of low samphire and saltbush shrubs, and from the surrounding ground. If flushed, departs with low, bouncy undulations when the yellowish rump is conspicuous against dark tail. **Variation:** 3 subspecies or forms. **Voice:** rapid, high twittering, almost tinkling, 'tsip-tsip-sip-sip-chip-chweeet, tsip-chip-', the occasional 'chweeet' louder, more musical. **Similar:** Buff-rumped Thornbill, but occurs in woodland. **Hab.:** the two western races inhabit saltbush and samphire flats ; the SE race lives in dense, low heath. **Status:** sedentary or nomadic; nominate race secure, common, others are smaller, confined populations, only locally common. **Br:** 391

Mountain Thornbill *Acanthiza katherina* 10 cm

Lives in mid and upper strata of rainforest; forages busily around outer foliage; it can be difficult to sight. Gregarious; gathers into small flocks after spring–summer breeding season. **Variation:** slight colour variation between birds of lowest altitude, around 400 m, and highest, 1200 m. **Voice:** intermingled quick, sharp, scratchy sounds and contrasting louder, mellow, musical, whistles: 'tsit-tsew-tseweit, tzit-tseet, toowheet, tsit-toowheet, tsit-tchip-tsew'. **Hab.:** mountain and high plateau rainforests above about 400 m altitude. **Status:** sedentary; moderately common within a restricted but secure range. **Br:** 390

Chestnut-rumped Thornbill *Acanthiza uropygialis* 9–10 cm

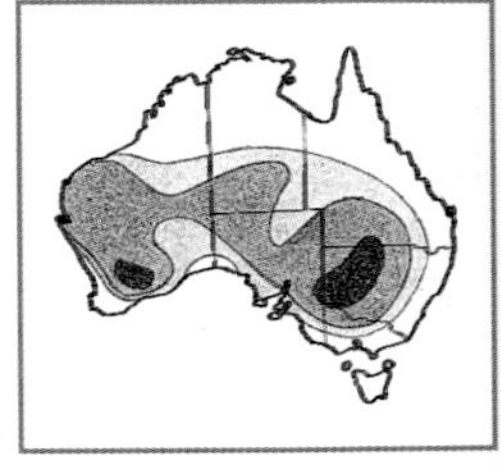

A small, busily active thornbill; seeks insects on foliage, branches, low shrubs and the ground; in pairs, singles or small parties; at times mixes with other small birds, often Weebills, pardalotes and other thornbills. **Imm.:** like adult, but dark eye. **Variation:** slightly darker backed, fawn flanked in SE; gradually slightly paler further into arid north–central regions. **Voice:** usual call is a high, peevish, squeaking 'tsee, tsee-tseep, tseep, tseeip'. Less often a whistled, clear, penetrating 'cheweep-cheweep-cheweep', or faster, lower, a mellow, musical warbling. **Similar:** Slaty-backed, Inland, Brown Thornbills. **Hab.:** drier eucalypt woodlands, mulga and similar acacia scrublands, mallee, casuarina, and, occasionally, low shrublands of saltbush, bluebush, lignum. **Status:** sedentary or seasonally locally nomadic; moderately common. **Br:** 391

Brown Thornbill *Acanthiza pusilla* 9.5–10.5 cm

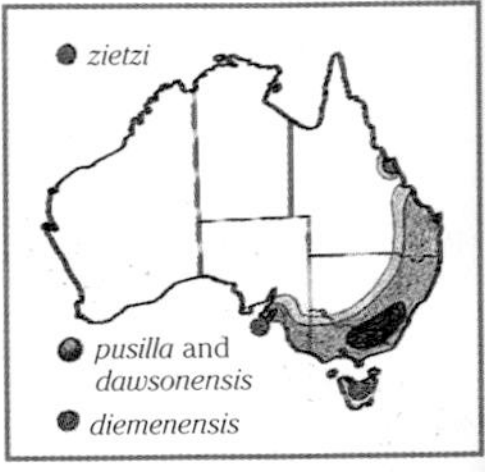

Actively seeks insects and and other small prey in undergrowth and lower foliage; rarely high in canopy or on ground. Alone or in pairs, rarely small family parties. Unlike Inland Thornbill, rarely holds tail upwards. **Variation:** northern race *dawsonensis*, mid E to SE Qld, has cream breast and belly. Race *pusilla* extends from SE Qld to the Mt Lofty Ra., SA; has dull creamy white to olive-grey breast and belly. Other races are *diemenensis* in Tas., *archbaldi* on King Is. and *zietzi* on Kangaroo Is. **Voice:** rich, musical with wide range of notes. Very high, clear whistles abruptly cascade in strong, rippling trills to notes unusually deep and mellow for so small a bird. Among these calls sharper scratchy sounds or mimicry of other birds may be inserted. In alarm, gives harsh, churring scoldings. **Hab.:** forests and woodlands with dense undergrowth, vegetation of creeks, rainforests, coastal dune thickets, mangroves. **Status:** sedentary; common. **Br:** 390

Inland Thornbill *Acanthiza apicalis* 9.5–10.5 cm

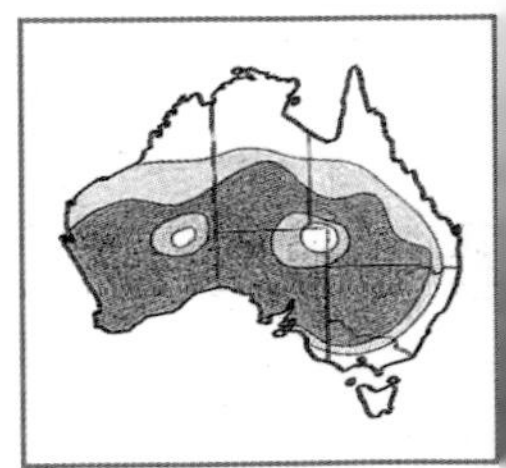

Forages briskly through shrubby lower growth rather than on ground or high in trees; at times in small flocks with other species of small birds. Unlike Brown Thornbill, often holds tail angled upwards, at times vertically. **Variation:** opinions differ whether variations are gradual (clinal) or sufficiently great or defined to justify races. Birds in SW of WA are more russet toned; on the E coast they have brighter rufous rumps: both grade gradually paler, greyer, further into drier interior habitats. **Voice:** lively, sharp 'tsip-chip-' calls; loud, clear, musically whistled 'chweeeip!' and deep, mellow, tremulously warbled 'quor-r-r-r-eip'; overall effect a varied 'tseeeip-tsip-tseep-chweeeeip, chip-tzzeep-tseep-quorr-r-reip! tsip seep-'. Also buzzing scoldings and passages of mimicry of other birds. **Similar:** Brown Thornbill, but rufous forehead, thinner tail band; Slaty-backed Thornbill, but has blue-grey upper parts; Chestnut-rumped Thornbill, but has white eye. **Hab.:** dry woodlands of eucalypts, acacias; in SW of WA where Brown Thornbill is absent, the Inland is also in wetter forests of karri and jarrah, and sandplain heaths both coastal and inland. **Status:** sedentary, seasonally locally nomadic; common. **Br:** 391

Three forms or races; differ slightly in colour; occur in separated localities and, in one instance, quite different habitat.
Eye pale, all races
— Forehead finely speckled
Bill small, slender
Pale olive-grey
Throat faintly mottled
Rump buffy white to pale yellow
— Race *iredalei*: under tail pale lemon yellow, but some almost white; flanks buffy white to buffy pale grey
Nominate race *iredalei* includes palest forms of this species.
Mottling heavier, flanks deeper olive
Back and wings darker olive-grey
Race *rosinae* has deepest rump colour.
Olive flanks
Tail, all races, very dark with pale tips
Race *rosinae* is darkest form; a third race, *hedleyi*, is intermediate in colour between the races shown.
The greenish grey back and faint yellowish tint to underparts are typical of this species at lower altitudes. Birds of the highest ranges tend to be grey backed and almost white beneath. Both sexes are similar; the immatures are relatively dull.
Back variable, olive-grey to grey-green
— Forehead noticeably speckled
Rump tawny rufous-buff
Eye white
Buff tint to sides of breast
The Mountain Thornbill is similar to the far more widespread Brown Thornbill, but obvious differences are the pale eye of this species, the song, and the nest, which is quite massive for the size of the bird.
Tail has broad black subterminal band, pale grey tip.
Slight buff tint to sides of flanks
Adult
The Chestnut-rumped Thornbill is distinguished from other reddish rumped species by the off-white eye of adults, its foraging on the ground as well as in foliage, and its choice of nest site (p 391).
Pale grey to brownish grey
Forehead has finely scalloped pattern
Eye stark creamy white; immatures are dark eyed.
Rump and base of tail are light, bright chestnut.
Underparts white with light grey on breast.
Tail blackish brown with pale central, white outer tips
Chestnut rump conspicuous when birds fly up from ground.
As name indicates, upper parts are warm grey-brown, but become more olive northwards.
— Rufous, pale tipped
Olive-brown
Streaked, all races
Grey-brown to olive-grey
Rump dull cinnamon
Rump and base of tail dull cinnamon
Southern forms, white to pale buff flanks
Race *dawsonensis*, E Qld, has deeper buffy yellow underparts.
Grey to white on outer tips
Tail brownish with narrow dark band
Tail may be held slightly to steeply upwards.
— Forehead is coarsely scalloped dark grey and white, unlike rufous tint on Brown Thornbill.
Overall slightly larger, more solidly built than the Inland and Chestnut-rumped Thornbills
Cinnamon rufous
Streaked, throat to lower breast
Eye red, all forms
Coarse black and white scalloping
Mid grey-brown
Central-western Qld form: pale grey backed, bright rufous rump, broad dark subterminal tail band (*cinerascens*)
Inland form (or, if a race, *whitlocki*) has pale upper parts, rufous on rump.
Flanks pale cinnamon
Heavy streaking on throat and over lower breast.
Strongly rufous
Rump quite bright rich rufous
Inland Qld–NSW form (*albiventris*): rump rich rufous, broad dark band above pale tip of tail
Flanks and under tail are whitish to pale cinnamon.
Belly white
Wide dark tail band
Flanks and under tail white

Slaty-backed Thornbill *Acanthiza robustirostris* 9–10 cm

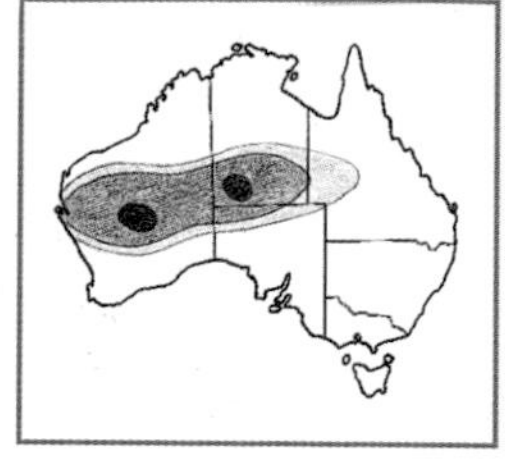

Back slaty grey, or perhaps brownish grey seen under warmer direct sun. Forages through foliage of low shrubs, and occasionally on ground. **Voice:** many calls: frequent 'tseeip' similar to the Chestnut-rumped and Inland Thornbills; some sounds may be mimicry of those species. Song varies—whistled, rich, musical notes interspersed with sharp, higher twittering: 'chwip-wip, chwip-wip, cheeowheeep, chwip, chwip-chwip, cheeowheeep, chwip, whip-e-chiew'. **Similar:** Chestnut-rumped and Inland Thornbills, but unstreaked crowns, eyes white or lighter red, finer bills. **Hab.:** arid mulga scrublands with shrub understorey; low shrubbery around claypans and salt lakes. **Status:** sedentary; generally uncommon. **Br:** 391

Tasmanian Thornbill *Acanthiza ewingii* 10 cm

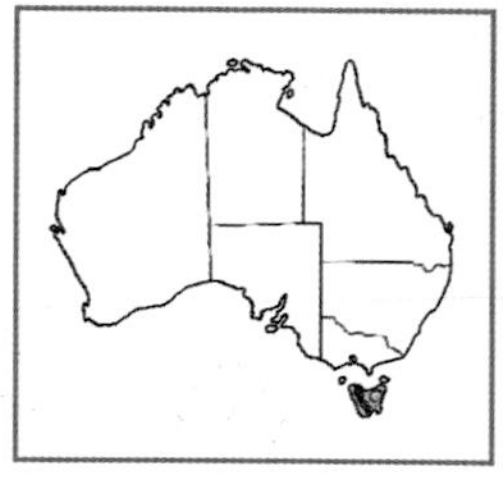

A very plain thornbill confined to the cool, wet rainforests of Tas. and Bass Str. islands. An arboreal species; forages through dense forest canopy and down into mid level vegetation, rarely very near or on the ground. **Voice:** contact calls are sharp, scratchy, twittering 'tszit, tszit'; in alarm gives harsh churrings and buzzing sounds. Other calls are richer, more musical, although disjointed rather than continuous, free-flowing song: 'Chip-cheerwheep. Chirr-chirrowp. Chir-r-r-r-owp!'. **Similar:** Brown Thornbill, a race of which is also in Tas., but has greyish outer edges to flight feathers, more heavily streaked throat, whiter flanks, and tends to keep to the drier forests. **Hab.:** cool temperate rainforests, wet eucalypt forests; moves to drier forests, woodlands through winter. **Status:** sedentary or with local seasonal movements; common. **Br:** 391

Southern Whiteface *Aphelocephala leucopsis* 10–12 cm

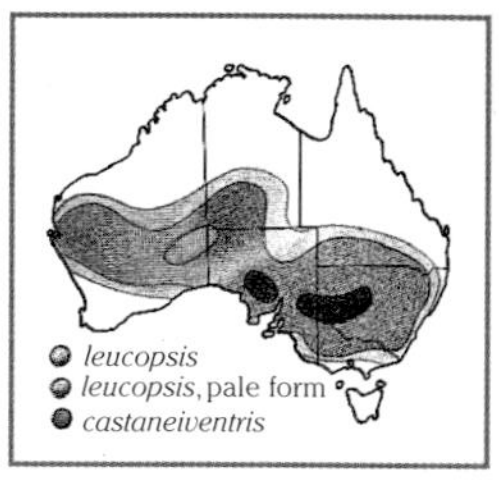

All species of whiteface have a white band across the forehead with a dark rear border down through the eye. A stubby, finch-like bill is indicative of a diet of seeds and insects. Lively, fossicks about on ground and low in shrubbery; often in small flocks that may include other species. **Variation:** *leucopsis* is paler, browner, in NW part of its range (formerly race *whitei*). **Voice:** a very rapid, scratchy, twittering 'tzip-tzip-tziptziptziptziptzip-' or 'tchip-tchip-chiptchipt-chipt-chip' and harsher 'tzzit, tzzit-tzzit, tzzit'. In alarm, a scolding 'kzzurrrk, kzzurrk-kzzurrk-kkzzurrk'. **Similar:** other whitefaces, but have black or chestnut breast-bands, chestnut on backs; several brownish, white-eyed thornbills, but these lack distinct white forehead patches. **Hab.:** semi-arid woodlands, mallee, mulga and similar dry-country scrublands. **Status:** sedentary; common. **Br:** 391

Chestnut-breasted Whiteface *Aphelocephala pectoralis* 10 cm

Rare whiteface; very restricted range. Often on ground in single-species groups or mixed flocks, seeking seeds and insects. Rather timid; when disturbed these birds travel some distance in typical whiteface undulating flight before dropping to ground again. **Juv.:** as adults, but breast-band narrower, dusky. **Voice:** rapid, weak trill often given as contact call while in flight, 'trit-trit-trt-trt-trt-trt-trt-trtt-rtt'; louder, plaintive, whistled song: 'wheeit, wheeit-weeeo'. **Similar:** Banded Whiteface, but much more clearly defined black breast-band; Southern Whiteface, but plain dull white breast, greyish back. **Hab.:** slightly elevated plains and low flat-topped mesa formations to W of Lake Eyre, largely gibber-stone surfaced with widely scattered mulga and very sparse low cover of shrubs, typically eremophilas, bluebush. **Status:** very confined distribution, difficult to locate; probably quite rare within that range; possibly nomadic. **Br:** 391

Banded Whiteface *Aphelocephala nigricincta* 10–11 cm

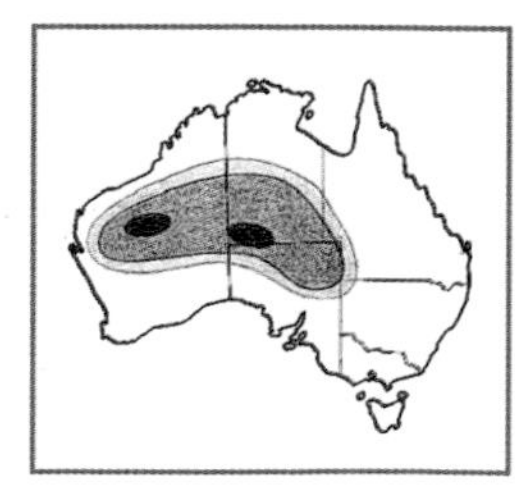

Inhabits some or most arid central regions, but well adapted—reportedly need not visit waterholes and wanders nomadically in search of seasonally favoured country. Gregarious except when nesting; in small flocks, sometimes in mixed foraging parties with other small birds. Fossicks busily over ground; quickly flies into bushes if disturbed. **Voice:** a mellow, musical succession of notes, rather bell-like, almost fast enough to be a trill, usually preceded by several longer, clearer notes, 'whh-wheeee-wheeee-wheeet-whit-whee-wheewhiwhitwhit-whit'; also slower, clear 'whit-whit-whit-whit-whit-whit-'. In alarm, a harsher, buzzing scolding. Breeding males give song-flights: rise steeply; flutter slowly, briefly, singing; then drop to bushes. **Similar:** other whitefaces, but either have plain breast or broad rufous band; White-fronted Chat, greyer, different face. **Hab.:** arid sparse mulga, eremophila, other stunted trees and shrubs; often scattered across spinifex dunes or saltbush flats. **Status:** seasonally nomadic; uncommon to locally quite common. **Br:** 391

The name of this species refers to 'slate grey', a slightly bluish grey common in slate, and typical of the dorsal colour of this thornbill.
Upper parts blue-grey with crown finely streaked black
Forehead, face, ear coverts have pale speckling.
Eye red-brown
Breast very pale grey
Rump dull pale chestnut
White with slight touch of buff along flanks
Tipped pale brown to dull white
Tail subterminal band broad, dark
Large thick bill
Flight feathers dark grey-brown
Flying up from low shrubbery or hovering shows rump colour.
Dark brown with black outer third
White edging to secondaries
Similar to Brown Thornbill with minor plumage variations; usually in different habitats
Mantle and shoulders olive, nape and back olive-brown (these more rufous on race ewingii, King Is.)
Rufous forehead
Outer vane of primaries is edged cinnamon-rufous.
Tail chestnut, blackish outer third; pale tips
Throat and breast are softly streaked dark grey and white, more heavily on male.
Tail edged cinnamon-rufous towards base
Flanks pale olive-grey
Cinnamon-rufous edges to flight feathers
Legs quite long, pinkish brown
Dark band, dull white tips
Forehead is white, shaped as two rounded patches, and with black rear edge that extends down to the eye.
Crown, ear coverts grey-brown
Stubby bill
While race leucopsis is palest in the far NW of its range, further westward the WA race castaneiventris is strongly mottled cinnamon-rufous along the flanks.
Grey-brown
Underparts greyish or off-white
W.A. race castaneiventris, flanks cinnamon-chestnut
Race leucopsis in the E of its range has mottled grey-brown flanks, gradually paler towards the NW: in WA, NT and NW of SA, flanks have but a faint smudge of brownish grey.
Flanks dull, pale grey-to brownish-grey
Tail dark, tipped white
Flanks mottled cinnamon- chestnut
Nominate race leucopsis
White forehead backed by dark band; merges into grey-brown crown, extends only slightly below eye.
Crown grey-brown, flecked darker
White forehead in two rounded patches
Forehead as horizontal band
Nape to rump rufous-chestnut
Distinguished by a broad rufous-chestnut breast-band; clearly defined only along its lower margin and blending softly upwards into buffy white of upper breast
Tail black, browner towards base and tipped white
Rufous patch down rear flanks
Southern Whiteface
Chestnut-breasted Whiteface
White forehead patch is backed by a blackish brown band that extends strongly well below the eye.
Forehead buffy white
Light cinnamon-rufous
Short, finch-bill shape
Pale margins
Flight feathers edged pale cinnamon, coverts tipped and finely edged off-white
Rump deeper chestnut
Brownish black breast-band sets this apart from other whiteface species, but gives some resemblance to White-fronted Chat.
Rufous scallopings along flanks
Conspicuous brownish-black breast-band across buffy white underparts
Tail dark brown, blackish nearer the buffy white tips

Yellow Wattlebird *Anthochaera paradoxa* 38–48 cm

A very large, gregarious honeyeater named for the long pendulous wattles hanging from the cheeks. Yellow Wattlebirds gather in flowering eucalypts, seeking both the nectar and the insects around the flowers, but will also peck into orchard fruit. Feeds mainly in forest canopy; moves among the trees with strong, undulating flight. Flocks wander from highlands to lowlands at end of summer; return as breeding pairs in spring. **Voice:** harsh, throaty, grating sounds, a gurgling 'grrowgk', 'chrrowk' or short, hollow 'klok' repeated irregularly. Also a duet; the female gives a gurgling sound, the male quickly responds with a harsh croak. **Hab.:** forests from subalpine to lowland; eucalypt and banksia woodlands and heathlands. **Status:** confined to Tas. and King Is., migratory or nomadic; common except W Tas. **Br:** 392

Red Wattlebird *Anthochaera carunculata* 32–36 cm Melbourne.

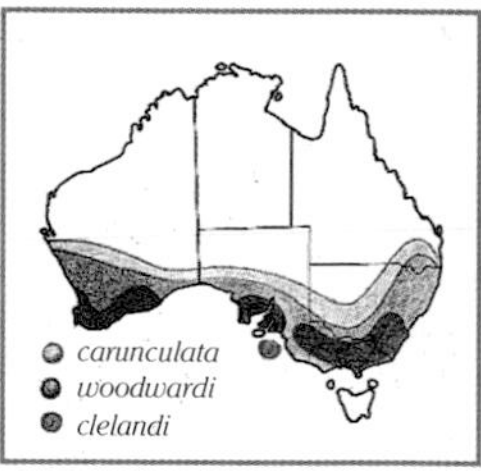

Large, noisy, aggressive honeyeater; widely distributed; feeds on flowers of eucalypts, other trees and shrubs; often in flocks, at times in large numbers. In flight, white tips of wings and tail are conspicuous. **Variation:** race *woodwardi*, WA to SA, intergrading gradually through SA with race *carunculata* of E Aust. On Kangaroo Is., race *clelandi*. **Voice:** harsh, grating cough or bark: 'hrarrark-hrak', 'hrrak-a-yak', 'hraak, hraka-yak'; slow, grating, growled 'graarrrrk'. Female gives a rather more pleasant, mellow 'kieuw-kieuw-kieuw', at times combined with harsher coughing noises contributed by the male. **Similar:** Little Wattlebird, but lacks red facial wattles and in flight shows chestnut in wings; Yellow Wattlebird, but it has a different range. **Hab.:** eucalypt forests and woodlands, mallee, gardens. **Status:** nomadic or sedentary; common **Br:** 392

Little Wattlebird *Anthochaera chrysoptera* 27–31 cm

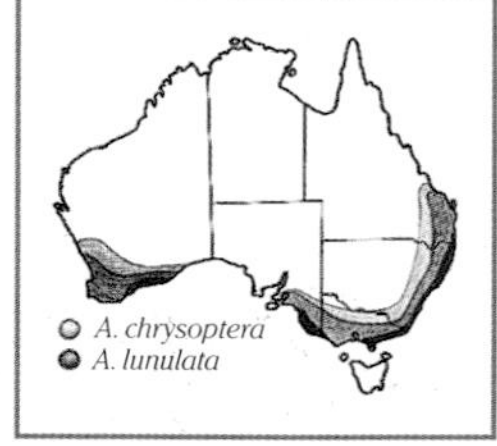

Large, rather plain grey-brown at a distance, intricately white streaked in close up. In flight, white tips to primaries mark path of wingbeats in conspicuous white arcs. Noisy, often aggressive; energetically seeks nectar, berries, insects from tree tops to low shrubbery; often in flocks. **Variation:** previously combined with Western Wattlebird, below. In Tas., race *tasmanica*; on Kangaroo Is., race *halmaturina*. **Voice:** varied; notes range from harsh to musical. Often a mellow 'kook-' alternating with a grating 'kraagk' as 'kook-kook-kraagk-kraagk, kook-kook-kraagk-kraagk' and 'kraagk-kook-kraagk-kook-'. Song of western race is higher, more musical, undulating: 'cook-cook-cup-hook, cook-cook-cup-hook'. All have a harsh, grating single 'graarrrk!'. **Similar:** Red Wattlebird, but has red wattles, yellow tint to belly, lacks chestnut in wing. **Hab.:** forests, woodlands, banksia heathlands, gardens. **Status:** migratory or nomadic; common. **Br:** 392

Western Wattlebird *Anthochaera lunulata* Was race *lunulata*; stronger plumage pattern, red eye. **Br:** 392

Spiny-cheeked Honeyeater *Acanthagenys rufogularis* 23–26 cm

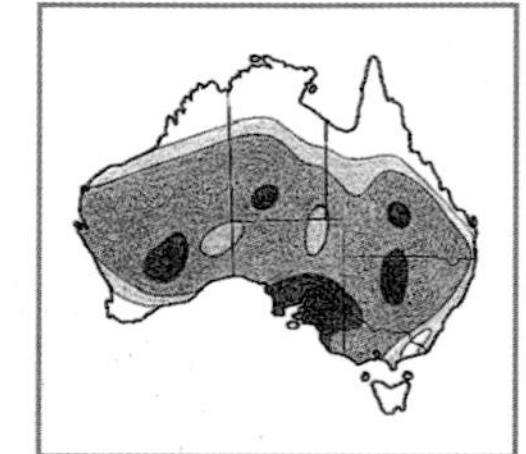

The clear, piping, musical yet melancholic notes that so often reveal this bird's presence always seem in keeping with the loneliness of its desert haunts and identify it unseen. Often calls from high bare perches and in high display flights. Feeds on insects and small native fruits as well as nectar; in pairs, alone or in small flocks. **Imm.:** as adult, bill paler pink, touch of yellow on 'spiny' cheeks. **Voice:** clear musical notes, but in most passages of song a bubbling character. Common sequences: 'quip-kpeeer-kpeeer-kpeeer-quipip-quipip-quipip'; 'chrriee-chrriee-chrriee'; 'whiteeer, whiteeer, whiteeer'. Also a gurgling, bubbling 'quorrok-quorrok-quorrok-quok'. **Hab.:** arid woodlands, mallee, mulga, often with groundcover of spinifex; also scrubby vegetation of drier parts of coast. **Status:** nomadic; generally common. **Br:** 392

Striped Honeyeater *Plectorhyncha lanceolata* 22–25 cm

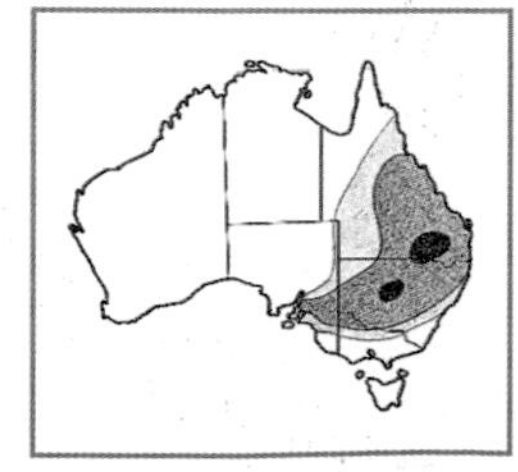

Although a honeyeater in origin, the Striped now uses many food sources including insects, seeds and fruits as well as nectar. The bill is short, sharply pointed and almost straight, more suited to prising insects from crevices than probing deep flower tubes; it has the honeyeater brush tongue and takes nectar at more shallow flowers such as eucalypts. Usually alone or in pairs, infrequently in family groups or small flocks. **Voice:** clear, rapid calls, rather more mellow than sharp, with slight variations giving a rollicking rise and fall: 'quirrip-quarreep-quirrip-quarreep-quirrip-quarreep'. **Hab.:** drier open forests, woodlands, mallee, mulga, also heathland and mangroves. **Status:** sedentary or nomadic in drier regions. **Br:** 392

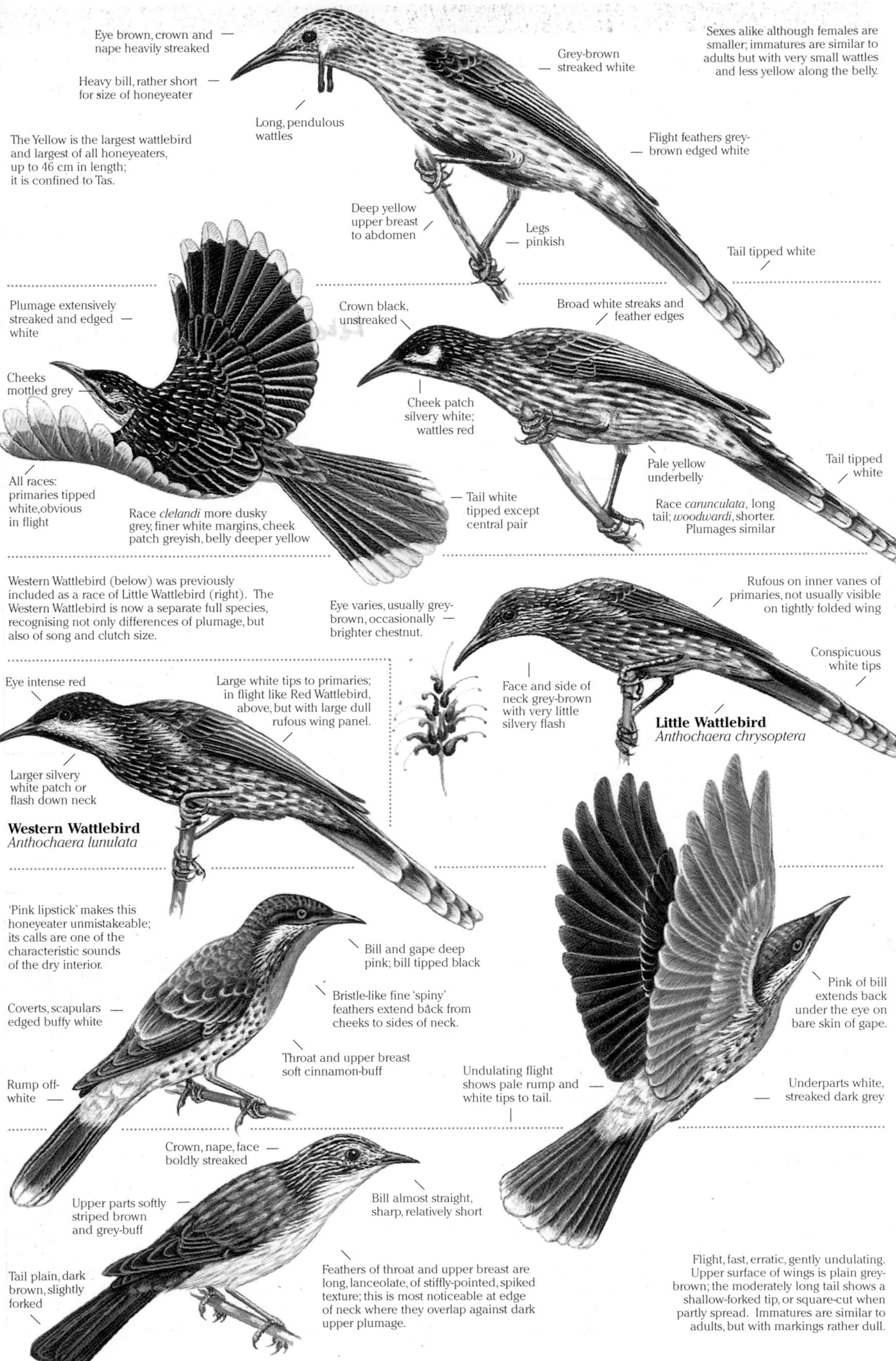

Eye brown, crown and nape heavily streaked
Heavy bill, rather short for size of honeyeater
Long, pendulous wattles
Grey-brown streaked white
Sexes alike although females are smaller; immatures are similar to adults but with very small wattles and less yellow along the belly.
The Yellow is the largest wattlebird and largest of all honeyeaters, up to 46 cm in length; it is confined to Tas.
Flight feathers grey-brown edged white
Deep yellow upper breast to abdomen
Legs pinkish
Tail tipped white
Plumage extensively streaked and edged white
Crown black, unstreaked
Broad white streaks and feather edges
Cheeks mottled grey
Cheek patch silvery white; wattles red
Pale yellow underbelly
Tail tipped white
All races: primaries tipped white, obvious in flight
Race clelandi more dusky grey, finer white margins, cheek patch greyish, belly deeper yellow
Tail white tipped except central pair
Race caruncalata, long tail; woodwardi, shorter. Plumages similar
Western Wattlebird (below) was previously included as a race of Little Wattlebird (right). The Western Wattlebird is now a separate full species, recognising not only differences of plumage, but also of song and clutch size.
Eye varies, usually grey-brown, occasionally brighter chestnut.
Rufous on inner vanes of primaries, not usually visible on tightly folded wing
Conspicuous white tips
Eye intense red
Large white tips to primaries; in flight like Red Wattlebird, above, but with large dull rufous wing panel.
Face and side of neck grey-brown with very little silvery flash
Little Wattlebird
Anthochaera chrysoptera
Larger silvery white patch or flash down neck
Western Wattlebird
Anthochaera lunulata
'Pink lipstick' makes this honeyeater unmistakeable; its calls are one of the characteristic sounds of the dry interior.
Bill and gape deep pink; bill tipped black
Pink of bill extends back under the eye on bare skin of gape.
Coverts, scapulars edged buffy white
Bristle-like fine 'spiny' feathers extend back from cheeks to sides of neck.
Throat and upper breast soft cinnamon-buff
Rump off-white
Undulating flight shows pale rump and white tips to tail.
Underparts white, streaked dark grey
Crown, nape, face boldly streaked
Upper parts softly striped brown and grey-buff
Bill almost straight, sharp, relatively short
Tail plain, dark brown, slightly forked
Feathers of throat and upper breast are long, lanceolate, of stiffly-pointed, spiked texture; this is most noticeable at edge of neck where they overlap against dark upper plumage.
Flight, fast, erratic, gently undulating. Upper surface of wings is plain grey-brown; the moderately long tail shows a shallow-forked tip, or square-cut when partly spread. Immatures are similar to adults, but with markings rather dull.

Helmeted Friarbird *Philemon buceroides* 32–37 cm

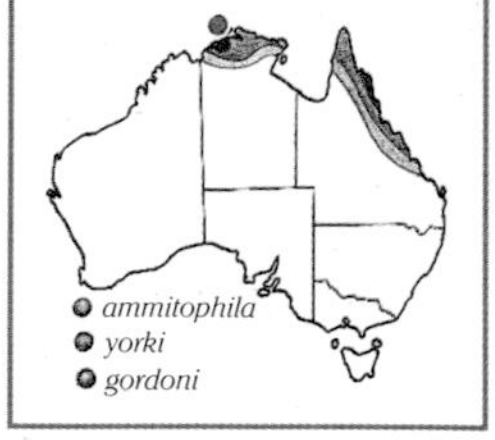

A very large honeyeater, largest of the friarbirds; aggressive toward most other birds around feeding sites; often boisterous and noisy within its foraging mobs. Feeds largely on nectar and fruit, but also insects from flowers or bark crevices, or caught in flight. **Variation:** three subspecies within Aust.: race *yorki* in NE Qld; race *ammitophila*, 'Sandstone Friarbird', Arnhem Land, NT; race *gordoni*, 'Melville Island Friarbird'. **Voice:** varied; includes harsh cackles, squawks and clanking 'chlank'. Some notes are quite musical, but are usually intermixed with harsh sounds: 'warruk, kuk-kaowww, warruk-kuk-kaowww'; 'kaowww-kuk, kaowww-kuk'; and other similar variations; also a higher 'kewik, kewik'. **Similar:** Silver-crowned Friarbird. **Hab.:** in Qld, margins of rainforests, monsoon forests, vine thickets, mangroves, eucalypt woodlands; also, in NT, sandstone escarpments. **Status:** nomadic, follows seasonal flowering; common. **Br:** 392

Silver-crowned Friarbird *Philemon argenticeps* 28–32 cm

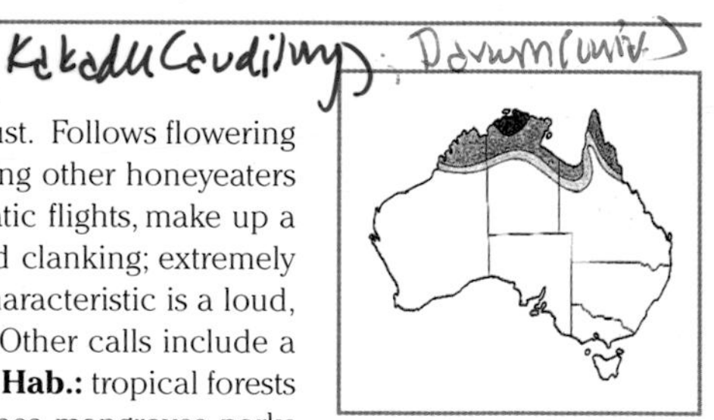

Large, conspicuous honeyeater; a characteristic species of tropical N Aust. Follows flowering of trees in noisy gatherings; aggressively dominates feeding sites, chasing other honeyeaters and lorikeets. Large insects of flowers and bark, taken in short acrobatic flights, make up a large part of the diet. **Voice:** loud, raucous, often rather metallic and clanking; extremely varied, prominent in dawn chorus. Most are typical friarbird noises; characteristic is a loud, rackety, raucous 'karrowk-kak-kuk, karrowk-kak-kuk, karrowk-kak-kuk'. Other calls include a clear, ringing, musical 'cherrowik, cherrowik' and sharp 'aik, aik, aik, aik'. **Hab.:** tropical forests and woodlands, usually of eucalypts and paperbarks; river-edge tree lines, mangroves, parks and gardens. **Status:** nomadic or locally migratory across N; common. **Br:** 392

Noisy Friarbird *Philemon corniculatus* 30–35 cm (L)

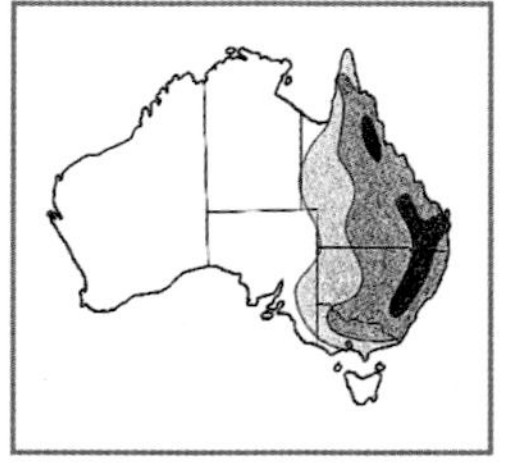

Noisy squabbling at flowering eucalypts makes any mob of Noisy Friarbirds conspicuous; they fight among themselves and aggressively harass other birds that might try to share their feeding territory. After summer, breeding flocks form, their wandering following the flowering of forests; they seek nectar, fruits and insects, and announce their presence with loud rollicking calls. Their flight may be direct or undulating. **Variation:** smaller, mid E Qld. northwards. **Voice:** commonly a deep, rollicking, goose-like honking, 'owk-orrok, owk-orrok, arrowk-kok-carrowk, kor-r-r-owk', parts of which can sound like 'tobacco' or 'four-o'-clock'. Also sharp, metallic 'owk!, owk!'. **Hab.:** open forests, woodlands, swamp-woodlands and watercourse trees; heathlands with banksia, gardens. **Status:** sedentary in N; migratory within S parts of range. **Br:** 392

Little Friarbird *Philemon citreogularis* 25–30 cm

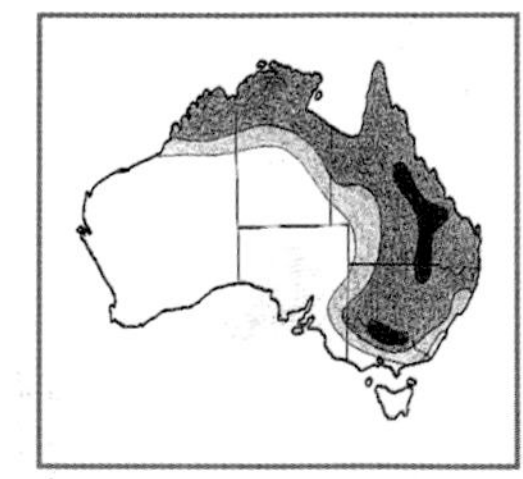

Much smaller than other friarbirds, but shares their pugnacious, noisy behaviour; often chases smaller birds from feeding sites. Nomadic flocks often follow flowering of trees; forages among flowers of outer foliage, taking insects on the wing. In flight shows white tipped tail, quick shallow wingbeats. **Voice:** varies from clear, musical notes, 'chip-wheeoo', 'chip-weik-weeoo' and 'chowk, chowk-ik', to loud, raucous, unpleasant sounds and quite guttural croaking noises. Often those higher, clear, liquid notes are combined with harsh 'scraarch' sounds: 'chwip-weik-skrarrrch-weeoo, chwip-weik-skraarrch-weeoo'; a common sequence is often paraphrased as 'rackety crook-shank'. **Hab.:** open forests, woodlands, river-edge and swamp woodlands, mangroves, gardens. **Status:** nomadic in N; summer migrant to SE Aust. **Br:** 392

Regent Honeyeater *Xanthomyza phrygia* 20–23 cm

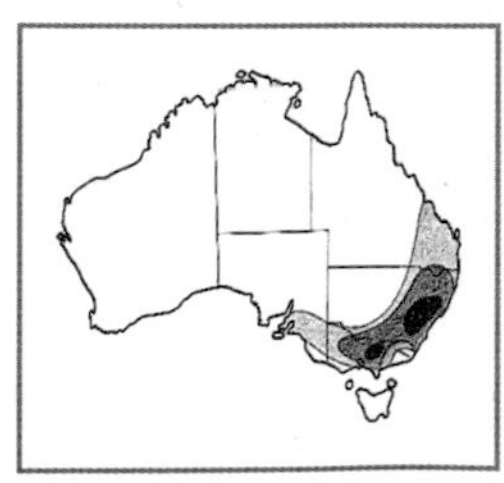

A brilliant black and yellow honeyeater; an extremely active bird of the upper foliage where it dominates its feeding sites and may drive away smaller honeyeaters, busily moving among trees with an audible flutter of wings. **Fem.:** slightly smaller. **Imm.:** browner on back, bill paler. **Voice:** clear bell-like notes, some quite sharp, others deep, rich, mellow, musical: 'quip-quorrip, quip-kip, quorrop-quip'; and sharper 'chlink, chlink'. Some mimicry of other birds. In aggression, makes harsh scolding noises. **Similar:** New Holland and White-cheeked Honeyeaters, but white feathered cheek patches; smaller. **Hab.:** a preference for ironbark forest, but also forests and woodlands of box, yellow gum, swamp mahogany and river oak. **Status:** greatly reduced in range and numbers; migratory with routine circuit of visits to forests as each comes into flower. Breeding migrant to SE. In flocks, formerly quite large, but now usually small. Scarce and endangered. **Br:** 392

Friarbirds are rather grotesque large honeyeaters distinguished by the extensive dark bare skin around the face and the knob-like structures on the top edge of the bill.
Silvery crown; nape has upturned white tufts.
Dark bare skin has rounded rear edge.
Eye bright red
Bill knob or casque is of variable shape and size, usually slopes back, gently rounded or slightly flat topped. Race *ammitophila*, 'Sandstone Friarbird', is larger, very little bill knob; race *gordoni*, 'Melville Island Friarbird', is smaller.
Adults are mottled silvery brown on throat and breast, merging into plainer fawny brown across remainder of underparts.
Back, wings and tail are brown, slightly darker than other species.
Knob very small, usually barely visible; eye brown
Adult
Immature has plainer breast, pale buffy scalloping on upper parts, throat lightly tinted yellow.
Tail plain mid brown without white at tips
Imm.
Both race *yorki*, north-eastern Qld
Bill knob rounded, rises gently from front, drops steeply to forehead. Bare blackish facial skin is shaped to a rearwards point towards the nape.
Forehead and crown silvery white, tufted on nape and hind neck. Throat feathers are grey-buff, silvery pointed.
Immatures have small or no knob on bill; eye is brown.
Back, wings, tail mid fawn-brown
Large friarbird with silvery white crown and prominent sloping casque
Underparts buffy white; faint yellow tint on throat
Isolated population on Cape York (possibly a second race, *kempi*) shows little difference—slightly larger with a casque that, on average, is a little smaller.
Tail plain mid brown, little if any white edging to tips
Adult
Imm.
Head bare, black skinned
Knob rounded, triangular
Mid fawn-brown
Silvery white plumes
Bare, entirely black head makes the Noisy Friarbird easily identifiable.
Two populations, provisionally two races: nominate *corniculatus* from Cape York to about 21° S; further southwards becomes a little larger and very slightly darker or browner, this possibly a second race, *monachus*.
Very small knob
Tail mid brown with obvious broad white tips
Underparts fawny white
Neck and breast are mottled or scalloped; back feathers are edged buff.
Imm.
Smallest, most widely distributed of the friarbirds
Crown plain grey-brown
Lacks knob on bill; bare facial skin is dark blue-grey.
Those from NW Qld through NT to Kimberleys may be a subspecies, *sordidus*, with upper parts and breast dull, mottled rather than plain grey-brown.
Grey-brown
Facial bare skin of immatures is paler, dull brownish; throat is tinted yellow.
Imm.
Tail mid-brown with narrow white tips
Adult
Wide, deep yellow bar across outer primaries; broad yellow outer margins to most flight feathers
Head black except for bare pinkish patch
Scalloped white and pale yellow
Large, boldly patterned honeyeater with bare, rough, salmon-pinkish facial skin
Bare, warty, salmon-pink facial skin
Diamond pattern, black margins
Most of flight feathers and tail edged yellow
White under tail
Sexes are alike, except female is slightly smaller.
Beneath tail, pale yellow
Tail broadly tipped and edged pale to deep yellow

Blue-faced Honeyeater *Entomyzon cyanotis* 25–31 cm

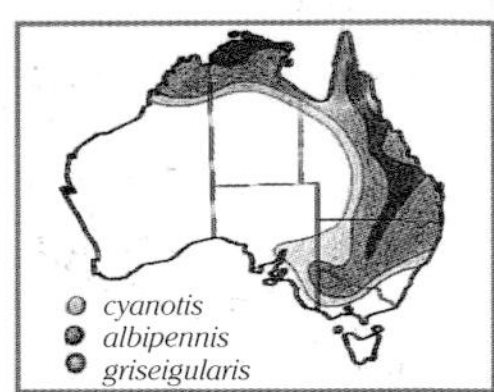

Large, gregarious, aggressive; feeds on nectar, fruits, insects; undulating flight. **Variation:** race *cyanotis*, E Aust.; race *griseigularis*, NE Qld, smaller; race *albipennis*, white underwings, NT, WA. **Voice:** noisy, often a loud, raucous, sharp yet husky 'kieeerk, kieeerk'; repetitive, sharp contact calls, 'whik, whik, whik, whik' and softer 'quorriek, quorriek'. Also a clear, whistled 'quorrieek'; begins low, finishes with loud, high shriek. **Hab.:** open forests and woodlands of eucalypts, paperbarks; river-edge vegetation, pandanus, plantations, gardens, edges of monsoon scrub, mangroves. **Status:** abundant, locally nomadic in N; migratory, uncommon in S of range. **Br:** 392

Bell Miner *Manorina melanophrys* 18–20 cm

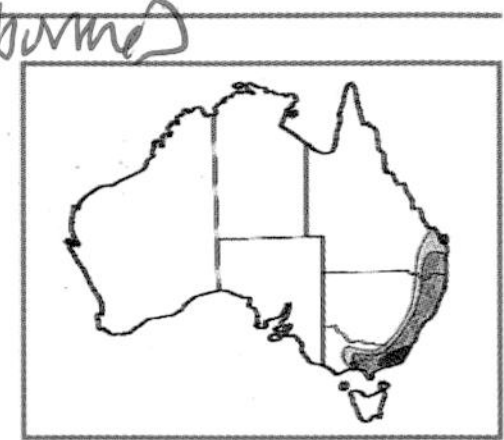

Famed for clear bell-like calls; difficult to sight in forest canopy. Feeds high, largely on lerps and other insects of leaves, flowers and bark. Usually rapid wingbeats, but in pugnacious displays often glides in with wings held up. **Voice:** abrupt, clear, penetrating 'peep'; becomes a 'tink' in the distance. From the hundreds of birds in a large colony, calls make a continual tinkling. Also a harsh 'chak, chak'. **Hab.:** dense, tall, wet eucalypt forests, rainforest margins, creek thickets. **Status:** usually sedentary; abundant. **Br:** 392

Yellow-throated Miner *Manorina flavigula* 26–28 cm

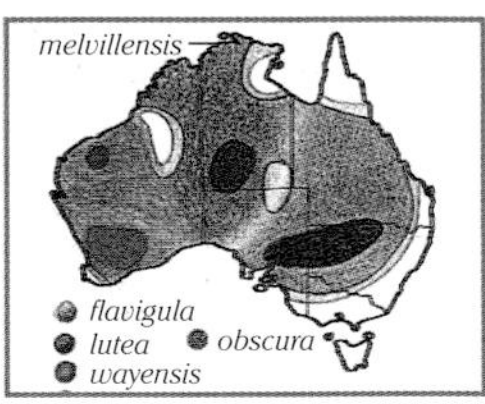

Noisy, gregarious, conspicuous; outside breeding season found in large pugnacious flocks; feeds more on insects taken from foliage, bark or ground than on nectar. **Variation:** mostly gradual, clinal differences. Opinions differ on number, names and boundaries of races; perhaps 4 to 6. **Voice:** loudest calls are in alarm or aggression; sharp nasal 'kiek-kiek-kierk', 'kweek-kewk' or 'kieerk-kieerk'. Also softer, plaintive contact 'chwip' and 'chwiep'. **Hab.:** woodlands, heaths, arid scrublands, grasslands. **Status:** locally nomadic; common. **Br:** 392

Black-eared Miner *Manorina melanotis* 26–28 cm

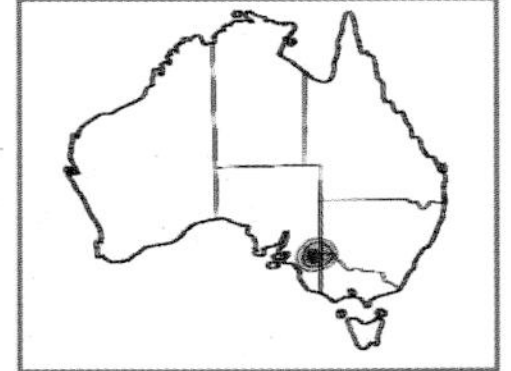

May not be a separate species, but a race, or darkest extreme, of a variable Yellow-throated Miner population; in any event, this dark form is distinctive and well known by its common name. It is endangered by partial clearing, allowing entry by Yellow-throated Miners. Interbreeding has reduced the population; now many hybrids exist. Positive field identification is difficult; reputedly very wary. **Voice:** more harsh, complaining than other miners. **Hab.:** dense mallee, remnant farmland mallee. **Status:** endangered, sedentary. **Br:** 392

Noisy Miner *Manorina melanocephala* 25–28 cm

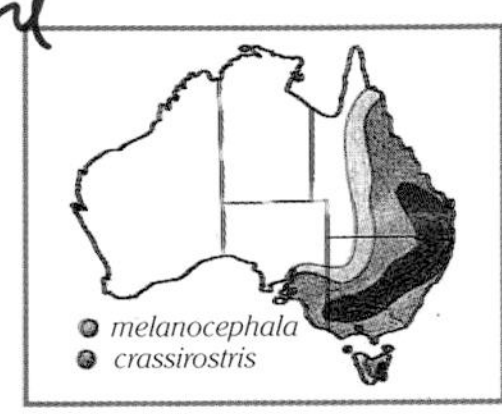

In colonies, often large. Forages treetops to ground; aggressive to other birds. **Variation:** race *crassirostris*, N of Brisbane, is paler. **Voice:** varied; some calls clear, mellow, quite musical, 'chwip-chwip, chwip-chwip', but others loud, raucous, complaining, noisy when taken up by others of flock: 'kairk-kairk-kairk'; harsher 'karrk-karrk-karrrk'. Territorial call, often given in flight, a musical 'tiew-tiew-tiew'. In breeding season, a clear musical song in dawn chorus. **Hab.:** open grassy forests and woodlands. **Status:** locally nomadic; common. **Br:** 392

Macleay's Honeyeater *Xanthotis macleayana* 19–21 cm

Inconspicuous in mid to upper canopy; usually solitary or in pairs; probes for insects or debris suspended in high vine tangles and epiphytes, rotting wood, bark crevices, foliage; nectar and fruits a smaller part of diet. **Voice:** clear, brisk, rather strident 'tsuweit-weet-weet, tsuweit-weet', undulating, with 'weit' each time louder, sharper, the following 'weet' lower, softer. **Similar:** Tawny breasted, paler, further N. **Hab.:** rainforests, monsoon and riverine forests, nearby woodlands, mangroves, gardens. **Status:** sedentary; common. **Br:** 392

Tawny-breasted Honeyeater *Xanthotis flaviventer* 18–21 cm (T)

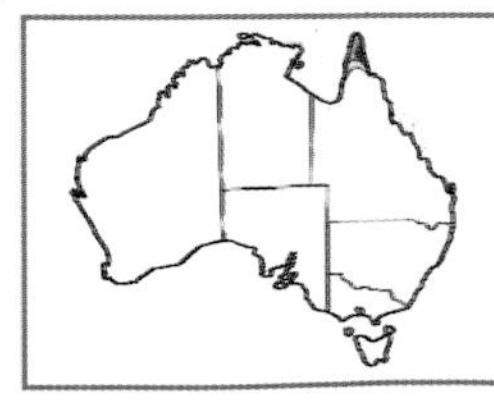

Uses high and mid levels of rainforests; finds insects among clumped epiphytes, tangled vines, bark crevices, but also takes nectar, some native fruits. Usually hard to see in high canopy, but undulating flight is often lower through the forest. **Variation:** many races in NG, race *filigera* on Cape York. **Voice:** series of strong whistles, the first loudest with longer, sharper ending followed by several lower, more abrupt, descending notes: 'WHEEEIT-whit-whut, WHEEEIT-whit-whut'; 'wht-wheeit-whit-wht'. **Hab.:** rainforests, gallery forests; rarely river-edge vegetation, mangrove and paperbark swamps, woodlands. **Status:** sedentary; uncommon. **Br:** 392

Large golden-olive honeyeater with unique bare blue face.
Golden-olive
Broken white nape crescent
Brownish olive, tipped white
Eastern coastal races *cyanotis* and *griseigularis* *both* have inconspicuous cinnamon-buff underwing patches
Adult
Imm.
Olive-green facial skin; or more yellow on juvenile
Bare skin around eye light to deep turquoise on race *albipennis*, slightly more sky blue to cobalt on eastern races *cyanotis* and *griseigularis*
Black 'bib'
Underwing coverts black
Kimberley–NT race *albipennis*, has large white underwing patch at base of primaries.
White tipped
Lores olive to yellow
Upper parts olive-green
Bill golden yellow
Brownish olive
Bare triangle behind eye is vermilion; immatures have paler orange.
Adult
Yellow behind eye and on gape under eye
Crown light grey; nape barred; variable yellow tint to sides of neck
Throat white with yellow tint
Race *flavigula*, rump white to off-white
Crown mid grey
WA race *obscura*, rump pale grey
Throat dark
Race *obscura*, dark mottled
Large whitish tips
Hind neck light grey, barred
In flight, pale rump shows.
Flight feathers edged citrine
Crown blackish grey
Yellow tint to forehead
Plain dark grey
Rump grey-brown, similar to back
Grey, mottled and barred darker grey
No distinct white band, but tips may appear faded pale brown.
Hind neck plain dark grey
Dark throat
Rump grey
Dark tail-tips
Finely barred
Grey-brown, mottled; immatures browner
Distinguished from other miners by dull white forehead merging or quite sharply joining black of crown
Bare skin behind eye is deep yellow with black surrounds from crown down to sides of throat.
Pale grey, barred darker
Rump, lower back, tail coverts are grey-brown.
Tail is dark grey tipped dull white.
A rather plump-bodied honeyeater with streaked plumage; immatures similar, but dull colours.
Black, nape streaked
Conspicuous rufous around eye
Heavily streaked
Breast deep grey, fine white streaks; much paler on immatures
Sides of breast light yellowish chestnut
Tail feathers edged yellow
Bare rufous-buff around eye, tufted yellowish ear to nape; dull on immatures
Off-white line of gape over yellowish chestnut malar streak
Ear coverts pale grey with yellowish or white tufts
Sides of neck grey
Distinctive stepped line through eye
Olive-brown
Upper parts brown with flight feathers, coverts and tail edged pale cinnamon-buff
Ear coverts dusky grey with tiny yellow ear-tuft
Distinctive bare buffy white line, curves under eye, back towards nape
Breast dull cinnamon-buff with fine, faint buffy white streaks
Buffy white
Tail without white tips
Sexes alike; immatures have dull, pale facial markings, greyer cinnamon underparts, finely scalloped buffy margins to mantle, coverts, wings.

Lewin's Honeyeater *Meliphaga lewinii* 19–22 cm
Typically in rainforests; forages actively in pairs or alone about the higher canopy; at times hovers to take insects from foliage, spiralling around trunks and branches, probing bark; also takes fruit and nectar. Undulating flight with audible wingbeat. **Variation:** colours darkest in S; lightest, Cape York population. Race *nea* is of doubtful status. **Voice:** usual call sharp, harshly chattered 'chak-ak-ak-ak-ak'; varies from slow, deliberate to rapid, machine-gun rattle; harsh raucous scold: 'kairrk'. **Hab.:** rainforests, but in far N usually above 200 m; elsewhere, coast to 1000 m and in wet eucalypt forests, woodlands, heaths. **Status:** abundant. **Br:** 393

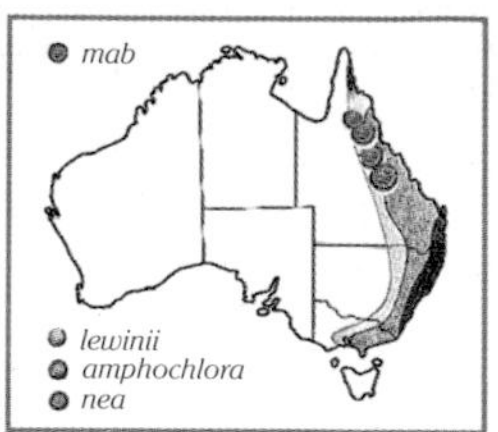

Yellow-spotted Honeyeater *Meliphaga notata* 17–19 cm
Intermediate-sized 'yellow-eared' honeyeater; tends to perch with body semi-horizontal. Noisy, aggressive; solitary, pairs or small parties; forages low to high among foliage. Flight direct, swift, undulating. **Variation:** nominate race *notata*, Cape York to Cooktown; race *mixta*, S from Cooktown, is slightly darker. **Voice:** loud, staccato 'plizk-plizk-plik-plik-plik-plik'; a plaintive, descending 'kiaak-kiaak-kiaak-kiaak'; harsher, descending 'kriaak-kriaak-kriaak'. **Hab.:** rainforests, coast and ranges, usually below 500 m altitude; also monsoon forests, vine scrubs, mangroves. **Status:** sedentary or localised wanderings; common. **Br:** 393

Graceful Honeyeater *Meliphaga gracilis* 14–17 cm
Most likely to be confused with Yellow-spotted Honeyeater; shares range and habitat. More slender, graceful; perches with more upright posture. Has a much more active jizz; restless, flicking wings, fluttering about the foliage seeking insects. Forages at all levels from near ground to high canopy. **Variation**: CY, *gracilis; imitatrix* darker. **Voice:** sharp, abrupt, musical 'tchip' repeated at intervals of several seconds. Similar, but clear, ringing, much faster 'whit-whit-whit-whit-' and weaker 'tsip-tsip-tsip'. **Hab.:** rainforests, lowlands and ranges to about 500 m; adjoining open forests, mangroves, gardens. **Status:** sedentary; common. **Br:** 393

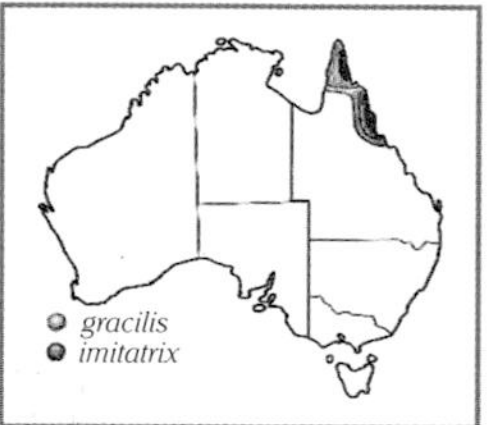

Bridled Honeyeater *Lichenostomus frenatus* 19–22 cm
Often in noisy, busy, pugnacious flocks in canopy of flowering trees. Attracted to nectar, but also seeks insects, often taken in flight, on bark and foliage. **Voice:** a loud, penetrating, rapid, slightly harsh 'chwip-a-whip, chwip-a-whip'. Also sharp, clear notes of its descending song ring in the rainforest, 'chiep-cheep-chierwiep-chier-chiew'. Scolding 'scrartchy-scrartchy'. **Similar:** Eungella Honeyeater, but different locality, gape not yellow. **Hab.:** mountain rainforests, descending to lowlands in winter; occasionally visits flowering trees of nearby eucalypt forests, woodlands, paperbarks. **Status:** locally nomadic; common. **Br:** 393

Eungella Honeyeater *Lichenostomus hindwoodi* 17–19 cm
Confined to small area of plateau rainforest in Clarke Ra., Qld. Noisy and aggressive around flowering trees and mistletoes, but otherwise may be rather wary and hard to locate in rainforest canopy. Flits about outer foliage and flowers to capture flying insects, and takes other small creatures from bark and foliage. **Voice:** extremely high, clear, sharp, almost scratchy, notes often preceded by low, abrupt, 'tchup' sounds: 'tchup, tchup, chwiep-chiep-cheep-chip-a-wiep'. **Hab.:** rainforest, lowest shrubs to canopy and outer margins; follows flowering of nearby riverine and open forests. **Status:** locally migratory; common. **Br:** 393

Yellow-faced Honeyeater *Lichenostomus chrysops* 16–18 cm
Busy, active forager in foliage of trees, shrubbery, low heath; feeds largely on manna flows through summer and autumn, nectar in spring, but always takes insects of foliage and bark. **Voice:** cheery ringing calls, 'whit, whit, whit' and 'chwikup, chwikup', these often combine in brisk, rollicking song: 'WHIT-chiwit-chiwickup, WHIT-chiwit-chiwickup, WHIT-chiwickup-chiwickup' with many variations. In flight, flock keeps together with soft 'clip, clip'; at other times use a rather peevish 'kreee' to maintain contact. **Variation:** probably insignificant. **Similar:** Varied, Singing and Mangrove Honeyeaters, but lack black beneath their yellow facial streaks; Bridled, but larger, basal third of bill yellow. **Hab.:** forests, woodlands, heath on high ranges and coastal lowlands; mangroves. **Status:** migratory; large flocks move N in autumn following Great Divide and coast—winter in northern NSW and SE Qld, return to breed in southern forests in spring. Common. **Br:** 393

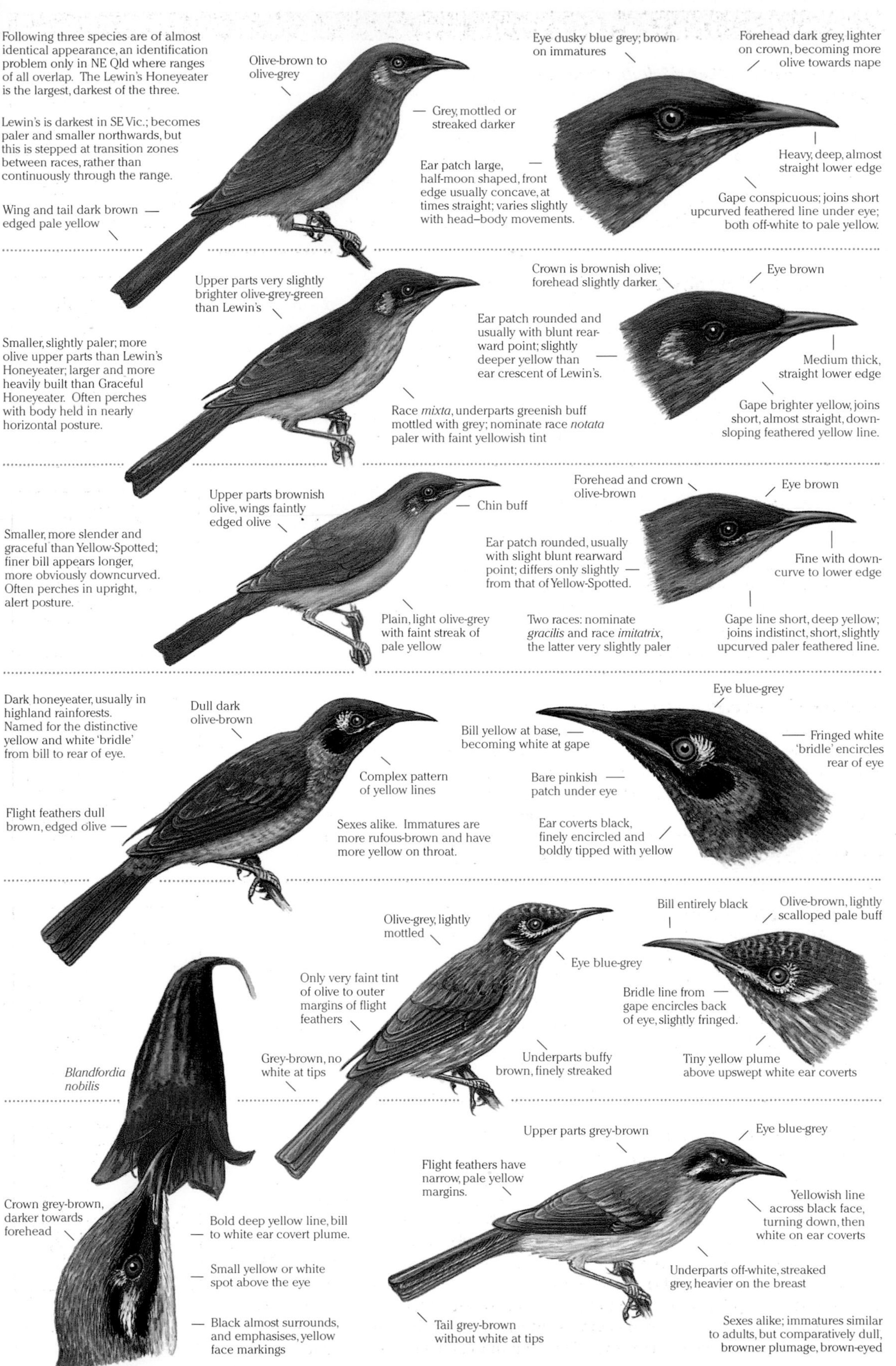
Following three species are of almost identical appearance, an identification problem only in NE Qld where ranges of all overlap. The Lewin's Honeyeater is the largest, darkest of the three.
Lewin's is darkest in SE Vic.; becomes paler and smaller northwards, but this is stepped at transition zones between races, rather than continuously through the range.
Wing and tail dark brown edged pale yellow
Olive-brown to olive-grey
Grey, mottled or streaked darker
Ear patch large, half-moon shaped, front edge usually concave, at times straight; varies slightly with head–body movements.
Eye dusky blue grey; brown on immatures
Forehead dark grey, lighter on crown, becoming more olive towards nape
Heavy, deep, almost straight lower edge
Gape conspicuous; joins short upcurved feathered line under eye; both off-white to pale yellow.
Upper parts very slightly brighter olive-grey-green than Lewin's
Smaller, slightly paler; more olive upper parts than Lewin's Honeyeater; larger and more heavily built than Graceful Honeyeater. Often perches with body held in nearly horizontal posture.
Race mixta, underparts greenish buff mottled with grey; nominate race notata paler with faint yellowish tint
Crown is brownish olive; forehead slightly darker.
Eye brown
Ear patch rounded and usually with blunt rearward point; slightly deeper yellow than ear crescent of Lewin's.
Medium thick, straight lower edge
Gape brighter yellow, joins short, almost straight, downsloping feathered yellow line.
Upper parts brownish olive, wings faintly edged olive
Chin buff
Smaller, more slender and graceful than Yellow-Spotted; finer bill appears longer, more obviously downcurved. Often perches in upright, alert posture.
Plain, light olive-grey with faint streak of pale yellow
Forehead and crown olive-brown
Eye brown
Ear patch rounded, usually with slight blunt rearward point; differs only slightly from that of Yellow-Spotted.
Fine with downcurve to lower edge
Two races: nominate gracilis and race imitatrix, the latter very slightly paler
Gape line short, deep yellow; joins indistinct, short, slightly upcurved paler feathered line.
Dark honeyeater, usually in highland rainforests. Named for the distinctive yellow and white 'bridle' from bill to rear of eye.
Dull dark olive-brown
Flight feathers dull brown, edged olive
Complex pattern of yellow lines
Sexes alike. Immatures are more rufous-brown and have more yellow on throat.
Eye blue-grey
Bill yellow at base, becoming white at gape
Fringed white 'bridle' encircles rear of eye
Bare pinkish patch under eye
Ear coverts black, finely encircled and boldly tipped with yellow
Olive-grey, lightly mottled
Eye blue-grey
Only very faint tint of olive to outer margins of flight feathers
Grey-brown, no white at tips
Underparts buffy brown, finely streaked
Bill entirely black
Olive-brown, lightly scalloped pale buff
Bridle line from gape encircles back of eye, slightly fringed.
Tiny yellow plume above upswept white ear coverts
Blandfordia nobilis
Crown grey-brown, darker towards forehead
Bold deep yellow line, bill to white ear covert plume.
Small yellow or white spot above the eye
Black almost surrounds, and emphasises, yellow face markings
Upper parts grey-brown
Eye blue-grey
Flight feathers have narrow, pale yellow margins.
Yellowish line across black face, turning down, then white on ear coverts
Underparts off-white, streaked grey, heavier on the breast
Tail grey-brown without white at tips
Sexes alike; immatures similar to adults, but comparatively dull, browner plumage, brown-eyed

White-lined Honeyeater *Meliphaga albilineata* 19–20 cm

Usually rather quiet, inconspicuous; at flowering trees can compete aggressively and noisily with other small birds for nectar, native fruits, insects. **Voice:** whistled calls, either slightly melancholic or brisk and cheery; sound often amplified by surrounding cliffs: 'tiew, tiew'; 'twieoo, tioo'; 'whee-a-whit, wheee-a-whit'; 'chwip, chwip'; 'whiew-wirrit-whirroo-wheeit-tiew'. **Hab.:** sandstone gorges, valleys, escarpments in watercourse, monsoon and paperbark vegetation, and in nearby eucalypt woodlands. **Status:** sedentary; common. **Br:** 393

White-gaped Honeyeater *Lichenostomus unicolor* 19–22 cm

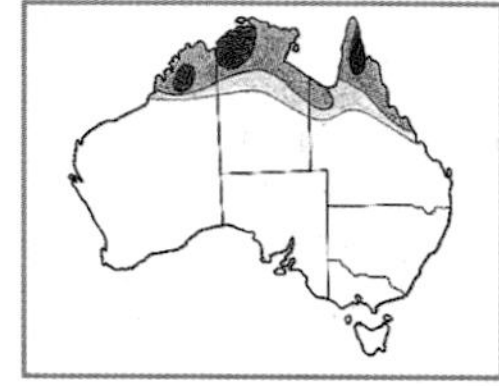

Often in dense canopy on limbs or suspended flood debris; seeks insects, spiders; briefly hovers near foliage, flowers. Alone or in very small groups, occasionally in larger gatherings of mixed species at flowering trees; then is more noisy, aggressive. **Voice:** clear, sharp whistle, 'tchiep, tchiep'; slightly husky 'wheit'; scratchy, scolding 'tzzeip'. Fast rollicking song, 'tchicka-tchicka-WHEIT-tchicka-tchicka-WHEIT'. **Hab.:** mangroves, paperbark swamps, riverine forest and pandanus, lush gardens. **Status:** common NT to Kimberley; sparse N Qld. **Br:** 393

Singing Honeyeater *Lichenostomus virescens* 18–22 cm

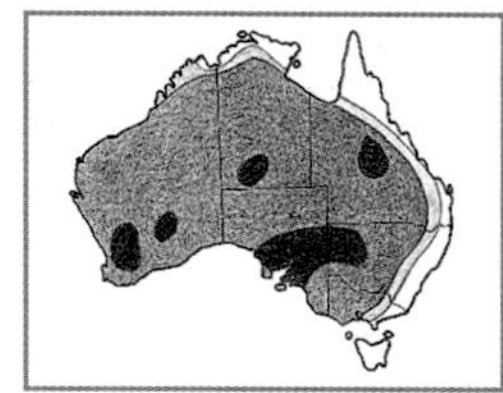

Widespread, common; in pairs or small parties, often solitary. Seeks berries, nectar, insects in shrubs, trees, on ground. **Voice:** typically an abrupt, musical 'prrip, prrip' at intervals of several seconds; faster if agitated, very fast at higher pitch when alarmed, 'prrrrrrrrt'. Song is best in dawn chorus; includes high, clear and deep, mellow notes, varied, 'cheewip-chip-quorricheep-quorit-chiwip'. **Hab.:** dry scrubby woodlands, mallee, mulga and other inland scrubs, sandplain and dune thickets, mangroves. **Status:** sedentary or nomadic; abundant. **Br:** 393

Mangrove Honeyeater *Lichenostomus fasciogularis* 18–21 cm

Congregate in noisy, frenzied activity at heavily flowering trees; aggressive toward smaller birds. Pairs or small parties forage for insects from upper foliage down to mangrove mud. **Variation:** may be combined with Varied Honeyeater; on present evidence retained separate. **Voice:** first note loud, clear. Following notes mellow, lower, 'CHWIP, chwiew, chwiuw, chwuu'. Repeated rapidly by many birds, becomes distinctive rollicking chorus. Softer 'chwiek, chwouk, chwouk'. **Similar:** Varied, but yellow underparts. **Hab.:** coast, offshore islands, usually in mangroves, but visits nearby eucalypt and paperbark woodlands, gardens. **Status:** common. **Br:** 393

Varied Honeyeater *Lichenostomus versicolor* 18–21 cm (T)

Roosts, nests and feeds in mangroves. Like previous species, descends to mangrove mud to take tiny crustaceans. Travels with swift, undulating flight to feed further afield, including offshore islands. Active, pugnacious in mixed gatherings at flowering trees. **Voice:** sharp, brisk, cheerful 'whitchu, whitchu'. Also piercing 'tchiew', often combined with deep, mellow chuckling: 'tchiew, quorrit-you, quorrit-you'. **Hab.:** mangroves, coastal woodlands, island coconut groves, gardens of coastal towns. **Status:** locally nomadic; common. **Br:** 393

White-eared Honeyeater *Lichenostomus leucotis* 19–22 cm

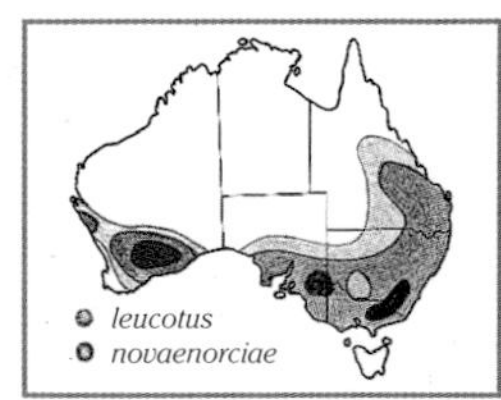

Actively searches for insects, spiders; often works over bark in treecreeper fashion. Prefers trees where peeling, flaking bark harbours insects. Nectar and fruits taken when available. Usually solitary, at times in small family groups. **Voice:** deep, mellow but slightly metallic 'chwok, chwok', 'choku-whit, choku-whit' and 'kwitchu, kwitchu'; very sharply scratchy, metallic 'chwik!'. **Hab.:** forests, woodlands, heathlands, mallee and dry inland scrublands. **Status:** sedentary, nomadic or local migratory movements; moderately common. **Br:** 393

Yellow Honeyeater *Lichenostomus flavus* 17–18 cm

Extensively yellow. Busy, almost constantly active in canopy and outer foliage; darts among trees; often in pairs or small groups, occasionally large or mixed-species flocks at flowering trees; audible wingbeat. **Variation:** smaller, brighter in N; probably clinal. **Voice:** high, yet slightly husky whistle, 'whee-whiew, whee-whiew'; scolding 'chzuk, chzuk, chuk-chuk'. **Hab.:** rainforest margins, riverine and swamp vegetation, eucalypt and paperbark forests, monsoon scrubs, mangroves, adjoining woodlands. **Status:** sedentary; locally common. **Br:** 394

The White-lined and the White-gaped (below) are two similar grey-brown northern honeyeaters, both have small whitish facial markings.
Upper parts grey-brown — mottled darker
Wings and tail faintly edged olive —
Pale grey to dull white, mottled brown
Eucalyptus phoenicea
Bill medium length, sturdy, downcurved
Eye blue-grey
Crown dark grey-brown
Bare gape is pale yellow. —
White upswept line under eye, usually around ear coverts
Plain grey-brown
Wide curve of bare gape is usually creamy white.
Eye grey to grey-brown
Immature has pale yellow skin on wide bare gape; creamy or off-white on adults.
Wings and tail faintly edged olive
Underparts grey-brown with very faint paler creamy mottling
Three species, far right and below, have long black masks underlined by yellow and white.
The Singing is one of the best known honeyeaters. It is widespread, common, and has clear, pleasant, easily remembered calls and bold, distinctive plumage markings. Variation gradual, apparently clinal, across the continent—darkest in humid environs, palest towards arid regions; large in S to small in N.
Bold black streak through eye and down onto sides of neck; underlined by yellow and white.
Crown and back vary from grey-brown in the SW to olive-brown in the SE of range.
Underparts dull off-white streaked grey-brown, often with a slight yellowish tint.
Wings and tail brown, edged yellow
Throat pale yellow or off-white, roughly barred or scaled dark grey
Black through eye down side of neck, underlined yellow or white; side of neck is white.
Blackish or grey-brown zone across breast
Wings and tail brown with narrow yellow margins
Olive-brown, softly mottled darker brown
Black and yellow across face
White streak down sides of yellow neck
Narrow brown streaks
Varied is similar to Mangrove and Singing, but has yellow across almost the entire underparts.
Immatures are like the adult; dull, paler streaks, brown eye.
Wings, tail edged and tinted yellow-olive
Mangrove and Varied overlap near Townsville, where some interbreeding occurs.
Crown dark grey, streaked black
White ear patch
Race *leucotis* of the E coast has intense green upper parts, and light greenish yellow on the belly. Inland, W of the Great Dividing Ra. these colours gradually become more dull olive, the birds smaller. This trend continues W until at Eyre Peninsula they are, in plumage and size, more like the WA race *novaenorcia*. There remains uncertainty where the boundary between races should be placed, and whether there are 2 or 3 races.
Upper parts, including wings and tail, are olive-green.
Black of face extends to upper breast
Rear of brow yellow
Upper parts olive-green, flight feathers brown, edged bright yellow
Adult
Race *novaenorciae*
Imm.
Darker through eye and gape. Line under eye is bright yellow.
Overall has rather plain yellowish, small-headed, plump-bellied jizz.
Underparts entirely pale yellow
Male and female are alike. Immature similar but lacks bold contrast of adult head pattern; crown is dull olive, black is replaced by dusky grey-brown, eye is brown.

Yellow-throated Honeyeater *Lichenostomus flavicollis* 18–21 cm

Confined to Tas., Bass Str. and islands. Exploits greatly diverse habitats, preferably where crevices and debris in rough or peeling tree bark provide abundant insects and other small prey. Scale insects are taken from foliage; flying insects are snapped up in brief aerial pursuits. Nectar and fruit appear make up a smaller part of the diet. Longer flights are swift, undulating, rather twisting or erratic. Solitary, in pairs, or, briefly after nesting, in small family parties. **Voice:** rich, mellow, rhythmic, musical 'whit-chorr, whit-chorr', 'chorr, chorr-chorr'; slightly higher 'whit-cheow'. Aggressive, rattling, sharper 'chop-chorr–chor-r-r-rr'. **Hab.:** wet forest to open woodlands, mallee and heaths. **Status:** sedentary; common. **Br:** 394

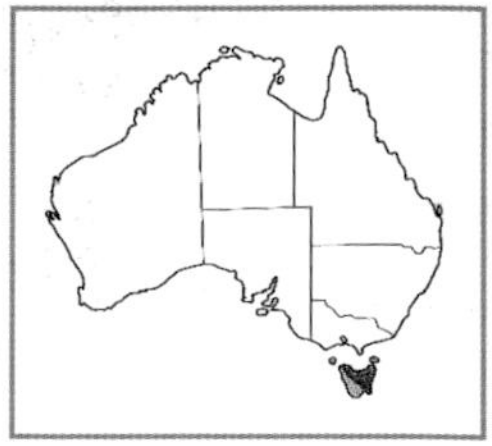

Yellow-tufted Honeyeater *Lichenostomus melanops* 17–23 cm

A noisy, active species; often in colonies of a few to a hundred or more; aggressive around flowering trees. Feeds on nectar, sap flows, fruit and insects from flowers or foliage, taken in flight. Often forages rather like treecreepers, over bark of trunks and limbs. Usually active in mid to upper storeys, occasionally in shrubs of understorey. **Voice:** whistled 'wheit, wheit' and 'tseit, tseit'; harsher, scolding 'chzeet'. Infrequent whistled song. **Hab.:** eucalypt forests and woodlands with shrub undergrowth; also mallee, brigalow, cypress. Race *cassidix*, 'Helmeted Honeyeater', in swamp-gum woodlands with melaleuca and tea-tree undergrowth. **Status:** nominate race nomadic, common; race *cassidix* rare and endangered. **Br:** 394

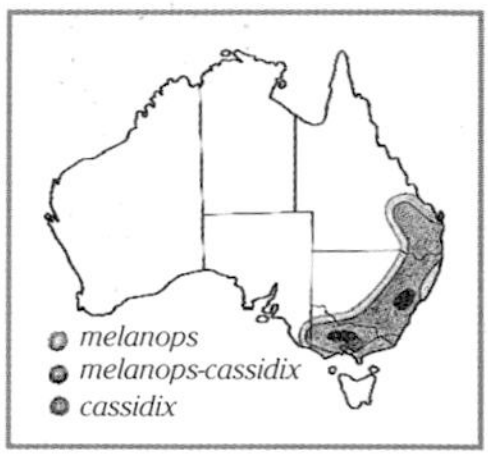

Purple-gaped Honeyeater *Lichenostomus cratitius* 16–19 cm

Rather shy, unobtrusive mallee honeyeater; usually alone or in pairs, at times in small parties at flowering trees. Forages in foliage of mallee and other vegetation for insects, spiders and nectar. **Voice:** some calls have distinctive, metallic, vibrant quality, especially a 'chairk, chairk, chairk', but evident in many notes. Also a clear, pleasant 'quitty' followed by contrasting louder, harsh, peevish calls: 'quitty-quitty-KAIRRK-KAIRRK'. In alarm or aggression, abrupt 'chek-chek-chekchekek'. **Similar:** Grey-fronted, Yellow-plumed, but both have streaked underparts, neither has lilac gape line. **Hab.:** mallee, open woodlands with shrubby understorey, broombush, heathlands. **Status:** locally nomadic; moderately common. **Br:** 394

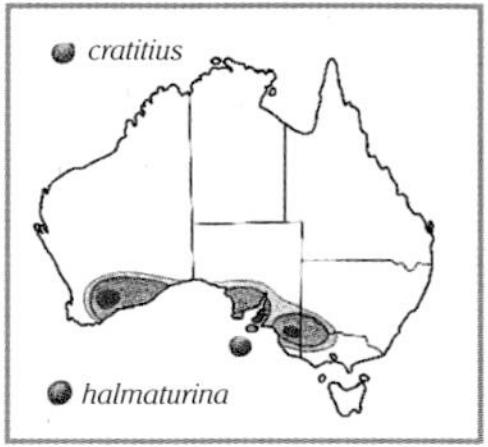

Grey-headed Honeyeater *Lichenostomus keartlandi* 15–16 cm

Northern arid regions, usually in ranges. Often gather in large numbers at sites of profuse flowering; many pairs take advantage of such favourable conditions, usually several months after heavy rains, to breed at almost any time of the year. A busy, active, noisy species; darts among flowering trees and shrubs. **Voice:** mellow, musical 'ka-towt, ka-towt'; clear, whistled 'whei-wheit-wheit-'; also a harsh, nasal 'kraak'; rapid rattling alarm trill. In flight, 'chek, chek'. **Hab.:** rocky gorges and escarpment of arid ranges with scattered trees, diverse shrubs, spinifex; less often seen in tree belts of larger rivers. **Status:** nomadic; common. **Br:** 394

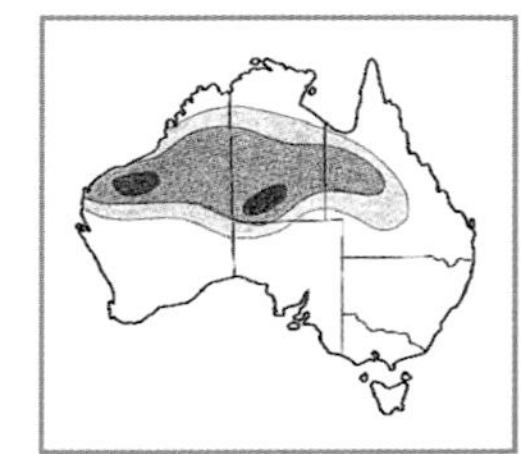

Yellow-plumed Honeyeater *Lichenostomus ornatus* 15–17 cm

A noisy, pugnacious species; its calls characterise habitats where it is common. Small flocks forage in foliage of trees and shrubs, finding insects, spiders and other small prey on leaves, bark and ground litter. Flocks quarrel noisily at flowering trees, resorting to physical attacks on each other and driving away other species. **Voice:** loud, clear, cheery, rollicking calls, 'WHIT-chier, WHIT-chier', 'whit-whit-CHIER, whit-whit-CHIER' and 'CHWIEP-ier, CHWIEP-ier'. Alarm call a loud, high, penetrating, musical trill; carries far through bushland, quickly taken up by others in group. Squabbling, a harsh, hard 'chzak-chzak-chzak'. **Hab.:** woodlands, open forests, mallee, heaths. **Status:** locally nomadic; common. **Br:** 394

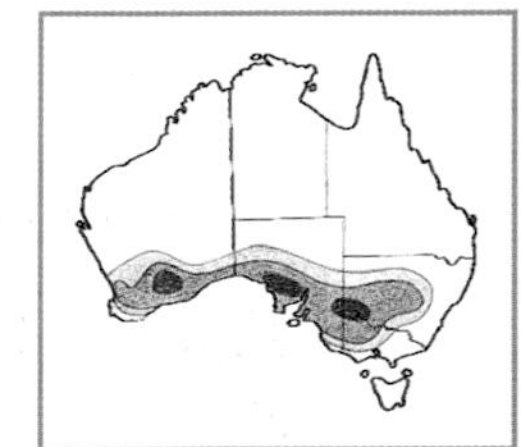

Grey-fronted Honeyeater *Lichenostomus plumulus* 15–17 cm

A busy, active honeyeater, seeking nectar; snatching insects in flight. In breeding season, males make erratic song flights, calling from high over mallee canopy. After breeding, often nomadic—small groups wander in search of flowering trees. Around flowers often aggressive towards other honeyeaters. **Voice:** a rapid, mellow, musical 'kwit, kwit-kweeit, kwit-kwit' and rather peevish 'queeeit-queeit-queeit'. Much sharper, more penetrating, loud, whistled 'peet-peet-peet'. Lower, persistent 'chwok, chwok'. **Hab.:** thickets of mallee; dry woodlands, mulga and other scrub; watercourse tree belts. **Status:** uncommon. **Br:** 394

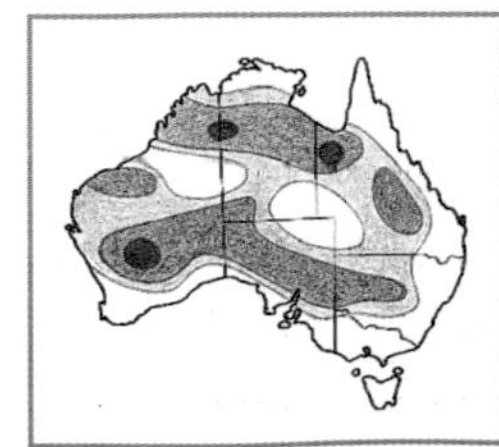

Widespread *melanops* intergrades with *cassidix* across Gippsland and SE of NSW. Population along the inland, W side of Great Divide may be a third race, *meltoni*.

Race *cratitius*, SW and SE mainland Aust., and race *halmaturina*, Kangaroo Is., SA: the latter is larger and darker. Opinion seems divided whether the latter population differs sufficiently to justify its recognition as a subspecies. If so, in naming this race (with Kangaroo Is. perhaps being the type locality for *cratitius*), the mainland population would take the next available name, *occidentalis*.

Male and female are alike. Immature plumage is dull; bare gape is yellowish.

The Grey-headed Honeyeater is similar to the Purpled-gaped, but occurs further N. Its range overlaps that of the Grey-fronted Honeyeater, which has heavier streaking of underparts and large yellow neck plume.

The Grey-fronted Honeyeater occupies a huge area of the semi-arid interior. Its variation across the continent may be clinal, gradual. If changing in recognisable steps, this could justify several races that have previously been proposed: *plumulus* across the central and western interior; the larger and darker *graingeri* in the mid-eastern interior, and *planasi*, with pale grey back and smaller patch of grey on the forehead, across the northern interior.

Sexes are alike; immatures have slightly dull colours, pale eye-ring.

Fuscous Honeyeater *Lichenostomus fuscus* 15–17 cm

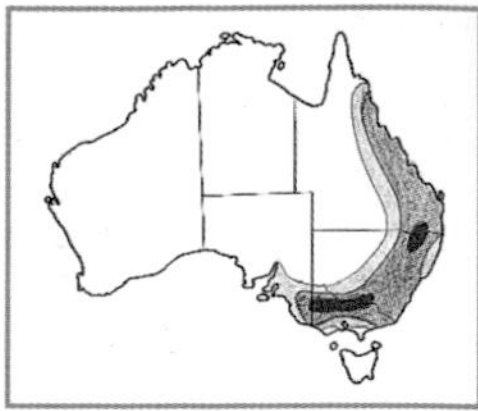

Drab grey-brown, dark-eyed, gregarious; in pairs or small to quite large flocks. **Br.:** bill entirely black, eye-ring dark. **Variation:** opinions vary: gradual, clinal variation, therefore no races; or 2 races—yellowish *subgermanus* in NE Qld S to Bowen–Mackay, thence dusky race *fuscus* to Vic, SA. **Voice:** calls include sharp 'twietch' followed by a rapid, hard rattle, 'twietch-chuka-chuka-chuka-', and sharp, abrupt 'tchwip, tchwip-tchwip, tchwip'. Songs include a ringing, clear, rhythmic 'teechu, teechu, tichu'; in song flight a faster, lilting, clear 'tichu-tichu-tichu'. Calls vary regionally. **Hab.:** forests, woodlands, rainforest margins, drier inland mallee and acacia scrubs, coastal heaths. **Status:** sedentary, locally nomadic, seasonal movement in SE; common. **Br:** 394

Yellow-tinted Honeyeater *Lichenostomus flavescens* 15–17 cm

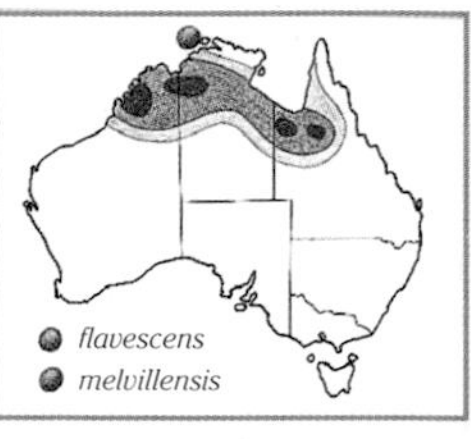

Overall yellowish, black diagonal at neck. Active, pugnacious, often in small flocks. Swift undulating flight; acrobatic pursuit of flying insects. **Variation:** nominate *flavescens* across N Aust.; race *melvillensis*, Melville Is., NT. Fuscous Honeyeater was previously included as a subspecies. **Voice:** contrasting calls: clear, high, even 'tiew-tiew, tiew-tiew'; harsh, rather raucous, complaining 'chaerp, chaerp'. More cheerful, a clear, high, whistled 'whi-whit, whi-whit' and quick, abrupt 'quorr-quorr, quorr-quorr-a-whit'. **Similar:** Fuscous, but olive-brown, range to E. **Hab.:** open forests and woodlands, favouring vicinity of rivers, watercourse tree belts, occasionally mangroves. **Status:** locally nomadic, follows flowering of trees; moderately common. **Br:** 394

White-plumed Honeyeater *Lichenostomus penicillatus* 15–17 cm

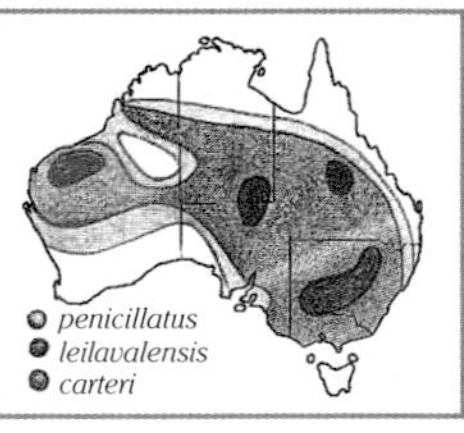

Widespread, conspicuous, noisy, gregarious; small loose flocks forage tree canopy to ground for nectar and insects. Feisty, alert; alarm calls ripple bird to bird at hint of danger from raptor or kookaburra. **Variation:** darker in SE; paler with more extensive yellow in arid and NW part of range. Opinions differ—subspecies or clines? How many? **Voice:** calls 'whitch-a-whee, whitch-a-whee-whit', 'whit-a-wheeit, whita-wheeit'. In song flight, clear, musical 'whee-whit-a-wheioo'. In alarm, raucous, nasal 'cak-ak-ak-ak-ak-ark'. **Hab.:** woodlands, mallee, tree-lined inland rivers. **Status:** usually sedentary; common. **Br:** 394

Green-backed Honeyeater *Glycichaera fallax* 11–12 cm

Active about the upper canopy, but small and inconspicuous; at times descends to lower shrubbery. More like flycatcher or white-eye than honeyeater; feeds largely on insects, less reliant on nectar. Clings among hanging twigs and leaves; flutters and hovers to snatch small prey from outer foliage or occasionally takes insects in aerial pursuit. Solitary, pairs, small groups or mixed parties with gerygones, fantails and other small birds. **Variation:** one race in Aust., others Aru Is., NG, and nearby islands. **Voice:** as contact, a high, weak 'piep'. A whistled 'whiet' is slightly husky yet quite sharp, often in brisk, even bursts of 4 to 10 notes: 'whiet-whiet-whiet-wheit'. **Similar:** Fairy Gerygone, but eye red, bill shorter, male brighter yellow. **Hab.:** rainforests, monsoon forests, vine scrubs, adjoining eucalypt woodlands. **Status:** sedentary; moderately common. **Br:** 394

Brown Honeyeater *Lichmera indistincta* 12–16 cm

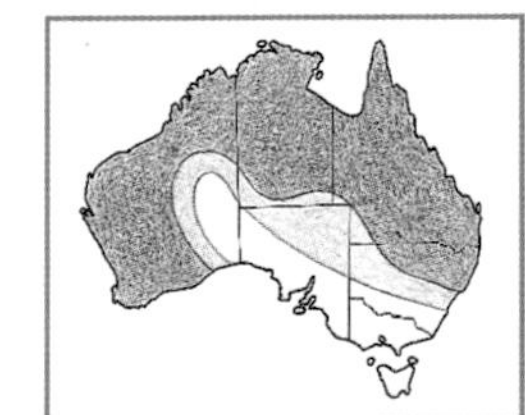

Principally a nectar-feeder; long bill well adapted to feed at deep flowers. Also takes insects in hovering and acrobatic pursuit. **Voice:** clear, ringing, musical, usually delivered in erratic bursts, 'whit, whit, whitchit'; hesitant 'whit,whit, whit, quorrit, quit'; vigorous 'quorrit-quorrit-quorrit'; rapidly trilled 'whitchit, whit, whit-whit-witiwitwitwit'; hesitant, disjointed pre-dawn version. **Hab.:** diverse—forests, woodlands, heaths, mulga, other arid scrubs, watercourse trees, mangroves, gardens. **Status:** sedentary, locally nomadic; common. **Br:** 395

Brown-backed Honeyeater *Ramsayornis modestus* 12–13 cm (T)

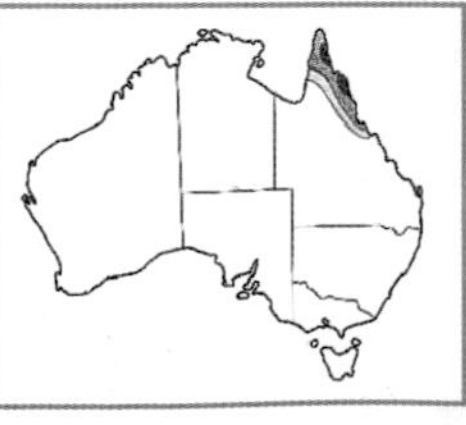

Often in loose flocks with other small honeyeaters; busily seeks nectar from canopy to low shrubbery. Captures insects around flowers, frequently on the wing; often on paperbark branches finding insects and spiders in bark crevices. **Imm.:** gape area buff, breast streaked. **Voice:** in contact, a sharp 'chwiet' and softer 'chwit' as flight call. Similar scratchy, sharp notes, slightly descending, slowing and fading: 'chwit-chwit-chwich-chwich-chwech-chwech'. **Similar:** immatures of Rufous-banded, Rufous-throated, female Scarlet Honeyeater, but no under-eye streak or breast bars. **Hab.:** swampy paperbark woodlands, river and rainforest edges, mangroves, nearby eucalypt woodlands. **Status:** nomadic or migratory; moderately common. **Br:** 395

Plume small, pale yellow, tucked behind blackish diagonal or point at rear edge of ear coverts
Back is dull grey-brown to olive-brown.
Wings and tail are olive–brown edged olive-yellow.
Slightly more yellow in N of range
Black
Dark eye-ring
Pale buffy-grey
Adult: breeding
Yellow base to bill
Fine yellow eye-ring
Immature has smaller, whiter plume at side of neck.
Nonbreeding adult similar, but less yellow on eye-ring gape and base of bill.
Flight feathers with yellow along outer margins
Yellow neck plume is not as obvious on races that have deeper rather than pale greyish plumage.
Tail edged yellow
Immatures are similar, yellowish base to bill, dull plumaged.
Head pale soft yellow, slightly greyer on crown
Upswept black line at rear of ear-coverts, with matching yellow plume
Nominate race flavescens has under-parts off-white to yellow-tinted, faintly streaked grey
Heavier yellow streaking
Race melvillensis
Western race: head yellow
Diagonal black line, white plume
Western race carteri has brighter, more extensive yellow extending as pale tint to underparts.
Back and wings of western and central Aust. races are lighter grey-brown.
Adult
Eastern race: crown grey, forehead yellow
Eastern race penicillatus has less yellow; underparts are pale grey-brown.
Imm. has yellow gape, and yellow on face around eye and on wing margins; remainder is pale grey-brown.
Back and wings dark olive-brown to olive-gey
Wings and tail edged yellow or olive, all races
Crown grey merging to olive-green on nape and back
Eye pale grey
Bill almost straight
Throat pale grey
A small honeyeater with almost straight bill; jizz and habits more like flycatcher or white-eye
Pale yellow softly mottled grey on breast
Wings grey-brown, flight feathers edged yellow
Adult
Immatures are like adults, but have dark brown eyes.
Bill quite long, downcurved, suited to feeding at deep tubular flowers
Small yellow triangle behind the eye
Anigozanthos manglesii
Upper parts olive-brown
Common, widespread, rather plain olive-brown honeyeater with long, rather heavy bill. Provisionally 3 races: ocularis Cape York S to NW of NSW and mid E coast Aust.; indistincta N Aust. and most of WA; melvillensis on Melville and Bathurst Islands
Flight feathers and tail edged yellow
Gape black on males, fine yellow line on females; immatures yellowish on and around gape, paler bill
Melaleuca viridiflora
Rufous-brown back appearing dark brown in shade; underparts white indistinctly barred and streaked brownish buff
Crown light brown with small darker flecks
Bill pinkish brown
Light to mid rufous-brown, softly mottled darker
Face darker, bill to ear coverts; white line under eye
Underparts dull white with faint, untidy brownish bars and streaks; immatures finely streaked
Tail plain brown
Legs pinkish brown

Bar-breasted Honeyeater *Ramsayornis fasciatus* 13–15 cm

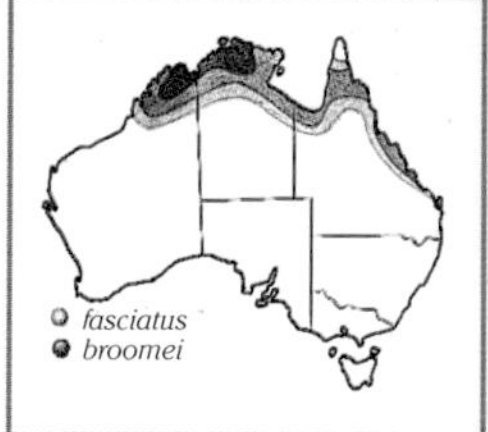

Typically found along tree-lined watercourses of ranges and floodplain paperbark swamps. Presence often revealed by large, domed nests of paperbark suspended from foliage, usually hanging quite low over the water; numerous along favoured waterways. Rather quiet, alone or in small groups busy in the canopy of flowering trees. Nectar is eagerly sought, but insects are also taken. Flight is swift, slightly undulating. **Voice:** sharp, slightly metallic, piercing 'chwit-chwit-chwit-chwit'; similar, faster 'chwee-chwee-chwee'. Also a lower pitch, rather raucous, nasal, rollicking yet brisk, cheerful: 'chawak-chawak-chawakety-chawak-chawakety'. **Hab.:** paperbark and eucalypt woodlands with a preference for vicinity of swamps, rivers, watercourses through ranges; also monsoon forests, mangroves. **Status:** locally nomadic; common. **Br:** 395

Rufous-banded Honeyeater *Conopophila albogularis* 12–13 cm (T)

These honeyeaters are widespread across the far N of the NT and Qld; draw attention with almost constant chatter as they forage for nectar and insects, working from canopy to low shrubs, often taking insects in agile aerial pursuits. Usually in pairs or small parties; often dart about the treetops chasing after each other or in territorial disputes with other honeyeaters. **Voice:** sharp, piercing, up-slurred 'shrieee' and 'szieep'. Similar notes of fluctuating strength are run together as an undulating song: 'shrip-shrip-SHRIEEE-shriee-shrip-SZRIEEE-shrip-SHRIEE-szriee'. In disputes, a scolding 'squaak' or 'scraach'. **Hab.:** melaleuca and riverine swamps, monsoon forests, mangroves; in dry season moves inland along waterways to escarpment gorges and nearby woodlands; returns to coastal swamps in wet season. **Status:** nomadic; common. **Br:** 395

Rufous-throated Honeyeater *Conopophila rufogularis* 13–14 cm

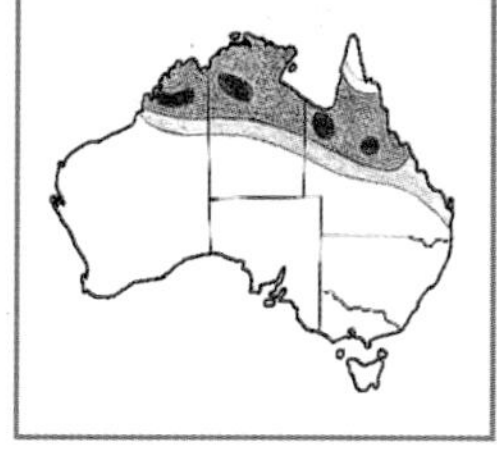

A small, quite bright, rufous-orange throat patch and rich, bright yellow edges to flight and tail feathers make adults of this otherwise dull plumaged species easy to identify. Often forages in flocks or mixed-species gatherings in flowering trees or low shrubs, aggressively trying to dominate. Insects are taken in addition to nectar, the agile birds darting out to take swarming termites, cicadas and other insects on the wing. **Voice:** as contact call, a sharp, slightly harsh 'chziep-chziep-chziep' or chzip-chzip-chzip'. Song a fast trill or twittering chatter, quite pleasant rather than harsh or scratchy, 'tzip-tzip-tsip-tip-tip'. Harsh, scolding 'scraak-scraak'. **Hab.:** prefers riverine paperbark forests and woodlands; also drier, open eucalypt woodlands, mangroves, gardens. **Status:** nomadic movements, migrations following tree flowerings. Common. **Br:** 395

Grey Honeyeater *Conopophila whitei* 11–12 cm

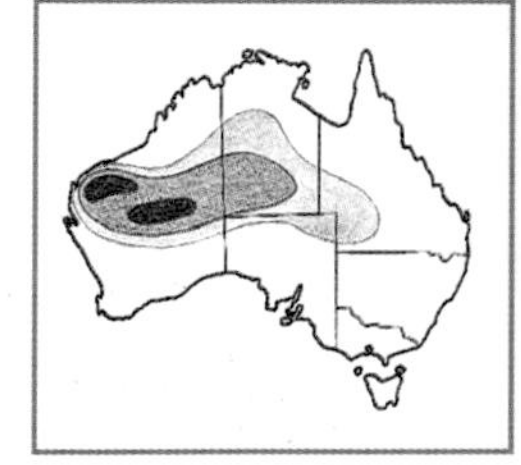

Almost straight bill, pale tipped tail; quiet, unobtrusive behaviour; less pugnacious than most other honeyeaters. Forages alone, in pairs or small groups; often in company with Western Gerygones, Yellow-rumped and probably other thornbill species; easily overlooked. Busily gleans surface of foliage, often hovering; captures flying insects, when white or pale tail-tips are most evident. Some deep, tubular flowers such as *eremophila* are pierced to reach nectar usually only available to the longer-billed honeyeaters; also takes nectar and berries such as mistletoe. **Voice:** piercing, metallic, quick double squeak, 'chirraWIEK-chirraWIEK' or 'tsi-WIEK-chirraWIEK'; weak, twittering, tinkling song. **Hab.:** mulga and other similar arid scrublands, arid woodlands. **Status:** not well known, probably nomadic; uncommon to rare. **Br:** 395

Tawny-crowned Honeyeater *Phylidonyris melanops* 16–18 cm

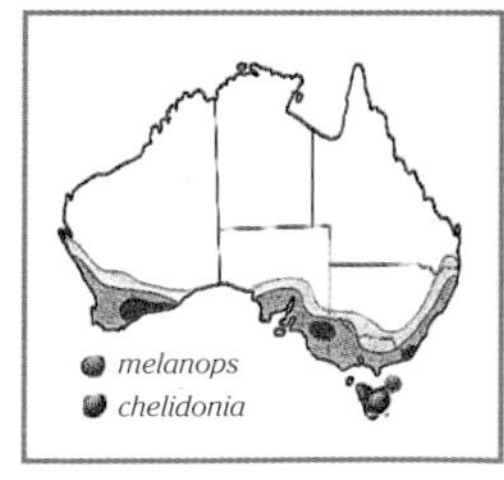

Inhabits heathlands; often present in extensive areas of sandplain and can also be found in small patches of low heathy vegetation within open woodlands. Flowering banksias, various bottlebrushes and other heath shrubs provide nectar, but the Tawny-crowned also hunts insects, often taken in flight. Usually alone or in pairs, occasionally larger numbers. Most likely to be noticed when breeding, as the clear lilting song, often given in aerial song flight, drifts across the heathlands. In display flights the bird rises steeply high above the low heath, spirals downwards while calling. **Voice:** clear, ringing, flute-like notes with wistful quality that seems appropriate to loneliness of windswept open heathlands. Usual call a high, clear, musical, mellow rather than sharp 'quip-peeer, pieer-pieer-piier', faster 'quip-pip-pip-pip'; many variations, from softly muted to loudly ringing. **Similar:** Eastern and Western Spinebills, but dark brown crowns and coloured throats; Crescent Honeyeater, but dark crown, bright yellow in wings. **Hab.:** heathlands with scattered banksias and other tall shrubs, open eucalypt-heath woodlands; sandplain heaths. **Status:** nomadic; patchy, common in extensive heaths. **Br:** 395

Heavy, black, uneven barring across white breast makes the Bar-breasted unique among Australian honey-eaters; its large hanging nests are conspicuous along tropical northern watercourses.
Kimberley population is perhaps larger and paler, and may have dark barring, mostly on the upper breast. These differences may not be sufficient to justify continued race status for the Kimberley population.
Mid brown, mottled and edged paler
Black, scalloped white, appearing finely speckled or barred
Bill pinkish brown
Fine black line
Scalloping narrow, not obvious
Black scalloping across breast and along flanks, usually forming rough barring. Kimberley form, shown, may tend to have the barring more confined to the upper breast.
Pinkish brown
Adult
Imm.
Immature is streaked rather than barred.
Head entirely and almost uniformly grey
Dull rufous–brown
Dull grey-brown
Clearly defined white chin and throat
Back chestnut-brown, flight feathers edged deep yellow
Broad cinnamon-rufous breast-band, paler tint along flanks
Underparts dull grey-buff with rufous-brown tone to breast
Lower breast to under-tail coverts white
Adult
Imm.
Light grey-brown
Flight feathers and the tail towards its base are conspicuously bright yellow along outer edges.
In flight the bold yellow outer webs of the flight feathers and tail retrices briefly add additional bright colour.
Grey-brown
Clearly defined rufous-orange patch, chin to throat
Throat same as rest of under-parts: white with faint grey-buff tint
Slight clinal variation: gradually trends paler and browner towards the W and inland. NT–WA form is shown.
Adult
Imm.
A small, very plain grey honeyeater without obvious distinguishing features. Appears unlike typical honeyeater in looks and activity.
Adult usually has a very faint pale eye-ring; can be entirely or partially hidden in shadow beneath brow.
Obvious pale eye-ring
Bill rather short, slightly downcurved, but not as tiny or straight as those of gerygones or thornbills.
Grey, becoming grey-brown in worn plumage
Chin to breast is grey to grey-brown; colour merges into similar tone of upper parts.
Weak buff tint on throat and cheeks
Pale tips
Similar in range and habitat: Western Gerygone, but far more white in tail and distinctly different song; several thornbills, but with blackish, pale-tipped tails, contrasting and often colourful rumps.
Buff edges to wing feathers
Yellowish tint to feather edges
Imm.
Crown tawny buff to cinnamon
Cinnamon crown and extended dark mask make this honeyeater distinctive. Two subspecies: nominate race *melanops* of Aust. mainland, Bass Str. islands and NE Tas.; race *chelidonia*, russet plumaged, NW and W coast of Tas.
Wing linings tinted pale cinnamon, adults and immatures
Bold black mask through eye and down sides of white throat and breast
Narrow buff or yellow margins to flight feathers
Adult
Underparts white, flanks softly streaked grey-brown
Tends to be secretive; keeps low in heath shrubbery; occasionally calls from higher perch; drops back to cover.
Tail dark grey-brown without white tips
Imm.: face mask blurred brownish rather than crisp black of adult

New Holland Honeyeater *Phylidonyris novaehollandiae* 17–19 cm

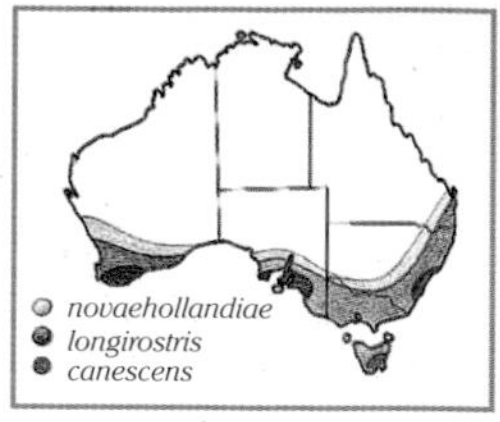

Bold, pugnacious, noisy; common in most Australian cities; attracts attention with sharp alarm calls; in flight, shows yellow in wings and tail, and white tail tips. Aggressively competes for nectar and insects, often taken in flight. **Variation:** 5 races include *caudata* on Bass Str. islands, *campbelli* on Kangaroo Is. **Voice:** abrupt, metallic, piercing 'tjik!', 'chwik!'; long, whistled 'tseeee' often in flight. As alarm, loud, piercingly sharp 'chwiep-chwiep-chwiep'; harsh, scolding 'tjuk, tjuk', slowly or in rapid bursts. **Hab.:** forests and woodlands with undergrowth, heaths, mallee, coastal thickets. **Status:** sedentary or locally nomadic. **Br:** 395

White-cheeked Honeyeater *Phylidonyris nigra* 16–19 cm

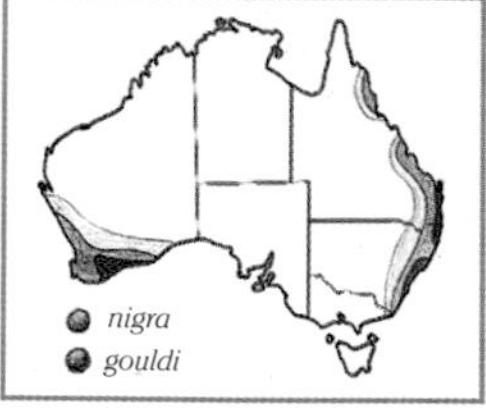

Often in groups foraging for nectar and insects; moves nomadically in search of flowering vegetation; busy and noisy, deep yapping calls. White cheeks and heavy bill separate from New Holland; seems less noisy and aggressive. **Variation:** two subspecies: nominate *nigra* has broad triangular white cheek patch; race *gouldi*, SW of WA, has a large oval white cheek patch. **Voice:** distinctive yapping call: 'chwikup, chwikup'; 'chwipit-chwipit'; 'chakup-chakup-chakup'. Song is a more musical, less sharp 'chwippy-choo, chwippy-choo' and lilting 'twee-tee, twee-tee, twee-tee', often given in display song flight when breeding. **Similar:** New Holland Honeyeater, but much smaller, white head markings, sharper calls. **Hab.:** forests, rainforest margins, heath undergrowth areas of open forests, woodlands, sandplains. **Status:** locally migratory or nomadic; common. **Br:** 395

Crescent Honeyeater *Phylidonyris pyrrhoptera* 15–16 cm

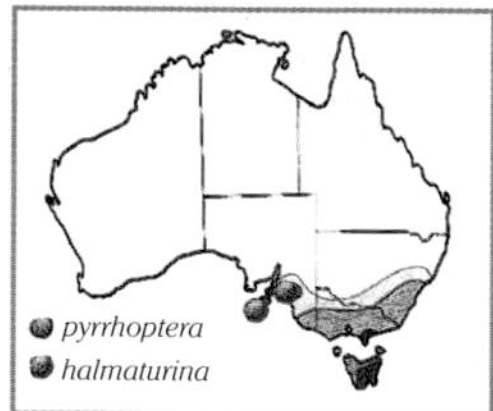

Uses wet forests of ranges; forages actively for nectar, sweet sap, insects; moves about forest in swift, undulating flight. Travels to lower altitudes in winter. **Variation:** nominate race *pyrrhoptera*, SE mainland; *halmaturina* on Mt Lofty Ra., SA, and nearby Kangaroo Island. **Voice:** loud, carrying 'tee-chiep'; variations include 'tee-cheop', 'ti-chi-cheop, ti-chi-cheop', 'tchop-tchop-tchip'; rollicking 'tchwop-it, tchwop-it, tchwop-it'. Higher, but not sharp, a musical 'tcheip-tcheip-'; sharp, rapid 'chip-chip-chip-' alarm. **Similar:** Tawny-crowned, but has buffy orange crown, lacks yellow wing panel. **Hab.:** wet eucalypt forests, alpine woodlands; prefers dense undergrowth including tall heaths. **Status:** locally migratory; common. **Br:** 396

White-fronted Honeyeater *Phylidonyris albifrons* 16–18 cm

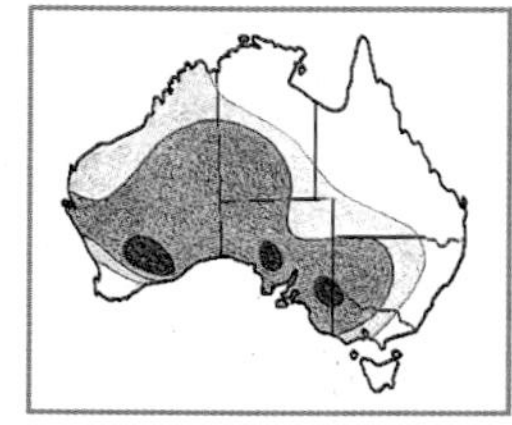

Superficially like the New Holland Honeyeater in appearance; also has a wide range, but through drier regions, mostly the interior. Wanders in search of the wealth of flowering that follows rain in such arid regions; then often gathers in large numbers, busily seeking nectar and insects, darting about the bushes with almost constant calling. When nesting, tends to be quiet, wary, skulking. **Juv.:** browner, wings edged dull yellow, throat mottled, lacks red eye spot. **Voice:** common call a quite high, nasal, metallic 'kzeeip' or rapid 'kzeip-kzeip-'. In song adds deep, mellow, musical notes, 'kzeip, chrrok-chrrok-chrrok' and higher 'pzeip-pzip, chrreik-chrreik'. **Similar:** New Holland Honeyeater, White-cheeked. **Hab.:** open dry eucalypt and acacia woodlands, scrublands, typically with scattered shrubs, eremophilas, grevilleas, hakeas. **Status:** nomadic; moderately common. **Br:** 395

Painted Honeyeater *Grantiella picta* 15.5–16 cm

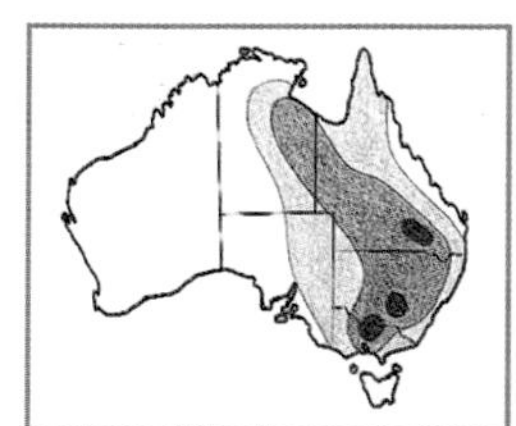

Bright pink bill and yellow feather margins justify the name 'Painted'. Atypical of honeyeaters in appearance and in its dependence on mistletoe berries, although nectar and insects are also taken. Alone or in groups; breeds in loose colonies. Often makes erratic, towering display flights from high perches. **Voice:** best-known call a loud, clear, far-carrying, rising, falling whistle 'wheeit-whiu, wheeit-whiu, wheeit-whiu' and, alternatively, 'whiu-wheeit.' Also a brisk, high 'chiewip-chiewip-' and fast 'whit-whew'. **Hab.:** forests, woodlands, dry scrublands, often with abundant mistletoe. **Status:** migratory, nomadic; uncommon. **Br:** 396

White-streaked Honeyeater *Trichodere cockerelli* 16–19 cm

Forages actively in crown and lower foliage of flowering paperbarks and other trees; solitary, pairs or small flocks; competes noisily with other small birds for nectar and insects. Flight swift, undulating, often in agile aerial pursuits of flying insects around flowers and foliage. **Voice:** call a harsh, grating 'scrairrk, scrairrk' and faster 'scraik-scraik-scraik'. Song a musical piping: 'chwip-cheweep-chwip-cheweep'; rapid 'wip-wipwipwip'. **Hab.:** monsoon and riverine rainforests, vine scrub margins, melaleuca swamps and heathlands, and adjoining eucalypt woodlands. **Status:** locally nomadic; moderately common. **Br:** 396

One of two similar species, but this with obviously much smaller white cheek patch than the White-cheeked Honeyeater (below)
Stark staring white eye
Bill long, quite slender
Head entirely black except white streaks and tufts, above and below the eye and at the side of neck.
Underparts streaked
Large golden wing panel
Tail yellow edged and tipped white
Imm: have black areas browner, ear tufts smaller, gape yellow, iris grey.
Conspicuous white eye
Flight and tail feathers have wide, deep yellow outer margins.
White tips to outer tail feathers show in flight.
Long tapering brow line
Eye brown, almost hidden in black
Western race shown; eastern race has broader, triangular cheek patch
Bill long, heavy
Very large white cheek patch;
Large golden wing panel
Imm.: gape and brow yellow, plumage dusky
Yellow-edged, no white on tips
Dark, rather cold grey upper parts emphasise the bright golden yellow of wing panels.
Thin white brow behind eye, both sexes
Female colours are dull, grey-brown with reduced yellow on wing.
Grey-black
Streaked
Epacris impressa
Crescent Honeyeater is named for the pattern of dark bands that, on both male and female, curve inwards, almost meeting to form a crescent, broken by a narrow gap at lower breast. Imm. is dull, crescent indistinct.
Bright yellow
Broken dark 'crescent'
♂
♀
Tail yellow on outer edge
On upper surface, white outer tips are visible on spread tail
Superficially like the New Holland honeyeater, it has a rather clown-like facial pattern and strange expression.
Tiny red wattle behind eye
White face mask
White malar line
Throat and breast black
Tail edged yellow
Flight feathers edged yellow
White 'front' divided down centre of forehead
Wings edged yellow and white
Upper parts are blackish; females and immatures are slightly lighter dusky brownish black.
Bill is deep pink, sharply triangular, almost straight.
♂
Short droplet shaped streaks or spots along flanks; females and immatures have plain white underparts.
♂
White tips are usually concealed when tail is folded; when it is spread, in flight and during display song flights, this white is conspicuous.
Confined to Cape York. Distinctive and, with no close relatives, it has been positioned as the sole species of the genus *Trichodere*.
Flight feathers edged yellow
Gape line pale blue-grey
Yellow-tufted ear coverts, fine yellow line under eye
Yellow towards base
Immature is similar to adult, but lacks streaky lanceolate plumes, throat to breast being plain with a faint yellow tint; gape yellow, eye red.
Adult
Breast feathers are finely, sharply pointed, appearing stiffly bristle-like with grey and white streaks.

Black-chinned Honeyeater *Melithreptus gularis* 15.5–17 cm

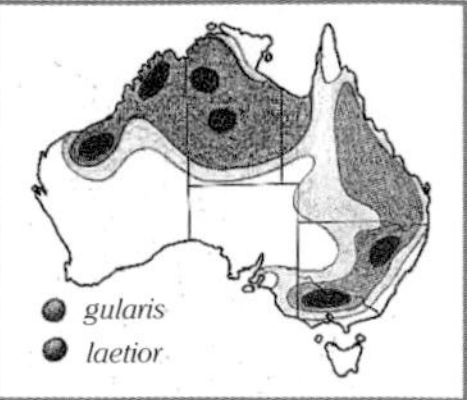

Active, noisy, forages in canopy of trees, usually in small parties; maintains contact with frequent calls; tends to work quickly through foliage, soon moving on with swift, direct flight, tree to tree across woodlands. **Variation:** race *gularis*, Black-chinned Honeyeater, E Aust.; race *laetior*, 'Golden-backed Honeyeater', NW Qld to Pilbara, WA. **Voice:** E race has vibrant, ringing calls, 'chi-chrrrip-chrrrip' and scolding 'quorrip-quorrip'. 'Golden-backed' has a clear, ringing, musical 'trrirrip-trrirrip-trrirrip-'; rapid 'trip-trip-trip-trip'. **Similar:** Brown-headed, White-naped, White-throated Honeyeaters. **Hab.:** forests, woodland of eucalypts, paperbarks; tree-lined watercourses of arid regions. **Status:** nomadic; common in NW Aust. to uncommon in SE. **Br:** 396

White-throated Honeyeater *Melithreptus albogularis* 13–15 cm

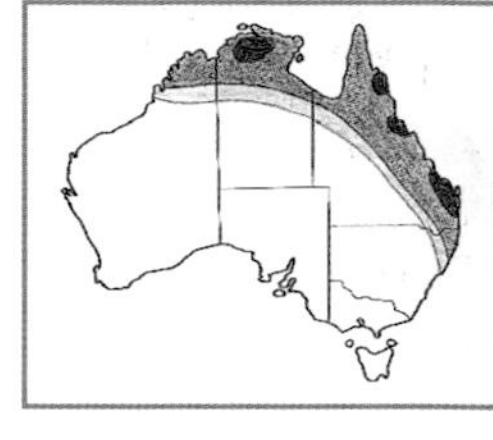

A small honeyeater of outer tree-foliage where its size and colour make it inconspicuous. Small parties climb among flowers and foliage attracting attention by calls and occasional sudden aerial pursuits. **Variation:** opinions differ: no races, or several. Probably race *albogularis* W of Atherton Tableland, NE Qld; race *inopinatus* to the SE. **Voice:** a sharp 'pit' or 'tsip' as contact, in trees and in flight; rasping, peevish 'querrk, querrk, querrk' and quick, higher 'quierk-quierk-quierk-quierk'. Song is a rapid, piping whistle, accelerating: 'pit, pit, pit pit-pit-pit-pitpit'. **Hab.:** paperbarks in forest and woodland, rainforest margins, riverside vegetation, mangroves, watercourse trees in arid regions. **Status:** sedentary, in some regions locally migratory or nomadic. **Br:** 396

Strong-billed Honeyeater *Melithreptus validirostris* 15–17 cm

The large, almost straight bill that tapers evenly in a sharply pointed triangular shape from a heavy, strong base plus a heavier build about the neck and shoulders are the most obvious feeding adaptations of this honeyeater, largest of its genus. As its searches for insects and other small prey it energetically probes, prises, rips bark from tree trunk and limbs; insects are rarely pursued in flight, and nectar appears to be of relatively little interest. **Voice:** a regular, sharp, treecreeper-like 'tip, tip, tip, tip' and rapid, sharp, scratchy 'whitch-whitch-whitch'; song a rather rollicking 'chwip-ckukachip-kukachip-kukachip'. Also a rather harsh 'kraach-kraach-'. **Hab.:** rainforests, eucalypt forests, woodlands, coastal heaths. **Status:** sedentary; common. **Br:** 396

White-naped Honeyeater *Melithreptus lunatus* 13–15 cm

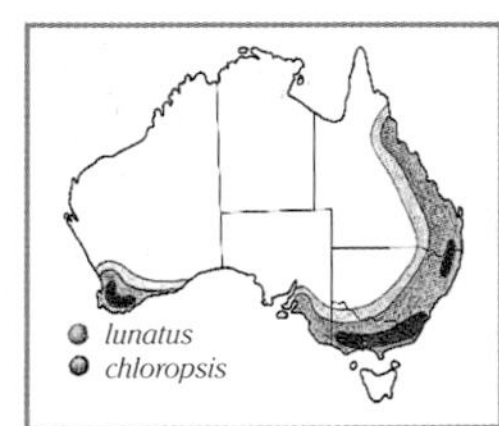

Distinctive husky contact calls call attention as individuals, groups or flocks forage in canopy for manna gum and honey-dew secretions of leaf insects; take nectar and insects as well. **Juv.:** dull, brown; eye-crescent dull tint of adult colour. **Variation:** nominate race *lunatus*, E Aust.; race *chloropsis*, drier parts of WA; race *whitlocki*, coastal SW of WA. **Voice:** contact call a husky 'shierk' or 'szerrk', abrupt in flight; may also be followed by a clear whistle: 'shierrk-WHIET'. Also loud, clear, musical 'wherrt-wherrt-wherrt'. In alarm, high, penetrating, rapid 'chwip-chwip'. **Similar:** others that have a white nape crescent. **Hab.:** eucalypt forests, woodlands. **Status:** usually sedentary, but, in SE Aust., conspicuous in migratory flocks; common. **Br:** 396

Brown-headed Honeyeater *Melithreptus brevirostris* 13–14 cm

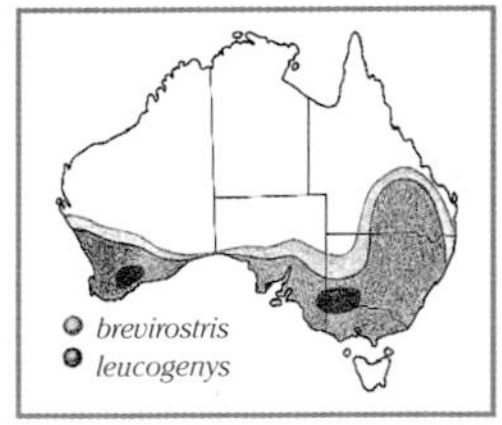

Forages in canopy around flowers, leaves, bark, finding insects, manna gum, pollen, nectar; in flocks or breeding communities of up to 20 birds. **Variation:** in SW Aust., race *leucogenys*. In SE, a darker coastal race, *brevirostris*; on Kangaroo Is., *magnirostris*. Also probably a valid subspecies is an inland, paler population, *pallidiceps*, and *wombeyi* in the Otway–Strzelecki Ra. **Voice:** sharp, scratchy 'chwik-chwik-chwik'; lively 'chak-chak-chak'. **Similar:** immatures of other white-naped species, but these have only partial-circle eye-rings. **Hab.:** forests, woodlands, mallee and heathlands. **Status:** locally nomadic; common. **Br:** 396

Black-headed Honeyeater *Melithreptus affinis* 13–15 cm

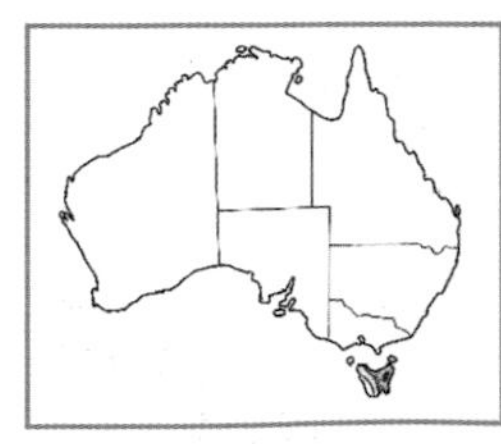

Busy, active, noisy; forages through upper foliage storey, small flocks moving tree to tree with swift undulating flight. **Voice:** harsh, rasping 'kherrk' like White-naped; sound becomes harsher, faster in aggression or alarm: 'scraak-scraak'. More distinctive is a high, sharp, clear whistle: 'tseip, tsheip, tseip'; similar, quick, double noted 'tsip-tsip, tsip-tsip, tsip-tsip' contact in flight. **Similar:** in Tas., only Strong-billed, but has white nape crescent. **Hab.:** eucalypt forests and woodlands, heath-woodlands; avoids wet rainforests and similar wet vegetation of SE Tas. **Status:** nomadic flocks after spring–summer breeding; common. **Br:** 396

Black-chinned Honeyeater,
nominate race *gularis*
White nape crescent
Partial white collar
Dull olive-green
Black centre line
Grey-buff tint
Adult
Nape crescent dull white
Blue eye-crescent
Crown and face brown
Dark throat
Imm
Yellowish eye crescent
Back to rump golden with slight olive tint
'Golden-backed Honeyeater' (race *laetior*) NW Qld to Pilbara, WA
Tail grey-brown
Wings dark grey-brown
Adult
Eye crescent pale blue or white
Nape crescent reaches close to eye.
White of chin extends up to gape.
Upper parts olive-green
Chin and entire underparts are white; black extends down sides of neck.
White-throated Honeyeater: closely resembles the Black-throated (above), but entirely white where that species has a black line, chin to throat.
Flight feathers grey edged olive-yellow
Immature: brownish crown, face, nape with white crescent at nape and over eyes
Bill brown with rufous-orange base
Dull olive-green
Entirely white
Tail olive
Eye-crescent pale blue
Bill long, almost straight, tapers evenly from a heavy base; small black chin patch
Rump rufous-olive
Underparts dull grey-green
Flanks olive-tan
Adult
Crown brownish black
Nape tinted yellow
Eye-ring and bill orange-buff
Yellow tint, throat to breast
Imm
In Tas., only one similar honeyeater, the Black-headed, but that species has a black hood without white at nape.
Eye crescent red
White of nape does not reach eye.
Olive-green
Black extends down slightly beneath bill, slightly onto chin.
Sides of neck black, throat to under-tail coverts white
Adult
Race: *lunatus*
Mottled olive, brown, black
Crescent dull white, small or absent
Eye crescent, bill base orange
Rump golden-olive
Entirely white
Imm: *lunatus*
Race *chloropsis* is slightly larger, has bolder nape crescent, longer bill.
Eye crescent greenish
Race *whitlocki* has white eye crescent.
Race *chloropsis*
Compared with others of the genus *Melithreptus*, the Brown-headed has a dull overall appearance: brown and grey where others are black and white, and more like juveniles of other white-naped species.
Immature is very dull plumaged, nape crescent indistinct, bill pinkish buff at base.
Eucalyptus pyriformis
Chin grey
Crown mid to dark brown. Dull white nape crescent meets a completely encircling, buffy white eye-ring.
Upper parts dull olive-grey with ochre tone to rump
Tail grey-brown with yellow ochre tint along margins
Dull grey-buff
Distinguished by the black that extends as a hood over the entire head and throat.
Does not have a white nape crescent.
Bare skin of eye crescent very pale turquoise
Olive-brown
Black bar down each side of neck
Rump paler rufous-olive
White shading to pale grey rearwards
Adult
Olive-brown mottled darker
Pale base to bill; yellow gape
Imm.
Very faint yellow tint
Confined to Tas. and Bass Str. islands; no subspecies

Eastern Spinebill *Acanthorhynchus tenuirostris* 14–16 cm

Colourful, fine-billed honeyeater of understorey; attracts attention with 'flop-flop' sound of wings, conspicuous flash of white tail-tips as it darts after insects, hovers around flowers, or rises high in display flights. Often at deeply tubular flowers such as those of *Grevillea* and *Epacris*, or probing the flower spikes of *Banksia* and *Callistemon*. **Variation:** probably 4 subspecies: *tenuirostris*, SE Qld to SE of SA; *cairnsensis*, NE Qld; *dubius* in Tas.; *halmaturinus*, Kangaroo Is. to Flinders Ra. **Voice:** sharp penetrating 'chip-chip-chip-chip-' in long sequences at constant pitch. Song a clear, high, sharply whistled 'cheerwhit-cheerwhit-cheerwhit', fast and cheerfully rollicking. **Similar:** Tawny-crowned Honeyeater, but cinnamon-buff crown, white throat; Crescent Honeyeater, but yellow wing panel, grey throat. **Hab.:** forests and woodlands with shrubby understorey, heathlands. **Status:** sedentary in N and coastal areas; in SE moves from ranges and coast in winter. **Br:** 396

Western Spinebill *Acanthorhynchus superciliosus* 12–15 cm

Lively, colourful and quite common small honeyeater unique to SW Aust. Attracts attention with clear, ringing distinctive calls and noticeable flop-flop-flop sound from wings. Alone or in pairs, seeks nectar and insects in low shrubbery or tree canopy; makes extensive use of the abundant wildflowers—the long, fine bill is well adapted to probing deep tubular flowers of kangaroo-paws and grevilleas. In spring, swarming termites are snatched from the air in acrobatic flight. **Imm.:** like female, but duller in colour; eye brown. **Voice:** sharp, abrupt, clear 'chwip, chwip, chwip, chwip' at slow regular intervals, or faster 'chip-chip-cherip'; usual song a rapid, very high, gradually descending, musical 'chri-chri-chri-chri-', often given in flight. **Similar:** in SW of WA, most alike is the Tawny-crowned Honeyeater with cinnamon crown, bars down sides of breast. **Hab.:** forests and woodlands with shrubby undergrowth; coastal, inland sandplain heaths with scattered banksias and large shrubs; parks and suburban gardens. **Status:** endemic to SW of WA, locally nomadic; common. **Br:** 396

Banded Honeyeater *Certhionix pectoralis* 12–14 cm

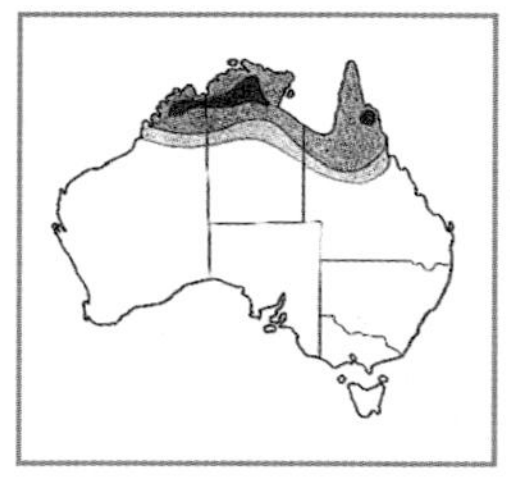

Nectar forms a large part of the diet of the Banded Honeyeater, although the bill appears rather straight and heavy compared with many other honeyeaters that rely largely on nectar. However, the Banded Honeyeater feeds in larger trees, the eucalypts and paperbarks that have cup- or brush-shaped rather than narrow tubular flowers. Except when nesting, these honeyeaters wander in pairs or small parties, or gather in great numbers when trees are heavily in flower, moving about the treetops in rather bouncy, undulating flight. **Voice:** loud metallic, vibrant 'tzziep, tzziep', higher buzzing 'chewip-chiewip-chrzieep'; song is a softer 'queeit-quitchie-quitchie'. **Similar:** Black Honeyeater, head entirely black; note its range. **Hab.:** eucalypt forests, woodlands, riverine forests, mangroves, dry pindan scrub. **Status:** nomadic; common. **Br:** 396

Black Honeyeater *Certhionix niger* 10–12 cm

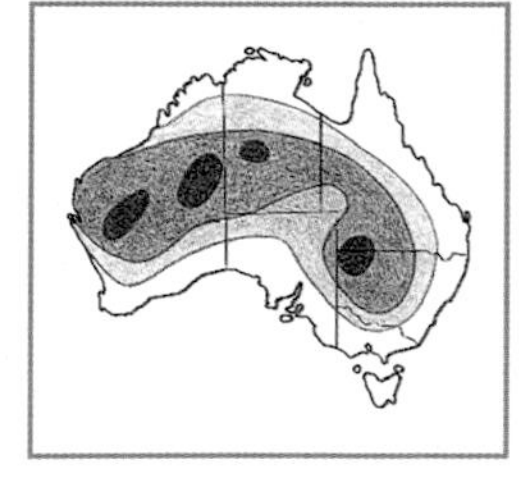

Very small, slender bodied, fine billed honeyeater of drier regions. Active, lively; takes insects in flight; hovers briefly about flowers and foliage; engages in conspicuous high song flights when breeding; collects charcoal after fires. Solitary, in pairs, small groups, when females and immatures often greatly outnumber adult males. Makes much use of flowers of eremophilas, as well as grevilleas, eucalypts and other wildflowers; moves nomadically in large numbers to localities where rains have induced flowering and prolific insect life. **Imm.:** like female, but mottled plumage, yellowish base to bill. **Voice:** soft, metallic 'chwit, chwit' and louder 'tieee', monotonously even pitch and spacing, at intervals of several seconds. In high song flights gives a double noted 'tieee-eee'. **Similar:** Pied Honeyeater, but white breast, white in wings. **Hab.:** rather dry and semi-arid regions, mostly inland; favours flowering regrowth. **Status:** nomadic; locally common. **Br:** 396

Pied Honeyeater *Certhionyx variegatus* 15–18 cm

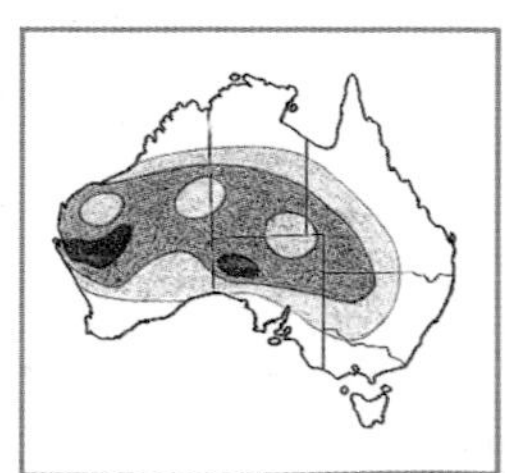

The highly nomadic Pied Honeyeater wanders wherever rains bring abundant flowers. In display flights, it rises almost vertically; shows white of wings and spread tail; gives its high, far-carrying calls; then drops vertically with wings closed. **Voice:** usually silent except in breeding season; then males seem to call incessantly in song flights—high, clear, extremely penetrating whistles at almost uniform pitch, 'tieeee, ti-tiee, tieeeee' or lengthened to 'tieee, ti-tieee, tieeee-tieeee-tieeee'. **Similar:** Hooded Robin, Black Honeyeater. **Hab.:** dry open woodlands with shrubby understorey; mallee, arid acacia scrublands, spinifex with scattered shrubs, drier heathlands. **Status:** nomadic; common but erratic occurrence. **Br:** 397

Spinebills, Eastern and, below, Western, are honeyeaters with very long, fine, strongly downcurved bills.
Upper surface of wings is dark blue-grey.
Hind neck rufous, blending into darker brown of back
Crown to sides of neck black
Bill fine, strongly downcurved
Throat chestnut
Breast white
Cinnamon-buff
Tail dark blue-grey; folded wings similar
Broad white band across outer tips
♂
Crown plain grey
Olive-brown
Indistinct, buffy white streak across lower face
Underparts pale dull olive
Female similar to male, slightly dulled colours, crown grey-brown
Tail blue-grey, white tipped
Imm.
Colourful small honeyeater with very long fine bill. Females are much less colourful than males; immatures are dull versions of female, but with brown eyes.
Bill long, fine, strongly downcurved
White brow
Bold white line under eye
Broad white band across outer tips
White and black breast-bands
♂
Nape bright rufous
Pale brow line and indistinct malar line in similar position to those of male
Underparts very light grey-buff, darker on breast
♀
Female very much less colourful than male, unlike the Eastern Spinebill, in which the female is almost identical to the male.
Bill appears heavy compared with many other small honeyeaters.
Black cap
White extends to chin and eye.
A small tropical honeyeater, adults crisply patterned black and white with neat, narrow black breast-band; male and female are alike.
Black breast-band merges into black of back and wings.
Rump white
Adult
Ear coverts tinted buff, tipped brown
Upper parts gingery brown, mottled or roughly streaked darker brown
Band paler or absent
Rump white
Juv.
Immatures become more like adults, but with some juvenile plumage; finally similar to adults but duller, back grey-brown
Bill very fine, strongly downcurved
Short pale brow
Male, has head, neck, back, wings, and breast black
Sexes unlike: male boldly pied; female dull grey-brown
White point or 'hook' at edge of breast; shape varies with bird's posture and angle of view.
♀
Black breast; continues down centre of breast in a long tapering point.
Rump and tail black
♂
Pied and Black Honeyeaters make great use of the many species of *eremophila* that flower strongly when rain reaches into arid regions.
Eremophila fraseri
Noticeably larger than the Black and Banded Honeyeaters; has white on wings, tail and rump.
Head and neck hooded black
Bill long, downcurved, heavier than that of Black Honeyeater
Mottled grey-brown
Blue-grey bare skin
Blue skin smaller
White bar along wing
Wing coverts, secondaries are edged creamy white.
Rump and upper-tail coverts white
♂
♀
Prominent white side panels
Immatures similar to female

Scarlet Honeyeater *Myzomela sanguinolenta* 10–11 cm

After breeding, flocks of Scarlet Honeyeaters wander nomadically seeking flowering trees. Species feeds in the outer canopy; the height, very small size of the birds and typical backlighting make viewing difficult. In some places such as heathlands, these birds descend to much lower flowering shrubs in their search for nectar and insects, often hovering around the flowers. **Imm.:** similar to females, but with a touch of yellow at base of the bill; males gradually develop scarlet feathering on chin, head, then shoulders and back, finally on breast. **Voice:** often a piercing whistle, 'tseeip, tsip, tseeip' and higher, weaker, squeaky 'seep-seep'. Male has a brisk, cheery song, given persistently in breeding season, a silvery, tinkling whistle, erratically rising and falling, 'chwip, swit-sweet-switty-swit-switty', at times including deeper, chattering 'chawak-chwakity' at end. **Hab.:** rainforest margins, eucalypt forests and woodlands, paperbark swamps, heath-woodlands, acacia scrubs. In NE Qld, mainly at higher altitudes; further south more common along coast. **Status:** sedentary in northern parts of range, erratic migratory visitor to southernmost parts of range, elsewhere nomadic; common. **Br:** 397

Red-headed Honeyeater *Myzomela erythrocephala* 11–13 cm (T)

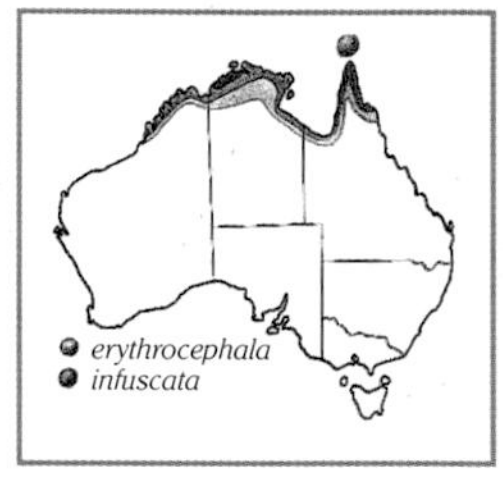

Small, agile; hovers around flowers and foliage to take nectar, and snatches insects in acrobatic aerial pursuits. Busily active, moving through mangroves in small parties, alone or in pairs, making longer trips between trees in undulating bouncy flight, or hovering to reach hanging foliage or flowers. **Variation:** 2 subspecies: *erythrocephala* N to NW mainland coast; race *infuscata* on islands of Torres Str., and as an intermediate form around N coast of Cape York. **Voice:** abrupt, sharp, scratchy, metallic 'tchwip-tchwip-tchwip-tchwip' or slightly softer 'swip-swip-swip-swip'. At times slightly harsher, more metallic, buzzing, emphatic 'tzzip, tzzip-tzzip, tzzip' and scolding 'charrk-charrk-'. **Similar:** Scarlet Honeyeater, but male has red also on breast, female a touch of red on forehead, a more dusky breast. **Hab.:** mangroves, swamp-paperbark forests, occasionally nearby tropical monsoon and riverine forests, woodlands. **Status:** locally nomadic; common. **Br:** 397

Dusky Honeyeater *Myzomela obscura* 12–15 cm (T)

Although of sombre plumage, can attract attention by its brisk foraging; flits about the upper storey, often hovering to reach nectar or insects. Often with other species of honeyeater, competing aggressively. **Variation:** *obscura* in NT; *harterti* in Qld; *fumata*, Torres Str. **Voice:** a quite musical 'whik-it, whik-it, whik-it', a uniform sequence, quality similar to voice of Noisy Pitta. May include richer, mellow notes rather like some of song of Brown Honeyeater, especially the pre-dawn calls of that species: 'whick-it, whik-a-chk, whik-it, whick-a-chuk, whikit'. Also various squeaks, harsh scolding. **Similar:** female of Red-headed Honeyeater, but has red on forehead, chin. **Hab.:** rainforests, monsoon scrubs, paperbarks, mangroves, watercourse thickets, nearby woodlands. **Status:** sedentary or locally nomadic; most common in N. **Br:** 397

Mistletoebird *Dicaeum hirundinaceum* 10–11 cm (See also pp. 338–39.)

The Mistletoebird is Australia's sole representative of a widespread family; superficially resembles the Red-headed and Scarlet Honeyeaters. It feeds almost exclusively on fruits of mistletoes and its digestive tract is modified for this diet. The outer skin of the berry is first removed, the fruit swallowed and the sweet jelly-like layer is digested. The seed, still very sticky, is defecated with a sideways bobbing action and wiped against the perch, which, if a living twig, will be penetrated by the tightly adhered germinating parasite, eventually giving rise to a new clump of mistletoe. Previously placed after honeyeaters and Yellow-bellied Sunbird in most references, but after finches on present classification. Comparative illustrations are positioned here with additional text. Main text and illustrations are on pp. 338–39. **Br:** 416

Yellow-bellied Sunbird *Nectarinia jugularis* 10–12 cm (See also pp. 338–39.)

Although superficially like Australian honeyeaters, family Meliphagidae, this sunbird is Australia's sole representative of another specialist nectar family, the Nectariniidae, which has more than a hundred species that inhabit tropical regions. The male sunbird is typically the brighter with iridescent, metallic colours. Both sunbirds and honeyeaters have adaptations for feeding at flowers, the most obvious being the long, fine, curved bill. The Yellow-bellied Sunbird is most like the small honeyeaters, and in earlier classifications was placed immediately after them. Some comparative illustrations are placed here in addition to the main entry in the current list position, pp. 338–39. **Br:** 416

Black wings add bold impact to the brilliant scarlet of this tiny honeyeater; adult males are not only spectacular, but unique.
Lores black
Head, upper body and breast are entirely glossy scarlet.
♂
Under-tail coverts to belly, dull white.
Wings black, flight feathers edged white
Males develop red first on head and back, finally on breast.
Black feathers gradually replace brown in wings, tail.
Imm. ♂
Immatures at first resemble adult females.
Pink tint to chin
Feathers of wings and tail are edged buff.
Females and some immatures are similar to those of the Red-headed Honeyeater, below.
♀
Lores black
Upper back and wings are brownish black.
Red tint to forehead, and slightly more on chin than female Scarlet
Red extends higher, on Torres Str race
Upper parts grey-brown
The northern Cape York population may be intermediate between the larger, more richly coloured Torres Str. race infuscata and the paler mainland race erythrocephala. The former has the red of rump always far up the mid back, and lower breast to under tail is uniformly dark dusky grey.
Lower back and rump are scarlet.
The scarlet of the male is glossy, reflecting fine highlight streaks under bright light.
On mainland race, blackish upper breast fades rearwards to pale grey
♂
Race erythrocephala (Aust. mainland form)
♀
Light grey
Immatures are similar to female, but with paler base to bill and olive-grey iris.
Flight feathers are edged dull white—conspicuous white under bright lighting.
Very dark brown, often with a slight scarlet undertone
Although it at first appears almost uniformly dull brown, the Dusky Honeyeater does have subtle variations of tone and detail under sunlight or other strong lighting.
Warm mid brown of breast pales rearwards to brownish cream.
Throat dark brown, often with a slight scarlet undertone
Three subspecies: obscura of NT; harterti of NE Qld; fumata on Torres Str. islands; the latter larger, darker.
Sexes are alike; immatures similar, slightly more uniform, yellowish gape and base of lower bill.
Red from chin to upper breast
Bill dark grey
Upper parts iridescent blue-black
Black central streak
Very pale creamy grey
Immatures are like females, but yellow at base of bill and gape.
Pink
♀
♂
♂
Eye is set in dark stripe between yellow lines.
Male is distinctive with iridescent blue-black from chin to upper breast, remainder of underparts bright yellow.
Iridescent deep blue-violet gorget
Female has underparts entirely yellow.
♂
Immature males are at first like females. Blue first appears down the centre of breast gorget, and gradually spreads to full width.
Often hovers at flowers to reach nectar, or to take small spiders from their webs or insects from leaves.
Widespread across islands to north; over 20 races, 1 in Aust.
♀

Crimson Chat *Epthianura tricolor* 11–12 cm

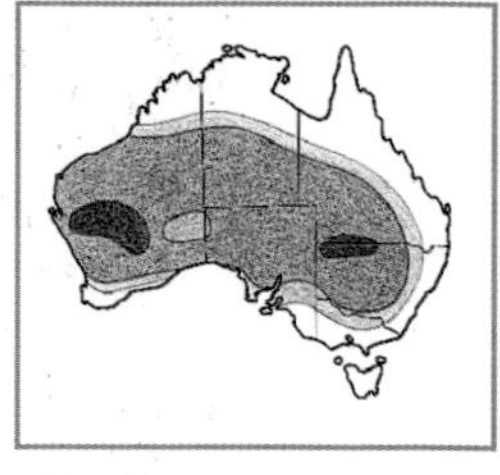

Usually in small flocks, working across herbage of open ground seeking small prey, perhaps occasionally nectar at low shrubs; flight bouncy-undulating. **Variation:** although it has an extremely wide distribution, the Crimson Chat has no subspecies; maintains uniformity by extensive nomadic wanderings in search of regions favoured by rain. **Voice:** when breeding the male often attracts attention with a high silvery even 'tseee-tseee-tseee' given in high song flight display over nest site, or from nearby high perch. As contact call in flight, 'tchek, tchek'; also a softer 'chikit, chikit'. **Similar:** Red-capped Robin, but a more plump silhouette, smaller red cap, and white wing streak; Mistletoe Bird, but lacks red crown, black ventral streak, strictly arboreal habits. **Hab.:** semi-arid mulga and other acacia scrublands, dry open woodlands, typically with low sparse ground cover of bluebush, samphire, other very low shrubs and tussocks, and often the very open surrounds of inland salt lakes. **Status:** nomadic and irruptive; generally common but vary locally by season, scarce to abundant. **Br:** 397

White-fronted Chat *Epthianura albifrons* 11–13 cm

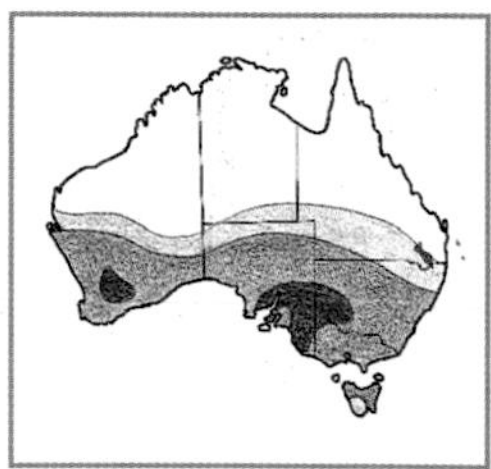

A familiar bird across southern Aust. where its use of open habitats has enabled it to persist or spread after substantial clearing for agriculture; the many local names reflect the call, 'Tin-tack', 'Tang'; or appearance, in 'Moonface', and 'Baldy-head'. Usually conspicuous, but very secretive in approach to nest; gives distraction displays. Highly gregarious; often nests in small colonies: after breeding, in small parties to large flocks. Seeks insects on ground and in low bushes. Flies with bouncy undulations, often rising high, calling. **Variation:** although spread across the continent, has no subspecies. **Voice:** a metallic, nasal 'tang' at wide irregular intervals, given in flight. **Hab.:** open country, often vicinity of inland salt lakes, coastal estuaries, saltmarshes with low and often sparse samphire; also swamp margins, coastal and inland open low heaths, remnant low vegetation on farmlands. **Status:** nomadic in arid parts, sedentary in wetter S of range. **Br:** 397

Orange Chat *Epthianura aurifrons* 10–12 cm

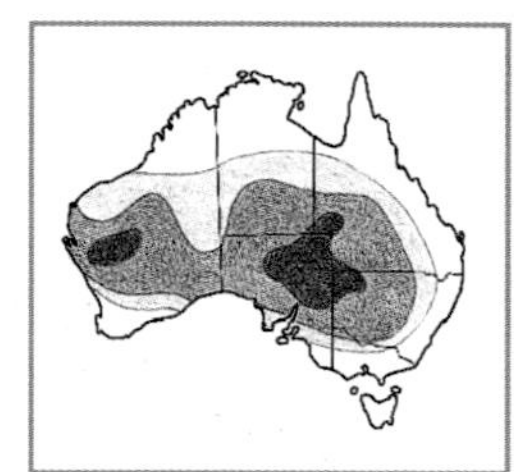

The first sighting of an Orange Chat is likely to be as a far distant, glowing golden speck, away across one of the wide samphire or saltbush flats that encircle salt lakes of the interior. This chat's habit of perching atop low vegetation on flat expanses of mudflats makes its sunlit gold visible from afar. Flight is undulating, quite often high. If approached, birds tend to be elusive, keeping low, popping up elsewhere in the distance. Especially wary when breeding; sneaks under cover to the nest. After breeding may be encountered in small groups. Insects and other tiny prey are taken from ground and vegetation. **Voice:** a vibrant, metallic 'tang', softer 'tchek, tchek'. **Similar:** Yellow Chat, smaller black throat crescent. **Hab.:** saltbush–samphire, sparsely vegetated gibber and similar plains. **Status:** nomadic; moderately common. **Br:** 397

Yellow Chat *Epthianura crocea* 10–12 cm

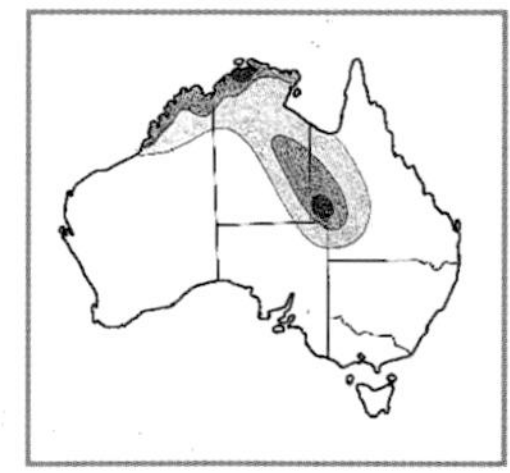

A rare chat inhabiting grassy swamplands along floodplains of several northern rivers, and natural springs and drains of Great Artesian Basin. Although each site is confined in area, these chats occur widely at suitable sites, with dispersive or nomadic movements across large parts of their range. Pairs or small parties seek insects in vegetation; occasionally 20 to 50 birds at a small but optimum habitat. **Voice:** a high, piping 'tee-tsue-tee' with the middle 'tsue' usually slightly lower. Males give displays: low undulating or dipping flights with metallic, piping 'tee-tee-tee-' of 2–15 notes. In alarm or threat near nest, low, harsh churring, scolding sounds. **Hab.:** coastal grassy swamps and lagoon margins with reeds, saltbush; inland around bore overflows, in swamp cane grass, cumbungi, lignum, bluebush. **Status:** nomadic, dispersive; rare. **Br:** 397

Gibberbird *Ashbyia lovensis* 12–13 cm

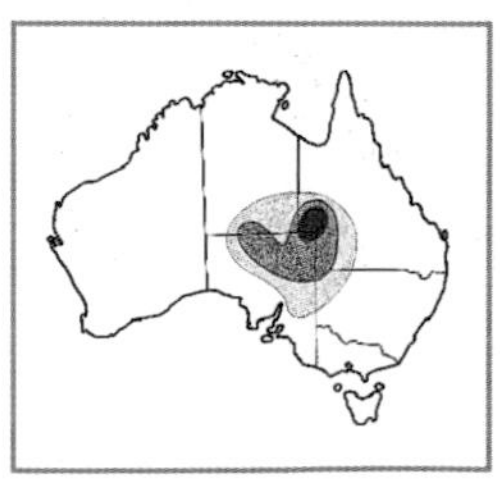

The distinctive character of the Gibberbird comes largely from its behaviour; far more terrestrial than chats, in many respects more like a small yellowish pipit, with similar upright stance, teetering and tail-wagging actions, abrupt short dashes across ground. When flushed, shows buffy brown rump and back, blackish white-tipped tail. Always lands on ground, perching on rocks rather than bushes. In pairs or small parties, occasionally larger flocks, at times with pipits; searches ground for small prey, occasionally flies up to snatch a flying insect. **Voice:** rather musical chatter, perhaps as contact call; in alarm, several piercingly sharp notes. Displays in steep, bouncy, high song-flight, giving high, clear 'tswee-tswee-tswee' calls. **Similar:** Yellow Wagtail, but long tail, dark grey or golden crown; female Orange Chat, but plumper, different behaviour. **Hab.:** waterless gibber-plains with extremely sparse vegetation. **Status:** nomadic; uncommon. **Br:** 397

Forehead and crown crimson
Throat white
Coverts, secondaries, edged buffy white
Breast to belly brilliant crimson
Rump crimson
♂
Underparts bright crimson except throat and under-tail coverts
Pale yellow eyes conspicuous in black face mask
Feathers of tail tipped white, conspicuous when tail is widely spread in flight
♂
Immatures are like female but lack pink, and have buff edged wing feathers.
Rump deep pink
Pale buffy brown
♀
Imm.
White encircled by black of hind crown, nape and breast-band
Eye yellowish
Back and rump silvery mid grey
Wings, upper tail coverts, dark brown
Underparts white with broad black breast-band
♂
White tips on dark brown, in flight conspicuous when tail is spread
Grey-brown
Dark breast band
♀
White tipped
Adult female dull grey-brown with indistinct markings, only the breast-band blackish, eye darker. Immatures similar to female, but breast-band very faint or missing
Male spectacular in deep, warm, cadmium yellow with orange overtone, strongest on crown and breast
Lores and throat black, eye deep red
Rump golden-orange
Tail finely tipped white
♂
Mottled grey-brown
Underparts softer fawny yellow
♀
In flight, shows golden orange rump and blackish tail narrowly tipped white. Immatures like adult female. No subspecies
Yellows will appear warmer in sun than in shade; late afternoon sun gives a warmer orange bias.
Male almost as brightly plumaged as the above species, but without such warm orange overtone, rather an intense, lemon yellow.
Head bright yellow, eye pale creamy grey to cream
Mottled grey
Female paler than male
Grey-brown; pale-margins to wings
Rump bright yellow
♂
Black breast crescent
♀
In flight, displays golden rump
Tail dark with white tips
Black breast mark can vary in shape.
♂
Although similar to the chats, the Gibberbird has sufficient differences to justify the separate genus *Ashbyia*.
Female similar to male, but with a buffy brown, smudgy band across breast, upper parts slightly browner
Black tail with white tips to inner webs of all except central pair
Upper parts softly mottled grey-brown
Rump buffy grey-brown
♂
Eye pale yellow
Light buffy grey-brown rump and back
Only slightly longer than the chats; more slender; often adopts a tall, upright posture.

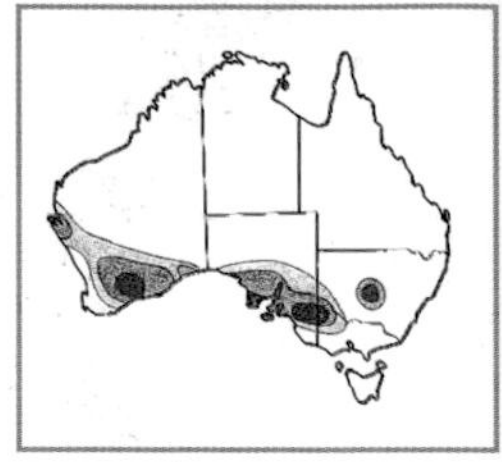

Southern Scrub-robin *Drymodes brunneopygia* 21–23 cm
An inhabitant of dense mallee and acacia scrubs. Attracts attention with persistent calls and song, but gives only fleeting glimpses, bouncing across the ground or inquisitively peering from a perch well screened by vegetation. Although appearing shy, may quietly sneak closer, scurrying across the ground, often scurrying behind shrub or debris, tail usually uplifted. The Southern Scrub-robin's colours blend into litter-layer tones—the browns and greys of dead twigs, buffy grey fallen leaves, dark shadows, reddish soil—making the bird difficult to see unless it moves. **Juv.:** overall grey-brown, streaked and mottled, throat feathers grey tipped, giving rather scalloped or roughly barred appearance. **Imm.:** similar to adults, slightly duller. **Variation:** slightly paler in SW of WA. **Voice:** a briskly whistled, cheery 'wheet-d-wheeeit', rising sharply. Also a brisk 'chwip-chWIPpee' and deeper, mellow 'chep-whep-wheeip'. Scolds with harsh, chattered 'scraach-chak-chak-chak'. **Similar:** Shy Heathwren, Chestnut-rumped Heathwren, but smaller, underparts heavily streaked, horizontal white brow line; Northern Scrub-robin, but far distant. **Hab.:** thickets of mallee eucalypts and similar low vegetation, including broombush, scrubby heaths and tea-tree thickets with a layer of leaf litter. **Status:** sedentary; patchy, uncommon. **Br:** 398

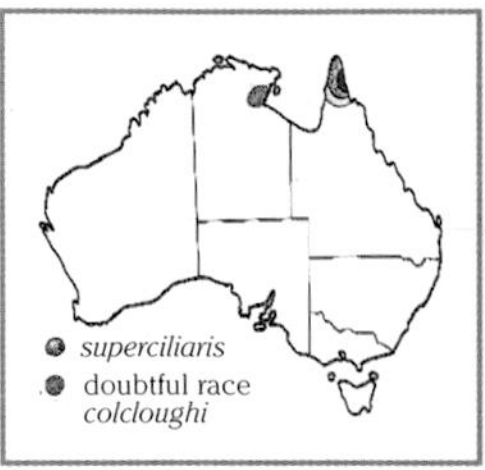

Northern Scrub-robin *Drymodes superciliaris* 21–23 cm
Forages on rainforest floor, scratching about vigorously with large, strong legs and feet, its colours matching the rufous, buff, grey and black tones of fallen leaf and twig debris. Solitary or in pairs, quiet, inconspicuous, yet inquisitive. May allow extended periods of observation, but, if alarmed, will bound away, relying first on long, strong legs to take it quickly into concealment in denser vegetation. Has a rather tall, upright stance, tail often uplifted. **Variation:** race *superciliaris*, Cape York. The apparently extinct NT race *colcloughi* is now thought not significantly different from some collected at Cape York; it also seems possible those few NT specimens may have come, incorrectly labelled, from Cape York. **Voice:** a strong, penetrating, drawn out whistle of several seconds duration, rising, 'wheeeeeiit', but may be steady, 'wheeeeet', undulating, 'wheeeur-eeeit', or interrupted, 'wheee-wheet'. Also often makes a hissing, rasping, scolding noise, a long drawn out, descending 'screeearch', level 'scrairch' or undulating 'screeeairarsch', similar to rise and fall of the whistled call. Has also a very harsh, scolding 'scraaach' as warning or alarm. **Hab.:** tropical rainforests and vine scrubs, typically where there is a deep litter layer, or thickets of tangled vines; often at the edge of the rainforest where the canopy slopes down densely to the ground. **Status:** sedentary; locally common, but doubtful NT subspecies is probably extinct. **Br:** 398

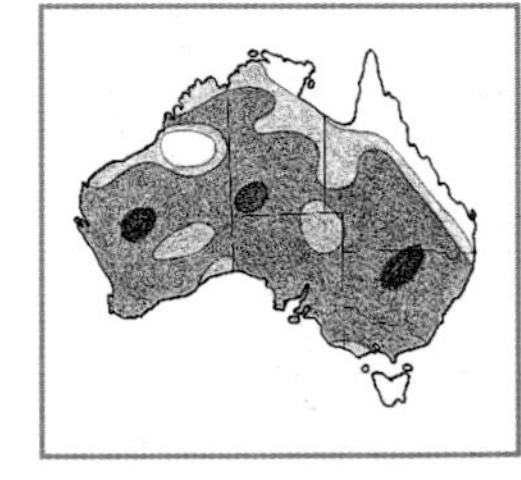

Hooded Robin *Melanodryas cucullata* 15–17 cm
Mature male's simple, clearcut, bold black and white plumage pattern is quite distinctive. Perches motionless, easily overlooked, for long periods as it scans the ground; occasionally darts down abruptly to take some small prey, momentarily displaying white bars in wings and conspicuous white panels in tail. Travels between trees with undulating flight. The female, although plain bodied in shades of grey-brown, has white like the male's in tail and wings. Usually in pairs or small family parties after breeding. **Juv.:** at first feathering and briefly after fledging, head and body are heavily and coarsely mottled brown and white. **Imm.:** like adult female, but rather dull with indistinct markings. Younger males, although breeding, may not have full intensity of black, and mottled rather than solid black on the breast. **Variation:** no subspecies, but size decreases slightly, gradually from S to N. **Voice:** a mellow series, first note far the highest, strongest, then descending, fading: 'CHIERP, chwep-chep-chep'. **Hab.:** drier and arid regions in open woodland of eucalypt, casuarina, pine, mallee, mulga; open banksia heathland of inland and drier parts of the coast; semi-cleared farmland. **Status:** adults sedentary except perhaps in very arid regions; immatures disperse. Moderately common. **Br:** 398

Dusky Robin *Melanodryas vittata* 16–17 cm
The sombre-plumaged Dusky Robin, confined to Tas. and islands of Bass Str., is most closely related to the Hooded Robin of the mainland. The Dusky and Hooded together make up the genus *Melanodryas*: although the males differ greatly in appearance, the female Hooded is more like the Dusky, a species in which male and female are similar. Even more than the Hooded, the Dusky sits motionless for long periods on a stump or low branch, occasionally dropping to the ground when it sees an insect or other small prey move. **Voice:** a plaintive, double-noted, carrying, mellow whistle, repeated monotonously, 'twei-twoo', the first part loud and clear, the second lower, muffled. The song is an excited, lively, cheery rush of abrupt notes, from low to very sharp and squeaky, which then falls away, 'chwot-chwert-chweet-chweeeit-chweeeit-chwert-chwot'. **Hab.:** diverse; rainforests to open woodlands, farmlands. **Status:** locally nomadic; common. **Br:** 398

Narrow white tips to outer feathers, often hidden with tail closed
Tail long, dull grey-brown, feathers edged rufous
Angle of tail varies; most typically is a position about mid-way between the two extremes shown.
Grey-brown
Face pale grey, lightest on chin and lores, white eye-ring
White tips to dark wing coverts form twin pale bars
Never really slender bodied, but also often fluffs out the feathers of breast and belly to appear quite rotund.
Adult
Distinctive, disruptive facial pattern with short dark line vertically through eye
Tail coverts and rump rufous to dull chestnut, merging into grey-brown of back
Legs and feet long, suited to foraging in ground debris
When on a perch clear of the ground, the tail tends to be lowered in line with the body; at other times slightly uplifted.
Long legged, long tailed, with head appearing relatively small. General appearance rather thrush-like, but most closely related to robins and flycatchers
A conspicuous identifying feature of this species is the bold black diagonal band down through the eye; it extends much further down onto the cheek than the similar mark on the Southern Scrub-robin.
Face off-white, merging to soft creamy buff on the breast
Male and female are alike; juveniles have throat and breast flecked darker brown, and plumage of crown and mantle edged dark brown.
Nape to back cinnamon-rufous, becoming brighter rufous on the rump and upper tail coverts
Flanks buff
Posssible race *colcloughi* (Roper R. NT), has been recorded as differing from those of Cape York in size, tail length, and with males more rufous.
Legs very long, feet large, an unusual, pale fleshy pink
Tail rufous nearest the base, each feather becoming gradually darker brown towards its white or pale tip
Black hood, unbroken by any markings
Upper parts mid grey-brown
Bold white shoulder bars
Hood extends downwards as rounded black bib on breast.
Bold white shoulder bar
White margins
White wing-bar and margins, but without male's curved white shoulder bar
Wing-bar
Outer four on each side form prominent white rectangles.
♂
Dull white, washed light to mid grey-brown across the breast
White bar along base of secondaries
♂
White on outer four on each side
♀
The sombre Dusky Robin is closely related to the Hooded, above; some of pale wing and tail markings are similar but far less conspicuous.
Upper parts dull olive-brown
Throat pale
Juvenile heavily mottled, brown on white, darker on upper parts
Weak wing-bar across base of secondaries
Tail dark olive-brown with pale, buffy white margins towards the base
White at bend of wing
Juv.
Juvenile plumage is soon replaced by immature: like that of adult, but with some streaking on wing coverts.
Adult
Pale wing-bar

Rose Robin *Petroica rosea* 10–12 cm

Forages actively in mid and upper levels of the forest, flitting about the foliage more like a flycatcher than a robin, the relatively long tail assisting the acrobatic chase for flying insects and hovering search in leaves. **Imm.:** like female, but no pink on breast. **Voice:** 'tick' in contact, but most distinctive is the male's repetitive song sequence, the initial notes sharp and clear, followed by vibrant, metallic, buzzing sounds: 'chwip-chwip-chwip-TZZEE-TZZEEE-TZZEEE'. **Similar:** Pink Robin. **Hab.:** rainforests, wet eucalypt forests with preference for understorey of large acacias, usually in gullies of ranges. **Status:** breeds SE ranges and coast; in autumn disperses out of high ranges; migrates W into SA and N into Qld. **Br:** 398

Pink Robin *Petroica rodinogaster* 11.5–13 cm

Takes prey on ground after watching patiently from perch; compared with similar Rose Robin, does not match that bird's energetic flycatcher-like activity about the higher canopy. **Juv.:** body mottled; wings and tail as adult female. **Imm.:** like adult female. **Voice:** cheery chattering trill, more like fairy-wren than robin; quite sharp, metallic, rapid, rattling, yet pleasant; has a barely discernible hesitation after the first note, then slightly descending, 'chwit, whit-wit-wit-wit-wit'; slower, more deliberate, 'whit, whit, whit, whit'. **Hab.:** gloomy tree-fern gullies and other dense undergrowth of rainforests and wet eucalypt forests in summer; moves to more open country in winter. **Status:** dispersive; uncommon. **Br:** 398

Flame Robin *Petroica phoenicea* 13–14 cm

Slender, rather small headed, long necked compared with plump, thick-necked build typical of robins; tends to perch in more upright posture on stumps, boulders; males spectacularly conspicuous. In open country takes prey on ground; also takes flying insects in heavy forest. Unique among Aust. robins in gathering into flocks in winter months; moves from high ranges to open lowland habitats. **Voice:** clear, piping trilling, sharp, penetrating, wildly fluctuating pitch, yet lilting, quite musical, 'chrip-a-chip, chrip-a-chip, chirripa-tirrrrrip'. Scolds quite harshly, a rapid 'scraach-scraach'. **Hab.:** in summer, eucalypt forests and woodland; in winter, open woodlands, farmlands. **Status:** migratory; common. **Br:** 398

Scarlet Robin *Petroica multicolor* 12–14 cm (N)

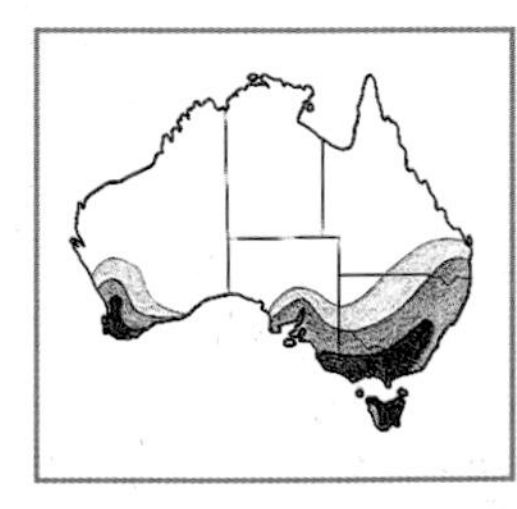

Best known of red robins; conspicuous in open and suburban habitats, its range includes all state capitals. Usually in pairs, waits patiently on low perch; drops to ground to take prey, displaying bold white markings in wings and tail, and again on flying back up to its perch. Longer flights direct, undulating. **Variation:** race *boodang* in SE Aust.; *campbelli* in SW Aust. Seem to have intermediates around Kangaroo Is., Fleurieu and Eyre Peninsulas, SA, however, position of boundary between races seems uncertain. SE race *boodang* includes a darker form in Tas., which may be a race (*leggi*), but these migrate and merge into the mainland population. **Voice:** song very high, whistled, reeling series of trills, not powerful, but, perhaps because of the shrill quality, travels a considerable distance. The call has a pleasant, cheery, rippling, quite musical quality; while each trill rises in pitch, the next begins lower, giving the song its ripple, 'tirrrit-tirrrit-tirrrit, tirrrit-tirrrit-tirrrit'. In aggression or defence of his territory, the male gives a rapid, aggressive, hard 'trrut! trrrut!'; contact call is a quiet, tapping 'tek'. **Similar:** other red robins within range. **Hab.:** forests, woodlands; heavier vegetation when breeding, more open and cleared in autumn and winter. **Status:** dispersive or locally migratory seasonal movements; common. **Br:** 398

Red-capped Robin *Petroica goodenovii* 11–12 cm

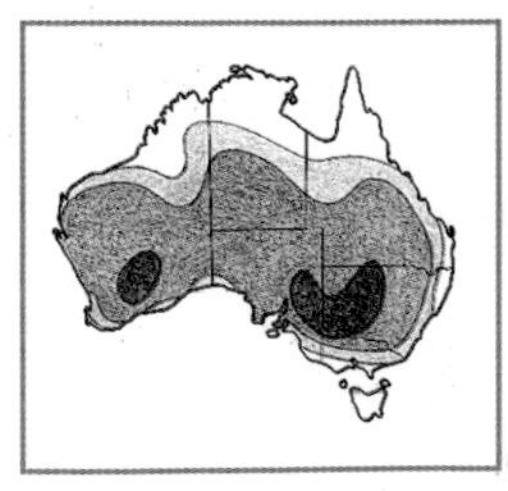

Smallest, most brightly coloured of red robins; intense crimson cap as well as crimson breast. This species's ticking trills are one characteristic sound of the drier regions, but, even when not calling, these birds are soon noticed—rarely still for long, they dart to ground, flying up to a new vantage point, only to duck to the ground again or away to try a new vantage point. While perched, often gives quick little flicks of tail and wings; fossicks actively on the ground, hopping and shuffling about the litter, half-spreading the wings with a sudden jerk as if to flush out insects. Usually in pairs or small family parties of three to five birds after breeding. Flight tends to be quite low, undulating. **Variation:** in central Aust., slightly paler; coastal populations have slight brown tint along flanks; no subspecies. **Voice:** cheery trill, metallic and rather insect-like, not loud but carries well through the quiet, open bushland: a rapid, ticking 'did-dit-d-wier, did-dit-d-wier, did-dit-d-wier'. As contact call, a hard, abrupt 'tchek, tchek'; rival males chase and threaten with scolding 'kek-kekekek-kekek'. **Hab.:** drier open woodlands, mallee; semi-arid mulga, similar scrublands, farms. **Status:** sedentary or nomadic; common. **Br:** 398

Head, back and throat slate grey, a slight blue bias
Wing-bar absent or very faint
Deep rose pink, not extending to belly
Belly all white
♂
Upper parts mid slaty grey
Often tinted pink
Pale buff wing-bar
Tail edged white
♀
Wing-bar absent or only a faint trace
Small white spot
Outer three mainly white
Coverts grey. Flight feathers grey-brown
Head, back, throat sooty brownish black
White forehead spot; buff on female
Faint tan-buff wing-bars
Magenta-pink
Pink extends far back on belly.
♂
Upper parts warm olive-brown
Cinnamon-buff; breast may have pink tint.
♀
Faint tan wing-bars
♂
Tail plain, dark brown
Wings and tail blackish brown; may have very faint tan or buff bars across flight feathers.
Upper parts dark slate grey
White bar along folded wing
Large forehead spot
Flame red
Flame colour extends to throat and belly.
♂
Usually has small buffy white spot over bill.
Uniform light brown
♀
White outer
Faint orange tint
Imm. ♂
Conspicuous white wing-bars
Outer feather white, both sexes
♂
Upper parts black
This species has largest white spot.
Throat black
Bold white bar across wing
Intense scarlet, barely reaches down to belly.
Outer two, each side, white
♂
Rather large head merges into plump body; wings are slightly drooped.
Always has large white spot.
Grey-brown
Variable orange tint
Buffy white double wing-bar
♀
Buffy wing-bars and coverts
Dark brown, streaked
Mottled brownish
Juv.
Prominent white wing-bars, male and female
Faint trailing-edge bar
Outer pair of feathers, each side, predominantly white
♂
Head, throat and upper parts brownish black
Unique with large crimson cap
Bold white bar across wing
White between crimson and black
White on outer feathers
♂
Smallest red robin, breast deepest crimson, but this only just reaches the belly. Wings are usually slightly drooped.
Light grey-brown
Faint dull red cap or forehead
Pale buffy grey
♀
Upper parts brown with fine white streaks
Buff bar
Mottled brown
Juv.
Immatures are like adult females, except young males have a touch of dull red on the breast.
White bars along coverts and flight feathers
Outer feathers white or white edged
♂

Grey-headed Robin *Heteromyias albispecularis* 16–18 cm

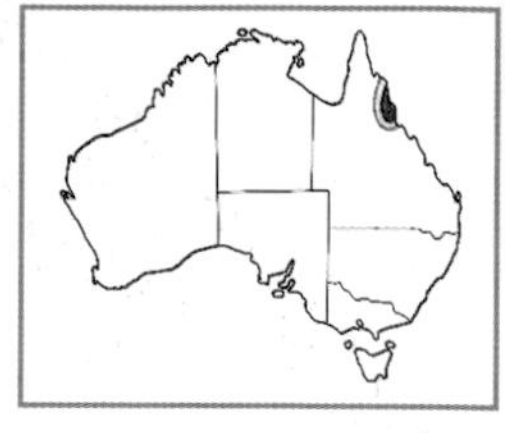

Softly colourful, unobtrusive robin of rainforest lower strata. Clings to trunks; drops to ground to take prey or forage in litter; flicks tail. **Juv.:** extensively rufous-brown, patchy whites; wing-bars like adult. **Voice:** soft piping whistle, rather mournful, single note or in long, leisurely, rather monotonous series of even pitch, occasional brief pauses, 'whiet-whiet-whiet', or high-low 'whiet-whit, whiet-whit'. Territorial song a long, high whistle followed by several lower, shorter notes. **Hab.:** rainforests of higher ranges, except a few situations where rainforest extends unbroken from ranges to coastal lowlands. **Status:** sedentary; common. **Br:** 399

White-browed Robin *Poecilodryas superciliosa* 15–17 cm

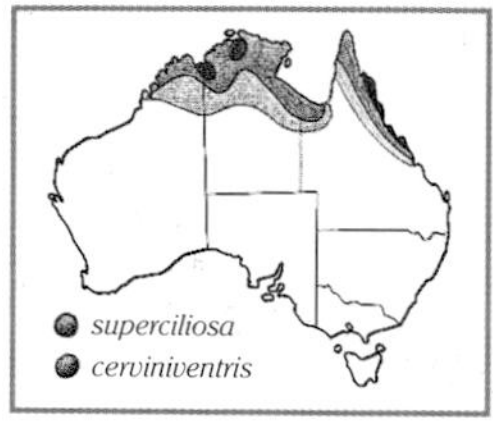

Boldly patterned; tail is often cocked upwards something like a scrub-wren, but the drooped wings and restless flicking of wings and tail are typically robin-like, as is their hunting, clinging to sides of trees, surveying ground, dropping to take prey from leaf litter or mangrove mud. **Juv.:** body more rufous. **Variation:** nominate race *superciliosa*, NE Aust.; race *cerviniventris*, 'Buff-sided Robin', W Kimberley to NW Qld. **Voice:** call of up to 5 high, clear, piping whistles, 'whiet-wheit-whiet', 'wheet-t-wit, whiew' or 'whieeeet-whiet'. Also a quick 'chokok, queitchiew' and 'chokoc-chiew'. As alarm, a harshly chattered 'chrok-chrorrok'. **Hab.:** rainforests, monsoon forests, vine thickets, mangroves. **Status:** sedentary; rather uncommon, uneven. **Br:** 399

Mangrove Robin *Eopsaltria pulverulenta* 15–16.5 cm

Conspicuous white on throat and rear flanks; in flight, bold contrast of white panels in dark tail. Unobtrusive but not shy inhabitant of mangroves; clings to trunks and branches, occasionally flicking wings; watches ground, dropping to probe the mud for tiny crabs and other creatures; also forages among foliage. **Voice:** most typical call a husky, mournful, descending 'whieer-wherrr…whurrr' and much lower, level 'whurrr-whurrr'. Song a higher, clearer 'whitch-a-whitchu' and lively chatterings. **Hab.:** mangroves, with preference for the shrubbier landward and seaward margins. **Status:** sedentary; moderately common. **Br:** 399

White-breasted Robin *Eopsaltria georgiana* 14.5–16 cm

Clings to tree trunks, low branches; occasionally drops to ground to take prey. **Imm.:** similar to adult, slight traces of juvenile streakings. **Variation:** isolated population on coast N of Perth; slightly smaller with warmer greys. **Voice:** attractive, mellow, liquid, abrupt 'wiCHWEK' with intervals of several seconds between calls; in distance, just 'chwek'. Song a higher, more lively 'tchiew-tchiew, whiet-siew, whiet-siew'; in alarm, hard 'tchek, tchek'. **Hab.:** in south, typically heavy forest in dense, tall, damp undergrowth, creek thickets; northern population in much drier coastal scrub. **Status:** sedentary, endemic to SW; common. **Br:** 399

Western Yellow Robin *Eopsaltria griseogularis* 15–16 cm

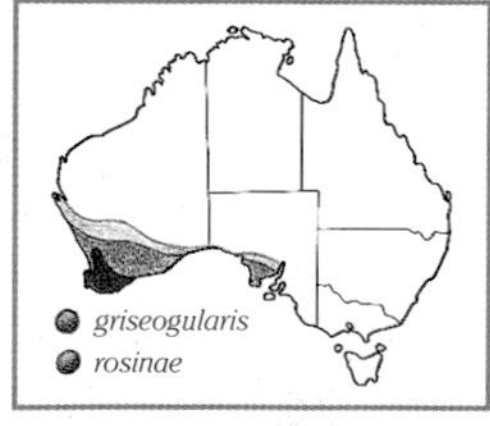

The western equivalent of the well-known Eastern Yellow Robin; differs most obviously in having a grey breast. Behaviour such as hunting is also similar; perches within several metres of ground, darting down to take prey. In pairs or family groups; unobtrusive but not shy. **Variation:** nominate race *griseogularis*, WA, bright yellow rump; race *rosinae*, SA, olive rump. **Voice:** most frequent call a long, even series of rapid, abrupt, clear piping whistles, 'whit-whit-whit-whit-'; then a slower, high 'chwip-chwip-chwip…'. Also a powerful, double noted call, delivered with force, slow and deliberate, and with whip-crack effect, 'chiOWP-chiOWP, chiOWP-chiOWP'. **Hab.:** diverse—forests, open woodlands, mallee, mulga. **Status:** local movements; common. **Br:** 399

Eastern Yellow Robin *Eopsaltria australis* 15–16 cm

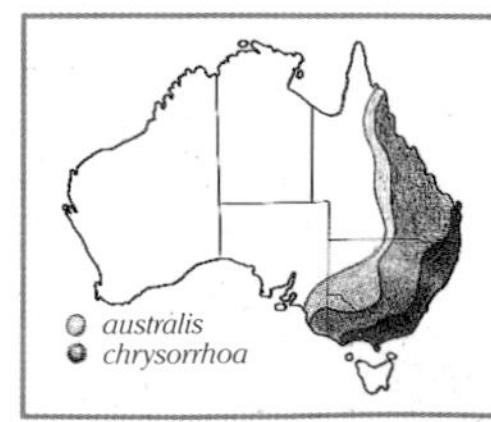

Usually pairs to small family groups; scans ground from a branch or while clinging to a tree trunk; watches patiently with an occasional flick of wings; darts perch to perch or drops to ground to take some small creature. Inquisitive, will approach squeaky sounds; soon becomes tame. Flights short, undulating, glides briefly to landing. **Variation:** nominate race SE Aust., race *chrysorrhoa* in N part of range. **Voice:** clear, even, piping whistle, 'tchiep-tchiep-tchiep'; double noted, loud 'tchweip-tchweip'. **Similar:** Western Yellow Robin, SA race *rosinae*, overlaps in SE of SA. **Hab.:** eucalypt and rainforests, woodlands, banksia heaths, drier mallee and acacia scrubs. **Status:** local seasonal movements; common. **Br:** 399

Crown and nape light grey, finely flecked darker
Heavy, pale tipped bill
Mantle olive-brown
Eye underlined white
Wing dark with short, uneven, dull white bar
Rufous-chestnut
Legs and feet large, solid, pale pink
Flanks buff
The Grey-headed Robin has a distinctive facial pattern: dark grey lores, dark olive-grey cheeks and ear coverts, white throat extending upwards to join a clear white underlining of the eye.
Conspicuous white wing-bar around base of flight feathers, faint trailing-edge wing-bar
Conspicuous, long, curved white brow; matching but inverted white line curves up from throat.
Bill long, thick
Secondaries tipped white
Both broad and fine wing-bars
Nominate race superciliosa
Crown and face darker
Rufous-buff
North-western race cerviniventris
Olive-brown
White band follows base of flight feathers
Secondaries tipped white
White tips
A second, faint bar along mid flight feathers
A slate-backed robin of coastal mangroves; its long straight bill, large for a robin, gives the head a rather whistler-like appearance.
Lores black; dark through eye to ear coverts
Bill relatively long, heavy
Conspicuous white throat
Underparts white with faint blue-grey wash to breast
Outer tail feathers white towards base
White
Upper parts mid to dark slaty grey
Faint pale wing-bar follows around base of flight feathers; hidden when wing is folded.
Large white panel each side of spread tail
Juveniles streaked brown above, mottled grey-brown beneath
Upper parts smoky mid grey blending softly into pale grey breast
Black lores
Wings and tail dark grey
Underparts white with pale grey across the breast
Outer tail tipped white
Plain grey and white robin, only in SW of WA; distinctive, almost a monochrome version of well-known yellow robins, below
Pale wing-bar around base of flight feathers; white underwing primary coverts
Juvenile streaked and mottled grey-brown; gape yellow
Upper wing has pale bar, hidden when wing is folded.
White tips to outer tail feathers
Grey back; only faint hint of wing-bar
White
Race: griseogularis, yellow rump
Grey breast
Bright yellow
Small white tips
Race rosinae, olive rump
Pale wing-bar visible in flight
Upper parts, face, chin grey
Wing-bar almost hidden in folded wing
Rump olive
Race australis
Juvenile brown, streaked white
Northern race chrysorrhoa, rump bright yellow
Race chrysorrhoa, 'Northern Yellow Robin', yellow rump, Qld and northern NSW
Chin pale grey
Concealed pale wing-bar around base of flight feathers
Pale matching wing-bar on undersurface

Pale-yellow Robin *Tregellasia capito* 12–13.5 cm

A small robin of rainforests that hunts more like a typical robin—clings to tree trunks, vines, dropping to ground to take prey—but often like small flycatchers, actively capturing insects in flight and from lower to mid-level foliage. **Voice:** bursts of 3 or 4 piercing, metallic squeaks, 'chweeik-chweeik-chweeik'; also harsh, scolding 'scraich-'. **Hab.:** rainforests, favouring parts with dense undergrowth, often lawyer vines; ventures into adjoining dense eucalypt forest. **Status:** sedentary; common. **Br:** 399

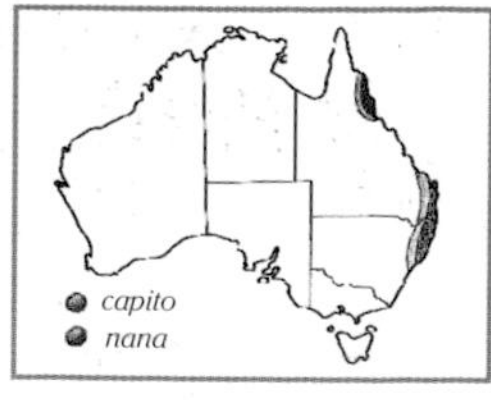

White-faced Robin *Tregellasia leucops* 12–13 cm

A black edged white face mask gives unique clown-like expression; sexes alike, immatures slightly duller; only the brown plumaged juvenile is significantly different. Forages in lower to mid-levels of rainforest; often perches across vertical trunks, vines; drops to take small prey on ground; pursues flying insects. **Voice:** song a cheery, rising whistle, 'whitia-whittik, whitia-whittik'. As contact, a soft 'tsip'; grating 'scraich' in alarm or aggression. **Hab.:** thickets of tropical rainforests and vine scrubs. **Status:** sedentary; moderately common. **Br:** 399

Lemon-bellied Flycatcher *Microeca flavigaster* 12–14 cm

Actively pursues insects in flight; flutters about the foliage, low to upper canopy, at times near and on the ground; often uses a perch in the open as a vantage point. Usually quiet and unobtrusive; alone, in pairs or small family parties. Performs song-flights when breeding; rises above treetops. **Juv.:** upper parts brown spotted cream; underparts white mottled brown. **Voice:** cheery, vigorous, musical whistling; notes range abruptly from squeaky, sharp to mellow, varied 'chiwi-chiwi-chweeip-chip-chrup' and 'quieee-chirrup-chi'. Also metallic, vibrant scolding sounds. **Hab.:** tropical savannah woodlands, rainforest margins, monsoon and vine scrubs, riverside vegetation, paperbark swamps, pandanus, mangroves. **Status:** sedentary; common. **Br:** 399

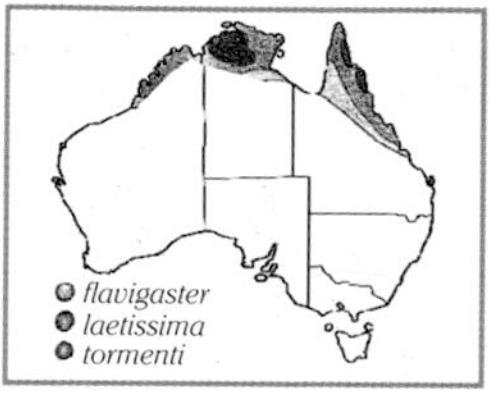

Yellow-legged Flycatcher *Microeca griseoceps* 11.5–12.5 cm

A restlessly active, very small flycatcher of upper and outer canopy of forests; often hovers to glean from foliage, darting after flying insects; when perched, often raises tail. Unobtrusive, in pairs or family groups. **Voice:** resonant, strong, whistled calls, but hesitant pauses between most notes, 'wheeit, wheeit, wheewit, whee–whi-wi-wi-whit, whit'; also high, clear, rapid, piping trill; buzzing 'scraich'. **Hab.:** tropical rainforests, favouring outer canopy; margins and edge of adjoining eucalypt woodlands; paperbark swamps. **Status:** sedentary; common. **Br:** 399

Jacky Winter *Microeca fascinans* 12–14 cm

Extremely widespread; well known in rural districts. Prefers open woodlands, farm paddocks with scattered trees. Uses open perches on branches, stumps or posts from which to scan the ground; drops on small prey more like a robin than flycatcher. More typical of flycatchers are its acrobatic pursuits of flying insects; unusual is its habit of hovering just above the grass, perhaps to flush out flying or hopping insects. On landing, wags tail side to side revealing the white—this is more obvious in flight. Alone, in pairs or family groups. **Variation:** three, perhaps four races; darkest in S, paler in far N and arid regions. **Voice:** carrying, clear, ringing, whistling call, a rapid 'chwit-chwit-chwit-queeter-queeter-queeter', slower 'cheweet, cheweet'. **Hab.:** open woodlands, mallee, mulga, cleared land with trees, stumps. **Status:** sedentary or nomadic; common. **Br:** 399

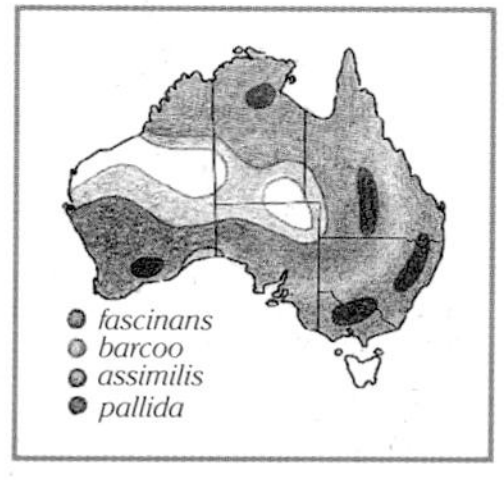

Narcissus Flycatcher *Ficedula narcissina* 13–14 cm

Breeds eastern Asian mainland, migrates in northern winter S to Java and Sumatra. Confirmed sightings, male only, on Barrow Is., NW Aust., late 1995 after a cyclone. Female more difficult to identify. **Voice:** usually silent on winter migration. **Similar:** Asian Tricoloured Flycatcher (not yet recorded in Aust.; see 'Similar', Blue-and-White Flycatcher). **Hab.:** usually woodlands, rainforest edges, gardens. On Barrow Is., acacia scrub with spinifex. **Status:** rare; vagrant.

Blue-and-White Flycatcher *Ficedula cyanomelana* 16–17 cm

Breeds China, Japan, winters S to Java. Sole Australian record an adult male dead on beach, NW coast, WA, 1995. **Variation:** Aust. record is nominate form; race *cumatilis* also possible. **Voice:** usually silent away from breeding range. **Similar:** this and Narcissus are distinctive; females most like small Petroicidae flycatchers. Both *Ficedula* flycatchers belong to family Muscicapidae (p.348), but placed here for comparison with other small flycatchers. **Status:** rare; vagrant.

Upper parts olive-grey; wings and tail grey-brown
Large white patch at lores joins white throat.
Pale yellow with olive-grey wash darkest on sides of breast
Nominate race *capito*, mid E coast
Race *nana*, NE Qld: pale rufous-buff on lores and eye-ring; slightly smaller
Juvenile rufous-brown, pale streaks on head; underparts paler, streaked on breast. Juvenile is like adult with touch of rufous on wing coverts.
Legs and feet dull pale pink, both races
Upper parts olive grey, rump olive
Head black, except white on forehead and chin, and around eye
Underparts yellow, olive tinge to breast
Juvenile rufous-chestnut, streaked white
The comic expression given by the white face mask is most evident in front view.
Small flycatcher with yellowish underparts, black legs
Pale eye-ring
White throat
Olive-brown
Pale lemon yellow
Black
Race *flavigaster*
Faint yellow tint
Deeper lemon yellow
Race *laetissima*
Olive-grey
Dull white
Race *tormenti*
Fine white wing-bar along coverts
Race *tormenti*
White tips, outer three
White edge to secondaries
A tiny flycatcher, unique among small flycatchers and robins for its bright orange-yellow legs and feet. Several others, including the White-faced Robin, have pink legs.
Head appears high, rounded compared with the sleek, low profile of the Lemon-bellied Flycatcher, above.
Juveniles brown, spotted buffy white on upper parts. Immatures like adults, but duller, pale-tipped wing coverts
Bill small but broad and flat, pale yellow lower base
White throat merges into light grey breast and pale lemon remainder of underparts.
Orange-yellow
SE race the darkest, deep grey-brown
Faint white brow line
Underparts white, faint brown wash over breast
Race *fascinans*
Outer feathers entirely white
Race *assimilis*, outermost feathers on each side white, becoming dark towards base of tail
Juvenile grey-brown, spotted off-white
Races *pallida*, N Aust., and *barcoo*, drier E interior regions, are paler; have white in outer tail. (White may not show in tightly folded tail.) *Barcoo* may be only a clinal variation within *pallida* rather than a race.
Male conspicuous black and yellow, female plain
White bar across coverts
Deep yellow
Long, wide brow is bright, deep yellow as also from chin to lower breast.
White belly
♂
Rump olive
Pale olive-buff, brownish wash over breast
♀
Crown bright cyan blue, darker on nape, encircles black face, breast.
Base of outer tail feathers white
Wings glossy blue, except dark brown primaries
♂
Lores, face, pale
Olive-brown
Wings and tail rufous-brown
♀
Imm. male is like the female, but with some blue showing on wings and tail.

Logrunner *Orthonyx temminckii* 18–21 cm

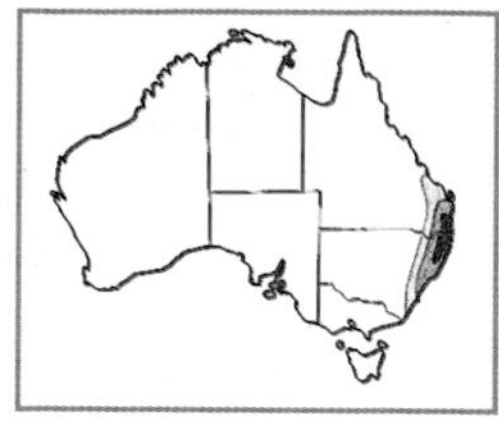

Forages in leaf litter on forest floor; scratches with large strong feet—props against broad stiff-tipped tail while throwing debris aside with one foot then the other, leaving a trail of small bare circles. Pairs, small family parties; unobtrusive, yet not shy; may approach closely while foraging. Walks rapidly away from danger; flies infrequently with whirring of short rounded wings. **Voice:** noisy at dawn, rapid, bubbly, sharp 'qwikit-qwikit-qwikit'; sharp, metallic 'queeik-queeik-queeik'; deeper, bubbly, mellow 'quokkit-quokkit-'. **Hab.:** subtropical rainforests, mainly on higher ranges. **Status:** sedentary; sparse in S, more common in N. **Br:** 400

Chowchilla *Orthonyx spaldingii* 27–29 cm

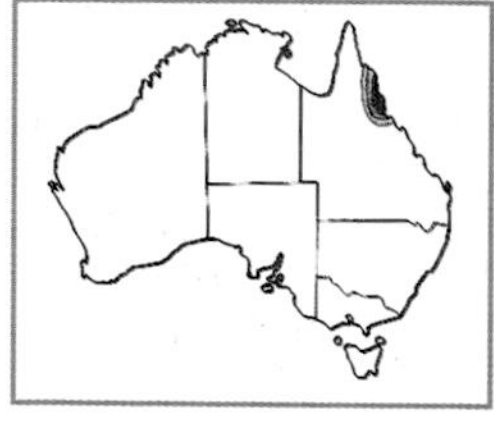

Blackish bird of rainforest floor; forages in leaf litter, propping against wide, stiff tail while raking debris out to one side then the other. Runs swiftly when alarmed; flies briefly on short, rounded, whirring wings. Inconspicuous, unobtrusive; presence may be revealed by rustling sounds or patches of soil bare of litter. Usually family groups, whose chorus gives a barrage of sound at dawn, dusk and occasionally through the day. **Voice:** calls usually in unison from several birds: one begins with rich, ringing, resonant, extremely loud 'chowk-chowk, chiowk-chiowk, chuck-chiow, chweik, chwieeik-chowk'; others simultaneously add softer musical warblings and low growlings. **Hab.:** tropical rainforests, NE Qld ranges. **Status:** sedentary; common. **Br:** 400

White-browed Babbler *Pomatostomus superciliosus* 19–22 cm

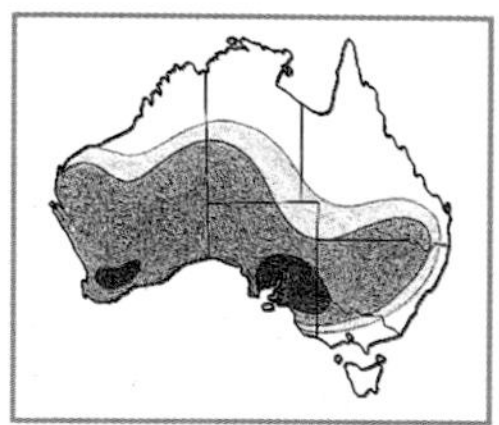

Noisy, gregarious behaviour has resulted in diverse descriptive names, e.g. 'Twelve Apostles'. Almost always found in mall family parties, with restless follow-the-leader foraging activity in shrubs, on ground, spread out or clustered. The passage of a family flock is typically marked by incessant churring and wheezing chatter. **Variation:** it is not clear whether the mostly small variations are smoothly clinal, or stepped so that several subspecies would probably be justified. **Voice:** varied nasal, squeaky and wheezy chatterings, cluckings and miaows: 'squarrk-squarrairk, wheeit-wheeit, chur-r-r-r-r'; clucking 'tchuk' contact calls; alarm whistles. **Hab.:** open forests and dry woodlands, mallee, tree-lined inland watercourses. **Status:** sedentary; common. **Br:** 400

Hall's Babbler *Pomatostomus halli* 22–24 cm

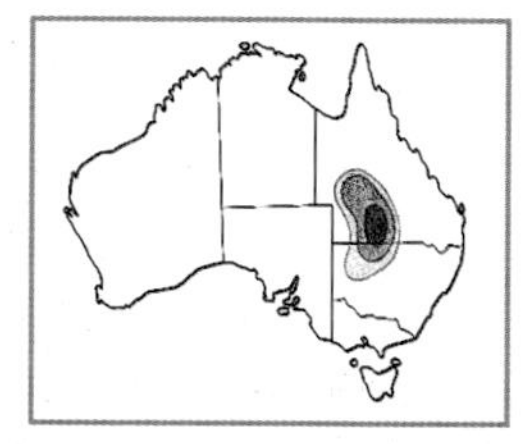

Superficially like White-browed Babbler in appearance, habits and calls. Not until 1963 was Hall's Babbler recognised as a separate species. In flocks of up to 20; forages mainly on ground. **Juv.:** as adults, but shorter bill and yellow gape at first. **Voice:** like that of White-browed, slightly deeper and more nasal, less wild variation; similar grating 'chweip-chweip-chur-r-r-r-'. **Similar:** White-browed Babbler, but much finer brow, central tail tipped white. **Hab.:** dry rocky slopes and ridges with mulga and other large shrubs, or open plains of sparse, stunted acacias or eucalypts. **Status:** locally nomadic; moderately common. **Br:** 400

Grey-crowned Babbler *Pomatostomus temporalis* 26–29 cm

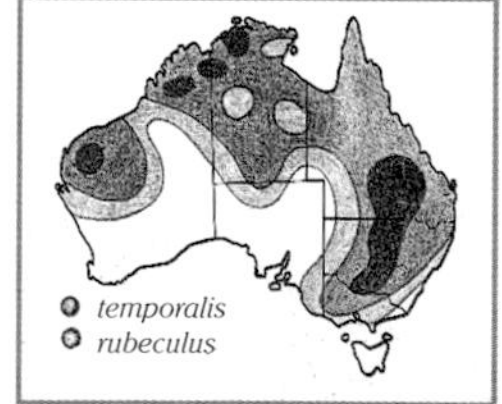

Highly gregarious; forms close, noisy family flocks to about 15 birds; energetic foraging from ground to high limbs. **Imm.:** brown-eyed to 2 or 3 years. **Voice:** diverse calls: most softer, more nasal than other babblers. Breeding adults give low–high duet: female a nasal 'awark'; male adds a high, clear 'tiew'; other birds of group join in—cacophonous calls, yet mellow, like distant yapping of many small dogs. Also a nasal 'chippa-wah, chippa-wah'. As contact, a soft, mellow 'tchuk'. **Hab.:** open forests, woodlands, road verges with grassy groundcover, sparse shrubbery. **Status:** sedentary; common in NW, uncommon in SE. **Br:** 400

Chestnut-crowned Babbler *Pomatostomus ruficeps* 21–23 cm

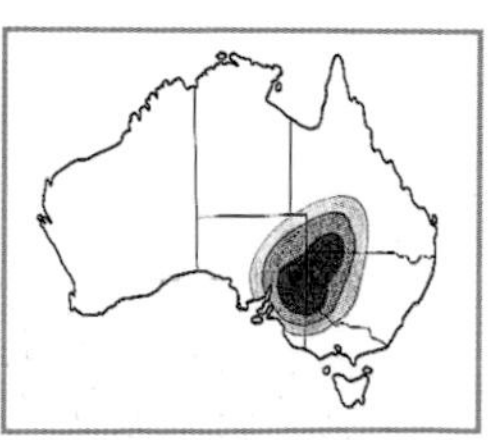

Close-knit groups bound energetically over ground, logs, through shrubs, onto tree trunks and limbs; while foraging keeps contact with low whistling calls. Probes ground litter and bark crevices with bill, seeking insects and their larvae. More shy than other babblers; they hop away behind cover, up into bushes, departing in long, low glides. When alarmed by potential predator, may huddle together under dense foliage. Like other babblers, roost together in disused nests. **Voice:** contact call a whistling 'tsee-tsee, tsee-tsee'; much chattering among members of family group; whistling calls intermixed with 'tchak-tchak-tchak', which becomes louder, more rapid in excited play, quarrels, alarm. Territorial song a rather strident piping. **Hab.:** sparse mulga, belah, mallee, lignum saltbush on plains and stony ranges. **Status:** sedentary; generally uncommon. **Br:** 400

Although both sexes are colourful, in the dim light of the rainforest the cinnamon-rufous of rump and flanks can appear dull brown unless caught in a patch of sunlight or strongly lit by camera flash.
Tail broad, dark, tipped with stiff feather-shaft 'spines'
Bold pattern of grey, black, orange
Female orange
The male, below, is less colourful than the female, at left.
Male white
A colourful species, but merges into the colours of the rainforest leaf litter: the golden, rufous and brown tones of fallen leaves and bark; and dark browns and black of moist earth and decaying leaves.
Immature is a darker, browner version of adult, but with underparts dull white, mottled or scalloped brown.
Tail very broad; dark with stiff, bare, feather-shaft spines
Short, rounded wings
Pale bluish white eye-ring
Male white, chin to belly
Flanks dark olive-rufous
Legs and feet large, black
Brownish black
Female orange, chin to upper breast
Male, left, is less colourful than female, right.
Immatures are overall dull rufous-brown finely flecked and scalloped with cinnamon.
Long, narrow, dark edged white brow, bill to nape
Upper parts, including crown, dusky mid to dark brown
Dark brown mask through dark eye
White tips to all except central all-dark pair
Throat and breast white, merging gradually to buffy brown flanks and belly
In flight, Hall's Babbler is darker blackish brown.
Bold white brow line extends full length of head, very broad and curving inwards at nape.
Dark blackish brown babbler with long and very broad brow line, conspicuous white bib and tail-tip.
Wide white brow
Black mask through dark eye
Tail tipped white, but narrowly on central pair
White bib has more sharply defined edge than on White-browed Babbler.
On spread tail of departing bird the white band continues narrowly around the central tail tip; (on the White-browed Babbler the all-dark central pair makes a dark break).
Long, wide, white brow lines separate narrow strip of grey on crown; dark face mask, eye yellow.
Nominate greyish race *temporalis* in SE Aust.; rufous-toned race *rubeculus*, in N and NW.
Dark mask across face
Wide white band around entire tail tip
Race *temporalis*, grey lower breast
In flight, chestnut across flight feathers, both races
Dark grey-brown
Bill long, heavy
Race *rubeculus*, 'Red-breasted Babbler', NW Aust. A clinal variation from greyish birds of SE to reddish NW birds
Tail has narrow white band around tip, including central pair
Distinctive slim, chestnut-crowned babbler, the only species to have white markings on the wings
Chestnut
White tips to some tertials
Brow line long, narrow, straighter than on other babblers
White bars across greater and median coverts; may also be a short bar on lesser coverts.
Adult
Large white tips to outer feathers; tapers to very narrow white on central pair.
Juv.
Juvenile like adult, but whites replaced by rufous-buff. Only briefly in this plumage

Western Whipbird *Psophodes nigrogularis* 22–24 cm

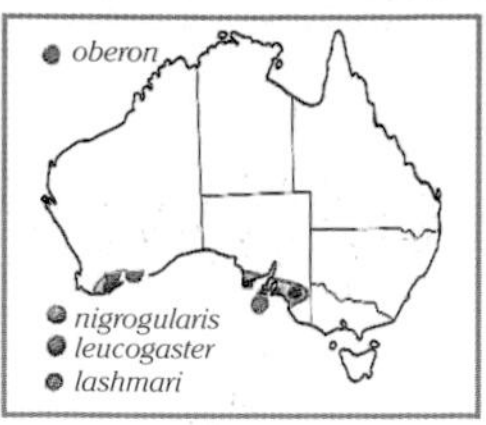

Rare, elusive, difficult to sight in dense scrub. Pairs forage in leaf and twig debris, seeking insects, seeds. If alarmed, vanishes under dense cover. **Voice:** not a whip-crack; has been likened to a squeaky cartwheel. While the whistled song does have a repetitive rotating rhythm, it is beautiful, musical, lilting, clear and liquid, 'WHIT-chee-a-WHEER-chwit': female often quickly answers 'chwik-it-up'. Also a clear, sharp, but downward, rattling finish, 'whit-chi-a-tr-r-r-r-r-t!'; as contact, 'chrrk!'. **Hab.:** dense heaths, mallee, broombush, but needs an accumulated litter layer. **Status:** locally common in protected sites, generally rare. **Br:** 401

Eastern Whipbird *Psophodes olivaceus* 25–30 cm

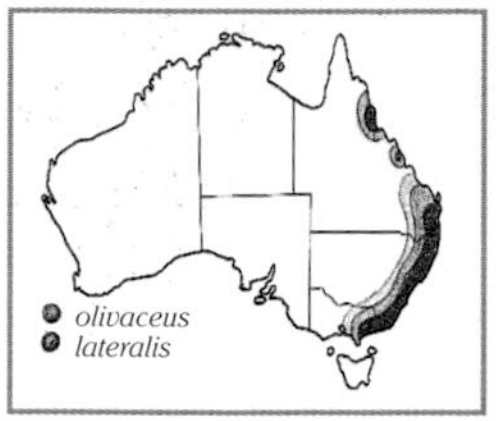

Loud ringing whip-crack call of this species is one of the most common, widely recognised eastern bird calls. Keeps to thickets, fossicking in ground litter; climbs through shrubbery; flutters weakly across open spaces. **Variation:** two subspecies, nominate *olivaceus*, coastal E Aust. except Atherton region, where race *lateralis* is smaller, olive-brown, with brownish tips to tail. **Voice:** from male, a long whistle building up to an explosive whip-crack; instantly answered with sharp 'tchew-tchew' from female. **Hab.:** thickets, margins of rainforests, wet eucalypt forests, heaths; dense regeneration. **Status:** sedentary; common. **Br:** 401

Chirruping Wedgebill *Psophodes cristatus* 19–21 cm

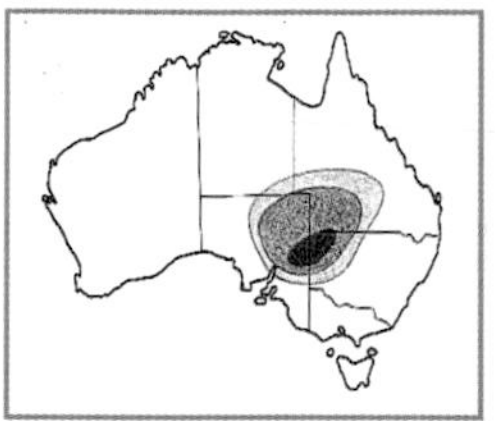

The two species of wedgebill appear almost identical, but differ in song and behaviour. This species less shy; allows closer approach to see a calling bird or watch a flock's activities. Small groups; run rather than hop. **Imm.:** like adult, but bill and legs brownish. **Voice:** a high squeak and strong, vibrating, descending trill, 'tsiep-TSIEEER', repeatedly, usually as male–female call–answer. **Hab.:** arid, open country, sparse acacias, eremophilas, saltbush, bluebush, lignum. **Status:** adults sedentary; common in N, rare in S of range. **Br:** 401

Chiming Wedgebill *Psophodes occidentalis* 20–22 cm

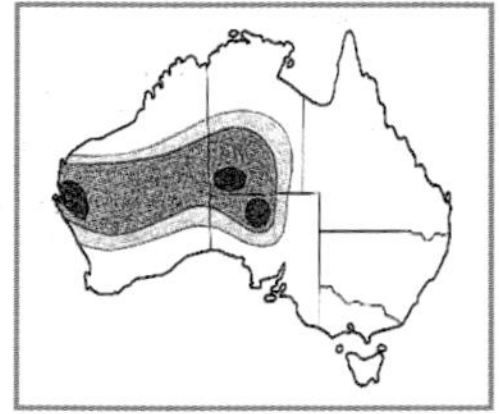

Extremely wary–if singing from a shrub-top perch will, at first distant sighting, become silent and drop to cover, or glide away out of sight. **Imm.:** like adult, but bill and legs brownish. **Voice:** far-carrying, beautiful, lonely notes; the first 3 high, ringing, metallic, the last much deeper, metallic, 'tchip-chipity-chiep, tchonk', repeatedly, vigorously. Also, high–low ringing, metallic 'WHI-whiet-WHI-wheit-'. **Hab.:** arid mulga and similar scrublands, broombush, mallee. **Status:** nomadic or dispersive; possibly most common near W coast **Br:** 401

Spotted Quail-thrush *Cinclosoma punctatum* 26–28 cm

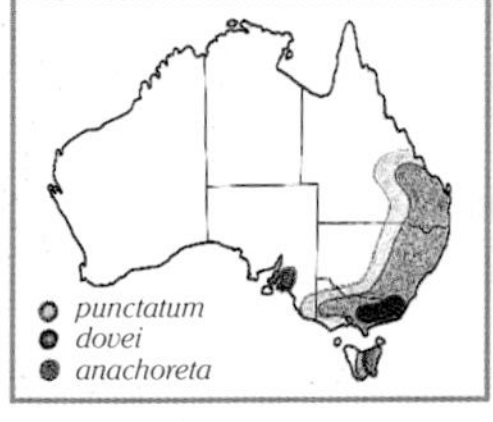

Largest quail-thrush, sole forest species. Alone or in pairs, forages on ground; crouches low, scuttling movements pecking at seeds, insects and other small prey. Very shy; initially relies on camouflage, until, almost underfoot, bursts up on whirring wings. Flight undulating; fanned tail displays white tips. Drops to cover 50 to 100 m away; then may be very timid, flushing at far greater distance. **Variation:** race *dovei*, Tas.; *anachoreta*, Mt Lofty Ra., SA. **Voice:** males call from low perch in trees. Song a penetrating double whistle, 'whee-it, whee-it, whee-it', the initial 'whee' with a mellow ringing quality, the final 'it' sharply higher, giving undulations throughout the long sequence. At times a faster, higher, 'sweeit-sweeit-sweeit'. Also a long, even whistler-like sequence, 'wheet-wheet-wheet' and extremely high twittering. **Hab.:** eucalypt forests, woodlands, favouring rock ridges with sparse shrubs, tussocky grass and abundant bark, leaf, twig litter, fallen logs. **Status:** sedentary; uncommon. **Br:** 401

Chestnut Quail-thrush *Cinclosoma castanotus* 22–26 cm

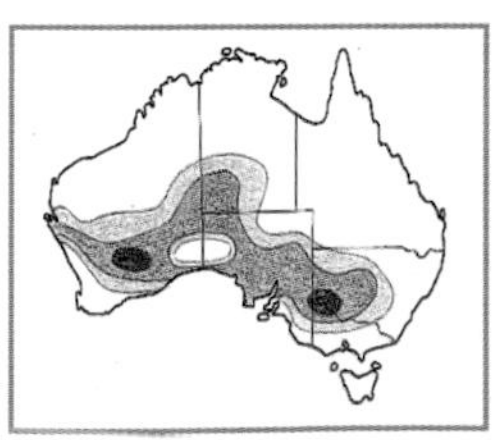

Wide range across Aust.; covers diverse habitats. In pairs or small family parties of 3 to 5, fossicking erratically among the leaf and twig ground litter. **Juv:** like female, but grey breast is scalloped. **Variation:** E of Spencer Gulf–Lake Torrens, SA, race *castanotus*. Opinions on those to the W varies: may be a clinal, more rufous continuation of *castanotus*, or may be a second race, *clarum*, or further split into coastal race, *fordianum*, and inland race, *clarum*. **Voice:** usual call a rapid, very high rush of squeaked whistles at uniform pitch, 'swit-swit-swit-swit-swit-swit'. Song by male, from a high perch, is an even series of rich, mellow whistles, 'wheit-wheit-wheit-wheit'; may be repeated at varied tempo and pitch, including a higher, quicker 'whit-whit-whit-whit-whit'. **Hab.:** open arid woodlands of eucalypts or cypress pine, mallee or mulga, with sparse shrub layer and litter debris; heathlands, coastal tea-tree thickets. **Status:** nomadic in arid regions, sedentary towards coasts; uncommon, patchy. **Br:** 401

Rare, elusive. Widely separated: race *nigrogularis*, 'Western Whipbird', SW of WA; race *leucogaster*, 'Mallee Whipbird', SA and Vic.
Small crest, often not erected
Male and female alike, male larger. Immature has dull indistinct face and throat markings.
Black line as edge to upper side of of white streak; black throat
Underparts uneven olive-grey
Lacks full black line along upper side of white streak.
Outer tips white
Dark subterminal band, outer tail
Race *nigrogularis*
White or pale central strip, lower breast and belly
Outer tips white
Race *leucogaster*
Crest and head black; eye deep red to ochre
Immature without white cheek streak until about second year; colours dull, greyer, mottled
Bold white streak, cheek and side of neck
Olive green
On tightly folded tails of whipbirds, white tips may be concealed from above by central feathers, which do not have white tips. Obvious on fanned tail, or from beneath
Black, with variable amount of patchy, streaky white
Adult
Imm.
When perched, appearance like Chiming Wedgebill except faintly streaked breast
Two almost identical species, the Chirruping Wedgebill, and the Chiming, below, can usually be identified by locality of sighting, (with reference to map) but also have differences in song and behaviour.
Breast white with faint indistinct darker flecks or streaks
Both Chirruping and Chiming Wedgebill often depart their shrub-top perch in a long low glide, giving a glimpse of widely-spread, conspicuously white-tipped tail.
Tall dark crest curves well forward.
Both wedgebills have a long tail that may be slightly or moderately uplifted, held straight, or lowered.
Bill rather stubby, almost finch-like
Adult
Breast plain, very pale buffy grey
White tipped
Adult
Lesser coverts tipped white
White tips to greater coverts give a thin white wing-bar.
Throat, face black, enclosing white patches
Sandy or rufous-brown, streaked black
Chestnut on tertial feathers
♂
Blue-grey, may be fluffed out hiding black coverts ♂
In the folded tail, white-tipped outer feathers are hidden under entirely grey central pair
All quail-thrush are long bodied, long tailed; rise quail-like on small whirring wings.
Juvenile, not shown, has upper parts olive-buff finely flecked black, underparts pale buff scalloped or mottled brown.
Flanks buff, with black streaks faint or absent down central belly
Buff
Race *punctatum*
♀
Outer tail feathers have white tips, conspicuous on take-off
Upper parts rufous and light brown
White brow and throat lines, grey breast. Imm: breast scalloped
Narrow white wing-bar
♀
Chestnut, brighter on western birds
Across southern inland has grey-brown upper parts (*castanotus*); gradually more rufous-backed towards N and W (possible race *clarum*). Opinion on SW population differs, *castanotus*, *clarum* or *fordianum*.
Black extends from chin to low on breast.
Dark brown, edged cinnamon
Inland-western form has chestnut more extensive, brighter
Grey on flanks between wing and black of breast
White tipped outer feathers hidden under plain central feathers
♂
White tips of dark outer feathers conspicuous when tail is spread

Cinnamon Quail-thrush *Cinclosoma cinnamomeum* 18–21 cm

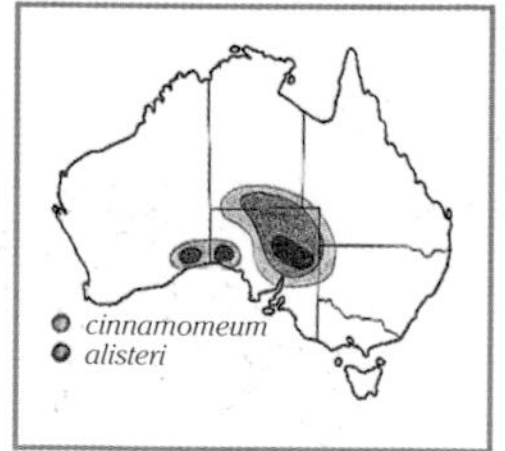

Lives on arid stony plains with sparse low shrub cover. Alone, in pairs or small family parties; wanders erratically among rocks and shrubs, maintaining contact with group by high faint whistles. Feeds exclusively on the ground, foraging birds wandering erratically, often hunched rather low, pecking at insects, seeds, occasionally moving small stones. If disturbed, will stand more upright; runs swiftly, squats low in cover of bush, rock or bare ground where the colour of a motionless bird renders it difficult to recognise. Put to flight, bursts upwards with quail-like take-off and whirring of relatively small wings; departs in undulating low flight, white tips of tail prominent; soon drops again to cover. **Variation:** 2 populations, each on stony plains, separated by unsuitable habitat. The nominate race *cinnamomeum* occupies a large area of arid central Aust. around the junction of SA, NT and Qld. Not as brightly coloured as the Nullarbor form, the cinnamon tone of upper parts is slightly greyed, the crown is brownish. The second race, the smaller Nullarbor form, is more brightly coloured. The northern half of the *cinnamomeum* population is paler; may justify another race, *tirariensis*. **Voice:** contact call a weak, high, piping whistle, uniform pitch, 'tsit, sit-sit'. Song, usually at dawn: 'seit, sit, sieee', 'tseit, see-seeeit' and tssi-sieee, seei-eit'. In each version the first note is very high, thin, even pitch, last part slightly lower and rather tremulous. **Similar:** Chestnut-breasted Quail-thrush, race *marginatum*, also has a black double band, but has chestnut in centre of breast and darker chestnut back. **Hab.:** sparsely vegetated shrub steppe country, usually dominated by bluebush and saltbush, on stony hard-pan or gibber-stone plains and low undulating ridges; vegetation along watercourses. **Status:** sedentary or locally nomadic. Nominate race quite common in localities of optimum habitat. Nullarbor race uncommon. **Br:** 401

'Nullarbor Quail-thrush' *Cinclosoma cinnamomeum alisteri* 18–20 cm

This distinctive race, at one time a full species known as the 'Nullarbor Quail-thrush', is slightly smaller and a brighter cinnamon-rufous on the upper parts, especially the crown. The most obvious difference on the male, however, is that the black of the breast is not broken into two bands by cinnamon or white as in the nominate race. The female is more like the nominate race, but smaller; upper parts are a lighter, brighter cinnamon, extending onto the crown. **Voice:** similar to nominate race. **Hab.:** confined to the Nullarbor on stony plains and grassland with sparse small shrubs of acacia, saltbush and bluebush. **Status:** sedentary or locally nomadic. Considered rare, possibly endangered by grazing and consequent alteration of the habitat that renders some areas unsuitable for the species. It remains common at some localities, but rare or absent in former habitat that has now become unsuitable.

Chestnut-breasted Quail-thrush *Cinclosoma castaneothorax* 20–24 cm

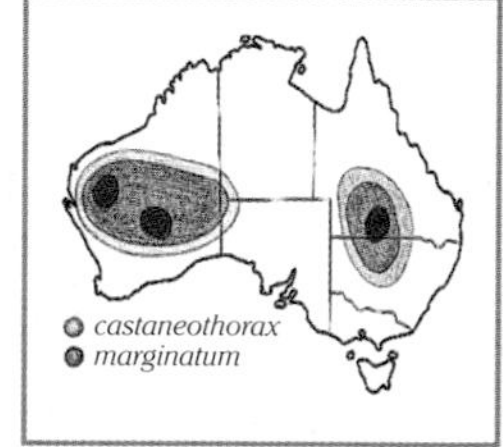

Although typically inconspicuous, elusive, shy of disturbance, this and other quail-thrush species may, if quietly approached, be observed at quite close range and their activities followed. Usually in pairs or small family groups, and, like other quail-thrushes, captures insects, typically grasshoppers, moths, beetles, ants, caterpillars and other small prey such as spiders and small skinks. Also eats large amounts of seed fallen to the ground from spinifex, acacias and other plants. Quail-thrushes walk rather deliberately with frequent rather erratic changes in direction, stopping to capture an insect or to probe with the bill into crevices, sand and under small rocks; apparently they do not use their feet to fossick or rake about in leaf and bark litter. If startled, they may squat where their colours blend with the surroundings, often to burst explosively into flight, quail-like, on small whirring wings, displaying prominent white tips of tail as they depart. More often they will run, skilfully using any vegetation as cover. **Variation:** two subspecies that have, in the past, been listed as full species. The nominate race *castaneothorax*, 'Chestnut-breasted Quail-thrush', inhabits the interior of eastern Aust. The brighter coloured race *marginatum*, 'Western Quail-thrush', occupies similar habitat through the central interior of WA and E to reach into SA and the NT. The eastern and western races have been separated by unsuitable desert sand dune country for a long period, perhaps since the development of those deserts, which has allowed their obvious differences of plumage to evolve. **Voice:** contact and alarm calls are high-pitched, weak, piping, insect-like squeaks, each with a slight uplift in pitch, 'tseit-tseit, tseit-tseit'. The song is a sequence of rather soft yet carrying whistles, tremulous, plaintive, and at times low-pitched compared with calls of other quail-thrushes, 'whit, whit-a-whittee'. This is repeated many times with the same pattern of notes, varying slightly, trending gradually lower, more tremulous, and much slower, 'whit, wheit, a-wheitee', and eventually an almost husky, soft, 'wheit, wht, a whteee'. **Similar:** Cinnamon Quail-thrush, eastern nominate race, but has centre of breast, between black of throat and breast-band, buffy white rather than cinnamon or chestnut. **Hab.:** mulga scrub on hard, stony, slightly elevated, dissected plateau country and low stony ranges; mallee, tea-tree on stony hills. **Status:** locally nomadic; nominate race in E Aust. uncommon. **Br:** 401

'Western Quail-thrush' *Cinclosoma castaneothorax marginatum* 21–25 cm

Western subspecies (formerly full species) is distinguished by more rufous upper parts: crown rufous-brown rather than olive-brown, breast paler cinnamon-buff. Black band across lower breast is much wider, heavier, more continuous in its extension rearwards along flanks. Behaviour and voice similar to nominate race. **Status:** locally nomadic; common.

Cinnamon Quail-thrush, nominate race *cinnamomeum*
Long down-curving white brow line
Crown Grey-brown
Dusky brownish cinnamon
Race *cinnamomeum*: upper parts, mantle to tail, and flanks, dusky cinnamon, not as bright as Nullarbor race *alisteri*
White-tipped black wing coverts, often hidden by long, upswept cinnamon feathers of breast and flanks
Cinnamon-white gap between black throat and breast-band
Breast grey
Black with white tips
♂
♀
Race *cinnamomeum*
Plain central tail feathers usually hide black, white-tipped feathers of outer tail.
Finely flecked brown
White tips to greater coverts form thin white wing band
Wing coverts buffy brown
♂
Juv.
Race *cinnamomeum*
Central tail feathers are plain dusky cinnamon, outers blackish, tipped white
Flight feathers brownish, with extensive edging of cinnamon or buff
Underparts entirely buff, streaked or mottled brown
Crown rufous
Upper parts brighter cinnamon than nominate race
Crown, face, cinnamon-grey
'Nullarbor Quail-thrush', race *alisteri*, upper parts entirely cinnamon-rufous, brighter than nominate race
Black continuous from chin to breast
Breast grey, flanks cinnamon
Concealed dark, white tipped outer feathers show when tail is spread.
♀ Race *alisteri*
White tipped black wing coverts are often concealed by long, upswept cinnamon feathers from sides of breast and flanks.
♂
Juvenile has underparts buff, flecked dark brown, wing coverts brown.
Race *alisteri*
Crown olive-brown
Creamy white brow, strongly downcurved; long, wide, pure white throat-edge streak
Rump, back, dull cinnamon merging to olive-brown on mantle, nape and crown
Black throat
Upper back grey-brown to olive-brown merging to chestnut on rump
Cinnamon-buff to chestnut
Brow and throat light buff
Narrow band
Breast grey-brown
Rufous-brown
White tail tips, usually hidden
♂
♀
Coverts black, tipped white
Pale cinnamon, lightly mottled brown
Nominate race *castaneothorax*, 'Chestnut-breasted Quail-thrush'
Broad cinnamon margins
Juv., both races
Outer tail brownish black, broad white tips
Undersurface pale buff, very lightly scalloped or flecked darker
Buff, finely scalloped or flecked brown
♂ Race *marginatum*
Brown with narrow, pale buff margins
Chestnut-breasted Quail-thrush, race *marginatum*, 'Western Quail-thrush', brighter rufous-chestnut
Brow and throat-streak white
Race *marginatum*
Black, white tipped shoulder often hidden beneath upswept cinnamon plumage of breast and flanks
Breast cinnamon-buff rather than chestnut
Rump rufous-chestnut, gradually darker along the tail; white tips usually hidden
Wider black breast-band, more solid black line back along flanks
♂
♀
Breast cinnamon-grey merging to cinnamon flanks; here partly covering dark shoulder

Varied Sittella *Daphoenositta chrysoptera* 11–13 cm

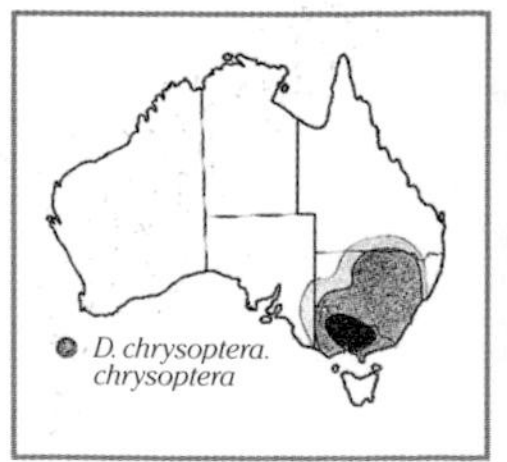

Very small birds of the upper limbs: would often be unnoticed but for the almost incessant sharp twittering calls, which seem loudest as the flocks (usually less than 10 birds, but up to 30) fly tree to tree. Flight is strongly undulating or jerky and rather erratic; the flock often abruptly changes direction, veering away to land in the top of a tree to one side or the other of their original course. The calls continue, maintaining contact, after the sittellas land in the crown of a tree, becoming slightly subdued as the group disperses through the upper twigs and branches. The birds work their way down, probing and prising with fine upturned bills into crevices of bark or timber, seeking spiders and insects. Their downward progress is erratic as individuals pause to pry out some small creature or pop food into chicks in an almost invisible nest before continuing nonchalantly downward as if no nest existed. Frequently they half-spread their wings, giving a glimpse of the wing-band colour. Reaching the lower limbs, the flock then flies noisily to the crown of another tree. Sittellas are strictly arboreal, and usually stay well above the ground. Through spring and summer the sittellas break into smaller breeding flocks, typically consisting of a breeding pair, several unmated adults, and two or three immatures of the previous season. All members seem to help build the nest, which is so sleekly blended and camouflaged that it appears a natural part of the branch, although usually in an open spot without foliage concealment. Feeding of young is also a combined effort. **Variation:** *Daphoenositta chrysoptera chrysoptera,* **'Varied Sittella'** and four other races, detailed and mapped below. These meet in intergrading populations that overlap in north–central Qld. All races have prominent wing bands of orange or white, mostly hidden while the wing is folded. **Voice:** a constant call-and-answer means that the sound is usually from many birds; gives the impression of very rapid calling. Contact call while foraging is a high, thin, metallic 'chwit'; suddenly louder, a harder and more metallic sound, yet with each note ending very sharply, 'tchweit-tchweit-tchweit'. Sounds agitated or urgent, with many birds calling together as a group departs for the next tree. The birds tend to travel in a compact if untidy group, and can be heard approaching from afar, simplifying location of their next landing place. **Similar:** treecreepers—like in showing orange or buff wing-bar in flight and foraging on bark, but unlike in descending to ground and foraging upward on limbs. **Hab.:** nominate race, SE Aust. N into central Qld in eucalypt forest and woodland, mallee, farm trees, shelter belts, roadside trees, and parks and gardens. While showing preference for rough-barked trees, these birds occur in most treed habitats except rainforest. **Status:** sedentary nearer coast, nomadic in arid regions. Common, except in regions of sparse tree cover. **Br:** 402

'Black-capped Sittella' *Daphoenositta chrysoptera pileata*

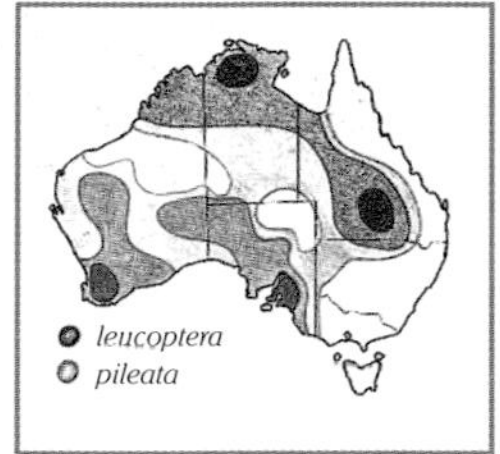

Has sharply defined black cap that extends below eye on female. Cap on male is much smaller than on nominate race, reduced by white brow line and white forehead, and does not extend far back onto nape; orange wing band. Underbody, both sexes, almost white without obvious streakings. **Hab.:** wide range across southern Aust. in diverse habitats from heavy wet coastal forests of SW of WA through open woodland to arid interior woodland, mallee, mulga, watercourse timber lines. **Status:** common.

'White-winged Sittella' *Daphoenositta chrysoptera leucoptera*

Well-defined blackish cap below eye on female; on male smaller and kept well above eye by white brow line; has white on forehead. Wing band is white; shows on folded wing only as white lines along primaries; white of secondaries hidden until wing is well extended. **Hab.:** north-central Qld westward through NT to Kimberley in WA; tropical open woodland on grassy plains and more stunted trees of ranges. Usually sedentary, perhaps locally nomadic; common.

'White-headed Sittella' *Daphoenositta chrysoptera leucocephala*

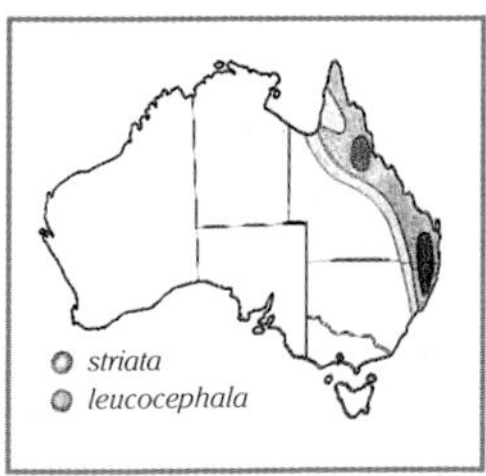

Combines white head with orange wing-bar; both sexes. Similar, but with combined white head and white wing-bar, is the doubtful race *albata*, 'Pied Sittella', known from only one specimen collected in 1864 near Bowen, Qld, and not recorded since, possibly from a hybrid population that has since died out. As with other sittellas, the wing-bars run the full length of the wing and are only partly visible, initially on the primaries, until the wing is fully spread. **Hab.:** forests and woodlands along the Great Divide, northern NSW to mid E coastal Qld, and inland on more open grassy woodland. **Status:** common.

'Striated Sittella' *Daphoenositta chrysoptera striata*

Heavily streaked underparts; on the female, more extensive dark about head with black on throat as well. Male has a smaller black cap; heavily streaked underbody extends to throat and chin; dark markings on lores, ear coverts. **Hab.:** a tropical race, Cape York S to Townsville, Qld, in eucalypt forest, woodland; like others, avoids rainforests. **Status:** sedentary, common. Sittellas were previously placed near treecreepers, most alike in ecology, and share the prominent wing stripe. Evidence now shows the sittella family not so closely related to the treecreepers. For comparison, a Rufous Treecreeper is shown with the sittellas, and also, the very different nests of the two families.

Tail short, dark, outer tips white, all races
Race *chrysoptera*
Rump white, all races
Streaked
Orange wing bar shows on primaries, but concealed on secondaries
Grey-brown
Legs and feet yellow, all races
♂
White or streaked
Yellow base
Pale orange wing patch is always visible to some extent on primaries, but only becomes exposed across secondaries as wing is extended.
Grey-black
♀
All races have yellow eye-ring. Race *chrysoptera*
Flight feathers and wing linings dark brown
Outer tips white, coverts barred, all races
White tipped coverts
White
Underbody whitish, streaked brown
Orange band translucent, intense if backlit
♀
Softly streaked light brown
Under-tail coverts barred, on all races
Race *pileata*, 'Black-capped Sittella', pale orange wing bar, mostly concealed until wing is fully spread
♀
Sharply defined black cap includes eyes and forehead.
Small blackish cap
Forehead white
♂
Male has dark cap separated from eye by white brow.
Dark band and central pair; white rump and outer tips, all races
Race *leucocephala*, 'White-headed Sittella', the one race combining white head and orange wing-bar
♂ ♀
Race *leucocephala*. 'White-headed Sittella, has combination of orange wing-bar with entirely white head.
White head and streaked underparts, both sexes
Race *striata*, 'Striated Sittella', dark head, underparts heavily streaked, white wing stripe
♀
♂ ♀
Black extends to throat; male has white brow, streaked throat.
♀
White bar
Race *leucoptera*, 'White-winged Sittella', white wing-bar, black cap and plain white underparts
Male differs with smaller dark cap separated from eye by white brow line.
White underparts
♂
Race *pileata*, 'Black-capped Sittella'
♀
Treecreepers (page 218) forage on tree bark, logs, ground, also show prominent wing-bars, but always work upwards, then glide to base of next tree. Unlike sitellas, all treecreepers nest in a tree hollow.
Nest is blended into a vertical fork, camouflaged with bark flakes. Birds spiral down; pause briefly to feed chicks; move on downwards, the whole sequence being little different to normal foraging activity.

Crested Bellbird *Oreoica gutturalis* 20–22 cm

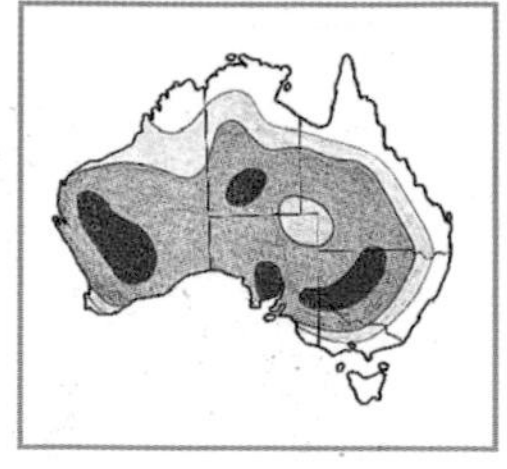

The deep, mellow, liquid and slightly melancholy note of the Crested Bellbird is one of the characteristic and, to many, nostalgic sounds of the Outback. The calling bird, usually the male, may be glimpsed in the distance, a rather upright shape, usually on an exposed perch above the general level of scrub. Flight is undulating; long downward glides from perches. Forages on ground, stumps, logs, in shrubs and lower parts of trees. In pairs during breeding season, then briefly in family groups, but solitary most of the year. **Variation:** slight, probably clinal—towards interior and north, gradually smaller and paler; these sometimes considered a race, *pallescens*. **Voice:** the song, usually begun by male with female contributing final sounds, often begins softly, builds louder, has a ventriloquial quality. Has a regular, strong rhythm, with unique soft, liquid 'ook' inserted among whistled notes. Variations include a low, husky sequence—rather slow rhythm, mellow whistles and deep waterdrop effect: 'dee, dee, dee-ook, dee, dee, dee-ook'; also higher, clearer, whistled 'whit, whit-whit-quiook', and low 'plonk-plon-plonka' **Similar:** wedgebills, but lack black markings. **Hab.:** arid mulga and similar scrublands, mallee, drier woodlands, farmlands. **Status:** through interior, nomadic, common; nearer coasts, sedentary and uncommon. **Br:** 402

Crested Shrike-tit *Falcunculus frontatus* 15–19 cm

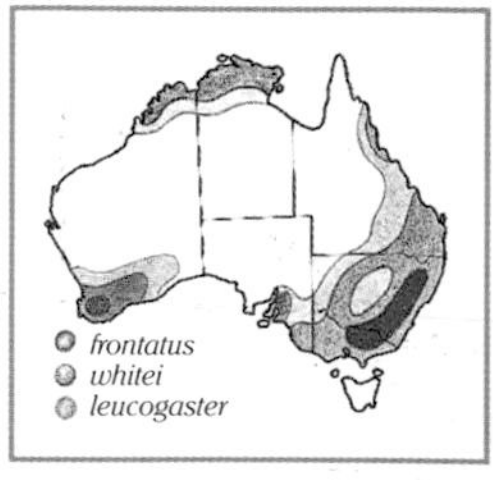

Although colourful, the Crested Shrike-tit is most unobtrusive; forages alone, in pairs or family groups, usually keeping quite high, but occasionally descending low on saplings. Unless attracting attention by calls, could easily be overlooked; on occasions reveals its presence through the sound of tearing bark. The function of the heavily built, deep bill is soon apparent if one of these birds is observed foraging. Strong pieces of bark are levered up by sliding the bill under, then rotating head and body to exert a twisting force through the flat blade of the bill. **Variation:** subspecies are widely separated, simplifying identification. The overall silhouette is distinctive, all races have large, crested head and deep bulky bill. The bold black-and-white head pattern is quite consistent between races, but colours of back, wings and underparts differ, while those birds from the far north are slightly smaller. Race *frontatus*, 'Eastern Shrike-tit'; race *whitei*, 'Northern Shrike-tit'; race *leucogaster*, 'Western Shrike-tit'. **Voice:** the song is a soft, rather mournful, undulating, piping whistle, often weak, seeming to come from a distance. The quality of voice in this call is mellow, slightly husky, almost nasal; tends to start softly, strengthen through several sequences of calls, with the last note of each set lowest, strongest and finally slightly up-slurred: 'whiert, whi-whit, wheeeir…whiert,whi-whit, wheeeir'. The northern race *whitei* varies this by alternating between lowering and rising sets of notes. Also a repeated, sharper, downwards call, 'peeeir, peeeir, peeeir'; scolds with a hard, harsh, rattling chatter. **Similar:** Golden Whistler, but white patch only on throat rather than above and below eye, and far louder, ringing, whip-crack calls; female whistler plain grey-brown except for yellow tint under tail. **Hab.:** open forests, woodlands, mallee, riverside and watercourse trees, stands of cypress pines, banksia woodlands. **Status:** sedentary or locally nomadic. Generally rare, but in some sites of optimum natural woodland habitat, may be quite common. **Br:** 402

Grey Whistler *Pachycephala simplex* 14–15 cm

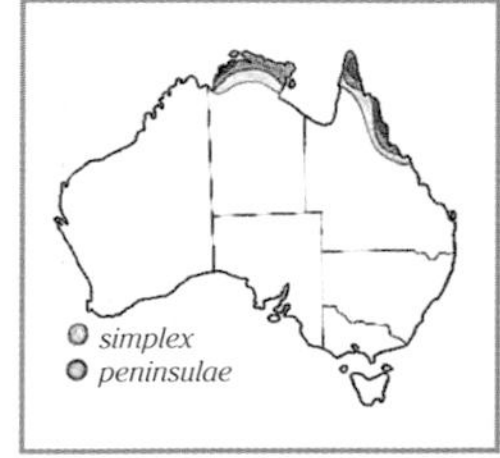

Not only far smaller than other whistlers, but more like a flycatcher in its active pursuit and gleaning of insects from foliage of mid to upper canopy; often in mixed foraging parties where its lack of distinctive markings makes it difficult to separate from slightly smaller flycatchers and gerygones. At other times, the whistler may sit inconspicuously for long periods among foliage. **Variation:** widely separated populations, previously full species, now subspecies. Race *simplex*, 'Brown Whistler', N coastal NT; race *peninsulae*, 'Grey Whistler', NE Qld. **Voice:** clear, sharp, penetrating, short whistles, often slow and deliberate, with a rise then fall, 'wheet, whit, wiet, wheat, whieew'; lacks final whip-crack. Can be a longer, vigorous rush of powerful, abrupt whistles. **Hab.:** rainforests, usually below 500 m for eastern race; also monsoon and riverine forests, paperbark swamps, mangroves and adjoining areas of eucalypts. **Status:** sedentary; common. **Br:** 402

Olive Whistler *Pachycephala olivacea* 20–22 cm

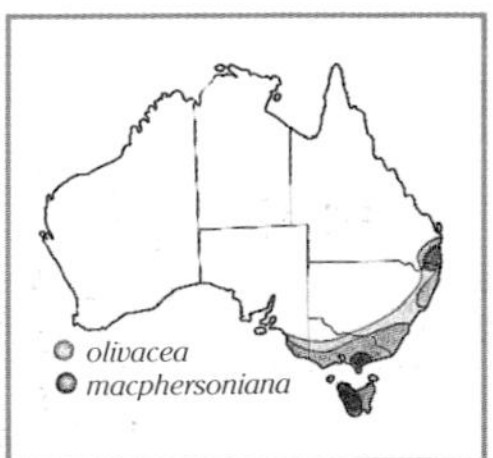

Large whistler plumaged in subdued rich colours; superb song. Keeps low, usually in undergrowth thickets, often on ground. Tends to be solitary much of year, but often in company with robins and scrubwrens. Retiring habits in dense vegetation and sombre colours would make the Olive Whistler easily overlooked but for loud, clear, liquid notes. **Variation:** two races: nominate *olivacea*, SE Aust., and Tas; race *macphersoniana*, NE NSW and SE Qld. **Voice:** strong whistles, high, sharp, deep and mellow. In south, whip-crack 'wheeeow, WHIT, whit, whit' and 'chee-o-whit, wheeeow'. Northern race high, thin, then abruptly low, 'whiee-whuooo', and low–high 'cheeowhit, wheee'. **Hab.:** N race above 400 m in rainforest; S race in undergrowth of eucalypt forests, tea-tree, heath. **Status:** sedentary or local movements; uncommon. **Br:** 402

Black-topped crest may be raised slightly higher than shown or tightly down as a black line.
Male has conspicuous and unique facial pattern, bright orange-red eye.
Combination of rounded black bib and overall sandy brown tone identify the male.
Female has only a black line as crest.
Eye dull chestnut, throat white
Underparts pale buff, slightly darker on breast
Male's crest folds flat; shows as black crown streak.
Slightly paler buffy margins
Immature is like female, but with buffy margins to wing feathers; male has mottled dark throat.
Tail slightly darker brown
Upper parts sandy brown to rufous-brown, primaries darker
Race frontatus, 'Eastern Shrike-tit', has mantle, back and rump olive-green to olive-brown.
Massive bill, shaped for levering and stripping bark
Males, all races, have throat black.
♂
Race frontatus, under-parts entirely yellow
Wings and tail vary, dark grey or dark grey-brown, fine pale margins, yellowish on race whitei.
All races have very broad white bands above and below the black eye-line.
Race frontatus, underparts yellow
All races have white lores joining across base of bill.
Cinnamon to olive-brown
Females, all races, have throat olive to olive-brown.
♀
Upper parts, including wing and tail edges, citrine yellow
Buffy white
Race leucogaster, 'Western Shrike-tit', white belly
♂
Juv.
♂
Race whitei, entirely yellow
Forehead grey-brown
Crown and nape grey
Eye dark brown, large, in keeping with shadowy habitats
Chin and throat white, very faintly streaked
Olive-brown
Fine white line under eye; faint white brow line extends forward to bill.
Breast and flanks tinted grey-buff
Pale streak at bend of wing, both races
Smallest of the whistlers, at 14 to 15 cm about the size of a yellow robin, and only very slightly larger than the Lemon-bellied Flycatcher
Adult
Belly and under tail white
Underparts pale yellow, darker grey-brown wash over breast
No white at tail-tip
Nominate race simplex
Race peninsulae ♂
Nominate race olivacea, darker
Eye brown
Northern race macphersoniana lighter colour, especially on crown; throat less heavily mottled
Dull olive-grey
Female has plain throat, overall dulled colours
Breast darker
Throat white, scalloped or mottled dark grey
Underparts mid tawny buff, or deeper on nominate race
♀
♂
Females slightly smaller. In Tas. have greyer head, red eye.
Immature has rufous-edged wing-coverts, reduced breast-band

Golden Whistler *Pachycephala pectoralis* 16–18 cm (L,N)

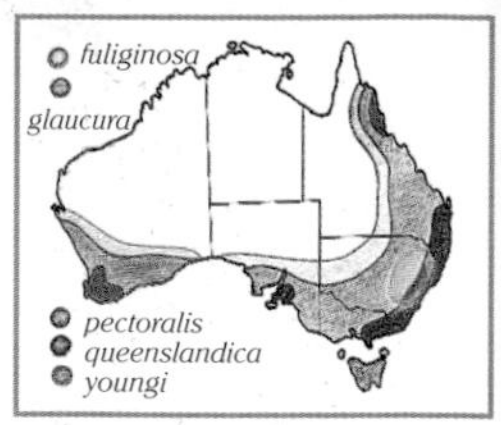

Conspicuous colourful males; strong ringing song in spring, with much posturing. Undulating flight between trees. Often solitary except when breeding; forages in trees, taller understorey shrubbery. **Variation:** in Aust., probably five races. **Voice:** persistent loud, ringing, whistled calls from both sexes, a long, rapid sequence swelling in volume: 'whit-whit-whit-whiet-whiet-wheet-quWHIT'; also often a repetitive short version, 'whit-whit-whew-WHIT', 'chwit-chwit-CHEW-WIT'. Contact a single, rising 'tseeip'. **Hab.:** rainforests, eucalypt forests, mallee, brigalow. **Status:** usually sedentary, some local migration in SE; common. **Br:** 403

Mangrove Golden Whistler *Pachycephala melanura* 15–17 cm

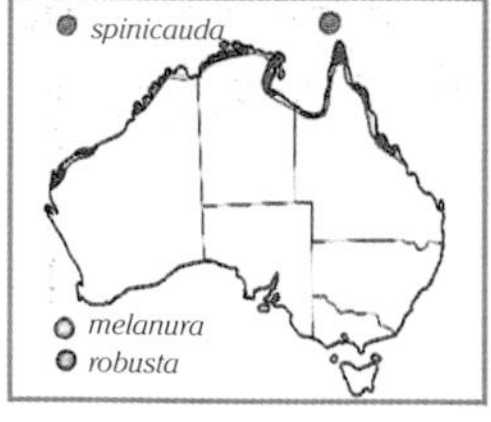

Loud calling when breeding; otherwise quiet and unobtrusive. Solitary or in pairs; forages on mangrove trees and mud. **Juv.:** like juvenile Golden Whistler. **Imm.:** like adult female, but in first year has rufous edged secondary coverts. **Variation:** nominate race *melanura*, NW WA; race *robusta*, W Kimberley to NE Qld. **Voice:** deeper, richer than Golden—throaty, low warblings that build up to clear whistles and sharp whip sounds which are more like the Golden: 'chwieop-cheiop-chweip-wheit-WHIT'. Frequent shorter call: 'whit-wheeta-WHIT' and single 'wheeeeit'. **Hab.:** mangroves; extends also into rainforest, monsoon forest and other dense vegetation adjoining mangrove swamps along coasts and rivers. **Status:** sedentary; moderately common. **Br:** 403

White-breasted Whistler *Pachycephala lanioides* 18–20 cm

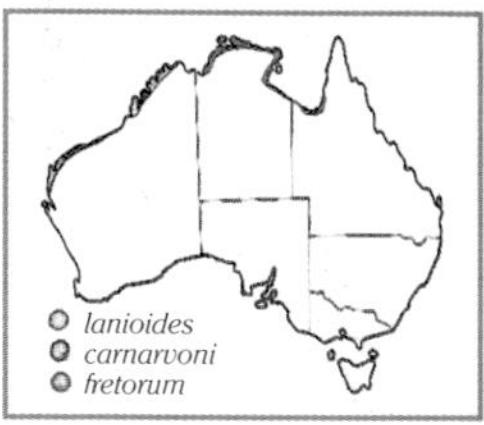

A large species with bill more typical of shrike-thrush than whistler. Usually solitary, quiet and unobtrusive; forages among foliage, stems and on mud at low tide when the large bill is used to take small crabs and other marine creatures. **Imm.:** like female, underparts deeper buff, more heavily streaked brown; wing feathers edged rufous. **Voice:** distinctive, more like the Rufous Whistler, most notes deep, mellow, somewhat husky-nasal quality, with frequent higher, clear, piercing whistles. **Similar:** Rufous Whistler, female, which is smaller than female of this species and lacks the long, heavy bill. **Hab.:** mangroves, usually thickets with low, dense canopy; also ventures into similar dense waterside vegetation. **Status:** sedentary; moderately common. **Br:** 403

Rufous Whistler *Pachycephala rufiventris* 16–17.5 cm

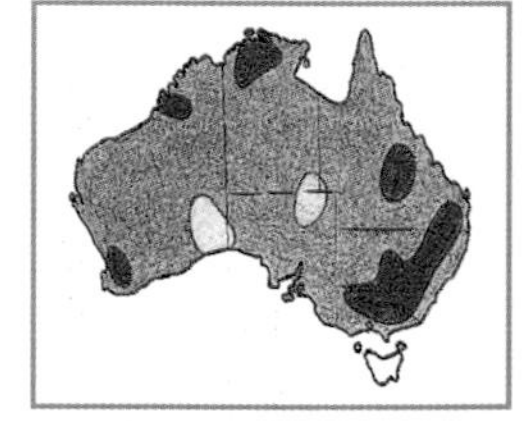

Through spring and early summer the loud outpouring of song makes this a most conspicuous bird; rival males bob, bow and chase. Forages in pairs or alone; moves rather sedately, methodically, through tree and shrub canopy; occasionally hovers in its search for insects and any other small edible creatures. **Juv.:** extensively rufous, underparts heavily streaked. **Imm.:** like female, but with cinnamon tint to inner flight margins and coverts. **Voice:** a long, loud, rapid succession of ringing notes, often 20 to 35 without pause: 'cheWIT-chWIT-chWIT-chWIT' or 'joey-joey-joey-'. Also a call with high, thin, drawn out beginning, powerful, ringing whip-crack finish: 'eeee–CHIEW!' and 'eee–CHONG!' **Hab.:** open eucalypt forest, woodland, mallee, mulga, watercourse vegetation in treeless regions, gardens. **Status:** migratory or sedentary; common. **Br:** 403

Red-lored Whistler *Pachycephala rufogularis* 19–21 cm

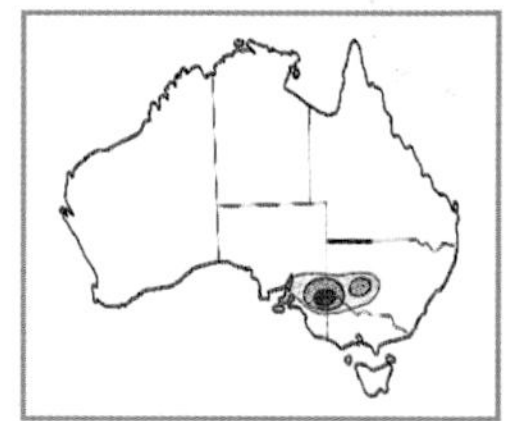

In dense low vegetation, elusive for most of year as it forages in low shrubbery and on ground; in breeding season is located by distinctive calls. **Voice:** series of deliberate, swelling yet abrupt whistles of even pitch and strength, 'wheit-wheit-wheit' or 'whieot-whieot-whieot-'; in distance like 'tchiopt-tchiop-tchiop'. Song a wistful, mellow, husky, almost scratchy sound, rising, falling, rising: 'wheit-chi-u, wheit-chi-u, wheit-chi-u'; 'wheit-chu, wheit-chu'. **Hab.:** dense low broombush or mallee heath, native pine and stringy bark–banksia heath with ground-cover of shrubs and spinifex. **Status:** usually sedentary; rare, confined. **Br:** 402

Gilbert's Whistler *Pachycephala inornata* 19–20 cm

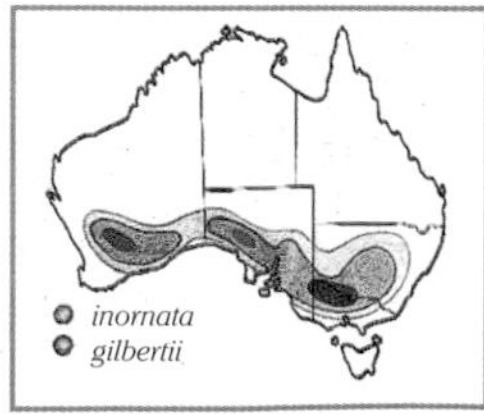

Usually inconspicuous, often in dense thickets; feeds low, in shrubs and on the ground. **Variation:** birds W of Flinders Ra. are often listed as *gilbertii*, but slight variations now appear clinal, leaving *inornata* an appropriate name across the entire range. **Voice:** a distinctive, far-carrying sequence of around 15 husky, mellow, slightly nasal, deep whistled calls, starting low, building up to a crescendo, each call with a sharper whip-crack effect: 'cheop-cheop-cheiop-CHEIOP-CHEEIOP-CHEEIOP'. Also a scratchy 'eechowk, eechOWK, eeCHOWK' and clear, strong, whistled 'eew-WHIT, ew-WHIT, ew-WHIT'. **Hab.:** dry woodlands, mallee, mulga and similar thickets with dense shrubby understorey, abundant litter. **Status:** sedentary or nomadic; uncommon. **Br:** 402

Head black, nape golden
Olive-green
Coverts and inner flight feathers edged olive
Rump olive, tail dark
White throat, black breast-band, golden underparts
♂
Grey with olive tint on shoulders
Under tail yellow-buff
♀
Imm.
Female and imm., light grey to buffy grey
Wings edged rufous
Head and body rufous
Pale gape
Wings dark grey; coverts, secondaries, edged rufous
Juv.
Golden collar broader at nape than on the Golden Whistler.
Bill longer, heavier
Black, slightly shorter than on Golden Whistler
Rump brighter olive-yellow
Slightly deeper yellow
Wings blackish, margins grey, or, on coverts, yellow
♂
Race *melanura*, back olive-brown
Olive-grey
Underparts buffy white; faint brown tint on breast
Yellow tint
♀
Race *robusta*, back olive
Extensive yellow tint
Black ♀
Male *robusta* like male *melanura*
Black crown edged rufous at nape
Grey-brown
Large, slightly hooked, appears heavy; more like shrike-thrush bill
White throat patch bordered by black; chestnut edged breast-band
Racial differences evident on female rather than male
♂
—Black, grey beneath
Varies slightly with race: *lanioides* grey to olive-grey, race *carnarvoni* grey-brown
Eye deep red, both sexes
Large, robust brownish bill
Underparts vary with race: pale buff on northern races *lanioides* and *fretorum*, deeper fawny buff on western race *carnarvoni*.
Wings dark brown with pale margins
♀
Crown grey; black more confined from lores through ear coverts to breast-band.
Back to tail mid grey
Wings grey-brown, pale margins
Southern form or race has deeper rufous underparts.
♂
♂
Pale inland form
Extremely wide distribution, but varies only slightly: from SE becomes gradually smaller and paler towards interior and N Aust.
Dark streaks
♀
Underparts pale buff to cinnamon
Plain cinnamon underparts, but lacks black band
Imm. ♂
Forehead, lores, face to throat orange-buff
Head and upper parts mid grey-brown
Wings dark grey-brown with fine olive margins to flight feathers
Grey breast-band, cinnamon rufous underparts
The Red-lored Whistler (far right) is one of three similar large and rather plump whistlers, the others being the Olive Whistler, and Gilbert's Whistler.
All have in common a rich yet subdued plumage with colours blending softly rather than being sharply defined. The Red-lored and Gilbert's have similar song and habitats.
Tail olive-grey to olive-brown
♂
Lores dark
Deep grey
Bill deep and stubby
Eye deep red, both sexes
Small, rich rufous-buff throat patch
Female throat pale
♀
Pale grey with light cinnamon tint
♂
Imm. similar, eye brown, flight feathers buff edged
Facial colours paler than male
Underparts paler cinnamon-buff
Red-lored Whistler ♀
Immature Red-lored is similar to female, except lores and throat are whitish, flight feathers are edged buff, eye is brown.

Little Shrike-thrush *Colluricincla megarhyncha* 17–19 cm

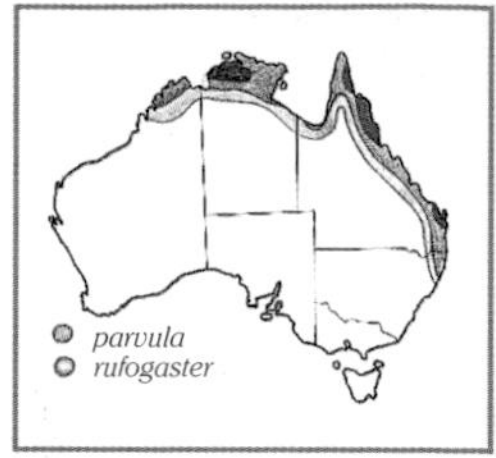

A widespread shrike-thrush, only slightly larger than Rufous Whistler, but, typical of shrike-thrushes, has a much longer, hook-tipped bill. Usually quiet, unobtrusive, wary; at times rather furtive. Alone or small groups; uses strong bill to tear bark, debris-entangled vines, epiphytes or ground litter. Also occurs in NG, with many races. **Variation:** populations in N are so different from those of the E that they have been previously listed as separate species; they are linked by several intermediate populations in a narrow corridor around the coast of the Gulf of Carpentaria. West of the Gulf is a single black-billed, brown-backed, sandy buff breasted race, *parvula*. To the E is a pale-billed, olive-backed, cinnamon-breasted race, *rufogaster*. The latter may be split in a number of very slightly different forms or races. **Voice:** often begins with high, drawn out, weak notes, then stronger, rollicking, deep, liquid-sounding notes: 'tseee-tsee-tsee, chok-ok, chok-a-wheet'; 'twheit-twheit-twhit-wheeee, chok-korr'; 'whicha-whieew' and other varied, clear, calls; harsh wheezy sounds. **Hab.:** monsoon and vine thickets, coastal woodlands, lowland rainforests, mangroves, swamps. **Status:** sedentary; common. **Br:** 403

Bower's Shrike-thrush *Colluricincla boweri* 19–20 cm

Usually confined to mountain rainforests above about 400 m altitude, Townsville north to Cooktown, inland to include the Atherton Tableland. Where rainforest continues unbroken to coastal lowlands, some may move to lower altitudes in winter. On higher parts, over about 600 m, it appears to displace the Little Shrike-thrush. Solitary or in pairs, rich song, but through winter quiet and unobtrusive. Rather slow, deliberate foraging; works middle and lower strata of forest, searching bark, foliage, entangled debris. Watches patiently from perch, occasionally dropping to ground. **Voice:** not powerful, but extremely varied with qualities reminiscent of well-known Grey Shrike-thrush. High, clear first notes are followed by deep, mellow, bubbling notes: 'trip-trip-shrieee, quorr-quorr-quorrot'. Other sequences include 'chuk-chuk-wheeeeit, quorr-quorr, quorreet-queeeit'; 'quorr-quorr-whitteet, quorr-quorr-whitteet'. Also loud 'chuk!', chirps, harsh scoldings. **Similar:** Little Shrike-thrush, but lighter build, longer tail, lighter streaking; upper parts usually more olive or brown. **Hab.:** rainforests, including their margins, clearings. **Status:** in some situations, altitudinal winter movements; common. **Br:** 403

Sandstone Shrike-thrush *Colluricincla woodwardi* 24–26 cm

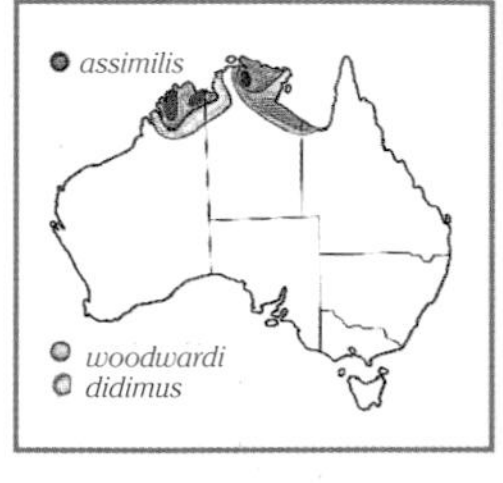

Typically seen at a distance on a vantage point of boulder or ledge, or occasionally a bare stick above sandstone escarpments and gorges; quick to hide among tumbled boulders or swoop away across a gorge to the opposite cliffs. Stays near the sheltering overhangs and crevices, darting on foot across the sandstone surfaces; usually visible only briefly, plumage similar in colour to the grey, buff and sandy buff tones of the stone. When fully revealed, the tail looks very long, emphasised by the slender, attenuated body shape. Alone or in pairs; methodically inspects rock face, probing crevices, finding small lizards, grasshoppers and spiders. **Variation:** 3 isolated populations, often listed as races; but current opinion seems to be that the differences are slight, probably clinal, therefore these are but variants of the nominate race, *woodwardi*. **Voice:** loud, ringing calls that reverberate from rock walls of gorges. Many variations, often beginning with high, clear, drawn out, rising whistles, quickly followed by abrupt, rich, mellow, at times bubbling notes. Among these: 'pieeer-pieeer, chokka-chok-chok'; 'wheeeit-wheeeit, quorr-quorr-quorr, queet-queet'; also, long, descending trill, single, quick 'cheewhip!' **Hab.:** escarpments, ravines of sandstone ranges, plateaux. **Status:** sedentary; common within specific habitat. **Br:** 403

Grey Shrike-thrush *Colluricincla harmonica* 22–25 cm

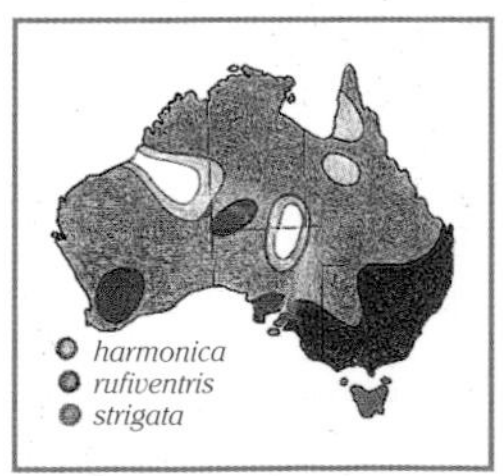

Known for its rich, varied song. Takes small insects, spiders, lizards, frogs, mice, nestlings and eggs of small birds; occasionally also seeds and fruit. Forages on tree trunks, branches and in foliage; finds prey under bark and in crevices; also often feeds on ground and around fallen logs. Flight swift, direct and undulating. **Male:** all races, black bill, pale lores, grey eye-ring. **Fem.:** pale-sided bill, greyish lores, more heavily streaked. **Variation:** many races and lesser variations, some gradual, or clinal. Distinctive races include *rufiventris*, from SA westwards, and *strigata* in Tas. The nominate *harmonica* is often split to recognise *brunnea* of Kimberley–NT and *superciliosa* on Cape York. **Voice:** a rich and varied repertoire of calls and songs—high, clear and often loudly ringing whistles intermingled with mellow, musical notes and deep, rich bubbly sounds. A common sequence is 'quorra-quorra-quorra, WHIEET-CHIEW', beginning with mellow throaty sounds and finishing with a high, clear, ringing whistle; similarly, 'wheit-wheit-quor-quor-quor, WHIEET-CHIEW'; 'wheit-wheit, WHIEET-chiew'. Calls include high, whistled 'whee-it' and 'whiet-whiet-wheeeit'. **Hab.:** widespread in extremely diverse habitats from coastal open forests, woodlands and gardens to tropical woodlands and the arid mallee and mulga of the interior. **Status:** sedentary; common. **Br:** 403

The Little Shrike-thrush has many variations, now grouped within two subspecies, *rufogaster* in E, *parvula* in NT and WA.
Head greyish; only slightly paler brow and lores
Bill large, hooked; pale pink along sides, brown above
Upper surfaces of wings and body plain brown. This subspecies, *parvula*, links to eastern *rufogaster* by intermediate populations around coast of Gulf of Carpentaria.
Olive-grey in SE of range, olive-brown towards NE; more russet brown on Cape York
Buffy white, faint streaks
Crown grey-brown to brown
Cinnamon; deeper rufous in NE Qld.
Adult
White brow and lores
Wing and tail feathers narrowly edged pale buff, or rufous in NE Qld form
Race *rufogaster*, 'Rufous Shrike-thrush'
Bill black
Juv.: like adults, but secondaries and their coverts are more broadly edged rufous.
Adult
Underparts pale fawny buff; throat white
Race *parvula*, 'Little Shrike-thrush', Top End of NT and Kimberley of WA
Slaty dark grey
Bill large, black or dark grey, hooked tip
Lores, brow, grey or rufous
Bower's Shrike-thrush is stoutly built, short tailed; gives impression of having a large head and heavy bill.
Cinnamon-buff, streaked grey
Confined within very restricted range, no subspecies
Wings and tail grey to grey-brown, edged lighter mid-brown or rufous
Olive-buff to deep rufous
More heavily streaked than the Little Shrike-thrush
Immature (not shown): like adult but breast more heavily streaked, back browner, rufous lores and brow, grey bill
Sexes are similar except male has grey lores and eye-ring, black bill; female has rufous lores and brow, dark grey bill.
Short tailed, heavy bodied appearance
Large, slender, long tailed; typically on or near sandstone cliffs; briefly perches in adjacent trees.
The Sandstone Shrike-thrush often bounces across rock face and ledges on foot, but also makes frequent short or long flights to higher rock or tree vantage points, across gorges, or from cliff tops gliding down to ravine boulders.
Lores pale, throat buffy white
Back sepia brown to olive-brown
Breast grey-buff, lightly streaked; lower underparts deeper buff
Grey crown merges to brownish back.
Immature has more rufous on wing coverts.
Bill long, with hooked tip
Slightly deeper rufous-buff in Kimberley population
Tail looks very long trailing back from slender body
Juv.: rufous brow
Male has black bill, white lores.
Race *rufiventris*, plain mid grey
Juv: dark streaks on underparts
Olive-brown back
Grey with light streaks on throat
Flight feathers darker grey-brown
Wings and tail grey-brown
Rufous-buff
Race *rufiventris*, 'Western Shrike-thrush'. Various splits have included *kolichisi* (NW Cape, WA), *whitei* (Eyre Pen. SA) and *murchisonii* (central Aust. to Pilbara of WA).
Nominate *harmonica*, often split into other local forms, which, from time to time, have been *brunnea* (Kimberley–NT), *superciliosa* (Cape York), *pallescens* (Qld), *halmaturina* (NSW)

Black-faced Monarch *Monarcha melanopsis* 16–20 cm

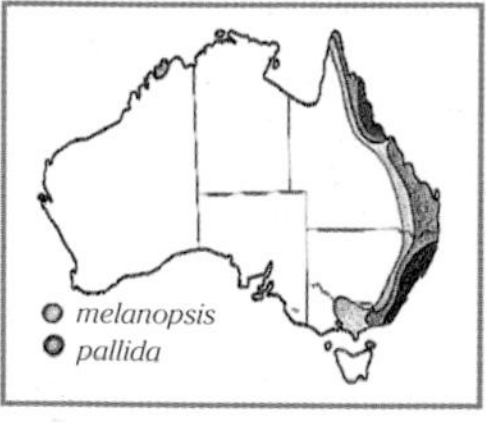

Monarch flycatchers are rather sedate and slow moving; unobtrusive, singly or pairs; take insects from foliage; forage in denser parts of mid-level forest; occasionally hawk for flying insects. **Voice:** deliberate whistles with distinctive mellowness in lower notes; slightly scratchy high whistles rise then fall—'wheech-iew, whieeeuw', 'whee-awhit, whieeeuw'; as single call, or widely spaced, 'wheit-chiew'. Also scratchy, husky, rising 'shreeeit, shreeeit, shreeeit' and harsh, scolding 'scraach'. **Hab.:** rainforests, mangroves, eucalypt forests and woodlands. **Status:** sedentary in NE, summer migrant southwards. **Br:** 404

Black-winged Monarch *Monarcha frater* 17–19 cm

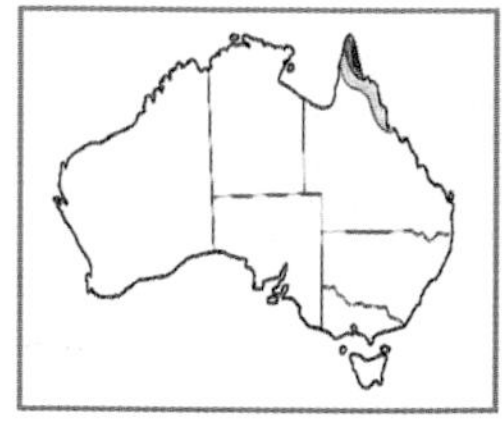

Active in mid to upper levels of the forest; gleans insects from foliage and from aerial flycatching; often found in mixed species feeding parties; attracts attention with noisy calls. **Voice:** similar to black-faced, but slightly more harsh, rising and falling 'whewheit-whieow, whewheit-whieow' and scratchy, husky, nasal 'wheit-chow, wheit-chow'. As a single, clearer, rising whistle at regular intervals: 'wheeeit, wheeeit, wheeeit'. **Hab.:** rainforests, mangroves, adjacent eucalypt woodlands. **Status:** summer migrant from NG; breeds Cape York where it is moderately common to uncommon; occasional vagrants as far S as Cairns area. **Br:** 404

Spectacled Monarch *Monarcha trivirgatus* 15–16.5 cm

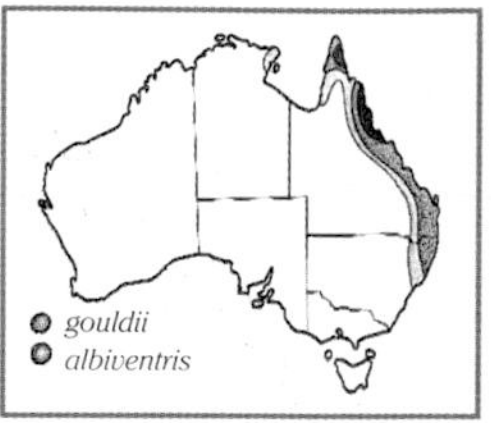

Slightly smaller, more active; darts and hovers about the foliage; hops about the branches in lively search for insects, with its tail often spread and waved to display the broad white tips. **Voice:** most distinctive call is a uniform series of strong whistles, beginning with a somewhat buzzing quality, each note drawn out and rising strongly to a clear, high finish: 'zreee-e-e-e, zreee-e-e-e, zreee-e-e-'. Also has a brief, squeaky warble, and scratchy chatterings and harsh scoldings. **Hab.:** usually rainforests, mangroves, but also often in moist gloomy gullies of dense wet eucalypt forests. **Status:** resident NE Qld, migrant to SE Qld and NE NSW. **Br:** 404

White-eared Monarch *Monarcha leucotis* 13–14 cm

Very active about the canopy and outer margins of rainforests; flutters among foliage to take insects in flight and from leaves. **Voice:** a loud, clear sequence of whistles, each rising, falling away, 'twei-tsieeew, twei-tsieeew, twei-tsieeew'; also has a series of long, piercing, descending whistles, 'tsieeeeuw, tsieeeeuw, tsieeeeuw'. Much softer is a musical, chattered song, brief, but rapidly repeated, 'chzit-chit-ch-weeeit, chzit-chit-ch-weeeit'. In each case the first notes are soft, slightly buzzing; the last much louder, higher, sharp and clear. Also harsh, low scoldings, buzzing chatterings. **Hab.:** rainforests, mainly coastal, also mangroves, swamps, watercourse thickets. **Status:** sedentary, some locally nomadic or migratory, probably altitudinal, movements; uncommon. **Br:** 404

Frilled Monarch *Arses telescophthalmus* 15–16 cm (T)

This species and the similar Pied are quite spectacular as they forage energetically in the rainforest, the males hopping and fluttering up tree trunks and limbs, wings and tail often partly spread, picking insects from fissures of bark; the females more often flutter about the foliage and pursue flying insects. In display, the white frill and crescent are widely spread. **Voice:** usual call a harsh, upward, rasping 'tzeeeeit, tzeeeeit' can be a single call, several times as slow, widely spaced calls, or extremely rapidly, in long series with resonant, buzzing, cicada-like effect, 'tzeet-zeet-zeet-zeetzeetzeet'. Also has harsh, squabbling scoldings or chatterings. **Hab.:** rainforest and adjoining eucalypt forest. **Status:** sedentary; common within restricted range. **Br:** 404

Pied Monarch *Arses kaupi* 14–15 cm

Like Frilled, forages with lively activity, fluttering and running up and down trunks of trees, treecreeper fashion, but more lively; often half-spreads wings, fans tail to dart out and take flying insects. At times displays in groups, perhaps 3 to 5 birds darting about, fluffing out white of nape and wings, much excited chasing, calling. **Imm.:** similar to Frilled, but with broad brown breast-band. **Voice:** a harsh 'tzeeeeeit', beginning as a low, rasping, buzzing sound, rising to a high, sharp finish; calls widely and unevenly spaced. Also has a long series of clearer, whistled notes, rising steadily higher throughout, 'wheit-wheit-wheit-wheit-' and 'whit-whit-whitwhitwhitwhit'. **Hab.:** rainforests, vine scrubs. **Status:** sedentary; moderately common. **Br:** 404

The Black-faced Monarch shows some variation: slightly paler, smaller northwards. Regarded by some as clinal (i.e. just one race, *melanopsis*); others take this variation as evidence of 2 races, *melanopsis* in S, *pallida* in N. There are other races in NG.
Head, back, most of folded wing and breast are mid blue-grey.
Pale bill
Dark bill
Black 'face', eye-ring
Face grey, pale lores
Rufous-buff wing linings
Flight feathers and spread tail are darker dusky grey.
Dusky mid grey
Rufous-buff
Tail without white at tips, blue-grey similar to rump and back
Adult
Imm.
The Black-winged Monarch differs from the Black-faced and Spectacled Monarchs most obviously in its much darker wings and entirely very dark tail.
Grey paler, more 'pearly'
Pale bill
Black slightly closer to eye, less on throat
Much less common than the Black-faced; usually confined to Cape York, occasionally seen around Cairns
Wings largely brownish black
Rufous-buff
Dark bill
Adult
Tail entirely plain grey-black
Imm.
Head entirely very light pearly grey
The Spectacled Monarch differs most obviously in its white belly and tail-tips, and black ear coverts
Black 'face' extends to ear coverts as well as throat.
Race *albiventris*, Cape York: rufous area reduced; has sharper boundary to white of flanks and lower breast.
Immature, race *gouldii*
Tail dark slate grey, outer 3 pairs with very large white tips
Race *gouldii*, S from Cooktown, has orange-buff breast and flanks merging softly into white belly.
Adult
Grey ear coverts and throat
Flanks cinnamon
Erectile feathers of forehead often give semi-crested high-crowned contour.
Dull, brownish cast to whites and blacks
White bars across coverts
White rump
White at ear, brow, under eye
Throat white or variably flecked with black
White across coverts, inner secondary margins
Rufous tint
Juv.
Female similar, slightly duller
Tail often slightly uplifted; broad white outer tips
Immature similar, less rufous than on juvenile
Broad white tips to outer tail
♂
Female, pale grey lores
Bill initially yellow or buff-brown
Across nape, broad fluffy looking ruff of erectile white feathers
Blue eye-ring, both sexes
Black chin
White chin; frill reduced, more freckled; may have faint buff tint at sides of breast.
White of scapulars and lower back form white crescent.
Juvenile brownish, dull whites, grey eye-ring; immature closer to adult
Both sexes have under-parts entirely white, unlike Pied, below.
Wings blackish-brown; rump to tail black
♂
♀
Incomplete collar
Erectile fluffy frill
In display the white parts of the plumage are fluffed out.
Lower back and rump white; tail coverts black
Thinner eye-ring
Black breast-band, which separates Pied Monarch from Frilled, becomes narrower towards the northern part of Pied's range.
Female usually has broader black band than male.
♂
♀
Frill is expanded, tail and wings are spread when excited or when calling or displaying.

Yellow-breasted Boatbill *Machaerirhynchus flaviventer* 11–12 cm

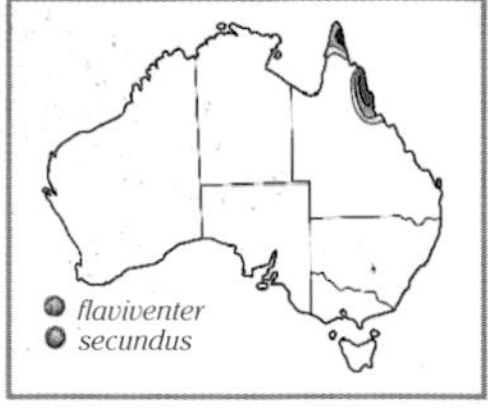

Forages actively in mid to upper strata of rainforest; darts out after flying insects; flutters around foliage to glean from leaves. Usually alone, otherwise in pairs. Occasionally holds tail partly cocked upward; rarely perches with body steeply upright. **Variation:** nominate race *flaviventer*, N Cape York, olive-back and secondary margins; race *secundus*, Cooktown S to Ingham, darker back, white wing margins. **Voice:** clear, sharp, rising, drawn out whistles, followed by much softer, rapid chuckling: 'wheee-wheeet, chuk-chuk-chuk'; similar 'wheit-wheit-wheit-wheit, tsi-tsi-tsi-tst-'. Also a softly trilled, musical, rhythmical 'whit-wh-wh-wh, wheeeee-whit'. **Hab.:** rainforests, vine and swamp thickets, nearby woodlands. **Status:** sedentary; uncommon. **Br:** 404

Broad-billed Flycatcher *Myiagra ruficollis* 15–16 cm (T)

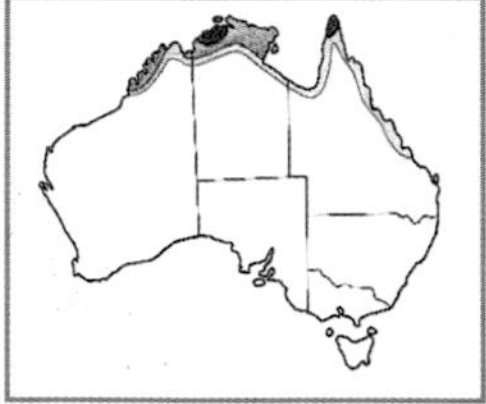

In pairs or solitary; actively forages about the crowns of mangroves or other dense vegetation; comes down to lower foliage around forest margins. Often darts out to take flying insects or flutters about the foliage. Rather quiet, but not shy; when perched rapidly quivers its tail. **Voice:** a clear, carrying, ringing, musical 'chiewip-chiwip-chwip-chwip', contrasting with harsh, grinding, buzzing 'tzzzeep, tzzeep, tzzeep'; abrupt, metallic, buzzing 'tzwip-tzwip-tzwip'; soft churring sounds. **Similar:** Leaden Flycatcher. **Hab.:** usually mangroves, but also monsoon forest, paperbark swamps, riverine forest. **Status:** sedentary; common. **Br:** 405

Leaden Flycatcher *Myiagra rubecula* 15–16 cm (L,T)

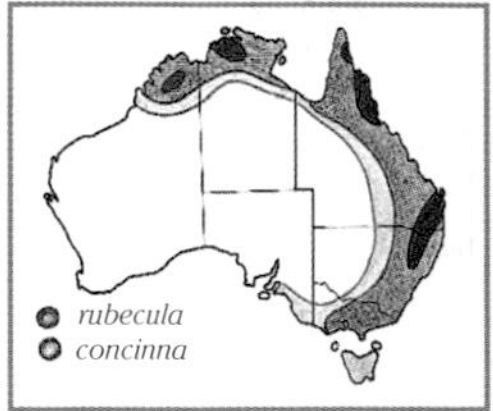

Busily active in mid to upper forest strata; hovers about the foliage; gleans small insects from leaves, snatching many in flight. Usually solitary, occasionally in pairs; quivers tail while perched; frequent calls attract attention. **Variation:** slight, 2 races. Or perhaps, tentatively, E coast race *rubecula* splits into 3: from S to N, *rubecula, yorki, okyri*. In Kimberley and NT, race *concinna* has a slightly narrower bill. **Voice:** clear, carrying 'whit-ee-eight, whit-ee-eight'; similar 'whee-ity, whee-ity','too-wheit, too-wheit'; 'wheeit-wheeit, wheeit-wheeit'. Also, harsh, nasal buzzing: 'tzzzeep'; 'scrzzarch'. **Hab.:** eucalypt forests, woodlands, scrubs, rainforest margins, mangroves, inland rivers. **Status:** sedentary and common in N; uncommon summer migrant to SE. **Br:** 405

Satin Flycatcher *Myiagra cyanoleuca* 15–17 cm (T)

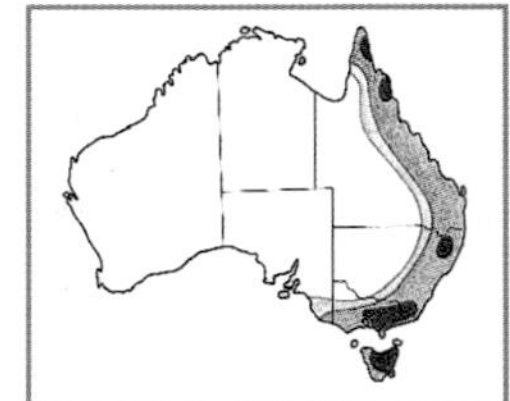

Energetically active in mid to upper levels of forest; darts out from perch to snap up flying insects; flutters about the foliage. On perches, restless with swaying, quivering tails; crests often erected. **Voice:** sharp, metallic 'chwee-wip, chwee-wip, chwee-wip' and faster, high, sharp 'cheeip-cheeip-cheeip-cheeip'. Harsh, grating grinding: 'grzzz-urk, grzz-urk'; 'gzzirk, gzzirk'. **Hab.:** forests and woodlands, mangroves, coastal heath scrubs, but avoids rainforests; in breeding season favours dense, wet gullies of heavy eucalypt forests. **Status:** summer breeding migrant into SE and Tas., winters NE Qld and NG; uncommon. **Br:** 405

Shining Flycatcher *Myiagra alecto* 16–18 cm (T)

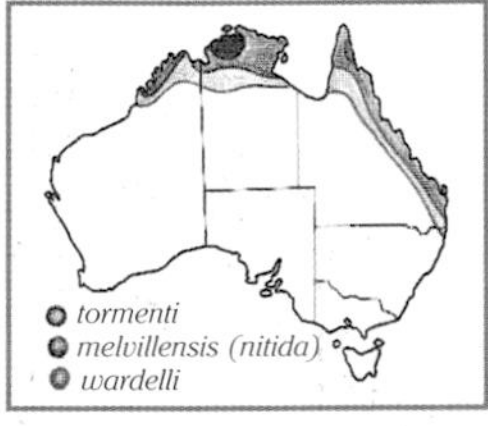

Typically near or over water, among stilt roots of mangroves or above a pool in rainforest; drops to take prey from mud or ground more than from foliage or in flight. When perched, swings and flicks tail, lifts crest. **Variation:** unsettled; race *melvillensis* is sometimes split into Kimberley race *tormenti* and NT race *nitida*. **Voice:** diverse calls—clear, musical whistles to frog-like buzzing, croaking sounds. Common call a rapid series of clear whistles—starts softly, increases in volume,'whit, whit, whit' or more rapid 'whit-whit-whi-'. A buzzing call begins low, increases suddenly,'zzzreEOW, zzzreEOW', both high, clear and low buzzing, 'kwit-zzzur, kwit-zzzur'. **Hab.:** mangroves, rainforests, pandanus creeks, paperbark swamps. **Status:** sedentary; common. **Br:** 405

Restless Flycatcher *Myiagra inquieta* 16–21 cm

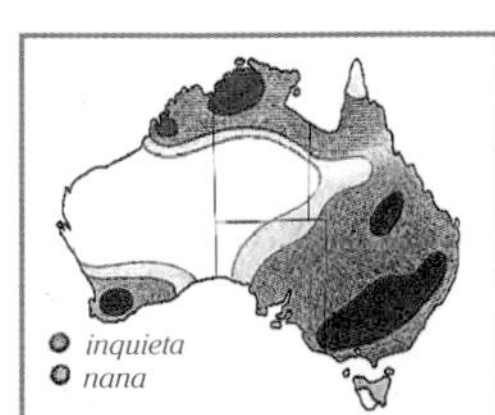

Hovers over grass, shrubbery, tree foliage, branches, logs, often with grinding calls; flies with buoyant action on slow moving wings. Restless when perched, tail waving; soon off after a flying insect or to a new perch. **Voice:** diverse and distinctive sounds. Common call a high, clear, musical 'toowhee-toowhee-toowhee', 'twheee-twheee' or, with slight buzzing, 'tzweet-tzweet'. In contrast, a grating, harsh 'grrzziek,'. Most remarkable is a call given while the bird is hovering: it is a rising series of metallic grinding noises that reverberate far through forest and woodland: 'kzowk! kzowk-kzowk, kziok-kziok, kzeek-kzeek, kzeik-kzeik-'. **Hab.:** open forests, woodlands, farmlands, inland scrubs. **Status:** sedentary or nomadic; common. **Br:** 405

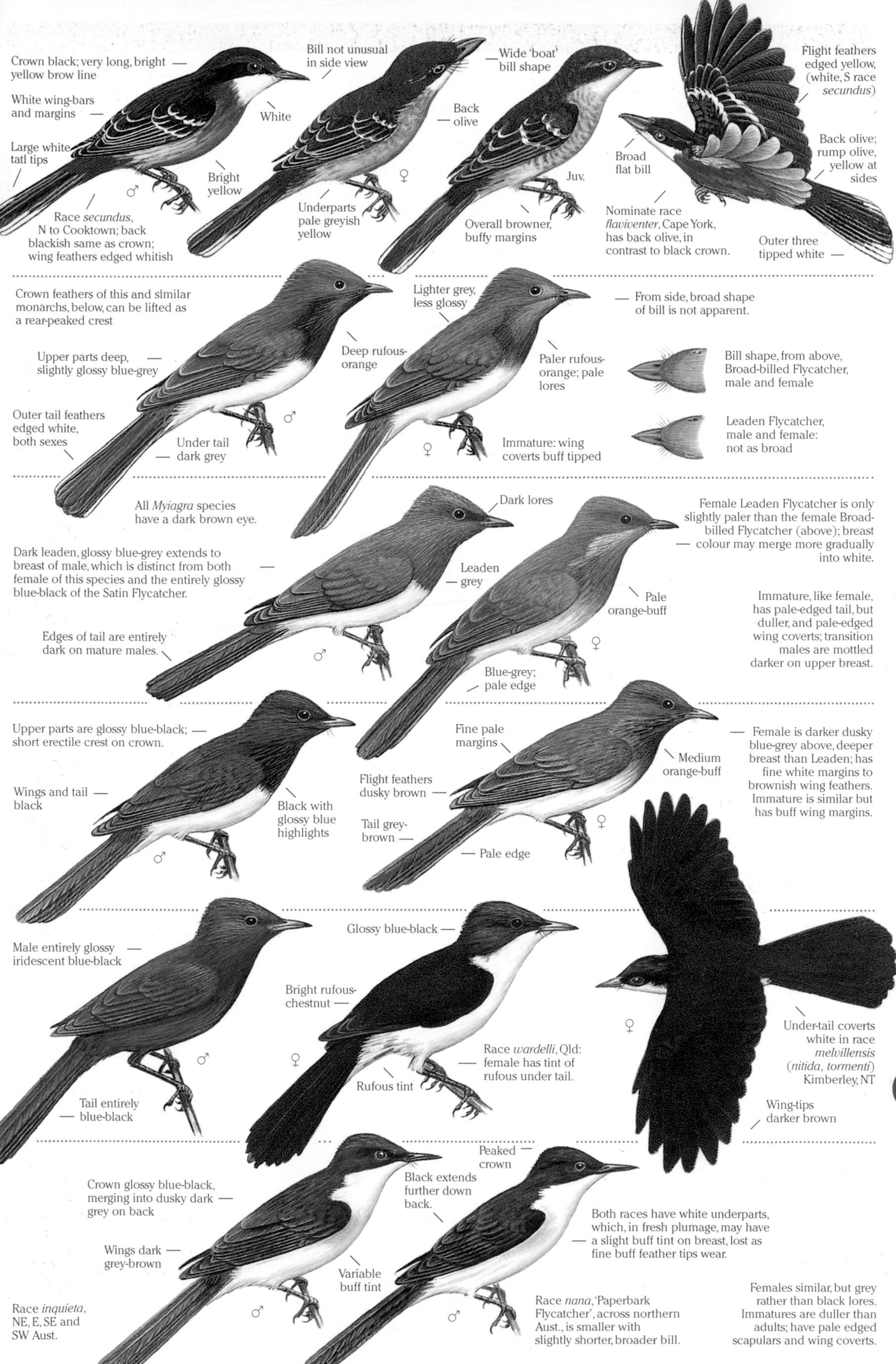

Crown black; very long, bright yellow brow line
White wing-bars and margins
Large white tatl tips
Race *secundus*, N to Cooktown; back blackish same as crown; wing feathers edged whitish
White
Bright yellow
♂
Bill not unusual in side view
Back olive
Underparts pale greyish yellow
♀
Wide 'boat' bill shape
Juv.
Overall browner, buffy margins
Broad flat bill
Nominate race *flaviventer*, Cape York, has back olive, in contrast to black crown.
Flight feathers edged yellow, (white, S race *secundus*)
Back olive; rump olive, yellow at sides
Outer three tipped white
Crown feathers of this and slmilar monarchs, below, can be lifted as a rear-peaked crest
Upper parts deep, slightly glossy blue-grey
Outer tail feathers edged white, both sexes
Under tail dark grey
♂
Deep rufous-orange
Lighter grey, less glossy
Paler rufous-orange; pale lores
♀
Immature: wing coverts buff tipped
From side, broad shape of bill is not apparent.
Bill shape, from above, Broad-billed Flycatcher, male and female
Leaden Flycatcher, male and female: not as broad
All *Myiagra* species have a dark brown eye.
Dark leaden, glossy blue-grey extends to breast of male, which is distinct from both female of this species and the entirely glossy blue-black of the Satin Flycatcher.
Edges of tail are entirely dark on mature males.
Dark lores
Leaden grey
♂
Female Leaden Flycatcher is only slightly paler than the female Broad-billed Flycatcher (above); breast colour may merge more gradually into white.
Pale orange-buff
♀
Blue-grey; pale edge
Immature, like female, has pale-edged tail, but duller, and pale-edged wing coverts; transition males are mottled darker on upper breast.
Upper parts are glossy blue-black; short erectile crest on crown.
Wings and tail black
Black with glossy blue highlights
♂
Fine pale margins
Flight feathers dusky brown
Tail grey-brown
Pale edge
Medium orange-buff
♀
Female is darker dusky blue-grey above, deeper breast than Leaden; has fine white margins to brownish wing feathers. Immature is similar but has buff wing margins.
Male entirely glossy iridescent blue-black
♂
Tail entirely blue-black
Glossy blue-black
Bright rufous-chestnut
♀
Rufous tint
Race *wardelli*, Qld: female has tint of rufous under tail.
♀
Under-tail coverts white in race *melvillensis* (*nitida, tormenti*) Kimberley, NT
Wing-tips darker brown
Crown glossy blue-black, merging into dusky dark grey on back
Wings dark grey-brown
Race *inquieta*, NE, E, SE and SW Aust.
Variable buff tint
♂
Peaked crown
Black extends further down back.
♂
Both races have white underparts, which, in fresh plumage, may have a slight buff tint on breast, lost as fine buff feather tips wear.
Race *nana*, 'Paperbark Flycatcher', across northern Aust., is smaller with slightly shorter, broader bill.
Females similar, but grey rather than black lores. Immatures are duller than adults; have pale edged scapulars and wing coverts.

Rufous Fantail *Rhipidura rufifrons* 15–16 cm

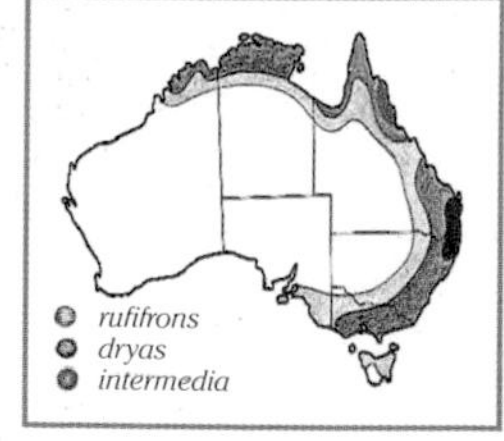

Flits and dances above the lower undergrowth and through mid level vegetation of shadowy forests, long tail often widely spread, or at times in sunlight that emphasises the bright orange-rufous of its tail **Variation:** race *intermedia*, mid or SE Qld to NE Qld (tail longer, tips paler northwards); race *rufifrons*, E and SE; these races intergrade mid to SE Qld. Both races winter N to Cape York and NG. Race *dryas*, with rufous part of tail smaller, and with larger white tail tips, may be a full species, 'Wood Fantail' or 'Arafura Fantail'. **Voice:** call is a very high, weak 'tsit' or 'tsi-tsit'. A song may follow on as a brisk, extremely high, squeaky, slightly descending, undulating tinkling, 'tsit, si-sit-tswit, tsit-tseit-tswit, tseit-tswit-tseit'; higher than notes of Grey Fantail. **Hab.:** rainforest, dense wet eucalypt and monsoon forests, paperbark and mangrove swamps, riverside vegetation; open country on migration. **Status:** in NE Aust., common; less common in SE, a summer breeding migrant. **Br.:** 406

Mangrove Grey Fantail *Rhipidura phasiana* 15–16 cm

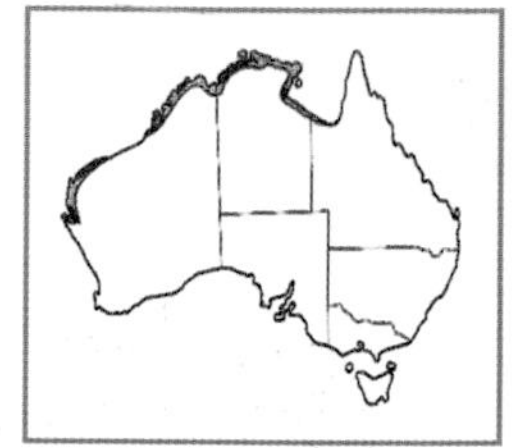

Paler, slightly smaller than the Grey Fantail. This species is confined to mangrove swamps; however, the Grey Fantail, also, often enters mangroves from adjoining woodland or forest. Agile and active in aerial pursuit of insects; flutters about the mangrove foliage, and low across mud and pools. **Voice:** contact call is an abrupt 'tek, tek'. As song, a series of short, twittering squeaks, 'tsit-tsit, tsit-chit-chit, chitty-chit,' distinctly different from the Grey's lively, cascading, silvery, squeaky song. **Hab.:** mangrove forests, tall to stunted, and adjoining samphire flats; occasionally acacia or melaleuca scrub nearby. **Status:** common. **Br:** 406

Grey Fantail *Rhipidura fuliginosa* 14–17 cm (L, N)

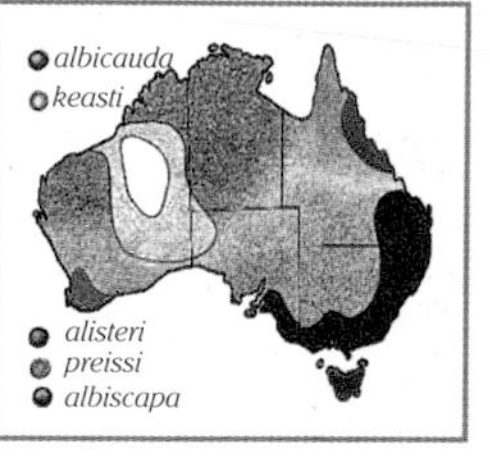

The twisting, turning aerial acrobatics of the Grey Fantail make it one of the best-known birds in Australian bush and gardens. It sometimes chase insects high above the vegetation, but more often flutters about the undergrowth and lower levels of tree foliage, rarely still, the long tail swinging side to side, frequently widely fanned. **Variation:** races *albicauda*, *keasti*, *alisteri*, *preissi* and *albiscapa*. Parts of southern-central populations migrate N in winter, spreading across most of interior and into N Aust., overlapping breeding ranges of central-northern races *albicauda* and *keasti*. Part of Tas. population migrates NE along mainland coast in winter. **Voice:** a cheery outpouring of scratchy, squeaky sounds, some so high that they verge on inaudible, mixed in seemingly haphazard manner with lower chatterings, 'twitch-twitchit, tsweeit-tseet, chit-twit, tswit-chat, tsweeit'. **Similar:** Mangrove Fantail, but wider white bars across wing coverts, faint dark breast-band; Northern Fantail, but tends to keep tail folded and lowered, song different. **Hab.:** found in most vegetation types from mangroves and rainforests to arid interior scrub; some races are restricted to just a few habitats. **Status:** southern and central Aust. populations migratory, breed in centre and S, wintering in centre and far N. Common. **Br.:** 406

Northern Fantail *Rhipidura rufiventris* 19–22 cm (T)

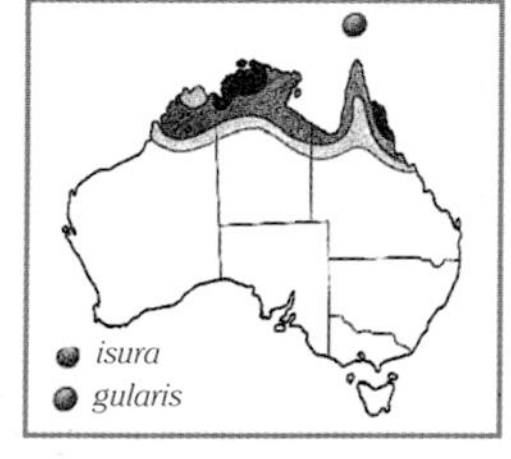

Rather sedate, even sluggish in its movements compared with the hyperactive Grey Fantail; usually keeps tail down, rarely fanned. Often waits quietly on a perch in slightly more upright posture, darts directly out to take insects that fly past. **Juv.:** is overall browner, buffy throat, and buff tint to wing margins. **Variation:** in Aust., one race, *isura*; many others on northern islands. A NG race, *gularis*, with upper parts darker, occurs on Boigu Is, Torres Str., Aust. **Voice:** calls distinctive, unlike any other fantail, resonant, metallic, yet rather mellow, 'kek', 'dek' or rather musical 'chunk' given repeatedly over long periods. An ascending, slightly buzzing call is often used: 'zziop–i-deet'. The song is similar to that of the Grey Fantail, but at lower pitch, lacking the highest squeaks of that species, yet the notes are sharp and clear, with descending, tinkling, musical trills, 'chi-pit, chwip-chit-chweip-chip-cheip-'. **Hab.:** margins of rainforests, monsoon and vine scrubs, paperbarks and mangroves, adjoining eucalypt forests, woodlands. **Status:** sedentary or locally nomadic; moderately common. **Br.:** 406

Willie Wagtail *Rhipidura leucophrys* 19–22 cm (T)

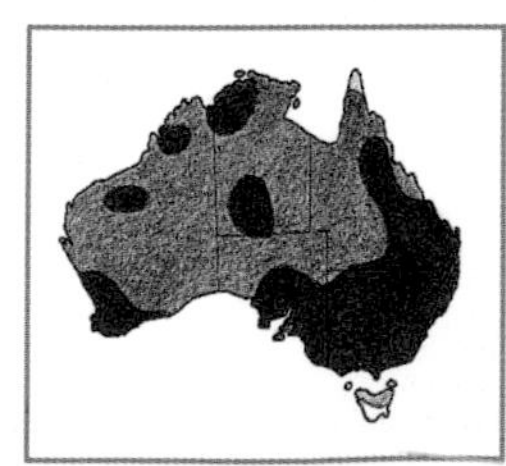

The name 'wagtail' is appropriate: the tail is waved incessantly rather than fanned. Always restless, swings body side to side, gives sudden flicks of the wings to flush hidden insects. Aggressive; near nest, attacks predators far larger than itself. **Voice:** brisk, strong, lively, a pleasant musical chatter, with sudden switches between sharp and low notes. Given repeatedly: 'whichity-wheit', 'whitch-i-wheit, whitchit'. Sings through still moonlit nights. In attack or defence gives harsh, loud, metallic, ratchetting chatter. **Hab.:** prefers open country, farms; avoids dense forests. **Status:** sedentary; widespread, common. **Br.:** 406

These are the most colourful of Australia's fan-tailed flycatchers. The tail shape and colour make this bird distinctive not only among fantails, but unique among all Australian birds.
Adult, race *rufifrons*
Tail base to lower back fiery orange-rufous
Rufous
Throat banded white and black, breast dappled
Does not have wing-bar.
Tips paler grey, more clearly defined, further N; lightest in NE Qld.
Wings entirely plain grey-brown
Tail is often spread in the insect-chasing, fluttering, acrobatic flight.
Race *rufifrons*
Tips light to mid dull grey, darkest least distinct, in S
Dull rufous
Buffy brown
Juv.
Wings edged rufous
Race *dryas*, smaller, rufous confined to base of tail, which otherwise is dark grey-brown with white tips that are largest on outer feathers (not shown; NT, Kimberleys)
The Mangrove Fantail was regarded as a race of the Grey Fantail (below) until detailed studies showed that it differed not only in details of appearance, but also had distinctive song, and differed in details of ecology and breeding.
Larger, pale buff or white brow line
Upper parts, crown to tail, paler than on Grey Fantail
Tail of Mangrove Fantail is proportionately shorter with broad white tips to outer 4 pairs.
Bill slightly longer than that of Grey Fantail
Throat bar very fine
White wing-bars
Adult
Tail feathers have white shafts; outer vane narrowly edged white.
Immatures have more light buff in the plumage than do adults.
Tips white, except central pair; outer edge white; all races
Race *alisteri*, upper parts mid grey
All races have white brow line and small streak behind the eye.
Often holds wings drooped.
Off-white to pale buff
Dark grey
Race *keasti*, upper parts very dark
All races have two wing-bars, usually fine; on some races indistinct.
All races have fine white shafts to all but central pair of tail feathers.
Central pair mid grey
Narrow grey outer edge, next 3 pairs of feathers
Less white at outer margin
Black
Pale buff underparts
Outer 2 pairs entirely white
Head and wing markings buff
All races have a grey breast-band, dark to pale, separating white throat from buff or creamy white underparts.
Overall slight to strong brownish tone; some similar to juvenile Rufous Fantail
Juv.
Mottled brown
Race *albicauda*, a most attractive white tailed form. It is so distinctive, in song as well as the tail, that it has at times been considered a separate species, the 'White-tailed Fantail'.
Breast white with faint or broken breast-band
Off-white
Other races: *preissi*, similar to *alisteri*, but mid grey breast-band and shorter bill; *albiscapa*, also like *alisteri*, but darker above, broader dark breast-band and less white on the tail
Immediately distinguished from other fantails by posture and behaviour, confirmed by plumage and calls or song. One race in Aust., others on northern islands
Upper parts dark grey, sides of face blackish
Tiny, dull brow mark
Wide, grey, white streaked breast-band
Tail typically slopes downwards, following the line of the body; rarely fanned.
Pale buff
Larger, more heavily built than the Grey Fantail
Tail edged and tipped white on outer feathers.
Tail edged white
Adult
Juveniles have rufous edges to wing feathers and brow line; tail initially short. Immatures are similar to adults but have pale tips to coverts.
White brow; size varies with level of excitement
Black throat
The long tail is entirely black, often fanned, and restlessly swung back and forth.
Wing entirely brownish black, lacking white or buff feather tips and margins shown by the smaller fantails
Widespread, generally familiar. Although a fantail, and so alike that it is placed with them in the genus *Rhipidura*, it is larger and of heavier build. The plumage pattern at throat and breast is a reversal of that of the small fantails: throat black, meeting the clean stark white of the underparts in a sharp breast-band line. This neat, crisp contrast of white against black is the essential element of the image of the 'Willie'.
Legs much longer than those of the other fantails, reflecting the inclusion of terrestrial hunting into the foraging activities of the Willie Wagtail.
One race only within Australia, as variation is negligible. Several races on northern islands

Spangled Drongo *Dicrurus bracteatus* 28–32 cm (T)

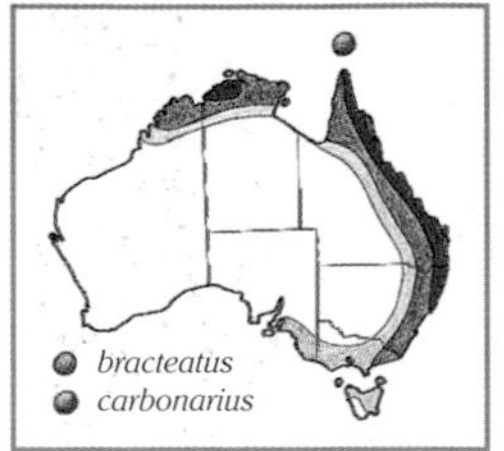

Unlike any other Australian bird, with long, outcurved, forked tail, iridescent black plumage and upright posture. The Drongo's habit of perching in the open, on bare limb or wire, acrobatic flights, loud, metallic tearing calls, all combine to make its presence obvious. Insects are taken in flight, often after much twisting, turning aerial pursuit, and brought back to the perch. May be solitary, or in pairs, chasing insects, ripping into bark for spiders, beetles and other small prey, or harassing other birds, including raptors. **Variation:** two Aust. races, may be split further. On mainland, *bracteatus*; in Torres Str.—Boigu, Saibai, Dauan—race *carbonarius*. **Voice:** noisy, harsh, tearing, sounds. Includes loud, rasping, brassy, nasal 'k-zark, kzairk-kzairk-', 'korrk-korrk–'; metallic 'wheit-', 'kairk-kiairk'; high 'kierk'; whistles, rackety chatterings. **Similar:** Metallic Starling, Trumpet Manucode. **Hab.:** open woodlands, often margins of lowland rainforests or vine scrubs, mangrove and paperbark swamps, riverside thickets, gardens. **Status:** in SE, a summer migrant from Cape York; in NE Aust., a summer migrant from NG. In the NT and Kimberley the species is present all year, but some may migrate N to Indonesia. **Br.:** 407

Magpie-lark *Grallina cyanoleuca* 26–30 cm (L)

Widespread, well-known magpie-like small bird. Distinctive, slow, buoyant flight with deep, uneven, lapping beats of broad, rounded wings; forages on the ground. **Variation:** across far N (NE Qld to Kimberley) has shorter tail and longer bill; may justify another race, *neglecta.* **Voice:** calls mellow, liquid yet clear; often a ringing, carrying 'tiu-weet, tiu-weet' and liquid 'cluip-cluip, cluip-cluip'; as alarm, gives a strident 'treee-treee-!'. Pairs, together give a closely synchronised duet, one a musical, mellow 'qwoo-zik', the other immediately following on with a sharp, yet rather harsh, 'wheeik'; in duet, 'qwoo-zik, wheeik…qwoo-zik-wheeik-', each call accompanied by lifting of wings. In flight together, a softer, liquid piping, 'qwoo-whik'. **Similar:** Australian Magpie, but has large, sharply triangular bill, swift swift flight; White-winged Triller, but black tail, and entirely white beneath. **Hab.:** diverse, coastal to semi-desert, almost anywhere with the water and trees needed for mud nest. Also requires open areas of bare soft ground for foraging, margins of lakes, rivers, swamps, temporary floodwaters. Has expanded range and numbers into more open environs where farmland has been cleared from forests and where dams provide permanent water. **Status:** nomadic; very common. Has adapted to use man-made environments. **Br.:** 405

White-winged Triller *Lalage tricolor (sueurii)* 16–19 cm

A well known, conspicuous migrant to southern Aust. Through mid spring into early summer, these trillers dominate bush sounds with the courting male's cheery and almost incessant loud trilling as he rises above his territory on rapidly fluttering wings. Insects are taken in the air, from foliage and on the ground. **Br.:** male, bold black and white. **Nonbr.:** similar to female, but wings black and rump grey like breeding male. **Variation:** *Lalage tricolor* is sometimes included within *Lalage sueurii* of Java and Lesser Sundas; only the nominate race in Aust. **Voice:** a vigorous chatter or trill, long sustained, loud, clear and carrying. The notes have intermingled sharp and mellow sounds, the overall quality cheerful, the tempo fast and rollicking: 'chwipa-wipa-wipa-wipa-wipa-'; 'chiffa-tiffa-tiffa-tiffa-'; often switches to a slightly softer but rapid, long trill, 'chif-chif-chif-chif-tif-tif-tif'. **Similar:** Varied Triller, but pale rufous under tail, underbody barred to variable extent, conspicuous brow line. **Hab.:** open forests, woodlands, vicinity of tree lines along rivers of semi-arid regions, nearby scrublands. **Status:** in N, sedentary or locally nomadic, common. In S, summer breeding migrant; uncommon. **Br.:** 408

Varied Triller *Lalage leucomela* 18–20 cm (T)

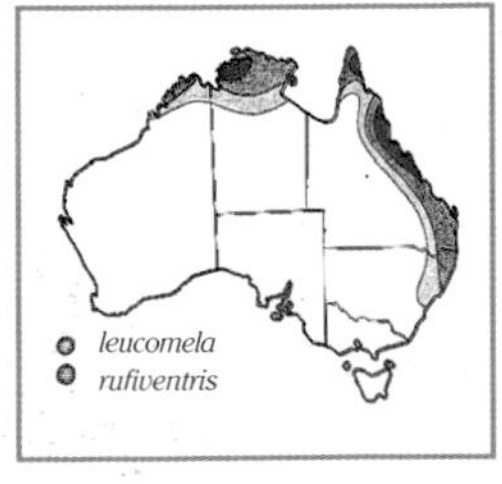

An inconspicuous inhabitant of dense and closed vegetation, where pairs or small parties move slowly, steadily, among the foliage in search of insects and fruits. Although not uncommon, may be overlooked unless calls attract attention. Varied Trillers forage at all levels of the forests, in rainforests from crown down to mid level, or where the habitat is more open, lower into undergrowth, occasionally on the ground. **Juv.:** like female, but flecked buff above, and with darker barring and streaking of underparts. **Variation:** nominate race *leucomela*, E coast; those of Cape York may prove to justify another race, *yorki*. Birds of the Top End, NT, and of NW Kimberley are race *rufiventris*; the latter may be a fourth race, *macrura*. **Voice:** while foraging, contact calls are given almost continuously. The sound is distinctive: a rolling, swelling and fading, mellow rattling, rising and strengthening for several seconds, 'trrreee-', then falling, '-iurr', giving an undulating 'trreee-iurr, trreee-iurr, trreee-iur-'. **Similar:** White-winged Triller, but lacks cinnamon tint and bars underneath, and has faint, indistinct brow line. **Hab.:** rainforests, monsoon forests, deciduous vine scrubs, paperbark swamps, tropical eucalypt forests and woodlands, river-edge thickets. Prefers the transitional environment between the closed and open forests. **Status:** sedentary; common in far N, uncommon to rare towards the SE. **Br.:** 408

Fine crest at nape, inconspicuous
Heavy bill with bristles around the base
Eye brilliant red
Brownish black
Eye brown
Range extends from India and China through SE Asia to NG. Many subspecies, but probably only one, *bracteatus*, on mainland Aust. Race *carbonarius* (smaller, bluish iridescence and upturned outer tail-tips) occurs on Australian islands in N Torres Str.
Plumage entirely black with a glossy greenish blue metallic sheen
Scattered, small, highly reflective, blue-green iridescent spots give a gleaming, glinting 'spangled' appearance.
Scattered white tips, mostly on breast feathers
Immatures are dusky black rather than the brownish black of juveniles and begin to show adult's iridescent greenish sheen. White tips are lost from breast, but are retained on under-tail coverts.
Legs and feet small, thin
Adult
Juv.
Unique among Australian birds is the distinctive, long, outward curving 'fishtail'.
Under-tail coverts retain pale tips on younger adults.
Under-tail coverts white tipped or white scalloped
Black subterminal band, much wider on central feathers
Rump white
Both the Drongo and the Magpie-lark are now placed in the monarch flycatcher family, Dicruridae.
Horizontal black line through eye; brow white
Black line down through eye; white brow and throat
♂
♀
White throat
Under-wing black with white coverts
Uneven white bar from shoulder back across the folded wing
In flight, large white wing patches very conspicuous across the broad black wings: from the shoulder, back across the coverts to the base of the inner flight feathers. Underwing coverts white
Tail white with wide black band
Juv-Imm.
Like adult male, imm. has a horizontal black eye line and white brow; but, like the female, has a vertical black band through eye, and white throat.
Legs and feet quite large, strong, reflecting terrestrial foraging
White-winged Triller, also in New Guinea, Indonesia. Widespread in Aust.; sole race *L. t. tricolor*
Narrow nape link, black of crown joins back
Brown in place of black
Like female; usually more white at sides of neck
Primaries black, with very fine white tips and outer margins.
Black down to lower rim of eye
Large white patch at side of neck
Bill brown at lower base
Faint pale brow, dark line through eye
Lower back and rump grey
Rump grey
White shoulder
Lightly mottled
Wings darker than on female
Underparts entirely white
♂
♂
♀
Juv. similar, but crown, back, scalloped
White margins widest on tertials, secondaries and their coverts
Coverts white
Tail black with white outer tips
Non-breeding plumage
Mid-brown
Trillers are small boldly-plumaged members of the cuckoo-shrike family. The Varied Triller has many subspecies on northern islands; in Aust. three, perhaps four races.
Crown to back, glossy black
Prominent white brow line
Black eye line
White shoulder bar
Breast to under tail pale buff with soft dusky barring
Lower back, rump, grey
Pale grey
Black, tipped white
Lightly barred
Cinnamon
Belly to vent, pale cinnamon
Barred grey-brown
♂
♂
♂
Female, race *leucomela*. NW races have dull colours, heavier barring.
♀
Race *yorki*
Race *rufiventris*
Nominate race: *leucomela*

Black-faced Cuckoo-shrike *Coracina novaehollandiae* 30–36 cm (T)

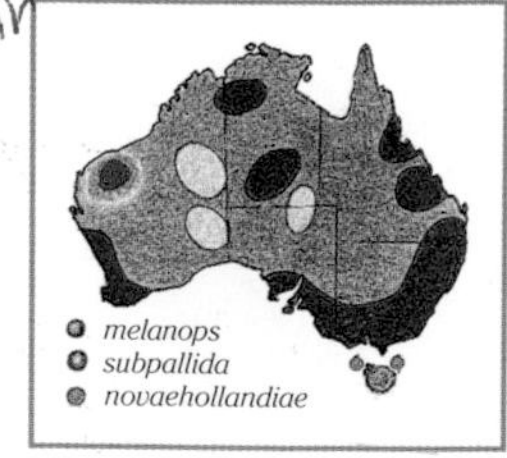

Found almost throughout Aust. in suitable habitats. Widespread familiarity has given rise to many local names including 'Blue Jay' and 'Shufflewing'. The restless lifting of the folded wings on landing (and in display), together with the strongly undulating flight, assist recognition. Its presence is made more conspicuous by its use of high, often dead, exposed limbs as vantage points from which to drop to the ground to take insects and other prey. Solitary, in pairs or small family parties, or in large flocks. **Voice:** a loud, sharp churring, harsh yet rather musical; often starts quite high then falls, 'churrieer', or a descending 'quarieer-quarieer-quarrieer '. Harsher sounds in aggression. **Hab.:** rainforests, eucalypt forests and woodlands, tree-lined rivers of interior, farms, gardens. **Status:** sedentary and nomadic; common. **Br.:** 407

White-bellied Cuckoo-shrike *Coracina papuensis* 26–28 cm (T)

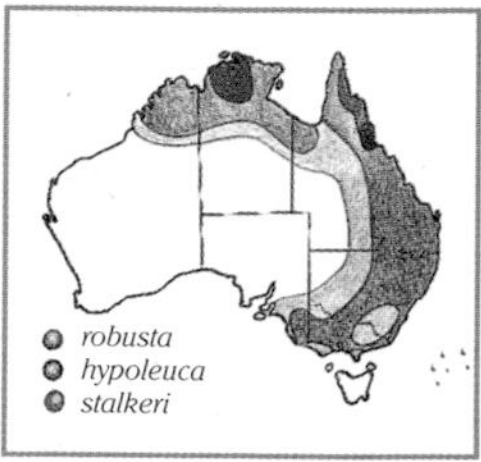

Flight undulating, long glides alternating with bursts of wingbeats. On landing, each wing briefly flicked up, then folded. In pairs or small family groups, forages for insects and fruit among foliage and branches. **Variation:** *hypoleuca*, 'White-breasted', NW Qld to Kimberley; race *stalkeri*, slightly darker, Cape York; *robusta*, largest and darkest, in SE. Melville–Bathurst islands birds may differ enough to justify subspecies *apsleyi*. **Voice:** distinctive, but reminiscent of some calls of Black-faced Cuckoo-shrike. Contact call, often while in flight, a scratchy, peevish, rather piercing, unmusical 'queeik-quisseeik', 'quirreeik, quirreeik'; also softer churrings. **Hab.:** rainforests, gallery forests, eucalypt forests, woodlands, mangroves, riverside tree belts. **Status:** locally nomadic; common in N, becoming sparse towards SE. **Br.:** 407

Barred Cuckoo-shrike *Coracina lineata* 24–26 cm

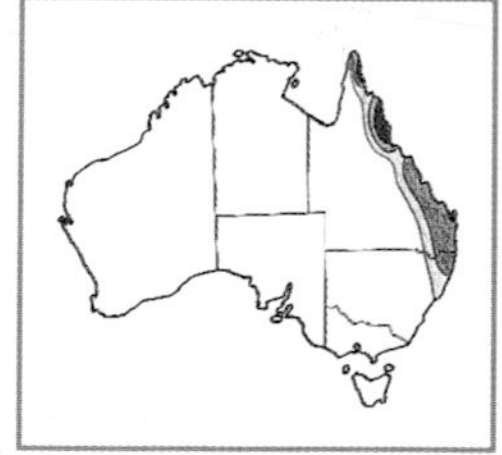

Also known as 'Yellow-eyed Cuckoo-shrike'. A strikingly plumaged cuckoo-shrike closely associated with rainforests where it feeds predominantly on native fruits. Gregarious, usually in flocks, often 10 to 20 birds, occasionally to 50. Flocks travel between food trees with characteristic undulating flight on rather downswept wings: bursts of rather fluttering action interspersed with glides; flicks wings. **Voice:** a sharp, brassy, toy trumpet sound, two or three notes, 'caw-airk-awk' and 'cwairk-awk', often given in flight. Also a single, nasal 'quairrn' and other soft chattering. **Similar:** Oriental Cuckoo, but has barred tail and underwings, different calls. **Hab.:** coastal rainforest and vine scrubs, nearby eucalypts, paperbarks, plantations and tropical gardens. **Status:** locally nomadic; common in far N, uncommon in S. **Br.:** 407

Ground Cuckoo-shrike *Coracina maxima* 34–37cm

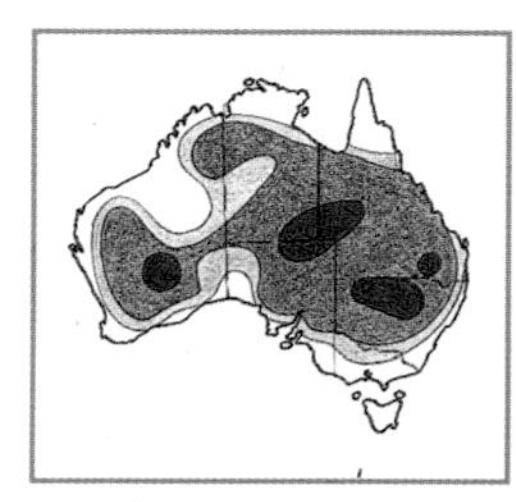

Flight is more direct, less undulating, and foraging is far more terrestrial than other cuckoo-shrikes. Usually in small flocks, covering the ground quickly with long legged, head bobbing strides, pausing to peck at small prey. When flushed, flies up to trees showing white rump between black wings. **Variation:** no subspecies, but in N may be paler. **Voice:** far-carrying contact call, usually in flight: a high, metallic, piercing, drawn out squeak following on from a short, soft note that may not be audible at a distance, 'chr-EEEIP' or 'gr-EEEIP'. Also a rapid sequence, 'weeip-weeip, weeip-weeip' and a musical, vibrating trill, 'tr-r-rweeip, tr-r-r-reeip, tr-r-r-r-rrrp'. **Similar:** Barred Cuckoo-shrike, but darker and heavier barring; usually in coastal forests. **Hab.:** sparse open woodlands, mulga and similar semi-arid scrublands; spinifex with scattered small trees; river and roadside tree belts; dry-land farms. **Status:** nomadic movements following rain; moderately common. **Br.:** 407

Cicadabird *Coracina tenuirostris* 24–26 cm (T)

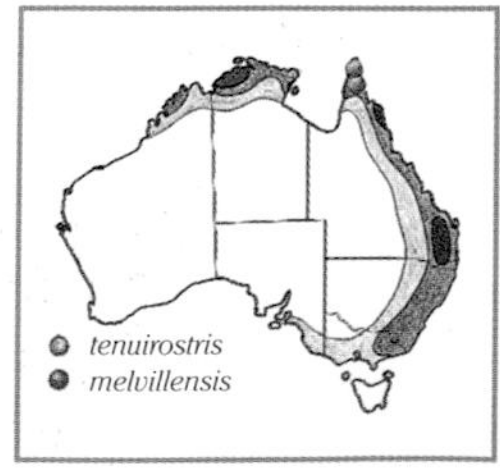

Feeding is arboreal; pairs work unobtrusively to find various insects. Can be hard to find unless it's calling, then elusive, easily put to flight: travels away swiftly in undulating flight through the treetops. **Variation:** widespread from Celebes to Solomon Is. Two subspecies in Aust. The race *tenuirostris* extends down the E coast of Aust.; *melvillensis* occurs across northern parts of the Kimberley, NT and Cape York. **Voice:** usually quiet, but through the breeding season males give a persistent, metallic buzzing: a long sequence of calls, even in pitch and volume. In far N, a drawn out 'trzeee, trzeee, trzeee-', repeated up to 20 times. In SE, call is much more rapid, high pitched, cicada-like, 'tzzeit-tzeit-tzeit-tzeit'. **Similar:** male unique. Female is like females of trillers, but those birds have wing feathers outlined in white in a prominent network of lines. **Hab.:** the foliage canopy of diverse forests and woodlands, mangrove and paperbark swamps. **Status:** in far N, sedentary and moderately common; in SE a migratory breeding visitor, Aug.–Oct. **Br.:** 407

The 4 species of Cuckoo-shrike often perch in slightly hump-backed posture, created by slightly raised, stiff feathers of the lower back
Delicate tints of grey and white are accentuated by black of face and wings.
Up-shuffled wing
Conspicuous black face and throat
Smaller area of black
Three subspecies: melanops over most of the mainland; novaehollandiae, Tas., darker, including mid grey breast, smaller bill; subpallida in Pilbara, WA, is paler.
Similar: dark morph of race robusta, White-bellied Cuckoo-Shrike (below)
Humped lower back
Light to mid grey
Fine dusky barring, mostly on the breast
Immediately after landing, shuffles wings up and down alternately before settling to resting position.
White
Adult
Race melanops
Juv.
Imm. is more like adult.
Head dusky black
Mottled and barred dusky black
Tail rounded and dark, tipped white
Small black mask from top of bill through lores and narrowly behind the fine white eye-ring
Darker grey
Dark morph of race robusta
Smaller than the Black-faced, above. Three Aust. subspecies. Palest is hypoleuca, slightly darker race stalkeri, and darkest, sometimes black-headed, robusta.
Light grey
Adult
Race hypoleuca: — white breasted
Adult
Black may extend back onto ear coverts.
Adult
Race robusta
Tail black, with outer feathers tipped white
Adult
Race stalkeri
Juvenile, robusta. Immature slightly more like adult
Breast faintly barred. Head, back-wing coverts mottled grey-brown
Eye bright light yellow
Flight feathers black, edged grey
Small black mask
Dark eye
In flight, typical long-tailed cuckoo-shrike shape; undulating travel
Distinctive with conspicuous, bright, light yellow eye and bold barring across the underparts. Only the nominate race lineata occurs within Aust.
Mid to dark grey
Bold fine bars
Imm.: faint barring. Juvenile is usually plain white.
Barred, breast to under-tail, and across underwing coverts
Tail black with grey base
Light grey; often white in sunlight
Mid grey
Eye bright, light yellow
Wings black
A large, pale cuckoo-shrike of the drier regions; more terrestrial than other species—legs are longer, feet are larger, heavier.
Fine, dark, scalloped barring
Eye dark
Crown, mantle, upper back silvery grey, or almost white under direct sunlight
Adult
Long, loose erectile feathers of lower back contribute to a humped, long-bodied, long-tailed appearance
More extensive brownish barring
Tail black; tip slightly forked
Lower back, rump and upper-tail coverts are white, finely barred brownish black.
Under-tail coverts white, extend well down beneath the long tail.
Juv.
Brown eye
Dark line through eye
Face black
Throat buffy white
Underwing coverts blue-grey
Wings black, with feathers edged grey
Wings dark, with feathers edged buff
Creamy buff barred dark brown
A small cuckoo-shrike, male entirely dark blue-grey. Two Aust. subspecies: nominate race tenuirostris, eastern coast, and melvillensis, across the Top End of the NT and Kimberley.
Legs and feet small, black
♂
Immature like female, but with pale-edged feathers of back giving scalloped pattern; heavier barring on underparts
♀
♂
Outer feathers black, edged grey
Primaries brownish black above and beneath

Yellow Oriole *Oriolus flavocinctus* 26–28 cm
Inconspicuous and rather elusive birds of the forest canopy, where their greenish coloration blends with surrounding foliage. Solitary or in pairs, they forage through mid and upper strata of gallery and monsoon forests, slowly and methodically seeking fruit; often spend hours in one tree, but eventually travel to the next in direct but deeply undulating flight. Most likely to be noticed when moving between trees, or when the typical bubbling calls are given. Males may call for long periods from one or other of trees within their territory, the easily recognisable calls revealing their approximate location, but these birds are always likely to be difficult to see if high in the canopy. **Variation:** regional differences subtle; opinions differ whether there is just 1 Aust. race, *flavocinctus*, or as many as 4; those on Cape York slightly brighter yellow. **Voice:** rich, mellow, liquid notes, delivered in rapid rollicking sequence, 'chiowk-chiowk-chorrok', a deeper 'chiok-chok-chorrok', and gurgling 'chorr-chor-chorr-chorrok'. Also a high, sharp, clear 'teoo-tieeew'. **Hab.:** tropical rainforests, monsoon and gallery forests, paperbark and mangrove swamps, lush gardens, plantations. **Status:** sedentary; common. **Br.:** 408

Olive-backed Oriole *Oriolus sagittatus* 25–28 cm
The most widely distributed of the three Aust. species of oriole, but the least colourful. As with the Yellow Oriole, the greenish and streaked plumage make this species difficult to sight while it is quietly foraging for fruit in the foliage of the tree canopy unless attention is drawn by the musical, bubbling calls or by flight between trees. Although the range of this and the Yellow Oriole overlap, the two species use different resources, the Olive-backed occupying open forests and woodlands, the Yellow Oriole the rainforests and similar closed tropical habitats. Usually solitary or in pairs, but is found in small, loose flocks through autumn and winter. **Fem.:** slightly paler, duller plumage, paler pink on the bill. **Variation:** 3 subspecies: the nominate *sagittatus* in the SE; race *affinis* in the NT; race *magnirostris* (or *grisescens*) on Cape York. **Voice:** calls are mellow and resonant, not the very deep bubbling of the Yellow Oriole, but mostly higher, clearer and more musical. A frequently repeated sequence is a rollicking 'orry-orry-orriole', quite clearly pronouncing itself to be an 'oriole'. The song is a prolonged version of the call, wandering through similar sequences of notes, often with mimicry of other birds, 'quiee-kwee-kworri-kworriole'; also varied querulous squawking and rasping sounds. **Hab.:** eucalypt forests, woodlands, mallee and taller inland scrubs; occasionally rainforests. **Status:** common in N; migrant to SE. **Br.:** 408

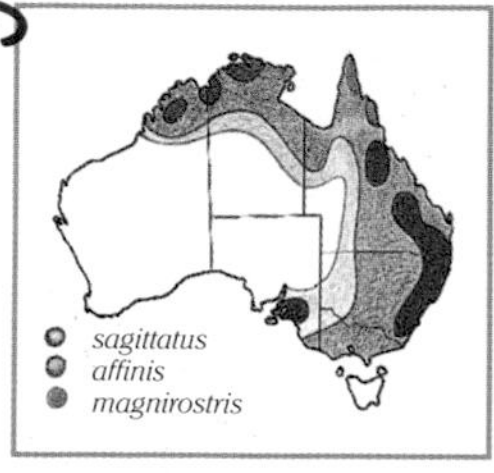

Figbird *Sphecotheres viridis* 28–29 cm
The Figbird, with the Yellow and Olive-backed Orioles, are Australia's representatives of a family that has many species from Africa and Eurasia to Aust. The Figbird, apart from appearance, differs from Yellow and Olive-backed Orioles in its much more gregarious behaviour. Figbirds are colonial, living in small groups through breeding months when several nests may be in quite close proximity. After breeding, Figbirds tend to congregate into larger flocks of 20 to 40, occasionally many more. Flocks gather to feed in various fruiting trees, showing a fondness for both wild and cultivated figs. In the treetops these birds clamber about in parrot-like manner, clinging and hanging among foliage and twigs to reach ripe fruits, insects and nectar. Unlike the often solitary and more subdued orioles, a flock of Figbirds raiding a tree will usually be noticed for their activity and constant chatter. **Male:** black head with rough-textured, deep buff, bare skin, which often reddens with excitement and is typically red in the breeding season. **Fem.:** upper parts olive brown streaked dark brown, underparts white with brown streakings; bare facial skin grey. **Variation:** about 6 subspecies through Indonesia and NG to Aust., where just 2 races occur. Race *flaviventris*, 'Yellow Figbird', has three populations, one on Cape York and NE coastal Qld, a second in the Top End of the NT, and the third in the NW Kimberley of WA. The NT and WA populations together are proposed as a possible third figbird race, *ashbyi*, which has males on average a brighter yellow. The southern race *vieilloti*, 'Green Figbird', extends from easternmost Vic. through coastal forests of NSW and Qld to the vicinity of Proserpine. Between ranges of *flaviventris* and *vieilloti* there is a broad zone of hybridisation, where the males show variable intermediate characteristics, mostly shown in the extent of yellow on the underbody. Females of these races are similar in appearance. **Voice:** calls are many and varied: commonly used is a rising and falling, loud, penetrating yet quite musical 'tsee-chieuw', or faster 'tsichiew'. In each case the beginning 'tsee' or 'tsi' is very brief and high, just a lead in to the long, descending, strongly whistled 'chieuw'. A few calls from within the group or flock are usually answered by others, the many whistles intermingling in a confusion of sound with the individual calls seeming to be given faster, the calling more excited: 'tsee-chieuw, tseechieuw, tsichiew-tchiew'. Also has various erratic notes both clear and harsh. The song is mellower but strong: 'too, too-heer, too-tooheer' and similar sequences including mimicry of other birds. **Similar:** the immatures are like those of the Olive-backed and, to a lesser extent, the Yellow Oriole. **Hab.:** diverse, including rainforest edges, mangrove and paperbark swamps, eucalypt forests and woodlands, orchards, tropical gardens. **Status:** immatures nomadic, adults sedentary; common. **Br.:** 408

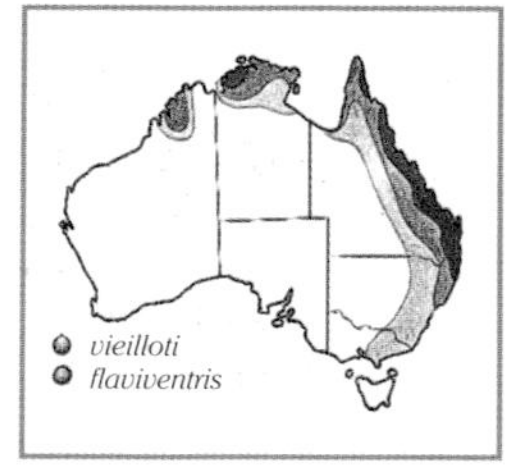

Australia's three orioles (including the Figbird), are birds of similar size and shape, all with olive or yellowish green in the plumage, and adults with red accent colour in eyes, or on face, bill or legs.
Yellowish olive or mustard yellow
Green less intense, heavier streaks
Yellowish olive like upper parts
Eye darker, bill brownish pink
Bill is deep salmon pink on adults of both sexes.
Mantle has thick black streaks, elsewhere fine.
Eye bright red
Wings dusky black with feathers edged buffy white to olive
Black
Grey
Underparts are brighter greenish yellow, streaks fine and sparse.
Imm. Female is similar, but with more intense eye. Bill and legs as for male
Tail dark with variable yellow tips
Wings dusky black edged buffy white to pale olive
Tail dark, edged olive, tipped yellow
♂
Widespread oriole with three subspecies in Aust., all with rather dull olive upper parts compared with the Yellow Oriole; birds of the SE are brighter olive, those from the NE appear more grey-brown on the back.
Olive to dull olive-grey
Bill varies, deep to dusky salmon pink.
Dusky brown
Eye dark
White, heavily streaked
Wing feathers edged rufous
Adult
Bill deep, dusky or pale salmon pink
Imm.
Streaked dark
Adult
Wings dusky dark grey, finely edged white
Tail mid grey tipped white
Feeds mainly on native forest fruits.
Race flaviventris, bright yellow
Outer tail feathers pale grey, whitish towards tips
Male: bare skin around the eye is buff to bright red.
Bare and warty around eye, usually deep buff; flushes red in excitement.
Primaries black, males of both races
Upper parts olive green, except black on head, primaries and tail
Throat bright yellow
Dull white
Legs deep pink
Chin to breast and collar, deep grey
Rump olive
Race vieilloti, grey throat
Race flaviventris, 'Yellow Figbird', tropical N Aust.
♂
Black, males of both races
White
♂
Race vieilloti
♂
Male, intergrade between flaviventris and vieilloti, has grey upper back, throat and upper breast; NE Qld.
Chin to breast grey
Tail is black with white tips larger until outermost is entirely white.
♂
Variable extent of yellow in hybrids
Bare skin around eye is grey.
Under-tail coverts white
Upper parts dull grey-brown with faint olive tint; pale scalloping more distinct
Imm.
Tail plain brown
♀
Female has olive brown upper parts, crown to tail; the latter are only faintly, narrowly, tipped white.
Females of races flaviventris and vieilloti are alike. Imm. are similar, but underbody streaks are more distinct, facial skin is brown.

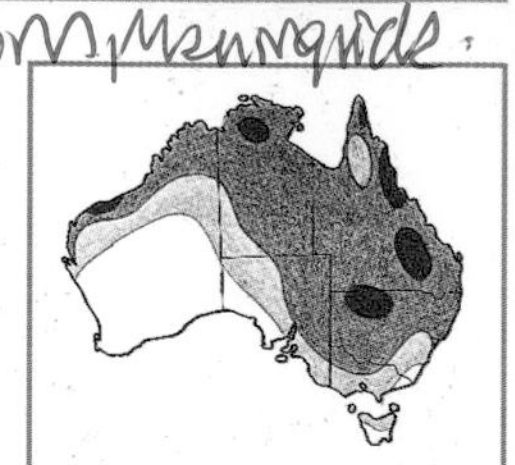

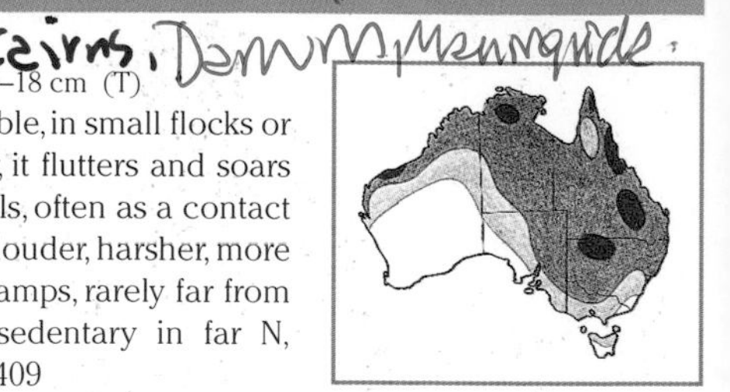

White-breasted Woodswallow *Artamus leucorynchus* 16–18 cm (T)
White breast, sharply defined at its lower edge; all-dark tail. Highly sociable, in small flocks or larger groups up to 50, occasionally several hundred. An aerial feeder, it flutters and soars over water or forest canopy in pursuit of flying insects. **Voice:** brisk calls, often as a contact chatter within the flock, a rather brassy 'aerk, aerk-aerk, aerk'. In alarm, louder, harsher, more strident. **Hab.:** above and around forests, paperbark and mangrove swamps, rarely far from water; extends far inland along pools of tree-lined rivers. **Status:** sedentary in far N, seasonally nomadic elsewhere, especially in S of range; common. **Br.:** 409

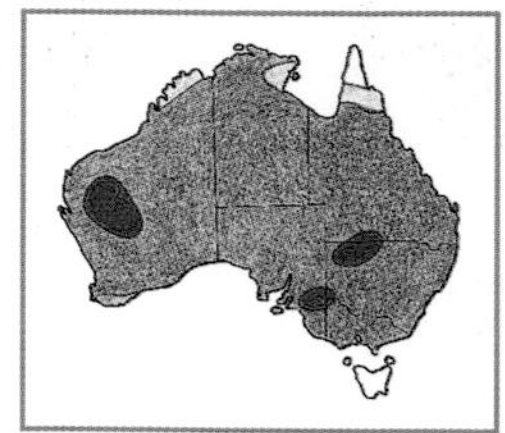

Masked Woodswallow *Artamus personatus* 19–20 cm (L, N)
Dusty tones of grey, black and white are typical of woodswallows, which are among the very few passerines that have powder down through the plumage, this giving the softly powdery appearance. Masked Woodswallows are gregarious, soaring and travelling high for hours in large flocks often intermixed with White-browed Woodswallows, hawking insects with flutter-and-glide wing action. **Juv.:** similar to female, slightly browner, flecked and mottled buff. **Voice:** a rather nasal, querulous 'chrrt' and 'chak'; soft 'chrrup', limited mimicry. **Hab.:** open forests, woodlands, heathlands, roadside and farm tree belts. **Status:** common. **Br.:** 409

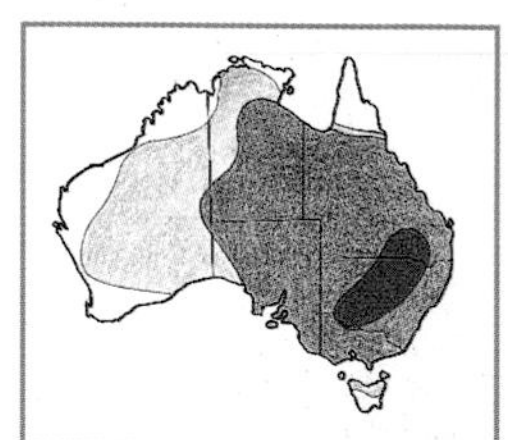

White-browed Woodswallow *Artamus superciliosus* 19–20 cm (N)
Often in high-flying flocks. Hundreds, sometimes thousands, of birds flutter, wheel and glide, showing chestnut underbody in colourful contrast against blue of sky and white of wings. Flocks stay aloft for hours; may descend suddenly to forage for nectar and insects among foliage. Flocks tend to travel N in autumn, wintering in cental Qld and NT, then go S in spring to breed in scattered colonies. **Voice:** contact call is a high, rather musical, descending 'tchip-tchep'. From large flocks this creates a constant yapping chatter. The song is a softer twittering that includes some mimicry; in alarm a harsh, scolding rattle. **Hab.:** forests, woodlands, heath, spinifex, farmland, suburbs. **Status:** highly nomadic; common. **Br.:** 409

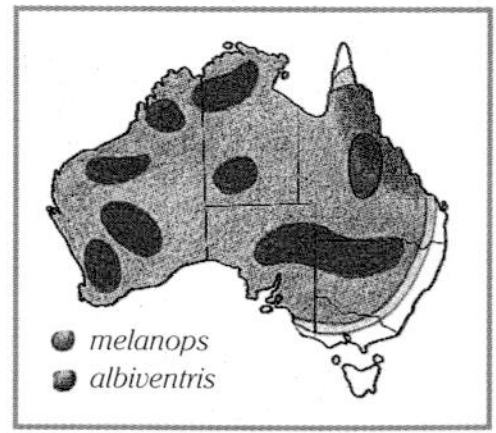

Black-faced Woodswallow *Artamus cinereus* 18–20 cm (L)
In small communal groups of several pairs, darts out from perches on bare limbs, fences or poles to chase insects in swift, powerful flight; soars briefly, circling, gliding, often showing white underwings. Occasionally hovers; drops to ground to take small prey. Also feeds among foliage, seeking nectar and insects. **Variation:** *melanops* (of which those in SW of WA may be *cinereus*); and *albiventris* (which alternatively may split as *normani* in N, *dealbatus* in S). Also in Timor (*cinereus*). **Voice:** calls subdued, slightly scratchy, 'tchif', 'tchif-tchiff' or 'tchif-tchap'. Song a brief, soft twittering; loud, harsh alarm call. **Hab.:** drier open country—woodlands, mulga, spinifex, gibber plains, samphire. **Status:** nomadic; common. **Br.:** 409

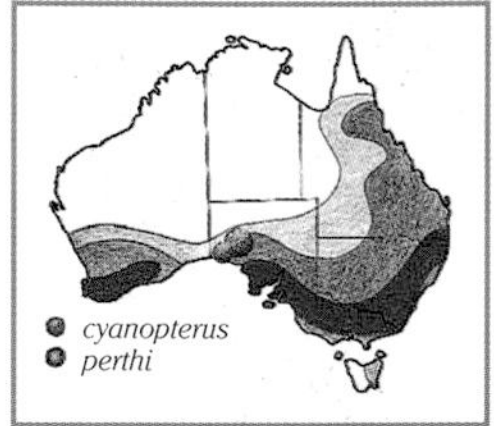

Dusky Woodswallow *Artamus cyanopterus* 17–18 cm
Lives in small flocks, hawking insects through clearings and above canopy of open forests and woodlands; flutters, soars, glides with white wing streak conspicuous. Between flights birds often gather closely along a limb, waving and rotating tails, preening. At night, members of flock cluster closely in a fork or shallow hollow. **Variation:** nominate race *cyanopterus* widespread in E; race *perthi* from SW of WA to SW of SA. **Voice:** contact calls, flight and while perched, brisk, vibrant 'tseit-tzeit', softer 'zut-zut'. A soft, quiet song often includes mimicry. In defence of nest, a harsh, scolding chatter. **Hab.:** drier open forests, woodlands, farmlands and roadsides with scattered trees. **Status:** migratory in SE, sedentary in SW; common. **Br.:** 409

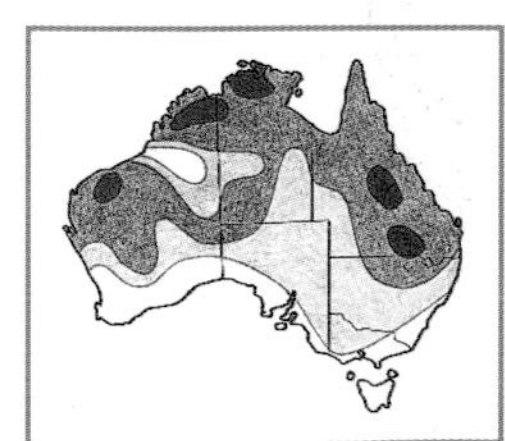

Little Woodswallow *Artamus minor* 12–14 cm
These small, dark woodswallows dart after insects, gliding, diving and hovering into the breeze, or catching wind uplift along the rims of cliffs and gorges of inland ranges. Groups often cluster on dead limbs, conspicuous in this bare habitat. Darting away after insects, they show all dark wings that lack the white streak of the similar Dusky Woodswallow. **Voice:** brisk, high, chirping 'peit-peit' as contact call, often given in flight. When perched, the song is a soft twittering. **Hab:** open tropical and semi-arid woodlands, acacia scrublands with spinifex groundcover, gorges and rocky outcrops of ranges with pools of water. **Status:** sedentary near gorges of ranges; nomadic on grassy plains.. **Br.:** 409

Blue-grey, all species
Sharp edge between throat and white breast
Adult
Adult
Upper parts dark brownish slate-grey, except white rump
Grey, speckled buff
Buffy white
Juv.
Rump white
Adult
Tail entirely dark; all other species have white tip
Adult
Square tipped
White except dark head and tail
Sharply defined, white edged, black facial mask
Silvery-grey
♂
Indistinct mask
Slightly browner
♀
White tip, slight 'V'
Adult
Masked Woodswallow, right, and White-browed, below, are the only species with marked difference between the sexes.
Metallic blue-grey
♂
Stark white brow line over blue-black face
Deep rich chestnut
Lighter grey with softly defined brow line
Paler cinnamon-buff
♀
Brow faint
Grey, mottled and streaked buff
Juv.
White tip to tail is smaller than on adults.
pale grey wings, richly coloured body
Tail-tip white, shallow fork
Adult
Pale smoky grey
Back ashy mid grey
Small, roughly triangular patch of black over the eyes
Pale grey
Wings grey-brown
Adult
Imm.: buff, heavily streaked brown, darkest on wings and white tipped tail
Juv.
Underwing greyish white, silvery in sunlight
Black chin and face
Tail brownish black, tipped white
Black under-tail coverts: *melanops* (and *cinereus*)
Underwing pale grey-brown
White under tail: *albiventris* (or, *normani* and *dealbatus*)
Dark lores, eyes
Adult
Smoky or dusky browns and greys
Unique white streak at edge of wing
Juvenile's body is brown streaked buff; wing as on adult.
Juv.
Smoky Blue-grey
White edges
Adult
Blue-black; white tips
In flight, body, head and tail are dark brown in contrast to the nearly white underwings
Adult
Dark tail with outer feathers white tipped
The Little Woodswallow is similar to the Dusky, above, but smaller and with body richer red-brown rather than smoky brown.
Lores dark
Adult
Wings plain blue-grey without any white edges to primaries
Adult
Tail blackish, outer feathers white
Adult
White-tipped, square-cut
Coverts lighter grey-brown, brighter if sunlit
Mid grey-brown, often appears darker against bright sky. Dusky, above, shows greater contrast between wings and body.
Wings are mid blue-grey without the white line along the edge that is shown by the similar Dusky Woodswallow.
Adult
Tail dark blue-grey, white tipped

Black Butcherbird *Cracticus quoyi* 38–44 cm (T)

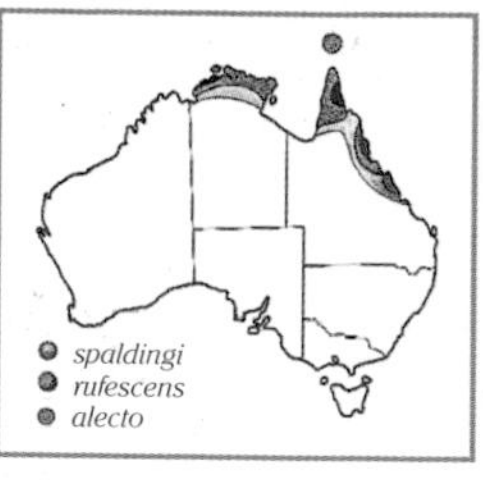

Inhabits dense, gloomy tropical forests; inconspicuous in dark plumage, wary and skulking. Flight between trees swift and direct; hunts from perches, at each site watching patiently for movement below. Takes prey—small reptile, rodent, frog or invertebrate—in swooping, downward dive, landing and almost simultaneously jabbing with powerful bill. Larger prey may be carried into trees, wedged in a narrow fork and ripped apart with the hooked tip of the bill. Occasionally hops on the ground to forage, but the short legs restrict this mode of hunting. Also finds prey in trees, including small birds, eggs and nestlings. **Variation:** Cape York form, smaller, may be a race, *jardini*. **Voice:** call is a deep, clonking 'kwok' or 'kwow'; in NT, currawong-like 'k-wonk'. Also a harsh 'carr-kark!'. Songs have typical butcherbird character, but deeper, abrupt. A musical four-note sequence has its first note highest, lilting, flute-like; two deep and mellow; the last rising: 'kowk, koork-koork, kowk'. Also gives a long, rapid, mellow, bubbling, yodelling 'kwowk-coor-coor, kwowk-coor-coor, kwowk-coor-'. **Hab.:** rainforests, monsoon forests, vine thickets, paperbarks, mangroves. **Status:** sedentary; quite common. **Br.:** 410

Grey Butcherbird *Cracticus torquatus* 26–30 cm

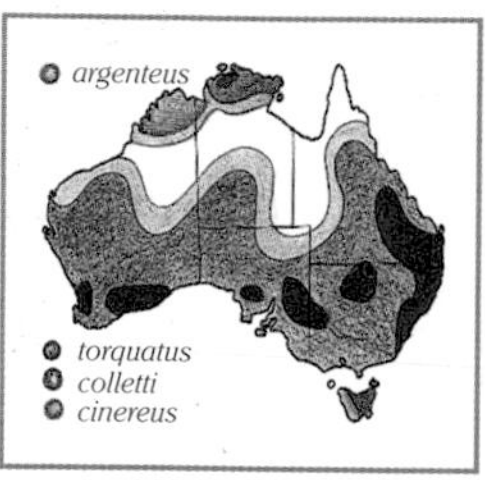

A very widespread butcherbird with minor variations of plumage distinguishing 4 or 5 subspecies; all have adults with a narrow partial collar; most races have a white spot ahead of the eye. Occurs alone, in pairs or small family groups; tends to sit quietly in ambush, occasionally dropping to the ground to take small prey; raids nests of other birds. Flight swift and direct, with rapid shallow wingbeats; glides with wings stiffly horizontal. Larger prey are often wedged into a narrow tree fork to assist dismembering and as short-term storage. **Voice:** rich and varied, including musical and mellow as well as harsh sounds; soft mimicry, strident shrieks in defence and aggression. Songs include slow, deep, mellow notes, 'quorrok-a-quokoo', and deep, bubbly to loud, clear 'kworrok-a-chowk-chowk-chowk-chowk'. Loudest, often given in wing-quivering flight, are vigorous piping notes that gradually develop a harsh, strident, scolding quality towards the end of the sequence, rapid and lively, 'quayk-quayk-quiak-qraik-qzaik–kzaaik-kzaaik-kzzaik'. **Similar:** other butcherbirds, but none has the narrow white part-collar added to the white throat and chin. **Hab.:** diverse, from margins of rainforests, monsoon and vine scrubs to paperbark swamps, eucalypt forests, woodlands, mallee, mulga and other acacia scrubland, partly cleared farmland, and suburban parks and gardens. **Status:** sedentary or locally nomadic; common. **Br.:** 410

Black-backed Butcherbird *Cracticus mentalis* 26–28 cm

Hunts from perches, dropping to ground to take small prey, but also searches branches and foliage for small reptiles, invertebrates, birds and their nests. Usual flight swift, direct, tree to tree. **Voice:** usually deeper, richer and slower than that of Grey Butcherbird; includes duets, mimicry. Often deep, mellow, throaty notes alternating with sharper sounds, 'whorrr-kwiiek, whorrr-kwiiek', and 'quorr-quorr-kweeik, quorr-quorr-kweeik'. Also a call similar to that of the Grey, a rapid, rollicking outpouring, slightly harsh, 'kwarr-kwiek, kwarr-kwiek, kwarr-kwiek', excitedly, as if territorial or warning. **Similar:** Pied Butcherbird, but has black throat; white collar continuous across nape, and the voice is quite different. Grey Butcherbird, but usually further south, narrow collar, greyed plumage tones. **Hab.:** tropical open forests, woodlands; becomes quite tame around homesteads. **Status:** sedentary, common, but of restricted range. **Br.:** 410

Pied Butcherbird *Cracticus nigrogularis* 33–36 cm

Black-hooded butcherbird with superb song. Hunts from bare limbs, poles or wires; often seen along roads; dives down to take prey that includes small reptiles, rodents, ground birds and large insects, all usually taken at first strike. Once on the ground, the butcherbird's short legs limit fast pursuits. Flight between trees is swift, low, undulating, with a swoop up to land, spread wings showing a bold, pied pattern. In small family parties to about six; often brownish immatures help feed the following year's nestlings. **Variation:** probably a single race, slight clinal variations. Alternatively, differences, although slight, may justify 2 races, meeting in NW Qld; those to W then race *picatus*. **Voice:** best heard at dawn during breeding season, or on still, moonlit nights. The sound combines the slow, rich, deep, mellow quality of butcherbirds with the flowing carolling of the Australian Magpie in fluted, far-carrying notes that seem to reflect the loneliness of its outback haunts. Often answered by others, near or distant, in duet or following in sequence, with notes that switch abruptly from high and clear to deep and mellow, but which always seem perfectly chosen. **Similar:** Black-backed Butcherbird, but has white throat and white collar that is broken at the nape; only on Cape York. **Hab.:** open country, woodlands, semi-arid acacia scrub, watercourse trees, spinifex, grasslands, farm and roadside trees. **Status:** sedentary; common. **Br:** 410

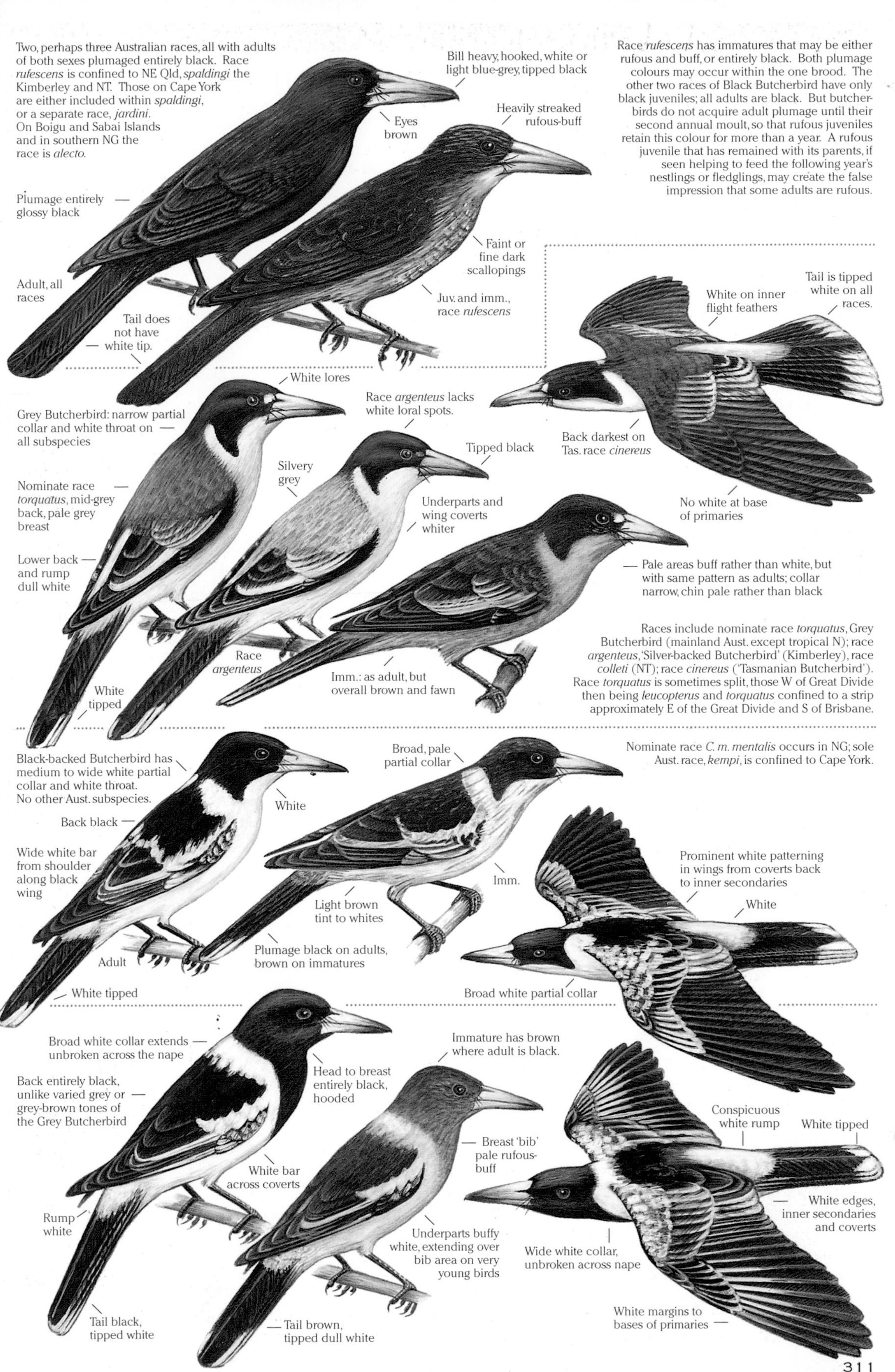
Two, perhaps three Australian races, all with adults of both sexes plumaged entirely black. Race *rufescens* is confined to NE Qld, *spaldingi* the Kimberley and NT. Those on Cape York are either included within *spaldingi*, or a separate race, *jardini*. On Boigu and Sabai Islands and in southern NG the race is *alecto*.
Bill heavy, hooked, white or light blue-grey, tipped black
Eyes brown
Heavily streaked rufous-buff
Race *rufescens* has immatures that may be either rufous and buff, or entirely black. Both plumage colours may occur within the one brood. The other two races of Black Butcherbird have only black juveniles; all adults are black. But butcherbirds do not acquire adult plumage until their second annual moult, so that rufous juveniles retain this colour for more than a year. A rufous juvenile that has remained with its parents, if seen helping to feed the following year's nestlings or fledglings, may create the false impression that some adults are rufous.
Plumage entirely glossy black
Faint or fine dark scallopings
Adult, all races
Juv. and imm., race *rufescens*
Tail does not have white tip.
White on inner flight feathers
Tail is tipped white on all races.
White lores
Grey Butcherbird: narrow partial collar and white throat on all subspecies
Race *argenteus* lacks white loral spots.
Back darkest on Tas. race *cinereus*
Tipped black
Silvery grey
Nominate race *torquatus*, mid-grey back, pale grey breast
Underparts and wing coverts whiter
No white at base of primaries
Lower back and rump dull white
Pale areas buff rather than white, but with same pattern as adults; collar narrow, chin pale rather than black
Races include nominate race *torquatus*, Grey Butcherbird (mainland Aust. except tropical N); race *argenteus*, 'Silver-backed Butcherbird' (Kimberley), race *colleti* (NT); race *cinereus* ('Tasmanian Butcherbird'). Race *torquatus* is sometimes split, those W of Great Divide then being *leucopterus* and *torquatus* confined to a strip approximately E of the Great Divide and S of Brisbane.
Race *argenteus*
Imm.: as adult, but overall brown and fawn
White tipped
Black-backed Butcherbird has medium to wide white partial collar and white throat. No other Aust. subspecies.
Broad, pale partial collar
Nominate race *C. m. mentalis* occurs in NG; sole Aust. race, *kempi*, is confined to Cape York.
White
Back black
Wide white bar from shoulder along black wing
Prominent white patterning in wings from coverts back to inner secondaries
Imm.
White
Light brown tint to whites
Adult
Plumage black on adults, brown on immatures
White tipped
Broad white partial collar
Broad white collar extends unbroken across the nape
Immature has brown where adult is black.
Head to breast entirely black, hooded
Back entirely black, unlike varied grey or grey-brown tones of the Grey Butcherbird
Conspicuous white rump
White tipped
Breast 'bib' pale rufous-buff
White bar across coverts
White edges, inner secondaries and coverts
Rump white
Underparts buffy white, extending over bib area on very young birds
Wide white collar, unbroken across nape
Tail black, tipped white
Tail brown, tipped dull white
White margins to bases of primaries

Australian Magpie *Gymnorhina tibicen* 37–44 cm

Widespread and conspicuous, this is one of Australia's best-known birds, common in natural bushland, cleared farmlands, country towns and suburbs of cities. The Australian Magpie is a large butcherbird with similar pied plumage and black tipped bill. The strong, direct flight of the butcherbirds is even more pronounced in the Magpie, its powerful, pointed wings driving it in fast, noisy flight, the rapid, lashing wingbeats audible at a distance, often attracting attention to these birds' aggressive chases. They often race far and high, then descend in long, fast glides on sleekly backswept wings. Butcherbirds hunt in trees and drop to the ground to take prey, but the Australian Magpie forages predominantly on the ground. Its longer, strong legs and feet enable it to run and bound across the ground in pursuit of prey. When foraging, the family party spreads out, usually still in sight of each other, certainly well within hearing range of any alarm call or softer call of anxiety. The birds stalk about the ground with bold, deliberate strides, occasionally running or bounding to a spot where movement was detected, then probing and jabbing into the ground or among leaf litter, logs or rocks. Also like the butcherbirds, the Australian Magpie has a rich, melodious song that is delivered with steeply uplifted bill, partly opened wings and fluffed-out plumage. Carolling duets and prolonged, powerful song battles with neighbouring magpie groups make these birds very conspicuous as the spring breeding season approaches. Unlike the butcherbirds, which are solitary, in pairs or temporarily in small family groups for the few months after the young have left the nest but are still dependent on the parents, the Australian Magpie lives within a complex social system. Family groups defend their territory very strongly against other magpies to ensure sufficient food and nesting sites for the group. Territorial transgressors are chased, often being attacked by the defending group, driven to the ground, grappled and pecked mercilessly. Each family flock consists of a dominant male, two or more females, often several subordinate males, and several newly fledged or juvenile birds that may be evicted from the territory when fully independent. Most matings are by the dominant male, which, with others of the flock, assists at the nest and strongly defends the site. Usually one of the females is dominant over the others. The flock's defence and support concentrates at her nest, and so nests of any other females are at risk of failure through distractions caused by assisting at and defending the dominant pair's nest and lack of support from others in the group. Groups ranging from 3 to 20 birds occupy areas of 2–40 ha. The smaller territories are sufficient in richer coastal environs, but far greater territories are needed to provide through the worst years in arid regions. The best sites, those with a balanced combination of productive open country together with trees suitable for nesting, are at a premium, needing strong defence. Weaker groups must make do with poorer country and in lean years may have little success in raising young. At the bottom of the ladder are displaced and mobile groups without permanent territory. Such groups feed and roost wherever they find temporary space. Among these birds are immatures and older birds whose territories have been invaded. But it is from these mobile groups that established groups may take recruits, perhaps after the death of an older member.

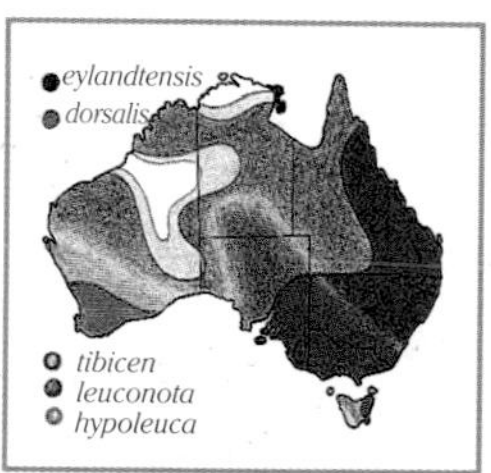

Male: in all races, bolder contrast, blacks richer, whites cleaner than on females. **Fem.:** in all races, white of nape has variable tint or mottling of grey; in white-backed races, the back lightly or heavily mottled with grey or black. **Voice:** strong, rich and varied carolling, with notes ranging from high and clear to deep and mellow. Often a second bird, or many of the group, join to construct a duet or chorus. This communal territory song is strongest in early to mid-spring months; often sings at night. Softer sub-song includes mimicry of other species. In alarm, loud, harsh 'quaark!'. **Hab.:** open country, scattered trees or clumps, usually eucalypt woodland, the vicinity of tree-lined rivers through drier inland regions, partly cleared farmlands, roadside tree belts, and, in suburbs, the more open gardens, parks and other areas with trees and grassy groundcover. **Status:** common to abundant in coastal and agricultural areas, sparse in most arid parts of the interior; sedentary or locally nomadic. **Variation:** 5 Aust. races and one NG race are usually recognised. **Br:** 410

'Black-backed Magpie', nominate race *tibicen*, wide range, coast to coast, but avoiding some wetter southern and tropical northern coastal country. Intergrades broadly with SE and SW white-backed forms; in SE with hybrids S to Melbourne.

'White-backed Magpie', race *leuconota*, occurs throughout Vic., most of SA except far W and NE, and into central and southern parts of the NT.

'Tasmanian Magpie', race *hypoleuca*, islands of Bass Str. and Tas., but absent from dense, wet forests of SW Tas.

'Western Magpie', race *dorsalis*, SW of WA from Shark Bay on the W coast to the vicinity of the WA–SA border on the Great Australian Bight.

Race *eylandtensis*, Groote Eylandt and adjacent coast of NE Arnhem Land, NT.

The widespread Black-backed *tibicen* may be split into 4 races. Most of Qld plus interior of NSW and perhaps parts of interior of Vic., western SA and NT would be the race *terraereginae*; the NT and Kimberley part of *tibicen* would be race *eylandtensis*; the Pilbara part of *tibicen* would be race *longirostris*. The part of race *tibicen* to retain that name would then be reduced to a comparatively small area—coastal NSW and SE Qld, E of the Great Dividing Range.

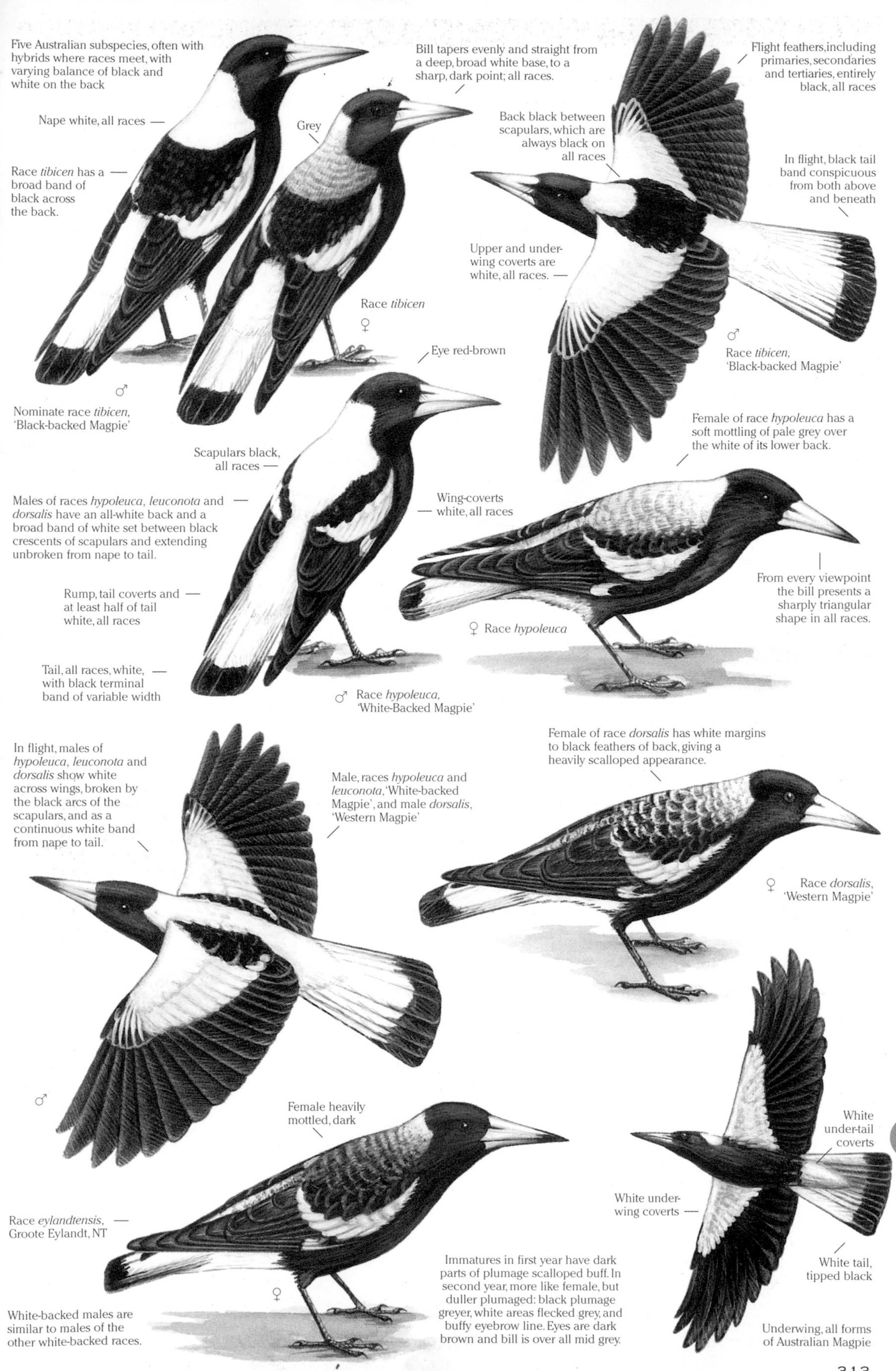
Five Australian subspecies, often with hybrids where races meet, with varying balance of black and white on the back
Nape white, all races
Race *tibicen* has a broad band of black across the back.
Grey
Nominate race *tibicen*, 'Black-backed Magpie'
♂
Race *tibicen*
♀
Bill tapers evenly and straight from a deep, broad white base, to a sharp, dark point; all races.
Flight feathers, including primaries, secondaries and tertiaries, entirely black, all races
Back black between scapulars, which are always black on all races
In flight, black tail band conspicuous from both above and beneath
Upper and underwing coverts are white, all races.
♂
Race *tibicen*, 'Black-backed Magpie'
Eye red-brown
Scapulars black, all races
Males of races *hypoleuca*, *leuconota* and *dorsalis* have an all-white back and a broad band of white set between black crescents of scapulars and extending unbroken from nape to tail.
Rump, tail coverts and at least half of tail white, all races
Tail, all races, white, with black terminal band of variable width
Wing-coverts white, all races
♂ Race *hypoleuca*, 'White-Backed Magpie'
Female of race *hypoleuca* has a soft mottling of pale grey over the white of its lower back.
From every viewpoint the bill presents a sharply triangular shape in all races.
♀ Race *hypoleuca*
In flight, males of *hypoleuca*, *leuconota* and *dorsalis* show white across wings, broken by the black arcs of the scapulars, and as a continuous white band from nape to tail.
Male, races *hypoleuca* and *leuconota*, 'White-backed Magpie', and male *dorsalis*, 'Western Magpie'
♂
Female of race *dorsalis* has white margins to black feathers of back, giving a heavily scalloped appearance.
♀ Race *dorsalis*, 'Western Magpie'
Female heavily mottled, dark
Race *eylandtensis*, Groote Eylandt, NT
White-backed males are similar to males of the other white-backed races.
♀
Immatures in first year have dark parts of plumage scalloped buff. In second year, more like female, but duller plumaged: black plumage greyer, white areas flecked grey, and buffy eyebrow line. Eyes are dark brown and bill is over all mid grey.
White under-tail coverts
White underwing coverts
White tail, tipped black
Underwing, all forms of Australian Magpie

Pied Currawong *Strepera graculina* 42–50 cm (L)

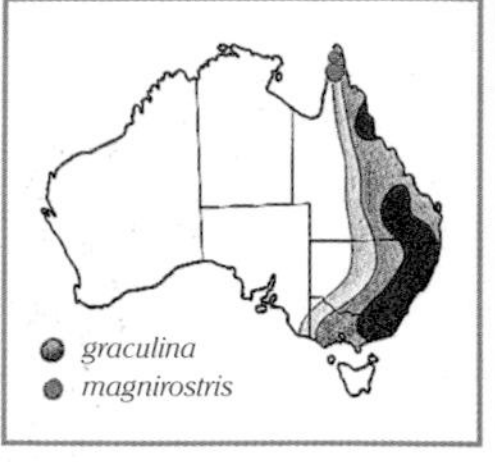

Conspicuous, often noisy; form large flocks that wander nomadically over large distances in autumn and winter. Prey includes insects, small reptiles, birds, carrion, berries; in suburbs, bold scavengers. In flight, a distinctive slow, uneven, rowing or lapping, semi-closing wing action, distinctly different from the steady, purposeful beat of crows and ravens. At dusk, often congregate in communal roosts. **Variation:** large-billed race *magnirostris*, Cape York; *graculina* from NE Qld S, with decreasing white in wings. NE Qld form of *graculina* may be another race, *robinsoni*; and S part, *nebulosa*. On Lord Howe Is., *crissalis*. **Voice:** a slow, rather rollicking series of mellow, often gurgling sounds: 'kurrok, kurrowk'; 'curra-currow-currowk'; 'carrow-carrow-currawowk'. In flight, a distinctive, wailing, raucous, descending 'kirrair-kirrair-kirrowk'. **Similar:** some forms of Grey Currawong, but none has the white crescent over base of the tail. **Hab.:** diverse, including woodlands, forests from coastal to alpine; rainforests, scrublands, farmlands, gardens. **Status:** seasonally nomadic; abundant. **Br.:** 410

Black Currawong *Strepera fuliginosa* 47–49 cm

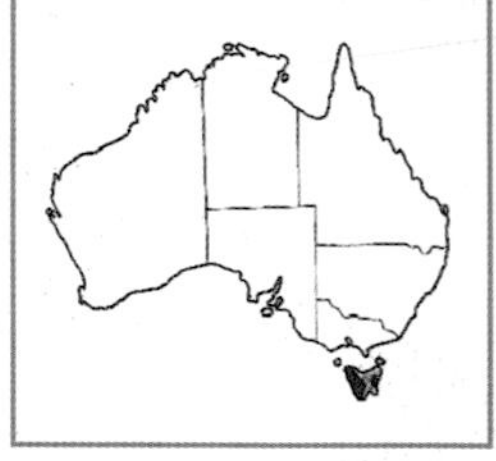

Most conspicuous in autumn and winter when flocks move from heavily forested ranges to more open lower woodlands and cleared country. Has loosely floating flight typical of currawongs; wingbeats slow, relaxed, with shallow, uneven, rowing, semi-closing action. Flocks, when travelling longer distances, keep high. Usually forages on the ground among leaf litter and low vegetation, and, on coasts and islands, fossicks on beaches among seaweed and rocks. Also, more than other currawongs, searches limbs of trees, probing crevices and ripping away bark. Takes small birds, reptiles, mice, insects, carrion, berries, orchard fruit. **Voice:** noisy, loud calls often given in flight; usually a long, rollicking, wailing yet rather musical 'kiarr-weeeik, weeeik-yarr'. Also shorter metallic and croaking noises. **Similar:** Grey Currawong, Tas. race *arguta*, but white in wing and under tail; Forest Raven, but no white, and has deep, regular wingbeats. **Hab.:** alpine moors, dense wet forests, drier open forests, woodlands, farms, orchards, suburbs, coasts, islands. **Status:** common. **Br.:** 410

Grey Currawong *Strepera versicolor* 45–50 cm

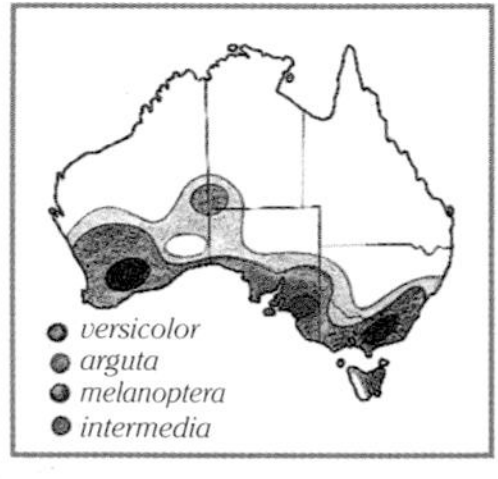

With a wide distribution that extends from SE to SW mainland Aust. and into the NT, the Grey Currawong shows considerable diversity of plumage. All Grey Currawongs, however, have in common a distinctive, higher pitched, clear, ringing, clinking call, and, at close range, a distinctive bill shape. Rather than the massive arched bills of Black and Pied species, with their heavier hooked tip, the Grey has straighter edges, tapering evenly to a somewhat lighter, finer pointed tip. Like other species, the Grey is omnivorous, finding small animals such as birds, their eggs and young, rodents, frogs, insects, carrion, and seeds and fruits. Forages on both the ground and trees. In Tas. appears more often to hunt on tree trunks and limbs, probing crevices of bark and timber. In flight has slow, languid, shallow, uneven, looping, semi-closing wing action. **Variation:** so diverse that some forms are more like other species, the Pied or Black Currawongs, than like other races of their own species, the Grey. There are probably 6 subspecies (listed below), some of which have in earlier times been considered separate full species, and have well-known common names. For observers in all but eastern localities, identification of currawongs is greatly simplified by the existence of only one species, the Grey Currawong, westwards of the extreme SE corner of SA. **Voice:** a loud, metallic 'kling-kling-kling' or 'chring-chring', and various softer mewings and squeaks. In SW Aust., a clear, ringing 'tiew-tiew-tiew', while in Tas. gives a metallic 'kier-kier-killink'. **Imm.:** paler dull grey-brown plumage, very faintly streaked on breast, pale yellow gape, dark brown eyes. **Similar:** Pied Currawong, but almost always has a white crescent over base of tail; deeper calls. In Tas., Black Currawong, but has no white on under-tail coverts. **Hab.:** diverse regions ranging from coastal to semi-desert. Includes forests, woodlands and mallee; larger, denser coastal thickets and heaths; remnant roadside and farm vegetation, orchards, suburbs. **Status:** common in optimum habitats, especially forested SE and SW; uncommon to rare in arid regions. **Br.:** 410

'Grey Currawong', nominate race *versicolor*: variable grey to grey-brown plumage, white tail-tip, under-tail coverts and across flight feathers.

'Clinking Currawong', race *arguta*, SE Aust.: a large, very dark form confined to Tas. and Bass Str. islands (large white patches on wings and under-tail coverts separate this race of Grey from an endemic Tas. species, the Black Currawong).

'Black-winged Currawong', race *melanoptera*: grey-black plumage without any white on wings, but retains white of under-tail coverts and tail-tip; mallee districts around Vic.–SA border.

'Brown Currawong', race *intermedia*: overall darker grey-brown plumage, white markings like nominate race; occurs on Eyre and Yorke Peninsulas, SA.

Race *plumbea*: probably distinct from, and replacing, nominate race in NW of SA, SW of NT and southern half of WA.

Race *halmaturina*: dark plumage, no white in wings and narrow white tail band; confined to Kangaroo Is., SA.

Bill large, straight, stout, hook tipped
Bills of Pied and Black Currawongs are similar to those of the butcherbirds and, to a lesser extent, the Australian Magpie, but are proportionately larger with strong, convex curvature to both upper and lower margins, and are entirely black, giving a heavy appearance.
White patches are largest in N Qld with clinal decrease to S and W; in extreme SW of range has little if any white at base of tail, and only small white patches in wings.
White panel across base of primaries, conspicuous in flight, accentuates the deep, uneven wing action.
Eye intense golden yellow
Variable white crescent around base of tail, wide to very narrow
Immatures of all species have dull plumage, pale yellow gape and dark eyes.
Body plumage predominantly black
All currawong species run, hop, bounce across ground on large, strong legs and feet.
Under-tail coverts white
White tips conspicuous
Black Currawong is confined to Tasmania.
Eye is intense golden yellow.
White tips to primaries
Tail white only at tip. Race *melanoptera* of Grey Currawong, below, similarly has black upper-tail base and lacks white wing panels, but differs in having white under-tail coverts.
Bill is massive with pronounced convex curvature to the upper edge and a deep base; its black silhouette merges smoothly to the black head, which seems relatively small. In most views the bill of the Black Currawong shows this character more strongly, and appears larger, heavier, than that of the Grey Currawong.
Under-tail coverts black
Small white tips to primaries; does not have white panel at base of primaries.
Bold white tips to tail feathers
Darkens to black around the yellow eye
Less massive; straighter taper to finer point
Race *versicolor* is mid grey in SE Aust., becoming gradually darker westwards with decreasing white in wing; very dark grey in SW of WA and in an isolated central Australian population.
Widespread, variable species, plumaged grey to dusky black; four races
Race *melanoptera*, 'Black-winged Currawong'; Vic.–SA mallee district
Nominate race *versicolor*, 'Grey Currawong'
No white in wing, or greatly reduced
White of wing patch may show in folded wing.
White across primaries
White
White-tipped flight feathers
White
White-tipped, all races
Race *arguta*, Tas.
Browner form of the Grey Currawong, 'Brown Currawong', race *intermedia*, Eyre and York Peninsulas, SA
Black around the eyes
White
All races have white under-tail coverts.
Less massive than bill of the Black Currawong
Has white in wing, unlike the other Tas. species, the Black Currawong.
Largest, darkest race, *arguta*, 'Clinking Currawong', confined to Tas. All races of the Grey Currawong have a white tipped tail.
White may show on wing, unlike Tasmanian Black Currawong, which has no white in wing.
Similar: White-winged Chough, but it has much larger white panels on wingtips, longer tail without white at tip, and more slender, strongly downcurved bill. (See p.320.)

Paradise Riflebird *Ptiloris paradiseus* 28–30 cm

The three Australian species of riflebird all have very similar plumages, differing in minor details that might not be evident when, from far below, one of these birds is seen high in a rainforest tree, a silhouette against the sky. Once the bird is known to be 'a riflebird', the locality of the sighting will establish which species. The long downcurved bill enables riflebirds to probe bark crevices, and leaf debris in tree forks and behind large ferns and other epiphytes, and to take rainforest fruits. Males perform elaborate, wing-clapping mating displays. **Voice:** a slow, drawn out rasp: starts low, builds to its loudest harsh, grating rasping; fades to a soft hiss, taking in all 3 to 4 sec. May be repeated quickly, but often several minutes apart: 'scraarsh, scraaarsh', or 'yaaarss'. Also long, upwards whistle; softer churring in displays. **Similar:** other riflebirds; Spangled Drongo, but short-billed, much longer outcurving tail. **Hab.:** subtropical and temperate rainforests, usually on ranges and nearby paperbark swamps, wet eucalypt forest; occasionally forages in adjoining eucalypt woodland. **Status:** sedentary; moderately common in SE Qld., less common towards southern end of range in NSW. **Br.:** 411

Victoria's Riflebird *Ptiloris victoriae* 23–25 cm

Closely resembles the Paradise Riflebird, but smaller. Has a wider and more intensely black band across the breast. Females have darker, deeper buff underparts than females of the Paradise Riflebird, and with different pattern of brown markings. Adult males, like those of the other riflebird species, hold territories dispersed through the forest, where several display sites are visited at close intervals through the breeding season. These are typically high exposed limbs above the rainforest canopy, or, lower, a tall stump. Here the male delivers loud rasping calls that carry far through the rainforest. The arrival of any female attracted to the calls triggers in the male an immediate transformation: his wings arch upwards, bill pointed vertically and gaping to show the yellowish lining. His display includes loud wing-claps and encircling the female in the arch formed by his wings. **Imm.:** like female; iridescent green feathers of immature males begin to show in second and third years. **Voice:** almost identical to calls of the Paradise Riflebird; a similarly harsh, rasping call, longer versions slightly undulating: 'scraark, scraar-aark'. **Similar:** Spangled Drongo, Shining Starling, but both have long tails and short bills. **Hab.:** tropical rainforest, mostly on ranges above 500 m. **Status:** sedentary; moderately common. **Br.:** 411

Magnificent Riflebird *Ptiloris magnificus* 28–33 cm

A NG species that in Aust., reaches only the NE parts of Cape York. In these dense, wet forests, the riflebird frequents the mid to upper strata, usually keeping to the tree trunks and thickest limbs, where, feeding among the clumps of ferns and mosses that festoon the bark, it is difficult to sight from below. At times the location will be revealed by the sounds of ripping bark as it searches for insects or probes into leaf and twig debris that has collected in a high fork or epiphyte. Even when it is calling loudly, the bird often cannot be detected, or only glimpsed as it flies with noisy rustling of feathers between high limbs. Display perches are often high, but some are quite low, like a fallen limb caught between trees that provides a long, thick, horizontal display perch. At times these birds come lower as they feed, when the male's magnificent rich iridescence will be appreciated. Movement changes the colours in the iridescent sheen of the plumage, especially the triangular shield across throat and upper breast. At one moment it is black, then iridescent turquoise or purple, often with intense yellow and bright green highlights, especially when the bird throws back its head to deliver the final, powerful 'whip' of its call. **Imm.:** like adult female; males acquire adult plumage after 2 years. **Variation:** the sole Australian race is *P.m. alberti*; 2 races in NG. **Voice:** males give a deep, powerful 'awoo-arr-WHEET', the sound swooping low through the 'woo', a growling 'arr', then rising to a powerful, sharp, drawn out, whipcrack 'WHEET'. **Hab.:** rainforests; monsoon and vine scrubs along rivers. **Status:** sedentary; moderately common but restricted range. **Br.:** 411

Trumpet Manucode *Manucodia keraudrenii* 28–32 cm (T)

Australia's fourth bird of paradise is placed in a genus apart from the riflebirds. It is not only different in appearance, but also in behaviour, with monogamous pairs sharing nesting responsibilities. Like other birds of paradise, however, it uses elaborate courtship displays. On extended chases, perch to perch, the male calls, bows, crouches, spreads or lifts his wings, plumage fluffed out. Manucodes may be alone, in pairs or small parties, feeding on fruits and berries high or low in the rainforest. **Voice:** a rasping, abrupt, gurgling and inwards gulping 'owwgk'. **Similar:** Spangled Drongo, but usually has a wide tipped tail. **Hab.:** tropical rainforests and vine scrubs. **Status:** sedentary; moderately common. **Br:** 411

Bill black with inside
of mouth yellow
Forehead, crown and
nape glossy, iridescent
blue-green
♂ Blue-black
with purplish
sheen
In flight, wings
of male make
quite a loud
rustling.
Tail very
short,
black
Green
margins
Triangular gorget of
iridescent metallic
green with gloss
of yellow and
blue-violet
Wide
black
band
Long brow line,
buffy white
Bill long, strongly
downcurved
Unmarked
off-white
Pale buff with
crescents of
dark brown
♀
Tail and wing rufous-
brown edged brighter
chestnut or orange
The Paradise and Victoria's Riflebirds are the most alike. When one of these is seen in sufficient detail to determine that it is a riflebird, the species identity can be established from the locality of the sighting. Intervening unsuitable habitat keeps the three sedentary riflebirds well apart. The Magnificent Riflebird of Cape York is the most distinctive.
Males use display perches, pointing
bill vertically, clapping wings.
Bill is black, long,
downcurved.
Inside of mouth
is yellow.
Black breast-band
is wider than on
the Paradise
Riflebird.
Velvety
black with
purplish
sheen
♂
Wider yellow-
green edging
on abdomen
and under
tail
Brow pale buff
Bill grey-black
Undersurface colour deepens rearwards. Off-white at chin and throat; from breast to under-tail coverts a deep buff or ochre with markings of dusky brown
Upper parts grey-brown;
flight feathers edged
brighter cinnamon
♀
Regular display sites
vary from stumps to
high, bare limbs.
♂
Head, sides of neck and nape
are velvety black except for
the iridescent blue-green
crown.
Unique to this
species, a colour
band, changing
with movement,
yellow, bronze
green or red
Fine, black
feathery
plumes
along
flanks.
♂
Every shift of position or tilt of
the head changes the iridescent
colours of the triangular
breast shield: deep blue
with yellow sheen,
then purple with
green highlights,
or partly black.
Breast to
under tail
is velvety
purplish
black.
Tail short; black except for
iridescent green on central
feathers of upper surface
Crown and mantle
cinnamon-grey, wings
brighter rufous
Long pale buffy white brow
Underparts off-white
with dense fine dark
speckling and barring
♀
Tail rufous-brown,
short, yet longer than
the folded wings
Wings are even shorter than tail
when folded; broadly rounded in
flight when the feathers of the
male's wings rustle loudly.
Largest and most
spectacular of the
Australian rifle-
birds; Cape York.
Long, fine rather glossy feathers
form clusters of plumes falling
back and down across nape,
neck and throat, giving streaked
blue-violet highlights that
change as the bird moves.
Plumage overall is black,
with an oily sheen that
changes with angle of
light and the position of
the bird, giving a dull
greenish overtone and
highlights to the black.
Immatures similar to
adults, but with little
gloss on the black
body plumage.
Tail long,
rounded
Adult
Adult
Eye brilliant red
Bill short,
straight, thick,
especially when
compared with the
other birds of
paradise, the riflebirds.
The Trumpet Manucode is, after the Magnificent Riflebird, the second bird of paradise on Cape York. As well as the obvious differences in appearance and behaviour, the male has a very long windpipe, coiled between the skin and the muscles of the breast, enabling him to produce his loud, remarkably deep, guttural, reverberating, trumpet-like blast of sound.
One subspecies in Australia, *M. k. gouldi*. About 8 races in all, others scattered through NG and nearby islands.

Australian Raven *Corvus coronoides* 48–54 cm (L)
Calls and behaviour are often best identifiers. In flight the Australian Raven appears rakish, backswept tapered wings that have a slow downstroke and quick upwards flick; glides with wings flat, tips widely fingered. Has a shallow, fluttering, 'returning to territory' flight with slow, tremulous descending calls. **Juv.:** eyes at first blue-grey, then brown; bare skin at sides of chin pinkish. **Variation:** gradually smaller W from Eyre Peninsula; either clinal (just one race across Aust.); or stepped (westwards, race *perplexus*). **Voice:** strong, first note of sequence rather high, loud, clear, then descending, fading to a deep, slow, muffled groan or gurgle: 'aairk, aark, aaarh, aargargh'; also gives high wailings. **Hab.:** open country, natural and cleared. **Status:** common. **Br.:** 411

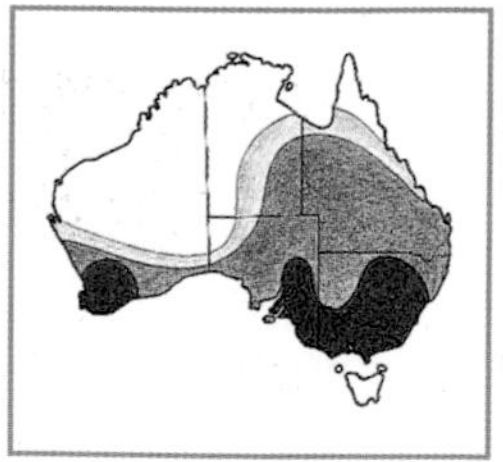

Forest Raven *Corvus tasmanicus* 51–54 cm
Largest raven. Neat, sleek plumage, heavy sluggish wingbeat, especially the short tailed southern race; glides on drooped, wide fingered, rounded wings. Gives a wing-flick flight display, but calls from perch without wing flips and with tail depressed. All 5 Aust. corvids give a 'returning to home territory' flight—a fluttering, shallow wingbeat accompanied by slower, drawn out, descending calls. Southern race forms flocks out of breeding season; may be seen flying in loose, unsynchronised flocks. **Juv.:** eye dark. **Variation:** nominate race *tasmanicus*, 'Forest Raven'; race *boreus*, 'Relict Raven', is confined to NE NSW. **Voice:** very deep, rough, bass 'karr, karr, kar-r-r-r', often with last note slow and falling away. 'Returning home' call is deeper, descending. **Hab.:** diverse—high snowfield plains and alpine forests of Tas. to islands, beaches, woodlands, tall, wet, dense eucalypt forests, open farmlands and coastal heaths. The sole corvid in Tas., but also occurs on mainland. Race *boreus* in forests of NE NSW. **Status:** common in Tas.; uncommon and patchy on mainland. **Br.:** 411

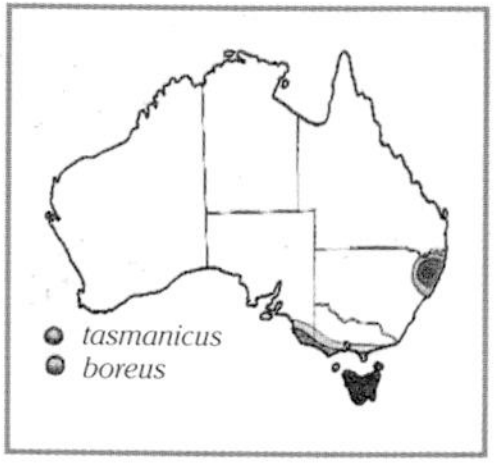

✓ **Little Raven** *Corvus mellori* 48–50 cm Melbourne
Flight agile; looks compact, wings tapered and only slightly fingered at tips. Only slightly smaller than the Australian Raven, with which it shares range and habitat, but has different behaviour. Breeds later, so competition for nest sites and food resources is reduced; defends a smaller territory. Gathers in tight flocks of several hundred that may attract attention by synchronised aerobatics and massed flights to and from roosting sites. **Juv.:** eyes brown, later hazel; fledglings have pink gape. **Voice:** a deep, guttural baritone, notes quick, clipped or abrupt, 'ok-ok-ok', then may fade away, 'ok-orhk-orrh', creakings. When perched, calls are accompanied by quick flicks of both wings together. The 'returning home' call, with wing-fluttering flight, is deep, croaking. **Hab.:** open plains, alpine woodlands. **Status:** common. **Br.:** 411

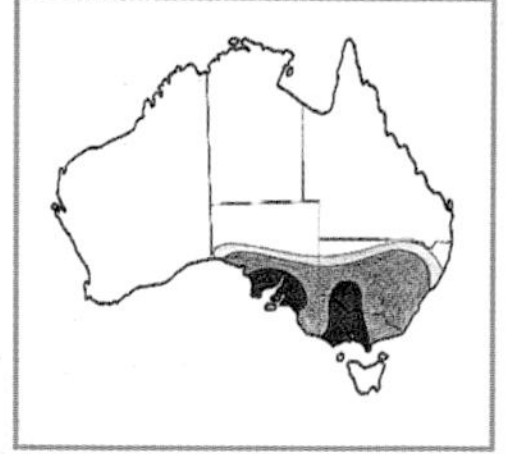

Little Crow *Corvus bennetti* 45–48 cm
Smallest Aust. corvid; plumage often rather dull black, perhaps from dust of terrestrial arid habitat. Agile in flight. Looks compact, rather blunt winged; soars with wings turned back at carpal joint; tail long and slender. This is the common corvid in bold gatherings about outback towns and stations. Much more gregarious than the Torresian Crow; flocks vary from small to very large, often performing spectacular aerial displays, turning, soaring and diving in unison. Gives only slight shuffle of wings after landing. **Juv.:** darker eyes, dull plumage. **Voice:** a rapid, deep, nasal baritone, 'nark-nark-nark', and slow creaking sounds. Sometimes gives a creaky call with fluttering 'returning home' or 'returning to flock' flight action. **Hab.:** open country, usually mulga, mallee, spinifex, arid and semi-arid regions. **Status:** abundant. **Br.:** 411

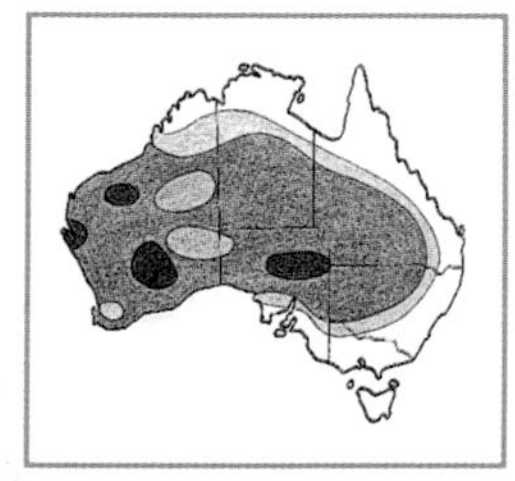

✓ **Torresian Crow** *Corvus orru* 48–53 cm (T) Darwin, Maningrida
In flight, broad tipped, blunt wings, broader tail than Little Crow and Australian Raven; more compact. Glides lazily on drooped wings. On arrival and in greeting, it rapidly and repeatedly flutters wings. In pairs, but in N and E Aust. may be more inclined to form flocks after breeding. **Juv.:** dark eye. **Variation:** in W of WA is slightly smaller, relatively larger bill, higher yodelling call. Race *orru*, Boigu and Sabai, Torres Str., and NG. **Voice:** nasal 'uk-uk-uk-uk-', often last notes slower. Also harsh, aggressive 'arrk-arrk-arrk-arrk, arrrrgk', evenly pitched and spaced until last deep gurgles. In 'returning home' flight, a long, nasal call, abrupt finish. **Hab.:** in coastal regions, open forests, woodlands, rainforest margins, beaches, mudflats, farms. In arid regions, sites with trees, usually along rivers, gorges. **Status:** common; adults sedentary, juveniles nomadic. **Br.:** 411

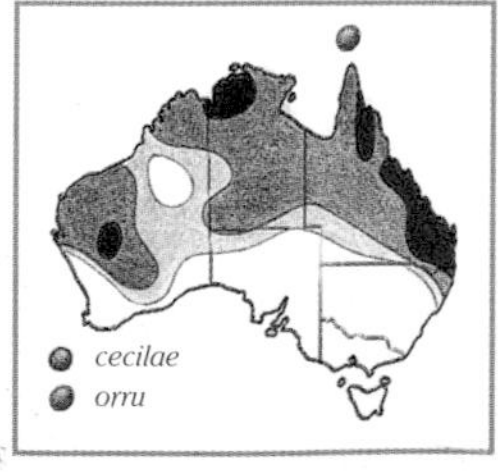

All three raven species have grey down at base of feathers of head and neck; hidden unless ruffled by wind.
Head slightly higher towards rear of crown
Eye is white in all adult Aust. corvids; juveniles change from blue-grey through brown to hazel.
Wings often held in slightly backswept, pointed shape
Large robust bill, slightly longer than head
Tail long, slender, round tipped
Shaggy, long, pointed hackles; fanned as baggy pouch and very conspicuous while calling
Long tail. Only Tas. race of Forest Raven has tail obviously shorter. Relative length of tail to folded wings too variable to reliably identify others
Perched territorial calling usually from a high tree; holds body horizontal, head low, and calls without any upwards flipping of wings.
Has wing-flick display while calling in flight.
Wings relatively longer, more slender than those of crows
Plumage black with purplish or greenish grey sheen
Massive bill, obviously heavier than that of any other Aust. corvid species
Rounded crown, shorter than very large bill
Heavy, laboured flight, but also gives wing-flick display while calling in flight.
Has only small bulge at throat; most obvious while calling; hackle tips forked.
Deep, intense, glossy black; hidden grey down at base of feathers
Race tasmanicus has shortest tail of all corvids relative to overall size of bird.
Race boreus, 'Relict Raven', has heavy bill, but less massive than that of nominate race.
Nominate race tasmanicus
Tail very short, broad
Broad, blunt tipped wings. Race boreus has longer wings and tail, more like those of Australian Raven.
Loud territorial calls are given from a high perch: body is more steeply upright and tail is held depressed.
Race boreus: tail longer
Only slightly smaller than the Australian Raven; glossy black plumage
Flight more agile, quick wing action
Tail relatively long, slender, but may be widely fanned while soaring.
Black with surface sheen; hidden grey base to feathers may show at nape if ruffled by wind.
Bill less massive, but still longer than head
While calling, shows only a small bulge of short, fork-tipped throat hackles.
Wings tapered, often backswept; but soars with wings held straight, tips broader.
When calling from a perch, flicks both of the closed wings simultaneously upwards.
Wings more slender and tapered than those of the Australian Raven
Smallest bill, shorter than head length
The 2 species of crows have white down concealed at the the base of the feathers; this white may show conspicuously if feathers are parted by wind.
From beneath, flight feathers slightly paler grey than underwing linings.
Shows only a slight bulge while calling; hackles very short, pointed, tips not forked.
Plumage is black with blue-violet sheen; but many rather dull.
Long, slender, round-tipped tail
Tail usually longer than folded wings, but varies, probably not reliable for identification.
When giving territorial flight calls, misses a beat by pulling wings close to body, giving an undulating motion like that of currawong or cuckoo-shrike.
Blunt wingtip; in flight appears comparatively small, compact.
When calling from a perch or ground, does not conspicuously flick the folded wings.
Torresian Crow, after landing, folds then lifts and shuffles wings. The one corvid occurring on Cape York, northernmost parts of the Kimberley and NT.
White down, hidden at base of feathers; may show if ruffled by wind.
When giving territorial flight call, pulls wings to body to miss a beat, giving an undulating motion.
Bill heavy, longer than head, and similar to the bill of the Australian Raven
Plumage glossy black with blue-violet sheen
Long, square tipped
Throat bulge not usually noticeable. But a smooth hackle bulge shows during extended calls while perched with body lowered to a nearly horizontal posture.
Wings appear blunt tipped compared with those of all ravens except southern race of Forest Raven.
Tail long, square cut, slightly broader than that of Australian Raven

House Crow *Corvus splendens* 42–44 cm

A very common crow from S Iran to W China; has colonised a wide region of SE Asia. Scavenges about towns and villages, occasionally transported by ships to other ports with occasional arrivals in Aust., most in WA. All reaching Aust. appear to have been destroyed, preventing establishment of a species that would prey on and displace small native birds, and become a damaging pest in agricultural districts. Where common, highly sociable; throughout the year large flocks gather in noisy masses on trees at regularly used roosting sites. **Voice:** rasping, high 'kza', downwards 'kzow', lower 'kiowk'. **Hab.:** agricultural and urban areas; mangroves; in Aust., ports of arrival. **Status:** occasional ship-assisted vagrant.

White-winged Chough *Corcorax melanorhamphos* 43–47 cm

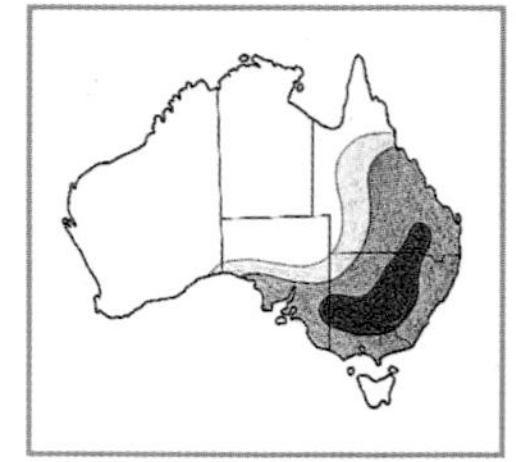

Superficially similar to ravens, crows and currawongs; but belonging to a different and distinctive family, the mud-nest builders, Corcoracidae, of which the only other member is the Apostlebird. Both the Chough and the Apostlebird build deep bowl nests of mud, and are highly gregarious in closely bonded family groups that share most activities, combining to forage as cooperative group, and sharing work of nest building, incubation, feeding of young; roost huddled closely together. The Chough flocks usually build up over several years from the offspring of one pair, and range from about 4 to 20 birds. Highly excitable. Events within the group or an external disturbance will often trigger displays with fanned tail, spread wings showing white patches, and outer eye-ring flushing brighter red. The usual soft chattering suddenly becomes a barrage of loud chattering and churring. The strongly terrestrial Choughs forage predominantly on the ground, moving forward in a scattered group with a head-swaying, tail-jerking swagger, stopping to probe and scratch through leaf and twig litter, seeking insects and their larvae, earthworms, snails and other small ground-dwelling prey. Any significant find is likely to be greeted with much excitement, and other members of the group run in for a share. Flight is used to travel between sites through the territory and to escape threatening disturbances. On these occasions the group keeps together; their slow, deep, flapping flight, interspersed with brief, flat-winged glides, shows the very large, conspicuous white patches of the broad, widely fingered outer wings. Even in flight, at some distance, the unusual downwards angle to the fine, sharp bill, which follows the downcurve of the forehead, gives a distinctive silhouette. **Variation:** has been considered negligible or very gradual, but population on Eyre Pa. and E to about Mt Lofty Ra., SA, may be sufficiently distinctive to form a separate race, *whiteae*, slightly smaller, bill more slender, faint pinkish tinge to white of wing patches. **Voice:** rather mournful, piping, whistling calls, beginning as a high, clear, musical whistle, descending and becoming more mellow; a pause, then rising high and clear again for the next long, downward slide: 't-i-e-e-e-ew, t-i-e-e-e-uw, t-i-e-e-uuw'. Usually given by many of a group together, calls intermingling. In alarm or aggression, harsh, rasping sound, 'ch-z-z-zark', which, from many birds together, becomes an almost continuous threatening barrage of unpleasant noise. **Hab.:** eucalypt woodlands and the drier, more open forests where there is abundant litter across relatively open ground. Includes mallee, mulga, timbered watercourse margins, cypress, pine plantations, farmlands. **Status:** locally nomadic; common but, in many localities, rather fragmented populations where habitat has been broken up and the under-tree litter layer is lost. **Br.:** 411

Apostlebird *Struthidea cinerea* 29–33cm

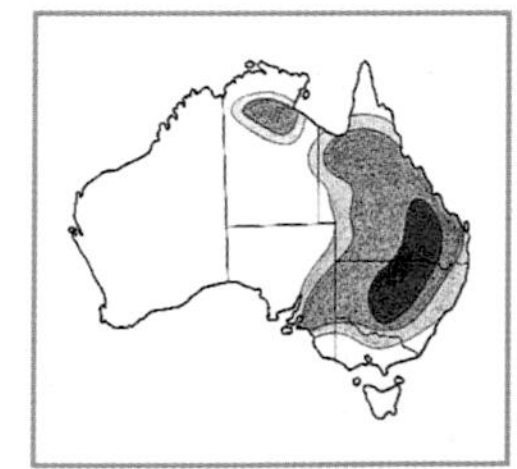

Conspicuous, gregarious family groups with a bustling, confident presence; usually noisy; often tame. One of the best-known birds throughout its range, its many names reflect these attributes—'Twelve Apostles', 'Happy Family', 'Grey Jumper'... Groups vary from 3 to about 20 birds, but typically 5 to 10. After the breeding season, groups may join to form larger flocks of up to 50 birds. On open ground birds tend to keep together as they search for seeds and insects; in woodlands, may become more scattered, but always within earshot of each other. Never far from water, visited several times daily by group in warm weather. A family group will hold a territory of 15–30 ha, defending it in the nesting season against other groups. Members of groups are highly sociable, often preening each other and roosting clustered closely together. **Juv:** fluffy plumage; both sexes brown-eyed at fledging; usually grey after first year. **Variation:** northernmost birds, NT to lower Cape York, may be very slightly, but sufficiently distinctive to be separated as a subspecies, *dalyi*—slightly larger but with smaller bill. **Voice:** varied harsh, discordant noises, usually taken up by many in a group. Includes a harsh, scratchy 'scrairch-scraach', grinding, grating sounds, louder 'tz-iew, tz-iew' and rough, abrupt warning calls; garrulous participation in any flock activities. Very noisy early in breeding season, with much loud chattering. **Hab.:** open and rather dry country, but never far from water of river or dam. Found in open forests, woodlands, river margin and roadside tree belts, remnant timber of paddocks and about homesteads. **Status:** common, but localised, patchy distribution. **Br.:** 411

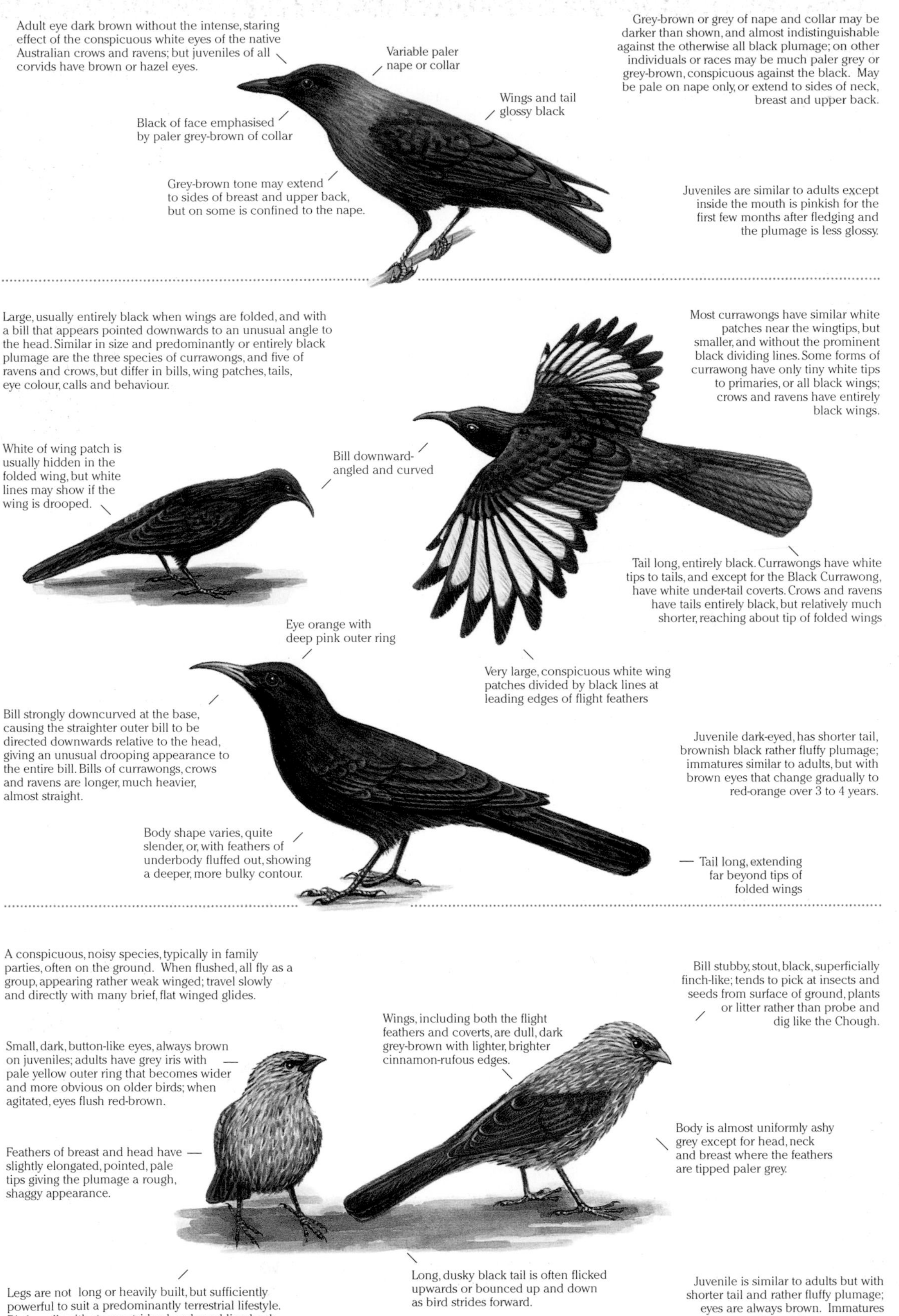
Adult eye dark brown without the intense, staring effect of the conspicuous white eyes of the native Australian crows and ravens; but juveniles of all corvids have brown or hazel eyes.
Variable paler nape or collar
Grey-brown or grey of nape and collar may be darker than shown, and almost indistinguishable against the otherwise all black plumage; on other individuals or races may be much paler grey or grey-brown, conspicuous against the black. May be pale on nape only, or extend to sides of neck, breast and upper back.
Wings and tail glossy black
Black of face emphasised by paler grey-brown of collar
Grey-brown tone may extend to sides of breast and upper back, but on some is confined to the nape.
Juveniles are similar to adults except inside the mouth is pinkish for the first few months after fledging and the plumage is less glossy.
Large, usually entirely black when wings are folded, and with a bill that appears pointed downwards to an unusual angle to the head. Similar in size and predominantly or entirely black plumage are the three species of currawongs, and five of ravens and crows, but differ in bills, wing patches, tails, eye colour, calls and behaviour.
Most currawongs have similar white patches near the wingtips, but smaller, and without the prominent black dividing lines. Some forms of currawong have only tiny white tips to primaries, or all black wings; crows and ravens have entirely black wings.
White of wing patch is usually hidden in the folded wing, but white lines may show if the wing is drooped.
Bill downward-angled and curved
Tail long, entirely black. Currawongs have white tips to tails, and except for the Black Currawong, have white under-tail coverts. Crows and ravens have tails entirely black, but relatively much shorter, reaching about tip of folded wings
Eye orange with deep pink outer ring
Very large, conspicuous white wing patches divided by black lines at leading edges of flight feathers
Bill strongly downcurved at the base, causing the straighter outer bill to be directed downwards relative to the head, giving an unusual drooping appearance to the entire bill. Bills of currawongs, crows and ravens are longer, much heavier, almost straight.
Juvenile dark-eyed, has shorter tail, brownish black rather fluffy plumage; immatures similar to adults, but with brown eyes that change gradually to red-orange over 3 to 4 years.
Body shape varies, quite slender, or, with feathers of underbody fluffed out, showing a deeper, more bulky contour.
Tail long, extending far beyond tips of folded wings
A conspicuous, noisy species, typically in family parties, often on the ground. When flushed, all fly as a group, appearing rather weak winged; travel slowly and directly with many brief, flat winged glides.
Bill stubby, stout, black, superficially finch-like; tends to pick at insects and seeds from surface of ground, plants or litter rather than probe and dig like the Chough.
Wings, including both the flight feathers and coverts, are dull, dark grey-brown with lighter, brighter cinnamon-rufous edges.
Small, dark, button-like eyes, always brown on juveniles; adults have grey iris with pale yellow outer ring that becomes wider and more obvious on older birds; when agitated, eyes flush red-brown.
Body is almost uniformly ashy grey except for head, neck and breast where the feathers are tipped paler grey.
Feathers of breast and head have slightly elongated, pointed, pale tips giving the plumage a rough, shaggy appearance.
Legs are not long or heavily built, but sufficiently powerful to suit a predominantly terrestrial lifestyle. Birds walk with strong strides, heads nodding back and forth, pausing to peck; often running, hopping.
Long, dusky black tail is often flicked upwards or bounced up and down as bird strides forward.
Juvenile is similar to adults but with shorter tail and rather fluffy plumage; eyes are always brown. Immatures are more like adults except males remain brown-eyed until second year.

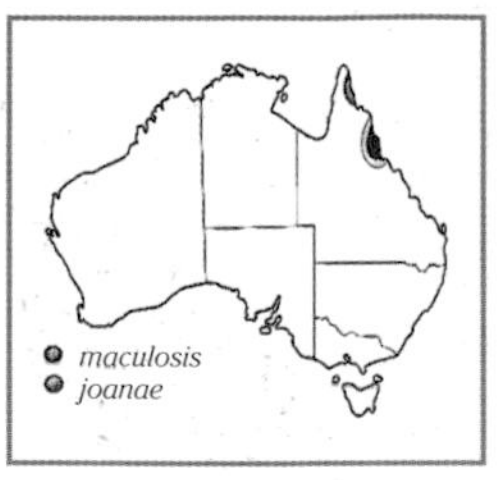

Spotted Catbird *Ailuroedus melanotis* 28–32 cm
Grating, wavering catbird calls become a part of northern Australian rainforests, especially in spring breeding months. In pairs or small groups through autumn and winter; they bounce limb to limb, mid to upper strata, probing bark and debris, finding fruits, buds, insects, small reptiles and nestlings. **Voice:** a cat-meowing, grating, wailing quality, similar to that of the Green Catbird, but probably more nasal. The 'here–I–are' call is slower, more drawn out, wavering and very grating: 'heeeir–I–aaa-arrr'. **Similar:** Green Catbird, but widely separated. **Hab.:** tropical rainforests and margins, monsoon and vine scrubs, riverine and paperbark forests. **Status:** sedentary or localised seasonal movements; common. **Br.:** 412

Green Catbird *Ailuroedus crassirostris* 24–33 cm
Unlike other members of the bowerbird family, the two species of catbird are monogamous, with a pair bonding that is maintained by the male feeding the female throughout the year, as well as by calls in duet. The sexes are alike in plumage, and both contribute to raising the young. They forage in trees, active and wary. **Voice:** similar to that of the Spotted Catbird, an undulating grating wail, which, with familiarity, becomes a welcome sound of the rainforest. The call varies from a short, recognisable 'heer–I–aar' to a long, drawn out, quavering version: 'heeeir–leee-aaa-aarr'. **Hab.:** subtropical and subtemperate rainforest and paperbarks, occasionally adjacent eucalypt forest. **Status:** sedentary; common. **Br.:** 412

Golden Bowerbird *Prionodura newtoniana* 23–25 cm
Unique to a small area of upland rainforest in NE Qld. Although small, this species builds the largest bower, a massive 'maypole' structure. During the breeding season the male is usually not very far from the bower, often in trees overhead, inconspicuous among foliage where many large rainforest leaves are yellowish. There he gives rattling calls that advertise the bower and entice females to approach. Forages in trees for fruits and insects; the male may hide some fruits in a cache near the bower. **Imm:** like female, eye brown. **Voice:** distinctive, including a rapid, pulsating, buzzing, metallic vibration: 'tzuz-uz-uz-uz-uz-'. Another noise is similar to the sound of rice grains shaken vigorously in a matchbox, a rapid 'chk-chk-chk-chk'. **Hab.:** tropical upland rainforests, lower on ranges in winter. **Status:** sedentary; moderately common. **Br.:** 412

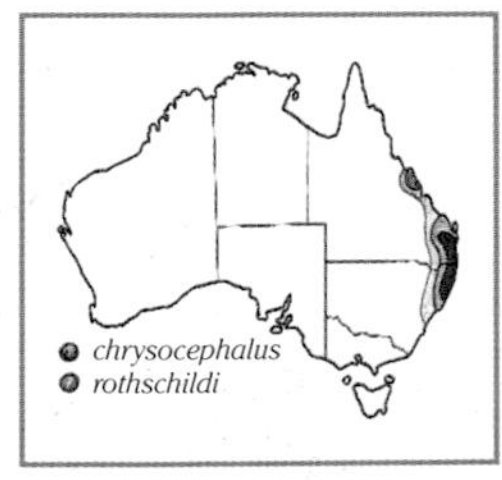

Regent Bowerbird *Sericulus chrysocephalus* 25–29 cm
Spectacular in flight through the rainforest, with the black of the body all but lost against the gloomy background, just the flickering gold of the wings, like a big, golden, tropical butterfly, commanding attention, identity unmistakeable. Only its quiet, retiring preference for dense thickets could make a sighting difficult. Females and young males (which take 5 years to acquire full colour) are far less conspicuous. **Variation:** northwards, gradually slightly smaller; perhaps an increase of orange tint to yellow of head and mantle. Northernmost population has been subspecies *rothschildi*, but possibly not distinctive enough. **Voice:** male in display at bower gives a husky, wheezy, grinding 'kzzzark-kzzaark, kzzzark-kzzaark, kzzzark', and low chattering. **Hab.:** from cool temperate rainforests of highest ranges down to coastal rainforests and scrubs; also in rainforest regrowth, dense thickets including blackberry and wild tobacco, densely planted gardens. **Status:** sedentary, or short seasonal movements from high to low country; generally scarce, locally common. **Br.:** 412

Satin Bowerbird *Ptilonorhynchus violaceus* 28–34 cm
With a range that includes the surrounds of major eastern cities, the Satin is the best-known of Australian bowerbirds, drawing attention with noisy wheezing buzzing sounds. Together with size and conspicuous location, this activity makes bowers easily found, and familiar. Forages widely, from treetops to ground, finding fruits, leaves, insects. Flight is swift, direct, with swooping undulations and slow, strong, deep wingbeats. In winter months forms flocks, usually 5 to 15, rarely over 100. Sedentary in breeding season, and most of the time solitary, except in encounters at the bower. **Imm.:** at first like adult female; in third year juvenile males develop richer green throat, in fourth year solid green band across breast, and through fifth and sixth years dark blue or black feathers begin to appear, with full overall blue-black adult plumage in the seventh year. **Voice:** noisy in breeding season, especially in vicinity of the bower, with loud, harsh, grinding churring and wheezy buzzing: rapid, rollicking 'tzzarr-tzzarr-tzzarr-tzzarrtz–tzzarr-tzzarrr', whistled 'whitchiew'; indulges in some mimicry. **Hab.:** rainforests, wet eucalypt forests. Through autumn and winter moves to more open woodlands, and occasionally to parks and gardens. **Status:** may be sedentary or locally nomadic; moderately common. **Br.:** 412

Two Australian subspecies are widely separated by unsuitable habitat. The differences are slight. Race *maculosus*, NE Qld, has brownish black head markings; race *joanae*, NE Cape York, has blacker markings about the head and underparts are deeper buff and more coarsely spotted. A population in the Cooktown–Atherton–Ingham district is intermediate in markings. Some authorities have suggested there be two Spotted Catbird species as well as the Green Catbird, below. Others have combined the Spotted, with all its slight variations, with the Green Catbird as a single species.

The Golden is the smallest and one of the most colourful of Australian bowerbirds. The yellow of the male's plumage has a gloss that reflects the direct light of sun or flash, especially backlighting, giving small, bright, almost sparkling highlights.

Two subspecies: nominate race *violaceus* from SE Qld to southern Vic. Race *minor*, NE Qld, is smaller and female is greener on the breast.

Brownish black ear coverts, crown; nape to upper back brownish, heavily spotted pale buff
Bill deep, short, pale pinkish buff
Lower back and wings are bright olive green. Coverts and secondaries may have spots at tips: very small, white, or pale buff, or without noticeably light tips.
White markings set in dark surrounds: white brow, partial white eye-ring and spotted crown
Throat to belly is brownish green streaked white.
Tail light olive
Adult
Tipped white
Iris bright red in direct sun or flash; white partial eye-ring
Slight grey tone to bill
Head is greenish brown mottled black and finely flecked pale buff.
Eye darker red for first 2 years. Eyes of adults can appear dark red under dull lighting conditions.
Back, wings and rump are brilliant emerald green with very conspicuous pure white spots at tips of tertiaries and secondaries, which, on tips of coverts, form two white wing-bars.
Greenish buff to dull emerald with short white streaks
Adult
Imm.
Plumage is like that of adults, but dull colours.
Tail brownish emerald with white tips
Buff
Able to lift a small glossy yellow crest at nape; mantle golden with glistening highlights.
Head olive brown with glossy, bright yellow crown, nape, throat; yellow eye
Olive brown
Wing and back plain yellowish olive-brown
Underparts golden yellow with glossy highlights
Entire underparts are light ashy grey.
Tail golden, but on the upper surface, central tail feathers are olive brown over yellow.
♂
♀
Deep golden yellow with an infusion of vermilion on forehead and faintly across the hind neck, mantle and shoulders
Black line over eye
Black nape
Pale, pointed streaks
Adult female has brownish yellow eyes. Immatures initially brown eyed; males become yellow eyed while still in female-like immature plumage.
Sooty black with slight purplish sheen
Distinctive gold and black face pattern
Extensive deep golden yellow on flight feathers
Two outer primaries black
Wide golden band across the wings
Broad black band across wings and back
♂
♀
♂
Black wing margin
Eye intense lilac to bright sapphire
Bill thick, quite short, pale, bluish white with yellow or green tint towards the tip
Entirely blue-black with a glossy sheen to the plumage giving diffuse, soft highlights of deep blue-violet.
Crown, mantle and upper back grey-green
Bill dark grey
Wings grey-brown with green tint to coverts; inner flight feathers sometimes tipped white
Eye bright lilac to turquoise
Legs and feet are stout; pale yellow ochre.
♂
♀
Tail plain grey-brown

Tooth-billed Bowerbird *Scenopoeetes dentirostris* 26–28 cm

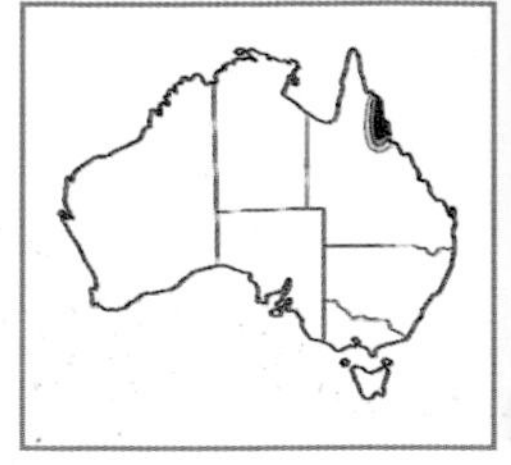

Dark bowerbird of upland rainforests at altitudes 500 to 1500 m; inconspicuous, quiet and wary outside breeding season, when male attracts attention with powerful calls from perches above his 'bower', a cleared circle of ground spread with upturned leaves. Solitary or in small parties, feeding on fruits, leaves, insects. **Voice:** extremely varied; switches abruptly from low, harsh, rasping noises to high, clear whistles and soft, mellow notes with passages recognisable from other birds' song such as the fine, clear notes of thornbill or honeyeater, or the more strident, harsh calls of a parrot. **Hab.:** tropical rainforests, vine scrubs, densely vegetated tropical gardens. **Status:** sedentary; common within restricted range and habitat. **Br.:** 412

Spotted Bowerbird *Chlamydera maculata* 25–31 cm

One of 4 similar grey or grey-brown *Chlamydera* bowerbirds that have adapted to open and mostly much drier habitats. Forages in trees, scrub thickets. Flight swift, direct, undulating, with long swoops on upcurved wings. Male maintains a display bower. **Fem.:** lilac nape crest smaller. **Imm.:** paler, pink nape tuft absent or just a few feathers. **Voice:** noisy near the bower with varied churring, grinding and metallic sounds. An accomplished mimic, imitates diverse bird calls and other sounds. **Similar:** Great Bowerbird, but greyish upper parts, pale fawny underparts. **Hab.:** woodland of eucalypts, cypress pine; brigalow scrub; around homesteads. **Status:** sedentary or locally nomadic; uncommon. **Br.:** 413

Western Bowerbird *Chlamydera guttata* 25–29 cm

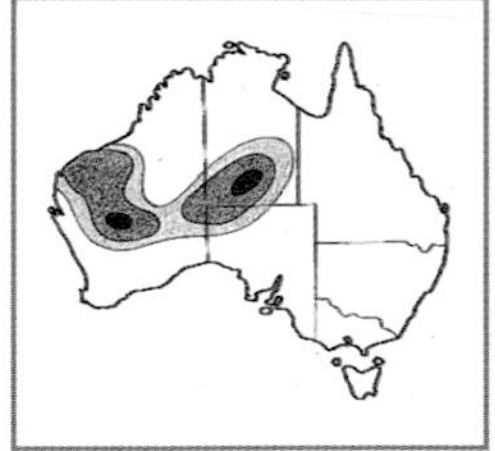

A slightly richer coloured version of Spotted Bowerbird, otherwise similar. **Variation:** at NW Cape, WA, smaller, more russet, perhaps justifies a race, *carteri*. **Voice:** loud, harsh calls echo far through rocky gorges. In display at bower, varied churring, grinding sounds; some like feral cat, perhaps mimicry. **Similar:** Spotted Bowerbird, but further east. **Hab.:** often in gorges of ranges in semi-arid regions, supported by pools and native figs that grow from cliff crevices. Also acacia scrub with dense, low thickets to conceal a bower. **Status:** sedentary in ranges near permanent water, nomadic in some localities; common. **Br.:** 413

Great Bowerbird *Chlamydera nuchalis* 33–38 cm

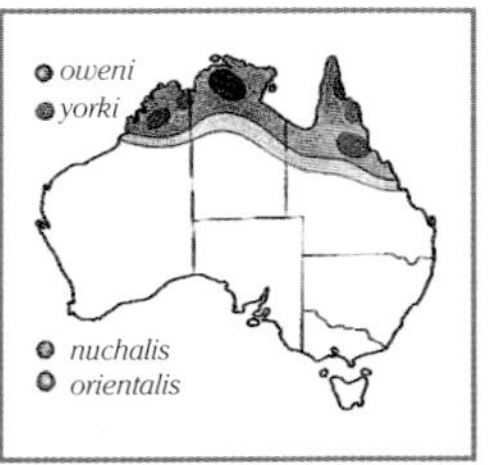

Large, fawny grey; male has pink crest. Constructs a twin-walled bower decorated with bleached bones, shells and green objects, including fruits. **Fem.:** often lacks crest, paler, smaller. **Imm.:** like female, but some barring of flanks and belly. **Variation:** 2, perhaps more races. Nominate *nuchalis* extends across N Aust. to NW Qld where it merges with race *orientalis* of N Qld. Race *yorki*, Cape York, differs only slightly, and is often included within *orientalis*. Likewise, in Kimberley, difference may be too slight to justify separation of race *oweni*, which then also would be within *nuchalis*. **Voice:** a mixture of strongly contrasting noises—wheezing, churring, raucous, harsh, grating, a rasping 'scraaach, graarrk'—intermingled with snatches of clear, piping whistles, often brief snatches of the clear song of a small bird, then perhaps switching to the demonic cackling of Blue-winged Kookaburras. **Hab.:** eucalypt woodlands, forests, favouring dense thickets, margins of monsoon and vine scrubs, mangroves, tropical gardens. **Status:** sedentary; common. **Br.:** 413

Fawn-breasted Bowerbird *Chlamydera cerviniventris* 27–30 cm

Overlaps range of Great Bowerbird in NE Cape York, but smaller and with plumage of richer colours, more like that of the Spotted Bowerbird. The bower is built on a quite massive platform that extends to give an elevated display area outside each end of the bower. Possibly this serves to raise the whole structure above flooded ground in the Cape York wet season. Although the male does not have a lilac-pink crest like other species of *Chlamydera*, it displays, as they do, by turning the head away to show to the female not an expanded crest, but the plain brown nape. **Imm.:** as adults, but with heavier white streakings on forehead and upper back. **Voice:** male is noisy in vicinity of bower during spring to summer breeding season, attracting attention with grating sounds, including long series of loud scratchy, rasping 'scaarrch-scarrch-scarrrch-', and longer 'sca-riarr-iarch'; almost explosive is a loud, harsh, abrupt 'tchuk,tchuk-,tchark. In contrast are whistled notes, perhaps mimicry of part of the the song of another bird. **Hab.:** coastal lowland fringes of tropical eucalypt and paperbark woodland, especially where there is undergrowth shrubbery; dense thickets along watercourses, margins of and clearings in rainforest, vine thickets and mangroves. **Status:** quite common within a very confined range. **Br.:** 413

A dull-plumaged bowerbird with robust 'toothed' bill used to snip and carry large leaves to maintain the display arena
Eye brown, encircled by a narrow, pale brown eye-ring
Bill deep, heavy with serrated cutting edge to both mandibles; grey-black
Upper parts dark olive brown including crown, back, wings and tail
Underparts off-white, heavily streaked dark brown
Immature is like adults, but has brown bill and eye is lighter brownish grey.
Display stage of green leaves, pale side up
Adult
The genus Chlamydera contains the only species of bowerbird adapted to dry open vegetation, especially the Spotted Bowerbird, which extends far into the semi-arid eastern interior, and the Western Bowerbird, found in deserts of central NT and NW of WA.
Pink at nape, usually just a small flat spot of colour, but in display these feathers lift, giving a wide, protruding lilac-pink crest.
Grey band, only on this species
Underparts buff to off-white with dusky brown mottling and barring
White stones, bleached bones and green berries decorate the bower floor.
Erectile lilac nuchal crest
Crown, nape, back, rump and wings are brown with feathers tipped and finely margined cinnamon-buff.
Under-parts are deep buff with brown scalloping, chin to breast.
Tail is dusky brown, lightly tipped buff or dull white.
Although both Spotted Bowerbird (above) and Western (left) share a similar plumage pattern, the brown of the Western is richer, with spots and underparts in strongly contrasting cinnamon or yellow ochre, while the Spotted has dull off-white underparts and spots of rufous-buff that are not as bold a contrast against a lighter brown.
In display at the bower, the male, like other Chlamydera bowerbirds, exhibits the pink nape crest by turning the head to face down and away.
Underparts are entirely soft pale grey or fawny grey.
Grey-brown tipped and finely edged white and pale grey
Has no crest.
Male builds avenue type of bower with twin walls to height around 30 cm.
Head is brown, finely flecked forehead to crown, and more heavily streaked on face, sides of neck and throat.
Upper parts grey-brown with prominent buffy white tips and fine, pale margins
Raised platform to 30 cm high
Softly streaked with brown
Breast to under-tail coverts is deep tawny buff.
Adult: male and female alike
Nest, built by female alone, p. 413

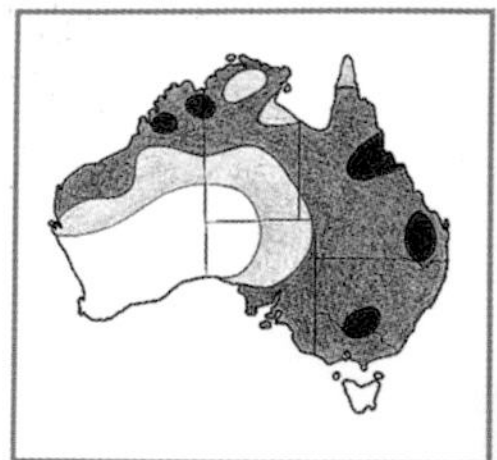

Singing Bushlark *Mirafra javanica* 12–15 cm
In flocks or solitary; forages on ground; runs without bobbing. If flushed, flutters low; short swoops make flight jerky; shows rufous in wings, white in tail. **Variation:** many local plumages, often matching soil colour; whether these represent few or very many subspecies seems unresolved. **Voice:** song varied, not as rich or strong as Skylark, interwoven with shrill trilling, rich, melodious sounds, mimicry. Given in flight or perched. Song flights in breeding season; rises high to hover while singing loudly. **Similar:** Australian (Richard's) Pipit, but larger, teeters and bobs; Common Skylark, larger, small crest. **Hab.:** open woodlands, tussock grasslands, saltbush. **Status:** migratory; locally abundant. **Br.:** 413

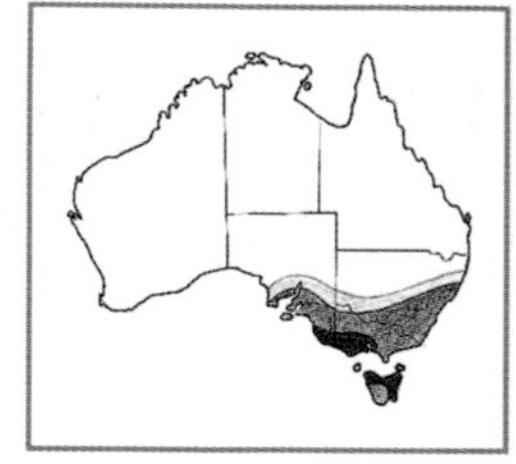

Skylark *Alauda arvensis* 17–19 cm
Introduced before 1850. One of Europe's best songsters, especially when the male rises high on fast, quivering wings, then undulates slowly down, pouring out a continuous stream of rippling and varied song. **Imm.:** smaller crest, reduced streaking on breast, paler edging to upper parts. **Voice:** lengthy outpourings of rich sound, a clear, very attractive and musical song that is often given during steep, upward flights; often gives a quick, mellow 'chirrup'. **Similar:** Australian Pipit, Singing Bushlark. **Hab.:** open country including pastures and other areas of short grass, heathlands, swamp margins, occasionally shores and coastal dunes. **Status:** introduced; nomadic or part-migratory; common in SE. **Br.:** 413

Australian (Richard's) Pipit *Anthus novaeseelandiae* 16–18 cm (C, L)
A strongly terrestrial bird that feeds, roosts and nests on the ground. Rather scattered flocks of up to 100 birds wander locally, then break into breeding pairs in spring. **Voice:** a brisk, cheery, rather abrupt 'chirrip' or 'ch'rip' and 'tsweip'. Courting male has a song flight from a low perch, undulating but gradually rising higher, with each downward dip accompanied by a quavering, trilled 'tiz-wee-ir'. **Similar:** Red-throated Pipit, larks. **Hab.:** grassland, forest clearings, grassy woodland, semi-open scrubland, beaches and hind-dunes, grassy roadsides; on outback tracks, often run out or flush from the grass beside and between wheel ruts. **Status:** usually sedentary or locally nomadic; common. **Br.:** 413

Red-throated Pipit *Anthus cervinus* 15–16 cm
Br. plumage: rusty red tone to the head extending down over throat and sides of neck to the breast. **Nonbr.:** traces of red about brow, face, throat. Underparts pale creamy buff, heavily streaked on upper and lower breast; heavy streaking extends further back down along the flanks than with Australian (Richard's) Pipit. **Fem.:** has slightly less red and heavier streaking on the breast. **Voice:** a distinctive, high pitched, piercing 'pseeiew', 'tsee-itz' or 'pseeeip'. **Similar:** Australian Pipit, but larger, taller, more upright stance, longer legs, little, if any, streaking on flanks; Pechora Pipit, but has prominent white streaks down each side of mantle, and whiter wing-bars. **Hab.:** within usual range, open, dry, bare or grassy sites, coastal to highlands, but also cultivated paddyfields and near water. **Status:** a migrant from Eurasia, S to Borneo and the Philippines, vagrant to Sulawesi and Christmas Is., extremely rare vagrant to N Aust.

Citrine Wagtail *Motacilla citreola* 16–18 cm
The wagtails are a widespread group within the family Motacillidae, which also contains the pipits. As a group, these wagtails are rather slender and long legged, and have long tails that are 'wagged' vertically. Unlike the Australian Willie Wagtail, which swings its long, broad tail through a horizontal arc, these northern wagtails bounce or teeter the rear end of the body, the movement being magnified at the tip of the long tail as an obvious and frequent tail wag. Forages for insects on wet grass, floating vegetation or wet ground, and occasionally springs up into the air in pursuit of a flying insect. Within usual range this species is gregarious, in small groups to large flocks. The breeding range of the Citrine extends through northern–central Asia with migration in the northern winter to China, India and SE Asia. It occasionally occurs in Indonesian islands, and, rarely, in Aust., where it has been recorded both in summer and winter. **Voice:** a husky, wheezy 'dzzeip' and, less frequently, a high, sharp, less rasping 'tsieeow'. **Similar:** Yellow Wagtail and Grey Wagtails, nonbreeding plumage, differs from Citrine's in having off-white rather than yellow breast, pale yellow rather than white under-tail area, and the dark band through the eye and ear coverts connects to the dark nape. Breeding plumages are distinctive and much less likely to be seen in Australia. **Hab.:** wet grasslands, muddy margins of wetlands, saltmarshes, floating vegetation. **Status:** the few Australian records have been scattered around the coastline at widely separated localities.

Five migrant or vagrant species of Eurasian and African wagtails of the family Motacillidae have been recorded in Australia; they are unrelated to the well-known resident Willie Wagtail, which is a fantail of the flycatcher family Muscicapidae.

Yellow Wagtail *Motacilla flava* 16–18 cm (C)
A summer migrant to SE Asia and Aust. Includes several subspecies, but most are either of the race *simillima*, which breeds in E Siberia, or race *taivana*, from the Asia to Japan region. Terrestrial in foraging, it walks briskly with nodding head, teetering tail, as it seeks insects, spiders and small molluscs. Flight is graceful, undulating, wings closing between beats. **Voice:** when flushed and in flight, a wheezy, drawn out 'chzeeip' or 'tsweeip', a call with each bounce of the undulating flight. **Similar:** nonbreeding and female plumages: Citrine Wagtail, but obvious white wing-bars and different face pattern; Grey Wagtail, but yellow on rump and upper-tail coverts, and white bar along base of flight feathers. **Hab.:** open habitats are usually preferred, often near water—swamp margins, salt marshes, sewage ponds, playing fields, grassed surrounds to airfields, extensive lawns, close-cut damp pastures, bare and ploughed ground; occasionally on drier inland plains. **Status:** a rare but regular migrant to parts of the Australian coast, especially in the NW between Broome and Darwin.

Grey Wagtail *Motacilla cinerea* 18–19 cm (C)
A vagrant that has, on a few occasions, been recorded on widely separated parts of the Australian coast, including NE Qld, the NT and near Adelaide, SA. Breeds across Europe and Asia; winters in Africa and SE Asia. This migration takes the Grey Wagtail as far S as Indonesia, and to NG where it is the most common of wintering wagtail species. Runs briskly, chasing small creatures of the stream edge or wetland margins, occasionally darting up to chase a flying insect. Throughout, the teetering movements constantly wag the long tail. Flight is undulating with alternate flapping and folding of wings. **Voice:** in flight, a sharp 'zichiep' or 'zittick'. **Similar:** Citrine and Yellow Wagtails. **Hab.:** in Aust., usually near fresh sandy or rocky streams, but also on mown grass, ploughed land, sewage ponds. **Status:** rare accidental vagrant.

Black-backed Wagtail *Motacilla lugens* 16–18 cm
This wagtail breeds in NE Asia, migrates to S Japan, S China and Taiwan, with some travelling much further south, having been recorded on several occasions on Australian coasts. It has at times been listed as a race, *lugens*, of the White Wagtail, but is now elevated to full species status. These species are not easily identified even in breeding plumages, which are more distinctive and boldly marked. Not only are there several races of the White Wagtail that may reach Australia, each with a slightly different pattern of black and white across head and breast, but these plumages vary with season, sex and age of birds. Shape and relative position of black and white areas may also change with the posture of the bird, with shapes of neck and throat patterns being distorted as the bird stands tall, turns or tucks the head down. **Voice:** like that of the White-backed Wagtail. **Hab.:** similar to the White-backed Wagtail. **Status:** rare accidental visitor; a probable adult female recorded Fraser Is., Qld, 1987, and a record from Derby, WA.

White Wagtail *Motacilla alba* 18–20 cm
White Wagtails spend much of the day foraging across open ground, busily active, running and walking briskly, head jerking back and forth, with frequent pauses when the tail wags up and down incessantly, and the bird pecks at insects on ground and grass. Within the normal range these birds are quite gregarious; in small parties or flocks, darting about grasslands or ploughed fields; roosts at night in large gatherings in dense tree foliage or reedbeds. Makes frequent brief or longer flights, the wings beating rapidly for a few seconds, then momentarily closing tightly to the body, giving a strongly undulating flight path. **Variation:** there are about 10 subspecies of the White Wagtail, of which the races *ocularis*, *leucopsis* and *baicalensis* have been, and seem most likely to be, sighted in Aust. The race *ocularis* breeds across northern Siberia, migrates S to winter from S China westwards to about Bangladesh. The race *leucopsis* breeds in far eastern Asia, mainly SE China, and winters from SW China to SE Asia. The third race, *baicalensis*, has its summer breeding grounds further to the SW, around Mongolia and Lake Baikal, and winters from SE China W to Iran. The usual southern limits of migration are the peninsulas and islands of SE Asia, about as far S as Borneo, where White Wagtails, race unknown, are reported as being quite common. Apparently very few reach Aust. Mostly seen down the W coast, but also sighted at Cape York and SW Vic. **Voice:** a sharp, rather hard and harsh, disyllabic 'chizzik' or 'tichizzik', given in flight. **Similar:** Black-backed Wagtail, but has white along the primaries in the folded wing, and an almost completely white wing in flight; black patch of breast extends up onto throat, but not to chin. **Hab.:** open country, often near water, paddyfields, open beaches, margins of effluent ponds and natural wetlands, and wet grasslands. **Status:** common within usual range; extremely rare vagrant to Aust.

Upper parts olive grey to olive green, most races; race *lutea* has an all-yellow head.
Crown grey, darker ear coverts; brow sharp-ended, stark white
Race *taivana*: crown to rump olive green, band through eye and ear coverts darker
White brow, dark ear coverts
Rump dark
♂ Breeding plumage
Yellow, chin to under-tail area
Underparts off-white
In breeding plumage, face and underparts entirely yellow
♀ and non-breeding ♂ race *simillina*
Weak yellow tint under tail
Legs black
Outer margins of tail white, all races
In flight, shows only thin, inconspicuous wing-bars.
Indistinct line from base of bill meets mottled, broken breast-band.
♂ Breeding plumage
May have faint yellow tint.
Imm.
Breeding male has unique combination of black throat and yellow underparts.
Throat white
Grey crown and back
Rump bright yellow
Throat white, breast yellow
May show one thin pale wing-bar.
♀ and non-breeding ♂
♀ and non-breeding male
Yellow above and beneath base of tail
Throat black, brow and cheek line white
Bold white bar along base of flight feathers; hidden in folded wing
Tail very long
♂ Breeding plumage
Legs pink or brown
Both back and breast are equally black, separated only by a narrow white strip of varying shape down each side of the neck, and between shoulder and breast.
Chin usually white, but throat to breast black
Crown darker
Wide, white band across folded wings. In flight the wings show predominantly white, in contrast to the comparatively dark wings of the White Wagtail, below, when outspread in flight.
♀ and non-breeding ♂
Grey from nape to tail coverts
Black reduced to band
♂ Breeding. Female pattern similar, but grey on back
Wing largely white
Black
Crown grey
The Black-backed Wagtail has a distinctive pattern of breeding plumage. The breeding female has a head to breast pattern similar to that of the breeding male.
♂ nonbreeding ♀ similar
Mottled breast-band
Dark wingtips
Imm.
Three races of White Wagtail possible in Australia: race *ocularis*, race *leucopsis* and race *baicalensis*
Forehead and face white
Breeding males, all races: similar wing pattern except darker on the lesser coverts, more white on median and secondary coverts, and on some secondary inner margins.
Black crown and eye-line
Large black bib includes chin
Crown mid to dark grey
Nape to rump mid grey, all races
Female and non-breeding
Shoulder white
White
♂ Breeding plumage
Smaller band
Underparts entirely white
Crown black fading to dark grey on mid back
Race *ocularis*, female, and male nonbreeding
Race *ocularis*
White wing stripe and large shoulder patch
Face and throat white; lacks eye-stripe.
Prominent white wing-bars
Mid grey
Face white; may have faint white eye-stripe.
♂ Breeding
White wing stripe, shoulder patch
Large , deep, parallel-sided breast patch
Race *leucopsis*. Female: similar but dull plumage
Outer margins always white, all races
♂ Race *baicalensis*
Female duller
Wings predominantly dark, in contrast to the largely white wings of the Black-backed Wagtail.

Zebra Finch *Taeniopygia guttata* 10 cm

Most widespread finch; seen in pairs, small parties and huge flocks, especially when coming to water; noisy with bouncy, undulating flight. Forages on ground, seeking fallen seeds rather than pulling down and stripping seeds from the heads; occasionally takes flying insects, especially when feeding nestlings. **Voice:** loud nasal or brassy twanging, 'tiarr', and abrupt 'tet, tet' from flocks in flight. **Hab.:** a bird of drier inland regions in spinifex, mulga and other semi-arid scrublands, grasslands, saltmarsh, open grassy woodlands, and cleared land, but never very far from water. **Status:** sedentary or seasonally nomadic; common. **Br.:** 414

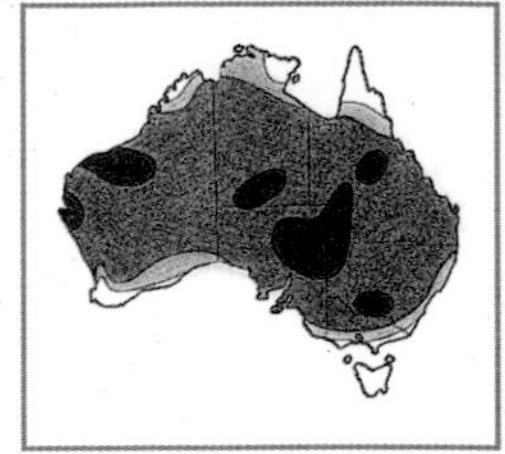

Double-barred Finch *Taeniopygia bichenovii* 10–11 cm

An 'owl-faced' finch; white face is encircled by a dark rim that continues across the throat as upper of the diagnostic twin bars. Typically in close flocks; feeds on ground under seeding grasses; bouncing, undulating flight. **Imm.:** dull, indistinct bars across fawn underparts. **Voice:** in contact, a brassy, drawn out 'tzeeaat, tzeeaat' like that of the Zebra Finch, but each call longer, more plaintive; similar sounds, weaker, may form a more continuous passage of song. In close contact, an abrupt, low 'tat, tat'. **Hab.:** dry grassy woodlands, open forests, grassy dry scrublands, farmlands, but never far from water. **Status:** nomadic; common, but more of an occasional visitor into extreme SE of range. **Br.:** 414

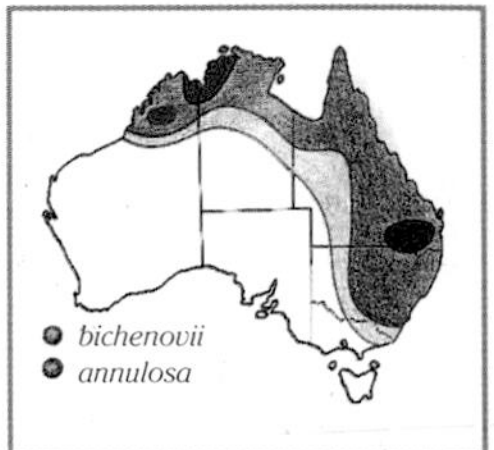

Long-tailed Finch *Poephila acuticauda* 15–16 cm

Also known as 'Blackheart' for the black throat patch. Strongly social, in pairs or flocks. Brief bobbing appeasement rituals maintain flock harmony. Forages on ground, may pull down grass to strip seed; insects are caught, sometimes in flight. **Variation:** bill colour red in W Qld, red to orange in NT, and red to yellow in Kimberley. May be colour forms (rather than subspecies where the red-billed are *P.a. hecki*). **Voice:** contact call, often in flight, a soft 'tek'; also a slow, rather mournful, descending 'whieeeuw' and loud, whistled 'whirrr'. **Hab.:** open woodland, grassland with scattered trees, shrubs and often pandanus; never far from water. **Status:** sedentary; common. **Br.:** 414

Black-throated Finch *Poephila cincta* 9–11 cm

Highly gregarious in small flocks of 10 to 30 birds; forages on ground for grass seed. **Variation:** race *cincta* from SE Qld northward; intergrades in NE Qld with *atropygialis* of Cape York Peninsula, the latter darker towards NW Cape York, either clinally, or as race *nigrotecta*. **Voice:** contact call a soft 'tek', often in flight; also a whistle similar to that of the Long-tailed Finch, but lower, more hoarse. **Hab.:** open woodland, scrubby plains, pandanus flats with deep cover of grasses; never far from water. **Status:** sedentary or locally nomadic; common in N of range, but southern white-rumped race declining and shrinking in range. **Br.:** 414

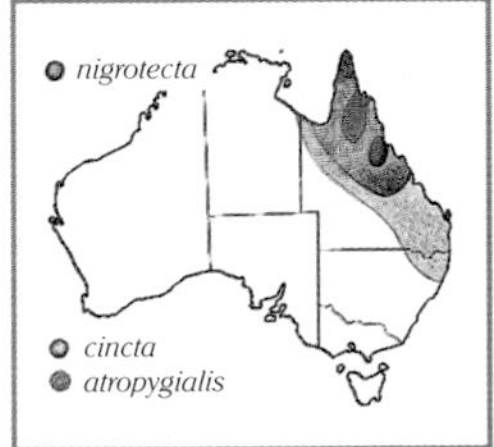

Masked Finch *Poephila personata* 12–14 cm

In pairs or small flocks forages on ground for fallen seeds of grasses. Flocks move to water morning and evening; at times thousands gather in trees around the water where the excited birds flick tails sideways, chatter and preen. **Juv.:** dull grey-brown without black mask; black bill. **Variation:** race *personata* from far NW Qld through NT to Kimberley, WA; race *leucotis*, 'White-eared Finch', lower and central Cape York. **Voice:** loud, brassy, nasal 'tziat' similar to Zebra Finch's call; nasal 'twet-twet' and soft chatterings in flocks. **Hab.:** grassy tropical woodland of eucalypts or paperbarks, grassland with scattered shrubs; never very far from water. **Status:** sedentary; common. **Br.:** 414

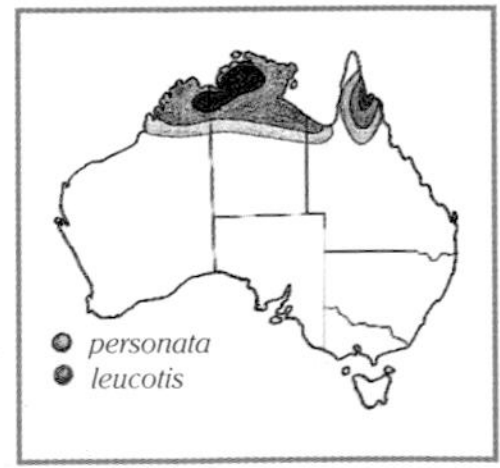

Pictorella Mannikin *Heteromunia pectoralis* 11–12 cm

Better adapted to the arid environment than other mannikins; highly nomadic, moving out from the coast to country that has received rain, then, with the return of arid condltions, moving again towards coastal grasslands. Although the Pictorellas visit water at least once a day, they will travel a considerable distance daily to water. In pairs to large flocks; seek seeds and insects. **Voice:** contact call within flocks, a low 'tsip'; identity call, a loud, sharp 'tliep' or longer 'tlee-ip'. **Hab.:** open country with sparse scattering of trees, groundcover of grass or spinifex, grassy flats along creeks. **Status:** nomadic; moderately common. **Br.:** 414

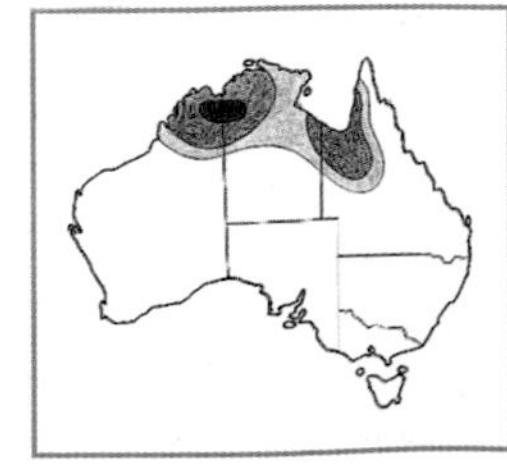

Australia has 18 native species of grass-finches, of which the Zebra Finch is the most widespread.
Crown blue-grey
Orange-red bill
White rump and heavily barred tail are conspicuous as finch flies up and away.
Rounded orange ear patch, black and white facial streaks
Flanks orange-chestnut, spotted white
♂
Female shares male's distinguishing facial streaks, white rump and barred tail.
Black
Juvenile like female, similar facial markings, but bill is black and eye darker.
♀
Juv.
Wings black with white spots heaviest on the coverts
White rump is conspicuous on eastern race.
Face and underparts white; two narrow black bands
Crown to lower back, cinnamon-brown
Black rump on western race
Bill always blue-grey
♂
♂
The two subspecies are identified by the rump colour. Nominate white-rumped race bichenovii to the east of the Gulf of Carpentaria from Cape York to SE Vic., and dark-rumped race annulosa W of the Gulf of Carpentaria through the NT into the Kimberley region of W.A.
Blue-grey
Yellow-billed form
Black bar continues down under wings, becoming a narrow point on flanks.
Soft tones of cinnamon
Red billed form
Tail coverts white
Black throat; slightly smaller on female
Dusky bill and throat
Black tail tapers to a long, fine point
Adult
Adult
White
Bright red
Grey to dull pink
Imm.
A plump, stocky finch; plumage pattern similar to that of the Long-tailed Finch
Head silvery and nape blue-grey
Throat and chin black, meeting black bill; lores also black
Unbroken black from lower back to tip of tail
Black bar continues down under wings, becoming a point on the flanks.
Cinnamon-fawn, slightly more cinnamon-brown on back and wings
Race nigrotecta (not shown) also has the black rump, but brown on the breast.
Upper-tail coverts white between black of upper rump, and tail
Adult
Race atropygialis
Legs red
Nominate race cincta
Adult
The only Peophila finch without a black throat marking; instead, larger areas of black on the lores, and very broad black patches on the flanks
Bill heavy, deep yellow
Black mask covers lores, encircles eyes.
Cinnamon-buff underparts become white towards the black mask.
Race personata
Race leucotis, 'White-eared Finch'
Tail sharply pointed, but far shorter than that of the Long-tailed Finch
Upper parts rufous-brown
Margin around mask varies: buff, cinnamon or orange.
Bill heavy, blue-grey
Black face mask includes throat and chin.
Breast scalloped black and white
In Australia there are three native and one introduced species of mannikin. These finches have relatively short, graduated tails and large stout bills. There are no subspecies of the Pictorella Mannikin, nor significant regional variation, but considerable variation among individuals.
Female is similar, but underparts are slightly paler, facial mask very dark brown rather than black, fewer white spots on wings, broader black barring on breast.
Juvenile is dull grey-brown above, face and throat dusky grey-brown, underparts paler pinkish buff, bill brownish.
Upper parts grey-brown; small spots across wings
Underparts, breast to under tail, cinnamon-buff with a slight violet overtone
Legs magenta

Crimson Finch *Neochmia phaeton* 13–14 cm

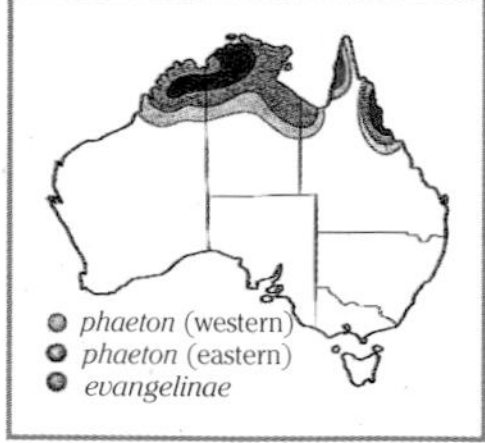

In its moist, lush habitats feeds only rarely on the ground, but usually higher, amongst seed heads of grass that may be well over a metre tall. Pairs or family parties, rather than large flocks, perch on strong grass stems, working at the large seed heads, fluttering between clumps. **Variation:** black-bellied race *phaeton*, W Kimberley through NT to NW Qld; then also, widely separated, another population on NE and E coasts of Qld (this previously a third race, *iredalei*). White-bellied race *evangelinae*, 'White-bellied Crimson Finch', Cape York. **Voice:** high, penetrating, squeaking 'tseit-tseit-tseit'; sharp, fine, single or double 'tsit-tsit' in contact; high 'tchieep' in alarm. When feeding birds are flushed, silvery tinkling contact calls. **Hab.:** waterside vegetation–pandanus, cane grass, paperbark, lush grasses, crops. **Status:** sedentary; common. **Br.:** 414

Plum-headed Finch *Neochmia modesta* 11–12 cm

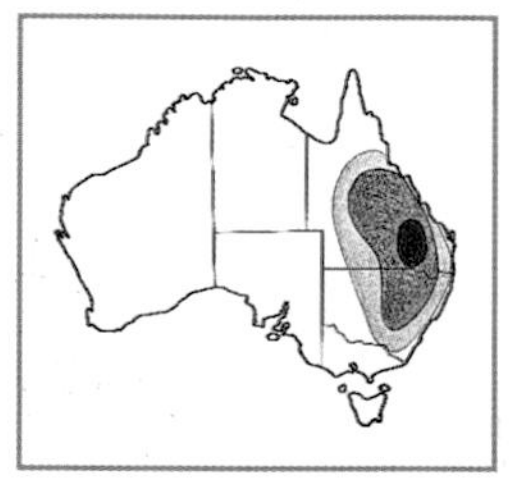

In flocks, often very large; feeds on the ground or climbs on taller grass to reach seed heads; drinks frequently. **Voice:** as close contact call, a soft 'tlip' or 'tleip'. In flight, a sharper 'tee-ip' or 'tleeip'; from a flock the calls combine to give a tinkling, quite loud. **Similar:** Nutmeg Mannikin, but underparts scalloped rather than barred, face and throat uniformly brown, no white spots on wings. **Hab.:** near swamps or rivers, in tall grass and shrubbery of nearby floodplains, lush grass of shallows, reedbeds and cumbungi. **Status:** migratory and locally nomadic; seasonally irruptive, then abundant. Generally rather uncommon. **Br.:** 414

Nutmeg Mannikin *Lonchura punctulata* 11–12 cm

Also known as Spice Finch; native to S Asia from India to S China, and S into the Philippines and Indonesia. Probably became established in Aust. when aviary escapees colonised the E coast, but kept to areas of human activity. Is seen in fast-flying flocks that make sharp turns in unison; drop suddenly into the grass. **Voice:** contact call is a soft 'tchp'; identity call is a high 'kh-teee', the first part very soft, the 'teee' penetrating; alarm call is a harsh 'krek-krek'. **Hab.:** grassland—prefers wetter areas, disturbed land, roadsides, crops on moist farmland, lantana. **Status:** introduced; common and expanding its range.

Chestnut-breasted Mannikin *Lonchura castaneothorax* 11–12 cm (T)

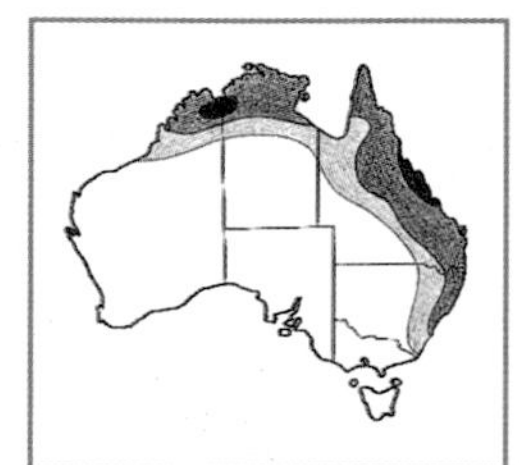

In small to large flocks, separating out in the breeding season; even then the flock keeps together, forming a loose breeding colony. When feeding, birds climb the strong grass stems to snip away the seed heads. **Voice:** as contact call, a clear, bell-like, quick 'tlit'; when used among a feeding flock, a slightly longer 'tleit'. The song is a long series of fine, clear notes in many variations: sounds such as 'twee, tee-oo, chie-ook, chee-ing'. **Hab.:** clumps of rank, tall grass in damp environs, wet grasslands, swamp margins and swampy heaths, mangroves, canefields, ricefields. **Status:** sedentary or seasonally nomadic; common. **Br.:** 414

Yellow-rumped Mannikin *Lonchura flaviprymna* 11–12 cm

Highly social; in flocks often intermixed with larger numbers of Chestnut-breasted Mannikins. Flocks are swift—birds, in unison, abruptly change direction before dropping back into tall grass. Individual birds in short, low flights have an undulating flight typical of grass-finches. **Voice:** like that of the Chestnut-breasted Mannikin, a sharp, clear, almost bell-like 'tseit' of varying length. **Hab.:** usually near water or damp situations, and woodlands with tall seeding grasses; also in reedbeds and mangroves. **Status:** locally nomadic; generally uncommon, but locally abundant, especially in irrigated crops. **Br.:** 414

Blue-faced Parrot-Finch *Erythrura trichroa* 11–12 cm

Unmistakeable finch, bright green with blue face patch. Usually in small parties, feeding largely in rainforest trees, finding seeds in lower to mid levels. **Variation:** about 10 races: islands of Indonesia, New Guinea, Solomons, Celebes, Micronesia. **Voice:** chattered, slightly harsh, 'chak' notes intermixed with extremely high, thin, insect-like, 'tsiep' squeaks: 'chak-chweip, tsiep, tak-chat, tsweip'. **Hab.:** lowlands to heights of ranges in grassy clearings and lush, low vegetation of rainforest margins and mangroves as well as open grassy areas of eucalypt forests. **Status:** seasonally nomadic or migratory; on highlands through summer, wintering on coastal lowlands. Easily overlooked, uncommon, possibly rare. **Br.:** 415

A finch of lush tropical habitats; the deep crimson plumage needing direct sunlight to reveal its full intensity of colour. Three races in Australia
Grey
Face crimson with matching bill
Olive grey with crimson wash
Dull red
Race *evangelinae*, 'White-bellied Crimson Finch', Cape York: white replaces black, belly to under tail.
Wings and back olive-grey with broad crimson margins
In sun, brilliant crimson, but dusky, dark red under dull lighting
Juvenile similar, bill brown
Crimson extends to the rump, upper-tail coverts and full length of tail
Black, central abdomen and under-tail coverts
Central belly and under-tail coverts white
♂
Race *phaeton*
♀
♂
Crown dark claret red or purplish brown, 'plum'; no white brow line
Fine white brow line
Upper parts, including crown, mid olive brown
Weak pale brow
Upper parts, wings, deep olive brown
Eyes brown, bill black
Throat to chin deep 'plum'
Without male's 'plum' patch under chin
Lacks any dark chin spot.
Wing coverts and inner flight feathers tipped white
Uneven, broken bars, heaviest along flanks
Dull brownish white with very weak barring
Tail black; coverts barred white
♂
♀
Juv.
Face and throat very dark brown
Entire upper parts chestnut-brown
Scalloped brown
Underparts plain brownish buff
Introduced species
Adult
Juv.
Colourful mannikin of lush environs of eastern and NE Aust. About 5 subspecies. Only one in Aust., the nominate race *castaneothorax*
♂
Bill silvery grey
Face and chin black
Olive brown
Underparts brownish buff, slightly deeper on breast
Has some similarities with the Yellow-rumped Manikin, which occupies drier inland regions. The two interbreed, producing intermediate forms.
Tail cinnamon and brown; under-tail coverts black
Breast orange-chestnut
Black barred flanks
Juv.
Adult male and female are alike, except that, in most pairs, seen together, the female may appear to have colours not quite as intense or bright.
Closely related to the Chestnut-breasted Mannikin. In regions of overlap, interbreeding occurs, producing hybrid birds of intermediate plumage.
Back bright chestnut
Head fawny white
Bill and legs pale blue-grey
Back and wings cinnamon-brown; flight feathers darker grey-brown
Lower rump to tail-tip soft warm ochre; under-tail coverts black
Fawny yellow
Adults; sexes alike
Juveniles have light cinnamon-brown upper parts, buffy white underparts, brownish-buff breast.
Forehead and face cobalt blue with mauve overtone
Upper parts dull grey-green
Flight feathers dusky rufous-brown with fine pale margins
Lacks blue face patch, rump dull brown.
Rump, upper-tail coverts and base of tail carmine or scarlet; outer tail dusky red
Plumage is predominantly bright grass green, slightly lighter on the underparts.
Female in similar but more subdued colours; blue face patch smaller
Pinkish brown
♂
Juv.
Bill is initially yellowish, gradually becomes black.

Diamond Firetail *Stagonopleura guttata* 12–13 cm

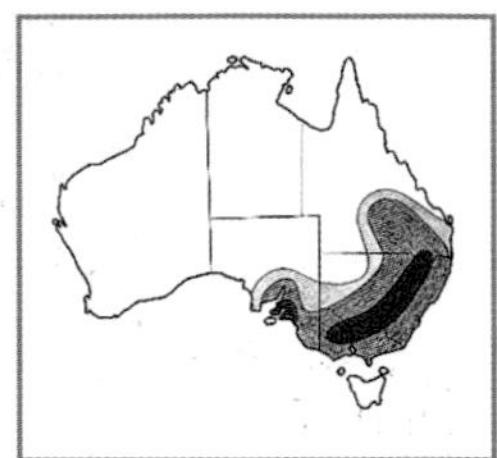

Usually in small flocks of 20 to 30 birds; tends to nest in loosely scattered colonies. In autumn and through winter, larger flocks may form. Feeds exclusively on the ground. Flight low with only slight undulations; large flocks travel in long lines. **Voice:** usual contact call is a rather plaintive, drawn out, whistled 'tioo-whieer' or 'tioo-wheee', the call of the female at a slightly higher pitch; male in display gives low, rasping sounds. Also soft, vibrating sounds when nest duties are exchanged. **Hab.:** localities with grassy groundcover underneath open forest; woodland, mallee, acacia scrublands, and timber belts along watercourses and roadsides. **Status:** sedentary or locally migratory; uncommon, patchy occurrence. **Br.:** 415

Beautiful Firetail *Stagonopleura bella* 11–13 cm

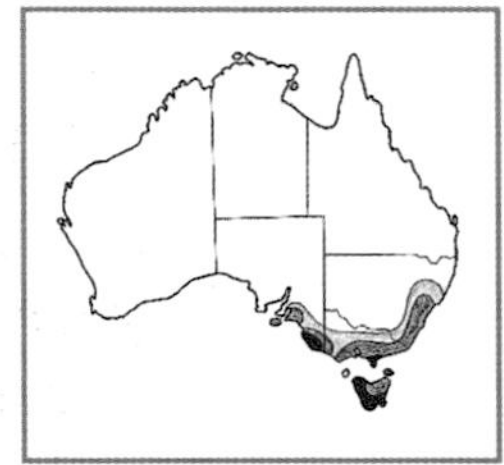

Usually in pairs or small family parties; forages on or near the ground, often concealed beneath overhanging grass or shrubbery, finding seeds of various grasses. Also seeks seeds and insects higher among foliage of shrubs and trees. Flight is low and direct on whirring wings with only very slight undulations. **Voice:** a single, mournful, undulating 'whee-ee-ee' as identification call; soft 'chrrit' as contact sound; in alarm gives abrupt 'tchup-tchup' sounds. **Hab.:** damp or swampy grassy spots in gullies, low-lying flats, woodland with dense, low undergrowth, dense thickets along creeks and rivers, coastal heaths; needs water nearby. **Status:** sedentary or locally nomadic; uncommon on mainland, more common Tas. **Br.:** 415

Red-eared Firetail *Stagonopleura oculata* 12–13 cm

In pairs or small family groups; secretive, low in dense undergrowth or on the ground seeking seeds of native grasses and sedges; bulky nests may reveal presence of the birds. **Juv.:** dull, lores dusky; lacks scarlet on ear coverts; bill initially dark. Although calls are soft, they help to locate this species in its dense habitats. **Voice:** a soft, rather mournful whistle, the shorter calls at an even, mid-level pitch, 'wheee'; longer calls slight, wavering 'wheeoo-ee'; also soft, murmured contact call. **Hab.:** in heavy forest, thickets along creeks, winter-wet paperbark flats; dense low S coastal heath. **Status:** sedentary; scarce except extreme SW. **Br.:** 415

Painted Finch *Emblema pictum* 11–12 cm

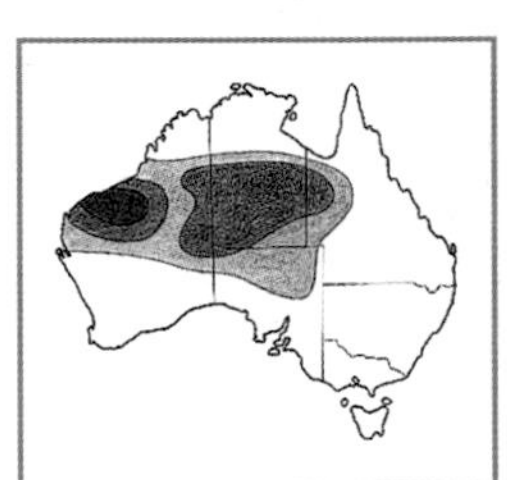

In flocks of typically 5 to 20 birds that travel with erratic, energetic, bouncing undulations. Noisy chatter, especially around waterholes. **Voice:** presence often announced by abrupt, slightly nasal 'chek' before birds on the ground, among spinifex, are seen. A flock, when flushed, passing overhead or arriving at water, creates a rapid, rather musical chattering: 'chek,chak-chek, chek' at random, with slightly different pitch of calls from individuals. **Hab.:** usually spinifex in rocky range country and vicinity of gorges where there are pools; also in sandplain spinifex if water is available. **Status:** highly nomadic; common. **Br.:** 415

Star Finch *Neochmia ruficauda* 10–12 cm

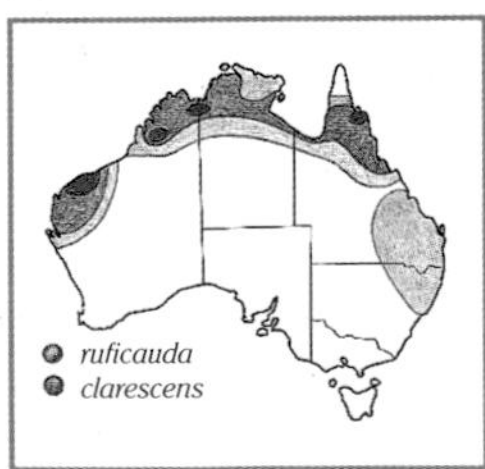

In pairs or small flocks up to 20 odd birds; feeds at seed heads in low vegetation and on ground. Flight strong, swift and erratic. **Variation:** nominate *ruficauda* (SE Qld, NSW); *clarescens* (N Qld, NT, WA). **Voice:** most noticeable and loudest call is a high, penetrating 'tseit, tseeit' given almost constantly in flight; from flock becomes a rather musical tinkling. Flocks feeding in grass keep in contact with an abrupt, softer 'tsit'. The male gives a weak twittering song. **Hab.:** lush green vegetation, tall, rank grass along watercourses, temporary or permanent; rushy margins of swamps, moist green crops. **Status:** locally common in NW in favourable seasons, but range and numbers of eastern race are greatly reduced. **Br.:** 415

Red-browed Finch *Neochmia temporalis* 11–12 cm

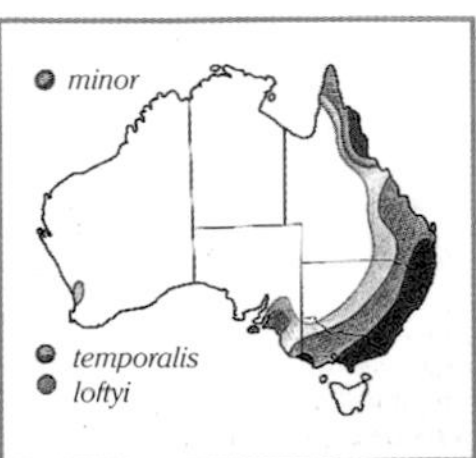

Highly sociable, in close-knit flocks; forages on the ground, but occasionally perching on grass stems to reach seed heads. When flushed, the flock departs with slightly undulating flight. **Variation:** nominate race *temporalis* (NE Qld down E coast through NSW and Vic. into SE of SA); race *minor* (Cape York); race *loftyi* (Mt Lofty–Kangaroo Is.), a doubtfully distinctive form. **Voice:** extremely high, almost inaudible, a drawn out squeak, 'tseee' and 'tseet'; in alarm a more abrupt 'tchip'. **Hab.:** undergrowth of forests; favours grassy clearings, coastal scrubs and heaths, mangroves, canefields. **Status:** sedentary or locally nomadic; common. **Br.:** 415

Named for the sharp, clear, white diamonds set in a band of black
Smoky grey
Black lores, conspicuous against the pale head, link the crimson-pink of the bill to these same colours in the eye and its encircling eye-ring.
Throat clear white
Rump and tail coverts crimson
Black breast-band extends along flanks displaying neat, sharply defined diamond spots of white.
White, mid breast to under tail
♂
Lores may be slightly darker than grey of head.
Bill mostly black; eye-ring and iris dark
Breast-band indistinct, flanks barred dusky grey-brown
Juv.
The bill of the juvenile is initially black, but soon begins to show red. The eye-ring encircling the dark eye later turns deep pink.
The adult female is almost identical to the male: the breast-band is slightly narrower, the lores are a less intense, slightly paler, brownish black.
Upper parts dark olive brown, at close range seen to be very finely and densely barred
Rump, upper-tail coverts and bases of darkly barred tail feathers are all glossy scarlet.
Black of lores also encircles the eyes and links across the base of the bill.
Underparts closely barred olive and black; boldest on the flanks
Blackish on central belly (hidden)
Adult
Crimson, like adult
Bill is initially black, begins to turn red at a very early age; so that some birds in this plumage are likely to have red on the bill. Lacks black on lores and surrounds to eyes.
Juvenile is even more uniformly dusky olive brown than adult, with finer barring that, at a distance, may appear to be an almost uniform olive grey-brown.
Immature
Ear patch scarlet, becoming deeper on breeding males
Bill scarlet
Olive grey to olive brown
Black of lores extends back around the eyes to meet the scarlet ear patch.
Rump, upper-tail coverts and base of tail feathers scarlet
Breast to under tail is black, heavily spotted white.
Adult
When flushed from low ground-cover, flying up and away, displays briefly the 'fire-tail' crimson.
Flight feathers grey-brown with widely spaced dusky brown bars
Adult in flight
Bill more slender
In its native haunts of rugged desert ranges and gorges—with sunlit colours of rust-red rock and brown clay, white flakes of quartz, black shadows—the Painted Finch, with intense, bold colours of plumage, becomes a part of the landscape.
Crimson face with splashes of colour down over velvety black of breast
Black underparts, crimson streak, heavily spotted flanks
♂
Upper parts, including wings, are cinnamon-brown.
Bill is red with black base to upper mandible; longer than that of other finches, it tapers to a finer point.
Less red on face, than male; none on throat
Juvenile is similar to adult female, but dull; red only on rump area.
♀
The Star Finch has two subspecies: nominate race *ruficauda* (E and SE of the Gulf of Carpentaria) and race *clarescens* (W of the Gulf through the NT and Kimberley to the vicinity of Shark Bay, WA).
Olive green to olive grey
Crown to throat and bill, light scarlet
Upper-tail coverts dull crimson with large white spots
Lemon, spotted white
Dull crimson on central tail feathers
Race *clarescens*
♂
Race *ruficauda*, paler under-parts
♂
Grey-brown, no red; eye dark
Soon turns red.
Pale, dull olive buff
Pale buff
Juv.
Female is similar to adult male, but red is confined close to forehead and eyes.
Possibly three races: *temporalis*, *minor* and *loftyi*
Rump and upper-tail coverts are scarlet, tail is grey-brown.
Olive-yellow
Brow long, red, wide at nape
Head and underparts grey
♂
Race *temporalis*
Red eyebrow line of the female tapers to a long, fine point at the nape, unlike the male's broad, square-cut brow.
Grey
Olive-yellow
Underparts white
♂
Race *minor*
Head, neck, underparts ashy brown
Juv.

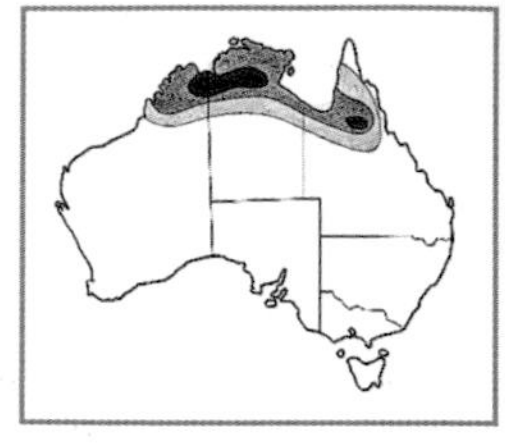

Gouldian Finch *Erythrura gouldiae* 12–14 cm
A well-known, brightly coloured finch, popular in aviculture, but endangered in the wild. Like many finches of open environs, this species is highly social; usually in flocks or small parties feeding in tall grass, rarely on the ground. Clings to strong stems of tropical grasses that may be 2 m tall, raiding seed heads. Insects, especially flying termites, are caught in flight. If flushed from their feeding in grass, the birds of a flock will fly together, their bright blue rumps conspicuous, to tops of nearby trees. **Variation:** there are no races, but rather three colour variations or morphs that occur throughout the range of the species. This variation is confined to a solid patch of colour that, like a mask, covers forehead, crown and face. While this does not cover the entire head, the colour forms are described as 'red-headed', 'golden-headed', 'black-headed'. The chin is always black, continuing as a black edge to red or yellow face mask; on the black morph it is an indistinguishable part of the mask. About 75% of individuals are black-headed, 24% red headed, and 0.1% to 0.5% have the golden yellow head. **Voice:** as contact call, a soft 'tsit' or 'ssit', sometimes a longer 'streee'. As alarm call, sharper, louder 'stret, stret-stret'. **Hab.:** open tropical woodlands with scattered trees and tall native grasses; spinifex with scattered shrubs. Never far from water, and often in vegetation along watercourses or edges of mangroves. When breeding, uses trees on low, stony ridges. **Status:** remains moderately common in parts of the Kimberley region of WA and Arnhem Land, but where formerly flocks of thousands came to waterholes, they now arrive in tens or hundreds. The Gouldian now occupies not much more than half its former range, a decline that may have been caused by trapping, frequent fires and disease. **Br.:** 415

Eurasian Tree Sparrow *Passer montanus* 13–15 cm (C)
The origins of the Australian population of the Tree Sparrow are uncertain. There may have been importations from both Europe and Japan. Often seen in company with the larger, bolder House Sparrow; both species forage in vicinity of human habitation. Occurs in small flocks even in the breeding season—nesting may be loosely colonial; at other times in larger flocks. **Juv.:** smaller, paler, markings indistinct. **Voice:** hard, metallic 'tchik', 'tchit-tchup' and high pitched chittering; in flight a softer 'tek'. **Similar:** House Sparrow. **Hab.:** cities, towns, suburbs, farms, keeping mostly to the vicinity of buildings and surrounding agricultural lands. **Status:** introduced species; generally sparse, but abundant in some localities.

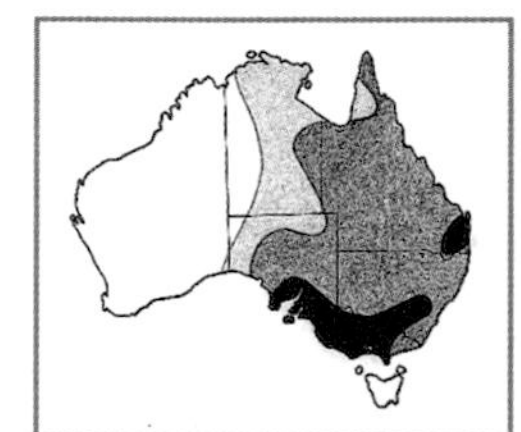

House Sparrow *Passer domesticus* 14–16 cm (N,T)
Introduced in the 1860s; now an abundant pest. Highly sociable and gregarious; usually in small colonies, but gathers into large flocks, at times of many thousands, after the breeding season. Feeds on grains, fruit and insects, and scavenges for food scraps. **Voice:** almost constant, monotonous 'chirrup' and 'chissik'; harsh chatterings. In alarm or excitement, a rattling 'kur-r-r-rit' and sharp 'treeee' alarm call. **Similar:** Tree Sparrow. **Hab.:** cities, country towns, farmlands and around farm buildings; needs holes in buildings or trees as nest sites. **Status:** introduced, widespread in mainland and Tas.; displaces native birds from nest sites.

European Goldfinch *Carduelis carduelis* 12–14 cm (M,N)
Widely distributed: Europe, central Asia, N Africa; introduced to Aust. in 1860s. In pairs, family groups through spring and summer; gathers in large flocks in autumn and winter. Feeds on seeds, climbing over plants, mostly introduced weeds, to raid the seed heads. If disturbed, flock flies up to perch on exposed top twigs of a tree. **Voice:** various rapid, tinkling, canary-like trills and chirpings, 'tswit-tsiewt-tswit-swit- tzwee-tzwee'; in flight a shrill 'tsieew, tseew'; tinkling sounds from feeding flocks. Also harsh, rasping, scolding sounds. **Hab.:** around towns and settlements, farms, roadsides and wastelands, usually where crops or weeds have replaced native vegetation. **Status:** locally nomadic; introduced species; common.

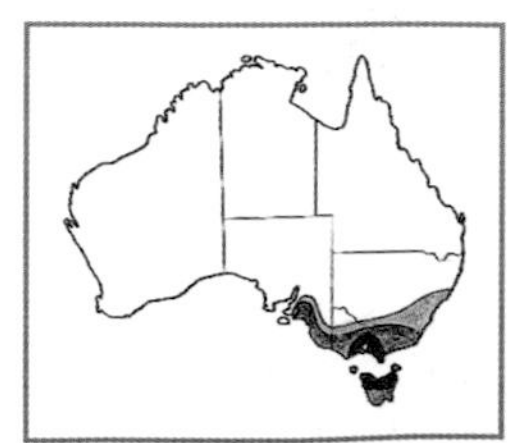

European Greenfinch *Carduelis chloris* 14–16 cm (N)
Introduced in 1880s. In small flocks, foraging for seeds on the ground; flight undulating. After the breeding season, gathers into much larger flocks. **Voice:** a rather harsh, chittering sound, 'chwitit', or run together as 'chwitchit-chit-chit-chit', or more abrupt and emphatic in alarm, 'zweet!'. In flight gives a nasal, buzzing 'tzwee-tzwee-tzwee'. **Hab.:** woodland in urban areas, parks, gardens, roadsides, farmlands and other areas not under native vegetation. **Status:** introduced species; sedentary, patchy, but tends to be most common near coast.

The Gouldian has no subspecies, but colour morphs that occur throughout the geographical range of the species. The large mask that covers forehead, crown and face may be red, black or yellow. Only this mask varies; other areas of the plumage remain unchanged among morphs.
Bright, light, grass green
Red morph: crown and face mask are scarlet.
Flanks and belly are deep golden yellow on all morphs.
Golden-headed morph is very rare; scattered through population.
Tail graduated, central pair have long, fine, black points.
♂
Colours slightly less intense
♀
Golden Headed morph: red and golden forms have a black border to the mask.
Turquoise border
♂
Black-headed morph: crown and face black. This is the most common variant.
Breast paler lilac
Under-tail coverts white, all races
Darker
Tail not as smoothly tapered if expanded in flight
Dull olive grey
Chin paler
Breast deep purple, males of all morphs
Female: shorter central tail feathers
Rump and upper-tail coverts cobalt blue to turquoise
Underparts dull pinkish buff to pale buff
Central tail feathers are not as extremely elongated on female, shorter again on the juvenile.
Under-tail coverts off-white
Juvenile, all morphs
Crown and nape chestnut; dark patch on white cheek
Streaked brown and black
Rump brownish yellow
Tail slightly forked
Adult
Small black throat patch; bill small, black
Two thin, weak wing-bars
Cheeks dull white without any dark mark in centre
Crown grey; nape chestnut
Head dull sandy brown, weak, pale brow line
Throat and breast black
Pale buff-brown
The crown, grey (male) or grey-brown (female), of the House Sparrow distinguishes it from the rich chestnut crown of both sexes of Tree Sparrow. The black throat of this species extends also to the breast, it has a single, very broad white wing-bar and an ear patch that is large, round, entirely white, not dark-centred.
Rump grey
Dark streaks
One wide, bold white wing-bar
Grey
Rather scruffy overall
♂
♀
Tail-tip notched
Tail quite long, rather thin, grey-brown
Juvenile male, in transition to adult plumage
Forehead, face and chin, deep scarlet broken by black lores
Wings like those of adults, but with feathers tipped buff rather than white
Wing black with wide, deep golden band
From side of head to throat white
Head and breast pale grey-brown
Plain off-white
Juv.
Upper-tail coverts white
Adult
Tail forked, black, tipped white
Outer primary entirely black
Predominantly green, light citrine to darker dull olive green
Primaries broadly edged bright yellow
Brownish buff on head and back
Wings like those of adults
Secondaries and tertials edged light to mid grey
Four subspecies of Greenfinch: Australian birds are of the nominate race, *chloris*, from Europe.
Female is similar to male in plumage; has narrower golden margins to outer flight feathers and less yellow on the outer tail feathers.
Central tail feathers are entirely black, outer feathers yellow on the basal half and black tipped.
♂
Underparts light brownish buff
Tail forked
Juv.

Yellow-bellied Sunbird *Nectarinia jugularis* 11–12 cm (T)
More than 20 subspecies across SE Asia and China. This species, with just one race, *frenata*, is the only Australian representative of the sunbird family, Nectariniidae, and an outlier of the NG population, where it is known as the 'Olive-backed Sunbird'. In NE Qld the Yellow-bellied Sunbird is common, in places abundant, and well known, not only as a bushland species but also in suburban gardens. This bird is conspicuous not only for its bright colours, but also for its bold raiding of nectar from garden flowers, the pugnacious defence of its territory against other sunbirds, and its use of verandas, from which it suspends nests. Often hovers while feeding from flowers and while taking spiders and insects. **Voice:** rather weak, squeaky, scratchy, metallic, rising 'tssee-ik', 'twieeik', stronger, harsher, aggressive 'tzzzeeik' and lively chattering. **Hab.:** rainforest, including its margins, clearings and regrowth; nearby plantations, gardens and lush watercourse vegetation; mangroves and coastal scrub. **Status:** sedentary; common. **Br.:** 416

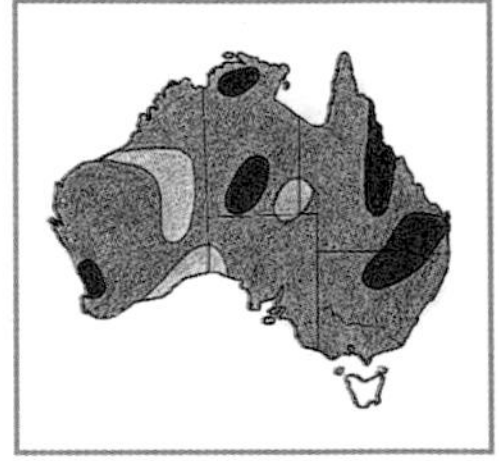

Mistletoebird *Dicaeum hirundinaceum* 10–11 cm
Usually alone or in pairs in foliage of tree or mistletoe clump; often high where it might not be noticed but for the distinctive calls. Flight is very fast, direct, with lively, sharp flight calls. Feeds on the berries of mistletoes; digests the soft, sweet under-skin layer, but not the large seed. When perched, twists one side to another, brushing the rear of its body across the twig or branch, ensuring that some of these seeds, still very sticky, are dropped or wiped onto the perch, so potentially starting a new mistletoe plant. The Mistletoebird is nomadic in most parts of its range; it follows the fruiting of various species of mistletoe. This may show as an increase or reduction in the local population. **Voice:** loud and spirited, carrying. The call is a sharp, squeaky 'tiech, tieech, ti-witch, tee-wietch-tieewietch-teewietch' and loud, sharp, rollicking 'kinzee-kinzee, perwita-perweeta-perweeta'. **Hab.:** wherever mistletoe occurs: dense wet forest of coastal ranges, mangroves, woodland, mallee, mulga and semi-arid spinifex country. **Status:** nomadic; common. **Br.:** 416

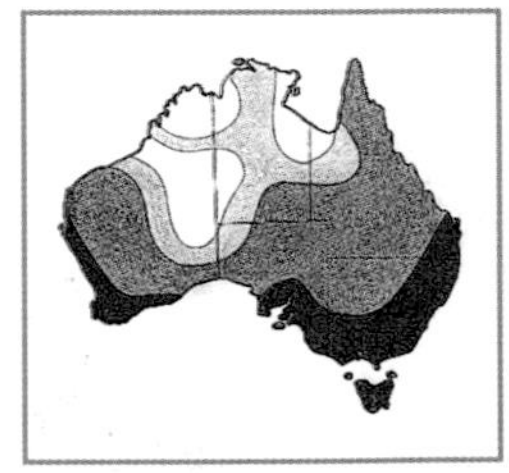

Welcome Swallow *Hirundo neoxena* 14–15 cm (L, M, N, T)
The slender, streamlined bodies and long, tapered wings of swallows and martins, family Hirundinidae, are easily recognised. Flight is fast, buoyant and acrobatic; they turn, bank, wheel and swoop to snap up flying insects with wide, bristle fringed bills. Dip low over pools of rivers or dams; momentarily touch the water. Often seen perched in rows along overhead wires, occasionally on the ground. The Welcome Swallow has adapted to the artificial environment of buildings, bridges and other structures that provide secure support for the mud nests. This has, in turn, made these birds very familiar part on farms, the verandas of country towns and homesteads. Pairs tend to return to the same nest site year after year. **Variation:** a gradual or clinal increase in size westwards: those from WA are slightly larger, possibly justifying a separate race, *carteri*, when compared with the smaller *neoxena* of E Aust. Those in Qld with less white in the tail have been described as a distinct race, *parsonsi*. **Voice:** as contact call in flight, an occasional single, squeaky, harsh 'tzeck' or 'tchek'. The song, usually from a perch, is a lively succession of cheery chattering and a twittering mixture of high squeaky and slightly harsh sounds. In alarm, a sharply whistled 'tseip-tseeeip'. **Similar:** Barn Swallow, but has black breast-band, longer tail streamers, white wing linings. **Hab.:** almost universal occurrence; tends to avoid very heavy forests and the most arid of desert country. **Status:** usually sedentary, nomadic or at times migratory; common. **Br.:** 416

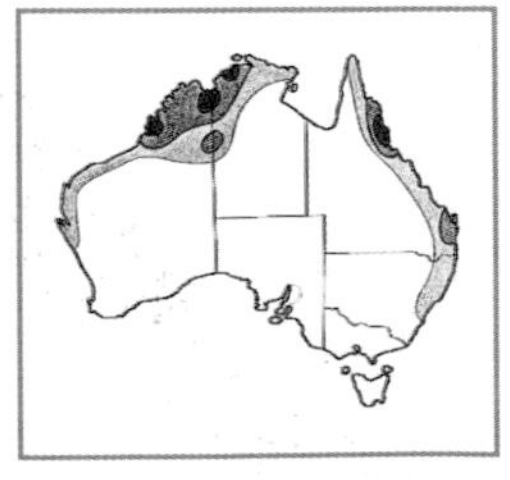

Barn Swallow *Hirundo rustica* 14–17 cm (C, K, T)
Similar to the Australian Welcome Swallow. Flight is swift; it chases insects with banking and turning, often swooping close to the ground, or skimming the surface of water. A very widespread swallow; breeds in Europe, Asia and North America. Those reaching Aust. appear to belong to the Asian race *gutturalis* that breeds from Japan through Korea to NE Burma. In the northern winter the Barn Swallow's migration route takes it southwards through India, the Malay Peninsula, Philippines, Celebes and Moluccas to Aru and western NG. In Aust., most arrivals tend to be in NE Qld around Innisfail and in the the far NW along the coast from Darwin to Broome. Before it used man-made structures, breeding Barn Swallows must have been restricted to cliffs or caves where their mud nests could be attached. With the spread of barns, houses, bridges and other artificial sites, this species has been able to extend its range where nest sites are provided. **Voice:** weak, twittering 'tsi-tswit-tsee-' and soft warblings. **Hab.:** open country, towns, often near water; tends to be noticed when perched on overhead wires in line with other swallows. **Status:** vagrant, but seen regularly in far N and NE.

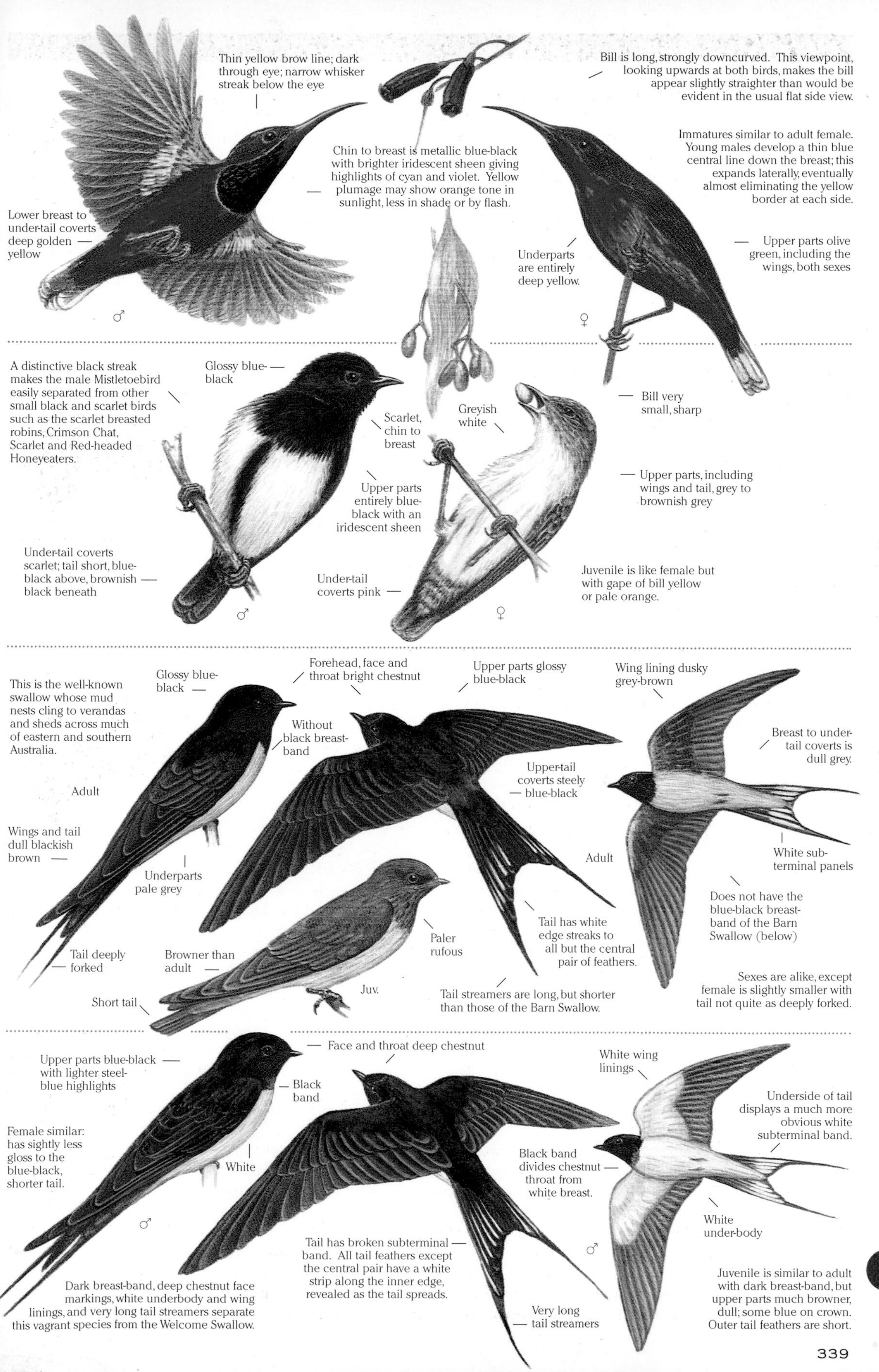
Thin yellow brow line; dark through eye; narrow whisker streak below the eye
Bill is long, strongly downcurved. This viewpoint, looking upwards at both birds, makes the bill appear slightly straighter than would be evident in the usual flat side view.
Immatures similar to adult female. Young males develop a thin blue central line down the breast; this expands laterally, eventually almost eliminating the yellow border at each side.
Chin to breast is metallic blue-black with brighter iridescent sheen giving highlights of cyan and violet. Yellow plumage may show orange tone in sunlight, less in shade or by flash.
Lower breast to under-tail coverts deep golden yellow
Underparts are entirely deep yellow.
Upper parts olive green, including the wings, both sexes
♂
♀
A distinctive black streak makes the male Mistletoebird easily separated from other small black and scarlet birds such as the scarlet breasted robins, Crimson Chat, Scarlet and Red-headed Honeyeaters.
Glossy blue-black
Scarlet, chin to breast
Greyish white
Bill very small, sharp
Upper parts entirely blue-black with an iridescent sheen
Upper parts, including wings and tail, grey to brownish grey
Under-tail coverts scarlet; tail short, blue-black above, brownish black beneath
Under-tail coverts pink
Juvenile is like female but with gape of bill yellow or pale orange.
♂
♀
This is the well-known swallow whose mud nests cling to verandas and sheds across much of eastern and southern Australia.
Glossy blue-black
Forehead, face and throat bright chestnut
Upper parts glossy blue-black
Wing lining dusky grey-brown
Without black breast-band
Breast to under-tail coverts is dull grey.
Upper-tail coverts steely blue-black
Adult
Wings and tail dull blackish brown
Underparts pale grey
Adult
White sub-terminal panels
Does not have the blue-black breast-band of the Barn Swallow (below)
Tail has white edge streaks to all but the central pair of feathers.
Paler rufous
Tail deeply forked
Browner than adult
Juv.
Short tail
Tail streamers are long, but shorter than those of the Barn Swallow.
Sexes are alike, except female is slightly smaller with tail not quite as deeply forked.
Face and throat deep chestnut
Upper parts blue-black with lighter steel-blue highlights
Black band
White wing linings
Underside of tail displays a much more obvious white subterminal band.
Female similar: has slightly less gloss to the blue-black, shorter tail.
White
Black band divides chestnut throat from white breast.
White under-body
♂
Tail has broken subterminal band. All tail feathers except the central pair have a white strip along the inner edge, revealed as the tail spreads.
♂
Juvenile is similar to adult with dark breast-band, but upper parts much browner, dull; some blue on crown. Outer tail feathers are short.
Dark breast-band, deep chestnut face markings, white underbody and wing linings, and very long tail streamers separate this vagrant species from the Welcome Swallow.
Very long tail streamers

Tree Martin *Hirundo nigricans* 12–13 cm (T)

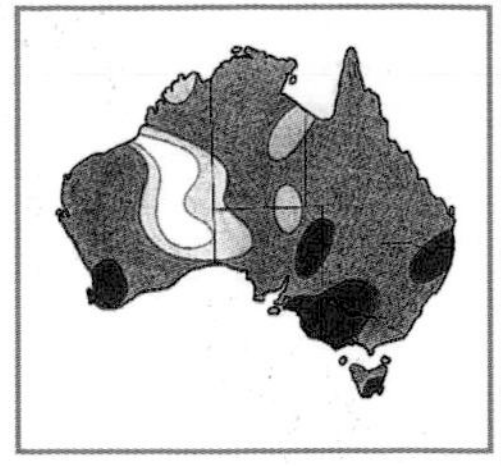

The Tree Martin's shallow, forked tail is shorter than its folded wings; it tends to look small, compact or stumpy when perched. It hunts insects in swift, twisting, turning flight around the canopy of trees, low over water of lakes and rivers, and above farm pastures. Often found in small colonies where there are large trees that have many small hollows in the upper limbs. **Variation:** has been divided into eastern race *nigricans*, western *neglecta*; alternative is that race *neglecta* is confined to mainland Aust.; the slightly larger, more rufous *nigricans* to Tas. through spring–summer breeding; after which, autumn and winter, it migrates N across the eastern mainland, some reaching NG. At that time, both races occur through eastern Aust. **Voice:** a slightly metallic 'tzweit' and musical, scratchy, squeaky chatter, 'chwip-chip-chzeit-chwip'. **Hab.:** open woodland and farmland with trees, not far from water. **Status:** nomadic or migratory over most of its range; common. **Br.:** 416

White-backed Swallow *Cheramoeca leucosternus* 14–15 cm

A graceful swallow with clearly defined black and white plumage. Usually in small flocks; tends to forage quite high; flight erratic, often swooping, fluttering with rapid wingbeats. At times skims across still water, bill touching the surface to drink. White-backed Swallows do not use man-made structures to support their nests and seem shy compared with the Welcome Swallow; they have not benefited from the spread of buildings. They are distinctive in excavating nest tunnels in sandy, loamy soil rather than building mud nests. Later, old nests are often used as overnight roosting places where many swallows gathering snugly for the night. **Voice:** a sharp, scratchy 'skiep, skiep, skiep'; from a flock, the very many calls combine as an almost continuous rattling chatter with varied pitch and strength; the overall effect is pleasant, cheerful, musical. **Hab.:** open country, woodland, semi-arid scrubland, heathland. Mainly inland, but coastal in drier regions, and, when breeding, sites with firm but soft, sandy, loamy earth into which nest tunnels can be drilled. **Status:** mostly sedentary with some localised nomadic wanderings or migratory movements. **Br.:** 417

Red-rumped Swallow *Hirundo daurica* 16–18 cm (C,T)

A medium to quite large swallow; migrates as far S as Borneo and NG, occasionally Aust. **Variation:** about 12 races, breeding eastern Asia to Spain; migrates into SE Asia, India, central Africa. Race *rufula* breeds in Europe; has a broad, rufous hind-neck collar. Races *daurica* from China and *japonica* from Japan have the rufous collar broken by blue where the nape joins the mantle; underbody colour also varies. The similar Striated Swallow, *H. striolata*, is sometimes included as a race of the Red-rumped. The Striated Swallow extends into the Indonesian chain of islands close to NW Aust., but these island races tend to be more sedentary. **Voice:** contact call, commonly given in flight, an abrupt, slightly nasal 'tweeit'; also a twittering, warbling song. **Hab.:** open country, coastal grasslands. **Status:** rare vagrant.

Fairy Martin *Hirundo ariel* 12–13 cm (L)

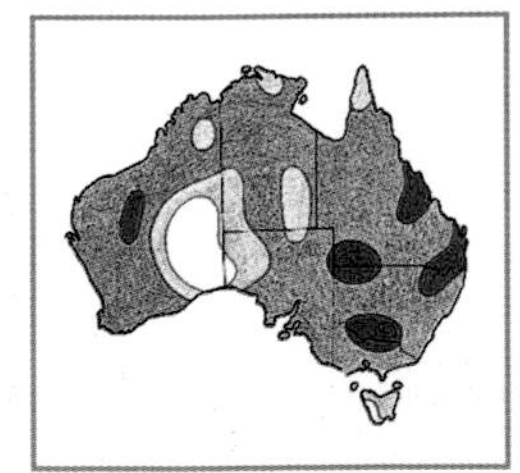

Perhaps best known as the builder of bottle-shaped nests that cluster densely on ceilings of shallow caves, under boulders and bridges. Nesting is colonial—from a few, to close-packed aggregations of hundreds. Small or large flocks hunt for airborne insects, high rather than close to ground. In breeding months, noisy flocks mass around the larger colonies. At other sites where there are only a few active nests, these birds tend to be inconspicuous. Fluttering, rather slow flight. **Juv.:** dull plumage, wing feathers tipped buff, crown pale. **Voice:** contact call, abrupt 'drrt, drrit'; also a feeble, twittering song. **Hab.:** open environs, often near water, cliffs, caves. **Status:** migratory moving N for winter, some to NG; uncommon. **Br.:** 416

Pacific Swallow *Hirundo tahitica* 13–14 cm

Widespread through islands of the South Pacific, W to Indonesia and NG. Banks, turns and glides in pairs to large flocks. **Variation:** there are about 8 subspecies of the Pacific Swallow; show a gradual, clinal lightening of colour westwards. The eastern birds of the Pacific islands have dark underparts, no white in tail; those of NG and westwards have pale underparts and white spots on inner webs of tail feathers that form a white subterminal bar. Within Aust. it is replaced by the similar Welcome Swallow. **Voice:** usual calls are a 'twzit-twzit' and longer 'tzitswee'; these calls may be run together in a twittering song. **Hab.:** usually near coasts, their cliffs and caves providing nest sites. **Status:** extremely rare vagrant.

Crown, nape and back metallic blue-black
Rump dull white with dark shaft streaks, upper-tail coverts slightly more heavily mottled
Adult
Tail shorter than wings; shallow fork at tip
Forehead chestnut
Off-white with fine dark streaks
Wings and tail dull blackish brown
Dull white
Juv.
Juvenile has grey-brown upper parts, less chestnut on forehead.
Underparts, including throat, off-white, streaked darker
Wing linings like flanks: off-white or with a slight rufous tint
Tail relatively short. When folded it has a shallow fork at the tip, but is likely to appear square tipped when the tail is partly or widely spread in flight.
Faint rufous tint along the flanks and on the under-tail coverts
Grey-brown flight feathers not in strong contrast to wing linings, unlike White-backed Swallow and Barn Swallow
Crown white with centre mottled or scaled brown
Chin to upper breast white
Clear, crisp, white mantle and upper back separates this species from Fairy and Tree Martins, which have dull white low on rump.
Lower back and rump black with a slight blue sheen
Long outer feathers of tail extend well beyond wingtips.
Adult
White back is conspicuous as the swallow turns and banks.
Buffy white
Dull dusky brown without blue sheen
Tail lacks streamers.
Juvenile
Black with blue sheen
White wing linings
Tail deeply forked
From beneath, a distinctive plumage pattern with white of wing linings linked by white of upper breast.
Some races have complete rufous collar around hind neck, others have crown and back linked by a narrow strip of blue down the back of the neck.
Whitish, no breast-band
Rump cinnamon to chestnut
Tail long, deeply forked
Upper parts blue-black with metallic blue sheen
Cinnamon-rufous rump is conspicuous as swallow banks and climbs.
Underwing coverts cinnamon to rufous
Underparts off-white to chestnut; plain, lightly or heavily streaked
Black under-tail coverts and under-wing covert colour confirm identity.
Has no white patches in tail.
Head cinnamon-rufous
Fine dark streaks
Rump white, tail quite short, slightly forked when folded
Underparts dull white
Mantle, back and scapulars are glossy blue-black.
Rump is off-white, but in sunlight appears bright white.
Slightly forked, dark
Under-wing coverts buffy white
Flight feathers grey-brown
Extremely rare visitor to N Aust. Subspecies from the Pacific are darker fronted; become increasingly pale westward through NG and Indonesia.
Crown, back of neck, mantle are black with a steely blue sheen.
Rufous-chestnut forehead, face, chin, throat
Does not have any black breast-band; blurred boundary, chestnut to white.
Rump and upper-tail coverts blue-black
Outer feathers of forked tail are approximately equal in length to tips of folded wings.
Dusky grey
In flight, underwing coverts and flight feathers are dusky brown. Upper surfaces of wings and tail are dull brownish black.
Under-tail coverts barred
White sub-terminal bar
Juvenile duller, browner, with less of the metallic blue sheen. Rufous of forehead and throat are less clearly defined; paler. Tertials and under-tail coverts are tipped buffy white.

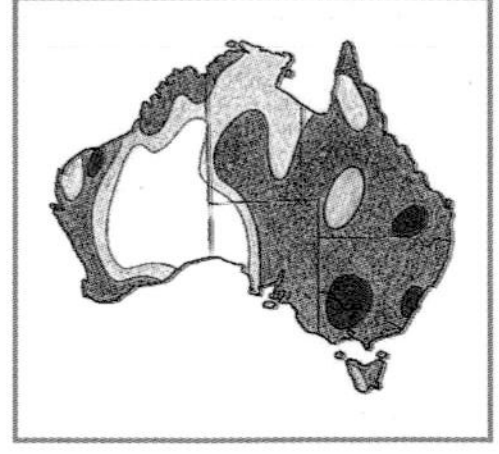

Clamorous Reed-Warbler *Acrocephalus stentoreus* 16–17 cm
The loud, spirited song of the Clamorous Reed-Warbler is likely to be heard from almost any reedbed, large or small, coastal or far inland. Disperses widely in its migration and seems to find most reedbeds, whether swamps, river pools, or gorge waterholes. Although distinctive song reveals its presence, the plain, olive brown bird is not easily observed unless it comes to the outside edge of the reeds as it climbs among their stems, or makes brief low flights across open water. **Voice:** the full song, given for long periods through the breeding season, is rich and powerful, often a metallic quality: 'cheewip-cheewip', 'quitt-quitt-quitt', 'kwitchy-kwitchy', 'kwarty-kwarty-warty' often flowing together in varied combinations, as long powerful passages of song—like a more powerful, richer version of Brown Honeyeater song. In alarm, abrupt 'tchuk, tchuk', 'kretch-kretch', scolding 'squarrk'. **Similar:** Oriental Reed-Warbler, but larger; rare vagrant. **Hab.:** wetlands; includes reedbeds of lakes, cumbungi, tall crops beside water, bamboo thickets, lantana beside water. **Status:** migratory; moves S into the SE and SW of its range to breed in summer; some remain through winter, but are quiet and often go unnoticed. **Br.:** 417

Oriental Reed-Warbler *Acrocephalus orientalis* 18–20 cm
Similar to Clamorous Reed-Warbler; identified within Aust. on rare occasions. Breeds Japan, China; winters S to Indonesia, NG. **Imm.:** like non-breeding adults. **Nonbr.:** compared with breeding plumage, may be slightly more rufous on upper parts, slightly stronger ochre tint to underparts. **Voice:** unlike the rich, musical song of the Clamorous Reed-Warbler, but instead gives loud metallic noises at lower pitch, commonly a grating 'krrek'; the song is a harsh, churring, chattering 'kratch-kraich, krok-krok-quorrark'. The voice, if the bird calls, will greatly assist in identification, but calls may be few in the nonbreeding season (winter in the northern hemisphere) **Hab.:** reeds, cane grass, cumbungi, sedges, waterside scrub and mangroves. **Status:** rare vagrant or migrant to coastal NW, N and NE Aust.

Arctic Warbler *Phylloscopus borealis* 11–12 cm
A small leaf-warbler that breeds across N Eurasia to NE Siberia and Alaska; winters S and SE Asia as far S as Philippines, Greater Sundas and islands of Indonesia. It occasionally reaches Ashmore Reef, and has been recorded from Scott Reef, Kimberley and Broome. Actively forages among foliage of trees and shrubs. **Voice:** includes a distinctive, harsh, scolding 'tzrick, tzrick' and more husky 'tszzic'; also has a reeling, buzzing trill. Begins calling from about March, before returning to breeding grounds. **Hab.:** open wooded country, forest edges, mangroves. **Status:** rare vagrant to W Kimberley and offshore reef islets.

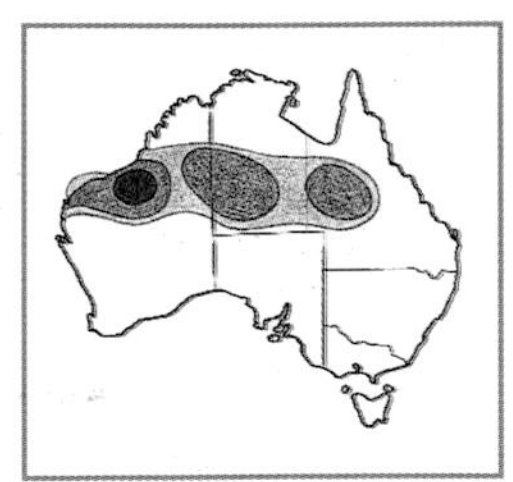

Spinifexbird *Eremiornis carteri* 14–16 cm
A long-tailed brownish bird concealed in large clumps of needle leaved spinifex, the sharp 'tjik' alarm calls likely to be heard before the bird is sighted. May then be seen moving about in the spinifex, or glimpsed fluttering away, low above the intervening patches of bare red earth, long, heavy tail trailing behind. Often one will climb a tall stem for a brief look at an intruder before scuttling down to vanish again in the spinifex. **Voice:** in alarm, 'tjik' and 'tjuk', like rocks being struck together. Frequently a high, quick 'cheeryit' or 'cheery-wheit', or slower downward 'cheer-y-a-roo'. **Hab.:** spinifex: favours areas where the clumps are large such as along watercourses, lower slopes of ranges; often intermixed with shrubs and small trees. Appears most common in rough, arid ranges. **Status:** sedentary; locally common. **Br.:** 417

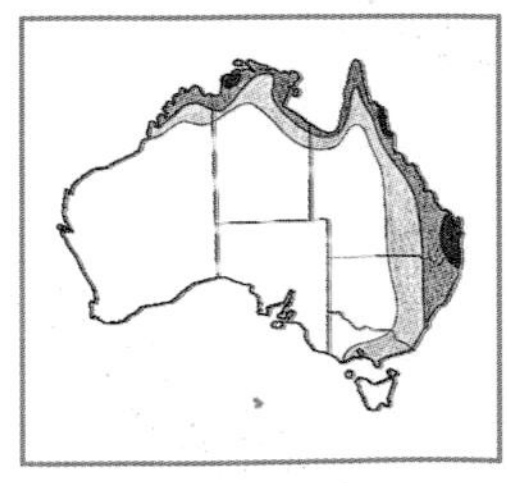

Tawny Grassbird *Megalurus timoriensis* 17–19 cm
A tawny rufous and brown bird of lush habitats; in pairs, parties, loose colonies. In breeding season, male is conspicuous and bold in display flights—flutters, almost hovering, tail down and spread while singing loudly. **Imm.:** slightly duller than adults and has finer streaking. **Variation:** one Australian race, *alisteri*, others in NG, and SE Asia. **Voice:** as alarm or warning, a single, sharp 'tjik'. Song is a loud, varied, initially high, squeaky sequence of descending, reeling notes, finishing with harsh, deep chuckling notes given in song flight or from a high perch; has some similarity to song of the Rufous Songlark. Calls mostly in early morning and towards evening. **Similar:** Little Grassbird, but smaller, has black-streaked crown and streaked underparts. **Hab.:** bullrushes and adjoining lush, wet grass—cane grass, cumbungi swamps, tall moist crops including sugar cane and corn. **Status:** nomadic and migratory; locally common but generally uncommon. **Br.:** 417

Crosses open water
between reedbeds
in low, swift, slightly
undulating flight.
When singing, the Reed-Warbler
lifts the feathers of the crown to
make a slight rear peak.
Slightly darker through
the eye, pale buff brow line.
Face, throat and most of
underparts are buffy white.
Underparts off-white
with buff or fawn tint
Tail quite
long; rounded tip
A very plain,
unstreaked, long tailed
inhabitant of dense,
wet reedbeds. Sexes
alike; juvenile has
slightly darker lores.
Clings to reed stems, often moving
down to water level, pecking at
tiny insects around the waterline,
or foraging over fallen, floating
debris and water plants.
Upper parts overall
olive brown, slightly
darker on the wings;
rump may be more
tawny or rufous.
Shorter outermost
primary gives wing a
more rounded shape
than wing of Oriental.
Primaries extend
further beyond tertials
than in the wing of the
Clamorous Reed-Warbler.
Primaries
Tertials
Upper parts are slightly
paler, less rufous, than
Clamorous Reed Warbler.
Flanks rufous
Bill relatively
heavier, shorter
In flight, wingtip is slightly less rounded, more
pointed, than is that of the Clamorous Reed-Warbler,
the outermost, primary being longer than the 6th
(numbered from innermost primary); this difference
would usually be evident only with the bird in hand.
Larger than Clamorous Reed-
Warbler and with proportionately
longer, more pointed wings
Paler; very fine
streaks beneath
Inside mouth is orange,
where Clamorous is yellow.
Long pale brow,
back to nape
Upper parts are soft,
mid olive grey.
Small white tips to greater coverts form a thin
pale wing-bar; a similar, fainter wing-bar may be
visible further forward on the median coverts.
Bill slender and dark with pale
orange-tan along the lower mandible
Thin dark line through eye joins
streaked grey ear coverts.
Underparts white to off-white
with faint dark streaks on breast;
flanks have slight buff-brown tint.
Legs pale, pinkish yellow
Tail is carried
lifted or lowered.
Upper parts unstreaked,
tawny brown to
grey-brown
Forehead, crown and
nape rufous
Fawny
white
Long tail coverts extend
unusually far down
the long tail.
Brow line off-
white, ear coverts
grey-brown
Long, broad, wedge-
tipped, dark
brown tail
Flutters low through
spinifex with drooped,
heavy looking tail.
Rufous
Crown rufous
In contrast to the
plain rufous tawny
crown and nape of
this species, the Little
Grassbird has crown
and nape streaked
black.
The Tawny Grassbird has upper
parts similar to the Little
Grassbird: tawny rufous
heavily streaked black
on back and wings.
Long coverts
Flight feathers,
especially the
tertials, are dark brown
edged rufous.
Underparts off-white
or fawny white; lack
the fine black streaks
that distinguish the
Little Grassbird.
Long, brown,
tapered tail
Long, graduated tail. Feathers are
dark brown edged cinnamon.

Little Grassbird *Megalurus gramineus* 13–15 cm

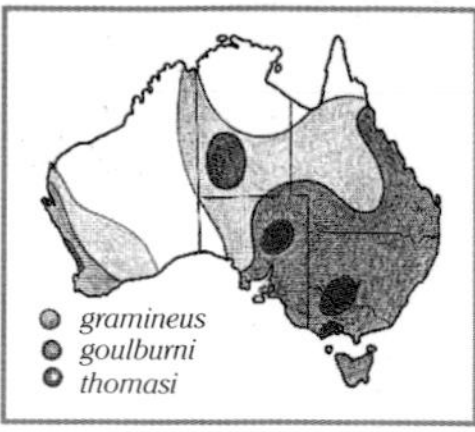

Secretive, skulking inhabitant of dense, wet vegetation; allows only brief glimpses; flutters away weakly through dense vegetation, tail trailing, often spread. **Imm.:** similar to adult; slightly greyer overall, more finely streaked, some almost plain on underparts. **Voice:** the call is often the only indication of the presence of the secretive grassbirds in their typically dense habitat; a mournful whistle of three notes, the first soft, low and brief, the second and third at high and even pitch, drawn out, 'whp-wheeee-wheeeee' and 'whp-whiooo'. In alarm, a harsh, rattling chatter. **Hab.:** wetlands with rushes, lignum swamps, cumbungi, cane grass, tidal marshes, mangroves. **Status:** sedentary; locally common, dispersive. **Br.:** 417

Golden-headed Cisticola *Cisticola exilis* 9–11 cm (T)

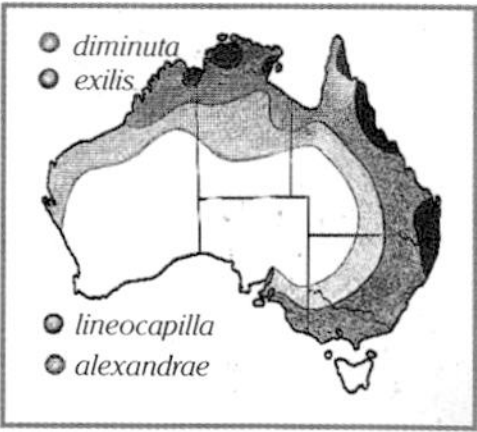

Breeding males attract attention with song flights and by calls from tall grass stems and fences; then drops suddenly to cover. In flight, the male's golden plumage in sunlight against background greenery is quite spectacular. It is known also as 'tailor-bird' for its stitched-leaf nests. **Variation:** about 9 races; 3 or 4 in Aust.; differences slight. **Voice:** spring and summer, calls almost incessantly. Drawn out, metallic, buzzing 'trzzzzeep' sounds are interspersed with quick, clear, musical 'teewip' calls; also has higher, faster, buzzing 'tizzzeip' and 'weezzz, whit-whit'. **Hab.:** wetlands, usually in lowland swamps, wet grass and rank herbage edges of swamps, rivers and irrigated pastures, samphire. **Status:** sedentary; common. **Br.:** 418

Zitting Cisticola *Cisticola juncidis* 9–11 cm

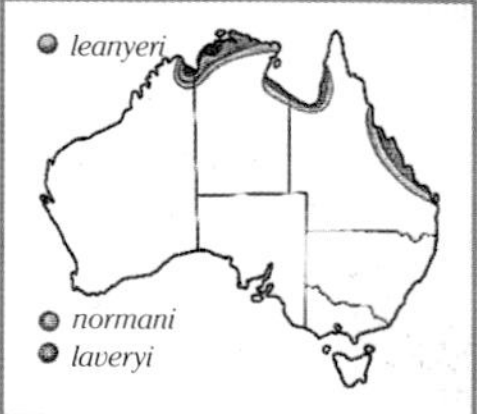

Behaviour is like that of the Golden-headed Cisticola. **Variation:** 3 races in Aust.: *leanyeri* of NT is, slightly, the brightest and most heavily striped; race *laveryi* of NE Qld has brighter cinnamon on tail edges; race *normani*, Gulf of Carpentaria, is palest. **Voice:** when breeding, the male has a persistent zitting call. Flying high in display, he almost hovers in slow, undulating flight into the wind, giving a 'tzip-zip, tzip-zip' call. **Hab.:** grassy swamps, seasonally damp or shallowly flooded, usually coastal plains or saltpan country dominated by sand-couch; margins of mangrove swamps. **Status:** migratory; uncommon. **Br.:** 418

Rufous Songlark *Cincloramphus mathewsi* 16–19 cm

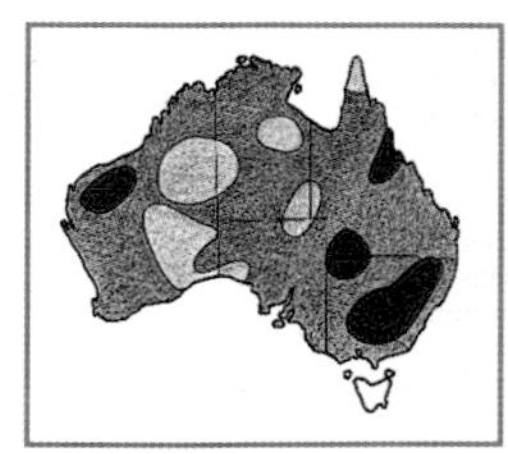

Attracts attention during the breeding months when the male calls almost continuously to attract females to his territory. His presence is proclaimed and advertised repeatedly in musical song from perches and in display flights in which he flies slowly, horizontally, between trees, pouring forth a succession of reeling, scratchy, musical notes. After mating, the female undertakes all nest building, incubation and care of the young. At other times of the year may form small parties, but quiet and inconspicuous. **Voice:** males are noisy with almost continuous song in the breeding season: a single ringing, almost whipcrack 'whitcher'; or joyful full song—a metallic, yet pleasant and musical, rollicking 'whitcher-whitcher, a-whitchy-wheedle-whitch'. **Similar:** Australian (Richard's) Pipit, Brown Songlark. **Hab.:** open grassland, grassy open woodland and mulga, farmland. **Status:** migratory: moves S to breed over summer; returns N in autumn. **Br:** 417

Brown Songlark *Cincloramphus cruralis* 18–19 cm

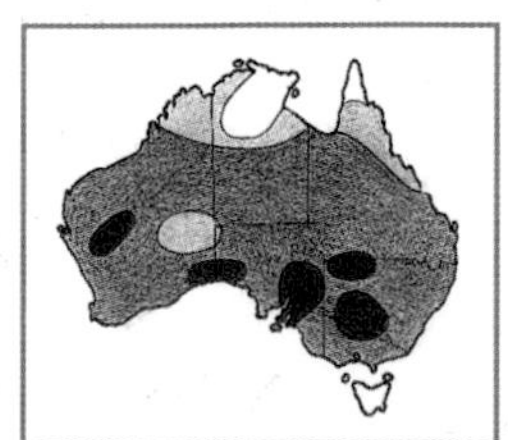

The male is conspicuous in the breeding season, plumaged in dusky cinnamon-brown, delivering his song from an exposed post or limb, or, most vigorously, in display flight. A displaying male rises steeply above his territory, pouring forth his song as the high climb levels out, and continues in full voice as he flutters and undulates downward with wings held upward. On alighting on his perch, he may continue the song. At other times the flight of this species, travelling between one patch of cover and another, is low. **Voice:** loud, clear; the tune or pattern of notes is jerky, erratic; the notes are often abrupt, varying from hard, metallic noises and buzzing sounds to clear and musical notes. Some of the sounds have been appropriately described as like the dry squeaking of a metal cartwheel turning slowly on a rusty axle. The overall effect, however, is cheerful, pleasant and distinctive. While the song is varied, patterns become recognisable. A small typical sample—'skzit-kotch-zzweiler, chweeip, kzzeech-kotch-aweiler'—finishes with a musical trill and whipcrack. This song has resulted in the Brown Songlark being known to children as the 'Skit, scot, a-wheeler'. **Hab.:** open grassland plain, woodland and grassy inland scrub, saltbush plain, paddocks. **Status:** highly nomadic, moving away from drought to regions that have received rain; fluctuates from locally scarce to abundant, but generally quite common. **Br.:** 417

Compared with Tawny Grassbird, upper parts are less rufous; black streaking includes head; underparts are also streaked.
Grey-buff, streaked black
Long whitish brow line
Forehead and crown dull rufous, increasingly grey towards back; streaked black
Underparts off-white with rufous tint towards rear; fine black streaks
Adult
Tail tapers to a point, dark brown.
Rump dull rufous, streaked
Tail dark, edged olive or buff; lacks any barring or white tips; in flight is often spread.
Three Australian races: nominate race gramineus, Tas., slightly darker overall; race goulburni, mainland except SW of WA; there race thomasi is slightly darker above and more heavily streaked.
Adult
Head golden buff, unstreaked, crested while calling
Paler buff, streaked
Crown dark streaked; face plain, deep buff
♂ Breeding
Light, plain golden buff
Heavily streaked
Underparts white, tinted buff, deeper on flanks and under tail
Rump golden buff
♀ and non-breeding male
Flying up, shows much golden cinnamon.
♂ Breeding
Tail dark, buff tipped and edged
Like the Golden-headed but lacks the rich gold on head and rump.
Female and non-breeding are more heavily streaked.
Crown heavily streaked
Streaked crown and nape are lighter in breeding plumage.
Back and wings buff-brown streaked black
Creamy white
Rump tawny buff, mottled brown
Cinnamon flanks
Tail dark brown, pale tips and edges
Inner flight feathers are dark, edged pale buff.
♂ Breeding
♀ and non-breeding male
♂ Breeding
Bill is flesh brown; becomes black on breeding male.
Brow pale, dark line through eye
Female is plumaged like the male, but slightly smaller, lacks dark eye line. Immature is like adults, slightly paler, weaker streaking to the breast.
Back mid brown
An unusual feature of this species is that, in the breeding season, the inside of the male's mouth is black, that of the female is pinkish; may be visible when bird is singing.
Rump and upper-tail coverts rufous
Reticulated pattern across wings with each dark brown feather edged in pale buff
Adult
Underparts greyish buff, slightly more dusky on the breast
Adult
Dark lores, bill paler
♂ Song flight
Nonbreeding male; juvenile male similar
Pale brow
Central strip dark brown
Lores and bill black
Lores, bill paler brown
In fresh plumage, fine white speckling on throat and breast, but it soon wears away.
Breeding plumage, singing posture: tail lifted and often with feathers fluffed out
Rump is dusky brown, unlike Rufous Songlark.
Underparts pale brownish white; belly centre very dark brown
This species is exceptional in that the male is much larger than the female.
♂ Overall sooty brown breeding plumage
Nonbreeding male is plumaged like the female, but retains the black lores and is obviously larger if seen with a female.
Female is smaller, more slender than the male.
♀

Pale White-eye *Zosterops citrinellus* 11–12 cm (T)

Within Aust. the Pale White-eye has been found only on small, well vegetated islands inside the Great Barrier Reef N of about Cooktown, and on small islands of Torres Str. The subspecies of Pale White-eye occurring on these islands is *Z. c. albiventris*. On islands N of Aust., through the Indonesian islands comprising Wallacia, this species is known as the Ashy-bellied White-eye. In pairs or small flocks it forages through foliage to seek native fruits, berries and insects. **Voice:** contact call is a quite loud chirp. The song is a rising and falling sequence of clear, sweetly warbled notes; overall less plaintive than that of the Silvereye. **Similar:** Yellow White-eye, but its entire underparts are pale yellow. **Hab.:** this white-eye appears to prefer small islands with cover of scrub thickets and trees; it has not established a presence on the mainland or the larger islands. **Status:** sedentary; locally quite common. **Br.:** 418

Yellow White-eye *Zosterops luteus* 11–12 cm

The Yellow White-eye is essentially a bird of the narrow mangrove belt that fringes most of Australia's NW to NE coast, although it does wander briefly into adjoining vegetation. In pairs to large flocks, forages through the yellow-olive foliage of the mangroves and occasionally on mud exposed at low tide. Between trees, uses a slow, fluttering flight, broken by short glides. Where there are long stretches of exposed rocky or sandy coast lacking sheltered bays with mudflats and mangroves, the substantial gaps in the mangrove belt create matching breaks in the distribution of mangrove birds. The isolated populations eventually become distinct races or species. **Variation:** possibly 2 subspecies: nominate race *luteus*, Qld and NT coasts, almost to WA border. Westwards along the Kimberley and Pilbara coasts, the race *balstoni* is slightly smaller, grey-green rather than olive green on upper parts, and has paler underparts. However, there may be insufficient difference in plumage or size to justify separate races. **Voice:** strong, high pitched, whistled contact or flocking calls, possibly louder than those of the Silvereye, and tinkling notes from foraging flocks. The song is strong, varied and tuneful with musical warbling and trills. **Similar:** Silvereye, especially the green backed western race, but it lacks the overall continuous yellow underbody; Pale White-eye, but it has white breast and belly, and very little, if any, overlap in range. **Hab.:** mangroves, but moves out to forage in adjoining monsoon scrub, dense river-edge vegetation, coastal scrub, paperbark swamps and coastal gardens. **Status:** sedentary, or very localised wanderings; common. **Br.:** 418

Silvereye *Zosterops lateralis* 11–13 cm (L, N)

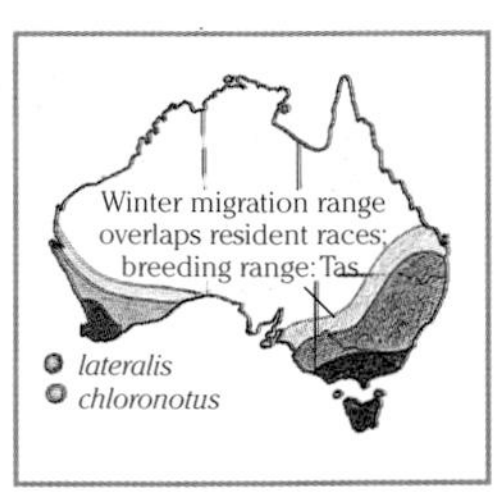

All races are gregarious; after breeding they gather into small parties and then into large flocks, these foraging through foliage of trees and undergrowth. Major migrations occur along Australia's E coast: the flocks form in late summer after the young have fledged, then begin to move northwards. The migration routes seem well established, following the eastern coastal plains where heathlands and gardens provide sustenance of wild and cultivated flowers, orchard and vineyard fruit. Longer flights appear to occur at night, but during the day the flocks work slowly through the treetops and undergrowth, feeding as they travel. Foraging Silvereyes are lively, busily active little birds, constantly on the move; they depart to the next patch of shrubbery with brisk, bouncy flight and much calling. The southernmost populations seem to undertake the longest migrations to escape the approaching southern winter. Large numbers, perhaps most, of the race *lateralis*, which breeds in Tas., travel across Bass Str., then disperse through Vic., NSW and into SE Qld. They tend to replace Silvereyes of those regions that have moved even further N. Silvereyes of the most northern regions and of the SW of WA tend to be locally nomadic rather than long-distance migrants. **Fem.:** sexes alike in all races, but the female in each pair tends to be slightly paler than the male. **Imm.:** similar to adult, but colours are not as bright. **Variation:** many subspecies scattered across islands of the SW Pacific and NZ. Possibly 6, but proposed 8 Aust. races. Renamed (with previous or alternative names in brackets) are races *cornwalli* (*familiaris*), SE Qld to S Vic.; *pinarochrous* (*halmaturina*), SW Vic. to Eyre Pa.; race *vegetus* (*ramsayi*) mid Qld coast N to Cape York; *chloronotus* (*gouldi*), southern WA to SW SA. Unchanged are nominate race *lateralis* (breeds in Tas.; migrates over SE and E mainland, overlapping both *cornwalli* and *pinarochrous*); *chlorocephalus*, on islands of the Bunker-Capricorn group; and *tephropleurus*, on Lord Howe Is. New would be: race *ochrochrous* on King Is., Bass Str.; *westernensis*, possible race replaces *cornwalli*, S and SE Vic. **Voice:** a clear, peevish 'tseeep'; song a rather slow, erratic series of clear, sharp, slightly peevish notes: 'tsweeip-cheeip, peeip-a-chweip, cheeip' with many slightly different variations in rambling sequence, often rather lengthy; there is a different, more hesitant version that is used at first light. In a softer, quieter sub-song, mimicry of some other birds may be included. Calls of western race, *gouldi*, are perhaps slightly more harsh. **Hab.:** diverse: eucalypt woodland, forest, coastal heath, mallee, mangroves and many other vegetation types; also can be found in gardens, orchards and vineyards. **Status:** migratory or nomadic; common to abundant. **Br.:** 418

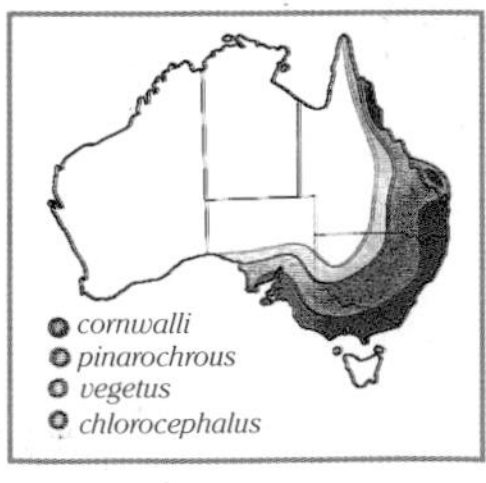

Three species of white-eye live in Australia. All have the characteristic narrow ring of white feathers surrounding the eye. The Pale White-eye has three races; only one in Aust. on Qld islands.
Forehead yellow; crown, nape and face olive yellow
Bill slightly more robust than on other white-eyes
Lores black, extending back to underline the white eye-ring
Throat to upper breast yellow
Upper parts pale olive-grey to pale yellowish olive
Rump and under-tail coverts pale yellow
Breast and belly off-white
The two species of white-eye and the Silvereye are the Australian representatives of the white-eye family, Zosteropidae, which is spread from Africa to the western Pacific. There is little variation within the family, to the extent that most of about 84 species are so alike that they have been grouped within a single genus, *Zosterops*.
Two races: nominate race *luteus* is slightly larger and brighter than NW race *balstoni*.
Forehead yellow
Upper parts mid yellowish olive
Thin dark line from lores through and underlining the eye
Flight feathers dark grey edged yellow
Underparts entirely bright yellow
Immatures are like adults, but colours not as bright.
Race *luteus*
Flanks slightly darker olive grey
Underside of flight feathers is plain grey-brown, but edged yellow on upper surface.
Head and nape olive green, all races
Back mid blue-grey
All races have the typical and conspicuous white eye-ring. The dark lores extend back as a grey or black shadow or black line under and behind the white eye-ring.
Back mid blue-grey
Rump olive yellow to olive green
Flanks vary, chestnut to dull cinnamon-brown.
White
Throat yellow
Under-tail coverts yellow
Race *chlorocephalus*, largest race, is like *vegetus*, but larger and has heavier bill.
Nominate race *lateralis*, in flight and perched
Back mid blue-grey
Throat white, sides of breast light grey
Pale blue-grey
Throat yellow
Flight feathers are dusky grey to grey-brown edged with yellow or citrine, all races.
Rump pale olive
Breast light grey
Flanks vary, chestnut to dull grey-brown.
Under-tail coverts white or greyish white
Yellow tint
Flanks rufous-buff to grey; belly off-white
Race *pinarochrous* (not illustrated) is like *lateralis* but off-white under tail, greyish buff throat, dull flanks.
Race *cornwalli*
Race *lateralis*
Race *vegetus* is like *cornwalli*, but has deeper colours.
Race *chlorocephalus* is like *vegetus*, but larger, with slightly heavier bill.
Back olive green rather than the grey of other silvereye races
Throat yellow
Throat yellow
Light Olive
Rump olive
Deep yellow
Yellow rump
Yellow
Olive
Pale yellow
Pale grey
Flanks pale cinnamon-grey
Race *vegetus*
Race *chlorocephalus*
Race *chloronotus*

Bassian Thrush *Zoothera lunulata* 27–29 cm

A secretive inhabitant of dense wet forests, typically in localities with dense overhead canopy and thick leaf-litter layer. Solitary, or in pairs or small parties, on the ground, scratching aside the accumulated debris to expose moist soil and the small creatures on which these 'Ground Thrushes' feed. May sometimes be detected in the quietness of the forest by rustling sounds as dry surface leaves and twigs are thrown aside. When disturbed, may run briefly, then stop, motionless, relying on the mottled brown and cinnamon plumage blending into background colours, a technique effective in the dim light at the forest floor. **Variation:** race *cuneata*, NE Qld; race *lunulata*, SE Qld to SW Vic.; those within SA will probably be recognised as a third race, *halmaturina*. **Voice:** usually silent, most likely to be heard at dawn or in dull weather. Song consists of three clear notes, the first level, the second rising briefly, the third again steady—'wheeer-aoo-whooo'—and may continue as a soft, tuneful sub-song. **Similar:** Russet-tailed Thrush. **Hab.:** heavy wet rainforests, eucalypt forests and woodlands with dense canopy and thick litter layer, heavily vegetated gullies, gardens. **Status:** sedentary or dispersive; common. **Br.:** 418

Russet-tailed Thrush *Zoothera heinei* 26–29 cm

A slightly more rufous 'ground' thrush species that inhabits mid to NE coastal forests and substantially overlaps the range of the Bassian Thrush, of which it was formerly a race, but now recognised as a separate species. Its name refers to its most obvious distinguishing feature, the overall colour, and especially that of the rump and upper-tail coverts, is slightly more rufous. A small patch of white near the tip to the outer tail feather is larger, whiter and more noticeable if the bird is flushed. In their habits the two species appear similar, but, in mountainous areas where both occur, the Bassian occupies the higher range tops, above 500 m altitude, while the Russet-tailed stays below 750 m. Where the two overlap there appears to be no hybrids. **Voice:** the call differs recognisably from that of the Bassian, described as two clear, whistled notes, the second lower, 'wheee-rooo'. **Hab.:** like Bassian Thrush, wet eucalypt and rainforest, but at lower altitude, at least in the breeding season. **Status:** local seasonal and in some places altitudinal movements; moderately common. **Br:** 418

Song Thrush *Turdus philomelos* 22–23 cm (L, M)

This European species was introduced into Aust. between 1850 and 1880, perhaps for its quite pleasant song that would have been a reminder of England. Forages on the ground finding insects, spiders and other small prey among leaf litter and surface soil; snails are broken against a rock. Flight is low and direct with little of glides or undulations; shows buff underwing coverts. **Voice:** male sings almost continuously through the breeding season, a clear, spirited sequence of notes; mimics native birds. In contact and in flight, a thin 'tsip'. **Hab.:** damp woodlands and forests, parks and gardens. **Status:** confined to the Melbourne–Geelong region; uncommon and of localised occurrence.

Blue Rock Thrush *Monticola solitarius* 22-23 cm

Breeds in S Europe, N Africa, central Asia, China, Japan and N Philippines. Closest to Aust. is the race *philippensis*, also known as the Red-bellied Rock Thrush. In the northern winter these birds migrate S into a region that includes Borneo, Sumatra, Sulawesi and the Moluccas, rarely to NG, and just one record for Aust., Oct. 1997, at Noosa Heads, Qld. This visitor was a male in the process of acquiring its first winter plumage. It was of the race *philippensis*, which winters closest to Aust., and has been recorded in NG. The Blue Rock Thrush tends to be solitary or only rarely in groups; it is shy, flushes easily, and is then quick to disappear into concealment of rock crevices and behind boulders. The Noosa individual kept to cliffs and shores of several headlands, usually on steep cliffs, steep slopes with sparse, scattered vegetation, rock platforms and stony slopes down to sea level. Data from its usual wintering territory, where it is present from early Sept. until early May, indicate that it typically perches on rocks and buildings, but will also use bare branches or wires, often adopting an upright posture, bobbing with body vertical and flicking tail up after landing. **Variation:** 2 subspecies: nominate race *solitarius* is entirely dusky blue with dark scallopings; race *philippensis* is brighter blue with chestnut from lower breast to under-tail coverts. **Voice:** low clucking sounds and a pleasant, reedy song. **Hab.:** rocky shorelines, cliffs, open stony country, coastal towns. **Status:** very rare vagrant or accidental occurrence in Aust.

Red-whiskered Bulbul *Pycnonotus jocosus* 20–22 cm

Conspicuous, distinctive species of southern Asian origin, introduced in Sydney suburbs about 1880, later to Melbourne. **Voice:** a trilled, fluid, musical whistle in contact and advertisement of presence or territory, and much lively chattering; harsh scolding sounds in alarm or warning. **Hab.:** in Aust. this species has kept within the alienated landscape and environs close around human habitation; it uses cover of shrubbery in parks, gardens, remnant and regrowth vegetation along roads and creeks. **Status:** introduced; locally common Sydney and Coffs Harbour, and in Qld around Mackay.

Previously the ground thrush in Aust. was thought to be the Eurasian *Zoothera dauma*. Australian birds are now separated and divided giving two endemic species: the Bassian Thrush, *Zoothera lunulata*, and the Russet-tailed Thrush, *Z. heinei*.
Brown, scalloped
Lores and eye-ring are off-white.
Northern race *cuneata* has slightly longer wings and bill.
Rufous-brown, scalloped
Bassian Thrush, race *lunulata*
Face, throat, upper breast pale buff; rest of underparts white, all with light to heavy black scalloping
Brown to olive brown heavily scalloped with black
Rump brown, scalloped
All forms have white panel in wing.
Slightly deeper buff on breast compared with race *lunulata*
Outermost feather on each side of tail has a small, dull white patch near its tip, smaller, duller than that of the Russet-tailed Thrush.
Bassian Thrush, race *cuneata*
Entire upper parts of the Russet-tailed Thrush are slightly brighter rufous compared with Bassian; this is more obvious on the rump and tail.
Russet-tailed Thrush
In both species male and female are alike. Immatures are similar to adults, but juveniles are duller, the black scallops are smaller, and the tail is tipped paler.
Outermost feather on each side of the tail has, near its tip, a spot larger and whiter than that of the Bassian Thrush.
Tail is slightly shorter than that of the Bassian Thrush.
The Russet-tailed Thrush has more white on the inner web of the outermost pair, which may not be visible in folded tail.
Plain rufous brown, crown to rump, wings and tail
Eye dark with white eye-ring, faint brow line
In flight, underside of flight feathers is brown, wing linings are buff.
Immatures are like adults, but the feathers of wing coverts are edged buff, and those of the mantle have fine, light buff shaft streaks.
Underparts are off-white, faintly buff on the breast and heavily streaked with dark arrow-shapes from throat almost to under-tail coverts.
Legs and feet are pinkish-brown and big, as suits this ground-foraging species.
Head and back blue: barred in the nonbreeding plumage, plain and brighter in breeding plumage
Nonbreeding: body plumage is barred black and grey, colours are dull compared with the unbarred breeding male.
Grey-brown mottled with white and black
Wings blackish
Throat and upper breast are slaty mid blue
Remainder of underparts are chestnut.
Rump and tail coverts blue
Pale eye-ring
♂
Buffy white, scalloped darker
♀
Tall, tapered black crest on crown.
Red spot behind the eye
Upper parts, including wings, are mid brown.
White cheek and throat are divided by a fine, black diagonal line.
Underparts pale fawny buff
Under-tail coverts red
The Red-whiskered Bulbul occurs naturally through India, Burma, Indo-China and the northern part of Malay Peninsula. There are about 9 races, of which one was introduced into Aust. It has also been introduced into the western islands of the Indonesian archipelago.
Male and female are alike. Immatures are similar, but with dark grey rather than black crown, smaller crest, lacks any red spot on the face, under-tail coverts pinkish rather than red, and overall rather dull coloration.
Tail dark brown above; paler on the underside; outer feathers have large white tips.

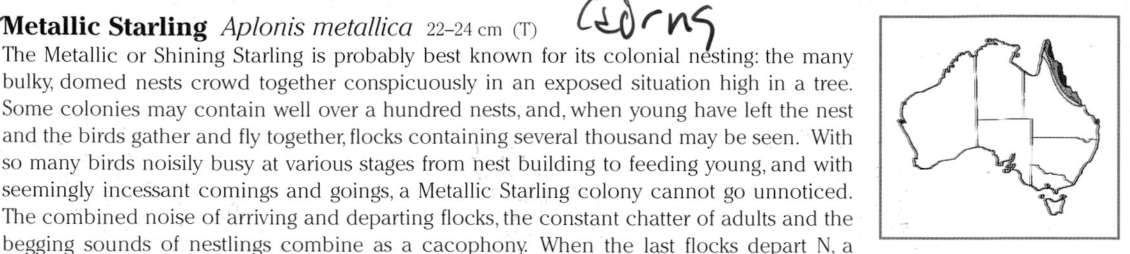

Metallic Starling *Aplonis metallica* 22–24 cm (T)
The Metallic or Shining Starling is probably best known for its colonial nesting: the many bulky, domed nests crowd together conspicuously in an exposed situation high in a tree. Some colonies may contain well over a hundred nests, and, when young have left the nest and the birds gather and fly together, flocks containing several thousand may be seen. With so many birds noisily busy at various stages from nest building to feeding young, and with seemingly incessant comings and goings, a Metallic Starling colony cannot go unnoticed. The combined noise of arriving and departing flocks, the constant chatter of adults and the begging sounds of nestlings combine as a cacophony. When the last flocks depart N, a peacefulness replaces the bedlam around the colony tree until they return. Flocks or groups forage together, flying out from the colony to surrounding rainforest to search through the canopy foliage seeking fruits and nectar as well as any insects about the fruit or flowers. In travelling from nest colonies to rainforest feeding sites, or between feeding grounds, the tight flocks hurtle low above the forest canopy, the sound of their calls and the wind over their wings attracting attention to their passage. Most migrate, leaving about Mar., probably for the islands to the N, and they return in Aug. **Variation:** The Metallic Starling occurs through most of the islands of the Moluccas and Lesser Sundas, NG and the Solomon Islands. **Voice:** various nasal and wheezing calls—from individuals, 'scriaarch, scraark, scraich-scraich-scraaairch, chrak-chrak-chrak' and similar sounds. From the flock, a jumbled racket of intermixed squawking and wheezing. **Similar:** Trumpet Manucode, but it has a blunt, rounded tail, different calls, and is comparatively secretive, being found alone or in small parties; Spangled Drongo, but it has outcurving tail. **Hab.:** tropical rainforests, nearby woodlands, coastal scrubs, mangroves, suburban gardens and parks of coastal towns in NE Qld. **Status:** migratory; common or locally abundant. **Br.:** 419

Common Starling *Sturnus vulgaris* 20–21 cm (L, M)
Introduced to Vic. in late 1850s, later in other eastern States. Climatic limitations prevent the species from spreading much N of Brisbane, while the barrier imposed by the Nullarbor, combined with patrols that seek and destroy any Common Starlings that cross the border, have kept WA free of this species, which displaces native bird species from nest hollows. **Variation:** across natural range, Europe to Russia, 11 subspecies; in Aust., one species introduced and established. **Voice:** most characteristic call a long downwards whistle, and 'chwee'; also high 'tizz-tzz', and diverse wheezing, rattling and clicking noises; includes mimicry. **Similar:** Common Blackbird, but plumage plain. **Hab.:** both urban and country areas, including woodlands, mallee, mulga, watercourse and roadside tree belts, reedbeds, cleared land, mallee, coastal, alpine, parks and gardens. **Status:** migratory, nomadic, dispersive; common, introduced species.

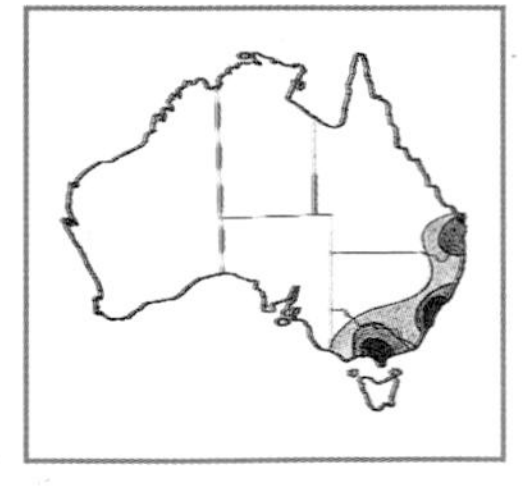

Common Myna *Acridotheres tristris* 23–25cm
First introduced to Melbourne in the mid to late 1800s, and in the late 1800s to N Qld. Seen in small to very large flocks and in mass gatherings when roosting. The Common Myna has proven to be a very successful colonist, scavenging around habitation, but also taking insects, fruits and young of other birds. Builds nests in crevices of buildings and in tree hollows. Undesirable because it damages orchards and market gardens. Nesting in cavities in walls and under roofs, it can create fire and vermin hazards. Away from town environs, it takes over hollows needed by native birds. **Voice:** a mix of loud, clear and mellow whistles, especially a high, clear and vigorous 'wheeoo, wheeoo, wheeoo' and contrasting harsh, rattling 'carrarrk, carrarrk' and sharp 'tseit-tseit'. Gives a clear, liquid note on take off. **Hab.:** urban areas; country towns, farms, agricultural and pastoral lands, canefields, roadside vegetation. **Status:** introduced species; abundant.

Common Blackbird *Turdus merula* 25–26 cm
Introduced at Melbourne around 1850 to 1860, and later in Adelaide, and has spread across most of SE Aust., occupying both coastal and inland regions. Native to Europe, N Africa and southern Asia. Unlike many other introduced species, the Blackbird has invaded natural bushland habitats in addition to alienated lands and the surrounds of towns, and it competes with native birds. A pest species in orchards, vineyards and market gardens. Mostly forages on the ground, probing, raking in litter, soil and lawns. **Voice:** in contact, a high, fine 'tseee', and harsh, almost screeching chatter of alarm when in flight. The song is a pleasant, mellow fluting sound with higher trills. **Hab.:** forests and woodlands, roadside and watercourse vegetation, coastal thickets and scrubs, parks and gardens. **Status:** introduced species; abundant.

The starling family, Sturnidae, is, in Aust., represented by a single species, the Metallic Starling. Worldwide there are more than 100 species in the family, which includes also the mynas. Two species, the Common Starling and the Common Myna, have been introduced; both are now widespread and common.
The Metallic Starling has two subspecies, the Australian population being of the race *circumscripta*, which occurs also in NG, and extends west to Tanimbar Is. in the Arafura Sea.
Nape and neck feathers are long and pointed.
Eye large, bulging, brilliant crimson, staring, obvious at a distance
Eye at first brown, soon becoming crimson
Plumage is entirely black with a glossy, metallic green and purple iridescence.
Adults, front and side views: male and female are similar.
Bill heavy, slightly downcurved, black
Underparts are white, streaked black. A black breast-band is the first stage in development of the all-black adult plumage.
Imm.
Adult
Tail, front view, is long and tapers to a point, differing from the shorter, thicker, square tipped tail of the Singing Starling, *Aplonis cantoroides*, of NG and Australian islands of northern Torres Str.
Tail, from side, is long; tapers evenly to a fine point.
The Metallic Starling is named for the glossy, oily, iridescent sheen on the surface of its black plumage, which gives multicoloured highlights and has some resemblance to the 'blued' finish given to some metal surfaces.
Long, pointed bill is dull brown except in breeding season.
Bill yellow through spring–summer breeding season
In autumn the new plumage has pointed buff and white tips and edges to the black feathers. These wear so that by the spring breeding season much of the plumage is plain glossy black.
By spring or early summer the plumage has worn to black. An iridescent sheen becomes evident as the pale tips are lost.
Legs reddish brown
Juveniles are plain grey-brown above, lighter grey beneath, and have a whitish throat.
Adult
Adult: summer breeding plumage
Head glossy black or blackish brown with greenish highlights
Bill deep yellow
Plain rufous-brown becoming darker blackish brown on head, wings and tail
Native to southern and SE Asia. Two races, one introduced to Aust.
Distinctive bare yellow skin around the eye tapers to a point behind the eye.
White shows in folded wing.
Tail blackish, tipped white
Tail long, dark, white-tipped
White
Legs and feet are large, strong, deep yellow.
Outer wing is blackish brown with a large white panel, conspicuous in flight.
Eye-ring is deep golden yellow, narrow.
Bill is dull brownish yellow.
Upper surfaces mid grey-brown, darker on wings and tail
Bill deep orange-yellow
Underparts light grey-brown mottled darker
Plumage is entirely black with a surface sheen.
About 16 races in Europe, southern Asia and Africa; one race in Aust.
Large, strong legs and feet; forages often on the ground, raking aside. litter and soil.
Juvenile like female, but with lighter underparts. Crown shows fine pale streaks of feather shafts.

Some of the most interesting accounts of nests and eggs (and other records of the avian fauna before the impact of European settlement) are those in the writings of those pioneers of ornithology, the egg collectors. The earliest ornithological publications, prior to the 1940s or 50s, the *Emu* and books such as North's *Nests and Eggs of Birds Breeding in Australia and Tasmania*, are filled with fascinating accounts of those very early years. Egg collecting—often along with the skins of the parent birds—was the driving force behind ornithology. Photography was primitive, but early pictures are now a valuable record and reference, especially of the habitats as they were.

There are photos of those early collectors at nests very high on branches or rope ladders—often in collar, tie and waistcoat! Very often Aboriginal guides found nests and went up very tall trees to collect the eggs. These skilled climbers are photographed clinging high on smooth tree trunks, tomahawk in hand, fingers and toes finding a grip in tiny notches.

Today we must observe and photograph rather than collect and shoot. (Except for scientific studies with permits, it is illegal to collect nests, eggs or birds, even if dead.) But we owe much to the early collectors who risked their lives to obtain the study specimens, notes and diaries of the birds, eggs and nests held in museum and other research collections.

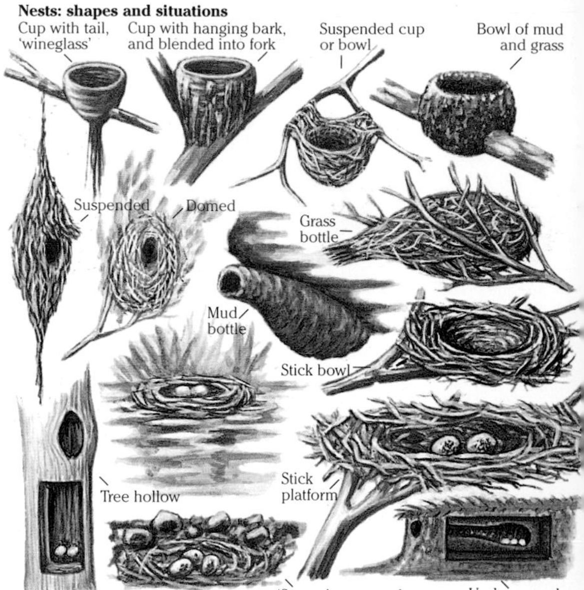

Emu Usually breeds Apr.–Jun.; often chooses an elevated site sheltered by a shrub or large clump of grass, but with a distant open view in at least one direction. Male makes a shallow scrape sparsely lined with grass, leaves, fine twigs. The 6–12 eggs (135 × 90 mm) are incubated about 60 days by the male. He may be rather torpid, temperature dropping several degrees, doing little more than incubate, turn the eggs and guard. The female may wander in the general area of the nest, mate with other males, or migrate. Eggs hatch over several days; the chicks stay with the male for up to 18 months. (p. 14)

Southern Cassowary Jun.–Dec.; female then less aggressive toward the male and pairs form; courtship initiated by either sex. Male prepares nest: leaves added to a scrape on ground, often at base of rainforest tree or in grass at forest margin. Female lays, then abandons eggs (140 × 90 mm); may pair up with other males. Male incubates clutch of 4; raises young; stays with them as family party for about 9 months. Green eggs, heavily granulated surface; incubation 47–53 days. (p. 14)

Malleefowl Aug.–Apr.; eggs incubated in mound of sand, sticks, fallen leaves and bark. Female lays eggs at intervals of 4 to 17 days. Usually 10–35 eggs (91 × 60 mm) in one season. In cooler spring months, heat for incubation is derived from fermentation of buried debris. Later, heat comes from the sun. Temperature is regulated almost constantly by the male, kept at 33° C by opening out, spreading or building up the sand. As chicks hatch, each after 50–90 days of incubation, they dig out, run into bush and fend for themselves. To breed, Malleefowl require mallee thickets or other dense, low vegetation with unburned litter accumulated over 20 or more years. (p. 16)

Orange-footed Scrubfowl Breeding and egg-laying extend about Aug.–Mar.; but the incubation mound is maintained throughout the year, the male doing most of the work in scraping up the debris and earth—it can be 3 m high and 10 m across, but averages 1.5 m high and 7 m across. Very small mounds would not generate the fermentation temperature (35–38° C) required to incubate the eggs. Several pairs may use a mound, females laying up to 15 eggs (90 × 50 mm) a season. The dark brown chicks struggle up through the mound's debris and run into the forest, independent from the beginning. (p. 16)

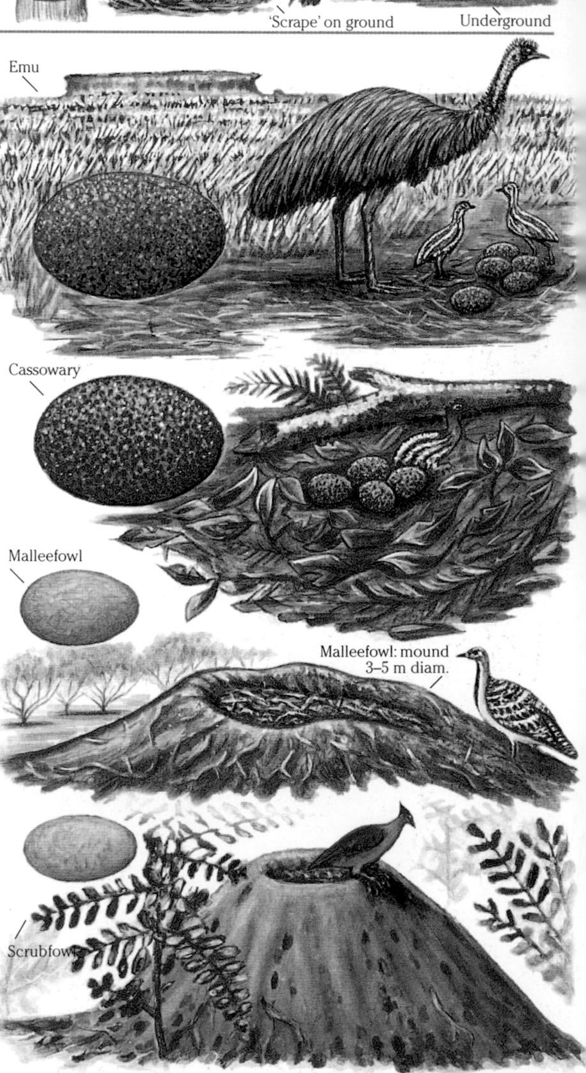

Orange-footed Scrubfowl

Eggs*: examples of shapes, variations

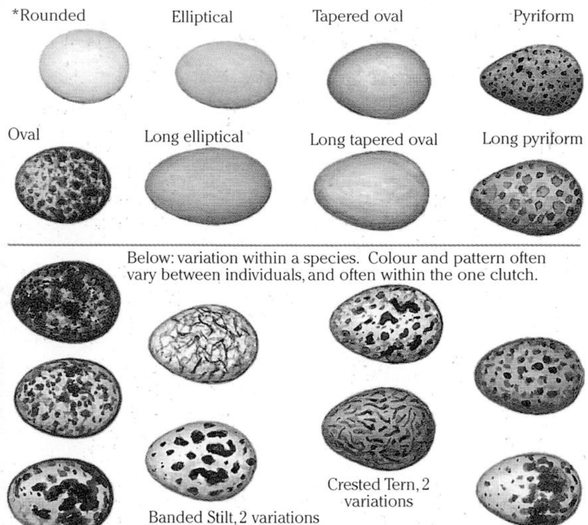

From the 1950s, as camera equipment improved, photography became a popular way to collect birds—the tree-climbing exploits of the early egg collectors were repeated in order to obtain photographs at nest. Most Australian birds had been captured on film at their nests by the 1980s. Not only did the photographers climb 20 to 30 m above ground, often with primitive and risky climbing methods, but they built hides on slender poles and with planks tied among branches overlooking the nests. Hours, days and nights were spent waiting in these treetop hides.

Now that almost all birds have been recorded at their nests, the trend is to record, on film, video, sound tape and notebook, the behaviour of birds, rather than the earlier portraits and feather-by-feather descriptions. Since the 1960s I have had the privilege of sharing the work, risks, excitement and rewards of treetop encounters with falcons, eagles and many other birds with several of these photographers. Some of the results are seen in the books listed on page 434.

For those interested in contributing their own nest sightings, the Royal Australian Ornithologist Union has a Nest Record Scheme. Breeding data is also needed for the new RAOU Atlas of Australian Birds: (03) 9882 2677, or e-mail to raou@raou.com.au.

(*White eggs not always shown; all egg sizes approximate.)

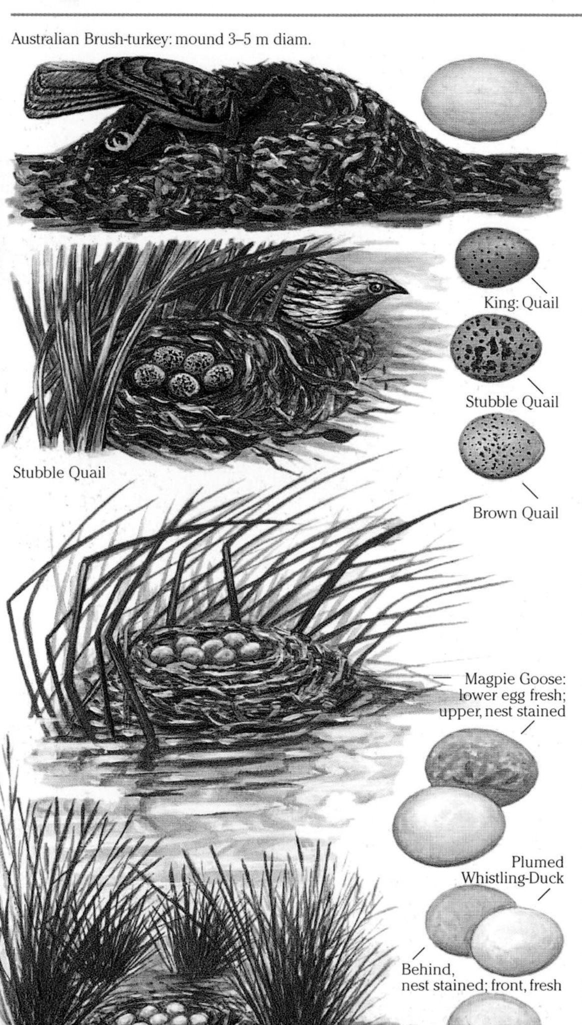

Australian Brush-turkey May–Dec.; males maintain, for up to 9 months each year, one or more huge leaf-litter mounds up to 5 m in diameter and 1.5 m in height. Fermentation of litter provides heat for incubation; solar heat helps at more open sites. A female may lay, perhaps into more than one mound, some 20–50 eggs (92 × 63 mm) a season. Incubation about 50 days; chicks dig out and depart unaided. (p. 16)

King Quail Throughout year; mainly Sep.–Mar. in S; Feb.–May in far N. Makes shallow depression in ground or in grass tussock; lined with grass; may be screened by overhanging plants. Both sexes incubate 4–10 eggs (25 × 20 mm) 18–20 days. (p.18)

Stubble Quail Breeds any time of year after rain,especially in drier regions; commonly Aug.–Mar. Often many broods. The nest, a scrape on the ground lined with grass, is well hidden under cover. Clutch 4–12 (29 × 22 mm); female incubates 21 days. Chicks leave nest site very soon after hatching. (p. 18)

Brown Quail Season variable, usually Aug.–May. Nest is a scrape in ground lined with grass; often among rank grass, not far from water, screened by overhanging plants. Clutch 7–10 (28 × 22 mm). Sexes share incubation, about 14 days, and care of young, which leave nest immediately after hatching. (p. 18)

Magpie Goose Breeds in large colonies after summer rains that turn river floodplains into swampy nesting habitat. Builds over shallow water; bends and tramples dense sedges or rushes to make a platform, to which other vegetation is added. Male often has several females; each clutch 6–9 (74 × 50 mm). Group shares incubation of combined clutch; 23–28 days. The eggs, at first white, become stained darker. (p. 20)

Plumed Whistling-Duck Breeds after rain; Feb.–May in N, often Aug.–Oct. in SE. Pairs disperse to build small, rough pads of grass on the ground among swamp-edge vegetation, sometimes away from water; use little down or soft lining. Clutch 9–12 (48 × 36 mm). Both sexes build nest and incubate. (p. 22)

Wandering Whistling-Duck Breeds with monsoon rain; usually Jan.–Apr. Pairs disperse, each to build a rough pad of grass in a small depression in the ground screened by vegetation. Often nests beside a swamp, but some well away from water. Clutch 7–10 (50 × 37 mm). Both sexes build nest and incubate; 28–30 days. (p. 22)

Blue-billed Duck One clutch each year; mainly Sep.–Dec., but appears to be an opportunistic breeder—takes advantage of good conditions almost any time of year. Probably responds to water levels in wetlands. Female builds in old, very dense, long-unburned typha reedbeds where masses of old dead leaves tangle near water level. Nest is a deep bowl with a domed hood of dead leaves pulled down and thickly interwoven, concealing the nest, which is indistinguishable from reed debris surrounds. At times builds into dense thicket or scrub of lignum or tea-tree low over water. Clutch 3–12, commonly 5–6; oval, palest green, soon stained (70 × 49 mm). Female incubates, about 24–26 days. Hatchlings downy, very dark grey. (p. 22)

Musk Duck Season varies; usually Jun.–Jan., but rain and rising water can trigger breeding at other times. Builds into old, dense reed clumps standing in water at least 1 m deep. Reeds are bent and trampled to make a water-level platform shaped on top as a shallow bowl and lined with grass and down. Reeds around and above are pulled down and woven to form a domed hood so that the nest appears a mass of dead reed debris typical of old, long-unburnt reedbeds. May also use a hollow in a log or stump over water, the cavity then lined with soft vegetation and down. Usual clutch 1–3, rarely 5 eggs, elongated oval, greenish white to pale olive-brown (80 × 55 mm). (p. 22)

Freckled Duck Usually Aug.–Dec.; will nest any time of year after flooding rain, especially in normally dry to arid regions. Pairs disperse to find a territory and nest site, which is usually the fork of a branch hanging low over water among flood debris, or in another water bird's old nest, but within 1 m of the surface. Clutch usually 5–8 (63 × 48 mm). Incubation takes about 26–28 days. (p. 26)

Black Swan Breeds any time of year after rain fills lakes and swamps; usually begins early–mid winter in S, summer–autumn in far N. Nest is a large pile of vegetation, largely reeds, in shallow water, on an island, or floating among reeds or other plants in deeper water. Clutch up to 10, usually 5–6 (105 × 66 mm). Incubation 35–45 days. Eggs, initially pale greenish cream, become stained darker greenish cream; are usually covered with damp vegetation when nest is unattended. Cygnets leave nest at 120–160 days. (p. 20)

Cape Barren Goose Leaves mainland to breed on islands Jun.–Oct. Pairs form and aggressively defend territories. Nest is on or close to ground, sheltered among tussock grass, shrubbery or boulders, often on a rocky headland on windward side of island; a slight hollow built up of grass and twigs, down lined. Eggs elliptical (83 × 55 mm), become stained. Usual clutch 3–5; female incubates alone; 34–37 days. (p. 20)

Australian Shelduck Appears to breed regularly once winter rains have brought fresh pastures; Jul.–Oct. Flocks disperse in pairs; coastal concentrations spread widely, often to well-watered farmland with pasture, dams, trees. Nest is a large hollow in a tree, usually high, often some distance from water, lined with grass and down; rarely on ground. Female incubates, male regularly escorts her to water where they swim, preen, mate. Clutch 5–14 (69 × 50 mm); incubation 30–35 days; ducklings then escorted to nearest water. (p. 26)

Radjah Shelduck Territories are occupied at start of wet season. Most eggs are laid Apr.–Jun. so that ducklings have optimum feeding areas as floodwaters recede. Nest is a large tree hollow chosen by the pair and lined with down; same site probably used each year. Clutch 6–12 (58 × 43 mm). After ducklings leave nest they are attended by both parents and remain together in a family group throughout the following dry season. (p. 26)

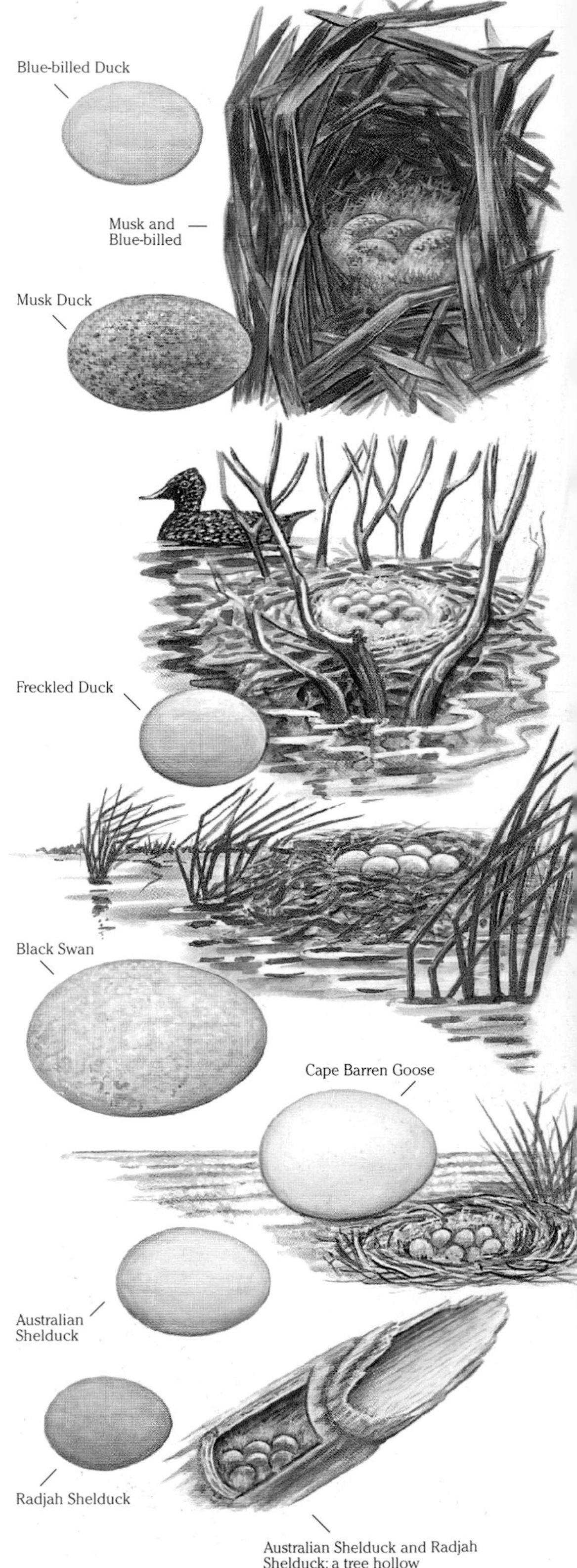

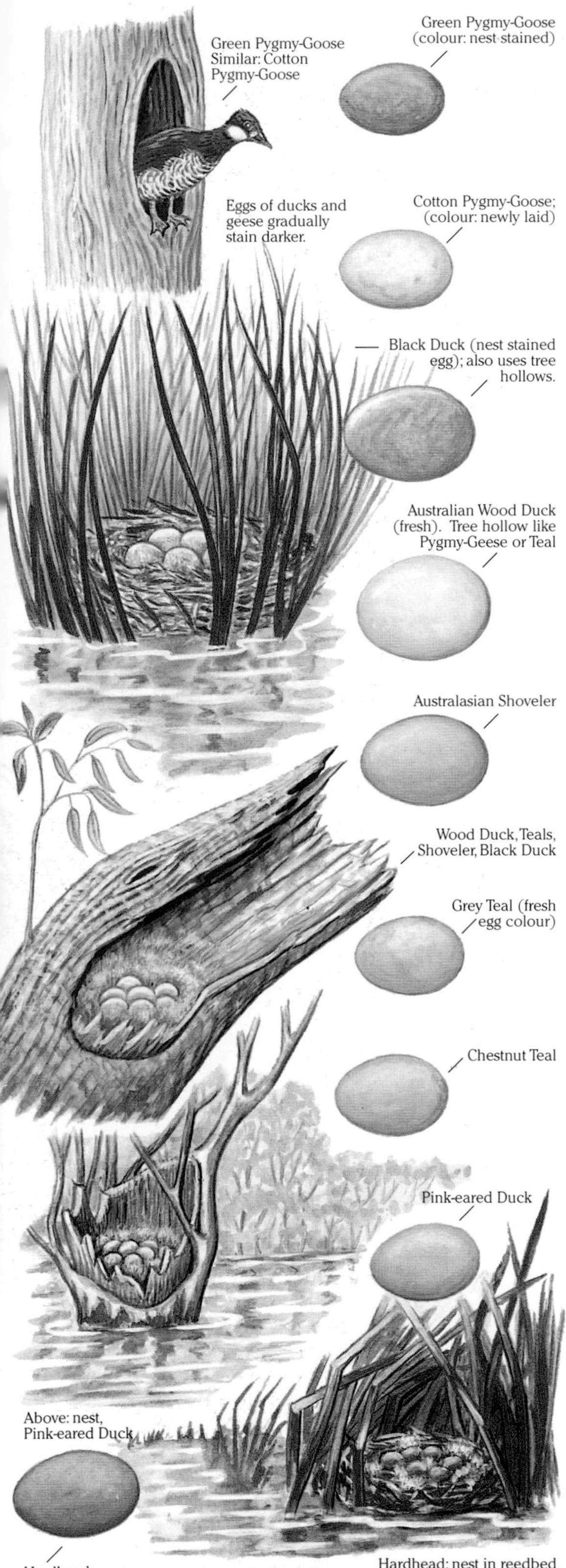

Green Pygmy-Goose Breeds in Wet, Dec.–Mar. Pair bond is strong, probably remain mated for life. Male flies to a tree hollow in wooded area of a swamp; female follows and inspects hollow. Many may be inspected before she decides on a suitable site, usually a hollow in a tree standing in water. Down provides the only lining. Clutch size 7–10 (46 × 33 mm); creamy white when first laid. (p. 28)

Cotton Pygmy-Goose Breeds in wet season, Dec.–Apr. Nest is a hollow, usually high, lined with a little down in the trunk of a dead tree standing in or near water. The 6–10 eggs (47 × 35 mm) are creamy white, gradually staining darker buffy or ochre tones as the 15–16 days of incubation proceeds. (p. 28)

Pacific Black Duck Jun.–Jan. in S; Jan.–Feb. in N. Nest sites vary depending on availability and competition from other pairs for the same sites. Tree hollows seem preferred, but will also use old nests of other waterbirds, stick nests of crows, or trample a cavity in dense, tangled undergrowth. May be close to water, or some distance away. Clutch 7–12 (57 × 42 mm). Down is used as a lining, more added throughout incubation, which lasts about 30 days; parental care by female. (p. 26)

Australian Wood Duck Nov. in southern Aust.; Jan.–Mar. in N Aust.; any time of the year after rain in semi-arid regions. Nests in a tree hollow, preferably over water, but some are distant from water; often high. A large cavity in a trunk or limb is thickly lined with down, usually in copious amounts that conceal the eggs. Clutch 8–14 (58 × 42 mm). Sexes share incubation of 28 days. Very soon after hatching the downy young jump from the hollow to be led to the security of water and pasture close by. (p. 24)

Australasian Shoveler Usually breeds Aug.–Nov., or after rain in semi-arid interior. Nest is a depression in the ground, on occasions a log or low, hollowed stump. It is lined with grass and down, and is usually well concealed, being either underneath or surrounded by long grass, thistles, clover or rushes. Clutch 9–11 (56 × 41 mm). (p. 28)

Grey Teal Responds swiftly to breed after substantial rain at any time of the year, but principally Jul.–Nov. in SW and SE Aust. where winter rainfall is reliable. Female establishes nest lined with down; incubates; cares for young. Not in colonies, although large numbers may use a favoured location. Usually in a hollow limb, but will nest on ground or in low vegetation or old nests; clutch 6–12 (50 × 36 mm). Incubation 24–26 days. (p. 24)

Chestnut Teal Breeds through year after rain, but usually Aug.–Nov. Preferred site is a tree hollow 6–10 m high, lower in mangroves. Alternatively it may be on the ground under dense grass near the water's edge. The nest is a mound of down confined within the hollow or by surrounding clumps of grass. Clutch 7–10 (53 × 38 mm). Male appears to give no assistance with incubation, but may help care for the young after hatching. Incubation about 26 days. (p. 24)

Pink-eared Duck Flooding may trigger breeding any time of year; nests as the floodwaters start to recede. Chooses a site over water in a tree hollow, on a stump or in the old nest of another waterbird. Nest is no more than a mass of down that surrounds and often covers the eggs. Clutch 6–8 (50 × 36 mm); incubation 26 days; female cares for young. (p. 24)

Hardhead Breeding dates probably influenced by rainfall; records from all seasons, mostly Aug.–Nov. in SE. Nest is a slightly hollowed platform of trampled reeds 10–90 cm above water level in very dense reeds, tea-tree, lignum, cumbungi or other thick vegetation. The nest is thickly lined with down; may be hidden by a canopy of overhead reeds. Female incubates the 9–13 eggs (57 × 42 mm) for about 30 days. (p. 26)

Australasian Grebe Breeding varies with climate and season: in S, regular winter rain, Sep.–Jan.; on tropical coasts, the Wet, Dec.–Apr.; in semi-arid districts, any time enough rain falls. Nest is a small, round pile of floating aquatic vegetation supported and anchored by reeds, waterlilies, partly submerged branches. Eggs are quickly covered with nest-edge plant debris whenever the incubating bird leaves; bluish or creamy white, they are increasingly stained brown by the wet nest vegetation. Clutch 4–5 (36 × 25 mm). Juveniles initially have striped head; follow adults for months. (p. 30)

Hoary-headed Grebe Breeds Aug.–Jan. Nest is a floating platform of water weeds, similar to that of Australasian Grebe, usually some distance out from shore among sparse reeds or other plants, anchored to and at least partly supported by them. A shallow depression on top is just slightly above water, so that the eggs are lying in water or damp. Eggs are concealed under pieces of wet vegetation when the incubating bird leaves so that the nest appears empty. Under full sun the covered eggs are warm and moist, and often may be left for some time. Eggs are oval, white; soon stain brownish ochre; clutch 2–5 (39 × 27 mm). Incubation 20–24 days. (p. 30)

Great Crested Grebe Oct.–Jan.; breeds in dispersed pairs on inland fresh waters. Many pairs can have territories close together, as loose colonies. Usually picks nest site near margins of open water, in water 1–2 m deep; nest is attached to, and partly supported by, aquatic plants like reeds, or tree branches and foliage that dip into the water. At times builds among floating vegetation and debris—sometimes far from shore. A typical nest is a large mound of pieces of aquatic plant that is built up from a shallow lake bottom or on a submerged log or other underwater supporting object. Clutch 2–4 (52 × 36 mm). Incubation 23–26 days. (p. 30)

Little Penguin Well known for the 'penguin parade' at Phillip Is., Vic. Breeds Aug.–Feb. in SE; Apr.–Dec. in WA. Nesting is usually colonial, but sometimes only a few pairs. The nest is a hole or tunnel scratched beneath sand-dune vegetation, or a space among boulders. Little Penguins come ashore at night, excavate burrows or find other sheltered sites where the 2 white eggs (55 × 42 mm) are laid. Incubation lasts about 35 days; young fledge at 7–10 weeks. A pair may have several broods in a season. (p. 34)

Great-winged Petrel Breeds May–Jul. in colonies of thousands of pairs on islands off S coast of WA. Nest is a burrow lined with vegetation. Clutch size 1, oval, smooth, dull white (66 × 47 mm). Both sexes incubate for about 56 days. At age 2–3 days, chick is visited at night to be fed, initially about once every 3 days, frequency decreasing to about one feed night in 7 after age 65 days; fledge at about 130 days. (p. 40)

Herald Petrel Breeds rather sporadically almost throughout year on Raine Is., islands of Coral Sea and more distant islands around southern oceans; possibly breeds on N Keeling Is. Forms small scattered colonies. Lays on ground or rock surface, usually under shelter of rock ledge, crevice beneath rocks, or a cliff ledge. Within Australasian region, the nests on Raine Is. are on the ground under cover of dense low shrubs, while suspected N Keeling breeding sites were beneath fallen palm fronds. The single egg is white (59 × 43 mm). Incubation shared by both adults, 51–53 days; fledge at about 14 weeks. (p. 44)

Black-winged Petrel Nov.–May. Seems to be expanding its breeding range: previously closest at Lord Howe and Norfolk Is., but possibly also on Capricorn Group, Qld, and Muttonbird Is., NSW. Nests colonially in burrows; also in crevices between rocks; 1 egg (66 × 47 mm). (p. 38)

Nests beneath low shrubs, fallen palm fronds, cavities beneath rocks.

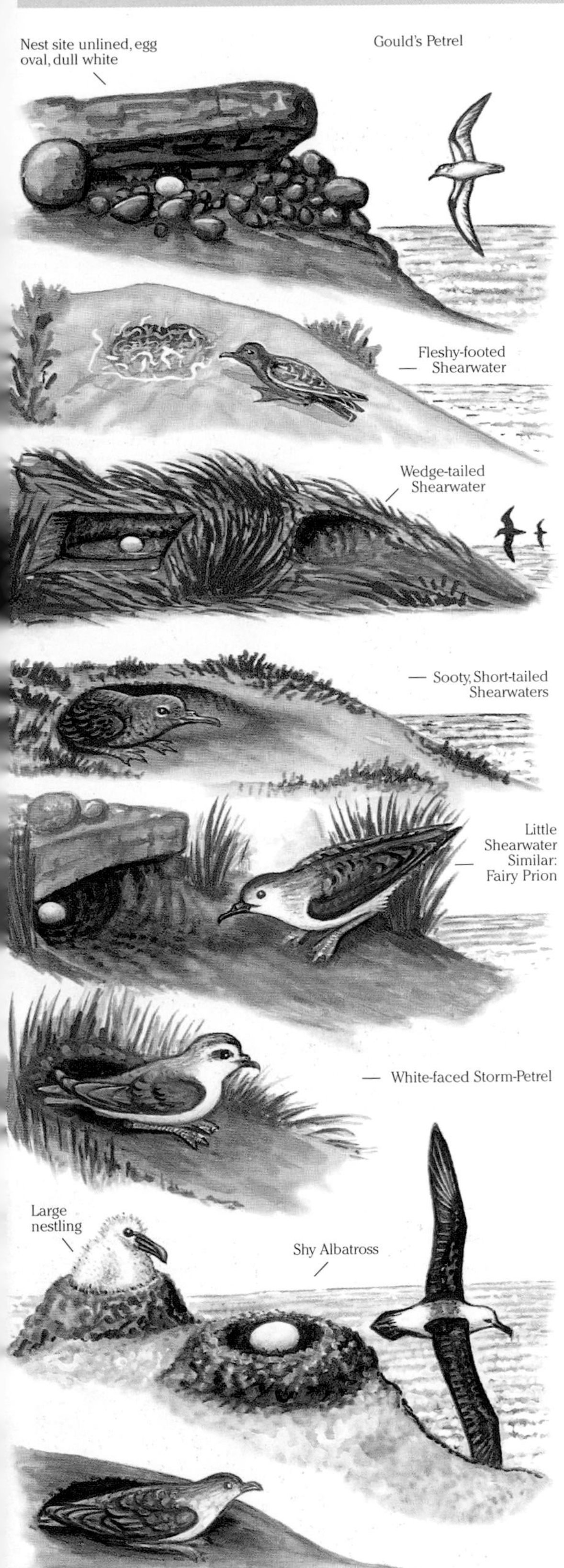

Gould's Petrel Oct.–Apr. or May. Nominate race breeds only on Cabbage Tree Is. off Port Stephens, NSW; disperses from this colony along the E coast to SE Qld—some as far as SW Aust. and Tas. Other races breed New Caledonia, but only rare visitors to Aust. seas. Nest is a space cleared among litter on ground under fallen palm fronds, or in a crevice under boulders. Clutch 1 (50 × 36 mm). Incubation by both sexes, 48–50 days. (p. 38)

Fairy Prion In Aust. waters, breeds on islands of Bass Str., around Tas. and S coast Vic.; Aug.–Feb. The nest is a burrow or rock crevice with some plant lining. Clutch of 1 egg, dull white (45 × 32 mm). Building, incubation and care of young shared by parents. Incubation about 56 days; young fledge 44–45 days after hatching. (p. 50)

Fleshy-footed Shearwater Breeds Sep.–May in 2 major groups: SW of WA; NZ–Lord Howe Is. Colonies on small offshore islands where burrows are dug 1–2 m into soft soil. Sexes share digging and incubation, 60–65 days. Clutch 1 (70 × 46 mm). All activity is at night; birds spend day in burrow or at sea. (p. 46)

Wedge-tailed Shearwater Oct.–May in densely crowded colonies on islands down WA coast from Montebello to Rottnest and Carnac near Perth. On E coast, from islands of Coral Sea S to Raine Is. and Montague Is., NSW. Builds sparse nest of grass and feathers in a burrow 2 m long, in sand, or a crevice under roots, rocks, vegetation or logs. Clutch 1 (63 × 40 mm). (p. 46)

Sooty Shearwater Sep.–Apr. in colonies shared with other more abundant species of shearwater. Prefers higher parts of islands, often on headlands. Digs a long burrow beneath forest canopy or under scrub cover, at times into quite hard ground; nest chamber is lined with grass. Clutch 1 (78 × 48 mm). Sexes share incubation, 58–60 days. (p. 46)

Short-tailed Shearwater Oct.–May in island colonies, often huge, Recherche Archipelago in W to Broughton Is., NSW. Concentrations of breeding colonies in Bass Str. and Spencer Gulf. Colonies are often densely packed with nest burrows. A tunnel of 1–2 m leads to a chamber lined with dry vegetation. Clutch 1, glossy white, often stained brown from soil. Incubation shared by sexes, 52–54 days; fledge at about 13 weeks. Burrow like that of Wedge-tailed and Sooty; egg similar—white, tapered oval (71 × 48 mm). (p. 46)

Little Shearwater Breeds Jun.–Nov. on islands of SW Aust.; also Norfolk, Lord Howe, other oceanic islands. Nest chamber sits at end of a tunnel up to 2 m long, or in cavity between rocks. Eggs laid Jun.–Jul.; clutch 1 (53 × 37 mm). Both parents incubate for 52–58 days. Nestling period 70–75 days. (p. 48)

White-faced Storm-Petrel Breeds Sep.–Mar. on islands of Aust. coast from Abrolhos, mid W coast, around S coast and Tas. to Broughton Is., NSW. Nest chamber is at the end of a narrow burrow, or in rock crevice, or under vegetation. Clutch 1, white, tapered oval (36 × 27 mm). Both parents incubate, about 55 days; young fledges at 53–67 days. (p. 62)

Shy Albatross Breeds Sep.–Apr. Only species of Albatross breeding in Aust. waters: in Bass Str. on Albatross Is., and off the SE Tas. coast on Mewstone and Pedra Branca. Nests on top of these almost inaccessible rocky peaks; constructs a bowl of mud reinforced with vegetation, feathers and bones. Clutch 1 (105 × 67 mm). Incubation by both sexes, 60–70 days; young fledge at 18–19 weeks. (p. 58)

Common Diving-Petrel Jul.–Dec. Several colonies on islands of Bass Str., including Seal Is. group; stronghold on islands around S Tas. Nest a short burrow, usually less than 0.5 m long. Often on steep islands under tussocks, among rocks or between tree roots. Clutch 1 (38 × 30 mm), white, oval. Incubation by both sexes; about 55 days. (p. 60)

Australian Pelican Breeding varies: Aug.–Nov. in S; but almost any time when heavy rain lifts water levels. Breeds in colonies on small islands or sand spits on remote lakes, estuaries. Nest is a ground scrape sparsely lined with debris. Clutch 2–3 (90 × 57 mm). Broods chick 6–8 days. Time between feeds gradually increases; chick fledges about 12 weeks. (p. 68)

Australasian Gannet Breeds Jul.–Dec.; few pairs or dense colonies of thousands on rocky offshore islands or headlands. Makes a compact mound of seaweed and other plant debris, soil and guano. A shallow bowl holds the single egg; chalky pale blue stains to creamy brownish white (77 × 47 mm). Both sexes incubate, 43–45 days; chick fledges at 100–110 days. (p. 68)

Great Frigatebird Season varies; protracted; Feb.–Nov. Breeding colonies form on downwind, protected side of island. Nest built of sticks, small twigs and leaves; top of tree or shrub, occasionally on ground or rock ledge. Single egg (68 × 47 mm); incubated alternately by parents for 55 days. Young is brooded for 14 days, guarded 14 days more; fledges about 170 days; depends on parents for another 220 days. (p. 68)

Lesser Frigatebird Breeding varies; prolonged; May–Dec. peak. Colonies on islands, often low cays. Builds platform of sticks, leaves and vines; top of tree or shrub, but often on ground if only low groundcover exists. Clutch 1 (64 × 45 mm); both sexes incubate, about 40 days. Young stays in nest 160 days or so; depends on adults long after that. (p. 68)

Red-tailed Tropicbird Varies; most months in tropics, Oct.–May in SW WA. Breeding colonies are sited on Qld islands from Raine Is. in the Coral Sea to several on the Barrier Reef; in WA, on islands from Ashmore Reef S to the SW corner—occasionally on mainland at Cape Naturaliste; also Christmas, Cocos-Keeling Is. Nests in cavity under rocks, bushes, in cliff crevice; a simple scrape in sand holds 1 egg (65 × 46 mm). Both sexes incubate. (p. 64)

White-tailed Tropicbird Nearest breeding colony at Ashmore Reef, NW from Kimberley coast; appears to breed throughout the year. Performs aerial display over site; groups of up to 10 birds fly in large circles at heights up to 100 m, tails drooped, long streamers trailing behind. Nests on the ground under an overhanging rock, shrub or grass, occasionally in a hollow limb; nest is usually no more than a slight scrape. Clutch 1 (54 × 35 mm); incubation 40–46 days. (p. 64)

Masked Booby Breeds most of year; peaks Sep.–Nov. Nests in colonies on islands far from mainland; sites are open and exposed to the prevailing winds. No nest built, at times a slight scrape, but a circle of debris and guano builds up around the sitting bird. Clutch 2–3 (65 × 44 mm). Incubation by both sexes, about 44 days; usually only one chick raised. Chicks fledge at about 4 months. (p. 64)

Red-footed Booby (Not illustrated.) Breeding season variable: usually Jul.–Oct. Uses oceanic islands near deep water: outer reef islands of NE Qld; Ashmore Reef, WA. Builds a substantial nest of sticks lined with finer twigs and leaves on trees or shrubs, occasionally on ground. The shallow central cup soon becomes heavily coated with guano. One white egg (60 × 41 mm). Incubation by both sexes, 45 days. (p. 64)

Brown Booby Breeds throughout year with peak activity Mar.–Apr. and Jul.–Oct. Colonies usually on oceanic islands or well out from continent. Uses islands of outer Barrier Reef, Qld; NW coast WA—Ashmore Reef, Adele, Lacepede and Bedout Is. Nests almost nonexistent on bare sandy cays, but some may be quite substantial platforms. New material frequently added by incubating bird. Both parents incubate, about 43 days; chick fledges at about 100 days. (p. 64)

Above: egg, nestling, nests of Masked Booby

Above: nestling and nest of Brown Booby

Black-faced Cormorant: nests on ledges of islands and mainland, usually close to the shore line.

Darter Irregular and erratic breeder, nesting at any time that water levels and food are suitable; more often spring and summer. Usually breeds Feb.–Apr. in tropical N; Aug.–Dec. in southern regions. Nests solitary or small colonies; often among cormorants, spoonbills, ibis. Nests are large, rough platforms of sticks, 40–50 cm diameter, well lined with leaves and built on tree limbs overhanging water. Clutch 2–6, usually 4, pale creamy green with chalky coating (58 × 37 mm). Incubation by both sexes, 28 days. Newly hatched chicks have flesh-pink heads, dark brown body; downy young are pale brown. Chicks first leave nest for water at about 40 days, often return to the nest or nest tree before finally flying at about 50 days. (p.66)

Great Cormorant Appears to breed any time of year given high enough water levels and adequate food supply. Nests in colonies, often with other cormorants, ibis, spoonbills, egrets. Although colonies continue to breed throughout the year, sometimes peaks of activity occur in autumn and, less often, spring. Varied sites: on trees, rocks, ledges, shrubs, reeds, the ground and man-made structures. Nest is a rough platform of sticks and plant debris, 45–55 cm diameter, with a shallow depression for the 3–5 faintly blue or greenish tinted white eggs (60 × 35 mm). Both parents incubate, 27–31 days. Chicks leave nest about 4 weeks of age; clamber around branches or rocks near nest; finally fledge at 7 weeks. (p.66)

Little Black Cormorant Breeds almost any time of year depending on local conditions—after the Wet in N, spring to summer in S. Colonies tend to be in well-vegetated swamps and lakes; birds nest on trees standing in water, preferably well out from land. Most are in fresh water, but some have been recorded in mangroves. Generally nests with other cormorants, darters, herons, ibis and spoonbills. Nest height varies: close to water level to tallest trees available. Nests are rough platforms of sticks, reeds, leaves; softer leaves, paperbark and feather lining. Clutch 3–5, dull white, faint blue-green tint (49 × 33 mm). (p.66)

Little Pied Cormorant Breeds in colonies on a variety of freshwater wetlands, large and small, but usually well vegetated. Nests are typically in trees overhanging lakes or rivers, or standing in swamps. The colonies are usually not large, often with Little Black and other cormorants, sometimes also ibis, herons and spoonbills. Nest platform is of sticks and bark, 30–35 cm diameter, its shallow depression lined with leaves. Clutch 3–5, white tinted light blue with white chalky coating (48 × 33 mm). Nestlings have bare reddish skin on head. (p.66)

Pied Cormorant Breeding season varies depending on conditions and locality. In some places colonies are only active late summer to autumn, while others also have birds breeding in spring. Probably those colonies on inland lakes have only an autumn nesting and those on marine coastal islands have the autumn and spring breeding. Coastal colony nests are close to ground on broken-down tops of low shrubbery or low on mangroves. On inland swamps, taller trees are used. At some sites, accompanies other cormorants, egrets, spoonbills, pelicans. Nests are substantial platforms built of plant stems, seaweed and debris. The eggs, in clutches of 2–4, have faintly blue shells with an overlying whitish chalky layer that soon becomes stained (60 × 40 mm). Incubation by both parents; nestlings have bare pinkish grey skin on the head. (p.66)

Black-faced Cormorant Breeding season varies; Sep.–Feb., probably also other months. Nests in pairs or colonies on bare, flat ledges of rocky islands or mainland coast, usually quite close to water's edge. Nest is a solidly constructed, large shallow bowl of driftwood, seaweed and other plant debris, 35–45 cm diameter. Clutch 2–3, elongated oval, faintly tinted blue-green (58 × 36 mm). (p.66)

Great-billed Heron Probably breeds most months, or triggered by monsoon rains. Solitary nesting, breeding pairs sited 1–5 km apart along rivers; remote sites, deep in mangroves. Builds in a horizontal fork of a large mangrove over river channel or tidal mudflat. A platform of sticks is lined with finer twigs and leaves, 1.2–1.4 m diameter. Clutch 2 (70 × 50 mm). (p. 72)

White-faced Heron Sep.–Dec. S Aust.; after monsoons in N tropics. Breeds in solitary pairs; rarely small, scattered colonies. Builds in a tall tree; lightly wooded country; can be 2–3 km from water. Stick nest is rough, flimsy, flat, 35–45 cm across; usually quite high on a horizontal fork. Clutch 3–5 (47 × 34 mm). Incubation shared, 21–24 days; young fly 40–45 days. (p. 72)

White-necked Heron Aug.–Feb. in S; Jan.–Apr. in N; after rain in semi-arid land. Usually in colonies, often with spoonbills, egrets, cormorants. Builds in tall, often dead trees standing in water; a wide, rough platform of sticks, 40–50 cm across. Clutch 4 (52 × 38 mm). Incubation probably by both parents, about 30 days. Young fly at 6–7 weeks. (p. 72)

Pied Heron Breeds Feb.–Apr. after floodwaters of Wet recede; in colonies, often with other herons, egrets, ibis. Builds 3–6 m above water in mangroves or other trees. Nest is a small, rough platform of sticks, 35–40 cm across, lined with fine twigs; slightly deeper, neater than other heron nests. Eggs are slightly stronger blue than other herons; clutch 3–4 (35–40 mm). (p. 72)

Eastern Reef Egret Some breeding throughout year, but mainly Sep.–Jan., Feb. Along N coast where species is common, nests in colonies on islands; further S, in pairs. Builds on forks or stilt roots of mangroves, pandanus and pisona, or on ledges, ground, and among shrubs and rocks. Platform is of sticks lined with seaweed; clutch 2–4 (46 × 35 mm). Both sexes incubate 25–28 days; young in nest 5 to 6 weeks. (p. 70)

Rufous (Nankeen) Night Heron In S, Sep.–Jan.; in far N, late wet season, Feb.–Apr. Nests in colonies, some of thousands, often with other herons, spoonbills and cormorants. Nest is a loosely constructed, almost flat platform of sticks on a horizontal fork in a tree standing in a swamp, lake shallows or mangrove swamp. Clutch 2–4, occasionally 5 (50 × 37 mm), pale blue-green. Both sexes incubate for 21–22 days. (p. 72)

Great Egret In N, late wet season, Mar.–May; in S, Oct.–Dec. In colonies with herons, ibis, spoonbills. Uses trees standing in water; builds in upper forks, often in topmost foliage, the nests frequently open to sky for unobstructed landing. Builds a rough, loose, shallow platform of sticks, diameter 50–70 cm. Eggs 2–5, pale blue-green (52 × 36 mm); both sexes incubate for 25–26 days; young fly at about 60 days. (p. 70)

Little Egret In S, Oct.–Jan.; in N, Feb.–Apr. In colonies, on fresh or saltwater wetlands; uses trees standing in water, often in mangroves, at times far from wetlands. Builds rough, shallow platform of sticks, 35–45 cm across. Clutch 4–6 (42 × 30 mm); incubation by both sexes, 20–25 days. Young scramble around branches near nest after 26 days; fly at 35–37 days. (p. 70)

Intermediate Egret Oct.–Jan. in S; Dec.–Mar. in tropical N; varies with rainfall, water levels. Breeds in colonies with other egrets, spoonbills, cormorants. Builds in trees over water, a rough shallow platform of sticks lined with a few leaves, 40–55 cm across. Clutch 3–4 (47 × 34 mm); both parents incubate, about 25 days. Young leave nest at about 63–70 days. (p. 70)

Cattle Egret Sep.–Oct. in SE; Mar.–May, or the end of the Wet, in the tropics. Breeds in colonies in clumps of trees or shrubs; often with other waterbirds. Most nests are 3–15 m high, each a rough, loose platform of sticks. Clutch 3–6 (45 × 34 mm); incubation by both parents, 24–25 days. Young leave nest aged 6 weeks, earlier if disturbed. (p. 70)

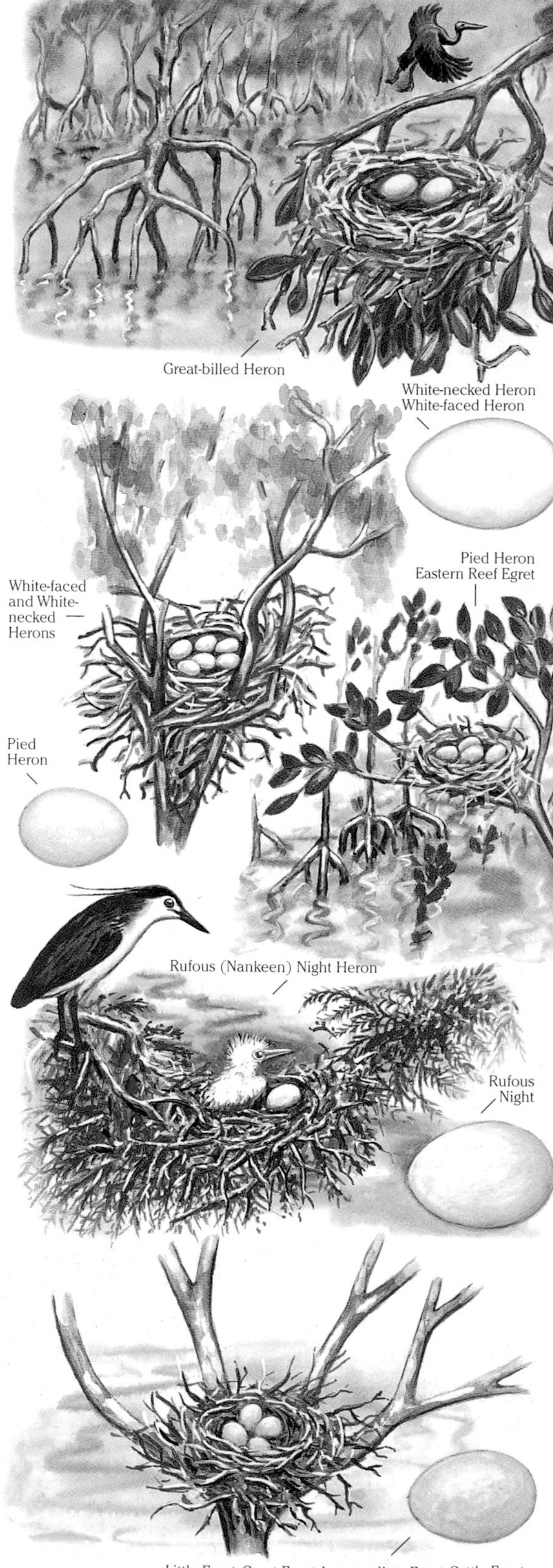

Little Egret, Great Egret, Intermediate Egret, Cattle Egret: nests and eggs similar, the latter vary mainly in size

Striated Heron Nov.–Apr. in N; Aug.–Dec. in SE. Solitary pairs build in a fork in a tree or shrub's dense foliage—mangrove, casuarina or melaleuca—in tidal or fresh swamps. Nest is a quite small, rough, loose platform of sticks lined with fine twigs. Eggs 3–4, very pale green (42 × 31 mm). Incubation 21–25 days. Chicks soon clamber about the branches near the nest. (p. 74)

Black Bittern Dec.–Mar. in N; Sep.–Dec. in S. Solitary pairs nest in secluded, densely vegetated wetlands; builds a substantial bowl of sticks, 35–45 cm diameter, up to 15 m high in a tree or large, dense shrub standing in or overhanging a swamp, lake or river. Clutch 3–5 (44 × 35 mm). (p. 74)

Little Bittern Dec.–Feb. in N; Oct.–Dec. in S. Solitary pairs. In optimum habitat, nests may be just 30–100 m apart. Builds a flimsy platform, 15–18 cm diameter, of reed stems, sedges and paperbark, usually 30–90 cm above water in a dense clump of reeds; occasionally in a densely foliaged tree overhanging water. Clutch 2–6 (33 × 25 mm). Incubation about 21 days, nestling in orange-buff down; young fly at 25–30 days. (p. 74)

Australasian Bittern Probably Sep.–Jan. throughout range. Solitary pairs breed in deep, densely vegetated swamps, building a large, rough platform of sticks and reeds just above water level, typically in shrubs standing in water within screening reed beds. Nest 45–70 cm diameter. Clutch 4–5, lustrous pale olive-green to light olive-brown (52 × 38 mm). (p. 74)

Royal Spoonbill Usually Mar.–May in N; Oct.–Mar. in SE and SW. In arid interior, any time after flooding rain. Often found with other waterbirds; colonies of 10–50 loosely scattered nests are often built in trees in or near water, in crowns and side branches. Others are low over water on reed beds, cumbungi or lignum. A large, substantial, dished platform built of sticks is lined with leaves and water plants; diameter 40–60 cm, similar to Yellow-billed. Clutch 2–3, white with fine red-brown speckling (69 × 42 mm). Incubation 20–25 days. (p. 76)

Yellow-billed Spoonbill Aug.–Mar. in S; Feb.–May in N; after heavy rain in semi-arid interior. Colonies nest in freshwater wetlands; build on trees, low shrubbery or reeds standing in water. Nest a wide, shallow platform of sticks lined with finer twigs, paperbark. Reed-bed nests are made of trampled reeds built up with added reeds and other plant debris. Clutch 2–3 (71 × 45 mm). Both sexes incubate, 27–30 days. Young fledge at about 5 weeks. (p. 76)

Straw-necked Ibis Mar. in N; Jul.–Dec. in S; throughout year or after heavy rain in some regions. Breeds in colonies, often with other ibis, egrets, spoonbills. Builds close together on dense reed beds or other vegetation—cumbungi, lignum, paperbarks or mangroves growing in water, occasionally on ground of wetland islands—a large, rough platform of interwoven, trampled plants lined with grass. Eggs dull, limy white, becoming dark stained; clutch 3–5 (65 × 44 mm). Both sexes incubate, 20–25 days; young fly after 30–40 days. (p. 76)

Australian White Ibis Feb.–May in N; Jun.–Nov. in S; after rain in arid regions. Breeds in large, closely packed colonies with herons, egrets and spoonbills. Rough, loose platforms are built in trees standing in water or close above water on trampled reeds, rushes, lignum or cumbungi. Nests are close-packed, sides can touch. Clutch 2–5 (65 × 45 mm). Both sexes incubate, 20–23 days. Young remain in nest about 48 days. (p. 76)

Glossy Ibis Oct.–Feb. in S; Feb.–Apr. in N; responds to rain and flooding, especially in semi-arid zones. In colonies, often with other species. Builds over water: platform of sticks in crown of a paperbark or mangrove; or dense, low trampled plants in water—reeds, lignum, cumbungi. Clutch 3–6 (52 × 35 mm). Both sexes incubate, about 21 days; young stay in nest 25 days. (p. 76)

Black-necked Stork (Jabiru) In N, probably breeds most months with a peak Mar.–May. In NSW, eggs in nests Aug.–Nov. Builds a massive stick nest, 1.2–1.6 m, even 2 m, diameter, up to 1 m deep, put as high as possible in a tree standing in or near water; usually lower when in the more stunted trees of mangrove swamps; occasionally on a low bush or stump when surrounded by water. Clutch 2–4 eggs, tapered oval in shape, shell coarse-grained, white, soon stained grey or brown (75 × 53 mm). Incubation by both adults, duration unknown. Nestlings initially in greyish down. Young remain in nest for 100–115 days. (p. 78)

Brolga During wet season in N, Jan.–Jun.; Aug.–Dec. in S. Solitary breeders; usually choose site with wide view of surrounds. A dry land nest may be no more than a bare patch of ground with a few sticks or other debris, just enough to prevent the eggs from rolling away. Other nests, usually those in a swamp, take the form of large mounds of trampled grass, reeds and other plant material, and are built up from the bottom in water less than about 30 cm deep, so that the eggs are above water level; sometimes such nests are semi-floating or on the ground on small islets. Slight depression on top holds the 2, very rarely 3, tapered oval eggs (94 × 60 mm). Incubation by both sexes, about 30 days. Young leave the nest 1 or 2 days after hatching and fly at about 14 weeks. (p. 78)

Sarus Crane Breeds early to late in the Wet, about Dec.–Mar. Builds nest on low vegetation in shallow water or beside water. Tends to use narrow wetlands between wooded ridges. The nest is a rough mound of grass and sedge dragged in from the water surrounding. Clutch size 2–3, white tinged blue (100 × 62 mm). Incubation by both sexes, about 30 days. (p. 78)

Osprey Breeds autumn to spring, egg laying as early as Apr. in N, as late as Oct. in S. Builds at sites with a commanding view—a high coastal headland, cliff top or offshore islands, especially small, high rock stacks that rise from the sea. Also uses trees, often dead, overlooking coasts and estuaries. Some nests are re-used for decades; new materials are added each year until it is a massive, flat-topped cone of sticks, bark, seaweed, driftwood, bones and other shoreline debris, both natural and man-made. Old nests, especially those on the ground, may eventually grow to be 2 m high and 2 m across. On the other hand, a new nest may be only a sparse collection of sticks. There are 2 or 3, rarely 4, eggs (60 × 45 mm) in a clutch, which is brooded by the female for some 33 days. The chicks initially have grey-brown down and are fed only by the female; the male does the hunting. Later both parents hunt to supply the nest. The nestling period extends for 50–60 days. (p. 80)

Brahminy Kite Breeds after wet season in N, Apr.–Jul.; spring months, Aug.–Nov., in S. Builds in a tree near or in water, typically at the seaward edge of a mangrove belt. The bulky mass of sticks, grass and seaweed is lined with leaves, high in a tree. Clutch consists of 1 or 2 eggs (52 × 48 mm). Both sexes incubate, but predominantly the female, for about 35 days. Downy nestlings are creamy on head, brownish on body; feathers show by 24 days. Young stay in nest 50–60 days. (p. 80)

Square-tailed Kite Breeds Aug.–Dec. in open forests or woodlands, usually near a watercourse in drier regions. Builds in the fork of a thick horizontal limb rather than in thin outer branches; may re-use old nests of other species. The nest is a substantial shallow bowl of large sticks lined with green eucalyptus leaves. Eggs 2–3, lightly to heavily spotted or blotched rust and brown (52 × 38 mm). Incubation mainly by female, about 40 days. Nestlings initially in white down; this is lost by 3 weeks of age. Female remains with young almost until they fly after about 9 weeks in the nest. (p. 82)

Black Kite In N, breeding recorded all months; further S, usually Jul.–Dec. The nest is a rough, slightly dished platform of large sticks high in the upper canopy of a tall tree, one of a clump or one in line along a watercourse; frequently among the bare branches of a baobab. Often re-uses nest of a crow or another raptor lined with various debris—dry leaves, wool, grass, bark. Clutch 2–3 (52 × 42 mm). Incubation about 35 days; young initially have grey down; fledge at about 40 days. (p. 80)

Whistling Kite In N Aust. breeds Feb.–Sep.; in S, Jul.–Nov. Uses tallest suitable tree available in woodland, near or standing in water, or beside a creek or dam, or, on coast, a tall mangrove; often a tree standing alone. Nest placed high, 10–30 m up in the tree. The bulky stick structure is lined with fresh leaves; the same one often re-used each year; grows gradually larger. Clutch 2–3, can be just 1 (55 × 43 mm). Incubation mostly by female; takes about 40 days. Hatchlings have short white down. (p.82)

Little Eagle Breeds May–Nov. in tropical N; in S, Aug.–Dec. Builds in a large woodland tree, often on a watercourse in scrub; as low as 5 m in stunted mallee, or as high as 45 m in coastal open forest. The nest is a platform of sticks some 70 cm in diameter that has a shallow depression lined with green leaves; often re-used. Eggs vary from almost white to finely spotted reddish brown and grey; markings are from quite strong to very faint. Usual clutch 2, occasionally 3 (56 × 45 mm). Incubation by both sexes, but predominantly female; about 35–40 days. Young hatch with pale, creamy white down; first pinfeathers show through at 3–4 weeks, down lost by 6–7 weeks; young finally fly at about 55–65 days. (p. 82)

Black-breasted Buzzard Breeds Jul.–Dec. Appears not to be influenced by rain—no evidence of unseasonal breeding after rain, but during prolonged, severe drought may not nest at all. Nests in semi-arid regions are typically built in a large tree beside a tree-lined watercourse, often in a dead tree or on a dead limb overhanging the watercourse, whether dry or containing pools. In the wetter tropics is more inclined to build in live trees, better screened by foliage, in open woodland. Nest is a large, shallow platform of sticks built on a stout, horizontal fork, kept lined with fresh eucalypt leaves; same nest, or another nearby, is used each year. Parents incubate equally for 36 days. Chick hatches in white down; fully feathered at 6 weeks, fledges at about 60 days. (p. 84)

Wedge-tailed Eagle Jun.–Nov. Appears not to breed in years of severe drought. Chooses a site with a clear view over a wide area, often in ranges overlooking plains, on side rather than top of a ridge or range, and often at the head of a gully. Also nests on open plains or in woodlands with restricted view—chooses one of the largest trees available with large lateral limbs or forks capable of holding the massive nest. In arid regions the nest may be within several metres of the ground, but usually as high as there are substantial limbs available. Nests are used over many years, growing as sticks are added annually. They begin as small as 1 m in diameter and 0.75 m deep, but commonly reach a diameter and depth of 2 m—nests of 2 m diameter and 4 m depth have even been seen. Clutch 1–3 (75 × 58 mm); both sexes incubate, 45 days. Hatchlings have pure white down. Usually only one, the oldest and strongest, survives. (p. 84)

White-bellied Sea-Eagle In N, breeds mid dry season, May–Sep.; in S, winter and spring (Jul.–Oct.). The nest is a huge pile of sticks on a cliff ledge or headland of remote coast or island, or high in a large tree near coast or river. Nests are usually within sight of water, re-used and enlarged each year until they reach over 2 m in diameter and almost as great a depth of accumulated sticks. Often a pair will use several nests randomly in rotation. Usual clutch 1–2 (75 × 59 mm). Both adults incubate, but predominantly the female, 35–40 days. (p.84)

Black-shouldered Kite Apr.–Dec. in S, also whenever prey is abundant; more erratic in semi-arid regions. May raise several broods in extended good seasons. Builds a stick nest lined with green leaves, bark, perhaps wool or fur; typically well hidden in dense leaves of top half of a tall tree; nest diameter 40–45 cm. Usual clutch 2–4 (41 × 32 mm); female incubates 30–34 days; male sits guard from nearby tree when not hunting. Nestlings first in sandy fawn down; turns grey by 9 days. Young fledge at about 35 days. (p. 88)

Letter-winged Kite Breeds whenever prey is abundant. Nests in loose colonies, 2–50 pairs, which keep breeding while season is good; many raise several broods, usually building a nest for each, often in the same tree. After breeding, the colony is abandoned and the kites disperse—can be over great distances. Nest an open shallow bowl, typically hidden in upper foliage; often of Beefwood, *Grevillea striata*. Clutch 3–6 (42 × 32 mm). Female incubates; male brings food and keeps guard. Young hatch in white down; fledge at about 35 days. (p. 88)

Pacific Baza Breeds Oct.–Feb. in a forest or woodland tree; prefers valleys with streams, or near pools or billabongs. Nest is usually concealed in dense outer foliage, large clump of regrowth or mistletoe in a horizontal fork towards the limb's outer extremity, 10–30 m up. The shallow, rather small bowl of sticks is well lined with green leaves. Clutch 3–4 (43 × 35 mm). Incubation by female; male brings food, briefly relieving her. Young hatch at 30–32 days; first down white; they stay in the nest for about 34 days. (p. 88)

Collared Sparrowhawk Breeds spring to early summer, Aug.–Dec. Nest is built in a fork of a tall tree, often hidden by foliage, mid level to high in the tree. Usually builds a new nest, occasionally refurbishes an old one of same or other species. The large, shallow bowl of sticks, about 50 cm diameter, is lined with green leaves. Leaves are added daily throughout the laying, incubation and nestling periods. Clutch 2–4 (40 × 31 mm); female incubates 35–37 days. After he brings food, male may brood eggs or small chicks while female leaves nest to feed and preen. The downy young are white. (p. 90)

Brown Goshawk Breeds Jul.–Dec. Builds in one of the largest trees available in locality, often near forest edge or stream, in a stand of trees on farmland, or beside a tree-lined inland watercourse. Nest is usually on a horizontal limb, high to very high, built of sticks and leaves, and kept lined with fresh eucalypt leaves. At times refurbishes another raptor's old nest. Clutch 2–4, usually 3 (45 × 36 mm). Female incubates, but male often takes over when she leaves nest. Young first in white down, replaced by salmon-brown after about 7 days. Parents defend nest aggressively; may attack humans nearby. (p.90)

Grey Goshawk Breeds regularly, Aug.–Dec. in S; in N, may be autumn also. Nest typically sited in heavily timbered country, often a gully or steep hillslope, where it is high in the canopy of a large forest tree; prefers spot open to morning sun, sheltered from strong wind. Nest is a large, shallow saucer of sticks lined with green leaves. Clutch 2–3 (47 × 37 mm). Female principal incubator; male relieves, but mostly hunts for both. Young hatch in white down, replaced by feathers between 14 and 21 days; leave nest at 40–45 days. (p. 90)

Red Goshawk Appears to have extended season through Dry; May–Dec. Builds high in live eucalypt or melaleuca on large limbs, often with view over surroundings. Nest is a large mass of sticks; the shallow bowl is lined with green leaves. Clutch 1–2 (54 × 43 mm). Incubation by female; male hunts for both, guards while female is off nest; about 40–43 days. Young first in white down; stay in nest about 50 days. (p. 90)

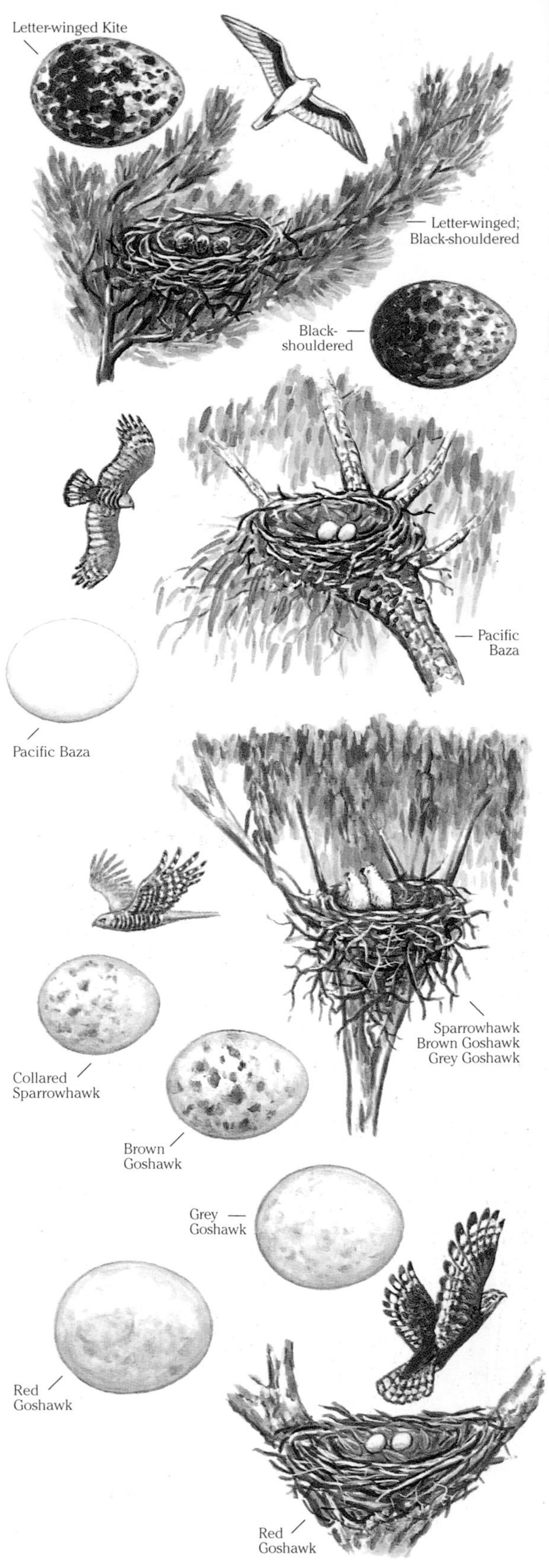

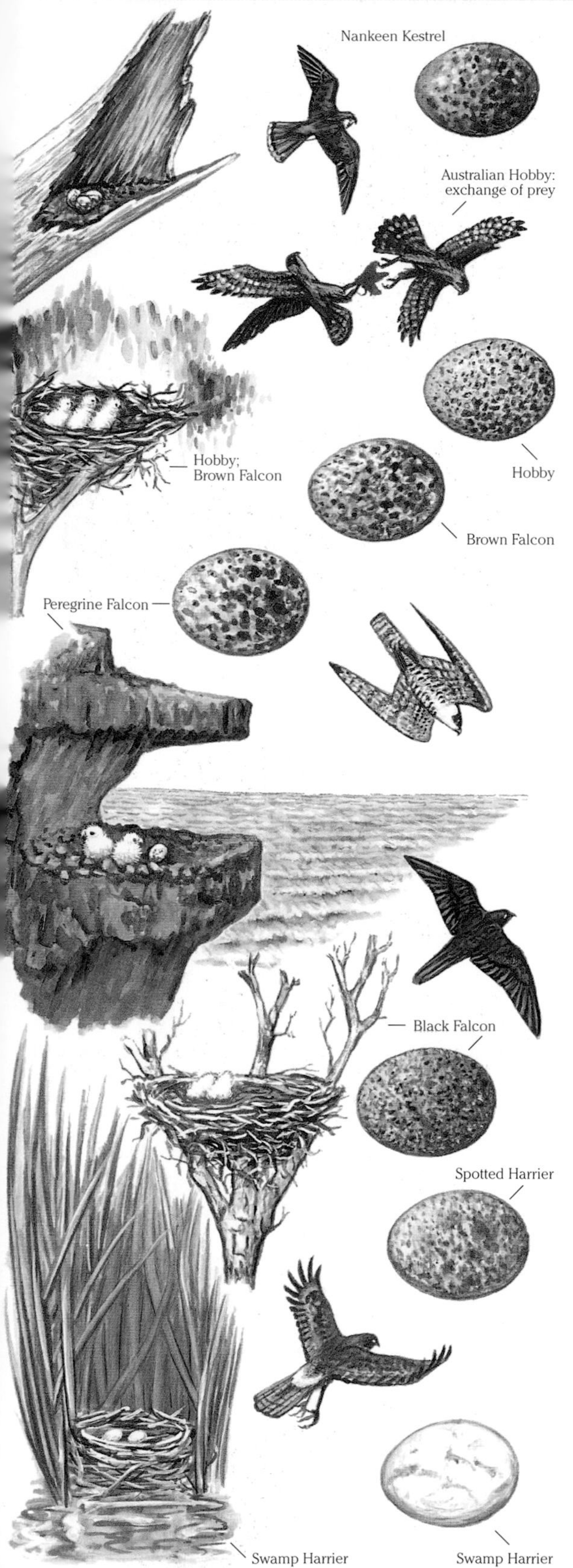

Nankeen Kestrel Most months; often Aug.–Dec. Nest is a tree hollow or cliff; otherwise a large, disused stick nest or man-made site such as a building ledge. Eggs are laid on wood dust in hollows or slight natural depression in cliff hole or ledge. Clutch 3–4 (37 × 30 mm). Incubation mostly by female; male occasionally is briefly on eggs, but mostly provides food during incubation, 28–29 days. Chicks first in white down; replaced by longer, creamy down; fledge at 28–31 days. (p. 92)

Australian Hobby Jul.–Nov. in N, Aug.–Jan. in S. Uses old nest of another raptor or corvid; picks one in highest part of tall tree on thin branches just below the canopy. Clutch 2–4 (44 × 33 mm). Incubation by female, except briefly by male after food delivery while female feeds nearby. Male does all hunting during incubation and for first 7–10 days after hatching. Young hatch in faintly pinkish white down; pale grey by 8 days; pale salmon-brown by 14 days; leave nest at 35–38 days. (p. 92)

Brown Falcon Apr.–Sep. in N; Aug.–Nov. in S. Uses old stick nest of raptor, crow or raven, babbler or magpie. No evidence of building own nest, but seems to reline nest with green leaves: picks one in the top of a tall tree, often on the sheltered side. Clutch 2–3 (49 × 38 mm). Incubation by female, 31–35 days; male brings all food. Young hatch in pale buff down; replaced by denser grey down within some 15 days. Tend to fly quite early, at about 30 days. (p. 94)

Peregrine Falcon Aug.–Nov. Builds no nest; uses ledge of cliff face; re-used many years; may rotate between several sites within territory. Peregrines will also use tree hollows (usually large, open, shallow like a cliff ledge); ledges of buildings in cities; large, high stick nest of other species. Clutch 2–3 (53 × 40 mm). Incubated by both sexes, about 33 days. Young in white down, creamy after 4 days; in nest about 40 days. (p. 92)

Black Falcon Jun.–Dec. Breeding appears largely confined to eastern interior, SW Qld to NW Vic. and adjoining NE of SA; sparse records elsewhere through interior. Uses an existing stick nest: usually disused nest of another raptor; may take over one in use. May scratch away accumulated debris to restore the slightly dished top, but does not restore the structure. Clutch 3–4 (42 × 32 mm). Female incubates, about 33 days. Hatchlings in buffy white down; fledge at about 40 days. (p. 94)

Grey Falcon Jul.–Oct. in S; Apr.–Jun. in N. Takes large, old stick nest of another raptor, crow or raven, such as that of an eagle or Black-breasted Buzzard, high in one of the tallest trees in the locality, usually near a river pool or dry watercourse. The old nest is scratched clean for the clutch of 2–3 eggs (52 × 39 mm). Incubation by female; male takes over while she feeds in a nearby tree after he brings her the prey. Young are hatched in off-white down. They are fully feathered at 40 days; fledge at 45–50 days. (p. 94)

Spotted Harrier Sep.–Dec. Usually in dense outer canopy foliage; usually high, occasionally quite low. Often constructed where a cluster of branchlets—regrowth on a broken branch or a large mistletoe clump—supports and conceals the nest, which is a large, rather flat, untidy platform of sticks lined with green leaves. Clutch 2–4, white or faintly blue (50 × 40 mm). Incubation approximately 33 days; chicks initially have off-white down and white facial discs. (p. 86)

Swamp Harrier Sep.–Dec. Builds nest of reeds and other plant material in dense reed beds of swamp or lake. May semi-float on reeds in water up to 1.5 m deep; or sit on dry ground in reeds, tall rank grass or crops. Clutch 2–5 (52 × 40 mm); female incubates, 32–34 days; hatch at intervals of 1–2 days. First down is white, second light grey. Usually only 1 or 2 oldest and largest chicks survive to fledge at 42–45 days. (p. 86)

Bush-hen Sep.–Apr.; may have 2 broods. Nest built in a large clump of grass or reeds, or centre of a grass-filled shrub—then it may be 2 m above ground; usually close beside or surrounded by shallow water. Constructs a shallow bowl of soft grass bent down from on-site grass, but with extra grass woven in. Other plants may be bent over to form a sparse canopy. Often has a ramp of vegetation sloping into the water. Clutch 3–6 (30 × 22 mm). Incubation by both sexes; takes about 18 days. (p. 98)

Purple Swamphen Breeds most months: mainly Jul.–Dec. in S; Dec.–Apr. in N; often several broods. Builds in a clump of reeds just above water level in a swamp or lake. The substantial nest bowl is formed of trampled rushes lined with finer, softer stems and grass. The 3–8 eggs (52 × 35 mm) are incubated by both sexes, often assisted by immatures. Incubation 23–29 days. Downy young are blackish with red-brown crown, pale grey bill and frontal shield, pink legs. (p. 96)

Dusky Moorhen Usually Sep.–Nov. in S; Jan.–Apr. in N. Builds a bulky bowl of sticks, reeds, bark and grass slightly above water level in a clump of reeds, tall rank grass, low-forked shrub, tree butt or log. Often nests are not well hidden. Clutch 4–9 (53 × 36 mm). Downy young are black with fine white tips to down of chin and throat; pinkish skin visible on head; bill and frontal shield red; bill tip yellowish. (p. 96)

Tasmanian Native-hen Group breeds Aug.–Dec.: usually two males, one female. May raise 2 or 3 broods in a season. Nest usually at or near banks of wetland, well hidden in dense clumps of overhanging sedges, reeds or ferns, or introduced blackberry thicket. A bulky cup of reeds and grass, lined with grass and with overhead canopy, is built; the entrance usually faces towards water. Usual clutch 5–8 (55 × 38 mm). Incubation by adults in the group, about 22 days. (p. 96)

Black-tailed Native-hen Opportunistic; in semi-arid regions breeds any time of year after heavy rain when it responds rapidly—may arrive in locality within a week and begin breeding. In SE and SW regions of regular winter rain, usually Aug.–Dec. Builds in dense ground vegetation; nest well hidden, cup-shaped; made of stems, reeds and leaves lined with feathers, grass, leaves. Clutch 5–7 (45 × 32 mm). Incubation 19–20 days. Downy young are greenish black. (p. 96)

Eurasian Coot Breeding season influenced by rain; usually Aug.–Mar. Nests in wetlands surrounded by water, often building up a heap of water weeds and other vegetation. Nests float, or are built from bottom of shallow water on a low or barely submerged stump or log; makes a neat, large bowl. The clutch is 4–12, usually 5–7, tapered, oval eggs (52 × 35 mm). Incubation by both sexes; lasts 22–26 days. Downy young are blackish with orange-pink skin on head and bill, and with red frontal shield and face. (p. 96)

White-browed Crake Sep.–Apr. Builds in dense reeds or grass in the shallows of a swamp or lake. The small, shallow bowl often has vegetation pulled over to form a rather sparse hood or canopy, and a ramp slopes down to the water. The usual clutch is 4–8, oval (28 × 20 mm). Incubation by both sexes, about 21 days. Down is black on newly hatched chicks; becomes rusty tan when they are half-grown. (p. 98)

Lewin's Rail Aug.–Jan. Builds in dense rushes, sedges or tall coarse grass surrounded by shallow water or mud—a small bowl of interwoven reeds and grass, slightly above water level with a runway, almost a tunnel, through the dense vegetation to the water. Overhead grass may form a canopy. The 4–6 eggs are oval and slightly lustrous (35 × 26 mm). Female incubates, about 20 days. Hatchlings have black down. Young of crakes and rails leave the nest within a few hours of hatching. (p. 98)

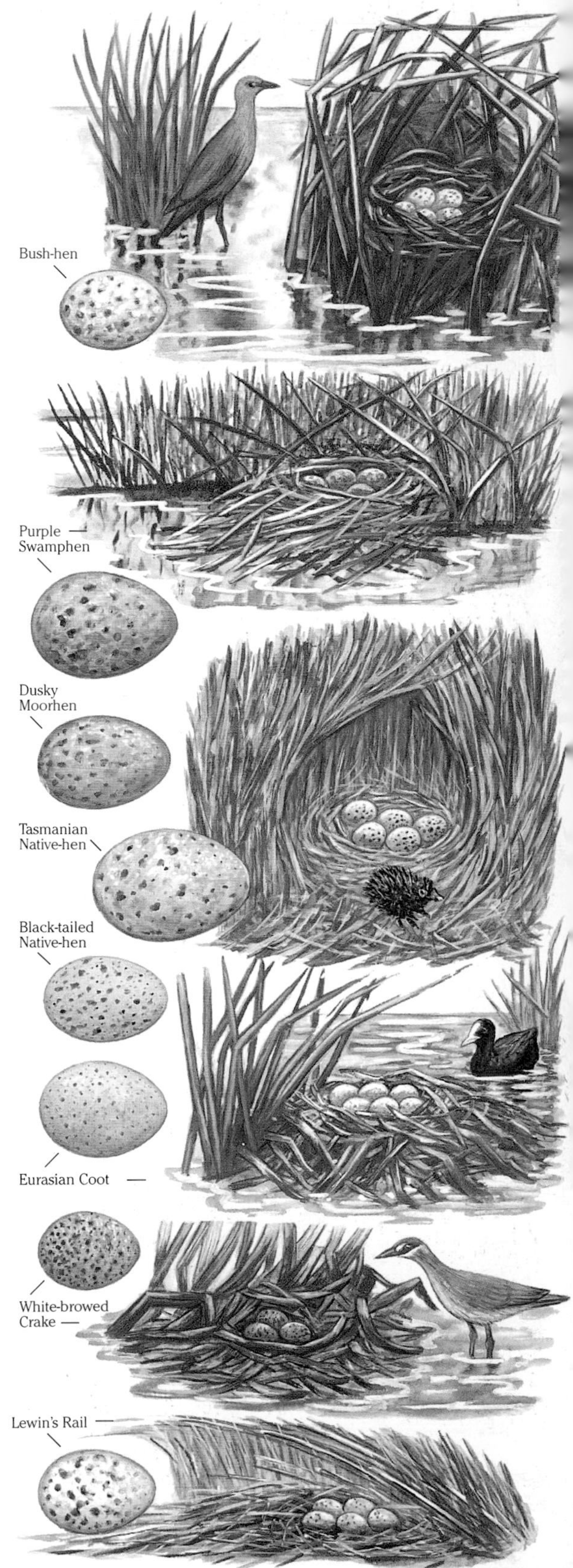

Baillon's Crake Aug.–Jan. in S; wet season, Dec.–Apr., in N. Nest is well hidden in dense vegetation above water level of a swamp or lake. The saucer of reed stems, grasses and aquatic plants often has an approach ramp sloping up from the water, and may have plants pulled close overhead as a rough canopy. Clutch 4–8 (28 × 20 mm); incubation by both sexes, 16–18 days. Downy young are greenish black with pale yellow bills. (p. 98)

Australian Spotted Crake Aug.–Feb. Nest like Baillon's Crake: usually a shallow bowl of reed stems, grass and aquatic plants hidden just above water level in a big, dense clump of reeds or grass. Other stems may be pulled over in a canopy. Clutch 3–6 (31 × 23 mm). Hatchlings have black down, white tipped on head and back; legs and feet are blue-black. (p. 98)

Spotless Crake Aug.–Jan. A bowl of reed stems, leaves and grass is hidden in a dense clump of swamp plants—grass, sedge or reeds; close to water level, up to 1 m above. An approach ramp or path of trampled plants slopes to the water. Overhead vegetation is often pulled over as a rough canopy. Clutch 3–6 (30 × 25 mm). Both sexes incubate for 19–22 days. Hatch in black down; bill black. (p. 98)

Red-necked Crake Dec.–Mar. Builds substantial bowl of leaves, twigs up to 2 m above ground in well-supported position —hollowed top of a stump, epiphytic bird's nest fern or crown of pandanus palm—but often just a shallow bowl of leaves on the ground beside buttress root or other shelter. Eggs glossy white, but are soon stained; clutch 3–5 (37 × 27 mm). (p. 100)

Buff-banded Rail Usually Aug.–Dec. in S; wet season in N; may have 2 broods some seasons. Nest closely hidden in dense vegetation, usually near water, but can be well away from it. The bowl of reeds or grass may be built low in a tussock, in a slight ground hollow, or under rocks or logs. Clutch 5–8, lustrous (36 × 28 mm). Incubation by both sexes, 18–19 days; young leave nest soon after all have hatched. (p. 100)

Chestnut Rail Sep.–Mar. Nest is built on lower branches or stilt roots of mangrove trees above reach of high tide, 1–3 m up; constructed of sticks, bark and seaweed. Apparently some nests are re-used each year after being refurbished. The parent rails reach the nest by climbing roots and branches rather than by flying. Clutch 3–4 (52 × 36 mm). (p. 100)

Australian Bustard Breeds Nov.–Jun. in N; Aug.–Nov. in S. Lays eggs on bare ground, often beside spinifex clump or low shrub. Clutch usually 1–2, rarely 3 (78 × 54 mm); incubation by female. Downy young buffy brown, striped and mottled darker brown. Young soon leave nest site; in alarm, they crouch motionless under spinifex or low bush where their mottled and striped pattern merges with the network of cast shadows. (p. 100)

Painted Snipe Dec.–May in N; Oct.–Dec. in S. Nest close to, often surrounded by, water—on an islet or dense, low samphire bushes in shallow water. May be a shallow, sparsely lined scrape on the ground or a substantial platform of vegetation, usually in low samphire bushes, grass or other partly concealing plants, some of which may be pulled over as a sparse hood. Clutch 4 (36 × 25 mm). Incubation by male, 19–20 days. (p. 106)

Comb-crested Jacana Sep.–May. Builds on water lilies or other floating plants in water a metre or more deep. The shallow platform of aquatic vegetation is low and wet, barely keeping the eggs above water. Clutch 3–4 (30 × 23 mm). Incubation by both sexes. If disturbed, the Jacanas may carry the eggs or chicks under their folded wings to a new site. Very soon after hatching the young are able to dive to escape danger. (p. 126)

Red-backed Button-quail Breeding varies with rainfall; usually Oct.–Jul. The nest is a small hollow lined with grass scratched in the dirt beside tall grass, often near flooded ground. May have a sparse overhead canopy. Clutch 2–4 (23 × 18 mm). At first chicks have dark brown down, two black stripes down back, white spots. Male incubates, cares for chicks. (p. 102)

Little Button-quail Any time after good rain. Nest mostly sheltered by grass tussock or low shrub; slight hollow in ground well lined with grass. Surrounding plants may be pulled over and interwoven with grass to form a domed hood, sparse or quite dense. Clutch 3–5 (27 × 20 mm); incubated by male, 13–15 days. Young hatch in dark brown down; male cares for chicks, which quickly leave nest site. (p. 102)

Red-chested Button-quail Variable; usually Sep.–Feb. in S, Feb.–Jul. in N. Nest is a small, grass-lined hollow in ground at base of a tussock; stems are pulled down, woven with more grass to form a hood. Clutch 2–4 (23 × 18 mm); incubated by male, 14–18 days; reach adult size in 6–8 weeks. (p. 102)

Chestnut-backed Button-quail Breeds through and following the Wet in N, usually Nov.–Jun. A small hollow under a grass tussock or a low bush is lined with grass and leaves that may be interwoven to form a partial or complete encircling wall, or a hood that will mask the nest from overhead. Clutch 2–4 (26 × 22 mm); incubated by male. (p. 102)

Buff-breasted Button-quail Wet season in the north, Dec.–May. Nest is a grass and leaf lined hollow in the ground under a grass tussock or low bush, often with a domed, screening hood. Clutch usually 3–4, pyriform oval (27 × 23 mm); male incubates and rears young. (p. 102)

Painted Button-quail Any month in N; Aug.–Feb. in S. Nest is a small hollow scratched in ground beside a grass tussock or low shrub. Lining of leaves and grass may extend upward as low walls or be interwoven with overhead grass to form a hood of variable extent and density. Usual clutch 3–4 (27 × 20 mm). Chicks in dark brown down with paler back stripes, head stripe; chin and throat white. (p. 104)

Black-breasted Button-quail Breeds most months in favourable conditions, but mostly Sep.–Mar. Nest is a shallow depression beside a grass tussock or low bush; lined with grass and leaves; may have surrounding vegetation pulled down and grass added to form a rough hood. Clutch 3–4 (28 × 21 mm). Eggs incubated by male, 15–16 days; male rears chicks, which are russet-brown with black and cream stripes. (p. 104)

Plains-wanderer May–Feb. if rainfall is sufficient; may not breed in drought years. Nest is a hollow about 3 cm deep scraped in ground among clumps of the sparse grass typical of habitat, lined with grass, often with some overhead screening of grass. Clutch 2–4 (30 × 23 mm). Incubation mostly by male, 23 days. Chicks first in creamy buff down with brown markings; underparts white and buff. (p. 104)

Bush Stone-curlew Aug.–Jan., but almost any time if conditions suit. Nest may be a shallow unlined scrape beside a low bush or log, but the well-camouflaged eggs are also laid on bare ground away from concealment. Clutch 2 (59 × 39 mm). Both sexes incubate for 28–30 days. Near time of hatching, sitting bird may lie flat when approached. Chicks walk within hours. Fully grown by 7 weeks. (p. 128)

Beach Stone-curlew Aug.–Feb.; usually Oct.–Dec. Nest is a shallow hollow scratched in sand or shingle beach, close above high-tide mark, often among seaweed and other wave-washed debris; at times by an estuary or creek close behind ocean beach. The single egg is 65 × 45 mm. Incubation by both sexes. Chick hatches in grey-buff down with black stripes. (p. 128)

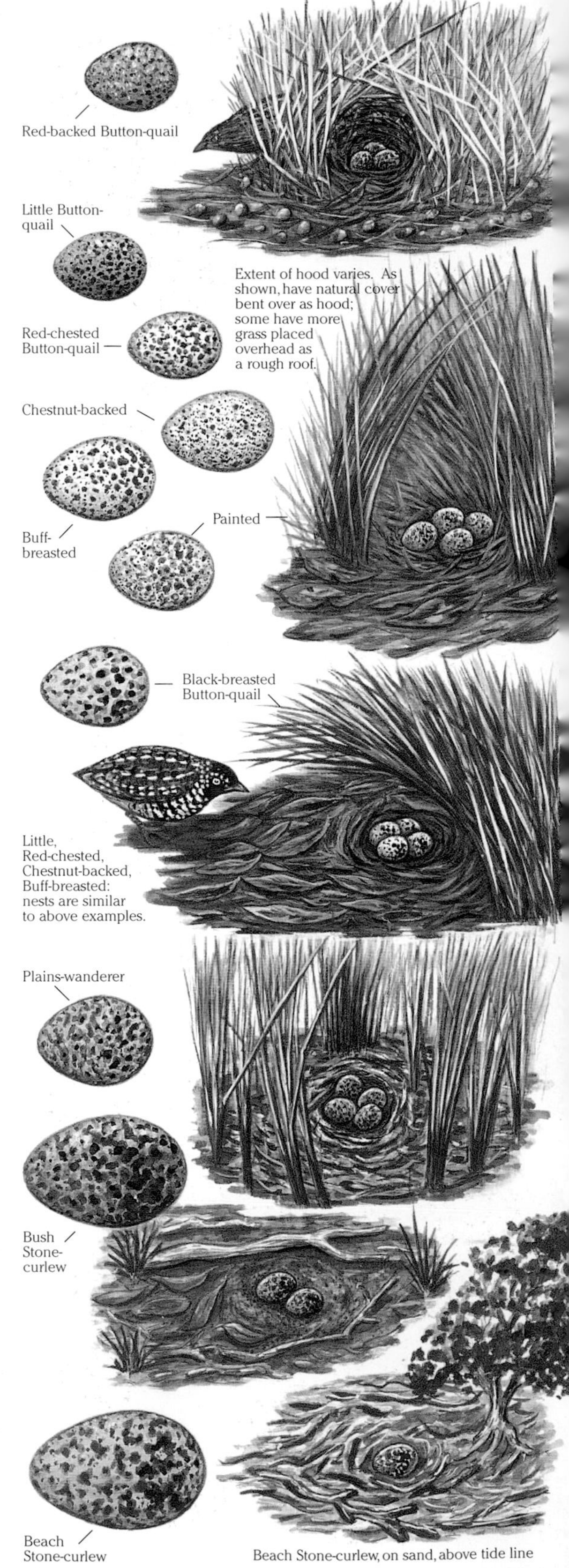

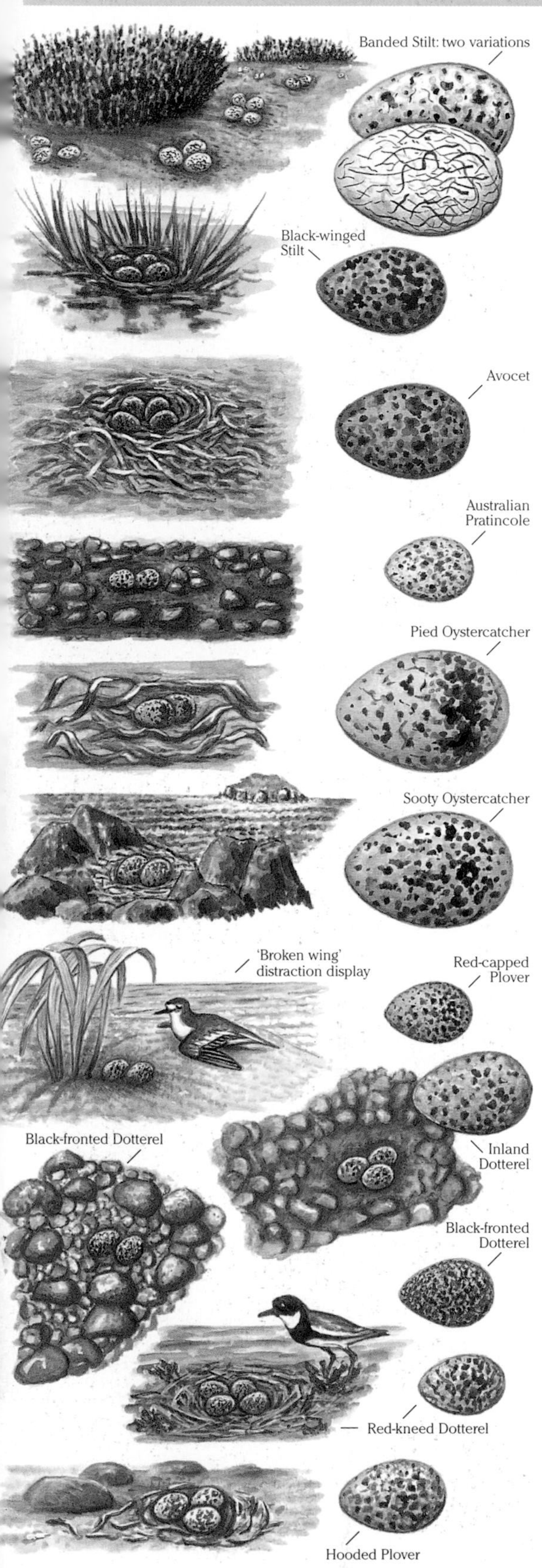

Banded Stilt Breeds almost any time after heavy rain fills lakes, but mostly Jun.–Nov. Nests in colonies, often thousands, on islands of temporary salt lakes, southern interior, WA and SA. Clutch 2–4 (54 × 40 mm). Eggs are laid in a slight hollow in soft ground; incubated by both sexes for 3–4 weeks. Young soon gather in large creches with adults on lakes. (p. 128)

Black-winged Stilt Usually Aug.–Nov.; inland and N, any month after substantial rain; solitary or small, scattered colonies. Nest is a small platform of plant material built up in shallows, or a shallow depression in ground of islet or beach. Clutch 3–4 (45 × 30 mm); both sexes incubate for 22–24 days. (p. 128)

Red-necked Avocet Aug.–Dec. or after good rain. In loose, small colonies beside water, or on islets of swamp or lake. Nest on ground: mound of sticks or slight depression lined with fine plant debris; around samphire or other small bushes. Clutch 3–5 (51 × 35 mm). Both sexes incubate, 3–4 weeks. (p. 128)

Australian Pratincole Any month after sufficient rain; usually Feb.–Aug. Eggs laid on earth—picks a flat spot on rough ground or small open area among gibberstones. May have a sparse ring of encircling stones or debris. Clutch 2 (32 × 23 mm); markings quite variable. Both sexes incubate, 20–21 days. Chicks leave nest for a more concealed, shaded hiding place very soon after hatching. (p. 130)

Pied Oystercatcher Usually Aug.–Jan., but early as Jun. in N. Nest is a shallow scrape in sand well above high-tide mark; often in seaweed, shells, small stones. Clutch 2–3 (60 × 41 mm); female is main incubator, 28–32 days. Young leave nest within several days. (p. 130)

Sooty Oystercatcher Jul.–Dec. in N; Sep.–Jan. in S: on off-shore islands, reefs above high-tide level. Nest is a shallow sand scrape near rocks or a slight depression on rock, sparsely lined with small pieces of plant (thicker layer on rocks). Clutch 2–3 (63 × 43 mm); both sexes incubate, 25–26 days. (p. 130)

Inland Dotterel Season varies; depends on rain; Feb.–Nov., mostly Aug.–Oct. Nest as pairs or in scattered colonies. Scrapes a hollow in bare open ground; rakes stones and debris into a circle, possibly adding material to encircle the eggs. Covers eggs with loose sand on leaving the nest. Incubation by both sexes. Clutch 3–5 (38 × 27 mm). (p. 136)

Red-capped Plover Most months, especially in N; usually Sep.–Dec. in S. Nest is a shallow scrape in sand on higher part of beach, or beside inland lake, flooded salt lake or claypan. Often partly sheltered by a small shrub or debris. May be a sparse circle of stones, debris or pebbles. Clutch 3–5 (37 × 27 mm). Both adults incubate 31–33 days and care for young. (p. 134)

Black-fronted Dotterel Aug.–Dec. in S; dry season in N. Makes a shallow scrape on claypan, gravel, sandbar, among riverbed stones or on open ground. Nests are usually very close to water; often among shoreline debris. Clutch is 1–3, usually 2 (29 × 21 mm). Incubation by both sexes, 25–26 days. (p. 138)

Hooded Plover Aug.–Jan. Nests on higher part of beach, above high-tide level, often partly sheltered or camouflaged by driftwood, sparse low vegetation or shells. Nests beside inland lakes tend to be closer to water level. Nest is a shallow scrape in sand or pebbles, sometimes encircled and lined with small stones, seaweed or debris. Clutch 2–3 (38 × 27 mm). (p. 138)

Red-kneed Dotterel Breeding season varies; follows rain, but usually Sep.–Dec. Scrapes a small hollow in sand or clay beside a lake, swamp or temporary inland claypan, often in the shelter of a clump of grass or small shrub; nest may be lined with fine twigs or other dry plant material. Clutch 2–3 (30 × 21 mm). Incubation by both sexes. (p. 138)

Masked Lapwing Jul.–Nov. in S; Nov.–May in N. Scrapes small hollow on bare ground; usually well away from cover. Unlined or sparsely lined with dry grass, rootlets or other plant debris; often several broods. Clutch 3–4 (43–49 × 31–36 mm). Sexes share nest preparation; incubation (about 28 days); care of young. Birds swoop aggressively to defend nest. (p. 138)

Banded Lapwing Jul.–Dec. in S; autumn in N; any month after rain in arid regions; pairs or small colonies. Nests are built in open, usually well away from shrubbery or other cover—low sparse pasture, playing fields, extensive bare stony ground, claypans. Scrapes a shallow hollow in the ground, sparsely lines it with local vegetation. Clutch 3–5 (42 × 31 mm); nest activities shared by sexes. Nest site defended strongly; also uses 'broken wing' distraction display. (p. 138)

Pacific Gull Sep.–Dec. in small scattered colonies or isolated pairs. Typical nest site is on higher part of islets or headlands overlooking coast. Makes a deep scrape in the sand, thickly lined with seaweed and plant debris. Pair shares nest activities. Clutch 2–3 (75 × 51 mm). Incubation 26–28 days. (p. 142)

Silver Gull Usually Aug.–Dec.; often in autumn; any month in N. Breeds in large colonies, very often dense, on sparsely vegetated small islands or sand spits; also on inland lakes, floodwaters. Typically makes a shallow scrape on the ground lined with vegetation, but often nests on low, dense shrubbery. Clutch 2–3 (53 × 38 mm). Both share nest preparation, incubation (22–26 days) and care of young. (p. 142)

Kelp Gull Sep.–Jan.; breeds in small colonies, widely spaced nests, or isolated pairs on a few islands off the coast, SE mainland and E Tas. The nest is a shallow depression lined with massed plant material; usually sheltered by a large tussock of dune grass, stunted shrub or rock. Clutch 2–3 (71 × 50 mm). Sexes share building, incubation and care of young, which fly at about 6 weeks. (p. 142)

Crested Tern Breeds Sep.–Dec. in S; Mar.–Jun. in N. Large, dense colonies form on small islands; thousands nest on sand or shingle among low vegetation behind the beaches. Nest is a small bowl-like scrape in the sand, unlined. Clutch usually 1, occasionally 2 (61 × 42 mm). Incubation, 25–26 days, and care of young are shared by both adults. (p. 146)

Lesser Crested Tern Mar.–Jun. on NW coast; Sep.–Dec. on E coast. Closely packed colonies, hundreds or thousands of birds, on low sand and coral cays of reefs, offshore islands or sandbars; often with other terns. Nest is a shallow, unlined scrape in the sand. Clutch 1 (50 × 36 mm). Sexes share incubation, 22–26 days, and care of the chick. (p. 146)

Caspian Tern Sep.–Dec. in S; almost any time of year in far N. Colonies may be small, occasionally large, densely packed. Commonly nest on offshore islands in shallow depressions in the sand, circled by flimsy collections of debris. Clutch 1–3, usually 2 (64 × 44 mm). Sexes share nest preparation. Incubation 20–22 days. (p. 146)

Gull-billed Tern Sep.–May, mostly Oct.–Dec. Forms large colonies to breed on islets in temporary lakes and saltmarshes of the interior and dry coastal regions. Nests on sand ridges and mud banks. Nest is a shallow depression scraped in sand or dried mud, lined with a few pieces of vegetation. Clutch 2–3 (50 × 37 mm). Incubation by both sexes, 22–23 days. (p. 146)

White-fronted Tern Breeds on islands of Furneaux Group, Bass Str.; possibly a fairly recent colonist from NZ. Oct.–Feb. on islets, rock stacks, beaches in small colonies. Each pair makes a shallow scrape in sand. Clutch 1–2 (48 × 32 mm). Sexes share nest building, incubation of 25–27 days and care of young, which fledge at about 30–35 days. (p. 148)

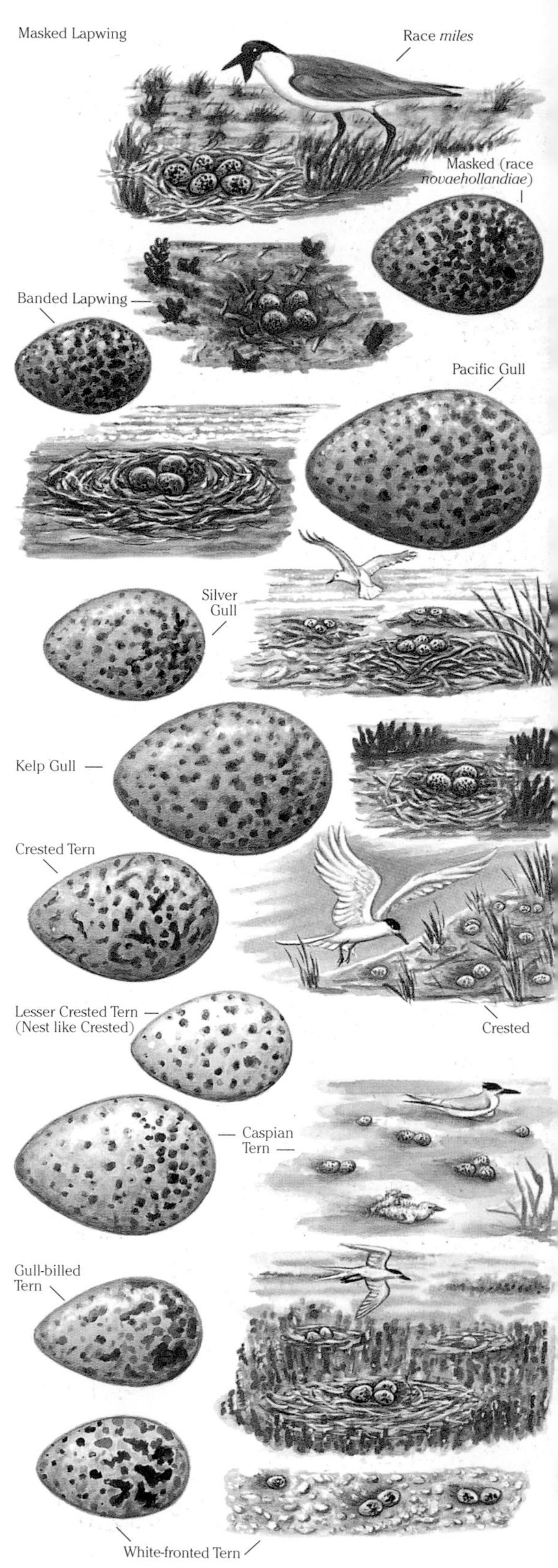

Lesser Noddy colony, Houtman Abrolhos
Similar: Black Noddy

Black-naped Tern Breeds Sep.–Jan. in small colonies on offshore coral islets, often with other terns. Nests are shallow scrapes in sand or coral debris not far above high-tide line; some nests are lined with small shell and coral fragments. Clutch 1–3, usually 2 (40 × 30 mm). Sexes share nest construction and incubation of about 25 days. (p. 148)

Little Tern Sep.–Mar. in SE; earlier in N. One brood a year. Small, rather scattered breeding colonies form along sandy beaches only slightly above high-tide mark where nests may be washed away, but people's activities more often disturb or destroy them. Camouflaged eggs and young are almost impossible to see once the adults have flown off. The nest is a small, inconspicuous, shallow, unlined depression in the sand. Clutch 2–3 (29 × 22 mm). Incubation about 19 or 20 days. (p. 150)

Roseate Tern Mar.–Jul. on islands off Kimberley; Sep.–Jan. in S. Breeding colonies form on offshore islands: may be small, or may have several thousand closely packed nests. Each is a shallow scrape in sand or a small cleared patch among rough coral debris. Clutch 1–3 (41 × 28 mm). Nest tasks are shared, including the 23–25 days of incubation. Young can fly at about 30 days. (p 148)

Fairy Tern Jul.–Nov. in NW; Sep.–Mar. in extreme SE of range. Nesting colonies are relatively small, a few pairs, usually up to 50, occasionally several hundred. Colonies typically are scattered along an open beach or the shores of islands, estuaries or coastal lakes. Each nest is a shallow scrape in the beach sand, often encircled by small pebbles and shell fragments. Clutch 1–3, rounded (35 × 25 mm). Both sexes incubate the eggs for about 18 days. (p. 150)

Whiskered Tern Breeding season variable; depends on rainfall and flooding of wetlands. Feb.–Apr. in N; Aug.–Nov. in S. Nests in colonies, often large, situated in shallow, often temporary, waters of inland swamps. The nest is a rough platform of vegetation built on samphire or other low shrubs standing in shallow water. Clutch 2–3, oval (37 × 27 mm). Each pair shares the nest building and incubation, which takes 18–20 days. The birds vigorously and noisily defend the colony by diving at intruders. (p. 152)

Sooty Tern May–Aug. in N; Sep.–Nov. in S. Forms huge colonies on favoured islands; tens of thousands of birds crowd together. Each nest is a slight hollow scraped in the sand, unlined except by pre-existing fine grass. Clutch 1 (52 × 36 mm). Sexes share nesting duties, 26–28 days. (p. 154)

Bridled Tern Sep.–Feb.; in Kimberley region often a second breeding period Mar.–Jun. Forms large, loose colonies on oceanic or continental-shelf islands. The eggs are laid on the sand or a stony surface under cover of a small shrub or rock overhang, usually not far from high-tide line. Clutch 1–3, usually 2 (56 × 32 mm). Both sexes incubate, 28–30 days. (p. 154)

Common Noddy May–Oct. in N; Aug.–Feb. in S. Forms huge colonies on some islands. A large platform of seaweed, leaves and other plant material is built on the dense top of a low shrub or on the ground. Clutch 1 (52 × 36 mm). Sexes share building and the 33–35 days of incubation. (p 154)

Black Noddy Breeding variable; mostly Jun.–Jan. in large, dense colonies; can be tens of thousands of birds. Builds a bulky nest of seaweed and other vegetation, securely plastered into the fork of a stunted tree or shrub. Clutch 1 (45 × 31 mm). (p. 154)

Lesser Noddy Aug.–Jan. Large dense colonies on islands of the Houtman Abrolhos, mid W coast of WA. Bulky seaweed and grass nests are cemented into the forks of small mangroves or shrubs. The single egg is 45 × 31 mm. Sexes share nest building and incubation. (p. 154)

Banded Fruit-Dove Breeds Apr.–Nov. Builds a very flimsy, sparse, almost flat platform of thin sticks across a horizontal fork in dense foliage, some in vines or creepers; egg often visible from below. Selects tree or shrub in monsoon thickets near cliffs of gorges or escarpments. Clutch 1 (36 × 26 mm). (p. 156)

White-headed Pigeon Sep.–Jan. Nest usually hidden high in rainforest canopy or other dense vegetation. A small but quite substantial dished platform of sticks and vine tendrils is built on a horizontal branch in dense foliage or tangled vines below the canopy, 4–20 m above ground. Clutch 1 (40 × 31 mm); incubation about 20 days. (p. 156)

Pied Imperial-Pigeon Aug.–Jan.; colonies in mangroves of offshore islands; a few in mainland rainforest. Builds a solid, small platform of sticks, often attaches green leaves, in mangrove fork 1–5 m up; higher in rainforest. Clutch 1 (45 × 31 mm). Both incubate 26–28 days; young fledge at 3–4 weeks. (p. 156)

Topknot Pigeon Jun.–Jan. Hidden in rainforest foliage; hard to spot; high, usually 20–40 m up. Builds substantial (but at times thin and flimsy) dished platform of vine tendrils and twigs. Clutch 1 (43 × 29 mm). Both sexes incubate, 22–24 days; young fledge at 3–4 weeks. (p. 156)

Crested Pigeon Breeds after rain in dry regions; several broods in best seasons; mostly Jul.–Dec.; all months in N. Nest is a rough, frail platform of twigs 15–30 cm across, built in dense foliage 1–6 m up. Clutch 2 (31 × 23 mm); both sexes incubate, 18–20 days. Young leave nest at about 3 weeks. (p. 158)

Common Bronzewing Breeds all year if conditions suit; mostly Aug.–Dec. Builds rough platform of slender twigs, usually flimsy; sometimes quite solid; 15–25 cm across. Sites range from very low to high, on solid, thick fork or branch, mistletoe clump, shallow hollow, stump top or in top of old nest of a magpie, crow or babbler. Clutch 2 (34 × 25 mm). (p. 158)

Brush Bronzewing Any month, mostly Oct.–Jan. Nest is a shallow saucer of twigs and rootlets; can be flimsy or quite substantial; in a dense, low shrub or tree, sometimes on ground in low cover or concealing debris. Structure is similar to other pigeon nests shown. Clutch 2, glossy, elliptical (34 × 25 mm), almost identical to Common Bronzewing eggs. Sexes share incubation, 16–18 days; young fledge by 3 weeks. Even more timid than Common Bronzewing: if flushed from nest, will readily desert nest and eggs. (p. 158)

Spinifex Pigeon Breeding determined by rainfall in drier regions; commonly Aug.–Dec. After summer wet season in tropics. Nest is a shallow scrape in bare, often rocky ground in the open or sheltered by spinifex or a low shrub; may be lined with grass or fine twigs. When on the nest, this pigeon's patterned cinnamon tones merge with the red earth, rocks and spinifex. The 2 eggs are white (27 × 20 mm). (p. 158)

Flock Bronzewing Breeding season varies with rainfall: Mar.–Sep. in N; Aug.–Dec. in S. The many breeding pairs in a flock nest close together, each making a shallow scrape, some sparsely lined with grass. Nest sites are usually sheltered by grass tussocks or low shrubs. Clutch of 2 creamy white, elliptical eggs (33 × 25 mm). Incubation by both sexes, female at night; takes 16 days; young wander from nest at about 1 week. (p. 158)

Squatter Pigeon Breeding varies: mostly early in dry season; later further S. Nest is a shallow scrape in the ground, sometimes sparsely lined with grass and often somewhat hidden by overhanging grass tussocks, low bushes or a log; is most like that of the Spinifex Pigeon, but usually among grass rather than spinifex. Clutch 2, creamy white, oval (30 × 23 mm). Incubation by both sexes, 16–17 days. (p. 160)

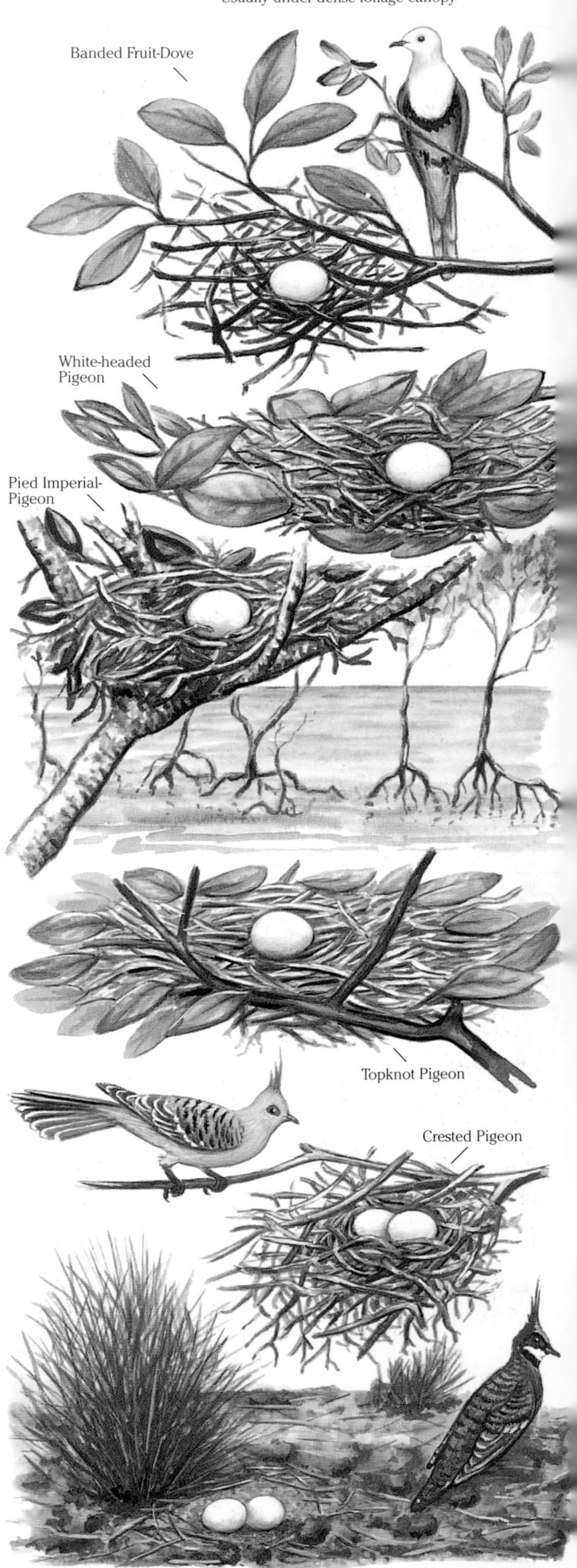

Spinifex Pigeon: usually on stony ground or rocky ranges among spinifex.
Similar: nests of Flock and Squatter, except usually on sandy, loamy, grassy plains

Pigeon tree nests differ only slightly. All are very rough, loosely built, shallow-dished platforms varying in size, some more fragile, others substantial. The location provides clues to identity—habitat, vegetation, concealment, height.

Partridge Pigeon Almost throughout year; peak activity Mar.–Aug. Makes a shallow scrape in sandy rather than rocky ground; sparsely lined with grass, leaves and semi-concealed by overhanging grass tussocks, bush, logs or among leaf-litter and plant debris of open grassy woodlands. Clutch 2, lustrous white (30 × 23 mm). Incubation about 17 days. (p. 160)

White-quilled Rock-Pigeon Throughout Dry; mostly Apr.–Nov. Rough sandstone country. Nest a sparse to substantial saucer of sticks on a rock ledge or in a cavity, often sheltered and shaded by overhanging rock. Clutch 2, lustrous creamy white (29 × 22 mm). Both sexes incubate for 17 days. (p. 160)

Chestnut-quilled Rock-Pigeon In Dry; peak Apr.–Nov. Cliffs and gorges of sandstone escarpments, western Arnhem Land. Nest is a quite substantial saucer of twigs, usually lined with fine spinifex, in a shaded crevice or under a rock overhang. Clutch 2 (28 × 22 mm). Both sexes incubate, about 16–18 days; young fledge at about 3 weeks. (p. 160)

Wompoo Fruit-Dove Jul.–Mar. A flimsy, untidy platform; small; twigs laid across a thin horizontal fork; outer branch of foliage in rainforest; 5–20 m up. Clutch 1 (42 × 28 mm). Sexes share incubation, 12–14 days. (p. 162)

Superb Fruit-Dove Jun.–Mar. Small, dished platform of thin sticks, mostly forked and interlocking; stronger than the flimsy, untidy appearance would suggest. Usually towards outer part of a slender branch from 5 to 30 m up in rainforest; occasionally in a shrub of nearby eucalypt woodland. Clutch 1 (30 × 20 mm); incubated by female at night, male by day, 14 days; young fledge after about 7 days. (p. 162)

Rose-crowned Fruit-Dove Aug–Feb., Mar. Nests in rainforest; occasionally in nearby mangroves or eucalypt woodland. Builds a small, untidy, flimsy, dished platform of thin twigs in foliage of tree, tall shrub or vine tangle. Clutch 1, glossy white, elliptical (32 × 21 mm); incubation by both sexes. Young fledge at 11–12 days. (p. 162)

Emerald Dove Sep.–Dec. on E coast; Jan.–Apr. in far N. Nest is a shallow saucer of fine twigs. Varies—thin and flimsy to thick, solidly built. Hidden in dense live or dead foliage: vine tangle, epiphytic fern, tree fern. Clutch 2, lustrous creamy white, oval (29 × 22 mm); both sexes incubate, 15–16 days. (p. 162)

Brown Cuckoo-Dove Jun.–Jan. A platform of sticks and vine tendrils; varies from flimsy to substantial; in tree fern or large epiphytic fern. Clutch 2 (35 × 24 mm). Incubated 16–18 days by both sexes; young fledge in 16 days. (p. 162)

Peaceful Dove Throughout year, but most Aug.–Jan.; often several broods in a season. Nest is a small, rough platform in various sites: often a horizontal, thick fork or dense clump of twigs 1–10 m up. Clutch 2 (21 × 17 mm). Both parents incubate, 16–17 days; young leave nest at 16–18 days. (p. 164)

Diamond Dove Sep.–Nov.; after rain in arid regions. Builds small bowl of fine twigs, grass, rootlets in a fork, occasionally on a stump; often quite low. Clutch 2 (20 × 15 mm). Both incubate, 13 days. Young fledge at about 14 days. (p. 164)

Bar-shouldered Dove Nov.–Jul. Builds small, rough bowl of fine twigs, vine tendrils 1–4 m up in dense foliage of shrub or lower part of a tree; in places uses cliff ledges. Eggs 2, lustrous white, oval (28 × 21 mm). Both sexes incubate, 14–16 days; young fledge at about 3 weeks. (p. 164)

Wonga Pigeon Sep.–Mar. Builds a dished platform of twigs; nests vary, solid to thin and flimsy. May be as low as 3 m in tree fern crown or vine tangle, but more often high, to 20 m, in a substantial fork of a tall tree, often at edge of rainforest. Clutch 2, smooth, white, elliptical (39 × 28 mm). (p. 164)

Palm Cockatoo Aug.–Feb. Nest is usually in an upward pointing, broken-off, dead top of a high, hollow tree trunk or stump. These sites are often found in open eucalypt forest near the edge of rainforest. The base of the hollow, about 0.5–1.5 m down, has a deep layer of twigs, possibly to keep the single egg and the young above the wet bottom of the hollow—the upward pointing spout would catch rain in the wet season. Hollows are usually re-used each year. The dull white egg is a tapered oval (50 × 36 mm). Incubation takes about 35 days; the young remains in the hollow for 3–4 months. (p. 166)

Red-tailed Black-Cockatoo Mar.–Jul. in N; both autumn and spring in S. Uses a large hollow trunk or limb, 20 cm or more at entrance, much larger at nest chamber. Often very high in tall coastal forests; may be low in stunted trees of arid inland. One egg (51 × 37 mm) is laid on a lining of woodchips chewed from walls inside the hollow. Male feeds female at nest during 30–31 day incubation; chick attended continuously for several weeks. Nestling at first in pale yellow down; leaves nest at 10–12 weeks. The hollow is likely to be re-used for many years. (p. 166)

Glossy Black-Cockatoo Mar.–Aug. Uses a large hollow, often in a dead tree; most are high. Floor of hollow is thick with woodchips and decayed wood on which the lone, dull white egg (45 × 35 mm) is laid. Female alone incubates, 29 days; is fed by male then and for the first week after the chick hatches. She then broods only through the nights. Sexes share feeding of young for next 9–10 weeks. (p. 166)

Yellow-tailed Black-Cockatoo Oct.–May in SE, Mar.–Aug. in N. Nests in a high hollow, usually in a massive limb, deep and wide, some 40 cm diameter at the nest floor. The bottom of the hollow is lined with woodchips, on which the eggs are laid. Clutch 2 (48 × 36 mm); incubation 28 days by female; male feeds her at nest. Usually only one chick survives; brooded by female for 20 days more, then fed by both parents. Leaves nest at about 11 weeks. (p 166)

Short-billed Black-Cockatoo May–Jan. Uses a deep hollow, usually 20 m or more high in a large, smooth-barked eucalypt, often a Salmon Gum. Uses gums as low as 3 m in inland. Entrance some 20 cm diameter; nest chamber at least 35 cm diameter. The eggs are laid on a layer of woodchips cut from the chamber's walls. Clutch 2 (46 × 35 mm). Incubation by female, 28 days; young fledge after 10 weeks; accompany adults a further 4–6 months. (p. 166)

Long-billed Black-Cockatoo Aug.–Jan. Breeds in heavy forests. Uses very large hollow, 35 cm or more in floor diameter, usually high in Karri or Marri tree. Clutch 1 or 2 (52 × 36 mm); laid on woodchips; incubated by female; brooded by her for first 3 weeks. Typically only one chick raised; fledges at 10–11 weeks; remains in the family group indefinitely. (p. 166)

Gang-Gang Cockatoo Aug.–Mar., mostly Oct.–Jan. Nests in the large hollow trunk or limb of a live eucalypt at heights of 12–27 m. Hollow is usually vertical or points steeply upward; occasionally horizontal. Clutch 2–3 (35 × 26 mm) laid on woodchips chewed from walls. Both incubate, female by night, 25–30 days. Young leave nest aged about 10 weeks; then accompany adults 4–6 weeks more. (p. 168)

Galah Jul.–Nov. in S; Feb.–May in N; several broods. Nests in hollows in limb or trunk of live or dead trees. Entrance up to 20 m high, usually much lower—down to 2 m in stunted inland trees. Bark chewed and worn away to create extensive bare wood around hollow and nearby limbs. Occasionally uses a cliff hole. Clutch 3–6 (36 × 26 mm). Floor of hollow lined with leafy green twigs; fresh leaves are added until the young fledge at 6–7 weeks. Young dependent until 3 months old. (p. 170)

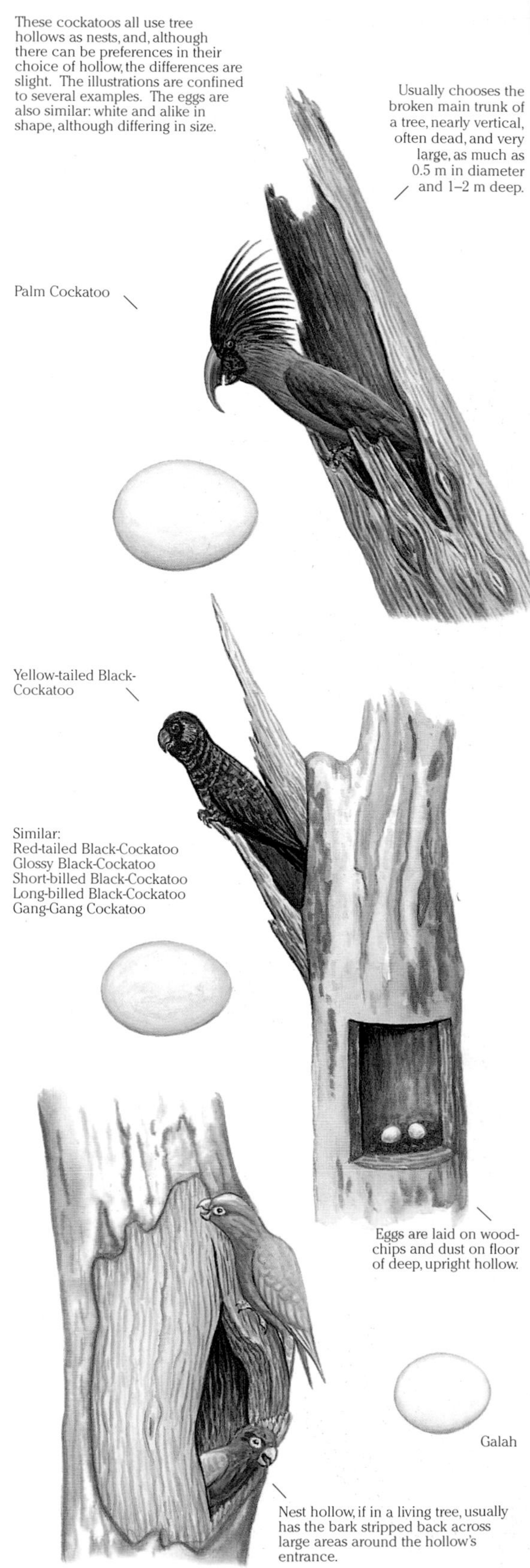

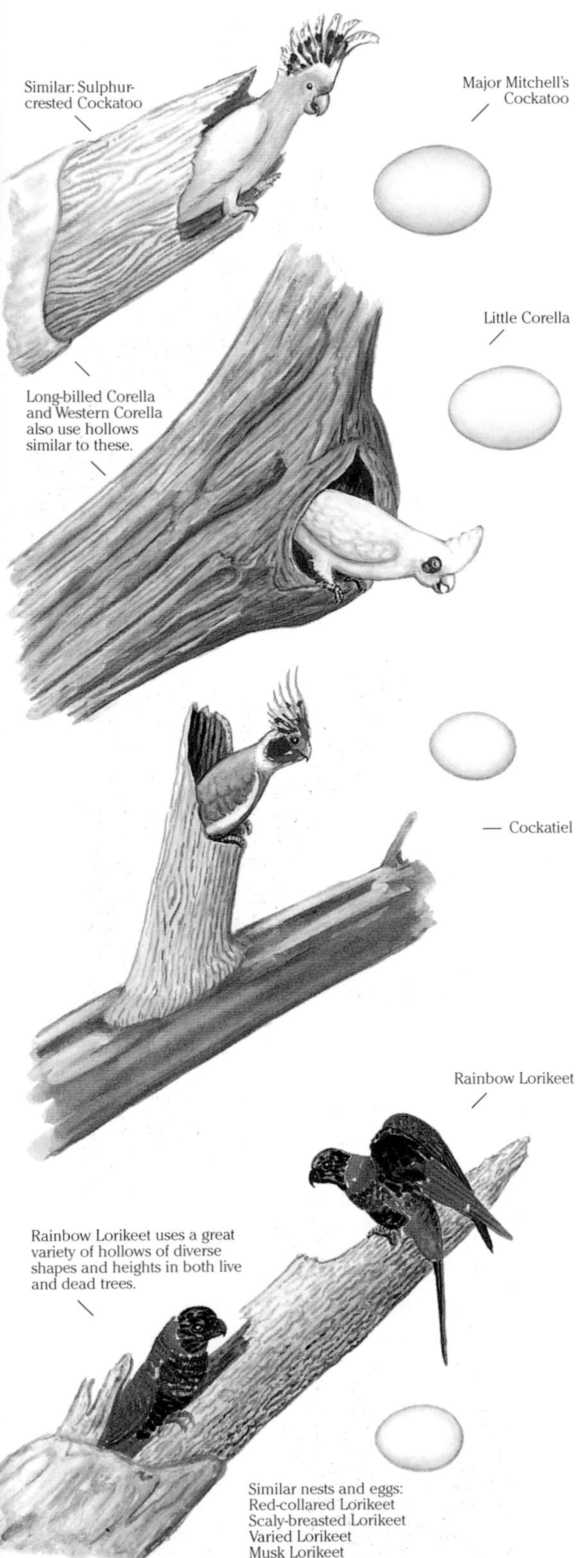

Major Mitchell's Cockatoo May–Sep. in N; Aug.–Dec. in S. The nest is a large, deep, often nearly vertical hollow in trunk or a major limb, typically in a live eucalypt or dead callitris. The floor is lined with wood dust and chips chewed from the walls. Clutch of 1–3 eggs, white with slight gloss (38 × 28 mm). Usually both sexes incubate for 25–26 days; some nests mostly by the female. Young initially in fine yellow down; fledge at 7–8 weeks; independent at 16 weeks. (p. 170)

Western Corella Aug.–Dec. Nests in big hollow: northern race uses Salmon Gum and Wandoo trees; southern race needs big old Jarrah or Marri trees. Clutch 1–3, long, oval, plain white (42 × 30 mm). Incubation 24–29 days by both sexes; female at night. Young are initially in fine yellow down; leave nest aged about 7 weeks; closely attended and fed by parents for a further month. Flocks wander widely after breeding. (p. 168)

Long-billed Corella Jul.–Dec.; varies with rainfall in drier regions. Uses a hollow high in tall Manna Gum or Sugar Gum, or, in drier regions, River Red Gums. Walls of hollow are chewed to bed the nest with chips, sometimes destroying the hollow after many years of use. Clutch 2–4, rounded oval eggs (38 × 28 mm). Incubation by both sexes, 25 days. (p. 168)

Little Corella Mar.–Aug. in N; Jul.–Nov. in S; influenced by rain in arid regions. Nests usually in large, nearly vertical hollows of trees on watercourses. Often use mangroves around N coast; occasionally holes in cliffs and tall termite mounds. Clutch 2–4, long oval, white (35 × 25 mm); laid on woodchips; incubated by both sexes, about 24 days. Chicks first clad in fine yellow down; leave the nest aged 6–8 weeks. (168)

Sulphur-crested Cockatoo Aug.–Jan. in S; May–Sep. in N. In courtship, male displays with crest raised, head bobbing. Nest is a large, nearly vertical hollow limb or trunk, often by water; can be a hole in a cliff. Clutch 2 (48 × 34 mm); incubated by both sexes, 28–30 days. Chicks initially in pale yellow down; fledge at 8 weeks; fed by adults for several months. (p. 170)

Cockatiel Breeds all year: mainly Aug.–Dec in S; Apr.–Aug. in N. Often picks a tree near or in water, frequently dead. The nest hollow may be in a dead limb or tall stump, small or large, and from 1–10 m above ground. Clutch 4–6 (27 × 19 mm); incubated by both sexes, 19–20 days; female at night. The chick is initially in fine yellow down. The young Cockatiel finally leaves the nests aged about 5 weeks. (p. 168)

Rainbow Lorikeet Jul.–Jan. in SE; Mar.–Jan. in N. Uses a large or smallish hollow limb or trunk in a smooth-barked tree, often near water and quite high. Clutch 2–3 (27 × 22 mm); eggs are laid on bare wood dust. Incubated by female, 25 days; young leave nest aged 7–9 weeks. (p. 172)

'Red-collared Lorikeet' Mar.–Jan. (Rainbow Lorikeet race; nest and eggs similar; probably a full species.) (p. 172)

Scaly-breasted Lorikeet Aug.–Dec. in S; May–Feb. in N. Uses a small, deep hollow, usually in a very high limb. The 1–3 eggs (26 × 20 mm) are laid on bare wood dust; female incubates for 23–25 days; sometimes 2 broods. Both parents feed chicks; they leave the nest aged about 6 weeks. (p. 172)

Varied Lorikeet Breeds anytime; mainly May–Sep. Nests in a small hollow, high limb, often near water. Chamber is lined with wood dust and chips, sometimes a few pieces of leaf. The 3 or 4 small, rounded, white eggs (24 × 20 mm) are incubated by the female for 21–22 days; young in nest 5–6 weeks. (p. 174)

Musk Lorikeet Breeds mainly Aug.–Jan. in a small hollow of a very high limb. Clutch 2 (24 × 20 mm); laid on bare wood dust; incubated 22 days by female. Male also roosts in the hollow at night. Young fledge at 5–6 weeks. (p. 174)

Purple-crowned Lorikeet Breeds Aug.–Dec.; found in large numbers where forests are flowering. Uses a small hollow in a trunk or limb: low in small inland trees; extremely high in tall Karri forests. Lays 3–4 eggs (20 × 17 mm). Incubation by female, 20 days. Young in nest are fed by both parents, finally leave nest aged 5–6 weeks. (p. 172)

Little Lorikeet Jul.–Jan. Nests in a hollow branch or knot-hole in a large limb or tree trunk; can be as low as 5 m, or very high; usually in a live eucalypt; often near water. Hollow cleaned and scratched out by both sexes. Clutch 3–5, usually 4, rounded, white (21 × 17 mm). Incubated by female, 22–23 days. Male feeds female at nest and shares care of young, which leave nest aged about 6 weeks. (p. 174)

Eclectus Parrot Jul.–Jan. Nest is a large and usually deep hollow, high in a rainforest tree, often overlooking a clearing or stream. The female incubates alone (26 days) and broods the young in their early stages; fed throughout by the male. Clutch 2, oval, white (42 × 33 mm). Several other Eclectus Parrots of either sex may help at the nest, these probably being immatures or young adults of broods from previous years. (p. 170)

Red-cheeked Parrot Aug.–Dec. Rather than use existing hollows, cuts into a softwood rainforest tree, often where a high broken limb has exposed its decaying centre. Inside, the nest hollow is lined with chewed wood. The 2–4 rounded white eggs (30 × 25 mm) are incubated by the female. The male feeds the female near the nest, later helps feed the young. (p. 170)

Double-eyed Fig-Parrot Aug.–Dec. The nest is a small tunnel chewed into the soft wood of a dead branch in the rainforest, usually high, but may be as low as 3 or 4 m. Clutch 2–3, rounded (22 × 19 mm); female incubates, about 18 days; she also broods the small young. The male feeds her at the nest or nearby. Both look after older young, which leave the nest hollow aged about 6 weeks. (p. 174)

Australian King-Parrot Sep.–Jan. Picks a deep hollow with entrance very high in a tall eucalypt. The nest chamber can be as far as 10 m down the hollow trunk: in some instances, almost back down to ground level. Clutch 4–6 (33 × 27 mm); takes about 20 days; the female incubates the eggs, and broods and feeds chicks for the first few weeks. She, in turn, is fed by the male. Thereafter both parents feed the young, which fly at about 5 weeks of age. (p. 176)

Red-winged Parrot Aug.–Jan. in S; May–Oct. in N. Picks a high, deep hollow in the trunk or limb of a tall, live eucalypt, often near water. Eggs are white, oval (31 × 26 mm). Incubation time for the clutch of 3–6 is 20 days; the female leaves the hollow only to be fed by the male. The young, when hatched, are fed by both parents and fly aged about 5 weeks. (p. 176)

Superb Parrot Sep.–Dec. Usually nests in a large River Red Gum, live or dead, near water. Prefers a very high, deep hollow in limb or trunk; tends to choose a dead one that overhangs a river pool. The nest's entrance is usually high, up to 20 m above ground; and the hollow inside may be equally deep. The 4–6 white eggs (30 × 24 mm) are incubated by the female for 20–21 days. The male visits the nest several times each day to feed the female. The young are later fed by both parents, finally fledging at about 5 weeks. (p. 176)

Regent Parrot Aug.–Dec. Chooses a high, deep hollow in a live or dead limb or tree, often a Wandoo or Salmon Gum in WA, or a River Red Gum in SE Aust. The female incubates alone for 21 days. Clutch of 4–6 rounded, white eggs (31 × 25 mm). The young leave the hollow aged 5 weeks, depend on their parents for another 3 weeks. (p. 176)

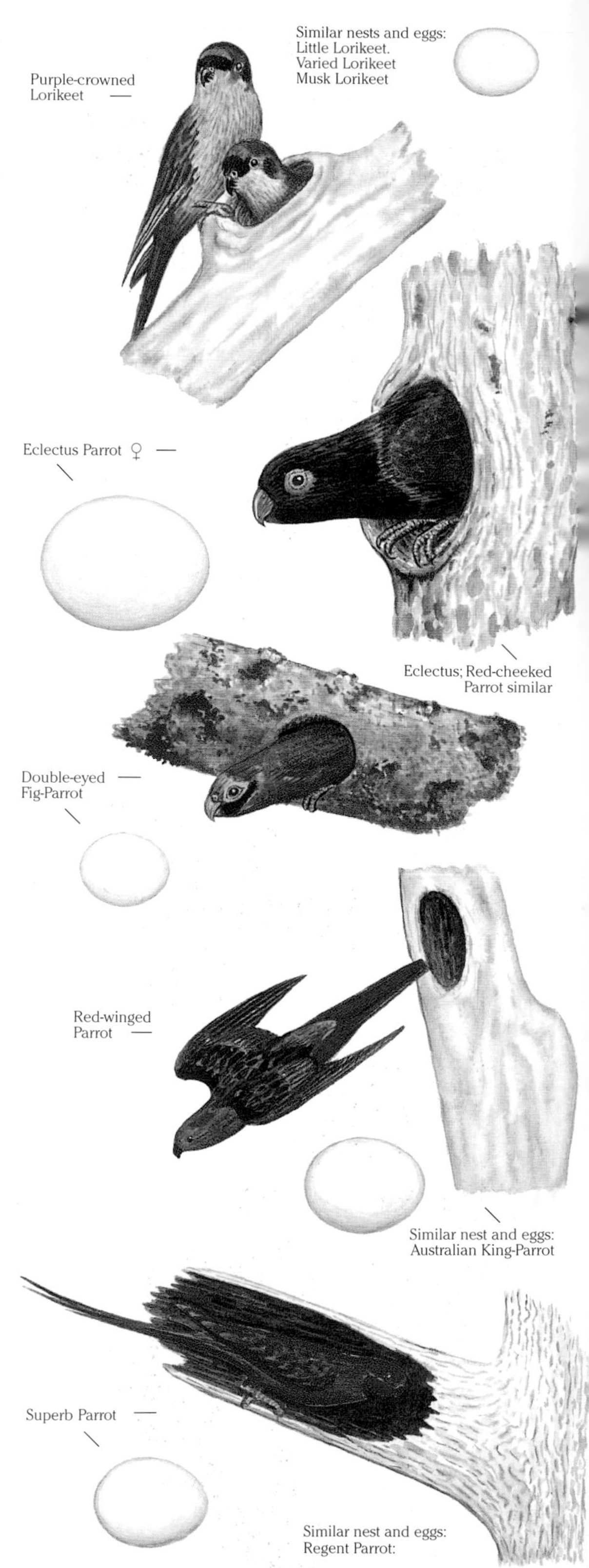

Princess Parrot Breeds Sep.–Dec.; dependent on rain in such an arid region. Often several pairs nest close together at a favoured locality, typically where trees are scattered along a watercourse; some trees may hold several nests. Hollows are likely to be relatively low in stunted eucalypts and casuarinas. The female incubates, about 20 days; 4–6 eggs (32 × 27 mm). During this time she is fed by the male near the nest. Later both sexes feed the young. (p. 176)

Green Rosella Sep.–Jan.; one brood a year. A very social species, maintains small localised flocks even through the breeding months; does not strongly defend any nest territory against others of the same species. Usually nests in a hollow high in a tall eucalypt. The creamy white, rounded eggs (30 × 24 mm) are laid on a floor of woodchips and dust. Incubation by the female, 20–22 days; young fly at about 5 weeks; are dependent on the parents for a further 4–6 weeks. (p. 178)

Crimson Rosella Sep.–Jan. Nests in hollow of live or dead tree; can be quite low or very high. The 4–8 white eggs (29 × 25 mm) are laid on wood dust. Female incubates, about 20 days; she leaves nest briefly several times a day to be fed by the male. Young leave the nest aged about 5 weeks and remain dependent on the adults for another 2 or 3 weeks. (p. 178)

Eastern Rosella Aug.–Dec. Nests in a deep, often narrow hollow in the trunk or limb of a live or dead tree; occasionally low in a stump or log, but usually quite high. Clutch size varies: usually 4–5, up to 8 round white eggs (27 × 22 mm). Incubation by female, 19–20 days. Male feeds her near nest and helps to feed the young when the female stops continuous brooding. Young leave nest aged about 5 weeks. (p. 180)

Pale-headed Rosella Aug.–Jan. and Mar.–Aug. after rain in drier parts of the bird's range. Nests in a high, often quite small hollow of a tree limb or trunk. Female alone incubates the clutch of rounded white eggs (27 × 22 mm) for 9–20 days. The male visits the nest to feed the female and later to help feed the young, which leave the hollow fully fledged at about 4 weeks of age. (p. 180)

Northern Rosella May–Nov.; mostly Jun.–Oct. The nest is a hollow in a tree, often one standing near water. The small entry leads into a limb or the upper trunk. Clutch 2–4, rounded, white (27 × 22 mm); incubated by the female, 19–20 days. She is fed near the nest by the male; later both parents feed larger young in the nest and after they fledge at about 5 weeks. (p. 180)

Western Rosella Aug.–Jan. Nests in a hollow branch or trunk, live or dead tree, 5–15 m above ground. Female incubates the clutch of 3–6 rounded white eggs (26 × 21 mm) for 19–20 days. The male feeds her near the nest and helps to feed the young when she has finished continuously brooding the very small chicks. The young rosellas leave the nest at about 5 weeks of age. (p. 180)

Australian Ringneck Feb.–Nov. in far N; Aug.–Dec. in S. Nests in a tree hollow 5–15 m up—occasionally within several metres of the ground in arid country with only small, stunted trees. The female incubates the clutch of rounded white eggs (30 × 24 mm) for 19–20 days. During incubation the male feeds her near the nest, and later helps to feed larger young, which leave the nest aged 5–6 weeks. (p. 182)

Red-capped Parrot Sep.–Dec. Nests in a small, deep hollow in a limb or hole in an upper trunk; commonly a Wandoo or Marri; usually high, 10–20 m up; occasionally as low as 3 m. Clutch 4–6 (30 × 24 mm); incubated by the female, who is fed by the male; he later helps to feed the chicks as they grow larger. The young leave the nest aged 5–6 weeks, around Nov. or Dec; they then call noisily for food. (p. 182)

Blue Bonnet Breeding is linked to rainfall in arid regions, but usually Aug.–Jan. Nests in a small but deep hollow no more than 5 or 6 m up, often much lower in the stunted trees of arid regions. Race *narethae* often nests in desert she-oaks. The 4–6 rounded white eggs (23 × 19 mm) are incubated by the female for 18–20 days. The male feeds her at the nest and helps feed the young, which leave the nest aged 4- 5 weeks. (p. 184)

Swift Parrot Sep.–Dec. These parrots nest in Tas. and some Bass Str. islands. In autumn (Mar.–May), after breeding, they migrate N to mainland Aust. They return to Tas. for the next breeding season in the early spring months. Gregarious, usually in small flocks, several nests may be in close vicinity; at times more than one in a tree if hollows are available. Typically choose a small hollow high in a tall eucalypt where the 3–5 eggs (26 × 21 mm) are laid on decayed wood dust. Female incubates for 20 days. The male helps feed the chicks, which leave the hollow aged about 6 weeks. (p. 186)

Red-rumped Parrot Jul.–Dec.; after rain in drier parts of range. Nests in a hollow in a live or dead tree or stump from 2 m to 15 m up, often near water. Clutch 4–7, usually 5, rounded, white (24 × 19 mm); incubated by the female, 18–20 days. The male feeds her and the chicks after they hatch. The young remain in the nest for 4 to 5 weeks. (p. 184)

Mulga Parrot The breeding of this bird of the dry, arid inland is influenced by rainfall, but is usually Jul.–Nov. In mulga or similar scrub it usually nests in bigger trees along water-courses, typically 3–5 m up, in a small, deep hollow in a branch or limb; usually in a live tree. In dry, treeless regions, it may nest in very low, cramped cavities in stunted trees, even holes in earth banks. The 4–5, rarely 6, rounded white eggs (24 × 19 mm) are incubated by the female, 18–20 days. She is attended at the nest by the male; young fledge at 4–5 weeks. (p. 184)

Golden-shouldered Parrot Mar.–Jun. Breeding begins as wet season grass begin to seed, but while the termite mounds used as nest sites are still damp and soft from the summer Wet and are easier to excavate. Tends to use flats that are boggy or under shallow water in the Wet. The nest tunnel is drilled, from an entrance hole positioned 1–2 m above ground, some 40 to 50 cm into the side of a tall spire or cone-shaped termite mound. This shape seems to be preferred to the thin, flat 'magnetic' shaped mounds. The 4–5 rounded white eggs (22 × 19 mm) are laid on the termite-earth floor of the nest chamber. The female incubates for 18–20 days; the male helps to feed the young, which leave the mound aged 4–5 weeks. In favourable seasons pairs may nest a second time in spring, Aug.–Nov. The same mound, or one nearby, may be used for later broods and over successive years. (p. 184)

Hooded Parrot Usually breeds after the wet season, Apr.–Aug., but some years continues through to about Oct. The nest is a chamber at the end of a tunnel, dug by both sexes, into a terrestrial termite mound. Uses tall, cylindrical, conical or rounded termitaria, rather than those that are the thin, flat, ribbed, 'magnetic' shape. The female incubates the eggs (22 × 19 mm), incubation taking 19–20 days. Meanwhile, the male feeds the incubating female, and later the chicks, which fledge when aged 4–5 weeks. (p. 184)

Paradise Parrot Almost certainly extinct, but records indicate nesting in spring to early summer. The nest was a tunnel dug into a large, but rather low, rounded, termite mound, unlike the tall tropical mounds used by the Golden-shouldered and Hooded Parrots. The clutch of rounded white eggs (21 × 17 mm) was probably brooded by the female, tended at the nest by the male, who probably also helped to rear the young. (p. 184)

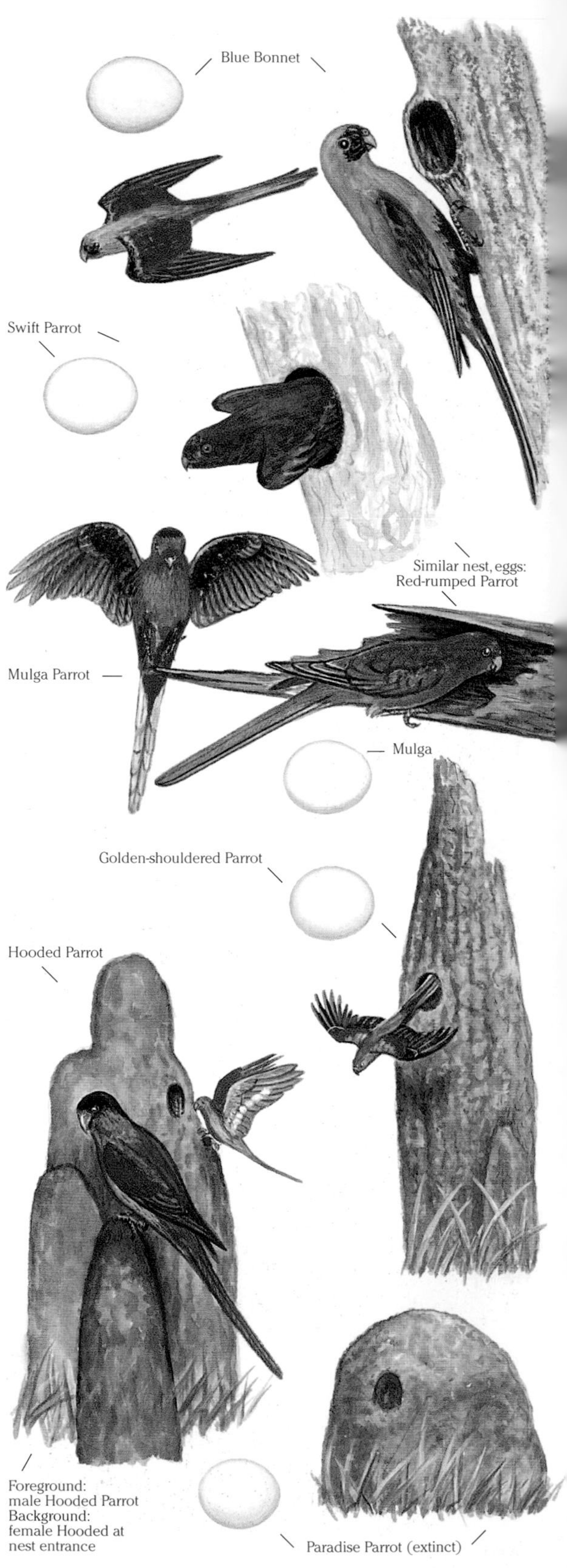

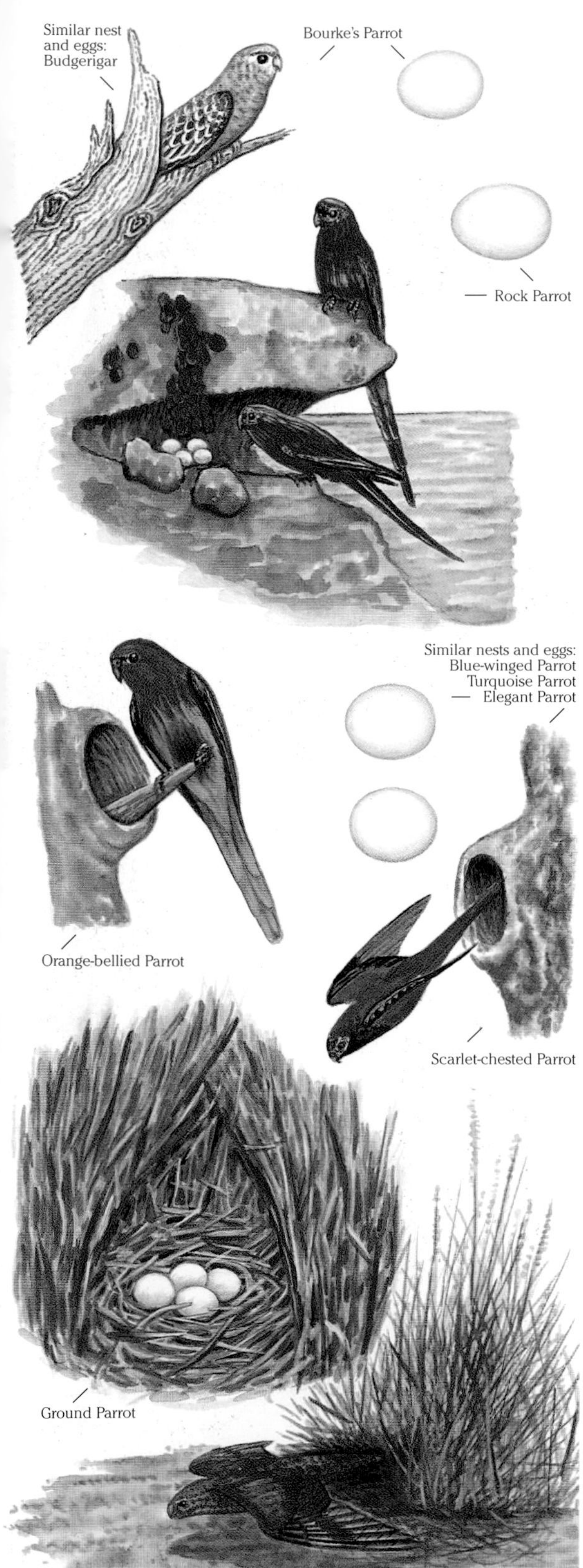

A Night Parrot is most likely to be found if one is flushed from its nest under the centre of a large clump of spinifex.

Budgerigar Usually Jun.–Jan.; any time after rain. Nests in a knothole or slender hollow branch of a tree trunk, 2–10 m high. When large numbers congregate to nest in a particularly favourable site, may be forced to use all available sites—holes in very low stumps, posts or logs. Clutch 4–8 (18 × 14 mm); incubated by female. Male feeds female, later helps to feed young, which fledge when aged 4–5 weeks. (p. 186)

Bourke's Parrot Jul.–Nov.; other months after good rain. The nest is in a small hollow branch, or eucalypt or desert oak trunk, 2–5 m up. In arid regions may use a small hollow very close to the ground in mulga or other such shrubs. Clutch 3–6 (21 × 18 mm); incubated by female 18–19 days; she is attended by the male. Chicks fledge aged about 4 weeks. (p. 186)

Blue-winged Parrot Oct.–Jan. Will nest in a small hollow high in a live or dead tree, or in a low stump, hollow fencepost or log. Clutch 4–6 (23 × 19 mm); incubated by the female, about 20 days; she leaves the hollow several times daily to be fed by the male at a nearby perch. The male helps care for the young; they fly aged 4–5 weeks. (p. 188)

Elegant Parrot Aug.–Dec. Nests in a small hollow limb or trunk, live or dead tree; usually very high in taller trees of coastal forests, quite low in small inland trees. Clutch 4–5 (21 × 19 mm); incubated by female, 18–19 days. Male regularly calls her from the hollow to be fed nearby; he feeds the chicks, which fledge aged 4–5 weeks. (p. 188)

Rock Parrot Jul.–Jan. Nests on offshore islets; occasionally on the mainland coast. Nest sites are close to the sea: misted by spray; under boulders; a cliff ledge or recess; screened by plants; in a disused seabird's nest tunnel. Clutch 4–5 (24 × 20 mm); incubated by the female, 18–19 days, during which time she is fed by the male. The young fledge at 4–5 weeks. (p. 188)

Orange-bellied Parrot Nov.–Jan. Breeds in SW Tas.; uses small trees in sparse clumps across open, wet coastal plains of grass and sedge. Nests in a small hollow; lays 4–6 white eggs (23 × 19 mm) on wood dust floor. Incubation by female, 20–21 days. Male feeds her; later the young, which fly at 4–5 weeks; all migrate to Vic. or SA in Mar.–Apr. (p. 188)

Turquoise Parrot Aug.–Dec. Typically picks a low hollow in a stump, post, fallen log or lower trunk of a dead tree. Often has a second brood in late spring to summer in good seasons. The 2–6 eggs (21 × 18 mm) are laid on a floor of wood dust. Female incubates; male feeds her frequently and later attends the young, which fly from the hollow aged 4–5 weeks. (p. 188)

Scarlet-chested Parrot Aug.–Dec.; other times after good rain; may have several broods in best seasons. Uses a hollow in the limb or trunk of a small tree, often dead. The 4–6 eggs are laid on wood dust on the hollow's floor. Incubation by female only, 18–21 days. Fledgling period 28–32 days. (p. 188)

Ground Parrot Sep.–Dec. Builds a shallow bowl of fine sticks and grass, well hidden under overhanging tall, coarse grass, sedge or low, heathy shrubbery; not a tree hollow. Screens nest from above and sides; often forms a tunnel in the surrounding dense, enclosing plants. The 3–4 white eggs (27 × 23 mm) have an incubation time of 21–24 days; young may leave the nest 3–4 weeks after hatching. (p. 186)

Night Parrot Probably after rain in its mostly arid habitat. Recorded as roosting and nesting in tunnel-like spaces inside dense clumps of spinifex: rough bowls of sticks and grass held 2–5 oval white eggs. While the few records are sketchy and uncertain, this rarely seen parrot probably has a nest similar to that of the Ground Parrot, but it would be very difficult to locate in the heart of a spinifex clump. (p. 186)

Pallid Cuckoo Aug.–Dec. Parasitises species with open cup nests and eggs similar to its own: Willie Wagtail, woodswallows, whistlers, robins, orioles, cuckoo-shrikes, honeyeaters. One host egg is replaced by cuckoo's. Clutch 1 per nest (24 × 17 mm). Incubation shorter than host species, 12–14 days: larger, stronger cuckoo chick ejects host's eggs or young. (p. 190)

Brush Cuckoo Sep.–Jan. in S; most months in N. Egg varies to match those of host group of species. (1) E Aust., uses open cup nests of *Myiagra* flycatchers. (2) N, NW, uses cup nests of fantails. (3) NE, uses domed nests of Bar-breasted and Brown-backed Honeyeaters. Clutch 1 per nest (19 × 15 mm). (p. 190)

Black-eared Cuckoo Jul.–Jan. in SE; after rain in semi-arid regions. In SE, often lays in Speckled Warbler's nest; elsewhere uses Redthroat's nest—it has similar eggs. Occasionally lays in other domed nests: those of thornbills, heathwrens, scrubwrens. Clutch 1 per nest (20 × 15 mm). (p. 190)

Chestnut-breasted Cuckoo Sep.–Feb. Little is known, but probably uses small domed nests. Unusual eggs have been reported in nests of Tropical Scrubwrens, and juveniles have been seen being fed by Tropical Scrubwrens; the Large-billed Scrubwren may also be a host. Eggs are likely to be similar to those of scrubwrens. (p. 192)

Fan-tailed Cuckoo Jul.–Feb. Parasitises small birds that build domed nests—thornbills, scrubwrens, heathwrens, Rock-warbler, Pilotbird and Redthroat. Only rarely uses cup nests such as that of Silvereye. Clutch 1 per nest (22 × 16 mm). Incubation 14–15 days; fledge in 16–17 days. (p. 192)

Horsfield's Bronze-Cuckoo Sep.–Jan.; after rain in arid regions. Exploits small birds with domed nests (fairy-wrens, thornbills) and those with open cup nests and speckled eggs that match (robins, honeyeaters, chats, whitefaces, scrubwrens). Lays 1 egg (18 × 13 mm). Incubation time usually shorter than that of the host; fledges in 16–18 days. (p. 192)

Shining Bronze-Cuckoo Aug.–Jan. Parasitises small birds, usually fairy-wrens and thornbills. The olive-brown egg (18 × 12 mm) differs from those of hosts, but is not obvious in the shadowy depths of their hooded nests. Young cuckoo outgrows the nest; leaves at 18–22 days. (p 192)

Gould's Bronze-Cuckoo Jul.–Mar. in N of range; Sep.–Jan. in S. Uses domed nests—usually Large-billed Gerygone, but also White-throated and other gerygones. Clutch 1 (18 × 12 mm). Dark brown egg is not obviously different in darkness of host's domed nest. (p. 192)

Little Bronze-Cuckoo Sep.–Feb.; mainly Oct.–Jan. Usually lays in domed nests of gerygones; 1 egg per nest (18 × 12 mm). Colour difference between eggs is unnoticeable in dim light inside a domed nest. Cuckoo usually hatches first; ejects host eggs or hatchlings. (p. 192)

Common Koel Sep.–Feb. A very large cuckoo. Parasitises much smaller birds (figbirds, friarbirds, Magpie-lark) with eggs roughly similar in markings, but much smaller than the Koel's (35 × 24 mm). The young Koel ejects host eggs or chicks; takes the entire food supply brought by the foster parents. (p. 194)

Channel-billed Cuckoo Jun.–Dec. A huge cuckoo; egg 43 × 30 mm. Currawongs are common hosts; also crows, ravens, Magpies, Magpie-lark, Chough, Sparrowhawk. The Channel-bill may break the host's eggs while laying. (p. 194)

Pheasant Coucal Aug.–Mar.; multiple broods. The only nonparasitic Australian cuckoo. Nest is a rough, trampled platform of vegetation, well hidden in dense, coarse grass, reeds, pandanus or sugar cane. Rounded oval (38 × 29 mm); both sexes incubate, about 15 days. (p. 194)

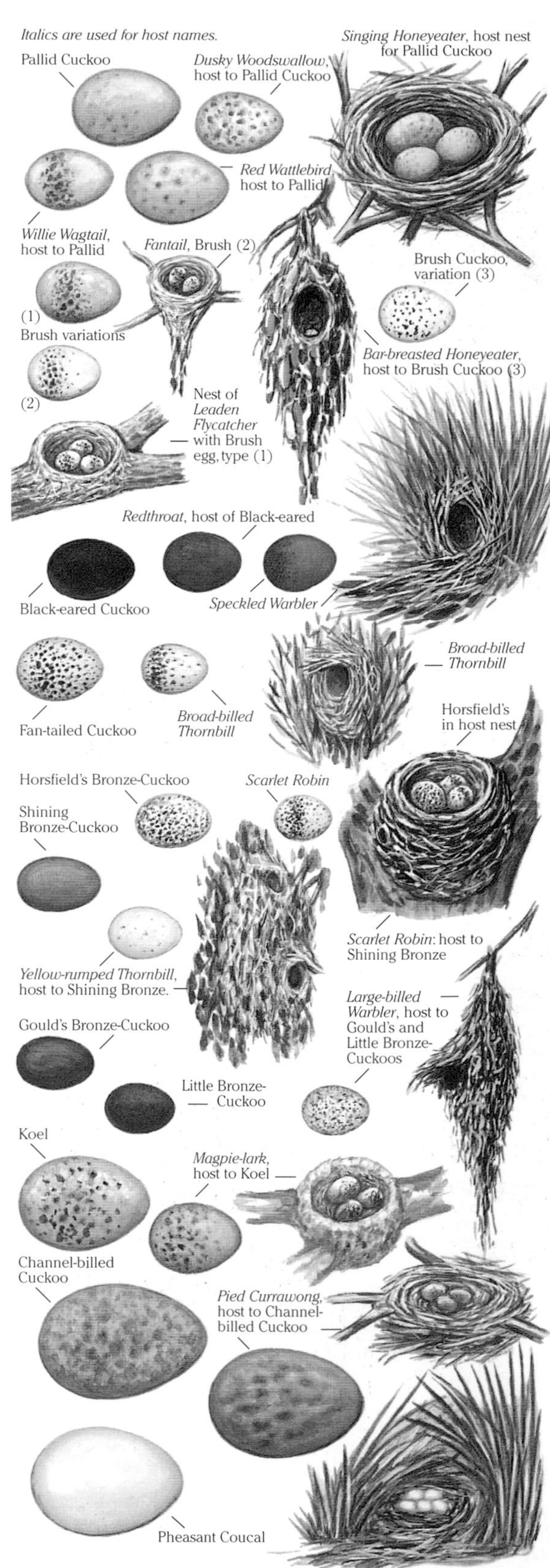

Coucal builds its own nest; incubates its own eggs.

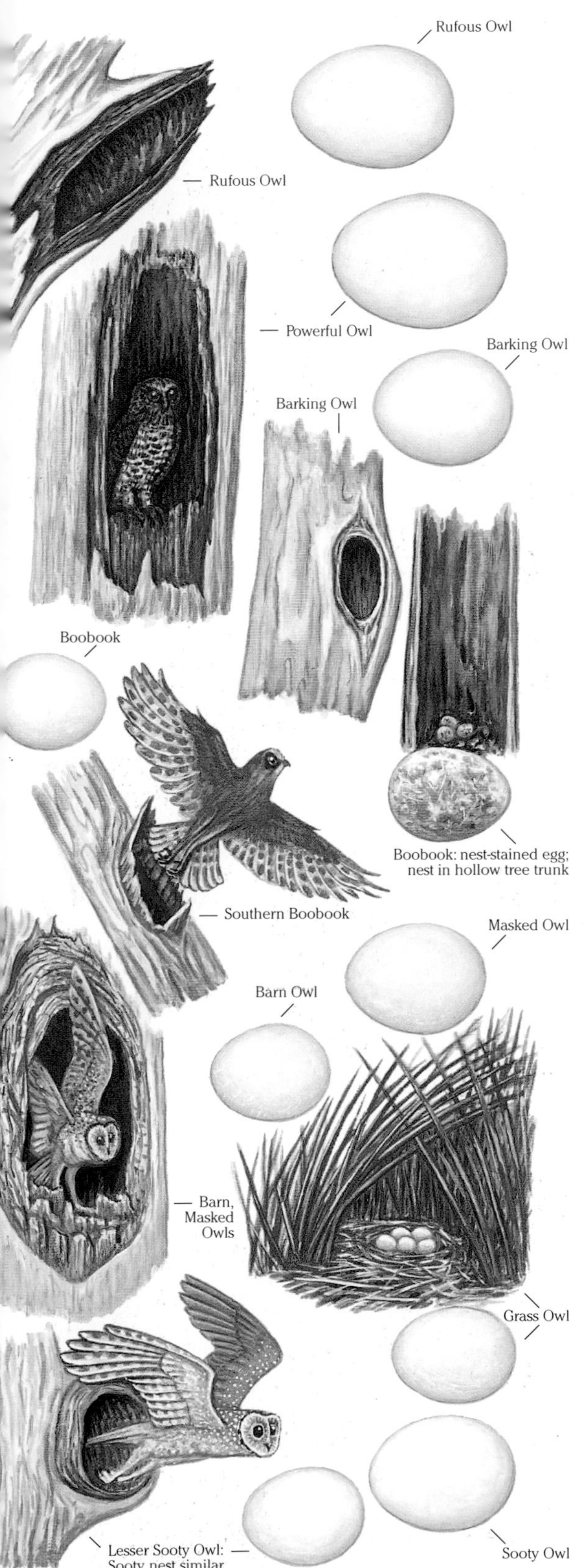

Rufous Owl Jun.–Aug. in N; Aug.–Dec. in E Qld. Likely to use the same hollow for many years. Male cleans out the hollow before female lays. No nest lining: eggs are laid on wood dust and fine debris on the hollow's floor. Clutch 1–2 (53 × 45 mm). Incubation by female, 36–38 days. Young in white down at first; feathered and leave hollow aged 6–8 weeks; depend on parents another 2 months. (p. 196)

Powerful Owl Apr.–Sep. Nest hollow typically in a massive forest tree, often on a hillside or head of a gully. Male prepares hollow; female lays 1–2 eggs (54 × 45 mm) on wood dust floor. Female fed at nest by male; incubation 36–38 days. Young in white down at first, soon discoloured in nest. First fly at about 8 weeks; depend on parents for several months more. (p. 196)

Barking Owl Jul.–Sep. in N; Aug.–Oct. in S. Uses big hollow, often not deep; typically in old eucalypts or melaleucas along rivers and gorges of drier regions. If suitable trees are scarce, may use a hole in a cliff or crevice among rocks. Hollow is unlined; a shallow depression is scratched in dusty floor. Clutch 2–3 (48 × 38 mm); incubated by female, 36 days. Young initially in white down; fledge at 5–6 weeks. (p. 196)

Southern Boobook Aug.–Dec.; little variation throughout range. Uses tree hollows ranging from low stumps to small holes in high limbs; at times a cliff hole or an old domed nest such as that of a babbler. Clutch 2–3 (43 × 35 mm); incubated by female, about 34 days; chicks' first down is white. Fledge at about 5 weeks; dependent on parents for several months more. (p. 196)

Barn Owl Breeds any time of year, commonly 2 broods unless food is scarce. In rodent plagues will brood successively while prey is abundant. Nest hollow typically a deep, cavernous space in trunk or major limb. May have an entrance 15m up, and be hollow right down to ground. Also uses cliff holes, caves, old mines and roof spaces in buildings. These sites are often occupied by Barn Owls for decades. Clutch 2–3 (43 × 32 mm). Incubation about 33 days; fledge at 7–9 weeks. (p. 198)

Masked Owl Usually Mar.–Jul.; but any time that prey is abundant. Nest is usually a cavernous hollow in the trunk or a main limb of a large tree in heavy forest, but often near open country over which the owls hunt. Clutch 2–4 (47 × 40 mm); incubated by female, 38–42 days. Nestlings first in white down; soon replaced by creamy buff. Young first leave hollow at 10–12 weeks; stay nearby to be fed for several months. (p. 198)

Grass Owl Breeds any time of year when prey is abundant, but mainly after Wet, Mar.–Jun. Makes a rough, thin platform of trampled grass on the ground, hidden under dense grass or sedge, with several access 'tunnels' through the surrounding grass; occasionally under a low dense shrub. The 3–8 eggs are dull white, tapered oval (42 × 31 mm); incubation 35–40 days. Chicks first in white down, then buff; fledge in adult plumage at about 2 months of age. (p. 198)

Lesser Sooty Owl Usually breeds Mar.–May, but breeding has been recorded most months. Nest is a large hollow in a tree trunk or big limb; varies from deep to quite shallow. The 1–2 eggs are dull white, rounded oval (42 × 39 mm); incubation 40–42 days. Chicks' first and second down is sooty grey. The young fledge in adult plumage aged about 7 weeks and stay with their parents until about 6 months old. (p. 198)

Sooty Owl Mostly Mar.–Jun.; also some breeding in spring. Nest is usually a cavernous hollow in a large, smooth-barked, live tree, often very high; occasionally a cave. Tree hollows vary 15–5 m depth. Eggs are 47 × 40 mm; incubation about 6 weeks. Chicks are first clad in sooty grey down. They leave the hollow aged about 6 weeks; stay in the vicinity being fed by the adults for several months. (p. 198)

Tawny Frogmouth Aug.–Dec.; after rain in arid regions. The nest is a very rough, untidy, loosely constructed platform of large sticks lined with a few leaves in a fork from 2 to 15 m up. Clutch of 2–5 white, slightly glossy eggs (42 × 29 mm). Male incubates by day and female at night, about 30 days. Young leave nest aged about 4 weeks. (p. 200)

Marbled Frogmouth Sep.–Dec. Builds a rough, dished platform of sticks and vine tendrils. It can be very high (40 m) in a big, old tree, or quite low—even on the ground. Often builds on a broad, horizontal fork, but also uses sites like the top of an ant plant (*Myrmecodia beccarii*) growing on a melaleuca's trunk. Clutch 1–2 (40 × 28 mm). (p. 200)

Papuan Frogmouth Aug.–Jan. The nest is a rough, untidy, flimsy and, for the size of the bird, small platform of sticks in a horizontal or upright fork, or on an ant plant, often in a paperbark or mangrove. Almost any height from very low to very high. The 1–2 eggs are white with a slight lustre (50 × 34 mm). (p. 200)

Australian Owlet-Nightjar Aug.–Dec. Nests and roosts in a hollow of a live or dead tree; ranges from high to very low; usually has a quite narrow entrance, but may be quite deep. In arid regions may be in a low stump, rock cavity or hole in an earth bank. A layer of leaves may be spread across the hollow's floor. The 2–5 eggs are glossy white, rounded oval (29 × 21 mm). Incubated by both sexes, about 25–27 days; young fledge aged around 22–28 days. (p. 202)

Large-tailed Nightjar Aug.–Jan.; one brood. No nest is constructed; eggs are laid onto leaf litter or on the ground. The bird's cryptic coloration makes it almost invisible when brooding. If the eggs are approached too closely or if there is excessive disturbance nearby, the bird may move them a short distance away. Clutch 2 (31 × 22 mm). The newly hatched young have fluffy, sandy grey down. (p. 202)

Spotted Nightjar Usually Sep.–Dec.; earlier in far N than in S. Does not build a nest; lays the single egg (35 × 24 mm) on the ground among litter of fallen leaves, bark, twigs and stones, often where lightly shaded by trees; the similar colours and tones of the sitting bird make it difficult to recognise. The egg is incubated about 30 days, probably by both sexes. (p. 202)

White-throated Nightjar Sep.–Jan. in N; Oct.–Feb. in S. The single egg (40 × 28 mm) is laid on the ground in litter of dead leaves, bark, twigs and stones, usually among fallen timber on rocky ground, and often at the top of a ridge. Both sexes incubate, 24–25 days. Chick walks within hours of hatching, and gradually wanders away from the nest site; the chick can fly by age 21 days. (p. 202)

White-rumped Swiftlet Jul.–Dec.; has between 2 and 4 broods each season. Breeds in colonies, attaching nests to the walls of caves and mine shafts, where they may be in total or semi-darkness. Nests are built of long, fine grass or casuarina leaves, feathers and moss, all cemented with dried saliva to form thin walled, half-cup shapes, so closely packed that many join to each other as well as adhering to the rockface. Clutch just 1 egg (19 × 14 mm). Incubation 22 days. (p. 204)

Azure Kingfisher Sep.–Feb. May have 2 broods. Nests in a narrow tunnel drilled in an earth bank beside or very close to water. Entrance is usually near the top of the bank. The tunnel extends 20–40 cm to a widened nest chamber; 4–6 white eggs (22 × 19 mm) are laid on bare earth. Incubation by both sexes, 20–22 days; young fly after 22–28 days. (p. 206)

Little Kingfisher Oct.–Mar. Nest is a small tunnel drilled in an earth bank, a soft, rotting stump, a mangrove root, a termite mound or low in a tree beside water. The tunnel, only 3–4 cm in diameter, extends 15–20 cm to a wider, unlined chamber where 4–5 white eggs (17 × 14 mm) are laid. Incubation details are unknown; both sexes feed the young. (p. 206)

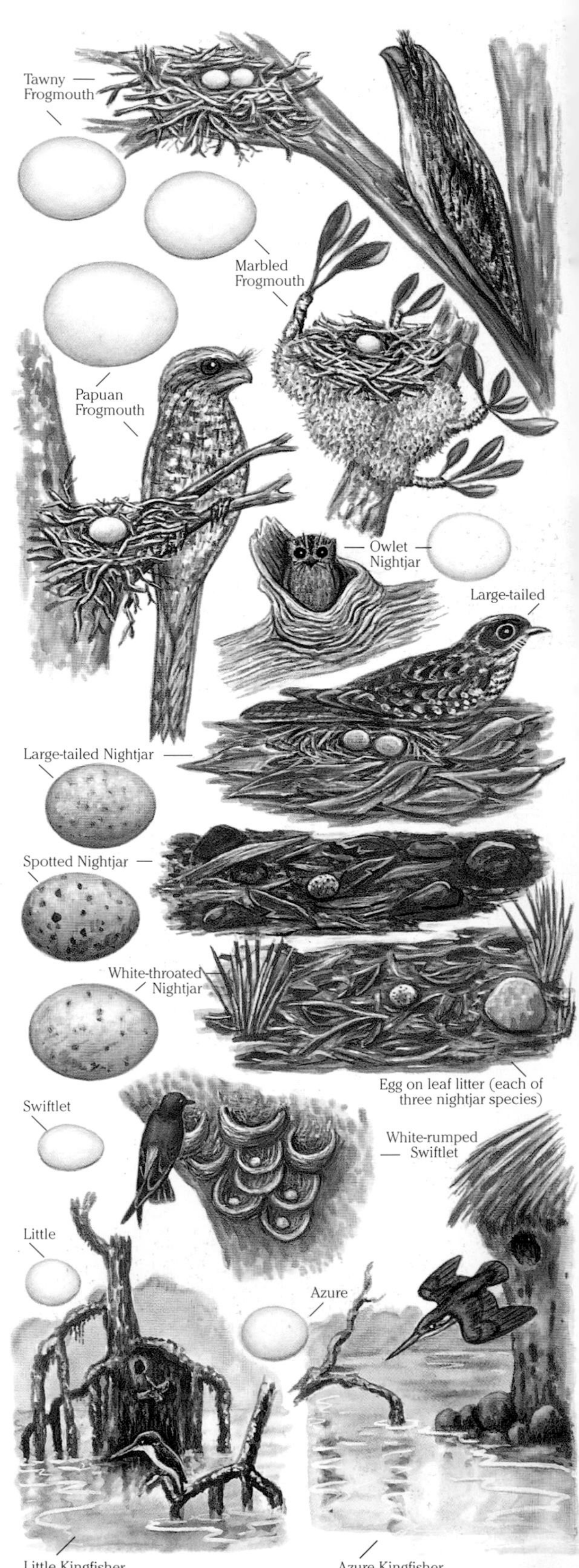

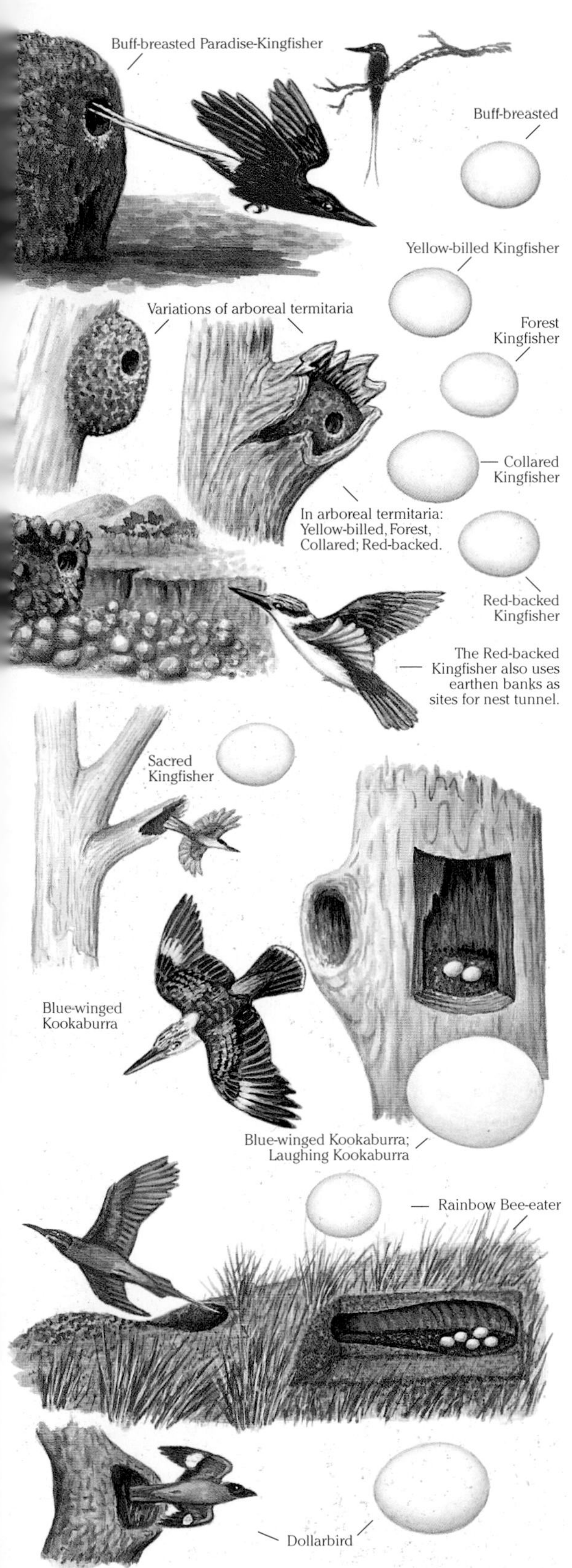

Buff-breasted Paradise-Kingfisher Oct.–Jan. Usually nests in a low, rounded termite mound rising from the rainforest floor. A tunnel, drilled 15–20 cm into the side of the mound, leads into an unlined nest chamber. The 3–4 lustrous, white, rounded eggs (25 × 22 mm) are laid on the bare termite-earth floor. Both parents incubate and feed the young, which leave the nest aged 22–25 days. (p. 206)

Yellow-billed Kingfisher Nov.–Feb. Drills into a termite mound attached to the side of a tree trunk or limb, or exposed within a hollow, or into soft, rotten wood in a tree; a hollow limb from which termite earth has been dug may be used. Nest is usually high. The tunnel extends 10–25 cm to the nest chamber where the eggs are laid on bare, termite-earth floor. Clutch 3–4, glossy white, rounded (26 × 23 mm). (p. 208)

Forest Kingfisher Aug.–Feb.; often 2 broods. The nest is drilled in an arboreal termite nest—can be 20 m up; occasionally dug in the termite-filled or soft, decaying centre of a limb, or firm earth of a creek bank. The entry tunnel leads to a deeper, wider, unlined chamber, where 3–6 glossy white, rounded eggs are laid (25 × 22 mm). Incubation 18–19 days; chicks fed by both parents; fledge aged about 23–25 days. (p. 208)

Collared Kingfisher Sep.–Mar.; often 2 broods. Digs a tunnel in an arboreal termite nest, tree hollow or, uncommonly, an earthen bank. The 2–3 eggs, glossy white (32 × 26 mm) are laid in an unlined chamber. Incubation and fledging times are probably similar to those of the Sacred Kingfisher. (p. 208)

Red-backed Kingfisher Sep.–Feb. Nest is a tunnel in any available earthen bank—the side of a watercourse or inside the shaft of an abandoned mine. May drill into an arboreal termite mound, or yet a tree hollow, especially if the inside is filled with termites. The tunnel leads to a wider, unlined nest chamber. Both sexes incubate the 3–5 slightly glossy, white eggs (26 × 22 mm), 17–19 days. Young fly aged about 21 days. (p. 208)

Sacred Kingfisher Aug.–Mar.; usually Sep.–Jan.; often two broods in the season. Nest is a tree hollow or a tunnel drilled in an arboreal termite nest or earth bank; in SW of WA, almost invariably uses a tree hollow. Tree nests are high, to about 25 m. The 3–6 glossy white eggs (25 × 22 mm) are incubated 17–18 days by both sexes; young leave nest aged 26–28 days. (p. 208)

Blue-winged Kookaburra Aug.–Feb., mostly Sep.–Jan. Nests in a tree hollow, hole cut in the soft trunk of a baobab tree or tunnel drilled in an arboreal termite mound; breeding pair is often helped by others in the family group. The 1–4 white, rounded eggs (45 × 36 mm) are laid in an unlined chamber. Both parents, and possibly others in family group, incubate for 23–25 days. Young fly at about 35 days. (p. 210)

Laughing Kookaburra Aug.–Jan.; most Sep.–Dec. Nest is unlined—usually in a tree hollow, can be a tunnel in an arboreal termite nest. Clutch of 2–4 eggs, white, semi glossy, rounded (45 × 36 mm). Incubated by female and others in family group, 23–25 days. Young fledge aged 32–37 days. (p. 210)

Rainbow Bee-eater In S, Nov.–Jan.; in N, before and after Wet, Aug.–Nov. and Apr.–Jun. Digs a long, narrow tunnel in soft, loamy soil of flat ground or a bank; extends almost a metre to a wider, unlined nest chamber. The 3–7 eggs, translucent white, glossy, rounded (24 × 18 mm) are incubated 21–24 days by both sexes, and at times by other birds from the communal group; the latter also help to excavate the tunnel and feed the young, which fledge at 28–31 days. (p. 210)

Dollarbird Oct.–Jan. Nests in a high, shallow, unlined tree hollow. The 3–5 eggs, glossy, slightly translucent white, rounded oval in shape (37 × 34 mm), are probably incubated by both sexes; times to hatching and fledging are unknown. (p. 210)

Red-bellied Pitta A summer, wet season migrant from PNG; arrives Oct.–Nov. for the breeding season, lasting until Jan.–Feb.; departs in Apr. The Red-bellied Pitta maintains a territory in dimly lit, dense, tropical monsoon forest where the nest is well concealed in litter and debris. The large, domed structure of sticks, dead leaves and mosses is on the ground at the base of a tree, or in a large fork among fallen debris, or in a tangled cluster of vines or pandanus foliage, or in the hollow top of a stump. Sticks and moss overhang the side entrance to form an awning; there is usually an entrance platform. The interior is lined with fine rootlets and bark fibre. The nest is very large, 25–35 cm, for the size of the bird. Clutch 2–4 (27 × 24 mm). While the female appears to undertake most of the incubation, the male may help occasionally. (p. 212)

Noisy Pitta Oct.–Jan. in S; in far N breeds through the later months of the wet season, Jan.–Apr. The Noisy Pitta defends a territory in rainforest or, in far N, monsoon forest. The nest is bulky, large for the size of the bird, and incorporates quite big sticks as well as dead leaves, bark and moss. It also has a debris-surrounded side entrance that is approached across a rough ramp of sticks and debris. The resulting appearance is almost indistinguishable from the natural accumulation of sticks and litter found between the buttresses of rainforest trees. Clutch 3–4 (32 × 25 mm). Incubation is by both sexes and probably takes some 16–18 days. (p. 212)

Rainbow Pitta Breeds through northern wet season, Nov.–Mar. The nest territory is within a dense patch of monsoon forest, vine thicket or stand of bamboo, where the overhead canopy creates a gloomy interior with an open, litter covered floor. Such habitats are usually restricted to damp sites, often quite small in area or in narrow strips along permanently damp see pages or creeks. The nest is a large sphere with a side entrance, but sometimes has a sparse overhead dome—it is then almost a large cup shape rather than a dome. Construction is of thick sticks, dead leaves, bark and grass lined with decayed wood and grass. Usually well hidden in a dense thicket or clump of vegetation such as vines or bamboo; occasionally is on the ground. Clutch 3–4 (30 × 24 mm). (p. 212)

Albert's Lyrebird Breeds May Oct. Males are well known for their song and display. Females seem to have their own separate territories, which partly overlap that of the male, and which they defend as feeding grounds rather than as the centre of a mating site. The female alone builds the nest, incubates the eggs and feeds the young. There is no evidence of any lasting pair-bond between the male and female. The nest is a big dome made of sticks and finer stems, rootlets, leaves and mosses. Common sites are in the crowns of tree ferns, on rock ledges and stumps, in forks of tree trunks, and, occasionally, on the ground. The overall appearance is rather like a pile of accumulated rainforest debris, which makes the nest quite inconspicuous. Clutch 1 (68 × 44 mm). Incubation takes about 6 weeks; the young fledge at 6 weeks. (p. 214)

Superb Lyrebird The breeding season is lengthy: the male renovates his forest floor display mound in Autumn, and the females begin building nests around May–Jun. This activity is probably timed to coincide with the effect of winter rain on the dense litter layer of the forest floor, when the supply of invertebrate fauna in the moist litter and topsoil increases. Only one clutch is reared by each female in each season. The nest is a bulky, untidy mound of sticks built into a rock crevice, large low fork or hollowed tree stump; its cave-like opening is in the exposed side. Despite its size, its resemblance to a mound of litter makes it inconspicuous. Clutch 1 (65 × 45 mm). Female incubates alone, 40–45 days, and cares for the young. (p. 214)

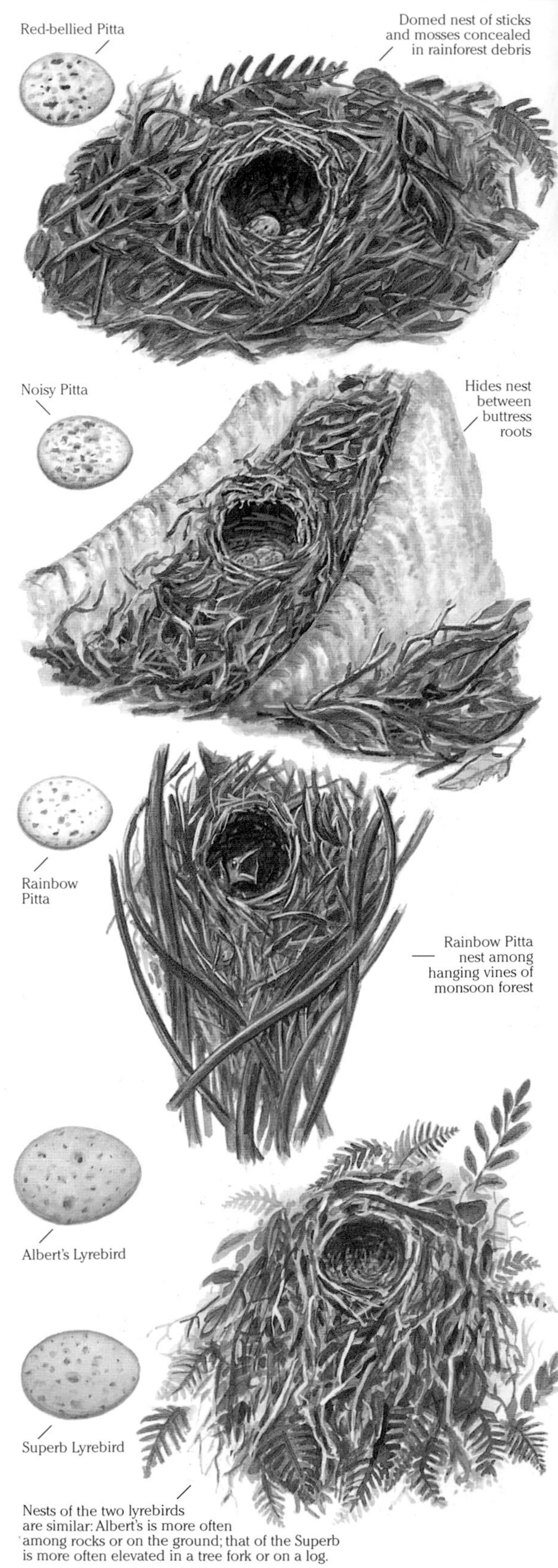

Nests of the two lyrebirds are similar: Albert's is more often among rocks or on the ground; that of the Superb is more often elevated in a tree fork or on a log.

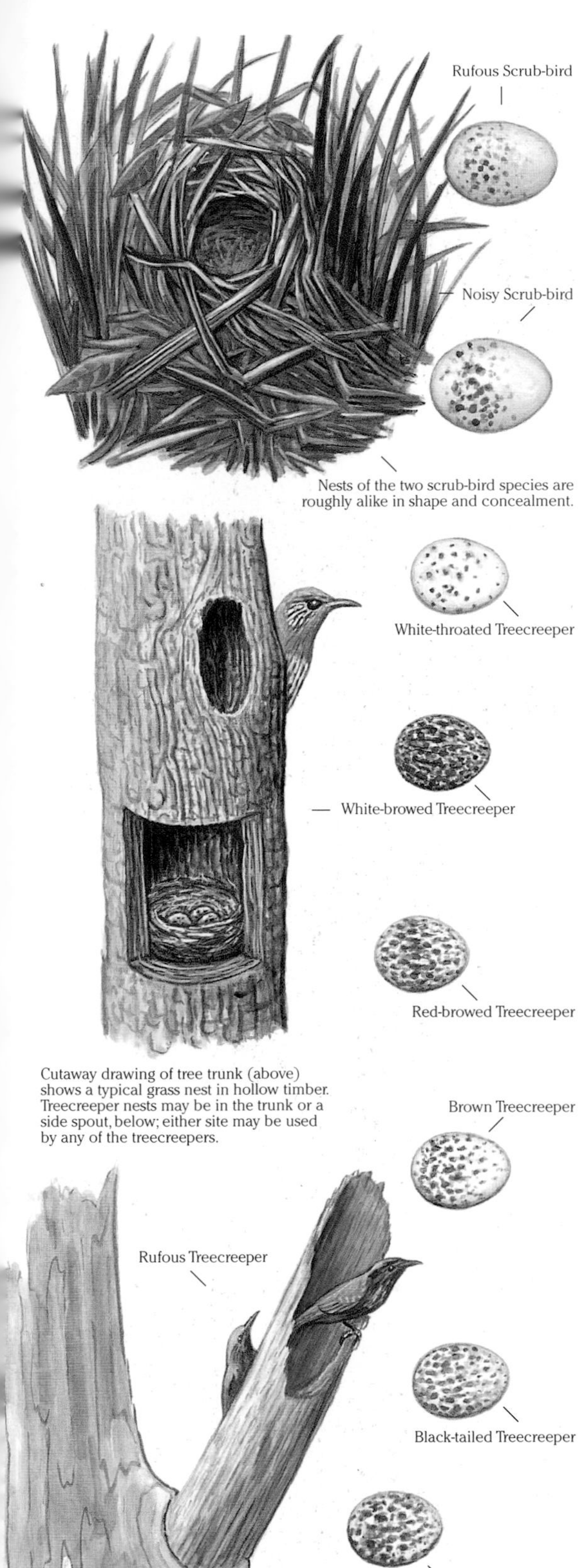

Nests of the two scrub-bird species are roughly alike in shape and concealment.

Cutaway drawing of tree trunk (above) shows a typical grass nest in hollow timber. Treecreeper nests may be in the trunk or a side spout, below; either site may be used by any of the treecreepers.

Rufous Scrub-bird Aug.–Jan. Secretive behaviour makes the female's territory, in which the nest would be built, very hard to locate—it is made more difficult by her reluctance to give any call other than a weak squeak if alarmed near the nest. A usual nest site is very close to the ground in or under a clump of rushes or other dense vegetation. The nest is quite large for the size of the bird, a thick-walled sphere about 15 cm in diameter, with a side entrance. The outside is woven of blady grass, rushes and ferns, while the inside is lined with of wet wood pulp and grass, which dries and sets to a cardboard-like consistency. Female incubates and rears young alone. Usual clutch 2 (29 × 20 mm); often only 1 hatches. (p. 216)

Noisy Scrub-bird May–Oct. Males begin calling extremely loudly as the winter breeding season approaches. The females, however, are not as easily located, and the nests, although large, are extremely well hidden in dense low undergrowth. The outside is built of rushes, leaves and grass, and looks like any of the surrounding rush clumps in the understorey in which leaf litter has accumulated. The nest has a side entrance; the inside is lined with wet, decaying rushes that dry to a cardboard-like layer. Lays a single egg (29 × 20 mm). (p. 216)

White-throated Treecreeper Aug.–Jan.; often 2 broods in a season. Lives in rainforest, wet eucalypt forest or tall, riverside trees. Nest hollows are high, up to 20 m above ground. A deep, narrow hollow in a tree trunk or limb is lined with bark, fur and grass to make a cup-shaped nest. Clutch 2–4 (23 × 17 mm). Nest preparation and incubation, which lasts 21–22 days, are by the female. The male helps to feed the female and the young, which fly aged about 25 days. (p. 218)

White-browed Treecreeper Aug.–Nov.; often 2 broods in favourable seasons. A cup of bark softly lined with grass, fur and hair is built into a small, deep hollow tree limb or trunk between 3 and 12 m up. In the mulga and casuarina woodland of the semi-arid inland, which comprises much of the habitat of this species, the trees are small and most hollows are quite low. Clutch 2 (23 × 16 mm). The female incubates alone for 22–23 days, but she is fed by the male. (p. 218)

Red-browed Treecreeper Aug.–Jan.; often 2 broods in a season. This is a strongly social species: the young of previous seasons, predominantly males, stay with the parent birds for several years forming a family group whose members help to build the nest and feed the dependent young. The cup nest of bark and fur is in a deep, narrow tree hollow at heights to 30 m (but some nests are as low as 5 m). Incubation of the 2–4 eggs (22 × 17 mm) takes about 18 days. (p. 218)

Brown Treecreeper May–Dec.; those of N and inland are earlier than birds of the SE. Bark and grass nest is built in hollow tree or stump. Clutch 2–3 (23 × 18 mm). Several independent young of previous years are part of the small family group that feeds the incubating female and the new chicks. Incubation takes 16–17 days; young fledge at 20–25 days. (p. 218)

Black-tailed Treecreeper Aug.–Jan.; probably also late in the summer wet season. Young males from previous years (more often than young females) seem to stay with the parents and help build a nest, and feed the incubating female and then the nestlings. Nest is of grass lined with feathers and is built in a small, deep hollow. Clutch 2–3 (25 × 20 mm). (p. 218)

Rufous Treecreeper Jun.–Jan.; 2 broods in some seasons. Young of previous years may help to build a nest and feed the nestlings. Nest is a deep, narrow hollow in a tree trunk or limb, occasionally in a fallen log, but usually 4–15 m up. The hollow's base is softly lined with bark fibre, feathers and wool. Female incubates. Clutch 2–3 (25 × 20 mm). (p. 218)

Purple-crowned Fairy-wren Nov.–Jul. Nest noticeably larger, made of coarser materials, than nests of other fairy-wrens. Paperbark, strips of pandanus leaf and grass are used. It is often placed low in a canegrass clump, sometimes at the centre of a pandanus. Clutch 2–4 (17 × 12 mm). (p. 220)

Superb Fairy-wren Jul.–Mar.; usually Sep.–Dec. A small sphere with a side entry is built of grass stems, rootlets, lined with feathers. Hidden in a grass clump, low dense shrub or bracken, 20–70 cm up. Clutch 3–4 (17 × 12 mm). Female incubates, 13–15 days. Rest of family group helps feed the chicks. (p. 220)

Splendid Fairy-wren Sep.–Jan. Often nests in open, upper part of undergrowth, 0.5–1.5 m up, where it is more easily seen. Domed nest is built of fine grass stems, well bound with webs and lined with feathers. Clutch 3–4 (17 × 13 mm); incubation by female, 14–16 days. Other family members, 'brown birds' of the group, help feed the nestlings. (p. 220)

Lovely Fairy-wren Throughout year; mostly Jul.–Dec. An upright, oval, domed nest is built low cover such as grass, less than 1 m up. Leaf skeletons added to the grass make the nest untidy, fragile. Clutch 2–3 (16 × 12 mm); incubated 13–15 days by female. Others in group help feed the young. (p. 222)

Variegated Fairy-wren Jul.–Feb.; in arid regions will breed in most months after rain. Nests often very low, 10–40 cm up in a clump of coarse grass or a small, dense shrub. Fine grass stems and bark fibre are used; webs lightly hold the structure together. Clutch 3–4 (17 × 13 mm). Incubated 14–16 days by female; young fly at 10–12 days. (p. 222)

Blue-breasted Fairy-wren Jul.–Dec.; mostly Aug.–Nov. Builds a small, neat, upright oval nest with side entrance from thin, dry grass stems and other fine material bound with webs, softly lined, and placed in undergrowth, 0.5–1 m up. Clutch 2–3 (16 × 13 mm). Female incubates. (p. 222)

Red-backed Fairy-wren Aug.–Jan. in S; Nov.–Apr. in tropical N. Builds a nest of grass and bark strips hidden very low in a clump of rank grass or a dense low shrub. Clutch 3–4 eggs (16 × 12 mm). Often found along watercourses; then nests tend to be placed higher, often in pandanus. (p. 224)

White-winged Fairy-wren Jul.–Sep.; other months after good rain. The small, upright oval nest, domed with side entrance, is built of grass, wool and fine rootlets bound with webs. Typically very low, hidden in stunted saltbush or samphire, or coarse grass. Clutch 3–4 (15 × 12 mm). (p. 220)

Red-winged Fairy-wren Sep.–Jan. Nests in undergrowth of moist valleys, creek-side thickets, swordgrass clumps and damp hill slopes, often where leaves of low *Xanthorrhoea* brush the ground. The domed nests are lower (10–30 cm up) and much better hidden than those of the Splendid or Blue-breasted. Decaying leaves are added to make a bulky, loosely bonded, fragile structure. Clutch 2–3 (16 × 13 mm). (p. 222)

Southern Emu-wren Aug.–Dec. Builds a domed, upright oval nest of grass, fine twigs and strips of sedge leaves; has a side entrance similar to those of the fairy-wrens. Is hidden close to ground in dense low heath or tussocks of rushes or grass. Clutch 2–4 (16 × 13 mm); female incubates, 10–12 days. (p. 224)

Mallee Emu-wren Sep.–Dec. Domed nest with side entry is made of fine twigs, grass stems and shreds of bark, bound with webs and softly lined. Nest blends into dead, grey-brown centre of a spinifex clump. Clutch 2–3 (16 × 12 mm). (p. 224)

Rufous-crowned Emu-wren Sep.–Dec.; almost any time after rain. Domed nest is built into a spinifex clump 15–30 cm above ground; is often visible, but hard to recognise. Clutch 2–3 (16 × 12 mm). (p. 224)

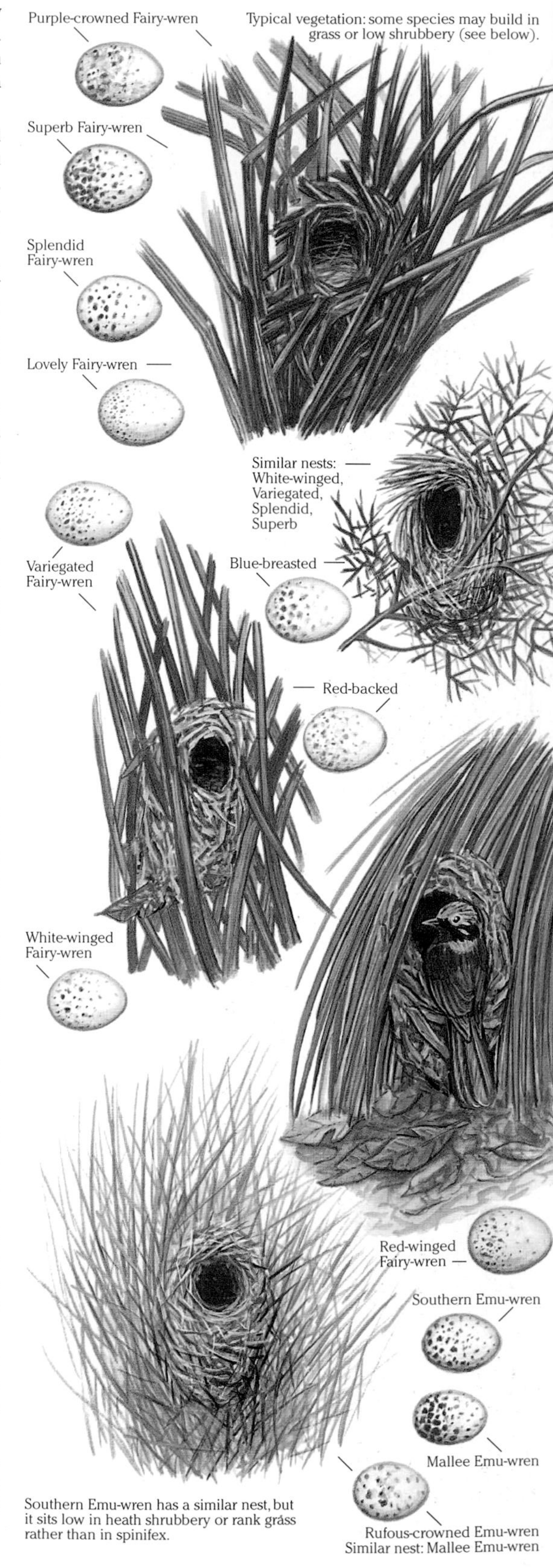

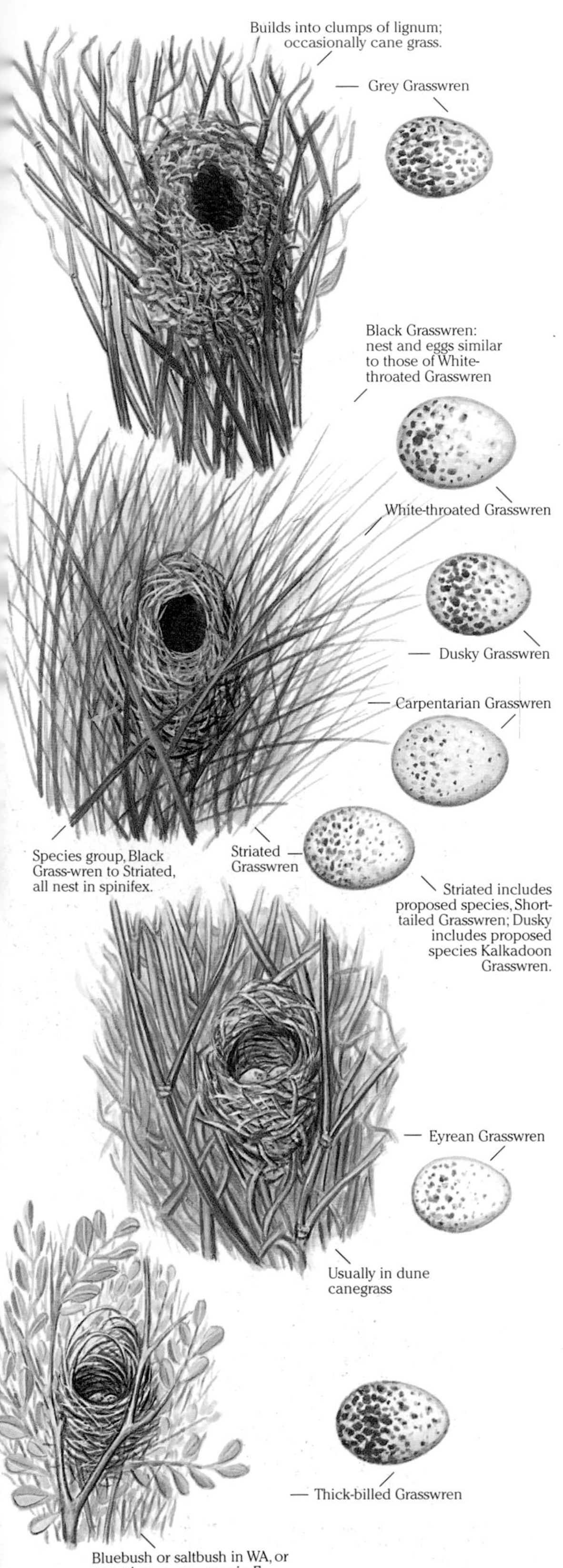

Grey Grasswren Jul.–Sep. Nests in dense swamp lignum thickets, less often cane grass, on floodplains of Cooper Creek, Bulloo and Lower Diamantina river systems. Before breeding, the normal foraging parties (15–20, occasionally 30 birds) break up and pairs form, each defending its own nest territory. Nest is usually quite low, 20–80 cm up, bulky, partially or fully hooded, built of fine stems and grass, and lined with small feathers and soft plant fibre. Clutch 2 (19 × 15 mm). (p. 226)

Black Grasswren Jan.–Mar. The few nests found were built in a large spinifex clump: bulky, domed, about 20 cm long, 14 cm high and wide; had a funnel entry at one end. Woven of spinifex stems and grass, the inside was lined with softer grass. Clutch 2 eggs, oval, slightly lustrous, white, sparsely spotted with fine grey, chestnut and dark brown (22 × 16 mm). Female incubates and broods the hatchlings; male brings food to the nest. (p. 226)

White-throated Grasswren Dec.–May. Nests in needle-pointed spinifex on sandstone escarpments of W Arnhem Land. Often visible, but the nest's top is rarely recognised—a mass of dead grass merges with the dead and disintegrating centre of the spinifex clump. Nest is larger, more substantial than that of the Striated Grasswren; fully domed, with side entry. Male defends the territory with song while the female builds the nest and incubates 2 eggs (23 × 16 mm). (p. 226)

Dusky Grasswren Aug.–Oct.; breeding recorded other times, after rain—may create a second breeding in late summer to autumn, enabling some pairs to raise two broods. Nest is a domed or partially hooded structure of grass and spinifex needles lined with softer, fine grass and rootlets. Most common location is within a clump of spinifex, but may be built into some other spiny, low, dense shrub; usually 10–50 cm from ground. Incubation is apparently entirely by the female, but both parents feed the young, both in the nest and after fledging. (p. 228)

Kalkadoon Grasswren Dusky, race *ballarae*, is proposed as a full species: similar breeding details, perhaps earlier. (p. 228)

Carpentarian Grasswren Nov.–Mar.; probably after wet season arrives. Bulky nest, about 12 cm diameter, has substantial domed roof; built of grass and spinifex needles with some dry eucalypt and acacia leaves. Placed in dry centre of a spinifex clump, where the colour matches dead foliage and accumulated leaf litter; usually 20–50 cm above ground. Incubation of the 2–3 eggs (23 × 16 mm) by female. (p. 226)

Striated Grasswren Aug.–Dec.; after summer cyclones or other rains in arid regions. Builds in the top of a spinifex clump where it is hard to recognise, even though barely concealed. Bulky, domed nest has side entrance; upper walls of interwoven dry spinifex stems, which are stiff and stick out roughly and untidily, merging with the centre of the spinifex. Base and lower walls include bark strips and grass; lining is thick with soft fibre, down and fur. Usual clutch of 2 eggs (21 × 16 mm) is incubated by the female, 13–14 days. (p. 228)

Short-tailed Grasswren Striated, race *merrotsyi*, is now proposed as a full species: similar breeding details. (p. 228)

Eyrean Grasswren Jul.–Sep.; other months after rain. Nest is compact, rounded; overhead hood complete, partially formed or sparse. It is thickly woven of grass, leaves, rootlets, lined with fine grass, vegetable down, spiders' cocoons. Usually hidden low in a tussock of cane grass. Clutch 2–3 (20 × 16 mm). (p. 226)

Thick-billed Grasswren Jul.–Oct.; Jan.–Apr. in years of abnormal summer rain. Nest is built in a low, dense bluebush or saltbush; in E, sometimes in cane grass. Nests are thin-walled. Most are fully domed—some flimsy or partially hooded. Female builds the nest, but the male, unlike other grasswrens, helps to incubate. Clutch 2–3 (21 × 12 mm). (p. 228)

Spotted Pardalote Jun.–Jan.; usually 2 broods in a season. A narrow tunnel is drilled in a steep or vertical bank of creek, roadside cutting or decayed stump. Entrance is close to top of bank, perhaps to avoid flooding. Entry tunnel is almost circular, about 30 mm diameter, and shows two grooves, like wheel tracks, made by the bird's feet. Extends 40–90 cm to an enlarged nest chamber where a spherical, domed nest with a side entrance is built of bark strips and lined with finer, softer material. Clutch 3–4 (16 × 13 mm). Nesting activities are shared; incubation, 14 days. The race *xanthopygus*, 'Yellow-rumped Pardalote', occupies inland mallee country with sandy soil where creek banks are rare; its tunnel is usually drilled into flat ground, often among mallee leaf litter; will use earth banks, especially of sandy road edges, where available. (p. 230)

Red-browed Pardalote Jun.–Oct. in S; Mar.–May in N; after rain in arid regions. Drills nest tunnel deep into a bank, often in very hard ground, or in a sand ridge in desert country. The tunnel and nest of bark and grass are similar to those of the Spotted Pardalote, above. Clutch 2–4 (16 × 13 mm). Both sexes build nest, incubate and care for young. (p. 230)

Forty-spotted Pardalote Aug.–Dec. Usually nests in a small, deep hollow from 1 to 20 m up in a tree limb or trunk; occasionally uses a hole in the ground. A nest of bark and grass is built, domed if space permits. Clutch 4 (17 × 13 mm). Sexes share incubation, 16 days, and both feed the young. (p. 230)

Striated Pardalote Jun.–Jan. Builds bark and grass nest, domed if space permits, in a tree hollow, or a tunnel in a creek bank, other bank or crevice in a building. In SW of WA, always in tree hollows. Clutch 3–4 (18 × 15 mm). Both sexes build nest, incubate and feed young. (p. 230)

Eastern Bristlebird Aug.–Jan.; mostly Sep.–Nov. Dense coastal heathland and scrub, creeks and swamps very effectively conceal the nest, which is rounded, domed and has a side opening. It is hidden close to the ground, usually in a clump of coarse sedge or grass, or in low shrubbery intermixed with grasses and litter. Coarse dry sedge leaves, grass stems, fine sticks and bark fibre form the outer walls; fine soft grass and paperbark form the lining. Clutch 2 (25 × 19 mm); incubated by female, about 21 days; usually only one egg is fertile. Both parents feed the young, which fledge at 18–22 days. (p. 216)

Western Bristlebird Aug.–Dec. The nest is built in a large clump of sedge, commonly known as Sword Grass. Nest situation and construction are very much like the Eastern Bristlebird; nest is large, loosely built, globular, has a side entry, is placed low. Clutch 2 (26 × 20 mm). Incubated by female, about 21 days; young fledge at 18–22days. (p. 216)

Rufous Bristlebird Sep.–Dec. Nest sites vary with diverse habitats. Most nest in sedge clumps, some in low shrubs in coastal scrub and along gullies of coastal forests. Nest is similar to but larger than those of Western and Eastern Bristlebirds; higher above ground, more exposed; it is put in top of sedge clumps rather than in the base. Clutch 2 (28 × 21 mm). Female builds nest; incubates eggs, 21 days. (p. 216)

Rockwarbler Aug.–Dec. The domed nest is suspended by a slender, rope-like attachment from the roof of a cave, rock overhang, mine shaft, bridge or building. Nest is a bulky, untidy, upright oval with side entrance, top tapered to domed roof and a slender tail. Clutch 3 (20 × 15 mm). (p. 232)

Pilotbird Aug.–Dec. Builds an untidy, globular nest with side entrance; makes it of bark, leaves, rootlets; lines it with feathers. It is well hidden: almost on the ground in dead ferns, fallen limbs and other forest litter, it looks like part of that debris. Clutch 2 (27 × 20 mm). Incubated by female, 20–22 days. (p. 232)

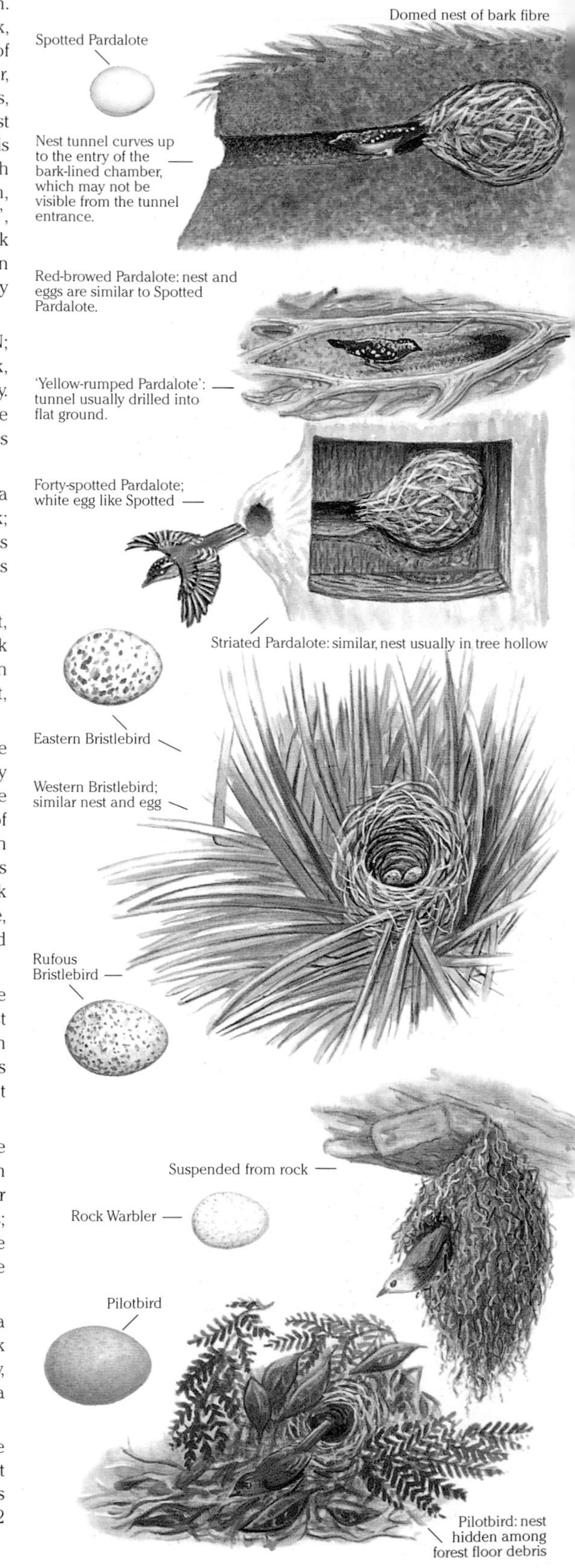

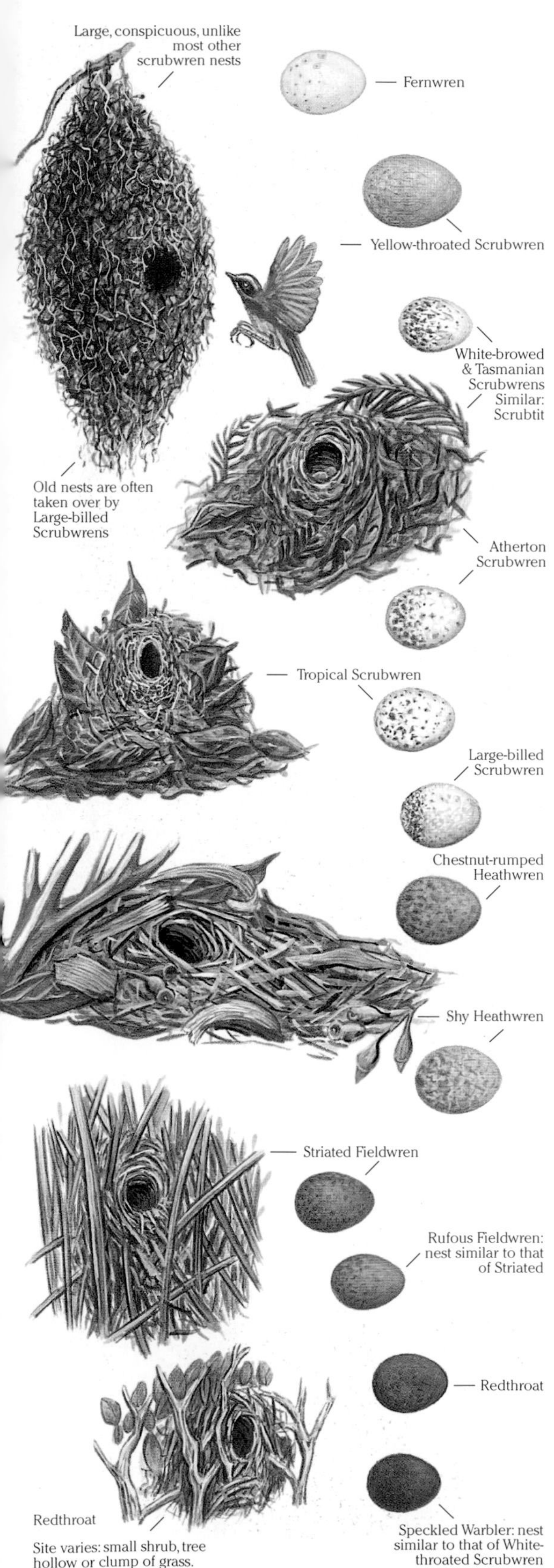

Fernwren May–Jan. Domed nest hidden under overhanging bank, fallen log or hollow log among forest debris, ferns, mosses. Large for bird's size, built of blackish rootlets, twigs, decaying leaves and dark green moss; difficult to recognise in its gloomy, cave-like setting. Clutch 2 (22 × 17 mm). (p.234)

Yellow-throated Scrubwren Jul.–Feb. Large, domed nest; often hangs above rainforest streams. Blackish rootlets and camouflage of moss, lichens and small ferns make the nest look like flood debris caught on branches. Clutch 2–3 (25 × 18 mm). The pair jointly builds the nest and incubates, 21 days. (p.234)

White-browed Scrubwren Jul.–Dec.; a domed nest of bark, grass, rootlets. The rough, loose exterior makes it look part of the forest debris among tangled low shrubbery, bark, fallen branches or logs on ground. Clutch 2–3 (19 × 15 mm). (p.234)

Tasmanian Scrubwren (Probably is a race of the White-browed.) Aug.–Dec. Domed nest of bark, grass, moss, softly lined with feathers, fur and soft plant fibre. Well hidden in dense groundcover, often in dead foliage, fern fronds, fallen bark and other debris. Clutch, eggs are like the White-browed. (p.234)

Atherton Scrubwren Oct.–Dec. Domed, globular nest of plant stems and fibre, decaying leaves and moss. Extremely well hidden, set in a hollow on a bank, or set on the ground among forest-floor debris, ferns and other low plants. The 2 eggs are sized and shaped like those of the White-browed. (p.232)

Tropical Scrubwren Oct.–Jan. Domed nest in dense foliage, debris, low vine tangles or exposed creek bank roots. Leaf skeletons, inbuilt and attached, are untidy, yet effectively disguise the shape. Clutch 2–3 (19 × 15 mm). (p.232)

Large-billed Scrubwren Jul.–Jan. Will take over nests, new or disused, commonly of the Yellow-throated. Otherwise, builds a nest similar to that of the White-browed: rough, loosely built and globular. Clutch 3–4 (19 × 15 mm). (p.232)

Scrubtit Sep.–Jan. Small, spherical nest of rootlets, bark, leaf skeletons and moss, near ground in dense, often dead debris of undergrowth; lined with fern rootlets, feathers, plant down. Often among dead tree-fern fronds. Clutch 3–4 (18 × 14 mm); white, fine reddish spots, mostly around the larger end. (p.232)

Chestnut-rumped Heathwren Jul.–Nov. Builds domed nest of dry grass stems, bark fibre and fine rootlets, very close to the ground in grass tussocks or sedge, or in dense, low shrubs on heathland. Clutch 3–4 (19 × 14 mm). Incubation apparently by the female. (p.236)

Shy Heathwren Aug.–Nov. Domed nest, hooded side entrance, almost on the ground in dense, low heathland; often at base of a mallee eucalypt among fallen branches, twigs, and strips of bark and leaves. Clutch 3–4 (20 × 15 mm). (p.236)

Striated Fieldwren Jul.–Dec. Spherical nest has a side entrance; placed on ground, tucked into a clump of sedge, grass or other dense, low plant; very well hidden. Lined inside with fine grass, fur and feathers. Clutch 3–4 (23 × 17 mm). (p.236)

Rufous Fieldwren Jul.–Nov. Domed nest set on ground, or in slight hollow under samphire, bluebush or other low, dense groundcover. Built of grass and stalks; lined with feathers, fur, plant down. Clutch 3–4 (21 × 16 mm). (p.236)

Redthroat Aug.–Nov. Domed nest, usually in low, dense bluebush, saltbush, coarse grass or sedge. Occasionally builds in a tree hollow or hole in the ground. Structure of twigs, bark, coarse grass. Clutch 3–4 (19 × 14 mm). (p.238)

Speckled Warbler Aug.–Jan. Rounded, domed nest built in slight hollow in ground or at base of low, dense plant. Often among fallen branches or other debris. Loosely, roughly built of dry grass, strips of bark. Clutch 3–4 (19 × 16 mm). (p.236)

Weebill Jul.–May. Hidden in dense, pendulous outer foliage of tree or shrub. Shape varies: domed with side entrance; but some globular, others tall, oval. Fine grass, bark and other plant fibre are bound with webs until soft and felty; spiders' white egg-sacs and brown caterpillar castings are often attached. Clutch 2–3 (15 × 11 mm). Incubated by female, 10–12 days. (p. 242)

Brown Gerygone Aug.–Feb. Nest domed with slight hood over side entrance. Roof tapers up to attach to a slender twig; beneath, narrows to a trailing tail. Built of long, fine bark fibre, grass and rootlets bound with webs; moss, lichens attached. Clutch 2–3 (17 × 13 mm). Incubation 12–14 days. (p. 238)

Dusky Gerygone Sep.–Jan. Suspended by domed roof in foliage of mangroves, 2–5 m up. Small, neat, compact; hooded side entrance; slender tail. Built of fine bark strips and fibre bound with webs and lined with feathers and plant down. Clutch 2–3 (18 × 12 mm). (p. 238)

Mangrove Gerygone Sep.–Feb.; in N, most of year. Neat, compact, upright oval with hooded side entrance and thin tail. Built of grass stems, rootlets, fine bark strips, closely bound with webs, decorated with pale cocoons and spiders' egg-sacs. Clutch 2–3 (18 × 12 mm). Incubated by female, 12–14 days. (p. 238)

Western Gerygone Aug.–Jan. Domed nest; slender top is attached to leafy twigs of outer foliage, 2–10 m up. Nest is woven of fine bark strips, dry grass stems, all matted with webs. Outside it is decorated with white egg cocoons and brown caterpillar castings. Clutch 2–3 (18 × 14 mm). (p. 238)

Large-billed Gerygone Sep.–Apr. Nest hangs on branch or vine, often over a stream or swamp. It is an elongated pear shape with extended ragged tail and hooded side entry to the egg chamber and it hangs down 50–80 cm from its support. The untidily attached pieces of bark, lichen and leaf make it look more like flood debris than an occupied nest. Clutch of 2–3 eggs (17 × 13 mm); incubation 12–14 days. (p. 240)

Fairy Gerygone Sep.–Mar. Nest hangs from a twig in outer foliage of a leafy tree. Spherical nest chamber tapers upward to a slender attachment; tail is beneath; has a side opening with deep hood. Built of bark, grass and other plant material decorated with lichens, moss and spiders' egg cocoons. Clutch 2–3 (18 × 13 mm). Incubation 12 days. (p. 240)

Green-backed Gerygone Oct.–Apr Globular nest with deeply hooded side entrance. Roof tapers up to attach to dense foliage, usually of mangrove, monsoon or paperbark forest; has a long, slender tail. Nest is bound with webs, which also attach many white spider egg-sacs. Clutch 2–3 (18 × 13 mm); incubated by female, about 12 days. (p. 240)

White-throated Gerygone Sep.–Nov.; Sep.–Jan. in N. Domed nest hangs from a slender support bound to a twig in the outer leaves, 3–15 m up; the nest's base tapers to a long tail. Nest is built of fine bark strips and fibres bound with webs. Clutch 2–3 (18 × 13 mm). Incubation about 12 days. (p. 240)

Mountain Thornbill Sep.–Jan. Bulky, tall, oval nest is built in a rainforest tree's outer foliage, 6–12 m up. A short attachment joins domed roof and supporting twig. The nest is large for so small a bird, and its outside is densely covered with brown and bright green moss, which trails in long strands below the nest. Usual clutch 2 (17 × 13 mm). (p. 244)

Brown Thornbill Aug.–Dec. Builds untidy, round, domed nest with side entrance from dry grass, fine bark strips and pieces of fern frond, lightly bound with webs. It is set among ferns or tussocks, low in the undergrowth; is often higher, in tall shrubs, in rainforest and in SE Qld. Clutch 2–3 (16 × 12 mm). Incubated by female, 17–20 days; young fly aged 15–16 days. (p. 244)

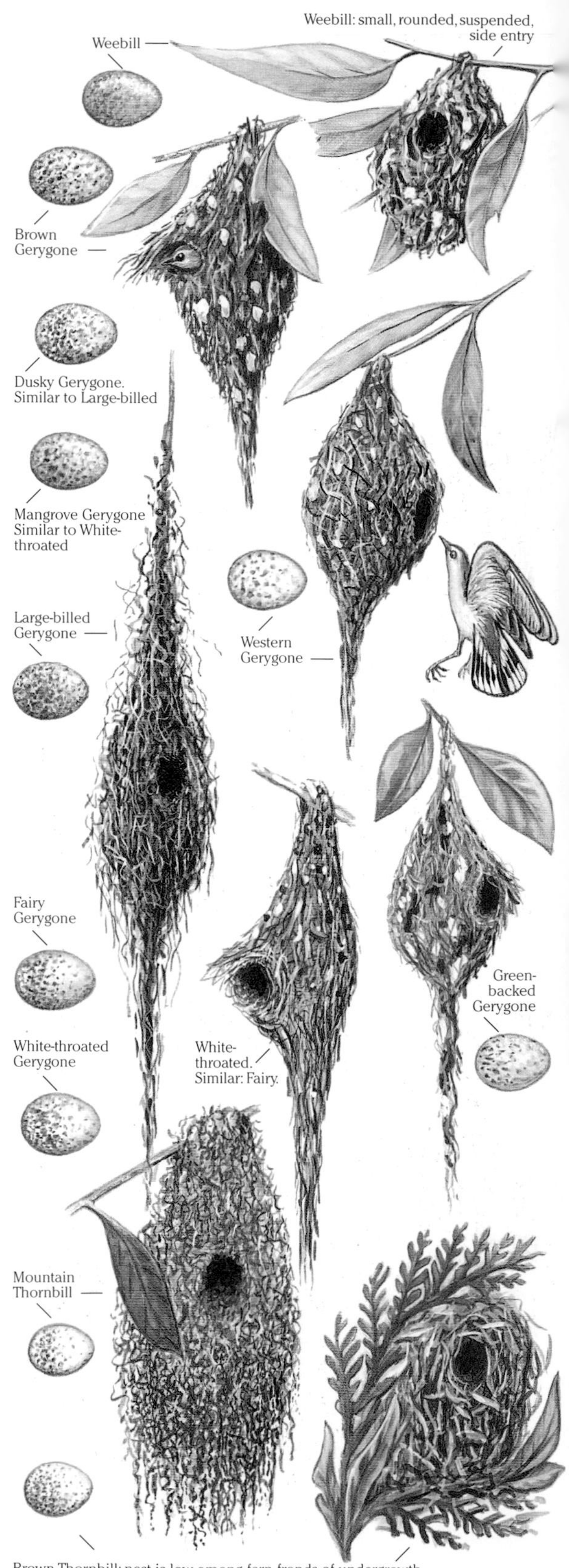

Brown Thornbill: nest is low among fern fronds of undergrowth.

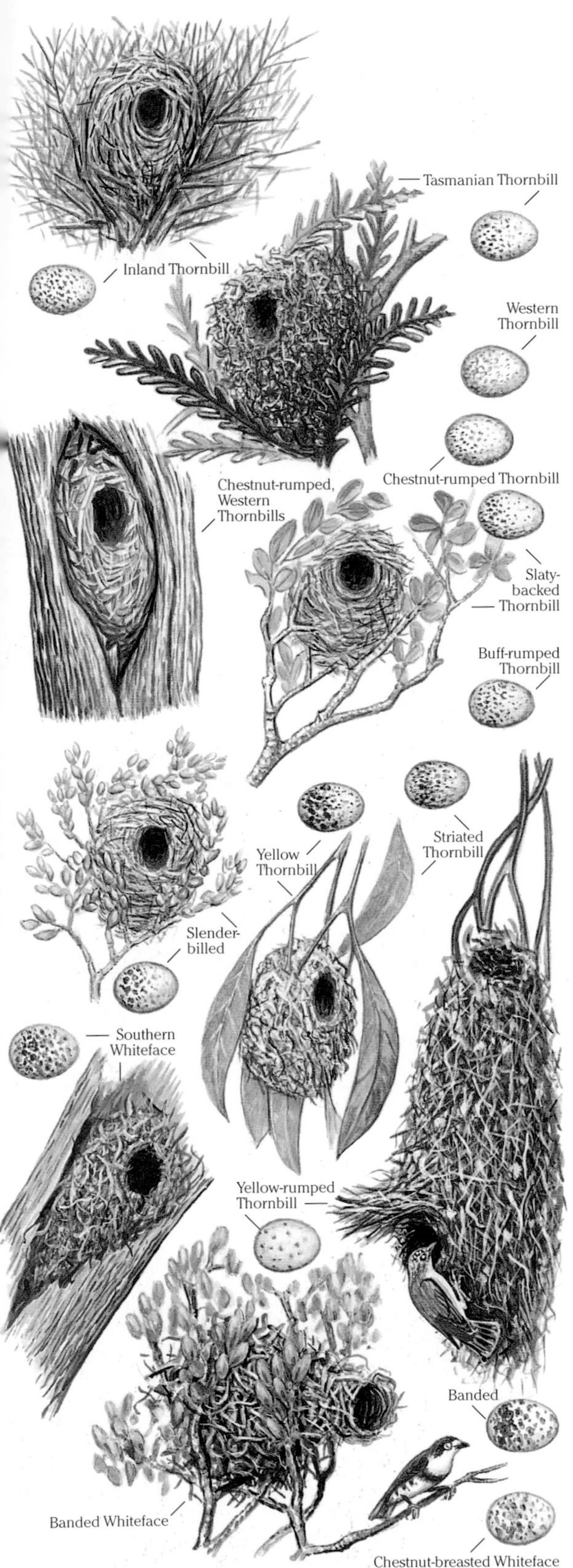

Inland Thornbill Jul.–Dec. A compact oval with hooded side entrance; in small shrubs within 0.5 m of ground. Built of fine twigs, dry grass neatly bound with webs, lined with feathers, wool. Clutch 3 (17 × 13 mm); incubated 19 days. (p. 244)

Tasmanian Thornbill Sep.–Jan. Domed nest is built of bark strips, fine grass bound with webs; decorated with dry brown fern tips, green moss; lined with feathers, hair and moss. Usually low in a shrub, among ferns or in a large grass tussock beneath shrubbery. Clutch 3–4 (18 × 13 mm). (p. 246)

Western Thornbill Aug.–Dec. Small, round, domed nest with side entrance built of bark fibre, grass; lined with feathers and fur. Built in tight spaces 0.5–2 m above ground—behind a sheet of peeling bark, in a knothole or in a dense plant such as a *xanthorrhoea* 'skirt' of leaves. Clutch 3 (16 × 13 mm). Incubated 19–21 days. (p. 242)

Chestnut-rumped Thornbill Jul.–Dec.; varies with rainfall; often 2 broods. Builds usual thornbill-type nest, but always in a small tree hollow, cavity behind bark, narrow fork like a hollow, or hollow fencepost. Made of bark and grass, lined with fur, feathers. Clutch 2–4 (16 × 12 mm). (p. 244)

Slaty-backed Thornbill Varies; almost anytime after rain; mostly Jul.–Nov. Builds small domed nest with side entrance in a low dense shrub. Made of dry grass bound with webs, lined with material such as hair. Clutch 3 (17 × 12 mm). (p. 246)

Buff-rumped Thornbill Aug.–Dec. Usually nests close to the ground among fallen twigs and leaves, clumps of sedge, grass, low dense shrubs, under logs or in hollows of fallen limbs. Builds a bulky, rough, loose nest of bark strips, grass stems and dry leaves; these help merge the nest into its surroundings. Clutch 3–5 (16 × 12 mm). (p. 242)

Yellow Thornbill Strictly arboreal—nests 3–15 m up. One of smallest thornbill nests; inconspicuous in outer foliage. Dry grass and bark fibre is bound with webs; decorated with green moss and spiders' egg-sacs. Clutch 3 (16 × 12 mm). (p. 242)

Striated Thornbill Jul.–Dec. Arboreal—relatively high nest sites. Small rounded nest hangs from a leafy twig 3–10 m up. Made of bark strips and fine grass lined with feathers and plant down. Usual clutch 3 (17 × 12 mm); incubated by female, 16–17 days. (p. 242)

Slender-billed Thornbill Any month after rain; mostly Jul.–Nov. Nests in dense, low samphire, saltbush and suchlike. Small, globular nest; grass and bark fibre bound with webs; lined with feathers, fur, plant down. (p. 244)

Yellow-rumped Thornbill Varies; mainly Jul.–Dec. Well-known, unusual double nest; a rough, open cup, false nest on top, the true nest chamber beneath—entry by a well-hidden, narrow funnel-hood. The grass and fibre structure, bound with webs, is built in a tree's drooping lower foliage or under a raptor's nest. Clutch 3–4 (18 × 13 mm). Female incubates, 18 days. (p. 242)

Southern Whiteface Jun.–Nov.; also after rain. Globular, side-entry nests may be built in hollows of trees or stumps, dense low shrubbery, or large sticks under nests of raptors; uses grass, bark lined with feathers, fur. Clutch 2–5 (19 × 14 mm). (p. 246)

Banded Whiteface Varies with rain; most Jul.–Oct. Unlike Southern Whiteface, nest is low in a shrub. Builds a bulky, untidy spherical nest lined with feathers, fur, wool; entry spout on one side. Clutch 2–4 (18 × 12 mm). (p. 246)

Chestnut-breasted Whiteface Breeds Aug.–Sept.; varies with rainfall. Nest is rarely seen: it has been described as large, roughly built of sticks; domed with a side entry, but no entry tunnel. Lined with feathers, fur, wool; sited in a low bush. Clutch 3 (18 × 13 mm). (p. 246)

Yellow Wattlebird Jul.–Apr.; 2 broods. In tree fork, 2–20 m up, large, rough cup shape of fine sticks, bark strips, lined with fine grass, plant down, feathers. Clutch 2–3 (36 × 24 mm). (p 248)

Red Wattlebird Most months; most Jul.–Dec. Large rough cup of twigs, bark, lined with soft grass, feathers, in vertical fork, 3–20 m. Clutch 2–3 (33 × 21 mm). Incubates 15–16 days. (p. 248)

Western Wattlebird (Is now the species *lunulata*, WA) Apr.–Jan.; most Jul.–Dec. Rough twig cup, smaller than Red Wattlebird's, soft fibre lining; built in dense shrub or tree, 1–10 m up. Clutch 1 (30 × 21 mm); incubates 14–15 days. (p. 248)

Little Wattlebird (SE Aust., Tas.) Jun.–Dec. Untidy bowl of sticks, bark, in fork, 1–10 m. Clutch 1–2 (2–3 in Tas.). (p. 248)

Spiny-cheeked Honeyeater Jul.–Jan.; after rain, arid land. Small, deep grass and web cup hangs by rim in leafy shrub, tree. Clutch 2–3 (24 × 18 mm); female incubates, 14 days. (p. 248)

Striped Honeyeater Aug.–Feb. Deep cup, soft fine grass, plant down, wool, webs; rim hangs from thin twigs, hidden by thick outer leaves, 2–5 m up. Clutch 3–4 (24 × 17 mm). (p. 248)

Noisy Friarbird Jul.–Feb. Large open cup—bark strips, dry grass bound with webs, lining of fine grass, often wool. Hangs by rim between slender branches of outer foliage, 2–20 m up. Clutch 2–4 (35 × 24 mm). Incubated by female. (p. 250)

Silver-crowned Friarbird Sep.–Jan., Feb. Bowl of bark strips, fibre and cobwebs, lining of finer grass, down. Usual site a leafy, drooping branch in outer foliage, 2–10 m above ground. Clutch 1–3 (30 × 20 mm). (p. 250)

Helmeted Friarbird Aug.–Apr. Deep, bulky cup of bark strips, fine twigs, lining of fine vine tendrils; hangs in outer foliage of tree or shrub, 2–15 m up. Clutch 2–5 (34 × 23 mm). (p. 250)

Little Friarbird Jul.–Mar.; 2 broods. Bark strips, grass, rootlets and cobwebs, lining of fine grass; hangs by rim in outer foliage 3–20 m up, often over water. Clutch 2–3 (27 × 20 mm); incubated by female. (p. 250)

Regent Honeyeater Aug.–Feb. Builds 2–20 m up in a thick horizontal or vertical fork or in a mistletoe clump. Makes a thick-walled bowl of bark strips lined with fine bark fibre and dry grass. Clutch 2–3 (24 × 18 mm); incubated by female. (p. 250)

Blue-faced Honeyeater Jul.–Jan; several broods. Often in loose colonies; bulky nest of bark, grass, rootlets, in tree fork, 3–20 m up. Will refurbish an old nest, often of a Grey-crowned Babbler. Clutch 2–3 (32 × 21 mm). (p. 252)

Bell Miner All year; several broods; breeds in colonies. Nest is a hanging cup of fine twigs, bark, grass decorated with lichens, insect cocoons, in fork 4–7 m up. Clutch 2–3 (23 × 16 mm). Others in colony may help feed the young. (p. 252)

Yellow-throated Miner Jul.–Dec; after rain, arid regions. Communal nesting; several broods in season. Builds a large, loose bowl of twigs and grass, well lined with feathers and wool in a tree fork, 2–6 m up. Clutch 3–4 (25 × 18 mm). Others in group often help. (p. 252)

Black-eared Miner (May be a race of Yellow-throated.) Aug.–Nov. Twig, grass bowl. Clutch 2–3 (25 × 18 mm). (p. 252)

Noisy Miner Jul.–Dec. Bowl of twigs, bark, leaves and cobwebs, lining of softer material, often flimsy. In tree or bush, 1–20 m up. Clutch 2–3 (25 × 20 mm). (p. 252)

Macleay's Honeyeater Oct.–Jan. Cup nest of bark and other fibre, bound with webs; hangs by rim in thin branches; mid level to high in forest. Clutch 2 (23 × 18 mm). (p. 252)

Tawny-breasted Honeyeater Nov.–Feb. Suspended cup; bark and webs, lining of fine rootlets. High in leaves of rain-forest, mangroves, woodlands. Clutch 2 (25 × 17 mm). (p. 252)

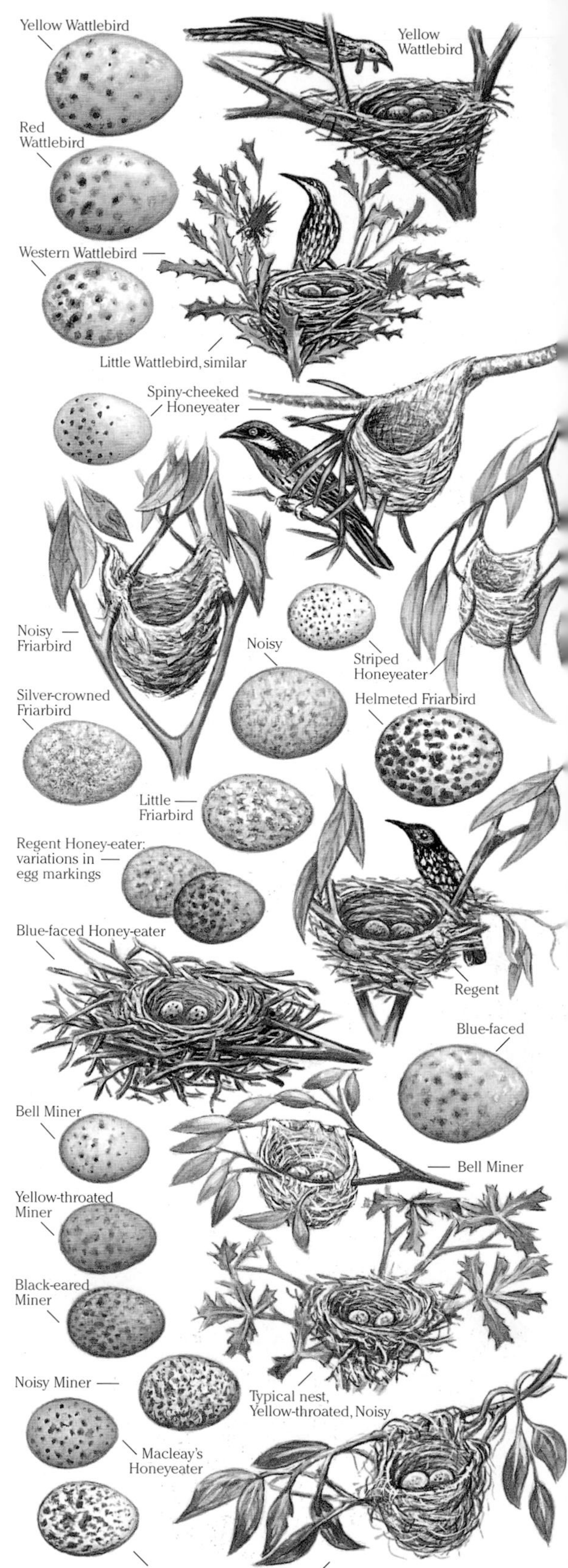

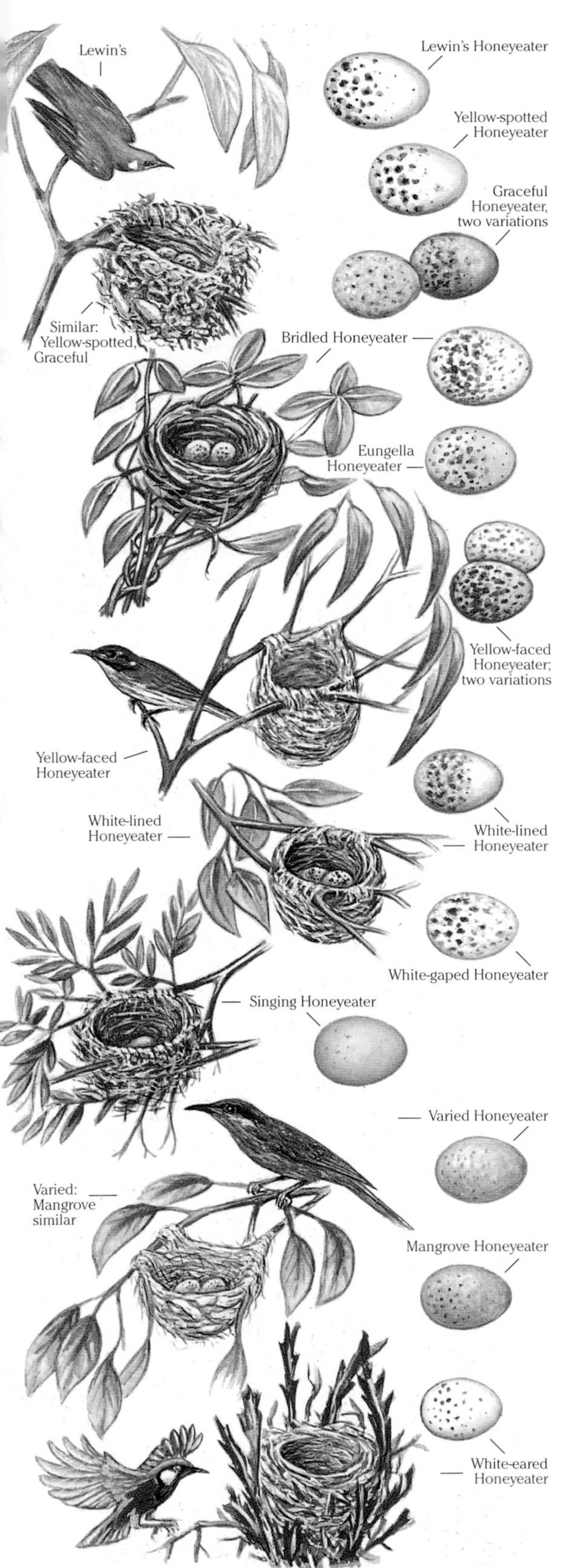

Lewin's Honeyeater Jul.–Mar.; mostly Aug.–Jan. Nest hangs by rim from twigs in dense foliage 2–5 m above ground; often in a gully or over water. A rough, bulky, cup, rather loosely built of bark strips, leaves and moss bound with webs and lined with soft down. Clutch 2–3 (24 × 17 mm). (p.254)

Yellow-spotted Honeyeater Aug.–Sep. to Jan–Feb. Deep cup, palm and bark fibre, and webs, hung by rim in foliage 2–4 m up, often edge of rainforest. Clutch 2–3 (23 × 16 mm). Female incubates, 15–16 days; chicks fledge about 15 days. (p.254)

Graceful Honeyeater Sep.–Oct. to Jan.–Feb. Deep cup, bark strips and green moss bound with webs, lined with fine grass or plant down. Nest hangs from its rim in a slender fork of outer foliage, 2–6 m up, often near rainforest edge or over water. Clutch 2 (20 × 15 mm). Incubation 14–15 days. (p.254)

Bridled Honeyeater Sep.–Jan., Feb. Builds cup shaped nest of vine tendrils, fine stems, ferns (all dry to a brownish tone) lined with soft plant fibre. Set among twigs or in vine tangle; usually quite low, but some 5 m or more above ground. Usual clutch 2 (25 × 15 mm). (p.254)

Eungella Honeyeater Sep.–Dec., Jan. Small, deep cup extensively decorated with grey-green moss, lined with fine plant fibre. Suspended at outer end of a slender, leafy branch, about 4 m up. Clutch 2 (22 × 16 mm). (p.254)

Yellow-faced Honeyeater Jul.–Jan. Small cup nest; hangs by the rim between slender leafy twigs from 2 to 6 m up. Deep, quite substantially built of grass and bark fibre well bound with webs and softly lined; sometimes has moss on exterior. Clutch 2–3 (21 × 14 mm). (p.254)

White-lined Honeyeater Aug.–Jan. Small cup; grey and brown twigs, fine dry stems, bound with webs, lined with very fine dry grass; hangs by the rim from a fork or between slender outer twigs of a leafy shrub or lower foliage of a tree; up to 5 m above ground. Usual clutch 2 (21 × 15 mm). (p.256)

White-gaped Honeyeater Almost throughout year, but usually Sep.–Feb. Nest is a deep cup of thin bark strips, fine grass and rootlets bound with webs, lined with finer fibre. It hangs by its rim in a horizontal fork of a shrub or tree between 2 and 20 m up, often over water. Clutch of 2 (22 × 18 mm); both sexes feed the young. (p.256)

Singing Honeyeater Jul.–Feb.; in arid regions, almost any time after rain. Builds a rather untidy cup of grass and thin plant stems, lightly bound with cobwebs and lined with finer rootlets, hair, grass and feathers. Hidden among dense twigs and leaves of a shrub or small tree; is placed from 0.5 to 5 m up. Usual clutch 2–3 (23 × 16 mm). (p.256)

Mangrove Honeyeater Aug.–Dec.; often 2 broods. Hangs a shallow, basket-like nest by its rim between branchlets, usually in a mangrove tree or shrub. Nest built of dry grass, sea-grasses, seaweed and spider's cocoons bound with webs, lined with finer grass, rootlets, soft plant down. Clutch 2 (24 × 17 mm). Female incubates; both sexes feed the young. (p.256)

Varied Honeyeater Some breeding almost all months, but most Sep.–Dec. Nest similar to Mangrove Honeyeater's, but built largely of seaweed; likely to be higher, at times above 4 m where mangroves are tall. Clutch 2 (23 × 17 mm). Female constructs nest, incubates; both sexes feed the young. (p.256)

White-eared Honeyeater Jul.–Aug. through to Feb.–Mar. Nest is built among tangled twigs and leaves, low in a small shrub, bracken or coarse grass from 0.5 m to 3 m up. Built of dry grass, fine stems, thin strips of bark and bound with webs, lined with soft plant down, feathers or fur. Clutch 2–3 (21 × 15 mm). Both sexes care for chicks. (p.256)

Yellow Honeyeater Aug.–Apr., but mainly Sep.–Feb. The small, cup shaped nest is suspended by its rim between slender twigs, in the foliage of a leafy shrub or tree. It is constructed of grass stems, thin strips of bark and, sometimes, palm fibre, all bound with webs and lined with fine dry grass and plant down. Clutch 2 (22 × 16 mm). (p. 256)

Yellow-throated Honeyeater Jul.–Jan; but mainly Aug.–Dec. The cup nest of bark and grass is lined with wool and hair, and is rather large, bulky. It is built into the dense inner part of a shrub or low mass of vegetation, live or dead—bracken or grass tussocks. Clutch 2–3 (23 × 17 mm). Incubated by both sexes for about 15 days. (p. 258)

Yellow-tufted Honeyeater Jul–Mar.; most Sep.–Jan.; often several broods. Makes a cup nest hung by its rim in low, dense shrubbery or regrowth among fallen or standing dead vegetation. Built of bark and grass bound with cobwebs, lined with fine grass, feathers, fur. Clutch 2–3 (23 × 17 mm). Incubated 14 days by both sexes; fledge at 13–15 days. (p. 258)

Purple-gaped Honeyeater Jul.–Jan., Feb. Both sexes build nest, a substantial cup of bark strips and grass, suspended by the rim in a shrub or low in a small tree, from 0.25 to 4 m up. Usual clutch 2 (20 × 15 mm). Adults share incubation for about 14 days; both feed young. (p. 258)

Grey-headed Honeyeater After rain; usually Jul.–Jan. Builds a small, neat cup of grass bound with webs, lined with rootlets and plant down, hung by the rim from twigs in a shrub 1–3 m high. Clutch 2 (20 × 14 mm). (p. 258)

Yellow-plumed Honeyeater Usually Jul.–Jan. Builds a small, neat, thin walled, suspended basket of fine dry grass and thin bark strips strongly bound with cobweb; attached by its rim to slender twigs of drooping foliage in a tree or hidden in a shrub, 1–3 m up. Clutch 2–3 (20 × 15 mm). (p. 258)

Grey-fronted Honeyeater Jul.–Jan.; after rain in arid regions. Builds a small, neat cup of fine grass and bark fibre bound with cobwebs, sometimes decorated externally with whitish spider's egg-sacs. The nest usually hangs from a slender fork in the drooping outer foliage of a small tree, 1–4 m up. Clutch 2–3 (20 × 15 mm). (p. 258)

Fuscous Honeyeater Jul.–Dec.; sometimes to Mar. Nest is a small cup of thin, dry grass stems and fine bark strips bound with cobwebs, and lined with finer grass, sometimes fur or wool. It hangs by the rim from thin twigs of the outer foliage of a tree or shrub, 1–15 m up. Clutch 2–3 (20 × 15 mm); both parents attend to young in nest and after they fledge. (p. 260)

Yellow-tinted Honeyeater Jul.–Mar. Constructs a small cup of dry grass stems and thin strips of bark, all bound together with cobwebs. But it has very little lining other than sparse, fine grass. Nest is suspended by its rim to slender branchlets of the lower outer foliage of a tree or shrub, typically 2–4 m above ground. Incubation 14 days. (p. 260)

White-plumed Honeyeater Breeds most months: often in response to rain in arid regions; usually Aug.–Dec. in S where rainfall is regular. A pair jointly builds a cup shaped nest of dry grass stems and fine bark strips, bound with webs and lined with feathers, wool or hair. It hangs in foliage at heights ranging from 1 to 9 m. Clutch 2–3 (20 × 15 mm). (p. 260)

Green-backed Honeyeater Habits are almost unknown; but sightings of young, out of nest birds suggest that the breeding season is in late summer. In NG, this species is reported to breed in late summer to mid winter. This description of eggs from NG possibly belong to it: light grey-brown with darker chocolate-brown at larger end. Clutch 2–3 (20 × 15 mm). (p. 260)

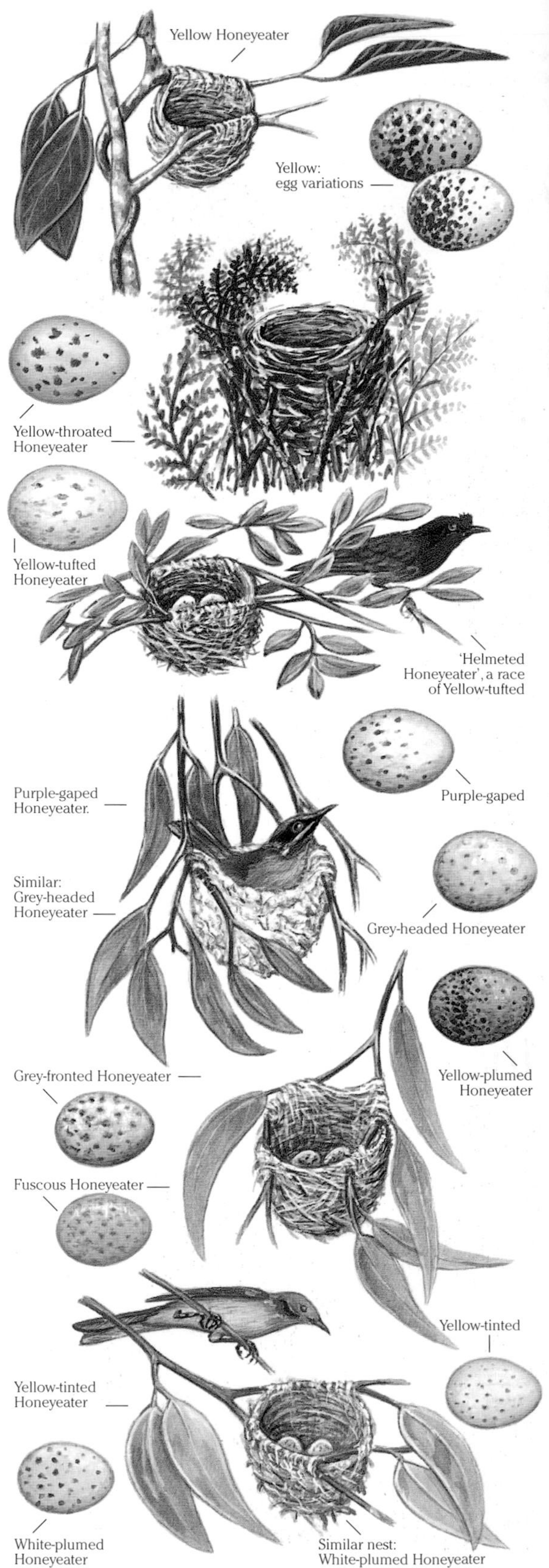

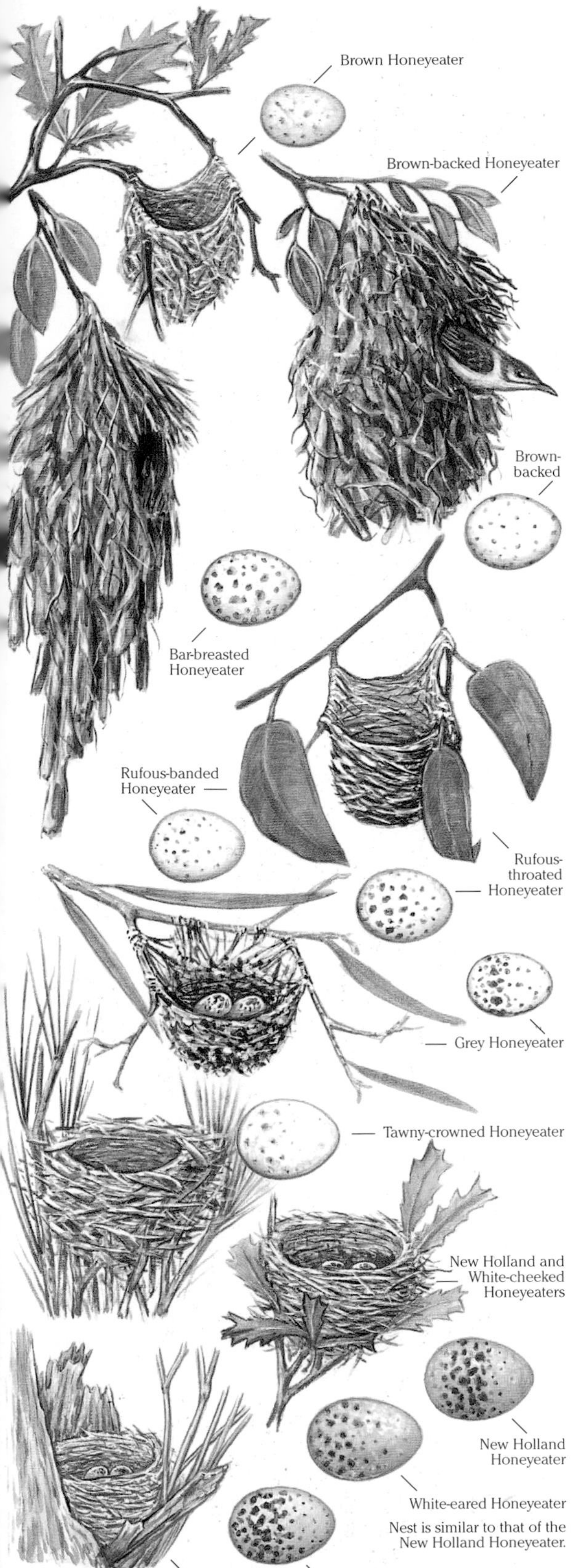

Brown Honeyeater Apr.–Nov. in N; Jun.–Feb in S. Small, neat cup of fine bark shreds, grass stems, plant down, heavily bound, matted with cobwebs, including egg-sacs; plant down as lining. Hangs from rim in a shrub, fern, bracken or tree; from very low to 5 m up. Clutch 2–3 (17 × 14 mm). (p. 260)

Brown-backed Honeyeater Aug.–Mar. Semi-colonial breeder in ideal habitat; isolated pairs otherwise. Large domed nest of paperbark strips and flakes; top of domed roof hangs from a thin, drooping, leafy outer branch; usually in wetlands over water or marshy ground, 2–4 m up. Clutch 2–3 (20 × 24 mm). Incubated by female, 15 days. (p. 260)

Bar-breasted Honeyeater Any time; mainly Sep.–Jan. Nest hangs in drooping foliage of a tree overhanging or standing in water. Many nests may be in close proximity over a favoured river pool or swamp. The large domed nest of paperbark strips has a hooded side entrance and tapers down to a long tail—it looks like a lump of flood debris caught in the foliage. Clutch 3–4 (21 × 14 mm). (p. 262)

Rufous-banded Honeyeater Aug.–Jun.; most Sep.–Mar.; 2 broods. Small, neat, deep basket hangs by rim from thin twigs; often one side is much higher: is well hidden in leafy, slender, drooping branch, 1–10 m up; often over water. Bark fibre and plant down are bound with cobwebs; the inside is lined with finer fibre. Clutch 2–3 (18 × 13 mm). Incubation about 12 days; young fledge at 11–13 days. (p. 262)

Rufous-throated Honeyeater Aug.–May; varies with season. Builds a deep, soft, suspended cup of fine grass stems, bark fibre and plant down bound with cobwebs. One side of rim is often much higher where supporting twigs are at different levels. Typically hidden in leafy outer foliage from 1 to 8 m above ground or water. Clutch 2–3 (18 × 13 mm). Incubation about 12 days; young fledge at 11–13 days. (p. 262)

Grey Honeyeater Aug.–Nov.; also Dec.–May after summer rain. Builds small, frail, shallow cup so thin and open that the eggs are visible from below; made of fine, dry grass stems, hair, plant down, bound with cobwebs; hangs by the rim between slender twigs, typically in outer, sparse foliage of a mulga shrub, 1–3 m up. Clutch 1–2 (17 × 13 mm). (p. 262)

Tawny-crowned Honeyeater Jul.–Feb.; most Sep.–Dec. Builds a sturdy, thick-walled cup of fine twigs, bark strips and grass; rough outside, but neat inside, softly, thickly lined with plant down, wool and fur. Is hidden in a low dense shrub or clump of grass. Adult birds sneak to and from nest through concealing plants, making it difficult to find. (p. 262)

New Holland Honeyeater Breeds any month that nectar plants flower; most Jul.–Dec. and Mar.–May. Has several broods in good seasons. The nest is a substantial cup of grass and bark, thickly lined with plant down or other soft fibre. It is built in dense foliage—a shrub or bushy tree—from 1 to 5 m up. Clutch 2–3 (21 × 15 mm). Incubate 14–15 days; young fly between 13 and 15 days. (p. 264)

White-cheeked Honeyeater Breeds any time; activity peaks Aug.–Nov. and Mar.–May; often nests in loose colonies. Cup nest is substantial, thick walled; made of bark strips, grass, leaves; thickly lined with softer material. Nest is supported and hidden by foliage of a low, dense shrub, clump of grass or reeds. Clutch 2–3 (21 × 15 mm). Female incubates, 15 days; both sexes feed young, which fly at 13–15 days. (p. 264)

White-fronted Honeyeater Aug.–Dec.; after rain in arid regions. Builds a cup of bark and grass with softer lining. Often hidden behind big bits of peeling bark around a fork in a mallee or other small tree or shrub; to 3 m up. Clutch 2–3 (20 × 15 mm). Incubation 14 days; fledge at 13–15 days. (p. 264)

Crescent Honeyeater Any time; most active Jul.–Dec., Mar.–Apr. Makes a thick walled cup of thin sticks, bark, grass; rough outside, neat and softly lined inside. Builds low in undergrowth in a shrub or tussock, usually below 2 m. Clutch 2–3 (20 × 15 mm); female incubates, 14 days. (p. 264)

Painted Honeyeater Aug.–Feb.; 2 broods. Shallow, flimsy looking basket of fine grass and rootlets bound with webs; hangs by the rim from twigs in the drooping outer foliage of a tree, 3–20 m up. Clutch 2 (20 × 15 mm). Both sexes build, incubate 15 days and care for the young. (p. 264)

White-streaked Honeyeater Jan.–May; sometimes Jun. Nest is a delicate looking, small basket of fine rootlets bound with cobwebs; so thin-walled that eggs or young can be seen from beneath; hangs by rim from twigs in the dense foliage of a low bush, 0.5–2 m up. Clutch 2 (16 × 14 mm). (p. 264)

Black-chinned Honeyeater Jul.–Dec.; often 2 broods. Small, neat, deep cup of fine bark strips and grass hangs by the rim in pendulous outer foliage 3–20 m up. Clutch 2–3 (22 × 15 mm); incubation 15 days. (p. 266)

White-throated Honeyeater throughout year; mostly Jul.–Dec. Small, deep, soft cup of fine grass and bark fibre bound with webs. Hangs by rim in pendulous outer foliage 5–15 m up. Clutch 2 (19 × 14 mm). (p. 266)

Strong-billed Honeyeater Aug.–Jan.; often small, loose colonies. Deep cup nest hangs by rim in fork of tree or shrub. Uses bark strips and grass bound with webs, softly lined; usually thicker walled than those of similar honeyeaters. Clutch 2–3 (22 × 16 mm). Sexes share nest building and incubation, 15 days; the young fledge at 14–16 days. (p. 266)

White-naped Honeyeater Jul.–Jan.; 2 broods. Small, deep, suspended cup of bark fibre and fine grass stems bound with cobwebs, softly lined, is hidden in drooping outer foliage, 3–15 m up. Clutch 2–3 (20 × 14 mm). Female incubates, 14 days; both parents grown young from previous nestings feed the nestlings, which fledge at 14–16 days. (p. 266)

Brown-headed Honeyeater Jul.–Jan. Small, deep cup of fine bark strips, grass, hair bound with cobwebs, softly lined, hangs by the rim in twigs of outer foliage, shrub or tree, 1–6 m up. Clutch 2–3 (16 × 13 mm). Both sexes feed nestlings, often being helped by other birds in the family group. (p. 266)

Black-headed Honeyeater Sep.–Jan. Both sexes make the nest, a deep cup of bark, fine dry grass and moss bound with webs; hangs in foliage of a tree or shrub. Clutch 2–3 (20 × 14 mm); incubated 16 days; young fledge aged 14–16 days. Adult young from previous broods may help. (p. 266)

Eastern Spinebill Aug.–Jan.; 2 broods. Makes small deep cup of grass and fine bark strips lined with feathers and soft plant fibre. Hangs by rim in a fork, outer foliage, 1–15 m up. Clutch 2–3 (17 × 13 mm). Both sexes incubate, 14 days. (p. 268)

Western Spinebill Jul.–Jan.; 2 broods. Builds small, neat cup of bark strips, fine grass and plant down bound with webs. Hangs by rim, or set in a tree fork or shrub, 1–7 m up. Clutch 1–2 (18 × 13 mm). Parents share incubation; both feed young on insects and nectar. (p. 268)

Banded Honeyeater Oct.–May. Makes deep, soft, thin walled cup of fine grass stems and bark fibre bound with webs; hangs it by the rim from slender twigs of outer foliage, tree or shrub, 1–5 m up. Clutch 2 (18 × 12 mm). (p. 268)

Black Honeyeater Sep.–Dec.; after rain in arid regions. Builds a very small cup of fine twigs and grass stems neatly lined with finer fibre. Outside is rough. Inconspicuous, even though often in an open, exposed position; usually below 2 m. Clutch 2–3 (16 × 12 mm). Shared incubation and feeding. (p. 268)

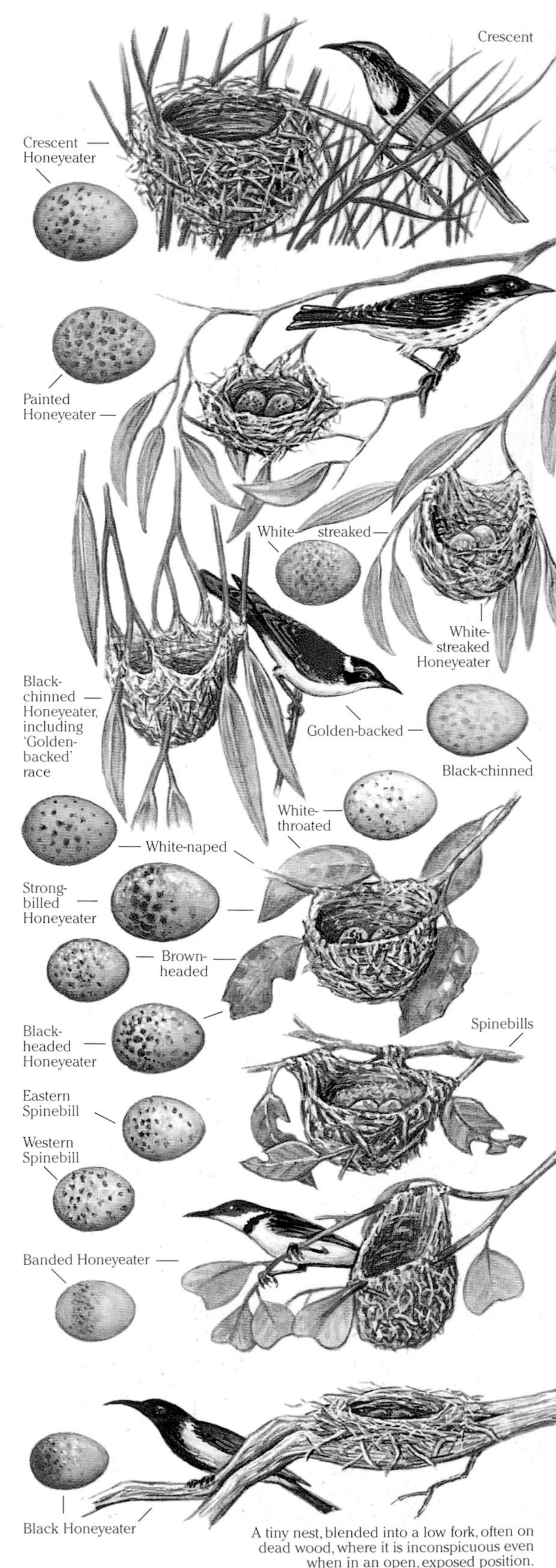

A tiny nest, blended into a low fork, often on dead wood, where it is inconspicuous even when in an open, exposed position.

Pied Honeyeater Jul.–Nov.; any other time after good rain. Small cup, rough outside, of fine twigs, grass, supported rather than hung in the fork of a shrub, 1–4 m up. Clutch 2–4 (22 × 16 mm). Sexes share nest building, incubation of 12–13 days, and feeding; young fledge aged 10–12 days. (p. 268)

Scarlet Honeyeater Jun.–Mar.; usually Jul.–Jan.; 2 broods. Female builds a small cup of fine bark strips, rootlets and grass bound with cobwebs, lined with fine grass. It hangs by the rim from thin twigs in dense outer foliage, 1–15 m up. Clutch 2 (16 × 12 mm). Both incubate, 11–12 days. Chicks are fed insects and nectar, collected by both adults. Young leave nest aged 11–13 days; parents feed them for another 10 days. (p. 270)

Red-headed Honeyeater Breeds any time; peaks Oct.–Jan. and perhaps Apr.–May; several broods in most seasons. Builds a small, thin walled cup of fine dry grass, bark fibre and rootlets bound with cobwebs, and lined with finer grass. The nest hangs by its rim from a horizontal fork in outer foliage, 1–10 m up; usually in or near mangroves. Clutch 2–3 (16 × 12 mm). Both sexes incubate and feed young. (p. 270)

Dusky Honeyeater Jul.–Mar.; most active Sep.–Dec. The small cup nest is of fine rootlets, bark fibre and thin dry grass bound with cobwebs. It hangs by the rim from a horizontal fork 2–15 m above ground, usually concealed by denser clumps of the outer foliage of a tree or shrub. Clutch 2 (17 × 14 mm). Incubation and fledging times are probably similar to those of Scarlet Honeyeater. (p. 270)

Crimson Chat Varies: Aug.–Oct. in S, winter rainfall. In dryer regions of N, almost any time after substantial rain. Flocks settle in favourable areas, typically plains with scattered, low, dense shrubs and carpeted with small ephemeral plants. A small cup nest of thin twigs and fine grass lined with rootlets is built in a supporting mass of twigs in a low, dense shrub or coarse grass tussock. Clutch 2–4 eggs (17 × 13 mm). Incubated by both sexes, 12–13 days; young fledge aged about 14 days. (p. 272)

Orange Chat All months after rain, but mostly Jul.–Dec.; often has several broods. The small, quite substantial cup nest is built very low in a small shrub, commonly saltbush or samphire, around a lake or claypan in the semi-arid regions. The site is usually where these small shrubs, which are mostly less than 30–40 cm high, are standing in very shallow water, 10–20 cm deep; this is often reduced to mud by the time the chicks have flown. Clutch 3–4 (17 × 14 mm). Incubated by both sexes, 12–13 days; fledge about 14 days. (p. 272)

White-fronted Chat Jul.–Jan. A substantial cup shaped nest, well concealed in the dense foliage of a low shrub, reeds or a grass tussock; tends to be low inside the vegetation, often almost on the ground, and usually lower than 1 m. The nest is constructed of sticks as well as grass and rootlets, which makes the outside rough and untidy, but the inside is neatly lined with fine grass and hair. Clutch 2–4 (17 × 14 mm). Incubated by both parents, 13–14 days; young fledge at 14–15 days. (p. 272)

Yellow Chat After rain, usually Sep.–Nov. in S; Mar.–Jun. in N. Builds small cup nest of plant stems, grass and feathers low in sedges, large grass tussocks, cane grass or a low shrub, often over shallow water in temporary or seasonal wetlands. Clutch 2–3 (17 × 13 mm). Incubation 12–13 days. (p. 272)

Gibberbird Breeds almost any time good rain, but usually Jun.–Dec. after winter rain. Nest is a small depression in ground, probably scratched out by the bird; may be sheltered by a bush or grass clump. The hollow is lined with sticks and grass to form a cup shaped nest; a surrounding platform is built of similar material. Clutch 2–4 (20 × 15 mm). Both sexes incubate; in hot weather brooding bird spreads wings to shade the nest. (p. 272)

Southern Scrub-robin Jul.–Jan.; 1 brood each season. Nests under screening mallee eucalypts, tea-tree thicket, broombush or coastal heath where debris is abundant. Nest is usually on the ground; rarely higher on a stump. A small bowl-shaped depression is scraped in the litter layer and lined with fine twigs and rootlets. Larger sticks and long strips of bark may be placed around the rim. Often the nest will be beside or between much larger sticks, and commonly close to a mallee, shrub, log or rock, blending into its surroundings exceptionally well. But when the single egg hatches, the naked chick is entirely black, conspicuous and vulnerable if not covered by the brooding adult. Clutch 1 (26 × 20 mm). (p. 274)

Northern scrub-robin Tropical wet season; Nov.–Feb. or Mar. Rakes aside leaf litter to scratch out a small, rounded depression, usually beside buttress roots of a tree, log or shrub; lines hollow with rootlets, fine twigs, bark strips and vine tendrils encircled by a rough rim of larger sticks. Clutch 2 (23 × 18 mm). Female probably builds and incubates. (p. 274)

Hooded Robin Jul.–Nov.; almost any time after rain in dry areas; often several broods. Builds a small, neat cup nest of bark and grass bonded and reinforced with webs, placed in a tree fork or crevice, 0.5–5 m above ground. The exterior is camouflaged with pieces of bark. Clutch 2–3 (22 × 16 mm). Incubated about 14 days by the female. (p. 274)

Dusky Robin Jul.–Jan.; up to 3 broods a year. Builds a small cup nest of fine twigs, rootlets, strips of bark and grass loosely held in place with web, rather untidy. Inside lined with hair, fine grass and other plant fibre. Rough outer usually not decorated with moss or lichen. Nest often in a fire-black, shallow hollow on a stump or tree, standing or fallen, or in an enclosed space—deep fork or between limbs, from ground level up to 5 m. Clutch 2–3 (22 × 16 mm). Female incubates, 14–15 days. (p. 274)

Rose Robin Sep.–Jan.; several broods. Nest is distinctive: thick walls, rolled inward towards the top so that the overall shape is almost spherical. The outside is very neatly bound with spiders' webs, soft yet resilient. Its camouflage varies: some have moss and lichen; others are entirely covered with lichen so that the underlying structure of moss and fine bark fibre is almost completely hidden. Placed in a fork 1–15 m up. Clutch 2–3 (18 × 14 mm). Incubated by female, 12–14 days. (p. 276)

Pink Robin Sep.–Jan.; 1–2 broods. Similar to Rose Robin's nest—slightly larger, not as neatly matted with webs, but has more green moss, especially around the rim where the Rose Robin's is closely covered with lichen flakes. Clutch 3–4 (18 × 14 mm). Probably incubated by female only, 13–14 days. (p. 276)

Flame Robin Aug.–Feb. Varied locations: tucked in behind bark; in burnt-out hollow of tree trunk or stump; deep fork; among exposed roots; cavity in earth bank—ground to 20 m up. Broad rather untidy bowl, mostly of bark loosely bound with web; little lichen or moss attached. Clutch 3–4 (18 × 14 mm). Female builds nest, incubates for about 14 days. (p. 276)

Scarlet Robin Jul.–Jan.; several clutches a season. Broad cup nest is usually in a fork, occasionally a cavity in a tree trunk or stump, 0.5–20 m up. Nest rather roughly built, mainly of bark lightly bound with webs; very little or no attached moss or lichen. Clutch 2–3 (18 × 14 mm). Female builds the nest and incubates alone, 14–16 days. (p. 276)

Red-capped Robin Jul.–Jan. Builds a tiny, neat, rounded cup nest of fine shreds of bark and grass bound with webs; often decorated with lichen; placed in vertical or horizontal fork. May be almost at ground level, or up to 10 m high, usually 1–3 m up. Clutch 2–4 (15 × 13 mm). Female builds the nest and incubates for 14–15 days; young fly aged about 14 days. (p. 276)

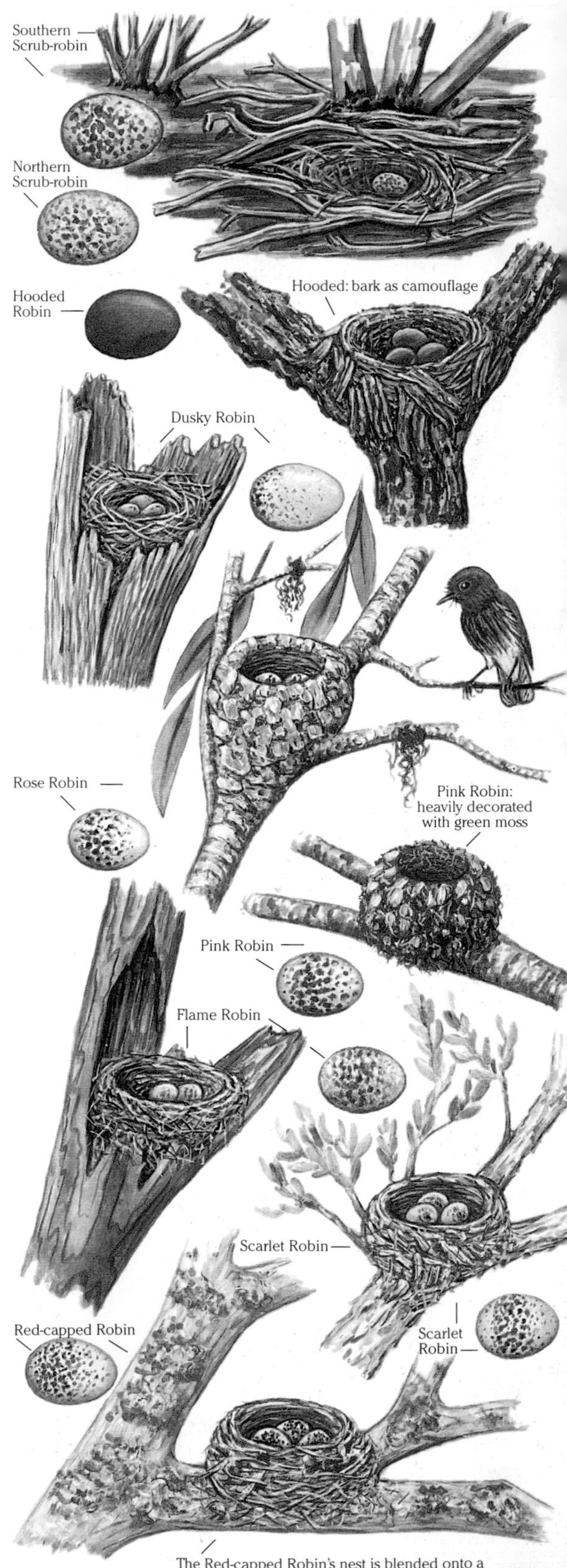

The Red-capped Robin's nest is blended onto a fork, and often camouflaged with pieces of lichen.

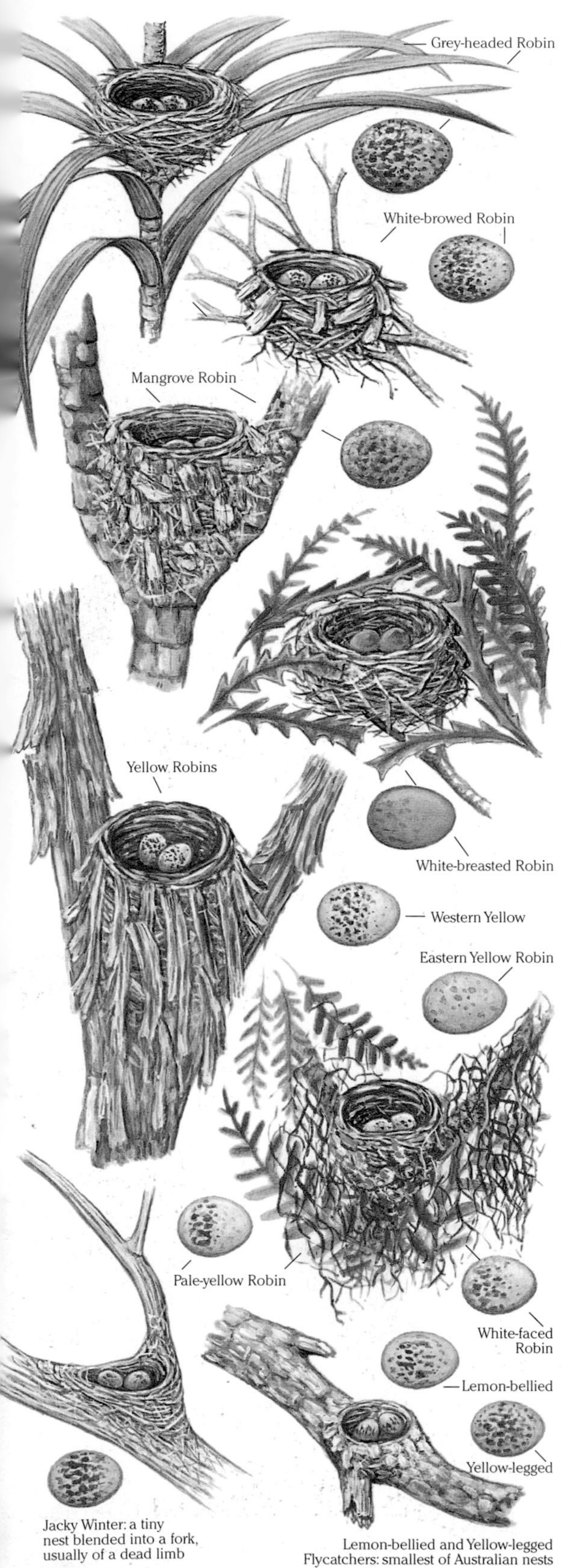

Jacky Winter: a tiny nest blended into a fork, usually of a dead limb

Lemon-bellied and Yellow-legged Flycatchers: smallest of Australian nests

Grey-headed Robin Aug.–Jan.; may be several broods in a season. Often builds on lower part of lawyer vine or palm, where leaves meet stem. Nest a rather rough cup of rootlets, fine twigs, decayed leaves; variable covering of green moss camouflage; lined with fine plant fibre. Placed 1–4 m above ground (some higher). Clutch 1–2 (25 × 19 mm). (p. 278)

White-browed Robin Aug.–Mar.; 1–3 broods, depending on the season. Builds a loosely woven cup of stems and bark fibre, dry and decaying leaves, lined with fine rootlets. Outside may be a few large thin flakes or strips of paperbark, attached by spider's web. Together with the rather rough, untidy exterior, this creates the appearance of flood debris caught on a branch 1–6 m up. Clutch 2 (19 × 15 mm). (p. 278)

Mangrove Robin Most months; mostly Aug.–Mar. Nest is a small, neat cup of fine bark strips, bound with webs and lined with fine rootlets. The outside is effectively camouflaged with strips of mangrove bark, which are attached with webs and hang vertically around the nest. Clutch 2–3 (19 × 14 mm). (p. 278)

White-breasted Robin Jul.–Dec. in N; Aug.–Jan in S. Builds in forks and twigs of undergrowth, shrubbery 0.5–1 m up. Nest is well camouflaged: this robin often chooses a dead shrub, putting the nest in tangled accumulated dead foliage such as dead bracken fronds, or perhaps an accumulation of peeling bark debris. Clutch 2 (20 × 15 mm). (p. 278)

Western Yellow Robin Aug.–Dec. Nest beautifully camouflaged; merges into a thick, rough-barked fork, usually in the open. Long pieces of bark are suspended vertically around the outside; the cup itself is built of finer strips of bark. Often the bark is grey externally but reddish where pulled from the tree, so that the whole outside is of rough-textured grey bark and the inside is smooth, often cinnamon-brown. Height varies from 0.5 to 7 m; most are 1–3 m. Clutch 2 (21 × 15 mm). (p. 278)

Eastern Yellow Robin Jul.–Jan.; several clutches. Nest like Western Yellow's, but of finer bark, grass; tends to have fewer pieces of hanging bark attached lower down. Rim and upper sides often decorated: scattered bits of pale turquoise lichen. Height 1–6 m, some 15 m. Clutch 2–3 (22 × 16 mm). (p. 278)

Pale-yellow Robin Jul.–Jan.; usually Aug.–Dec. Deep, neat cuplet built in a thin vertical fork, rainforest shrub, sparse foliage, or higher in a small tree, 1–10 m. Often built on lawyer vine—sharp spines, hooks protect it. Outside camouflaged with lots of thread-like green moss. Clutch 2 (20 × 16 mm). (p. 280)

White-faced Robin Sep.–Feb.; most Nov.–Jan. Builds small neat cup on a rainforest lawyer vine or in the fork of a sapling at heights of 1–8 m. The exterior is sparsely decorated with tangled threads of green moss and flakes of pale bark. Usual clutch two, (19 × 15 mm). (p. 280)

Lemon-bellied Flycatcher Jul.–Mar.; may have 2 broods. Smallest nest of any Aust. bird—a tiny shallow rim of fine bark strips tightly bound with webs, in a horizontal fork. Exterior is camouflaged with flakes of bark to match the limb. Nest may be invisible from below; from side, it is but a slight bump. Often over water, 1–10 m up. Clutch 1 (16 × 13 mm). (p. 280)

Yellow-legged Flycatcher Nov.–Feb. Has a nest almost identical to that of Lemon-bellied, but not quite as small; perhaps slightly deeper, of fine fibre, and with camouflage of paperbark flakes and lichen. On horizontal branch, 5–15 m up. Clutch 2, (15 × 12 mm). (p. 280)

Jacky Winter Jul.–Dec. and after rain. A tiny nest of grass, bark fibre and fine rootlets heavily matted with web. The outside has no added decoration. Placed at heights of 0.5–20 m. Usual clutch 2 (19 × 14 mm). (p. 280)

Logrunner Mar.–Oct.; most Jun.–Sep.;often 2 broods. Nest is usually on ground or very low on a fallen log or stump; favoured site is a tangled mass of lawyer vine—nest may be 2 m above ground. Often near tree's base, between buttress roots or against a log. Female constructs the nest alone: first lays a platform of thick, short sticks; then builds up to meet overhead as a roof. This is covered with dry leaves and green moss, which overhang the side entry. The interior remains dry even in heavy downpours. Leaf litter gathers on the roof, so the nest looks like any other accumulation of forest debris. Clutch 2 (28 × 21 mm). The female alone incubates for 22–25 days; feeds the young with small prey brought by the male. (p. 282)

Chowchilla Breeds most months; mainly Mar.–Aug. Nest is a domed sphere of sticks, many of them large for the size of bird. Moss is used extensively through the structure and as lining. The nest is usually at heights up to 1 m, occasionally as high as 3 m; built in a dense tangle of lawyer vine, staghorn or bird's nest fern, or on top of a wide, low stump or log—moss and litter on top help to hide the nest. Clutch 1 (36 × 26 mm). (p. 282)

White-browed Babbler Jun.–Dec.; 1–2 broods. Nest is built among vertical forks of a large shrub or bushy tree, 1–3 m up, some to 6 m; lower nests are often in dense bushes with thorny or prickly branches. Nest is spherical, domed with side entrance; 30–40 cm diameter. There may be a hood over the entrance and some have a small entry platform. Outside is very roughly built: many long straggly sticks project out haphazardly. The egg chamber is thickly lined with fine dry grass, feathers and wool. Old disused nests become compacted, flattened. Clutch 2–3 (25 × 17 mm). While building, the female is helped by others in the group, who also feed her while she is incubating and the young after they have hatched. (p. 282)

Hall's Babbler Jun.–Oct. Builds in vertical multiple fork of a large acacia, other shrub or small tree, 3–8 m up; mulga is commonly used. The domed nest is smaller than the White-browed Babbler's, which it most closely resembles. However, finer sticks are used, giving a neater outside. Small, if any, hood and platform at the side entrance. The inside is deeply lined with fine dry grass, fur and feathers. Clutch 2 (23 × 17 mm). The female builds and incubates, fed by others in the family group or troop. Occasionally several females will lay in one nest. These birds, like other babblers, are communal breeders; others of the troop feed and preen the incubating female between her shifts on the eggs, and the troop defends the nest territory. The troop builds many nests in their territory, apparently for roosting rather than for eggs; these usually have little or no lining. (p. 282)

Grey-crowned Babbler Jul.–Feb. Builds very big domed nest (40–50 cm diameter); uses thicker sticks than the White-browed, giving the exterior a heavier, rougher texture. Sticks project above and below the side entrance to make a hood and platform or an entry tunnel. Tend to be in taller trees and higher, 4–7 m, built into an upward-sloping or horizontal, multiple forked branch in the tree's upper outer foliage. Communal nesting by troop: several females may lay in one nest; brooding female is fed by others. Clutch 2–3 (28 × 21 mm). (p. 282)

Chestnut-crowned Babbler Jul.–Dec.; may have several broods. Builds a very large and conspicuous nest about 50 cm in diameter and almost 1 m from top to bottom. The typical site is an upright fork of mallee, mulga or some other tree or large shrub from 4 to 10 m above ground. Long sticks are used, finer than those in the nest of the White-browed Babbler; the nest looks neater with fewer sticks protruding. Clutch 3–5 (27 × 19 mm). The Chestnut-crowned Babbler is a communal breeding species; the entire troop helps to build, feed the incubating female, and defend the territory. (p. 282)

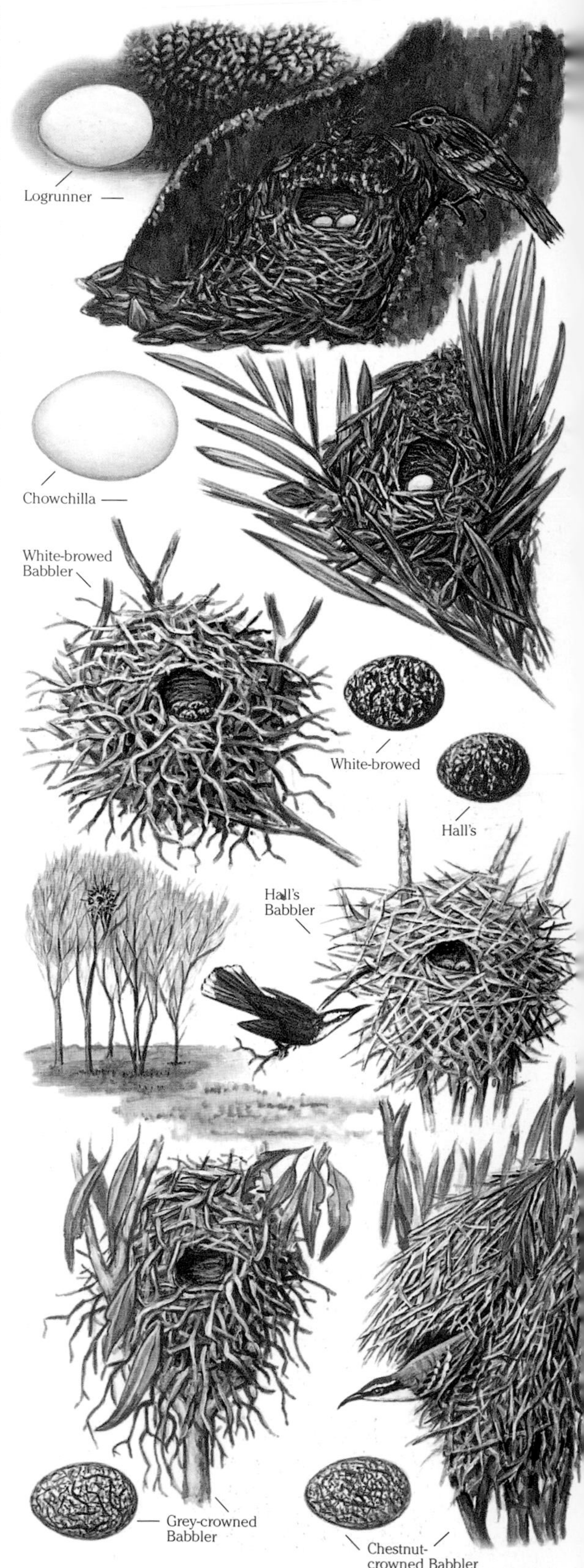

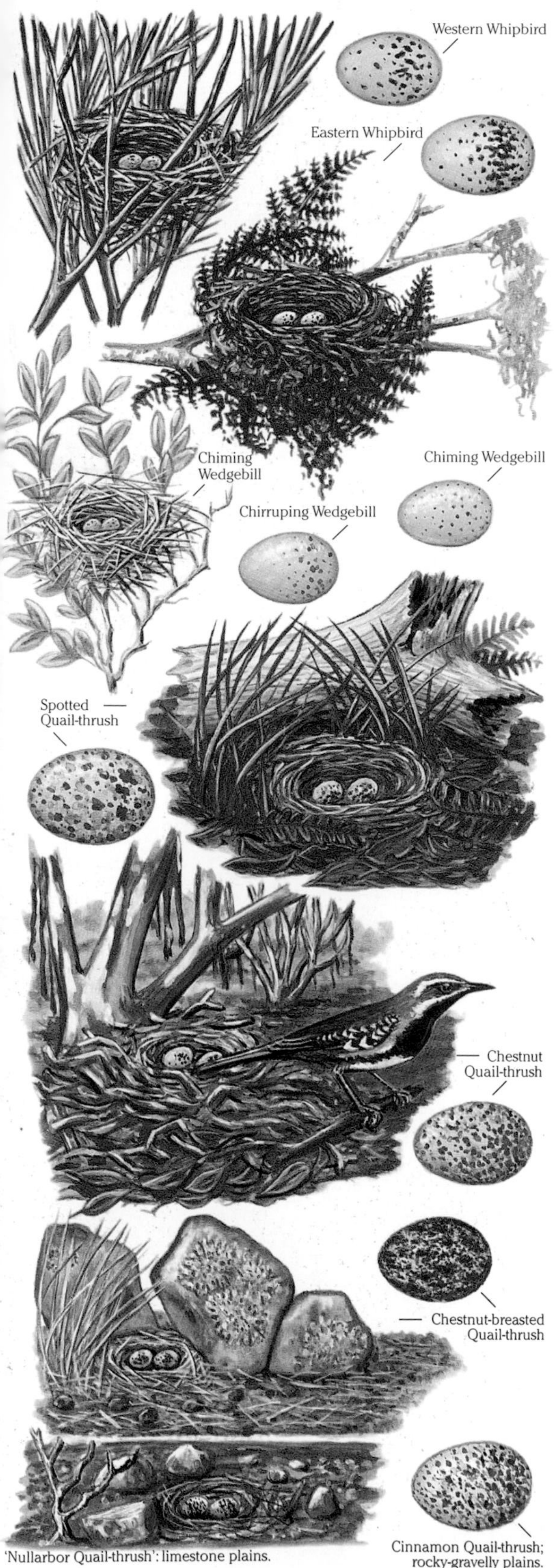

'Nullarbor Quail-thrush': limestone plains.
Cinnamon Quail-thrush; rocky-gravelly plains.

Western Whipbird Jul.–Oct., Nov. Territorial song strongest in midwinter, early in breeding season. Builds in low, dense heath scrub or sedge clumps. In SW of WA, often in prickly *Banksia cayleyi* and low shrubs of *Hakea*, *Agonis* and *Davesia*. Nest is extremely well hidden in old foliage and accumulated litter, low in the densest inner part of the shrub. The cup nest has thick outer walls of fine twigs, bark strips, dry grass, strands of sedge leaf; the outside is rough and untidy, matching its surrounds. Inside is neatly lined with fine, dry native grass. Clutch 2 (25 × 19 mm). Incubated by both parents, 21 days; young leave nest aged 12–14 days. (p. 284)

Eastern Whipbird Jul.–Dec. May have 2 broods. Nest site varies greatly with diverse habitats. Usually 0.5–2 m above ground, may nest in a low but open shrub in rainforest undergrowth, entangled thickets of wet eucalypt forest, or well hidden, low in very dense coastal heath. The nest is bulky, its outside loosely woven of fine twigs, bark fibre and vine tendrils, the inside much more neatly lined with fine rootlets and plant fibre. Clutch 2–3 (27 × 20 mm). The female builds the nest and incubates 17–18 days; the male helps feed the young. (p. 284)

Chiming Wedgebill Aug.–Nov; other months after good rain, often Feb.–May. The nest is hidden deep inside one of the denser shrubs in the territory—a clump of saltbush or a mass of mistletoe; up to 4 m above ground in mulga. The nest is small but strong, its outside of interlocking twigs, the cup lined with fine dry grass. This species is extremely wary: unlikely to be seen near its nest; sneaks away low through the shrubbery. It is unlikely to be seen returning while a human, or any source of potential danger, is nearby. Clutch 2–3 (24 × 17 mm). (p. 284)

Chirruping Wedgebill Aug.–Nov.; after rain, Mar.–Apr. Occupies more open habitat than the Chiming; favours low thickets of acacia, box-thorn and nitre-bush. However, the nests of both look identical. Clutch 2–3 (24 × 17 mm). This wedgebill is neither as shy nor as secretive as the Chiming; breeding behaviour is probably similar to that species. (p. 284)

Spotted Quail-thrush Jul.–Feb.; most Aug.–Dec.; up to 3 broods in good seasons. The nest is built on the ground close to a fallen log, large stone, base of a tree or a shrub. An open cup of grass, leaves and bark is formed in a slight depression in the ground or leaf debris. It is so loosely constructed and fragile that it would disintegrate if not frequently pushed back into shape around the incubating bird's body. Clutch 2–3 (32 × 23 mm). Nest is totally hidden by the incubating adult, which is so confident of its camouflaged plumage that it can almost be walked on, before flushing. Downy chicks are dark brown; fledged young are as perfectly disguised as the adults. (p. 284)

Chestnut Quail-thrush Jul.–Dec., but varies with rain; up to 3 broods. The nest is typically beneath a low bush, or clump of spinifex or grass, or beside rocks or a fallen limb. It is but a slight depression in the ground or leaf litter, which is lined with a sparse and fragile layer of leaves, bark strips and grass. Clutch 2–3 (30 × 21 mm). (p. 284)

Chestnut-breasted Quail-thrush Any month; depends on rain; often Jul.–Oct. The nest is a scrape in litter and soil beside a log, rock or shrub, lined with a loose layer of dry leaves, bark and grass. Clutch 2–3 (28 × 20 mm). Incubated by female. Behaviour at nest, visibility of adults, chicks and eggs are similar to Spotted Quail-thrush. (p. 286)

Cinnamon Quail-Thrush Breeds any month after rain; mostly Jul.–Sep. Shy and unobtrusive. Makes a slight depression in stony, open ground, including the Nullarbor. The nest is lined with grass, bark strips and long, narrow, flat leaves. Clutch 1–2 (29 × 22 mm). Female builds nest and incubates eggs. (p. 286)

Varied Sittella Sep.–Dec. in S; Aug.–Oct. in N, often also Feb.–Mar. Sittellas forage on rough-barked trees, spiralling head-first downward around branches and trunk, pausing frequently to probe into crevices in search of insects. Their nest is so well hidden on an exposed limb that, when the bird pauses to put food into a tiny open bill, it looks as if the bird is simply stopping to peck into the bark for its prey. To find the nest, watch the birds with binoculars. When a bird is seen carrying insects or building material in its bill, watch for where that insect etc. disappears during the bird's downwards travel. That may be the site of the nest, to be confirmed if the parent birds visit the same spot repeatedly. The nest is a small cone in shape, built into and blended to the sides of the tree fork. Construction is of bark and other fibre bound with webs and camouflaged externally with flakes of bark that match the tree in which it is built. The nest may be in a vertical or sloping fork, often a dead limb, and usually high, 5–25 m above ground. From below, it looks to be no more than a slight bump in the fork or on the limb. Clutch 2–3 (17 × 13 mm). Incubated by female, 18–20 days. (p. 288)

Crested Shrike-tit Aug.–Jan. Attaches its small cup nest to the highest leafy twigs of a eucalypt sapling where the supporting branchlets are far too slender to carry the weight of any predator that might try to climb up from below. The nest's height ranges from about 6 m to 30 m. It is a neat, deep cup or cone, narrowed at the rim, and built of fine grass stems and bark fibre, heavily bound with webs to give a soft, flexible structure; it may be covered with pale grey-green lichen, heavily coated with spiders' webs and egg-sacs. Clutch 2–3 (24 × 15 mm). Nests of 'Western' and 'Northern' shrike-tits usually lack the lichen decoration. Incubation 15–16 days, mainly by female; young fledge aged 14–17 days. (p. 290)

Crested Bellbird Aug.–Mar.; varies with rain. Builds a deep, substantial cup, large for the size of the bird. Outer structure is of sticks, often quite thick; inner walls are strips of bark lined with fine bark and grass. Often placed in a deep fork where multiple stout stems or branches meet, or between clustered thin branches and main trunk, 1–3 m up. Occasionally it is built in the top of a stump or cavity in the side of a trunk. Clutch 3–4 (26 × 20 mm). Sometimes puts live hairy caterpillars, paralysed by pinching, around the nest rim. (p. 290)

Olive Whistler Aug.–Feb. Nests in gullies of ranges where undergrowth is thickest and intermixed with long grass; builds into a shrub, 1–3 m up. The large, solidly built, deep open cup has its outside roughly constructed of quite coarse twigs and strips of bark. Inside the nest cup is neatly lined with fine rootlets and dry grass. Clutch 2–3 (27 × 19 mm). (p. 290)

Red-lored Whistler Sep.–Jan. In broombush and similar mallee. Nests may be hidden low in dense, tangled creepers or among the long, twisted ribbons of bark that accumulate in the forks of mallee trees. Some nests are built in clusters of old grey banksia cones. Most of these sites match the nest's rough exterior of grey twigs and bark strands. The inside is neatly lined with fine rootlets. Clutch 2–3 (23 × 17 mm). (p. 292)

Gilbert's Whistler Sep.–Dec. Builds a deep, thick walled cup of twigs, fine fibre such as bark, fur or wool loosely bound with webs. Chooses a site very much like the Red-lored, but also uses disused babbler's nests. Clutch 2–4 (23 × 17 mm). Sexes share building, incubation and feeding of young. (p. 292)

Grey Whistler All months; most often Sep.–Mar. A small whistler, rather flycatcher-like in appearance and behaviour, and in having a small, neat nest high in the forest canopy Nest is built in a fork among foliage or in a high tangle of vines. The cup nest is made of dry leaves, bark and vine tendrils bound with webs and lined with finer fibre. Clutch 2 (20 × 16 mm). (p. 290)

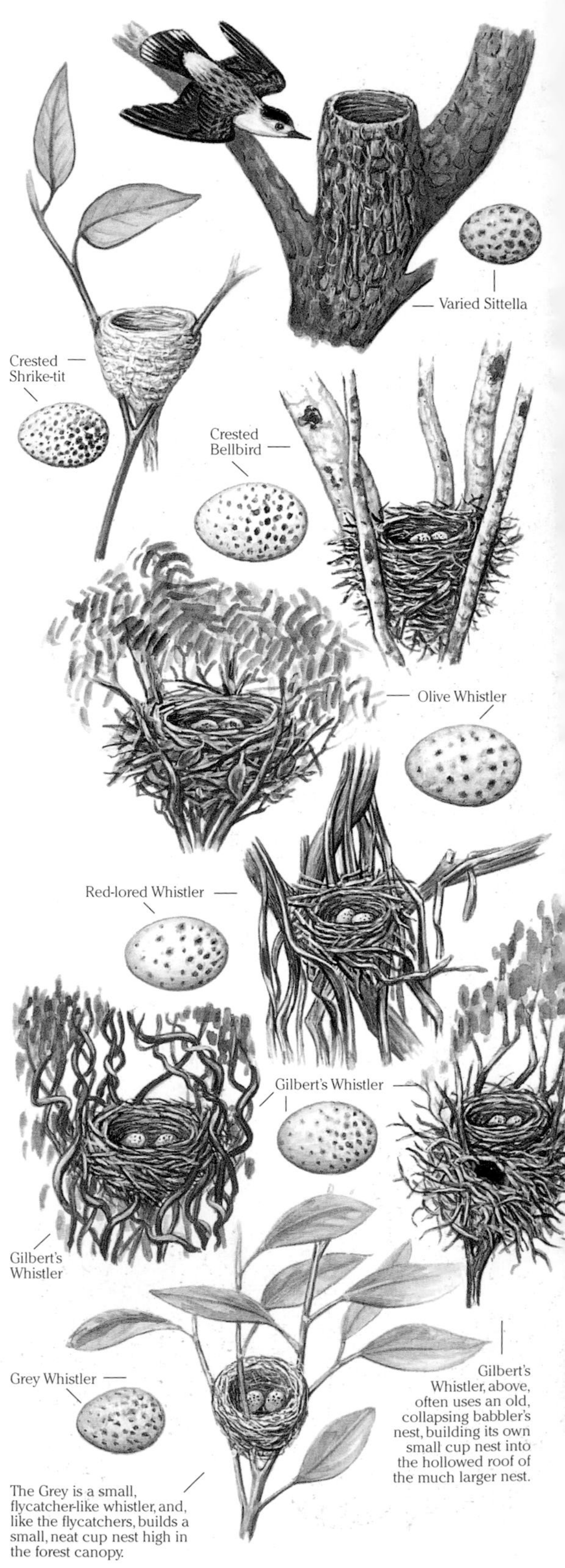

Gilbert's Whistler, above, often uses an old, collapsing babbler's nest, building its own small cup nest into the hollowed roof of the much larger nest.

The Grey is a small, flycatcher-like whistler, and, like the flycatchers, builds a small, neat cup nest high in the forest canopy.

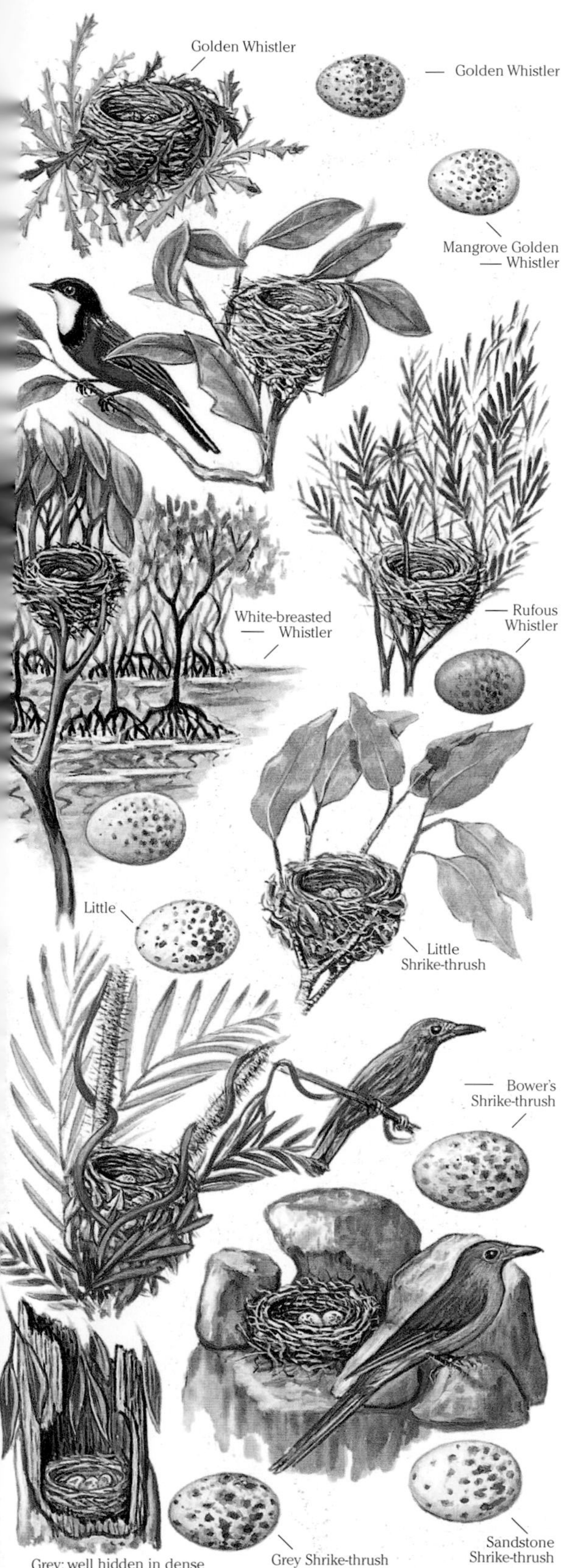

Grey: well hidden in dense vegetation or shallow hollow

Golden Whistler Aug.–Jan. This widespread whistler nests in an open cup. Outside material varies with locality: some are woven of fine casuarina leaves, some use plant stems and rootlets, while others use pieces of dry fern frond or skeletons of leaves. The material is held together with a light binding of cobweb. The nest is lined with finer fibre, and its height varies greatly, from 0.5 m up to 6 m; it is built in a fork, or tangled stems or vine, or among debris collected in a shrub or tree. Clutch 2 (23 × 17 mm). Incubated by both sexes; male often on the nest by day. Incubation 14–17 days. (p. 292)

Mangrove Golden Whistler Sep.–Feb. Usually builds in a mangrove tree or shrub, occasionally in nearby coastal rainforest or monsoon forest. Nest is placed in an upright leafy fork or tangle of stems and vines, 1–5 m up. Externally the nest is composed of thin twigs, stems and grasses, and lined with fine hair-like rootlets; almost identical to the nest of the Golden Whistler. Clutch 2–3 (21 × 16 mm). Both sexes build the nest, incubate the eggs and care for young. (p. 292)

White-breasted Whistler Mar.–Nov.; occasionally other months. Builds in an upright mangrove fork, 1–5 m up; often near the high-tide line, perhaps because the frequently exposed mud abounds with small crabs and other prey. Nest is a rather flimsy, shallow bowl of interwoven thin twigs and fine roots lined with very fine rootlets. Clutch 1–2 (27 × 19 mm). The female undertakes nest building and incubation. (p. 292)

Rufous Whistler Sep.–Feb.; sometimes 2 broods. Builds an open cup nest, rather flimsy, of long, fine twigs lined with finer fibre. Some are so thin that the eggs can be seen from below through the floor. Nest is usually supported by an upright fork or multi-stemmed fork 1–8 m up in a tree or tall shrub. Clutch 2–4 (23 × 17 mm); incubation by both sexes, 14–15 days. (p. 292)

Little Shrike-thrush Sep.–Feb. The nest is well hidden among foliage, built in a deep upright fork, crown of pandanus or other palm, or a tangled mass of lawyer vine, 0.5–10 m up. It is a bowl of twigs, bark strips and leaves bound with cobweb and vine tendrils; lining is of fine rootlets and other soft dry fibre. Clutch 2–3 (25 × 19 mm). (p. 294)

Bower's Shrike-thrush Oct.–Feb., Mar.; 1 brood only. Builds in an upright fork of a shrub or tangle of stems and limbs in rainforest, monsoon forest, vine thicket or paperbark swamp. Often hidden in spiny canes of lawyer vine or base of radiating, saw-edged pandanus leaves, 1–8 m up, usually below 5 m. Nest is a bowl of small, thin paperbark sheets and many dry leaves that give the nest a golden-buff colour. These materials are bound in place with thin vine stems, tendrils and spiders' webs. The interior is lined with fine vine tendrils and rootlets. Clutch 2–3 (26 × 19 mm). (p. 294)

Sandstone Shrike-thrush Oct.–Feb., Mar.; 1 brood. Nests on a ledge of sandstone cliff or escarpment, or in a crevice between sandstone boulders. Nest is usually positioned with a clear outwards view, often where it is protected by overhanging rock. Builds a bulky bowl of neatly interwoven, thin, crinkly roots, mostly those of spinifex, but also includes shreds of bark. The lining is of finer rootlets and grasses. Often the nest is distorted to a rough semicircle, oval or triangle to match the cavity or space in the sandstone into which it has been built. Clutch 2–3 (28 × 21 mm). (p. 294)

Grey Shrike-thrush Jul.–Mar.; after rain in arid regions. Nests in various spots, mostly enclosed—a hollow or crevice in a tree trunk or stump, a rock ledge or crevice, dense tangles of vegetation, a tree-fork or accumulated debris on forest floor. Builds bowl of bark strips, grass lined with fine fibre. Clutch 2–4 (29 × 21 mm). Male helps incubate, 17–18 days. (p. 294)

Yellow-breasted Boatbill Jul.–Mar.; probably 1 brood a season. Nest is a small, shallow, saucer shaped basket, hanging by the rim in a thin, narrow-angled, horizontal fork in a rainforest tree's outer foliage. With external diameter about 70 mm and depth only some 40 mm, it is often smaller than the leaves among which it is built. The outer support is an open net of dry brown vine tendrils and thin plant stems, through which the sparse lining of fine tendrils and grass stems is clearly visible. The structure is bound together and the rim attached on each side with spider's webs. It is so thin, lace-like, that the eggs can be seen from below; if the bird is sitting, its yellow breast can be glimpsed through the fabric. Clutch 2 (17 × 13 mm). Built mostly by the male; incubated by both. (p. 298)

Black-faced Monarch Sep.–Feb.; most Oct.–Dec. During the breeding season they move to secluded damp gullies and brush covered creek margins, especially where there are palms, tree ferns and luxuriant undergrowth. Calling begins in spring as nesting activity starts. The beautifully constructed nest is built in an upright or horizontal fork of an understorey shrub or sapling, 2–5 m up, usually in an open site with little or no screening foliage. It may be blended into the branch and is constructed of long, fine, needle-like leaves of casuarina intermixed with green moss, which the birds form into a thick, bright green layer that covers the whole of the nest's exterior. The nest's interior is lined with a thin layer of fine, black, thread-like fibre. Clutch 2–3 eggs (23 × 17 mm). (p. 296)

Black-winged Monarch Nov.–Apr. Nest resembles that of the Black-faced Monarch, but it has more bright green moss attached, and has many conspicuous silky-white spiders' egg-sacs decorating the exterior. It is also much more likely to be in a vertical fork; this species shows a preference for large-leaved trees or saplings. Clutch 2–3 (23 × 16 mm). (p. 296)

Spectacled Monarch Sep.–Feb. Breeds in dense scrub in gullies of coastal ranges where its nest is often built in a slender, upright fork or between the upright stems of a vine, 1–4 m up. The nest is a deep cup shape of thin strips of bark, leaf skeletons and moss bound with cobweb; the outside is decorated with an overall light covering of bright green moss and white silken spiders' cocoons. The nest is lined with black thread-like fibre or fine, dry grasses. Clutch 2 (22 × 16 mm). (p. 296)

White-eared Monarch Sep.–Feb. Builds high in upper canopy of the rainforest, 20–35 m up, where, being so small and among screening foliage, the nest is very hard to find. Rarely it is in a lower fork, 10–20 m up, below the canopy where it can be observed. More than 70 years passed between description of the bird and, in 1923, the finding of the first nest, probably because the lower nests are built in the southern part of its range where the bird is rare. The nest is a cup of grass, bark, leaves and moss bound with cobweb, lined with palm fibre, and decorated with moss and white cocoons. Clutch 2 (19 × 13 mm). The male helps build nest, incubate eggs and feed young. (p. 296)

Frilled Monarch Sep.–Mar. The cup-shaped nest is slung hammock-like between slender, vertical stems of a vine, 3–10 m up. The rim is attached at several points with cobwebs. The delicate, open textured, loosely woven basket is built of vine tendrils and dark rootlets bound with webs and lined with fine, glossy black fibre. Overall, the outside is dark brown, but is decorated with flakes of pale green and white lichen, which are attached with webs. Clutch 2 (20 × 15 mm). (p. 296)

Pied Monarch Sep.–Apr. Builds a small, hammock-like basket bound and attached with webs, woven of brown vine tendrils and decorated with pale lichen. It is almost identical to that of the Frilled, perhaps slightly smaller, more flimsy. Clutch usually 2 (19 × 14 mm). (p. 296)

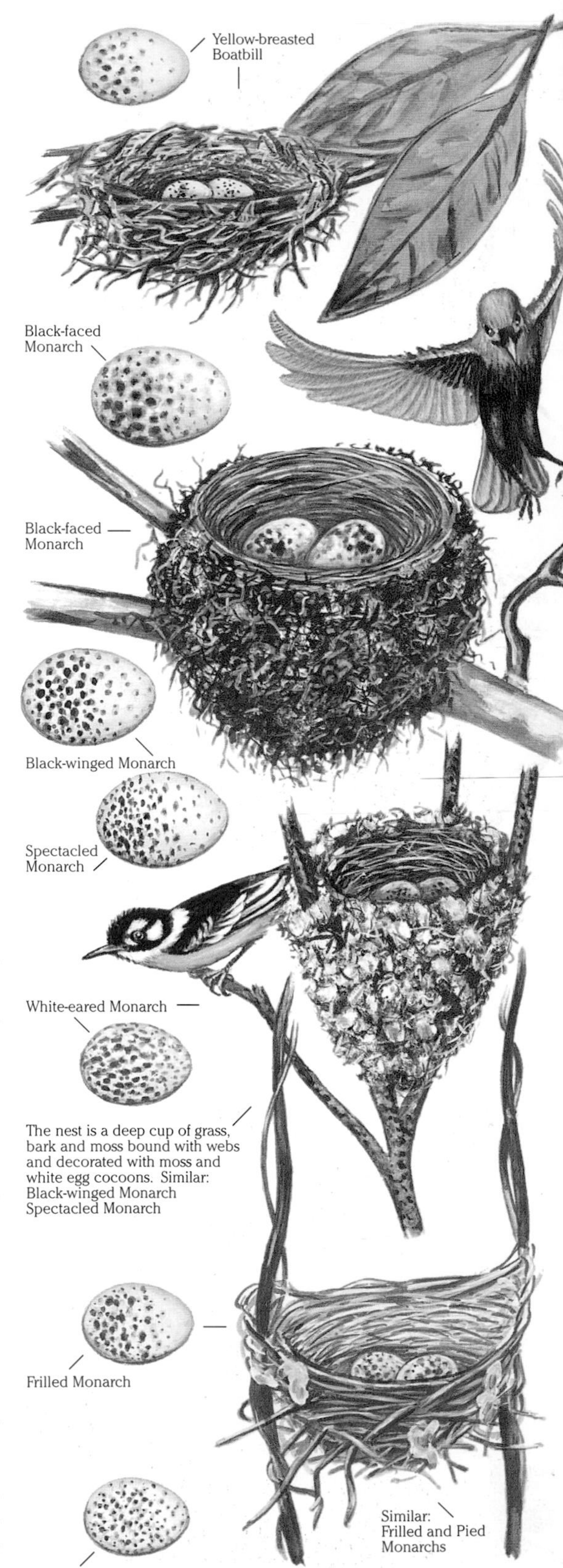

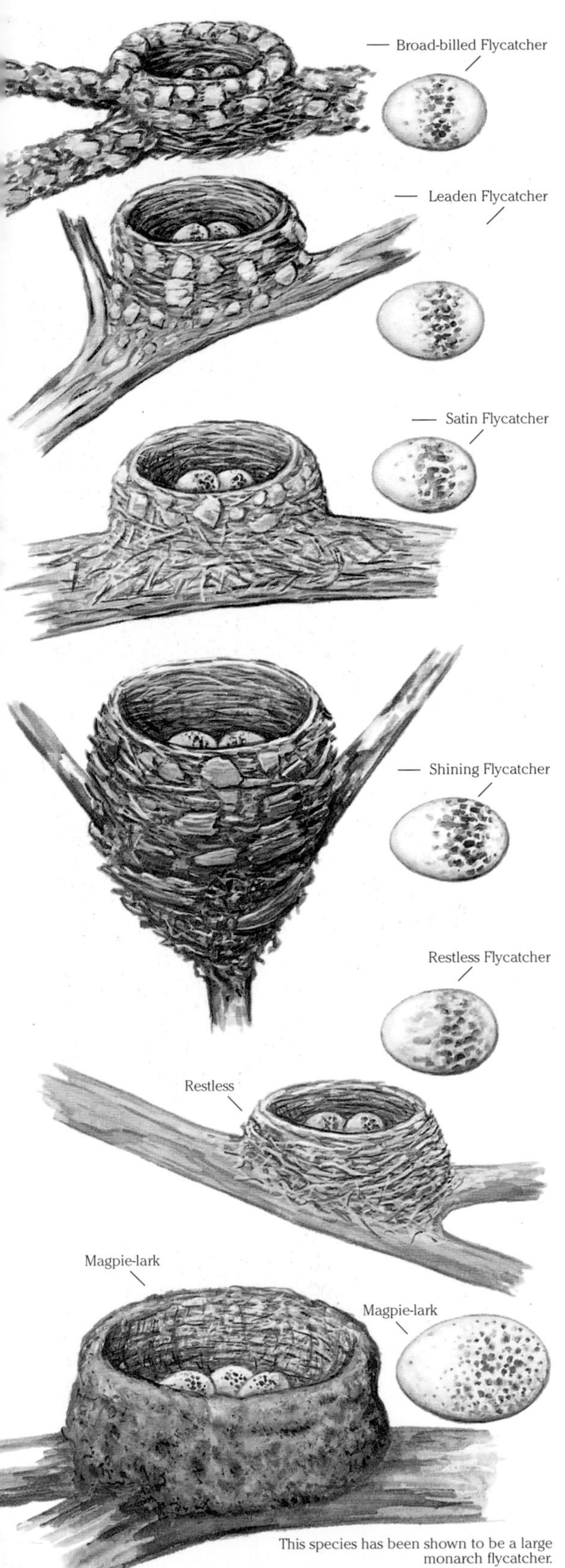

This species has been shown to be a large monarch flycatcher.

Broad-billed Flycatcher Oct.–Mar; 1 brood each season. The nest is built in an upright mangrove fork, 0.5–2 m above high water level, often overhanging a deep channel. It is a small, cup-shaped nest, the walls curving in slightly towards the rim; the base is shaped to blend smoothly onto the contours of the supporting fork. Thin strips of bark, vine tendrils and other plant fibres are heavily bound with cobweb to form a strong, neat, flexible structure. Numerous flakes of pale greenish or nearly white lichen are attached to the outside, mostly to the upper walls and rim. Clutch usually 2 (18 × 14 mm). Male and female share nest construction. (p. 298)

Leaden Flycatcher Sep.–Nov. in SE; Aug.–Feb. in far N. Builds in a forest or woodland tree, choosing a site on a limb extending well out from the main trunk yet short of the outer foliage, 6–20 m up. The nest is open and exposed, and would be conspicuous but for its superb camouflage. Often some protection is given by placement just below a large leafy branch. The nest is an open cup of thin bark strips bound with webs and lined with fine rootlets. The exterior is skilfully and effectively disguised with strips or flakes of bark and sometimes lichen taken from elsewhere on the host tree. The nest then matches its supporting branch almost perfectly in colour and texture, appearing as a bump like a knothole or knob on the limb. Clutch 2–3 eggs (18 × 14 mm). All nest duties, including 14–15 days of incubation, are shared. (p. 298)

Satin Flycatcher Sep.–Mar.; most Oct.–Jan.; 1 brood each season. There may be several nest territories close nearby, perhaps 30–50 m apart, in very small, loose associations. Nest is built on a horizontal fork of a small dead branch, but close under a large live limb and its green foliage, 5–25 m above ground. It is built of strips of bark bound with cobweb and lined with fine roots. Bark strips blended onto the branch with lavish use of spiders' webs form the outside. In most instances the nest's natural ochre and pale cinnamon tones match the branch, but some birds add pieces of lichen to the rim and upper sides. Clutch 3 (20 × 15 mm). (p. 298)

Shining Flycatcher Aug.–Mar. Breeds in mangrove or paperbark swamps; favours gloomy spots where a dense canopy shuts out most light. Uses an upright fork or vines, usually of a mangrove or paperbark standing in water, or a limb reaching out over a watercourse; is placed 1–4 m above high-water mark. The nest is a rounded cup shape of bark strips bound with spiders' webs and disguised on the outside with pieces of ochre-coloured paperbark and pale blue lichen. Clutch size 2–3 (20 × 15 mm). (p. 298)

Restless Flycatcher Aug.–Jan. in S; Aug.–Apr. in far N. Builds on a sloping or horizontal fork, well out from the trunk. The nest is a small shallow cup; the lower walls often taper outwards to merge onto the branch. It is made of dry grass and thin strips of bark bound with cobweb and lined with fine rootlets, fur and dry grass. It may be disguised by having small pieces of bark, lichen or spiders' egg-sacs attached to it. Clutch 3–4 (20 × 16 mm). Male helps to build nest and incubate eggs, 14–15 days; young fledge at 13–16 days. (p. 298)

Magpie-lark Aug.–Feb.; inland, most months after rain; several broods. DNA comparison has shown this species to be a very large monarch flycatcher. Its well-known mud nest is a larger version, similar in shape and situation, to the nest of the Restless Flycatcher, above, but the embedded grass and other plant material are simply bound with copious dollops of mud rather than cobweb. It is lined thickly with grass, fur and feathers, and its diameter is 150 mm. Nests or nearby sites may be re-used annually; usually near water. Clutch 3–5 (29 × 20 mm). Incubated by both parents, 18 days. (p. 302)

Rufous Fantail Sep.–Apr.; mostly Oct.–Jan.; often several broods. In the N of its range the Rufous ('Arafura') Fantail is a year-round resident, but those of S Qld, NSW and Vic. migrate northwards in Autumn to N Qld and NG. In spring, Sep. and Oct., the movement is reversed as the Rufous Fantail population spreads southward again, pairs returning to their SE breeding territories. Although this fantail is seen in open country while on migration, it breeds in dense wet forests—rainforests, mangroves, the wet fern gullies in eucalypt forests and other dense vegetation. The nest is attached to a thin sloping or horizontal branch, usually at a fork; some are attached to vines, but large-leaved trees or shrubs seem to be preferred. The nest is usually low, 0.5–3 m up, but some are as high as 7 m. Sites along creek banks seem favoured, some nests being on branches hanging over the water. A very small cup of grass and soft pale bark is built, heavily bound with spiders' webs, moulded and blended to the contours of the branch, then tapering below to a long tail. Clutch 2–3 (17 × 14 mm). Parent birds share nest construction and incubation. (p. 300)

'Arafura Fantail' Proposed species, presently race *dryas* of Rufous. Nest and eggs similar to those of Rufous Fantail. (p. 300)

Grey Fantail Jul.–Mar.; most Aug.–Dec.; several broods per season. The nests of all the races of this widespread species vary only slightly, merely reflecting the available materials; shape differs negligibly across the continent. The nest is built on a thin horizontal or sloping branch, usually, but not always, where a fork gives added support; height varies 1–7 m. Dead branches away from screening foliage are often chosen, yet rarely is there any attempt to disguise the exterior. The nest is very neat, elegant, being shaped like a slender wineglass. The composition of its walls varies, but is usually fine shreds of soft inner bark, bound and matted with spiders' webs. Some nests in moist mountain valleys use the reddish down of new tree fern fronds, while others add thistle down. However, they all use webs extensively, making the walls flexible. Colour varies only slightly, warm grey to soft ochre. The nest's tail may be up to 100 mm, tending to be strips of bark, loosely rather than tightly bound with webs. Clutch 2–3 (16 × 13 mm). Male and female share nest building, incubation, 14–15 days, and feeding of young. (p. 300)

Mangrove Grey Fantail Sep.–Feb.; most Oct.–Jan. Builds in mangroves fringing the far N and NW coast, 1–4 m above high tide on a slender or sloping branch or fork. Nest is a tiny cup of plant fibre similar to that of the Grey Fantail, above. It is heavily bound with spider's webs and tapers to a long, slender, loose tail; the inside is lined with fine, soft fibre. Except for its being slightly smaller and the use of a mangrove shrub or tree as the site, the nest is probably indistinguishable from that of the Grey Fantail. Clutch 2–3 (16 × 13 mm). (p. 300)

Northern Fantail Aug.–Jan.; 1 or 2 broods in a season. The nest is built on a thin sloping or horizontal branch, sometimes at a fork, 2–20 m up. Usually in open, away from concealing foliage, in a tree or sapling in woodland, or in fringes of rainforest, or monsoon or other closed forest. Nest is slightly larger than that of the Grey Fantail; tail of paperbark is perhaps more bulky. Clutch 2 (17 × 15 mm). Sexes share building, incubation and care of young. (p. 300)

Willie Wagtail Throughout year in good times, but mostly Aug.–Dec. One of the best-known Aust. nests, especially in rural areas—most places have a Willie nesting near the homestead or shearing shed. The nest is a small cup of grass bound with webs like those of the fantails, but lacking the tail. It is built on a horizontal branch, 1–15 m up; can overhang water. Clutch 2–4 (20 × 15 m). Nest and its vicinity are defended vigorously. Incubated 14–15 days by both sexes. (p. 300)

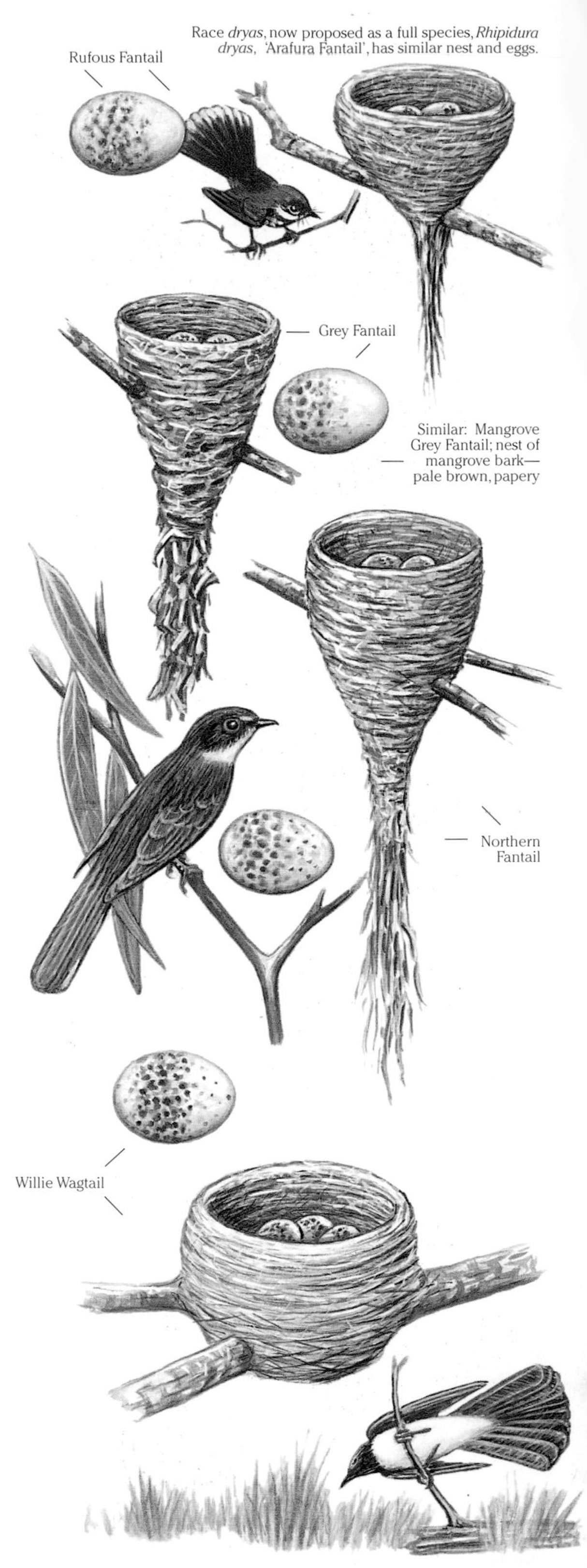

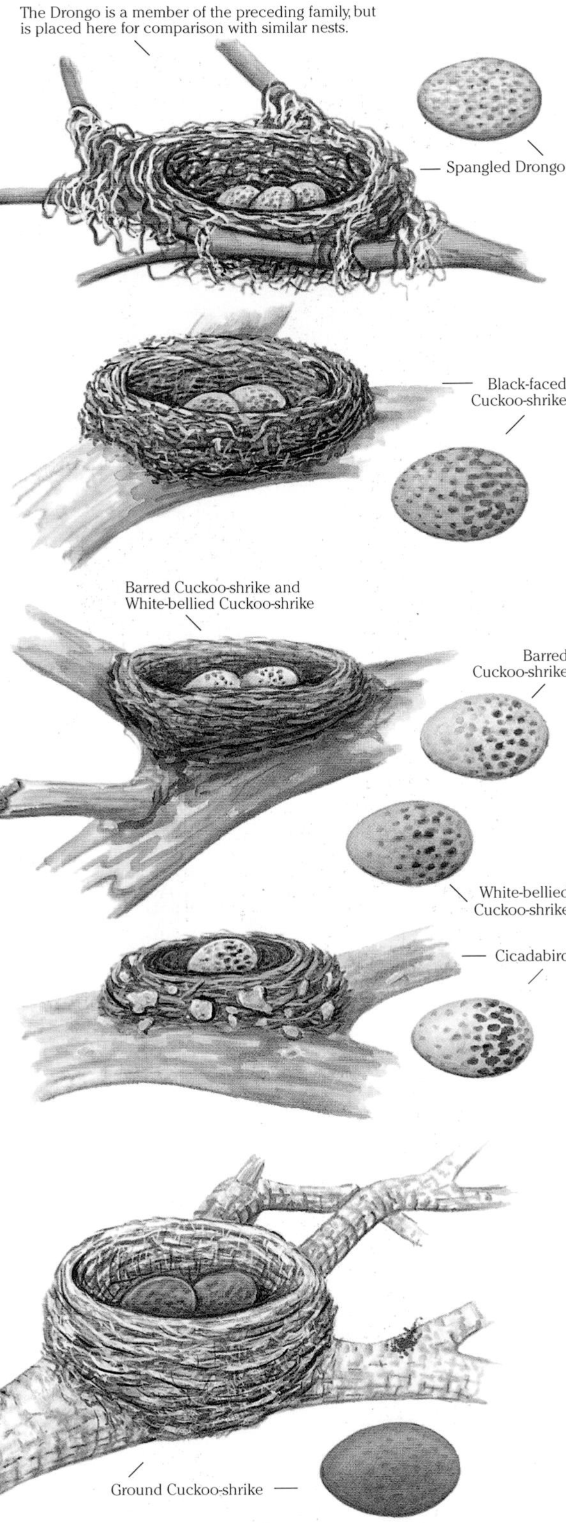

The Drongo is a member of the preceding family, but is placed here for comparison with similar nests.

Barred Cuckoo-shrike and White-bellied Cuckoo-shrike

The broad bowl nest of the Ground Cuckoo-shrike has the appearance and soft, resilient feel of the Willie Wagtail's, but is almost twice the diameter.

Spangled Drongo Aug.–May; mostly Oct.–Jan. The nest hangs in a slender fork in dense foliage where it is usually difficult to see, at a height of 10–25 m above ground; often in a tree near the rainforest margin. Commonly used trees are Blackbutt, Moreton Bay Ash, Box and Dwarf White Gum. The Drongo's nest is a hammock-like basket attached by the rim or sides to leafy twigs at the extremity of the outspread branches, surrounded and hidden by leaves. It is constructed almost entirely of long, curling, pliant vine tendrils interwoven to form the mesh-like fabric of the basket. The shape of the eggs can sometimes be seen from below through this open network. Clutch 3–5 (30 × 20 mm). Both adults incubate, brood and feed the young. (p. 302)

Black-faced Cuckoo-shrike Throughout the year; most Aug.–Jan. This Cuckoo-shrike occurs almost throughout Aust., occupying exceptionally diverse habitats from semi-desert scrub to the tall, wet forests of the SE and Tas. The nest may be only several metres up in stunted trees of arid regions, or more than 20 m high in tall coastal trees. It is small for the size of the bird, and is built on an open, exposed part of a limb. Blended into the angle of a fork, usually of quite thick limbs, it is very hard to see: from below it is either invisible, or just a slight bump on the upper side of the branch. The nest's construction and appearance are common among small flycatchers, but unusual among larger birds. Fine rootlets, bark fibre and casuarina leaves are heavily bound and matted with spiders' webs to give a soft, yet strong, resilient structure—small, shallow and saucer shaped. Clutch 2 (31 × 22 mm). Male shares building and incubation. (p. 304)

Barred Cuckoo-shrike Sep.–Mar. Nest is slightly smaller than the Black-faced Cuckoo-shrike's, but quite similar—a shallow saucer of fine twigs, casuarina needles and rootlets heavily bound with webs. It is built in a broad fork of a thick branch, between 15 and 30 m above ground, and is very difficult to recognise, even though usually in an exposed position. Clutch 2 (30 × 21 mm). Both male and female join in nest construction, incubation and care of young. (p. 304)

White-bellied Cuckoo-shrike Jul.–Mar.; most Aug.–Jan. The nest is most like that of the Black-faced Cuckoo-Shrike, but is slightly smaller. The rim is so low that the shallow saucer-shaped depression that holds the eggs seems almost flat. The nest is composed of bark strips bound with cobwebs and is built in a thick fork, with no concealing foliage, 6–10 m above ground. From a distance, its size, shape and colour make it hard to recognise as a nest. In NE Qld, Bloodwood and Beefwood trees seem preferred. Clutch of 2–3 (30 × 22 mm). Sexes share nest building and the incubation of 21–22 days; young leave the nest aged about 21 days. (p. 304)

Cicadabird Sep.–Jun.; most Nov.–Feb. Builds in a fork of a sloping or horizontal limb; seems to prefer rough-barked trees, at 4–30 m above ground. Although often high, the nest, as typical of the cuckoo-shrikes, is built below the canopy on a limb about the thickness of the nest's width, which hides it from below. The nest is a very shallow, low-rimmed saucer constructed of short, thin stems and casuarina needles mixed with lichen and well bound with spiders' webs. The rim is more heavily decorated with lichen. Clutch just a single egg (30 × 21 mm); incubated by female, 22 days. (p. 304)

Ground Cuckoo-shrike Jul.–Feb.; in arid regions, after rain. Nest is built on a stout, nearly horizontal fork, 2–25 m up. Grass, bark fibre and roots are interwoven with softer plant down, fur, feathers or wool, all bound with webs into a soft yet strong structure; the soft lining is of loose fur, feathers or wool. Larger and far more obvious than nests of other cuckoo-shrikes. Clutch 2–3 (34 × 24 mm). Both sexes build and incubate. (p. 304)

White-winged Triller Recorded nesting most of year, but usually Sep.–Dec. Breeding male sings almost incessantly, but most strongly during song-flights, rising high and circling above the nest sites, pouring forth loud, spirited, trilling, chattering notes that may be continued from a high perch, and even while sitting on the nest. Breeding tends to be semi-colonial: several pairs nest within 20 m; sometimes in the one tree. However, each pair has its own foraging area, apparently some 10–15 ha in extent. Male and female combine to choose a nest site and build; the male contributes nearly constant trilling and chattering during these activities. Nest is a neat, shallow cup, quite small for the size of the bird, made of fine grass and rootlets heavily bound with webs. It is smoothly blended into a thick fork, horizontal or sloping, between 2 and 10 m up. Although exposed rather than hidden in foliage, it is hard to see unless the incubating or feeding birds are present—from below, if visible at all, it will be only a slight bump on top of the limb. Clutch 2–3 (21 × 15 mm). Both sexes incubate, about 14 days, and the young are in the nest for a further 13–15 days. (p. 302)

Varied Triller Usually Sep.–Dec.; some breeding almost throughout the year. Nest is very small for the size of the bird—the smallest, proportionately, among the Australian cuckoo-shrikes and trillers. It is built of very fine twigs and vine tendrils bound with spiders' webs, most heavily on the outside, giving a whitish colour and a neat exterior that blends smoothly to the thick-stemmed, usually horizontal fork in which it is built. A typical site is among upper branches toward the tree's canopy. The Varied Triller's single egg is large for the size of both it and its nest (26 × 18 mm). Male and female jointly build nest, incubate eggs and feed young. (p. 302)

Yellow Oriole Jul.–Mar.; usually Sep.–Jan. The nest territory can be revealed by the male's almost incessant bubbling calls, sometimes continuing for several hours without obvious pause. However, while indicating a territory, the calls may not prove an active nest; a male may call sporadically throughout the year. Nest is a deep cup-shaped basket of long, thin, interwoven strips of flexible bark, often paperbark, and vine tendrils; lining is of finer, softer fibre. It is slung from the rim in a slender horizontal fork, in or slightly beneath the foliage canopy, from 2 to 15 m up. Clutch 2–3 (33 × 23 mm). (p. 306)

Olive-backed Oriole Breeding recorded most months; most activity Sep.–Jan. Nest is a deep, rather untidy cup; hangs by the rim. It is a larger version of the type of nest used by many small honeyeaters. The materials used include strips of bark, grass, wool and leaves bound with webs to a form a compact yet strong, soft and flexible nest. The rim is tied with fine fibres and webs into a slender, horizontal fork in a tree or tall shrub's outer foliage canopy, 2–15 m up. Clutch 2–4 (33 × 22 mm). Female builds and incubates; the male makes only brief visits during this time, but takes an equal share in feeding nestlings. If, in a favourable season, there is a second brood, he may continue to feed the young after they leave the nest while the female builds a new nest for a second clutch. Incubation about 18 days; young fledge at 16–17 days. (p. 306)

Figbird Usually Oct.–Jan.; some breeding Jul.–Mar. Nest is cup shaped, made of fine twigs and vine tendrils, rather loosely woven, not as lavishly bound with webs, nor as deep or well-built as those of the orioles. It hangs by the rim in a slender horizontal fork, usually quite high in the leafy canopy of a tree in or near rainforest or tropical monsoon forest. Clutch 2–3 (34 × 24 mm). Figbirds gather in flocks out of the breeding season, and carry this social behaviour through into breeding; they nest in small, semi-colonial groups, the nests being no great distance apart. Incubation by both sexes. (p. 306)

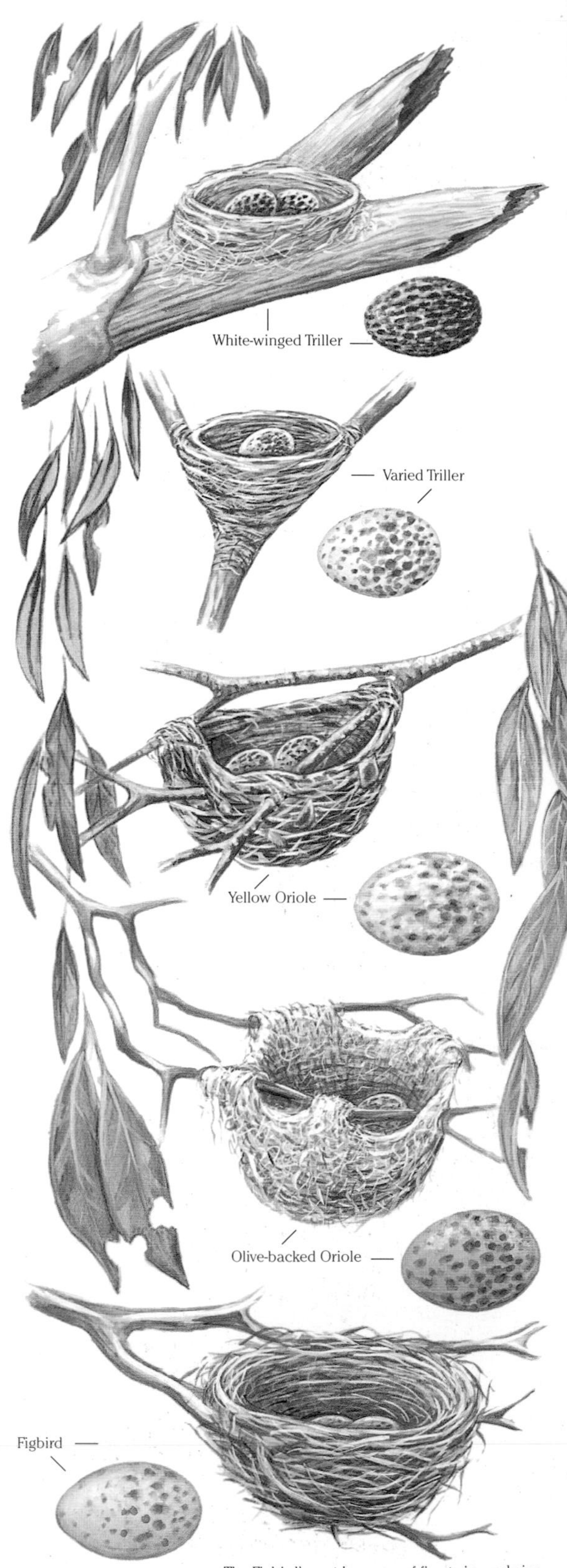

The Figbird's nest is woven of fine twigs and vine tendrils; looser-knit than those of orioles.

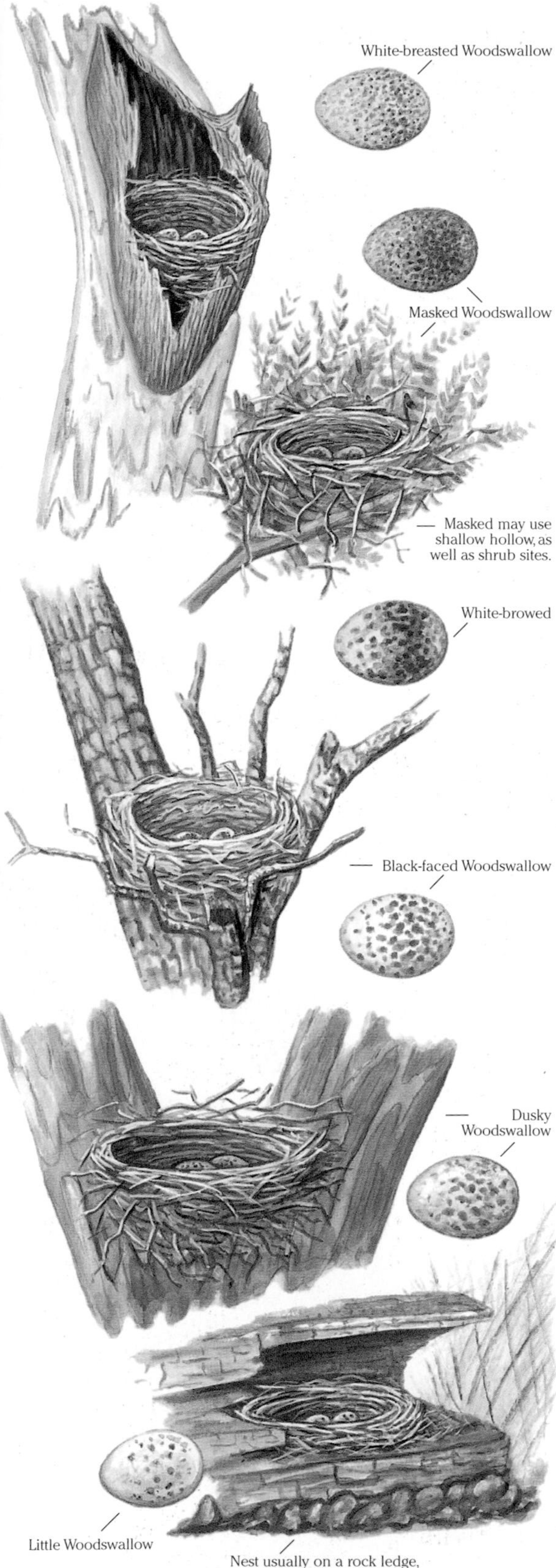

White-breasted Woodswallow Throughout year; most Aug.–Jan. Nest is a bowl of dry grass, rootlets and twigs placed in diverse positions. Its location is often betrayed by the birds' behaviour: if disturbed while on the nest, they will fly away with plaintive calls; then, with no disguise of their approach, will soon return. Nest sites range in height, 10–30 m. Where large trees are available, uses shallow spouts or hollows where a limb has snapped or ripped from the trunk forming a recess into which the grass bowl can be built. Also uses forks, preferring the deeply enclosed spaces between large diverging trunks or limbs. On the coast where large trees are absent, their nests are built among the spine-edged leaves of a pandanus. Occasionally build in the disused nest of a Magpie-lark. Clutch 3–4 (23 × 17 mm). Sexes share building and incubation. (p. 308)

Masked Woodswallow Aug.–Dec.; most months after substantial rains; 1 or 2 clutches. Builds a bowl of fine sticks that point untidily in all directions on the outside. The interior of the cup, however, is neatly lined with fibrous rootlets and grass. It is usually quite low, 1–3 m above ground, but some can be 6 m up. A dense bush or sapling is often used, the nest supported by the surrounding mass of stiff twigs. Also common is a multiple fork site where many thin branches have suckered out from the trunk of a mallee so that the nest can be tucked in between these and the trunk, often hidden by the long, twisted ribbons of bark that tend to hang over such forks. On occasions, builds in the top of a hollowed stump, and in old nests of other larger birds, including the Australian Magpie. Clutch 2–3 (22 × 17 mm). Male assists in building and incubation. (p. 308)

White-browed Woodswallow Throughout year; most Aug.–Dec. Often in large, highly nomadic flocks that congregate in green areas after good rain or where insects are abundant. The breeding of so many birds in a confined region results in nests being placed in every possible site, often very exposed—in shrubs, multiple forks of trees such as mallees, and hollow stumps and posts. The nest is bowl shaped—the outside is very roughly formed of sticks, the inside is neatly lined with fine rootlets and dry grass, and it usually sits 0.5–3 m above ground. Clutch 2–3 (21 × 16 mm). Incubation shared. (p. 308)

Black-faced Woodswallow Breeds any month that conditions suit; usually Oct.–Dec. The Black-faced Woodswallow is usually sedentary, even in arid regions. Small flocks, 5–20 birds, are locally nomadic; usually nest in the same spot year after year. Builds an open, cup-shaped nest of thin twigs and dry plant stalks lined with fine, fibrous rootlets. Usually placed in a low, dense shrub, less than 1 m above ground; some use the top of a hollow stump. Clutch 3–4 (22 × 17 mm). Sexes combine to build, incubate and feed the young. (p. 308)

Dusky Woodswallow Some breeding occurs throughout the year; most Aug.–Dec.; often 2 broods. Nest is bowl shaped, of twigs, rather rough and dishevelled looking outside, but neatly shaped and lined inside. Usually built on a thick, forked limb of shrub or tree, or tucked behind a slab of bark that has separated from the trunk. Also in hollow tree trunks, stumps or posts. Clutch 3–4 (24 × 17 mm). Some pairs are helped to build and feed nestlings by young of previous years. (p. 308)

Little Woodswallow Aug.–Jan.; most other months after good rain through arid regions. Little Woodswallow lives in rocky ranges and gorges in groups of 10–20, sometimes flocks to 200 birds. The nest is put on a sheltered rock ledge of a cliff or in a crevice among boulders; occasionally built in the top of a dead stump: a small bowl of twigs, stems and grass lined with rootlets. In tight hollows, the lining may be sparse. Clutch 2–3 (21 × 14 mm). Both parent birds build the nest, incubate the eggs and feed the young. (p. 308)

Black Butcherbird Sep.–Feb.; mainly Oct.–Jan. Builds in a large, often vertical tree fork, 5–10 m up, or in a mangrove about 2 m above high-water level. The nest is a large externally untidy bowl of sticks that may be lined with fine rootlets. Usual clutch 3–4 (35 × 25 mm). Black Butcherbirds are bold and aggressive in defence of their young, which are fed for several months after fledging, then disperse. Some populations have rufous-plumaged juveniles; these retain the rufous until their second moult. Some rufous-plumaged yearlings are still with their parents, helping to feed the next season's brood; some still have rufous in the plumage in their first breeding season. (p.310)

Grey Butcherbird Jul.–Jan.; usually Aug.–Dec. The nest is a bowl of sticks, untidy outside, but with a neat nest cavity smoothly lined with fine grass, rootlets and other soft fibre. A vertical fork in a tree or sapling is often used, 3–10 m up. Usual clutch 3–5 (31 × 23 mm). Incubated 24–26 days by female; fed at the nest by the male. Both feed the young, and fiercely defend the nest by diving at intruders. The young fledge aged 28–30 days, and often stay in the territory a considerable time, helping feed the following year's young. (p 310)

Black-backed Butcherbird Jul.–Dec. Nest like that of Grey Butcherbird, in vertical fork of a sapling or tree, to about 5 m up in open tropical eucalypt woodland. Clutch 3–4 (30 × 27 mm). Male feeds the female in courtship, while she builds the nest, and also during incubation. Both adults feed the young and aggressively defend the nest. (p.310)

Pied Butcherbird In months following winter rain in S; May–Nov. in the far N; almost any time after rain in arid regions. The nest is a quite large, externally rough bowl of sticks, neatly lined with fine dry grass and other soft fibre. The usual site is a vertical tree fork, 3–15 m up. Usual clutch 3–5 (32 × 24 mm). Incubated by the female for 20–21 days; young fledge aged about 30–32 days. (p.310)

Australian Magpie Jun.–Dec.; usually Aug.–Oct. Builds in upright forks of slender upper branches, in or near foliage of crown, occasionally in a dead tree or on a power pole. Height is determined by size of trees—usually 5–20 m up. The nest is a bulky, untidy mass of sticks, sometimes pieces of wire, with a neat, shallow egg cavity lined with grass and feathers or other soft material. Clutch 1–6, most 3–5 (38 × 25 mm). Incubated by female, 20 days. Young are fed by female; the male helps occasionally, but is often preoccupied with flock and territorial defence. Young fledge aged about 28 days. (p.312)

Pied Currawong Jul.–Nov. in N of range; Sep.–Jan. in S. Builds in slender fork of upper or outer leafy canopy; height ranges 5–25 m, determined by size of available trees. The nest is a bulky but shallow, untidy bowl of sticks, some quite large. The egg cavity is lined with finer grass, bark strips and rootlets. Clutch 2–4 (41 × 30 mm). Male feeds female while she builds, and through the 21 days of incubation. Both feed young in the nest and for several months after fledging. (p.314)

Black Currawong Aug.–Dec. Builds in a fork of a tree's upper canopy, usually high. The rough, bulky mass of sticks has a shallow nest cavity lined with finer rootlets, grass and bark. Clutch 2–4 (40 × 29 mm). Incubation and fledging details are not known, but probably similar to the Pied Currawong. Defends young in nest against close human approach. (p.314)

Grey Currawong Aug.–Dec. Makes a large, rough, shallow bowl of sticks, similar to the nests of other currawongs in a horizontal fork of slender branches in a tree's outer leafy canopy, usually quite high, 5–15 m. The egg cavity is lined with fine rootlets and grass. Usual clutch 2–3 (42 × 30 mm). Incubated by female; both sexes attend young. (p.314)

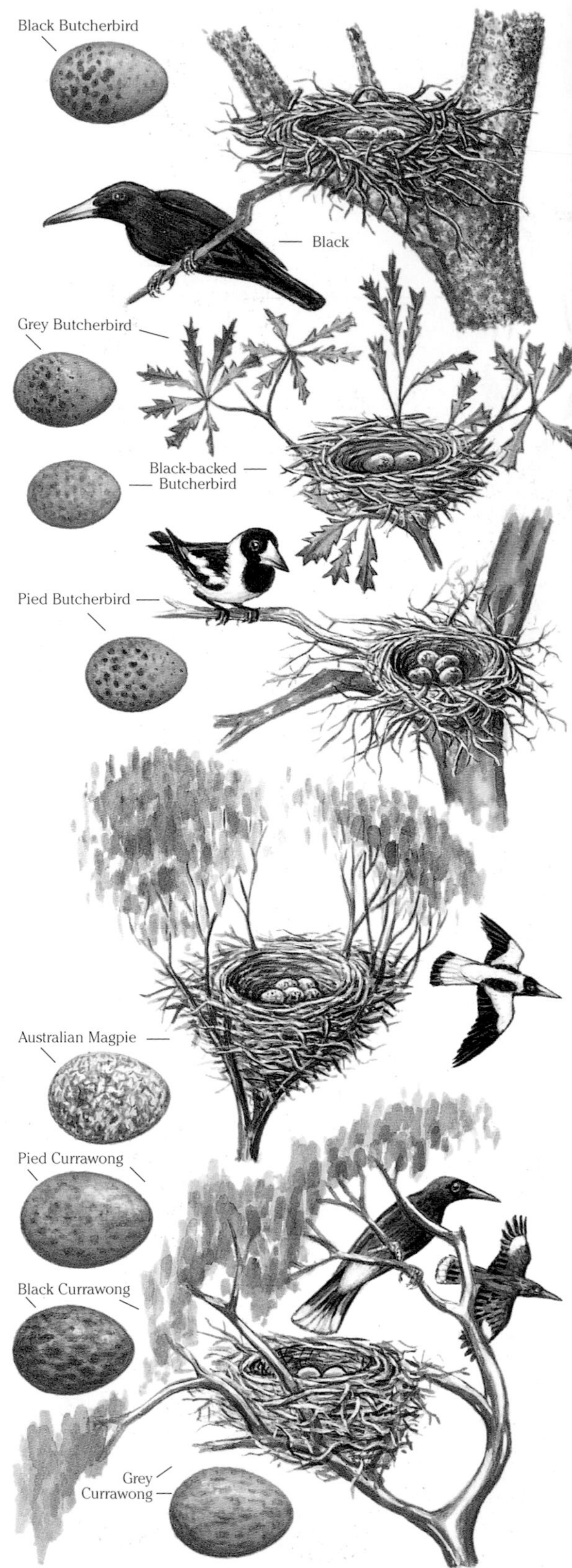

Nests of currawong species are similar: bowls of sticks among slender branches of the upper canopy in one of the tallest trees available.

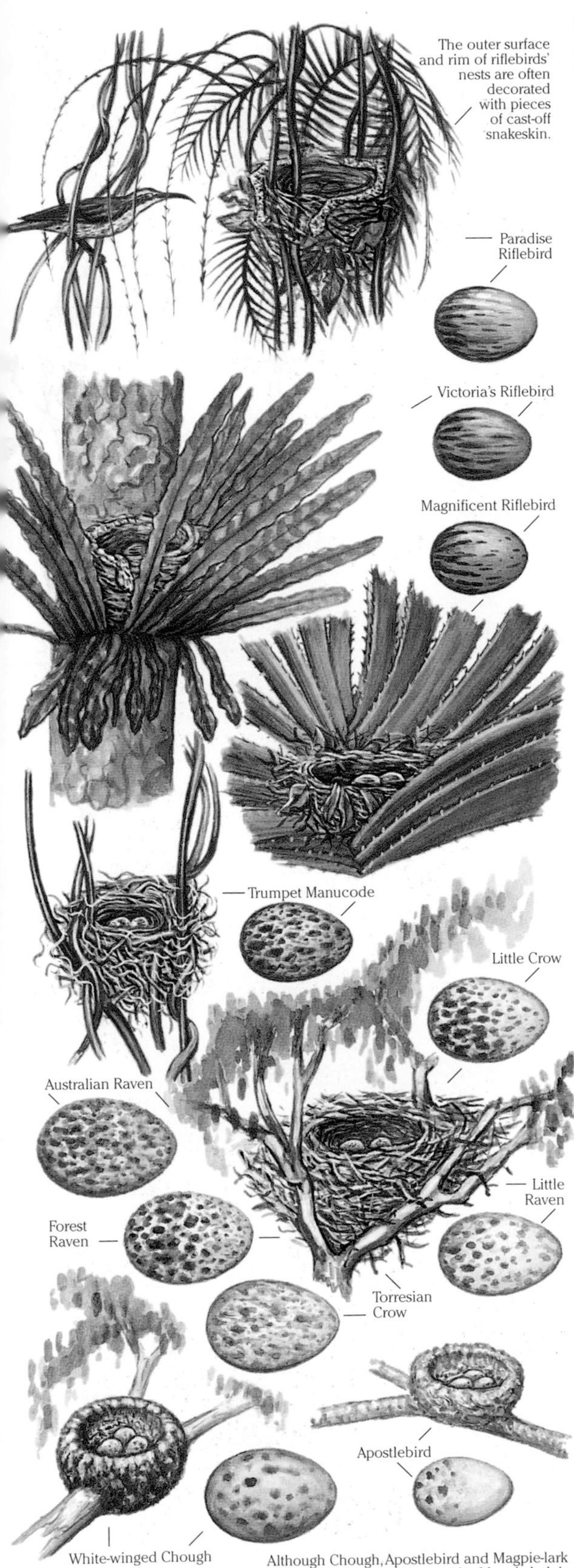

Although Chough, Apostlebird and Magpie-lark all build mud bowl nests, the Magpie-lark is more closely related to monarch flycatchers.

Paradise Riflebird Sep.–Feb. Unlike spectacular, noisy male riflebirds, the brownish females are inconspicuous. They build the nest, and incubate and raise the young without help from the male. The nest is a bulky, deep bowl of vine stems, palm fronds, leaves and twigs, bound with fine vine tendrils; hidden in dense foliage, often in tangled masses of vines or mistletoe, usually 5–35 m up. Clutch 2 (36 × 24 mm). Incubation 15–16 days; young fledge at 4 weeks. (p.316)

Victoria's Riflebird Sep.–Feb. Males are noisy show-offs; the quiet brownish females build the nest, incubate and raise the young alone. The nest is well hidden in foliage such as the centre of a dense, high tangled mass of vines, the crown of a tall palm, or tucked into a large epiphytic fern on a tall rainforest tree. It is built of rootlets, leaves and vine tendrils lined with leaves and fine rootlets. Clutch usually 2 (33 × 23 mm). (p.316)

Magnificent Riflebird Sep.–Feb.; possibly other months. Female builds the nest alone—fragile loose mass of dry leaves and vine tendrils. The crown of a pandanus palm is probably the most common site, but will also use the top of a stump amid foliage or a dense tangle of vines and other vegetation. Clutch 2 (34 × 24 mm). Incubation 15–16 days. (p.316)

Trumpet Manucode Oct.–Feb. Nest is a small bowl of interwoven vine tendrils lined with finer tendrils, built in a vertical or horizontal fork or among slender twigs in the upper foliage. Usually positioned in the mid to upper levels of the forest canopy, 15–25 m up, or occasionally lower in the foliage of a smaller tree, down to 5 m. Clutch 2 (37 × 25 mm). Incubation and care of young undertaken by both sexes. (p.316)

Australian Raven Jul.–Oct. Builds a large, rough bowl of sticks lined with soft grass, wool and fur. Very high in a vertical tree fork of the canopy, or on a power pole or other structure. Clutch 4–5 (45 × 35 mm). Female incubates for 20 days; fed at nest by male, who helps in building, feeding young. (p.318)

Forest Raven Aug.–Dec. Builds a bulky, rough bowl of sticks; egg cavity lined with bark strips, fur or wool. Typically in an upright fork among slender branches near the crown of a tall forest tree. Clutch usually 4 (45 × 32 mm). Both build nest and feed young; female incubates, about 20 days. (p.318)

Little Raven Apr.–Dec. Builds a bulky stick nest like other corvids, but often semi-colonial; nests much lower, under 10 m, in smaller inland trees; occasionally on ground in treeless areas. Clutch 4–5 (44 × 41 mm). Sexes share building and feeding of young; female incubates, 19–20 days. (p.318)

Little Crow Feb.–May in N; Aug–Sep. in S; after rain in arid regions. A rough bulky bowl, some mud among sticks of lower parts, is lined with wool, fur, emu feathers. Built in tree fork near upper canopy; often low in arid land. Clutch 3–6 (39 × 26 mm). Female incubates, 19–20 days; both feed young. (p.318)

Torresian Crow Aug.–Mar.; 1 brood. Rough bowl of sticks, sparsely grass lined, built in a tall tree's leafy canopy. Clutch 3–5 (45 × 30 mm). Incubated by female, fed on nest by the male, 20 days; young fledge aged about 40 days. (p.318)

White-winged Chough Aug.–Dec.; after rain. A large, deep bowl of mud is reinforced with bark and grass; lined with soft dry grass and other fibre. Most of family flock, 5–10 birds, combine to build nest, share incubation, 19 days. Clutch 3–5 (40 × 31 mm). Several females may lay in one nest. (p.320)

Apostlebird Aug.–Mar.; after rain. Builds deep mud bowl reinforced and lined with grass on a horizontal branch, 2–20 m up. Family party shares building, incubation (18 days) and care of young. Clutch 3–4 (30 × 22 mm). Several females may lay in one nest to make a larger, combined clutch. (p.320)

Spotted Catbird Sep.–Mar.; just 1 brood per season. Unlike other Australian members of the bowerbird family, both catbirds are monogamous. The female builds the nest, incubates the eggs and broods the young while they are very small. The male defends the nest site and its surrounding territory, and helps to feed the nestlings. Nest is like that of the Green Catbird: bulky, outside of quite large sticks, and with a layer of rotted wood beneath the inner lining of sticks. This bulk is supported by multiple forks and stems in the outer foliage of a small tree in the rainforest's lower canopy; otherwise is tucked in the centre of a low palm, 2–5 m up. Clutch 2–3 (40 × 24 mm). (p. 322)

Green Catbird Usually Oct.–Jan., but a few earlier, in Sep., or as late as Mar. The nest is large, bulky, thick walled; the exterior is built of quite large sticks bound with vine tendrils and with an inner lining of dry leaves. An unusual feature of its construction is the thick layer of wet, soft decayed wood beneath the lining of fine twigs and leaves. This layer adds to the size of the nest, already large for the size of the bird. The nest is placed in the dense crown of a low understorey tree where there are many twigs and stems to support its bulk, or in thorny or stinging trees, the crown of a tree fern, or in the upwardly cupped fronds of a bird's nest fern. Clutch 2–3 (44 × 31 mm). Incubated by female for 23–24 days; male assists in rearing young through the 20–22 days to fledging. (p. 322)

Golden Bowerbird Oct.–Feb. Nest is built by female; a deep, bulky, cup shape, mostly of dry leaves on a foundation of sticks, and lined with fine vine tendrils. Most nests are built in a deep, narrow crevice of a tree; strong preference for those that close inward overhead, as a 'roof'. Such sites occur occasionally in buttressed tree, where two large buttresses that meet the trunk close together, forming a narrow space between. Narrow cavities are also formed within the curtain of hanging roots of strangler figs as they thicken to become like multiple, vertical trunks. Closely clustered saplings sometimes create a suitable enclosed space, as may a rotted-out cavity in a dead stump. These sites are usually within 150 m of a bower in active use by an adult male, and tend to be re-used annually. Clutch 1–2 (35 × 25 mm). Incubation of 21–22 days, and care of young in nest, 17–20 days, by the female. (p. 322)

Regent Bowerbird Oct.–Feb.; probably only 1 brood. The female builds a flimsy nest of sticks, stems and thin roots, lining it with finer twigs: does not have the massive bulk of the catbird nests, nor as solidly built as the Golden Bowerbird's nest. It is built among dense foliage—a tangle of vines, or where multiple forks and leafy twigs give both support and concealment. Clutch 2–3 (38 × 27 mm). As well as receiving no help from the male in nest construction (the male being fully preoccupied with activities around his bower) the female incubates, 20–21 days, and cares for nestlings by herself. (p. 322)

Satin Bowerbird Sep.–Oct.; most Oct.–Feb.; usually only 1 brood per season. The nest is a shallow bowl of sticks lined with dry leaves, positioned in upright forks in the foliage of a tree's crown, or in a dense mistletoe clump from 1 to 30 m up. The male takes no part in nest building, or other nest duties: he is too much occupied some distance away at the bower. Clutch 2–3 (44 × 30 mm). Incubation takes 21–22 days; young fledge when aged about 21 days. (p. 322)

Tooth-billed Bowerbird Nov.–Jan.; 1 brood per season. Nest is a rather flimsy bowl of fine sticks and vine tendrils built in tangled vines or twigs of the tree canopy, 4–30 m up. Clutch 1–2 (42 × 29 mm). Like other bowerbirds, the males maintain a site to which females are attracted by song and display, a court of upturned green leaves. (p. 324)

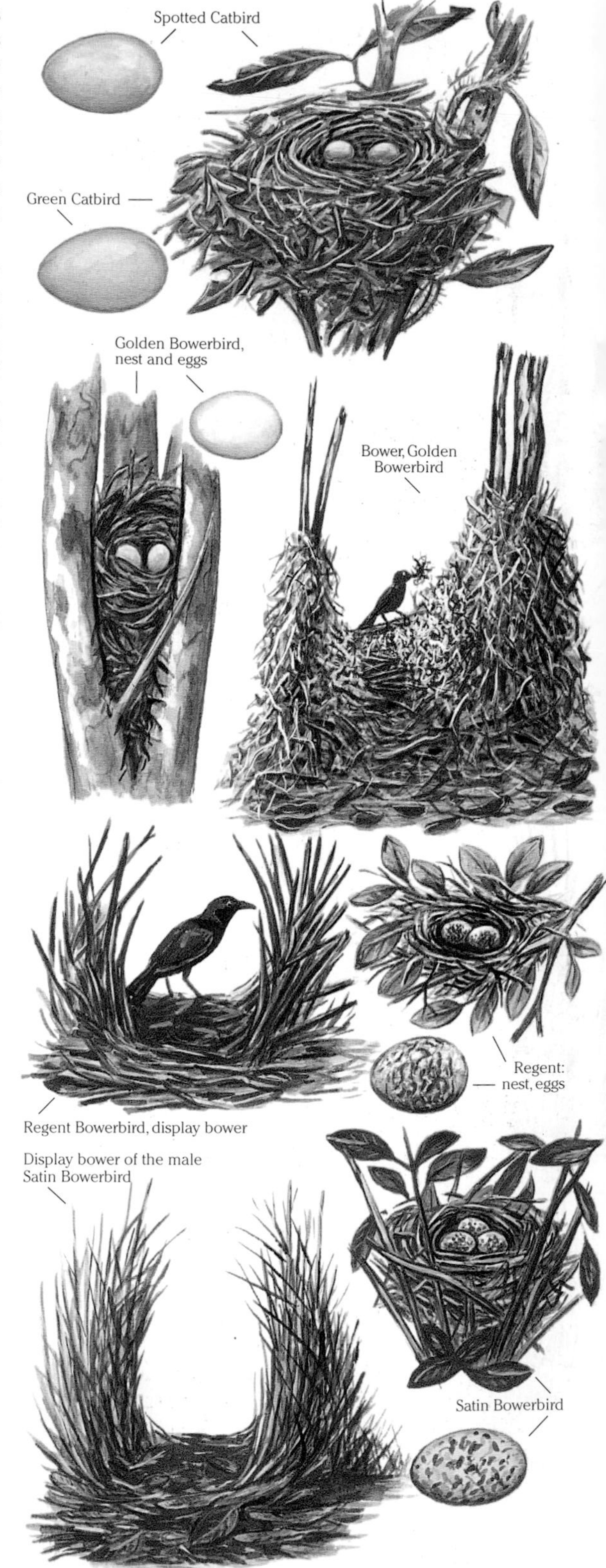

Tooth-billed, see illustrations at top of p. 413.

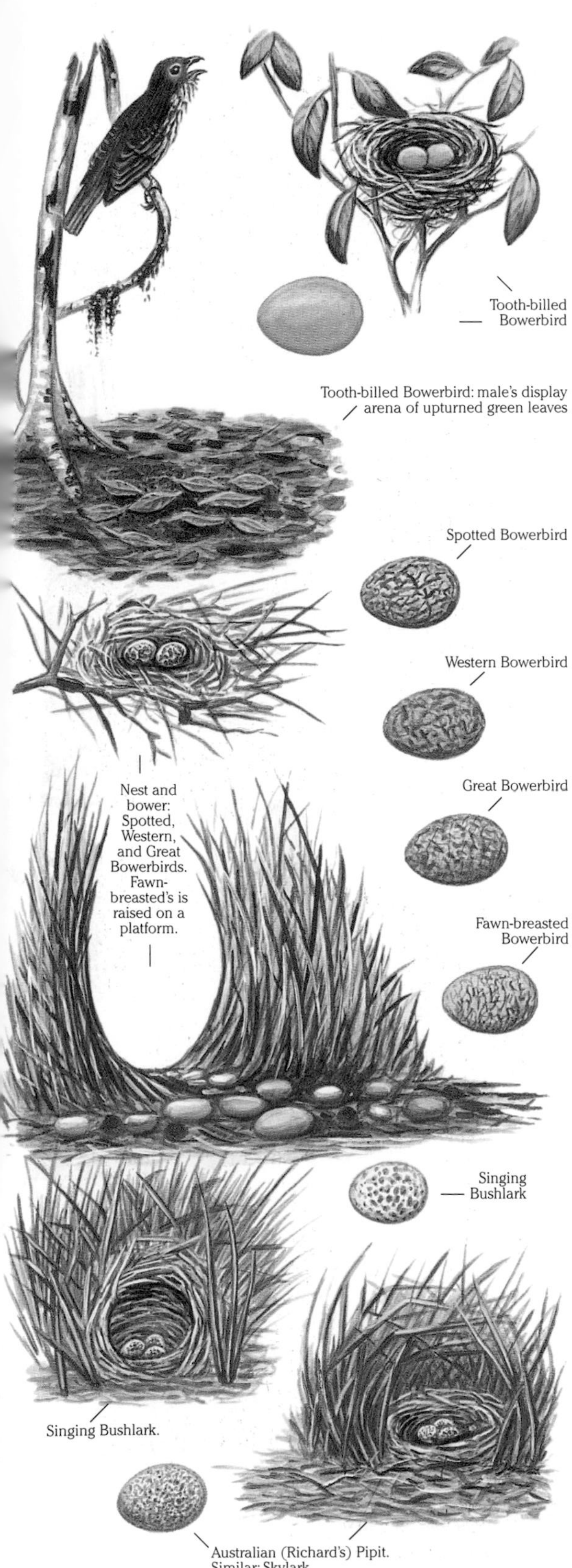

Spotted Bowerbird Aug.–Jan.; usually Oct.–Dec. Like the other Australian bowerbirds, unlike the monogamous catbirds, the male Spotted Bowerbird is entirely preoccupied with enticing any passing female of his species into his bower. With the need to maintain the bower and all its ornaments in the pristine perfection that seems a prerequisite for his conquests, the male has no time to help with any of the domestic duties that follow a successful courtship display. Alone, the female builds the nest, incubates the eggs and feeds the young. The Spotted Bowerbird's nest is a rough, shallow cup of twigs built into a tangle of forks and branchlets within the outer canopy of a shrub or tree, mistletoe clump or tangle of vines and stems, at heights ranging from 2 to 15 m, usually 50 m or more from any active bower. Clutch 1–2 (38 × 28 mm). Incubation 21–22 days; the young spend a similar period in the nest. (p. 324)

Western Bowerbird Usually Aug.–Dec. Builds a rough, shallow saucer of twigs lined with finer twigs and, in some instances, casuarina needles. It is typically situated near the top of a casuarina, mulga or other large shrub or stunted tree, from 2 to 6 m above ground. Clutch is usually 2 (33 × 26 mm). The eggs are incubated by the female, the period probably being 21–22 days, like that of the Spotted Bowerbird, and the time to fledging is probably similar—around 21–22 days. (p. 324)

Great Bowerbird Aug.–Feb.; most Oct.–Jan. Female tends to build her nest well away from the male's bower. Nest is a rough, shallow but substantial bowl of sticks lined with finer twigs and built in dense plants where the many stiff branchlets and forks conceal and support the rather loose nest structure. Height is from 2 to 10 m above ground. Studies have shown that the male's display period extends to include females that mate again later if eggs or young are lost by predation or other mishap. Clutch 1–2 (40 × 28 mm). Incubation by the female takes approximately 21–22 days, and the young stay in the nest for a similar period. (p. 324)

Fawn-breasted Bowerbird Usually Sep.–Dec., but some breeding activity spreads May–Mar., that is, for most of the year. The female alone builds the rough, shallow bowl of sticks lined with finer twigs and rootlets; it is built in a dense, bushy part of a shrub or low tree where the mass of forks and twigs hold the loose structure, giving it support and protective concealment. Clutch 1–2 (40 × 28 mm). The nest tends to be closer to an active bower than are the other *Chlamydera* species, sometimes no more than 20 m. Incubated by female only, 21–22 days; young fledge at around 21 days. (p. 324)

Singing Bushlark Usually Sep.–Dec.; any time of year after good rains. The nest is well hidden between tussocks of grass, and set in a slight depression in the ground. A thick walled cup of dry grass is made, its rim slightly above ground level, but much higher on one side; this partial hood is usually where clumps of grass overhang. Bare or lightly vegetated ground on the opposite side gives the bird clear access to the nest and a good line of sight from it. Clutch 2–4 (15 × 13 mm). (p. 326)

Skylark Aug.–Feb., but mainly between Sep. and Dec. Builds a cup-shaped nest of grass tucked into a small depression in the ground under an overhanging clump of grass. Clutch of 3–5 eggs (24 × 17 mm). Incubation 12–14 days; young leave the nest aged about 12 days. (p. 326)

Australian Pipit (Former common name 'Richard's Pipit') Recorded most months; mainly Aug.–Dec.; may have 2–3 clutches a season. Builds in a depression sheltered by a rock or clump of vegetation. A deep grass cup is lined with soft grass. Clutch 2–4 (23 × 17 mm); incubated by female (male keeps watch) about 13 days; young in nest about 14 days. (p. 326)

Zebra Finch Usually breeds Oct.–Apr.; in the drier regions is quick to nest any time after good rain. Female picks nest site: usually quite low in a dense, often thorny shrub or tree, occasionally in a tree hollow. Nest is a bulky, untidy, rounded dome with a deeply hooded side entrance and lining of soft feathers and fur. Clutch 4–6 (16 × 11 mm). Both sexes incubate, 12–16 days, and feed the nestlings. (p. 330)

Long-tailed Finch Jan.–Jun.; may be several broods. Male picks the nest site in dense foliage of eucalypt, mistletoe or pandanus; female builds while the male collects material. Nest is large, domed, oval bottle shape with side entry extended as a hooded tunnel, built of long dry grass stems, lined with feathers, plant down and bits of charcoal. Clutch 4–7 (16 × 12 mm). Incubation about 13 days; young in nest 22 days. (p. 330)

Black-throated Finch Aug.–Apr. Male initiates; female chooses final nest site. Often outer foliage of tree to 10 m, but can be built in base of a raptor's nest or in hollow tree or termite mound. Nest domed with entrance spout. Charcoal included in otherwise soft lining. Clutch 4–7 (17 × 13 mm). Incubated 12–14 days; young fledge aged about 21 days. (p. 330)

Masked Finch Mar.–Jun. and Aug.–Dec.; usually 1 brood. Nest sits on the ground in grass, beside a log, or low in a bush or tree, tree hollow or termite mound, a domed oval of grass blades and stems with a side opening and landing platform. Clutch 4–6 (17 × 12 mm); incubation 14–16 days. Sociable finches: flock and feed together; often nest in scattered colonies. (p. 330)

Crimson Finch On N coast of WA and NT, Jan.–Apr.; in NE Qld, Sep.–May. Often builds into junction of serrated leaves of pandanus; occasionally in cavities of buildings. Nest is a round, domed shape with side entrance but no entry tunnel, built of coarse grass, bark and leaves and lined with feathers. Clutch 5–8 (14 × 12 mm). Incubated by both sexes; young are attended by both parents, fledge at 20–21 days. (p. 332)

Double-barred Finch In N and inland, after rain. The nest is a rough sphere with side opening and short entry tunnel, lined with finer grass, feathers and plant down. Built 1–5 m above ground in pandanus, shrub, old babbler's nest, a hollow, under eaves of a building, and, very often, close to an active wasps' nest. Clutch 4–7 (16 × 11 mm). Incubated by both sexes, 12–14 days; young fledge at about 21 days. (p. 330)

Plum-headed Finch Aug.–May; variable. Builds a small, domed nest, slightly taller than wide; lacks entrance tunnel; made of green grasses, sometimes lined with feathers. Built low in grass, often of living grass interwoven, or slightly higher in a dense shrub. Clutch 4–6 (17 × 12 mm). Incubated by both, 12–14 days; young fly aged about 21 days. (p. 332)

Chestnut-breasted Mannikin Breeds in Wet in coastal N; inland, any time after good rain. Highly social, often in large flocks; nests in colonies. Small, tall, domed ovoid of grass, fine grass lining, up to 2 m from ground in tall grass, reeds, scrub, sugarcane or some other tall crop. Clutch 4–7 (17 × 12 mm). Incubation 12–13 days; young fledge at 21–22 days. (p. 332)

Pictorella Mannikin Jan.–Apr.; can extend Nov.–May. Builds a rough, bulky, domed nest with no entrance spout, lined with feathers and placed low in grass, spinifex or small dense shrub. Clutch 4–6 (16 × 12 mm). Incubated by both sexes, period unknown; fledge at 20–23 days. (p. 330)

Yellow-rumped Mannikin Dec.–Apr.; can be 2 broods. Bulky, spherical nest of dry grass lined with fine grass and feathers; side entrance but no entry tunnel. Usually quite low in reeds or grass, some of which is woven into the nest. Incubated 12–13 days by both sexes. Young are fed by both adults, leave nest aged 20–22 days. (p. 332)

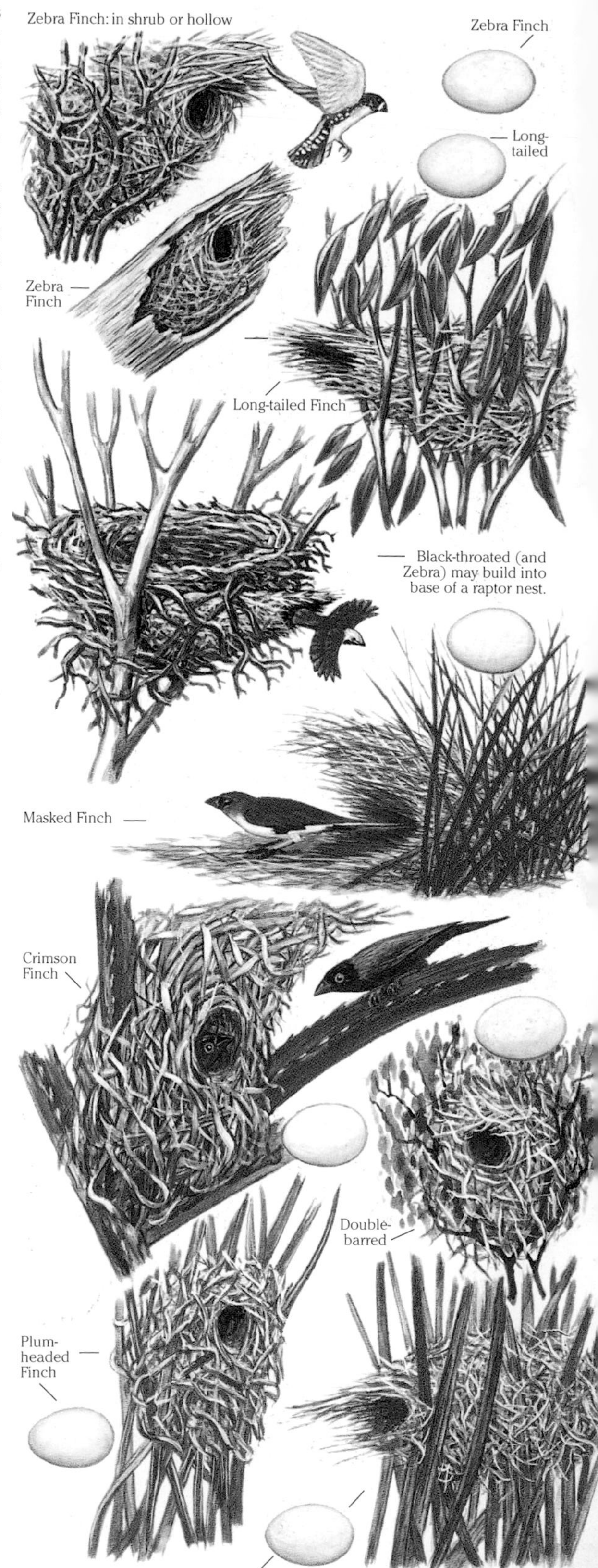

The three Mannikins have similar nests and eggs: Chestnut-breasted; Pictorella; Yellow-rumped

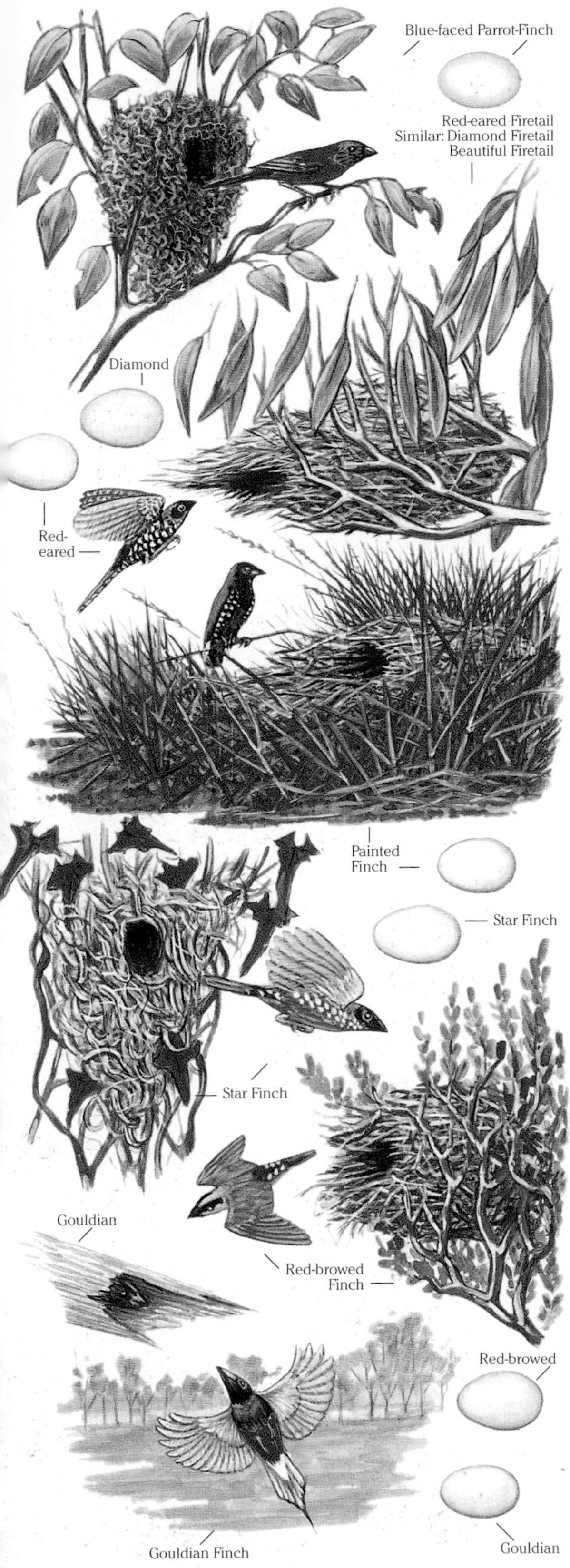

Blue-faced Parrot-Finch Nov.–Apr. Nest is bulky oval to pear shaped; has an entry tunnel at one end. This rainforest species has a choice of building materials—vine shreds and tendrils and green moss, as well as grass. A typical site is in, or at the edge of the forest, sheltered by the foliage canopy in a vertical fork of tree or sapling, 1–7 m above ground. Clutch 3–6 (15 × 10 mm). Incubated by both sexes, 12–14 days; young fledge after about 22 days. (p. 332)

Red-eared Firetail Sep.–Jan.; usually Oct.–Nov.; may rear 2 broods in a season. Male, in his courtship display, holds a very long, fine piece of grass, up to 0.5 m long, hanging from his bill, and gives a soft, undulating nest-site call. Similar long pieces of grass are used to build the large, long, oval, domed nest. In forests where nests are high up, these finches can be seen slowly towing long grass stems upward. The large nest is well hidden in dense foliage 2–15 m above ground in forest country; lower in the heathland of the S coast. The nest has a long entrance tunnel that slopes upward into it. Clutch 4–6 (18 × 12 mm). Both sexes incubate for 12–14 days, and feed chicks until they fledge at 21–24 days. (p. 334)

Diamond Firetail Aug.–Jan.; mostly Oct.–Dec. Uses long grass stems and blades to build a horizontal, bottle-shaped nest with entry tunnel at one end. Inside it is lined with fine grass and white feathers. The nest is well concealed in dense foliage of a tree, shrub or mistletoe clump at heights up to 10 m. Clutch 4–6 (18 × 13 mm). Incubated by both sexes, 14–15 days; young fledge aged 23 days. (p. 334)

Beautiful Firetail Sep.–Jan. A bulky, horizontal bottle-shape is built of green grass and leaves lined with finer grass and feathers; a long entrance tunnel is at one end. It is usually hidden in dense foliage of a shrub or tree 1–5 m above ground. Clutch 4–5 (18 × 12 mm). Incubated by both sexes for 13–15 days; young fly aged about 23 days. (p. 334)

Painted Finch Any time of year after good rain. Nest is built on a platform of bark, sticks or stones on top of a low clump of spinifex, usually where the centre is dead so that the rough nest of grass and sticks matches the grey-brown debris typical of the heart of an old ring of spinifex. Clutch 3–5 (14 × 11 mm). Incubation 12–14 days; fledge at 20–24 days. (p. 334)

Star Finch Dec.–May in far N; after heavy rain in arid regions. Builds a round nest of grass; has a side entry but no entrance tunnel. Lined with feathers, it is hidden in tall, rank grass or a low shrub in tall grass, often beside a watercourse that flooded during preceding rain, promoting the growth of lush grass. Clutch 3–6 (14 × 12 mm). Incubated by both sexes, 12–14 days; fledging period 15–17 days. (p. 334)

Red-browed Finch Jan.–Apr. in N; Sep.–Dec. in S; often several broods. A large, untidy, horizontal, oval bottle nest. The entrance may have a small overhanging hood, sometimes extended as a short funnel. Green and dry grass is used for the body of the nest; the interior is softly lined with small feathers. Usual site is a dense bush or tree, often thorny, 1–3 m up, rarely to 8 m. Clutch 5–8 (16 × 12 mm). Incubated by both sexes, 13–15 days; young fledge by 24 days. (p. 334)

Gouldian Finch Nov.–Apr.; mostly Dec.–Mar.; 1 or more broods annually. Most nests are in small, low tree hollows where a sparse lining of dry grass forms a globular nest with a short entrance spout. Rarely, perhaps only when hollows are not available, a pair will build a domed nest in a shrub. Clutch 4–8 (17 × 13 mm). May nest in small colonies: several pairs in close proximity—in one tree that has multiple hollows; two pairs will even occasionally share one hollow. Incubation by both sexes, 12–13 days; young fledge at 21 days. (p. 336)

Yellow-bellied Sunbird ('Olive-backed Sunbird' in NG.) Breeds throughout the year; mostly Oct.–Feb. Bark strips, fine rootlets, grass and other plant fibre are strongly bound with cobwebs and decorated with small flakes of golden buff paperbark, white spiders' cocoons and other buff and chestnut toned material. The side entrance has a small hood and the egg chamber is lined with plant down. The total length of the suspended domed nest and its long tail is 30–50 cm; it hangs by a slender attachment to a thin branch of a shrub or tree, usually quite low and sometimes over water. It is very often fixed to some part of a building, commonly beams of a verandah, rafters in an open shed or the underside of a bridge. The female completes most of the construction and all of the incubation. Meanwhile the male pugnaciously defends the nest territory. Clutch 2–3 (17 × 12 mm). Incubation time is 15 days; the young are in the nest for about another 15 days. (pp. 270, 338)

Mistletoebird The suspended nest is tiny, neat, soft and pear shaped, made of plant down densely bound with webs to create soft, felt-like walls. The nest's bottom has no tail, but is smoothly rounded. Its shape and the soft thin walls, like fine woollen knitting, have led many to describe the nest as being like a baby's bootee. The top of the domed roof is attached to a slender twig; the nest is well hidden in a tree or shrub's outer foliage, and is unlikely to be seen unless the birds are spotted busily carrying mistletoe berries to the nestlings. Clutch 2–3 (17 × 11 mm). The nest is built and the eggs are incubated by the female; both parents feed the nestlings, which fly from the nest aged about 15 days. (pp. 270, 338)

Welcome Swallow Aug.–Dec. The Welcome Swallow's mud cup must be one of the most familiar of all nests, especially in country areas. These birds nest under the verandah rooves of almost every homestead and country town shop, in shearing sheds, under bridges, in mine shafts and culverts. Where no such structure is available, the swallows must resort to natural sites: the underside of overhanging cliffs, and inside large tree hollows. The nest is a semicircular bowl of mud: hundreds, perhaps thousands of pellets of mud are attached to a wall of rock, brick or timber, and then softly lined with grass and feathers. Usual clutch 4–6 (19 × 13 mm). Sexes share both building and the incubation of 14–16 days. (p. 338)

Tree Martin Recorded nesting all months, but are most active during spring and summer, Aug.–Dec. Uses small hollows, mostly in the upper limbs of trees: a single big, old dead tree can provide nest sites for many Tree Martins. The nests are easily and often seen: the martins, made bold by the height of their hollows, dart in and out continuously. The hollow is lined with grass and leaves—those that are too big may be reduced with pellets of mud. The same hollow may be re-used year after year. Clutch 3–5 (18 × 13 mm). Sexes share incubation of 15–16 days; both feed the nestlings. (p. 340)

Fairy Martin These birds' presence is often revealed by their long necked, bottle-shaped mud nests attached under rock overhangs, shallow caves, bridges, culverts, abandoned buildings and large pipes. Often the nests show that Fairy Martins have been in the locality, even though the breeding season has long gone and the birds have moved on. This species is more wary of human activity than the Welcome Swallow and usually does not use parts of inhabited buildings as nest sites. Each nest is a flask-shaped chamber of mud pellets with a long, narrow entry tunnel. Always nests in colonies of few to prolific nests, these often clustered so densely that they join, overlap and are buried by later additions, the protruding entry tunnels being the only visible part. Clutch 4–5 (17 × 12 mm). Sexes share in building and incubation of 14–16 days. (p. 340)

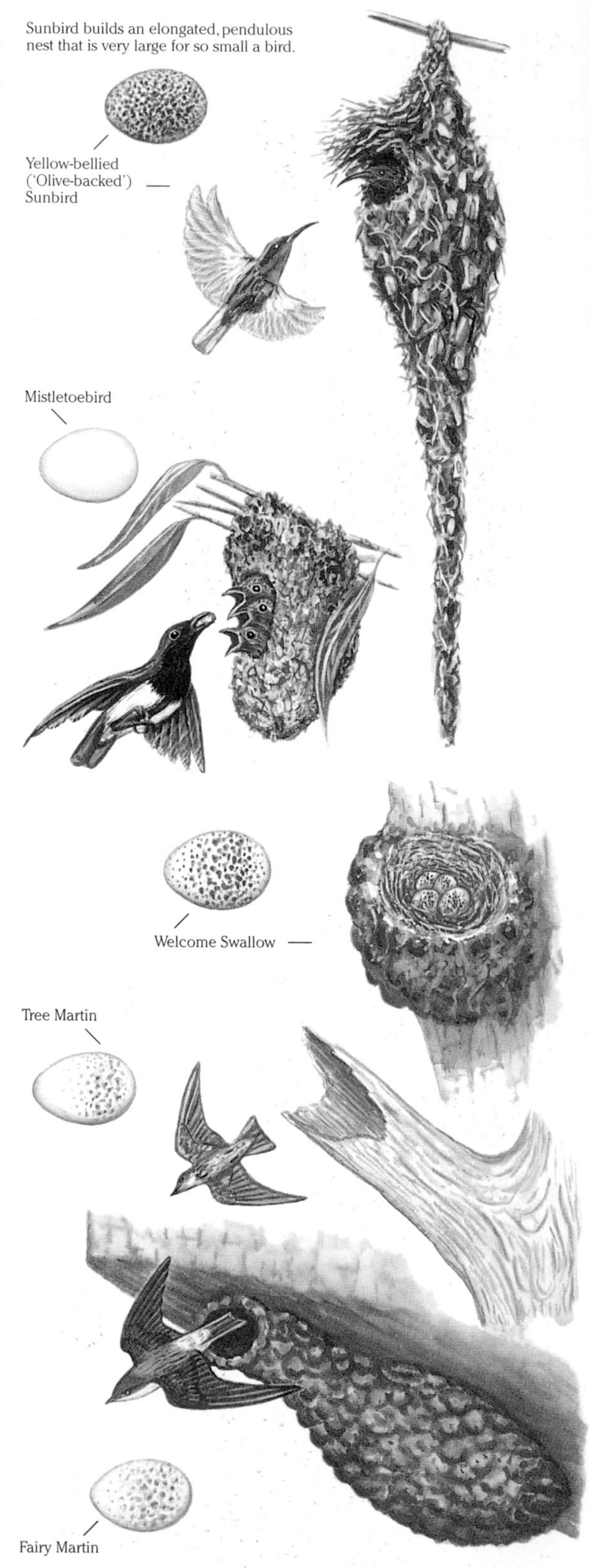

Sunbird builds an elongated, pendulous nest that is very large for so small a bird.

These tunnel-like bottle nests, when not used by the martins, may be taken over by other birds, including the hollow-nesting Striated Pardalote.

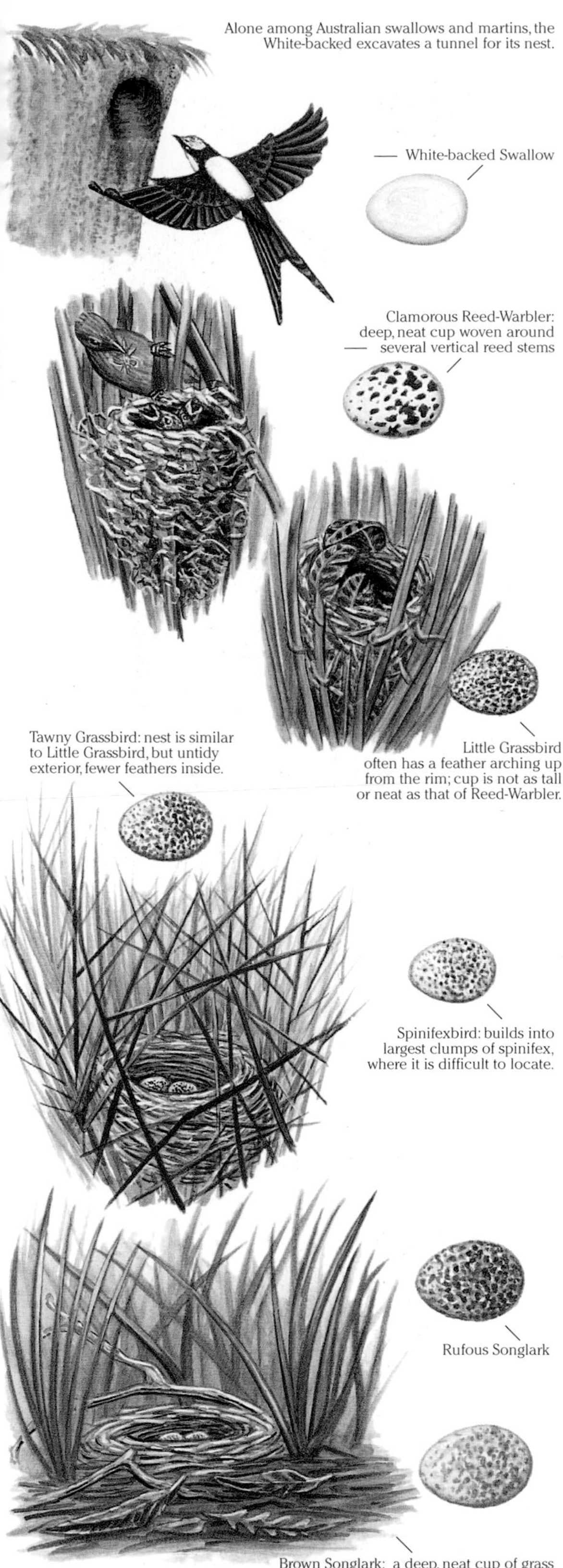

White-backed Swallow Nests recorded most months; usually Jun.–Dec. Uniquely, nest is a tunnel in a bank of soft earth; some extend almost a metre to the wider nest chamber in which is built a bowl of grass, rootlets and feathers. Is alone or in a colony with a few or many other tunnels nearby. Uses various sites, man-made or natural—roadside cuttings, sand pits, watercourse banks. Clutch 4–6 (17 × 12 mm). Incubated by both sexes, probably, 14–16 days. (p. 340)

Clamorous Reed-Warbler Sep.–Feb. In dense reeds or rushes, a small, very deep cup is attached to several stems, usually 0.5–1 m above water level. The nest bowl narrows towards the top opening so that, even should the reeds bend far over in a strong wind, the eggs would not roll out. The wet, decayed water weeds included in the nest bond the structure when they dry. The attachment to the reeds is cleverly devised: fine strips of reed or other fibre are wrapped and tied around the smooth, round stems in a way that prevents the nest sliding down when the reeds sway in the wind. Nests are usually in dense clumps of reeds in a lake or swamp, but the birds also use bamboos, willows, cumbungi, and other dense, slender stemmed plants in or over water. Clutch 3–4 (20 × 14 mm). Incubated by female, 14–15 days; young leave nest aged 14–16 days. (p. 342)

Little Grassbird Most nest Aug.–Jan.; a few begin as early as Jul.; others as late as May. Nest is supported by the density of vegetation beneath it rather than by being firmly tied to reed stems. The cup is very deep; large curved feathers arch upward in a delicate, semi-domed roof that screens the interior. May refurbish or build onto an abandoned Reed-Warbler's nest. Clutch 2–3 (20 × 15 mm). (p. 344)

Tawny Grassbird Aug.–Apr. Builds a deep cup shape that narrows slightly at the rim. The outer structure is made of coarse grass or the flat blades of sedges; the inside is lined with finer grass. The nest is extremely well hidden; its untidy grass outside is almost indistinguishable from tangles of brown and green stems and leaves found at the centre of most large clumps of grass, which are typical sites. Clutch 2–3 (20 × 15 mm). The male gives a hint of the nest site's approximate location when early in the breeding season he makes conspicuous display flights above his territory. (p. 342)

Spinifexbird Aug.–Nov.; after rain. The nest is an open cup, quite thick walled, built of grass and other herbage, and lined with fine roots. It is typically hidden inside a large clump of spinifex, very close to the ground. Clutch 2–3 (19 × 13 mm). These birds, like other low-nesting inhabitants of spinifex, grassland and heath, rarely give away their nest's location, rather they sneak in, beneath and through the concealing vegetation. Prefers large, often massive, clumps of spinifex a metre or more in height and 3–4 m in diameter—dense and with needle-like leaves that deter entry. (p. 342)

Rufous Songlark Aug.–Feb. A deep cup of grass lined with finer grass sits in dense, low grass or other vegetation; often beside a tall clump of grass or log that shelters one side. The nest is on, rather than dug into, the ground. While the male loudly proclaims his territory in high song flights, the female secretively builds the nest, incubates the eggs and feeds the young, keeping out of sight as she creeps through the screening plants. Clutch 3–4 (23 × 17 mm). (p. 344)

Brown Songlark Sep.–Feb. in S; any time of year after good rain in drier mid-northern regions. The nest is built in a small depression in the ground, usually hidden in a clump of grass or other groundcover. As with the Rufous Songlark, the male sings while the female builds the nest, incubates the eggs and rears the young. Clutch 3–4 (23 × 17 mm). (p. 344)

Zitting Cisticola Breeds through tropical wet season; usually Dec.–Mar., when males are conspicuous with display flights and calls over their territories. The females quietly and secretively build nests, incubate the eggs and rear the young. The nest is an open cup shape with leaves of grass sewn together overhead to provide a screen for the contents. The plant material in the walls is bound with cobwebs, and the inside is warmly lined with soft plant down. Incubation and feeding of young by the female. Clutch 3–4 (15 × 12 mm). (p.344)

Golden-headed Cisticola Breeds Sep.–Mar., before and through the summer wet season. The nest is a rounded dome with entrance in the side, towards the top. Its neat, softly textured walls are made of fine grass, plant down, spiders' webs and cocoons, and, where they can be pulled and stitched into position, large leaves of the plant the nest sits in. It is assumed that the inclusion of living leaves in the external surface is a form of camouflage. Clutch 3–4 (16 × 13 mm). The male helps in nest building, but not in incubation. (p.344)

Zitting Cisticola, similar nest: but of cup shape with a frail open canopy of grass leaves drawn together overhead.

Pale White-eye Nesting probably extends throughout the year; may, like the Yellow Silvereye, have a peak just or late in the wet season, Jan.–Jun. Very few nests have been found. One record describes a small, neat cup-shaped nest of dry leaf skeletons bound together with cobwebs, and decorated or camouflaged on the outside with semi-transparent, whitish flakes or strips of paperbark. The lining is of fine, dry, wiry grass. The nest is suspended, mostly by the rim, from several thin twigs in a horizontal fork of a tree's outer foliage. Clutch 2–4 (17 × 13 mm). Incubation is probably around 10 days, with young leaving the nest aged 9–11 days. (p.346)

Yellow White-eye Breeds throughout year. In NT, eggs are laid in every month except Nov.; nesting activity peaks through Sep.–Oct. Most nests are built in mangroves, typically in the upper foliage of low mangroves, at an average height of about 2 m. The nest is suspended by the rim in a horizontal or diagonal fork; it is built of grass and other plant matter, bound with cobwebs and lined with finer grass. Clutch of 2–3 eggs (17 × 13 mm). Incubation by both sexes; takes a very short 9–11 days; young are in the nest for a very short period, about 10 days. The white-eyes generally seem to have exceptionally short incubation and fledging times. (p.346)

Silvereye Sep.–Jan. Builds a small, deep cup of fine grass and other plant matter bound with webs. In drier regions the nests are the pale golden buff of freshly-dried grass; in more humid coastal regions such as Tasmania, and elsewhere that such materials are available, the nest may be quite green with fine threads of moss. The nest is suspended by the rim from slender twigs, usually well hidden in the dense foliage of a shrub or a tree's lower canopy. Usual clutch 2–4 eggs (18 × 13 mm). Both parents share the nest building, incubation of eggs (10 days), and feeding of the young. (p.346)

Bassian Thrush Aug.–Dec.; also recorded Jul.–Mar. The nest is a large, deep cup shape made of bark strips, leaves and rootlets, camouflaged outside with green mosses and lichens. Typical nest sites are the shallow depression on top of a stump, the cavity formed between a tree trunk and a major limb, the space between several closely packed trunks, or between a tree trunk and a large slab of peeling bark. Height varies from very low up to about 15 m. Clutch 2–3 (35 × 23 mm). The young are attended by both parents. (p.348)

Russet-tailed Thrush Aug.–Dec. Builds a bowl of bark strips, leaves and grass, camouflaged with moss, lined with fine rootlets, and placed in a large fork or on a stump; can be up to 15 m above ground. Clutch 2–3 (31 × 22 mm). The young are attended by both parents. (p.348)

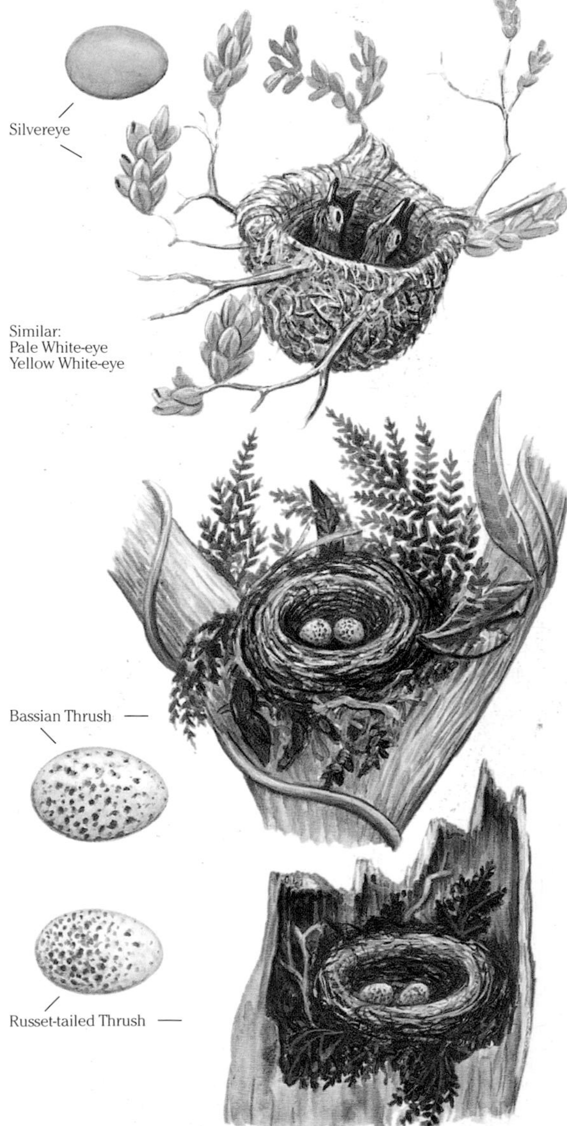

The difference between thrushes may be slight, but by about 1900 egg collectors had recognised that the eggs of the Russet-tailed are smaller and more finely speckled.

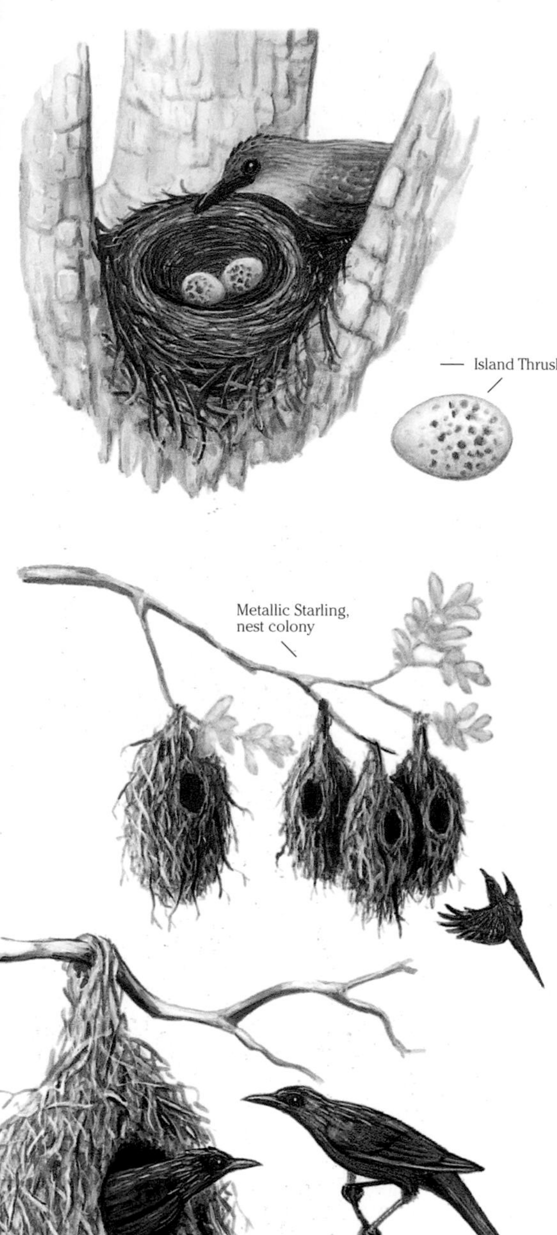

The similar Singing Starling (*Aplonis cantoroides*), a species widespread in NG, occurs within Australian territory on Boigu, and possibly also on Saibai and Dauan islands, all in Torres Str., and may breed on one or more of those islands. The nests of the Singing Starling are quite unlike those of the Metallic Starling, shown above. Rather than the conspicuous cluster of hanging nests, it builds a grass nest in a tree hollow or cavity, among rocks, or in house rooves. Where cavity space is available, there may be small colonies. This and the Metallic will happily share the one tree, one in hanging nests, the other in hollows.

Island Thrush (Christmas Island) Summer wet season, Nov.–Apr.; usually has several broods. A widespread species with many races, native from Indonesia through New Guinea to the Pacific Islands, including Norfolk and Lord Howe Islands; includes the Christmas Island Thrush. Visitors to that island will often meet this common species in the forests, where it is remarkably tame and tolerant, even at its nest; in fact it will often be the bird, bold and curious, that makes the close approach. Nests, whether currently in use or old, are often seen; those in more slender forks and open situations are quite obvious. The deep, bulky cup nest is blended neatly into the fork of a tree, often quite low and in the open. (p. 426)

Metallic Starling Aug.–Jan. Breeds in large colonies where great numbers of nests are clustered in a tree's highest branches. Each nest is a bulky, deep oval with a side entrance that is sometimes extended to form a hood or short entrance tunnel. The nest can be up to 0.5 m high, built of curly vine tendrils and lined with fine strips of palm leaves. Each nest is attached at various points on its domed top, often with many nests clustered closely together. Some colonies can number many hundreds, even thousands, of birds, the entire crown of a large rainforest tree being festooned with these rather untidy nests. Clutch 2–4 (30 × 21 mm). (p. 350)

Nest Site Precautions: When you observe birds at nests, avoid those with newly hatched, naked chicks. These need the adult's constant attention and warmth; without it, they will soon weaken, be unable to take food and perish. Better that they are a few days or a week old, feathered and lively. If putting up a hide, set it up well back and wait in the far distance to be sure that the bird confidently returns to the nest. When moving the hide closer, wait each time to be certain that the bird accepts it as harmless, that it is not keeping the bird away from the nest. If the bird makes several attempts to return to the nest, but baulks each time and turns back, move the hide further away. Be prepared to abandon observation or photography rather than endanger the nest. When finished, ensure that the nest is well screened by foliage, at least as effectively as it was when found.

As a general rule, it is best to stay well back from any nest that is in use; watch proceedings through binoculars. Some birds tolerate careful human intrusion; with others, there is some risk that prolonged stress will cause them to desert the nest. The safe distance from the nest increases with the size of the bird. A robin will soon come to the nest with observers 20 m away, and usually is quick to accept a much closer approach. A parrot might still be wary at three or even five times that distance, while an eagle might stay away while a human is present within several kilometres of its nest.

Crows, butcherbirds and kookaburras are keen watchers of humans' movements, if only to scavenge the scraps so often left behind; they are likely to pay a visit to any place that they have seen receiving human attention. Touching a nest may also leave scents that invite closer inspection by other predators such as feral cats, foxes or goannas. Camping close to a nest may on rare occasions be another cause of disaster. Campers may be quite unaware that a pair of parrots is sitting for hours in a distant tree, waiting to return to a nest nearby where eggs are becoming cold or the chicks are hungry and cold. Be sensitive to the possibility that birds hanging about, perhaps circling from tree to tree around a campsite, might be trying to summon up courage to reach a tiny nest in the foliage or in a tree hollow, perhaps just above someone's car or tent. If, through fear, the bird cannot reach its nest within a reasonable time—an hour or thereabouts, depending on air temperature and time of day—only a shift of camp will avert disaster.

Many birds of Australia's island territories are also on the Australian mainland bird list; others are subspecies that differ only slightly from Australian species. Some, not on the Australian list, are shared with nearby islands or continents—New Guinea, Indonesia, New Zealand or Antarctica. Newly explored by some dedicated birdwatchers are the islands of Torres Strait, especially Saibai and Boigu Islands, both very close to the New Guinea coast. Christmas Island shares many species with nearby Indonesia. On the following pages are birds not already covered in the preceding Australian list, mostly those that differ from Australian birds at the species level. Some subspecies or morphs are included, typically those that are well known by common name and distinctive to the island (for example, the 'Christmas Island Goshawk' and the 'Golden Bosunbird'). Birds that occur on any of these islands as well as in the Australian list (pp. 14–351) are marked in that list by a (T–Torres), (C–Christmas), (K–Cocos-Keeling), (N–Norfolk), (L–Lord Howe), (H–Heard) or (M–Macquarie) in the heading for the species. Recent unconfirmed sightings and species believed to be extinct are listed on pages 432–33.

TORRES STRAIT ISLANDS, ASHMORE REEF (T)

Grey-headed Goshawk *Accipiter poliocephalus* 32–38 cm
Saibai Island. Usually alone, inconspicuous; hunts insects and small reptiles among foliage and branches; appears not to soar above the canopy. **Voice:** like calls of Australian Brown Goshawk, but more rapid, very high and thin, 'ki-kikikiki–'. **Similar:** white phase of Grey Goshawk (NG Variable Goshawk), but has white upper parts. **Hab.:** rainforests, their margins and regrowth; gallery forests; probably in mangroves. **Status:** quite common through NG and nearby islands, including Saibai. **Br.:** compact, solidly built nest of sticks and leafy twigs; similar to those of Aust. goshawks; may be positioned at least as high as 30 m. Recorded at nests, Aug.–Mar. Eggs white, extensively blotched olive and buff.

Gurney's Eagle *Aquila gurneyi* Boigu and Saibai. See page 84, where placed for comparison.

Brown Hawk-Owl *Ninox scutulata* Ashmore Reef. See page 196, where placed for comparison.

Rufous-bellied Kookaburra *Dacelo gaudichaud* 27–29 cm
Saibai Island. Deep cinnamon underparts, broad, completely encircling white collar, white throat and chin, and black head make this large kingfisher unmistakeable. Males have dark blue tails; females' cinnamon-rufous. Heard more often than seen—its loud and recognisable calls carry far through the rainforest, marking its presence even though dense vegetation hides the calling bird. Hunts from perches in mid and lower levels of the forest, dropping down to take small terrestrial creatures; probably also raids nests of small birds. Strongly territorial, defends the nest. **Voice:** a loud, barking, chopping 'tchk, tchok, tchk' and descending, rattling 'tok-tok-tok-', sometimes given by two or more in duet; also varied shrieks and rasping sounds. Some calls like those of Blue-winged Kookaburra. **Hab.:** rainforests and margins; also monsoon forests, mangroves. **Status:** in Australian territory, Saibai Is., the southern edge of a wide range through PNG and nearby small islands. **Br.:** nests in a tunnel with unlined chamber drilled into an arboreal termite nest.

Red-capped Flowerpecker *Dicaeum geelvinkianum* 8.5–9.5 cm
Saibai Island. A bird of canopy foliage, often high; heard more often than seen, usually alone, occasionally in pairs; calls distinctive once learnt. Flight like Mistletoebird, swift and darting, restless, much fluttering about foliage and especially epiphytic ant-house plants. Feeds on mistletoe berries and other fruit, spiders and insects. **Voice:** a high, whistled, drawn out, rather buzzing 'zzweeeit' and a short, sharp, upwards note similar to that of the Sunbird. **Similar:** Mistletoebird, but neither sex has the red cap; male has black breast streak; upper parts much darker blue-black. **Hab.:** foliage canopy and secondary growth, gallery woodlands, mangroves, gardens, from sea level to 2000 m. **Status:** Saibai Is., probably Boigu Is. Throughout PNG and nearby small islands, sea level up to 2000 m.; common. **Br.:** nest like Mistletoebird, tiny, softly felted consistency, hooded, hangs by the roof from leafy twig in foliage canopy.

Singing Starling *Aplonis cantoroides* 19–21 cm
Boigu and Saibai Islands. Black with dull green sheen; short, square tipped tail. Solitary, in pairs or, after breeding, gathers in large flocks. Forages for fruits; sometimes takes flying insects; perches conspicuously on exposed limbs. **Voice:** a pleasant, somewhat drawn out, downwards whistle, similar to call of Bronze Cuckoo, repeated irregularly, 'tsieeeuw'; also a shorter, musical, rather bell-like double note, 'tsu-wei'. **Similar:** Metallic Starling, which also occurs on the northernmost island of Torres Str. Metallic has conspicuous nest colonies; differs in appearance and calls. **Hab.:** forests, coasts, mangroves, offshore islands, some towns. **Status:** Saibai and Boigu Is. (also NG and its surrounding islands). **Br.:** nests in a tree hole, recess among palm fronds, hole in cliff or building, or among clustered nests of a colony of Metallic Starlings; colonial where there are numerous hollows or other suitable sites close together.

Pechora Pipit *Anthus gustavi* 14–15 cm
Ashmore Reef. Sighted 1984 on Middle Islet. Solitary, in pairs or small parties; unobtrusive and rather wary, keeps to cover of low bushes. **Voice:** usually silent on migration; has an abrupt 'prtt' and 'tzeeip'. **Similar:** Australian Pipit, lighter breast streakings, plain or only faintly streaked, larger; Red-throated Pipit, heavily streaked underparts but more prominent facial (malar) streak, deeper buff in double wing-bar. **Hab.:** surrounds of swamps, low scrub, forest margins and regrowth. **Status:** vagrant; breeds in N Asia; migrates as far S as Indonesia, including W Timor. **Br.:** nests NE Asia, May–Jul.

Boigu, Saibai and Dauan Islands are just 5–10 km S of the PNG coast. Boigu and Saibai are flat and have extensive mangroves; Dauan is mostly a pyramid peak of boulders. All require access permission. Christmas Island, about 1400 km NW from the Australian mainland, and less than 350 km S of Java, is a 360 m high, cliff-edged, rainforested plateau. The Cocos-Keeling Islands, a further 900 km SW, are low, flat coral cays. East of the NSW coast are two popular tourist islands, Lord Howe and Norfolk Islands, both with birdlife links to NZ. Heard, 53° S, and Macquarie, about 54° S, Islands are the Australian territories nearest to Antarctica.

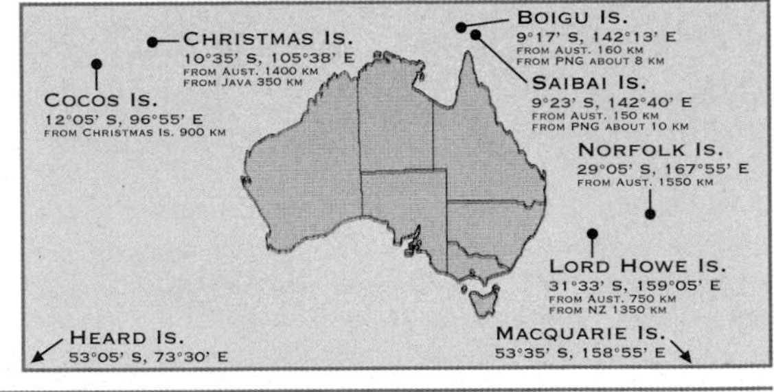

Grey-headed Goshawk

A distinctive, grey and white, medium sized goshawk, rather lightly built; conspicuously deep orange cere, eye-ring, legs and feet. The female is larger than the male. In flight, wing quills and tail are darker grey; underwing coverts are white. Immatures have dark longitudinal streaks, and under tail and underside of primaries are barred; darker upper parts.

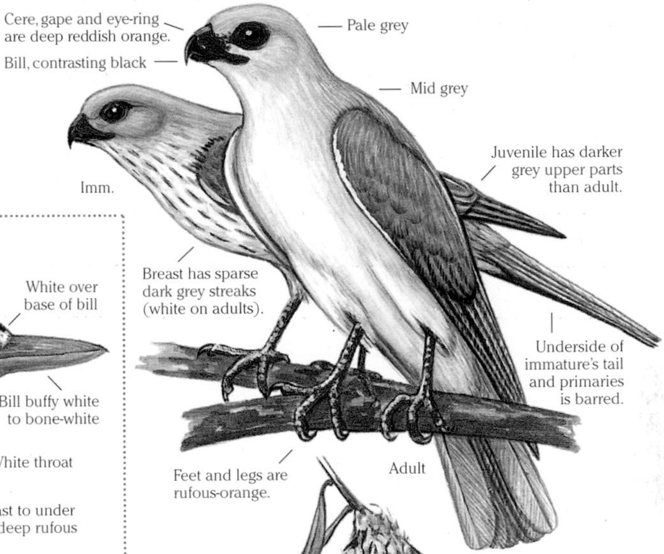

Cere, gape and eye-ring are deep reddish orange.
Bill, contrasting black
Pale grey
Mid grey
Juvenile has darker grey upper parts than adult.
Imm.
Breast has sparse dark grey streaks (white on adults).
Underside of immature's tail and primaries is barred.
Feet and legs are rufous-orange.
Adult

Rufous-bellied Kookaburra

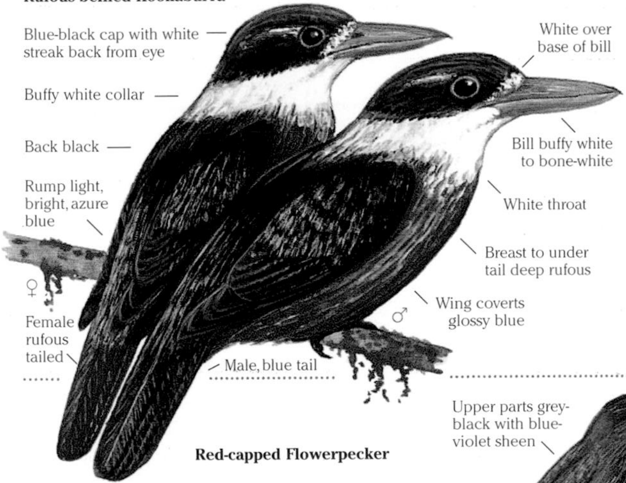

Blue-black cap with white streak back from eye
Buffy white collar
Back black
Rump light, bright, azure blue
♀ Female rufous tailed
Male, blue tail
White over base of bill
Bill buffy white to bone-white
White throat
Breast to under tail deep rufous
♂ Wing coverts glossy blue

Red-capped Flowerpecker

This tiny flowerpecker is similar to Australia's Mistletoebird (of which there is a doubtful record for the Gulf of Papua). The Red-capped Flowerpecker is sometimes regarded as a subspecies of the Papuan Flowerpecker, *Dicaeum pectorale*, which occurs on westernmost Irian Jaya, has no red on cap or rump, and has a white chin.

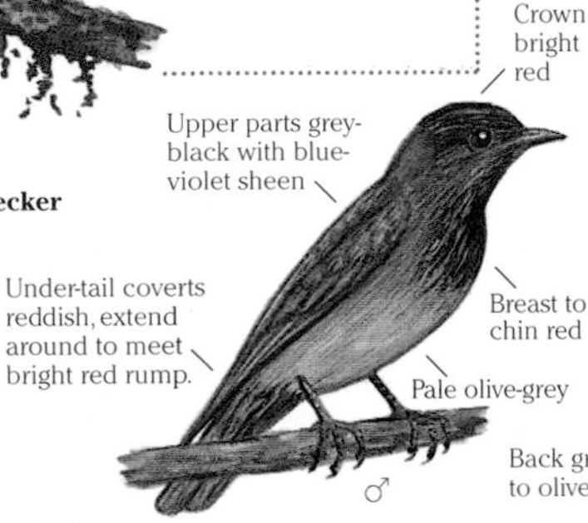

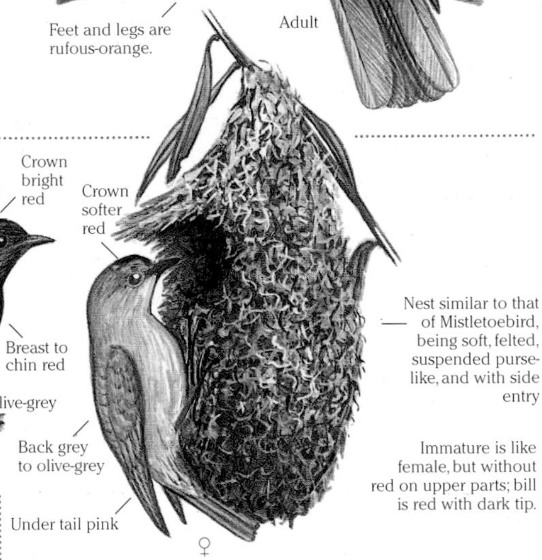

Crown bright red
Upper parts grey-black with blue-violet sheen
Under-tail coverts reddish, extend around to meet bright red rump.
Breast to chin red
Pale olive-grey
♂
Crown softer red
Back grey to olive-grey
Under tail pink
♀
Nest similar to that of Mistletoebird, being soft, felted, suspended purse-like, and with side entry
Immature is like female, but without red on upper parts; bill is red with dark tip.

Singing Starling

Metallic Starling is similar, but has longer, tapering tail, and, on the black plumage, a more glossy, metallic green sheen; has violet on crown and nape.

Tail is relatively short and has a square tip.
Eye bulging, bright red
Bill slightly heavier than that of Metallic Starling
Plumage overall black with dull oily green sheen
Immatures have upper parts grey-brown; underparts buffy white with fine dark streaks.
Adult

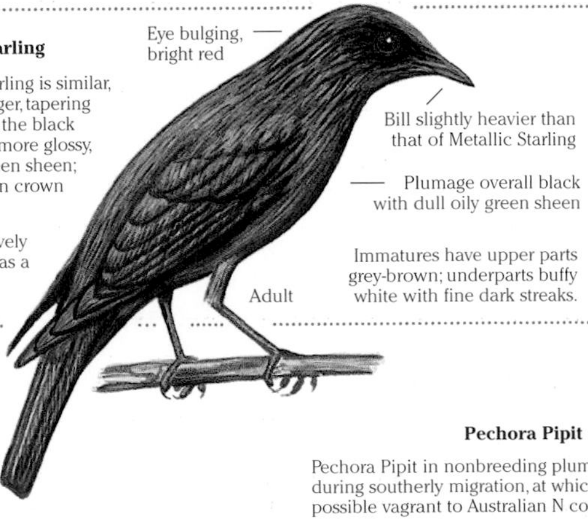

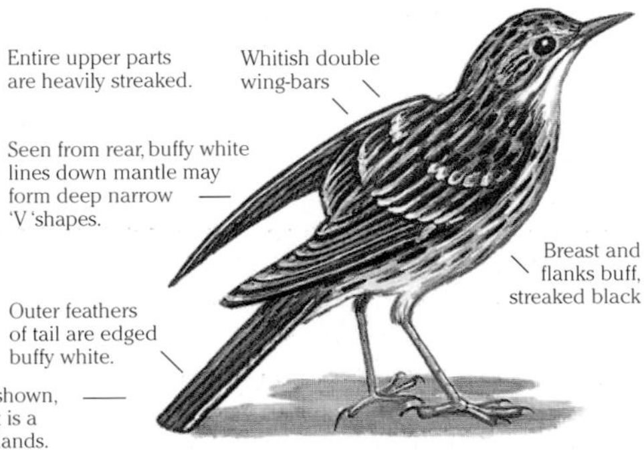

Pechora Pipit

Entire upper parts are heavily streaked.
Whitish double wing-bars
Seen from rear, buffy white lines down mantle may form deep narrow 'V' shapes.
Breast and flanks buff, streaked black
Outer feathers of tail are edged buffy white.

Pechora Pipit in nonbreeding plumage, as shown, during southerly migration, at which time it is a possible vagrant to Australian N coast or islands.

'Golden Bosunbird' *Phaethon lepturus* about 90 cm including tail streamers of 40–45 cm (C)
Golden-tinted morph of White-tailed Tropicbird; sometimes listed as a race, *fulvus*, of the White-tailed. The golden morph is rare except on this island and the surrounding tropical seas. The white morph is uncommon on Christmas Is., predominant elsewhere. Is conspicuous around coast and settlement, soaring and fluttering above and along the limestone cliffs, long streamers trailing. Seen from beneath, against a blue sky, or from the clifftops looking down, against the deep blue and turquoise of the ocean, their sunlit, golden tinted plumage makes one of the most spectacularly beautiful bird sights on the island. Along the limestone cliffs it intermingles with the slightly larger 'Silver Bosunbird' (local name for Red-tailed Tropicbird). **Voice:** a harsh, croaking, grating 'grrr-uk' in display flights. **Hab.:** open ocean for plunge-dive fishing; cliffs and forest canopy of island for rest and shelter, breeding display, nesting. **Status:** abundant within the confined island habitat; total population relatively small, 5000–10 000 pairs. **Br.:** all months; nests in holes and crevices of the coastal limestone cliffs, and in hollows of large plateau-top trees. Lays just one egg per brood.

Abbott's Booby *Papasula abbotti* 78–80 cm (C)
Unique to Christmas Is. and its surrounding seas; little is known of its movements away from the island. A large, slender, very long winged booby; dives for fish, plummeting bill-first into the ocean; includes squid and flying fish. This booby's flight is distinctive: slow, relaxed wing-flaps and glides, a more leisurely action than other smaller, shorter winged boobies. Its long necked body and the great reach of its narrow wings give a distinctive, 'flying cross' shape. **Voice:** has a deep bellow; gives brief male–female duets. Young have monotonous, quavering, begging calls. **Hab.:** Christmas Is. plateau and high limbs of rainforest canopy for resting and nesting; open ocean for fishing. **Status:** endemic species, threatened, endangered; only 2500–3500 pairs. Breeding habitat restricted to island's high plateau between 160 and 260 m above sea level. Parts of this habitat were destroyed by guano strip mining, but are now being replanted and included in national park. Those birds forced by mining to move to less preferred forest have lower breeding success. Nests on trees near cleared areas are more likely to be exposed or damaged by wind. Any young booby that launches itself on its first flight from a treetop nest or nearby limb must get a clear glide across the trees and out over the ocean. If forced down by wind or other misadventure to fall through the canopy to the ground, it is likely to perish—the wings are superbly designed for open ocean, but are useless among forest trunks, limbs and undergrowth. **Br.:** Apr.–Nov. Builds a large nest of sticks and leaves high up on a horizontal fork of a large rainforest tree; clutch, just a single egg.

Javan Pond-Heron *Ardeola speciosa* 44–46 cm (C)
Christmas Is. Within usual range (Indonesia and SE Asia) is common; seen alone or in small flocks. On wetlands, often stands motionless, neck folded, bill ready to spear down. May be with other herons or egrets. In evening, small groups fly to communal roosts; wingbeats shallow, rapid. **Voice:** when disturbed, harsh or creaky 'craar' or 'craark'. **Hab.:** swamps, ponds, lakes, paddy fields, mangroves, coastal mudflats. **Status:** possible rare vagrant; presence as yet unconfirmed.

Black-crowned Night-Heron *Nycticorax nycticorax* 55–60 cm (K)
Widely distributed species almost everywhere except Australia, where only recorded on Cocos-Keeling Is. and Ashmore Reef, WA; an uncommon nonbreeding vagrant to Indonesia. Mostly nocturnal, in small groups; roosts by day in dense foliage; flies out at dusk to feed; takes off with clatter of wings. In flight, a compact, stocky shape; the trailing legs are quite short. **Voice:** in flight, harsh 'kowark'; in alarm, a deep 'kwok'. **Hab.:** within usual range (closest part being Timor) hunts in shallow, fresh water—mangroves, swamps, paddy fields. Prefers swamps with some plant shelter, but open rather than very dense. **Status:** Cocos-Keeling, vagrant, several records; Ashmore Reef, NW from Kimberley coast of WA, 1 record.

Purple Heron *Ardea purpurea* 75–90 cm (C)
Christmas Island. Within usual range, often solitary; in shallow water, stalks slowly or stands hunched or tall; in flight has slow, deep wingbeat. **Voice:** a harsh, rather nasal croak, usually given in flight. **Similar:** in Java, Sumatra and other parts of usual range, the Grey Heron. **Hab.:** more often in freshwater wetlands of swamps, streams; less often brackish waters, mudflats, mangroves. **Status:** widespread in Eurasia, Africa, Asia, SE Asia to Indonesia; probable vagrant to Christmas Is.

Malayan Night-Heron *Gorsachius melanolophus* 48–50 cm. (C)
Christmas Island. Occasionally sighted, apparently mostly immatures. Wary and elusive nature reduces chances of being seen; the presence of this inconspicuous night-heron is unlikely to be noticed except by chance; its actual frequency of occurrence may be significantly higher than on record. **Hab.:** usually forages along streams and swamps in forests; Christmas Is. habitat occurs on terraces around coast where there are spring-fed pools beneath rainforest and freshwater mangroves. Has been seen on shore terraces below Ross Hill Springs and forests of island. **Voice:** most distinctive is the dawn and dusk calling from high in the forest canopy—a series of 10–15 deep 'hoo, hoo, hoo–' notes at intervals of 1–2 sec. Also a 'quorr-quorr-' and a rasping 'arrr-arrr-arrr'; in flight, a rasping croak. **Status:** vagrant.

Christmas Frigatebird *Fregata andrewsi* (C) Breeds only on Christmas Island. Placed with Great and Lesser Frigatebirds (p. 68) for comparison, as any of these may be seen together both around Christmas Is. and on Aust. seas.

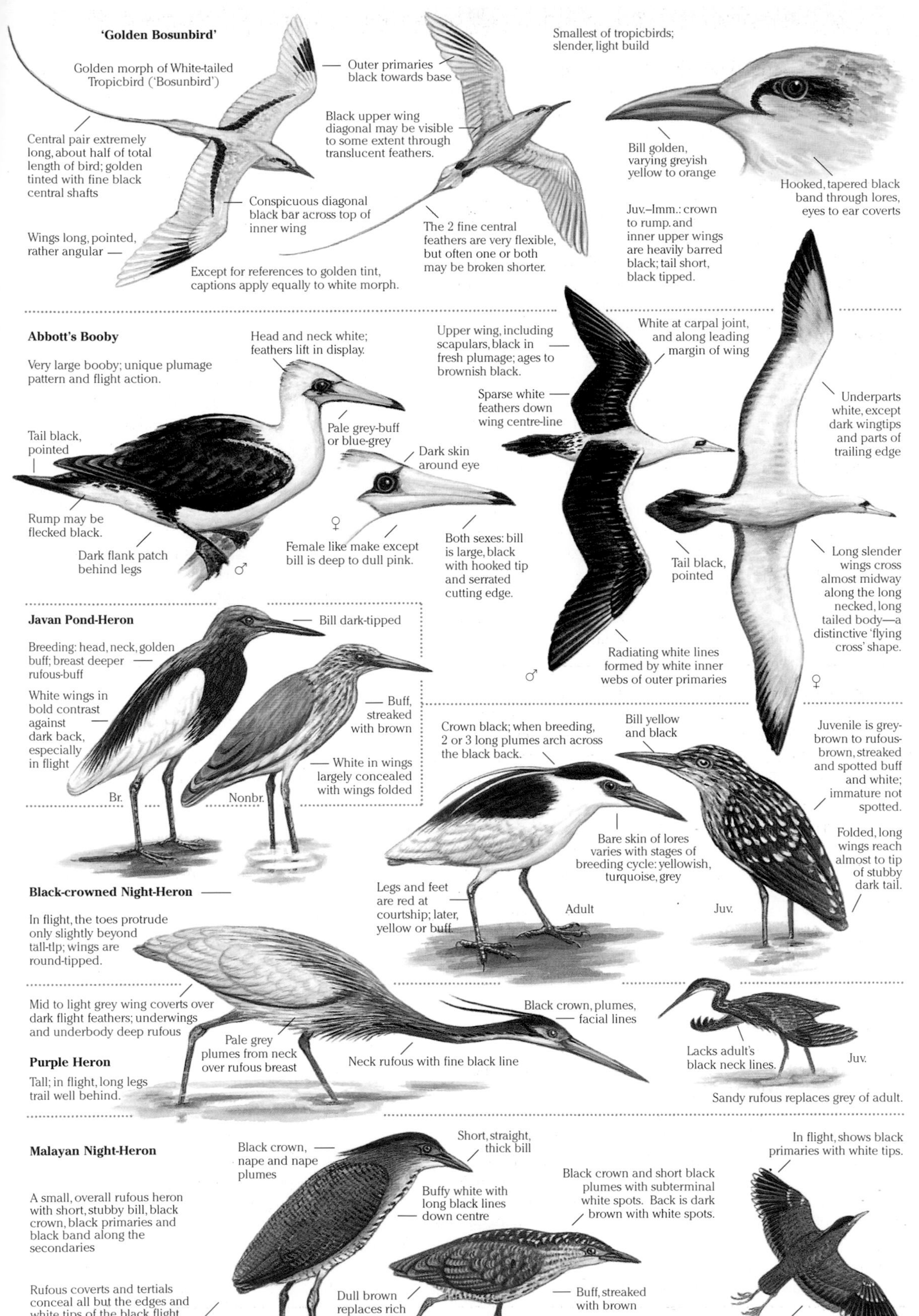
'Golden Bosunbird'
Golden morph of White-tailed Tropicbird ('Bosunbird')
Outer primaries black towards base
Smallest of tropicbirds; slender, light build
Central pair extremely long, about half of total length of bird; golden tinted with fine black central shafts
Black upper wing diagonal may be visible to some extent through translucent feathers.
Bill golden, varying greyish yellow to orange
Hooked, tapered black band through lores, eyes to ear coverts
Conspicuous diagonal black bar across top of inner wing
Wings long, pointed, rather angular
The 2 fine central feathers are very flexible, but often one or both may be broken shorter.
Juv.–Imm.: crown to rump, and inner upper wings are heavily barred black; tail short, black tipped.
Except for references to golden tint, captions apply equally to white morph.
Abbott's Booby
Very large booby; unique plumage pattern and flight action.
Head and neck white; feathers lift in display.
Upper wing, including scapulars, black in fresh plumage; ages to brownish black.
White at carpal joint, and along leading margin of wing
Sparse white feathers down wing centre-line
Underparts white, except dark wingtips and parts of trailing edge
Tail black, pointed
Pale grey-buff or blue-grey
Dark skin around eye
Rump may be flecked black.
Dark flank patch behind legs
♂
♀
Female like make except bill is deep to dull pink.
Both sexes: bill is large, black with hooked tip and serrated cutting edge.
Tail black, pointed
Long slender wings cross almost midway along the long necked, long tailed body—a distinctive 'flying cross' shape.
Radiating white lines formed by white inner webs of outer primaries
♂
♀
Javan Pond-Heron
Bill dark-tipped
Breeding: head, neck, golden buff; breast deeper rufous-buff
White wings in bold contrast against dark back, especially in flight
Buff, streaked with brown
White in wings largely concealed with wings folded
Br.
Nonbr.
Crown black; when breeding, 2 or 3 long plumes arch across the black back.
Bill yellow and black
Juvenile is grey-brown to rufous-brown, streaked and spotted buff and white; immature not spotted.
Bare skin of lores varies with stages of breeding cycle: yellowish, turquoise, grey
Folded, long wings reach almost to tip of stubby dark tail.
Black-crowned Night-Heron
Legs and feet are red at courtship; later, yellow or buff.
Adult
Juv.
In flight, the toes protrude only slightly beyond tail-tip; wings are round-tipped.
Mid to light grey wing coverts over dark flight feathers; underwings and underbody deep rufous
Black crown, plumes, facial lines
Pale grey plumes from neck over rufous breast
Neck rufous with fine black line
Lacks adult's black neck lines.
Juv.
Purple Heron
Tall; in flight, long legs trail well behind.
Sandy rufous replaces grey of adult.
Malayan Night-Heron
Black crown, nape and nape plumes
Short, straight, thick bill
In flight, shows black primaries with white tips.
Black crown and short black plumes with subterminal white spots. Back is dark brown with white spots.
A small, overall rufous heron with short, stubby bill, black crown, black primaries and black band along the secondaries
Buffy white with long black lines down centre
Rufous coverts and tertials conceal all but the edges and white tips of the black flight feathers. These in turn almost reach the tip of the stubby tail.
Dull brown replaces rich chestnut of adult.
Buff, streaked with brown
Adult
Imm.
Broad white bar along underwing

'Christmas Island Goshawk' *Accipiter fasciatus natalis* 35–45 cm (C)
Christmas Island only. Often known by local common name. Much smaller than Aust. Brown Goshawk, distinctive shape, colour. Unusually tame; allows close approach; often seen on roads and tracks. **Voice:** hard, rapid 'kyek-kek-kek-kek-'. **Similar:** Nankeen Kestrel; Japanese Sparrowhawk (possible rare vagrant). **Hab.:** forests. **Status:** moderately common.

White-breasted Waterhen *Amaurornis phoenicurus* 29–31 cm (C)
Widespread: Asia, S through Indonesia to Christmas Is. Often skulks near dense vegetation; not very wary—quite often emerges from cover. **Voice:** noisy, may call at night with nasal squawking, chuckling, screaming; squabbling noises if several together. **Hab.:** tropical wetlands, swamps, tall grass, forest margins. **Status:** former vagrant, probably established.

Ruddy-breasted Crake *Porzana fusca* 20–22 cm (C)
Christmas Island. Forages around outer margins of dense concealment; active in half-light of dawn and dusk. **Voice:** a regular, soft 'kiewk', often followed by a descending, squeaky, bubbling burst of sound; also a knocking 'kok, kok, kok-kokkok'. **Hab.:** swampy sites—mangroves, forest pools, wet grasslands. **Similar:** Red-legged Crake. **Status:** vagrant from Indonesia; elusive, even where, as in Java, it is locally common; may be more frequent than sightings suggest.

Watercock *Gallicrex cinerea* 38–42 cm (C)
Christmas Island. Usually shy, wary; a skulking inhabitant of dense, wet places. Largely nocturnal, but may be seen at dusk or dawn, or in cloudy, dull weather. Solitary, pairs or small groups. **Voice:** silent except in summer breeding range, then deep booming. **Hab.:** reedy swamps and close surrounding thickets. **Status:** vagrant; usually winters S to Java, Bali.

Greater Flamingo *Phoenicopterus ruber* 1.5–2 m (K)
Cocos-Keeling Island is site of sole record for Australasia, Apr.–Jun. 1988. **Status:** rare vagrant; nearest usually E India.

Christmas Island Imperial-Pigeon *Ducula whartoni* 38–40 cm (C)
Endemic species, similar to Pied Imperial-Pigeon, but softer tones, purplish grey upper parts blending to dusky grey-brown underparts. Flocks often descend to feed in Japanese Cherry bushes used to replant phosphate mine sites. **Voice:** strong, drawn out 'coo' or 'crew', a characteristic sound of the island's rainforests. **Hab.:** canopy of tall primary rainforest on island's plateau and regrowth on old mine sites. **Similar:** Emerald Dove, *Chalcophaps indica natalis*, a subspecies unique to this island (the only other pigeon on the island) but much smaller with softly iridescent green wings and back, pink bill. **Status:** common. **Br.:** wet season, Oct.–Apr.; a rough stick nest in tree or shrub; 2–3 white eggs.

Christmas Island Hawk-Owl *Ninox natalis* 26–29 cm (C)
Only owl on island; previously a race, *natalis*, of the Moluccan Hawk-Owl, *Ninox squamipila*. Recent DNA studies gave the Christmas Island Hawk-Owl full species status—an endemic species for this island. These owls often perch near street lights taking large insects from the ground, occasionally in flight. Elsewhere they may be tracked down by their carrying calls. **Voice:** loud, double hooted 'boo-book', first note higher; also a soft 'porr-porr-'. **Similar:** no similar owl recorded on Christmas Is. **Hab.:** dense closed canopy rainforest on island heights, tall coastal deciduous forest. **Status:** small population confined to Christmas Is. **Br.:** nests in a tree hollow, often in rainforest; eggs white, rounded.

Savanna Nightjar *Caprimulgus affinis* 23–25 cm (C, K?)
Christmas Is., possibly Cocos-Keeling. A tiny nightjar, most active dusk and dawn; hawks insects 10–40 m up among trees, over canopy. **Voice:** loud, vibrant 'tzweeip'. **Hab.:** grassland, savanna-grassland with scattered trees, forest clearings, margins of forests, mangroves; open surrounds of airstrip are a favoured site. **Status:** probable resident.

Christmas Island Glossy Swiftlet *Collocalia esculenta natalis* 9.5–10 cm (C)
Subspecies *natalis* of widespread Glossy Swiftlet; unique to this island. Many other races through SE Asia, Indonesia, NG. This swiftlet is tiny; most easily seen in flight, hawking for insects, threading on fluttering wings through rainforest, low into open spaces, close above the forest canopy. **Voice:** faint twittering. **Hab.:** forests, open areas nearby. **Status:** common. **Br.:** wet season, in semi-darkness of island's caves; small half-bowl nests on rock ceiling of cave.

Edible-nest Swiftlet *Collocalia fuciphaga* **Black-nest Swiftlet** *Collocalia maxima* Both 12 cm (C)
Edible-nest, Black-nest Swiftlets almost identical except in hand or at nest; both fly strongly, high, both possible vagrants.

Asian House Martin *Delichon dasypus* 12–13 cm (C)
Christmas Is. Sightings reported since 1996. Feeds high in slow, agile, fluttery flight; swoops, glides, often with other high-feeding martins or swiftlets. **Voice:** twitter. **Hab.:** above gorges, coastal cliffs, buildings. **Status:** unconfirmed vagrant.

Christmas Island Goshawk: a distinctive, smaller, endemic race of the Australian Brown Goshawk
Upper parts dark grey rather than the brownish grey of the mainland Brown Goshawk
Chin rufous-buff (male white); throat rufous mottled grey
Rufous bars of belly become broader upwards over lower breast until solid rufous on upper breast.
Juvenile has upper parts scalloped rufous.
♀
Underwing barred brown, heavier on coverts (darker in cast shadow)
♀
This race has more rounded wingtips.
White-breasted Waterhen
Dark grey-green
Throat and breast white; face and forehead white or mottled with black
Rufous-buff
Under tail and belly barred
White
Dull chestnut
Ruddy-breasted Crake: similar, but larger, are the Red-legged and Red-necked Crakes, Indonesian species that could reach Christmas Is.
Watercock: very large, distinctive rail
Darker crown
♂ Nonbreeding, and ♀
Frontal shield red (breeding)
Brown, each feather buff edged
♂ Breeding
Buff with brown barring
Black with brown edges to feathers of back, wings, belly
Underwing similar to breast; usually appears darker in shadow against bright sky.
Red-brown
Forehead paler grey
White spot
Dark purplish grey
Upper parts have greenish iridescence.
Dark grey with variable iridescence of green, purple or brown
Christmas Island Imperial-Pigeon
Adult
Red-brown
Juv.: eye brown, legs grey, plumage dull brownish
Christmas Island Hawk-Owl
White X between eyes, onto brows
Entirely barred rufous
Adult: Juv. is downy, paler, with indistinct barring.
Slightly smaller than Australia's Southern Boobook; but barred and brighter rufous
Savanna Nightjar
Upper parts rather uniform mottled greys
White patch in wing
Outer 2 pairs mostly white
♂
Small whitish neck patches, rather than part-collar
Folded wings almost reach tail-tip.
♂
Female: no white in tail; wing spot buff
Outer 2 white
Similar: most likely on this island are Large-tailed and Grey Nightjars, both larger and have less than half length of outer tail feathers white.
Greater Flamingo
(Shown about half size compared with others on this page)
Distinctive down-angled, black tipped bill; filters water and mud to obtain plankton.
In flight, both neck and legs are outstretched so that the feet trail well behind the short tail. Within its usual range, usually in flocks, very rarely alone.
Almost entirely pink, becoming crimson on upper wing coverts and black on flight feathers.
Wades in shallows, head upside-down to scoop water through the bill.
Christmas Island Glossy Swiftlet
Glossy black with green sheen
Square to shallow fork
White
Slender back-curved wings, longer than body
Grey throat
Edible-nest, Black-nest Swiftlets
In usual range, in flocks. Both nest in caves, making white or black edible cup nests on rock walls.
Edible: medium to shallow notch
Black-nest: shallow to square tip
Variable rump, dark to pale grey
Adult
Underparts entirely mid to dark grey-brown, both species
Black-nest Swiftlet has slightly larger, heavier build; in hand or at nest, the heavy feathering of lower leg (tarsus) can be seen—leg of Edible-nest Swiftlet is bare or lightly-feathered.
Asian House-Martin
Upper parts black with steely blue sheen
Underparts dusky white, becoming pale grey across upper breast
White rump
Shallow fork

CHRISTMAS ISLAND (C)

Christmas Island White-eye *Zosterops natalis* 11–12 cm
This large white-eye is unique to Christmas Island; not conspicuously different from Aust. Silvereye. Common and obvious, often in flocks, sometimes in very large numbers. **Voice:** a high, squeaky song, more often a sharp squeak. **Hab.:** natural forests, settlement gardens. **Status:** endemic; abundant. **Br.:** Nov.–Feb., small open cup, 2–3 blue-green eggs.

Island Thrush *Turdus poliocephalus erythropleurus* 20 cm
Widespread, SE Asia and Pacific islands; this Christmas Island population is an endemic race, *erythropleurus.* Common, conspicuous, approachable; has no fear of humans; often on ground in clearings and thickets of forest. **Voice:** clear notes, at first slowly alternating between high and low, then faster, stronger. Alarm call loud, rattling. **Hab.:** forest and settlement. **Status:** endemic; common. **Br.:** nests quite low, often noticed in tree fork or building; neat cup, 2–3 eggs (p. 419).

Java Sparrow *Padda oryzivora* 16 cm
Often large flocks; forages in grass, shrubbery, on ground. **Voice:** soft 'tchup'; sharper chirps, trill ending with long whistles. **Hab.:** on Christmas Island, around settlement, gardens and roadsides. **Status:** introduced from Java; locally common.

Purple-backed Starling *Sturnus sturninus* 19 cm
Forages on ground in open areas. **Voice:** harsh squawks and whistles typical of starlings. **Hab.:** coastal areas, forest edges and clearings, regrowth, gardens. **Status:** rare vagrant; breeds central–eastern Asia, winters SE Asia including Java.

Brown Shrike *Lanius cristatus* 20–22 cm
Hunts from low perches; drops to ground to take small reptiles, insects. **Voice:** often silent; harsh chatterings. **Hab.:** open areas, forest edges, regrowth, gardens. **Status:** uncommon visitor S to Indonesia; rare vagrant to Christmas Island.

NORFOLK ISLAND (N)

South Island Pied Oystercatcher *Haematopus finschi* 43–47 cm
The pied oystercatchers of Norfolk Island were assumed to be vagrants of the Australian species, *haematopus longirostris*, but are now found to be the South Island Pied Oystercatcher, *H. finschi.* The rare sightings of oystercatchers on Lord Howe Is. are probably also this New Zealand species. The South Island Pied Oystercatcher is a known, regular, long-distance migrant, and New Zealand is closer to Lord Howe. On the other hand, the Australian species typically only undertakes short coastal travels. **Voice:** a loud, sharp piping, given in flight and from beach. **Similar:** Australian Pied Oystercatcher. **Hab.:** beaches and mudflats, usually of estuaries, occasionally reefs, paddocks. **Status:** uncommon vagrant.

Norfolk Red-crowned Parakeet *Cyanoramphus novaezelandiae cookii* 21–26 cm
Likely to be elevated to species status. In flocks or pairs, forages at all levels from canopy to ground level; takes seeds, nectar, invertebrates. **Voice:** a rattling 'ki-ki-ki-ki', 'chik-chik-chik' and slower, harsher chatter. **Hab.:** on Norfolk Island, in tall rainforest remnants; also in eucalypt plantations and gardens. **Status:** endemic to Norfolk Is.; endangered; Lord Howe Island race *subflavescens* is extinct. **Br.:** earth depression in tunnel or hollow tree or stump. Eggs 4–9, rounded, white.

Norfolk Island Gerygone *Gerygone modesta* 10–11 cm
Rather dusky toned gerygone. **Voice:** not as strong, clear as other gerygones; final notes rather slurred. **Hab.:** most areas of forest, pasture with trees, gardens. **Status:** common. **Br.:** Oct.–Dec. Domed nest made of fine dry grass, rootlets, moss, a side entrance near the top, suspended among foliage. The 3 eggs are white with brown and reddish spots.

Slender-billed White-eye *Zosterops tenuirostris* 13.5–14.0 cm
Usually in flocks; often foraging on branches and trunks, probing bark crevices. **Hab.:** rainforest and tall regrowth; tends to avoid Silvereye habitat of thickets and garden shrubbery. **Status:** common. (The Robust White-eye, formerly on Lord Howe Is., is now extinct.) **Br.:** Oct.–Dec. Small cup of grass and webs suspended at rim among foliage; 2–4 eggs.

White-chested White-eye *Zosterops albogularis* 14–14.5 cm
Largest white-eye. Olive upper parts, distinctive russet flanks, whitish throat. **Hab.:** remnant forest. **Status:** endangered; confined to Mt Pitt Reserve and adjoining vegetation. **Br.:** Oct.–Dec. Cup-shaped nest suspended by rim.

Norfolk Island Whistler *Pachycephala pectoralis xanthoprocta* 16,5–17 cm Not illustrated.
Unlike Aust. Golden Whistler, males closely resemble females. Neither has the bold black breast-band, both have grey heads; male distinguished by slightly deeper yellow underparts. Very 'tame' in human presence. **Status:** endangered.

Christmas Island White-eye
Widespread and common throughout forests and gardens of the island. Usually in flocks, busy, conspicuous, quite noisy. Large, with long tail and rounded wingtips.
White eye-ring
Dull olive green
Black lores extending back under eye
Dull white with pale grey flanks
Yellow tint
Island Thrush
('Christmas Island Thrush')
Conspicuous eye-ring
Bill yellow-orange
Upper parts mid to dark grey-brown
Pale Grey
Cinnamon-buff
Legs and feet large, deep yellow
Java Sparrow
Head and neck black
Bill deep pink to dusky crimson
Conspicuous oval white cheek patch
Pearl grey
Grey-brown
Dull blue-grey
Black
Dusky pink to salmon-pink
Adult
Imm.
Brownish tint
Brown Shrike
Thick black eye-stripe
Underparts and face white; flanks barred
Bill stout
Upper parts rufous-brown
Immature: colour dull; underparts more extensively barred
Adult
Superciliosus, most common race in Java, nearest part of usual range of this species
Purple-backed Starling
Back iridescent purple
Black nape spot
Wings iridescent dark green
Rump buffy white
Iridescent green
White wing-bars
Female: ashy grey above; nape spot brown, wings and tail black
♂
Immature: brownish above; underparts mottled brown
South Island Pied Oystercatcher
White of wing stripe obvious along bottom edge of folded wing (barely visible on Pied Oystercatcher)
Wings not quite to tail tip
Very long, straight red bill; proportionately longer than that of Australian Pied Oystercatcher
Similar to Australia's Pied (p. 130)
Legs short, stout. South Island Pied has legs 1/3 shorter than those of Australia's Pied, giving comparatively low, short legged, dumpy look.
Immature has brownish black plumage, dull red bill and legs.
South Island Pied has underwing almost entirely white.
Black upper parts, except white of wing-bars, upper tail coverts, rump and lower back.
White bar extends to inner secondaries, reaches trailing edge, unlike the narrow tapering bar of Australia's Pied.
White far up back, in 'V' shape between wings
Norfolk Red-crowned Parrakeet
Red forehead and crown (forehead only on immature)
Red through lores to ear coverts (on female and immature, thinner, or reduced to a small ear spot)
Light, bright yellow-green
Red patch each side of rump
Upper surface of wing bright green, except blue outer primaries and their coverts
Slender-billed White-eye
Closely resembles the well-known Australian Silvereye, but rather olive-brown above and yellowish beneath.
Black of lores extends back to encircle the white-ringed eyes.
Upper parts olive-green, slightly more brownish on lower back and rump
Longer, more slender than bill of Silvereye
Throat to belly yellow
Under-tail coverts brighter yellow
White-chested White-eye
Lores black; also under, and behind, the white eye-ring
Throat to belly white
Brownish green
Flanks dull rufous-brown
Coverts yellow-green
Norfolk Island Gerygone
Broken white orbital ring
Dusky grey-green
Rump and upper tail olive-brown
Grey face, darker lores
Flanks pale grey, belly dull white
Broad dark mid-tail band
Dull white band near tail-tip across all except several central feathers
Juvenile: upper parts darker olive-brown; pale underparts; flanks and underwing coverts have light olive-yellow tint.

LORD HOWE ISLAND (L)

Canada Goose *Branta canadensis* Smallest race about 55 cm; largest about 100 cm
Many races occur in N America; vary greatly in size; body and wing colour grey to grey-brown and light to dark in tone. Introduced to NZ, probably race *maxima*; well established on S Island, scattered records across N Island; vagrant to Lord Howe Island; usually in flocks. **Voice:** loud honking before and during flight. **Hab.:** in NZ, on pasture, crops, shallows of wetlands. **Status:** rare vagrant to Lord Howe, common in parts of NZ. **Br.:** grass bowl on ground, in grass.

Paradise Shelduck *Tadorna variegata* 63–65 cm
Usually in pairs or groups; the females' colour combination of bright chestnut body and white head is distinctive. In distance on water, the stout build and rather short neck are evident. **Voice:** male has a deep, honking 'zzonk', female a higher 'zeeik', often as a duet while in flight. **Hab.:** in NZ, hilly farmland, dams, tussock grasslands. **Status:** endemic to NZ. Vagrant to Lord Howe Island—a small flock arrived in 1950; 4 were caught, fed, ringed and released.

Lord Howe Woodhen *Gallirallus sylvestris* 32–42 cm
A moderately large, heavily built, flightless rail; body almost entirely dull brownish olive, wings rufous-brown, barred. May be alone, in pairs or small groups; rather wary but not extremely timid; curious and will approach to investigate unusual sounds. Runs if alarmed; jumps with wings flapping, almost flying for several metres. Forages in wet leaf litter, mossy rotten logs, fallen palm fronds on forest floor; always requires water nearby. **Voice:** territorial call of loud piercing notes; also low grunting contact calls. **Hab.:** moss forests of highest summit are this species's stronghold. Also in lowland vegetation of palms, figs, margins of forests adjoining pasture and gardens; only rarely in rainforest. **Status:** endemic species, now rare; reduced by loss of habitat, attacks of introduced rats and owls. **Br.:** nest is a shallow depression in the ground lined with grass and leaves. The 4 eggs are white, blotched and finely spotted grey and chestnut.

Lord Howe White-eye *Zosterops lateralis tephropleurus* 12–13 cm
At times regarded as a full species, more recently a subspecies of the Silvereye of mainland; differs in being slightly larger, proportionately longer bill and tail, and rounded wingtip. Forages in foliage for insects; takes nectar at flowers; exploits cultivated fruits. **Voice:** high pitched, squeaky peeps and twitters similar to those of other white-eyes. **Hab.:** throughout island in most types of vegetation, including gardens. **Status:** endemic species or race; abundant. (A second Lord Howe *Zosterops* species, the Robust White-eye, *Z. tenuirostris strenuus*, is now extinct; its nominate race *Z. tenuirostris tenuirostris* survives on Norfolk Island.) **Br.:** small cup of fine fibre bound with cobwebs and suspended by rim, usually in lower foliage of shrubbery. The 3 eggs are uniformly pale blue-green.

Common Chaffinch *Fringilla coelebs* 11–12 cm
European species introduced to NZ, and vagrant or introduced to many islands of region. Often in flocks, mostly feeding on ground; flight undulating. **Voice:** song 'tsip-tsip-tsell-sell-tsee-chie-weeoo'. **Hab.:** farms, gardens. **Status:** common.

HEARD (H) AND MACQUARIE (M) ISLANDS

Emperor Penguin *Aptenodytes forsteri* 1–1.35 m (H)
Circumpolar, rarely N of 60° S. Breeds around coast and adjacent islands; ventures further afield to islands further N, including Heard Island. **Voice:** most at breeding site. Both sexes give loud, trumpeting calls in contact and courtship. **Hab.:** usually in areas of pack-ice where there is 40–60% open water between ice slabs. **Status:** stable, around 450 000 individuals; some colonies fluctuate. **Br.:** winter, in colonies on sea-ice, near glacier or iceberg.

Imperial Shag *Leucocarbo atriceps* 72–76 cm (H, M)
A large marine species found around rocky coasts of Macquarie and Heard Islands, nearby seas; vagrant to adjacent small islands. **Voice:** male noisy in colonies, approaching nest, threatening, fighting. Barks, gargles, growls; female hisses. **Hab.:** marine; keeps to shallower inshore waters; seeks mostly bottom-dwelling fish. **Status:** race (or species) on Heard Is. is *nivalis* ('Heard Island Shag'); on Macquarie, local race is *purpurascens* ('Macquarie Shag'). Both have small breeding populations; presently secure. **Br.:** small colonies on rocks and cliffs; shallow bowl of vegetation, 1–3 pale blue eggs.

Black-faced Sheathbill *Chionis minor* 38–41 cm (H)
A common shoreline bird of subantarctic islands of Indian Ocean; in small noisy flocks. Inquisitive, almost tame. In flight, strong, direct wingbeats; travels direct, pigeon-like. **Voice:** harsh 'kek-kek-'. **Hab.:** shoreline, feeding on algae, marine invertebrates; also scavenges for carrion around colonies of seabirds and elephant seals. **Status:** breeding resident.

Common Redpoll *Carduelis flammea* 11.5–12.5 cm (M, L) Not illustrated.
Introduced to NZ, then to surrounding islands. **Status:** introduced, common, breeding. Also vagrant to Lord Howe Island.

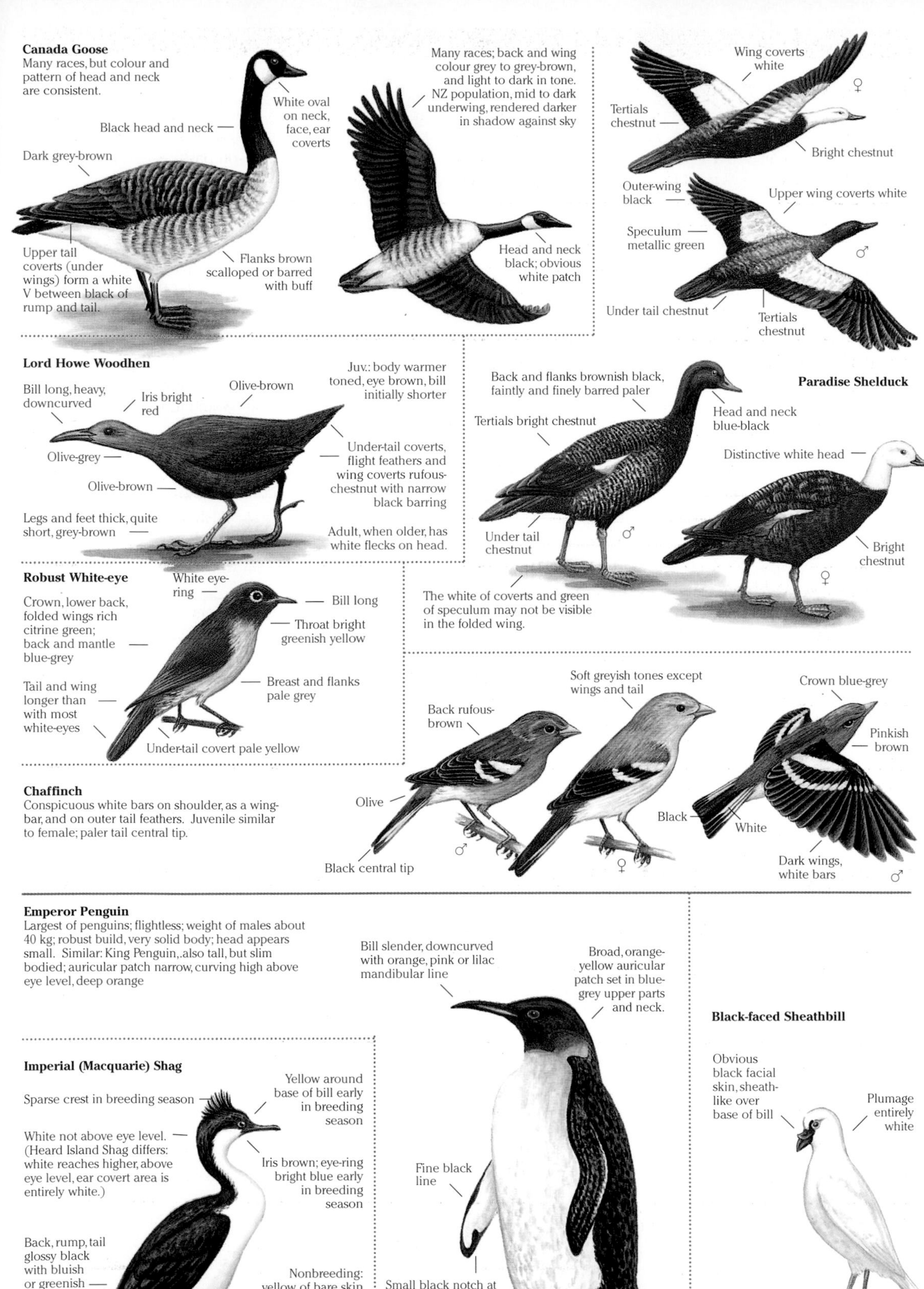
Canada Goose
Many races, but colour and pattern of head and neck are consistent.
White oval on neck, face, ear coverts
Black head and neck
Dark grey-brown
Upper tail coverts (under wings) form a white V between black of rump and tail.
Flanks brown scalloped or barred with buff
Many races; back and wing colour grey to grey-brown, and light to dark in tone. NZ population, mid to dark underwing, rendered darker in shadow against sky
Head and neck black; obvious white patch
Wing coverts white
Tertials chestnut
Bright chestnut
Outer-wing black
Upper wing coverts white
Speculum metallic green
Under tail chestnut
Tertials chestnut
Lord Howe Woodhen
Bill long, heavy, downcurved
Iris bright red
Olive-brown
Juv.: body warmer toned, eye brown, bill initially shorter
Olive-grey
Olive-brown
Under-tail coverts, flight feathers and wing coverts rufous-chestnut with narrow black barring
Legs and feet thick, quite short, grey-brown
Adult, when older, has white flecks on head.
Back and flanks brownish black, faintly and finely barred paler
Paradise Shelduck
Tertials bright chestnut
Head and neck blue-black
Distinctive white head
Under tail chestnut
Bright chestnut
The white of coverts and green of speculum may not be visible in the folded wing.
Robust White-eye
White eye-ring
Bill long
Crown, lower back, folded wings rich citrine green; back and mantle blue-grey
Throat bright greenish yellow
Breast and flanks pale grey
Tail and wing longer than with most white-eyes
Under-tail covert pale yellow
Soft greyish tones except wings and tail
Crown blue-grey
Back rufous-brown
Pinkish brown
Chaffinch
Conspicuous white bars on shoulder, as a wing-bar, and on outer tail feathers. Juvenile similar to female; paler tail central tip.
Olive
Black
White
Black central tip
Dark wings, white bars
Emperor Penguin
Largest of penguins; flightless; weight of males about 40 kg; robust build, very solid body; head appears small. Similar: King Penguin,.also tall, but slim bodied; auricular patch narrow, curving high above eye level, deep orange
Bill slender, downcurved with orange, pink or lilac mandibular line
Broad, orange-yellow auricular patch set in blue-grey upper parts and neck.
Black-faced Sheathbill
Imperial (Macquarie) Shag
Yellow around base of bill early in breeding season
Sparse crest in breeding season
White not above eye level. (Heard Island Shag differs: white reaches higher, above eye level, ear covert area is entirely white.)
Iris brown; eye-ring bright blue early in breeding season
Fine black line
Obvious black facial skin, sheath-like over base of bill
Plumage entirely white
Back, rump, tail glossy black with bluish or greenish sheen
Nonbreeding: yellow of bare skin under base of bill and blue of eye are brightest at courtship, then fade towards grey. Yellow on base of upper bill becomes dull.
Small black notch at lower edge near tip. (On King Penguin, black patch is much larger, on centre and upper edge rather than lower edge.)
Stout legs and feet, pinkish grey. Toes are not webbed.

As well as the birds listed in the nests and eggs pages are many that visit Australia regularly but breed elsewhere. Large numbers of these are migratory waders that escape the ice and snow of the far northern winter by following the sun to southern shores.

These waders make this long return journey each year, travelling a migration route staged with various feeding grounds. At these sites (tidal mudflats or beaches that terminate each long flight on the journey), the birds feed and rest—waves of birds of many species arrive, recuperate, then move on across land or ocean on the next leg.

Extensive mudflats and beaches on shallow, protected coasts of islands or continents make suitable feeding sites, and loss of just a few major sites could break the flocks' vital migration chain by creating an impossibly long flight distance between stops for rest and refuelling. Any partial loss of feeding area or loss of smaller sites to building or agricultural use, or to pollution damage from nearby activities, could reduce numbers of birds able to undertake the migration.

RAMSAR, an international convention, was set up to protect wader sites and the East Asian–Australian Shorebird Reserve Network was launched at the March 1996 RAMSAR Convention. Known as the Brisbane Initiative, it aims to identify and protect the network of wetlands that supports the waders' migratory path. This flight path extends from the Arctic coasts of Europe, Asia and Alaska south through central and southern Asia, Japan, the far western Pacific islands and south-east Asian peninsula to the island chains of Indonesia and New Guinea, northern coastal Australia, and, finally, southern Australia and New Zealand. Eight nations within this fly-way, all supporters of RAMSAR, have nominated wetlands within their territories that are important to migratory shorebirds.

The northernmost, in Russia, is the Moroshechnaya Estuary; in China lie the Shuangtaizi Estuary, Yellow River Delta and Hong Kong's Mai Po marshes. Japan nominated the Yatsu tidal flats and Yoshino Estuary while, in the SE Asian island chain, Olango Island in the Philippines and Indonesia's Wasur National Park were listed. In New Zealand the sites are Farewell Spit and the Firth of Thames, and Australia has listed nine sites. Globally, the East Asian network supports the Western Hemisphere Shorebird Reserve Network, an agreement that covers the flight paths of African–Asian Migratory Waterbirds, and the North American Waterfowl Management Plan.

Some major and many minor sites are spread around Australia's northern coast, these being vital to the shorebirds on arrival and departure. Here they feed and rest, building up their reserves for the next long flight—southward across or around the continent's huge arid interior in spring, or north across the ocean to Asian island or continental sites in autumn.

Some crucial touchdown sites in Australia are Roebuck Bay and Eighty-mile Beach in the Kimberley, and the south-eastern corner of the Gulf of Carpentaria. From these, birds move to Australia's south, spreading out over multitudes of sites; they reach the southern coasts of Western Australia, South Australia, Victoria and Tasmania. The waders arrive in Australia's north between September and October.

On arrival, some retain part or most of their colourful breeding plumage, but most are in transition to their relatively dull winter or nonbreeding plumage. They reach the most southerly sites around November, by which time most are in soft greys and browns, often without prominent markings to help identification.

The return migration begins in the autumn; the waders begin to move out of the southernmost sites in mid February. Huge numbers of them congregate in the north on the main feeding grounds such as the Eighty Mile Beach before finally departing for the breeding grounds in the far north in March and April. By then, many have begun to acquire their colourful breeding plumage.

Lake Argyle's southern and eastern shallows support about 200 000 waterbirds, including waders in passage in spring and autumn; among these are the Long-toed Stint and Wood Sandpiper. Downstream on the Ord River floodplain are sites at Lake Kununurra, Packsaddle Swamp, and, nearer Wyndham, Marlgu Billabong. Species here include Oriental Pratincole, Swinhoe's Snipe, Little Curlew, Wood Sandpiper and Pectoral Sandpiper.

Eighty Mile Beach and Roebuck Bay: Each year about 300 000 waders use the Eighty Mile Beach, with some 170 000 at Roebuck Bay, site of the Broome Bird Observatory (where the total bird list exceeds 330 species, about 50 of shorebirds). Waders pass through on migration in late September to early October and again in March to early April. At Roebuck Bay around 14 000 Bar-tailed Godwits, 13 000 Great Knots, and many thousands each of Greater Sand Plovers, Curlew Sandpipers, Red-necked Stints and Sharp-tailed Sandpipers have been recorded. Contact the observatory on 08 929333; E-mail bbo@tpgi.com.au.

Shark Bay: The extensive shallow bays, mangroves, mudflats, beaches and salt lakes (Lake McLeod) of this flat coast attract waders; offshore islands are sites of seabird colonies; total bird list around 120 species.

Forrestdale and Thompson's Lakes: On the Swan coastal plain great numbers of birds, mostly waders, sometimes gather in early to mid-summer as the water becomes shallow over wide expanses of mud. Some 20 000 have been recorded on Forrestdale Lake, among them the rare Long-toed Stint and, at the other extreme, up to 3000 Banded Stilts.

Lake Dumbleyung: This large, deep but rarely full lake is a refuge during drought for some 40 waterbird species, including 12 shorebird species.

Lake Warden: North of Esperance, this is one of a chain of lakes where waders congregate, as do big flocks of Banded Stilts.

Peel-Yalgorup is comprised of the Peel and Harvey Estuaries with Lakes Clifton and Preston, and the nearby Creery Marshes. The extensive shallows on the east and south of Peel Inlet are a major feeding site for waders (counts around 20 000); with the other birds on the lakes, the total sometimes exceeds 40 000.

Vasse and Wonnerup Estuaries near Busselton form an important waterbird and wader habitat in the south-west. Waders include the Long-tailed Stint, Wood Sandpiper, Marsh Sandpiper and Pectoral Sandpiper.

Eyre Bird Observatory: At the western end of the Great Australian Bight, the bird list exceeds 200 bushland and coastal species. Influx of migratory waders peaks in Oct.–Nov. and Mar.–May. Species include Oriental Plover, Pectoral Sandpiper, Buff-breasted Sandpiper, Baird's Sandpiper.

Albany coast: Extensive tidal mudflats in the landlocked bays and estuaries, including the south-western shores of Princess Royal Harbour, the south-west flats in Oyster Harbour, and the Lower King and Lower Kalgan Rivers, support many summer waders.

Most waders that reach Australia breed across northern and central Asia in the northern summer, but just a few are from northern Europe, and Alaska and Canada in the east. While most of those from Asia breed in the region marked with darker shading, some breed further south in Mongolia, northern India, southern China and Japan. These waders spend the winter in southern India, South-East Asia (including the Philippines and Indonesia), New Guinea, Australia's mainland, Tasmania and New Zealand.

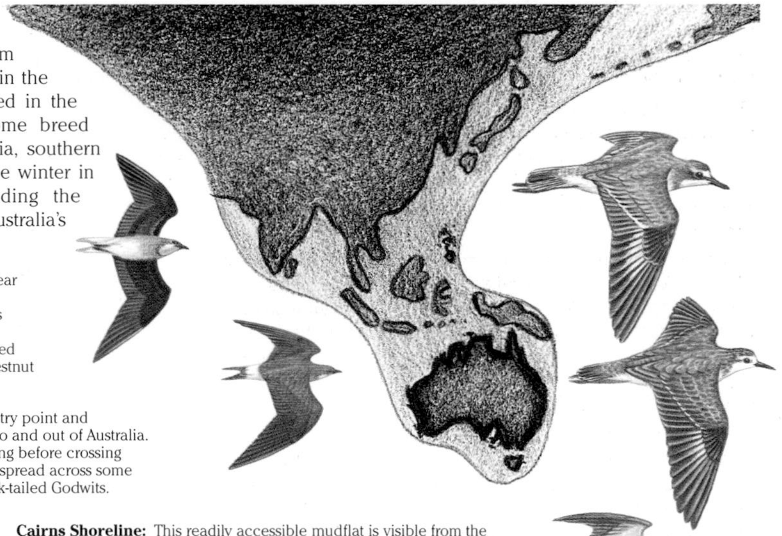

Darwin shoreline: Mudflats and beaches in and near Darwin, including Lee Point, Buffalo Creek, Leanyer Swamp and Sanderson Sewerage Ponds, have waders such as Redshank, Pacific Golden Plover, Pectoral Sandpiper, Caspian Plover, Long-toed Stint, Little Ringed Plover, Common Greenshank and Ruff, as well as Chestnut Rail and Great-billed Heron.

SE Gulf of Carpentaria: An important entry point and recuperation site for waders on passage into and out of Australia. As many as 250 000 birds, feeding and resting before crossing desert to the south or sea to the north, can spread across some 5400 km². These include some 50 000 Black-tailed Godwits.

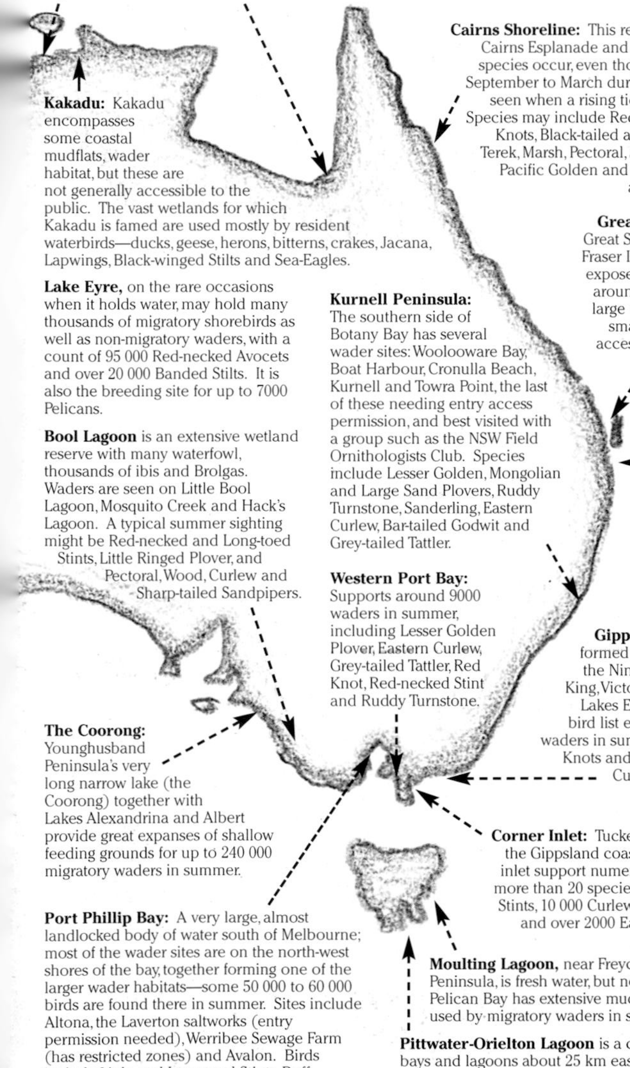

Kakadu: Kakadu encompasses some coastal mudflats, wader habitat, but these are not generally accessible to the public. The vast wetlands for which Kakadu is famed are used mostly by resident waterbirds—ducks, geese, herons, bitterns, crakes, Jacana, Lapwings, Black-winged Stilts and Sea-Eagles.

Lake Eyre, on the rare occasions when it holds water, may hold many thousands of migratory shorebirds as well as non-migratory waders, with a count of 95 000 Red-necked Avocets and over 20 000 Banded Stilts. It is also the breeding site for up to 7000 Pelicans.

Bool Lagoon is an extensive wetland reserve with many waterfowl, thousands of ibis and Brolgas. Waders are seen on Little Bool Lagoon, Mosquito Creek and Hack's Lagoon. A typical summer sighting might be Red-necked and Long-toed Stints, Little Ringed Plover, and Pectoral, Wood, Curlew and Sharp-tailed Sandpipers.

The Coorong: Younghusband Peninsula's very long narrow lake (the Coorong) together with Lakes Alexandrina and Albert provide great expanses of shallow feeding grounds for up to 240 000 migratory waders in summer.

Port Phillip Bay: A very large, almost landlocked body of water south of Melbourne; most of the wader sites are on the north-west shores of the bay, together forming one of the larger wader habitats—some 50 000 to 60 000 birds are found there in summer. Sites include Altona, the Laverton saltworks (entry permission needed), Werribee Sewage Farm (has restricted zones) and Avalon. Birds include Little and Long-toed Stints, Ruff, Greater Sand Plover, and White-rumped, Buff-breasted, Terek, Wood, Pectoral and Broad-billed Sandpipers. Enquiries to Birds Australia or BOC.

Kurnell Peninsula: The southern side of Botany Bay has several wader sites: Woolooware Bay, Boat Harbour, Cronulla Beach, Kurnell and Towra Point, the last of these needing entry access permission, and best visited with a group such as the NSW Field Ornithologists Club. Species include Lesser Golden, Mongolian and Large Sand Plovers, Ruddy Turnstone, Sanderling, Eastern Curlew, Bar-tailed Godwit and Grey-tailed Tattler.

Western Port Bay: Supports around 9000 waders in summer, including Lesser Golden Plover, Eastern Curlew, Grey-tailed Tattler, Red Knot, Red-necked Stint and Ruddy Turnstone.

Cairns Shoreline: This readily accessible mudflat is visible from the Cairns Esplanade and has diverse waders; regularly more than 26 species occur, even though total numbers (several thousand from September to March during peak use) are not huge. Best views are seen when a rising tide pushes the birds closer to the foreshore. Species may include Red-necked Stints, Greenshank, Great and Red Knots, Black-tailed and Bar-tailed Godwits, Curlew, Broad-billed, Terek, Marsh, Pectoral, Sharp-tailed and Common Sandpipers, and Pacific Golden and Grey Plovers. The mangroves at Ellie Point and around Trinity Inlet are also good sites.

Great Sandy Coast: Hervey Bay south through the Great Sandy Strait to Tin Can Bay is sheltered between Fraser Island and the coast, and extensive mudflats are exposed at low tide. Counts for this region have been around 14 000 birds, including 2000 Eastern Curlews, large numbers of plovers and Bar-tailed Godwits, and smaller numbers of other wader species. Points of access include Woodgate, Burrum Heads, Mary River Heads, Boonooroo and Tin Can Bay.

Moreton Bay: Here the Brisbane coast is enclosed by a north-south line of islands—Bribie, Moreton, North Stradbroke and South Stradbroke. Within this large expanse lie many smaller sites. South of the mouth of the Brisbane River, Raby Bay has extensive mudflats frequented in summer by many waders: Red and Great Knots, Red-necked Stint, Broad-billed, Sharp-tailed, Terek and Curlew Sandpipers, and Lesser and Greater Sand Plovers. In summer waders can also feed on the exposed flats of the tidal seagrass shallows off Moreton Island's southern tip.

Gippsland Lakes: Inlets, coastal lakes and swamps have formed behind the long sweep of sandy coast and dunes of the Ninety Mile Beach—the long, narrow Lake Reeve, Lakes King, Victoria, Wellington and Tyers, and the waterways around Lakes Entrance—some parts are within national parks. The bird list exceeds 170 species, including some 15 000 to 20 000 waders in summer. Among these are Red-necked Stints, 5000 Red Knots and 1800 Curlew Sandpipers. Other species are Eastern Curlew, Greenshank, Sharp-tailed, Common and Curlew Sandpipers, Bar-tailed Godwit, and Latham's Snipe.

Corner Inlet: Tucked between Wilson's Promontory and the Gippsland coast, the extensive mudflats of this large inlet support numerous waders—around 45 000 birds of more than 20 species, including some 20 000 Red-necked Stints, 10 000 Curlew Sandpipers, 7000 Bar-tailed Godwits and over 2000 Eastern Curlews, have been on record.

Moulting Lagoon, near Freycinet Peninsula, is fresh water, but nearby Pelican Bay has extensive mudflats used by migratory waders in summer.

Pittwater-Orielton Lagoon is a complex of bays and lagoons about 25 km east of Hobart, and has several wader sites. Near Sorell waders use Orielton Lagoon, Pittwater and Barilla Bay; on the South Arm peninsula, Ralph's Bay, and Pipeclay, Rushy and Clear Lagoons.

BIRDS BELIEVED EXTINCT: Excluded from previous pages were the following:

Australian mainland and offshore islands: King Island Emu, *Dromaius ater*, Kangaroo Island Emu, *D. baudinianus*.

Lord Howe Island: White Gallinule, *Porphyrio albus*; Tasman Booby, *Sula tasmani*; White-throated Pigeon, *Columba vitiensis godmanae*; Lord Howe race of Red-fronted Parakeet, *Cyanoramphus novaezelandiae subflavescens*; Lord Howe Boobook, *Ninox novaeseelandiae albaria*; Vinous-tinted Blackbird, *Turdus xanthopus vinitinctus*; Lord Howe Fantail, *Rhipidura cervina*; Lord Howe Gerygone, *Gerygone insularis*; Robust White-eye, *Zosterops strenuus*; Lord Howe Starling, *Aplonis fuscus hullianus*.

Norfolk Island: Tasman Booby, *Sula tasmani*; Norfolk Island Kaka, *Nestor productus*; Norfolk Island Ground-Dove, *Gallicolumba norfolciensis*; New Zealand Pigeon, *Hemiphaga novaeseelandiae*.

NEW DISCOVERIES The following species are recent sightings, or reports awaiting confirmation. Many of the additions are from Australia's island territories, which have only recently received close attention. The Torres Strait islands, close to New Guinea, and Christmas and Cocos-Keeling are giving successive new observations of birds that are island residents, migrants or vagrants from SE Asia. New discoveries can be followed through such publications as *Wingspan*, *Australian Birding* and the *Australian Bird Watcher*. Now there are also web pages devoted to Australian birds. In time, the author hopes to provide a web page that will update and extend this guide, and show previous and forthcoming bird books: try www.michaelmorcombe.com.au, or search on michael morcombe or michaelmorcombe.

SPECIES	UPDATE NOTES
Green Junglefowl *Gallus varius* male 60 cm; female 45 cm (K) Cocos Is. list. Introduced to West Island about 1880, not reported between about 1981 and 1993, then several large flocks observed. Males greenish blue-black with a large, rounded, red tipped, unserrated red comb, red face, and yellow tipped, red throat wattle. The Green Junglefowl's neck is entirely blue and green, in contrast to the Red Junglefowl's orange-red neck. **Status:** feral population.	
Chukar Partridge *Alectoris chukar* 30–35 cm Australian list. A plump ground bird; white face with black line through the eye and encircling the white throat. Flanks buff, heavily barred. Usually runs; flies only if forced. **Voice:** drumming and wailing sounds. **Status:** feral; released for sporting shooting near Gulong, NSW.	
Helmeted Guinea Fowl *Numida meleagris* 55–65 cm Australian list. Plump, blackish, speckled, short legged fowl; a proportionately tiny head that carries bony red casque; red wattles hang from bare bluish face. Runs; flies if forced. **Voice:** varied loud cackles, cluckings, clickings and thin wailing. **Status:** several small feral populations; some on Barrier Reef islands, including Heron Is., Qld.	
Little Grebe *Tachybaptus ruficollis* 24–25 cm (See also pp. 30, 31) Recently reported and photographed near Darwin. It is superficially like the Australian Grebe, except in breeding plumage: throat, cheeks rufous-red rather than black; sides and front of neck rufous not rufous and black; flanks dark, almost black, rather than pale rufous; much smaller white patch under tail; eye red rather than yellow. In flight, the Little Grebe shows little or no white wing-bar, a conspicuous feature on Australian and Hoary-headed Grebes. In nonbreeding and immature plumages, it is harder to separate species: wing-bar difference still evident; eye dark rather than pale; side of head slightly deeper buff rather than buffy white. **Voice:** sharply chattered 'ki-ki-ki-' and softer churring sounds. **Hab.:** in NT, a sewage farm near Darwin, Sep.–Nov. 1999. **Status:** rare vagrant; probably one of several SE Asian races of a species with very wide range in Eurasia and Africa.	
Newell's Shearwater *Puffinus (auricularis) newelli* 30–38 cm Possible new species for Australian list. Similar to Townsend's Shearwater, *P. auricularis* (with which it may be combined as a race), and the Manx Shearwater, *P. puffinus*. Of these, only Newell's has completely white under-tail coverts. All 3 have a very small, although slightly varying, notch of white, upward into the lower edge of the black cap just behind the ear coverts. Recorded from Phillip Is., Norfolk Island group, Dec. 1977.	

Abbott's Booby *Papasula abbotti* (see also Christmas Is. birds, pp. 422, 423)
Breeds only on Christmas Is.; feeds and disperses across surrounding seas. First Australian record 17 Dec. 1999 at Broome, W Kimberley coast of WA. This solitary vagrant was found soon after 2 tropical cyclones passed; both had originated in the Indian Ocean, and one from near Christmas Is. This bird, an adult male in good condition, was flown back to Christmas Is. for release on 24 Dec. 1999. Reference: Boyle, A., *Western Australian Bird Notes* (RAOU, WA Group), no. 93, March 2000.

Japanese Sparrowhawk *Accipiter gularis* 22–32 cm (C)
Christmas Is. list; unconfirmed. A small sparrowhawk, male with breast barred cinnamon-rufous, upper parts blue-grey; underside of wings, including coverts, barred; tail has 4 dark bands. Female larger, dusky barring, grey-brown upper parts. **Hab.:** forest margins, open woodlands. **Status:** reaches S to Indonesia and Timor in winter.

Buffy Fish-Owl *Ketupa ketupu* 45–50 cm (K)
Cocos Is., 1941. A large owl with obvious, long, perhaps drooping, tufts out from the eyes over the ear coverts. Rich buff plumage: upper parts broadly streaked dark brown; underparts have fine, dark streaks. Eyes golden. **Voice:** loud, high, shrill 'kootookoo' and a soft, musical 'tiewee, tiewee'. **Hab.:** forests, especially along streams. **Status:** widespread in SE Asia, S to Sumatra, Java and Bali.

Northern Wheatear *Oenanthe oenanthe* 14–15 cm (C)
A small chat-like bird seen usually on the ground where, in typically upright posture, it often perches on a rock or mound. Bird has a white rump and white base of outer tail. **Voice:** abrupt, hard 'chak' or 'chuk', and 'chak-wheit'. **Hab.:** open, rather bare land. **Status:** breeds in Palaearctic regions of northern hemisphere; migrates to India during N winter; vagrant further S to Philippines, Borneo and, probably only very rarely, reaches Christmas Is.

Sooty (Dark-sided) Flycatcher *Muscicapa sibirica* 12–13 cm
Occurs Borneo, N Sumatra; a rare visitor to Java. One Aust. record in Pilbara region of WA at the Shay Gap sewage farm, Oct. 1983. Overall sooty grey-brown, upper parts very dark. A conspicuous white collar encircles all but the rear of the neck; on the throat it meets a broad white streak that extends from chin through centre of breast and belly to under-tail coverts. Has whitish eye-ring, and buffy white wing-bar and edges to some flight feathers. The immatures are spotted buffy white on face and back. **Voice:** abrupt 'tzi-up, tzi-up', repeated rapidly in flight. **Hab.:** undergrowth and mid levels of forests; darts out to snap up passing insects.

Streak-headed Mannikin *Lonchura tristissima* 9–10 cm (T)
Recorded on Sabai Is., Torres Strait, 1998. The only member of the family in this region that has a pale rump in otherwise dark plumage. (If the race *leucosticta* is separated as a full species, this sighting may be of the species *leucosticta*, the White-spotted Mannikin.) **Voice:** abrupt nasal buzz repeated rapidly in flight. **Hab.:** damp grassy stream edges, forest clearings, regrowth, bamboo clumps in forest.

FURTHER NOTES

Accidental: straying beyond usual range or migratory path.

Adult: a bird in its final plumage excluding seasonal changes.

Breeding plumage: acquired for duration of courting and nesting.

Carpal (joint): the wrist joint—in flight, this is the back-angled wrist joint, becoming the forward-most bend or point in the folded wing.

Casque: helmet-like tall ridge or shield on skull or bill.

Cheek: area of plumage or skin under the eye from gape of bill back to ear coverts.

Chevrons: broad V-shaped markings on plumage, arrowheads.

Cline: a gradual or graded series of small changes in plumage or another characteristic across a geographical area.

Colonial: many birds roosting together; nests close together, from touching to just a few metres apart.

Cosmopolitan: occurring through most of the world.

Coverts: small covering feathers in rows that conceal bases of larger feathers and, finally, of the largest flight feathers and tail feathers.

Endemic: found only in a locality, region or continent.

Eye-ring: a row of tiny feathers or area of skin that forms a distinct encircling ring around the eye.

Facial disc: the pale facial feathers, heart-shaped or rounded, defined by a dark rim that is typical of many owls and, less obviously, harriers.

Form: a broad term that covers all the variants of a species, whether subspecies or morphs or others that are distinguishable or distinctive.

Frontal shield: a fleshy or horny oval or rectangular patch that lacks feathers, on the forehead area.

Gape: corner of mouth, a fleshy rim, often yellow, and most conspicuous on young birds.

Genus: a group of closely related species, the genus name being the first of the two names given each species, and capitalised.

Graduated: indicating a tapered shape, the term is often used to describe the tail: a graduated tail has the outer feathers shortest and each pair inwards being longer.

Gregarious: in groups or flocks; suggests some degree of social interaction among members.

Hackles: longer, pointed and stiff-looking feathers of neck or throat.

Immature: yet to reach maturity; between juvenile and mature.

Juvenile: young bird; usually describes the first plumage after naked nestling or natal down stage; often acquired while still in the nest.

Lamellae: comb-like fringe to inner edge of bill; used to sieve fine food particles from water.

Lobes: wide flat fringes along sides of toes of water birds; an alternative to webbing between the toes in making the feet more effective as paddles.

Lores: small area of plumage or skin between base of bill and eye.

Nomadic: bird species that undertake wandering travels of irregular pattern in timing, direction or distance.

Nuchal: area of the neck.

Orbital ring: circle of tiny feathers or bare skin around the eye.

Order: grouping of families; subdivision of class.

Passerine: belonging to by far the largest bird group, the Order Passeriformes, song or perching birds, characterised by perching with three toes forward, one back.

Phase: of plumage—a difference between individuals that is not related to race or age; sometimes confined to one sex.

Primaries: outermost very large flight feathers of the wing, attached to the 'hand' portion of the wing.

Wattles: fleshy, usually colourful growths of crown, face or neck.

Wing coverts: rows of small feathers covering bases of larger feathers, including base of flight feathers.

Bibliography

Christides, L. & W.E. Boles. 1994. *The Taxonomy and Species of Birds of Australia and its Territories*. RAOU Monograph 2. RAOU: Hawthorn East.

Beehler, B.M., et al. 1986. *Birds of New Guinea*. Princeton.

Beruldsen, G. 1980. *A Field Guide to Nests and Eggs of Australian Birds*. Rigby: Adelaide.

Beruldsen, G. 1995. *Raptor Identification*. G. Beruldsen: Kenmore Hills.

Blakers, M., et al. 1984. *The Atlas of Australian Birds*. Melbourne University Press: Carlton.

Serventy, D.L. 1977. *Distribution of Birds on the Australian Mainland*. Prepared by J.R. Busby & S.J.J.F. Davies. CSIRO.

Coates, B.J. 1985–90. The Birds of Papua New Guinea. Vols. 1, *Non-Passerines*, and 2, *Passerines*. Dove Publications: Alderley.

Coates, B.J. & K.D. Bishop. 1997. *A Guide to the Birds of Wallacia*. Dove Publications: Alderley.

Emison, W.B., et al. 1987. *Atlas of Victorian Birds*. Department of Conservation, Forests and Lands, and RAOU: Melbourne.

Frith, H.J. 1982. *Pigeons and Doves of Australia*. Rigby: Adelaide.

Forshaw, J. 1981. *Australian Parrots*. 2nd ed. Landsowne Editions: Melbourne.

Fry, C. H., K. Fry & A. Harris. 1992. *Kingfishers, Bee-eaters and Rollers*. Christopher Helm, A & C Black: London.

Harrison, P. 1985. *Seabirds: An Identification Guide*. Revised ed. Croome Helm: Beckenham.

Hayman, P., J. Marshal & T. Prater. 1986. *Shorebirds: an Identification Guide to the Waders of the World*. Croome Helm: Beckenham.

Hollands, D. 1984. *Eagles, Hawks and Falcons of Australia*. Nelson: Melbourne.

Hollands, D. 1991. *Birds of the Night: Owls, Frogmouths and Nightjars of Australia*. Reed: Sydney.

Hollands, D. 1999. *Kingfishers and Kookaburras: Jewels of the Australian Bush*. New Holland: Frenchs Forest.

Hutton, I. 1950. *Birds of Lord Howe Island, Past and Present*. Ian Hutton: Coffs Harbour.

Johnstone, R.E. 1990. 'Mangroves and mangrove birds of Western Australia'. *Rec. West. Aust. Mus.* Suppl. 32: 1–120.

Johnstone, R.E. & L.A. Smith. 1981. *Birds of Mitchell Plateau and Adjacent Coasts etc., Kimberley, W.A.* W.A. Museum: Perth.

King, B., M. Woodcock & E.C. Dickinson. 1975. *A Field Guide to the Birds of South East Asia*. Collins: London.

Lekagul, B. & P.D. Round. 1991. *A Guide to the Birds of Thailand*. Saha Karn Bhaet: Bangkok.

MacKinnon, J. & K. Phillipps. 1993. *A Field Guide to the Birds of Borneo, Sumatra, Java and Bali*. OUP: Oxford.

Marchant, S. & P.J. Higgins (Eds.). *Handbook of Australian, New Zealand and Antarctic Birds*. Vols. 1–4. OUP: Melbourne.

Morcombe, M. 1986. *The Great Australian Birdfinder*. Lansdowne Press: Sydney.

National Photographic Index of Australian Wildlife: Birds. Vols. 1–10. 1982–1994. Angus & Robertson: Sydney.

North, A.J. 1901–14. *Nests and Eggs of Birds Found Breeding in Australia and Tasmania*. Australian Museum: Sydney.

Pizzey, G. 1997. *Field Guide to the Birds of Australia*. Angus & Robertson: Sydney.

Reville, B.J. 1993. *A Visitor's Guide to the Birds of Christmas Island, Indian Ocean*. 2nd ed. Christmas Island Natural History Association: Christmas Island.

Schodde, R. 1982. *The Fairy-wrens: A Monograph of the Maluridae*. Lansdowne: Melbourne.

Schodde, R. & I.J. Mason. 1999. *The Directory of Australian Birds: Passerines*. CSIRO Publishing: Collingwood.

Serventy, D.L. & J. Warham. 1971. *The Handbook of Australian Seabirds*. AH & AW Reed: Sydney.

Serventy, D.L. & H.M. Whittell. 1976. *Birds of Western Australia*. 5th ed. University of Western Australia Press: Perth.

Slater, P., P. Slater & R. Slater. 1986. *The Slater Field Guide to Australian Birds*. Rigby: Sydney.

Storr, G.M. & R.E. Johnstone. 1985. *A Field Guide to the Birds of Western Australia*. 2nd ed. Western Australian Museum: Perth.

Viney, C. & K. Phillipps. 1988. *Birds of Hong Kong*. 4th ed. Government Printer: Hong Kong.

Magazines

The publications of the many bird groups listed on the following page are also of great value, especially for nests and eggs, and particularly older volumes of the *EMU*, from the early 1900s to around the 1940s.

Australian Birding, an independent publication conceived, edited and published through its early years by David Andrew, has reported on new discoveries, often from the island territories, and has specialised in identification within bird groups. It is now published by Andrew Isles Natural History Books (03 9510 5750; aislesbs@anzaab.com.au).

Birds Australia: The Royal Australasian Ornithologists Union was established in 1901 and has as its aim the conservation of Australian birds and their habitat; it is a non-profit conservation-based organisation. The organisation is member-based, giving bird information, access to activities, free entry to the country's reserves and observatories, discounted optical equipment and subscription to *Wingspan*, a quarterly magazine; it also publishes a journal of ornithological research. More than 6000 volunteers Australia-wide take part in research and field activities that will help conserve our native bird populations. This contribution by the members would be worth many millions annually.

Because there is great strength in numbers, membership of Birds Australia is one of the best ways to join the fight to save our unique native bird life. To become a member, simply visit the website: www.birdsaustralia.com.au

or contact the national office at 415 Riversdale Road, Hawthorn East, Vic. 3123; phone 03 9882 2622.

On average, Birds Australia is involved in about 30 conservation-related projects each year. Among these are:

The Bird Habitat Acquisition Fund: The importance of land acquisition for the protection of threatened species and their habitats is now well understood. Following the successful purchase and establishment of the Birds Australia Gluepot Reserve in 1997, the Birds Australia Council have established the Bird Habitat Acquisition Fund (formerly the Birds Australia Land Fund). This fund will receive land purchase donations and bequests so that the purchase and management of significant habitat reserves for protection of threatened species can continue. Newhaven Station will be the first acquisition from this fund. Should Birds Australia receive more donations than required to buy and establish Newhaven, remaining funds will be retained for subsequent land purchases.

The Australian Nest Record Scheme: This scheme (NRS) has been running since 1964 to gather information and increase knowledge of the breeding biology of Australian birds, a goal achieved by collecting information on nest building, clutch size, fledging success and the timing and location of breeding. The scheme now has close to 90 000 records from approximately 1500 contributors; at any one time about 100 people are active contributors. To participate, contact the office of Birds Australia to receive data sheets and instructions. Further enquiries to: Nest Record Scheme, Birds Australia, 415 Riversdale Road, Hawthorn East, Vic. 3123; phone 03 9882 2622; fax 03 9882 2677. I hope that the nests and eggs section (pages 352–419) may help those participating to locate and identify nests.

Other Birds Australia offices:

Sydney: 2/399 Pacific Hwy, Crows Nest, NSW 2065
Ph.: 02 9436 0388 Webpage: http://web.one.net.au/~rosella/

WA: Perry House, 71 Oceanic Drive, Floreat, WA 6014
Ph.: 08 9383 7749 e-mail: birdswa@starwon.com.au

Regional groups are established in Victoria, southern NSW–ACT; northern NSW, WA, north Queensland and Tasmania. Contact information: RAOU offices and *Wingspan*.

Bird observatories:

Barren Grounds Bird Observatory, PO Box 3, Jamberoo, NSW 2533; 02 4236 0195; www.users.bigpond.com/barren.grounds

Broome Bird Observatory, PO Box 1313, Broome, WA 6725; 08 9193 5600; e-mail bbo@tpgi.com.au

Eyre Bird Observatory, Cocklebiddy via Norseman, WA 6443; 08 9039 3450

Rotamah Island Bird Observatory, PO Box 75, Paynesville, Vic. 3880; e-mail rotamah@i-o.net.au

Reserves: Birds Australia Gluepot Reserve, PO Box 345, Waikerie, SA 5330; 08 8892 9600 (6–8 pm); e-mail condor@riverland.net.au

Special interest groups: These include Australasian Raptor Association (ARA); Australasian Wader Study Group (AWSG); Australasian Seabird Group (ASG); Birds Australia Parrot Association. Contact details from RAOU offices.

Bird groups, publications and addresses

Australian Bird Study Association—*Corella*, PO Box A313, Sydney South, NSW 1235

Avicultural Society of Australia—*Australian Aviculture*, 52 Harris Road, Elliminyt, Vic. 3249

Bird Observers' Association of Tasmania—*Tasmanian Bird Report*, GPO Box 68A, Hobart, Tas. 7000

Cumberland Bird Observers' Club, PO Box 550, Baulkham Hills, NSW 2153

Bird Observers' Club of Australia—*The Australian Bird Watcher* and *The Bird Observer*, PO Box 185, Nunawading, Vic. 3131

Canberra Ornithologists' Group—*Canberra Bird Notes*, PO Box 301, Civic Square, ACT 2608

Illawarra Bird Observers' Club, PO Box 56, Fairy Meadow, NSW 2519

Northern Field Naturalists' Club—*Nature Territory*, PO Box 39565, Winnelli, NT 0821

NSW Field Ornithologists' Group—*Australian Birds*, PO Box Q277, QVB PO, Sydney, NSW

Queensland Ornithological Society—*The Sunbird*, *QOS Newsletter*, PO Box 97, St Lucia, Qld 4067

South Australian Ornithologists' Union—*South Australian Ornithologist*, SA Museum, North Terrace, Adelaide, SA 5000

Toowoomba Bird Observer, PO Box 4730, Toowoomba East, Qld 4350.

Western Australian Naturalists' Club—*The Western Australian Naturalist*, PO Box 8257, Perth Business Centre, WA 6849

Photography, sound recordings, scientific collecting

At times concern is expressed because scientific collecting, photography, recording or even close observation might cause losses to nesting birds. While taking all precautions to prevent any detrimental interference (it would be unthinkable to do otherwise) the reality is that any harm to the bird population caused by these or similar activities is insignificant compared with the permanent losses, especially of habitat and hollows in old trees, caused by land clearing, clearfelling of old-growth forest, wildfires or feral predators.

One, at the most several, of the millions of feral cats and foxes that live on wildlife in the bush and suburbs would account for more losses each year than all the scientific collecting, unintentional and accidental losses (for example, a crow finding a nest left exposed, or a bird deserting its nest because a photographic hide was placed too close) caused by all Australian birdwatchers, photographers and museum collectors combined. Likewise, so would one or two Goshawks, Sparrowhawks or Square-tailed Kites that feed on small birds' nestlings. Spring forestry burns and land clearing would each year account for far greater losses.

On the credit side, the recorded observations of birds at nests, photographs and study skins have made possible all the books, documentaries and articles that have built public interest in, and appreciation and awareness of, the need to conserve large areas of all types of habitat.

Coinciding with the increase in such publications since 1960 has been a growth in the demands from an increasingly appreciative but alarmed public who want more protection of habitat and funding for scientific research to discover optimum means of conservation, whether of endangered species or habitat. Any small losses caused by scientific collecting, photography and the like have been far more than offset by the explosion of public awareness and interest, and the demand for more national parks; these have forced a political response—budget allocations to protect the natural environment and to provide for sustainable ecotourism have increased.

But any loss, however small, could be significant to the survival of very rare or endangered species like the Helmeted Honeyeater or the Noisy Scrub-bird; none of these should be put at risk. However, the populations of most birds are limited by much broader constraints: confines of suitable habitat; available food resources; effects of fire, drought or extensive clearing; predator pressure; or gradual contamination or degradation of habitat.

R

S

The numbers of pages on which species are illustrated are set in **bold type**. The names given in quotation marks are widely used names for a species or subspecies that are not the preferred RAOU common name for that species.